ULTRASONIC NONDESTRUCTIVE TESTING OF MATERIALS

THEORETICAL FOUNDATIONS

ULTRASONIC NONDESTRUCTIVE TESTING OF MATERIALS

THEORETICAL FOUNDATIONS

Karl-Jörg Langenberg, René Marklein,
and **Klaus Mayer**

CRC Press is an imprint of the
Taylor & Francis Group, an **informa** business

Theoretische Grundlagen der zerstörungsfreien Materialprüfung mit Ultraschall
2009 Oldenbourg Wissenschaftsverlag GmbH
Rosenheimer Straße 145, D-81671 München
Telefon. (089) 45051-0
Oldenbourg.de

Authorized translation from German language edition published by Oldenbourg Wissenschaftsverlag GmbH

CRC Press
Taylor & Francis Group
6000 Broken Sound Parkway NW, Suite 300
Boca Raton, FL 33487-2742

First issued in paperback 2017

Version Date: 20120125

ISBN 13: 978-1-138-07596-2 (pbk)
ISBN 13: 978-1-4398-5588-1 (hbk)

Visit the Taylor & Francis Web site at
http://www.taylorandfrancis.com

and the CRC Press Web site at
http://www.crcpress.com

Contents

Preface

The present book stands between the fundamental elaborations on elastic waves in solids (e.g., Achenbach 1973; Auld 1973; de Hoop 1995) and emerging applications for ultrasonic nondestructive testing (e.g., Rose 1999; Schmerr 1998; Schmerr and Song 2007). The latter emphasize engineering viewpoints in contrast to the more physical and mathematical elastic wave propagation theory. As a consequence, we consider the following chapters to be a missing link, on one hand elaborating on the physics and mathematics of ultrasound propagation in solids and on the other hand exemplifying it on standard nondestructive testing problems. As a typical example, worldwide, engineers tend to argue with plane wave knowledge, speaking of longitudinal and transverse ultrasonic beams, thus ignoring that plane waves represent rather idealized and artificial solutions of wave equations exhibiting their polarization as a consequence of their respective wave speeds and their underlying physical nature as pressure and shear waves, and they are not beams! These arise from an approximate solution of wave equations or as exact solutions of approximated wave equations. Of course, ultrasonic nondestructive testing may often be roughly understood in terms of plane waves and beams; yet, the key issues are transducer radiation, defect scattering and imaging, respectively, and this has to be thoroughly formulated with the help of—physically spoken—point source synthesis or—mathematically spoken—utilizing representation integrals with Green functions. Again recognizing the book as a missing link, we introduce Green functions for the simplest scalar acoustic case, always accompanied by intuitive interpretations, and approach the relevant tensorial elastodynamic case step by step. Apropos tensors: We avoid the often used but somewhat confusing index notation and rely on our electromagnetic heritage of a coordinate free formulation as introduced by Chen (1983). Yet, we do not leave the reader alone; we provide the rules of this calculus as a mathematical introduction.

Another specific feature of this book comprises the utilization of numerical computational tools to explain specific wave propagation phenomena and to compare the results with those obtained by—mostly approximate—analytical formulations.

Finally, we are thankful to the late Paul Höller, founding director of the Fraunhofer Institute for Nondestructive Testing in Saarbrücken, Germany, who was responsible for our switch from electromagnetics to elastodynamics, as well as many colleagues from the German Society for Nondestructive

Testing (DGZfP) and the Federal Institute for Materials Research and Testing (BAM), Berlin, Germany, for their continuous stimulation, support, and interest.

Karl-Jörg Langenberg

Authors

Prof. Dr. rer. nat. Karl-Jörg Langenberg was educated in physics at the University of the Saarland, Saarbrücken, Germany, where he earned his Doctor of Natural Sciences (Dr. rer. nat.) and his *venia legendi.* A subsequent three-year period as principal scientist at the Fraunhofer Institute for Nondestructive Testing, Saarbrücken, Germany, ended with the acceptance of the Chair for Electromagnetic Theory at the University of Kassel, Germany. Dr. Langenberg is a Fellow of IEEE.

Priv.-Doz. Dr.-Ing. René Marklein was educated in electrical engineering at the University of Kassel, Germany, where he earned his Doctor of Engineering Sciences (Dr.-Ing.), *venia legendi,* and his Private Docent (Priv.-Doz.). A subsequent period as principal engineer at the University of Kassel ended when he founded his own engineering office. Beside this, he works presently for the Fraunhofer Institute for Wind Energy and Energy System Technology (IWES) and teaches as a private lecturer at the University of Kassel, both in Kassel, Germany. Dr. Marklein is a recipient of the Berthold-Prize of the German Society for Nondestructive Testing, and received the European NDT Innovation Diploma in the category of basic research.

Dr.-Ing. Klaus Mayer was educated in electrical engineering at the University of the Saarland, Saarbrücken, Germany. He received his Doctor of Engineering Sciences (Dr.-Ing.) from the Department of Electrical Engineering at the University of Kassel, Germany, where, since then, he holds a graduate occupation. Dr. Mayer is a recipient of the Berthold-Prize of the German Society for Nondestructive Testing.

1

Contents

1.1 Introduction

Ultrasonic nondestructive testing (US-NDT) relies on the excitation, propagation, and scattering of elastic waves in solids; this topic is absolutely nontrivial, regarding neither its physics nor its mathematical formulation. One of the reasons is that elastic waves occur in two modes (in isotropic materials): pressure and shear waves (longitudinal and transverse waves) with different wave speeds. This fact considerably complicates the interpretation of ultrasonic signals and makes a "common sense interpretation" often impossible; the support of mathematical and numerical modeling of elastic wave propagation is definitely required. Then, heuristically introduced concepts such as "ultrasonic beams" or "reflector" can be precisely defined.

For example, US-NDT uses the term "pressure" being measured and displayed as an A-scan; as a matter of fact, the real meaning of it is the field quantity $p(\underline{\mathbf{R}}, t)$ at a given location characterized by the vector of position $\underline{\mathbf{R}}$ as a function of time t. This pressure is a scalar quantity that is uniquely determined by only *one* number (with a physical unit). Yet the fundamental physical field quantity of *elastic waves* is the vector displacement $\underline{\mathbf{u}}(\underline{\mathbf{R}}, t)$ with, in general, *three* scalar components that defines the symmetric deformation tensor $\underline{\underline{\mathbf{S}}}(\underline{\mathbf{R}}, t)$ with six scalar components, and the latter one is related to the symmetric stress tensor $\underline{\underline{\mathbf{T}}}(\underline{\mathbf{R}}, t)$ through Hooke's law as constitutive equation. This reveals that the physics of elastic waves has to be described by a theory of space- and time-dependent scalar, vector, and tensor fields. In the following chapters, we will outline this theory with relevance to US-NDT and we will illustrate it by examples. Therefore, a certain amount of mathematical calculus is necessary, but we will always try to depict the meaning of abstract formalisms.

At first, we define spatially dependent scalar, vector, and tensor fields and their algebraic conjunction; we continue to talk about space and time variations of these fields, in particular about gradients, divergence, and curl densities. The time variable t is opposed by the (circular) frequency variable ω as a conjugate variable; the frequency "content" of a pulsed signal, its spectrum, is quantified by the Fourier transform. To describe an ultrasonic beam, we additionally need the Fourier transform with regard to spatial (Cartesian) coordinates. Elastic ultrasonic waves are excited by "transducers"; the relation between a sound field and its sources is given by Green functions that are nothing more than the respective fields from idealized point sources. Hence, for a mathematical formulation, such point sources have to be thoroughly defined, which introduces Dirac's delta function; yet, this "function" is not a function at all but a distribution that requires a closer explanation.

Having provided these mathematical tools, we turn our attention to four fundamental NDT-relevant problems; propagation of elastic waves in isotropic and anisotropic materials—idealized as plane waves and elementary waves from point sources—radiation from volume and surface sources, scattering by material inhomogeneities and imaging of those, say: material defects.

In the following, we discuss the flow chart of Figure 1.1 that guides us through the subsequent chapters like a thread.

1.2 Contents Flow Chart

Linear elastodynamics is based upon the Newton–Cauchy equation of motion—relating the time variation of the linear momentum density with the source density of the stress tensor and prescribed force densities—and, additionally, the deformation rate equation as definition of the time derivative of the deformation tensor through the symmetric part of the gradient dyadic of the particle velocity; the prescribed source of that equation is the injected deformation rate. At material jump discontinuities, both equations reduce to transition conditions for the components of the particle velocity and the vector traction as projection of the stress tensor to the surface normal vector. Because both equations contain different field quantities, they cannot be immediately combined: Material properties have to be introduced before that relate field quantities in terms of constitutive equations and those do not follow from the governing equations, they have to be postulated instead knowing the physical properties of the underlying materials. As a consequence of the constitutive equations, we obtain elastodynamic governing equations as a coupled system of first-order partial differential equations. Nevertheless, constitutive equations must not violate basic physical principles, for example, elastodynamic energy conservation; as a result, the involved material tensors like the forth rank stiffness (compliance) tensor in the linear nondissipative Cauchy–Hooke law have to satisfy various symmetries. According to the requirements of US-NDT, the resulting governing equations of elastodynamics have now to be solved; the closer the actual model problem is to reality—for example, defect imaging in a dissimilar weld—the more unlikely a solution with "paper and pencil," that is, to say with analytical methods, is at hand. A request for numerical methods arises that could either be utilized after some preparatory analytical work or be directly operated on the governing differential equations. Our own numerical tool of the second category is called elastodynamic finite integration

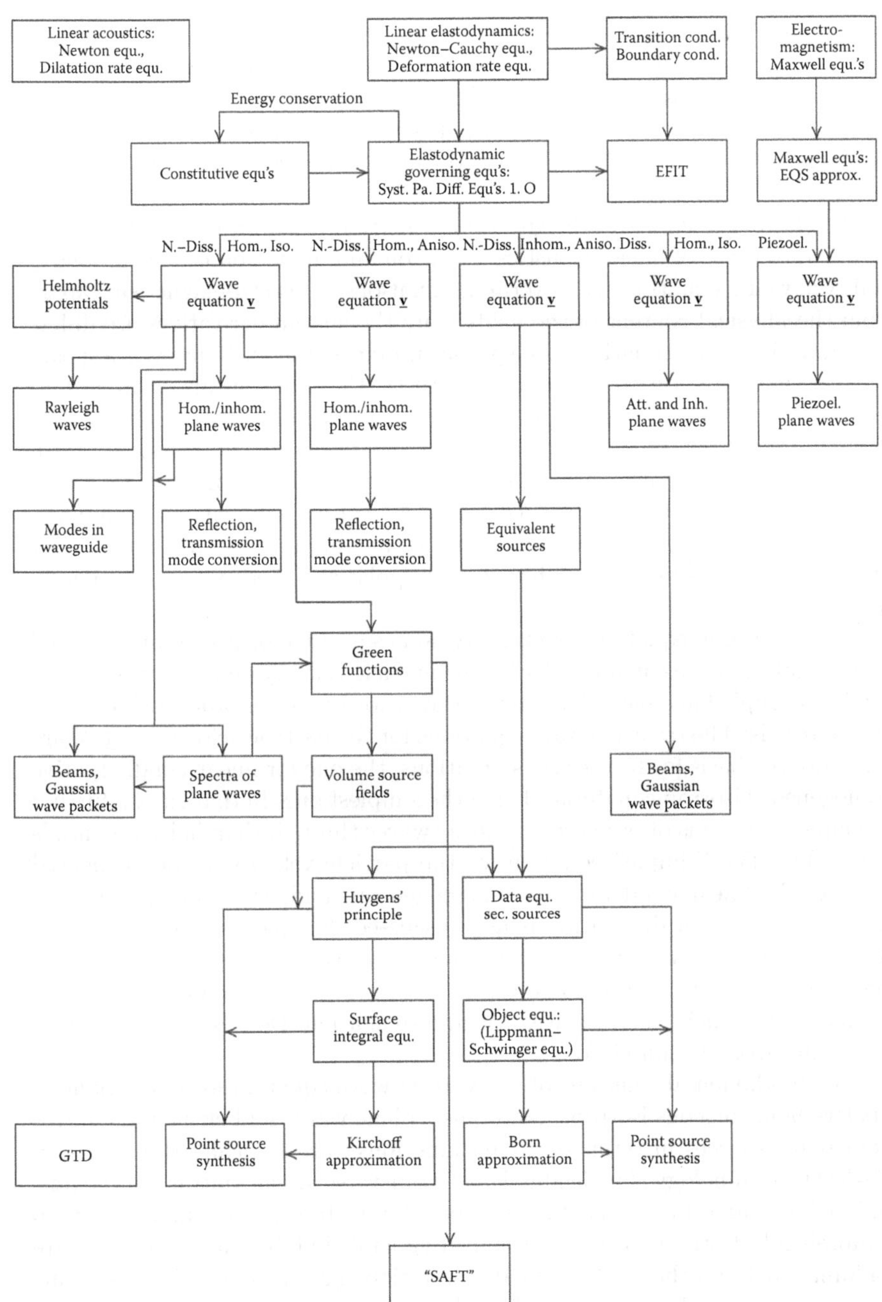

FIGURE 1.1
Contents flow chart.

technique (EFIT) that needs nothing but the governing equations, prescribed sources, given materials, and transition conditions. In the chapters to follow, we will often discuss results obtained with the EFIT-code, yet we will not go into the details of the method because it is well documented in the literature (Fellinger 1991; Fellinger et al. 1995; Marklein 1997, 2002; Langenberg et al. 2002; Bihn 1998).

The formal structure of the governing equations of linear elastodynamics is identical to those of linear acoustics—Newton and dilatation rate equations—and Maxwell's equations as governing equations of electromagnetism; apart from the physical content of the fields, only the spatial derivatives are different, and Maxwell's equations are special insofar as the curl-operator appears in both equations. Based on these similarities, the solutions also exhibit similarities and, therefore, we find it appropriate to include chapters on the fundamental solutions of acoustics and electromagnetism: plane waves, Green functions, and Huygens' principle. That is not only of interest for NDT applications; moreover, scalar acoustic fields often serve us to "simplify" the somewhat complex vector and tensor calculus of elastodynamics. Last but not least: two-dimensional horizontally polarized shear waves *are* strictly scalar.

The governing equations of elastodynamics as a coupled system of partial differential equations of first order for the particle velocity and the stress tensor can be decoupled in a partial differential equation of second order each for various materials: The resulting wave equations for the particle velocity $\underline{\mathbf{v}}(\underline{\mathbf{R}}, t)$ are generally the basis for further considerations, the one for linear nondissipative homogeneous isotropic materials being the simplest one. In that case, a further decoupling in terms of pressure and shear waves through Helmholtz potentials is possible. The "simple" equation for the particle velocity is complemented by those for linear nondissipative inhomogeneous and/or anisotropic materials; the influence of dissipation is also discussed. For piezoelectric materials, the equations of elastodynamics are coupled to Maxwell's equations, resulting in a piezoelectric wave equation for elastic waves if Maxwell's equations are reduced to their electroquasistatic approximation. Details can be found in Marklein's dissertation (1997).

The fundamental solutions of the $\underline{\mathbf{v}}(\underline{\mathbf{R}}, t)$ wave equation for linear nondissipative homogeneous isotropic materials—plane waves and elementary waves from point sources in terms of Green functions—are basically the source of US-NDT terminology: longitudinal and transverse waves, ultrasonic beams, and point source synthesis. Plane waves, for instance, are fundamental to comprehend elastic wave propagation in general, but beyond that, they are building blocks for the mathematical description of realistic sound fields. Plane waves are the simplest solutions of the homogeneous wave equation, the wave equation containing no given sources; with the *ansatz* "plane waves with planar phase fronts," this equation looks like an eigenvalue problem: their phase velocities are the eigenvalues and their polarization vectors are the eigenvectors. For isotropic materials, two of the three eigenvalues coincide, they refer to secondary plane waves, while the remaining eigenvalue stands for primary waves, because they arrive first at a specific point of observation due to their larger phase velocity. The polarization of *plane* primary waves is longitudinal whereas the polarization of both *plane* secondary waves is independently transverse to the direction of propagation because of the coinciding eigenvalues. Therefore, the US-NDT terminology "longitudinal/transverse" can be synonymously used to "primary/secondary" as long as *plane* waves are under concern. But even more general due to its applicability to sound fields of transducers is the terminology "pressure/shear" because primary waves are always pressure waves and secondary waves are always shear waves in homogeneous isotropic materials. For nondissipative homogeneous isotropic materials, phase and amplitude fronts of plane waves either coincide for homogeneous plane waves or they are orthogonal to each other for inhomogeneous plane waves. The latter ones appear when plane waves are reflected at planar boundaries of elastic half-spaces, they represent evanescent surface waves.

Reflection, mode conversion, and transmission of elastic plane waves—either pressure or shear waves with vertical or horizontal polarization—at planar boundaries of nondissipative homogeneous isotropic half-spaces is an important analytically solvable canonical US-NDT problem that is extensively discussed in the respective chapter; moreover, it is an example of the decoupling of the two shear wave polarizations in two spatial dimensions. Using this opportunity, we will critically emblaze the term "sound pressure", even though it can be properly defined via the stress tensor, it is by no means a field quantity that satisfies boundary or transition conditions at jump discontinuities of material properties.

A "finely tuned" superposition of evanescent pressure and shear waves on the stress-free boundary of an elastic half-space yields Rayleigh surface waves as special solutions of the homogeneous wave equation.

We refer to the literature regarding modal propagation of horizontally polarized shear or Lamb waves in wave guides (Rose 1999).

Due to their infinitely extended phase and amplitude fronts, plane waves contain infinite elastodynamic energy; they are physically nonrealizable. Nevertheless, they are useful to model realistic sound fields in terms of spatial plane wave spectra, in particular, if one does not know Green's functions analytically, for example, in case of the nondissipative homogeneous isotropic half-space with stress-free surface: Only Green's functions spatial elastic plane wave spectra can be analytically derived to be evaluated with the method of stationary phase yielding the Miller–Pursey factors for far-field computations of piezoelectric transducer sound fields.

Application of the paraxial approximation to spatial plane wave spectra leads us to the mathematical representation of Gaussian beams—at first scalar—*ultrasound beams*. Yet, the generalization to pulsed Gaussian beams or Gaussian wave packets runs into problems: It is better to solve a parabolically approximated time domain wave equation exactly. Such beam solutions can also be found for weekly inhomogeneous and even anisotropic materials, and they replace no longer existent plane waves; being mathematically more complex, they give more physical insight.

Until now, no sources of elastic waves have been taken into account, the keyword "Green functions" brings them into play. Yet before we explicitly introduce them we refer to various wave equations for more complex materials. Nondissipative homogeneous anisotropic materials also allow for plane wave solutions, removing the degeneracy of coinciding shear wave eigenvalues: One finds three independent wave modes with different phase velocities. The pertinent polarizations are no longer longitudinal or transverse but quasilongitudinal and quasitransverse, forming an orthogonal trihedron with a uniquely defined orientation that is prescribed by the anisotropy under concern. Additionally, the physical property of pressure and shear waves is also lost: Quasilongitudinal plane waves are only quasipressure waves—for weakly anisotropic materials, they degrade into pressure waves—and quasitransverse plane waves are only quasishear waves that also degrade into shear waves for weak anisotropy. The most important consequences of anisotropy for US-NDT are the following ones: The phase velocities of the three wave modes depend upon the propagation direction of the phase and the direction of energy propagation—defined by the elastodynamic Poynting vector yielding the energy velocity vector—does no longer coincide with the direction of phase propagation, that is to say the energy velocity is no longer orthogonal to the phase front. A 45° shear wave transducer designed for isotropic—ferritic—steel radiates into a completely different direction in an anisotropic material! As a consequence, the magnitude and direction of the energy velocity is primary for the propagation velocity of an ultrasonic impulse. We discuss details of anisotropy consequences only for the simplest case, i.e., materials that are anisotropic in a direction orthogonal to an isotropy plane, the so-called transversely isotropic materials. These are approximately realized by austenitic steel and carbon fiber reinforced composites. We present results even for reflection, mode conversion, and transmission of plane waves at the planar boundary between isotropic and transversely isotropic half-spaces. Existence of physically possible wave modes is always verified with the energy velocity diagram, *not* with the slowness diagram as in the isotropic case; evanescence of inhomogeneous waves is also defined with respect to the energy velocity direction.

Wave equations for nondissipative *inhomogeneous* materials, either isotropic or anisotropic, exhibit an additional complexity as spatial derivatives—del operator calculations—have also to be applied to the material parameters; these are Lamé constants for isotropic materials and the stiffness tensor for anisotropic materials. The *ansatz* of plane waves is no longer working! Initially, one tries a generalization in terms of *locally* plane waves with nonlinear spatially dependent phase and amplitude. To avoid the laborious vector and tensor calculus for elastic waves, it is advisable to investigate first a similar *ansatz* for scalar acoustic waves in materials with a spatially varying sound velocity; if the material properties are only slowly changing within a wavelength, a differential equation for the phase—the eikonal equation—and a differential equation for the amplitude—the transport equation—can be derived. Solutions of the eikonal equation constitute the nonplanar surfaces of constant phase with the orthogonal phase vector that defines a ray trajectory. It is noteworthy that a Taylor expansion of the phase yields a beam propagating along the ray trajectory. Now one can move on to *elastic* wave rays and beams. There are parallels to plane waves in homogeneous materials: Inhomogeneous isotropic materials support the independent propagation of longitudinal and transverse beams along primary and secondary ray trajectories; inhomogeneous anisotropic materials require the addition "quasi" to the polarizations, and propagation occurs along the ray trajectory for the energy. The pertinent pulsed solutions are (Gaussian) wave packets.

The partial derivatives of the material parameters in the wave equation for inhomogeneous (an)isotropic materials "disappear" if the inhomogeneity has compact support, i.e., is restricted to a finite volume: In that case, all these terms can be collected on the right-hand side of the equation, where the prescribed (primary) sources reside anyway, thus defining secondary sources that replace the inhomogeneity, they are equivalent to it, hence the terminology *equivalent* sources is introduced. As a consequence, the field scattered by the inhomogeneity can be formally calculated in the same manner as the one for the primary sources. This solution is *formal* in the sense that the equivalent sources depend on their own scattered field, which is not explicitly known; therefore, the equivalent sources must first be calculated as solutions of integral equations.

The consideration of dissipation is achieved via the "design" of appropriate constitutive equations. Yet basic physical principles must not be violated; for instance, causality directly implies the frequency dependence of the material parameters and, hence, dispersion of pulsed waves. Surfaces of constant phase and amplitude of plane waves in homogeneous dissipative materials may coincide, accounting for an attenuation in propagation direction, or they may include an arbitrary angle not equal to 90°. These inhomogeneous plane waves are excited in dissipative half-spaces by plane waves under arbitrary angles of incidence: The attenuation is orthogonal to the half-space surface and *not* in the direction of propagation as it is true for homogeneous plane waves.

Up to now, we only considered idealized solutions of homogeneous wave equations, but we came already close to the description of radiated sound fields introducing the concept of ultrasonic beams. Yet, ultrasound must be excited; therefore, the mathematical dependence between prescribed sources—force densities and deformations rates—and their pertinent radiation field is required. (We always allow for *both* sources to prepare for a consistent derivation of Huygens' principle.) Again, the answer is in terms of idealized solutions of inhomogeneous wave equations: *Point*-sources are prescribed and their radiation field is calculated as so-called Green functions; based on the linearity of the elastodynamic governing equations, Green functions constitute a point source synthesis for the radiation field of spatially extended sources, i.e., extended sources are broken apart into point-sources and their respective fields are superimposed. Physically, Green functions of elastodynamics are nothing but elastodynamic elementary waves emanating from point-like force densities and deformation rates; because compact support inhomogeneities can be replaced by secondary or equivalent sources, it is anticipated that scattered fields can also be calculated utilizing point source synthesis; hence, it turns out that Green functions ultimately constitute the mathematical building blocks of two fundamental problems of US-NDT: the radiation and the scattering problem. This underlines their eminent importance, and therefore we discuss Green functions in thorough detail.

To calculate elastodynamic Green functions explicitly, the vector differential operators applied to the particle velocity in terms of wave equations have to be inverted in order to formally relate the particle velocity to the given sources on the right-hand side of the wave equations. This task is split into the inversion of the derivatives and the inversion of the vector operators. The latter are not present in the pressure wave equation for scalar acoustic waves, and therefore it might be wise to calculate and discuss the scalar Green function at first. As a matter of fact, this is tackled for time harmonic and pulsed point sources in two and three spatial dimensions, because an actual US problem can often be modeled two dimensionally. With the resulting scalar point source synthesis, we are ready to turn to elastodynamics: It is exposed that the scalar Green function is again the key concept, we "simply" have to bring the inverted vector operations into play and we have to account (in homogeneous isotropic materials) for the excitation of primary pressure and secondary shear waves emanating from point sources, that is to say, we need *two* scalar Green functions for elementary waves with different wave speeds. Resulting are tensor Green functions differing whether we want to calculate the particle velocity originating from a point force density or from a point deformation rate.

In the first case, a second rank Green tensor is required and in the second case a third rank Green tensor; both contain a pressure and a shear term. The different tensor operations on the primary and secondary scalar Green functions determine the spatially dependent amplitudes of the—inhomogeneous isotropic materials spherical—elementary waves, i.e., their far-field point characteristics. We explicitly point out that pressure elementary waves are only longitudinal in the far-field and the same is true for shear elementary transverse waves. If there is a request not only for the point source synthesis of the particle velocity from given sources but also for the stress tensor, the third rank Green tensor to calculate the contribution from force densities is needed once more, but additionally, a forth rank tensor representing the contribution from deformation rates has to be introduced. Based on the knowledge of the mathematical structure of elastodynamic elementary waves in terms of Green functions, we can now formulate the point source synthesis of primary volume sources; in homogeneous isotropic materials of infinite extent, the result is a volume integral extending over the sources multiplied by "matching" Green functions. These representations give rise to far-field approximations in two and three spatial dimensions defining elastodynamic radiation patterns. Surface sources are special cases of volume sources, and if they are residing in an infinite elastic space, they come already close to the US-NDT aperture radiator (close: because, in reality, the aperture—the piezoelectric transducer—sits on a stress-free surface); therefore, several examples of that kind will be discussed and, as always in this book, also for pulsed excitation.

Green functions represent physical wave fields satisfying a homogeneous wave equation in a half-space that does *not* contain the point source. We already mentioned that such wave fields can be decomposed into spatial spectra of plane waves yielding spatial plane wave spectra even for spherical elementary waves; mathematically spoken, Green functions have representations in terms of two-dimensional inverse Fourier integrals (Weyl integral representations). Those will be extremely useful to calculate sound fields in elastic half-spaces with planar stress-free surfaces, i.e., transducer radiation fields.

As a matter of fact, the generally applied radiating sources for US-NDT are aperture radiators residing on the surfaces of components being usually considered as stress free. Therefore, the sound fields of such transducers have to satisfy an appropriate boundary condition that is not inherent in our previous point source synthesis because it implies Green functions of infinite space; Green functions have to be found that are compatible with the boundary condition! For scalar acoustic waves, the solution is comparatively simple: The adequate Green functions, at least for planar surfaces, can be calculated imaging a point source at the surface. Due to mode conversion, this is not possible for elastic waves and, hence, explicit analytic expressions for the relevant Green tensors are not available: Only Weyl-type integral equations can be developed! Utilizing the method of stationary phase, these integral representations are evaluated in the far-field, yielding the Miller–Pursey point source characteristics; thus, an approximate point source synthesis can be constituted being also applicable to calculate the near-field of aperture radiators on stress-free surfaces. A fundamental task of US-NDT has been solved! What remains is the computation of fields scattered by finite volume inhomogeneities, i.e., defects in the widest sense; again, elementary waves described by Green functions prove to be essential to formulate Huygens' principle for scattered fields.

Interesting enough, a radiation or scattered field can even be calculated if the primary or secondary sources are not explicitly known; instead, the field on an arbitrary surface enclosing the sources is known. Christiaan Huygens has formulated his principle in the 17th century: Elementary waves—spherical waves—emanate form each point on such a surface being weighted with the pertinent amplitude; the field *outside* the closed surface is composed by the envelope of all elementary waves. In addition, Huygens claims that the elementary waves superimpose to a null field in the *interior*. We state that: Elementary waves are given by Green functions, and, therefore, a wave equation-based theory must exist to derive Huygens' principle mathematically. This theory is advantageously shaped for scalar acoustic waves first, inspiring physical meaning to Green's second formula: Outside the closed surface containing the sources in the interior, the field is found as an integral over the surface, extending the "principle" of Huygens in the sense that not only isotropic spherical waves but also dipole waves—elementary waves with a dipole radiation characteristics—have to be accounted for. For scalar acoustic waves, the first ones are weighted with the normal derivative of the pressure and the second ones with the pressure itself. Surprisingly, this integral yields zero values in the interior of the surface as has been heuristically claimed by Huygens. Huygens' principle initially constitutes an equivalence principle: The surface integration over field values is equivalent to the integration over sources. Yet, its real value is getting obvious if scattered fields have to be computed; if the Huygens-surface encloses a scatterer where the surface field has to satisfy certain boundary conditions—the scatterer surface may be sound soft or rigid—the Huygens-surface is contracted to the scatterer surface, inserting explicitly the boundary conditions and thus canceling either the pressure or its normal derivative in the integral. The remaining term can then be considered as an equivalent (secondary) source of the scattered field. Yet, as it is true for equivalent volume sources, this equivalent surface source depends upon the (scattered) field itself, requiring its calculation at first. This is achieved if the Huygens integral *representation* is again subject to the boundary condition, resulting in an integral *equation* for the equivalent source. Having solved it, the field can be calculated with the Huygens integral anywhere outside the scatterer. The inherent surface integral equation in Huygens' principle is obviously prestage to a point-source synthesis of scattered fields, satisfying boundary conditions on scatterers.

Now we take considerable advantage from having consequently considered *both* sources of elastodynamic fields—forces and deformation rates—because they appear simultaneously as field-dependent equivalent surface sources multiplied by the pertinent Green tensors in the elastodynamic version of Huygens' principle, revealing that these tensors represent the elementary waves of elastodynamics. Modeling crack scattering for US-NDT purposes often allows us to postulate stress-free surfaces, canceling the equivalent source "surface force density" and leaving us with the surface deformation rate as a source of the elementary wave related to the third rank Green tensor. For this remaining equivalent source, we obtain surface integral equations. Note: The radiation field of a piezoelectric transducer is modeled through specification of the primary surface source force density on a stress-free surface whereas the field scattered by an inhomogeneity of finite volume with a stress-free surface is modeled through calculation of Huygens-type equivalent surface deformation rates, which means that radiation and scattered fields are composed of completely different elementary waves; for radiation fields, it is the second rank Green tensor, and for scattered fields, it is the third rank Green tensor!

As an example, we derive the surface integral equation for a two-dimensional crack model and present results of a numerical solution.

To reduce the numerical cost for calculation of fields scattered by arbitrary geometries (with stress-free surfaces), we discuss a widely spread approximation for determination of equivalent sources, namely, the Kirchhoff approximation originating from electromagnetics. The scatterer surface is subdivided into patches, and the pertinent secondary deformation rate source is obtained from reflection and mode conversion of plane waves at planar boundaries. Based explicitly on this Kirchhoff approximation, we formulate a standard system model for US-NDT: transmitting transducer, scatterer, and receiving transducer.

In case the finite inhomogeneity is neither a crack nor a void but an inclusion—potentially inhomogeneous anisotropic—it must be modeled as a penetrable scatterer. Looking at our flow chart, we discover that equivalent *volume* sources serve this purpose. Even though they are field dependent similar to the Huygens' surface equivalent sources they can, nevertheless, be inserted into a volume source integral, yielding a data equation likewise to the Huygens' integral. In a second step, the equivalent sources must be calculated; the volume integral is written down for observation points in the interior of the scatterer, resulting in an object equation, the so-called Lippmann–Schwinger integral equation—a volume integral equation—complementing the surface integral equations for stress-free scatterers. The object equation must generally be solved numerically; yet, an approximation is also at hand, the so-called Born approximation: The initially unknown field in the interior of the scatterer is replaced by the known incident field being certainly permissible for weak scatterers. Even for penetrable scatterers embedded in inhomogeneous isotopic materials, a point source synthesis to calculate scattered fields can be derived.

A limited number of canonical scattering geometries (with stress-free surfaces)—cylinder and sphere—allow for an analytical solution of the underlying surface integral equation utilizing a matching coordinate system and solving the wave equation in terms of eigenfunctions—cylindrical and spherical functions. We will carry out this solution and discuss numerical results. Such analytical solutions for canonical problems—implying scattering by a wedge—can be utilized to calculate scattered fields by superposition of the fields coming from an ensemble of characteristic scattering centers into which the scatterer has been decomposed; this is often possible for high frequencies, the pertinent technique is called geometric theory of diffraction (GTD) that is well documented in the literature (Achenbach et al. 1982).

It remains to discuss the intrinsic problem of US-NDT: imaging material defects. The synthetic aperture focusing technique (SAFT) is established as a solution using heuristic arguments that can nevertheless be embedded into a thorough inverse scattering theory yielding simultaneously effective algorithmic alternatives—FT-SAFT applying mostly Fourier transforms. Again, Green functions turn out to be the basic principle!

It is obvious that plane waves and Green functions provide the theoretical fundament for US-NDT.

2

Mathematical Foundations

2.1 Scalar, Vector, and Tensor Fields

2.1.1 Vector of position

To characterize a specific point in space, for example, on the surface of a specimen, we necessitate coordinates; the simplest ones are Cartesian coordinates "length, width, height" being denoted by x, y, z or x_1, x_2, x_3 (x_i, $i = 1, 2, 3$), respectively. Figure 2.1 shows a (right-handed[1]) Cartesian coordinate system with the particular coordinates x_0, y_0, z_0 of a spatial point $P_0(x_0, y_0, z_0)$. The location of that point is known if the three figures x_0, y_0, and z_0 are known under the assumption of an arbitrary but fixed coordinate origin and the arbitrary but fixed orientation of the coordinate axes. Figure 2.1 also displays that P_0 can be equally characterized by the knowledge of cylindrical r_0, φ_0, z_0 or spherical coordinates $R_0, \vartheta_0, \varphi_0$. The following coordinate transforms are immediately obvious:

$$\begin{aligned} x_0 &= r_0 \cos \varphi_0, \\ y_0 &= r_0 \sin \varphi_0, \\ z_0 &= z_0; \end{aligned} \tag{2.1}$$

$$\begin{aligned} x_0 &= R_0 \sin \vartheta_0 \cos \varphi_0, \\ y_0 &= R_0 \sin \vartheta_0 \sin \varphi_0, \\ z_0 &= R_0 \cos \vartheta_0. \end{aligned} \tag{2.2}$$

Given the coordinate origin O—compare Figure 2.2—we can equally specify the location of P_0 through the direction and length of the so-called vector of position[2] $\underline{\mathbf{R}}$. This is graphically descriptive, yet the question arises how to characterize $\underline{\mathbf{R}}$ mathematically. We consider Figure 2.3, where the directions

[1]Heinrich Hertz (1890) writes that: We assume that the coordinate system of the x, y, z is of the kind that, if the direction of the positive x is towards you and the direction of the positive z is upward, then the y grow from left to right.

[2]To distinguish them from scalars, vectors are denoted by fat characters with a single underline; that way, we have the possibility to denote tensors of second and higher rank by underlining fat characters according to the tensor rank; consequently, the second rank deformation tensor reads $\underline{\underline{\mathbf{S}}}$ and the forth rank stiffness tensor $\underline{\underline{\underline{\underline{\mathbf{c}}}}}$.

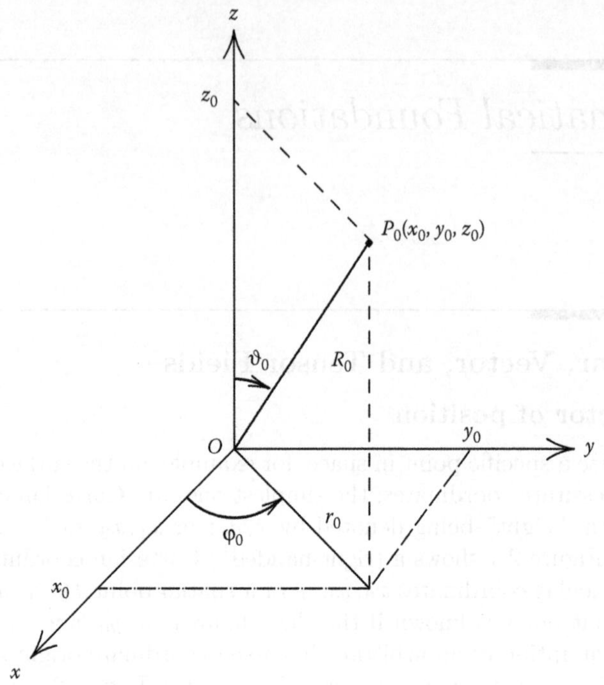

FIGURE 2.1
Cartesian coordinates; cylinder and spherical coordinates.

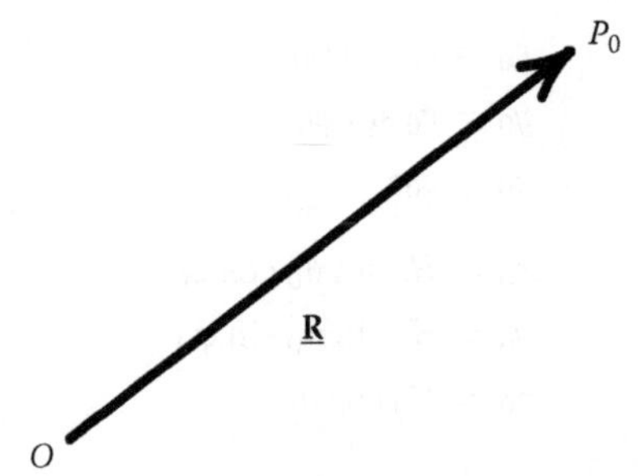

FIGURE 2.2
Vector of position $\underline{\mathbf{R}}$.

of Cartesian coordinate axes are given by three orthogonal unit vectors[3] $\underline{\mathbf{e}}_x$, $\underline{\mathbf{e}}_y$, $\underline{\mathbf{e}}_z$; per definition, a unit vector has length one. This system of three unit vectors is called an orthonormal trihedron. By drawing the vector of position $\underline{\mathbf{R}}$

[3]Except for some standard unit vectors (e.g., $\underline{\mathbf{e}}_x, \underline{\mathbf{e}}_y, \underline{\mathbf{e}}_z, \underline{\mathbf{n}}$), we characterize them by a hat, hence, for example, $\hat{\underline{\mathbf{R}}} = \underline{\mathbf{R}}/R$.

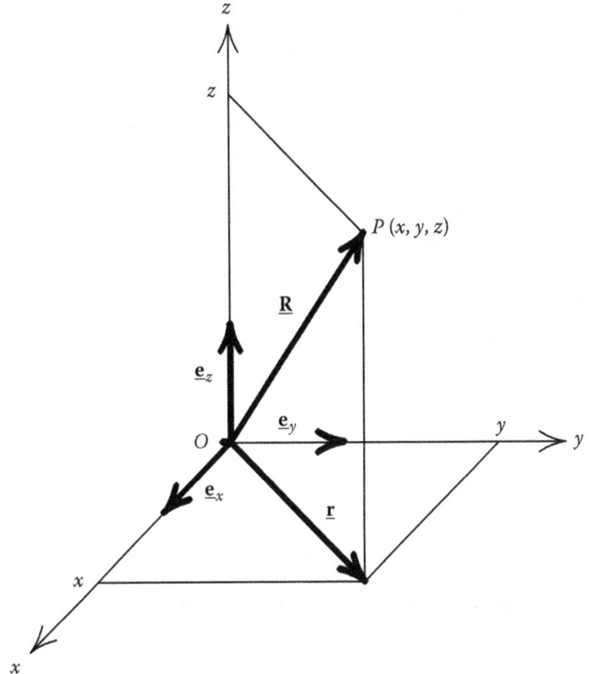

FIGURE 2.3
Vector of position $\underline{\mathbf{R}}$ in Cartesian coordinates.

to the point $P(x, y, z)$ with coordinates x, y, z, we immediately recognize that the projections of $\underline{\mathbf{R}}$ on the pertinent coordinate axes are equal to the coordinates of the point: The (scalar) components of the vector of position are *coordinates* of that point which it characterizes. Now we construct the three vectors $x\underline{\mathbf{e}}_x$, $y\underline{\mathbf{e}}_y$, and $z\underline{\mathbf{e}}_z$; they are directed as the orthonormal trihedron; therefore, they are equally orthogonal but no longer normalized to one, their lengths being[4] x, y, and z. Defining the addition of two vectors $\underline{\mathbf{R}}_1$ and $\underline{\mathbf{R}}_2$ as in Figure 2.4, we find that the vector $\underline{\mathbf{r}}$ as depicted in Figure 2.3 is obviously given as

$$\underline{\mathbf{r}} = x\underline{\mathbf{e}}_x + y\underline{\mathbf{e}}_y \tag{2.3}$$

and, hence, the vector of position $\underline{\mathbf{R}}$ as

$$\begin{aligned}\underline{\mathbf{R}} &= \underline{\mathbf{r}} + z\underline{\mathbf{e}}_z \\ &= x\underline{\mathbf{e}}_x + y\underline{\mathbf{e}}_y + z\underline{\mathbf{e}}_z;\end{aligned} \tag{2.4}$$

[4]We implicitly assumed that x, y, z are greater than zero; for example, if we had $x < 0$, the length of $x\underline{\mathbf{e}}_x$ would be $|x|$.

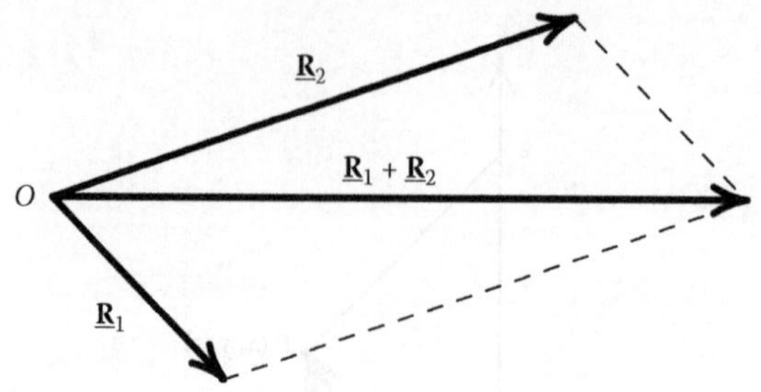

FIGURE 2.4
Addition of two vectors.

x, y, z are scalar components and $x\underline{\mathbf{e}}_x, y\underline{\mathbf{e}}_y, z\underline{\mathbf{e}}_z$ are vector components of $\underline{\mathbf{R}}$. Talking subsequently about "components," we always mean scalar components. According to Pythagoras' theorem, we obtain the length r—the magnitude of $r = |\underline{\mathbf{r}}|$—according to

$$r = \sqrt{x^2 + y^2} \tag{2.5}$$

and the length R—the magnitude $R = |\underline{\mathbf{R}}|$—of the vector of position $\underline{\mathbf{R}}$ according to

$$\begin{aligned} R &= \sqrt{r^2 + z^2} \\ &= \sqrt{x^2 + y^2 + z^2}. \end{aligned} \tag{2.6}$$

The length—the magnitude—of a vector is always denoted by the same character, yet not fat. The magnitude $|\underline{\mathbf{R}}|$ of the vector of position for the point P is obviously identical with the radial spherical coordinate R of P.

The theory of elastic waves often requires to distinguish between *two* vectors of position $\underline{\mathbf{R}}$ and $\underline{\mathbf{R}}'$ (Figure 2.5); $\underline{\mathbf{R}}'$, the vector of position for the source point Q (also denoted by $\underline{\mathbf{R}}_Q$), varies in a source volume where the forces and deformation rates radiating elastic waves are nonzero, and $\underline{\mathbf{R}}$, the vector of position for the observation point P (also denoted by $\underline{\mathbf{R}}_P$), is that point where the pertinent elastic wave is currently observed. In homogeneous isotropic materials, the Green function relating the source density at $\underline{\mathbf{R}}'$ and the particle velocity at $\underline{\mathbf{R}}$ depends only on the distance between $\underline{\mathbf{R}}$ and $\underline{\mathbf{R}}'$. Allocating

$$\underline{\mathbf{R}}' = x'\underline{\mathbf{e}}_x + y'\underline{\mathbf{e}}_y + z'\underline{\mathbf{e}}_z \tag{2.7}$$

with x', y', z' as coordinates to the source point $\underline{\mathbf{R}}'$, we obtain

$$\underline{\mathbf{R}} - \underline{\mathbf{R}}' = (x - x')\underline{\mathbf{e}}_x + (y - y')\underline{\mathbf{e}}_y + (z - z')\underline{\mathbf{e}}_z \tag{2.8}$$

and consequently applying Pythagoras' theorem

$$|\underline{\mathbf{R}} - \underline{\mathbf{R}}'| = \sqrt{(x - x')^2 + (y - y')^2 + (z - z')^2} \tag{2.9}$$

analogous to (2.6).

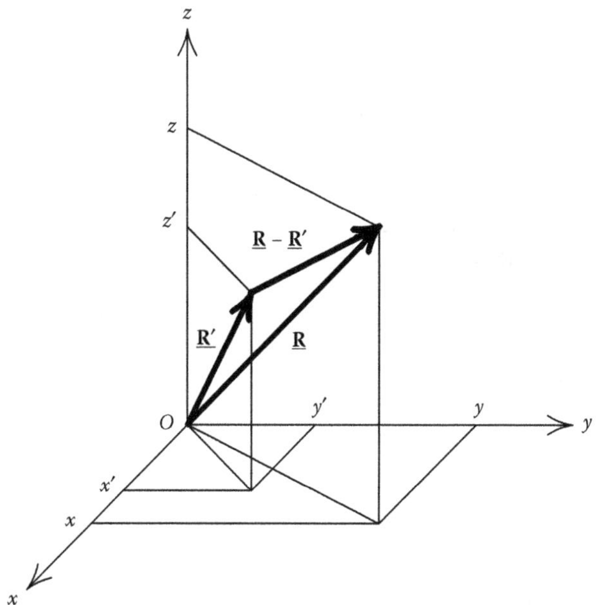

FIGURE 2.5
Distance between observation point $\underline{\mathbf{R}}$ and source point $\underline{\mathbf{R}}'$.

2.1.2 Scalar and vector fields

In Figure 2.6, a measurement point $P_M(\underline{\mathbf{R}}) = P_M(x, y, z)$ on the surface of a specimen is given by the pertinent vector of position $\underline{\mathbf{R}}$; Figure 2.7 sketches an A-scan, for instance, the sound pressure[5] $p(\underline{\mathbf{R}}, t)$ measured at P_M as a function of time t. We consider $p(\underline{\mathbf{R}}, t)$ as a scalar field quantity and consulting (2.4), we know that this function depends on the three spatial coordinates and time, it is a function of *four* variables. The detailed notation would be $p(x, y, z, t)$, yet in short-hand, we write $p(\underline{\mathbf{R}}, t)$.

An elastic wave in a solid primarily consists of displacements of infinitesimally small volume elements, the so-called displacement $\underline{\mathbf{u}}(\underline{\mathbf{R}}, t)$ at the point $\underline{\mathbf{R}}$ and time t; it is by nature a vector field quantity because the volume elements are displaced in terms of direction and magnitude. Figure 2.8 illustrates such a displacement. To define the (scalar) Cartesian components u_x, u_y, u_z of $\underline{\mathbf{u}}$, we draw a displaced (dashed) coordinate system with origin at the position vector $\underline{\mathbf{R}}$ and project $\underline{\mathbf{u}}$ on the respective coordinate axes. The resulting

[5]Section 9.1.1 reveals that a something like a sound pressure in a solid with $\mu \neq 0$ (λ, μ: Lamé constants) can only properly be defined for plane waves, hence, strictly speaking, it cannot be measured.

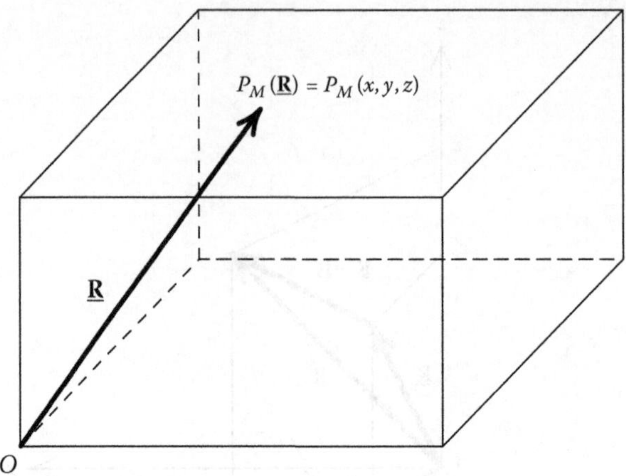

FIGURE 2.6
Measurement point on the surface of a specimen.

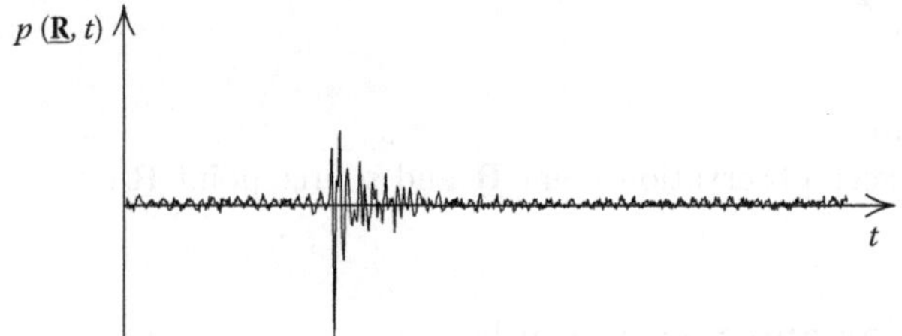

FIGURE 2.7
Measured sound pressure A-scan.

(Cartesian) component representation reads similar to (2.4):

$$\begin{aligned}\underline{\mathbf{u}}(\underline{\mathbf{R}}, t) &= \underline{\mathbf{u}}(x, y, z, t)\\ &= u_x(\underline{\mathbf{R}}, t)\,\underline{\mathbf{e}}_x + u_y(\underline{\mathbf{R}}, t)\,\underline{\mathbf{e}}_y + u_z(\underline{\mathbf{R}}, t)\,\underline{\mathbf{e}}_z\\ &= u_x(x, y, z, t)\,\underline{\mathbf{e}}_x + u_y(x, y, z, t)\,\underline{\mathbf{e}}_y + u_z(x, y, z, t)\,\underline{\mathbf{e}}_z.\end{aligned} \tag{2.10}$$

We abide by the following: Each (cartesian) component of the vector displacement depends upon each (cartesian) coordinate. It is this property of vector fields that requires the definition of certain differential operators—gradient, divergence, and curl—to calculate physically meaningful spatial variations of fields (Section 2.2).

The magnitude $u(\underline{\mathbf{R}}, t)$ of $\underline{\mathbf{u}}(\underline{\mathbf{R}}, t)$ is obtained similar to (2.6):

$$u(\underline{\mathbf{R}}, t) = \sqrt{u_x^2(\underline{\mathbf{R}}, t) + u_y^2(\underline{\mathbf{R}}, t) + u_z^2(\underline{\mathbf{R}}, t)}. \tag{2.11}$$

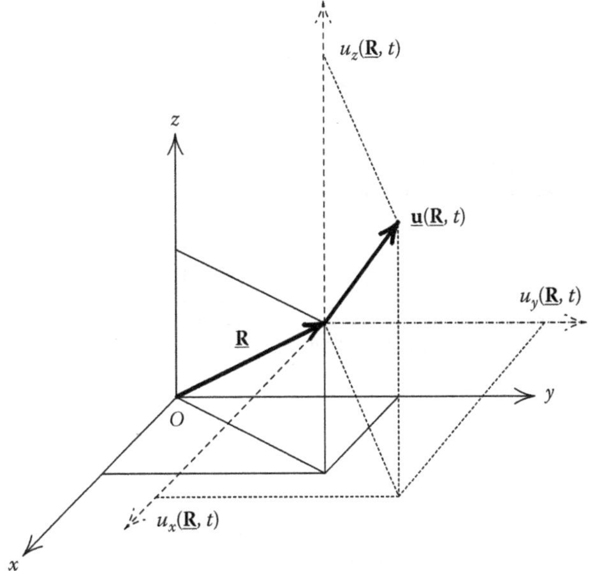

FIGURE 2.8
Particle displacement vector.

2.1.3 Vector products

We distinguish three different product of vectors named according to the respective result:

- Scalar product
- Vector product
- Dyadic product.

Scalar product: The scalar (dot) product $\underline{\mathbf{A}} \cdot \underline{\mathbf{B}}$ of two vectors $\underline{\mathbf{A}}$ and $\underline{\mathbf{B}}$ is denoted by a dot and it can be intuitively illustrated. Figure 2.9 depicts a vector $\underline{\mathbf{A}}$ being projected onto a unit vector $\underline{\hat{\mathbf{e}}}$ with the result

$$\underline{\mathbf{A}} \cdot \underline{\hat{\mathbf{e}}} = A \cos \phi, \tag{2.12}$$

if ϕ is the angle between $\underline{\mathbf{A}}$ and $\underline{\hat{\mathbf{e}}}$. Replacing $\underline{\hat{\mathbf{e}}}$ by a vector $\underline{\mathbf{B}}$ with magnitude B, the generalization of (2.12) reads as

$$\begin{aligned} \underline{\mathbf{A}} \cdot \underline{\mathbf{B}} &= \underline{\mathbf{B}} \cdot \underline{\mathbf{A}} \\ &= AB \cos \phi \end{aligned} \tag{2.13}$$

and defines the (commutative) scalar product $\underline{\mathbf{A}} \cdot \underline{\mathbf{B}}$.

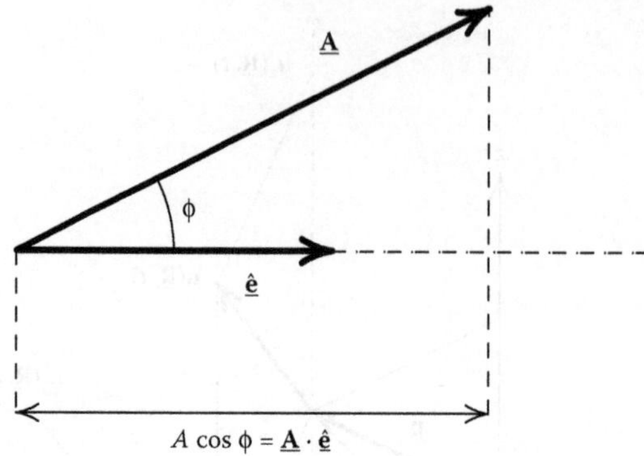

FIGURE 2.9
Illustration of the scalar product.

We obviously have $\underline{\mathbf{A}} \cdot \underline{\mathbf{B}} = 0$ if $\underline{\mathbf{A}}$ and $\underline{\mathbf{B}}$ are orthogonal to each other; consequently, the orthogonality of two vectors is guaranteed finding the value zero of their scalar product.

The orthonormal trihedron of cartesian coordinates has the property:

$$\begin{aligned} \underline{\mathbf{e}}_x \cdot \underline{\mathbf{e}}_y &= 0, \\ \underline{\mathbf{e}}_x \cdot \underline{\mathbf{e}}_z &= 0, \\ \underline{\mathbf{e}}_y \cdot \underline{\mathbf{e}}_z &= 0; \end{aligned} \tag{2.14}$$

$$\begin{aligned} \underline{\mathbf{e}}_x \cdot \underline{\mathbf{e}}_x &= 1, \\ \underline{\mathbf{e}}_y \cdot \underline{\mathbf{e}}_y &= 1, \\ \underline{\mathbf{e}}_z \cdot \underline{\mathbf{e}}_z &= 1. \end{aligned} \tag{2.15}$$

Numbering cartesian coordinates according to x_i, $i = 1, 2, 3$, with the trihedron $\underline{\mathbf{e}}_{x_i}$, $i = 1, 2, 3$, and utilizing the Kronecker symbol

$$\delta_{ij} = \begin{cases} 1 & \text{for } i = j \\ 0 & \text{for } i \neq j, \end{cases} \tag{2.16}$$

we can write the six equations of (2.14) and (2.15) as a single equation:

$$\underline{\mathbf{e}}_{x_i} \cdot \underline{\mathbf{e}}_{x_j} = \delta_{ij} \text{ for } i, j = 1, 2, 3. \tag{2.17}$$

The scalar product is useful to calculate the components of a vector $\underline{\mathbf{A}}$, for example, in Cartesian coordinates; with (2.12), it follows per definition

$$\begin{aligned} A_x &= \underline{\mathbf{A}} \cdot \underline{\mathbf{e}}_x, \\ A_y &= \underline{\mathbf{A}} \cdot \underline{\mathbf{e}}_y, \\ A_z &= \underline{\mathbf{A}} \cdot \underline{\mathbf{e}}_z. \end{aligned} \tag{2.18}$$

Now we calculate

$$\underline{\mathbf{A}} \cdot \underline{\mathbf{B}} = (A_x\, \underline{\mathbf{e}}_x + A_y\, \underline{\mathbf{e}}_y + A_z\, \underline{\mathbf{e}}_z) \cdot (B_x\, \underline{\mathbf{e}}_x + B_y\, \underline{\mathbf{e}}_y + B_z\, \underline{\mathbf{e}}_z) \tag{2.19}$$

with the (Cartesian) component representation of $\underline{\mathbf{A}}$ and $\underline{\mathbf{B}}$ and formally find by distributive multiplication and utilization of (2.17) observing the commutative property of the scalar product

$$\underline{\mathbf{A}} \cdot \underline{\mathbf{B}} = A_x B_x + A_y B_y + A_z B_z. \tag{2.20}$$

That way, we have the possibility to find the value of the scalar product if Cartesian *components* of the respective vectors are given. Similarly, the angle between two vectors with nonzero magnitudes is obtained as

$$\begin{aligned} \cos\phi &= \frac{\underline{\mathbf{A}} \cdot \underline{\mathbf{B}}}{AB} \\ &= \frac{A_x B_x + A_y B_y + A_z B_z}{\sqrt{A_x^2 + A_y^2 + A_z^2}\sqrt{B_x^2 + B_y^2 + B_z^2}}. \end{aligned} \tag{2.21}$$

The square root of the scalar product $\underline{\mathbf{A}} \cdot \underline{\mathbf{A}}$ obviously yields the magnitude of $\underline{\mathbf{A}}$:

$$\begin{aligned} A &= \sqrt{\underline{\mathbf{A}} \cdot \underline{\mathbf{A}}} \\ &= \sqrt{A_x^2 + A_y^2 + A_z^2}\,; \end{aligned} \tag{2.22}$$

in addition, we obtain

$$\begin{aligned} \hat{\underline{\mathbf{A}}} &= \frac{\underline{\mathbf{A}}}{A} \\ &= \frac{\underline{\mathbf{A}}}{\sqrt{\underline{\mathbf{A}} \cdot \underline{\mathbf{A}}}} \\ &= \frac{A_x}{A}\,\underline{\mathbf{e}}_x + \frac{A_y}{A}\,\underline{\mathbf{e}}_y + \frac{A_z}{A}\,\underline{\mathbf{e}}_z \end{aligned} \tag{2.23}$$

as the unit vector $\hat{\underline{\mathbf{A}}}$ in the direction of $\underline{\mathbf{A}}$. If applied to the vector of position, (2.23) provides

$$\hat{\underline{\mathbf{R}}} = \frac{x}{\sqrt{x^2 + y^2 + z^2}}\,\underline{\mathbf{e}}_x + \frac{y}{\sqrt{x^2 + y^2 + z^2}}\,\underline{\mathbf{e}}_y + \frac{z}{\sqrt{x^2 + y^2 + z^2}}\,\underline{\mathbf{e}}_z. \tag{2.24}$$

We quote another two—abbreviated—notations for the scalar product. The serially numbered version of (2.18)

$$A_{x_i} = \underline{\mathbf{A}} \cdot \underline{\mathbf{e}}_{x_i} \text{ for } i = 1, 2, 3 \tag{2.25}$$

and equally for $\underline{\mathbf{B}}$ results in

$$\underline{\mathbf{A}} \cdot \underline{\mathbf{B}} = \sum_{i=1}^{3} A_{x_i} B_{x_i} \tag{2.26}$$

instead of (2.19). If we agree that the x_i—as in this case—are cartesian coordinates, we can continue, according to $A_{x_i} \Longrightarrow A_i$, $B_{x_i} \Longrightarrow B_i$, abbreviating (2.26):

$$\underline{\mathbf{A}} \cdot \underline{\mathbf{B}} = \sum_{i=1}^{3} A_i B_i. \tag{2.27}$$

Einstein's summation convention goes even further omitting the summation sign in (2.27):

$$\underline{\mathbf{A}} \cdot \underline{\mathbf{B}} = A_i B_i. \tag{2.28}$$

Equation 2.28 is translated as: If an index on one side of an equation—in this case i—appears at least twice and is not found on the other side, a summation from $i = 1$ to $i = 3$ is understood, the index is contracted;[6] if the index also appears on the other side, it is not contracted. This summation convention is extensively applied in the literature on elastodynamics (e.g., Achenbach 1973; de Hoop 1995); nevertheless, we generally prefer the coordinate-free representation $\underline{\mathbf{A}} \cdot \underline{\mathbf{B}}$ instead of (2.28), because it is much more practical for analytical derivations; yet, in case numbers are requested as a result of a physical problem, one must rely on coordinates.

Once again, we consider a specimen as in Figure 2.6 and imagine that a point-like piezoelectric "transducer" at the measurement point $P_M(\underline{\mathbf{R}})$ exclusively measures the component of the particle displacement $\underline{\mathbf{u}}(\underline{\mathbf{R}}, t)$ normal to the surface (Figure 2.10). To characterize this "normal component" $u_n(\underline{\mathbf{R}}, t)$

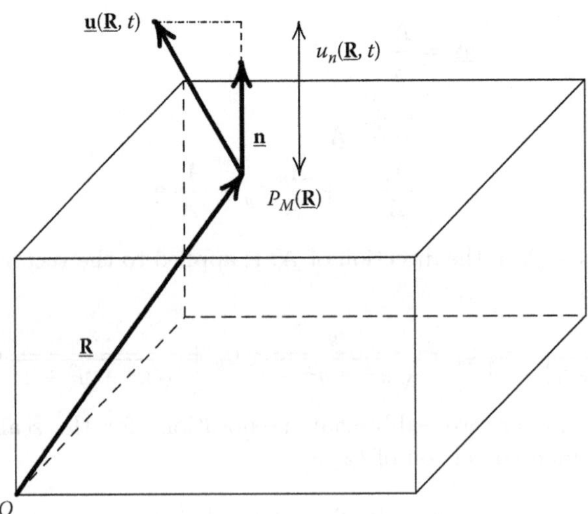

FIGURE 2.10
Normal component of the particle displacement.

[6]Therefore, a dot product (scalar product) $\underline{\mathbf{A}} \cdot \underline{\mathbf{B}}$ implies contraction of adjacent indices of the scalar components of the vectors in the immediate neighborhood of the dot.

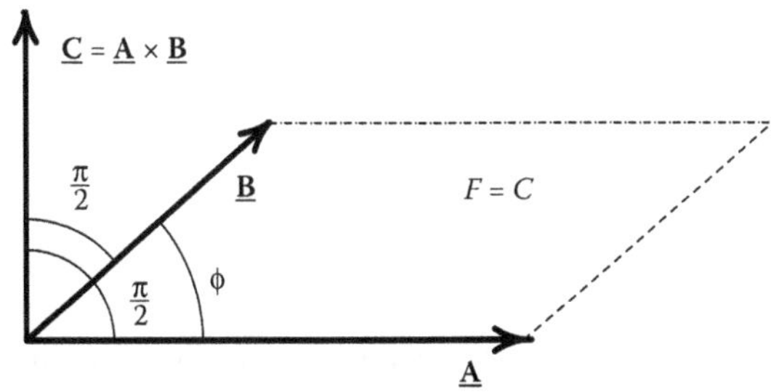

FIGURE 2.11
Definition of the vector product.

mathematically, we define a unit vector $\underline{\mathbf{n}}$ being orthogonal to the surface of the specimen.[7] Per definition, we have

$$u_n(\underline{\mathbf{R}}, t) = \underline{\mathbf{u}}(\underline{\mathbf{R}}, t) \cdot \underline{\mathbf{n}} = \underline{\mathbf{n}} \cdot \underline{\mathbf{u}}(\underline{\mathbf{R}}, t). \tag{2.29}$$

Being difficult to simultaneously measure the tangential components of $\underline{\mathbf{u}}(\underline{\mathbf{R}}, t)$, the normal component $u_n(\underline{\mathbf{R}}, t)$ for "all" points $\underline{\mathbf{R}}$ on a measurement surface S_M and all times t is generally the maximum obtainable information in US-NDT. In connection with imaging methods, we will learn how to process it.

Vector product: The definition of the vector product $\underline{\mathbf{A}} \times \underline{\mathbf{B}}$—that is, $\underline{\mathbf{A}}$ cross $\underline{\mathbf{B}}$—is illustrated in Figure 2.11. Two vectors $\underline{\mathbf{A}}$ and $\underline{\mathbf{B}}$ span a rhomboid with the area

$$F = AB \sin \phi; \tag{2.30}$$

the vector $\underline{\mathbf{C}}$ with magnitude F being right-handed[8] orthogonal to the rhomboid area is called the vector product

$$\underline{\mathbf{C}} = \underline{\mathbf{A}} \times \underline{\mathbf{B}} \tag{2.31}$$

of $\underline{\mathbf{A}}$ and $\underline{\mathbf{B}}$. Because of its definition implying right-handedness, the vector product is not commutative; we rather have[9]

$$\underline{\mathbf{B}} \times \underline{\mathbf{A}} = -\underline{\mathbf{A}} \times \underline{\mathbf{B}}. \tag{2.32}$$

---

[7]This unit vector multiply appears with the same meaning, hence, the hat is omitted. To calculate it, the surface must be suitably parameterized.

[8]The orthogonality of $\underline{\mathbf{C}}$ to the rhomboid area only defines the shaft of the arrow representing $\underline{\mathbf{C}}$. With regard to the tip, there is the choice "upward" or "downward." The *arbitrary* decision is "up" specified by the right-hand rule: If the cranked fingers of the right hand point from $\underline{\mathbf{A}}$ to $\underline{\mathbf{B}}$, the vector product $\underline{\mathbf{C}} = \underline{\mathbf{A}} \times \underline{\mathbf{B}}$ should point into the direction of the thumb of the right hand. Because of this choice, the vector product yields a so-called axial or pseudo-vector.

[9]Compare Footnote 8.

Obviously, two vectors are parallel or antiparallel if their cross product vanishes. It follows:

$$\begin{aligned} \underline{\mathbf{e}}_x \times \underline{\mathbf{e}}_x &= \underline{\mathbf{0}}, \\ \underline{\mathbf{e}}_y \times \underline{\mathbf{e}}_y &= \underline{\mathbf{0}}, \\ \underline{\mathbf{e}}_z \times \underline{\mathbf{e}}_z &= \underline{\mathbf{0}}. \end{aligned} \tag{2.33}$$

The symbol $\underline{\mathbf{0}}$ denotes the null vector, that is to say a vector with zero cartesian scalar components. We immediately verify

$$\begin{aligned} \underline{\mathbf{e}}_y \times \underline{\mathbf{e}}_z &= \underline{\mathbf{e}}_x, \\ \underline{\mathbf{e}}_x \times \underline{\mathbf{e}}_y &= \underline{\mathbf{e}}_z, \\ \underline{\mathbf{e}}_z \times \underline{\mathbf{e}}_x &= \underline{\mathbf{e}}_y. \end{aligned} \tag{2.34}$$

Distributive multiplication of the component representations of $\underline{\mathbf{A}}$ and $\underline{\mathbf{B}}$ utilizing (2.33) and (2.34) yields

$$\begin{aligned} \underline{\mathbf{C}} &= (A_x\,\underline{\mathbf{e}}_x + A_y\,\underline{\mathbf{e}}_y + A_z\,\underline{\mathbf{e}}_z) \times (B_x\,\underline{\mathbf{e}}_x + B_y\,\underline{\mathbf{e}}_y + B_z\,\underline{\mathbf{e}}_z) \\ &= (A_yB_z - A_zB_y)\,\underline{\mathbf{e}}_x + (A_zB_x - A_xB_z)\,\underline{\mathbf{e}}_y + (A_xB_y - A_yB_x)\,\underline{\mathbf{e}}_z \end{aligned} \tag{2.35}$$

for the components of $\underline{\mathbf{C}}$.

Orthogonality of the cross product to its vector factors has as a consequence

$$\begin{aligned} \underline{\mathbf{A}} \cdot (\underline{\mathbf{A}} \times \underline{\mathbf{B}}) &= 0, \\ \underline{\mathbf{B}} \cdot (\underline{\mathbf{A}} \times \underline{\mathbf{B}}) &= 0. \end{aligned} \tag{2.36}$$

The product

$$\underline{\mathbf{A}} \cdot (\underline{\mathbf{B}} \times \underline{\mathbf{C}}) = \underline{\mathbf{C}} \cdot (\underline{\mathbf{A}} \times \underline{\mathbf{B}}) = \underline{\mathbf{B}} \cdot (\underline{\mathbf{C}} \times \underline{\mathbf{A}}) \tag{2.37}$$

is nothing but the volume of the parallelepiped spanned by $\underline{\mathbf{A}}, \underline{\mathbf{B}}, \underline{\mathbf{C}}$.

The relation

$$\underline{\mathbf{n}} \times \underline{\mathbf{u}}(\underline{\mathbf{R}}, t) = \underline{\mathbf{u}}_{\tan}(\underline{\mathbf{R}}, t) \tag{2.38}$$

defines the vector of the particle displacement tangential to the surface being characterized by the normal vector $\underline{\mathbf{n}}$, i.e., its tangential "component." For instance, electromagnetic ultrasonic transducers or laser vibrometers are able to measure this particular component. Note: The vector tangential component $\underline{\mathbf{u}}_{\tan}(\underline{\mathbf{R}}, t)$ is orthogonal to $\underline{\mathbf{n}}$ and $\underline{\mathbf{u}}(\underline{\mathbf{R}}, t)$, it is not in the plane spanned by $\underline{\mathbf{n}}$ and $\underline{\mathbf{u}}(\underline{\mathbf{R}}, t)$ as it is true for the vector tangential component $\underline{\mathbf{u}}_t(\underline{\mathbf{R}}, t)$ (Equation 2.97; Figure 2.12).

Dyadic product: Now, we define a dyadic product of two vectors where the intuitive interpretation only follows after its definition and application, hence we proceed formally and put two vectors adjacent to each other without dot or cross in terms of their cartesian component representation:

$$\underline{\mathbf{A}}\,\underline{\mathbf{B}} = (A_x\,\underline{\mathbf{e}}_x + A_y\,\underline{\mathbf{e}}_y + A_z\,\underline{\mathbf{e}}_z)(B_x\,\underline{\mathbf{e}}_x + B_y\,\underline{\mathbf{e}}_y + B_z\,\underline{\mathbf{e}}_z). \tag{2.39}$$

Distributive multiplication produces the pertinent dyadic products of the unit vectors:

$$\begin{aligned}\underline{\mathbf{A}}\,\underline{\mathbf{B}} &= A_xB_x\,\underline{\mathbf{e}}_x\underline{\mathbf{e}}_x + A_xB_y\,\underline{\mathbf{e}}_x\underline{\mathbf{e}}_y + A_xB_z\,\underline{\mathbf{e}}_x\underline{\mathbf{e}}_z\\ &\quad + A_yB_x\,\underline{\mathbf{e}}_y\underline{\mathbf{e}}_x + A_yB_y\,\underline{\mathbf{e}}_y\underline{\mathbf{e}}_y + A_yB_z\,\underline{\mathbf{e}}_y\underline{\mathbf{e}}_z\\ &\quad + A_zB_x\,\underline{\mathbf{e}}_z\underline{\mathbf{e}}_x + A_zB_y\,\underline{\mathbf{e}}_z\underline{\mathbf{e}}_y + A_zB_z\,\underline{\mathbf{e}}_z\underline{\mathbf{e}}_z \qquad (2.40)\\ &= \sum_{i=1}^{3}\sum_{j=1}^{3} A_{x_i}B_{x_j}\,\underline{\mathbf{e}}_{x_i}\underline{\mathbf{e}}_{x_j} \qquad (2.41)\\ &= A_{x_i}B_{x_j}\,\underline{\mathbf{e}}_{x_i}\underline{\mathbf{e}}_{x_j} \quad \text{(summation convention).} \qquad (2.42)\end{aligned}$$

Summation convention means that summation from 1 to 3 on the right-hand side affects the indices i and j appearing twice on that side.

The vector with the component representation

$$\underline{\mathbf{A}} = A_x\,\underline{\mathbf{e}}_x + A_y\,\underline{\mathbf{e}}_y + A_z\,\underline{\mathbf{e}}_z \tag{2.43}$$

can be written as a single-column matrix (column vector)

$$\underline{\mathbf{A}} = \begin{pmatrix} A_x \\ A_y \\ A_z \end{pmatrix} \tag{2.44}$$

or as a single-row matrix (row vector)

$$\underline{\mathbf{A}}^{\mathrm{T}} = \begin{pmatrix} A_x & A_y & A_z \end{pmatrix}, \tag{2.45}$$

being the transpose—indicated by the upper index T—of the single-column matrix.[10] The unit vectors in (2.43) refer to the position of the scalar component in the pertinent matrix scheme. Similarly, we can choose the scheme

$$\underline{\mathbf{A}}\,\underline{\mathbf{B}} = \begin{pmatrix} A_xB_x & A_xB_y & A_xB_z \\ A_yB_x & A_yB_y & A_yB_z \\ A_zB_x & A_zB_y & A_zB_z \end{pmatrix} \tag{2.46}$$

of a 3×3-matrix for the dyadic product (2.40)—the *dyadic* $\underline{\mathbf{A}}\,\underline{\mathbf{B}}$. Obviously, the dyadic products $\underline{\mathbf{e}}_{x_i}\underline{\mathbf{e}}_{x_j}$, $i, j = 1, 2, 3$, indicate the position of the element $A_{x_i}B_{x_j}$ in the matrix if we agree upon the choice of the first index as row index and the second index as column index. We adhere that in this sense a dyadic possesses nine scalar components in contrast to the three scalar

[10] We only must know the coordinate system for the components.

components of a vector; nevertheless, in the present case, the nine components are determined by the six vector components of the two vectors forming the dyadic product. From the definition of the dyadic product, we deduce that it is not commutative:

$$\underline{\mathbf{A}}\,\underline{\mathbf{B}} \neq \underline{\mathbf{B}}\,\underline{\mathbf{A}}. \tag{2.47}$$

The dyadic product yields a descriptive meaning when applied via a dot product (contraction) from left or right to a vector. Hence, we try to interpret the operation

$$\underline{\mathbf{A}}\,\underline{\mathbf{B}} \cdot \underline{\mathbf{C}} \tag{2.48}$$

or

$$\underline{\mathbf{C}} \cdot \underline{\mathbf{A}}\,\underline{\mathbf{B}} \tag{2.49}$$

writing $\underline{\mathbf{A}}\,\underline{\mathbf{B}} \cdot \underline{\mathbf{C}}$ in components

$$\underline{\mathbf{A}}\,\underline{\mathbf{B}} \cdot \underline{\mathbf{C}} = \sum_{i=1}^{3}\sum_{j=1}^{3} A_{x_i} B_{x_j}\, \underline{\mathbf{e}}_{x_i}\underline{\mathbf{e}}_{x_j} \cdot \sum_{k=1}^{3} C_{x_k}\, \underline{\mathbf{e}}_{x_k} \tag{2.50}$$

and using (2.17) to calculate

$$\begin{aligned}
\underline{\mathbf{A}}\,\underline{\mathbf{B}} \cdot \underline{\mathbf{C}} &= \sum_{i=1}^{3}\sum_{j=1}^{3}\sum_{k=1}^{3} A_{x_i} B_{x_j} C_{x_k}\, \underline{\mathbf{e}}_{x_i}\underline{\mathbf{e}}_{x_j} \underbrace{\underline{\mathbf{e}}_{x_j} \cdot \underline{\mathbf{e}}_{x_k}}_{=\,\delta_{jk}} \\
&= \sum_{i=1}^{3}\sum_{k=1}^{3} A_{x_i} B_{x_k} C_{x_k}\, \underline{\mathbf{e}}_{x_i} \\
&\Uparrow \\
&\quad \text{due to } \delta_{jk} \text{ only the term } j = k \\
&\quad \text{remains from the } j\text{-summation} \\
&= \underbrace{\sum_{i=1}^{3} A_{x_i}\, \underline{\mathbf{e}}_{x_i}}_{=\,\underline{\mathbf{A}}} \underbrace{\sum_{k=1}^{3} B_{x_k} C_{x_k}}_{=\,\underline{\mathbf{B}} \cdot \underline{\mathbf{C}}} \\
&= \underline{\mathbf{A}}(\underline{\mathbf{B}} \cdot \underline{\mathbf{C}}) \\
&= (\underline{\mathbf{B}} \cdot \underline{\mathbf{C}})\underline{\mathbf{A}}.
\end{aligned} \tag{2.51}$$

The left-sided contraction of a dyadic product with a vector is nothing but the contraction of the indices of the adjacent vectors—in this case, $\underline{\mathbf{B}}$ and $\underline{\mathbf{C}}$; the scalar product $\underline{\mathbf{B}} \cdot \underline{\mathbf{C}}$ shows up as a *scalar* factor of the remaining vector $\underline{\mathbf{A}}$, the left factor of the dyadic product.

In complete analogy, we compute

$$\underline{\mathbf{C}} \cdot \underline{\mathbf{A}}\,\underline{\mathbf{B}} = (\underline{\mathbf{C}} \cdot \underline{\mathbf{A}})\underline{\mathbf{B}}, \tag{2.52}$$

and obviously we find

$$\underline{\mathbf{A}}\,\underline{\mathbf{B}} \cdot \underline{\mathbf{C}} \neq \underline{\mathbf{C}} \cdot \underline{\mathbf{A}}\,\underline{\mathbf{B}}. \tag{2.53}$$

The dyadic *operator* $\underline{\mathbf{A}}\,\underline{\mathbf{B}}$ rotates the vector $\underline{\mathbf{C}}$ into the direction of the vector $\underline{\mathbf{A}}$ according to $\underline{\mathbf{A}}\,\underline{\mathbf{B}} \cdot \underline{\mathbf{C}}$ and the vector $\underline{\mathbf{C}}$ into the direction of the vector $\underline{\mathbf{B}}$ according to $\underline{\mathbf{C}} \cdot \underline{\mathbf{A}}\,\underline{\mathbf{B}}$.

Commercially available shear wave transducers radiate transverse waves under various angles applying normal forces to surfaces: The related particle displacement as a vector has quite different directions that do not comply with the normal to the surface. Therefore, the transformation force $\Longrightarrow$ wave must be mathematically procured by a dyadic operator; in the case of point-like forces, it is just Green's dyadic. Its explicit mathematical structure is required to model sound fields of piezoelectric transducers.

Utilizing the matrix representations (2.46) and (2.44) of $\underline{\mathbf{A}}\,\underline{\mathbf{B}}$ and $\underline{\mathbf{C}}$, we find $\underline{\mathbf{A}}\,\underline{\mathbf{B}} \cdot \underline{\mathbf{C}}$ as a single-column matrix resulting from matrix multiplication:

$$\begin{aligned}
\begin{pmatrix} A_xB_x & A_xB_y & A_xB_z \\ A_yB_x & A_yB_y & A_yB_z \\ A_zB_x & A_zB_y & A_zB_z \end{pmatrix}\begin{pmatrix} C_x \\ C_y \\ C_z \end{pmatrix} &= \begin{pmatrix} (B_xC_x + B_yC_y + B_zC_z)A_x \\ (B_xC_x + B_yC_y + B_zC_z)A_y \\ (B_xC_x + B_yC_y + B_zC_z)A_z \end{pmatrix} \\
&= (B_xC_x + B_yC_y + B_zC_z)\begin{pmatrix} A_x \\ A_y \\ A_z \end{pmatrix}.
\end{aligned} \tag{2.54}$$

Analogously, we find $\underline{\mathbf{C}} \cdot \underline{\mathbf{A}}\,\underline{\mathbf{B}}$ as a single-row matrix:

$$\begin{aligned}
&(C_x, C_y, C_z)\begin{pmatrix} A_xB_x & A_xB_y & A_xB_z \\ A_yB_x & A_yB_y & A_yB_z \\ A_zB_x & A_zB_y & A_zB_z \end{pmatrix} \\
&\quad = (C_xA_x + C_yA_y + C_zA_z)\begin{pmatrix} B_x & B_y & B_z \end{pmatrix}.
\end{aligned} \tag{2.55}$$

The explicit calculation of $\underline{\mathbf{A}}\,\underline{\mathbf{B}} \cdot \underline{\mathbf{C}}$ (or $\underline{\mathbf{C}} \cdot \underline{\mathbf{A}}\,\underline{\mathbf{B}}$) becomes most obvious utilizing the summation convention

$$\begin{aligned}
\underline{\mathbf{A}}\,\underline{\mathbf{B}} \cdot \underline{\mathbf{C}} &= A_iB_j\, \underline{\mathbf{e}}_{x_i}\underline{\mathbf{e}}_{x_j} \cdot C_k\, \underline{\mathbf{e}}_{x_k} \\
&= A_iB_jC_k\, \underline{\mathbf{e}}_{x_i}\, \underbrace{\underline{\mathbf{e}}_{x_j} \cdot \underline{\mathbf{e}}_{x_k}}_{=\,\delta_{jk}} \\
&= A_i\, \underline{\mathbf{e}}_{x_i}\, B_kC_k.
\end{aligned} \tag{2.56}$$

We nicely see that the dot product contracts adjacent indices, i.e., one index—in this case, j—disappears.

It is evident that

$$\underline{\mathbf{A}}\,\underline{\mathbf{B}} \times \underline{\mathbf{C}} \tag{2.57}$$

and

$$\underline{\mathbf{C}} \times \underline{\mathbf{A}}\,\underline{\mathbf{B}} \tag{2.58}$$

become meaningful through (2.51): $\underline{\mathbf{A}}\,\underline{\mathbf{B}} \times \underline{\mathbf{C}}$ is the dyadic (!) product of the vector $\underline{\mathbf{A}}$ with the (axial) vector $\underline{\mathbf{B}} \times \underline{\mathbf{C}}$, and $\underline{\mathbf{C}} \times \underline{\mathbf{A}}\,\underline{\mathbf{B}}$ is the dyadic $\underline{\mathbf{D}}\,\underline{\mathbf{B}}$ mit $\underline{\mathbf{D}} = \underline{\mathbf{C}} \times \underline{\mathbf{A}}$.

Linear independence: Three vectors $\underline{\mathbf{A}}_1, \underline{\mathbf{A}}_2, \underline{\mathbf{A}}_3$ are linearly independent if

$$\alpha_1\,\underline{\mathbf{A}}_1 + \alpha_2\,\underline{\mathbf{A}}_2 + \alpha_3\,\underline{\mathbf{A}}_3 = \underline{\mathbf{0}} \tag{2.59}$$

only holds for $\alpha_1 = \alpha_2 = \alpha_3 = 0$. Therefore, linear dependence implies that the three vectors span a triangle.

Complex valued vectors: The frequency spectrum[11] $\underline{\mathbf{u}}(\underline{\mathbf{R}}, \omega)$ of the time-dependent particle displacement $\underline{\mathbf{u}}(\underline{\mathbf{R}}, t)$ apparently is a vector field

$$\underline{\mathbf{u}}(\underline{\mathbf{R}}, \omega) = u_x(\underline{\mathbf{R}}, \omega)\,\underline{\mathbf{e}}_x + u_y(\underline{\mathbf{R}}, \omega)\,\underline{\mathbf{e}}_y + u_z(\underline{\mathbf{R}}, \omega)\,\underline{\mathbf{e}}_z, \tag{2.60}$$

whose components are frequency spectra of the components of $\underline{\mathbf{u}}(\underline{\mathbf{R}}, t)$ (Equation 2.10). Yet, frequency spectra generally are complex valued functions of the (real) variable ω (Section 2.3) with consequences regarding algebraic operations like, for instance, computing the magnitude of $\underline{\mathbf{u}}(\underline{\mathbf{R}}, \omega)$. If we calculate

$$\underline{\mathbf{u}}(\underline{\mathbf{R}}, \omega) \cdot \underline{\mathbf{u}}(\underline{\mathbf{R}}, \omega) = u_x^2(\underline{\mathbf{R}}, \omega) + u_y^2(\underline{\mathbf{R}}, \omega) + u_z^2(\underline{\mathbf{R}}, \omega), \tag{2.61}$$

the single terms

$$\begin{aligned} u_{x_i}^2(\underline{\mathbf{R}}, \omega) &= u_{x_iR}^2(\underline{\mathbf{R}}, \omega) - u_{x_iI}^2(\underline{\mathbf{R}}, \omega) + 2\mathrm{j}u_{x_iR}(\underline{\mathbf{R}}, \omega)u_{x_iI}(\underline{\mathbf{R}}, \omega), \\ & i = 1, 2, 3, \end{aligned} \tag{2.62}$$

are *complex* numbers with real

$$u_{x_iR}(\underline{\mathbf{R}}, \omega) = \Re\{u_{x_i}(\underline{\mathbf{R}}, \omega)\}, \quad i = 1, 2, 3, \tag{2.63}$$

and imaginary part

$$u_{x_iI}(\underline{\mathbf{R}}, \omega) = \Im\{u_{x_i}(\underline{\mathbf{R}}, \omega)\}, \quad i = 1, 2, 3, \tag{2.64}$$

of $u_{x_i}(\underline{\mathbf{R}}, \omega)$. As a consequence, (2.61) is no longer the square of the "length" of the complex valued vector $\underline{\mathbf{u}}(\underline{\mathbf{R}}, \omega)$. However, if we investigate the so-called Hermite product

$$\underline{\mathbf{u}}(\underline{\mathbf{R}}, \omega) \cdot \underline{\mathbf{u}}^*(\underline{\mathbf{R}}, \omega) = |u_x(\underline{\mathbf{R}}, \omega)|^2 + |u_y(\underline{\mathbf{R}}, \omega)|^2 + |u_z(\underline{\mathbf{R}}, \omega)|^2, \tag{2.65}$$

[11]For physical quantities, we use the same character $\underline{\mathbf{u}}$ for the (spatially dependent) time function $\underline{\mathbf{u}}(\underline{\mathbf{R}}, t)$ and for the (spatially dependent) spectrum $\underline{\mathbf{u}}(\underline{\mathbf{R}}, \omega)$ and distinguish them through explicit indication of the variable t or ω, respectively; often one finds $\hat{\underline{\mathbf{u}}}$, $\tilde{\underline{\mathbf{u}}}$, $\bar{\underline{\mathbf{u}}}$, $\underline{\mathbf{U}}$ for the spectrum. Note that the physical dimension of $\underline{\mathbf{u}}(\underline{\mathbf{R}}, \omega)$ is equal to the physical dimension of $\underline{\mathbf{u}}(\underline{\mathbf{R}}, t)$ multiplied by the physical dimension "time."

where $\underline{\mathbf{u}}^*(\underline{\mathbf{R}}, \omega)$ has the complex conjugate components of $\underline{\mathbf{u}}(\underline{\mathbf{R}}, \omega)$, then the magnitudes of the complex numbers $u_{x_i}(\underline{\mathbf{R}}, \omega)$ appearing in (2.64)

$$|u_{x_i}(\underline{\mathbf{R}}, \omega)| = \sqrt{\Re\{u_{x_i}(\underline{\mathbf{R}}, \omega)\}^2 + \Im\{u_{x_i}(\underline{\mathbf{R}}, \omega)\}^2}, \quad i = 1, 2, 3, \tag{2.66}$$

are real valued.

Generalizing (2.22), we define the real positive length of a complex vector $\underline{\mathbf{C}}$ according to

$$|\underline{\mathbf{C}}| = \sqrt{\underline{\mathbf{C}} \cdot \underline{\mathbf{C}}^*}. \tag{2.67}$$

2.1.4 Tensor fields

Tensor components: Compared to a vector $\underline{\mathbf{A}}$ with single index components A_{x_i} a dyadic $\underline{\mathbf{A}}\,\underline{\mathbf{B}}$ has doubly indexed components $A_{x_i}B_{x_j}$; therefore, we stipulate the notation

$$\underline{\underline{\mathbf{D}}} = \underline{\mathbf{A}}\,\underline{\mathbf{B}} \tag{2.68}$$

with two underlines for $\underline{\underline{\mathbf{D}}}$. In lieu of characterizing $\underline{\underline{\mathbf{D}}}$ by the doubly indexed matrix elements $D_{x_ix_j} = A_{x_i}B_{x_j}$ as a dyadic product A_{x_i} and B_{x_j} according to (2.40), we may generalize

$$\underline{\underline{\mathbf{D}}} = \begin{pmatrix} D_{xx} & D_{xy} & D_{xz} \\ D_{yx} & D_{yy} & D_{yz} \\ D_{zx} & D_{zy} & D_{zz} \end{pmatrix}; \tag{2.69}$$

that way, we interpret the matrix elements $D_{xx}, D_{xy}, D_{xz}, \ldots$—the nonreducible components $D_{x_ix_j}$, $i, j = 1, 2, 3$—as components of a tensor of second rank:[12]

$$\begin{aligned} \underline{\underline{\mathbf{D}}} &= \sum_{i=1}^{3}\sum_{j=1}^{3} D_{x_ix_j}\, \underline{\mathbf{e}}_{x_i}\underline{\mathbf{e}}_{x_j} \\ &= D_{x_ix_j}\, \underline{\mathbf{e}}_{x_i}\underline{\mathbf{e}}_{x_j} \text{ (summation convention)} \\ &= D_{ij}\, \underline{\mathbf{e}}_{x_i}\underline{\mathbf{e}}_{x_j} \text{ (summation convention).} \end{aligned} \tag{2.70}$$

Contraction of tensors with vectors: Let us perform some calculus, for instance, the left-sided contraction of the vector $\underline{\mathbf{C}}$ with the tensor $\underline{\underline{\mathbf{D}}}$

$$\underline{\mathbf{C}} \cdot \underline{\underline{\mathbf{D}}} = \underline{\mathbf{E}}, \tag{2.71}$$

that is to say, we search the components of the resulting vector $\underline{\mathbf{E}}$. With the components C_{x_k} of $\underline{\mathbf{C}}$ and the components of $D_{x_ix_j}$ of $\underline{\underline{\mathbf{D}}}$, we obtain

[12] A second rank tensor may be represented by a matrix of its (scalar) components, yet a matrix must not necessarily be a tensor: Tensor components transform like vector components in a prescribed manner if the coordinate system is changed (e.g., Morse and Feshbach 1953; Chen 1983)!

$$
\begin{aligned}
\underline{\mathbf{C}} \cdot \underline{\underline{\mathbf{D}}} &= \sum_{k=1}^{3} C_{x_k}\, \underline{\mathbf{e}}_{x_k} \cdot \sum_{i=1}^{3}\sum_{j=1}^{3} D_{x_i x_j}\, \underline{\mathbf{e}}_{x_i}\underline{\mathbf{e}}_{x_j} \\
&= \sum_{k=1}^{3}\sum_{i=1}^{3}\sum_{j=1}^{3} C_{x_k} D_{x_i x_j}\, \underbrace{\underline{\mathbf{e}}_{x_k} \cdot \underline{\mathbf{e}}_{x_i}}_{=\,\delta_{ki}}\, \underline{\mathbf{e}}_{x_j} \\
&= \sum_{j=1}^{3}\left(\sum_{i=1}^{3} C_{x_i} D_{x_i x_j}\right) \underline{\mathbf{e}}_{x_j} \\
&= C_i D_{ij}\, \underline{\mathbf{e}}_{x_j} \quad \text{(summation convention).}
\end{aligned} \tag{2.72}
$$

With the summation convention notation, $\underline{\mathbf{E}}$ has the components[13]

$$E_j = C_i D_{ij}; \tag{2.73}$$

the dot in $\underline{\mathbf{C}} \cdot \underline{\underline{\mathbf{D}}}$ contracts adjacent indices, namely the index $\underline{\mathbf{C}}$ with the first index of $\underline{\underline{\mathbf{D}}}$. Analogously, we calculate—this time exclusively utilizing the summation convention—

$$
\begin{aligned}
\underline{\underline{\mathbf{D}}} \cdot \underline{\mathbf{C}} &= D_{ij}\, \underline{\mathbf{e}}_{x_i}\underline{\mathbf{e}}_{x_j} \cdot C_k\, \underline{\mathbf{e}}_{x_k} \\
&= D_{ij} C_k\, \underline{\mathbf{e}}_{x_i} \underbrace{\underline{\mathbf{e}}_{x_j} \cdot \underline{\mathbf{e}}_{x_k}}_{=\,\delta_{jk}} \\
&= D_{ik} C_k\, \underline{\mathbf{e}}_{x_i};
\end{aligned} \tag{2.74}
$$

the dot in $\underline{\underline{\mathbf{D}}} \cdot \underline{\mathbf{C}}$ again contracts adjacent indices, yet this time the second index of $\underline{\underline{\mathbf{D}}}$ with the $\underline{\mathbf{C}}$-index. Therefore, the product $\underline{\underline{\mathbf{D}}} \cdot \underline{\mathbf{C}}$ has $\sum_{k=1}^{3} D_{ik} C_k$ as i-component. We state that in general

$$\underline{\underline{\mathbf{D}}} \cdot \underline{\mathbf{C}} \neq \underline{\mathbf{C}} \cdot \underline{\underline{\mathbf{D}}} \tag{2.75}$$

holds. Equation 2.75 comes with an equality sign *only if* the contractions from right and from left, namely over the first and over the second index of $\underline{\underline{\mathbf{D}}}$, are equal; this implies

$$D_{ik} = D_{ki}, \tag{2.76}$$

because then (summation convention understood)

$$D_{ik} C_k = D_{ki} C_k = C_k D_{ki} \tag{2.77}$$

holds.

Symmetric tensors: A tensor with the property (2.76) is symmetric: The matrix representation does not change mirroring components at the main

[13]On the right-hand side of (2.73), summation over i is understood but not over j because j appears on both sides: j counts the components of $\underline{\mathbf{E}}$ resulting in three ($j = 1, 2, 3$) equations hidden in (2.73).

diagonal—inverting rows and columns, i.e., inverting indices. Indicating the mirroring by an upper index T for "transpose," we have

$$\underline{\underline{\mathbf{D}}} = \underline{\underline{\mathbf{D}}}^{\mathrm{T}} \tag{2.78}$$

for a symmetric tensor $\underline{\underline{\mathbf{D}}}$. Mirroring at the main diagonal of the tensor matrix

$$\underline{\underline{\mathbf{D}}} = D_{ij}\, \underline{\mathbf{e}}_{x_i} \underline{\mathbf{e}}_{x_j} \text{ (summation convention)} \tag{2.79}$$

implies in components inverting places according to

$$\underline{\underline{\mathbf{D}}}^{21} = D_{ij}\, \underline{\mathbf{e}}_{x_j} \underline{\mathbf{e}}_{x_i} \text{ (summation convention).} \tag{2.80}$$

The former second vector $\underline{\mathbf{e}}_{x_j}$ in the dyadic product $\underline{\mathbf{e}}_{x_i} \underline{\mathbf{e}}_{x_j}$ appears now in first place and vice versa, indicated by $\underline{\underline{\mathbf{D}}}^{21}$. Renaming (2.80) according to $i \Longrightarrow j$, $j \Longrightarrow i$, results in

$$\begin{aligned} \underline{\underline{\mathbf{D}}}^{21} &= D_{ji}\, \underline{\mathbf{e}}_{x_i} \underline{\mathbf{e}}_{x_j} \text{ (summation convention)} \\ &= \underline{\underline{\mathbf{D}}}^{\mathrm{T}}, \end{aligned} \tag{2.81}$$

meaning that the former places $\underline{\mathbf{e}}_{x_i} \underline{\mathbf{e}}_{x_j}$ in the component scheme (2.79) contain the mirrored matrix elements D_{ji}. Therefore, symmetry of a tensor equally implies

$$\underline{\underline{\mathbf{D}}} = \underline{\underline{\mathbf{D}}}^{21}. \tag{2.82}$$

Inverting component places with the upper index notation is conveniently applied for tensors of higher order when the transpose is meaningless.

Simultaneously, transposing $\underline{\underline{\mathbf{D}}}$ obviously allows for the exchange of the contractions

$$\underline{\underline{\mathbf{D}}} \cdot \underline{\mathbf{C}} = \underline{\mathbf{C}} \cdot \underline{\underline{\mathbf{D}}}^{\mathrm{T}} = \underline{\mathbf{C}} \cdot \underline{\underline{\mathbf{D}}}^{21}. \tag{2.83}$$

Therefore, a symmetric tensor is characterized by

$$\underline{\underline{\mathbf{D}}} \cdot \underline{\mathbf{C}} = \underline{\mathbf{C}} \cdot \underline{\underline{\mathbf{D}}}. \tag{2.84}$$

An arbitrary tensor $\underline{\underline{\mathbf{D}}}$ can be used to construct a symmetric tensor through

$$\underline{\underline{\mathbf{D}}}_{\mathrm{s}} = \frac{1}{2}(\underline{\underline{\mathbf{D}}} + \underline{\underline{\mathbf{D}}}^{21}); \tag{2.85}$$

$\underline{\underline{\mathbf{D}}}_{\mathrm{s}}$ is the symmetric part of $\underline{\underline{\mathbf{D}}}$. Because of $(\underline{\underline{\mathbf{D}}}^{21})^{21} = \underline{\underline{\mathbf{D}}}$—twofold mirroring at the main diagonal—the symmetry of $\underline{\underline{\mathbf{D}}}_{\mathrm{s}}$ is obvious. The factor 1/2 can be understood postulating *a priori* symmetry of $\underline{\underline{\mathbf{D}}}$.

Antisymmetric tensors: The antisymmetric part $\underline{\underline{\mathbf{D}}}_{\mathrm{a}}$ of $\underline{\underline{\mathbf{D}}}$ with the property

$$\underline{\underline{\mathbf{D}}}_{\mathrm{a}}^{21} = -\underline{\underline{\mathbf{D}}}_{\mathrm{a}} \tag{2.86}$$

is obtained according to

$$\underline{\underline{\mathbf{D}}}_{\mathrm{a}} = \frac{1}{2}(\underline{\underline{\mathbf{D}}} - \underline{\underline{\mathbf{D}}}^{21}). \tag{2.87}$$

Note: The requirement (2.86) implies zero values of all main diagonal elements of $\underline{\underline{\mathbf{D}}}_{\mathrm{a}}$.

The cartesian components of the rotation vector $\langle \underline{\underline{\mathbf{D}}} \rangle$ of a second rank tensor is defined as

$$\langle \underline{\underline{\mathbf{D}}} \rangle = D_{ij}\, \underline{\mathbf{e}}_{x_i} \times \underline{\mathbf{e}}_{x_j} \text{ (summation convention)}, \tag{2.88}$$

where the appliance to a dyadic product

$$\begin{aligned} \langle \underline{\mathbf{A}}\,\underline{\mathbf{B}} \rangle &= A_i B_j\, \underline{\mathbf{e}}_{x_i} \times \underline{\mathbf{e}}_{x_j} \text{ (summation convention)} \\ &= \underline{\mathbf{A}} \times \underline{\mathbf{B}} \end{aligned} \tag{2.89}$$

explains the terminology. Through calculation of the cartesian components, we can show that the construction of—$\underline{\underline{\mathbf{I}}}$ is the unit tensor; see below—

$$\frac{1}{2}\langle \underline{\underline{\mathbf{D}}}^{21} \rangle \times \underline{\underline{\mathbf{I}}} = \underline{\underline{\mathbf{D}}}_{\mathrm{a}} \tag{2.90}$$

always yields the antisymmetric part $\underline{\underline{\mathbf{D}}}_{\mathrm{a}}$ of $\underline{\underline{\mathbf{D}}}$. As a matter of fact, (2.90) is the most general representation of an antisymmetric tensor. Applied to a dyadic product $\underline{\mathbf{A}}\,\underline{\mathbf{B}}$, we have

$$\begin{aligned} \frac{1}{2}\langle \underline{\mathbf{A}}\,\underline{\mathbf{B}} \rangle \times \underline{\underline{\mathbf{I}}} &= \underline{\mathbf{A}}\,\underline{\mathbf{B}}_{\mathrm{a}} \\ &= \frac{1}{2}(\underline{\mathbf{B}} \times \underline{\mathbf{A}}) \times \underline{\underline{\mathbf{I}}}; \end{aligned} \tag{2.91}$$

its antisymmetric part is equal to the null tensor if and only if $\underline{\mathbf{A}}$ and $\underline{\mathbf{B}}$ are parallel.

Tensor fields: The symmetric deformation tensor $\underline{\underline{\mathbf{S}}}(\underline{\mathbf{R}}, t)$ and the symmetric stress tensor $\underline{\underline{\mathbf{T}}}(\underline{\mathbf{R}}, t)$ represent important tensor fields in NDT, because spatial and time variations of the particle velocity characterize the space- and time-dependent deformation state of a solid, and sources of stresses result in acceleration of volume elements, that is to say second time derivatives of particle velocities and, hence, waves.

Unit tensor of second rank: The unit matrix of matrix calculus corresponds to the unit tensor of second rank ($\underline{\underline{\mathbf{I}}}$dentity tensor):

$$\begin{aligned} \underline{\underline{\mathbf{I}}} &= \sum_{i=1}^{3}\sum_{j=1}^{3} \delta_{ij}\, \underline{\mathbf{e}}_{x_i}\underline{\mathbf{e}}_{x_j} \\ &= \delta_{ij}\, \underline{\mathbf{e}}_{x_i}\underline{\mathbf{e}}_{x_j} \text{ (summation convention)} \\ &= \underline{\mathbf{e}}_{x_i}\underline{\mathbf{e}}_{x_i} \text{ (summation convention)} \\ &= \underline{\mathbf{e}}_{x}\underline{\mathbf{e}}_{x} + \underline{\mathbf{e}}_{y}\underline{\mathbf{e}}_{y} + \underline{\mathbf{e}}_{z}\underline{\mathbf{e}}_{z}; \end{aligned} \tag{2.92}$$

obviously, in the matrix representation of $\underline{\underline{\mathbf{I}}}$, only the main diagonal elements are nonzero and all are equal to one:

$$\underline{\underline{\mathbf{I}}} = \begin{pmatrix} 1 & 0 & 0 \\ 0 & 1 & 0 \\ 0 & 0 & 1 \end{pmatrix} . \tag{2.93}$$

Accordingly, in circular cylindrical and spherical coordinates, we have

$$\underline{\underline{\mathbf{I}}} = \underline{\mathbf{e}}_r \, \underline{\mathbf{e}}_r + \underline{\mathbf{e}}_\varphi \, \underline{\mathbf{e}}_\varphi + \underline{\mathbf{e}}_z \, \underline{\mathbf{e}}_z , \tag{2.94}$$

$$\underline{\underline{\mathbf{I}}} = \underline{\mathbf{e}}_R \, \underline{\mathbf{e}}_R + \underline{\mathbf{e}}_\vartheta \, \underline{\mathbf{e}}_\vartheta + \underline{\mathbf{e}}_\varphi \, \underline{\mathbf{e}}_\varphi . \tag{2.95}$$

The symmetric unit tensor has the property

$$\underline{\underline{\mathbf{I}}} \cdot \underline{\mathbf{A}} = \underline{\mathbf{A}} \cdot \underline{\underline{\mathbf{I}}} = \underline{\mathbf{A}}. \tag{2.96}$$

Vector components tangential to a surface: With $\underline{\mathbf{n}}$, we denote the normal to a surface at point $\underline{\mathbf{R}}$; then

$$\underline{\mathbf{u}}_t(\underline{\mathbf{R}}, t) = (\underline{\underline{\mathbf{I}}} - \underline{\mathbf{n}}\,\underline{\mathbf{n}}) \cdot \underline{\mathbf{u}}(\underline{\mathbf{R}}, t) \tag{2.97}$$

apparently yields a vector component of $\underline{\mathbf{u}}(\underline{\mathbf{R}}, t)$ tangential to the surface (Figure 2.12) because

$$\begin{aligned} \underline{\mathbf{n}} \cdot \underline{\mathbf{u}}_t(\underline{\mathbf{R}}, t) &= \underline{\mathbf{n}} \cdot (\underline{\underline{\mathbf{I}}} - \underline{\mathbf{n}}\,\underline{\mathbf{n}}) \cdot \underline{\mathbf{u}}(\underline{\mathbf{R}}, t) \\ &= (\underline{\mathbf{n}} - \underline{\mathbf{n}} \cdot \underline{\mathbf{n}}\,\underline{\mathbf{n}}) \cdot \underline{\mathbf{u}}(\underline{\mathbf{R}}, t) \\ &= 0. \end{aligned} \tag{2.98}$$

The tangential component $\underline{\mathbf{u}}_{tan}(\underline{\mathbf{R}}, t)$ defined according to (2.38) relates to $\underline{\mathbf{u}}_t(\underline{\mathbf{R}}, t)$ in the following way (Figure 2.12):

$$\underline{\mathbf{u}}_t(\underline{\mathbf{R}}, t) = -\underline{\mathbf{n}} \times \underline{\mathbf{u}}_{tan}(\underline{\mathbf{R}}, t), \tag{2.99}$$

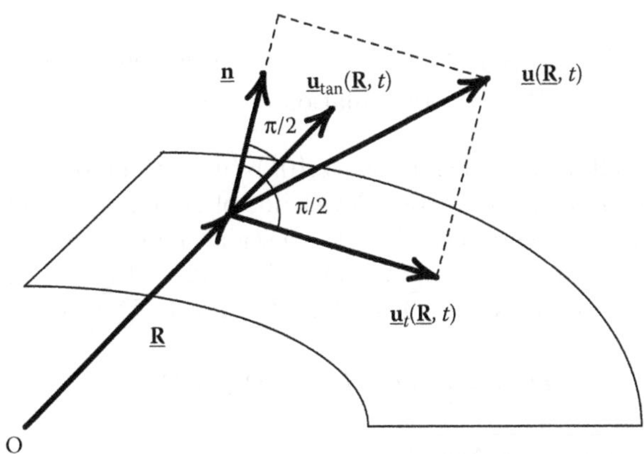

FIGURE 2.12
Vector components tangential to a surface.

$$\underline{\mathbf{u}}_{tan}(\underline{\mathbf{R}}, t) = \underline{\mathbf{n}} \times \underline{\mathbf{u}}_t(\underline{\mathbf{R}}, t)$$
$$= \underline{\mathbf{n}} \times \underline{\mathbf{u}}(\underline{\mathbf{R}}, t). \quad (2.100)$$

The vector tangential component $\underline{\mathbf{u}}_t(\underline{\mathbf{R}}, t)$ lies in the plane spanned by $\underline{\mathbf{n}}$ and $\underline{\mathbf{u}}(\underline{\mathbf{R}}, t)$, whereas $\underline{\mathbf{u}}_{tan}(\underline{\mathbf{R}}, t)$ is orthogonal to it.

Contraction of tensors with tensors: Certainly, contractions of adjacent indices of two tensors (second or higher rank) can be equally accomplished: for instance,

$$\underline{\underline{\mathbf{D}}} \cdot \underline{\underline{\mathbf{C}}} = \underline{\mathbf{e}}_{x_i} D_{ij} C_{jk} \, \underline{\mathbf{e}}_{x_k} \quad \text{(summation convention)}$$
$$= D_{ij} C_{jk} \, \underline{\mathbf{e}}_{x_i} \underline{\mathbf{e}}_{x_k} \quad \text{(summation convention)} \quad (2.101)$$

yields another tensor of second rank with $(x_i x_k)$-components $\sum_{j=1}^{3} D_{ij} C_{jk}$: The second index of $\underline{\underline{\mathbf{D}}}$ is contracted with the first index of $\underline{\underline{\mathbf{C}}}$, and the resulting tensor has the first index of $\underline{\underline{\mathbf{D}}}$ as first index and the second index of $\underline{\underline{\mathbf{C}}}$ as second index. In contrast, calculating $\underline{\underline{\mathbf{C}}}^{21} \cdot \underline{\underline{\mathbf{D}}}^{21}$ also implies contraction of the first index of $\underline{\underline{\mathbf{C}}}$ with the second index of $\underline{\underline{\mathbf{D}}}$, yet the resulting tensor has the second index of $\underline{\underline{\mathbf{C}}}$ as first index and the first index of $\underline{\underline{\mathbf{D}}}$ as second index. Consequently,

$$(\underline{\underline{\mathbf{C}}}^{21} \cdot \underline{\underline{\mathbf{D}}}^{21})^{21} = \underline{\underline{\mathbf{D}}} \cdot \underline{\underline{\mathbf{C}}} \quad (2.102)$$

or

$$(\underline{\underline{\mathbf{D}}} \cdot \underline{\underline{\mathbf{C}}})^{21} = \underline{\underline{\mathbf{C}}}^{21} \cdot \underline{\underline{\mathbf{D}}}^{21}. \quad (2.103)$$

Double contraction of tensors with tensors: Double contractions are performed subsequently with adjacent indices:[14]

$$\underline{\underline{\mathbf{D}}} : \underline{\underline{\mathbf{C}}} = D_{ij} C_{jk} \, \underline{\mathbf{e}}_{x_i} \cdot \underline{\mathbf{e}}_{x_k} \quad \text{(summation convention)}$$
$$= D_{ij} C_{ji} \quad \text{(summation convention)}, \quad (2.104)$$

that is to say, after the contraction (2.101) (upper dot of :) the adjacent indices i and k are also contracted (lower dot of :). The result of the double contraction of two tenors of second rank is a scalar, namely, the "double scalar product" of the second index of $\underline{\underline{\mathbf{D}}}$ with the first index of $\underline{\underline{\mathbf{C}}}$ and the first index of $\underline{\underline{\mathbf{D}}}$ with the second index of $\underline{\underline{\mathbf{C}}}$. Therefore, $\underline{\underline{\mathbf{D}}} : \underline{\underline{\mathbf{I}}}$ or $\underline{\underline{\mathbf{I}}} : \underline{\underline{\mathbf{D}}}$, respectively, double contracts to

$$\underline{\underline{\mathbf{D}}} : \underline{\underline{\mathbf{I}}} = \underline{\underline{\mathbf{I}}} : \underline{\underline{\mathbf{D}}} = D_{xx} + D_{yy} + D_{zz}; \quad (2.105)$$

---

[14] Another definition used by Auld (1973) reads as

$$\underline{\underline{\mathbf{D}}} : \underline{\underline{\mathbf{C}}} = D_{ij} D_{kl} \, \underline{\mathbf{e}}_{x_i} \underline{\mathbf{e}}_{x_j} : \underline{\mathbf{e}}_{x_k} \underline{\mathbf{e}}_{x_l}$$
$$= D_{ij} C_{kl} \, (\underline{\mathbf{e}}_{x_i} \cdot \underline{\mathbf{e}}_{x_k})(\underline{\mathbf{e}}_{x_j} \cdot \underline{\mathbf{e}}_{x_l})$$
$$= D_{ij} C_{ij}.$$

the sum of the main diagonal elements of the matrix representation of $\underline{\underline{\mathbf{D}}}$ is called the trace of the tensor $\underline{\underline{\mathbf{D}}}$:

$$\underline{\underline{\mathbf{D}}} : \underline{\underline{\mathbf{I}}} = \text{trace}\, \underline{\underline{\mathbf{D}}}. \tag{2.106}$$

With the scalar product $\underline{\mathbf{A}} \cdot \underline{\mathbf{A}}$, we define the magnitude—the length—of a vector that is intuitively independent of the choice of the coordinate system (the components of $\underline{\mathbf{A}}$ are not!); similarly, the trace of a tensor is not coordinate dependent.

Unit tensors of rank four: The elastic properties of solid materials must be characterized by constitutive equations; these do not follow from the governing equations of elastodynamics, they have to be postulated based on experimental results and physical considerations. One of the most important constitutive equations is Hooke's law relating the deformation state of a solid with its stress state in a *linear* way; to achieve this, a tensor of rank four—the compliance or, alternatively, the stiffness tensor—is required (Section 4.2). Per definitionem, for isotropic materials, the components of this fourth rank tensor must be independent of the coordinate system.[15] The most general isotropic tensor of rank four is constructed as follows:

$$\underline{\underline{\underline{\underline{\mathbf{I}}}}} = \alpha_1 \underline{\underline{\underline{\underline{\mathbf{I}}}}}^{\delta} + \alpha_2 \underline{\underline{\underline{\underline{\mathbf{I}}}}}^{+} + \alpha_3 \underline{\underline{\underline{\underline{\mathbf{I}}}}}^{-}, \tag{2.107}$$

where

$$\begin{aligned}
\underline{\underline{\underline{\underline{\mathbf{I}}}}}^{\delta} &= \underline{\underline{\mathbf{I}}}\,\underline{\underline{\mathbf{I}}} \\
&= \delta_{ij}\delta_{kl}\, \underline{\mathbf{e}}_{x_i}\underline{\mathbf{e}}_{x_j}\underline{\mathbf{e}}_{x_k}\underline{\mathbf{e}}_{x_l} && \text{(summation convention)} \\
&= \underline{\mathbf{e}}_{x_i}\underline{\mathbf{e}}_{x_i}\underline{\mathbf{e}}_{x_k}\underline{\mathbf{e}}_{x_k} && \text{(summation convention)}, \qquad (2.108)
\end{aligned}$$

$$\underline{\underline{\underline{\underline{\mathbf{I}}}}}^{+} = \frac{1}{2}(\underline{\underline{\mathbf{I}}}\,\underline{\underline{\mathbf{I}}}^{1342} + \underline{\underline{\mathbf{I}}}\,\underline{\underline{\mathbf{I}}}^{1324}), \tag{2.109}$$

$$\underline{\underline{\underline{\underline{\mathbf{I}}}}}^{-} = \frac{1}{2}(\underline{\underline{\mathbf{I}}}\,\underline{\underline{\mathbf{I}}}^{1342} - \underline{\underline{\mathbf{I}}}\,\underline{\underline{\mathbf{I}}}^{1324}); \tag{2.110}$$

$$\begin{aligned}
\underline{\underline{\mathbf{I}}}\,\underline{\underline{\mathbf{I}}}^{1342} &= \delta_{ij}\delta_{kl}(\underline{\mathbf{e}}_{x_i}\underline{\mathbf{e}}_{x_j}\underline{\mathbf{e}}_{x_k}\underline{\mathbf{e}}_{x_l})^{1342} && \text{(summation convention)} \\
&= \delta_{ij}\delta_{kl}\, \underline{\mathbf{e}}_{x_i}\underline{\mathbf{e}}_{x_k}\underline{\mathbf{e}}_{x_l}\underline{\mathbf{e}}_{x_j} && \text{(summation convention)} \\
&= \underline{\mathbf{e}}_{x_i}\underline{\mathbf{e}}_{x_j}\underline{\mathbf{e}}_{x_j}\underline{\mathbf{e}}_{x_i} && \text{(summation convention)}, \qquad (2.111)
\end{aligned}$$

$$\begin{aligned}
\underline{\underline{\mathbf{I}}}\,\underline{\underline{\mathbf{I}}}^{1324} &= \delta_{ij}\delta_{kl}(\underline{\mathbf{e}}_{x_i}\underline{\mathbf{e}}_{x_j}\underline{\mathbf{e}}_{x_k}\underline{\mathbf{e}}_{x_l})^{1324} && \text{(summation convention)} \\
&= \delta_{ij}\delta_{kl}\, \underline{\mathbf{e}}_{x_i}\underline{\mathbf{e}}_{x_k}\underline{\mathbf{e}}_{x_j}\underline{\mathbf{e}}_{x_l} && \text{(summation convention)} \\
&= \underline{\mathbf{e}}_{x_i}\underline{\mathbf{e}}_{x_j}\underline{\mathbf{e}}_{x_i}\underline{\mathbf{e}}_{x_j} && \text{(summation convention)}; \qquad (2.112)
\end{aligned}$$

[15]The electromagnetic properties of materials may be characterized by permittivity and permeability tensors of second rank; in that case, isotropy prevails if these constitutive tensors are proportional to the unit tensor $\underline{\underline{\mathbf{I}}}$ because $\underline{\underline{\mathbf{I}}}$ has the representation (2.92) as sum of the dyadic products of the orthonormal trihedron vectors in any coordinate system (Chen 1983; Equations 2.94 and 2.95).

the α_i, $i = 1, 2, 3$, denote arbitrary constants. The tensors $\underline{\underline{\mathbf{I}}}\,\underline{\underline{\mathbf{I}}}^{1342}$, $\underline{\underline{\mathbf{I}}}\,\underline{\underline{\mathbf{I}}}^{1324}$, $\underline{\underline{\underline{\underline{\mathbf{I}}}}}^{\delta}$, $\underline{\underline{\underline{\underline{\mathbf{I}}}}}^{+}$, $\underline{\underline{\underline{\underline{\mathbf{I}}}}}^{-}$ have the following properties:

$$\underline{\underline{\mathbf{I}}}\,\underline{\underline{\mathbf{I}}}^{1342} : \underline{\underline{\mathbf{D}}} = \underline{\underline{\mathbf{D}}} : \underline{\underline{\mathbf{I}}}\,\underline{\underline{\mathbf{I}}}^{1342} = \underline{\underline{\mathbf{D}}}, \tag{2.113}$$

$$\underline{\underline{\mathbf{I}}}\,\underline{\underline{\mathbf{I}}}^{1324} : \underline{\underline{\mathbf{D}}} = \underline{\underline{\mathbf{D}}} : \underline{\underline{\mathbf{I}}}\,\underline{\underline{\mathbf{I}}}^{1324} = \underline{\underline{\mathbf{D}}}^{21}, \tag{2.114}$$

$$\underline{\underline{\underline{\underline{\mathbf{I}}}}}^{\delta} : \underline{\underline{\mathbf{D}}} = \underline{\underline{\mathbf{D}}} : \underline{\underline{\underline{\underline{\mathbf{I}}}}}^{\delta} = \underline{\underline{\mathbf{I}}}\,\mathrm{trace}\,\underline{\underline{\mathbf{D}}}, \tag{2.115}$$

$$\underline{\underline{\underline{\underline{\mathbf{I}}}}}^{+} : \underline{\underline{\mathbf{D}}} = \underline{\underline{\mathbf{D}}} : \underline{\underline{\underline{\underline{\mathbf{I}}}}}^{+} = \underline{\underline{\mathbf{D}}}_{\mathrm{s}}, \tag{2.116}$$

$$\underline{\underline{\underline{\underline{\mathbf{I}}}}}^{-} : \underline{\underline{\mathbf{D}}} = \underline{\underline{\mathbf{D}}} : \underline{\underline{\underline{\underline{\mathbf{I}}}}}^{-} = \underline{\underline{\mathbf{D}}}_{\mathrm{a}}. \tag{2.117}$$

Inverse, adjoint, and determinant of a second rank tensor: As mentioned earlier, the second rank Green tensor rotates the direction of a point force (density) at the source point $\underline{\mathbf{r}}'$ into the direction of the particle velocity at the observation point $\underline{\mathbf{R}}$. Therefore, Green's tensor must be inverted—apart from the wave propagation from source to observation point—to calculate the particle velocity originating from a force density: We face the fundamental problem of NDT generalizing inversion to scattering of ultrasonic waves by material inhomogeneities (Chapter 16).

The inverse (second rank) tensor $\underline{\underline{\mathbf{D}}}^{-1}$ of a second rank tensor $\underline{\underline{\mathbf{D}}}$, if existing, has the property

$$\underline{\underline{\mathbf{D}}} \cdot \underline{\underline{\mathbf{D}}}^{-1} = \underline{\underline{\mathbf{D}}}^{-1} \cdot \underline{\underline{\mathbf{D}}} = \underline{\underline{\mathbf{I}}}. \tag{2.118}$$

That way the relation

$$\underline{\underline{\mathbf{D}}} \cdot \underline{\mathbf{A}} = \underline{\mathbf{B}} \tag{2.119}$$

can be inverted according to

$$\underline{\mathbf{A}} = \underline{\underline{\mathbf{D}}}^{-1} \cdot \underline{\mathbf{B}}. \tag{2.120}$$

$\underline{\underline{\mathbf{D}}}^{-1}$ can be calculated in terms of (Chen 1983)

$$\underline{\underline{\mathbf{D}}}^{-1} = \frac{\mathrm{adj}\,\underline{\underline{\mathbf{D}}}}{\det \underline{\underline{\mathbf{D}}}}, \tag{2.121}$$

where $\mathrm{adj}\,\underline{\underline{\mathbf{D}}}$ denotes the adjoint tensor of $\underline{\underline{\mathbf{D}}}$ whose matrix representation reads as follows:[16]

$$\mathrm{adj}\,\underline{\underline{\mathbf{D}}} = \begin{pmatrix} D_{yy}D_{zz} - D_{yz}D_{zy} & D_{zy}D_{xz} - D_{zz}D_{xy} & D_{xy}D_{yz} - D_{xz}D_{yy} \\ D_{zx}D_{yz} - D_{yx}D_{zz} & D_{xx}D_{zz} - D_{xz}D_{zx} & D_{xz}D_{yx} - D_{xx}D_{yz} \\ D_{yx}D_{zy} - D_{zx}D_{yy} & D_{xy}D_{zx} - D_{xx}D_{zy} & D_{xx}D_{yy} - D_{xy}D_{yx} \end{pmatrix}; \tag{2.122}$$

[16] The coordinate-free representation of the adjoint tensor and its components utilizes the completely antisymmetrical third rank permutation tensor by Levi–Cività (Chen 1983; de Hoop 1995); yet in the present elaboration, it is not urgently needed.

with

$$\det \underline{\underline{\mathbf{D}}} = D_{xx}D_{yy}D_{zz} + D_{xy}D_{yz}D_{zx} + D_{yx}D_{zy}D_{xz} - D_{xz}D_{yy}D_{zx} - D_{xy}D_{yx}D_{zz} - D_{xx}D_{zy}D_{yz}, \tag{2.123}$$

we refer to the determinant of $\underline{\underline{\mathbf{D}}}$. Obviously, inversion of a tensor necessarily requires $\det \underline{\underline{\mathbf{D}}} \neq 0$; if this is not true, the tensor is singular. Chen (1983) gives many formulas to calculate determinants and adjoints of tensors with given algebraic structure; we cite them in our Appendix "Collection of Formulas."

Complex valued tensors: A tensor $\underline{\underline{\mathbf{D}}}$ is complex valued if its components are complex numbers; this is generally true for the Fourier spectra of tensor fields, for instance, the Fourier spectrum $\underline{\underline{\mathbf{T}}}(\underline{\mathbf{R}}, \omega)$ of the stress tensor $\underline{\underline{\mathbf{T}}}(\underline{\mathbf{R}}, t)$. The Hermite-conjugate tensor $\underline{\underline{\mathbf{D}}}^+$ is obtained via transposition and simultaneous insertion of complex-conjugate components :

$$\begin{aligned} \underline{\underline{\mathbf{D}}}^+ &= D_{ij}^* \, \underline{\mathbf{e}}_{x_j} \underline{\mathbf{e}}_{x_i} \quad \text{(summation convention)} \\ &= D_{ji}^* \, \underline{\mathbf{e}}_{x_i} \underline{\mathbf{e}}_{x_j} \quad \text{(summation convention).} \end{aligned} \tag{2.124}$$

A complex valued tensor is called Hermitian if[17]

$$\underline{\underline{\mathbf{D}}}^+ = \underline{\underline{\mathbf{D}}}. \tag{2.125}$$

Analogously to (2.67), we define the "magnitude" of a complex valued (second rank) tensor:

$$|\underline{\underline{\mathbf{D}}}| = \sqrt{\underline{\underline{\mathbf{D}}} : \underline{\underline{\mathbf{D}}}^+}. \tag{2.126}$$

Then

$$\hat{\underline{\underline{\mathbf{D}}}} = \frac{\underline{\underline{\mathbf{D}}}}{|\underline{\underline{\mathbf{D}}}|} \tag{2.127}$$

turns out to be a "unit tensor" with magnitude 1.

Eigenvalue problems: Phase velocities of elastic plane waves in isotropic and anisotropic materials result as eigenvalues from an eigenvalue problem that originates from the time and space Fourier transformed wave equation; the longitudinal polarization of primary plane pressure and the transverse polarization of secondary plane shear waves in isotropic materials are consequences of the orientation of the eigenvectors of the eigenvalue problem.

Eigenvalues α of a second rank tensor $\underline{\underline{\mathbf{D}}}$ are defined as[18] those factors of a vector $\underline{\mathbf{A}}$ if the rotation $\underline{\underline{\mathbf{D}}} \cdot \underline{\mathbf{A}}$ exceptionally results in a (may be complex valued) length change of $\underline{\mathbf{A}}$: One states the eigenvalue problem

$$\underline{\underline{\mathbf{D}}} \cdot \underline{\mathbf{A}} = \alpha \underline{\mathbf{A}} \tag{2.128}$$

[17]The main diagonal elements are real valued, the off-diagonal elements are complex conjugate.

[18]Eigenvalue problems are also formulated for $n \times n$-matrices with $n > 3$.

and understands "exceptional" in the sense that this can only be true for selected vectors $\underline{\mathbf{A}}$—the eigenvectors—that are allocated to the pertinent eigenvalues. Eigenvalues and eigenvectors are solutions of (2.128).

By writing (2.128) according to

$$(\underline{\underline{\mathbf{D}}} - \alpha\underline{\underline{\mathbf{I}}}) \cdot \underline{\mathbf{A}} = \underline{\mathbf{0}}, \tag{2.129}$$

we see that the components of eventually existing eigenvectors must be solutions of the system of homogeneous equations

$$\begin{aligned} (D_{xx} - \alpha)A_x + D_{xy}A_y + D_{xz}A_z &= 0, \\ D_{yx}A_x + (D_{yy} - \alpha)A_y + D_{yz}A_z &= 0, \\ D_{zx}A_x + D_{zy}A_y + (D_{zz} - \alpha)A_z &= 0 \end{aligned} \tag{2.130}$$

with the coefficient matrix $\underline{\underline{\mathbf{D}}} - \alpha\underline{\underline{\mathbf{I}}}$. Systems of homogeneous equations only have nontrivial—nonzero—solutions if and only if the determinant of the coefficient matrix vanishes. Hence, we require

$$\det(\underline{\underline{\mathbf{D}}} - \alpha\underline{\underline{\mathbf{I}}}) = 0. \tag{2.131}$$

With (2.123), we find the explicit representation of (2.131) as a third-degree polynomial for the eventually existing eigenvalues; Chen (1983) gives the following short-hand notation:

$$\alpha^3 - \alpha^2 \operatorname{trace} \underline{\underline{\mathbf{D}}} + \alpha \operatorname{trace} \operatorname{adj} \underline{\underline{\mathbf{D}}} - \det \underline{\underline{\mathbf{D}}} = 0. \tag{2.132}$$

This so-called characteristic polynomial (characteristic for $\underline{\underline{\mathbf{D}}}$) exhibits (implying real valued components of $\underline{\underline{\mathbf{D}}}$)

- Either three not necessarily different real valued zeroes[19]
- Or one real valued and two complex conjugate zeroes.

For real symmetric (and complex Hermitian) tensors, only the first alternative is true: Their eigenvalues are always real valued! If the tensor is additionally positive definite, the eigenvalues are positive. The tensor $\underline{\underline{\mathbf{D}}}$ is positive definite if the quadratic form $\underline{\mathbf{R}} \cdot \underline{\underline{\mathbf{D}}} \cdot \underline{\mathbf{R}}$ is greater than zero for $R > 0$ and zero only if $R = 0$ holds.

In order to predict properties of eigenvectors $\underline{\mathbf{A}}_i$, $i = 1, 2, 3$, belonging to the eigenvalues α_i, $i = 1, 2, 3$, results concerning the structure of tensor adjoints for vanishing tensor determinants are required. We cite Chen (1983): For $\det(\underline{\underline{\mathbf{D}}} - \alpha_i \underline{\underline{\mathbf{I}}}) = 0$ either $\underline{\underline{\mathbf{D}}} - \alpha_i \underline{\underline{\mathbf{I}}}$ or $\operatorname{adj}(\underline{\underline{\mathbf{D}}} - \alpha_i \underline{\underline{\mathbf{I}}})$ is the dyadic product of two vectors, a so-called linear tensor; in the first case, $\operatorname{adj}(\underline{\underline{\mathbf{D}}} - \alpha_i \underline{\underline{\mathbf{I}}})$ is the null tensor. If $\underline{\underline{\mathbf{D}}} - \alpha_i \underline{\underline{\mathbf{I}}}$ is a dyadic, any vector orthogonal to the right factor of this dyadic is an eigenvector to the eigenvalue α_i of $\underline{\underline{\mathbf{D}}}$, and if $\operatorname{adj}(\underline{\underline{\mathbf{D}}} - \alpha_i \underline{\underline{\mathbf{I}}})$ is a dyadic (and not the null tensor), the eigenvector to the eigenvalue α_i is proportional to the left factor of that dyadic.

[19] The eigenvalue $\alpha = 0$ only exists for $\det \underline{\underline{\mathbf{D}}} = 0$, which means that the noninvertibility of a tensor (a matrix) can also be recognized by a vanishing eigenvalue.

For real valued symmetric tensors $\underline{\underline{\mathbf{D}}}$, the eigenvectors are real valued, and if they belong to different (real valued) eigenvalues, they are orthogonal to each other. It is exactly this result that we meet when we calculate the phase velocities and polarizations of plane waves in anisotropic materials (Section 8.3): The wave tensor is real valued and symmetric, its eigenvalues—the phase velocities—are real valued and distinct, the eigenvectors are real valued and orthogonal to each other. For isotropic materials (Section 8.1), we face a so-called degeneracy: Two eigenvectors are equal, hence *any* vector orthogonal to the eigenvector belonging to the third eigenvalue is an eigenvector to the identical eigenvalues. In the terminology of NDT: The polarization of transverse waves is arbitrary with regard to the polarization of longitudinal waves.

2.2 Vector and Tensor Analysis

Propagation of elastic waves implies the variation of vector fields—e.g., $\underline{\mathbf{u}}(\underline{\mathbf{R}}, t)$—and tensor fields—e.g., $\underline{\underline{\mathbf{S}}}(\underline{\mathbf{R}}, t)$, $\underline{\underline{\mathbf{T}}}(\underline{\mathbf{R}}, t)$—in space and time. What are the possibilities to forge appropriate mathematical equations for these physical variations? Fortunately, we are no longer in the situation of Isaac Newton who had to invent the necessary calculus beforehand; we can shop for vector and tensor analysis.

2.2.1 Del-operator: Gradient dyadic, gradient, divergence, and curl

Gradient dyadic: The variation of a scalar function $f(x)$ with x is characterized by its derivative[20]

$$
\begin{aligned}
f'(x) &= \lim_{\Delta x \to 0} \frac{f(x + \Delta x) - f(x)}{\Delta x} \\
&\stackrel{\text{def}}{=} \frac{\mathrm{d}f(x)}{\mathrm{d}x}.
\end{aligned}
\tag{2.133}
$$

Yet, the vector field quantity $\underline{\mathbf{u}}(\underline{\mathbf{R}}, t)$ has three scalar components each depending on three coordinates (Equation 2.10); therefore, in total, nine so-called partial derivatives can be calculated:

$$
\begin{array}{lll}
\dfrac{\partial u_x(x,y,z,t)}{\partial x}, & \dfrac{\partial u_y(x,y,z,t)}{\partial x}, & \dfrac{\partial u_z(x,y,z,t)}{\partial x}; \\
\dfrac{\partial u_x(x,y,z,t)}{\partial y}, & \dfrac{\partial u_y(x,y,z,t)}{\partial y}, & \dfrac{\partial u_z(x,y,z,t)}{\partial y}; \\
\dfrac{\partial u_x(x,y,z,t)}{\partial z}, & \dfrac{\partial u_y(x,y,z,t)}{\partial z}, & \dfrac{\partial u_z(x,y,z,t)}{\partial z}.
\end{array}
\tag{2.134}
$$

[20]As far as the mathematical conditions for the existence of derivatives are concerned, we refer to the literature (e.g.: Burg et al. 1990).

Additionally, three partial derivatives of the components with regard to time may be under concern:

$$\frac{\partial u_x(x,y,z,t)}{\partial t}, \quad \frac{\partial u_y(x,y,z,t)}{\partial t}, \quad \frac{\partial u_z(x,y,z,t)}{\partial t}. \tag{2.135}$$

The three time derivatives can be organized as a vector—the vector $\underline{\mathbf{v}}(\underline{\mathbf{R}},t)$ of the particle velocity—

$$\frac{\partial \underline{\mathbf{u}}(\underline{\mathbf{R}},t)}{\partial t} \stackrel{\text{def}}{=} \frac{\partial u_x(\underline{\mathbf{R}},t)}{\partial t}\underline{\mathbf{e}}_x + \frac{\partial u_y(\underline{\mathbf{R}},t)}{\partial t}\underline{\mathbf{e}}_y + \frac{\partial u_z(\underline{\mathbf{R}},t)}{\partial t}\underline{\mathbf{e}}_z \tag{2.136}$$

$$= \underline{\mathbf{v}}(\underline{\mathbf{R}},t) \tag{2.137}$$

and the nine spatial derivatives constitute a second rank tensor

$$\nabla\underline{\mathbf{u}}(\underline{\mathbf{R}},t) \stackrel{\text{def}}{=} \begin{pmatrix} \frac{\partial u_x(\underline{\mathbf{R}},t)}{\partial x} & \frac{\partial u_y(\underline{\mathbf{R}},t)}{\partial x} & \frac{\partial u_z(\underline{\mathbf{R}},t)}{\partial x} \\ \frac{\partial u_x(\underline{\mathbf{R}},t)}{\partial y} & \frac{\partial u_y(\underline{\mathbf{R}},t)}{\partial y} & \frac{\partial u_z(\underline{\mathbf{R}},t)}{\partial y} \\ \frac{\partial u_x(\underline{\mathbf{R}},t)}{\partial z} & \frac{\partial u_y(\underline{\mathbf{R}},t)}{\partial z} & \frac{\partial u_z(\underline{\mathbf{R}},t)}{\partial z} \end{pmatrix} \tag{2.138}$$

with the above matrix representation. Comparing this matrix representation with the one for the dyadic product (Equation 2.46), the interpretation $\nabla\underline{\mathbf{u}}(\underline{\mathbf{R}},t)$ as a dyadic product of the vector differential operator[21]

$$\nabla = \underline{\mathbf{e}}_x\frac{\partial}{\partial x} + \underline{\mathbf{e}}_y\frac{\partial}{\partial y} + \underline{\mathbf{e}}_z\frac{\partial}{\partial z} \tag{2.139}$$

with the vector $\underline{\mathbf{u}}(\underline{\mathbf{R}},t)$ is self-evident. Because of the upside down Greek letter capital delta ∇ is called del-operator; it is not underlined due to the fact that it is not a vector but a vector operator. The product $\nabla\underline{\mathbf{u}}(\underline{\mathbf{R}},t)$ is called gradient dyadic of $\underline{\mathbf{u}}(\underline{\mathbf{R}},t)$. The notion "gradient" is immediately plausible if we tentatively apply ∇ to a scalar field quantity, for instance, the acoustic pressure $p(\underline{\mathbf{R}},t)$.

Gradient: Applying the del-operator to $p(\underline{\mathbf{R}},t)$, we formally receive

$$\nabla p(\underline{\mathbf{R}},t) = \underline{\mathbf{e}}_x\frac{\partial p(\underline{\mathbf{R}},t)}{\partial x} + \underline{\mathbf{e}}_y\frac{\partial p(\underline{\mathbf{R}},t)}{\partial y} + \underline{\mathbf{e}}_z\frac{\partial p(\underline{\mathbf{R}},t)}{\partial z}, \tag{2.140}$$

[21] We purposely write the differential operator components *behind* the vectors of the orthonormal trihedron, because it is coercively necessary in other than cartesian coordinates; formally, we obtain, for instance, the $(x_i x_j)$-component of $\nabla\underline{\mathbf{u}}(\underline{\mathbf{R}},t)$ as

$$\underline{\mathbf{e}}_{x_i}\frac{\partial}{\partial x_i}(u_{x_j}\underline{\mathbf{e}}_{x_j}) = \underline{\mathbf{e}}_{x_i}\frac{\partial u_{x_j}}{\partial x_i}\underline{\mathbf{e}}_{x_j} = \frac{\partial u_{x_j}}{\partial x_i}\underline{\mathbf{e}}_{x_i}\underline{\mathbf{e}}_{x_j},$$

where the first equality sign holds because of the coordinate independence of the vectors of the orthonormal trihedron. However, for non-cartesian coordinates, the vector components of $\underline{\mathbf{u}}$ and therefore the pertinent j-unit vectors have to be differentiated with the ith coordinate (Section 2.2.4).

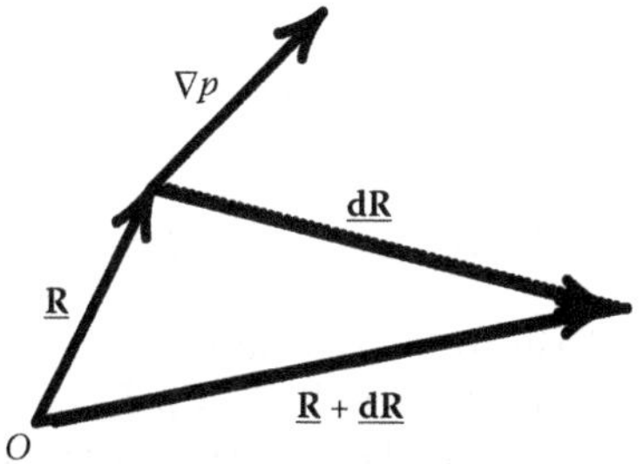

FIGURE 2.13
Definition of the gradient.

i.e., a vector. Evidently, this vector can be calculated for any spatial point $\underline{\mathbf{R}}$ (and any time t); it has a magnitude and a direction as sketched in Figure 2.13. Let us now consider a close-by spatial point $\underline{\mathbf{R}} + \underline{\mathbf{dR}}$ that is dislodged from $\underline{\mathbf{R}}$ by an infinitesimal vector $\underline{\mathbf{dR}}$; generally, the field quantity p will then have changed by the infinitesimal value

$$\mathrm{d}p = p(\underline{\mathbf{R}} + \underline{\mathbf{dR}}, t) - p(\underline{\mathbf{R}}, t). \tag{2.141}$$

This change $\mathrm{d}p$ can be calculated as total differential—sum of products of p-changes in the respective coordinate directions with the infinitesimal coordinate changes—

$$\mathrm{d}p = \frac{\partial p(\underline{\mathbf{R}}, t)}{\partial x}\,\mathrm{d}x + \frac{\partial p(\underline{\mathbf{R}}, t)}{\partial y}\,\mathrm{d}y + \frac{\partial p(\underline{\mathbf{R}}, t)}{\partial z}\,\mathrm{d}z, \tag{2.142}$$

which can be written as

$$\mathrm{d}p = \boldsymbol{\nabla} p(\underline{\mathbf{R}}, t) \cdot \underline{\mathbf{dR}}, \tag{2.143}$$

with $\mathrm{d}x, \mathrm{d}y, \mathrm{d}z$ denoting the components of $\underline{\mathbf{dR}}$. Combining (2.143) with (2.141) yields

$$p(\underline{\mathbf{R}} + \underline{\mathbf{dR}}, t) = p(\underline{\mathbf{R}}, t) + \boldsymbol{\nabla} p(\underline{\mathbf{R}}, t) \cdot \underline{\mathbf{dR}}. \tag{2.144}$$

Now we choose two particular spatial directions $\underline{\mathbf{dR}}$:

- $\underline{\mathbf{dR}}$ orthogonal to the vector $\boldsymbol{\nabla} p(\underline{\mathbf{R}}, t)$: The scalar product $\boldsymbol{\nabla} p(\underline{\mathbf{R}}, t) \cdot \underline{\mathbf{dR}}$ is equal to zero, that is to say, the variation of $p(\underline{\mathbf{R}}, t)$ orthogonal to $\boldsymbol{\nabla} p(\underline{\mathbf{R}}, t)$ is zero!

- $\underline{\mathbf{dR}}$ parallel to the vector $\boldsymbol{\nabla} p(\underline{\mathbf{R}}, t)$: The scalar product $\boldsymbol{\nabla} p(\underline{\mathbf{R}}, t) \cdot \underline{\mathbf{dR}}$ is maximum, that is to say, the variation of $p(\underline{\mathbf{R}}, t)$ in the direction of $\boldsymbol{\nabla} p(\underline{\mathbf{R}}, t)$ is maximum!

We conclude that: In any spatial point, the vector $\nabla p(\underline{\mathbf{R}}, t)$ points into that direction which coincides with the strongest variation of the field quantity; hence, it is called the gradient of $p(\underline{\mathbf{R}}, t)$ with the occasional notation

$$\nabla p(\underline{\mathbf{R}}, t) = \operatorname{grad} p(\underline{\mathbf{R}}, t). \tag{2.145}$$

Let S_g be a—really existing or mathematically virtual—closed surface with outward normal $\underline{\mathbf{n}}$; projecting the gradient $\nabla p(\underline{\mathbf{R}}, t)$ as calculated at a particular point on this surface onto the direction of the normal, we obtain the so-called normal derivative

$$\underline{\mathbf{n}} \cdot \nabla p(\underline{\mathbf{R}}, t) \stackrel{\text{def}}{=} \frac{\partial p(\underline{\mathbf{R}}, t)}{\partial n}, \quad \underline{\mathbf{R}} \in S_g, \tag{2.146}$$

of the scalar field $p(\underline{\mathbf{R}}, t)$.

The normal derivative plays an important role in Huygens' principle for scalar fields (Section 15.1.2) because its knowledge on the total surface S_g together with the knowledge of $p(\underline{\mathbf{R}}, t)$ on S_g is sufficient for the knowledge of $p(\underline{\mathbf{R}}, t)$ interior or exterior of S_g depending on whether the sources of the field $p(\underline{\mathbf{R}}, t)$ are located interior or exterior of S_g.

Apparently, the matrix scheme (2.138) of the gradient dyadic of a vector field exhibits the gradients of the scalar components of the field as column vectors:

$$\nabla \underline{\mathbf{u}}(\underline{\mathbf{R}}, t) = \left(\nabla u_x(\underline{\mathbf{R}}, t) \quad \nabla u_y(\underline{\mathbf{R}}, t) \quad \nabla u_z(\underline{\mathbf{R}}, t) \right), \tag{2.147}$$

so that we have a dyadic $\nabla \underline{\mathbf{u}}(\underline{\mathbf{R}}, t)$ at hand that contains the complete information about the variation of the vector field $\underline{\mathbf{u}}(\underline{\mathbf{R}}, t)$ at any spatial point and for any time. As a matter of fact, the symmetric part of the gradient dyadic $\nabla \underline{\mathbf{u}}(\underline{\mathbf{R}}, t)$ defines the deformation tensor $\underline{\underline{\mathbf{S}}}(\underline{\mathbf{R}}, t)$ (Section 3.1).

Divergence: The gradients of the scalar components of $\underline{\mathbf{u}}(\underline{\mathbf{R}}, t)$ originate from the respective aggregation of the components of the gradient dyadic $\nabla \underline{\mathbf{u}}(\underline{\mathbf{R}}, t)$. There are two other possibilities to combine components of $\nabla \underline{\mathbf{u}}(\underline{\mathbf{R}}, t)$ in a way that the resulting expressions give evidence of the physical properties of the vector field $\underline{\mathbf{u}}(\underline{\mathbf{R}}, t)$. The first possibility yields the divergence (source density) of the vector field. We compose the trace of the gradient dyadic according to

$$\operatorname{trace} \nabla \underline{\mathbf{u}}(\underline{\mathbf{R}}, t) = \frac{\partial u_x(\underline{\mathbf{R}}, t)}{\partial x} + \frac{\partial u_y(\underline{\mathbf{R}}, t)}{\partial y} + \frac{\partial u_z(\underline{\mathbf{R}}, t)}{\partial z} \tag{2.148}$$

and state that we can write it formally as a contraction of the del-operator with $\underline{\mathbf{u}}(\underline{\mathbf{R}}, t)$:

$$\nabla \cdot \underline{\mathbf{u}}(\underline{\mathbf{R}}, t) = \frac{\partial u_x(\underline{\mathbf{R}}, t)}{\partial x} + \frac{\partial u_y(\underline{\mathbf{R}}, t)}{\partial y} + \frac{\partial u_z(\underline{\mathbf{R}}, t)}{\partial z}. \tag{2.149}$$

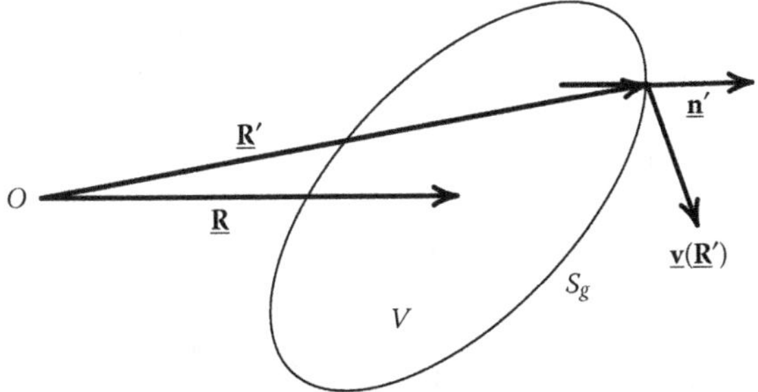

FIGURE 2.14
Definition of the divergence of a vector field.

One calls

$$\nabla \cdot \underline{\mathbf{u}}(\underline{\mathbf{R}}, t) = \operatorname{div} \underline{\mathbf{u}}(\underline{\mathbf{R}}, t) \tag{2.150}$$

divergence of the vector field; the reason will be made plausible below.

In Figure 2.14, a closed surface S_g is sketched that encloses the volume V; the vector $\underline{\mathbf{R}}$ indicates the center of gravity of this volume. The outward normal of S_g at the point $\underline{\mathbf{R}}'$ is given by $\underline{\mathbf{n}}'$. We consider a (stationary) vector field $\underline{\mathbf{v}}$ that might represent the flow velocity of an incompressible fluid. At $\underline{\mathbf{R}}'$ on S_g, $\underline{\mathbf{v}}(\underline{\mathbf{R}}')$ should have the direction as indicated, that is to say, the flow exhibits a component parallel and a component orthogonal to the surface. Evidently, the orthogonal component[22]—the normal component $\underline{\mathbf{n}}' \cdot \underline{\mathbf{v}}(\underline{\mathbf{R}}')$—measures the flux through the surface; summation—i.e., integration—of this local flux over S_g yields the total flux of the vector field through S_g:[23]

$$\text{flux of } \underline{\mathbf{v}} \text{ through } S_g = \int\!\!\int_{S_g} \underline{\mathbf{n}}' \cdot \underline{\mathbf{v}}(\underline{\mathbf{R}}')\, \mathrm{d}S'; \tag{2.151}$$

here, $\mathrm{d}S'$ denotes the infinitesimal surface element of S_g at $\underline{\mathbf{R}}'$. This flux of $\underline{\mathbf{v}}$ through S_g is a number that may be positive, negative, or zero: If positive, we observe a loss of fluid in the volume V, and due to the incompressibility of the stationary flow, this must be equivalent to the existence of a (net-)source in the interior of S_g that exhibits the same intensity. Accordingly, a negative flux is tantamount to a (net-)drain (sink) in the interior of S_g, and a vanishing flux means that there are neither sources nor sinks in V or, equivalently, sources and sinks cancel each other, and the outward and inward fluxes are equal.

With the flux, we define a global property of a vector field; with the divergence, we do that locally. The global definition of the flux would change into a local definition via a series of volumes contracting to the point $\underline{\mathbf{R}}$;

[22]The parallel component is subsequently considered to define the curl of a vector field.
[23]Concerning the explicit calculation of such a surface integral, we refer to the literature (e.g., Burg et al. 1990; Langenberg 2005).

yet, in the limit of an arbitrarily small volume, the flux integral (2.151) would yield a zero value because of the arbitrarily small integration surface. As a remedy, we normalize the flux to the respective volume—thus defining a flux density—yielding the indefinite expression "zero over zero" that might have a finite value; in that case, it defines the positive or negative source density—the divergence—of the vector field locally at $\underline{\mathbf{R}}$:

$$\operatorname{div}\underline{\mathbf{v}}(\underline{\mathbf{R}}) = \lim_{V\to 0}\frac{1}{V}\int\!\!\int_{S_g}\underline{\mathbf{n}}'\cdot\underline{\mathbf{v}}(\underline{\mathbf{R}}')\,\mathrm{d}S'. \tag{2.152}$$

The above recipe to calculate a divergence is rather intuitive yet it lacks practicability: It would be better to do it based on the components of $\underline{\mathbf{v}}$! To achieve this, the series of integrals is evaluated for a cubic volume fitting into a cartesian coordinate system; then, the limit is calculated with the help of the midpoint theorem (Burg et al. 1990): We find

$$\begin{aligned}\operatorname{div}\underline{\mathbf{v}}(\underline{\mathbf{R}}) &= \frac{\partial v_x(\underline{\mathbf{R}})}{\partial x} + \frac{\partial v_y(\underline{\mathbf{R}})}{\partial y} + \frac{\partial v_z(\underline{\mathbf{R}})}{\partial z}\\ &= \boldsymbol{\nabla}\cdot\underline{\mathbf{v}}(\underline{\mathbf{R}}).\end{aligned} \tag{2.153}$$

The "generalization" to time-dependent vector fields is given by equation (2.149).

The mathematical evaluation of the physically significant divergence definition as contraction of the del-operator with a vector[24] immediately allows for generalizations, for instance, the divergence of the tensor according to

$$\begin{aligned}\boldsymbol{\nabla}\cdot\underline{\underline{\mathbf{T}}}(\underline{\mathbf{R}},t) &\stackrel{\text{def}}{=} \operatorname{div}\underline{\underline{\mathbf{T}}}(\underline{\mathbf{R}},t)\\ &= \frac{\partial T_{x_ix_j}(\underline{\mathbf{R}},t)}{\partial x_i}\,\underline{\mathbf{e}}_{x_j}\ \text{(summation convention)}\\ &= \left(\frac{\partial T_{xx}(\underline{\mathbf{R}},t)}{\partial x} + \frac{\partial T_{yx}(\underline{\mathbf{R}},t)}{\partial y} + \frac{\partial T_{zx}(\underline{\mathbf{R}},t)}{\partial z}\right)\underline{\mathbf{e}}_x\\ &\quad + \left(\frac{\partial T_{xy}(\underline{\mathbf{R}},t)}{\partial x} + \frac{\partial T_{yy}(\underline{\mathbf{R}},t)}{\partial y} + \frac{\partial T_{zy}(\underline{\mathbf{R}},t)}{\partial z}\right)\underline{\mathbf{e}}_y\\ &\quad + \left(\frac{\partial T_{xz}(\underline{\mathbf{R}},t)}{\partial x} + \frac{\partial T_{yz}(\underline{\mathbf{R}},t)}{\partial y} + \frac{\partial T_{zz}(\underline{\mathbf{R}},t)}{\partial z}\right)\underline{\mathbf{e}}_z.\end{aligned} \tag{2.154}$$

If $\underline{\underline{\mathbf{T}}}(\underline{\mathbf{R}},t)$ denotes the stress tensor $\operatorname{div}\underline{\underline{\mathbf{T}}}(\underline{\mathbf{R}},t)$ defines, according to Cauchy, a force density inherent to the solid that is the origin, according to Newton, for particle accelerations: We found the Newton–Cauchy governing equation of elastodynamics!

[24]Obviously, this contraction does not commute like a scalar product because $\underline{\mathbf{u}}(\underline{\mathbf{R}},t)\cdot\boldsymbol{\nabla}$ is meaningless.

Curl: Apparently, after the contraction application of the del-operator, it is only a small step to the cross product application according to

$$\begin{aligned}\nabla \times \underline{\mathbf{u}}(\underline{\mathbf{R}}, t) = &\left(\frac{\partial u_z(\underline{\mathbf{R}}, t)}{\partial y} - \frac{\partial u_y(\underline{\mathbf{R}}, t)}{\partial z}\right) \underline{\mathbf{e}}_x \\ &+ \left(\frac{\partial u_x(\underline{\mathbf{R}}, t)}{\partial z} - \frac{\partial u_z(\underline{\mathbf{R}}, t)}{\partial x}\right) \underline{\mathbf{e}}_y \\ &+ \left(\frac{\partial u_y(\underline{\mathbf{R}}, t)}{\partial x} - \frac{\partial u_x(\underline{\mathbf{R}}, t)}{\partial y}\right) \underline{\mathbf{e}}_z. \end{aligned} \tag{2.155}$$

At first, we observe that we have "discovered" another possibility to combine the elements of the gradient dyadic $\nabla \underline{\mathbf{u}}(\underline{\mathbf{R}}, t)$: The elements directly below the main diagonal are subtracted from their mirror elements and declared as z- and x-components, respectively, of a vector, and subtraction of the upper right corner element from the lower left corner element yields the missing y-component of this so-called rotation (curl) vector of the gradient dyadic [compare (2.89)]:

$$\langle \nabla \underline{\mathbf{u}}(\underline{\mathbf{R}}, t) \rangle = \nabla \times \underline{\mathbf{u}}(\underline{\mathbf{R}}, t). \tag{2.156}$$

The rotation vector $\langle \nabla \underline{\mathbf{u}}(\underline{\mathbf{R}}, t) \rangle$ even has the physical meaning[25] of a rotation or curl density of $\underline{\mathbf{u}}(\underline{\mathbf{R}}, t)$. We recall Figure 2.14, to illustrate the source density $\operatorname{div} \underline{\mathbf{v}}(\underline{\mathbf{R}})$ of a stationary incompressible fluid, we added up the normal components of $\underline{\mathbf{v}}(\underline{\mathbf{R}})$ in terms of a flux integral; now we integrate the tangential components according to (2.38) to define a "curl" integral:

$$\text{curl of } \underline{\mathbf{v}} \text{ on } S_g = \int\!\!\int_{S_g} \underline{\mathbf{n}}' \times \underline{\mathbf{v}}(\underline{\mathbf{R}}') \, \mathrm{d}S'. \tag{2.157}$$

Normalization to the volume V and performance of the limit $V \longrightarrow 0$ intuitively yields a local curl density of $\underline{\mathbf{v}}(\underline{\mathbf{R}})$:

$$\operatorname{curl} \underline{\mathbf{v}}(\underline{\mathbf{R}}) = \lim_{V \to 0} \frac{1}{V} \int\!\!\int_{S_g} \underline{\mathbf{n}}' \times \underline{\mathbf{v}}(\underline{\mathbf{R}}') \, \mathrm{d}S'. \tag{2.158}$$

Calculation of the integral and the limit in Cartesian coordinates for a cubic volume actually provides (Burg et al. 1990)

$$\operatorname{curl} \underline{\mathbf{v}}(\underline{\mathbf{R}}) = \nabla \times \underline{\mathbf{v}}(\underline{\mathbf{R}}), \tag{2.159}$$

namely (2.155) according to

$$\nabla \times \underline{\mathbf{u}}(\underline{\mathbf{R}}, t) = \operatorname{curl} \underline{\mathbf{u}}(\underline{\mathbf{R}}, t), \tag{2.160}$$

---

[25]With

$$\nabla \underline{\mathbf{u}}_{\mathrm{a}} = -\frac{1}{2} (\nabla \times \underline{\mathbf{u}}) \times \underline{\underline{\mathbf{I}}},$$

we can calculate the antisymmetric part $\nabla \underline{\mathbf{u}}_{\mathrm{a}}$ of the gradient dyadic $\nabla \underline{\mathbf{u}}$ according to (2.91).

"generalizing" once again to a spatially and time-dependent vector field. Since we add up vectors in terms of tangential components in the integral (2.158), it is obvious that curl $\underline{\mathbf{v}}(\underline{\mathbf{R}})$ is a vector; we can envisage this vector just like the angular momentum of a small $\underline{\mathbf{v}}$-vortex, which, as it is well-known, is oriented orthogonally to the vortex surface.[26]

2.2.2 Application of the del-operator to products of field quantities, chain rules, delta-operator

Application of the del-operator to products of field quantities: Many possibilities exist to construct products of field quantities and, depending on the version of the result—scalar, vector, and tensor—the del-operator can be applied as gradient, divergence, or curl. Often needed results can be found in the formula collections of respective books; a very comprehensive collection is appended to this elaboration.

Some examples particularly useful for the derivation of plane wave solutions of the governing equations of elastodynamics and for the mathematical formulation of Huygens' principle, and the energy conservation theorem will be explicitly discussed.

The simplest product is the product of two scalar field quantities $\Phi(\underline{\mathbf{R}}, t)$ and $\Psi(\underline{\mathbf{R}}, t)$; We investigate the gradient of this product:[27]

$$\nabla(\Phi\Psi) = \Psi\nabla\Phi + \Phi\nabla\Psi. \tag{2.161}$$

To prove this equation, the product rule of differential calculus is applied to the partial derivatives as contained in ∇; afterward, the single terms are combined to ∇-operations.

We consider Ψ in (2.161) as ith scalar component of a vector $\underline{\mathbf{A}}$, calculate

$$\nabla(\Phi\Psi_i) = \Psi_i\nabla\Phi + \Phi\nabla\Psi_i, \quad i = 1, 2, 3, \tag{2.162}$$

and combine the three vector equations to the dyadic

$$\nabla(\Phi\underline{\mathbf{A}}) = (\underline{\mathbf{A}}\nabla\Phi)^{21} + \Phi\nabla\underline{\mathbf{A}}, \tag{2.163}$$

bearing in mind that the first index must be the ∇-index.

Again, we replace Ψ in (2.161) by a vector $\underline{\mathbf{A}}$, this time calculating the divergence using the summation convention:

---

[26]This is exceptionally descriptive with Ampère's theorem stating the following for magnetic fields $\underline{\mathbf{H}}(\underline{\mathbf{R}})$ of stationary current densities $\underline{\mathbf{J}}(\underline{\mathbf{R}})$:

$$\operatorname{curl}\underline{\mathbf{H}}(\underline{\mathbf{R}}) = \underline{\mathbf{J}}(\underline{\mathbf{R}});$$

an infinitely long current carrying wire is surrounded by circular magnetic field lines: The curl density curl $\underline{\mathbf{H}}(\underline{\mathbf{R}})$ of the magnetic field is oriented in the direction of the current density $\underline{\mathbf{J}}(\underline{\mathbf{R}})$.

[27]We ignore the arguments to enhance the facility of inspection.

$$\begin{aligned}
\boldsymbol{\nabla} \cdot (\Phi \underline{\mathbf{A}}) &= \underline{\mathbf{e}}_{x_i} \cdot \frac{\partial}{\partial x_i}(\Phi A_{x_j} \underline{\mathbf{e}}_{x_j}) \\
&= \underbrace{\underline{\mathbf{e}}_{x_i} \cdot \underline{\mathbf{e}}_{x_j}}_{= \delta_{ij}} \left(\frac{\partial \Phi}{\partial x_i} A_{x_j} + \Phi \frac{\partial A_{x_j}}{\partial x_i} \right) \\
&= \frac{\partial \Phi}{\partial x_i} A_{x_i} + \Phi \frac{\partial A_{x_i}}{\partial x_i} \\
&= (\boldsymbol{\nabla} \Phi) \cdot \underline{\mathbf{A}} + \Phi \boldsymbol{\nabla} \cdot \underline{\mathbf{A}}.
\end{aligned} \tag{2.164}$$

Without the summation convention, this result is also found, only the number of symbols to write is larger. Up to the last but one line a calculus as above is more or less trivial, and only the combination to explicit del-operations as in the last line requires some thinking.

The curl of the product $\Phi\underline{\mathbf{A}}$ is taken from the collection of formulas:[28]

$$\boldsymbol{\nabla} \times (\Phi \underline{\mathbf{A}}) = \Phi \boldsymbol{\nabla} \times \underline{\mathbf{A}} - \underline{\mathbf{A}} \times \boldsymbol{\nabla} \Phi. \tag{2.165}$$

We continue with the gradient of the scalar product of two vectors:

$$\begin{aligned}
\boldsymbol{\nabla}(\underline{\mathbf{A}} \cdot \underline{\mathbf{B}}) &= \underline{\mathbf{e}}_{x_i} \frac{\partial}{\partial x_i}(A_{x_j} B_{x_j}) \\
&= \underline{\mathbf{e}}_{x_i} \left(\frac{\partial A_{x_j}}{\partial x_i} B_{x_j} + A_{x_j} \frac{\partial B_{x_j}}{\partial x_i} \right) \\
&= \underline{\mathbf{e}}_{x_i} \frac{\partial A_{x_j}}{\partial x_i} B_{x_j} + \underline{\mathbf{e}}_{x_i} \frac{\partial B_{x_j}}{\partial x_i} A_{x_j} \\
&= (\boldsymbol{\nabla} \underline{\mathbf{A}}) \cdot \underline{\mathbf{B}} + (\boldsymbol{\nabla} \underline{\mathbf{B}}) \cdot \underline{\mathbf{A}};
\end{aligned} \tag{2.166}$$

obviously, the gradient dyadics of the respective vectors appear. Writing down the last line of (2.166), we have to be careful with the contraction of the correct indices of the gradient dyadic: In $\boldsymbol{\nabla}(\underline{\mathbf{A}} \cdot \underline{\mathbf{B}})$, the vector index is the index of $\boldsymbol{\nabla}$ and this must also be true for the final result.

The divergence of the dyadic product of two vectors is calculated as follows:

$$\begin{aligned}
\boldsymbol{\nabla} \cdot (\underline{\mathbf{A}}\,\underline{\mathbf{B}}) &= \underline{\mathbf{e}}_{x_i} \cdot \frac{\partial}{\partial x_i}(A_{x_j} B_{x_k} \underline{\mathbf{e}}_{x_j} \underline{\mathbf{e}}_{x_k}) \\
&= \underbrace{\underline{\mathbf{e}}_{x_i} \cdot \underline{\mathbf{e}}_{x_j}}_{= \delta_{ij}} \underline{\mathbf{e}}_{x_k} \left(\frac{\partial A_{x_j}}{\partial x_i} B_{x_k} + A_{x_j} \frac{\partial B_{x_k}}{\partial x_i} \right) \\
&= \underline{\mathbf{e}}_{x_k} \left(\frac{\partial A_{x_i}}{\partial x_i} B_{x_k} + A_{x_i} \frac{\partial B_{x_k}}{\partial x_i} \right) \\
&= \frac{\partial A_{x_i}}{\partial x_i} B_{x_k} \underline{\mathbf{e}}_{x_k} + A_{x_i} \frac{\partial B_{x_k}}{\partial x_i} \underline{\mathbf{e}}_{x_k} \\
&= (\boldsymbol{\nabla} \cdot \underline{\mathbf{A}}) \underline{\mathbf{B}} + \underline{\mathbf{A}} \cdot \boldsymbol{\nabla} \underline{\mathbf{B}}.
\end{aligned} \tag{2.167}$$

[28]Without the Levi–Cività tensor, the calculation is somewhat circumstantial.

A last example, the vector $\underline{\mathbf{S}}(\underline{\mathbf{R}}, t)$ of the elastodynamic energy flow density—the elastodynamic counterpart to the electromagnetic Poynting vector (Section 4.3)—is defined as such $\underline{\mathbf{S}}(\underline{\mathbf{R}}, t) = -\underline{\mathbf{v}}(\underline{\mathbf{R}}, t) \cdot \underline{\underline{\mathbf{T}}}(\underline{\mathbf{R}}, t)$; according to the law of energy conservation, its (positive or negative) local source density—its divergence—must be equivalent to the local increase or loss of elastodynamic energy density. Providently, we calculate

$$
\begin{aligned}
\nabla \cdot (\underline{\mathbf{A}} \cdot \underline{\underline{\mathbf{D}}}) &= \underline{\mathbf{e}}_{x_i} \cdot \frac{\partial}{\partial x_i} (A_{x_j} D_{x_j x_k} \underline{\mathbf{e}}_{x_k}) \\
&= \underbrace{\underline{\mathbf{e}}_{x_i} \cdot \underline{\mathbf{e}}_{x_k}}_{= \delta_{ik}} \left(\frac{\partial A_{x_i}}{\partial x_i} D_{x_j x_k} + A_{x_j} \frac{\partial D_{x_j x_k}}{\partial x_i} \right) \\
&= \frac{\partial A_{x_j}}{\partial x_k} D_{x_j x_k} + A_{x_j} \frac{\partial D_{x_j x_k}}{\partial x_k} \\
&= (\nabla \underline{\mathbf{A}}) : \underline{\underline{\mathbf{D}}} + \underline{\mathbf{A}} \cdot \nabla \cdot \underline{\underline{\mathbf{D}}}^{21}.
\end{aligned} \tag{2.168}
$$

Chain rules for gradient, divergence, and curl: The mathematical representation of a time harmonic plane wave contains the function

$$
\mathrm{e}^{\mathrm{j}k \hat{\underline{\mathbf{k}}} \cdot \underline{\mathbf{R}}} \tag{2.169}
$$

with $k > 0$ being a constant and $\hat{\underline{\mathbf{k}}}$ a unit vector. Based on the time harmonic scalar Green function

$$
\frac{\mathrm{e}^{\mathrm{j}k|\underline{\mathbf{R}} - \underline{\mathbf{R}}'|}}{4\pi|\underline{\mathbf{R}} - \underline{\mathbf{R}}'|}, \tag{2.170}
$$

the dyadic Green function of elastodynamics is derived. Both examples are functions—exponential function e^{ϕ_1} and hyperbolic function $1/\phi_2$—whose arguments ϕ_1, ϕ_2 are functions of the vector of position. Gradient calculation of (2.169) and (2.170), therefore, requires an "interior derivative"; the counterpart of differential calculus is the chain rule, and here, we present the chain rules for gradient, divergence, and curl:

$$
\nabla \Phi[\phi(\underline{\mathbf{R}})] = \frac{\partial \Phi(\phi)}{\partial \phi} \nabla \phi(\underline{\mathbf{R}}); \tag{2.171}
$$

$$
\nabla \cdot \underline{\mathbf{A}}[\phi(\underline{\mathbf{R}})] = \frac{\partial \underline{\mathbf{A}}(\phi)}{\partial \phi} \cdot \nabla \phi(\underline{\mathbf{R}}); \tag{2.172}
$$

$$
\nabla \times \underline{\mathbf{A}}[\phi(\underline{\mathbf{R}})] = - \frac{\partial \underline{\mathbf{A}}(\phi)}{\partial \phi} \times \nabla \phi(\underline{\mathbf{R}}); \tag{2.173}
$$

$$
\nabla \underline{\mathbf{A}}[\phi(\underline{\mathbf{R}})] = \left[\frac{\partial \underline{\mathbf{A}}(\phi)}{\partial \phi} \nabla \phi(\underline{\mathbf{R}}) \right]^{21}. \tag{2.174}
$$

With the short-hand notation $\phi(\underline{\mathbf{R}}) = \mathrm{j}k \, \hat{\underline{\mathbf{k}}} \cdot \underline{\mathbf{R}}$, we calculate with the help of the summation convention:

$$
\begin{aligned}
\nabla \mathrm{e}^{\mathrm{j}k \hat{\underline{\mathbf{k}}} \cdot \underline{\mathbf{R}}} &= \nabla \mathrm{e}^{\phi(\underline{\mathbf{R}})} \\
&= \mathrm{e}^{\phi(\underline{\mathbf{R}})} \nabla \phi(\underline{\mathbf{R}})
\end{aligned}
$$

$$
\begin{aligned}
&= \mathrm{e}^{\mathrm{j}k\,\hat{\underline{\mathbf{k}}}\cdot\underline{\mathbf{R}}}\,\mathrm{j}k\,\nabla(\hat{\underline{\mathbf{k}}}\cdot\underline{\mathbf{R}}) \\
&= \mathrm{e}^{\mathrm{j}k\,\hat{\underline{\mathbf{k}}}\cdot\underline{\mathbf{R}}}\,\mathrm{j}k\,\underline{\mathbf{e}}_{x_i}\frac{\partial}{\partial x_i}(\hat{k}_{x_j}x_j) \\
&= \mathrm{e}^{\mathrm{j}k\,\hat{\underline{\mathbf{k}}}\cdot\underline{\mathbf{R}}}\,\mathrm{j}k\,\hat{k}_{x_j}\underline{\mathbf{e}}_{x_i}\underbrace{\frac{\partial x_j}{\partial x_i}}_{=\,\delta_{ij}} \\
&= \mathrm{e}^{\mathrm{j}k\,\hat{\underline{\mathbf{k}}}\cdot\underline{\mathbf{R}}}\,\mathrm{j}k\,\hat{k}_{x_i}\underline{\mathbf{e}}_{x_i} \\
&= \mathrm{j}k\,\hat{\underline{\mathbf{k}}}\,\mathrm{e}^{\mathrm{j}k\,\hat{\underline{\mathbf{k}}}\cdot\underline{\mathbf{R}}}.
\end{aligned}
\tag{2.175}
$$

To assess the gradient of the scalar Green function (2.170), we utilize the gradient product rule (2.161) according to

$$
\begin{aligned}
\nabla\frac{\mathrm{e}^{\mathrm{j}k|\underline{\mathbf{R}}-\underline{\mathbf{R}}'|}}{4\pi|\underline{\mathbf{R}}-\underline{\mathbf{R}}'|} &= \frac{1}{4\pi|\underline{\mathbf{R}}-\underline{\mathbf{R}}'|}\nabla\mathrm{e}^{\mathrm{j}k|\underline{\mathbf{R}}-\underline{\mathbf{R}}'|} \\
&\quad + \mathrm{e}^{\mathrm{j}k|\underline{\mathbf{R}}-\underline{\mathbf{R}}'|}\,\nabla\frac{1}{4\pi|\underline{\mathbf{R}}-\underline{\mathbf{R}}'|},
\end{aligned}
\tag{2.176}
$$

introduce $\phi(\underline{\mathbf{R}}) = |\underline{\mathbf{R}}-\underline{\mathbf{R}}'|$, and find with the gradient chain rule:

$$
\begin{aligned}
\nabla\frac{\mathrm{e}^{\mathrm{j}k|\underline{\mathbf{R}}-\underline{\mathbf{R}}'|}}{4\pi|\underline{\mathbf{R}}-\underline{\mathbf{R}}'|} &= \frac{1}{4\pi\phi(\underline{\mathbf{R}})}\nabla\mathrm{e}^{\mathrm{j}k\phi(\underline{\mathbf{R}})} + \mathrm{e}^{\mathrm{j}k\phi(\underline{\mathbf{R}})}\,\nabla\frac{1}{4\pi\phi(\underline{\mathbf{R}})} \\
&= \frac{\mathrm{j}k}{4\pi\phi(\underline{\mathbf{R}})}\,\mathrm{e}^{\mathrm{j}k\phi(\underline{\mathbf{R}})}\,\nabla\phi(\underline{\mathbf{R}}) - \mathrm{e}^{\mathrm{j}k\phi(\underline{\mathbf{R}})}\frac{1}{4\pi\phi^2(\underline{\mathbf{R}})}\nabla\phi(\underline{\mathbf{R}}) \\
&= \mathrm{j}k\,\frac{\underline{\mathbf{R}}-\underline{\mathbf{R}}'}{4\pi|\underline{\mathbf{R}}-\underline{\mathbf{R}}'|^2}\,\mathrm{e}^{\mathrm{j}k|\underline{\mathbf{R}}-\underline{\mathbf{R}}'|} - \frac{\underline{\mathbf{R}}-\underline{\mathbf{R}}'}{4\pi|\underline{\mathbf{R}}-\underline{\mathbf{R}}'|^3}\,\mathrm{e}^{\mathrm{j}k|\underline{\mathbf{R}}-\underline{\mathbf{R}}'|} \\
&= \frac{\underline{\mathbf{R}}-\underline{\mathbf{R}}'}{|\underline{\mathbf{R}}-\underline{\mathbf{R}}'|}\,\frac{\mathrm{e}^{\mathrm{j}k|\underline{\mathbf{R}}-\underline{\mathbf{R}}'|}}{4\pi|\underline{\mathbf{R}}-\underline{\mathbf{R}}'|}\left(\mathrm{j}k - \frac{1}{|\underline{\mathbf{R}}-\underline{\mathbf{R}}'|}\right),
\end{aligned}
\tag{2.177}
$$

because

$$
\begin{aligned}
\nabla\phi(\underline{\mathbf{R}}) &= \nabla|\underline{\mathbf{R}}-\underline{\mathbf{R}}'| \\
&= \frac{\underline{\mathbf{R}}-\underline{\mathbf{R}}'}{|\underline{\mathbf{R}}-\underline{\mathbf{R}}'|},
\end{aligned}
\tag{2.178}
$$

as it is readily computed in cartesian coordinates using (2.9).[29] Two facts are worth being noticed:

- The gradient of the magnitude of the distance between source and observation point $|\underline{\mathbf{R}}-\underline{\mathbf{R}}'|$ is the unit vector in $(\underline{\mathbf{R}}-\underline{\mathbf{R}}')$-direction.

[29]Note: Due to the use of the gradient product and chain rules, we need coordinates to calculate $\nabla|\underline{\mathbf{R}}-\underline{\mathbf{R}}'|$.

- The gradient of the scalar Green function—the elastodynamic Green dyadic results from a double gradient of the scalar Green function!—reproduces the scalar Green function, and it contains an additional term proportional to the inverse distance between source and observation point: If this distance is very large (whatever that means at the moment), this term may be eventually disregarded. To calculate far-fields of transducers, this simplification is tremendously useful.

Delta-operator: We consider the gradient $\nabla p(\underline{\mathbf{R}}, t)$ of a scalar field and compute the divergence:

$$\begin{aligned}\nabla \cdot \nabla p(\underline{\mathbf{R}}, t) &= \frac{\partial}{\partial x}\left(\frac{\partial p(\underline{\mathbf{R}}, t)}{\partial x}\right) + \frac{\partial}{\partial y}\left(\frac{\partial p(\underline{\mathbf{R}}, t)}{\partial y}\right) + \frac{\partial}{\partial z}\left(\frac{\partial p(\underline{\mathbf{R}}, t)}{\partial z}\right) \\ &= \frac{\partial^2 p(\underline{\mathbf{R}}, t)}{\partial x^2} + \frac{\partial^2 p(\underline{\mathbf{R}}, t)}{\partial y^2} + \frac{\partial^2 p(\underline{\mathbf{R}}, t)}{\partial z^2} \\ &\stackrel{\text{def}}{=} \Delta p(\underline{\mathbf{R}}, t). \end{aligned} \tag{2.179}$$

A scalar differential operator Δ results that contains double partial derivatives with regard to x, y, z:

$$\Delta = \frac{\partial^2}{\partial x^2} + \frac{\partial^2}{\partial y^2} + \frac{\partial^2}{\partial z^2}; \tag{2.180}$$

it is called "delta- or Laplace operator." In connection with the second time derivative of $p(\underline{\mathbf{R}}, t)$ according to

$$\Delta p(\underline{\mathbf{R}}, t) - \frac{1}{c^2}\frac{\partial^2 p(\underline{\mathbf{R}}, t)}{\partial t^2} = 0, \tag{2.181}$$

it constitutes an important term in any wave equation, here: a scalar wave equation for the acoustic pressure $p(\underline{\mathbf{R}}, t)$, which contains the constant c as (phase-)velocity of acoustic waves.

However, in contrast to acoustics, elastic waves are vector waves in terms of the particle velocity $\underline{\mathbf{u}}(\underline{\mathbf{R}}, t)$; therefore, we try to apply the delta-operator to a vector field according to

$$\Delta \underline{\mathbf{u}}(\underline{\mathbf{R}}, t) = \underline{\mathbf{e}}_x\, \Delta u_x(\underline{\mathbf{R}}, t) + \underline{\mathbf{e}}_y\, \Delta u_y(\underline{\mathbf{R}}, t) + \underline{\mathbf{e}}_z\, \Delta u_z(\underline{\mathbf{R}}, t) \tag{2.182}$$

and state the vector $\Delta \underline{\mathbf{u}}(\underline{\mathbf{R}}, t) = \nabla \cdot \nabla \underline{\mathbf{u}}(\underline{\mathbf{R}}, t)$—the divergence of the gradient dyadic of $\underline{\mathbf{u}}(\underline{\mathbf{R}}, t)$—as result, whose three (Cartesian) components

$$\underline{\mathbf{e}}_x \cdot \Delta \underline{\mathbf{u}}(\underline{\mathbf{R}}, t) = \Delta u_x(\underline{\mathbf{R}}, t), \tag{2.183}$$

$$\underline{\mathbf{e}}_x \cdot \Delta \underline{\mathbf{u}}(\underline{\mathbf{R}}, t) = \Delta u_y(\underline{\mathbf{R}}, t), \tag{2.184}$$

$$\underline{\mathbf{e}}_z \cdot \Delta \underline{\mathbf{u}}(\underline{\mathbf{R}}, t) = \Delta u_z(\underline{\mathbf{R}}, t) \tag{2.185}$$

are applications of the delta-operator to scalar field quantities. We emphasize that the component representation (2.183) through (2.185) is only correct in Cartesian coordinates (Section 2.2.4)!

Apart from the fact that $\underline{\mathbf{u}}(\underline{\mathbf{R}}, t)$ satisfies a vector wave equation, we will ascertain (Chapter 7) that the additional term $\boldsymbol{\nabla}\boldsymbol{\nabla} \cdot \underline{\mathbf{u}}(\underline{\mathbf{R}}, t)$ (gradient of the divergence of $\underline{\mathbf{u}}(\underline{\mathbf{R}}, t)$) with second spatial derivatives appears.[30] We calculate its components:

$$\begin{aligned}\boldsymbol{\nabla}\boldsymbol{\nabla} \cdot \underline{\mathbf{u}}(\underline{\mathbf{R}}, t) = \underline{\mathbf{e}}_x & \left(\frac{\partial^2 u_x(\underline{\mathbf{R}}, t)}{\partial x^2} + \frac{\partial^2 u_y(\underline{\mathbf{R}}, t)}{\partial x \partial y} + \frac{\partial^2 u_z(\underline{\mathbf{R}}, t)}{\partial x \partial z}\right) \\ + \underline{\mathbf{e}}_y & \left(\frac{\partial^2 u_x(\underline{\mathbf{R}}, t)}{\partial y \partial x} + \frac{\partial^2 u_y(\underline{\mathbf{R}}, t)}{\partial y^2} + \frac{\partial^2 u_z(\underline{\mathbf{R}}, t)}{\partial y \partial z}\right) \\ + \underline{\mathbf{e}}_z & \left(\frac{\partial^2 u_x(\underline{\mathbf{R}}, t)}{\partial z \partial x} + \frac{\partial^2 u_y(\underline{\mathbf{R}}, t)}{\partial z \partial y} + \frac{\partial^2 u_z(\underline{\mathbf{R}}, t)}{\partial z^2}\right). \end{aligned} \tag{2.186}$$

The two differential operators $\boldsymbol{\nabla}\boldsymbol{\nabla} \cdot \underline{\mathbf{u}}(\underline{\mathbf{R}}, t)$ and $\boldsymbol{\nabla} \cdot \boldsymbol{\nabla}\underline{\mathbf{u}}(\underline{\mathbf{R}}, t) = \Delta\underline{\mathbf{u}}(\underline{\mathbf{R}}, t)$ can be combined to a single vector differential operator:

$$\boldsymbol{\nabla}\boldsymbol{\nabla} \cdot \underline{\mathbf{u}}(\underline{\mathbf{R}}, t) - \boldsymbol{\nabla} \cdot \boldsymbol{\nabla}\underline{\mathbf{u}}(\underline{\mathbf{R}}, t) = \boldsymbol{\nabla} \times \boldsymbol{\nabla} \times \underline{\mathbf{u}}(\underline{\mathbf{R}}, t). \tag{2.187}$$

Two identities of multiple del-operator applications deserve particular attention:[31] The curl of a gradient field is always equal to the null vector, and the divergence of a curl field is always equal to zero:

$$\boldsymbol{\nabla} \times (\boldsymbol{\nabla}\Phi) \equiv \underline{\mathbf{0}}, \tag{2.188}$$

$$\boldsymbol{\nabla} \cdot (\boldsymbol{\nabla} \times \underline{\mathbf{A}}) \equiv 0. \tag{2.189}$$

Nota bene: These two equations hold for any scalar field $\Phi(\underline{\mathbf{R}}, t)$ and any vector field $\underline{\mathbf{A}}(\underline{\mathbf{R}}, t)$.

2.2.3 Gauss' theorem, Gauss' integral theorems, Green's formulas

Gauss' theorem: With (2.152), we defined a local source density of a vector field via the limit of a normalized flux integral. For a small but still finite volume V, we can write (2.152) according to

$$V \operatorname{div} \underline{\mathbf{v}}(\underline{\mathbf{R}}) \simeq \iint_{S_g} \underline{\mathbf{n}}' \cdot \underline{\mathbf{v}}(\underline{\mathbf{R}}') \, \mathrm{d}S'; \tag{2.190}$$

The flux of $\underline{\mathbf{v}}$ through S_g is proportional to an averaged source density multiplied by the volume. Even not a proof of Gauss' theorem, it is intuitively clear that, for an arbitrary volume V, the flux through its surface S_g equals the (net-)source density of $\underline{\mathbf{v}}$ in V, namely, the added up positive and negative "divergences" $\underline{\mathbf{v}}$:

$$\iiint_V \operatorname{div} \underline{\mathbf{v}}(\underline{\mathbf{R}}) \, \mathrm{d}V = \iint_{S_g} \underline{\mathbf{n}} \cdot \underline{\mathbf{v}}(\underline{\mathbf{R}}) \, \mathrm{d}S; \tag{2.191}$$

[30] In contrast to scalar acoustics, we expect pressure and shear waves.

[31] The physical terms "pressure" and "shear" waves become plausible that way (Section 7.2).

the distinction of the integration variables in the surface and the volume integral is no longer necessary because the integration itself defines the regime of variation of $\underline{\mathbf{R}}$. Equation 2.191 is the Gauss theorem holding for any vector field $\underline{\mathbf{v}}(\underline{\mathbf{R}})$ that satisfies the respective mathematical assumptions (Burg et al. 1990).

Gauss' integral theorems: Writing (2.191) according to

$$\iiint_V \nabla \cdot \underline{\mathbf{v}}(\underline{\mathbf{R}})\, \mathrm{d}V = \iint_{S_g} \underline{\mathbf{n}} \cdot \underline{\mathbf{v}}(\underline{\mathbf{R}})\, \mathrm{d}S, \qquad (2.192)$$

the formal content of this equation is revealed: Replace the operation $\nabla\cdot$ in the volume integral by $\underline{\mathbf{n}}\cdot$ in the surface integral. Formulated as such, the following pendants to Gauss' theorem—Gauss' integral theorems—are immediately at hand:

$$\iiint_V \nabla \Phi(\underline{\mathbf{R}})\, \mathrm{d}V = \iint_{S_g} \underline{\mathbf{n}}\, \Phi(\underline{\mathbf{R}})\, \mathrm{d}S; \qquad (2.193)$$

$$\iiint_V \nabla \underline{\mathbf{v}}(\underline{\mathbf{R}})\, \mathrm{d}V = \iint_{S_g} \underline{\mathbf{n}}\, \underline{\mathbf{v}}(\underline{\mathbf{R}})\, \mathrm{d}S; \qquad (2.194)$$

$$\iiint_V \nabla \times \underline{\mathbf{v}}(\underline{\mathbf{R}})\, \mathrm{d}V = \iint_{S_g} \underline{\mathbf{n}} \times \underline{\mathbf{v}}(\underline{\mathbf{R}})\, \mathrm{d}S; \qquad (2.195)$$

$$\iiint_V \nabla \cdot \underline{\underline{\mathbf{D}}}(\underline{\mathbf{R}})\, \mathrm{d}V = \iint_{S_g} \underline{\mathbf{n}} \cdot \underline{\underline{\mathbf{D}}}(\underline{\mathbf{R}})\, \mathrm{d}S. \qquad (2.196)$$

By the way, the integral theorem (2.195) has already been used to define the curl integral,[32] and the theorem (2.193) similarly serves to define the gradient. The integral theorems (2.194) and (2.196) are required to transform the differential style of the governing equations of elastodynamics into an integral style, thus providing the basis for the EFIT as a numerical method to compute elastodynamic fields (Fellinger 1991; Marklein 1997).

Stokes' integral theorem: Gauss' theorem is complemented by Stokes' theorem (in a similar way: Stokes' integral theorems):

$$\int_{C_g} \underline{\mathbf{v}}(\underline{\mathbf{R}}) \cdot \underline{\mathbf{dR}} = \iint_S \underline{\mathbf{n}} \cdot \operatorname{rot} \underline{\mathbf{v}}(\underline{\mathbf{R}})\, \mathrm{d}S. \qquad (2.197)$$

Here, $\underline{\mathbf{dR}}$ denotes the infinitesimal vector tangential to the arbitrary closed integration path C_g, that is to say, the line integral adds up all tangential components of the vector field $\underline{\mathbf{v}}$ along the integration path. If such an integral is nonzero, the vector field exhibits vortices, and, as a matter of fact, Stokes' theorem claims that the result of this integration exactly equals the

[32] Or: The integral definition of the curl yields an intuitive explanatory statement of the integral theorem (2.195).

surface integral of the curl density of $\underline{\mathbf{v}}$ where S is the membrane surface spanned by C_g (it may be arbitrarily distorted). Ampère's law (Footnote 26) provides a physically intuitive example for Stoke's theorem, and it is indeed particularly useful for the theory of electromagnetic fields.

Green's integral formulas: Gauss' theorem (2.192) serves to derive the first and the second Green formulas; Green's second formula is the basis for the mathematical specification of Huygens' principle for scalar fields (Section 15.1.2). As a consequence, Huygens' principle is no longer a principle but an implication of the wave equation.

We specify

$$\underline{\mathbf{v}}(\underline{\mathbf{R}}) = \Phi(\underline{\mathbf{R}})\boldsymbol{\nabla}\Psi(\underline{\mathbf{R}}) \tag{2.198}$$

in (2.192); here, $\Phi(\underline{\mathbf{R}})$ and $\Psi(\underline{\mathbf{R}})$ denote arbitrary scalar functions. We utilize (2.164) and calculate

$$\begin{aligned}\boldsymbol{\nabla}\cdot\underline{\mathbf{v}}(\underline{\mathbf{R}}) &= \boldsymbol{\nabla}\cdot[\Phi(\underline{\mathbf{R}})\boldsymbol{\nabla}\Psi(\underline{\mathbf{R}})]\\ &= \boldsymbol{\nabla}\Phi(\underline{\mathbf{R}})\cdot\boldsymbol{\nabla}\Psi(\underline{\mathbf{R}}) + \Phi(\underline{\mathbf{R}})\,\Delta\Psi(\underline{\mathbf{R}}).\end{aligned} \tag{2.199}$$

Insertion into (2.192) yields Green's first formula:

$$\iiint_V [\Phi(\underline{\mathbf{R}})\,\Delta\Psi(\underline{\mathbf{R}}) + \boldsymbol{\nabla}\Phi(\underline{\mathbf{R}})\cdot\boldsymbol{\nabla}\Psi(\underline{\mathbf{R}})]\,\mathrm{d}V = \iint_{S_g} \Phi(\underline{\mathbf{R}})\frac{\partial\Psi(\underline{\mathbf{R}})}{\partial n}\,\mathrm{d}S. \tag{2.200}$$

We have used (2.146) for $\underline{\mathbf{n}}\cdot\boldsymbol{\nabla}\Psi(\underline{\mathbf{R}})$.

Green's second formula is obtained if the above procedure is applied to

$$\underline{\mathbf{v}}(\underline{\mathbf{R}}) = \Psi(\underline{\mathbf{R}})\boldsymbol{\nabla}\Phi(\underline{\mathbf{R}}), \tag{2.201}$$

subtracting the result from (2.200):

$$\begin{aligned}&\iiint_V [\Phi(\underline{\mathbf{R}})\,\Delta\Psi(\underline{\mathbf{R}}) - \Psi(\underline{\mathbf{R}})\,\Delta\Phi(\underline{\mathbf{R}})]\,\mathrm{d}V\\ &\quad= \iint_{S_g}\left[\Phi(\underline{\mathbf{R}})\frac{\partial\Psi(\underline{\mathbf{R}})}{\partial n} - \Psi(\underline{\mathbf{R}})\frac{\partial\Phi(\underline{\mathbf{R}})}{\partial n}\right]\mathrm{d}S.\end{aligned} \tag{2.202}$$

Now, we simply have to provide a physical meaning for the fields Φ and Ψ and to interpret (2.202) in terms of wave theory to obtain Huygens' principle as a mathematical formulation: It is the Δ-operator appearing in the wave equation (2.181) and in both Green formulas suggesting this.

2.2.4 Cylindrical and spherical coordinates

In isotropic materials, phase surfaces of waves emanating from a point source are spherical; in general, the amplitude is direction dependent. Insofar, the mathematical characterization of these wave fronts does not fit into the cartesian coordinate system that has only been used until now; the utilization of spherical coordinates is mandatory (transducer sound fields originate from the superposition of spherical waves)! Additionally, cylindrical coordinates are often useful, for instance, to characterize a specimen like a pipe mathematically. Therefore, we briefly refer to essential differences of such orthogonal curvilinear coordinates as compared to cartesian coordinates.

Circular cylindrical coordinates r, φ, z are nothing but polar coordinates r, φ in the xy-plane combined with the cartesian component z. Cartesian coordinates are spanned by a trihedron of orthogonal unit vectors $\underline{\mathbf{e}}_x, \underline{\mathbf{e}}_y, \underline{\mathbf{e}}_z$; (scalar) vector components result from the projection (scalar products) of a vector to the orthonormal trihedron vectors, and therefore the definition of a similar orthonormal trihedron for cylindrical coordinates is appropriate. We refer to Figure 2.15: For simplicity, we only sketch the xy-plane—the unit vector $\underline{\mathbf{e}}_z$ characterizes the cylinder coordinate z—and identify a point • in this plane through the radial coordinate r and the angular coordinate φ, counted from the x-axis; we have $0 \leq r < \infty$ and $0 \leq \varphi \leq 2\pi$. The relation between r, φ and x, y is given by coordinate transform equations (2.1):

$$x = r\cos\varphi, \tag{2.203}$$

$$y = r\sin\varphi. \tag{2.204}$$

The cartesian x- and y-coordinates are spanned by $\underline{\mathbf{e}}_x$ and $\underline{\mathbf{e}}_y$, and because the pertinent x- and y-coordinate lines are straight, the unit vectors $\underline{\mathbf{e}}_x$ and $\underline{\mathbf{e}}_y$

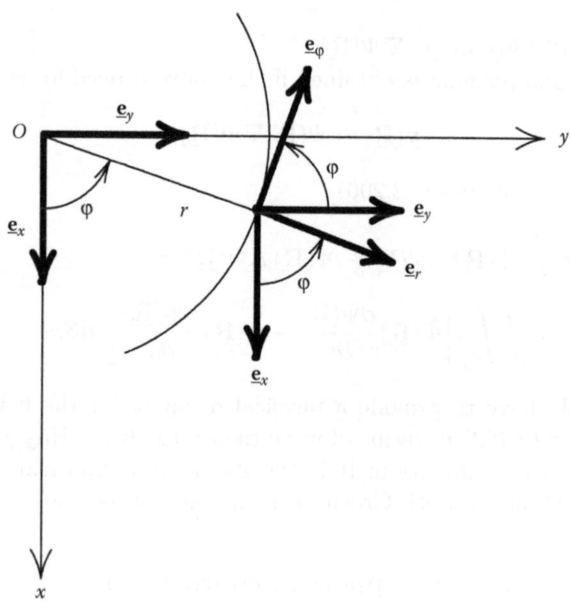

FIGURE 2.15
Orthogonal unit vectors for circular cylindrical coordinates.

always have the same direction, which also holds for the point with coordinates r, φ: $\underline{\mathbf{e}}_x$ and $\underline{\mathbf{e}}_y$ indicate for any point in the xy-plane the direction of variation of the respective coordinate. Unit vectors $\underline{\mathbf{e}}_r$ and $\underline{\mathbf{e}}_\varphi$ for the cylindrical coordinates r and φ similarly should point into those directions of pertinent coordinate variations. Consequently, $\underline{\mathbf{e}}_r$ points into radial direction and $\underline{\mathbf{e}}_\varphi$ into the direction tangential to a circle with radius r, of course in the direction of increasing φ. We stated that: A vector is defined by its length and direction, both parameters can be computed given the cartesian components of the vector; this must also be true for the unit vectors $\underline{\mathbf{e}}_r$ and $\underline{\mathbf{e}}_\varphi$. We obtain their cartesian components through projection to the unit vectors $\underline{\mathbf{e}}_x$ and $\underline{\mathbf{e}}_y$:

$$\underline{\mathbf{e}}_r = (\underline{\mathbf{e}}_r \cdot \underline{\mathbf{e}}_x)\,\underline{\mathbf{e}}_x + (\underline{\mathbf{e}}_r \cdot \underline{\mathbf{e}}_y)\,\underline{\mathbf{e}}_y, \tag{2.205}$$

$$\underline{\mathbf{e}}_\varphi = (\underline{\mathbf{e}}_\varphi \cdot \underline{\mathbf{e}}_x)\,\underline{\mathbf{e}}_x + (\underline{\mathbf{e}}_\varphi \cdot \underline{\mathbf{e}}_y)\,\underline{\mathbf{e}}_y. \tag{2.206}$$

In Figure 2.15, we immediately read off these projection:[33]

$$\underline{\mathbf{e}}_r = \cos\varphi\,\underline{\mathbf{e}}_x + \sin\varphi\,\underline{\mathbf{e}}_y, \tag{2.207}$$

$$\underline{\mathbf{e}}_\varphi = -\sin\varphi\,\underline{\mathbf{e}}_x + \cos\varphi\,\underline{\mathbf{e}}_y, \tag{2.208}$$

if we assume per definitionem that $\underline{\mathbf{e}}_r$ and $\underline{\mathbf{e}}_\varphi$ are unit vectors; yet, with (2.23), we immediately prove this fact. The calculation of

$$\underline{\mathbf{e}}_r \cdot \underline{\mathbf{e}}_\varphi = 0 \tag{2.209}$$

confirms orthogonality of $\underline{\mathbf{e}}_r$ and $\underline{\mathbf{e}}_\varphi$; trivially, $\underline{\mathbf{e}}_z$ is orthogonal to both. Apparently, with $\underline{\mathbf{e}}_r, \underline{\mathbf{e}}_\varphi, \underline{\mathbf{e}}_z$, we have found the right-handed orthonormal trihedron for circular cylindrical coordinates! The spatial dependence of this trihedron, in this case, the dependence on φ, represents the essential difference with regard to cartesian coordinates.

With $\underline{\mathbf{e}}_r, \underline{\mathbf{e}}_\varphi, \underline{\mathbf{e}}_z$, the components A_r, A_φ, A_z of a vector $\underline{\mathbf{A}}$ in cylindrical coordinates can be defined:

$$\underline{\mathbf{A}} = A_r\,\underline{\mathbf{e}}_r + A_\varphi\,\underline{\mathbf{e}}_\varphi + A_z\,\underline{\mathbf{e}}_z, \tag{2.210}$$

where

$$A_r = \underline{\mathbf{A}} \cdot \underline{\mathbf{e}}_r,$$

$$A_\varphi = \underline{\mathbf{A}} \cdot \underline{\mathbf{e}}_\varphi, \tag{2.211}$$

$$A_z = \underline{\mathbf{A}} \cdot \underline{\mathbf{e}}_z. \tag{2.212}$$

With $\underline{\mathbf{A}} = A_x\,\underline{\mathbf{e}}_x + A_y\,\underline{\mathbf{e}}_y + A_z\,\underline{\mathbf{e}}_z$ and (2.207) and (2.208), we immediately obtain equations to transform cartesian components A_x, A_y, A_z into circular cylindrical components A_r, A_φ, A_z:

$$\begin{aligned} A_r &= A_x \cos\varphi + A_y \sin\varphi, \\ A_\varphi &= -\,A_x \sin\varphi + A_y \cos\varphi, \\ A_z &= A_z, \end{aligned} \tag{2.213}$$

[33]Clearly, Equations 2.207 and 2.208 can be formally derived from the coordinate transform equations (2.203) and (2.204) (Langenberg 2005).

and the matrix notation of these equations

$$\begin{pmatrix} A_r \\ A_\varphi \\ A_z \end{pmatrix} = \begin{pmatrix} \cos\varphi & \sin\varphi & 0 \\ -\sin\varphi & \cos\varphi & 0 \\ 0 & 0 & 1 \end{pmatrix} \begin{pmatrix} A_x \\ A_y \\ A_z \end{pmatrix} \tag{2.214}$$

directly reveals how to obtain the transform of circular cylindrical components A_r, A_φ, A_z into cartesian components A_x, A_y, A_z; the coefficient matrix has to be inverted. The property of orthogonality of this matrix yields the inverse to be equal to the transpose:

$$\begin{pmatrix} \cos\varphi & \sin\varphi & 0 \\ -\sin\varphi & \cos\varphi & 0 \\ 0 & 0 & 1 \end{pmatrix}^{-1} = \begin{pmatrix} \cos\varphi & \sin\varphi & 0 \\ -\sin\varphi & \cos\varphi & 0 \\ 0 & 0 & 1 \end{pmatrix}^{\mathrm{T}} = \begin{pmatrix} \cos\varphi & -\sin\varphi & 0 \\ \sin\varphi & \cos\varphi & 0 \\ 0 & 0 & 1 \end{pmatrix}. \tag{2.215}$$

With the help of this matrix, we can also show that the value of the scalar product of two vectors $\underline{\mathbf{A}}$ and $\underline{\mathbf{B}}$ is independent of the coordinate system:

$$A_r B_r + A_\varphi B_\varphi + A_z B_z = A_x B_x + A_y B_y + A_z B_z. \tag{2.216}$$

The elastodynamic energy densities are defined as scalar product of two vectors and the double contraction of two second rank tensors, respectively (Section 4.3), and therefore their independence from the coordinate system is ensured. Here, we meet the cue: tensors in other than cartesian coordinates. For example, the $r\varphi$-component of a tensor of second rank $\underline{\underline{\mathbf{D}}}$ is defined by:[34]

$$\begin{aligned} D_{r\varphi} &= \underline{\mathbf{e}}_r \cdot \underline{\underline{\mathbf{D}}} \cdot \underline{\mathbf{e}}_\varphi \\ &= \underline{\underline{\mathbf{D}}} : \underline{\mathbf{e}}_\varphi \underline{\mathbf{e}}_r; \end{aligned} \tag{2.217}$$

as a consequence, the following transform equation corresponding to (2.214) is obtained:

$$\begin{pmatrix} D_{rr} & D_{r\varphi} & D_{rz} \\ D_{\varphi r} & D_{\varphi\varphi} & D_{\varphi z} \\ D_{zr} & D_{z\varphi} & D_{zz} \end{pmatrix} = \begin{pmatrix} \cos\varphi & \sin\varphi & 0 \\ -\sin\varphi & \cos\varphi & 0 \\ 0 & 0 & 1 \end{pmatrix} \begin{pmatrix} D_{xx} & D_{xy} & D_{xz} \\ D_{yx} & D_{yy} & D_{yz} \\ D_{zx} & D_{zy} & D_{zz} \end{pmatrix} \begin{pmatrix} \cos\varphi & -\sin\varphi & 0 \\ \sin\varphi & \cos\varphi & 0 \\ 0 & 0 & 1 \end{pmatrix}. \tag{2.218}$$

Applying the summation convention to (2.218), we can rapidly show that the double contraction of two second rank tensors is also independent of the coordinate system (the double contraction is, just like the scalar product, only a number).

---

[34]Numbering cylindrical coordinates r, φ, z in terms of ξ_i, $i = 1, 2, 3$, we obtain all tensor components as

$$D_{\xi_i \xi_j} = \underline{\mathbf{e}}_{\xi_i} \cdot \underline{\underline{\mathbf{D}}} \cdot \underline{\mathbf{e}}_{\xi_j}, \quad i, j = 1, 2, 3;$$

the short-hand notation $D_{\xi_i \xi_j} = D_{ij}$ requires the understanding of the underlying coordinate system.

In principle, all facts are at hand to investigate consequences of coordinate changes for the analysis of scalar, vector, and tensor fields. The essential tool of this analysis is the del-operator whose components possess a physical dimension, namely the unit m^{-1}, under the assumption that x, y, z are (cartesian) coordinates with unit m (meter) (Equation 2.139). In case of cylindrical coordinates, $\partial/\partial r$ and—of course—$\partial/\partial z$ exhibit this unit, yet $\partial/\partial \varphi$ does not. Therefore, we must supply the unit m to the differential variation along the φ-coordinate line, replacing $\partial\varphi$ by the differential arc length variation $\partial s = r\partial\varphi$ on a circle with radius r. Consequently, the del-operator in circular cylindrical coordinates reads as

$$\boldsymbol{\nabla} = \underline{\mathbf{e}}_r \frac{\partial}{\partial r} + \underline{\mathbf{e}}_\varphi \frac{1}{r}\frac{\partial}{\partial \varphi} + \underline{\mathbf{e}}_z \frac{\partial}{\partial z}. \tag{2.219}$$

As a matter of fact, the same representation is mathematically obtained if the so-called scale factors of the orthogonally curvilinear cylindrical coordinates are introduced.[35] With (2.219) and (2.210), it is finally clear what we have to cope with doing analysis in other than cartesian coordinates; for instance, calculation of the divergence of a vector field $\underline{\mathbf{A}}(\underline{\mathbf{R}}) = \underline{\mathbf{A}}(r, \varphi, z)$ in cylindrical coordinates requires the computation of

$$\begin{aligned}
\boldsymbol{\nabla}\cdot\underline{\mathbf{A}}(\underline{\mathbf{R}}) &= \left(\underline{\mathbf{e}}_r \frac{\partial}{\partial r} + \underline{\mathbf{e}}_\varphi \frac{1}{r}\frac{\partial}{\partial \varphi} + \underline{\mathbf{e}}_z \frac{\partial}{\partial z}\right)\cdot \\
&\quad \Big[A_r(r,\varphi,z)\,\underline{\mathbf{e}}_r(\varphi) + A_\varphi(r,\varphi,z)\,\underline{\mathbf{e}}_\varphi(\varphi) + A_z(r,\varphi,z)\,\underline{\mathbf{e}}_z\Big] \\
&= \frac{\partial A_r(r,\varphi,z)}{\partial r} + \underline{\mathbf{e}}_\varphi \cdot \frac{1}{r}\frac{\partial}{\partial\varphi}\Big[A_r(r,\varphi,z)\,\underline{\mathbf{e}}_r(\varphi)\Big] \\
&\quad + \underline{\mathbf{e}}_\varphi \cdot \frac{1}{r}\frac{\partial}{\partial\varphi}\Big[A_\varphi(r,\varphi,z)\,\underline{\mathbf{e}}_\varphi(\varphi)\Big] + \frac{\partial A_z(r,\varphi,z)}{\partial z} \\
&= \frac{\partial A_r(r,\varphi,z)}{\partial r} + \frac{A_r(r,\varphi,z)}{r}\,\underline{\mathbf{e}}_\varphi \cdot \underbrace{\frac{\partial \underline{\mathbf{e}}_r(\varphi)}{\partial\varphi}}_{=\underline{\mathbf{e}}_\varphi} + \frac{1}{r}\frac{\partial A_\varphi(r,\varphi,z)}{\partial\varphi} \\
&\quad + \frac{A_\varphi(r,\varphi,z)}{r}\,\underline{\mathbf{e}}_\varphi \cdot \underbrace{\frac{\partial \underline{\mathbf{e}}_\varphi(\varphi)}{\partial\varphi}}_{=-\underline{\mathbf{e}}_r} + \frac{\partial A_z(r,\varphi,z)}{\partial z} \\
&= \frac{\partial A_r(r,\varphi,z)}{\partial r} + \frac{A_r(r,\varphi,z)}{r} + \frac{1}{r}\frac{\partial A_\varphi(r,\varphi,z)}{\partial\varphi} + \frac{\partial A_z(r,\varphi,z)}{\partial z} \\
&= \frac{1}{r}\frac{\partial r A_r(r,\varphi,z)}{\partial r} + \frac{1}{r}\frac{\partial A_\varphi(r,\varphi,z)}{\partial\varphi} + \frac{\partial A_z(r,\varphi,z)}{\partial z}
\end{aligned} \tag{2.220}$$

---

[35] For circular cylindrical coordinates, the scale factors read as

$$\begin{aligned} h_r &= 1, \\ h_\varphi &= r, \\ h_z &= 1. \end{aligned}$$

according to (2.150)—we explicitly refer to the dependence of the unit vectors $\underline{\mathbf{e}}_r(\varphi), \underline{\mathbf{e}}_\varphi(\varphi)$ upon φ which, therefore, must be differentiated too—: It is often mentioned that the divergence-$\boldsymbol{\nabla}$-operator is written as

$$\boldsymbol{\nabla} = \underline{\mathbf{e}}_r \frac{1}{r}\frac{\partial}{\partial r} r + \underline{\mathbf{e}}_\varphi \frac{1}{r}\frac{\partial}{\partial \varphi} + \underline{\mathbf{e}}_z \frac{\partial}{\partial z}; \tag{2.221}$$

yet, this is only true if it is agreed upon that (2.221) is only applied to the scalar components A_r, A_φ, A_z. With that in mind, it is correct to state that the gradient-$\boldsymbol{\nabla}$-operator exhibits a different representation than (2.221), and the curl-$\boldsymbol{\nabla}$-operator does not at all have a component representation in other than cartesian coordinates. Yet, consequently staying with (2.219) thus always agreeing to differentiate the vector components—compare (2.210)—we even obtain

$$\begin{aligned} \boldsymbol{\nabla} \times \underline{\mathbf{A}}(\underline{\mathbf{R}}) = &\left(\underline{\mathbf{e}}_r \frac{\partial}{\partial r} + \underline{\mathbf{e}}_\varphi \frac{1}{r}\frac{\partial}{\partial \varphi} + \underline{\mathbf{e}}_z \frac{\partial}{\partial z} \right) \\ &\times \left[A_r(r,\varphi,z)\,\underline{\mathbf{e}}_r(\varphi) + A_\varphi(r,\varphi,z)\,\underline{\mathbf{e}}_\varphi(\varphi) + A_z(r,\varphi,z)\,\underline{\mathbf{e}}_z \right] \end{aligned} \tag{2.222}$$

and correct results for all other $\boldsymbol{\nabla}$-applications. Corresponding formulas are listed in the Appendix.

Spherical coordinates: As already mentioned, ultrasonic radiation fields exhibit demonstrative features only in spherical coordinates. As it is obvious from the simpler example of cylindrical coordinates, it is basically sufficient to know the coordinate transform equations and, already derived from them, the cartesian component representation of the orthonormal trihedron. Coordinate transform equations can be taken from Figure 2.16: The polar coordinate r in the xy-plane depends on the magnitude of the vector of position, the spherical coordinate R, via

$$r = R \sin\vartheta, \tag{2.223}$$

where ϑ denotes the coordinate "polar angle"; in connection with (2.203), (2.204), and another look at Figure 2.16, we obtain

$$\begin{aligned} x &= R \sin\vartheta \cos\varphi, \\ y &= R \sin\vartheta \sin\varphi, \\ z &= R \cos\vartheta; \end{aligned} \tag{2.224}$$

the spherical coordinate φ is called "azimuth angle". The orientation of the right-handed orthonormal trihedron ordered according to $\underline{\mathbf{e}}_R, \underline{\mathbf{e}}_\vartheta, \underline{\mathbf{e}}_\varphi$ can also be extracted from Figure 2.16, as well as the projections to cartesian coordinates:

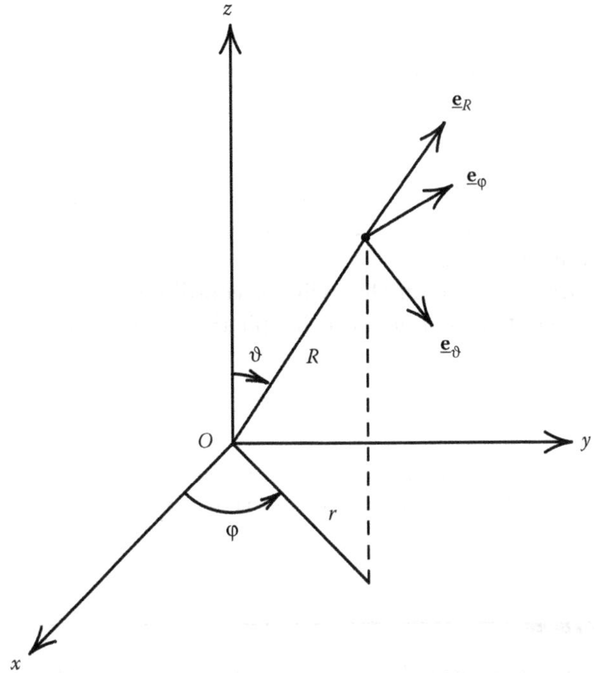

FIGURE 2.16
Orthonormal trihedron of spherical coordinates R, ϑ, φ.

$$\begin{aligned}
\underline{\mathbf{e}}_R &= (\underline{\mathbf{e}}_R \cdot \underline{\mathbf{e}}_x)\, \underline{\mathbf{e}}_x + (\underline{\mathbf{e}}_R \cdot \underline{\mathbf{e}}_y)\, \underline{\mathbf{e}}_y + (\underline{\mathbf{e}}_R \cdot \underline{\mathbf{e}}_z)\, \underline{\mathbf{e}}_z \\
&= \sin\vartheta \cos\varphi\, \underline{\mathbf{e}}_x + \sin\vartheta \sin\varphi\, \underline{\mathbf{e}}_y + \cos\vartheta\, \underline{\mathbf{e}}_z, \\
\underline{\mathbf{e}}_\vartheta &= (\underline{\mathbf{e}}_\vartheta \cdot \underline{\mathbf{e}}_x)\, \underline{\mathbf{e}}_x + (\underline{\mathbf{e}}_\vartheta \cdot \underline{\mathbf{e}}_y)\, \underline{\mathbf{e}}_y + (\underline{\mathbf{e}}_\vartheta \cdot \underline{\mathbf{e}}_z)\, \underline{\mathbf{e}}_z \\
&= \cos\vartheta \cos\varphi\, \underline{\mathbf{e}}_x + \cos\vartheta \sin\varphi\, \underline{\mathbf{e}}_y - \sin\vartheta\, \underline{\mathbf{e}}_z, \\
\underline{\mathbf{e}}_\varphi &= (\underline{\mathbf{e}}_\varphi \cdot \underline{\mathbf{e}}_x)\, \underline{\mathbf{e}}_x + (\underline{\mathbf{e}}_\varphi \cdot \underline{\mathbf{e}}_y)\, \underline{\mathbf{e}}_y + (\underline{\mathbf{e}}_\varphi \cdot \underline{\mathbf{e}}_z)\, \underline{\mathbf{e}}_z \\
&= -\sin\varphi\, \underline{\mathbf{e}}_x + \cos\varphi\, \underline{\mathbf{e}}_y.
\end{aligned} \tag{2.225}$$

We explicitly refer to

$$\underline{\mathbf{e}}_R = \hat{\underline{\mathbf{R}}}, \tag{2.226}$$

that is to say, the vector of position has the component representation

$$\underline{\mathbf{R}} = R \sin\vartheta \cos\varphi\, \underline{\mathbf{e}}_x + R \sin\vartheta \sin\varphi\, \underline{\mathbf{e}}_y + R \cos\vartheta\, \underline{\mathbf{e}}_z \tag{2.227}$$

in the cartesian orthonormal trihedron.

The system of Equations 2.225 defines the transform matrix for vector and tensor components, i.e., the transformation of the cartesian components A_x, A_y, A_z of a vector $\underline{\mathbf{A}}$ into its spherical components

$$\underline{\mathbf{A}} = A_R\, \underline{\mathbf{e}}_R + A_\vartheta\, \underline{\mathbf{e}}_\vartheta + A_\varphi\, \underline{\mathbf{e}}_\varphi \tag{2.228}$$

according to:

$$\begin{pmatrix} A_R \\ A_\vartheta \\ A_\varphi \end{pmatrix} = \begin{pmatrix} \sin\vartheta\cos\varphi & \sin\vartheta\sin\varphi & \cos\vartheta \\ \cos\vartheta\cos\varphi & \cos\vartheta\sin\varphi & -\sin\vartheta \\ -\sin\varphi & \cos\varphi & 0 \end{pmatrix} \begin{pmatrix} A_x \\ A_y \\ A_z \end{pmatrix}. \tag{2.229}$$

Again, the inverse of the transform matrix is equal to its transpose, immediately yielding the inversion of (2.229) and the transform equation for tensor components similar to (2.218).

The same arguments as in the cylinder coordinate paragraph lead us to the representation of the del-operator in spherical coordinates:[36]

$$\nabla = \underline{\mathbf{e}}_R \frac{\partial}{\partial R} + \underline{\mathbf{e}}_\vartheta \frac{1}{R}\frac{\partial}{\partial \vartheta} + \underline{\mathbf{e}}_\varphi \frac{1}{R\sin\vartheta}\frac{\partial}{\partial \varphi}; \tag{2.230}$$

single and multiple gradients, divergences, and curls can then be calculated; the respective formulas may be taken from the Appendix.

2.3 Time and Spatial Spectral Analysis with Fourier Transforms

The so-called kernel of the Fourier transform

$$\begin{aligned} F(\omega) &= \mathcal{F}\{f(t)\} \\ &= \int_{-\infty}^{\infty} f(t)\, \mathrm{e}^{\mathrm{j}\omega t}\, \mathrm{d}t \end{aligned} \tag{2.231}$$

of a time function $f(t)$ into a spectrum $F(\omega)$ is an exponential function $\mathrm{e}^{\mathrm{j}\omega t}$ with imaginary argument,[37] that is to say, the spectrum is generally complex. Therefore, we include a brief discussion of complex numbers before we turn to the Fourier transform.

---

[36] The scale factors in spherical coordinates read as

$$\begin{aligned} h_R &= 1, \\ h_\vartheta &= R, \\ h_\varphi &= R\sin\vartheta. \end{aligned}$$

[37] In communication theory, the Fourier transform is often defined with the complex conjugate kernel $\mathrm{e}^{-\mathrm{j}\omega t}$; yet, the theory of acoustic, elastic, and electromagnetic waves prefers the above *ansatz* because the respective Green function (2.170) then appears with the positive sign in the exponent. For real-valued time functions, the wave theoretical and communication theoretical spectra are apparently complex conjugate to each other. Caution is appropriate if mapping equations are under concern that explicitly contain the imaginary unit.

2.3.1 Complex numbers and complex valued functions of a complex variable

The equation

$$x^2 + 1 = 0 \tag{2.232}$$

does not have a solution in the space of real numbers x; therefore, we define solutions[38]

$$x_{1/2} = \pm \mathrm{j} = \sqrt{-1} \tag{2.233}$$

with the imaginary unit[39] j. Now j is utilized as building block for complex numbers

$$z = x + \mathrm{j}y, \tag{2.234}$$

which are attributed a real part with the real valued number x

$$\Re z = x \tag{2.235}$$

and an imaginary part with the real valued number y

$$\Im z = y; \tag{2.236}$$

the imaginary part counts the imaginary units j as "imaginary part" of z. With (2.234), a complex number has two "components" in a xy-"coordinate system," that is called the complex Gauss plane that exhibits a phasor (Figure 2.17), pointing from the origin to the complex number z under the phase angle φ . The complex number

$$z^* = x - \mathrm{j}y \tag{2.237}$$

is called conjugate complex to z, its phase angle is $-\varphi$ or $2\pi - \varphi$, respectively.

Addition and subtraction of two complex numbers $z_1 = x_1 + \mathrm{j}y_1$, $z_2 = x_2 + \mathrm{j}y_2$ is trivially defined as

$$z_1 \pm z_2 = x_1 \pm x_2 + \mathrm{j}(y_1 \pm y_2). \tag{2.238}$$

Their multiplication is easily calculated noting $\mathrm{j}^2 = -1$:

$$z_1 z_2 = x_1 x_2 - y_1 y_2 + \mathrm{j}(x_1 y_2 + x_2 y_1); \tag{2.239}$$

special cases are obtained as

$$z^2 = x^2 - y^2 + 2\mathrm{j}xy \tag{2.240}$$

and

$$zz^* = x^2 + y^2. \tag{2.241}$$

[38]As soon as we define the nth root of a complex number, we find that the square root $\sqrt{-1}$ has always two values, namely +j and −j; both are solutions of Equation 2.232.

[39]In the engineering sciences, in particular in electrical engineering, the notation j is commonly used whereas in physics, it is called i; to distinguish the imaginary unit j from the counting index j, we use a *roman character*.

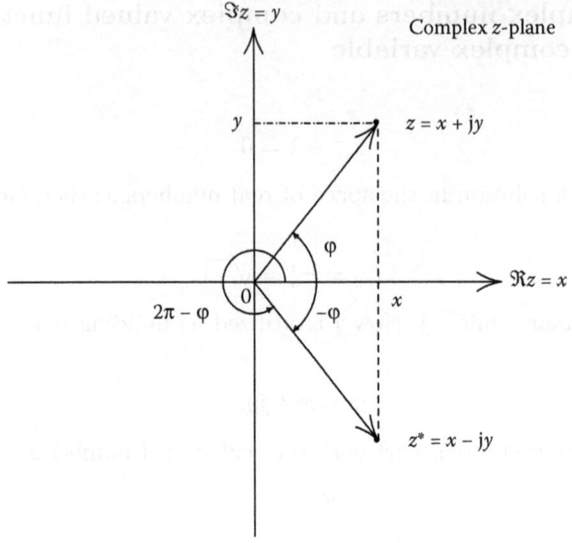

FIGURE 2.17
Complex number z and complex conjugate number z^* in the Gauss plane with real axis $\Re z$ and imaginary axis $\Im z$.

Evidently, the product zz^* is real valued; Figure 2.17 reveals that the magnitude $|z|$ of z according to

$$|z| = \sqrt{zz^*} \tag{2.242}$$

is exactly the length of the phasor. That way, $|z|$ and φ may serve as "polar coordinates" for z, supplying the complementary representation[40]

$$z = |z| \cos \varphi + \mathrm{j}\, |z| \sin \varphi. \tag{2.243}$$

The magnitude calculation according to (2.242) is also utilized to base the division of two complex numbers on something well-known:

$$\begin{aligned} \frac{z_1}{z_2} &= \frac{z_1 z_2^*}{z_2 z_2^*} \\ &= \frac{x_1 x_2 + y_1 y_2 + \mathrm{j}(y_1 x_2 - x_1 y_2)}{x_2^2 + y_2^2}. \end{aligned} \tag{2.244}$$

---

[40]Obviously, we have $\tan\varphi = y/x$, providing a way to calculate φ from the real and imaginary part via the inverse function of the tangent; yet, the arc tangent is multivalued, requesting case distinctions with regard to the signs of x:

$$\varphi = \begin{cases} PV \arctan y/x & \text{for } x > 0 \\ PV \arctan y/x + \pi & \text{for } x < 0 \end{cases}.$$

"PV" stands for principal value; we have $-\pi/2 < \varphi < \pi/2$.

With (2.240) all powers z^n are defined allowing for the immediate construction of a polynomial

$$P(z) = a_0 + a_1 z + a_2 z^2 + a_3 z^3 + \cdots + a_n z^n \tag{2.245}$$

(eventually with complex coefficients) as a function of a complex variable. The next step leading to a "complex analysis"—the theory of complex functions of a complex variable—is the power series:

$$f(z) = \sum_{n=0}^{\infty} a_n z^n; \tag{2.246}$$

immediately, the question arises for those values of z ensuring convergence of the series. This question has a very general answer[41] (Behnke and Sommer 1965) that will not be discussed in detail. We rather present the way to obtain a complex valued pendant of the real valued power series expansions for examples like the exponential, sine, and cosine functions: We replace the real variable x through the complex variable z! Hence:

$$\mathrm{e}^x = \sum_{n=0}^{\infty} \frac{x^n}{n!} \Longrightarrow \mathrm{e}^z = \sum_{n=0}^{\infty} \frac{z^n}{n!}, \tag{2.247}$$

$$\sin x = \sum_{n=0}^{\infty} (-1)^n \frac{x^{2n+1}}{(2n+1)!} \Longrightarrow \sin z = \sum_{n=0}^{\infty} (-1)^n \frac{z^{2n+1}}{(2n+1)!}, \tag{2.248}$$

$$\cos x = \sum_{n=0}^{\infty} (-1)^n \frac{x^{2n}}{(2n)!} \Longrightarrow \cos z = \sum_{n=0}^{\infty} (-1)^n \frac{z^{2n}}{(2n)!}; \tag{2.249}$$

all these power series converge in the open z-plane, namely for all values of z except[42] $z = \infty$. From these power series, the following relations are deduced:

$$\mathrm{e}^{\pm \mathrm{j}z} = \cos z \pm \mathrm{j} \sin z, \tag{2.250}$$

$$\cos z = \frac{\mathrm{e}^{\mathrm{j}z} + \mathrm{e}^{-\mathrm{j}z}}{2}, \tag{2.251}$$

$$\sin z = \frac{\mathrm{e}^{\mathrm{j}z} - \mathrm{e}^{-\mathrm{j}z}}{2\mathrm{j}}, \tag{2.252}$$

$$\cos \mathrm{j}z = \frac{\mathrm{e}^{z} + \mathrm{e}^{-z}}{2} \stackrel{\text{def}}{=} \cosh z, \tag{2.253}$$

$$\frac{1}{\mathrm{j}} \sin \mathrm{j}z = \frac{\mathrm{e}^{z} - \mathrm{e}^{-z}}{2} \stackrel{\text{def}}{=} \sinh z. \tag{2.254}$$

All the relations that we know from the real valued functions—addition theorems, derivatives, etc.—can be transferred to the complex regime; among

[41]The power series (2.246) converges in the z-plane within the largest circle around $z = 0$, exhibiting no singularity of the function that it represents.

[42]As a matter of fact, in the theory of complex functions of a complex variable the *single point* ∞ is defined.

others, we find the following separations of $\sin z$ and $\cos z$ into real and imaginary parts:

$$\begin{aligned}\sin z &= \sin(x + \mathrm{j}y)\\ &= \sin x \cosh y + \mathrm{j} \cos x \sinh y, \qquad (2.255)\\ \cos z &= \cos(x + \mathrm{j}y)\\ &= \cos x \cosh y - \mathrm{j} \sin x \sinh y. \qquad (2.256)\end{aligned}$$

The reflection of plane elastic waves at the plane boundary between two materials with different elastic properties exhibits critical angles of incidence if the sine of the transmission angle of the longitudinal wave and the transverse wave, respectively, gets larger than 1, being possible via analytic continuation of this angle into a complex plane and utilization of the above relations.

The complex exponential function is 2π-periodic on the imaginary axis:

$$\mathrm{e}^{z+\mathrm{j}2\pi k} = \mathrm{e}^z, \quad k = 0, \pm 1, \pm 2, \ldots. \qquad (2.257)$$

From (2.253) and (2.254), we have

$$\mathrm{e}^{\pm z} = \cos \mathrm{j}z \mp \mathrm{j} \sin \mathrm{j}z, \qquad (2.258)$$

that is to say,

$$\mathrm{e}^{\pm \mathrm{j}\varphi} = \cos \varphi \pm \mathrm{j} \sin \varphi, \qquad (2.259)$$

finally, yielding the representation

$$z = |z|\, \mathrm{e}^{\mathrm{j}\varphi} \qquad (2.260)$$

of a complex number in terms of magnitude $|z|$ and phase (argument) $\arg z = \varphi$ if we observe (2.243). Then, the construction of the integer nth power

$$z^n = |z|^n\, \mathrm{e}^{\mathrm{j}n\varphi} \qquad (2.261)$$

is exceptionally simple as it is true for multiplication and division:

$$z_1 z_2 = |z_1||z_2|\, \mathrm{e}^{\mathrm{j}(\varphi_1+\varphi_2)}, \qquad (2.262)$$

$$\frac{z_1}{z_2} = \frac{|z_1|}{|z_2|}\, \mathrm{e}^{\mathrm{j}(\varphi_1-\varphi_2)}. \qquad (2.263)$$

Besides integer powers, we can also construct nth roots

$$\sqrt[n]{z} = \sqrt[n]{|z|}\, \mathrm{e}^{\mathrm{j}\frac{\varphi}{n}} \qquad (2.264)$$

observing (2.260); however, we have to recognize nonuniqueness because of the 2π-periodicity of the exponential function; this will be explained for the square root. Certainly,

$$\sqrt{z} = \sqrt{|z|}\, \mathrm{e}^{\mathrm{j}\frac{\varphi}{2}} \qquad (2.265)$$

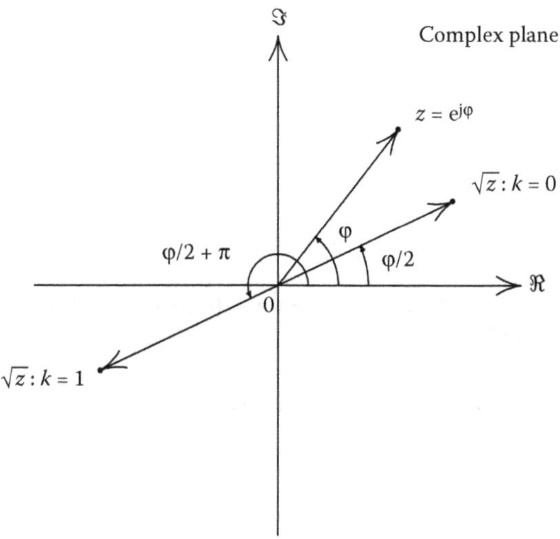

FIGURE 2.18
Nonuniqueness of the complex square root.

is the square root of the the complex number z, because the square of (2.265) again yields z; $\sqrt{z}$ according to (2.265) is a complex number with the argument $0 \leq \arg \sqrt{z} \leq \pi$, because it results from bisection of the phase angle $0 \leq \varphi \leq 2\pi$ of z. Yet, the bisection also maps the periodicity interval $2\pi \leq \varphi + 2\pi \leq 4\pi$ of the exponential function $e^{j\varphi}$ into the basic interval from 0 to 2π, namely into the part $\pi \leq \arg \sqrt{z} \leq 2\pi$. In other words: Besides (2.265)

$$\begin{aligned}\sqrt{z} &= \sqrt{|z|}\, e^{j\frac{\varphi}{2}+j\pi} \\ &= -\sqrt{|z|}\, e^{j\frac{\varphi}{2}}\end{aligned} \tag{2.266}$$

is a square root of z too. Both square roots—(2.265) and (2.266)—can be reconciled according to

$$\sqrt{z} = \sqrt{|z|}\, e^{j\frac{\varphi}{2}+j\pi k}, \quad k = 0, 1. \tag{2.267}$$

For $k = 0$, we obtain (2.265) and for $k = 1$ (2.266). Figure 2.18 illustrates the nonuniqueness[43] of the square root. Accordingly, the nth root is n-fold nonunique; in real valued space, this is not "visible," because the exponential function is only periodic on the imaginary axis.

In dissipative materials, elastic waves experience attenuation in propagation direction that is characterized by the imaginary part of the complex wave number k; yet, the square of the wave number $z = k^2$ is related to the material

[43]One half of a $\sqrt{z}$-plane already originates from a whole z-plane, thus forcing the whole $\sqrt{z}$-plane to supply from two z-planes and this is nonunique. Uniqueness of the square root is ensured if we precisely agree how to combine the two z-planes in terms of a so-called Riemann plane.

parameters, thus requiring the real and imaginary part of $\sqrt{z} = k$ as a function of $x = \Re k^2$ and $y = \Im k^2$ if real and imaginary parts of $z = x + \mathrm{j}y$ are given. We find

$$\Re k = \pm\sqrt{\frac{1}{2}\left(\sqrt{x^2+y^2}+x\right)}, \tag{2.268}$$

$$\Im k = \pm\sqrt{\frac{1}{2}\left(\sqrt{x^2+y^2}-x\right)}, \tag{2.269}$$

where the signs have to be chosen based on physical arguments—attenuation in propagation direction.

2.3.2 Time domain spectral analysis

Certainly, the Fourier integral (2.231) does not exist for arbitrary time functions $f(t)$; a sufficient condition is the absolute integrability of $f(t)$. But this does not imply that the spectrum $F(\omega)$ is absolutely integrable with the consequence that the inverse Fourier integral

$$\begin{aligned} f(t) &= \mathcal{F}^{-1}\{F(\omega)\} \\ &= \frac{1}{2\pi}\int_{-\infty}^{\infty} F(\omega)\,\mathrm{e}^{-\mathrm{j}\omega t}\,\mathrm{d}\omega \end{aligned} \tag{2.270}$$

may not exist. Yet, *if* a time function $f(t)$ yields an existing Fourier transform $F(\omega)$ leading again to the respective time function via inversion according to (2.270)—the mathematical assumptions are detailed, for instance, by Doetsch (1967)—we call it a correspondence and write

$$f(t) \circ\!\!-\!\!\bullet\ F(\omega). \tag{2.271}$$

A brief remark regarding negative frequencies: For real valued time functions $f(t)$—components of physical wave fields *are* real valued—we have

$$F(-\omega) = F^*(\omega), \tag{2.272}$$

i.e., negative frequencies do not contain any new spectral information. Measurements (or calculations) of a spectrum for positive frequencies yields real valued time functions after Fourier inversion complementing the spectrum according to (2.272).

The general properties of spectra are best discussed with examples.

Examples

1. Rectangular impulse $q_T(t)$:

The Fourier transform of the rectangular impulse of duration $2T$ symmetric to the origin (Figure 2.19)

$$q_T(t) = \begin{cases} 1 & \text{for } |t| < T \\ 0 & \text{for } |t| > T \end{cases} \tag{2.273}$$

is calculated via elementary integration (Figure 2.19):

$$\mathcal{F}\{q_T(t)\} = \frac{2\sin T\omega}{\omega}. \tag{2.274}$$

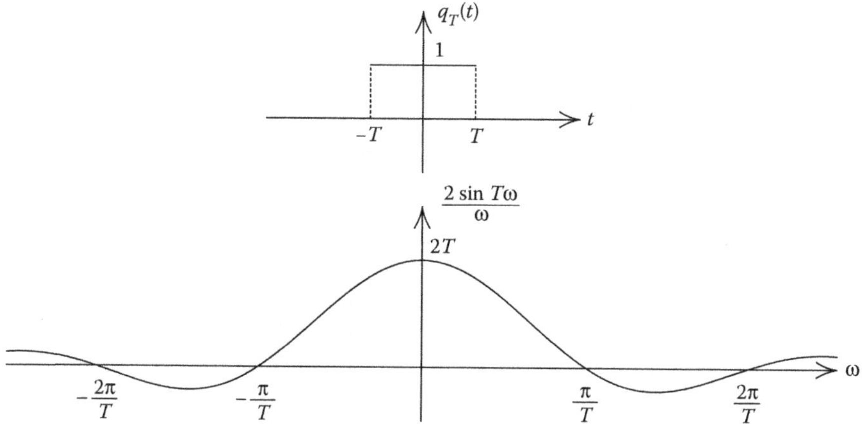

FIGURE 2.19
Rectangular impulse symmetric to the origin and spectrum.

We regognize the following:

- A time function of limited duration has an infinitely broadband spectrum; this—and the opposite—is always true.
- The so-called sinc-function (2.274) exhibits zeroes at $\omega = \pm n\pi/T$, $n = 1, 2, 3, \ldots$. Thus we can define a bandwidth B (of the "main lobe") through $B = 2\pi/T$; with decreasing T, this bandwidth B increases and vice versa: Long impulses have a small bandwidth, short impulses have a large bandwidth, which is called the uncertainty relation.
- The sinc-function (2.274) is not absolutely integrable; therefore, the inversion integral has to be defined as a Cauchy principle value, the result is not (2.273) but a rectangular impulse possessing the value 1/2 for $t = \pm T$, whereas (2.273) is not defined for these times.

2. RC$N(t)$-impulse:
To visualize simulation results, we often use a standard impulse that is called RC$N(t)$-impulse; RC stands for *raised cosine* and N for the number of oscillations of duration T_0 corresponding to the carrier (circular) frequency ω_0 according to $T_0 = 2\pi/\omega_0$:

$$\mathrm{RC}N(t) = \begin{cases} \underbrace{\left(1 + \cos\frac{\omega_0}{N}t\right)}_{= eN(t)} \cos\omega_0 t & \text{for} - N\frac{\pi}{\omega_0} \leq t \leq N\frac{\pi}{\omega_0} \\ 0 & \text{else} \end{cases} . \tag{2.275}$$

Obviously, the total duration of RC$N(t)$ is NT_0, the amplitudes of the ω_0-oscillations are modulated according to a raised cosine, the envelope $eN(t)$. Figure 2.20 illustrates an RC2(t)- and an RC4(t)-impulse together with the

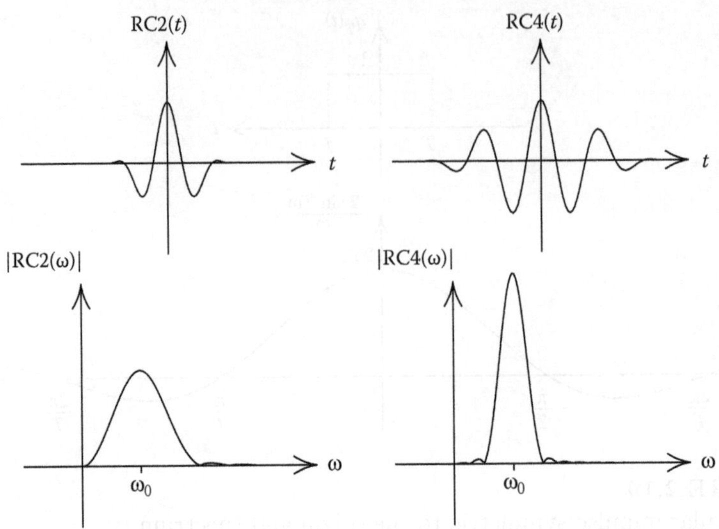

FIGURE 2.20
RC2(t)- and RC4(t)-impulses together with the magnitudes of their spectra for $\omega > 0$. (For $\omega < 0$ the spectra have to be symmetrically complemented.)

magnitudes of their spectra (we use the same letter for the spectra distinguishing them from the time functions by the argument)

$$
\begin{aligned}
\mathrm{RC}N(\omega) &= (-1)^{N+1} \sin\left(N\frac{\pi}{\omega_0}\omega\right) \\
&\times \left[\frac{\omega}{\omega^2 - \left(\frac{N+1}{N}\right)^2 \omega_0^2} - \frac{2\omega}{\omega^2 - \omega_0^2} + \frac{\omega}{\omega^2 - \left(\frac{N-1}{N}\right)^2 \omega_0^2}\right]
\end{aligned}
\tag{2.276}
$$

calculated via elementary evaluation of the Fourier integral recognizing (2.251) and $\mathrm{e}^{\pm \mathrm{j}N\pi} = (-1)^N$. Apparently, the spectra magnitudes are maximum at $\omega = \omega_0$; increasing the number of oscillations of the RC$N(t)$-impulse keeping T_0 constant decreases the spectral bandwidth (uncertainty relation!).

3. Exponential function symmetric to the origin:
In connection with the reflection of pulsed plane SV-waves at the stress-free boundary of a half-space, we need the inverse Fourier transform of $\mathrm{e}^{-\alpha z|\omega|}$ for $\alpha > 0$, $z > 0$ (Section 9.1.2); we calculate

$$
\begin{aligned}
\mathcal{F}^{-1}\{\mathrm{e}^{-\alpha z|\omega|}\} &= \frac{1}{2\pi}\int_{-\infty}^{\infty} \mathrm{e}^{-\alpha z \mathrm{sign}(\omega)\omega}\, \mathrm{e}^{-\mathrm{j}\omega t}\, \mathrm{d}\omega \\
&= \frac{1}{2\pi}\int_{-\infty}^{0} \mathrm{e}^{\alpha z\omega - \mathrm{j}\omega t}\, \mathrm{d}\omega + \frac{1}{2\pi}\int_{0}^{\infty} \mathrm{e}^{-\alpha z\omega - \mathrm{j}\omega t}\, \mathrm{d}\omega \\
&= \frac{1}{\pi}\frac{\alpha z}{\alpha^2 z^2 + t^2}.
\end{aligned}
\tag{2.277}
$$

Therefore,

$$\frac{1}{\pi}\frac{\alpha z}{\alpha^2 z^2 + t^2} \circ\!\!-\!\!\bullet\ \mathrm{e}^{-\alpha z|\omega|}. \tag{2.278}$$

In Section 9.1.2, the limit $z \longrightarrow 0$ is of interest:

$$\lim_{z\to 0}\frac{1}{\pi}\frac{\alpha z}{\alpha^2 z^2 + t^2} = \delta(t). \tag{2.279}$$

4. Gaussian impulse:
We cite the correspondence

$$\mathrm{e}^{-\alpha t^2} \circ\!\!-\!\!\bullet \sqrt{\frac{\pi}{\alpha}}\,\mathrm{e}^{-\frac{\omega^2}{4\alpha}} \tag{2.280}$$

as an example for time functions and spectra being of the same type: a Gaussian function. Again, this correspondence reflects the uncertainty relation.

As far as further correspondences are of interest, we refer to tables (Doetsch 1967; Erdélyi 1954). Sometimes, the relation

$$F(t) \circ\!\!-\!\!\bullet\ 2\pi f(-\omega) \tag{2.281}$$

is useful; it results from the symmetry of Fourier and inverse Fourier transform: Consider a given spectrum $F(\omega)$ of a time function $f(t)$ as a time function and it follows that its Fourier transform is equal to the original time function with the argument $-\omega$ (times 2π).

Standard functions in field theory—for example, the unit-step function $u(t)$, the sign function $\mathrm{sign}(t) = 2u(t) - 1$, the complex exponential function $\mathrm{e}^{\pm \mathrm{j}\omega_0 t}$, and the hyperbolic function t^{-1}—are not absolutely integrable; hence, Fourier transforms can only be defined in the space of (tempered) distributions (Doetsch 1967). We anticipate (Section 2.4.3) that:

$$\mathcal{F}\{u(t)\} = \pi\delta(\omega) + \mathrm{j}\,\mathrm{PV}\frac{1}{\omega}, \tag{2.282}$$

$$\mathcal{F}\{\mathrm{sign}(t)\} = 2\mathrm{j}\,\mathrm{PV}\frac{1}{\omega}, \tag{2.283}$$

$$\mathcal{F}\{\mathrm{e}^{\pm \mathrm{j}\omega_0 t}\} = 2\pi\delta(\omega \pm \omega_0), \tag{2.284}$$

$$\mathcal{F}\{\cos\omega_0 t\} = \pi[\delta(\omega - \omega_0) + \delta(\omega + \omega_0)], \tag{2.285}$$

$$\mathcal{F}\{\sin\omega_0 t\} = \mathrm{j}\pi[\delta(\omega - \omega_0) - \delta(\omega + \omega_0)], \tag{2.286}$$

$$\mathcal{F}\left\{\pi\delta(t) - \mathrm{jPV}\frac{1}{t}\right\} = 2\pi u(\omega), \tag{2.287}$$

$$\mathcal{F}\left\{\mathrm{PV}\frac{1}{t}\right\} = \mathrm{j}\pi\,\mathrm{sign}(\omega); \tag{2.288}$$

here, $\delta(\omega)$ is the delta-"function" (delta-distribution), and PV means computation of the inverse Fourier integral of ω^{-1} (or the Fourier integral of t^{-1}) in the sense of Cauchy's principal value.

Fourier transformation rules are of tremendous importance for our applications.

2.3.3 Fourier transformation rules

Certain operations on time functions—for instance, time shift, differentiation, and convolution—often have simpler counterparts in the spectral domain. In the following, we cite these so-called transformation rules.

Similarity rule: Be $F(\omega)$ the Fourier transform of $f(t)$; then $F(\omega/a)/|a|$ is the Fourier transform of $f(at)$, where $a \neq 0$ is a real valued parameter. We write

$$f(at) \circ\!\!-\!\!\bullet \frac{1}{|a|} F\left(\frac{\omega}{a}\right). \tag{2.289}$$

Therefore, measuring a time axis with the unit μs instead of s, we must change the dimension of the frequency axis from Hz to MHz: a is equal to 10^{-6}.

Shifting rule: Shifting a time function by $\pm t_0$ on the t-axis yields a modulation of the spectrum with $\mathrm{e}^{\mp \mathrm{j} t_0 \omega}$:

$$f(t \pm t_0) \circ\!\!-\!\!\bullet \mathrm{e}^{\mp \mathrm{j} t_0 \omega} F(\omega). \tag{2.290}$$

Note: Former real valued spectra (e.g., Equation 2.274) turn into complex spectra, that is to say, the rectangular impulse $q_T(t - T)$ starting at the origin has the complex spectrum $2\mathrm{e}^{\mathrm{j}T\omega} \sin T\omega/\omega$.

Modulation rule: The symmetry between Fourier transform and inverse Fourier transform generally brings symmetric transformation rules, i.e., the modulation of a time function results in a spectral shift with the modulation frequency:

$$f(t)\, \mathrm{e}^{\pm \mathrm{j}\omega_0 t} \circ\!\!-\!\!\bullet F(\omega \pm \omega_0). \tag{2.291}$$

Differentiation rule: Governing equations of any wave phenomena are partial differential equations in space and time, they exhibit spatial and time derivatives of field quantities. In case of linear governing equations, time derivatives can be advantageously eliminated transforming field quantities into their pertinent Fourier spectra applying the differentiation rule. Under certain assumptions (the time function and its derivatives must vanish for $t \longrightarrow \pm\infty$ (Doetsch 1967)), we have:

$$f^{(n)}(t) \circ\!\!-\!\!\bullet (-\mathrm{j}\omega)^n F(\omega), \quad n = 1, 2, 3, \ldots. \tag{2.292}$$

Integration rule: For $n = 1$, the "inversion" of (2.292) reads as

$$\int_{-\infty}^{t} f(\tau)\, \mathrm{d}\tau \circ\!\!-\!\!\bullet \frac{F(\omega)}{(-\mathrm{j}\omega)}, \tag{2.293}$$

where the assumption $F(0) = 0$, i.e., the zero average of $f(t)$ has to be guaranteed (Doetsch 1967).

Convolution rule: The convolution integral

$$\begin{aligned} f(t) &= g(t) * h(t) \\ &= h(t) * g(t) \\ &= \int_{-\infty}^{\infty} h(t-\tau)g(\tau)\,\mathrm{d}\tau \end{aligned} \tag{2.294}$$

of two time functions $g(t)$ and $h(t)$ is mapped into the product of the spectra $G(\omega)$ and $H(\omega)$ through the Fourier transform:

$$g(t) * h(t) \circ\!\!-\!\!\bullet\; G(\omega)H(\omega). \tag{2.295}$$

Spectral convolution rule: The spectral convolution rule symmetric to (2.295) reads as

$$g(t)h(t) \circ\!\!-\!\!\bullet\; \frac{1}{2\pi}G(\omega) * H(\omega), \tag{2.296}$$

where

$$\begin{aligned} G(\omega) * H(\omega) &= H(\omega) * G(\omega) \\ &= \int_{-\infty}^{\infty} H(\omega-\omega')G(\omega')\,\mathrm{d}\omega'. \end{aligned} \tag{2.297}$$

2.3.4 Analytic signal and Hilbert transform

The Hilbert transform (Doetsch 1967; with different signs: Hahn 1997)

$$\begin{aligned} f(t) &= \mathcal{H}\{g(\tau)\} \\ &= -\frac{1}{\pi}\,\mathrm{PV}\int_{-\infty}^{\infty} \frac{g(\tau)}{t-\tau}\,\mathrm{d}\tau \end{aligned} \tag{2.298}$$

is an integral transform with convolution kernel; the inverse integral

$$\begin{aligned} g(\tau) &= \mathcal{H}^{-1}\{f(t)\} \\ &= \frac{1}{\pi}\,\mathrm{PV}\int_{-\infty}^{\infty} \frac{f(t)}{\tau-t}\,\mathrm{d}t \end{aligned} \tag{2.299}$$

looks completely similar apart from the sign; again PV stands for principal value, namely Cauchy's principal value. The Hilbert transform plays an important role in the field of the Fourier transform.

The convolution kernel of the Hilbert transform implies

$$\begin{aligned}\mathcal{F}\{f(t)\} &= F(\omega)\\ &= \mathcal{F}\{\mathcal{H}\{g(\tau)\}\}\\ &= -\frac{1}{\pi}\mathcal{F}\left\{\mathrm{PV}\frac{1}{t} * g(t)\right\}\\ &= -\frac{1}{\pi}\mathcal{F}\left\{\mathrm{PV}\frac{1}{t}\right\} G(\omega)\\ &= -\mathrm{j}\,\mathrm{sign}(\omega)G(\omega) \end{aligned} \tag{2.300}$$

in connection with the convolution rule (2.295) of the Fourier transform and the correspondence (2.288). For the Fourier spectra of two Hilbert transforms, the Hilbert transform reveals itself as a filter with the frequency response $-j\,\mathrm{sign}(\omega)$, that is to say, $G(\omega)$ is multiplied by $-j$ for positive frequencies and by j for negative frequencies. The relation (2.300) can be utilized to calculate a pair of Hilbert transforms analytically or numerically:

$$f(t) = \mathcal{F}^{-1}\{-\mathrm{j}\,\mathrm{sign}(\omega)\mathcal{F}\{g(t)\}\}. \tag{2.301}$$

With (2.300) and the correspondences (2.285) and (2.286), we immediately show

$$\mathcal{H}\{\sin\omega_0\tau\} = \cos\omega_0 t, \tag{2.302}$$

$$\mathcal{H}\{\cos\omega_0\tau\} = -\sin\omega_0 t, \tag{2.303}$$

hence,

$$\mathrm{e}^{-\mathrm{j}\omega_0 t} = \cos\omega_0 t + \mathrm{j}\,\mathcal{H}\{\cos\omega_0\tau\}. \tag{2.304}$$

If $a(t) \geq 0$ is defined as amplitude modulation $a(t)\cos\omega_0 t$ of a real valued carrier oscillation with (circular) frequency ω_0 and bandlimited spectrum $A(\omega) \equiv 0$ for[44] $|\omega| > \omega_{\max} < \omega_0$, we can even prove

$$\mathcal{H}\{a(\tau)\,\sin\omega_0\tau\} = a(t)\,\cos\omega_0 t, \tag{2.305}$$

$$\mathcal{H}\{a(\tau)\,\cos\omega_0\tau\} = -\,a(t)\,\sin\omega_0 t \tag{2.306}$$

and therefore

$$a(t)\,\mathrm{e}^{-\mathrm{j}\omega_0 t} = a(t)\,\cos\omega_0 t + \mathrm{j}\,\mathcal{H}\{a(\tau)\,\cos\omega_0\tau\}. \tag{2.307}$$

The amplitude modulated real valued carrier oscillation turns into a complex valued carrier oscillation through the imaginary complement of its Hilbert transform exhibiting the modulation—the envelope—as magnitude.

---

[44]Even though the RC$N(t)$-impulse is time limited, we can approximately take the assumptions concerning $a(t)$ for $eN(t)$ as granted.

If $a(t)$ is not bandlimited, the Hilbert transform of, for example, $a(t)\sin\omega_0 t$ shows up correction terms

$$\mathcal{H}\{a(\tau)\sin\omega_0\tau\} = a(t)\cos\omega_0 t - \mathcal{F}^{-1}\{A(\omega-\omega_0)u(-\omega)\} \\ - \mathcal{F}^{-1}\{A(\omega+\omega_0)u(\omega)\} \tag{2.308}$$

that refer to the nonvanishing spectral parts of $A(\omega-\omega_0)$ for negative frequencies and of $A(\omega+\omega_0)$ for positive frequencies.

We generalize (2.307) in terms of the so-called analytic signal (Gabor 1946)

$$f_+(t) = f(t) + \mathrm{j}\,\mathcal{H}\{f(\tau)\} \tag{2.309}$$

and define $|f_+(t)|$ as envelope of $f(t)$. This is advantageously utilized for ultrasonic signal processing with imaging algorithms (Langenberg et al. 1993).

We already mentioned that real valued time functions have spectra with no additional information for negative frequencies; what would happen if this information is completely deleted? We consider spectra $F(\omega)$ that are equal to zero for $\omega < 0$ according to the identity

$$F(\omega) = F(\omega)u(\omega). \tag{2.310}$$

Formal Fourier transform of (2.310) applying the convolution rule

$$f(t) = f(t) * \mathcal{F}^{-1}\{u(\omega)\} \tag{2.311}$$

yields

$$f(t) = \frac{1}{2}\,f(t) - \frac{\mathrm{j}}{2\pi}\,f(t) * \mathrm{PV}\frac{1}{t} \tag{2.312}$$

together with (2.287) and (2.361), hence

$$f(t) = -\mathrm{j}\frac{1}{\pi}\,\mathrm{PV}\int_{-\infty}^{\infty}\frac{f(\tau)}{t-\tau}\,\mathrm{d}\tau \tag{2.313}$$

as a Hilbert transformation rule. In (2.313), $f(t)$ cannot be real valued why we separate into real and imaginary parts:

$$\Re\{f(t)\} = \frac{1}{\pi}\,\mathrm{PV}\int_{-\infty}^{\infty}\frac{\Im\{f(\tau)\}}{t-\tau}\,\mathrm{d}\tau, \tag{2.314}$$

$$\Im\{f(t)\} = -\frac{1}{\pi}\,\mathrm{PV}\int_{-\infty}^{\infty}\frac{\Re\{f(\tau)\}}{t-\tau}\,\mathrm{d}\tau. \tag{2.315}$$

Therefore, we have

$$f(t) = \Re\{f(\tau)\} + \mathrm{j}\,\mathcal{H}\{\Re\{f(\tau)\}\} \tag{2.316}$$

or

$$f_+(t) \stackrel{\mathrm{def}}{=} f(t) + \mathrm{j}\mathcal{H}\{f(\tau)\} \tag{2.317}$$

with $f(t)$ as real valued time function. Time functions with zero spectra for $\omega < 0$ are complex valued, yet the imaginary part is not independent from the real part, it is its Hilbert transform: $f_+(t)$ according to (2.317) denotes an analytic signal, the spectrum $\mathcal{F}\{f_+(t)\}$ is constrained to positive frequencies, whence the notation $f_+(t)$ comes from.

How does the spectrum of $f_+(t)$ depend on $F(\omega)$, the spectrum of $f(t)$? We calculate

$$\begin{aligned}
\mathcal{F}\{f_+(t)\} &= \mathcal{F}\{f(t)\} + \mathrm{j}\mathcal{F}\left\{-\frac{1}{\pi}\int_{-\infty}^{\infty}\frac{f(\tau)}{t-\tau}\,\mathrm{d}\tau\right\} \\
&= \mathcal{F}\{f(t)\} + \mathrm{j}\mathcal{F}\left\{-\frac{1}{\pi}f(t) * \mathrm{PV}\frac{1}{t}\right\} \\
&\stackrel{(2.288)}{=} F(\omega) + F(\omega)\mathrm{sign}\,(\omega) \\
&= \begin{cases} 2F(\omega) & \text{for } \omega > 0 \\ 0 & \text{for } \omega < 0. \end{cases}
\end{aligned} \tag{2.318}$$

We analytically obtain the inverse Fourier transform

$$\begin{aligned}
f_+(t) &= \frac{1}{\pi}\int_0^{\infty} F(\omega)\,\mathrm{e}^{-\mathrm{j}\omega t}\,\mathrm{d}\omega \\
&= \mathcal{F}_+^{-1}\{F(\omega)\}.
\end{aligned} \tag{2.319}$$

Disregarding negative frequencies in (2.319) even permits to define $f_+(t)$ for complex times with negative imaginary part, because the integral (2.319) is an analytic function of t, whence the terminology *analytic* signal comes from. In the limit of real t-values, we have the representation (2.317) for $f_+(t)$. For example, (2.319) allows for the definition of $f_+(t - \mathrm{j}\gamma)$ with $\gamma > 0$ according to

$$f_+(t-\mathrm{j}\gamma) = \frac{1}{\pi}\int_0^{\infty} F(\omega)\mathrm{e}^{-\gamma\omega}\mathrm{e}^{-\mathrm{j}\omega t}\,\mathrm{d}\omega \tag{2.320}$$

$$\begin{aligned}
&= \Re\left\{\frac{1}{\pi}\int_0^{\infty} F(\omega)\mathrm{e}^{-\gamma\omega}\mathrm{e}^{-\mathrm{j}\omega t}\,\mathrm{d}\omega\right\} + \mathrm{j}\Im\left\{\frac{1}{\pi}\int_0^{\infty} F(\omega)\mathrm{e}^{-\gamma\omega}\mathrm{e}^{-\mathrm{j}\omega t}\,\mathrm{d}\omega\right\} \\
&\stackrel{\mathrm{def}}{=} f_\gamma(t) + \mathrm{j}\mathcal{H}\{f_\gamma(\tau)\}
\end{aligned} \tag{2.321}$$

with the real valued time function $f_\gamma(t)$ because Fourier inversion of a spectrum $F(\omega)\mathrm{e}^{-\gamma\omega}$ without negative frequencies defines an analytic signal, in this case, the separation of $f_+(t - \mathrm{j}\gamma)$ into real and imaginary parts. Utilizing $f_+(t) = \mathrm{e}^{-\mathrm{j}\omega_0 t}$ with $\omega_0 > 0$ as an analytic signal, we obtain $F(\omega) = \pi\delta(\omega - \omega_0)$ and consequently

$$\begin{aligned}
f_+(t-\mathrm{j}\gamma) &= \int_0^{\infty} \delta(\omega-\omega_0)\mathrm{e}^{-\gamma\omega}\mathrm{e}^{-\mathrm{j}\omega t}\,\mathrm{d}\omega \\
&= \mathrm{e}^{-\gamma\omega_0}\mathrm{e}^{-\mathrm{j}\omega_0 t},
\end{aligned} \tag{2.322}$$

$$f_\gamma(t) = \mathrm{e}^{-\gamma\omega_0}\cos\omega_0 t, \tag{2.323}$$

$$\mathcal{H}\{f_\gamma(\tau)\} = -\,\mathrm{e}^{-\gamma\omega_0}\sin\omega_0 t, \tag{2.324}$$

a result that we could have simply found by insertion in this particular case. Yet, (2.320) even holds for arbitrary $F(\omega)$, which will become important in connection with pulsed ultrasonic beams (Section 12.2).

The symmetry between Fourier and inverse Fourier transform suggests that causal time functions—time functions with zero values for $t < 0$—possess spectra with real and imaginary parts being Hilbert transforms of each other. As a matter of fact, we immediately find through Fourier transform of the "causality condition"

$$f(t) = f(t)u(t), \tag{2.325}$$

application of the spectral convolution rule, utilization of the correspondence (2.282), and the convolution relation (2.361)

$$\Re\{F(\omega)\} = -\frac{1}{\pi}\,\mathrm{PV}\int_{-\infty}^{\infty} \frac{\Im\{F(\omega')\}}{\omega - \omega'}\,\mathrm{d}\omega', \tag{2.326}$$

$$\Im\{F(\omega)\} = \frac{1}{\pi}\,\mathrm{PV}\int_{-\infty}^{\infty} \frac{\Re\{F(\omega')\}}{\omega - \omega'}\,\mathrm{d}\omega'. \tag{2.327}$$

As far as the mathematical assumptions are concerned, which have to be satisfied to ensure validity of (2.314), (2.315) and (2.326), (2.327), respectively, we again refer to Doetsch (1967); essentially, time and spectral causality have to be complemented by quadratic integrability (finite energy) of the time function and its spectrum.

If $F(\omega)$ represents the complex valued spectrum of a material parameter in a linear constitutive equation, Equations 2.326 and 2.327 are called Kramers–Kronig relations (Langenberg 2005). As a consequence, the phase velocity and the attenuation of a wave are not independent of each other: Materials without losses are basically not existent.

We often meet (complex) spectra $F(\omega)$ with $F(-\omega) = F^*(\omega)$ in the theory of wave propagation that are multiplied by a frequency-independent factor according to $F(\omega)\mathrm{e}^{\mathrm{j}\varphi}$ (Section 9.1.2). Yet, if a real valued time function should correspond to $F(\omega)\mathrm{e}^{\mathrm{j}\varphi}$, we have to complement $F(\omega)\mathrm{e}^{\mathrm{j}\varphi\,\mathrm{sign}(\omega)}$ for negative frequencies with the outcome

$$\mathcal{F}^{-1}\{\mathrm{e}^{\mathrm{j}\varphi\,\mathrm{sign}(\omega)}\,F(\omega)\} = \cos\varphi\, f(t) - \sin\varphi\,\mathcal{H}\{f(\tau)\}; \tag{2.328}$$

the resulting real valued time function also contains the Hilbert transform of $f(t) = \mathcal{F}^{-1}\{F(\omega)\}$.

2.3.5 Spatial domain spectral analysis

Evidently, the spelling of the Fourier variable is arbitrary why the notation

$$F(K_x) = \int_{-\infty}^{\infty} f(x)\,\mathrm{e}^{-\mathrm{j}K_x x}\,\mathrm{d}x, \tag{2.329}$$

$$f(x) = \frac{1}{2\pi}\int_{-\infty}^{\infty} F(K_x)\,\mathrm{e}^{\mathrm{j}K_x x}\,\mathrm{d}K_x \tag{2.330}$$

is also permitted. Utilization of x as original space variable suggests a cartesian spatial coordinate, thus allowing for the interpretation of (2.329) as spatial spectrum of the spatially dependent function $f(x)$; therefore, K_x has the dimension of a reciprocal length, hence the unit m^{-1}. Note: Additionally exploiting the arbitrariness of the sign in the kernel of Fourier and inverse Fourier transform, we simultaneously changed it with regard to (2.231) and (2.270); there are good reasons for that, yet, at the moment, we will not discuss them. We only refer to the fact that utilization of the complex conjugate kernels has consequences for the transformation rules of the spatial Fourier transform: In (2.290), (2.291), and (2.292), we have to replace j by −j.

In wave theory, spatial functions are functions of three cartesian coordinates; therefore, we can Fourier transform $\phi(x, y, z)$ subsequently with regard to x, y, and z denoting the Fourier variables by K_x, K_y and K_z:

$$\Phi(K_x, K_y, K_z) = \int_{-\infty}^{\infty} \left[\int_{-\infty}^{\infty} \left[\int_{-\infty}^{\infty} \phi(x, y, z)\, \mathrm{e}^{-\mathrm{j}K_x x}\, \mathrm{d}x \right] \times \mathrm{e}^{-\mathrm{j}K_y y}\, \mathrm{d}y \right] \mathrm{e}^{-\mathrm{j}K_z z}\, \mathrm{d}z, \tag{2.331}$$

$$\phi(x, y, z) = \frac{1}{2\pi} \int_{-\infty}^{\infty} \left[\frac{1}{2\pi} \int_{-\infty}^{\infty} \left[\frac{1}{2\pi} \int_{-\infty}^{\infty} \Phi(K_x, K_y, K_z)\, \mathrm{e}^{\mathrm{j}K_x x}\, \mathrm{d}K_x \right] \times \mathrm{e}^{\mathrm{j}K_y y}\, \mathrm{d}K_y \right] \mathrm{e}^{\mathrm{j}K_z z}\, \mathrm{d}K_z. \tag{2.332}$$

Combining x, y, z to the vector of position

$$\underline{\mathbf{R}} = x\underline{\mathbf{e}}_x + y\underline{\mathbf{e}}_y + z\underline{\mathbf{e}}_z \tag{2.333}$$

allows for a similar procedure for the Fourier variables K_x, K_y, K_z:

$$\underline{\mathbf{K}} = K_x\underline{\mathbf{e}}_x + K_y\underline{\mathbf{e}}_y + K_z\underline{\mathbf{e}}_z; \tag{2.334}$$

this leads to the short-hand notation of (2.331) and (2.332):

$$\Phi(\underline{\mathbf{K}}) = \mathcal{F}_{3\mathrm{D}}\{\phi(\underline{\mathbf{R}})\} = \int_{-\infty}^{\infty}\int_{-\infty}^{\infty}\int_{-\infty}^{\infty} \phi(\underline{\mathbf{R}})\, \mathrm{e}^{-\mathrm{j}\underline{\mathbf{K}}\cdot\underline{\mathbf{R}}}\, \mathrm{d}^3\underline{\mathbf{R}}, \tag{2.335}$$

$$\phi(\underline{\mathbf{R}}) = \mathcal{F}_{3\mathrm{D}}^{-1}\{\Phi(\underline{\mathbf{K}})\} = \frac{1}{(2\pi)^3}\int_{-\infty}^{\infty}\int_{-\infty}^{\infty}\int_{-\infty}^{\infty} \Phi(\underline{\mathbf{K}})\, \mathrm{e}^{\mathrm{j}\underline{\mathbf{K}}\cdot\underline{\mathbf{R}}}\, \mathrm{d}^3\underline{\mathbf{K}} \tag{2.336}$$

as three-dimensional Fourier and inverse Fourier transform with the vector variables $\underline{\mathbf{R}}$ and $\underline{\mathbf{K}}$; the image space of the spatial Fourier transform is called $\underline{\mathbf{K}}$-space.

Bracewell (1978) cites some correspondences for (2.335) and (2.336); in particular, the three-dimensional Fourier transform of the "ball" $u(a - R)$ of radius R is given by the three-dimensional generalization of the sinc function:

$$\mathcal{F}_{3\mathrm{D}}\{u(a - R)\} = 4\pi\, \frac{\sin aK - aK\cos aK}{K^3}. \tag{2.337}$$

In addition, we have in the sense of distributions

$$\mathcal{F}_{3D}\{e^{\pm j\underline{\mathbf{k}}\cdot\underline{\mathbf{R}}}\} = (2\pi)^3\delta(\underline{\mathbf{K}} \mp \underline{\mathbf{k}}) \tag{2.338}$$

as generalization of (2.284). Anticipating that $e^{j\underline{\mathbf{k}}\cdot\underline{\mathbf{R}}}$ (together with the time function $e^{-j\omega_0 t}$) is a time harmonic plane wave propagating in $\underline{\mathbf{k}}$-direction (Section 8.1.2), we know that, consulting (2.338), its spatial spectrum is given by a δ-singularity at $\underline{\mathbf{K}} = \underline{\mathbf{k}}$ in $\underline{\mathbf{K}}$-space: According to (2.284), the time harmonic oscillation $e^{-j\omega_0 t}$ is assigned one spectral line at the circular oscillation frequency $\omega = \omega_0$, and according to (2.338), the spatially harmonic oscillation $e^{j\underline{\mathbf{k}}\cdot\underline{\mathbf{R}}}$ is assigned a δ-point spectrum at the (vectorial) spatial frequency $\underline{\mathbf{K}} = \underline{\mathbf{k}}$; the Fourier vector $\underline{\mathbf{K}}$ points into the direction of the phase propagation vector $\underline{\mathbf{k}}$ of the plane wave with the length $k = \omega/c$. Therefore, varying the propagation direction at fixed frequency varies $\underline{\mathbf{K}}$ on the so-called Ewald sphere $K = k$.

In the following, we cite the three-dimensional versions of relevant transformation rules.

Shifting rule

$$\phi(\underline{\mathbf{R}} \pm \underline{\mathbf{R}}') \overset{3D}{\circ\!\!-\!\!\bullet} \Phi(\underline{\mathbf{K}})\, e^{\pm j\underline{\mathbf{K}}\cdot\underline{\mathbf{R}}'}. \tag{2.339}$$

Modulation rule

$$\phi(\underline{\mathbf{R}})\, e^{\pm j\underline{\mathbf{k}}\cdot\underline{\mathbf{R}}} \overset{3D}{\circ\!\!-\!\!\bullet} \Phi(\underline{\mathbf{K}} \mp \underline{\mathbf{k}}). \tag{2.340}$$

Differentiation rule

$$\nabla\phi(\underline{\mathbf{R}}) \overset{3D}{\circ\!\!-\!\!\bullet} j\underline{\mathbf{K}}\Phi(\underline{\mathbf{K}}). \tag{2.341}$$

Convolution rule

$$\phi(\underline{\mathbf{R}}) *** \psi(\underline{\mathbf{R}}) \overset{3D}{\circ\!\!-\!\!\bullet} \Phi(\underline{\mathbf{K}})\Psi(\underline{\mathbf{K}}), \tag{2.342}$$

where

$$\begin{aligned}
&\phi(\underline{\mathbf{R}}) *** \psi(\underline{\mathbf{R}}) \\
&\quad = \int_{-\infty}^{\infty}\int_{-\infty}^{\infty}\int_{-\infty}^{\infty} \phi(\underline{\mathbf{R}} - \underline{\mathbf{R}}')\psi(\underline{\mathbf{R}}')\,d^3\underline{\mathbf{R}}' \\
&\quad = \int_{-\infty}^{\infty}\int_{-\infty}^{\infty}\int_{-\infty}^{\infty} \phi(x - x', y - y', z - z')\psi(x', y', z')\,dx'dy'dz'.
\end{aligned} \tag{2.343}$$

Wave field quantities are often vector or even tensor functions; their three-dimensional Fourier transform, for instance,

$$\underline{\mathbf{V}}(\underline{\mathbf{K}}) = \int_{-\infty}^{\infty}\int_{-\infty}^{\infty}\int_{-\infty}^{\infty} \underline{\mathbf{v}}(\underline{\mathbf{R}})\, e^{-j\underline{\mathbf{K}}\cdot\underline{\mathbf{R}}}\,d^3\underline{\mathbf{R}}, \tag{2.344}$$

consequently must be understood component wise.

2.4 Delta Function

2.4.1 Delta function as distribution

The physicist Paul Dirac was hoping to "invent" a function $\delta(x)$ that is zero everywhere except for $x = 0$ in a sense of being "strongly infinite" for $x = 0$ in order to sift the value $\phi(0)$ from the integral

$$\int_{-\infty}^{\infty} \delta(x)\phi(x)\,\mathrm{d}x = \phi(0). \tag{2.345}$$

The relation (2.345) is called the sifting property of the delta function (Dirac function, Dirac impulse). Yet, integration calculus tells us that a function with the property (2.345) does not exist with the consequence of defining either a distribution space to which δ belongs, thus providing a strictly mathematical sense to a functional like (2.345) (Doetsch 1967), or we symbolically understand (2.345) in an intuitive engineering sense according to

$$\int_{-\infty}^{\infty} \delta(x)\phi(x)\,\mathrm{d}x \stackrel{\mathrm{s}}{=} \phi(0) \tag{2.346}$$

trying to define rules of calculation—algebra and analysis—based on (2.346) for an appropriate δ-"function" (Dudley 1994; Langenberg 2005). If the resulting properties of δ comply with the mathematical theory of distributions, we do not have arguments against (2.346). We then symbolically speak of the delta function even though we have the delta *distribution* in mind.

It is already clear from (2.345), and in particular from (2.346), that we cannot allocate a value to the delta function for $x = 0$. This exhibits similarities to the analysis of functions that are not differentiable at jump discontinuities; their derivatives do not have a value at those points. For example, let us consider the unit-step function $u(x)$ (Figure 2.21); for $x \neq 0$, we have $u'(x) = 0$, and for $x = 0$, the discontinuous function $u(x)$ is not differentiable, that is to say, $u'(x)$ does not have a value for $x = 0$ (Figure 2.21). The rule of partial integration

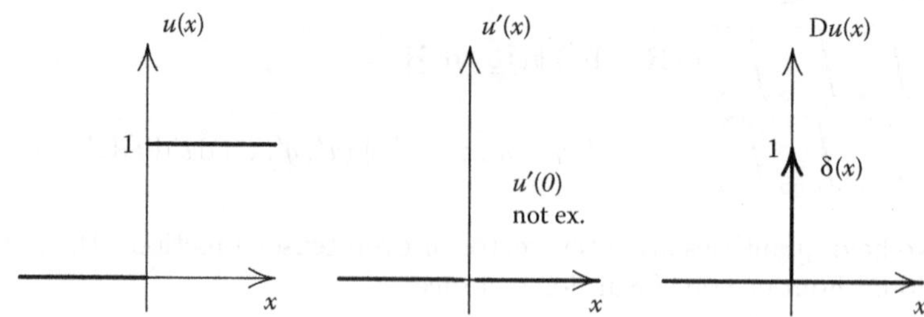

FIGURE 2.21
Unit-step function, derivative, and distributional derivative.

$$\int_a^b f'(x)g(x)\,\mathrm{d}x = f(x)g(x)\Big|_a^b - \int_a^b f(x)g'(x)\,\mathrm{d}x \tag{2.347}$$

tells us that, in particular for a vanishing integrated part $f(x)g(x)|_a^b$, we can transfer the differentiation from $f(x)$ onto $g(x)$; advantage: A nondefined operation on $f(x)$ may be absolutely permitted on $g(x)$. Substantiating this idea for the unit-step function, we first choose "test functions" $g(x) \Longrightarrow \phi(x)$ yielding a vanishing integrated part, in particular for infinite integration limits due to $\phi(\pm\infty) = 0$; the resulting relation

$$\int_{-\infty}^{\infty} u'(x)\phi(x) \stackrel{\mathrm{s}}{=} -\int_{-\infty}^{\infty} u(x)\phi'(x)\,\mathrm{d}x \tag{2.348}$$

can obviously only be understood in the above-mentioned symbolic sense because—we may turn it over and over again—the left-hand side of (2.348) does not exist. Nevertheless, it can be assigned a meaning through the right-hand side:

$$\int_{-\infty}^{\infty} u'(x)\phi(x)\,\mathrm{d}x \stackrel{\mathrm{s}}{=} \phi(0), \tag{2.349}$$

because

$$\begin{aligned} -\int_{-\infty}^{\infty} u(x)\phi'(x)\,\mathrm{d}x &= -\int_0^{\infty} \phi'(x)\,\mathrm{d}x \\ &= -\phi(x)\Big|_0^{\infty} \\ &= \phi(0). \end{aligned} \tag{2.350}$$

This new meaning is expressed by the notation

$$\int_{-\infty}^{\infty} \mathrm{D}u(x)\,\phi(x)\,\mathrm{d}x \stackrel{\mathrm{s}}{=} \phi(0) \tag{2.351}$$

of the distributional derivative $\mathrm{D}u(x)$ von $u(x)$. This distributional derivative of $u(x)$ is not the (conventional) derivative of $u(x)$, but its symbolic (distributional) generalization. Since $\phi(x)$ denotes an arbitrary test function (a member of the well-defined space of test functions), the comparison of (2.351) with (2.346) reveals that apparently we have

$$\delta(x) = \mathrm{D}u(x). \tag{2.352}$$

The symbolic graphical representation of (2.352) can also be found in Figure 2.21: $\delta(x)$ as an arrow with a unit "amplitude."

2.4.2 Delta distribution calculus

Computational rules for $\delta(x)$ (and other distributions) are always found following the above scheme: Transfer of a nondefined operation onto the test

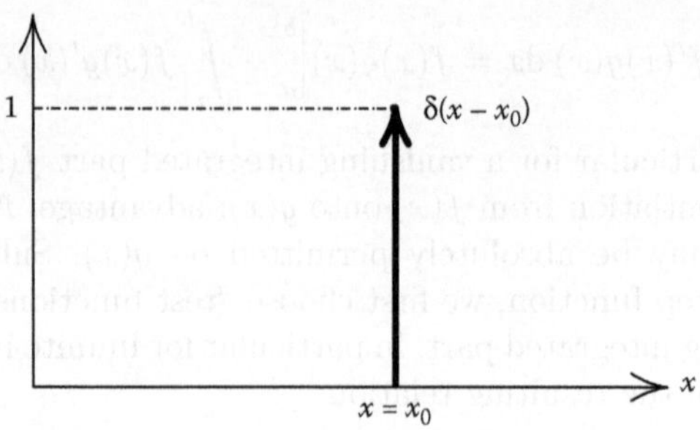

FIGURE 2.22
Illustration of the δ-function.

function. Intuitively, $\delta(x - x_0)$ is displayed as a "unit arrow" at $x = x_0$ (Figure 2.22); symbolically calculating, we find that:

$$\begin{aligned}\int_{-\infty}^{\infty} \delta(x - x_0)\phi(x)\,\mathrm{d}x &\stackrel{\mathrm{s}}{=} \int_{-\infty}^{\infty} \delta(x)\phi(x + x_0)\,\mathrm{d}x \\ &\stackrel{\mathrm{s}}{=} \phi(x_0). \end{aligned} \tag{2.353}$$

Analogously, we can show that $\alpha\delta(x)$ is illustrated by a δ-arrow with α-"amplitude" according to

$$\int_{-\infty}^{\infty} \alpha\delta(x)\phi(x)\,\mathrm{d}x \stackrel{\mathrm{s}}{=} \alpha\phi(0). \tag{2.354}$$

For real valued $a \neq 0$, we find

$$\delta(ax) = \frac{1}{|a|}\,\delta(x), \tag{2.355}$$

indicating that $\delta(ax)$ with dimensionless variable x has the reciprocal unit of the dimension of the parameter a, in other words, $\delta(t)$ has the unit s^{-1} if t denotes time, and $\delta(x)$ has the unit m^{-1} if x is a spatial coordinate. In addition, (2.355) implies

$$\delta(-x) = \delta(x). \tag{2.356}$$

The $\delta(x - x_0)$-distribution can be utilized to sample a function $\alpha(x)$:

$$\begin{aligned}\int_{-\infty}^{\infty} \alpha(x)\delta(x - x_0)\phi(x)\,\mathrm{d}x &\stackrel{\mathrm{s}}{=} \int_{-\infty}^{\infty} \delta(x - x_0)\alpha(x)\phi(x)\,\mathrm{d}x \\ &\stackrel{\mathrm{s}}{=} \alpha(x_0)\phi(x_0) \\ &\stackrel{\mathrm{s}}{=} \alpha(x_0)\int_{-\infty}^{\infty} \delta(x - x_0)\phi(x)\,\mathrm{d}x \\ &\stackrel{\mathrm{s}}{=} \int_{-\infty}^{\infty} \alpha(x_0)\delta(x - x_0)\phi(x)\,\mathrm{d}x; \end{aligned} \tag{2.357}$$

consequently,[45]

$$\alpha(x)\delta(x - x_0) = \alpha(x_0)\delta(x - x_0) \tag{2.358}$$

yielding

$$(x - x_0)\,\delta(x - x_0) = 0. \tag{2.359}$$

The δ-distribution can also be differentiated in the distributional sense:

$$\int_{-\infty}^{\infty} \mathrm{D}\delta(x)\phi(x)\,\mathrm{d}x \stackrel{\mathrm{s}}{=} -\phi'(0). \tag{2.360}$$

For simplicity, we write $\mathrm{D}\delta(x) = \delta'(x)$.

Even the convolution of $\delta(x)$ with a function $\alpha(x)$ can be symbolically calculated with the sifting property:[46]

$$\begin{aligned}\delta(x) * \alpha(x) &\stackrel{\mathrm{s}}{=} \int_{-\infty}^{\infty} \delta(x - x')\alpha(x')\,\mathrm{d}x' \\ &= \alpha(x);\end{aligned} \tag{2.361}$$

further, we have

$$\alpha(x) * \delta(x - x_0) = \alpha(x - x_0). \tag{2.362}$$

We are now getting rather bold if we put $\alpha(x) = \delta(x - x_0)$ in (2.361) claiming that

$$\delta(x) * \delta(x - x_0) = \delta(x - x_0) \tag{2.363}$$

holds and even

$$\delta(x - x_1) * \delta(x - x_2) = \delta(x - x_1 - x_2). \tag{2.364}$$

Of course, the above relations can be mathematically proven using distribution theory (Doetsch 1967).

2.4.3 Delta function and Fourier transform

Similar to (2.361), we can define the Fourier transform of the δ-function using the sifting property for plausibility [consult Doetsch (1967) for Fourier transform of distributions]:

$$\int_{-\infty}^{\infty} \delta(t - t_0)\,\mathrm{e}^{\mathrm{j}\omega t}\,\mathrm{d}t \stackrel{\mathrm{s}}{=} \mathrm{e}^{\mathrm{j}t_0\omega}; \tag{2.365}$$

[45]The following is not correct: $\alpha(x)\delta(x - x_0) = \alpha(x_0)$.

[46]Since $\alpha(x)$ must not necessarily be a test function, we have circumvented the distributional path [look at Doetsch (1967) for a correct calculation].

it follows:

$$\delta(t) \circ\!\!-\!\!\bullet\ 1, \tag{2.366}$$

$$\delta(t \pm t_0) \circ\!\!-\!\!\bullet\ \mathrm{e}^{\mp \mathrm{j} t_0 \omega}. \tag{2.367}$$

If, on the other hand, inverse Fourier transforming $2\pi\delta(\omega \pm \omega_0)$ yields

$$\frac{1}{2\pi}\int_{-\infty}^{\infty} 2\pi\delta(\omega \pm \omega_0)\,\mathrm{e}^{-\mathrm{j}\omega t}\,\mathrm{d}\omega \stackrel{\mathrm{s}}{=} \mathrm{e}^{\pm \mathrm{j}\omega_0 t}, \tag{2.368}$$

accordingly, in the distributional sense

$$\mathrm{e}^{\pm \mathrm{j}\omega_0 t} \circ\!\!-\!\!\bullet\ 2\pi\delta(\omega \pm \omega_0) \tag{2.369}$$

should hold. Utilizing Euler's formulas (2.251) and (2.252), we obtain the correspondences (2.285) and (2.286) and

$$1 \circ\!\!-\!\!\bullet\ 2\pi\delta(\omega). \tag{2.370}$$

Similar to (2.369) and (2.370), the Fourier transforms of $u(t)$ and $\mathrm{sign}(t)$ only exist symbolically or in the distributional sense. To find them, we first calculate $\mathcal{F}\{t^{-1}\}$ accounting for the singularity of t^{-1} in terms of a Cauchy principal value of the integral:

$$\begin{aligned}
\mathcal{F}\left\{\frac{1}{t}\right\} &\stackrel{\mathrm{def}}{=} \mathcal{F}\left\{\mathrm{PV}\frac{1}{t}\right\} \\
&= \mathrm{PV}\int_{-\infty}^{\infty} \frac{1}{t}\,\mathrm{e}^{\mathrm{j}\omega t}\,\mathrm{d}t \\
&\stackrel{\mathrm{def}}{=} \lim_{\epsilon \to 0}\left(\int_{\epsilon}^{\infty} \frac{1}{t}\,\mathrm{e}^{\mathrm{j}\omega t}\,\mathrm{d}t + \int_{-\infty}^{-\epsilon} \frac{1}{t}\,\mathrm{e}^{\mathrm{j}\omega t}\,\mathrm{d}t\right) \\
&= \lim_{\epsilon \to 0}\int_{\epsilon}^{\infty} \frac{1}{t}\left(\mathrm{e}^{\mathrm{j}\omega t} - \mathrm{e}^{-\mathrm{j}\omega t}\right)\mathrm{d}t \\
&= 2\mathrm{j}\lim_{\epsilon \to 0}\int_{\epsilon}^{\infty} \frac{\sin \omega t}{t}\,\mathrm{d}t \\
&= 2\mathrm{j}\int_{0}^{\infty} \frac{\sin \omega t}{t}\,\mathrm{d}t \\
&= 2\mathrm{j}\begin{cases} \dfrac{\pi}{2} & \text{for } \omega > 0 \\ -\dfrac{\pi}{2} & \text{for } \omega < 0 \end{cases} \\
&= \mathrm{j}\pi\,\mathrm{sign}(\omega).
\end{aligned} \tag{2.371}$$

The last but one equality sign of (2.371) holds on behalf of the definition of the sine integral. With $\mathrm{sign}(\omega) = 2u(\omega) - 1$, it follows from (2.371)

$$\mathcal{F}\left\{\frac{1}{2}\,\delta(t) - \frac{\mathrm{j}}{2\pi}\,\mathrm{PV}\frac{1}{t}\right\} = u(\omega). \tag{2.372}$$

With the symmetry relation (2.281), we further obtain from (2.371):

$$\mathrm{sign}(t) \circ\!\!-\!\!\bullet\; 2\mathrm{j}\,\mathrm{PV}\frac{1}{\omega}, \tag{2.373}$$

$$u(t) \circ\!\!-\!\!\bullet\; \pi\delta(\omega) + \mathrm{j}\,\mathrm{PV}\frac{1}{\omega}. \tag{2.374}$$

2.4.4 Three-dimensional delta function

If the variable in $\delta(x)$ corresponds to a cartesian coordinate, an enhancement to $\delta(y)$ and $\delta(z)$ is immediately at hand to define a spatial point source with the three-dimensional sifting property:

$$\int_{-\infty}^{\infty}\int_{-\infty}^{\infty}\int_{-\infty}^{\infty} \delta(x)\delta(y)\delta(z)\phi(x,y,z)\,\mathrm{d}x\mathrm{d}y\mathrm{d}z \stackrel{\mathrm{s}}{=} \phi(0,0,0). \tag{2.375}$$

Nota bene: Because each integral in (2.375) must be considered symbolically (in the distributional sense: as a functional), $\delta(x)\delta(y)\delta(z)$ is only a symbolical product of δ-functions. With $\delta(\underline{\mathbf{R}}) = \delta(x)\delta(y)\delta(z)$, we introduce the short-hand notation

$$\int_{-\infty}^{\infty}\int_{-\infty}^{\infty}\int_{-\infty}^{\infty} \delta(\underline{\mathbf{R}})\phi(\underline{\mathbf{R}})\,\mathrm{d}^3\underline{\mathbf{R}} \stackrel{\mathrm{s}}{=} \phi(\underline{\mathbf{0}}) \tag{2.376}$$

and generalize to

$$\int_{-\infty}^{\infty}\int_{-\infty}^{\infty}\int_{-\infty}^{\infty} \delta(\underline{\mathbf{R}}-\underline{\mathbf{R}}')\phi(\underline{\mathbf{R}})\,\mathrm{d}^3\underline{\mathbf{R}} \stackrel{\mathrm{s}}{=} \phi(\underline{\mathbf{R}}'). \tag{2.377}$$

It is important for the sifting property (2.377) that $\underline{\mathbf{R}}'$ is located within the integration region, which is always true for infinite integration limits, but not if the integration extends only over a finite region V of $\mathbb{R}^3$:

$$\iiint_V \delta(\underline{\mathbf{R}}-\underline{\mathbf{R}}')\phi(\underline{\mathbf{R}})\,\mathrm{d}^3\underline{\mathbf{R}}' \stackrel{\mathrm{s}}{=} \begin{cases} \phi(\underline{\mathbf{R}}') & \text{for } \underline{\mathbf{R}}' \in V \\ 0 & \text{for } \underline{\mathbf{R}}' \notin V. \end{cases} \tag{2.378}$$

Ultimately, the extinction theorem of Helmholtz' integral formulation of Huygens' principle turns out to be a consequence of (2.378) (Section 15.1.2).

With (2.377), the correspondences

$$\delta(\underline{\mathbf{R}}) \stackrel{\mathrm{3D}}{\circ\!\!-\!\!\bullet}\; 1, \tag{2.379}$$

$$\delta(\underline{\mathbf{R}} \pm \underline{\mathbf{R}}') \stackrel{\mathrm{3D}}{\circ\!\!-\!\!\bullet}\; \mathrm{e}^{\pm\mathrm{j}\underline{\mathbf{K}}\cdot\underline{\mathbf{R}}'}, \tag{2.380}$$

$$\mathrm{e}^{\pm\mathrm{j}\underline{\mathbf{k}}\cdot\underline{\mathbf{R}}} \stackrel{\mathrm{3D}}{\circ\!\!-\!\!\bullet}\; (2\pi)^3\delta(\underline{\mathbf{K}} \mp \underline{\mathbf{k}}) \tag{2.381}$$

of the three-dimensional Fourier transform are rapidly plausible.

As far as the representation of $\delta(\underline{\mathbf{R}})$ in other than cartesian coordinates is concerned, we refer to Langenberg (2005).

2.4.5 Singular function of a surface

The singular function $\gamma(\underline{\mathbf{R}})$ of a closed surface S has the multidimensional sifting property to reduce a volume integral to a surface integral (Bleistein 1984; Bamler 1989):

$$\int_{-\infty}^{\infty}\int_{-\infty}^{\infty}\int_{-\infty}^{\infty} \gamma(\underline{\mathbf{R}})\phi(\underline{\mathbf{R}})\,\mathrm{d}V = \int\!\!\int_S \phi(\underline{\mathbf{R}})\,\mathrm{d}S. \tag{2.382}$$

To make that plausible and to define $\gamma(\underline{\mathbf{R}})$, we recall that the distributional derivative of the unit-step function has the same sifting property as the δ-function (Equation 2.352). We start to consider an example of a closed surface, a sphere of radius a; the interior of the sphere is given by its characteristic function

$$u(a-R) = \begin{cases} 1 & \text{for } R < a \\ 0 & \text{for } R > a \end{cases} \tag{2.383}$$

with a dependence of the radial variable R as displayed in Figure 2.23. Obviously,

$$\underline{\mathbf{e}}_R \cdot \boldsymbol{\nabla} u(a-R) = -\delta(a-R) \tag{2.384}$$

yields a δ-function, being singular on the whole surface of the sphere; this singular function $\gamma_a(\underline{\mathbf{R}}) = \delta(a-R)$ of the sphere has the desired property

$$\begin{aligned}\int_0^{2\pi}\int_0^{\pi}\int_0^{R} \delta(a-R)\phi(R,\vartheta,\varphi)\,\underbrace{R^2\sin\vartheta\,\mathrm{d}R\mathrm{d}\vartheta\mathrm{d}\varphi}_{=\,\mathrm{d}V} \\ = \int_0^{2\pi}\int_0^{\pi} \phi(a,\vartheta,\varphi)\,\underbrace{a^2\sin\vartheta\,\mathrm{d}\vartheta\mathrm{d}\varphi}_{=\,\mathrm{d}S}\end{aligned} \tag{2.385}$$

according to (2.382).

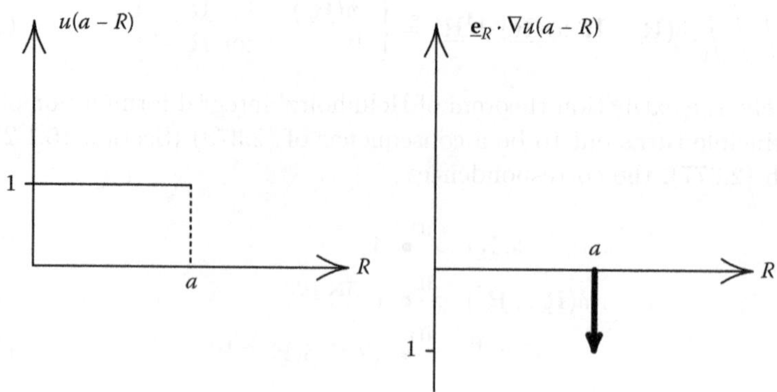

FIGURE 2.23
Illustration of the singular function of a spherical surface.

We generalize as follows; at first, we define

$$\Gamma(\underline{\mathbf{R}}) = \begin{cases} 1 & \text{for } \underline{\mathbf{R}} \in V \\ 0 & \text{for } \underline{\mathbf{R}} \notin V \end{cases} \tag{2.386}$$

as a characteristic function of a volume V with surface S and outer normal $\underline{\mathbf{n}}$; then, we construct

$$\gamma(\underline{\mathbf{R}}) = -\underline{\mathbf{n}} \cdot \boldsymbol{\nabla}\Gamma(\underline{\mathbf{R}}) \tag{2.387}$$

the singular function of S as generalization of (2.384); it definitely has the property (2.382). With

$$\underline{\boldsymbol{\gamma}}(\underline{\mathbf{R}}) = -\boldsymbol{\nabla}\Gamma(\underline{\mathbf{R}}), \tag{2.388}$$

we denote the vector singular function of S.

3

Governing Equations of Elastodynamics

Maxwell's equations of electromagnetism cannot be proven in terms of a theoretical derivation from even more fundamental equations; they compose the brilliant design of a theory that must, of course, describe experimental observations without contradictions, and as such, they are axiomatically put on top in order to draw conclusions to be validated experimentally.[47] In a similar sense, we put the governing equations of elastodynamics (3.1) and (3.2)—Newton–Cauchy's equation and the deformation rate equation—axiomatically at the beginning, but we present physical arguments in the subsequent section because, different from Maxwell's equations, they can be deduced from the physical laws of mechanics under the continuum hypothesis, the geometric linearization of small particle displacements, and the neglect of products of field quantities, that is to say, in a *linear* approximation.

3.1 Newton–Cauchy Equation of Motion and Deformation Rate Equation in the Time and Frequency Domain

We write the governing equations of elastodynamics in the following form:

$$\frac{\partial \underline{\mathbf{j}}(\underline{\mathbf{R}}, t)}{\partial t} = \nabla \cdot \underline{\underline{\mathbf{T}}}(\underline{\mathbf{R}}, t) + \underline{\mathbf{f}}(\underline{\mathbf{R}}, t), \tag{3.1}$$

$$\frac{\partial \underline{\underline{\mathbf{S}}}(\underline{\mathbf{R}}, t)}{\partial t} = \frac{1}{2}\left\{\nabla \underline{\mathbf{v}}(\underline{\mathbf{R}}, t) + [\nabla \underline{\mathbf{v}}(\underline{\mathbf{R}}, t)]^{21}\right\} + \underline{\underline{\mathbf{h}}}(\underline{\mathbf{R}}, t). \tag{3.2}$$

They linearly relate the subsequent elastodynamic field quantities:

- Linear momentum vector $\underline{\mathbf{j}}(\underline{\mathbf{R}}, t)$,
- Symmetric stress tensor $\underline{\underline{\mathbf{T}}}(\underline{\mathbf{R}}, t)$ of second rank,
- Symmetric deformation tensor $\underline{\underline{\mathbf{S}}}(\underline{\mathbf{R}}, t)$ of second rank,
- Particle velocity vector $\underline{\mathbf{v}}(\underline{\mathbf{R}}, t)$;

[47]The classical example is Heinrich Hertz' experiment to excite the electromagnetic waves that are predicted by Maxwell's theory.

the quantities

- Force density $\underline{\mathbf{f}}(\underline{\mathbf{R}}, t)$ and
- Injected deformation rate $\underline{\underline{\mathbf{h}}}(\underline{\mathbf{R}}, t)$ as a symmetric second rank tensor

denote prescribed[48] volume sources that are the physical origin of the elastodynamic field as inhomogeneities of the governing equations.

To solve (3.1) and (3.2), we need physically based mathematical relations between the field quantities, so-called constitutive equations (Chapter 4). This is similarly true for Maxwell's equations except in vacuum.

With the definition of Fourier spectra of field and source quantities according to

$$\underline{\mathbf{j}}(\underline{\mathbf{R}}, \omega) = \int_{-\infty}^{\infty} \underline{\mathbf{j}}(\underline{\mathbf{R}}, t)\, \mathrm{e}^{\mathrm{j}\omega t}\, \mathrm{d}t, \tag{3.3}$$

$$\underline{\underline{\mathbf{T}}}(\underline{\mathbf{R}}, \omega) = \int_{-\infty}^{\infty} \underline{\underline{\mathbf{T}}}(\underline{\mathbf{R}}, t)\, \mathrm{e}^{\mathrm{j}\omega t}\, \mathrm{d}t, \tag{3.4}$$

$$\underline{\underline{\mathbf{S}}}(\underline{\mathbf{R}}, \omega) = \int_{-\infty}^{\infty} \underline{\underline{\mathbf{S}}}(\underline{\mathbf{R}}, t)\, \mathrm{e}^{\mathrm{j}\omega t}\, \mathrm{d}t, \tag{3.5}$$

$$\underline{\mathbf{v}}(\underline{\mathbf{R}}, \omega) = \int_{-\infty}^{\infty} \underline{\mathbf{v}}(\underline{\mathbf{R}}, t)\, \mathrm{e}^{\mathrm{j}\omega t}\, \mathrm{d}t, \tag{3.6}$$

$$\underline{\mathbf{f}}(\underline{\mathbf{R}}, \omega) = \int_{-\infty}^{\infty} \underline{\mathbf{f}}(\underline{\mathbf{R}}, t)\, \mathrm{e}^{\mathrm{j}\omega t}\, \mathrm{d}t, \tag{3.7}$$

$$\underline{\underline{\mathbf{h}}}(\underline{\mathbf{R}}, \omega) = \int_{-\infty}^{\infty} \underline{\underline{\mathbf{h}}}(\underline{\mathbf{R}}, t)\, \mathrm{e}^{\mathrm{j}\omega t}\, \mathrm{d}t, \tag{3.8}$$

we move to the governing equations of elastodynamics

$$-\mathrm{j}\omega\, \underline{\mathbf{j}}(\underline{\mathbf{R}}, \omega) = \nabla \cdot \underline{\underline{\mathbf{T}}}(\underline{\mathbf{R}}, \omega) + \underline{\mathbf{f}}(\underline{\mathbf{R}}, \omega), \tag{3.9}$$

$$-\mathrm{j}\omega\, \underline{\underline{\mathbf{S}}}(\underline{\mathbf{R}}, \omega) = \frac{1}{2}\left\{\nabla \underline{\mathbf{v}}(\underline{\mathbf{R}}, \omega) + [\nabla \underline{\mathbf{v}}(\underline{\mathbf{R}}, \omega)]^{21}\right\} + \underline{\underline{\mathbf{h}}}(\underline{\mathbf{R}}, \omega) \tag{3.10}$$

in the frequency domain. Note: The spectral quantities contain the factor "second" in their unit, in contrast to the time domain quantities; nevertheless, we stick to the physical terminology of the time domain quantities; instead of calling $\underline{\underline{\mathbf{h}}}(\underline{\mathbf{R}}, \omega)$ the Fourier transformed injected deformation rate, we stay with the term injected deformation rate.

With the solutions of (3.9) and (3.10) for $\omega \geq 0$ and the continuation relation (2.272) for negative frequencies, we retrieve the real valued quantities in the time domain as solutions of (3.1) and (3.2)

[48] If, for example, $\underline{\mathbf{f}}(\underline{\mathbf{R}}, t)$ denotes the Lorentz force density produced by an EMAT, it is not really "prescribed," but it must be calculated with Maxwell's equations, which means that this step has already been performed.

$$\underline{\mathbf{j}}(\underline{\mathbf{R}}, t) = \frac{1}{2\pi} \int_{-\infty}^{\infty} \underline{\mathbf{j}}(\underline{\mathbf{R}}, \omega) \, \mathrm{e}^{-\mathrm{j}\omega t} \, \mathrm{d}t, \tag{3.11}$$

$$\underline{\underline{\mathbf{T}}}(\underline{\mathbf{R}}, t) = \frac{1}{2\pi} \int_{-\infty}^{\infty} \underline{\underline{\mathbf{T}}}(\underline{\mathbf{R}}, \omega) \, \mathrm{e}^{-\mathrm{j}\omega t} \, \mathrm{d}t, \tag{3.12}$$

$$\underline{\underline{\mathbf{S}}}(\underline{\mathbf{R}}, t) = \frac{1}{2\pi} \int_{-\infty}^{\infty} \underline{\underline{\mathbf{S}}}(\underline{\mathbf{R}}, \omega) \, \mathrm{e}^{-\mathrm{j}\omega t} \, \mathrm{d}t, \tag{3.13}$$

$$\underline{\mathbf{v}}(\underline{\mathbf{R}}, t) = \frac{1}{2\pi} \int_{-\infty}^{\infty} \underline{\mathbf{v}}(\underline{\mathbf{R}}, \omega) \, \mathrm{e}^{-\mathrm{j}\omega t} \, \mathrm{d}t, \tag{3.14}$$

$$\underline{\mathbf{f}}(\underline{\mathbf{R}}, t) = \frac{1}{2\pi} \int_{-\infty}^{\infty} \underline{\mathbf{f}}(\underline{\mathbf{R}}, \omega) \, \mathrm{e}^{-\mathrm{j}\omega t} \, \mathrm{d}t, \tag{3.15}$$

$$\underline{\underline{\mathbf{h}}}(\underline{\mathbf{R}}, t) = \frac{1}{2\pi} \int_{-\infty}^{\infty} \underline{\underline{\mathbf{h}}}(\underline{\mathbf{R}}, \omega) \, \mathrm{e}^{-\mathrm{j}\omega t} \, \mathrm{d}t \tag{3.16}$$

via inverse Fourier integrals. Therefore, it is our choice whether we work in the spectral or in the time domain, a fact that is extensively exploited in the present elaboration; roughly speaking, it is often easier to calculate in the frequency domain and to interpret results in the time domain.

Alternative to the Fourier transform of (3.1) and (3.2), we can make the *ansatz* of real valued time harmonic fields with circular frequency $\omega_0 > 0$, for instance, using the example of the momentum density:

$$\underline{\mathbf{j}}(\underline{\mathbf{R}}, t) \Longrightarrow \underline{\mathbf{j}}(\underline{\mathbf{R}}, t, \omega_0) = \Re\left\{\underline{\mathbf{j}}(\underline{\mathbf{R}}, \omega_0) \, \mathrm{e}^{-\mathrm{j}\omega_0 t}\right\}. \tag{3.17}$$

Here, following the terminology of electrical engineering, $\underline{\mathbf{j}}(\underline{\mathbf{R}}, \omega_0)$ is called the (complex valued) phasor. This results in the governing equations

$$-\mathrm{j}\omega_0 \underline{\mathbf{j}}(\underline{\mathbf{R}}, \omega_0) = \nabla \cdot \underline{\underline{\mathbf{T}}}(\underline{\mathbf{R}}, \omega_0) + \underline{\mathbf{f}}(\underline{\mathbf{R}}, \omega_0), \tag{3.18}$$

$$-\mathrm{j}\omega_0 \, \underline{\underline{\mathbf{S}}}(\underline{\mathbf{R}}, \omega_0) = \frac{1}{2} \left\{\nabla \underline{\mathbf{v}}(\underline{\mathbf{R}}, \omega_0) + [\nabla \underline{\mathbf{v}}(\underline{\mathbf{R}}, \omega_0]^{21}\right\} + \underline{\underline{\mathbf{h}}}(\underline{\mathbf{R}}, \omega_0) \tag{3.19}$$

for the phasors. However, if we select two spectral lines $\omega = \pm\omega_0$, $\omega_0 > 0$ out of the Fourier spectrum of $\underline{\mathbf{j}}(\underline{\mathbf{R}}, \omega)$ and combine them according to

$$\begin{aligned}\underline{\mathbf{j}}(\underline{\mathbf{R}}, \omega, \omega_0) &\stackrel{\text{def}}{=} \pi \underline{\mathbf{j}}(\underline{\mathbf{R}}, \omega) \delta(\omega - \omega_0) + \pi \underline{\mathbf{j}}^*(\underline{\mathbf{R}}, \omega) \delta(\omega + \omega_0) \\ &= \pi \underline{\mathbf{j}}(\underline{\mathbf{R}}, \omega_0) \delta(\omega - \omega_0) + \pi \underline{\mathbf{j}}^*(\underline{\mathbf{R}}, \omega_0) \delta(\omega + \omega_0),\end{aligned} \tag{3.20}$$

we obviously obtain the real valued time harmonic field quantity via Fourier inversion with the correspondence (2.284):

$$\begin{aligned}\mathcal{F}^{-1}\left\{\underline{\mathbf{j}}(\underline{\mathbf{R}}, \omega, \omega_0)\right\} &= \frac{1}{2} \underline{\mathbf{j}}(\underline{\mathbf{R}}, \omega_0) \, \mathrm{e}^{-\mathrm{j}\omega_0 t} + \frac{1}{2} \underline{\mathbf{j}}^*(\underline{\mathbf{R}}, \omega_0) \, \mathrm{e}^{\mathrm{j}\omega_0 t} \\ &= \Re\left\{\underline{\mathbf{j}}(\underline{\mathbf{R}}, \omega_0) \, \mathrm{e}^{-\mathrm{j}\omega_0 t}\right\} \\ &= \underline{\mathbf{j}}(\underline{\mathbf{R}}, t, \omega_0);\end{aligned} \tag{3.21}$$

ω_0-phasors and conjugate complex ω_0-phasors (multiplied by π) are nothing but the amplitudes of spectral lines of real valued time harmonic field quantities at $\omega = \pm\omega_0$, and as such, they represent a discrete sample out of the continuous Fourier spectrum.

3.2 Physical Foundations

3.2.1 Mass conservation

To physically justify the governing equations of elastodynamics (3.1) and (3.2), we essentially follow de Hoop (1995), yet we do not use the index notation with summation convention but the coordinate-free notation, which, according to our opinion, is more transparent.

De Hoop starts with the continuum hypothesis that a particle distribution at $\underline{\mathbf{R}}$ for time t—$\underline{\mathbf{R}}$ is the vector of position in a fixed reference system—can be described through a (piecewise) *continuous* particle *density* $n(\underline{\mathbf{R}}, t)$ that is defined as the number of particles $N_\epsilon(\underline{\mathbf{R}}, t)$ per (small) reference volume $V_\epsilon(\underline{\mathbf{R}})$. The macroscopic (particle) drift velocity $\underline{\mathbf{v}}(\underline{\mathbf{R}}, t)$ is introduced as average value of the velocity vectors of single particles in $V_\epsilon(\underline{\mathbf{R}})$, so to average out chaotic (thermal) contributions. The calculation of the time variation of the total number $N(t)$ of particles in a volume $V(t)$ on the basis of

$$N(t) = \iiint_{V(t)} n(\underline{\mathbf{R}}, t)\,\mathrm{d}V \tag{3.22}$$

immediately yields the conservation theorem

$$\iiint_{V(t)} \frac{\partial n(\underline{\mathbf{R}}, t)}{\partial t}\,\mathrm{d}V + \iint_{S(t)} n(\underline{\mathbf{R}}, t)\,\underline{\mathbf{v}}(\underline{\mathbf{R}}, t) \cdot \underline{\mathbf{dS}} = 0 \tag{3.23}$$

for the particle flow $n(\underline{\mathbf{R}}, t)\underline{\mathbf{v}}(\underline{\mathbf{R}}, t)$ provided particles are neither created nor annihilated;[49] here, $S(t)$ is the surface[50] of $V(t)$ and $\underline{\mathbf{dS}}$ its vector differential surface element. Evidently, $n(\underline{\mathbf{R}}, t)\underline{\mathbf{v}}(\underline{\mathbf{R}}, t) \cdot \underline{\mathbf{dS}}\,\Delta t$ is the (average) number of particles passing $\mathrm{d}S$ during the time interval Δt, thus changing the particle

[49]Otherwise, the right-hand side of (3.23) would not be zero but the difference between creation and annihilation rates (de Hoop 1995).

[50]The time dependence of $V(t)$ and $S(t)$ is understood as follows: During the time interval Δt, the surface $S(t)$ changes to $S(t + \Delta t)$ according to $\underline{\mathbf{v}}(\underline{\mathbf{R}}, t)\Delta t$, where $\underline{\mathbf{v}}(\underline{\mathbf{R}}, t)$ is the vector drift velocity $\underline{\mathbf{v}}(\underline{\mathbf{R}}, t)$ of the particle density of each surface point of $S(t)$. If particles are neither created nor annihilated, particle conservation mean that the time variation of the total number of particles within this time-dependent volume $V(t)$ is equal to zero. The conservation theorem (3.23) expresses this fact in the following way: If the particle density $n(\underline{\mathbf{R}}, t)$ does not change in the volume $V(t)$ for fixed time t, it must be compensated by the particle flow $n(\underline{\mathbf{R}}, t)\underline{\mathbf{v}}(\underline{\mathbf{R}}, t)$ normal to $S(t)$.

density in $V(t)$. Because (3.23) must hold for each volume $V(t)$, application of Gauss' theorem yields the continuity equation

$$\frac{\partial n(\underline{\mathbf{R}}, t)}{\partial t} + \nabla \cdot [n(\underline{\mathbf{R}}, t)\, \underline{\mathbf{v}}(\underline{\mathbf{R}}, t)] = 0 \tag{3.24}$$

for the particle flow.

In elastodynamics, we are particularly interested in the material properties of the particles,[51] hence de Hoop now defines the (volume) mass density $\rho(\underline{\mathbf{R}}, t)$ as volume average of the single particle masses (of an arbitrary particle type). If the single masses (of this type) are all equal to m, it follows

$$\rho(\underline{\mathbf{R}}, t) = m\, n(\underline{\mathbf{R}}, t) \tag{3.25}$$

for the mass density and

$$\underline{\mathbf{j}}(\underline{\mathbf{R}}, t) = \rho(\underline{\mathbf{R}}, t)\underline{\mathbf{v}}(\underline{\mathbf{R}}, t) \tag{3.26}$$

for the mass flow density (momentum density of linear momentum: drift momentum as average of the particle momentum times particle density) that originates as a macroscopic quantity via averaging from the microscopic vector particle momentum. Where required, we have to sum over the various particle types to find the total mass density and the total mass flow density.

From particle conservation according to (3.23), we immediately deduce mass conservation

$$\iiint_{V(t)} \frac{\partial \rho(\underline{\mathbf{R}}, t)}{\partial t}\, \mathrm{d}V + \iint_{S(t)} \underline{\mathbf{j}}(\underline{\mathbf{R}}, t) \cdot \underline{\mathbf{dS}} = 0, \tag{3.27}$$

and from the continuity equation (3.24) for the particle flow, we obtain the continuity equation for the mass flow (of the particle type under consideration):

$$\frac{\partial \rho(\underline{\mathbf{R}}, t)}{\partial t} + \nabla \cdot \underline{\mathbf{j}}(\underline{\mathbf{R}}, t) = 0. \tag{3.28}$$

If $\Psi(\underline{\mathbf{R}}, t)$ denotes any function attributed to a particle, de Hoop calculates the total time variation of $\Psi(\underline{\mathbf{R}}, t)$ in $V(t)$ according to[52]

$$\begin{aligned} \frac{\mathrm{d}}{\mathrm{d}t} \iiint_{V(t)} \Psi(\underline{\mathbf{R}}, t)\, \mathrm{d}V = & \iiint_{V(t)} \frac{\partial \Psi(\underline{\mathbf{R}}, t)}{\partial t}\, \mathrm{d}V \\ & + \iint_{S(t)} \Psi(\underline{\mathbf{R}}, t)\underline{\mathbf{v}}(\underline{\mathbf{R}}, t) \cdot \underline{\mathbf{dS}}, \end{aligned} \tag{3.29}$$

[51] In electromagnetism, it is the electric and magnetic properties.

[52] Note: Following Footnote 50, $V(t)$ is a very special time-dependent volume whose time variation must be differentiated on the left-hand side of (3.29); how is told by the right-hand side of (3.29).

assuming vanishing particle creation and annihilation. Equation 3.29, modified according to

$$\frac{\mathrm{d}}{\mathrm{d}t}\iiint_{V(t)} \Psi(\underline{\mathbf{R}},t)\,\mathrm{d}V = \iiint_{V(t)} \left\{ \frac{\partial \Psi(\underline{\mathbf{R}},t)}{\partial t} + \boldsymbol{\nabla}\cdot[\underline{\mathbf{v}}(\underline{\mathbf{R}},t)\Psi(\underline{\mathbf{R}},t)] \right\} \mathrm{d}V, \tag{3.30}$$

with the help of Gauss' theorem is called Reynold's transport theorem. If an operator $\delta/\delta t$ according to

$$\frac{\mathrm{d}}{\mathrm{d}t}\iiint_{V(t)} \Psi(\underline{\mathbf{R}},t)\,\mathrm{d}V = \iiint_{V(t)} \frac{\delta \Psi(\underline{\mathbf{R}},t)}{\delta t}\,\mathrm{d}V \tag{3.31}$$

is introduced [de Hoop writes $\dot{\Psi}(\underline{\mathbf{R}},t)$], (3.30) implies the definition

$$\frac{\delta \Psi(\underline{\mathbf{R}},t)}{\delta t} = \frac{\partial \Psi(\underline{\mathbf{R}},t)}{\partial t} + \boldsymbol{\nabla}\cdot[\underline{\mathbf{v}}(\underline{\mathbf{R}},t)\Psi(\underline{\mathbf{R}},t)]. \tag{3.32}$$

3.2.2 Convective time derivative

Now, $\Psi(\underline{\mathbf{R}},t)$ should denote any (scalar) macroscopic physical quantity (e.g., mass, density, cartesian linear or angular momentum components, and kinetic energy) that is attributed to a particular particle type; the time variation of $\Psi(\underline{\mathbf{R}},t)$-total of all particles contained in $V(t)$ results from Reynold's transport theorem (3.29) explicitly incorporating the particle conservation law:

$$\begin{aligned}\frac{\mathrm{d}}{\mathrm{d}t}\iiint_{V(t)} n(\underline{\mathbf{R}},t)\Psi(\underline{\mathbf{R}},t)\,\mathrm{d}V &= \iiint_{V(t)} \frac{\partial}{\partial t}[n(\underline{\mathbf{R}},t)\Psi(\underline{\mathbf{R}},t)]\,\mathrm{d}V \\ &\quad + \iint_{S(t)} n(\underline{\mathbf{R}},t)\Psi(\underline{\mathbf{R}},t)\underline{\mathbf{v}}(\underline{\mathbf{R}},t)\cdot\underline{\mathbf{dS}}.\end{aligned} \tag{3.33}$$

Applying Gauss' theorem, evaluating the time and spatial derivatives, and recognizing the particle continuity equation, (3.24) yields

$$\frac{\mathrm{d}}{\mathrm{d}t}\iiint_{V(t)} n(\underline{\mathbf{R}},t)\Psi(\underline{\mathbf{R}},t)\,\mathrm{d}V = \iiint_{V(t)} n(\underline{\mathbf{R}},t)\frac{\mathrm{D}\Psi(\underline{\mathbf{R}},t)}{\mathrm{D}t}\,\mathrm{d}V, \tag{3.34}$$

where the derivative operator $\mathrm{D}/\mathrm{D}t$ stands as a short-hand notation for

$$\frac{\mathrm{D}}{\mathrm{D}t} = \frac{\partial}{\partial t} + \underline{\mathbf{v}}(\underline{\mathbf{R}},t)\cdot\boldsymbol{\nabla}. \tag{3.35}$$

With $\underline{\mathbf{dR}} = \underline{\mathbf{v}}(\underline{\mathbf{R}},t)\mathrm{d}t$ and the truncated Taylor expansion

$$\begin{aligned}\Psi(\underline{\mathbf{R}}+\underline{\mathbf{dR}},t+\mathrm{d}t) &\simeq \Psi(\underline{\mathbf{R}},t) + \left[\frac{\partial \Psi(\underline{\mathbf{R}},t)}{\partial t} + \underline{\mathbf{v}}(\underline{\mathbf{R}},t)\cdot\boldsymbol{\nabla}\Psi(\underline{\mathbf{R}},t)\right]\mathrm{d}t \\ &\simeq \Psi(\underline{\mathbf{R}},t) + \frac{\mathrm{D}\Psi(\underline{\mathbf{R}},t)}{\mathrm{D}t}\,\mathrm{d}t\end{aligned} \tag{3.36}$$

in space and time, we immediately recognize that $\mathrm{D}/\mathrm{D}t$ has the meaning of a convective derivative: It denotes the time variation for an observer simultaneously traveling with the drift velocity.

We subsequently apply the calculation instruction (3.34) to different realizations of $\Psi(\underline{\mathbf{R}}, t)$.

Particle mass: $\Psi(\underline{\mathbf{R}}, t) = m$: Because $\mathrm{D}m/\mathrm{D}t \equiv 0$, we have

$$\frac{\mathrm{d}}{\mathrm{d}t}\iiint_{V(t)} n(\underline{\mathbf{R}}, t)\, m \,\mathrm{d}V = \frac{\mathrm{d}}{\mathrm{d}t}\iiint_{V(t)} \rho(\underline{\mathbf{R}}, t)\,\mathrm{d}V = 0, \tag{3.37}$$

and this is nothing but the mass conservation (3.27) in a different notation.[53]

Cartesian component of the particle linear momentum: $\Psi(\underline{\mathbf{R}}, t) = m\,\underline{\mathbf{v}}(\underline{\mathbf{R}}, t)\cdot\underline{\mathbf{e}}_{x_i}$:

$$\begin{aligned}\frac{\mathrm{d}}{\mathrm{d}t}&\iiint_{V(t)} n(\underline{\mathbf{R}}, t)\, m\,\underline{\mathbf{v}}(\underline{\mathbf{R}}, t)\cdot\underline{\mathbf{e}}_{x_i}\,\mathrm{d}V\\ &= \iiint_{V(t)} n(\underline{\mathbf{R}}, t)\, m\,\frac{\mathrm{D}\underline{\mathbf{v}}(\underline{\mathbf{R}}, t)\cdot\underline{\mathbf{e}}_{x_i}}{\mathrm{D}t}\,\mathrm{d}V;\end{aligned} \tag{3.38}$$

combining all three components to the momentum vector results in the calculation instruction for the time variation of the total momentum of the volume $V(t)$:

$$\frac{\mathrm{d}}{\mathrm{d}t}\iiint_{V(t)} \rho(\underline{\mathbf{R}}, t)\,\underline{\mathbf{v}}(\underline{\mathbf{R}}, t)\,\mathrm{d}V = \iiint_{V(t)} \rho(\underline{\mathbf{R}}, t)\,\frac{\mathrm{D}\underline{\mathbf{v}}(\underline{\mathbf{R}}, t)}{\mathrm{D}t}\,\mathrm{d}V. \tag{3.39}$$

Cartesian component of the particle angular momentum: $\Psi(\underline{\mathbf{R}}, t) = m\,\underline{\mathbf{R}}\times\underline{\mathbf{v}}(\underline{\mathbf{R}}, t)\cdot\underline{\mathbf{e}}_{x_i}$: We immediately write down the vector combination of the components:

$$\frac{\mathrm{d}}{\mathrm{d}t}\iiint_{V(t)} \rho(\underline{\mathbf{R}}, t)\underline{\mathbf{R}}\times\underline{\mathbf{v}}(\underline{\mathbf{R}}, t)\,\mathrm{d}V = \iiint_{V(t)} \rho(\underline{\mathbf{R}}, t)\frac{\mathrm{D}[\underline{\mathbf{R}}\times\underline{\mathbf{v}}(\underline{\mathbf{R}}, t)]}{\mathrm{D}t}\,\mathrm{d}V. \tag{3.40}$$

On behalf of

$$\frac{\mathrm{D}[\underline{\mathbf{R}}\times\underline{\mathbf{v}}(\underline{\mathbf{R}}, t)]}{\mathrm{D}t} = \frac{\partial[\underline{\mathbf{R}}\times\underline{\mathbf{v}}(\underline{\mathbf{R}}, t)]}{\partial t} + \underline{\mathbf{v}}(\underline{\mathbf{R}}, t)\cdot\boldsymbol{\nabla}[\underline{\mathbf{R}}\times\underline{\mathbf{v}}(\underline{\mathbf{R}}, t)], \tag{3.41}$$

we must calculate

$$\frac{\partial\underline{\mathbf{R}}\times\underline{\mathbf{v}}(\underline{\mathbf{R}}, t)}{\partial t} = \underbrace{\frac{\partial\underline{\mathbf{R}}}{\partial t}}_{=\underline{\mathbf{0}}}\times\underline{\mathbf{v}}(\underline{\mathbf{R}}, t) + \underline{\mathbf{R}}\times\frac{\partial\underline{\mathbf{v}}(\underline{\mathbf{R}}, t)}{\partial t} \tag{3.42}$$

---

[53] With (3.25) and (3.22), the left-hand side of (3.37) yields the time variation of the total mass in the time varying volume $V(t)$; according to Footnote 50, this volume exactly moves with the mass flow on $S(t)$, whence, recognizing mass conservation, the time variation of the total mass in the time varying volume must be zero.

and[54]

$$\nabla[\underline{\mathbf{R}} \times \underline{\mathbf{v}}(\underline{\mathbf{R}}, t)] = (\nabla\underline{\mathbf{R}}) \times \underline{\mathbf{v}}(\underline{\mathbf{R}}, t) - [\nabla\underline{\mathbf{v}}(\underline{\mathbf{R}}, t)] \times \underline{\mathbf{R}}$$
$$= \underline{\underline{\mathbf{I}}} \times \underline{\mathbf{v}}(\underline{\mathbf{R}}, t) - [\nabla\underline{\mathbf{v}}(\underline{\mathbf{R}}, t)] \times \underline{\mathbf{R}}, \tag{3.43}$$

respectively,

$$\underline{\mathbf{v}}(\underline{\mathbf{R}}, t) \cdot \nabla[\underline{\mathbf{R}} \times \underline{\mathbf{v}}(\underline{\mathbf{R}}, t)] = \underline{\mathbf{v}}(\underline{\mathbf{R}}, t) \cdot [\underline{\underline{\mathbf{I}}} \times \underline{\mathbf{v}}(\underline{\mathbf{R}}, t)] - \underline{\mathbf{v}}(\underline{\mathbf{R}}, t) \cdot [\nabla\underline{\mathbf{v}}(\underline{\mathbf{R}}, t)] \times \underline{\mathbf{R}}$$
$$= \underbrace{\underline{\mathbf{v}}(\underline{\mathbf{R}}, t) \times \underline{\mathbf{v}}(\underline{\mathbf{R}}, t)}_{=\underline{\mathbf{0}}} + \underline{\mathbf{R}} \times [\underline{\mathbf{v}}(\underline{\mathbf{R}}, t) \cdot \nabla]\underline{\mathbf{v}}(\underline{\mathbf{R}}, t). \tag{3.44}$$

Note that the $(\times\underline{\mathbf{R}})$-vector product in (3.43) refers to the right-factor $\underline{\mathbf{v}}(\underline{\mathbf{R}}, t)$ of the gradient dyadic $\nabla\underline{\mathbf{v}}(\underline{\mathbf{R}}, t)$; therefore, in (3.44), we may write $\underline{\mathbf{R}}\times$ (accordingly changing the sign), because only the first index of $\nabla\underline{\mathbf{v}}(\underline{\mathbf{R}}, t)$ is used up by the $[\underline{\mathbf{v}}(\underline{\mathbf{R}}, t)]$-scalar product, whence $\underline{\mathbf{R}}\times$ can only refer to $\underline{\mathbf{v}}(\underline{\mathbf{R}}, t)$ in $\nabla\underline{\mathbf{v}}(\underline{\mathbf{R}}, t)$. Consequently, we find

$$\frac{\mathrm{D}[\underline{\mathbf{R}} \times \underline{\mathbf{v}}(\underline{\mathbf{R}}, t)]}{\mathrm{D}t} = \underline{\mathbf{R}} \times \frac{\partial\underline{\mathbf{v}}(\underline{\mathbf{R}}, t)}{\partial t} + \underline{\mathbf{R}} \times [\underline{\mathbf{v}}(\underline{\mathbf{R}}, t) \cdot \nabla]\underline{\mathbf{v}}(\underline{\mathbf{R}}, t)$$
$$= \underline{\mathbf{R}} \times \frac{\mathrm{D}\underline{\mathbf{v}}(\underline{\mathbf{R}}, t)}{\mathrm{D}t}, \tag{3.45}$$

and that is why

$$\frac{\mathrm{d}}{\mathrm{d}t} \iiint_{V(t)} \rho(\underline{\mathbf{R}}, t)\, \underline{\mathbf{R}} \times \underline{\mathbf{v}}(\underline{\mathbf{R}}, t)\, \mathrm{d}V = \iiint_{V(t)} \rho(\underline{\mathbf{R}}, t)\, \underline{\mathbf{R}} \times \frac{\mathrm{D}\underline{\mathbf{v}}(\underline{\mathbf{R}}, t)}{\mathrm{D}t}\, \mathrm{d}V \tag{3.46}$$

results as time variation of the total angular momentum associated with $V(t)$.

3.2.3 Linear momentum conservation: Newton–Cauchy equation of motion

Newton says "force is equal to mass times acceleration," therefore, the time variation of the total momentum of all particles in $V(t)$ is equal to the sum of all forces acting on the particles. Figure 3.1 depicts a solid body volume V_M with surface S_M—a specimen or part—from which we select a partial volume $V(t)$ with surface $S(t)$. Contact forces of the surrounding material act on $S(t)$, for example, the surface force density $\mathrm{d}\underline{\mathbf{F}}_S$ on $\mathrm{d}S$; according to

$$\underline{\mathbf{t}} = \frac{\mathrm{d}\underline{\mathbf{F}}_S}{\mathrm{d}S}, \tag{3.47}$$

[54]Similar product rules are collected in the Appendix; they are proven via calculation in cartesian coordinates with the recommendation to use the summation convention and the Levi–Cività tensor.

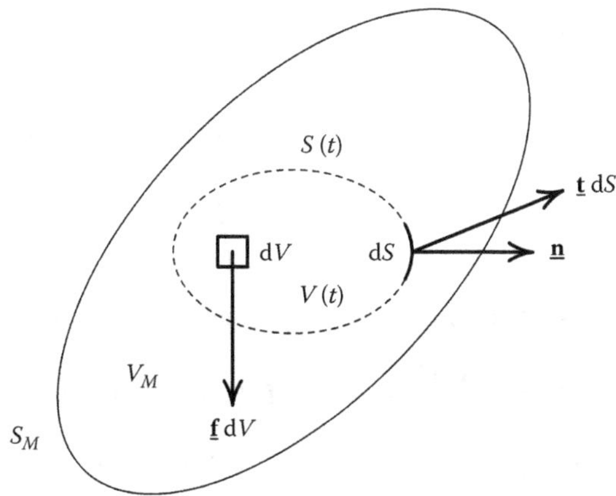

FIGURE 3.1
Newton–Cauchy equation of motion.

we define the traction vector $\underline{\mathbf{t}}$ as surface force *density* (dimension: force/surface) for $\mathrm{d}S \longrightarrow 0$. Simultaneously, volume forces $\mathrm{d}\underline{\mathbf{F}}_V$ may act on volume elements $\mathrm{d}V$ that may be assigned to volume force densities (dimension: force/volume)

$$\underline{\mathbf{f}} = \frac{\mathrm{d}\underline{\mathbf{F}}_V}{\mathrm{d}V} \tag{3.48}$$

for $\mathrm{d}V \longrightarrow 0$.

The *collective* motion of *all* particle types with the *same* drift velocity is characteristic for a solid; hence, the total momentum of $V(t)$, as it appears on the left-hand side of Newton's law

$$\frac{\mathrm{d}}{\mathrm{d}t} \iiint_{V(t)} \rho(\underline{\mathbf{R}}, t)\, \underline{\mathbf{v}}(\underline{\mathbf{R}}, t)\, \mathrm{d}V = \iint_{S(t)} \underline{\mathbf{t}}(\underline{\mathbf{R}}, t)\, \mathrm{d}S + \iiint_{V(t)} \underline{\mathbf{f}}(\underline{\mathbf{R}}, t)\, \mathrm{d}V \tag{3.49}$$

is understood as the sum of the single momentum of each particle type; on the right-hand side stands the sum of all forces. With (3.39), we finally obtain the version

$$\iiint_{V(t)} \rho(\underline{\mathbf{R}}, t)\, \frac{\mathrm{D}\underline{\mathbf{v}}(\underline{\mathbf{R}}, t)}{\mathrm{D}t}\, \mathrm{d}V = \iint_{S(t)} \underline{\mathbf{t}}(\underline{\mathbf{R}}, t)\, \mathrm{d}S + \iiint_{V(t)} \underline{\mathbf{f}}(\underline{\mathbf{R}}, t)\, \mathrm{d}V \tag{3.50}$$

of the momentum conservation law. To assume a common integrand of all integrals, as it was true for the particle and mass conservation, [(3.23) $\Longrightarrow$ (3.24), (3.27) $\Longrightarrow$ (3.28)], we must relate the traction $\underline{\mathbf{t}}(\underline{\mathbf{R}}, t)$ depending on the orientation of $\mathrm{d}S$ to the outer normal $\underline{\mathbf{n}}$ expressing this orientation; we postulate the linear (Cauchy) relation

$$\underline{\mathbf{t}}(\underline{\mathbf{R}}, t) = \underline{\mathbf{n}} \cdot \underline{\underline{\mathbf{T}}}(\underline{\mathbf{R}}, t) \tag{3.51}$$

via a tensor of second rank $\underline{\underline{\mathbf{T}}}(\underline{\mathbf{R}}, t)$—the stress tensor—to obtain the integral version Newton–Cauchy equation of motion

$$\iiint_{V(t)} \rho(\underline{\mathbf{R}}, t) \frac{\mathrm{D}\underline{\mathbf{v}}(\underline{\mathbf{R}}, t)}{\mathrm{D}t} \,\mathrm{d}V = \iiint_{V(t)} \nabla \cdot \underline{\underline{\mathbf{T}}}(\underline{\mathbf{R}}, t) \,\mathrm{d}V + \iiint_{V(t)} \underline{\mathbf{f}}(\underline{\mathbf{R}}, t) \,\mathrm{d}V \tag{3.52}$$

after applying Gauss' theorem; accordingly, the differential form reads

$$\rho(\underline{\mathbf{R}}, t) \frac{\mathrm{D}\underline{\mathbf{v}}(\underline{\mathbf{R}}, t)}{\mathrm{D}t} = \nabla \cdot \underline{\underline{\mathbf{T}}}(\underline{\mathbf{R}}, t) + \underline{\mathbf{f}}(\underline{\mathbf{R}}, t), \tag{3.53}$$

because (3.52) must hold for each arbitrary volume $V(t)$.

In general and in particular in ultrasonic NDT, the volume forces $\underline{\mathbf{f}}(\underline{\mathbf{R}}, t)$ are considered as outer—prescribed—forces that are independent of the stress and motion field and, therefore, appear as inhomogeneities in the Newton–Cauchy equation of motion.

The notation

$$\frac{\mathrm{d}}{\mathrm{d}t} \iiint_{V(t)} \underline{\mathbf{j}}(\underline{\mathbf{R}}, t) \,\mathrm{d}V \tag{3.54}$$

for the time variation of the total linear momentum of a volume $V(t)$ in Newton's law

$$\frac{\mathrm{d}}{\mathrm{d}t} \iiint_{V(t)} \underline{\mathbf{j}}(\underline{\mathbf{R}}, t) \,\mathrm{d}V = \iint_{S(t)} \underline{\mathbf{t}}(\underline{\mathbf{R}}, t) \,\mathrm{d}S + \iiint_{V(t)} \underline{\mathbf{f}}(\underline{\mathbf{R}}, t) \,\mathrm{d}V \tag{3.55}$$

yields the generalization

$$\frac{\mathrm{d}}{\mathrm{d}t} \iiint_{V(t)} \underline{\mathbf{j}}(\underline{\mathbf{R}}, t) \,\mathrm{d}V = \iiint_{V(t)} \left[\nabla \cdot \underline{\underline{\mathbf{T}}}(\underline{\mathbf{R}}, t) + \underline{\mathbf{f}}(\underline{\mathbf{R}}, t)\right] \,\mathrm{d}V \tag{3.56}$$

of the Newton–Cauchy equation of motion (3.52), because, for instance, in geophysics, applications exist where the reduction of $\underline{\mathbf{j}}(\underline{\mathbf{R}}, t)$ to a mass flow density according to (3.26) is not adequate to describe macroscopic physical phenomena.

With (3.31) and (3.32), Equation 3.56 reads in differential form

$$\frac{\delta \underline{\mathbf{j}}(\underline{\mathbf{R}}, t)}{\delta t} = \nabla \cdot \underline{\underline{\mathbf{T}}}(\underline{\mathbf{R}}, t) + \underline{\mathbf{f}}(\underline{\mathbf{R}}, t) \tag{3.57}$$

as generalization of (3.53).

3.2.4 Angular momentum conservation: Stress tensor symmetry

The moment $\underline{\mathbf{N}}$ of a force $\underline{\mathbf{F}}$ is defined as

$$\underline{\mathbf{N}} = \underline{\mathbf{R}} \times \underline{\mathbf{F}}; \tag{3.58}$$

hence, conservation of angular momentum for the total number of particles (sum over all particle types) in $V(t)$ is written according to recognizing (3.46)

$$
\begin{aligned}
\frac{\mathrm{d}}{\mathrm{d}t}\iiint_{V(t)} \rho(\underline{\mathbf{R}},t)\,\underline{\mathbf{R}}\times\underline{\mathbf{v}}(\underline{\mathbf{R}},t)\,\mathrm{d}V &= \iint_{S(t)} \underline{\mathbf{R}}\times\underline{\mathbf{t}}(\underline{\mathbf{R}},t)\,\mathrm{d}S \\
&\quad + \iiint_{V(t)} \underline{\mathbf{R}}\times\underline{\mathbf{f}}(\underline{\mathbf{R}},t)\,\mathrm{d}V \\
&= \iiint_{V(t)} \rho(\underline{\mathbf{R}},t)\,\frac{\mathrm{D}[\underline{\mathbf{R}}\times\underline{\mathbf{v}}(\underline{\mathbf{R}},t)]}{\mathrm{D}t}\,\mathrm{d}V.
\end{aligned}
\tag{3.59}
$$

In the following calculation, we change the order of vectors in all vector products of (3.59)—it is getting more obvious that way—the resulting minus signs being canceled. In the surface integral of (3.59), we replace $\underline{\mathbf{t}}(\underline{\mathbf{R}},t)$ by (3.51) and then we have to calculate $\nabla\cdot[\underline{\underline{\mathbf{T}}}(\underline{\mathbf{R}},t)\times\underline{\mathbf{R}}]$ after applying Gauss' law:[55]

$$
\begin{aligned}
\nabla\cdot[\underline{\underline{\mathbf{T}}}(\underline{\mathbf{R}},t)\times\underline{\mathbf{R}}] &= [\nabla\cdot\underline{\underline{\mathbf{T}}}(\underline{\mathbf{R}},t)]\times\underline{\mathbf{R}} + \underline{\underline{\mathbf{T}}}^{21}(\underline{\mathbf{R}},t)\,\dot{\times}\,\nabla\underline{\mathbf{R}} \\
&= [\nabla\cdot\underline{\underline{\mathbf{T}}}(\underline{\mathbf{R}},t)]\times\underline{\mathbf{R}} + \underline{\underline{\mathbf{T}}}^{21}(\underline{\mathbf{R}},t)\,\dot{\times}\,\underline{\underline{\mathbf{I}}} \\
&= [\nabla\cdot\underline{\underline{\mathbf{T}}}(\underline{\mathbf{R}},t)]\times\underline{\mathbf{R}} + \langle\,\underline{\underline{\mathbf{T}}}^{21}(\underline{\mathbf{R}},t)\,\rangle \\
&= [\nabla\cdot\underline{\underline{\mathbf{T}}}(\underline{\mathbf{R}},t)]\times\underline{\mathbf{R}} - \langle\underline{\underline{\mathbf{T}}}(\underline{\mathbf{R}},t)\rangle.
\end{aligned}
\tag{3.60}
$$

Equation 3.59 changes into

$$
\begin{aligned}
&\iiint_{V(t)} \rho(\underline{\mathbf{R}},t)\,\frac{\mathrm{D}\underline{\mathbf{v}}(\underline{\mathbf{R}},t)}{\mathrm{D}t}\times\underline{\mathbf{R}}\,\mathrm{d}V \\
&\quad = \iiint_{V(t)} [\nabla\cdot\underline{\underline{\mathbf{T}}}(\underline{\mathbf{R}},t) + \underline{\mathbf{f}}(\underline{\mathbf{R}},t)]\times\underline{\mathbf{R}}\,\mathrm{d}V \\
&\quad\quad - \iiint_{V(t)} \langle\underline{\underline{\mathbf{T}}}(\underline{\mathbf{R}},t)\rangle\,\mathrm{d}V,
\end{aligned}
\tag{3.61}
$$

---

[55]The expression $\underline{\underline{\mathbf{A}}}\,\dot{\times}\,\underline{\underline{\mathbf{B}}}$ is understood as (summation convention notation!)

$$
\begin{aligned}
\underline{\underline{\mathbf{A}}}\,\dot{\times}\,\underline{\underline{\mathbf{B}}} &= A_{ij}B_{kl}\,(\underline{\mathbf{e}}_{x_i}\times\underline{\mathbf{e}}_{x_l})\,\underline{\mathbf{e}}_{x_j}\cdot\underline{\mathbf{e}}_{x_k} \\
&= A_{ik}B_{kl}\,\underline{\mathbf{e}}_{x_i}\times\underline{\mathbf{e}}_{x_l},
\end{aligned}
$$

i.e., so to speak, the "vector product" of the first index of $\underline{\underline{\mathbf{A}}}$ with the second index of $\underline{\underline{\mathbf{B}}}$ after contraction of the second index of $\underline{\underline{\mathbf{A}}}$ with the first one of $\underline{\underline{\mathbf{B}}}$. With $\underline{\underline{\mathbf{B}}}=\underline{\underline{\mathbf{I}}}$, we obtain

$$
\begin{aligned}
\underline{\underline{\mathbf{A}}}\,\dot{\times}\,\underline{\underline{\mathbf{I}}} &= A_{il}\,\underline{\mathbf{e}}_{x_i}\times\underline{\mathbf{e}}_{x_l} \\
&= \langle\,\underline{\underline{\mathbf{A}}}\,\rangle;
\end{aligned}
$$

$\langle\underline{\underline{\mathbf{A}}}\rangle$ denotes the so-called rotation vector of the second rank tensor $\underline{\underline{\mathbf{A}}}$ (Equation 2.88). It follows:

$$
\langle\underline{\underline{\mathbf{A}}}^{21}\rangle = \underline{\underline{\mathbf{A}}}^{21}\,\dot{\times}\,\underline{\underline{\mathbf{I}}} = A_{li}\,\underline{\mathbf{e}}_{x_i}\times\underline{\mathbf{e}}_{x_l} = -A_{li}\,\underline{\mathbf{e}}_{x_l}\times\underline{\mathbf{e}}_{x_i} = -\langle\underline{\underline{\mathbf{A}}}\rangle.
$$

and considering the momentum conservation law (3.50), we finally obtain

$$\langle \underline{\underline{\mathbf{T}}}(\underline{\mathbf{R}}, t) \rangle \equiv \underline{\mathbf{0}} \tag{3.62}$$

as a consequence of the angular momentum conservation law.

According to (2.85) and (2.87), we decompose $\underline{\underline{\mathbf{T}}}$ into a symmetric and antisymmetric part

$$\begin{aligned} \underline{\underline{\mathbf{T}}} &= \underline{\underline{\mathbf{T}}}_{\mathrm{s}} + \underline{\underline{\mathbf{T}}}_{\mathrm{a}} \\ &= \frac{1}{2}\left(\underline{\underline{\mathbf{T}}} + \underline{\underline{\mathbf{T}}}^{21}\right) + \frac{1}{2}\left(\underline{\underline{\mathbf{T}}} - \underline{\underline{\mathbf{T}}}^{21}\right); \end{aligned} \tag{3.63}$$

with Footnote 55, it follows

$$\langle \underline{\underline{\mathbf{T}}} \rangle = \langle \underline{\underline{\mathbf{T}}}_{\mathrm{a}} \rangle, \tag{3.64}$$

because $\langle \underline{\underline{\mathbf{T}}}_{\mathrm{s}} \rangle \equiv \underline{\mathbf{0}}$; yet, the rotation vector of an antisymmetric tensor $\underline{\underline{\mathbf{T}}}_{\mathrm{a}}$ is only identically zero, if $\underline{\underline{\mathbf{T}}}_{\mathrm{a}}$ is the null tensor $\underline{\underline{\mathbf{0}}}$: $\underline{\underline{\mathbf{T}}}$ must be symmetric!

If the angular momentum conservation law (3.59) basically contains prescribed volume moments, the symmetry of $\underline{\underline{\mathbf{T}}}$ is accordingly broken (Auld 1973); this is true for electrically or magnetically prepolarized materials (ferroelectrica or ferromagnetica) in electric or magnetic fields (Langenberg 2005). Within the frame of linear elastodynamics, such effects may be neglected, thus approximately keeping the symmetry of the stress tensor.

The generalization of (3.56) of Newton–Cauchy's equation of motion does not immediately exhibit the symmetry of $\underline{\underline{\mathbf{T}}}$, why de Hoop makes it "visible" through

$$\begin{aligned} &\frac{\mathrm{d}}{\mathrm{d}t} \iiint_{V(t)} \underline{\mathbf{j}}(\underline{\mathbf{R}}, t)\, \mathrm{d}V \\ &\quad = \iiint_{V(t)} \left\{ \frac{1}{2} \nabla \cdot \left[\underline{\underline{\mathbf{T}}}(\underline{\mathbf{R}}, t) + \underline{\underline{\mathbf{T}}}^{21}(\underline{\mathbf{R}}, t) \right] + \underline{\mathbf{f}}(\underline{\mathbf{R}}, t) \right\} \mathrm{d}V. \end{aligned} \tag{3.65}$$

The main diagonal elements $T_{x_i x_i}$ of the (symmetric) stress tensor are called normal stresses and the off diagonal elements $T_{x_i x_j}$, $i \neq j$ are called shear stresses.

3.2.5 Deformation rate equation

Using the concept of the drift velocity, we calculate the relative change of position of two mass points within the time interval Δt; this will result in the concept of the (linear) deformation rate of a solid. In Figure 3.2, the two points are denoted by P_R and P_Q; during the time interval Δt, they move with their drift velocities $\underline{\mathbf{v}}(\underline{\mathbf{R}}, t)$ and $\underline{\mathbf{v}}(\underline{\mathbf{Q}}, t)$ to $P_{R'}$ and $P_{Q'}$, respectively, thus relating the new vectors of position $\underline{\mathbf{R}}'$ and $\underline{\mathbf{Q}}'$ at time $t + \Delta t$ with the old vectors of position $\underline{\mathbf{R}}$ and $\underline{\mathbf{Q}}$ at time t through

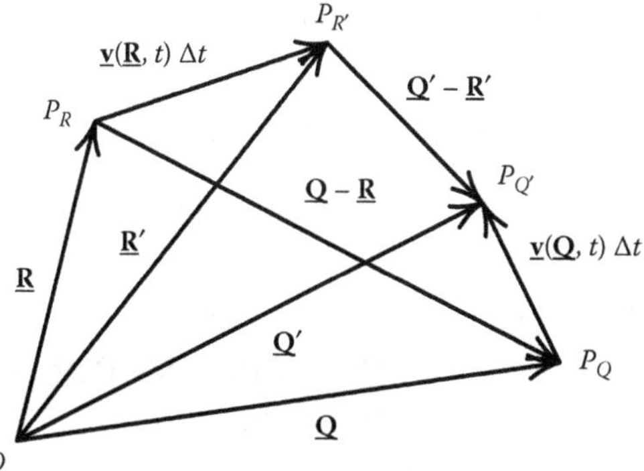

FIGURE 3.2
Relative change of position of two mass points P_R and P_Q during the time interval Δt.

$$-\underline{\mathbf{R}}' \simeq \underline{\mathbf{R}} + \underline{\mathbf{v}}(\underline{\mathbf{R}}, t)\,\Delta t, \tag{3.66}$$

$$\underline{\mathbf{Q}}' \simeq \underline{\mathbf{Q}} + \underline{\mathbf{v}}(\underline{\mathbf{Q}}, t)\,\Delta t; \tag{3.67}$$

the error associated with these truncated Taylor expansions tends to zero with $\Delta t \longrightarrow 0$. On the other hand, we can calculate $\underline{\mathbf{v}}(\underline{\mathbf{Q}}, t)$ from $\underline{\mathbf{v}}(\underline{\mathbf{R}}, t)$ for small values of $|\underline{\mathbf{Q}} - \underline{\mathbf{R}}|$ according to

$$\underline{\mathbf{v}}(\underline{\mathbf{Q}}, t) \simeq \underline{\mathbf{v}}(\underline{\mathbf{R}}, t) + (\underline{\mathbf{Q}} - \underline{\mathbf{R}}) \cdot \boldsymbol{\nabla}\underline{\mathbf{v}}(\underline{\mathbf{R}}, t); \tag{3.68}$$

the next term in the Taylor expansion (3.68) would be quadratic in $|\underline{\mathbf{Q}} - \underline{\mathbf{R}}|$. The velocity of the relative change of the distance $\underline{\mathbf{Q}} - \underline{\mathbf{R}}$ of the mass points—the deformation rate—is consequently obtained in a *linear* approximation:

$$\begin{aligned} \lim_{\Delta t \to 0} \frac{(\underline{\mathbf{Q}}' - \underline{\mathbf{R}}') - (\underline{\mathbf{Q}} - \underline{\mathbf{R}})}{\Delta t} &= \underline{\mathbf{v}}(\underline{\mathbf{Q}}, t) - \underline{\mathbf{v}}(\underline{\mathbf{R}}, t) \\ &= (\underline{\mathbf{Q}} - \underline{\mathbf{R}}) \cdot \boldsymbol{\nabla}\underline{\mathbf{v}}(\underline{\mathbf{R}}, t). \end{aligned} \tag{3.69}$$

A linear velocity of the solid common to both points is canceled that way. Within the linear approximation (3.69), the "deformation rate"—we use quotation marks because the "real" deformation rate is even defined without the rotation velocity of the solid[56]—is obviously completely specified by the gradient dyadic $\boldsymbol{\nabla}\underline{\mathbf{v}}(\underline{\mathbf{R}}, t)$ of the drift velocity. In the following, we want to show that $\boldsymbol{\nabla}\underline{\mathbf{v}}(\underline{\mathbf{R}}, t)$ actually contains the rotation of P_R around O and how we can "subtract" it.

[56]As far as soccer is concerned: We are neither interested in the drive nor in the spin of the ball.

The track velocity $\underline{\mathbf{w}}(\underline{\mathbf{R}}, t)$ of the point P_R rotating around O with spatially constant angular velocity $\underline{\mathbf{\Omega}}(t)$ is given by

$$\underline{\mathbf{w}}(\underline{\mathbf{R}}, t) = \underline{\mathbf{\Omega}}(t) \times \underline{\mathbf{R}}; \tag{3.70}$$

in the present connection, we are interested how to calculate $\underline{\mathbf{w}}(\underline{\mathbf{R}}, t)$ from $\nabla\underline{\mathbf{w}}(\underline{\mathbf{R}}, t)$. Therefore, we investigate[57]

$$\begin{aligned}\nabla \times \underline{\mathbf{w}}(\underline{\mathbf{R}}, t) &= \nabla \times [\underline{\mathbf{\Omega}}(t) \times \underline{\mathbf{R}}]\\ &= -\,\underline{\mathbf{\Omega}}(t) \cdot \nabla\underline{\mathbf{R}} + (\nabla \cdot \underline{\mathbf{R}})\underline{\mathbf{\Omega}}(t)\\ &= -\,\underline{\mathbf{\Omega}}(t) + 3\underline{\mathbf{\Omega}}(t)\\ &= 2\underline{\mathbf{\Omega}}(t)\end{aligned} \tag{3.71}$$

and insert it into (3.70):

$$\begin{aligned}\underline{\mathbf{w}}(\underline{\mathbf{R}}, t) &= \frac{1}{2}\,[\nabla \times \underline{\mathbf{w}}(\underline{\mathbf{R}}, t)] \times \underline{\mathbf{R}}\\ &= \frac{1}{2}\,\{[\underline{\mathbf{w}}(\underline{\mathbf{R}}, t)\nabla] \cdot \underline{\mathbf{R}} - [\nabla\underline{\mathbf{w}}(\underline{\mathbf{R}}, t)] \cdot \underline{\mathbf{R}}\}\\ &= \frac{1}{2}\,\underline{\mathbf{R}} \cdot \{\nabla\underline{\mathbf{w}}(\underline{\mathbf{R}}, t) - [\nabla\underline{\mathbf{w}}(\underline{\mathbf{R}}, t)]^{21}\}\\ &= \underline{\mathbf{R}} \cdot [\nabla\underline{\mathbf{w}}(\underline{\mathbf{R}}, t)]_{\mathrm{a}};\end{aligned} \tag{3.72}$$

it follows: After a similar projection as in (3.69), the *antisymmetric* part of $\nabla\underline{\mathbf{w}}(\underline{\mathbf{R}}, t)$ is responsible for the rotation track velocity that does not yield a local deformation. Therefore, we define the symmetric part $[\nabla\underline{\mathbf{v}}(\underline{\mathbf{R}}, t)]_{\mathrm{s}} = \frac{1}{2}\{\nabla\underline{\mathbf{v}}(\underline{\mathbf{R}}, t) + [\nabla\underline{\mathbf{v}}(\underline{\mathbf{R}}, t)]^{21}\}$ as second rank deformation rate tensor:

$$\underline{\underline{\mathbf{D}}}(\underline{\mathbf{R}}, t) = \frac{1}{2}\,\{\nabla\underline{\mathbf{v}}(\underline{\mathbf{R}}, t) + [\nabla\underline{\mathbf{v}}(\underline{\mathbf{R}}, t)]^{21}\}\,. \tag{3.73}$$

In the Newton–Cauchy equation of motion the right-hand side $\nabla \cdot \underline{\underline{\mathbf{T}}}(\underline{\mathbf{R}}, t) + \underline{\mathbf{f}}(\underline{\mathbf{R}}, t)$ is the *origin* of the time variation $\delta/\delta t$ of the linear momentum density $\underline{\mathbf{j}}(\underline{\mathbf{R}}, t)$; *here*, the deformation *rate* $\underline{\underline{\mathbf{D}}}(\underline{\mathbf{R}}, t)$ is the origin of the time variation $\delta/\delta t$ of the accordingly defined *deformation* $\underline{\underline{\mathbf{S}}}(\underline{\mathbf{R}}, t)$ that turns out to be a symmetric second rank tensor. This deformation is causally induced by stresses; in addition, we can introduce a source term[58] $-\underline{\underline{\mathbf{h}}}(\underline{\mathbf{R}}, t)$

[57]The direct calculation

$$\begin{aligned}\nabla\underline{\mathbf{w}}(\underline{\mathbf{R}}, t) &= \nabla[\underline{\mathbf{\Omega}}(t) \times \underline{\mathbf{R}}]\\ &= -\,(\nabla\underline{\mathbf{R}}) \times \underline{\mathbf{\Omega}}(t)\\ &= -\,\underline{\underline{\mathbf{I}}} \times \underline{\mathbf{\Omega}}(t)\\ &= -\,\underline{\mathbf{\Omega}}(t) \times \underline{\underline{\mathbf{I}}}\end{aligned}$$

with $\underline{\mathbf{\Omega}}(t) \times \underline{\underline{\mathbf{I}}}$ yields the general representation of an antisymmetric tensor where $\underline{\mathbf{\Omega}}$ can be calculated according to $\underline{\mathbf{\Omega}} = -\frac{1}{2}\langle\nabla\underline{\mathbf{w}}^{21}\rangle = \frac{1}{2}\langle\nabla\underline{\mathbf{w}}\rangle$. This is the same result as (3.71).

[58]In contrast to de Hoop, we formally choose $-\underline{\underline{\mathbf{h}}}(\underline{\mathbf{R}}, t)$ as source term in order to have the source terms $\underline{\mathbf{f}}$ and $\underline{\underline{\mathbf{h}}}$ on the right-hand sides with the same sign as it is true for Maxwell's equations (6.1) and (6.2).

as prescribed—symmetric—deformation rate tensor finally resulting in the deformation rate equation:

$$\frac{\delta \underline{\underline{\mathbf{S}}}(\underline{\mathbf{R}}, t)}{\delta t} = \frac{1}{2} \left\{ \nabla \underline{\mathbf{v}}(\underline{\mathbf{R}}, t) + [\nabla \underline{\mathbf{v}}(\underline{\mathbf{R}}, t)]^{21} \right\} + \underline{\underline{\mathbf{h}}}(\underline{\mathbf{R}}, t). \tag{3.74}$$

Utilizing the $\underline{\underline{\underline{\underline{\mathbf{I}}}}}^+$-tensor of rank four according to (2.109), we can write (3.74) in short-hand notation:

$$\frac{\delta \underline{\underline{\mathbf{S}}}(\underline{\mathbf{R}}, t)}{\delta t} = \underline{\underline{\underline{\underline{\mathbf{I}}}}}^+ : \nabla \underline{\mathbf{v}}(\underline{\mathbf{R}}, t) + \underline{\underline{\mathbf{h}}}(\underline{\mathbf{R}}, t). \tag{3.75}$$

3.2.6 Linear elastodynamics: Newton–Cauchy equation of motion and deformation rate equation

The continuum hypothesis and the particle conservation law guided us to Reynold's transport theorem (3.30); its utilization in Newton's conservation law for the linear momentum resulted in the Newton–Cauchy equation of motion (3.56) after introduction of the stress tensor instead of the traction. A similar formulation of the angular momentum conservation law together with the constitutive equation (3.26) provided the symmetry of the stress tensor. For the physical justification of the deformation rate equation (3.74), we needed the *geometric* linearization of the deformation rate (3.69) and the elimination of the rotation of the solid with the consequence of the symmetry of the deformation rate tensor. Writing the governing equations of elastodynamics (3.56) and (3.74) explicitly utilizing the operator (3.32) according to

$$\frac{\partial \underline{\mathbf{j}}(\underline{\mathbf{R}}, t)}{\partial t} + \nabla \cdot [\underline{\mathbf{v}}(\underline{\mathbf{R}}, t)\underline{\mathbf{j}}(\underline{\mathbf{R}}, t)] = \nabla \cdot \underline{\underline{\mathbf{T}}}(\underline{\mathbf{R}}, t) + \underline{\mathbf{f}}(\underline{\mathbf{R}}, t), \tag{3.76}$$

$$\frac{\partial \underline{\underline{\mathbf{S}}}(\underline{\mathbf{R}}, t)}{\partial t} + \nabla \cdot [\underline{\mathbf{v}}(\underline{\mathbf{R}}, t)\underline{\underline{\mathbf{S}}}(\underline{\mathbf{R}}, t)] = \frac{1}{2} \left\{ \nabla \underline{\mathbf{v}}(\underline{\mathbf{R}}, t) + [\nabla \underline{\mathbf{v}}(\underline{\mathbf{R}}, t)]^{21} \right\} + \underline{\underline{\mathbf{h}}}(\underline{\mathbf{R}}, t), \tag{3.77}$$

we nicely recognize the nonlinearity of these equations regarding the elastodynamic field quantities. In NDT with ultrasound, the amplitudes of field quantities are generally rather small, allowing for the neglect of the relevant terms. This approximation results in the linear equations

$$\frac{\partial \underline{\mathbf{j}}(\underline{\mathbf{R}}, t)}{\partial t} = \nabla \cdot \underline{\underline{\mathbf{T}}}(\underline{\mathbf{R}}, t) + \underline{\mathbf{f}}(\underline{\mathbf{R}}, t), \tag{3.78}$$

$$\frac{\partial \underline{\underline{\mathbf{S}}}(\underline{\mathbf{R}}, t)}{\partial t} = \frac{1}{2} \left\{ \nabla \underline{\mathbf{v}}(\underline{\mathbf{R}}, t) + [\nabla \underline{\mathbf{v}}(\underline{\mathbf{R}}, t)]^{21} \right\} + \underline{\underline{\mathbf{h}}}(\underline{\mathbf{R}}, t) \tag{3.79}$$

of elastodynamics that are—as already introduced as Equations 3.1 and 3.2—the basis of further evaluations. The subsequent step now consists in the combination of these equations: The keywords are "constitutive equations" (Chapter 4).

3.3 Transition and Boundary Conditions

3.3.1 Discontinuous material properties: Homogeneous and inhomogeneous transition conditions

Even *without* any knowledge of the precise elastic properties of materials, we are able to specify the conditions for elastodynamic fields at jump discontinuities of materials: These conditions immediately follow from the governing Equations 3.1 and 3.2.

We refer to the sketch in Figure 3.3: The homogeneous or inhomogeneous, isotropic or anisotropic, dissipative or nondissipative material (1) contains an "inclusion" V with material properties (2) that may equally be arbitrary than those of material (1), they should just vary discontinuously on the surface S of the inclusion; $\underline{\mathbf{n}}$ denotes the outer normal[59] on S. Now we select a "very small" piece ΔS on S—it should be considered as planar—and coat it with a volume V_i with surface S_i and outer normal $\underline{\mathbf{n}}_i$; V_i simultaneously contains material (1) as well as material (2) (Figure 3.3a). In the following, we investigate the volume integrals

$$\iiint_{V_i} \frac{\partial \underline{\mathbf{j}}(\underline{\mathbf{R}}, t)}{\partial t}\, \mathrm{d}V = \iiint_{V_i} \nabla \cdot \underline{\underline{\mathbf{T}}}(\underline{\mathbf{R}}, t)\, \mathrm{d}V + \iiint_{V_i} \underline{\mathbf{f}}(\underline{\mathbf{R}}, t)\, \mathrm{d}V, \quad (3.80)$$

$$\iiint_{V_i} \frac{\partial \underline{\underline{\mathbf{S}}}(\underline{\mathbf{R}}, t)}{\partial t}\, \mathrm{d}V = \frac{1}{2} \iiint_{V_i} \{\nabla \underline{\mathbf{v}}(\underline{\mathbf{R}}, t) + [\nabla \underline{\mathbf{v}}(\underline{\mathbf{R}}, t)]^{21}\}\, \mathrm{d}V + \iiint_{V_i} \underline{\underline{\mathbf{h}}}(\underline{\mathbf{R}}, t)\, \mathrm{d}V \quad (3.81)$$

of the governing equations of elastodynamics (3.1) and (3.2) having in mind the limit $i \longrightarrow \infty$ of a series of volumes V_i similar to the transition from

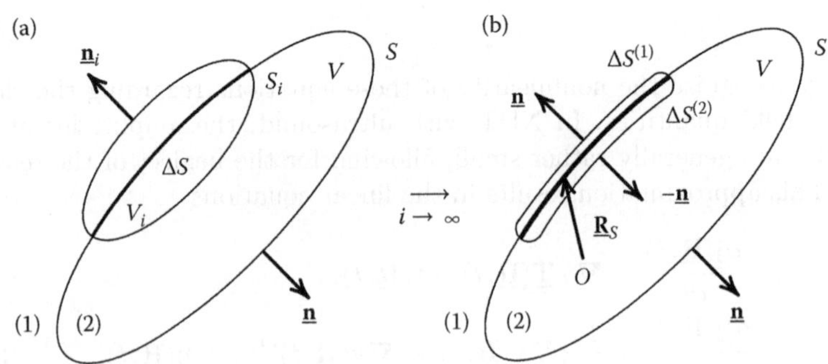

FIGURE 3.3
Derivation of transition conditions.

[59]We postulate that S exhibits only "rounded" edges and corners with an existing normal. Furthermore, we assume particle motions on S so small that S can be considered as time invariant.

Figure 3.3a to b; in this limit, the volumes should approach the geometry of a flat box adapting more and more to ΔS from both sides, finally resulting in an outer surface $\Delta S^{(1)}$ and an inner surface $\Delta S^{(2)}$ whose pertinent (outer) normals $\underline{\mathbf{n}}$ and $-\underline{\mathbf{n}}$ originate from $\underline{\mathbf{n}}_i$ for $i \longrightarrow \infty$. Applying Gauss' theorems to the first integrals on the right-hand sides of (3.80) and (3.81), we have to evaluate the limit $i \longrightarrow \infty$ in the equations

$$\iiint_{V_i} \frac{\partial \underline{\mathbf{j}}(\underline{\mathbf{R}}, t)}{\partial t}\, \mathrm{d}V = \iint_{S_i} \underline{\mathbf{n}}_i \cdot \underline{\underline{\mathbf{T}}}(\underline{\mathbf{R}}, t)\, \mathrm{d}S + \iiint_{V_i} \underline{\mathbf{f}}(\underline{\mathbf{R}}, t)\, \mathrm{d}V, \tag{3.82}$$

$$\begin{aligned}\iiint_{V_i} \frac{\partial \underline{\underline{\mathbf{S}}}(\underline{\mathbf{R}}, t)}{\partial t}\, \mathrm{d}V = &\frac{1}{2} \iint_{S_i} \left[\underline{\mathbf{n}}_i \underline{\mathbf{v}}(\underline{\mathbf{R}}, t) + \underline{\mathbf{v}}(\underline{\mathbf{R}}, t) \underline{\mathbf{n}}_i\right] \mathrm{d}S \\ &+ \iiint_{V_i} \underline{\underline{\mathbf{h}}}(\underline{\mathbf{R}}, t)\, \mathrm{d}V.\end{aligned} \tag{3.83}$$

Let us first consider the volume integrals of elastodynamic fields on the left-hand side: If the fields are "physically reasonable," i.e., without mathematical singularities, the integrals tend to be zero with vanishing integration volume.[60]

The surface integrals in (3.82) and (3.83) tend to integrals over $\Delta S^{(1)}$ and $\Delta S^{(2)}$ for $i \longrightarrow \infty$, where the normal $-\underline{\mathbf{n}}$ on $\Delta S^{(2)}$ accounts for the negative sign:

$$\lim_{i \to \infty} \iint_{S_i} \underline{\mathbf{n}}_i \cdot \underline{\underline{\mathbf{T}}}(\underline{\mathbf{R}}, t)\, \mathrm{d}S = \iint_{\Delta S^{(1)}} \underline{\mathbf{n}} \cdot \underline{\underline{\mathbf{T}}}(\underline{\mathbf{R}}, t)\, \mathrm{d}S - \iint_{\Delta S^{(2)}} \underline{\mathbf{n}} \cdot \underline{\underline{\mathbf{T}}}(\underline{\mathbf{R}}, t)\, \mathrm{d}S, \tag{3.84}$$

$$\begin{aligned}\lim_{i \to \infty} \frac{1}{2} \iint_{S_i} \left[\underline{\mathbf{n}}_i \underline{\mathbf{v}}(\underline{\mathbf{R}}, t) + \underline{\mathbf{v}}(\underline{\mathbf{R}}, t) \underline{\mathbf{n}}_i\right] \mathrm{d}S = &\frac{1}{2} \iint_{\Delta S^{(1)}} \left[\underline{\mathbf{n}}\, \underline{\mathbf{v}}(\underline{\mathbf{R}}, t) + \underline{\mathbf{v}}(\underline{\mathbf{R}}, t) \underline{\mathbf{n}}\right] \mathrm{d}S \\ &- \frac{1}{2} \iint_{\Delta S^{(2)}} \left[\underline{\mathbf{n}}\, \underline{\mathbf{v}}(\underline{\mathbf{R}}, t) + \underline{\mathbf{v}}(\underline{\mathbf{R}}, t) \underline{\mathbf{n}}\right] \mathrm{d}S.\end{aligned} \tag{3.85}$$

According to the mean value theorem of integral calculus (Burg et al. 1990) we always find a vector of position $\underline{\mathbf{R}}_S$ on ΔS—it equally resides on $\Delta S^{(1)}$ and $\Delta S^{(2)}$ due to the adaptation of $\Delta S^{(1)}$ and $\Delta S^{(2)}$ to ΔS—which satisfies

$$\iint_{\Delta S^{(j)}} \underline{\mathbf{n}} \cdot \underline{\underline{\mathbf{T}}}(\underline{\mathbf{R}}, t)\, \mathrm{d}S = \underline{\mathbf{n}} \cdot \underline{\underline{\mathbf{T}}}^{(j)}(\underline{\mathbf{R}}_S, t) \Delta S, \tag{3.86}$$

$$\frac{1}{2} \iint_{\Delta S^{(j)}} \left[\underline{\mathbf{n}}\, \underline{\mathbf{v}}(\underline{\mathbf{R}}, t) + \underline{\mathbf{v}}(\underline{\mathbf{R}}, t) \underline{\mathbf{n}}\right] \mathrm{d}S = \frac{1}{2} \left[\underline{\mathbf{n}}\, \underline{\mathbf{v}}^{(j)}(\underline{\mathbf{R}}_S, t) + \underline{\mathbf{v}}^{(j)}(\underline{\mathbf{R}}_S, t) \underline{\mathbf{n}}\right] \Delta S, \quad j = 1, 2; \tag{3.87}$$

with $\underline{\underline{\mathbf{T}}}^{(j)}(\underline{\mathbf{R}}_S, t)$, $\underline{\mathbf{v}}^{(j)}(\underline{\mathbf{R}}_S, t)$ we denote the limits of field quantities $\underline{\underline{\mathbf{T}}}(\underline{\mathbf{R}}, t)$, $\underline{\mathbf{v}}(\underline{\mathbf{R}}, t)$ if $\underline{\mathbf{R}}$ tends to $\underline{\mathbf{R}}_S$ in material (1) or (2).

What remains is the investigation of the V_i-integrals over the prescribed sources $\underline{\mathbf{f}}(\underline{\mathbf{R}}, t)$, $\underline{\underline{\mathbf{h}}}(\underline{\mathbf{R}}, t)$: In the following, we distinguish two cases.

[60] There is nothing to accumulate (integrate) in a zero volume.

Homogeneous transition conditions: Continuity of the traction vector, the surface deformation rate tensor, and the particle displacement vector: The given source functions $\underline{\mathbf{f}}(\underline{\mathbf{R}}, t)$, $\underline{\underline{\mathbf{h}}}(\underline{\mathbf{R}}, t)$ should represent *volume* sources without singularities; then they do not contribute to the limit of Equations (3.80) and (3.81) for $i \longrightarrow \infty$, and hence the governing equations of elastodynamics (3.82) and (3.83) are reduced to the *homogeneous* transition conditions due to (3.84), (3.85), (3.86), and (3.87):

$$\underline{\mathbf{n}} \cdot \underline{\underline{\mathbf{T}}}^{(1)}(\underline{\mathbf{R}}_S, t) - \underline{\mathbf{n}} \cdot \underline{\underline{\mathbf{T}}}^{(2)}(\underline{\mathbf{R}}_S, t) = \underline{\mathbf{0}}, \quad \underline{\mathbf{R}}_S \in S, \tag{3.88}$$

$$\underline{\mathbf{n}}\,\underline{\mathbf{v}}^{(1)}(\underline{\mathbf{R}}_S, t) + \underline{\mathbf{v}}^{(1)}(\underline{\mathbf{R}}_S, t)\underline{\mathbf{n}} - \underline{\mathbf{n}}\,\underline{\mathbf{v}}^{(2)}(\underline{\mathbf{R}}_S, t) - \underline{\mathbf{v}}^{(2)}(\underline{\mathbf{R}}_S, t)\underline{\mathbf{n}} = \underline{\underline{\mathbf{0}}}, \quad \underline{\mathbf{R}}_S \in S; \tag{3.89}$$

we could divide by the small but finite surface element ΔS unfolding the independence of the resulting equations from the arbitrary partial surface ΔS of S ensuring that $\underline{\mathbf{R}}_S$ in (3.88) and (3.89) may finally be a vector of position of any point on S. The homogeneous transition conditions (3.88) and (3.89) therefore require the *continuity* of the traction vector $\underline{\mathbf{n}} \cdot \underline{\underline{\mathbf{T}}}(\underline{\mathbf{R}}, t)$ as surface traction density and the tensor $\underline{\mathbf{n}}\,\underline{\mathbf{v}}(\underline{\mathbf{R}}, t) + \underline{\mathbf{v}}(\underline{\mathbf{R}}, t)\underline{\mathbf{n}}$ as surface deformation rate if $\underline{\mathbf{R}}$ moves from one side of S in material (1) to the other side of S in material (2), *even* if the material properties exhibit a jump discontinuity on S. The governing elastodynamic equations do not tell anything regarding other field vector and tensor components.

The homogeneous transition condition (3.89) can even be simplified. We write (3.89) short-hand

$$\underline{\mathbf{n}}\,\underline{\mathbf{v}} + \underline{\mathbf{v}}\,\underline{\mathbf{n}} = \text{continuous} \tag{3.90}$$

and take subsequent projections of this tensor equation into the direction of the normal on S and tangential to S. Hence:

$$\begin{aligned} \underline{\mathbf{n}} \cdot (\underline{\mathbf{n}}\,\underline{\mathbf{v}} + \underline{\mathbf{v}}\,\underline{\mathbf{n}}) &= \underline{\mathbf{v}} + \underline{\mathbf{v}} \cdot \underline{\mathbf{n}}\,\underline{\mathbf{n}} \\ &= \underline{\mathbf{v}}_t + 2\underline{\mathbf{v}} \cdot \underline{\mathbf{n}}\,\underline{\mathbf{n}} = \text{continuous}; \end{aligned} \tag{3.91}$$

we have replaced $\underline{\mathbf{v}}$ by the sum $\underline{\mathbf{v}} = \underline{\mathbf{v}}_t + \underline{\mathbf{v}}_n$ of the tangential vector

$$\underline{\mathbf{v}}_t = (\underline{\underline{\mathbf{I}}} - \underline{\mathbf{n}}\,\underline{\mathbf{n}}) \cdot \underline{\mathbf{v}}$$

and the normal vector

$$\underline{\mathbf{v}}_n = \underline{\mathbf{v}} \cdot \underline{\mathbf{n}}\,\underline{\mathbf{n}}.$$

Then, we calculate the projection

$$(\underline{\underline{\mathbf{I}}} - \underline{\mathbf{n}}\,\underline{\mathbf{n}}) \cdot (\underline{\mathbf{n}}\,\underline{\mathbf{v}} + \underline{\mathbf{v}}\,\underline{\mathbf{n}}) = \underbrace{(\underline{\mathbf{v}} - \underline{\mathbf{v}} \cdot \underline{\mathbf{n}}\,\underline{\mathbf{n}})}_{= \underline{\mathbf{v}}_t}\,\underline{\mathbf{n}} = \text{continuous}; \tag{3.92}$$

requiring the continuity of $\underline{\mathbf{v}}_t$; therefore, the continuity of $\underline{\mathbf{v}}_n$ is required in combination with (3.91). Both facts result in the homogeneous transition condition

$$\underline{\mathbf{v}}^{(1)}(\underline{\mathbf{R}}_S, t) - \underline{\mathbf{v}}^{(2)}(\underline{\mathbf{R}}_S, t) = \underline{\mathbf{0}}, \quad \underline{\mathbf{R}}_S \in S, \tag{3.93}$$

namely, the continuity of the particle velocity vector. To deduce the continuity of the particle *displacement* vector, we need an additional argument (de Hoop 1995):

Due to the relation

$$\underline{\mathbf{v}}(\underline{\mathbf{R}}, t) = \frac{\partial \underline{\mathbf{u}}(\underline{\mathbf{R}}, t)}{\partial t} \tag{3.94}$$

between particle velocity and particle displacement, the transition condition (3.93) is equivalent to

$$\frac{\partial \underline{\mathbf{u}}^{(1)}(\underline{\mathbf{R}}_S, \tau)}{\partial \tau} = \frac{\partial \underline{\mathbf{u}}^{(2)}(\underline{\mathbf{R}}_S, \tau)}{\partial \tau}, \tag{3.95}$$

therefore, time integration yields

$$\begin{aligned} \int_0^t \frac{\partial \underline{\mathbf{u}}^{(1)}(\underline{\mathbf{R}}_S, \tau)}{\partial \tau} \, d\tau &= \underline{\mathbf{u}}^{(1)}(\underline{\mathbf{R}}_S, t) + \underline{\mathbf{u}}^{(1)}(\underline{\mathbf{R}}_S, 0) \\ &= \underline{\mathbf{u}}^{(2)}(\underline{\mathbf{R}}_S, t) + \underline{\mathbf{u}}^{(2)}(\underline{\mathbf{R}}_S, 0). \end{aligned} \tag{3.96}$$

It makes sense to postulate that elastodynamic fields are "switched on" at a certain time instant being identically zero for smaller times; consequently, we choose the time origin as far in the past that $\underline{\mathbf{u}}^{(1)}(\underline{\mathbf{R}}_S, 0) = \underline{\mathbf{u}}^{(2)}(\underline{\mathbf{R}}_S, 0) \equiv \underline{\mathbf{0}}$ holds, i.e., we deal with *causal* fields. According to (3.96), we conclude the continuity of the particle displacement vector for *those* fields:

$$\underline{\mathbf{u}}^{(1)}(\underline{\mathbf{R}}_S, t) - \underline{\mathbf{u}}^{(2)}(\underline{\mathbf{R}}_S, t) = \underline{\mathbf{0}}, \quad \underline{\mathbf{R}}_S \in S. \tag{3.97}$$

Of course, the homogeneous transition conditions (3.88) and (3.97) also hold for the Fourier spectra:[61]

$$\underline{\mathbf{n}} \cdot \underline{\underline{\mathbf{T}}}^{(1)}(\underline{\mathbf{R}}_S, \omega) - \underline{\mathbf{n}} \cdot \underline{\underline{\mathbf{T}}}^{(2)}(\underline{\mathbf{R}}_S, \omega) = \underline{\mathbf{0}}, \quad \underline{\mathbf{R}}_S \in S, \tag{3.98}$$

$$\underline{\mathbf{u}}^{(1)}(\underline{\mathbf{R}}_S, \omega) - \underline{\mathbf{u}}^{(2)}(\underline{\mathbf{R}}_S, \omega) = \underline{\mathbf{0}}, \quad \underline{\mathbf{R}}_S \in S. \tag{3.99}$$

Inhomogeneous transition conditions: Definition of surface source densities: As announced, for the second case, we allow for the existence of

[61] At first sight, it looks as if (3.99) follows from the Fourier transformed equation (3.93) *without* any further assumptions; yet (3.93) leads to the Fourier transformed equation

$$\omega \left[\underline{\mathbf{u}}^{(1)}(\underline{\mathbf{R}}_S, \omega) - \underline{\mathbf{u}}^{(2)}(\underline{\mathbf{R}}_S, \omega) \right] = \underline{\mathbf{0}},$$

and the conclusion can only read

$$\underline{\mathbf{u}}^{(1)}(\underline{\mathbf{R}}_S, \omega) - \underline{\mathbf{u}}^{(2)}(\underline{\mathbf{R}}_S, \omega) = \underline{\mathbf{u}}_0(\underline{\mathbf{R}}_S)\delta(\omega)$$

with an arbitrary vector $\underline{\mathbf{u}}_0(\underline{\mathbf{R}}_S)$ because $\omega\delta(\omega) = 0$. An inverse Fourier transform and the comparison with (3.96) reveals that $\underline{\mathbf{u}}_0(\underline{\mathbf{R}}_S)/2\pi = \underline{\mathbf{u}}^{(1)}(\underline{\mathbf{R}}_S, t = 0) - \underline{\mathbf{u}}^{(2)}(\underline{\mathbf{R}}_S, t = 0)$ so that only causal fields in the time domain yield $\underline{\mathbf{u}}_0(\underline{\mathbf{R}}_S) \equiv \underline{\mathbf{0}}$.

prescribed *surface* source densities on S besides singularity-free *volume* source densities. In terms of mathematics surface source *densities* can be considered as "amplitudes" of δ-singular volume source densities on S[62] according to—we use the singular function $\gamma_S(\underline{\mathbf{R}})$ of the surface S:

$$\underline{\mathbf{f}}_S(\underline{\mathbf{R}}, t) = \underline{\mathbf{t}}(\underline{\mathbf{R}}, t)\gamma_S(\underline{\mathbf{R}}), \tag{3.100}$$

$$\underline{\underline{\mathbf{h}}}_S(\underline{\mathbf{R}}, t) = \underline{\underline{\mathbf{g}}}(\underline{\mathbf{R}}, t)\gamma_S(\underline{\mathbf{R}}), \tag{3.101}$$

because, only in that case, the V_i-volume integration of $\underline{\mathbf{f}}_S$ and $\underline{\underline{\mathbf{h}}}_S$ yields a finite value:

$$\begin{aligned}\iiint_{V_i} \underline{\mathbf{f}}_S(\underline{\mathbf{R}}, t)\,\mathrm{d}V &= \iiint_{V_i} \underline{\mathbf{t}}(\underline{\mathbf{R}}, t)\gamma_S(\underline{\mathbf{R}})\,\mathrm{d}V \\ &= \iint_{\Delta S} \underline{\mathbf{t}}(\underline{\mathbf{R}}, t)\,\mathrm{d}S \\ &= \underline{\mathbf{t}}(\underline{\mathbf{R}}_S, t)\Delta S;\end{aligned} \tag{3.102}$$

the last sign of equality implies the application of the mean value theorem of integral calculus. Similarly, we obtain

$$\iiint_{V_i} \underline{\underline{\mathbf{h}}}_S(\underline{\mathbf{R}}, t)\,\mathrm{d}V = \underline{\underline{\mathbf{g}}}(\underline{\mathbf{R}}_S, t)\Delta S. \tag{3.103}$$

With (3.102), (3.103), and (3.104) through (3.107), the *inhomogeneous* transition conditions

$$\underline{\mathbf{n}} \cdot \underline{\underline{\mathbf{T}}}^{(1)}(\underline{\mathbf{R}}_S, t) - \underline{\mathbf{n}} \cdot \underline{\underline{\mathbf{T}}}^{(2)}(\underline{\mathbf{R}}_S, t) = -\underline{\mathbf{t}}(\underline{\mathbf{R}}_S, t), \quad \underline{\mathbf{R}}_S \in S, \tag{3.104}$$

$$\begin{aligned}\frac{1}{2}\Big[\underline{\mathbf{n}}\,\underline{\mathbf{v}}^{(1)}(\underline{\mathbf{R}}_S, t) + \underline{\mathbf{v}}^{(1)}(\underline{\mathbf{R}}_S, t)\underline{\mathbf{n}} - \underline{\mathbf{n}}\,\underline{\mathbf{v}}^{(2)}(\underline{\mathbf{R}}_S, t) - \underline{\mathbf{v}}^{(2)}(\underline{\mathbf{R}}_S, t)\underline{\mathbf{n}}\Big] \\ = -\underline{\underline{\mathbf{g}}}(\underline{\mathbf{R}}_S, t), \quad \underline{\mathbf{R}}_S \in S,\end{aligned} \tag{3.105}$$

for the traction vector and the tensor of the surface deformation rate are obtained provided surface sources on S are—no matter how—prescribed. Such prescribed sources yield a discontinuity of the field quantities involved.

We might read the inhomogeneous transition conditions (3.104) and (3.105) from left to right: If the traction vector $\underline{\mathbf{n}} \cdot \underline{\underline{\mathbf{T}}}(\underline{\mathbf{R}}, t)$ and the tensor $\underline{\mathbf{n}}\,\underline{\mathbf{v}}(\underline{\mathbf{R}}, t) + \underline{\mathbf{v}}(\underline{\mathbf{R}}, t)\underline{\mathbf{n}}$ are—for any reasons—discontinuous on a surface S, such a discontinuity *defines* surface source densities. This interpretation will be extremely helpful to understand Huygens' principle in elastodynamics (Section 15.1.3).

The spectral versions of (3.104) and (3.105) apparently read as

$$\underline{\mathbf{n}} \cdot \underline{\underline{\mathbf{T}}}^{(1)}(\underline{\mathbf{R}}_S, \omega) - \underline{\mathbf{n}} \cdot \underline{\underline{\mathbf{T}}}^{(2)}(\underline{\mathbf{R}}_S, \omega) = -\underline{\mathbf{t}}(\underline{\mathbf{R}}_S, \omega), \quad \underline{\mathbf{R}}_S \in S, \tag{3.106}$$

[62] The dimension (of the components) of $\underline{\mathbf{t}}$ is force/area and the dimension (of the components) of $\underline{\underline{\mathbf{g}}}$ is length/second because the dimension of γ_S is length^{-1}.

$$\frac{1}{2}\left[\underline{\mathbf{n}}\,\underline{\mathbf{v}}^{(1)}(\underline{\mathbf{R}}_S,\omega)+\underline{\mathbf{v}}^{(1)}(\underline{\mathbf{R}}_S,\omega)\underline{\mathbf{n}}-\underline{\mathbf{n}}\,\underline{\mathbf{v}}^{(2)}(\underline{\mathbf{R}}_S,\omega)-\underline{\mathbf{v}}^{(2)}(\underline{\mathbf{R}}_S,\omega)\underline{\mathbf{n}}\right]$$
$$=-\underline{\underline{\mathbf{g}}}(\underline{\mathbf{R}}_S,\omega),\quad \underline{\mathbf{R}}_S\in S; \tag{3.107}$$

the Fourier spectrum of the surface deformation tensor $(\underline{\mathbf{n}}\,\underline{\mathbf{u}}+\underline{\mathbf{u}}\,\underline{\mathbf{n}})/2$ is obtained from (3.106):

$$\frac{1}{2}\left[\underline{\mathbf{n}}\,\underline{\mathbf{u}}^{(1)}(\underline{\mathbf{R}}_S,\omega)+\underline{\mathbf{u}}^{(1)}(\underline{\mathbf{R}}_S,\omega)\underline{\mathbf{n}}-\underline{\mathbf{n}}\,\underline{\mathbf{u}}^{(2)}(\underline{\mathbf{R}}_S,\omega)-\underline{\mathbf{u}}^{(2)}(\underline{\mathbf{R}}_S,\omega)\underline{\mathbf{n}}\right]$$
$$=-\frac{\mathrm{j}}{\omega}\,\underline{\underline{\mathbf{g}}}(\underline{\mathbf{R}}_S,\omega),\quad \underline{\mathbf{R}}_S\in S. \tag{3.108}$$

The "simple version" (3.99) does no longer exist in the case of inhomogeneous transition conditions.

3.3.2 Infinite discontinuity of material properties: Boundary conditions

Vacuum is infinitely compliable regarding its elastic properties; therefore, it does not allow for the propagation of elastic waves. The same is true for idealized materials with infinite mass density. We refer to the terminology of acoustics and speak of perfectly soft and perfectly rigid materials. If our inclusion V is supposed to be made of such a material, $\underline{\mathbf{v}}^{(2)}(\underline{\mathbf{R}},t)$ and $\underline{\underline{\mathbf{T}}}^{(2)}(\underline{\mathbf{R}},t)$ are identically zero in V. As a stress-free boundary condition, the perfectly soft material consequently enforces its surface to be free of stresses—more precisely: free of tractions—

$$\underline{\mathbf{n}}\cdot\underline{\underline{\mathbf{T}}}(\underline{\mathbf{R}}_S,t)=\underline{\mathbf{0}},\quad \underline{\mathbf{R}}_S\in S, \tag{3.109}$$
$$\frac{1}{2}\left[\underline{\mathbf{n}}\,\underline{\mathbf{v}}(\underline{\mathbf{R}}_S,t)+\underline{\mathbf{v}}(\underline{\mathbf{R}}_S,t)\underline{\mathbf{n}}\right] = -\,\underline{\underline{\mathbf{g}}}(\underline{\mathbf{R}}_S,t),\quad \underline{\mathbf{R}}_S\in S, \tag{3.110}$$

because the infinitely compliable surface allows for deformations (surface deformation rates), yet, it does not support tractions (surface force densities). Complementary to the boundary of a perfectly soft material, a perfectly rigid material yields the boundary condition of a surface free of deformation rates:[63]

$$\underline{\mathbf{n}}\cdot\underline{\underline{\mathbf{T}}}(\underline{\mathbf{R}}_S,t) = -\,\underline{\mathbf{t}}(\underline{\mathbf{R}}_S,t),\quad \underline{\mathbf{R}}_S\in S, \tag{3.111}$$
$$\underline{\mathbf{v}}(\underline{\mathbf{R}}_S,t)=\underline{\mathbf{0}},\quad \underline{\mathbf{R}}_S\in S. \tag{3.112}$$

In Figure 3.4, the two perfect boundary conditions are compared to each other. It is quite clear that the stress-free boundary condition is particularly relevant for NDT because it simulates the surface of parts or specimens (in vacuum)

[63]For $\underline{\underline{\mathbf{g}}}(\underline{\mathbf{R}}_S,t)\equiv\underline{\mathbf{0}}$, the sum of the dyadic products $\underline{\mathbf{n}}\,\underline{\mathbf{v}}$ and $\underline{\mathbf{v}}\,\underline{\mathbf{n}}$ is equal to zero if $\underline{\mathbf{v}}$ is equal to zero.

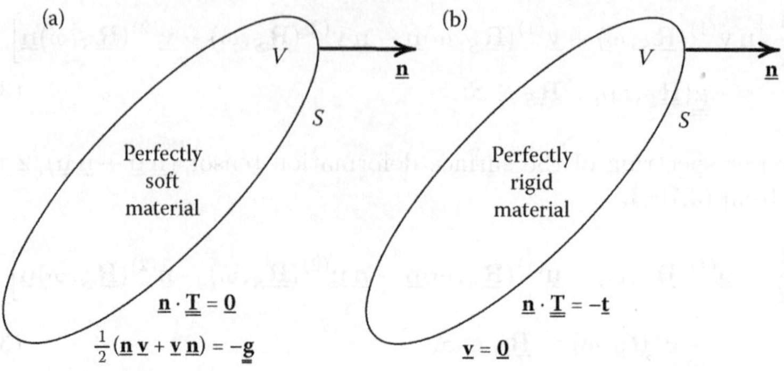

FIGURE 3.4
Boundary conditions on the surface of perfectly soft (a) and perfectly rigid (b) materials.

and—approximately—air-filled inclusions and perfect cracks (for infinitely flat volumes).

Evidently, the earlier discussed boundary conditions consist of one homogeneous and one inhomogeneous condition. It is by no means a formal question whether *both* boundary conditions may be homogeneous, e.g., whether we can arbitrarily prescribe $\underline{\underline{\mathbf{g}}}(\underline{\mathbf{R}}_S, t)$ in (3.110) and $\underline{\mathbf{t}}(\underline{\mathbf{R}}_S, t)$ in (3.111), respectively, even assuming them to be zero. The answer is no! A stress-free surface *must* be deformed, and a deformation-free surface *must* exhibit stresses provided elastic waves are present in the material outside V, because, via the enforcement of the boundary conditions, the waves induce the surface sources. Otherwise spoken, exclusively homogeneous boundary conditions are only possible for elastodynamic fields that are zero in entire infinite space. This is an immediate consequence of Huygens' principle for elastodynamic waves as a mathematical solution of the governing equations (Section 15.1.3).

3.3.3 Boundary between elastic and fluid materials: Homogeneous and inhomogeneous transition conditions

US-NDT often applies fluid immersed transducers, thus accounting for a boundary between an elastic and a fluid material. How should we modify the transition conditions (3.88), (3.97) and (3.104), (3.105), respectively, in this case? Figure 3.5 displays on the left-hand side the relevant governing equations for the solid material (1) and the fluid material (2) [Equations (3.1), (3.2) and (5.1), (5.2)]. We formally remove the incompatibility between both systems of equations introducing a stress tensor $\underline{\underline{\mathbf{T}}}(\underline{\mathbf{R}}, t) = -p(\underline{\mathbf{R}}, t)\,\underline{\underline{\mathbf{I}}}$ for the fluid as well as taking the trace $S(\underline{\mathbf{R}}, t) = \mathrm{trace}\,\underline{\underline{\mathbf{S}}}(\underline{\mathbf{R}}, t)$ of the tensor deformation rate equation of the solid and defining $h(\underline{\mathbf{R}}, t) = \mathrm{trace}\,\underline{\underline{\mathbf{h}}}(\underline{\mathbf{R}}, t)$ because the

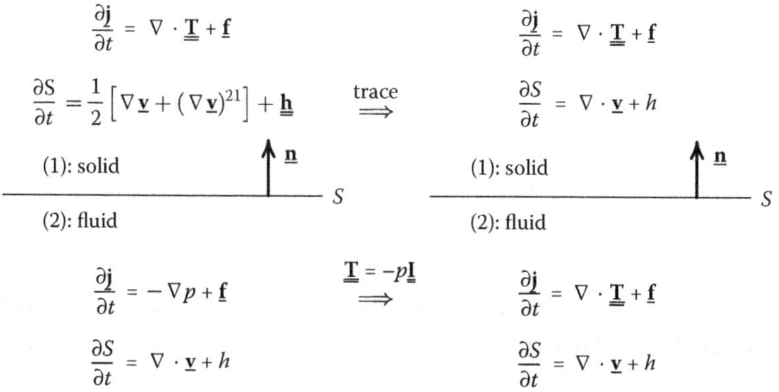

$$\frac{\partial \underline{\mathbf{j}}}{\partial t} = \nabla \cdot \underline{\underline{\mathbf{T}}} + \underline{\mathbf{f}} \qquad\qquad \frac{\partial \underline{\mathbf{j}}}{\partial t} = \nabla \cdot \underline{\underline{\mathbf{T}}} + \underline{\mathbf{f}}$$

$$\frac{\partial \underline{\underline{\mathbf{S}}}}{\partial t} = \frac{1}{2}\left[\nabla \underline{\mathbf{v}} + (\nabla \underline{\mathbf{v}})^{21}\right] + \underline{\underline{\mathbf{h}}} \quad \overset{\text{trace}}{\Longrightarrow} \quad \frac{\partial S}{\partial t} = \nabla \cdot \underline{\mathbf{v}} + h$$

(1): solid $\uparrow \underline{\mathbf{n}}$ S (1): solid $\uparrow \underline{\mathbf{n}}$ S

(2): fluid (2): fluid

$$\frac{\partial \underline{\mathbf{j}}}{\partial t} = -\nabla p + \underline{\mathbf{f}} \quad \overset{\underline{\underline{\mathbf{T}}} = -p\underline{\underline{\mathbf{I}}}}{\Longrightarrow} \quad \frac{\partial \underline{\mathbf{j}}}{\partial t} = \nabla \cdot \underline{\underline{\mathbf{T}}} + \underline{\mathbf{f}}$$

$$\frac{\partial S}{\partial t} = \nabla \cdot \underline{\mathbf{v}} + h \qquad\qquad \frac{\partial S}{\partial t} = \nabla \cdot \underline{\mathbf{v}} + h$$

FIGURE 3.5
Derivation of transition conditions for a boundary between elastic and fluid materials.

respective equation for the fluid only has information about the cubic dilatation[64] $S(\underline{\mathbf{R}}, t)$. That way, we have arrived at the right-hand part of Figure 3.5 and can now proceed as in Figure 3.3. We immediately obtain the following inhomogeneous transition conditions [also compare Schmerr (1998)]:

$$\underline{\mathbf{n}} \cdot \underline{\underline{\mathbf{T}}}^{(1)}(\underline{\mathbf{R}}_S, t) + p^{(2)}(\underline{\mathbf{R}}_S, t)\underline{\mathbf{n}} = -\underline{\mathbf{t}}(\underline{\mathbf{R}}_S, t), \quad \underline{\mathbf{R}}_S \in S, \tag{3.113}$$

$$\underline{\mathbf{n}} \cdot \underline{\mathbf{v}}^{(1)}(\underline{\mathbf{R}}_S, t) - \underline{\mathbf{n}} \cdot \underline{\mathbf{v}}^{(2)}(\underline{\mathbf{R}}_S, t) = -h(\underline{\mathbf{R}}_S, t), \quad \underline{\mathbf{R}}_S \in S, \tag{3.114}$$

because $\underline{\mathbf{n}} \cdot \underline{\underline{\mathbf{T}}}^{(2)}(\underline{\mathbf{R}}_S, t) = -p^{(2)}(\underline{\mathbf{R}}_S, t)\underline{\mathbf{n}} \cdot \underline{\underline{\mathbf{I}}} = -p^{(2)}(\underline{\mathbf{R}}_S, t)$; $\underline{\mathbf{t}}(\underline{\mathbf{R}}_S, t)$ and $h(\underline{\mathbf{R}}_S, t)$ represent prescribed tractions and surface dilatation rates, respectively. The vector equation (3.113) is appropriately separated into normal and tangential components relative to S:

$$\underline{\mathbf{n}} \cdot \left[\underline{\mathbf{n}} \cdot \underline{\underline{\mathbf{T}}}^{(1)}(\underline{\mathbf{R}}_S, t)\right] + p^{(2)}(\underline{\mathbf{R}}_S, t) = -\underline{\mathbf{n}} \cdot \underline{\mathbf{t}}(\underline{\mathbf{R}}_S, t), \quad \underline{\mathbf{R}}_S \in S, \tag{3.115}$$

$$(\underline{\underline{\mathbf{I}}} - \underline{\mathbf{n}}\,\underline{\mathbf{n}}) \cdot \left[\underline{\mathbf{n}} \cdot \underline{\underline{\mathbf{T}}}^{(1)}(\underline{\mathbf{R}}_S, t)\right] = -\underline{\mathbf{t}}_t(\underline{\mathbf{R}}_S, t), \quad \underline{\mathbf{R}}_S \in S, \tag{3.116}$$

where $\underline{\mathbf{t}}_t = (\underline{\underline{\mathbf{I}}} - \underline{\mathbf{n}}\,\underline{\mathbf{n}}) \cdot \underline{\mathbf{t}}$ denotes the tangential part of the prescribed traction.

If there are no surface sources prescribed on the boundary, we obtain the homogeneous transition conditions

$$\underline{\mathbf{n}} \cdot \left[\underline{\mathbf{n}} \cdot \underline{\underline{\mathbf{T}}}^{(1)}(\underline{\mathbf{R}}_S, t)\right] + p^{(2)}(\underline{\mathbf{R}}_S, t) = 0, \quad \underline{\mathbf{R}}_S \in S, \tag{3.117}$$

$$(\underline{\underline{\mathbf{I}}} - \underline{\mathbf{n}}\,\underline{\mathbf{n}}) \cdot \left[\underline{\mathbf{n}} \cdot \underline{\underline{\mathbf{T}}}^{(1)}(\underline{\mathbf{R}}_S, t)\right] = \underline{\mathbf{0}}, \quad \underline{\mathbf{R}}_S \in S, \tag{3.118}$$

$$\underline{\mathbf{n}} \cdot \underline{\mathbf{u}}^{(1)}(\underline{\mathbf{R}}_S, t) - \underline{\mathbf{n}} \cdot \underline{\mathbf{u}}^{(2)}(\underline{\mathbf{R}}_S, t) = 0, \quad \underline{\mathbf{R}}_S \in S, \tag{3.119}$$

[64]One defines $\frac{1}{3}\underline{\underline{\mathbf{I}}}\,\mathrm{trace}\,\underline{\underline{\mathbf{S}}}(\underline{\mathbf{R}}, t)$ as (isotropic) dilatation (de Hoop 1995).

where we switched again to the particle displacement vector as in the transition from (3.93) to (3.97). Actually, the relations (3.117) through (3.119) separate into homogeneous *transition* conditions for the normal components of the vectors $\underline{\mathbf{u}}(\underline{\mathbf{R}}_S, t)$ and $\underline{\mathbf{n}} \cdot \underline{\underline{\mathbf{T}}}(\underline{\mathbf{R}}_S, t)$ and one *boundary* condition for the vector tangential component $\underline{\mathbf{n}} \cdot \underline{\underline{\mathbf{T}}}(\underline{\mathbf{R}}_S, t)$.

The time harmonic version of (3.117) through (3.119) looks formally similar.

3.3.4 Boundary between two elastic materials with fluid coupling: Homogeneous and inhomogeneous transition conditions

According to (3.118), shearing forces on the surface of an elastic material are not transmitted into the adjacent fluid; therefore, the transition conditions for fluid coupled elastic materials should account for it. Figure 3.6 illustrates such a coupling with a fluid layer (f): We postulate homogeneous transition conditions (3.117) through (3.119) for both boundaries S_1 and S_2, where $\underline{\mathbf{R}}_{S_1} \in S_1$ and $\underline{\mathbf{R}}_{S_2} \in S_2$. For a very thin fluid layer $\underline{\mathbf{R}}_{S_1} \simeq \underline{\mathbf{R}}_S$, $\underline{\mathbf{R}}_{S_2} \simeq \underline{\mathbf{R}}_S$, $\underline{\mathbf{n}}_1 = \underline{\mathbf{n}}$, $\underline{\mathbf{n}}_2 = -\underline{\mathbf{n}}$ holds, reducing both transition systems to a single one via elimination of $p^{(f)}(\underline{\mathbf{R}}_S, t)$ and $\underline{\mathbf{n}} \cdot \underline{\mathbf{u}}^{(f)}(\underline{\mathbf{R}}_S, t)$:

$$\underline{\mathbf{n}} \cdot \left[\underline{\mathbf{n}} \cdot \underline{\underline{\mathbf{T}}}^{(1)}(\underline{\mathbf{R}}_S, t)\right] - \underline{\mathbf{n}} \cdot \left[\underline{\mathbf{n}} \cdot \underline{\underline{\mathbf{T}}}^{(2)}(\underline{\mathbf{R}}_S, t)\right] = 0, \quad \underline{\mathbf{R}}_S \in S, \tag{3.120}$$

$$\underline{\mathbf{n}} \cdot \underline{\mathbf{u}}^{(1)}(\underline{\mathbf{R}}_S, t) - \underline{\mathbf{n}} \cdot \underline{\mathbf{u}}^{(2)}(\underline{\mathbf{R}}_S, t) = 0, \quad \underline{\mathbf{R}}_S \in S, \tag{3.121}$$

$$(\underline{\underline{\mathbf{I}}} - \underline{\mathbf{n}}\,\underline{\mathbf{n}}) \cdot \left[\underline{\mathbf{n}} \cdot \underline{\underline{\mathbf{T}}}^{(1)}(\underline{\mathbf{R}}_S, t)\right] = \underline{\mathbf{0}}, \quad \underline{\mathbf{R}}_S \in S, \tag{3.122}$$

$$(\underline{\underline{\mathbf{I}}} - \underline{\mathbf{n}}\,\underline{\mathbf{n}}) \cdot \left[\underline{\mathbf{n}} \cdot \underline{\underline{\mathbf{T}}}^{(2)}(\underline{\mathbf{R}}_S, t)\right] = \underline{\mathbf{0}}, \quad \underline{\mathbf{R}}_S \in S. \tag{3.123}$$

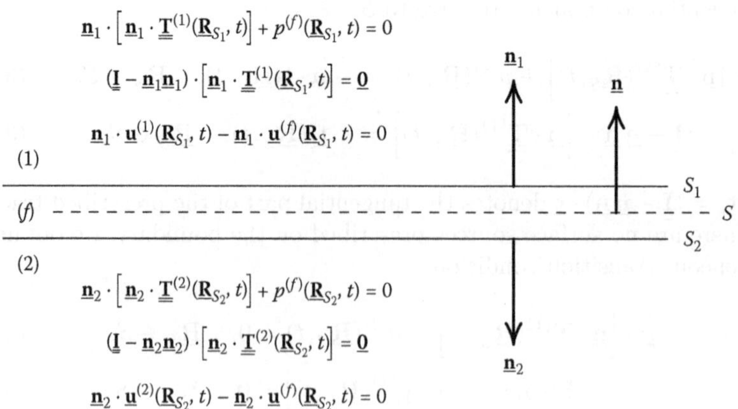

FIGURE 3.6
Derivation of (homogeneous) transition conditions for fluid coupled elastic materials.

The normal component of the traction vector $\underline{\mathbf{t}} = \underline{\mathbf{n}} \cdot \underline{\underline{\mathbf{T}}}$ is continuously transmitted as well as the normal component of the displacement $\underline{\mathbf{u}}$, whereas the shear components of the traction vector are indeed also continuous, yet they undergo a zero pass. The independence of Equations 3.122 and 3.123 is particularly obvious if, for instance, in material (1), tangential surface force densities are prescribed according to (3.116): They are not transmitted into material (2). Therefore, the inhomogeneous version of (3.120) through (3.123) should suggestively read as

$$\underline{\mathbf{n}} \cdot \left[\underline{\mathbf{n}} \cdot \underline{\underline{\mathbf{T}}}^{(1)}(\underline{\mathbf{R}}_S, t)\right] - \underline{\mathbf{n}} \cdot \left[\underline{\mathbf{n}} \cdot \underline{\underline{\mathbf{T}}}^{(2)}(\underline{\mathbf{R}}_S, t)\right] = -\underline{\mathbf{n}} \cdot \underline{\mathbf{t}}(\underline{\mathbf{R}}_S, t), \quad \underline{\mathbf{R}}_S \in S, \tag{3.124}$$

$$\underline{\mathbf{n}} \cdot \underline{\mathbf{u}}^{(1)}(\underline{\mathbf{R}}_S, t) - \underline{\mathbf{n}} \cdot \underline{\mathbf{u}}^{(2)}(\underline{\mathbf{R}}_S, t) = -h(\underline{\mathbf{R}}_S, t), \quad \underline{\mathbf{R}}_S \in S, \tag{3.125}$$

$$(\underline{\underline{\mathbf{I}}} - \underline{\mathbf{n}}\,\underline{\mathbf{n}}) \cdot \left[\underline{\mathbf{n}} \cdot \underline{\underline{\mathbf{T}}}^{(1)}(\underline{\mathbf{R}}_S, t)\right] = \underline{\mathbf{0}}, \quad \underline{\mathbf{R}}_S \in S, \tag{3.126}$$

$$(\underline{\underline{\mathbf{I}}} - \underline{\mathbf{n}}\,\underline{\mathbf{n}}) \cdot \left[\underline{\mathbf{n}} \cdot \underline{\underline{\mathbf{T}}}^{(2)}(\underline{\mathbf{R}}_S, t)\right] = \underline{\mathbf{0}}, \quad \underline{\mathbf{R}}_S \in S, \tag{3.127}$$

where $\underline{\mathbf{n}} \cdot \underline{\mathbf{t}}$ and h could be differences of prescribed surface source densities on S_1 or S_2, respectively.

Sometimes the transition conditions as discussed here are "simply written down"; yet, we emphasized that they may not only but also must be derived from the governing equations.

4

Constitutive Equations, Governing Equations, Elastodynamic Energy Conservation

4.1 Constitutive Equations

The governing equations of elastodynamics, be they linear or not, express facts concerning the time and spatial variations of field quantities (Equations 3.57 and 3.75):

$$\frac{\delta \underline{\mathbf{j}}(\underline{\mathbf{R}}, t)}{\delta t} = \nabla \cdot \underline{\underline{\mathbf{T}}}(\underline{\mathbf{R}}, t) + \underline{\mathbf{f}}(\underline{\mathbf{R}}, t), \tag{4.1}$$

$$\frac{\delta \underline{\underline{\mathbf{S}}}(\underline{\mathbf{R}}, t)}{\delta t} = \underline{\underline{\underline{\underline{\mathbf{I}}}}}^{+} : \nabla \underline{\mathbf{v}}(\underline{\mathbf{R}}, t) + \underline{\underline{\mathbf{h}}}(\underline{\mathbf{R}}, t). \tag{4.2}$$

Apparently, Newton–Cauchy's equation of motion (4.1) contains field quantities different from the deformation rate equation (4.2) requesting, in the most general form, composition operators $\underline{\mathsf{j}}$, $\underline{\underline{\mathsf{S}}}$ according to

$$\frac{\delta \underline{\mathbf{j}}(\underline{\mathbf{R}}, t)}{\delta t} = \underline{\mathsf{j}}\left[\underline{\mathbf{v}}(\underline{\mathbf{R}}, t), \underline{\underline{\mathbf{T}}}(\underline{\mathbf{R}}, t)\right], \tag{4.3}$$

$$\frac{\delta \underline{\underline{\mathbf{S}}}(\underline{\mathbf{R}}, t)}{\delta t} = \underline{\underline{\mathsf{S}}}\left[\underline{\mathbf{v}}(\underline{\mathbf{R}}, t), \underline{\underline{\mathbf{T}}}(\underline{\mathbf{R}}, t)\right], \tag{4.4}$$

the so-called constitutive equations (de Hoop 1995). They have to be based on physical arguments, in particular, they do not follow from the governing equations. Yet, modeling a solid should satisfy the criteria "close to reality" and "simplicity." Due to the latter, the dependence of the operators $\underline{\mathsf{j}}$ and $\underline{\underline{\mathsf{S}}}$ on *both* field quantities is usually sacrificed. We approximate

$$\frac{\delta \underline{\mathbf{j}}(\underline{\mathbf{R}}, t)}{\delta t} = \underline{\mathsf{j}}\left[\underline{\mathbf{v}}(\underline{\mathbf{R}}, t)\right], \tag{4.5}$$

$$\frac{\delta \underline{\underline{\mathbf{S}}}(\underline{\mathbf{R}}, t)}{\delta t} = \underline{\underline{\mathsf{S}}}\left[\underline{\underline{\mathbf{T}}}(\underline{\mathbf{R}}, t)\right]. \tag{4.6}$$

Considering (3.39), we specify

$$\frac{\delta \underline{\mathbf{j}}(\underline{\mathbf{R}}, t)}{\delta t} = \underline{\underline{\boldsymbol{\rho}}}(\underline{\mathbf{R}}) \cdot \frac{\mathrm{D}\underline{\mathbf{v}}(\underline{\mathbf{R}}, t)}{\mathrm{D}t}, \tag{4.7}$$

$$\frac{\delta \underline{\underline{\mathbf{S}}}(\underline{\mathbf{R}},t)}{\delta t} = \underline{\underline{\underline{\underline{\mathbf{s}}}}}(\underline{\mathbf{R}}) : \frac{\mathrm{D}\underline{\underline{\mathbf{T}}}(\underline{\mathbf{R}},t)}{\mathrm{D}t} \tag{4.8}$$

and linearize according to $\delta/\delta t \Longrightarrow \partial/\partial t$, $\mathrm{D}/\mathrm{D}t \Longrightarrow \partial/\partial t$

$$\frac{\partial \underline{\mathbf{j}}(\underline{\mathbf{R}},t)}{\partial t} = \underline{\underline{\boldsymbol{\rho}}}(\underline{\mathbf{R}}) \cdot \frac{\partial \underline{\mathbf{v}}(\underline{\mathbf{R}},t)}{\partial t}, \tag{4.9}$$

$$\frac{\partial \underline{\underline{\mathbf{S}}}(\underline{\mathbf{R}},t)}{\partial t} = \underline{\underline{\underline{\underline{\mathbf{s}}}}}(\underline{\mathbf{R}}) : \frac{\partial \underline{\underline{\mathbf{T}}}(\underline{\mathbf{R}},t)}{\partial t} \tag{4.10}$$

with the consequence

$$\underline{\mathbf{j}}(\underline{\mathbf{R}},t) = \underline{\underline{\boldsymbol{\rho}}}(\underline{\mathbf{R}}) \cdot \underline{\mathbf{v}}(\underline{\mathbf{R}},t), \tag{4.11}$$

$$\underline{\underline{\mathbf{S}}}(\underline{\mathbf{R}},t) = \underline{\underline{\underline{\underline{\mathbf{s}}}}}(\underline{\mathbf{R}}) : \underline{\underline{\mathbf{T}}}(\underline{\mathbf{R}},t). \tag{4.12}$$

The constitutive equations (4.7) and (4.8) and the linear constitutive equations (4.11) and (4.12), define a second rank mass density tensor $\underline{\underline{\boldsymbol{\rho}}}(\underline{\mathbf{R}})$ and the forth rank compliance tensor $\underline{\underline{\underline{\underline{\mathbf{s}}}}}(\underline{\mathbf{R}})$. Both tensors characterize a time invariant instantaneously reacting inhomogeneous locally reacting anisotropic material: time invariant, because they do not explicitly depend on time, and instantaneously reacting, because $\underline{\mathbf{j}}(\underline{\mathbf{R}},t)$ and $\underline{\underline{\mathbf{S}}}(\underline{\mathbf{R}},t)$ depend on $\underline{\mathbf{v}}(\underline{\mathbf{R}},t)$ and $\underline{\underline{\mathbf{T}}}(\underline{\mathbf{R}},t)$, respectively, only at the same time t. In a similar sense, the material (4.11) and (4.12) is spatially invariant (inhomogeneous) and locally reacting: inhomogeneous, because $\underline{\underline{\boldsymbol{\rho}}}(\underline{\mathbf{R}})$ and $\underline{\underline{\underline{\underline{\mathbf{s}}}}}(\underline{\mathbf{R}})$ depend on the vector of position $\underline{\mathbf{R}}$ and locally reacting, because $\underline{\mathbf{j}}(\underline{\mathbf{R}},t)$ and $\underline{\underline{\mathbf{S}}}(\underline{\mathbf{R}},t)$ at point $\underline{\mathbf{R}}$ depend on $\underline{\mathbf{v}}(\underline{\mathbf{R}},t)$ and $\underline{\underline{\mathbf{T}}}(\underline{\mathbf{R}},t)$, respectively, only at the same point. The material is anisotropic because the variation of one (cartesian) component of $\underline{\mathbf{v}}$ and $\underline{\underline{\mathbf{T}}}$ yields variations of all other components of $\underline{\mathbf{j}}$ and $\underline{\underline{\mathbf{S}}}$, that is to say, the relative orientation, for example, of $\underline{\mathbf{v}}$ and $\underline{\mathbf{j}}$, depends on the direction of $\underline{\mathbf{v}}$: The material exhibits a macroscopic inner structure.

Specializations of (4.11) and (4.12) are homogeneity and isotropy (Section 4.2.2) of the material. Other important generalizations comprise noninstantaneously reacting materials for a mathematical description of the physical phenomenon of dissipation (Section 4.4).

4.2 Linear Nondissipative Materials: Cauchy–Hooke Law

4.2.1 Anisotropic materials, Voigt notation, transversely isotropic materials

Anisotropic materials: Symmetries of the compliance tensor: Even though there are several reasons in geophysics to introduce a mass density tensor (de Hoop 1995), we disregard this in our further elaboration of wave propagation and consider the mass density $\rho(\underline{\mathbf{R}})$ as a scalar quantity:

$$\underline{\mathbf{j}}(\underline{\mathbf{R}}, t) = \rho(\underline{\mathbf{R}})\underline{\mathbf{v}}(\underline{\mathbf{R}}, t). \tag{4.13}$$

The actual Hooke law is the linear relation between the elongation of a spring and the applied weight. Here, we denote each linear relation between stress tensor and deformation tensor (or vice versa) as (Cauchy–) Hooke's law:

$$\underline{\underline{\mathbf{S}}}(\underline{\mathbf{R}}, t) = \underline{\underline{\underline{\underline{\mathbf{s}}}}}(\underline{\mathbf{R}}) : \underline{\underline{\mathbf{T}}}(\underline{\mathbf{R}}, t). \tag{4.14}$$

The constitutive equations (4.13) and (4.14) characterize a linear time invariant instantaneously reacting inhomogeneous anisotropic locally reacting material. Due to the symmetry of $\underline{\underline{\mathbf{T}}}(\underline{\mathbf{R}}, t)$, the compliance $\underline{\underline{\underline{\underline{\mathbf{s}}}}}(\underline{\mathbf{R}})$ must be symmetric with regard to the last two indices and due to the symmetry of $\underline{\underline{\mathbf{S}}}(\underline{\mathbf{R}}, t)$, it must be symmetric with regard to the first two indices:

$$\underline{\underline{\underline{\underline{\mathbf{s}}}}}^{1234} = \underline{\underline{\underline{\underline{\mathbf{s}}}}}^{1243} = \underline{\underline{\underline{\underline{\mathbf{s}}}}}^{2143} = \underline{\underline{\underline{\underline{\mathbf{s}}}}}^{2134}; \tag{4.15}$$

the index notation of (4.15) reads

$$s_{ijkl} = s_{ijlk} = s_{jilk} = s_{jikl}. \tag{4.16}$$

In Section 4.3.1, we will show that the elastodynamic energy conservation for instantaneously reacting (nondissipative) materials additionally enforces the symmetry

$$\underline{\underline{\underline{\underline{\mathbf{s}}}}}^{1234} = \underline{\underline{\underline{\underline{\mathbf{s}}}}}^{3412} \Longleftrightarrow s_{ijkl} = s_{klij}. \tag{4.17}$$

Stiffness tensor: Very often, the stiffness tensor $\underline{\underline{\underline{\underline{\mathbf{c}}}}}(\underline{\mathbf{R}})$ is used instead of the compliance tensor; it is defined through inversion of (4.14):

$$\underline{\underline{\mathbf{T}}}(\underline{\mathbf{R}}, t) = \underline{\underline{\underline{\underline{\mathbf{c}}}}}(\underline{\mathbf{R}}) : \underline{\underline{\mathbf{S}}}(\underline{\mathbf{R}}, t); \tag{4.18}$$

due to (2.116) and the symmetry of $\underline{\underline{\mathbf{S}}}$ and $\underline{\underline{\mathbf{T}}}$, we must have

$$\underline{\underline{\underline{\underline{\mathbf{c}}}}}(\underline{\mathbf{R}}) : \underline{\underline{\underline{\underline{\mathbf{s}}}}}(\underline{\mathbf{R}}) = \underline{\underline{\underline{\underline{\mathbf{s}}}}}(\underline{\mathbf{R}}) : \underline{\underline{\underline{\underline{\mathbf{c}}}}}(\underline{\mathbf{R}}) = \underline{\underline{\underline{\underline{\mathbf{I}}}}}^{+}. \tag{4.19}$$

Obviously, $\underline{\underline{\underline{\underline{\mathbf{c}}}}}$ has to satisfy the same symmetries as $\underline{\underline{\underline{\underline{\mathbf{s}}}}}$.

Voigt notation: A forth rank tensor has 81 components represented by a 3×3-matrix whose nine elements are 3×3-matrices themselves. The symmetry of $\underline{\underline{\underline{\underline{\mathbf{s}}}}}$ and $\underline{\underline{\underline{\underline{\mathbf{c}}}}}$ with regard to the two first indices reduces the number of independent components to 54, the additional symmetry with regard to the last two indices to 36. Due to the symmetry (4.17), there remain only 21 independent components. Those can be inserted—for example, for the stiffness tensor—into a 6×6-matrix according to:[65]

$$\underline{\underline{C}} = \begin{pmatrix} c_{11} & c_{12} & c_{13} & c_{14} & c_{15} & c_{16} \\ c_{12} & c_{22} & c_{23} & c_{24} & c_{25} & c_{26} \\ c_{13} & c_{23} & c_{33} & c_{34} & c_{35} & c_{36} \\ c_{14} & c_{24} & c_{34} & c_{44} & c_{45} & c_{46} \\ c_{15} & c_{25} & c_{35} & c_{45} & c_{55} & c_{56} \\ c_{16} & c_{26} & c_{36} & c_{46} & c_{56} & c_{66} \end{pmatrix} . \tag{4.20}$$

This is called the Voigt notation of the stiffness tensor (similarly: of the compliance tensor). Note that $\underline{\underline{C}}$ is a matrix, not a tensor! The symmetry of $\underline{\underline{\mathbf{S}}}$ and $\underline{\underline{\mathbf{T}}}$ leaves six independent components to both tensors that can be numbered according to

$$\begin{pmatrix} T_{xx} & T_{xy} & T_{xz} \\ T_{xy} & T_{yy} & T_{yz} \\ T_{xz} & T_{yz} & T_{zz} \end{pmatrix} \Longrightarrow \begin{pmatrix} T_1 & T_6 & T_5 \\ T_6 & T_2 & T_4 \\ T_5 & T_4 & T_3 \end{pmatrix} \Longrightarrow \begin{pmatrix} T_1 \\ T_2 \\ T_3 \\ T_4 \\ T_5 \\ T_6 \end{pmatrix} = \underline{T}, \tag{4.21}$$

$$\begin{pmatrix} S_{xx} & S_{xy} & S_{xz} \\ S_{xy} & S_{yy} & S_{yz} \\ S_{xz} & S_{yz} & S_{zz} \end{pmatrix} \Longrightarrow \begin{pmatrix} S_1 & \frac{1}{2}S_6 & \frac{1}{2}S_5 \\ \frac{1}{2}S_6 & S_2 & \frac{1}{2}S_4 \\ \frac{1}{2}S_5 & \frac{1}{2}S_4 & S_3 \end{pmatrix} \Longrightarrow \begin{pmatrix} S_1 \\ S_2 \\ S_3 \\ S_4 \\ S_5 \\ S_6 \end{pmatrix} = \underline{S} \tag{4.22}$$

and combined to 6×1-matrices (column "vectors"). Then, Hooke's law reads[66]

$$\begin{pmatrix} T_1 \\ T_2 \\ T_3 \\ T_4 \\ T_5 \\ T_6 \end{pmatrix} = \begin{pmatrix} c_{11} & c_{12} & c_{13} & c_{14} & c_{15} & c_{16} \\ c_{12} & c_{22} & c_{23} & c_{24} & c_{25} & c_{26} \\ c_{13} & c_{23} & c_{33} & c_{34} & c_{35} & c_{36} \\ c_{14} & c_{24} & c_{34} & c_{44} & c_{45} & c_{46} \\ c_{15} & c_{25} & c_{35} & c_{45} & c_{55} & c_{56} \\ c_{16} & c_{26} & c_{36} & c_{46} & c_{56} & c_{66} \end{pmatrix} \begin{pmatrix} S_1 \\ S_2 \\ S_3 \\ S_4 \\ S_5 \\ S_6 \end{pmatrix}$$

$$\begin{aligned} \underline{T}(\underline{\mathbf{R}}, t) &= \underline{\underline{C}}(\underline{\mathbf{R}})\underline{S}(\underline{\mathbf{R}}, t), \\ \Longleftrightarrow T_\alpha(\underline{\mathbf{R}}, t) &= C_{\alpha\beta}(\underline{\mathbf{R}})S_\beta(\underline{\mathbf{R}}, t), \\ \alpha, \beta &= 1, \ldots, 6. \end{aligned} \tag{4.23}$$

Yet, in this elaboration, we prefer the tensor version of (4.18) and (4.14), respectively, because it can immediately be written in coordinates via projection onto an orthonormal trihedron of any coordinate system.

---

[65] The explicit transformation of the c_{ijkl}, $i, j, k, l = 1, 2, 3$, into $C_{\alpha\beta}$, $\alpha, \beta = 1, \ldots, 6$, can be found in Helbig (1994).

[66] The summation convention is generalized insofar as summation from 1 to 6 is performed over Greek indices appearing twice.

As references for stiffness anisotropy, we mention Auld (1973), Ben-Menahem and Singh (1981), Royer and Dieulesaint (2000), and Helbig (1994).

Transversely isotropic materials: According to the crystal symmetries of solids, models for anisotropy with increasing complexity can be formulated. The simplest one[67] accounting for crystals with hexagonal symmetry is the model of transverse isotropy perpendicular to a preference direction $\underline{\hat{\mathbf{a}}}$, where five elastic constants $\lambda_\perp, \lambda_\|, \mu_\perp, \mu_\|, \nu$ (instead of the 21 for the general case) are involved (Spies 1992, 1994):

$$\begin{aligned}\underline{\underline{\underline{\underline{\mathbf{c}}}}}^{\text{triso}}(\underline{\mathbf{R}}) &= \lambda_\perp \underline{\underline{\underline{\underline{\mathbf{I}}}}}^\delta + 2\mu_\perp \underline{\underline{\underline{\underline{\mathbf{I}}}}}^+ \\ &\quad + [\lambda_\perp + 2\mu_\perp + \lambda_\| + 2\mu_\| - 2(\nu + 2\mu_\|)]\underline{\hat{\mathbf{a}}}\,\underline{\hat{\mathbf{a}}}\,\underline{\hat{\mathbf{a}}}\,\underline{\hat{\mathbf{a}}} \\ &\quad + (\nu - \lambda_\perp)(\underline{\underline{\mathbf{I}}}\,\underline{\hat{\mathbf{a}}}\,\underline{\hat{\mathbf{a}}} + \underline{\hat{\mathbf{a}}}\,\underline{\hat{\mathbf{a}}}\,\underline{\underline{\mathbf{I}}}) \\ &\quad + (\mu_\| - \mu_\perp)(\underline{\underline{\mathbf{I}}}\,\underline{\hat{\mathbf{a}}}\,\underline{\hat{\mathbf{a}}}^{1324} + \underline{\hat{\mathbf{a}}}\,\underline{\hat{\mathbf{a}}}\,\underline{\underline{\mathbf{I}}}^{1324} + \underline{\underline{\mathbf{I}}}\,\underline{\hat{\mathbf{a}}}\,\underline{\hat{\mathbf{a}}}^{1342} + \underline{\hat{\mathbf{a}}}\,\underline{\hat{\mathbf{a}}}\,\underline{\underline{\mathbf{I}}}^{1342}). \end{aligned} \tag{4.24}$$

The inhomogeneity of $\underline{\underline{\underline{\underline{\mathbf{c}}}}}^{\text{triso}}(\underline{\mathbf{R}})$ may show up in the elastic constants $\lambda_\perp(\underline{\mathbf{R}})$, $\mu_\perp(\underline{\mathbf{R}})$, $\lambda_\|(\underline{\mathbf{R}})$, $\mu_\|(\underline{\mathbf{R}})$, $\nu(\underline{\mathbf{R}})$ and in the spatial dependence of the preference direction $\underline{\hat{\mathbf{a}}}(\underline{\mathbf{R}})$. An example for spatially independent constants yet a spatially dependent preference direction is the crystal orientation within an austenitic weld (Langenberg et al. 2000); by the way, in that case, only four independent constants are required because $\nu = \lambda_\perp - \mu_\perp + \mu_\|$ must hold (Neumann et al. 1995). For $\underline{\hat{\mathbf{a}}} = \underline{\mathbf{e}}_x$—thus disregarding the spatial dependence of the preference direction—the stiffness tensor (4.24) reads in Voigt notation:

$$\begin{aligned}&\underline{\underline{\mathit{C}}}^{\text{triso}}(\underline{\mathbf{R}}) \\ &= \begin{pmatrix} \lambda_\|(\underline{\mathbf{R}}) + 2\mu_\|(\underline{\mathbf{R}}) & \nu(\underline{\mathbf{R}}) & \nu(\underline{\mathbf{R}}) & 0 & 0 & 0 \\ \nu(\underline{\mathbf{R}}) & \lambda_\perp(\underline{\mathbf{R}}) + 2\mu_\perp(\underline{\mathbf{R}}) & \lambda_\perp(\underline{\mathbf{R}}) & 0 & 0 & 0 \\ \nu(\underline{\mathbf{R}}) & \lambda_\perp(\underline{\mathbf{R}}) & \lambda_\perp(\underline{\mathbf{R}}) + 2\mu_\perp(\underline{\mathbf{R}}) & 0 & 0 & 0 \\ 0 & 0 & 0 & \mu_\perp(\underline{\mathbf{R}}) & 0 & 0 \\ 0 & 0 & 0 & 0 & \mu_\|(\underline{\mathbf{R}}) & 0 \\ 0 & 0 & 0 & 0 & 0 & \mu_\|(\underline{\mathbf{R}}) \end{pmatrix};\end{aligned} \tag{4.25}$$

apparently, the relation of the Lamé parameters with the Voigt parameters depends on the coordinate system because, for $\underline{\hat{\mathbf{a}}} = \underline{\mathbf{e}}_z$ we obtain

$$\begin{aligned}&\underline{\underline{\mathit{C}}}^{\text{triso}}(\underline{\mathbf{R}}) \\ &= \begin{pmatrix} \lambda_\perp(\underline{\mathbf{R}}) + 2\mu_\perp(\underline{\mathbf{R}}) & \lambda_\perp(\underline{\mathbf{R}}) & \nu(\underline{\mathbf{R}}) & 0 & 0 & 0 \\ \lambda_\perp(\underline{\mathbf{R}}) & \lambda_\perp(\underline{\mathbf{R}}) + 2\mu_\perp(\underline{\mathbf{R}}) & \nu(\underline{\mathbf{R}}) & 0 & 0 & 0 \\ \nu(\underline{\mathbf{R}}) & \nu\underline{\mathbf{R}}) & \lambda_\|(\underline{\mathbf{R}}) + 2\mu_\|(\underline{\mathbf{R}}) & 0 & 0 & 0 \\ 0 & 0 & 0 & \mu_\|(\underline{\mathbf{R}}) & 0 & 0 \\ 0 & 0 & 0 & 0 & \mu_\|(\underline{\mathbf{R}}) & 0 \\ 0 & 0 & 0 & 0 & 0 & \mu_\perp(\underline{\mathbf{R}}) \end{pmatrix};\end{aligned} \tag{4.26}$$

[67]Even simpler is a hypothetic uniaxial model (Lindell and Kiselev 2000):

$$\underline{\underline{\underline{\underline{\mathbf{c}}}}}^{\text{uni}} = \alpha \underline{\underline{\underline{\underline{\mathbf{I}}}}}^\delta + \beta \underline{\underline{\underline{\underline{\mathbf{I}}}}}^+ + \gamma \underline{\hat{\mathbf{a}}}\,\underline{\hat{\mathbf{a}}}\,\underline{\hat{\mathbf{a}}}\,\underline{\hat{\mathbf{a}}}.$$

Further coordinate-free representations of stiffness tensors for higher degrees of anisotropy of different crystal classes (cubic, orthorhombic, and tetragonal) including their Voigt notations can be found in Marklein (1997).

4.2.2 Isotropic materials

By definition, the stiffness tensor for isotropic materials must have a representation that does not exhibit any macroscopic structural parameters; furthermore, its double contraction with the symmetric deformation tensor must yield a symmetric stress tensor. This is generally achieved with the forth rank tensor $\underline{\underline{\underline{\underline{\mathbf{I}}}}}$ according to (2.107) with $\alpha_3 = 0$; typically, we write

$$\begin{aligned}\underline{\underline{\underline{\underline{\mathbf{c}}}}}^{\text{iso}}(\underline{\mathbf{R}}) &= \lambda(\underline{\mathbf{R}})\,\underline{\underline{\underline{\underline{\mathbf{I}}}}}^{\delta} + 2\mu(\underline{\mathbf{R}})\,\underline{\underline{\underline{\underline{\mathbf{I}}}}}^{+} \\ &= \lambda(\underline{\mathbf{R}})\,\underline{\underline{\mathbf{I}}}\,\underline{\underline{\mathbf{I}}} + \mu(\underline{\mathbf{R}})(\underline{\underline{\mathbf{I}}}\,\underline{\underline{\mathbf{I}}}^{1342} + \underline{\underline{\mathbf{I}}}\,\underline{\underline{\mathbf{I}}}^{1324})\end{aligned} \tag{4.27}$$

with the Lamé constants $\lambda(\underline{\mathbf{R}})$ and $\mu(\underline{\mathbf{R}})$. Consequently, Hooke's law reads as

$$\begin{aligned}\underline{\underline{\mathbf{T}}}(\underline{\mathbf{R}},t) &= \underline{\underline{\underline{\underline{\mathbf{c}}}}}^{\text{iso}}(\underline{\mathbf{R}}) : \underline{\underline{\mathbf{S}}}(\underline{\mathbf{R}},t) \\ &= \lambda(\underline{\mathbf{R}})\,\underline{\underline{\mathbf{I}}}\,\operatorname{trace}\underline{\underline{\mathbf{S}}}(\underline{\mathbf{R}},t) + 2\mu(\underline{\mathbf{R}})\underline{\underline{\mathbf{S}}}(\underline{\mathbf{R}},t).\end{aligned} \tag{4.28}$$

The compliance tensor $\underline{\underline{\underline{\underline{\mathbf{s}}}}}^{\text{iso}}(\underline{\mathbf{R}})$ has a structure analogous to (4.27):

$$\underline{\underline{\underline{\underline{\mathbf{s}}}}}^{\text{iso}}(\underline{\mathbf{R}}) = \Lambda(\underline{\mathbf{R}})\underline{\underline{\underline{\underline{\mathbf{I}}}}}^{\delta} + 2M(\underline{\mathbf{R}})\underline{\underline{\underline{\underline{\mathbf{I}}}}}^{+}, \tag{4.29}$$

where[68] (de Hoop 1995)

$$\Lambda(\underline{\mathbf{R}}) = -\frac{\lambda(\underline{\mathbf{R}})}{2\mu(\underline{\mathbf{R}})[3\lambda(\underline{\mathbf{R}}) + 2\mu(\underline{\mathbf{R}})]}, \tag{4.30}$$

$$M(\underline{\mathbf{R}}) = \frac{1}{4\mu(\underline{\mathbf{R}})}. \tag{4.31}$$

The stiffness tensor (4.27) is written as a Voigt matrix as follows:

$$\underline{\underline{C}}^{\text{iso}}(\underline{\mathbf{R}}) = \begin{pmatrix} \lambda(\underline{\mathbf{R}})+2\mu(\underline{\mathbf{R}}) & \lambda(\underline{\mathbf{R}}) & \lambda(\underline{\mathbf{R}}) & 0 & 0 & 0 \\ \lambda(\underline{\mathbf{R}}) & \lambda(\underline{\mathbf{R}})+2\mu(\underline{\mathbf{R}}) & \lambda(\underline{\mathbf{R}}) & 0 & 0 & 0 \\ \lambda(\underline{\mathbf{R}}) & \lambda(\underline{\mathbf{R}}) & \lambda(\underline{\mathbf{R}})+2\mu(\underline{\mathbf{R}}) & 0 & 0 & 0 \\ 0 & 0 & 0 & \mu(\underline{\mathbf{R}}) & 0 & 0 \\ 0 & 0 & 0 & 0 & \mu(\underline{\mathbf{R}}) & 0 \\ 0 & 0 & 0 & 0 & 0 & \mu(\underline{\mathbf{R}}) \end{pmatrix}. \tag{4.32}$$

[68]These formulas are reciprocal:

$$\lambda(\underline{\mathbf{R}}) = -\frac{\Lambda(\underline{\mathbf{R}})}{2M(\underline{\mathbf{R}})[3\Lambda(\underline{\mathbf{R}}) + 2M(\underline{\mathbf{R}})]},$$

$$\mu(\underline{\mathbf{R}}) = \frac{1}{4M(\underline{\mathbf{R}})}.$$

4.2.3 Elastodynamic governing equations

With the constitutive equations (4.11) and (4.12), we obtain the elastodynamic governing equations:

$$\underline{\underline{\rho}}(\underline{\mathbf{R}}) \cdot \frac{\partial \underline{\mathbf{v}}(\underline{\mathbf{R}}, t)}{\partial t} = \boldsymbol{\nabla} \cdot \underline{\underline{\mathbf{T}}}(\underline{\mathbf{R}}, t) + \underline{\mathbf{f}}(\underline{\mathbf{R}}, t), \tag{4.33}$$

$$\underline{\underline{\underline{\underline{\mathbf{s}}}}}(\underline{\mathbf{R}}) : \frac{\partial \underline{\underline{\mathbf{T}}}(\underline{\mathbf{R}}, t)}{\partial t} = \frac{1}{2}\left\{\boldsymbol{\nabla}\underline{\mathbf{v}}(\underline{\mathbf{R}}, t) + [\boldsymbol{\nabla}\underline{\mathbf{v}}(\underline{\mathbf{R}}, t)]^{21}\right\} + \underline{\underline{\mathbf{h}}}(\underline{\mathbf{R}}, t). \tag{4.34}$$

They describe the propagation of elastic waves in linear time invariant instantaneously and locally reacting inhomogeneous anisotropic nondissipative materials.

4.3 Elastodynamic Energy Conservation Theorem for Nondissipative Materials in the Time and Frequency Domains

4.3.1 Elastodynamic Poynting vector in the time domain

Convincing reasons exist in elastostatics to define (Ben-Menahem and Singh 1981)

$$w(\underline{\mathbf{R}}) = \frac{1}{2}\,\underline{\underline{\mathbf{S}}}(\underline{\mathbf{R}}) : \underline{\underline{\mathbf{T}}}(\underline{\mathbf{R}}) \tag{4.35}$$

as a potential deformation energy density that is locally contained in a static deformation-stress field $\underline{\underline{\mathbf{S}}}(\underline{\mathbf{R}}), \underline{\underline{\mathbf{T}}}(\underline{\mathbf{R}})$. To generalize it to elastodynamics, we tentatively use Equation 4.35 for time-dependent deformations and stresses $\underline{\underline{\mathbf{S}}}(\underline{\mathbf{R}}, t), \underline{\underline{\mathbf{T}}}(\underline{\mathbf{R}}, t)$; in addition, we have to account for the kinetic energy density of the time varying motion of the material particles, finally leading to the Hamiltonian expression as an *ansatz* for the elastodynamic energy density (Ben-Menahem and Singh 1981):

$$w_{\mathrm{el}}(\underline{\mathbf{R}}, t) = \frac{1}{2}\,\underline{\mathbf{j}}(\underline{\mathbf{R}}, t) \cdot \underline{\mathbf{v}}(\underline{\mathbf{R}}, t) + \frac{1}{2}\,\underline{\underline{\mathbf{S}}}(\underline{\mathbf{R}}, t) : \underline{\underline{\mathbf{T}}}(\underline{\mathbf{R}}, t). \tag{4.36}$$

An energy (conservation) law expresses a balance for the energy density: If the latter locally changes with time, energy is either flowing or created/annihilated. Therefore, we investigate the time derivative of (4.36):

$$\begin{aligned}\frac{\partial w_{\mathrm{el}}(\underline{\mathbf{R}}, t)}{\partial t} = {} & \frac{1}{2}\,\frac{\partial \underline{\mathbf{j}}(\underline{\mathbf{R}}, t)}{\partial t} \cdot \underline{\mathbf{v}}(\underline{\mathbf{R}}, t) + \frac{1}{2}\,\underline{\mathbf{j}}(\underline{\mathbf{R}}, t) \cdot \frac{\partial \underline{\mathbf{v}}(\underline{\mathbf{R}}, t)}{\partial t} \\ & + \frac{1}{2}\,\frac{\partial \underline{\underline{\mathbf{S}}}(\underline{\mathbf{R}}, t)}{\partial t} : \underline{\underline{\mathbf{T}}}(\underline{\mathbf{R}}, t) + \frac{1}{2}\,\underline{\underline{\mathbf{S}}}(\underline{\mathbf{R}}, t) : \frac{\partial \underline{\underline{\mathbf{T}}}(\underline{\mathbf{R}}, t)}{\partial t}.\end{aligned} \tag{4.37}$$

Now we should bring the elastodynamic governing equations into play to account for elastodynamics, i.e., for the time and spatial variations of elastic fields; after all, we want to characterize the time and spatial energy flow. In fact, the (linearized) governing equations immediately tell us something about $\partial\mathbf{j}/\partial t$ and $\partial\underline{\underline{\mathbf{S}}}/\partial t$, yet the same is only true for $\partial\underline{\mathbf{v}}/\partial t$ and $\partial\underline{\underline{\mathbf{T}}}/\partial t$ if we add constitutive equations. With

$$\mathbf{j}(\underline{\mathbf{R}}, t) = \rho(\underline{\mathbf{R}})\underline{\mathbf{v}}(\underline{\mathbf{R}}, t), \tag{4.38}$$

$$\underline{\underline{\mathbf{S}}}(\underline{\mathbf{R}}, t) = \underline{\underline{\underline{\underline{\mathbf{s}}}}}(\underline{\mathbf{R}}) : \underline{\underline{\mathbf{T}}}(\underline{\mathbf{R}}, t), \tag{4.39}$$

we postulate a linear time invariant instantaneously and locally reacting nondissipative (nondispersive) inhomogeneous anisotropic material to obtain

$$\begin{aligned} w_{\text{el}}(\underline{\mathbf{R}}, t) &= \frac{1}{2}\,\rho(\underline{\mathbf{R}})\underline{\mathbf{v}}(\underline{\mathbf{R}}, t) \cdot \underline{\mathbf{v}}(\underline{\mathbf{R}}, t) + \frac{1}{2}\,\underline{\underline{\underline{\underline{\mathbf{s}}}}}(\underline{\mathbf{R}}) : \underline{\underline{\mathbf{T}}}(\underline{\mathbf{R}}, t) : \underline{\underline{\mathbf{T}}}(\underline{\mathbf{R}}, t) \\ &= \frac{1}{2}\,\rho(\underline{\mathbf{R}})|\underline{\mathbf{v}}(\underline{\mathbf{R}}, t)|^2 + \frac{1}{2}\,\underline{\underline{\underline{\underline{\mathbf{s}}}}}(\underline{\mathbf{R}}) : \underline{\underline{\mathbf{T}}}(\underline{\mathbf{R}}, t) : \underline{\underline{\mathbf{T}}}(\underline{\mathbf{R}}, t) \end{aligned} \tag{4.40}$$

instead of (4.36) and, therefore, instead of (4.37):

$$\begin{aligned} \frac{\partial w_{\text{el}}(\underline{\mathbf{R}}, t)}{\partial t} &= \rho(\underline{\mathbf{R}})\,\frac{\partial \underline{\mathbf{v}}(\underline{\mathbf{R}}, t)}{\partial t} \cdot \underline{\mathbf{v}}(\underline{\mathbf{R}}, t) + \frac{1}{2}\,\underline{\underline{\underline{\underline{\mathbf{s}}}}}(\underline{\mathbf{R}}) : \frac{\partial \underline{\underline{\mathbf{T}}}(\underline{\mathbf{R}}, t)}{\partial t} : \underline{\underline{\mathbf{T}}}(\underline{\mathbf{R}}, t) \\ &\quad + \frac{1}{2}\,\underline{\underline{\underline{\underline{\mathbf{s}}}}}(\underline{\mathbf{R}}) : \underline{\underline{\mathbf{T}}}(\underline{\mathbf{R}}, t) : \frac{\partial \underline{\underline{\mathbf{T}}}(\underline{\mathbf{R}}, t)}{\partial t}, \end{aligned} \tag{4.41}$$

where we have combined the first two terms of (4.37). Yet, the last two terms can only be combined if the commutation

$$\underline{\underline{\underline{\underline{\mathbf{s}}}}}(\underline{\mathbf{R}}) : \frac{\partial \underline{\underline{\mathbf{T}}}(\underline{\mathbf{R}}, t)}{\partial t} : \underline{\underline{\mathbf{T}}}(\underline{\mathbf{R}}, t) = \underline{\underline{\underline{\underline{\mathbf{s}}}}}(\underline{\mathbf{R}}) : \underline{\underline{\mathbf{T}}}(\underline{\mathbf{R}}, t) : \frac{\partial \underline{\underline{\mathbf{T}}}(\underline{\mathbf{R}}, t)}{\partial t} \tag{4.42}$$

is allowed, and this requests the symmetry

$$\underline{\underline{\underline{\underline{\mathbf{s}}}}}^{1234} = \underline{\underline{\underline{\underline{\mathbf{s}}}}}^{3412} \Longleftrightarrow s_{ijkl} = s_{klij} \tag{4.43}$$

of the compliance tensor. With the symmetry (4.43), Equation 4.41 reads as

$$\frac{\partial w_{\text{el}}(\underline{\mathbf{R}}, t)}{\partial t} = \rho(\underline{\mathbf{R}})\,\frac{\partial \underline{\mathbf{v}}(\underline{\mathbf{R}}, t)}{\partial t} \cdot \underline{\mathbf{v}}(\underline{\mathbf{R}}, t) + \underline{\underline{\underline{\underline{\mathbf{s}}}}}(\underline{\mathbf{R}}) : \frac{\partial \underline{\underline{\mathbf{T}}}(\underline{\mathbf{R}}, t)}{\partial t} : \underline{\underline{\mathbf{T}}}(\underline{\mathbf{R}}, t). \tag{4.44}$$

Final insertion of the governing equations (4.33) and (4.34) leads us to—we utilize the symmetry of $\underline{\underline{\mathbf{T}}}(\underline{\mathbf{R}}, t)$—

$$\begin{aligned} \frac{\partial w_{\text{el}}(\underline{\mathbf{R}}, t)}{\partial t} &= \nabla \cdot \underline{\underline{\mathbf{T}}}(\underline{\mathbf{R}}, t) \cdot \underline{\mathbf{v}}(\underline{\mathbf{R}}, t) + \nabla \underline{\mathbf{v}}(\underline{\mathbf{R}}, t) : \underline{\underline{\mathbf{T}}}(\underline{\mathbf{R}}, t) \\ &\quad + \underline{\mathbf{f}}(\underline{\mathbf{R}}, t) \cdot \underline{\mathbf{v}}(\underline{\mathbf{R}}, t) + \underline{\underline{\mathbf{h}}}(\underline{\mathbf{R}}, t) : \underline{\underline{\mathbf{T}}}(\underline{\mathbf{R}}, t). \end{aligned} \tag{4.45}$$

The last two terms on the right-hand side of the above equation are identified as time variation of an energy density

$$\frac{\partial w_Q(\underline{\mathbf{R}},t)}{\partial t} = \underline{\mathbf{f}}(\underline{\mathbf{R}},t)\cdot\underline{\mathbf{v}}(\underline{\mathbf{R}},t) + \underline{\underline{\mathbf{h}}}(\underline{\mathbf{R}},t) : \underline{\underline{\mathbf{T}}}(\underline{\mathbf{R}},t), \tag{4.46}$$

that is locally "injected" into the stress-motion field $\underline{\underline{\mathbf{T}}}(\underline{\mathbf{R}},t), \underline{\mathbf{v}}(\underline{\mathbf{R}},t)$ by prescribed force densities $\underline{\mathbf{f}}$ and deformation rates $\underline{\underline{\mathbf{h}}}$; consequently, the first two terms on the right-hand side of Equation 4.45 must have the meaning of an energy density flow. To make it obvious, we combine them as a divergence of a vector[69]

$$\underline{\mathbf{S}}(\underline{\mathbf{R}},t) = -\underline{\mathbf{v}}(\underline{\mathbf{R}},t)\cdot\underline{\underline{\mathbf{T}}}(\underline{\mathbf{R}},t) \tag{4.47}$$

according to

$$-\nabla\cdot\underline{\mathbf{S}}(\underline{\mathbf{R}},t) = \nabla\underline{\mathbf{v}}(\underline{\mathbf{R}},t) : \underline{\underline{\mathbf{T}}}(\underline{\mathbf{R}},t) + \nabla\cdot\underline{\underline{\mathbf{T}}}(\underline{\mathbf{R}},t)\cdot\underline{\mathbf{v}}(\underline{\mathbf{R}},t), \tag{4.48}$$

exploiting the symmetry of $\underline{\underline{\mathbf{T}}}$. The result is the energy conservation law of elastodynamics:

$$\frac{\partial w_{\mathrm{el}}(\underline{\mathbf{R}},t)}{\partial t} = -\nabla\cdot\underline{\mathbf{S}}(\underline{\mathbf{R}},t) + \frac{\partial w_Q(\underline{\mathbf{R}},t)}{\partial t} \tag{4.49}$$

for nondissipative materials. The minus sign in (4.47) is based on the following argument: If the vector $\underline{\mathbf{S}}(\underline{\mathbf{R}},t)$ should represent a physical energy flow density, e.g., an energy per time and per area: a surface power density, a locally positive divergence of $\underline{\mathbf{S}}(\underline{\mathbf{R}},t)$ refers to an "escape" of energy, that is to say, $\partial w_{\mathrm{el}}(\underline{\mathbf{R}},t)/\partial t$ must be negative if vanishing energy delivery is assumed: $\partial w_Q(\underline{\mathbf{R}},t)/\partial t \equiv 0$; correspondingly, a locally negative divergence of $\underline{\mathbf{S}}(\underline{\mathbf{R}},t)$ for $\partial w_Q(\underline{\mathbf{R}},t)/\partial t \equiv 0$ results in a local increase of energy density. The vector $\underline{\mathbf{S}}(\underline{\mathbf{R}},t)$ is the elastodynamic analogon to the Poynting vector for electromagnetic waves; hence, it is sometimes called the elastodynamic Poynting vector.

The above derivation of the energy conservation law starts from a physically plausible definition of elastodynamic energy density utilizing the governing equations of elastodynamics together with specially selected constitutive equations; as a consequence, a physically meaningful definition of the elastodynamic Poynting vector $\underline{\mathbf{S}}(\underline{\mathbf{R}},t)$ according to (4.47) arises; yet, $\underline{\mathbf{S}}(\underline{\mathbf{R}},t)$ is not uniquely defined that way because the curl of any arbitrary vector could be added without changing the energy conservation law. Nevertheless, the definition (4.47) of $\underline{\mathbf{S}}(\underline{\mathbf{R}},t)$ has always proved of value.

Proceeding conversely to the above derivation, the symmetry stipulation (4.43) for the compliance tensor is *mandatory* in order to formulate the

---

[69]We have

$$\nabla\cdot(\underline{\mathbf{v}}\cdot\underline{\underline{\mathbf{T}}}) = \nabla\underline{\mathbf{v}} : \underline{\underline{\mathbf{T}}} + \nabla\cdot\underline{\underline{\mathbf{T}}}^{21}\cdot\underline{\mathbf{v}}.$$

elastodynamic energy density consistently: With (4.48), we take the negative divergence of the elastodynamic Poynting vector $\underline{\mathbf{S}}(\underline{\mathbf{R}}, t)$ as postulated as an energy flow density, write $\nabla\underline{\mathbf{v}} : \underline{\underline{\mathbf{T}}}\ \frac{1}{2}[\nabla\underline{\mathbf{v}} + (\nabla\underline{\mathbf{v}})^{21}] : \underline{\underline{\mathbf{T}}}$ relying on the symmetry of $\underline{\underline{\mathbf{T}}}$, and insert the governing equations (3.1) and (3.2) without specifying constitutive equations:

$$-\nabla \cdot \underline{\mathbf{S}}(\underline{\mathbf{R}}, t) = \underbrace{\frac{\partial \underline{\underline{\mathbf{S}}}(\underline{\mathbf{R}}, t)}{\partial t} : \underline{\underline{\mathbf{T}}}(\underline{\mathbf{R}}, t) + \frac{\partial \underline{\mathbf{j}}(\underline{\mathbf{R}}, t)}{\partial t} \cdot \underline{\mathbf{v}}(\underline{\mathbf{R}}, t)}_{\stackrel{!}{=} \frac{\partial w_{\mathrm{el}}(\underline{\mathbf{R}}, t)}{\partial t}} \underbrace{- -\underline{\underline{\mathbf{h}}}(\underline{\mathbf{R}}, t) : \underline{\underline{\mathbf{T}}}(\underline{\mathbf{R}}, t) - \underline{\mathbf{f}}(\underline{\mathbf{R}}, t) \cdot \underline{\mathbf{v}}(\underline{\mathbf{R}}, t)}_{= -\frac{\partial w_Q(\underline{\mathbf{R}}, t)}{\partial t}}. \tag{4.50}$$

That way, $-\nabla \cdot \underline{\mathbf{S}}(\underline{\mathbf{R}}, t)$ defines the time variation of the elastodynamic energy density in a conservation law but not—like (4.36)—the energy density itself. To be consistent with (4.37) following from Equation 4.36, we have to claim the symmetry (4.43) under the assumption of the special constitutive equations (4.38) and (4.39). We emphasize that: The elastodynamic energy conservation law in the time domain enforces the symmetry $\underline{\underline{\underline{\underline{\mathbf{s}}}}}(\underline{\mathbf{R}})^{1234} = \underline{\underline{\underline{\underline{\mathbf{s}}}}}(\underline{\mathbf{R}})^{3412}$ of the compliance tensor and the respective symmetry $\underline{\underline{\underline{\underline{\mathbf{c}}}}}(\underline{\mathbf{R}})^{1234} = \underline{\underline{\underline{\underline{\mathbf{c}}}}}(\underline{\mathbf{R}})^{3412}$ of the stiffness tensor for nondissipative (time invariant instantaneously reacting) materials. For dissipative materials, (4.36) must indeed be modified.

4.3.2 Complex valued elastodynamic Poynting vector in the frequency domain

The elastodynamic Poynting vector in the time domain being defined as a product of two (real valued) time functions corresponds to a convolution integral in the frequency domain, more precisely: three convolution integrals for the three components of $\underline{\mathbf{S}}$:

$$\underline{\mathbf{S}}(\underline{\mathbf{R}}, \omega) = -\frac{1}{2\pi}\int_{-\infty}^{\infty} \underline{\mathbf{v}}(\underline{\mathbf{R}}, \omega') \cdot \underline{\underline{\mathbf{T}}}(\underline{\mathbf{R}}, \omega - \omega')\, \mathrm{d}\omega'. \tag{4.51}$$

Realizing that real valued time harmonic time functions have δ-functions as spectral lines and, hence, that the above convolution of δ-functions again results in δ-functions, we expect the Poynting vector of time harmonic fields to be equally time harmonic since Fourier inversion of a δ is time harmonic; the resulting phasor should then be proportional to the product of the phasors of $\underline{\mathbf{v}}(\underline{\mathbf{R}}, t)$ and $\underline{\underline{\mathbf{T}}}(\underline{\mathbf{R}}, t)$.

Therefore, we put [compare (3.17)]

$$\underline{\mathbf{v}}(\underline{\mathbf{R}}, t, \omega_0) = \Re\left\{\underline{\mathbf{v}}(\underline{\mathbf{R}}, \omega_0)\, \mathrm{e}^{-\mathrm{j}\omega_0 t}\right\}, \tag{4.52}$$

$$\underline{\underline{\mathbf{T}}}(\underline{\mathbf{R}}, t, \omega_0) = \Re\left\{\underline{\underline{\mathbf{T}}}(\underline{\mathbf{R}}, \omega_0)\, \mathrm{e}^{-\mathrm{j}\omega_0 t}\right\} \tag{4.53}$$

and calculate the spectra

$$\underline{\mathbf{v}}(\underline{\mathbf{R}}, \omega, \omega_0) = \pi\, \underline{\mathbf{v}}(\underline{\mathbf{R}}, \omega_0)\delta(\omega - \omega_0) + \pi\, \underline{\mathbf{v}}^*(\underline{\mathbf{R}}, \omega_0)\delta(\omega + \omega_0), \quad (4.54)$$

$$\underline{\underline{\mathbf{T}}}(\underline{\mathbf{R}}, \omega, \omega_0) = \pi\, \underline{\underline{\mathbf{T}}}(\underline{\mathbf{R}}, \omega_0)\delta(\omega - \omega_0) + \pi\, \underline{\underline{\mathbf{T}}}^*(\underline{\mathbf{R}}, \omega_0)\delta(\omega + \omega_0). \quad (4.55)$$

With (2.364), we compute according to (4.51):

$$\begin{aligned}
\underline{\mathbf{S}}(\underline{\mathbf{R}}, \omega, \omega_0) &= -\frac{1}{2\pi}\int_{-\infty}^{\infty} \underline{\mathbf{v}}(\underline{\mathbf{R}}, \omega', \omega_0) \cdot \underline{\underline{\mathbf{T}}}(\underline{\mathbf{R}}, \omega - \omega', \omega_0)\, d\omega' \\
&= -\frac{\pi}{2}\, \underline{\mathbf{v}}(\underline{\mathbf{R}}, \omega_0) \cdot \underline{\underline{\mathbf{T}}}(\underline{\mathbf{R}}, \omega_0)\, \underbrace{\delta(\omega - \omega_0) * \delta(\omega - \omega_0)}_{= \delta(\omega - 2\omega_0)} \\
&\quad - \frac{\pi}{2}\, \underline{\mathbf{v}}(\underline{\mathbf{R}}, \omega_0) \cdot \underline{\underline{\mathbf{T}}}^*(\underline{\mathbf{R}}, \omega_0)\, \underbrace{\delta(\omega - \omega_0) * \delta(\omega + \omega_0)}_{= \delta(\omega)} \\
&\quad - \frac{\pi}{2}\, \underline{\mathbf{v}}^*(\underline{\mathbf{R}}, \omega_0) \cdot \underline{\underline{\mathbf{T}}}(\underline{\mathbf{R}}, \omega_0)\, \underbrace{\delta(\omega + \omega_0) * \delta(\omega - \omega_0)}_{= \delta(\omega)} \\
&\quad - \frac{\pi}{2}\, \underline{\mathbf{v}}^*(\underline{\mathbf{R}}, \omega_0) \cdot \underline{\underline{\mathbf{T}}}^*(\underline{\mathbf{R}}, \omega_0)\, \underbrace{\delta(\omega + \omega_0) * \delta(\omega + \omega_0)}_{= \delta(\omega + 2\omega_0)}, \quad (4.56)
\end{aligned}$$

and indeed obtain three spectral lines at $\omega = 0, \pm 2\omega_0$. Consequently, the Poynting vector for time harmonic fields reads as

$$\begin{aligned}
\underline{\mathbf{S}}(\underline{\mathbf{R}}, t, \omega_0) &= \Re\left\{-\frac{1}{2}\, \underline{\mathbf{v}}(\underline{\mathbf{R}}, \omega_0) \cdot \underline{\underline{\mathbf{T}}}^*(\underline{\mathbf{R}}, \omega_0)\right\} \\
&\quad + \Re\left\{-\frac{1}{2}\, \underline{\mathbf{v}}(\underline{\mathbf{R}}, \omega_0) \cdot \underline{\underline{\mathbf{T}}}(\underline{\mathbf{R}}, \omega_0)\, e^{-2j\omega_0 t}\right\}. \quad (4.57)
\end{aligned}$$

In case of time averaging

$$\frac{1}{T_0}\int_0^{T_0} \underline{\mathbf{S}}(\underline{\mathbf{R}}, t, \omega_0)\, dt = \Re\left\{-\frac{1}{2}\, \underline{\mathbf{v}}(\underline{\mathbf{R}}, \omega_0) \cdot \underline{\underline{\mathbf{T}}}^*(\underline{\mathbf{R}}, \omega_0)\right\}, \quad (4.58)$$

the term oscillating with $2\omega_0$ in (4.57) vanishes because

$$\int_0^{T_0} e^{-2j\omega_0 t}\, dt = 0, \quad (4.59)$$

and only the dc-term prevails. The result of time averaging is the real part of the phasor

$$\underline{\mathbf{S}}_{\mathrm{K}}(\underline{\mathbf{R}}, \omega_0) = -\frac{1}{2}\, \underline{\mathbf{v}}(\underline{\mathbf{R}}, \omega_0) \cdot \underline{\underline{\mathbf{T}}}^*(\underline{\mathbf{R}}, \omega_0) \quad (4.60)$$

as product of phasors that can obviously be regarded as a spectral alternative to (4.47), allowing ω_0 to be an arbitrary frequency[70] ω if the phasors are identified as spectral amplitudes according to (3.20). We call

$$\underline{\mathbf{S}}_{\mathrm{K}}(\underline{\mathbf{R}}, \omega) = -\frac{1}{2}\,\underline{\mathbf{v}}(\underline{\mathbf{R}}, \omega) \cdot \underline{\underline{\mathbf{T}}}^*(\underline{\mathbf{R}}, \omega) \tag{4.61}$$

the complex elastodynamic Poynting vector, and we know with (4.58) that its real part gives us the time averaged energy density flow of real valued time harmonic fields.

To formulate an energy conservation law for time averages of real valued time harmonic fields, we first have to define the corresponding time harmonic energy density of time harmonic fields on the basis of (4.57) analogous to (4.57)

$$\begin{aligned} w_{\mathrm{el}}(\underline{\mathbf{R}}, t, \omega_0) &= \Re\left\{\frac{1}{4}\,\underline{\mathbf{j}}(\underline{\mathbf{R}}, \omega_0) \cdot \underline{\mathbf{v}}^*(\underline{\mathbf{R}}, \omega_0) + \frac{1}{4}\,\underline{\underline{\mathbf{S}}}(\underline{\mathbf{R}}, \omega_0) : \underline{\underline{\mathbf{T}}}^*(\underline{\mathbf{R}}, \omega_0)\right\} \\ &\quad + \Re\left\{\left[\frac{1}{4}\,\underline{\mathbf{j}}(\underline{\mathbf{R}}, \omega_0) \cdot \underline{\mathbf{v}}(\underline{\mathbf{R}}, \omega_0) + \frac{1}{4}\,\underline{\underline{\mathbf{S}}}(\underline{\mathbf{R}}, \omega_0) : \underline{\underline{\mathbf{T}}}(\underline{\mathbf{R}}, \omega_0)\right] \mathrm{e}^{-2\mathrm{j}\omega_0 t}\right\}. \end{aligned} \tag{4.62}$$

Taking the time derivative of (4.62) and subsequently the time average

$$\frac{1}{T_0} \int_0^{T_0} \frac{\partial w_{\mathrm{el}}(\underline{\mathbf{R}}, t, \omega_0)}{\partial t}\, \mathrm{d}t \equiv 0, \tag{4.63}$$

we find it to be always zero, independent of any postulated constitutive equations. For vanishing phasors of the volume force density $\underline{\mathbf{f}}(\underline{\mathbf{R}}, \omega_0)$ and the deformation rate $\underline{\underline{\mathbf{h}}}(\underline{\mathbf{R}}, \omega_0)$, averaging the energy conservation law (4.49) yields

$$\nabla \cdot \Re\left\{\underline{\mathbf{S}}_{\mathrm{K}}(\underline{\mathbf{R}}, \omega_0)\right\} = 0 \tag{4.64}$$

for time harmonic fields. This is surprising because the material could be dissipative, and this should have a locally negative divergence of the time averaged energy flow density as consequence. Yet, the result (4.64) is definitely plausible for nondissipative materials as described by the constitutive equations (4.38) and (4.39). It follows that: The definition (4.62)—and (4.36)—is by no way the ultimate wisdom for dissipative materials.

[70]In electrical engineering, the factor 1/2 is sometimes deleted defining effective values of phasors, e.g.,

$$\underline{\mathbf{v}}_{\mathrm{eff}}(\underline{\mathbf{R}}, \omega_0) = \frac{\underline{\mathbf{v}}(\underline{\mathbf{R}}, \omega_0)}{\sqrt{2}}.$$

By the way, starting from $\underline{\mathbf{S}}(\underline{\mathbf{R}}, t) = -\underline{\underline{\mathbf{T}}}(\underline{\mathbf{R}}, t) \cdot \underline{\mathbf{v}}(\underline{\mathbf{R}}, t)$ would have resulted in $\underline{\mathbf{S}}_{\mathrm{K}}(\underline{\mathbf{R}}, \omega_0) = -\underline{\underline{\mathbf{T}}}(\underline{\mathbf{R}}, \omega_0) \cdot \underline{\mathbf{v}}^*(\underline{\mathbf{R}}, \omega_0)/2$, the complex conjugate value; yet, the physically meaningful real part of $\underline{\mathbf{S}}_{\mathrm{K}}(\underline{\mathbf{R}}, \omega_0)$ remains unchanged.

As before in the time domain, we start from (4.50) and calculate the mean values for time harmonic fields:

$$
\begin{aligned}
&-\nabla\cdot\Re\{\underline{\mathbf{S}}_{\mathrm{K}}(\underline{\mathbf{R}},\omega_0)\}\\
&=\Re\left\{-\frac{\mathrm{j}\omega_0}{2}\,\underline{\underline{\mathbf{S}}}(\underline{\mathbf{R}},\omega_0):\underline{\underline{\mathbf{T}}}^*(\underline{\mathbf{R}},\omega_0)-\frac{\mathrm{j}\omega_0}{2}\,\underline{\mathbf{j}}(\underline{\mathbf{R}},\omega_0)\cdot\underline{\mathbf{v}}^*(\underline{\mathbf{R}},\omega_0)\right\}\\
&\quad-\Re\left\{\frac{1}{2}\,\underline{\underline{\mathbf{h}}}(\underline{\mathbf{R}},\omega_0):\underline{\underline{\mathbf{T}}}^*(\underline{\mathbf{R}},\omega_0)+\frac{1}{2}\,\underline{\mathbf{f}}(\underline{\mathbf{R}},\omega_0)\cdot\underline{\mathbf{v}}^*(\underline{\mathbf{R}},\omega_0)\right\},
\end{aligned}
\tag{4.65}
$$

where we compute the phasor, for example, of $\partial\underline{\underline{\mathbf{S}}}(\underline{\mathbf{R}},t,\omega_0)/\partial t$, according to

$$
\frac{\partial\underline{\underline{\mathbf{S}}}(\underline{\mathbf{R}},t,\omega_0)}{\partial t}=\Re\left\{-\mathrm{j}\omega_0\,\underline{\underline{\mathbf{S}}}(\underline{\mathbf{R}},\omega_0)\,\mathrm{e}^{-\mathrm{j}\omega_0 t}\right\};
\tag{4.66}
$$

for the sake of completeness, we also take prescribed force densities and deformation rates into account. Yet, even for $\underline{\mathbf{f}}=\underline{\mathbf{0}}$, $\underline{\underline{\mathbf{h}}}=\underline{\underline{\mathbf{0}}}$, there may be "something left" on the right-hand side of Equation 4.65. Yet, the constitutive equations

$$
\underline{\mathbf{j}}(\underline{\mathbf{R}},\omega_0)=\rho(\underline{\mathbf{R}})\underline{\mathbf{v}}(\underline{\mathbf{R}},\omega_0),
\tag{4.67}
$$

$$
\underline{\underline{\mathbf{S}}}(\underline{\mathbf{R}},\omega_0)=\underline{\underline{\underline{\underline{\mathbf{s}}}}}(\underline{\mathbf{R}}):\underline{\underline{\mathbf{T}}}(\underline{\mathbf{R}},\omega_0)
\tag{4.68}
$$

of a nondissipative material at frequency ω_0 should yield—as in Equation 4.64—

$$
\Re\left\{-\frac{\mathrm{j}\omega_0}{2}\,\underline{\underline{\underline{\underline{\mathbf{s}}}}}(\underline{\mathbf{R}}):\underline{\underline{\mathbf{T}}}(\underline{\mathbf{R}},\omega_0):\underline{\underline{\mathbf{T}}}^*(\underline{\mathbf{R}},\omega_0)-\frac{\mathrm{j}\omega_0}{2}\,\rho(\underline{\mathbf{R}})\underline{\mathbf{v}}(\underline{\mathbf{R}},\omega_0)\cdot\underline{\mathbf{v}}^*(\underline{\mathbf{R}},\omega_0)\right\}=0.
\tag{4.69}
$$

Since $\underline{\mathbf{v}}(\underline{\mathbf{R}},\omega_0)\cdot\underline{\mathbf{v}}^*(\underline{\mathbf{R}},\omega_0)=|\underline{\mathbf{v}}(\underline{\mathbf{R}},\omega_0)|^2$ is always real, we only have to check the first term. We can write

$$
\Re z=\frac{1}{2}(z+z^*)
\tag{4.70}
$$

for the real part of a complex number, and therefore we should have

$$
-\frac{\mathrm{j}\omega_0}{4}\,\underline{\underline{\underline{\underline{\mathbf{s}}}}}(\underline{\mathbf{R}}):\underline{\underline{\mathbf{T}}}(\underline{\mathbf{R}},\omega_0):\underline{\underline{\mathbf{T}}}^*(\underline{\mathbf{R}},\omega_0)+\frac{\mathrm{j}\omega_0}{4}\,\underline{\underline{\underline{\underline{\mathbf{s}}}}}(\underline{\mathbf{R}}):\underline{\underline{\mathbf{T}}}^*(\underline{\mathbf{R}},\omega_0):\underline{\underline{\mathbf{T}}}(\underline{\mathbf{R}},\omega_0)=0,
\tag{4.71}
$$

and this is true if $\underline{\underline{\underline{\underline{\mathbf{s}}}}}(\underline{\mathbf{R}})$ satisfies the symmetry

$$
\underline{\underline{\underline{\underline{\mathbf{s}}}}}(\underline{\mathbf{R}})^{1234}=\underline{\underline{\underline{\underline{\mathbf{s}}}}}(\underline{\mathbf{R}})^{3412}.
\tag{4.72}
$$

Under the assumption (4.72), the elastodynamic energy conservation law for real valued time harmonic fields and for nondissipative materials results as a time average:

$$
\nabla\cdot\Re\{\underline{\mathbf{S}}_{\mathrm{K}}(\underline{\mathbf{R}},\omega_0)\}=\Re\left\{\frac{1}{2}\,\underline{\mathbf{f}}(\underline{\mathbf{R}},\omega_0)\cdot\underline{\mathbf{v}}^*(\underline{\mathbf{R}},\omega_0)+\frac{1}{2}\,\underline{\underline{\mathbf{h}}}(\underline{\mathbf{R}},\omega_0):\underline{\underline{\mathbf{T}}}^*(\underline{\mathbf{R}},\omega_0)\right\}.
\tag{4.73}
$$

For the time averaged energy density

$$\frac{1}{T_0}\int_0^{T_0} w_{\rm el}(\underline{\mathbf{R}}, t, \omega_0)\,{\rm d}t = \langle w_{\rm el}(\underline{\mathbf{R}}, t, \omega_0)\rangle \tag{4.74}$$

of real valued time harmonic fields, we consequently obtain

$$\begin{aligned}\langle w_{\rm el}(\underline{\mathbf{R}}, t, \omega_0)\rangle &= \frac{\rho(\underline{\mathbf{R}})}{4}\,|\underline{\mathbf{v}}(\underline{\mathbf{R}}, \omega_0)|^2 + \frac{1}{4}\,\underline{\underline{\underline{\underline{\mathbf{s}}}}}(\underline{\mathbf{R}}) : \underline{\underline{\mathbf{T}}}(\underline{\mathbf{R}}, \omega_0) : \underline{\underline{\mathbf{T}}}^*(\underline{\mathbf{R}}, \omega_0)o\\ &= \frac{\rho(\underline{\mathbf{R}})}{4}\,|\underline{\mathbf{v}}(\underline{\mathbf{R}}, \omega_0)|^2 + \frac{1}{4}\,\underline{\underline{\mathbf{S}}}(\underline{\mathbf{R}}, \omega_0) : \underline{\underline{\underline{\underline{\mathbf{c}}}}}(\underline{\mathbf{R}}) : \underline{\underline{\mathbf{S}}}^*(\underline{\mathbf{R}}, \omega_0),\end{aligned} \tag{4.75}$$

because under the assumption (4.72), the expression $\underline{\underline{\underline{\underline{\mathbf{s}}}}}(\underline{\mathbf{R}}) : \underline{\underline{\mathbf{T}}}(\underline{\mathbf{R}}, \omega_0) : \underline{\underline{\mathbf{T}}}^*(\underline{\mathbf{R}}, \omega_0)$ and, hence, $\underline{\underline{\mathbf{S}}}(\underline{\mathbf{R}}, \omega_0) : \underline{\underline{\underline{\underline{\mathbf{c}}}}}(\underline{\mathbf{R}}) : \underline{\underline{\mathbf{S}}}^*(\underline{\mathbf{R}}, \omega_0)$ are always real.

Replacing phasors by spectral amplitudes, we have (4.73) for the Fourier spectra:

$$\nabla \cdot \Re\{\underline{\mathbf{S}}_{\rm K}(\underline{\mathbf{R}}, \omega)\} = \Re\left\{\frac{1}{2}\,\underline{\mathbf{f}}(\underline{\mathbf{R}}, \omega) \cdot \underline{\mathbf{v}}^*(\underline{\mathbf{R}}, \omega) + \frac{1}{2}\,\underline{\underline{\mathbf{h}}}(\underline{\mathbf{R}}, \omega) : \underline{\underline{\mathbf{T}}}^*(\underline{\mathbf{R}}, \omega)\right\}. \tag{4.76}$$

4.4 Linear Dissipative Materials

Several models for elastodynamic dissipation are discussed in the literature (e.g., Auld 1973; Ben-Menahem and Singh 1981; de Hoop 1995). Here, we do not aim at their physical basis, but we especially discuss the consequences of any kind of dissipation for the propagation of elastic waves.

4.4.1 Maxwell model

Maxwell model: As a complete formal analogon to the conductivity energy losses of electromagnetic fields—for instance, as given by Ohm's law: $\underline{\mathbf{J}}_l(\underline{\mathbf{R}}, t) = \underline{\underline{\sigma}}_{\rm e}(\underline{\mathbf{R}}) \cdot \underline{\mathbf{E}}(\underline{\mathbf{R}}, t)$—we postulate respective losses of elastodynamic energy by the additional linear Maxwell terms $\underline{\underline{\mathbf{K}}}(\underline{\mathbf{R}}) \cdot \underline{\mathbf{v}}(\underline{\mathbf{R}}, t)$ and $\underline{\underline{\underline{\underline{\Gamma}}}}(\underline{\mathbf{R}}) : \underline{\underline{\mathbf{T}}}(\underline{\mathbf{R}}, t)$ in the constitutive equations (4.9) and (4.10) (Ben-Menahem and Singh 1981; de Hoop 1995):

$$\frac{\partial \underline{\mathbf{j}}(\underline{\mathbf{R}}, t)}{\partial t} = \underline{\underline{\rho}}(\underline{\mathbf{R}}) \cdot \frac{\partial \underline{\mathbf{v}}(\underline{\mathbf{R}}, t)}{\partial t} + \underline{\underline{\mathbf{K}}}(\underline{\mathbf{R}}) \cdot \underline{\mathbf{v}}(\underline{\mathbf{R}}, t), \tag{4.77}$$

$$\frac{\partial \underline{\underline{\mathbf{S}}}(\underline{\mathbf{R}}, t)}{\partial t} = \underline{\underline{\underline{\underline{\mathbf{s}}}}}(\underline{\mathbf{R}}) : \frac{\partial \underline{\underline{\mathbf{T}}}(\underline{\mathbf{R}}, t)}{\partial t} + \underline{\underline{\underline{\underline{\Gamma}}}}(\underline{\mathbf{R}}) : \underline{\underline{\mathbf{T}}}(\underline{\mathbf{R}}, t). \tag{4.78}$$

Here, $\underline{\underline{\mathbf{K}}}(\underline{\mathbf{R}})$ is the second rank tensor coefficient of a friction force and $\underline{\underline{\underline{\underline{\Gamma}}}}(\underline{\mathbf{R}})$ the forth rank tensor coefficient of an inverse viscosity called inviscidness that—per definitionem—must satisfy the symmetries $\underline{\underline{\underline{\underline{\Gamma}}}}^{1234} = \underline{\underline{\underline{\underline{\Gamma}}}}^{1243} = \underline{\underline{\underline{\underline{\Gamma}}}}^{2143} = \underline{\underline{\underline{\underline{\Gamma}}}}^{2134}$; $\underline{\underline{\mathbf{K}}}$ must not satisfy any symmetry conditions.

Governing equations of elastodynamics; attenuation and dispersion of plane waves: With the constitutive equations (4.77) and (4.78), the elastodynamic governing equations read as:

$$\underline{\underline{\rho}}(\underline{\mathbf{R}}) \cdot \frac{\partial \underline{\mathbf{v}}(\underline{\mathbf{R}}, t)}{\partial t} = \nabla \cdot \underline{\underline{\mathbf{T}}}(\underline{\mathbf{R}}, t) - \underline{\underline{\mathbf{K}}}(\underline{\mathbf{R}}) \cdot \underline{\mathbf{v}}(\underline{\mathbf{R}}, t) + \underline{\mathbf{f}}(\underline{\mathbf{R}}, t), \tag{4.79}$$

$$\underline{\underline{\underline{\underline{\mathbf{s}}}}}(\underline{\mathbf{R}}) : \frac{\partial \underline{\underline{\mathbf{T}}}(\underline{\mathbf{R}}, t)}{\partial t} = \underline{\underline{\underline{\underline{\mathbf{I}}}}}^{+} : \nabla \underline{\mathbf{v}}(\underline{\mathbf{R}}, t) - \underline{\underline{\underline{\underline{\Gamma}}}}(\underline{\mathbf{R}}) : \underline{\underline{\mathbf{T}}}(\underline{\mathbf{R}}, t) + \underline{\underline{\mathbf{h}}}(\underline{\mathbf{R}}, t), \tag{4.80}$$

where it is clear that $\underline{\underline{\mathbf{K}}} \cdot \underline{\mathbf{v}}$ is in fact a field-induced force density term that, due to the negative sign, counters the prescribed driving force density, i.e., it decelerates. Similarly, $\underline{\underline{\underline{\underline{\Gamma}}}} : \underline{\underline{\mathbf{T}}}$ is directed opposite to the prescribed deformation rate.

Another time derivative of (4.79) and subsequent insertion of (4.80) yields Equation 7.3 augmented by dissipation terms:

$$\begin{aligned}
&\nabla \cdot \underline{\underline{\underline{\underline{\mathbf{c}}}}}(\underline{\mathbf{R}}) : \nabla \underline{\mathbf{v}}(\underline{\mathbf{R}}, t) - \underline{\underline{\rho}}(\underline{\mathbf{R}}) \cdot \frac{\partial^2 \underline{\mathbf{v}}(\underline{\mathbf{R}}, t)}{\partial t^2} - \underline{\underline{\mathbf{K}}}(\underline{\mathbf{R}}) \cdot \frac{\partial \underline{\mathbf{v}}(\underline{\mathbf{R}}, t)}{\partial t} \\
&\quad - \nabla \cdot \underline{\underline{\underline{\underline{\mathbf{c}}}}}(\underline{\mathbf{R}}) : \underline{\underline{\underline{\underline{\Gamma}}}}(\underline{\mathbf{R}}) : \underline{\underline{\mathbf{T}}}(\underline{\mathbf{R}}, t) \\
&\quad = -\frac{\partial \underline{\mathbf{f}}(\underline{\mathbf{R}}, t)}{\partial t} - \nabla \cdot \underline{\underline{\underline{\underline{\mathbf{c}}}}}(\underline{\mathbf{R}}) : \underline{\underline{\mathbf{h}}}(\underline{\mathbf{R}}, t).
\end{aligned} \tag{4.81}$$

To survey the consequences of dissipation terms for the propagation of plane waves, we consider the homogeneous equation (4.81) simplifying it by putting the $\underline{\underline{\underline{\underline{\Gamma}}}}$-term equal to zero:

$$\nabla \cdot \underline{\underline{\underline{\underline{\mathbf{c}}}}}(\underline{\mathbf{R}}) : \nabla \underline{\mathbf{v}}(\underline{\mathbf{R}}, t) - \underline{\underline{\rho}}(\underline{\mathbf{R}}) \cdot \frac{\partial^2 \underline{\mathbf{v}}(\underline{\mathbf{R}}, t)}{\partial t^2} - \underline{\underline{\mathbf{K}}}(\underline{\mathbf{R}}) \cdot \frac{\partial \underline{\mathbf{v}}(\underline{\mathbf{R}}, t)}{\partial t} = \underline{\mathbf{0}}; \tag{4.82}$$

after Fourier transforming with regard to t, we can combine $\underline{\underline{\rho}}(\underline{\mathbf{R}})$ and $\underline{\underline{\mathbf{K}}}(\underline{\mathbf{R}})$ to a complex (frequency-dependent) material tensor $\underline{\underline{\rho}}_c(\underline{\mathbf{R}})$:

$$\nabla \cdot \underline{\underline{\underline{\underline{\mathbf{c}}}}}(\underline{\mathbf{R}}) : \nabla \underline{\mathbf{v}}(\underline{\mathbf{R}}, \omega) + \omega^2 \underbrace{\left[\underline{\underline{\rho}}(\underline{\mathbf{R}}) + \mathrm{j} \frac{\underline{\underline{\mathbf{K}}}(\underline{\mathbf{R}})}{\omega} \right]}_{= \underline{\underline{\rho}}_c(\underline{\mathbf{R}})} \cdot \underline{\mathbf{v}}(\underline{\mathbf{R}}, \omega) = \underline{\mathbf{0}} \tag{4.83}$$

We specialize to a homogeneous isotropic dissipative material according to[71]

[71] In (4.85), K is not the magnitude of the Fourier vector $\underline{\mathbf{K}}$ but the scalar friction coefficient $\underline{\underline{\mathbf{K}}} = K \, \underline{\underline{\mathbf{I}}}$.

$$\underline{\underline{\underline{\underline{\mathbf{c}}}}}(\mathbf{R}) = \lambda\,\underline{\underline{\underline{\underline{\mathbf{I}}}}}^{\delta} + 2\mu\,\underline{\underline{\underline{\underline{\mathbf{I}}}}}^{+}, \tag{4.84}$$

$$\begin{aligned} \underline{\underline{\rho}}_c(\mathbf{R}) &= \left(\rho + \mathrm{j}\frac{K}{\omega}\right)\underline{\underline{\mathbf{I}}} \\ &= \rho_c\underline{\underline{\mathbf{I}}} \end{aligned} \tag{4.85}$$

and choose, as in Section 8.1, the z-axis as propagation direction of a plane wave with $\partial/\partial x \equiv \partial/\partial y \equiv 0$; taking components of (4.83) yields

$$(\lambda + 2\mu)\frac{\mathrm{d}^2 v_z(z,\omega)}{\mathrm{d}z^2} + \omega^2\rho_c v_z(z,\omega) = 0, \tag{4.86}$$

$$\mu\,\frac{\mathrm{d}^2 v_{x,y}(z,\omega)}{\mathrm{d}z^2} + \omega^2\rho_c v_{x,y}(z,\omega) = 0 \tag{4.87}$$

as generalization of (8.9) and (8.35) to dissipative materials. The definition of complex wave numbers

$$\begin{aligned} k_{\mathrm{P}c}(\omega) &= \omega\sqrt{\frac{\rho_c}{\lambda + 2\mu}} \\ &= \Re k_{\mathrm{P}c}(\omega) + \mathrm{j}\Im k_{\mathrm{P}c}(\omega), \end{aligned} \tag{4.88}$$

$$\begin{aligned} k_{\mathrm{S}c}(\omega) &= \omega\sqrt{\frac{\rho_c}{\mu}} \\ &= \Re k_{\mathrm{S}c}(\omega) + \mathrm{j}\Im k_{\mathrm{S}c}(\omega) \end{aligned} \tag{4.89}$$

reveals that plane ω_0-time harmonic primary and secondary waves [compare (8.24) and (8.45)]

$$v_z(z,t) = v_z(\omega_0)\,\mathrm{e}^{\mathrm{j}z\Re k_{\mathrm{P}c}(\omega_0)}\mathrm{e}^{-z\Im k_{\mathrm{P}c}(\omega_0)}\mathrm{e}^{-\mathrm{j}\omega_0 t}, \tag{4.90}$$

$$v_{x,y}(z,t) = v_{x,y}(\omega_0)\,\mathrm{e}^{\mathrm{j}z\Re k_{\mathrm{S}c}(\omega_0)}\mathrm{e}^{-z\Im k_{\mathrm{S}c}(\omega_0)}\mathrm{e}^{-\mathrm{j}\omega_0 t} \tag{4.91}$$

propagating, for example, in $+z$-direction are now exponentially attenuated in propagation direction for $z > 0$ according to the imaginary parts of the complex wave numbers (provided the correct—positive!—sign of imaginary parts has been chosen). The physical origin of the attenuation is the friction coefficient K in ρ_c, it is responsible for dissipation.

Besides dissipation, K is also conveyed to dispersion because the complex wave numbers are no longer proportional to frequency as it was true for the lossless case; the phase velocities

$$c_{\mathrm{P,S}}(\omega_0) = \frac{\omega_0}{\Re k_{\mathrm{P,S}c}(\omega_0)} \tag{4.92}$$

become frequency dependent in dissipative materials! Consequently, an impulse traveling in a dissipative material experiences a distortion with increasing[72] z. Apropos impulse propagation: To apply an inverse Fourier

[72] The numerical example for electromagnetic Maxwell-dispersion in Langenberg (2005) can be quantitatively assigned to the present elastodynamic case.

transform leading into the time domain, we consider frequency dependent functions, for example $\rho_c(\omega)$, as complex spectra; yet, their time functions are only physically realistic if they are causal and contain finite energy. According to Section 2.3.4, we must assume that $\Re\rho_c(\omega)$ and $\Im\rho_c(\omega)$ fulfill Kramers–Kronig relations, i.e., they should be Hilbert transform pairs. As a matter of fact, this is not true for the Maxwell model (4.85) (Langenberg 2005), hence it may only be used within limited frequency ranges.

4.4.2 Elastodynamic energy conservation law: Dissipation energy

Elastodynamic energy conservation law in the time domain: Elastodynamic dissipation must reflect itself in an elastodynamic energy conservation law. We refer to (4.50) and insert the Maxwell constitutive equations (4.77) and (4.78) [we immediately exploit the symmetry (4.15) of $\underline{\underline{\underline{\underline{\mathbf{s}}}}}$ and $\underline{\underline{\underline{\underline{\Gamma}}}}$]:

$$
\begin{aligned}
-\nabla\cdot\underline{\mathbf{S}}(\underline{\mathbf{R}},t) = {} & \underbrace{\underline{\underline{\mathbf{T}}}(\underline{\mathbf{R}},t):\underline{\underline{\underline{\underline{\mathbf{s}}}}}(\underline{\mathbf{R}}):\frac{\partial\underline{\underline{\mathbf{T}}}(\underline{\mathbf{R}},t)}{\partial t}+\underline{\mathbf{v}}(\underline{\mathbf{R}},t)\cdot\underline{\underline{\boldsymbol{\rho}}}(\underline{\mathbf{R}})\cdot\frac{\partial\underline{\mathbf{v}}(\underline{\mathbf{R}},t)}{\partial t}}_{=\frac{\partial w_{\mathrm{el}}(\underline{\mathbf{R}},t)}{\partial t}}\\
&+\underbrace{\underline{\underline{\mathbf{T}}}(\underline{\mathbf{R}},t):\underline{\underline{\underline{\underline{\Gamma}}}}(\underline{\mathbf{R}}):\underline{\underline{\mathbf{T}}}(\underline{\mathbf{R}},t)+\underline{\mathbf{v}}(\underline{\mathbf{R}},t)\cdot\underline{\underline{\mathbf{K}}}\cdot\underline{\mathbf{v}}(\underline{\mathbf{R}},t)}_{=\frac{\partial w_{\mathrm{d}}(\underline{\mathbf{R}},t)}{\partial t}}\\
&\underbrace{-\,\underline{\underline{\mathbf{h}}}(\underline{\mathbf{R}},t):\underline{\underline{\mathbf{T}}}(\underline{\mathbf{R}},t)-\underline{\mathbf{f}}(\underline{\mathbf{R}},t)\cdot\underline{\mathbf{v}}(\underline{\mathbf{R}},t)}_{=-\frac{\partial w_Q(\underline{\mathbf{R}},t)}{\partial t}}.
\end{aligned}
\tag{4.93}
$$

The first bottom bracket in (4.93) defines the time variation of the (nondissipative) instantaneous elastodynamic energy density[73] and the second bottom bracket apparently defines the time variation of the elastodynamic dissipation energy density. Combining terms according to

$$
\frac{\partial w(\underline{\mathbf{R}},t)}{\partial t}=\frac{\partial w_{\mathrm{el}}(\underline{\mathbf{R}},t)}{\partial t}+\frac{\partial w_{\mathrm{d}}(\underline{\mathbf{R}},t)}{\partial t}
\tag{4.94}
$$

as the time variation of the total energy density, we obtain the concise formulation of the energy conservation law:

$$
\frac{\partial w_Q(\underline{\mathbf{R}},t)}{\partial t}-\nabla\cdot\underline{\mathbf{S}}(\underline{\mathbf{R}},t)=\frac{\partial w(\underline{\mathbf{R}},t)}{\partial t}.
\tag{4.95}
$$

[73]That can be calculated via time derivation of the (nondissipative) instantaneous elastodynamic energy density itself (Equation 4.40), considering the symmetry of $\underline{\underline{\boldsymbol{\rho}}}$ and the symmetry (4.43) of $\underline{\underline{\underline{\underline{\mathbf{s}}}}}$.

The sum of a positive local energy flow and the time variation of energy induced from outside results in a local increase of the total energy density, namely the increase of kinetic and deformation energy density as well as the replacement of losses due to dissipation. As a dissipation term, $\partial w_{\mathrm{d}}(\underline{\mathbf{R}}, t)/\partial t \geq 0$ must be positive; this yields the requirement of non-negative definiteness of the tensors $\underline{\underline{\mathbf{K}}}$ and $\underline{\underline{\underline{\underline{\Gamma}}}}$ and [not the symmetries $\underline{\underline{\mathbf{K}}} = \underline{\underline{\mathbf{K}}}^{21}$ and $\underline{\underline{\underline{\underline{\Gamma}}}}^{1234} = \underline{\underline{\underline{\underline{\Gamma}}}}^{3412}$ as claimed by Auld (1973)].

From (4.94) follows (4.93) for causal fields via time integration, recognizing the symmetries $\underline{\underline{\boldsymbol{\rho}}} = \underline{\underline{\boldsymbol{\rho}}}^{21}$, $\underline{\underline{\underline{\underline{\mathbf{s}}}}}^{1234} = \underline{\underline{\underline{\underline{\mathbf{s}}}}}^{3412}$:

$$
\begin{aligned}
w(\underline{\mathbf{R}}, t) = {} & \frac{1}{2}\, \underline{\mathbf{v}}(\underline{\mathbf{R}}, t) \cdot \underline{\underline{\boldsymbol{\rho}}}(\underline{\mathbf{R}}) \cdot \underline{\mathbf{v}}(\underline{\mathbf{R}}, t) + \frac{1}{2}\, \underline{\underline{\mathbf{T}}}(\underline{\mathbf{R}}, t) : \underline{\underline{\underline{\underline{\mathbf{s}}}}}(\underline{\mathbf{R}}) : \underline{\underline{\mathbf{T}}}(\underline{\mathbf{R}}, t) \\
& + \int_0^t \left[\underline{\underline{\mathbf{T}}}(\underline{\mathbf{R}}, \tau) : \underline{\underline{\underline{\underline{\Gamma}}}}(\underline{\mathbf{R}}) : \underline{\underline{\mathbf{T}}}(\underline{\mathbf{R}}, \tau) + \underline{\mathbf{v}}(\underline{\mathbf{R}}, \tau) \cdot \underline{\underline{\mathbf{K}}}(\underline{\mathbf{R}}) \cdot \underline{\mathbf{v}}(\underline{\mathbf{R}}, \tau) \right] \mathrm{d}\tau.
\end{aligned}
\tag{4.96}
$$

That is to say, the elastodynamic energy density (4.40) as instantaneous energy density of nondissipative materials must be amended by the dissipation energy density for dissipative materials; the definition (4.96) replaces the definition (4.36) for actual constitutive equations of dissipative materials. As a consequence, the contradiction formulated with (4.63) is resolved writing (4.96) for real time harmonic fields with subsequent time averaging.

Elastodynamic conservation law in the frequency domain: For real time harmonic fields, the time averaging of Poynting's theorem (4.93) results in [compare (4.65)]

$$
\begin{aligned}
& -\nabla \cdot \Re\{\underline{\mathbf{S}}_{\mathrm{K}}(\underline{\mathbf{R}}, \omega_0)\} \\
& = \Re\left\{ -\frac{\mathrm{j}\omega_0}{2}\, \underline{\underline{\mathbf{T}}}(\underline{\mathbf{R}}, \omega_0) : \underline{\underline{\underline{\underline{\mathbf{s}}}}}(\underline{\mathbf{R}}) : \underline{\underline{\mathbf{T}}}^*(\underline{\mathbf{R}}, \omega_0) - \frac{\mathrm{j}\omega_0}{2}\, \underline{\mathbf{v}}(\underline{\mathbf{R}}, \omega_0) \cdot \underline{\underline{\boldsymbol{\rho}}}(\underline{\mathbf{R}}) \cdot \underline{\mathbf{v}}^*(\underline{\mathbf{R}}, \omega_0) \right\} \\
& \quad + \Re\left\{ \frac{1}{2}\, \underline{\underline{\mathbf{T}}}(\underline{\mathbf{R}}, \omega_0) : \underline{\underline{\underline{\underline{\Gamma}}}}(\underline{\mathbf{R}}) : \underline{\underline{\mathbf{T}}}^*(\underline{\mathbf{R}}, \omega_0) + \frac{1}{2}\, \underline{\mathbf{v}}(\underline{\mathbf{R}}, \omega_0) \cdot \underline{\underline{\mathbf{K}}}(\underline{\mathbf{R}}) \cdot \underline{\mathbf{v}}^*(\underline{\mathbf{R}}, \omega_0) \right\} \\
& \quad - \Re\left\{ \frac{1}{2}\, \underline{\underline{\mathbf{h}}}(\underline{\mathbf{R}}, \omega_0) : \underline{\underline{\mathbf{T}}}^*(\underline{\mathbf{R}}, \omega_0) + \frac{1}{2}\, \underline{\mathbf{f}}(\underline{\mathbf{R}}, \omega_0) \cdot \underline{\mathbf{v}}^*(\underline{\mathbf{R}}, \omega_0) \right\}
\end{aligned}
\tag{4.97}
$$

for the phasors and the Fourier spectra, respectively. On behalf of the symmetries $\underline{\underline{\boldsymbol{\rho}}} = \underline{\underline{\boldsymbol{\rho}}}^{21}$, $\underline{\underline{\underline{\underline{\mathbf{s}}}}}^{1234} = \underline{\underline{\underline{\underline{\mathbf{s}}}}}^{3412}$, the first term on the right-hand side of (4.97) is equal to zero leading to the elastodynamic energy conservation law in the frequency domain:

$$
\begin{aligned}
& \frac{1}{2}\, \Re\left\{ \underline{\underline{\mathbf{h}}}(\underline{\mathbf{R}}, \omega) : \underline{\underline{\mathbf{T}}}^*(\underline{\mathbf{R}}, \omega) + \underline{\mathbf{f}}(\underline{\mathbf{R}}, \omega) \cdot \underline{\mathbf{v}}^*(\underline{\mathbf{R}}, \omega) \right\} - \nabla \cdot \Re\{\underline{\mathbf{S}}_{\mathrm{K}}(\underline{\mathbf{R}}, \omega)\} \\
& \quad = \frac{1}{2}\, \Re\left\{ \underline{\underline{\mathbf{T}}}(\underline{\mathbf{R}}, \omega) : \underline{\underline{\underline{\underline{\Gamma}}}}(\underline{\mathbf{R}}) : \underline{\underline{\mathbf{T}}}^*(\underline{\mathbf{R}}, \omega) + \underline{\mathbf{v}}(\underline{\mathbf{R}}, \omega) \cdot \underline{\underline{\mathbf{K}}}(\underline{\mathbf{R}}) \cdot \underline{\mathbf{v}}^*(\underline{\mathbf{R}}, \omega) \right\}.
\end{aligned}
\tag{4.98}
$$

Evidently, the equation

$$\left\langle \frac{\partial w(\underline{\mathbf{R}}, t, \omega_0)}{\partial t} \right\rangle = \left\langle \frac{\partial w_{\mathrm{d}}(\underline{\mathbf{R}}, t, \omega_0)}{\partial t} \right\rangle \tag{4.99}$$

resolves the contradiction (4.63).

4.4.3 Rayleigh and Kelvin–Voigt model

The (linear) Rayleigh model postulates dissipative constitutive equations

$$\underline{\mathbf{j}}(\underline{\mathbf{R}}, t) = \underline{\underline{\boldsymbol{\rho}}}(\underline{\mathbf{R}}) \cdot \underline{\mathbf{v}}(\underline{\mathbf{R}}, t) + \underline{\underline{\mathbf{K}}}(\underline{\mathbf{R}}) \cdot \underline{\mathbf{u}}(\underline{\mathbf{R}}, t), \tag{4.100}$$

$$\underline{\underline{\mathbf{T}}}(\underline{\mathbf{R}}, t) = \underline{\underline{\underline{\underline{\mathbf{c}}}}}(\underline{\mathbf{R}}) : \underline{\underline{\mathbf{S}}}(\underline{\mathbf{R}}, t) + \underline{\underline{\underline{\underline{\boldsymbol{\eta}}}}}(\underline{\mathbf{R}}) : \frac{\partial \underline{\underline{\mathbf{S}}}(\underline{\mathbf{R}}, t)}{\partial t}, \tag{4.101}$$

where $\underline{\underline{\mathbf{K}}}(\underline{\mathbf{R}})$, as for the Maxwell model, is a tensor friction coefficient and $\underline{\underline{\underline{\underline{\boldsymbol{\eta}}}}}(\underline{\mathbf{R}})$ a tensor viscosity that satisfies the symmetries $\underline{\underline{\underline{\underline{\boldsymbol{\eta}}}}}^{1234} = \underline{\underline{\underline{\underline{\boldsymbol{\eta}}}}}^{1243} = \underline{\underline{\underline{\underline{\boldsymbol{\eta}}}}}^{2143} = \underline{\underline{\underline{\underline{\boldsymbol{\eta}}}}}^{2134}$ because $\underline{\underline{\mathbf{S}}}$ and $\underline{\underline{\mathbf{T}}}$ are symmetric. If we write (4.101) according to the "normal form"

$$\underline{\underline{\mathbf{S}}}(\underline{\mathbf{R}}, t) = \underline{\underline{\underline{\underline{\mathbf{s}}}}}(\underline{\mathbf{R}}) : \underline{\underline{\mathbf{T}}}(\underline{\mathbf{R}}, t) - \underline{\underline{\underline{\underline{\boldsymbol{\tau}}}}}(\underline{\mathbf{R}}) : \frac{\partial \underline{\underline{\mathbf{S}}}(\underline{\mathbf{R}}, t)}{\partial t} \tag{4.102}$$

of a constitutive equation, the relaxation tensor arises (its elements have the dimension of time)

$$\underline{\underline{\underline{\underline{\boldsymbol{\tau}}}}}(\underline{\mathbf{R}}) = \underline{\underline{\underline{\underline{\mathbf{s}}}}}(\underline{\mathbf{R}}) : \underline{\underline{\underline{\underline{\boldsymbol{\eta}}}}}(\underline{\mathbf{R}}) \tag{4.103}$$

with the symmetries $\underline{\underline{\underline{\underline{\boldsymbol{\tau}}}}}^{1234} = \underline{\underline{\underline{\underline{\boldsymbol{\tau}}}}}^{1243} = \underline{\underline{\underline{\underline{\boldsymbol{\tau}}}}}^{2143} = \underline{\underline{\underline{\underline{\boldsymbol{\tau}}}}}^{2134}$, suggesting to call (4.102) a Kelvin–Voigt relaxation model (Ben-Menahem and Singh 1981).

With the definition of the deformation rate $\partial \underline{\underline{\mathbf{S}}}(\underline{\mathbf{R}}, t)/\partial t = \underline{\underline{\underline{\underline{\mathbf{I}}}}}^{+} : \boldsymbol{\nabla} \underline{\mathbf{v}}(\underline{\mathbf{R}}, t)$, we obtain the governing equations

$$\underline{\underline{\boldsymbol{\rho}}}(\underline{\mathbf{R}}) \cdot \frac{\partial \underline{\mathbf{v}}(\underline{\mathbf{R}}, t)}{\partial t} = \boldsymbol{\nabla} \cdot \underline{\underline{\mathbf{T}}}(\underline{\mathbf{R}}, t) - \underline{\underline{\mathbf{K}}}(\underline{\mathbf{R}}) \cdot \underline{\mathbf{v}}(\underline{\mathbf{R}}, t) + \underline{\mathbf{f}}(\underline{\mathbf{R}}, t), \tag{4.104}$$

$$\underline{\underline{\underline{\underline{\mathbf{s}}}}}(\underline{\mathbf{R}}) : \frac{\partial \underline{\underline{\mathbf{T}}}(\underline{\mathbf{R}}, t)}{\partial t} = \underline{\underline{\underline{\underline{\mathbf{I}}}}}^{+} : \boldsymbol{\nabla} \underline{\mathbf{v}}(\underline{\mathbf{R}}, t) + \underline{\underline{\underline{\underline{\boldsymbol{\tau}}}}}(\underline{\mathbf{R}}) : \boldsymbol{\nabla} \frac{\partial \underline{\mathbf{v}}(\underline{\mathbf{R}}, t)}{\partial t} + \underline{\underline{\mathbf{h}}}(\underline{\mathbf{R}}, t) \tag{4.105}$$

for Rayleigh-dissipative materials (4.100) and (4.102). Inserting the constitutive equations into (4.50) defines the time variation of the dissipation energy density analogously to (4.93)

$$\frac{\partial w_{\mathrm{d}}(\underline{\mathbf{R}}, t)}{\partial t} = -\underline{\underline{\mathbf{T}}}(\underline{\mathbf{R}}, t) : \underline{\underline{\underline{\underline{\boldsymbol{\tau}}}}}(\underline{\mathbf{R}}) : \boldsymbol{\nabla} \frac{\partial \underline{\mathbf{v}}(\underline{\mathbf{R}}, t)}{\partial t} + \underline{\mathbf{v}}(\underline{\mathbf{R}}, t) \cdot \underline{\underline{\mathbf{K}}}(\underline{\mathbf{R}}) \cdot \underline{\mathbf{v}}(\underline{\mathbf{R}}, t). \tag{4.106}$$

Insertion of the governing equations into each other generates the differential equation

$$\begin{aligned}
&\nabla \cdot \underline{\underline{\underline{\underline{c}}}}(\underline{\mathbf{R}}) : \nabla \underline{\mathbf{v}}(\underline{\mathbf{R}}, t) + \nabla \cdot \underline{\underline{\underline{\underline{\eta}}}}(\underline{\mathbf{R}}) : \nabla \frac{\partial \underline{\mathbf{v}}(\underline{\mathbf{R}}, t)}{\partial t} - \underline{\underline{\mathbf{K}}}(\underline{\mathbf{R}}) \cdot \frac{\partial \underline{\mathbf{v}}(\underline{\mathbf{R}}, t)}{\partial t} \\
&\quad - \underline{\underline{\rho}}(\underline{\mathbf{R}}) \cdot \frac{\partial^2 \underline{\mathbf{v}}(\underline{\mathbf{R}}, t)}{\partial t^2} \\
&\quad = -\frac{\partial \underline{\mathbf{f}}(\underline{\mathbf{R}}, t)}{\partial t} - \nabla \cdot \underline{\underline{\underline{\underline{c}}}}(\underline{\mathbf{R}}) : \underline{\underline{\mathbf{h}}}(\underline{\mathbf{R}}, t)
\end{aligned} \tag{4.107}$$

for the particle velocity that has the formal advantage over the special Kelvin–Voigt relaxation (4.81) to be decoupled from $\underline{\underline{\mathbf{T}}}(\underline{\mathbf{R}}, t)$. The Fourier transform with regard to t

$$\begin{aligned}
&\nabla \cdot \left[\underline{\underline{\underline{\underline{c}}}}(\underline{\mathbf{R}}) - \mathrm{j}\omega \underline{\underline{\underline{\underline{\eta}}}}(\underline{\mathbf{R}}) \right] : \nabla \underline{\mathbf{v}}(\underline{\mathbf{R}}, \omega) + \omega^2 \left[\underline{\underline{\rho}}(\underline{\mathbf{R}}) + \mathrm{j} \frac{\underline{\underline{\mathbf{K}}}(\underline{\mathbf{R}})}{\omega} \right] \cdot \underline{\mathbf{v}}(\underline{\mathbf{R}}, \omega) \\
&\quad = \mathrm{j}\omega \underline{\mathbf{f}}(\underline{\mathbf{R}}, \omega) - \nabla \cdot \underline{\underline{\underline{\underline{c}}}}(\underline{\mathbf{R}}) : \underline{\underline{\mathbf{h}}}(\underline{\mathbf{R}}, \omega)
\end{aligned} \tag{4.108}$$

reveals once again that dissipation terms yield complex frequency-dependent material parameters. Consequence: Elastic waves in Rayleigh-dissipative materials experience attenuation and dispersion. Marklein (1997) offers a detailed elaboration of plane waves with homogeneous isotropic Kelvin–Voigt dissipation ($\underline{\underline{\mathbf{K}}} = \underline{\underline{\mathbf{0}}}$).

4.4.4 Relaxation models

It was Boltzmann who already proposed relaxation models

$$\frac{\partial \underline{\mathbf{j}}(\underline{\mathbf{R}}, t)}{\partial t} = \underline{\underline{\rho}}(\underline{\mathbf{R}}) \cdot \frac{\partial \underline{\mathbf{v}}(\underline{\mathbf{R}}, t)}{\partial t} + \int_0^t \underline{\underline{\mu}}(\underline{\mathbf{R}}, t - \tau) \cdot \frac{\partial \underline{\mathbf{v}}(\underline{\mathbf{R}}, \tau)}{\partial \tau} \, \mathrm{d}\tau, \tag{4.109}$$

$$\frac{\partial \underline{\underline{\mathbf{S}}}(\underline{\mathbf{R}}, t)}{\partial t} = \underline{\underline{\underline{\underline{s}}}}(\underline{\mathbf{R}}) : \frac{\partial \underline{\underline{\mathbf{T}}}(\underline{\mathbf{R}}, t)}{\partial t} + \int_0^t \underline{\underline{\underline{\underline{\chi}}}}(\underline{\mathbf{R}}, t - \tau) : \frac{\partial \underline{\underline{\mathbf{T}}}(\underline{\mathbf{R}}, \tau)}{\partial \tau} \, \mathrm{d}\tau \tag{4.110}$$

as dissipation terms in linear constitutive equations (Ben-Menahem and Singh 1981; de Hoop 1995); here, $\underline{\underline{\mu}}(\underline{\mathbf{R}}, t)$ denotes the tensor inertia and $\underline{\underline{\underline{\underline{\chi}}}}(\underline{\mathbf{R}}, t)$ the tensor compliance kernel within the respective convolution integral; for $\underline{\underline{\underline{\underline{\chi}}}}$, the usual commutation of the first and the last two indices holds. Note: Postulating the reasonable causality of fields and relaxation kernels, the limits of the convolution integrals become 0 and t, that is to say, $\partial \underline{\mathbf{j}}(\underline{\mathbf{R}}, t)/\partial t$ and $\partial \underline{\underline{\mathbf{S}}}(\underline{\mathbf{R}}, t)/\partial t$ depend at time t only upon the past of the fields and not upon the future. In contrast to the instantaneous terms, the relaxation terms in (4.109) and (4.110) represent a noninstantaneously reacting material, yet it is still time invariant because the relaxation kernels are functions of $t - \tau$ and not of "t comma τ". Insertion of (4.109) and (4.110) into (4.50) identifies the noninstantaneous reaction as dissipation yielding the nonnegative definiteness of the tensors $\underline{\underline{\mu}}$ and $\underline{\underline{\underline{\chi}}}$.

Causal relaxation kernels $\underline{\underline{\mu}}(\underline{\mathbf{R}}, t)$, $\underline{\underline{\underline{\chi}}}(\underline{\mathbf{R}}, t)$ with finite energy (Section 2.3.4) must have complex spectra $\underline{\underline{\mu}}(\underline{\mathbf{R}}, \omega)$, $\underline{\underline{\underline{\chi}}}(\underline{\mathbf{R}}, \omega)$, real and imaginary parts of the tensor components must even be Hilbert transform pairs, Kramers–Kronig relations must hold. Therefore, the relaxation models again yield complex (frequency-dependent) material parameters to describe dissipation (and dispersion) in the frequency domain, yet in contrast to the Maxwell and Rayleigh–Kelvin–Voigt models, they are physically consistent.

Starting from basic physical principles—for instance, causality and finite energy—Karlsson and Kristensson (1992) were able to prove for the electromagnetic case that relaxation models represent the most general form of linear dissipative constitutive equations and they derive the properties of the respective material tensor functions.

4.5 Piezoelectricity and Magnetostriction

4.5.1 Piezoelectricity

Piezoelectric effect: A rather extensive elaboration of elastic–electromagnetic (piezoelectric) waves in piezoelectric materials together with numerous references can be found in Marklein (1997).

A crystal shows the piezoelectric effect if exterior mechanical stresses resulting in deformations create electric charge densities. A precondition is that the crystal does not exhibit a symmetry center (Figure 4.1); therefore, it is always anisotropic.

Consequently, the creation of electrical stresses resulting in deformations through electrical forces—electrical field strengths—is called the inverse piezoelectric effect.

Piezoelectric governing and constitutive equations: The piezoelectric effect relates mechanical and electrical field quantities, resulting in governing equations as a combination of the elastodynamic governing equations (3.1) and (3.2) as well as Maxwell's equations (6.1) through (6.4); always concentrating on time-dependent phenomena, we have to account for the complete Maxwell equations, we only put magnetic source terms equal to zero:

$$\frac{\partial \underline{\mathbf{j}}(\underline{\mathbf{R}}, t)}{\partial t} = \nabla \cdot \underline{\underline{\mathbf{T}}}(\underline{\mathbf{R}}, t) + \underline{\mathbf{f}}(\underline{\mathbf{R}}, t), \tag{4.111}$$

$$\frac{\partial \underline{\underline{\mathbf{S}}}(\underline{\mathbf{R}}, t)}{\partial t} = \underline{\underline{\underline{\underline{\mathbf{I}}}}}^{+} : \nabla \underline{\mathbf{v}}(\underline{\mathbf{R}}, t) + \underline{\underline{\mathbf{h}}}(\underline{\mathbf{R}}, t), \tag{4.112}$$

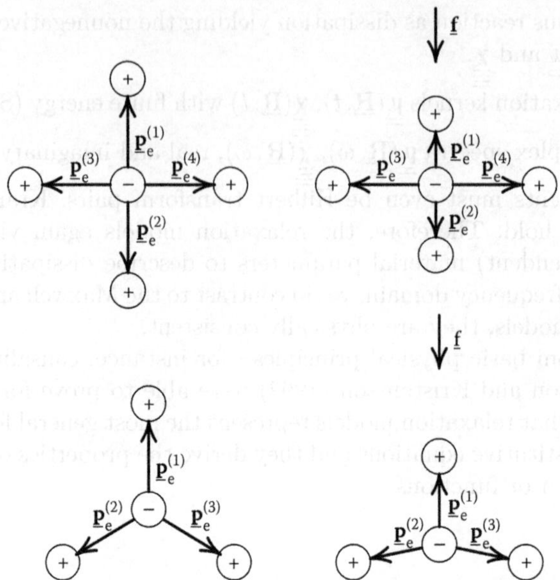

FIGURE 4.1
Crystal models with (top) and without (bottom) symmetry center. (With a center, the sum of all electric dipole moments $\underline{\mathbf{p}}_e^{(i)}$ with and without mechanical stresses is always zero; without a symmetry center, a finite dipole moment may result from mechanical stresses.)

$$\frac{\partial \underline{\mathbf{D}}(\underline{\mathbf{R}}, t)}{\partial t} = \nabla \times \underline{\mathbf{H}}(\underline{\mathbf{R}}, t) - \underline{\mathbf{J}}_e(\underline{\mathbf{R}}, t), \tag{4.113}$$

$$\frac{\partial \underline{\mathbf{B}}(\underline{\mathbf{R}}, t)}{\partial t} = -\nabla \times \underline{\mathbf{E}}(\underline{\mathbf{R}}, t), \tag{4.114}$$

$$\nabla \cdot \underline{\mathbf{D}}(\underline{\mathbf{R}}, t) = \varrho_e(\underline{\mathbf{R}}, t), \tag{4.115}$$

$$\nabla \cdot \underline{\mathbf{B}}(\underline{\mathbf{R}}, t) = 0. \tag{4.116}$$

As usual, a system of governing equations has to be complemented by constitutive equations that should, in the case of piezoelectricity, result in a coupling of elastic and electromagnetic waves: We are talking about piezoelectric waves.

For a physical justification of piezoelectric constitutive equations, we briefly refer to the physical background of the constitutive equation $\underline{\mathbf{D}}(\underline{\mathbf{R}}, t) = \epsilon_0 \underline{\underline{\epsilon}}_r(\underline{\mathbf{R}}) \cdot \underline{\mathbf{E}}(\underline{\mathbf{R}}, t)$ (Equation 6.51); it expresses the electric polarizability of matter via electric Coulomb forces (Equation 6.16). The result of this microscopic polarizability, that is to say, the mutual displacement of (positive) ions and electrons or the orientation of already existing statistically distributed electric dipole moments, is the macroscopic polarization vector $\underline{\mathbf{P}}_e(\underline{\mathbf{R}}, t)$ with polarization volume charges created by Coulomb forces as negative sources:

$$\nabla \cdot \underline{\mathbf{P}}_e(\underline{\mathbf{R}}, t) = -\varrho_{\mathrm{Pol}}(\underline{\mathbf{R}}, t). \tag{4.117}$$

In vacuum, the electric field strength $\underline{\mathbf{E}}(\underline{\mathbf{R}}, t)$ and the electric flux density $\underline{\mathbf{D}}(\underline{\mathbf{R}}, t)$ solely express different concepts to characterize the presence of electric charges (Sommerfeld 1964); therefore, they are related by the electric field constant of vacuum ϵ_0 with the physical dimension charge per voltage per length (Equation 6.10). However, in an electrically polarizable material, we define

$$\underline{\mathbf{D}}(\underline{\mathbf{R}}, t) = \epsilon_0 \underline{\mathbf{E}}(\underline{\mathbf{R}}, t) + \underline{\mathbf{P}}_{\mathrm{e}}(\underline{\mathbf{R}}, t); \tag{4.118}$$

with the linear *ansatz*

$$\underline{\mathbf{P}}_{\mathrm{e}}(\underline{\mathbf{R}}, t) = \epsilon_0 \underline{\underline{\boldsymbol{\chi}}}_{\mathrm{e}}(\underline{\mathbf{R}}) \cdot \underline{\mathbf{E}}(\underline{\mathbf{R}}, t) \tag{4.119}$$

of a time invariant instantaneously and locally reacting electrically polarizable inhomogeneous anisotropic material, we obtain the permittivity tensor $\underline{\underline{\boldsymbol{\epsilon}}}_r(\underline{\mathbf{R}})$ according to

$$\underline{\underline{\boldsymbol{\epsilon}}}_r(\underline{\mathbf{R}}) = \underline{\underline{\mathbf{I}}} + \underline{\underline{\boldsymbol{\chi}}}_{\mathrm{e}}(\underline{\mathbf{R}}) \tag{4.120}$$

resulting from the electric susceptibility $\underline{\underline{\boldsymbol{\chi}}}_{\mathrm{e}}(\underline{\mathbf{R}})$.

Corresponding to the direct piezoelectric effect, additional electric volume charges $\varrho_{\mathrm{Pi}}(\underline{\mathbf{R}}, t)$ can be created by exterior deformations (stresses) representing negative sources of the piezoelectric polarization vector:

$$\boldsymbol{\nabla} \cdot \underline{\mathbf{P}}_{\mathrm{Pi}}(\underline{\mathbf{R}}, t) = -\varrho_{\mathrm{Pi}}(\underline{\mathbf{R}}, t). \tag{4.121}$$

With the repeated linear *ansatz*

$$\underline{\mathbf{P}}_{\mathrm{Pi}}(\underline{\mathbf{R}}, t) = \underline{\underline{\underline{\mathbf{e}}}}(\underline{\mathbf{R}}) : \underline{\underline{\mathbf{S}}}(\underline{\mathbf{R}}, t) \tag{4.122}$$

of a time invariant instantaneously and locally piezoelectric polarizable inhomogeneous anisotropic material, the constitutive equation of the direct piezoelectric effect results:

$$\underline{\mathbf{D}}(\underline{\mathbf{R}}, t) = \epsilon_0 \underline{\underline{\boldsymbol{\epsilon}}}_r^{\underline{\underline{\mathbf{S}}}}(\underline{\mathbf{R}}) \cdot \underline{\mathbf{E}}(\underline{\mathbf{R}}, t) + \underline{\underline{\underline{\mathbf{e}}}}(\underline{\mathbf{R}}) : \underline{\underline{\mathbf{S}}}(\underline{\mathbf{R}}, t). \tag{4.123}$$

Here, the third rank tensor $\underline{\underline{\underline{\mathbf{e}}}}(\underline{\mathbf{R}})$ contains the (adiabatic) piezoelectric stress constants that, on behalf of the symmetry of $\underline{\underline{\mathbf{S}}}$, satisfies the symmetry relation

$$\underline{\underline{\underline{\mathbf{e}}}}(\underline{\mathbf{R}}) = \underline{\underline{\underline{\mathbf{e}}}}^{132}(\underline{\mathbf{R}}); \tag{4.124}$$

the first index of $\underline{\underline{\underline{\mathbf{e}}}}(\underline{\mathbf{R}})$ is sort of the electric index. The permittivity tensor $\underline{\underline{\boldsymbol{\epsilon}}}_r^{\underline{\underline{\mathbf{S}}}}(\underline{\mathbf{R}})$ in (4.123) receives the upper index $\underline{\underline{\mathbf{S}}}$: It is measured for constant deformation tensor (and constant entropy: adiabatic).

As an alternative to (4.123), we can consider the exterior stresses as origin of the direct piezoelectric effect:

$$\underline{\mathbf{D}}(\underline{\mathbf{R}}, t) = \epsilon_0 \underline{\underline{\boldsymbol{\epsilon}}}_r^{\underline{\underline{\mathbf{T}}}}(\underline{\mathbf{R}}) \cdot \underline{\mathbf{E}}(\underline{\mathbf{R}}, t) + \underline{\underline{\underline{\mathbf{d}}}}(\underline{\mathbf{R}}) : \underline{\underline{\mathbf{T}}}(\underline{\mathbf{R}}, t), \tag{4.125}$$

where the third rank tensor $\underline{\underline{\underline{\mathbf{d}}}}(\underline{\mathbf{R}})$ contains the (adiabatic) piezoelectric strain constants; it satisfies the symmetry

$$\underline{\underline{\underline{\mathbf{d}}}}(\underline{\mathbf{R}}) = \underline{\underline{\underline{\mathbf{d}}}}^{132}(\underline{\mathbf{R}}). \tag{4.126}$$

The permittivity tensor $\underline{\underline{\boldsymbol{\epsilon}}}_r^{\mathbf{T}}(\underline{\mathbf{R}})$ must be measured for constant stress tensor (and constant entropy).

Postulating Hooke's law (4.14) between $\underline{\underline{\mathbf{T}}}$ and $\underline{\underline{\mathbf{S}}}$ for vanishing electric field strength, we have:

$$\underline{\underline{\underline{\mathbf{d}}}}(\underline{\mathbf{R}}) = \underline{\underline{\underline{\mathbf{e}}}}(\underline{\mathbf{R}}) : \underline{\underline{\underline{\underline{\mathbf{s}}}}}^{\mathbf{E}}(\underline{\mathbf{R}}) \tag{4.127}$$

or, respectively,

$$\underline{\underline{\underline{\mathbf{e}}}}(\underline{\mathbf{R}}) = \underline{\underline{\underline{\mathbf{d}}}}(\underline{\mathbf{R}}) : \underline{\underline{\underline{\underline{\mathbf{c}}}}}^{\mathbf{E}}(\underline{\mathbf{R}}). \tag{4.128}$$

Therefore, the requirement of constant electric field strength to measure the compliance tensor $\underline{\underline{\underline{\underline{\mathbf{s}}}}}^{\mathbf{E}}(\underline{\mathbf{R}})$ is a consequence of the "piezoelectrically augmented" Hooke law

$$\underline{\underline{\mathbf{S}}}(\underline{\mathbf{R}}, t) = \underline{\underline{\underline{\underline{\mathbf{s}}}}}^{\mathbf{E}}(\underline{\mathbf{R}}) : \underline{\underline{\mathbf{T}}}(\underline{\mathbf{R}}, t) + \underline{\underline{\underline{\mathbf{d}}}}^{231}(\underline{\mathbf{R}}) \cdot \underline{\mathbf{E}}(\underline{\mathbf{R}}, t) \tag{4.129}$$

as a formulation of the inverse piezoelectric effect that also exhibits the piezoelectric strain constants; however, the "electric index" of $\underline{\underline{\underline{\mathbf{d}}}}$ has to be contracted with $\underline{\mathbf{E}}$. The "stress tensor alternative" to (4.129) reads as

$$\underline{\underline{\mathbf{T}}}(\underline{\mathbf{R}}, t) = \underline{\underline{\underline{\underline{\mathbf{c}}}}}^{\mathbf{E}}(\underline{\mathbf{R}}) : \underline{\underline{\mathbf{S}}}(\underline{\mathbf{R}}, t) - \underline{\underline{\underline{\mathbf{e}}}}^{231}(\underline{\mathbf{R}}) \cdot \underline{\mathbf{E}}(\underline{\mathbf{R}}, t), \tag{4.130}$$

where the minus sign becomes plausible if (4.130) is inserted into (4.129) and if (4.127) and the symmetries of $\underline{\underline{\underline{\mathbf{e}}}}$, $\underline{\underline{\underline{\underline{\mathbf{s}}}}}^{\mathbf{E}}$, and $\underline{\underline{\underline{\underline{\mathbf{c}}}}}^{\mathbf{E}}$ as well as (4.19) are utilized.

With the constitutive equation (4.129) of the inverse and the constitutive equation (4.123) of the direct piezoelectric effect as well as

$$\underline{\mathbf{j}}(\underline{\mathbf{R}}, t) = \rho(\underline{\mathbf{R}})\underline{\mathbf{v}}(\underline{\mathbf{R}}, t), \tag{4.131}$$

$$\underline{\mathbf{B}}(\underline{\mathbf{R}}, t) = \mu_0 \underline{\mathbf{H}}(\underline{\mathbf{R}}, t), \tag{4.132}$$

the piezoelectric governing equations read

$$\rho(\underline{\mathbf{R}}) \frac{\partial \underline{\mathbf{v}}(\underline{\mathbf{R}}, t)}{\partial t} = \nabla \cdot \underline{\underline{\mathbf{T}}}(\underline{\mathbf{R}}, t) + \underline{\mathbf{f}}(\underline{\mathbf{R}}, t), \tag{4.133}$$

$$\underline{\underline{\underline{\underline{\mathbf{s}}}}}^{\mathbf{E}}(\underline{\mathbf{R}}) : \frac{\partial \underline{\underline{\mathbf{T}}}(\underline{\mathbf{R}}, t)}{\partial t} = \underline{\underline{\underline{\underline{\mathbf{I}}}}}^{+} : \nabla \underline{\mathbf{v}}(\underline{\mathbf{R}}, t) - \underline{\underline{\underline{\mathbf{d}}}}^{231}(\underline{\mathbf{R}}) \cdot \frac{\partial \underline{\mathbf{E}}(\underline{\mathbf{R}}, t)}{\partial t} + \underline{\underline{\mathbf{h}}}(\underline{\mathbf{R}}, t), \tag{4.134}$$

$$\epsilon_0 \underline{\underline{\boldsymbol{\epsilon}}}_r^{\mathbf{S}}(\underline{\mathbf{R}}) \cdot \frac{\partial \underline{\mathbf{E}}(\underline{\mathbf{R}}, t)}{\partial t} = \nabla \times \underline{\mathbf{H}}(\underline{\mathbf{R}}, t) - \underline{\underline{\underline{\mathbf{e}}}}(\underline{\mathbf{R}}) : \frac{\partial \underline{\underline{\mathbf{S}}}(\underline{\mathbf{R}}, t)}{\partial t} - \underline{\mathbf{J}}_{\mathrm{e}}(\underline{\mathbf{R}}, t), \tag{4.135}$$

$$\mu_0 \frac{\partial \underline{\mathbf{H}}(\underline{\mathbf{R}}, t)}{\partial t} = -\nabla \times \underline{\mathbf{E}}(\underline{\mathbf{R}}, t), \tag{4.136}$$

$$\epsilon_0 \nabla \cdot \left[\underline{\underline{\epsilon}}_r^{\mathbf{S}}(\underline{\mathbf{R}}) \cdot \underline{\mathbf{E}}(\underline{\mathbf{R}}, t) \right] = -\nabla \left[\underline{\underline{\mathbf{e}}}(\underline{\mathbf{R}}) : \underline{\underline{\mathbf{S}}}(\underline{\mathbf{R}}, t) \right] + \varrho_e(\underline{\mathbf{R}}, t), \tag{4.137}$$

$$\nabla \cdot \underline{\mathbf{H}}(\underline{\mathbf{R}}, t) = 0. \tag{4.138}$$

Evidently, this is a coupled elastodynamic electromagnetic system of governing equations.

Homogeneous piezoelectric wave equations: The coupling term $\underline{\underline{\underline{\mathbf{d}}}}^{231}(\underline{\mathbf{R}}) \cdot \partial \underline{\mathbf{E}}(\underline{\mathbf{R}}, t)/\partial t$ in (4.134) may be interpreted as electromagnetically induced negative source of deformation rate; equally, $\underline{\underline{\underline{\mathbf{e}}}}(\underline{\mathbf{R}}) \cdot \partial \underline{\underline{\mathbf{S}}}(\underline{\mathbf{R}}, t)/\partial t$ is an elastodynamically induced electric current density with a divergence that actually is, according to a continuity equation, equal to the negative time derivative of the elastodynamically induced electric volume charge density appearing in (4.137).

To define piezoelectric plane waves, we consider a homogeneous anisotropic piezoelectric material; with the above cited physical interpretation of the coupling terms in (4.134) and (4.135), we can immediately refer to (7.15) and (6.75) to write down the following homogeneous wave equations:

$$\nabla \cdot \underline{\underline{\underline{\underline{\mathbf{c}}}}}^{\mathbf{E}} : \nabla \underline{\mathbf{v}}(\underline{\mathbf{R}}, t) - \rho \frac{\partial^2 \underline{\mathbf{v}}(\underline{\mathbf{R}}, t)}{\partial t^2} = \nabla \cdot \underline{\underline{\underline{\mathbf{e}}}}^{231} \cdot \frac{\partial \underline{\mathbf{E}}(\underline{\mathbf{R}}, t)}{\partial t}, \tag{4.139}$$

$$-\nabla \times \nabla \times \underline{\mathbf{E}}(\underline{\mathbf{R}}, t) - \epsilon_0 \mu_0 \underline{\underline{\epsilon}}_r^{\mathbf{S}} \cdot \frac{\partial^2 \underline{\mathbf{E}}(\underline{\mathbf{R}}, t)}{\partial t^2} = \mu_0 \underline{\underline{\underline{\mathbf{e}}}} : \underline{\underline{\underline{\underline{\mathbf{I}}}}}^{+} : \nabla \frac{\partial \underline{\mathbf{v}}(\underline{\mathbf{R}}, t)}{\partial t}; \tag{4.140}$$

we have used $\underline{\underline{\underline{\mathbf{e}}}}^{231} = \underline{\underline{\underline{\underline{\mathbf{c}}}}}^{\mathbf{E}} : \underline{\underline{\underline{\mathbf{d}}}}^{231}$, necessitating (4.128) and the symmetries of $\underline{\underline{\underline{\mathbf{e}}}}$, $\underline{\underline{\underline{\mathbf{d}}}}$, and $\underline{\underline{\underline{\underline{\mathbf{c}}}}}^{\mathbf{E}}$. The system of coupled wave equations (4.139) and (4.140) has now to be solved.

Electroquasistatic approximation of the piezoelectric wave equations: Elastic and electromagnetic (monochromatic) waves exhibit rather different wavelengths for the same frequency because the phase velocities differ by several orders of magnitude. However, the dimensions of piezoelectric devices (piezoelectric transducers) are matched to elastic waves; hence, they are generally much smaller than the wavelengths of electromagnetic waves; for this reason, Equations 4.139 and 4.140 are advantageously solved with the electroquasistatic (EQS) approximation neglecting the time derivative of the vector potential in the representation (6.112) of the electric field strength:

$$\underline{\mathbf{E}}(\underline{\mathbf{R}}, t) = -\nabla \Phi(\underline{\mathbf{R}}, t). \tag{4.141}$$

Therefore, the source-free divergence equation (4.137) immediately yields via time derivation:

$$\epsilon_0 \nabla \cdot \underline{\underline{\epsilon}}_r^{\mathbf{S}} \cdot \nabla \frac{\partial \Phi(\underline{\mathbf{R}}, t)}{\partial t} = \nabla \cdot \underline{\underline{\underline{\mathbf{e}}}} : \underline{\underline{\underline{\underline{\mathbf{I}}}}}^{+} : \nabla \underline{\mathbf{v}}(\underline{\mathbf{R}}, t); \tag{4.142}$$

within the frame of the EQS approximation, this equation is equivalent to equation (4.140) if the divergence of the latter is calculated and a time integration (of causal fields) is performed. Note: Equation 4.142 is Poisson's equation of electrostatics (Blume 1991) for the time derivative of a time-dependent scalar potential.

Equation 4.139 within the quasistatic approximation

$$\nabla \cdot \underline{\underline{\underline{\underline{\mathbf{c}}}}}^{\mathbf{E}} : \nabla \underline{\mathbf{v}}(\underline{\mathbf{R}}, t) - \rho \frac{\partial^2 \underline{\mathbf{v}}(\underline{\mathbf{R}}, t)}{\partial t^2} = -\nabla \cdot \underline{\underline{\underline{\mathbf{e}}}}^{231} \cdot \nabla \frac{\partial \Phi(\underline{\mathbf{R}}, t)}{\partial t} \tag{4.143}$$

represents—together with (4.142)—the electroelastically coupled system of differential equations of EQS-approximated piezoelectric waves. As usual, the Fourier transform with respect to time leads to the corresponding time harmonic equations

$$\nabla \cdot \underline{\underline{\underline{\underline{\mathbf{c}}}}}^{\mathbf{E}} : \nabla \underline{\mathbf{v}}(\underline{\mathbf{R}}, \omega) + \rho\omega^2 \underline{\mathbf{v}}(\underline{\mathbf{R}}, \omega) = \mathrm{j}\omega \nabla \cdot \underline{\underline{\underline{\mathbf{e}}}}^{231} \cdot \nabla \Phi(\underline{\mathbf{R}}, \omega), \tag{4.144}$$

$$-\mathrm{j}\omega\epsilon_0 \nabla \cdot \underline{\underline{\epsilon}}_r^{\mathbf{S}} \cdot \nabla \Phi(\underline{\mathbf{R}}, \omega) = \nabla \cdot \underline{\underline{\underline{\mathbf{e}}}} : \underline{\underline{\underline{\underline{\mathbf{I}}}}}^{+} : \nabla \underline{\mathbf{v}}(\underline{\mathbf{R}}, \omega). \tag{4.145}$$

Piezoelectric plane waves: Piezoelectrically stiffened stiffness tensor: In Section 8, we take the particle displacement as the basis to derive plane elastic waves; the corresponding Equations 4.144 and 4.145 read as

$$\nabla \cdot \underline{\underline{\underline{\underline{\mathbf{c}}}}}^{\mathbf{E}} : \nabla \underline{\mathbf{u}}(\underline{\mathbf{R}}, \omega) + \rho\omega^2 \underline{\mathbf{u}}(\underline{\mathbf{R}}, \omega) = -\nabla \cdot \underline{\underline{\underline{\mathbf{e}}}}^{231} \cdot \nabla \Phi(\underline{\mathbf{R}}, \omega), \tag{4.146}$$

$$\epsilon_0 \nabla \cdot \underline{\underline{\epsilon}}_r^{\mathbf{S}} \cdot \nabla \Phi(\underline{\mathbf{R}}, \omega) = \nabla \cdot \underline{\underline{\underline{\mathbf{e}}}} : \underline{\underline{\underline{\underline{\mathbf{I}}}}}^{+} : \nabla \underline{\mathbf{u}}(\underline{\mathbf{R}}, \omega). \tag{4.147}$$

The solution *ansatz* of homogeneous plane waves (Sections 8.1.2 and 8.3)

$$\underline{\mathbf{u}}(\underline{\mathbf{R}}, \omega) \Longrightarrow \underline{\mathbf{u}}(\underline{\mathbf{R}}, \omega, \hat{\underline{\mathbf{k}}}) = \underline{\mathbf{u}}(\omega, \hat{\underline{\mathbf{k}}})\, \mathrm{e}^{\pm \mathrm{j} \frac{\hat{\underline{\mathbf{k}}} \cdot \underline{\mathbf{R}}}{c(\hat{\underline{\mathbf{k}}})} \omega} \tag{4.148}$$

for the frequency spectrum of the elastodynamic particle displacement with the (phase) propagation direction $\hat{\underline{\mathbf{k}}}$ and the phase velocity $c(\hat{\underline{\mathbf{k}}})$ transforms (4.146) and (4.147) into

$$\left[\frac{1}{c^2(\hat{\underline{\mathbf{k}}})} \hat{\underline{\mathbf{k}}} \cdot \underline{\underline{\underline{\underline{\mathbf{c}}}}}^{\mathbf{E}} \cdot \hat{\underline{\mathbf{k}}} - \rho \underline{\underline{\mathbf{I}}}\right] \cdot \omega^2 \underline{\mathbf{u}}(\underline{\mathbf{R}}, \omega, \hat{\underline{\mathbf{k}}}) = \nabla \cdot \underline{\underline{\underline{\mathbf{e}}}}^{231} \cdot \nabla \Phi(\underline{\mathbf{R}}, \omega) \tag{4.149}$$

$$\epsilon_0 \nabla \cdot \underline{\underline{\epsilon}}_r^{\mathbf{S}} \cdot \nabla \Phi(\underline{\mathbf{R}}, \omega) = \frac{\omega^2}{c^2(\hat{\underline{\mathbf{k}}})} \hat{\underline{\mathbf{k}}} \cdot \underline{\underline{\underline{\mathbf{e}}}} : \underline{\underline{\underline{\underline{\mathbf{I}}}}}^{+} : \hat{\underline{\mathbf{k}}}\, \underline{\mathbf{u}}(\underline{\mathbf{R}}, \omega, \hat{\underline{\mathbf{k}}}). \tag{4.150}$$

Applying a three-dimensional spatial Fourier transform, we obtain

$$\epsilon_0 \underline{\mathbf{K}} \cdot \underline{\underline{\epsilon}}_r^{\mathbf{S}} \cdot \underline{\mathbf{K}}\, \Phi(\underline{\mathbf{R}}, \omega) = \frac{\omega^2}{c^2(\hat{\underline{\mathbf{k}}})} \hat{\underline{\mathbf{k}}} \cdot \underline{\underline{\underline{\mathbf{e}}}} : \underline{\underline{\underline{\underline{\mathbf{I}}}}}^{+} : \hat{\underline{\mathbf{k}}}\, \underline{\mathbf{u}}(\omega, \hat{\underline{\mathbf{k}}})(2\pi)^3 \delta \left[\underline{\mathbf{K}} \mp \frac{\omega}{c(\hat{\underline{\mathbf{k}}})} \hat{\underline{\mathbf{k}}}\right] \tag{4.151}$$

from (4.150); Fourier inversion yields

$$\Phi(\underline{\mathbf{R}},\omega) = \frac{1}{c^2(\hat{\underline{\mathbf{k}}})}\hat{\underline{\mathbf{k}}}\cdot\underline{\underline{\underline{\mathbf{e}}}}:\underline{\underline{\underline{\mathbf{I}}}}^+:\hat{\underline{\mathbf{k}}}\,\underline{\mathbf{u}}(\omega,\hat{\underline{\mathbf{k}}})\int_{-\infty}^{\infty}\int_{-\infty}^{\infty}\int_{-\infty}^{\infty}\delta\left[\underline{\mathbf{K}}\mp\frac{\omega}{c(\hat{\underline{\mathbf{k}}})}\hat{\underline{\mathbf{k}}}\right] \\ \times\frac{\mathrm{e}^{\,\mathrm{j}\underline{\mathbf{K}}\cdot\underline{\mathbf{R}}}}{\epsilon_0\frac{1}{\omega}\underline{\mathbf{K}}\cdot\underline{\underline{\boldsymbol{\epsilon}}}_r^{\mathbf{S}}\cdot\frac{1}{\omega}\underline{\mathbf{K}}}\,\mathrm{d}^3\underline{\mathbf{K}}. \tag{4.152}$$

The factor ω^2 in (4.151) has been purposely distributed between the $\underline{\mathbf{K}}$-factors of the integration denominator (4.152) to indicate that we do not have a problem to apply the sifting property of the δ-Distribution even for $\omega = 0$:[74]

$$\Phi(\underline{\mathbf{R}},\omega)\Longrightarrow\Phi(\underline{\mathbf{R}},\omega,\hat{\underline{\mathbf{k}}}) = \underbrace{\frac{\hat{\underline{\mathbf{k}}}\cdot\underline{\underline{\underline{\mathbf{e}}}}:\underline{\underline{\underline{\mathbf{I}}}}:\hat{\underline{\mathbf{k}}}\,\underline{\mathbf{u}}(\omega,\hat{\underline{\mathbf{k}}})}{\epsilon_0\hat{\underline{\mathbf{k}}}\cdot\underline{\underline{\boldsymbol{\epsilon}}}_r^{\mathbf{S}}\cdot\hat{\underline{\mathbf{k}}}}}_{\stackrel{\text{def}}{=}\,\Phi(\omega,\hat{\underline{\mathbf{k}}})}\,\mathrm{e}^{\pm\mathrm{j}\frac{\hat{\underline{\mathbf{k}}}\cdot\underline{\mathbf{R}}}{c(\hat{\underline{\mathbf{k}}})}\omega}, \tag{4.153}$$

that is to say, the scalar electric potential behaves just like a plane wave with the same phase velocity as the elastic wave (4.148): It is enforced by the elastic wave through the piezoelectric effect. With the symmetry (4.124) of $\underline{\underline{\underline{\mathbf{e}}}}$, we can write $\Phi(\underline{\mathbf{R}},\omega,\hat{\underline{\mathbf{k}}})$ as follows:

$$\Phi(\underline{\mathbf{R}},\omega,\hat{\underline{\mathbf{k}}}) = \frac{\hat{\underline{\mathbf{k}}}\cdot\underline{\underline{\underline{\mathbf{e}}}}\cdot\hat{\underline{\mathbf{k}}}}{\epsilon_0\hat{\underline{\mathbf{k}}}\cdot\underline{\underline{\boldsymbol{\epsilon}}}_r^{\mathbf{S}}\cdot\hat{\underline{\mathbf{k}}}}\cdot\underline{\mathbf{u}}(\underline{\mathbf{R}},\omega,\hat{\underline{\mathbf{k}}}). \tag{4.154}$$

The electric field strength related to the potential results from the negative gradient

$$\underline{\mathbf{E}}(\underline{\mathbf{R}},\omega,\hat{\underline{\mathbf{k}}}) = \mp\mathrm{j}\frac{\omega}{c(\hat{\underline{\mathbf{k}}})}\hat{\underline{\mathbf{k}}}\frac{\hat{\underline{\mathbf{k}}}\cdot\underline{\underline{\underline{\mathbf{e}}}}\cdot\hat{\underline{\mathbf{k}}}}{\epsilon_0\hat{\underline{\mathbf{k}}}\cdot\underline{\underline{\boldsymbol{\epsilon}}}_r^{\mathbf{S}}\cdot\hat{\underline{\mathbf{k}}}}\cdot\underline{\mathbf{u}}(\underline{\mathbf{R}},\omega,\hat{\underline{\mathbf{k}}}) \tag{4.155}$$

as a longitudinal enforced field strength; it does not exist without the "elastic companion."

To calculate the phase velocity $c(\hat{\underline{\mathbf{k}}})$ and the polarization $\hat{\underline{\mathbf{u}}}(\hat{\underline{\mathbf{k}}})$ of the piezoelectric wave according to the factorization $\underline{\mathbf{u}}(\omega,\hat{\underline{\mathbf{k}}}) = u(\omega)\hat{\underline{\mathbf{u}}}(\hat{\underline{\mathbf{k}}})$ as postulated in Sections 8.1.2 and 8.3, we take advantage of (4.149) with (4.154); analogous to (8.61) and (8.204), we obtain

$$\left[\frac{1}{\rho}\hat{\underline{\mathbf{k}}}\cdot\underline{\underline{\underline{\underline{\mathbf{c}}}}}^{\mathbf{E}}\cdot\hat{\underline{\mathbf{k}}} - c^2(\hat{\underline{\mathbf{k}}})\,\underline{\underline{\mathbf{I}}}\right]\cdot\underline{\mathbf{u}}(\underline{\mathbf{R}},\omega,\hat{\underline{\mathbf{k}}}) = -\frac{1}{\rho}\hat{\underline{\mathbf{k}}}\cdot\frac{\underline{\underline{\underline{\mathbf{e}}}}^{231}\cdot\hat{\underline{\mathbf{k}}}\hat{\underline{\mathbf{k}}}\cdot\underline{\underline{\underline{\mathbf{e}}}}}{\epsilon_0\hat{\underline{\mathbf{k}}}\cdot\underline{\underline{\boldsymbol{\epsilon}}}_r^{\mathbf{S}}\cdot\hat{\underline{\mathbf{k}}}}\cdot\hat{\underline{\mathbf{k}}}\cdot\underline{\mathbf{u}}(\underline{\mathbf{R}},\omega,\hat{\underline{\mathbf{k}}}), \tag{4.156}$$

[74]Since $\underline{\underline{\boldsymbol{\epsilon}}}_r^{\mathbf{S}}$ must be positive-definite due to the expression (6.57) for the electromagnetic instantaneous energy density, it follows that $\underline{\mathbf{K}}\cdot\underline{\underline{\boldsymbol{\epsilon}}}_r^{\mathbf{S}}\cdot\underline{\mathbf{K}}$ is larger than zero and zero only for $K = 0$; yet, $K = 0$ corresponds to $\omega = 0$.

and this is the conventional eigenvalue problem

$$\underline{\underline{\mathbf{D}}}^{\text{piezo}}(\hat{\underline{\mathbf{k}}}) \cdot \hat{\underline{\mathbf{u}}}(\hat{\underline{\mathbf{k}}}) = c^2(\hat{\underline{\mathbf{k}}})\,\hat{\underline{\mathbf{u}}}(\hat{\underline{\mathbf{k}}}) \tag{4.157}$$

for the real valued symmetric tensor

$$\underline{\underline{\mathbf{D}}}^{\text{piezo}}(\hat{\underline{\mathbf{k}}}) = \frac{1}{\rho}\hat{\underline{\mathbf{k}}} \cdot \left(\underline{\underline{\underline{\underline{\mathbf{c}}}}}^{\mathbf{E}} + \frac{\underline{\underline{\underline{\mathbf{e}}}}^{231} \cdot \hat{\underline{\mathbf{k}}}\,\hat{\underline{\mathbf{k}}} \cdot \underline{\underline{\underline{\mathbf{e}}}}}{\epsilon_0 \hat{\underline{\mathbf{k}}} \cdot \underline{\underline{\boldsymbol{\epsilon}}}_r^{\mathbf{S}} \cdot \hat{\underline{\mathbf{k}}}} \right) \cdot \hat{\underline{\mathbf{k}}} \tag{4.158}$$

with the real valued eigenvalues $c^2(\hat{\underline{\mathbf{k}}})$ and the real valued eigenvectors $\hat{\underline{\mathbf{u}}}(\hat{\underline{\mathbf{k}}})$. In (4.158), it appears

$$\underline{\underline{\underline{\underline{\mathbf{c}}}}}^{\text{piezo}} = \underline{\underline{\underline{\underline{\mathbf{c}}}}}^{\mathbf{E}} + \frac{\underline{\underline{\underline{\mathbf{e}}}}^{231} \cdot \hat{\underline{\mathbf{k}}}\,\hat{\underline{\mathbf{k}}} \cdot \underline{\underline{\underline{\mathbf{e}}}}}{\epsilon_0 \hat{\underline{\mathbf{k}}} \cdot \underline{\underline{\boldsymbol{\epsilon}}}_r^{\mathbf{S}} \cdot \hat{\underline{\mathbf{k}}}} \tag{4.159}$$

as the piezoelectrically stiffened stiffness tensor.

For piezoelectric crystals with transverse isotropy and the preference direction $\hat{\underline{\mathbf{a}}}$, i.e., for the stiffness tensor (4.24), the permittivity tensor (Equation 6.53),

$$\underline{\underline{\boldsymbol{\epsilon}}}_r^{\text{e}} = \epsilon_{\perp}^{\mathbf{S}} \underline{\underline{\mathbf{I}}} + (\epsilon_{\parallel}^{\mathbf{S}} - \epsilon_{\perp}^{\mathbf{S}})\,\hat{\underline{\mathbf{a}}}\,\hat{\underline{\mathbf{a}}} \tag{4.160}$$

and the piezoelectric coupling tensor

$$\underline{\underline{\underline{\mathbf{e}}}} = \eta_1 \hat{\underline{\mathbf{a}}}\,\underline{\underline{\mathbf{I}}} + \eta_2(\underline{\underline{\mathbf{I}}}\,\hat{\underline{\mathbf{a}}} + \underline{\underline{\mathbf{I}}}\,\hat{\underline{\mathbf{a}}}^{132}) + \eta_3 \hat{\underline{\mathbf{a}}}\,\hat{\underline{\mathbf{a}}}\,\hat{\underline{\mathbf{a}}}, \tag{4.161}$$

the phase and energy velocity diagrams have been calculated by Marklein (1997). Since the resulting wave tensor $\underline{\underline{\mathbf{W}}}^{\text{piezo}}(\hat{\underline{\mathbf{k}}}, c^2)$ exhibits the same mathematical structure as in the transverse isotropic case without piezoelectricity (Equation 8.247), we can stick to the (orthogonal) polarizations SH, qP, and qSV (Section 8.3).

4.5.2 Magnetostriction

The magnetization of a ferromagnet may yield strains of a crystal and vice versa; it is called linear magnetostriction—similar to the electrostriction, the real magnetostriction is a nonlinear effect—or piezomagnetism (Landau et al. 1984; Auld 1973). Analogous to (4.123) and (4.130), we postulate piezomagnetic constitutive equations (Wilbrand 1989; IEEE Committee 1973):

$$\underline{\mathbf{B}}(\underline{\mathbf{R}}, t) = \mu_0 \underline{\underline{\mu}}_r^{\mathbf{S}}(\underline{\mathbf{R}}) \cdot \underline{\mathbf{H}}(\underline{\mathbf{R}}, t) + \underline{\underline{\underline{\mathbf{m}}}}(\underline{\mathbf{R}}) : \mathbf{S}(\underline{\mathbf{R}}, t), \tag{4.162}$$

$$\mathbf{T}(\underline{\mathbf{R}}, t) = \underline{\underline{\underline{\underline{\mathbf{c}}}}}^{\mathbf{H}}(\underline{\mathbf{R}}) : \mathbf{S}(\underline{\mathbf{R}}, t) - \underline{\underline{\underline{\mathbf{m}}}}^{231}(\underline{\mathbf{R}}) \cdot \underline{\mathbf{H}}(\underline{\mathbf{R}}, t). \tag{4.163}$$

The term $-\nabla \cdot [\underline{\underline{\underline{\mathbf{m}}}}^{231}(\underline{\mathbf{R}}) \cdot \underline{\mathbf{H}}(\underline{\mathbf{R}}, t)]$ turns out to be a magnetostrictive volume force density depending on the magnetic field that additionally appears as an inhomogeneity in the elastodynamic governing equations: The magnetoelastic coupling of magnetostriction is able to create elastic waves in ferromagnets. This applies to the construction of so-called electromagnetic-acoustic transducers (EMATs); however, the impact of Lorentz forces is superimposed that are otherwise solely utilized to excite ultrasound without mechanical contact in nonferromagnetic materials.

5

Acoustics

Before we draw conclusions for the propagation of elastic waves from the elastodynamic governing equations augmented by constitutive relations, we will refer to the governing equations of acoustics and electromagnetics (Chapter 6) complemented by plane waves as well as source field and scattered field representations. On one hand, US-NDT likes to think in the terminology of (scalar) acoustics, and on the other hand, coupling mechanisms between elastic and electromagnetic phenomena—electromagnetic-acoustic transducers, piezoelectric transducers, and laser excitation of ultrasound—are of fundamental importance. In addition, microwave methods gain more and more attention—e.g., the application of ground probing radars for NDT of concrete (Krieger et al. 1998; Mayer et al. 2003)—suggesting a comparison of the theoretical foundations of electromagnetic and elastic waves. Yet, as already mentioned, we will only cite and not derive facts, eventually providing plausible arguments.

5.1 Governing Equations of Acoustics

Based on the mechanical physical properties of fluids and gases, the governing equations of acoustics (acoustodynamics) are derived within certain approximations (Morse and Ingard 1968; de Hoop 1995):

$$\frac{\partial \underline{\mathbf{j}}(\underline{\mathbf{R}}, t)}{\partial t} = -\nabla p(\underline{\mathbf{R}}, t) + \underline{\mathbf{f}}(\underline{\mathbf{R}}, t), \tag{5.1}$$

$$\frac{\partial S(\underline{\mathbf{R}}, t)}{\partial t} = \nabla \cdot \underline{\mathbf{v}}(\underline{\mathbf{R}}, t) + h(\underline{\mathbf{R}}, t). \tag{5.2}$$

Formally, these equations result from (3.1) and (3.2) if the stress tensor is replaced by the isotropic pressure tensor according to[75] $\underline{\underline{\mathbf{T}}}(\underline{\mathbf{R}}, t)$

[75]We decompose $\underline{\underline{\mathbf{T}}}$ according to

$$\begin{aligned}\underline{\underline{\mathbf{T}}} &= \frac{1}{3}\underline{\underline{\mathbf{I}}}\,\mathrm{trace}\,\underline{\underline{\mathbf{T}}} + \left(\underline{\underline{\mathbf{T}}} - \frac{1}{3}\underline{\underline{\mathbf{I}}}\,\mathrm{trace}\,\underline{\underline{\mathbf{T}}}\right)\\ &= \sigma\underline{\underline{\mathbf{I}}} + \left(\underline{\underline{\mathbf{T}}} - \frac{1}{3}\underline{\underline{\mathbf{I}}}\,\mathrm{trace}\,\underline{\underline{\mathbf{T}}}\right)\end{aligned}$$

into the isotropic stress tensor $\sigma\underline{\underline{\mathbf{I}}}$ and the deviatoric stress tensor and neglect the latter; then we put $p = -\sigma$.

$$\underline{\underline{\mathbf{T}}}(\underline{\mathbf{R}}, t) \Longrightarrow \underline{\underline{\mathbf{P}}}(\underline{\mathbf{R}}, t) = -p(\underline{\mathbf{R}}, t)\,\underline{\underline{\mathbf{I}}} \tag{5.3}$$

and bringing the scalar cubic dilatation $S(\underline{\mathbf{R}}, t)$ and the scalar injected dilatation rate $h(\underline{\mathbf{R}}, t)$ into play via the trace of (3.2):

$$\underline{\underline{\mathbf{S}}}(\underline{\mathbf{R}}, t) \Longrightarrow S(\underline{\mathbf{R}}, t) = \operatorname{trace} \underline{\underline{\mathbf{S}}}(\underline{\mathbf{R}}, t), \tag{5.4}$$

$$\underline{\underline{\mathbf{h}}}(\underline{\mathbf{R}}, t) \Longrightarrow h(\underline{\mathbf{R}}, t) = \operatorname{trace} \underline{\underline{\mathbf{h}}}(\underline{\mathbf{R}}, t). \tag{5.5}$$

Calculating the acoustic Poynting vector (the acoustic energy density flow)

$$\underline{\mathbf{S}}(\underline{\mathbf{R}}, t) = \underline{\mathbf{v}}(\underline{\mathbf{R}}, t) p(\underline{\mathbf{R}}, t) \tag{5.6}$$

and inserting the acoustic governing equations (5.1) and (5.2) yield the acoustic energy conservation law

$$\frac{\partial w_{\mathrm{ak}}(\underline{\mathbf{R}}, t)}{\partial t} = -\boldsymbol{\nabla} \cdot \underline{\mathbf{S}}(\underline{\mathbf{R}}, t) + \frac{\partial w_Q(\underline{\mathbf{R}}, t)}{\partial t}, \tag{5.7}$$

where the time derivative of the acoustic energy density

$$\frac{\partial w_{\mathrm{ak}}(\underline{\mathbf{R}}, t)}{\partial t} = \frac{\partial \underline{\mathbf{j}}(\underline{\mathbf{R}}, t)}{\partial t} \cdot \underline{\mathbf{v}}(\underline{\mathbf{R}}, t) - \frac{\partial S(\underline{\mathbf{R}}, t)}{\partial t} p(\underline{\mathbf{R}}, t) \tag{5.8}$$

for the constitutive equations (5.22) and (5.23) of linear nondissipative acoustic "materials" coincides with the time derivative of the acoustic energy density

$$w_{\mathrm{ak}}(\underline{\mathbf{R}}, t) = \frac{1}{2} \underline{\mathbf{j}}(\underline{\mathbf{R}}, t) \cdot \underline{\mathbf{v}}(\underline{\mathbf{R}}, t) - \frac{1}{2} p(\underline{\mathbf{R}}, t) S(\underline{\mathbf{R}}, t); \tag{5.9}$$

regarding dissipative materials, we refer to the elastodynamic case in Section 4.4 and to Marklein (1997). The term

$$\frac{\partial w_Q(\underline{\mathbf{R}}, t)}{\partial t} = \underline{\mathbf{f}}(\underline{\mathbf{R}}, t) \cdot \underline{\mathbf{v}}(\underline{\mathbf{R}}, t) - h(\underline{\mathbf{R}}, t) p(\underline{\mathbf{R}}, t) \tag{5.10}$$

in (5.7) denotes the time variation of the energy density injected from exterior.

5.2 Transition and Boundary Conditions

The inhomogeneous transition conditions

$$\underline{\mathbf{n}}\,[p^{(1)}(\underline{\mathbf{R}}_S, t) - p^{(2)}(\underline{\mathbf{R}}_S, t)] = \underline{\mathbf{t}}(\underline{\mathbf{R}}_S, t), \quad \underline{\mathbf{R}}_S \in S, \tag{5.11}$$

$$\underline{\mathbf{n}} \cdot [\underline{\mathbf{v}}^{(1)}(\underline{\mathbf{R}}_S, t) - \underline{\mathbf{v}}^{(2)}(\underline{\mathbf{R}}_S, t)] = -g(\underline{\mathbf{R}}_S, t), \quad \underline{\mathbf{R}}_S \in S, \tag{5.12}$$

for the boundary $\underline{\mathbf{R}} = \underline{\mathbf{R}}_S \in S$, representing a jump discontinuity of material properties immediately follow from the governing equations (5.1) and (5.2); $\underline{\mathbf{t}}(\underline{\mathbf{R}}_S, t)$ and $g(\underline{\mathbf{R}}_S, t)$ are prescribed surface force densities and prescribed surface dilatation rates. For vanishing prescribed surface sources, we obtain the homogeneous transition conditions

$$p^{(1)}(\underline{\mathbf{R}}_S, t) - p^{(2)}(\underline{\mathbf{R}}_S, t) = 0, \quad \underline{\mathbf{R}}_S \in S, \tag{5.13}$$

$$\underline{\mathbf{n}} \cdot [\underline{\mathbf{v}}^{(1)}(\underline{\mathbf{R}}_S, t) - \underline{\mathbf{v}}^{(2)}(\underline{\mathbf{R}}_S, t)] = 0, \quad \underline{\mathbf{R}}_S \in S, \tag{5.14}$$

of pressure continuity and continuity of the normal component of the particle velocity. From the arguments based on Equations 3.95 and 3.96, we conclude the continuity of the normal component of the displacement

$$\underline{\mathbf{n}} \cdot [\underline{\mathbf{u}}^{(1)}(\underline{\mathbf{R}}_S, t) - \underline{\mathbf{u}}^{(2)}(\underline{\mathbf{R}}_S, t)] = 0, \quad \underline{\mathbf{R}}_S \in S \tag{5.15}$$

from Equation 5.14.

In case the material (2) does not allow for the propagation of acoustic waves, Equations 5.11 and 5.12 reduce to the (perfectly) soft (Dirichlet) boundary condition

$$p(\underline{\mathbf{R}}_S, t) = 0, \quad \underline{\mathbf{R}}_S \in S, \tag{5.16}$$

with the consequence of definition of an induced surface dilatation rate according to

$$\underline{\mathbf{n}} \cdot \underline{\mathbf{v}}(\underline{\mathbf{R}}_S, t) = -g(\underline{\mathbf{R}}_S, t), \quad \underline{\mathbf{R}}_S \in S; \tag{5.17}$$

alternatively, a (perfectly) rigid boundary condition

$$\underline{\mathbf{n}} \cdot \underline{\mathbf{v}}(\underline{\mathbf{R}}_S, t) = 0, \quad \underline{\mathbf{R}}_S \in S, \tag{5.18}$$

with the consequence of definition of an induced surface force density

$$\underline{\mathbf{n}}\, p(\underline{\mathbf{R}}_S, t) = \underline{\mathbf{t}}(\underline{\mathbf{R}}_S, t), \quad \underline{\mathbf{R}}_S \in S, \tag{5.19}$$

or

$$p(\underline{\mathbf{R}}_S, t) = \underline{\mathbf{n}} \cdot \underline{\mathbf{t}}(\underline{\mathbf{R}}_S, t), \quad \underline{\mathbf{R}}_S \in S, \tag{5.20}$$

respectively, can be defined. Differentiating (5.18) with respect to time and utilization of the homogeneous equation (5.11) (vanishing prescribed surface source densities), Equation 5.18 is written as Neumann's boundary condition for the pressure:

$$\underline{\mathbf{n}} \cdot \nabla p(\underline{\mathbf{R}}, t)\Big|_{\underline{\mathbf{R}}=\underline{\mathbf{R}}_S} = 0, \quad \underline{\mathbf{R}}_S \in S. \tag{5.21}$$

5.3 Wave Equations in the Time and Frequency Domains

With constitutive equations[76]

$$\underline{\mathbf{j}}(\underline{\mathbf{R}}, t) = \rho(\underline{\mathbf{R}})\underline{\mathbf{v}}(\underline{\mathbf{R}}, t), \tag{5.22}$$

$$S(\underline{\mathbf{R}}, t) = -\kappa(\underline{\mathbf{R}})p(\underline{\mathbf{R}}, t) \tag{5.23}$$

for linear inhomogeneous nondissipative acoustic "materials"—$\kappa(\underline{\mathbf{R}})$ is the (adiabatic) compressibility—we obtain the acoustic complements of (4.33) and (4.34):

$$\rho(\underline{\mathbf{R}}) \frac{\partial \underline{\mathbf{v}}(\underline{\mathbf{R}}, t)}{\partial t} = -\nabla p(\underline{\mathbf{R}}, t) + \underline{\mathbf{f}}(\underline{\mathbf{R}}, t), \tag{5.24}$$

$$-\kappa(\underline{\mathbf{R}}) \frac{\partial p(\underline{\mathbf{R}}, t)}{\partial t} = \nabla \cdot \underline{\mathbf{v}}(\underline{\mathbf{R}}, t) + h(\underline{\mathbf{R}}, t); \tag{5.25}$$

through mutual insertion:

$$\nabla \cdot \left[\frac{1}{\rho(\underline{\mathbf{R}})} \nabla p(\underline{\mathbf{R}}, t)\right] - \kappa(\underline{\mathbf{R}}) \frac{\partial^2 p(\underline{\mathbf{R}}, t)}{\partial t^2} = \nabla \cdot \left[\frac{1}{\rho(\underline{\mathbf{R}})} \underline{\mathbf{f}}(\underline{\mathbf{R}}, t)\right] + \frac{\partial h(\underline{\mathbf{R}}, t)}{\partial t}, \tag{5.26}$$

respectively, through explicit differentiation

$$\begin{aligned} &\Delta p(\underline{\mathbf{R}}, t) - \rho(\underline{\mathbf{R}})\kappa(\underline{\mathbf{R}}) \frac{\partial^2 p(\underline{\mathbf{R}}, t)}{\partial t^2} - [\nabla \ln \rho(\underline{\mathbf{R}})] \cdot \nabla p(\underline{\mathbf{R}}, t) \\ &= -[\nabla \ln \rho(\underline{\mathbf{R}})] \cdot \underline{\mathbf{f}}(\underline{\mathbf{R}}, t) + \nabla \cdot \underline{\mathbf{f}}(\underline{\mathbf{R}}, t) + \rho(\underline{\mathbf{R}}) \frac{\partial h(\underline{\mathbf{R}}, t)}{\partial t} \end{aligned} \tag{5.27}$$

or even repeated utilization of (5.24)

$$\begin{aligned} &\Delta p(\underline{\mathbf{R}}, t) - \rho(\underline{\mathbf{R}})\kappa(\underline{\mathbf{R}}) \frac{\partial^2 p(\underline{\mathbf{R}}, t)}{\partial t^2} + \frac{\partial \underline{\mathbf{v}}(\underline{\mathbf{R}}, t)}{\partial t} \cdot \nabla \rho(\underline{\mathbf{R}}) \\ &= \nabla \cdot \underline{\mathbf{f}}(\underline{\mathbf{R}}, t) + \rho(\underline{\mathbf{R}}) \frac{\partial h(\underline{\mathbf{R}}, t)}{\partial t} \end{aligned} \tag{5.28}$$

[76]Formally, Equation 5.23 results from Hooke's law $\underline{\underline{\mathbf{S}}}(\underline{\mathbf{R}}, t) = \underline{\underline{\underline{\underline{\mathbf{s}}}}}(\underline{\mathbf{R}}) : \underline{\underline{\mathbf{T}}}(\underline{\mathbf{R}}, t)$ through calculation of the trace—trace $\underline{\underline{\mathbf{S}}} = \underline{\underline{\mathbf{I}}} : \underline{\underline{\mathbf{S}}}$—and neglect of the deviatoric stress tensor; then, the compressibility $\kappa(\underline{\mathbf{R}})$ is given according to

$$\kappa(\underline{\mathbf{R}}) = \underline{\underline{\mathbf{I}}} : \underline{\underline{\underline{\underline{\mathbf{s}}}}}(\underline{\mathbf{R}}) : \underline{\underline{\mathbf{I}}}$$

with the compliance tensor $\underline{\underline{\underline{\underline{\mathbf{s}}}}}(\underline{\mathbf{R}})$. Hence, the minus sign in (5.23) has nothing to do with the compressibility but with the relation stress versus pressure.

as well as[77]

$$\nabla \left[\frac{1}{\kappa(\underline{\mathbf{R}})} \nabla \cdot \underline{\mathbf{v}}(\underline{\mathbf{R}}, t) \right] - \rho(\underline{\mathbf{R}}) \frac{\partial^2 \underline{\mathbf{v}}(\underline{\mathbf{R}}, t)}{\partial t^2} = -\frac{\partial \underline{\mathbf{f}}(\underline{\mathbf{R}}, t)}{\partial t} - \nabla \left[\frac{1}{\kappa(\underline{\mathbf{R}})} h(\underline{\mathbf{R}}, t) \right], \tag{5.29}$$

respectively, through explicit differentiation

$$\begin{aligned} &\nabla\nabla \cdot \underline{\mathbf{v}}(\underline{\mathbf{R}}, t) - \rho(\underline{\mathbf{R}})\kappa(\underline{\mathbf{R}}) \frac{\partial^2 \underline{\mathbf{v}}(\underline{\mathbf{R}}, t)}{\partial t^2} - [\nabla \ln \kappa(\underline{\mathbf{R}})] \nabla \cdot \underline{\mathbf{v}}(\underline{\mathbf{R}}, t) \\ &\quad = [\nabla \ln \kappa(\underline{\mathbf{R}})] h(\underline{\mathbf{R}}, t) - \nabla h(\underline{\mathbf{R}}, t) - \kappa(\underline{\mathbf{R}}) \frac{\partial \underline{\mathbf{f}}(\underline{\mathbf{R}}, t)}{\partial t} \end{aligned} \tag{5.30}$$

or even repeated utilization of (5.25)

$$\begin{aligned} &\nabla\nabla \cdot \underline{\mathbf{v}}(\underline{\mathbf{R}}, t) - \rho(\underline{\mathbf{R}})\kappa(\underline{\mathbf{R}}) \frac{\partial^2 \underline{\mathbf{v}}(\underline{\mathbf{R}}, t)}{\partial t^2} + \frac{\partial p(\underline{\mathbf{R}}, t)}{\partial t} \nabla \kappa(\underline{\mathbf{R}}) \\ &\quad = -\nabla h(\underline{\mathbf{R}}, t) - \kappa(\underline{\mathbf{R}}) \frac{\partial \underline{\mathbf{f}}(\underline{\mathbf{R}}, t)}{\partial t}. \end{aligned} \tag{5.31}$$

The differential operators for $p(\underline{\mathbf{R}}, t)$ in (5.27) or (5.28), respectively, and for $\underline{\mathbf{v}}(\underline{\mathbf{R}}, t)$ in (5.30) or (5.31), respectively, contain "extra terms" $\nabla\rho(\underline{\mathbf{R}})$ or $\nabla \ln \rho(\underline{\mathbf{R}})$ and $\nabla\kappa(\underline{\mathbf{R}})$ or $\nabla \ln \kappa(\underline{\mathbf{R}})$ [the operators in (5.28) and (5.31) are not even decoupled] that are considerably annoying solving the differential equations. Therefore, to find arguments for appropriate approximations (Born approximation), they are transferred to the right-hand sides as additional inhomogeneities in terms of equivalent sources (Section 5.6).

For homogeneous materials, the differential equations (5.28) and (5.31) immediately decouple:

$$\Delta p(\underline{\mathbf{R}}, t) - \kappa\rho \frac{\partial^2 p(\underline{\mathbf{R}}, t)}{\partial t^2} = \nabla \cdot \underline{\mathbf{f}}(\underline{\mathbf{R}}, t) + \rho \frac{\partial h(\underline{\mathbf{R}}, t)}{\partial t}, \tag{5.32}$$

$$\nabla\nabla \cdot \underline{\mathbf{v}}(\underline{\mathbf{R}}, t) - \kappa\rho \frac{\partial^2 \underline{\mathbf{v}}(\underline{\mathbf{R}}, t)}{\partial t^2} = -\nabla h(\underline{\mathbf{R}}, t) - \kappa \frac{\partial \underline{\mathbf{f}}(\underline{\mathbf{R}}, t)}{\partial t}. \tag{5.33}$$

With (2.187), we alternatively obtain regarding (5.33):

$$\Delta \underline{\mathbf{v}}(\underline{\mathbf{R}}, t) + \nabla \times \nabla \times \underline{\mathbf{v}}(\underline{\mathbf{R}}, t) - \kappa\rho \frac{\partial^2 \underline{\mathbf{v}}(\underline{\mathbf{R}}, t)}{\partial t^2} = -\nabla h(\underline{\mathbf{R}}, t) - \kappa \frac{\partial \underline{\mathbf{f}}(\underline{\mathbf{R}}, t)}{\partial t}. \tag{5.34}$$

[77]Obviously, Equation 5.29 also results specializing (7.3) to

$$\underline{\underline{\underline{\mathbf{c}}}}(\underline{\mathbf{R}}) = \kappa^{-1}(\underline{\mathbf{R}})\underline{\underline{\mathbf{I}}}\,\underline{\underline{\mathbf{I}}}$$

and

$$\underline{\underline{\mathbf{h}}}(\underline{\mathbf{R}}, t) = \frac{1}{3} h(\underline{\mathbf{R}}, t)\underline{\underline{\mathbf{I}}}.$$

From the governing equation (5.24) for $\rho(\underline{\mathbf{R}}) = \rho$, we gain the so-called compatibility relation[78] via calculation of the curl:

$$\nabla \times \frac{\partial \underline{\mathbf{v}}(\underline{\mathbf{R}}, t)}{\partial t} = \frac{1}{\rho} \nabla \times \underline{\mathbf{f}}(\underline{\mathbf{R}}, t) \tag{5.35}$$

transferring the double curl term in (5.34) into an inhomogeneity on the right-hand side:

$$\Delta \underline{\mathbf{v}}(\underline{\mathbf{R}}, t) - \kappa\rho \frac{\partial^2 \underline{\mathbf{v}}(\underline{\mathbf{R}}, t)}{\partial t^2} = -\nabla h(\underline{\mathbf{R}}, t) - \kappa \frac{\partial \underline{\mathbf{f}}(\underline{\mathbf{R}}, t)}{\partial t} - \frac{1}{\rho} \nabla \times \int_0^t \underline{\mathbf{f}}(\underline{\mathbf{R}}, \tau)\, \mathrm{d}\tau. \tag{5.36}$$

The differential equations for the pressure (5.32) and the particle velocity vector (5.36) emerge as d'Alembert wave equations or, via Fourier transform

$$\Delta p(\underline{\mathbf{R}}, \omega) + \omega^2 \kappa\rho\, p(\underline{\mathbf{R}}, \omega) = \nabla \cdot \underline{\mathbf{f}}(\underline{\mathbf{R}}, \omega) - \mathrm{j}\omega\rho\, h(\underline{\mathbf{R}}, \omega), \tag{5.37}$$

$$\begin{aligned} \Delta \underline{\mathbf{v}}(\underline{\mathbf{R}}, \omega) + \omega^2 \kappa\rho\, \underline{\mathbf{v}}(\underline{\mathbf{R}}, \omega) = -\nabla h(\underline{\mathbf{R}}, \omega) - \frac{1}{\mathrm{j}\omega\rho} \big[& \omega^2 \kappa\rho\, \underline{\mathbf{f}}(\underline{\mathbf{R}}, \omega) \\ & - \nabla \times \underline{\mathbf{f}}(\underline{\mathbf{R}}, \omega) \big] \end{aligned} \tag{5.38}$$

as Helmholtz equations (reduced wave equations). We complement (5.38) with the Fourier transformed differential equation (5.33) as alternative:

$$\nabla\nabla \cdot \underline{\mathbf{v}}(\underline{\mathbf{R}}, \omega) + \omega^2 \kappa\rho\, \underline{\mathbf{v}}(\underline{\mathbf{R}}, \omega) = -\nabla h(\underline{\mathbf{R}}, \omega) + \mathrm{j}\kappa\omega \underline{\mathbf{f}}(\underline{\mathbf{R}}, \omega). \tag{5.39}$$

5.4 Solutions of the Homogeneous Acoustic Wave Equations in Homogeneous Materials: Plane Longitudinal Pressure Waves

The homogeneous acoustic Helmholtz equations read

$$\Delta p(\underline{\mathbf{R}}, \omega) + k^2 p(\underline{\mathbf{R}}, \omega) = 0, \tag{5.40}$$

$$\Delta \underline{\mathbf{v}}(\underline{\mathbf{R}}, \omega) + k^2 \underline{\mathbf{v}}(\underline{\mathbf{R}}, \omega) = \underline{\mathbf{0}}, \tag{5.41}$$

where the wave number k according to

$$k = \omega\sqrt{\kappa\rho} \tag{5.42}$$

has been introduced as short-hand notation. One special solution of (5.40) is obtained as:

$$p(\underline{\mathbf{R}}, \omega) = p(\omega)\, \mathrm{e}^{\pm \mathrm{j}\underline{\mathbf{k}} \cdot \underline{\mathbf{R}}}, \tag{5.43}$$

[78]For (causal) solutions of the homogeneous wave equation, we have $\nabla \times \underline{\mathbf{v}}(\underline{\mathbf{R}}, t) = \underline{\mathbf{0}}$; as a consequence, plane acoustic waves are longitudinally polarized.

provided the wave number vector $\underline{\mathbf{k}}$ satisfies the dispersion relation

$$\underline{\mathbf{k}} \cdot \underline{\mathbf{k}} = k^2 = \omega^2 \kappa \rho \tag{5.44}$$

or the slowness vector $\underline{\mathbf{s}} = \underline{\mathbf{k}}/\omega$, respectively, the dispersion relation

$$\underline{\mathbf{s}} \cdot \underline{\mathbf{s}} = \kappa \rho. \tag{5.45}$$

Choosing particularly

$$\underline{\mathbf{k}} = k \hat{\underline{\mathbf{k}}} \tag{5.46}$$

with arbitrary real valued unit vector $\hat{\underline{\mathbf{k}}}$, the phase propagation vector, the Fourier inversion of (5.43) yields

$$p(\underline{\mathbf{R}}, t) = p\left(t \mp \frac{\hat{\underline{\mathbf{k}}} \cdot \underline{\mathbf{R}}}{c}\right) \tag{5.47}$$

with

$$p(t) = \mathcal{F}^{-1}\{p(\omega)\} \tag{5.48}$$

a homogeneous[79] real valued plane pressure wave, provided the initially arbitrary spectral amplitude $p(\omega)$ is constrained to

$$p(-\omega) = p^*(\omega); \tag{5.49}$$

with

$$c = \frac{1}{\sqrt{\kappa \rho}}, \tag{5.50}$$

we have denoted the phase velocity of this wave.[80] The ($\pm$)-sign in the phase of (5.43) or the ($\mp$)-sign in the phase of (5.47), respectively, determines the propagation direction: The negative (positive) sign in (5.47) results in a propagation in positive (negative) $\hat{\underline{\mathbf{k}}}$-direction; based on our choice of the sign in the kernel of the Fourier transform, the positive (negative) sign in (5.43) results in a propagation in the positive (negative) $\hat{\underline{\mathbf{k}}}$-direction.

Inserting (5.43) into the Fourier transformed governing equation (5.24) yields the particle velocity for the pressure wave

$$\underline{\mathbf{v}}(\underline{\mathbf{R}}, \omega) = \underbrace{\pm \frac{1}{Z} p(\omega)}_{= v(\omega)} \, \mathrm{e}^{\pm \mathrm{j} k \hat{\underline{\mathbf{k}}} \cdot \underline{\mathbf{R}}} \, \hat{\underline{\mathbf{k}}} \tag{5.51}$$

[79]The planes of constant phase and constant amplitude coincide and are orthogonal to the phase propagation direction $\hat{\underline{\mathbf{k}}}$.

[80]The derivation and interpretation details can be found in Sections 8.1.1 and 8.1.2.

as a special solution of (5.41), where

$$Z = \rho c \tag{5.52}$$

denotes the acoustic impedance—the wave impedance—of the ρc-material. Apparently, the pressure wave is longitudinally polarized in the particle velocity. Contracting with $\hat{\underline{\mathbf{k}}}$, we obtain from (5.51)

$$p(\underline{\mathbf{R}}, \omega) = \pm Z\, \underline{\mathbf{v}}(\underline{\mathbf{R}}, \omega) \cdot \hat{\underline{\mathbf{k}}}, \tag{5.53}$$

thus relating the Fourier transformed pressure and the Fourier transformed scalar particle motion velocity $\underline{\mathbf{v}}(\underline{\mathbf{R}}, \omega) \cdot \hat{\underline{\mathbf{k}}}$ of a plane wave. Turning to the Fourier transformed particle displacement (Equation 3.94) according to $\underline{\mathbf{v}}(\underline{\mathbf{R}}, \omega) = -\mathrm{j}\omega\, \underline{\mathbf{u}}(\underline{\mathbf{R}}, \omega)$, Equation 5.53 reads as

$$\begin{aligned} p(\underline{\mathbf{R}}, \omega) &= \mp \mathrm{j}\omega Z\, \underline{\mathbf{u}}(\underline{\mathbf{R}}, \omega) \cdot \hat{\underline{\mathbf{k}}}, \\ &= \mp \mathrm{j}\omega Z\, u(\omega)\, \mathrm{e}^{\pm \mathrm{j}k \hat{\underline{\mathbf{k}}} \cdot \underline{\mathbf{R}}}, \end{aligned} \tag{5.54}$$

where $-\mathrm{j}\omega u(\omega) = v(\omega) = p(\omega)/Z$. Equation 5.54 represents—apart from the factor $\mp \mathrm{j}$—the scalar pressure displacement relation as cited by Krautkrämer and Krautkrämer (1986). Yet this factor is important if we cross over to the real valued field quantities in the space time domain:

$$\underline{\mathbf{v}}(\underline{\mathbf{R}}, t) = \pm \frac{1}{Z} p \left(t \mp \frac{\hat{\underline{\mathbf{k}}} \cdot \underline{\mathbf{R}}}{c} \right) \hat{\underline{\mathbf{k}}}, \tag{5.55}$$

$$p(\underline{\mathbf{R}}, t) = \pm Z\, \underline{\mathbf{v}}(\underline{\mathbf{R}}, t) \cdot \hat{\underline{\mathbf{k}}}, \tag{5.56}$$

$$p(\underline{\mathbf{R}}, t) = \pm Z\, \frac{\partial \underline{\mathbf{u}}(\underline{\mathbf{R}}, t)}{\partial t} \cdot \hat{\underline{\mathbf{k}}}. \tag{5.57}$$

A relation similar to (5.53) holds for the far-field of acoustic source fields (Section 13.1.4).

5.5 Acoustic Source Fields in Homogeneous Materials: Point Source Synthesis with Green Functions

5.5.1 Green functions for pressure sources

To solve the inhomogeneous—scalar—Helmholtz equation (5.37) in a homogeneous infinitely extended $\kappa\rho$-material, we confine ourselves at first to a unit-point source located at the arbitrarily chosen point $\underline{\mathbf{R}}'$, the so-called source point, that is to say, we consider the Helmholtz equation

$$\Delta G(\underline{\mathbf{R}}, \underline{\mathbf{R}}', \omega) + k^2 G(\underline{\mathbf{R}}, \underline{\mathbf{R}}', \omega) = -\delta(\underline{\mathbf{R}} - \underline{\mathbf{R}}') \tag{5.58}$$

for the Fourier spectrum of the three-dimensional scalar Green function $G(\underline{\mathbf{R}}, \underline{\mathbf{R}}', \omega)$ of homogeneous infinite space. The δ-function (δ-distribution) represents the mathematical model of a point source, and the minus sign is just convention. The solution of (5.58) relevant to us is the time harmonic outward bound spherical wave (Section 13.1; Langenberg 2005; Becker 1974; de Hoop 1995; King and Harrison 1969; DeSanto 1992)

$$G(\underline{\mathbf{R}}, \underline{\mathbf{R}}', \omega) = G(\underline{\mathbf{R}} - \underline{\mathbf{R}}', \omega) = \frac{\mathrm{e}^{\mathrm{j}k|\underline{\mathbf{R}}-\underline{\mathbf{R}}'|}}{4\pi|\underline{\mathbf{R}} - \underline{\mathbf{R}}'|}. \tag{5.59}$$

With (5.59), the solution of (5.37) can immediately be written down:[81]

$$p(\underline{\mathbf{R}}, \omega) = \int_{-\infty}^{\infty}\int_{-\infty}^{\infty}\int_{-\infty}^{\infty} \left[-\nabla' \cdot \underline{\mathbf{f}}(\underline{\mathbf{R}}', \omega) + \mathrm{j}\omega\rho\, h(\underline{\mathbf{R}}', \omega)\right] G(\underline{\mathbf{R}} - \underline{\mathbf{R}}', \omega)\, \mathrm{d}^3\underline{\mathbf{R}}', \tag{5.60}$$

because the application of the $(\Delta + k^2)$-operator onto $p(\underline{\mathbf{R}}, \omega)$ with regard to $\underline{\mathbf{R}}$ can be pulled under the integral, yet it only applies to the variable $\underline{\mathbf{R}}$ in Green's function, and with (5.58) as well as the sifting property of the delta-distribution, we actually obtain (5.37). The physical interpretation (Section 13.1) of this mathematical representation of the pressure source field turns out to be a $[-\nabla' \cdot \underline{\mathbf{f}}(\underline{\mathbf{R}}', \omega) + \mathrm{j}\omega\rho\, h(\underline{\mathbf{R}}', \omega)]$-weighted synthesis of $\underline{\mathbf{R}}'$-point sources:[82] From each source point, $\underline{\mathbf{R}}'$ a $[-\nabla' \cdot \underline{\mathbf{f}}(\underline{\mathbf{R}}', \omega) + \mathrm{j}\omega\rho\, h(\underline{\mathbf{R}}', \omega)]$-weighted time harmonic elementary spherical wave emerges whose amplitudes and phases are superimposed for each observation point $\underline{\mathbf{R}}$. The travel time $t(\underline{\mathbf{R}}, \underline{\mathbf{R}}') = |\underline{\mathbf{R}} - \underline{\mathbf{R}}'|/c$ of the elementary spherical waves from the source point $\underline{\mathbf{R}}$ to the observation point $\underline{\mathbf{R}}'$ only depends on the magnitude of their mutual distance.

In general, the source volume V_Q is finite—the sources are equal to zero outside V_Q—yielding a finite integration volume V in (5.60) that completely contains V_Q in its interior[83] $(V \supset V_Q)$:

$$p(\underline{\mathbf{R}}, \omega) = \iiint_{V \supset V_Q} \left[-\nabla' \cdot \underline{\mathbf{f}}(\underline{\mathbf{R}}', \omega) + \mathrm{j}\omega\rho\, h(\underline{\mathbf{R}}', \omega)\right] G(\underline{\mathbf{R}} - \underline{\mathbf{R}}', \omega)\, \mathrm{d}^3\underline{\mathbf{R}}'. \tag{5.61}$$

---

[81]In cartesian coordinates, this is a three-dimensional convolution integral.

[82]The method to calculate source fields with Green functions is a point source synthesis.

[83]An alternative distributional calculation goes as follows: With the characteristic function $\Gamma_Q(\underline{\mathbf{R}})$ of V_Q, we explicitly confine $\underline{\mathbf{f}}(\underline{\mathbf{R}}', \omega) \Longrightarrow \underline{\mathbf{f}}(\underline{\mathbf{R}}', \omega)\Gamma_Q(\underline{\mathbf{R}}')$ on V_Q with the consequence having to differentiate in the distributional sense according to

$$\nabla' \cdot \underline{\mathbf{f}}(\underline{\mathbf{R}}', \omega) \Longrightarrow [\nabla' \cdot \underline{\mathbf{f}}(\underline{\mathbf{R}}', \omega)]\Gamma_Q(\underline{\mathbf{R}}') - \underline{\mathbf{n}}'_Q \cdot \underline{\mathbf{f}}(\underline{\mathbf{R}}', \omega)\gamma_Q(\underline{\mathbf{R}}'),$$

hence the singular function $\gamma_Q(\underline{\mathbf{R}})$ of the surface S_Q of V_Q emerges. Consequently, intergration over infinite space yields

$$\int_{-\infty}^{\infty}\int_{-\infty}^{\infty}\int_{-\infty}^{\infty} [-\nabla' \cdot \underline{\mathbf{f}}(\underline{\mathbf{R}}', \omega)]\, G(\underline{\mathbf{R}} - \underline{\mathbf{R}}', \omega)\, \mathrm{d}^3\underline{\mathbf{R}}' \Longrightarrow$$

$$\iiint_{V_Q} [-\nabla' \cdot \underline{\mathbf{f}}(\underline{\mathbf{R}}', \omega)]\, G(\underline{\mathbf{R}} - \underline{\mathbf{R}}', \omega)\, \mathrm{d}^3\underline{\mathbf{R}}' + \iint_{S_Q} \underline{\mathbf{n}}'_Q \cdot \underline{\mathbf{f}}(\underline{\mathbf{R}}', \omega)\, G(\underline{\mathbf{R}} - \underline{\mathbf{R}}', \omega)\, \mathrm{d}S';$$

Equation 5.62 and the application of Gauss' theorem to V_Q finally results in the respective term in (5.63).

With

$$\nabla' \cdot [\underline{\mathbf{f}}(\underline{\mathbf{R}}', \omega) G(\underline{\mathbf{R}} - \underline{\mathbf{R}}', \omega)] \\ = [\nabla' \cdot \underline{\mathbf{f}}(\underline{\mathbf{R}}', \omega)] G(\underline{\mathbf{R}} - \underline{\mathbf{R}}', \omega) + \underline{\mathbf{f}}(\underline{\mathbf{R}}', \omega) \cdot \nabla' G(\underline{\mathbf{R}} - \underline{\mathbf{R}}', \omega), \qquad (5.62)$$

we can express $[\nabla' \cdot \underline{\mathbf{f}}]G$ by $\underline{\mathbf{f}} \cdot \nabla' G$. The integral over $\nabla' \cdot [\underline{\mathbf{f}}G]$ can be transformed into a surface integral over the surface S of V with Gauss' theorem producing $\underline{\mathbf{n}}' \cdot [\underline{\mathbf{f}}G] = [\underline{\mathbf{n}}' \cdot \underline{\mathbf{f}}]G$; according to our assumption, we have $\underline{\mathbf{f}} \equiv \underline{\mathbf{0}}$ on S (also holding for the normal components of $\underline{\mathbf{f}}$), hence this integral vanishes. The remaining volume integrals over V can equally be extended over V_Q:

$$p(\underline{\mathbf{R}}, \omega) = \iiint_{V_Q} [\mathrm{j}\omega\rho\, h(\underline{\mathbf{R}}', \omega) G(\underline{\mathbf{R}} - \underline{\mathbf{R}}', \omega) \\ + \underline{\mathbf{f}}(\underline{\mathbf{R}}', \omega) \cdot \nabla' G(\underline{\mathbf{R}} - \underline{\mathbf{R}}', \omega)]\, \mathrm{d}^3\underline{\mathbf{R}}'. \qquad (5.63)$$

In this integral representation of the pressure source field, the sources $h(\underline{\mathbf{R}}', \omega)$ and $\underline{\mathbf{f}}(\underline{\mathbf{R}}', \omega)$ appear *explicitly*. One says that (de Hoop 1995): The inhomogeneity $h(\underline{\mathbf{R}}', \omega)$ of the pressure rate equation (5.25) requires the scalar Green function $G^{ph}(\underline{\mathbf{R}} - \underline{\mathbf{R}}', \omega) = \mathrm{j}\omega\rho G(\underline{\mathbf{R}} - \underline{\mathbf{R}}', \omega)$, whereas the inhomogeneity $\underline{\mathbf{f}}(\underline{\mathbf{R}}', \omega)$ of the equation of motion (5.24) requires the vector Green function $\underline{\mathbf{G}}^{pf}(\underline{\mathbf{R}} - \underline{\mathbf{R}}', \omega) = \nabla' G(\underline{\mathbf{R}} - \underline{\mathbf{R}}', \omega) = -\nabla G(\underline{\mathbf{R}} - \underline{\mathbf{R}}', \omega)$ (compare Figure 5.1). The point source synthesis defined as such superimposes spherical waves $G(\underline{\mathbf{R}} - \underline{\mathbf{R}}', \omega)$ with direction-independent amplitude and phase and spherical waves $\nabla' G(\underline{\mathbf{R}} - \underline{\mathbf{R}}', \omega)$ with direction-dependent amplitude and phase, the so-called "dipole waves" (Langenberg 2005).

By the way, the integral representation (5.63) holds for all observation points, either in the exterior or in the interior of V_Q: For $\underline{\mathbf{R}} \in V_Q$, a convergent improper integral emerges (Martensen 1968).

5.5.2 Green functions for velocity sources

Evidently, a solution with structure (5.60) of the vector Helmholtz equation (5.38) utilizing the scalar Green function $G(\underline{\mathbf{R}}, \underline{\mathbf{R}}', \omega)$ can be written down for each scalar component. Yet, to arrive at a representation equivalent to (5.63), it is recommended to define a second rank Green tensor of the vector equation (5.39) according to

$$\nabla\nabla \cdot \underline{\underline{\mathbf{G}}}_v(\underline{\mathbf{R}}, \underline{\mathbf{R}}', \omega) + k^2 \underline{\underline{\mathbf{G}}}_v(\underline{\mathbf{R}}, \underline{\mathbf{R}}', \omega) = -\underline{\underline{\mathbf{I}}}\,\delta(\underline{\mathbf{R}} - \underline{\mathbf{R}}'). \qquad (5.64)$$

As usual (Section 13.1), it is advisable to calculate the solution of (5.64) with the help of the three-dimensional spatial Fourier transform; we immediately obtain

$$(\underline{\mathbf{K}}\underline{\mathbf{K}} - k^2\, \underline{\underline{\mathbf{I}}}) \cdot \underline{\underline{\tilde{\mathbf{G}}}}_v(\underline{\mathbf{K}}, \underline{\mathbf{R}}', \omega) = \underline{\underline{\mathbf{I}}}\, \mathrm{e}^{-\mathrm{j}\underline{\mathbf{K}} \cdot \underline{\mathbf{R}}'}, \qquad (5.65)$$

and the application of two Chen formulas for the calculation of det $(\underline{\mathbf{K}}\,\underline{\mathbf{K}} - k^2 \underline{\underline{\mathbf{I}}})$ and adj $(\underline{\mathbf{K}}\,\underline{\mathbf{K}} - k^2 \underline{\underline{\mathbf{I}}})$ (Chen 1983; Appendix "Formula Collection") results in

$$\tilde{\underline{\underline{\mathbf{G}}}}_v(\underline{\mathbf{K}}, \underline{\mathbf{R}}', \omega) = -\frac{1}{k^2}\left(\underline{\underline{\mathbf{I}}} - \frac{\underline{\mathbf{K}}\,\underline{\mathbf{K}}}{K^2 - k^2}\right) \mathrm{e}^{-\mathrm{j}\underline{\mathbf{K}}\cdot\underline{\mathbf{R}}'}. \tag{5.66}$$

Fourier inversion with twofold application of the differentiation theorem (2.341) yields [compare de Hoop (1995)]

$$\begin{aligned}\underline{\underline{\mathbf{G}}}_v(\underline{\mathbf{R}}, \underline{\mathbf{R}}', \omega) &= \underline{\underline{\mathbf{G}}}_v(\underline{\mathbf{R}} - \underline{\mathbf{R}}', \omega) \\ &= -\frac{1}{k^2}\underline{\underline{\mathbf{I}}}\,\delta(\underline{\mathbf{R}} - \underline{\mathbf{R}}') - \frac{1}{k^2}\nabla\nabla\frac{\mathrm{e}^{\mathrm{j}k|\underline{\mathbf{R}}-\underline{\mathbf{R}}'|}}{4\pi|\underline{\mathbf{R}} - \underline{\mathbf{R}}'|} &(5.67)\\ &= -\frac{1}{k^2}\underline{\underline{\mathbf{I}}}\,\delta(\underline{\mathbf{R}} - \underline{\mathbf{R}}') - \frac{1}{k^2}\nabla'\nabla'\frac{\mathrm{e}^{\mathrm{j}k|\underline{\mathbf{R}}-\underline{\mathbf{R}}'|}}{4\pi|\underline{\mathbf{R}} - \underline{\mathbf{R}}'|}. &(5.68)\end{aligned}$$

Two important remarks regarding the mathematical structure of (5.67) are appropriate:

- For $\underline{\mathbf{R}} \neq \underline{\mathbf{R}}'$, the "strange" δ-term[84] is irrelevant and the $\nabla\nabla$-Differentiation applied to the nonsingular scalar Green function for $\underline{\mathbf{R}} \neq \underline{\mathbf{R}}'$ does not cause any problems (for simplicity, we choose $\underline{\mathbf{R}}' = \underline{\mathbf{0}}$):

$$\begin{aligned}\underline{\underline{\mathbf{G}}}_v^{(0)}(\underline{\mathbf{R}}, \omega) &\stackrel{\text{def}}{=} -\frac{1}{k^2}\nabla\nabla\frac{\mathrm{e}^{\mathrm{j}kR}}{4\pi R}, \quad R \neq 0 &(5.69)\\ &= \left[\hat{\underline{\mathbf{R}}}\,\hat{\underline{\mathbf{R}}} - \frac{\mathrm{j}}{kR}(\underline{\underline{\mathbf{I}}} - 3\hat{\underline{\mathbf{R}}}\,\hat{\underline{\mathbf{R}}}) + \frac{1}{k^2R^2}(\underline{\underline{\mathbf{I}}} - 3\hat{\underline{\mathbf{R}}}\,\hat{\underline{\mathbf{R}}})\right]\frac{\mathrm{e}^{\mathrm{j}kR}}{4\pi R}.\end{aligned}$$

- In the resulting source representation[85]

$$\begin{aligned}\underline{\mathbf{v}}(\underline{\mathbf{R}}, \omega) &= \iiint_{V\supset V_Q} \underline{\underline{\mathbf{G}}}_v(\underline{\mathbf{R}} - \underline{\mathbf{R}}', \omega)\cdot\left[\nabla' h(\underline{\mathbf{R}}', \omega) - \mathrm{j}\omega\kappa\,\underline{\mathbf{f}}(\underline{\mathbf{R}}', \omega)\right]\mathrm{d}^3\underline{\mathbf{R}}' \\ &= \iiint_{V\supset V_Q}\left[\nabla' h(\underline{\mathbf{R}}', \omega) - \mathrm{j}\omega\kappa\,\underline{\mathbf{f}}(\underline{\mathbf{R}}', \omega)\right]\cdot\underline{\underline{\mathbf{G}}}_v(\underline{\mathbf{R}} - \underline{\mathbf{R}}', \omega)\,\mathrm{d}^3\underline{\mathbf{R}}'\end{aligned} \tag{5.70}$$

  of the particle velocity, we can basically choose source points exterior and interior of the finite source volume. For source points in the exterior, the above item is relevant, yet for source points in the interior, the δ-term in (5.67) is relevant on one hand, and on the other hand, the $\nabla\nabla$-differentiation of the singular scalar Green function for

[84] We present arguments for its necessity in the last paragraph of this subsection.

[85] We apply the differential operator (5.33) to the first row of (5.70), shift it under the integral—it applies only to $\underline{\mathbf{R}}$—and use (5.64). Since $\underline{\underline{\mathbf{G}}}_v$ is symmetric, we can interchange the factors in the integrand.

$\underline{\mathbf{R}}' = \underline{\mathbf{R}}$ causes problems: We conclude from Equation 5.58 that a twofold $\boldsymbol{\nabla}$-differentiation of the singular scalar Green function has to be interpreted in a distributional sense, it might yield δ-terms. In addition, the last term in (5.69) is obviously $\sim R^{-3}$, and the integral (5.70) over such a (hyper)singular term does not converge in the usual sense (Martensen 1968), yet under certain assumptions[86] (Langenberg 2005) in the sense of a Cauchy PV. A detailed investigation[87] of this singularity topic leads to

$$\begin{aligned}\underline{\underline{\mathbf{G}}}_v(\underline{\mathbf{R}} - \underline{\mathbf{R}}', \omega) &= -\frac{1}{k^2}\underline{\underline{\mathbf{I}}}\,\delta(\underline{\mathbf{R}} - \underline{\mathbf{R}}') \\ &\quad + \mathrm{PV}\,\underline{\underline{\mathbf{G}}}_v^{(0)}(\underline{\mathbf{R}} - \underline{\mathbf{R}}', \omega) + \frac{1}{3k^2}\underline{\underline{\mathbf{I}}}\,\delta(\underline{\mathbf{R}} - \underline{\mathbf{R}}') \\ &= -\frac{2}{3k^2}\underline{\underline{\mathbf{I}}}\,\delta(\underline{\mathbf{R}} - \underline{\mathbf{R}}') + \mathrm{PV}\,\underline{\underline{\mathbf{G}}}_v^{(0)}(\underline{\mathbf{R}} - \underline{\mathbf{R}}', \omega). \end{aligned} \tag{5.71}$$

To transform (5.70) into a structure comparable to (5.63), we advantageously utilize the three-dimensional Fourier transform—we have $\underline{\underline{\tilde{\mathbf{G}}}}_v(\underline{\mathbf{K}}, \omega) = \underline{\underline{\tilde{\mathbf{G}}}}_v(\underline{\mathbf{K}}, \underline{\mathbf{R}}' = \underline{\mathbf{0}}, \omega)$—:

$$\begin{aligned}\underline{\tilde{\mathbf{v}}}(\underline{\mathbf{K}}, \omega) = \underbrace{\mathrm{j}\underline{\mathbf{K}}\,\tilde{h}(\underline{\mathbf{K}}, \omega) \cdot \underline{\underline{\tilde{\mathbf{G}}}}_v(\underline{\mathbf{K}}, \omega)}_{=\tilde{h}(\underline{\mathbf{K}}, \omega)\frac{\mathrm{j}\underline{\mathbf{K}}}{K^2 - k^2}} - \mathrm{j}\omega\kappa\,\underline{\tilde{\mathbf{f}}}(\underline{\mathbf{K}}, \omega) \cdot \underline{\underline{\tilde{\mathbf{G}}}}_v(\underline{\mathbf{K}}, \omega). \end{aligned} \tag{5.72}$$

It follows:[88]

$$\begin{aligned}\underline{\mathbf{v}}(\underline{\mathbf{R}}, \omega) = \iiint_{V_Q} \big[&- \mathrm{j}\omega\kappa\,\underline{\mathbf{f}}(\underline{\mathbf{R}}', \omega) \cdot \underline{\underline{\mathbf{G}}}_v(\underline{\mathbf{R}} - \underline{\mathbf{R}}', \omega) \\ &- h(\underline{\mathbf{R}}', \omega)\boldsymbol{\nabla}' G(\underline{\mathbf{R}} - \underline{\mathbf{R}}', \omega)\big]\, \mathrm{d}^3\underline{\mathbf{R}}'. \end{aligned} \tag{5.73}$$

This integral representation of the particle velocity field again explicitly exhibits the sources $\underline{\mathbf{f}}(\underline{\mathbf{R}}', \omega)$ and $h(\underline{\mathbf{R}}', \omega)$. One says that (de Hoop 1995): The inhomogeneity $\underline{\mathbf{f}}(\underline{\mathbf{R}}', \omega)$ of the equation of motion (5.24) requires the dyadic Green function $\underline{\underline{\mathbf{G}}}^{vf}(\underline{\mathbf{R}} - \underline{\mathbf{R}}', \omega) = -\mathrm{j}\omega\kappa\,\underline{\underline{\mathbf{G}}}_v(\underline{\mathbf{R}} - \underline{\mathbf{R}}', \omega)$ and the inhomogeneity $h(\underline{\mathbf{R}}', \omega)$ of the pressure equation (5.25) requires the vector Green function $\underline{\mathbf{G}}^{vh}(\underline{\mathbf{R}} - \underline{\mathbf{R}}', \omega) = -\boldsymbol{\nabla}' G(\underline{\mathbf{R}} - \underline{\mathbf{R}}', \omega) = \boldsymbol{\nabla} G(\underline{\mathbf{R}} - \underline{\mathbf{R}}', \omega) = -\underline{\mathbf{G}}^{pf}(\underline{\mathbf{R}} - \underline{\mathbf{R}}', \omega)$. Figure 5.1 displays this assignment graphically.

[86]The (small) exclusion volume around the singularity point $\underline{\mathbf{R}}' = \underline{\mathbf{R}} \in V_Q$ must be spherical. If not, a new definition of the PV is required (van Bladel 1991). This is particularly important if (5.70) must be numerically calculated via a discretization of the source volume because the geometry of the underlying voxels becomes important.

[87]In 1961, van Bladel referred to this problem for the first time while investigating the dyadic Green function for electromagnetic waves presenting a heuristic solution (van Bladel 1961). In the meantime, this result has been multiply assured.

[88]The minus sign in $-h\boldsymbol{\nabla}' G$ originates from the application of the differentiation and convolution theorems of the three-dimensional Fourier transform to the bracketed term in (5.72) yielding $\boldsymbol{\nabla}[h(\underline{\mathbf{R}}, \omega) * G(\underline{\mathbf{R}}, \omega)]$ at first.

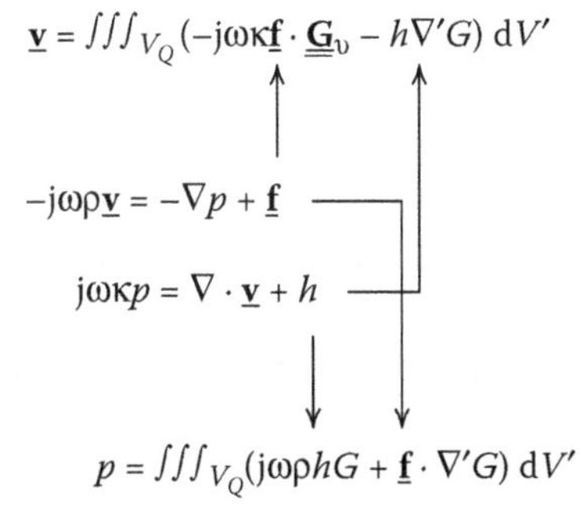

$$\underline{\mathbf{v}} = \iiint_{V_Q} (-\mathrm{j}\omega\kappa \underline{\mathbf{f}} \cdot \underline{\underline{\mathbf{G}}}_{\upsilon} - h\nabla' G)\, \mathrm{d}V'$$

$$-\mathrm{j}\omega\rho\underline{\mathbf{v}} = -\nabla p + \underline{\mathbf{f}}$$

$$\mathrm{j}\omega\kappa p = \nabla \cdot \underline{\mathbf{v}} + h$$

$$p = \iiint_{V_Q} (\mathrm{j}\omega\rho h G + \underline{\mathbf{f}} \cdot \nabla' G)\, \mathrm{d}V'$$

FIGURE 5.1
Assignment of Green functions for homogeneous isotropic acoustic materials to the source terms $\underline{\mathbf{f}}$ and h.

For source points in the exterior of V_Q, we can also use the time harmonic version

$$\underline{\mathbf{v}}(\underline{\mathbf{R}}, \omega) = \frac{1}{\mathrm{j}\omega\rho} \nabla p(\underline{\mathbf{R}}, \omega) \tag{5.74}$$

of (5.24) to calculate $\underline{\mathbf{v}}(\underline{\mathbf{R}}, \omega)$ from (5.63) via application of the gradient: For $\underline{\mathbf{R}} \neq \underline{\mathbf{R}}'$, the gradient can be shifted under the integral without convergence problems, thus transforming it into $-\nabla'$. The result is (5.73).

Green functions can be equally defined for inhomogeneous $\rho(\underline{\mathbf{R}})\kappa(\underline{\mathbf{R}})$-materials; nevertheless, analytical expressions are only available for special cases: A typical example is the one-dimensionally layered material (Chew 1990). This is the reason that the practical point source synthesis computation of source fields is usually constrained to homogeneous $\rho\kappa$-materials.

5.5.3 Justification of the distributional term appearing in the second rank Green tensor of acoustics

Footnote 84 already announced that there are arguments to justify the δ-term in (5.67). For $\underline{\mathbf{R}} \in V_Q$,

$$\underline{\mathbf{v}}(\underline{\mathbf{R}}, \omega) = \frac{1}{\mathrm{j}\omega\rho} \nabla p(\underline{\mathbf{R}}, \omega) - \frac{1}{\mathrm{j}\omega\rho} \underline{\mathbf{f}}(\underline{\mathbf{R}}, \omega) \tag{5.75}$$

must hold. With the source field representation (5.63) of the pressure that is also valid for $\underline{\mathbf{R}} \in V_Q$, we obtain for $\underline{\mathbf{v}}(\underline{\mathbf{R}}, \omega)$ according to (5.75):

$$\underline{\mathbf{v}}(\underline{\mathbf{R}}, \omega) = \frac{1}{\mathrm{j}\omega\rho} \nabla \iiint_{V_Q} \big[\nabla' G(\underline{\mathbf{R}} - \underline{\mathbf{R}}', \omega) \cdot \underline{\mathbf{f}}(\underline{\mathbf{R}}', \omega) + \mathrm{j}\omega\rho h(\underline{\mathbf{R}}', \omega) G(\underline{\mathbf{R}} - \underline{\mathbf{R}}', \omega)\big]\, \mathrm{d}^3\underline{\mathbf{R}}' - \frac{1}{\mathrm{j}\omega\rho} \underline{\mathbf{f}}(\underline{\mathbf{R}}, \omega). \tag{5.76}$$

Definitely realizing that the convergence of the V_Q-integral over the resulting $\nabla'\nabla' G$-term must be carefully investigated, we shift $\nabla \Longrightarrow -\nabla'$ under the integral:

$$\underline{\mathbf{v}}(\underline{\mathbf{R}}, \omega) = \frac{1}{\mathrm{j}\omega\rho} \iiint_{V_Q} \big[- \nabla'\nabla' G(\underline{\mathbf{R}} - \underline{\mathbf{R}}', \omega) \cdot \underline{\mathbf{f}}(\underline{\mathbf{R}}', \omega) - \mathrm{j}\omega\rho h(\underline{\mathbf{R}}', \omega) \nabla' G(\underline{\mathbf{R}} - \underline{\mathbf{R}}', \omega) \big] \, \mathrm{d}^3\underline{\mathbf{R}}' - \frac{1}{\mathrm{j}\omega\rho} \underline{\mathbf{f}}(\underline{\mathbf{R}}, \omega). \tag{5.77}$$

Considering

$$\iiint_{V_Q} \delta(\underline{\mathbf{R}} - \underline{\mathbf{R}}') \underline{\underline{\mathbf{I}}} \cdot \underline{\mathbf{f}}(\underline{\mathbf{R}}', \omega) \, \mathrm{d}^3\underline{\mathbf{R}}' = \underline{\mathbf{f}}(\underline{\mathbf{R}}, \omega) \tag{5.78}$$

for $\underline{\mathbf{R}} \in V_Q$, we can add the isolated $\underline{\mathbf{f}}$-Term in (5.77) according to

$$\underline{\mathbf{v}}(\underline{\mathbf{R}}, \omega) = \frac{1}{\mathrm{j}\omega\rho} \iiint_{V_Q} \Big\{ \underbrace{\big[- \nabla'\nabla' G(\underline{\mathbf{R}} - \underline{\mathbf{R}}', \omega) - \delta(\underline{\mathbf{R}} - \underline{\mathbf{R}}') \underline{\underline{\mathbf{I}}} \big]}_{= k^2 \underline{\underline{\mathbf{G}}}_v(\underline{\mathbf{R}} - \underline{\mathbf{R}}', \omega)} \cdot \underline{\mathbf{f}}(\underline{\mathbf{R}}', \omega) - \mathrm{j}\omega\rho \, h(\underline{\mathbf{R}}', \omega) \nabla' G(\underline{\mathbf{R}} - \underline{\mathbf{R}}', \omega) \Big\} \, \mathrm{d}^3\underline{\mathbf{R}}' \tag{5.79}$$

to a second rank Green tensor that reveals itself as $k^2 \underline{\underline{\mathbf{G}}}_v(\mathbf{R}'', \omega)$ in comparison to (5.73).

The Green functions in (5.63) do not contain an additional δ-term; why, because for $\underline{\mathbf{R}} \in V_Q$, we have

$$p(\underline{\mathbf{R}}, \omega) = \frac{1}{\mathrm{j}\omega\kappa} \nabla \cdot \underline{\mathbf{v}}(\underline{\mathbf{R}}, \omega) + \frac{1}{\mathrm{j}\omega\kappa} h(\underline{\mathbf{R}}, \omega) \tag{5.80}$$

analogous to (5.75). If we equally insert (5.73), the pendant to (5.77) reads as

$$p(\underline{\mathbf{R}}, \omega) = \iiint_{V_Q} \Big[\nabla' \cdot \underline{\underline{\mathbf{G}}}_v(\underline{\mathbf{R}} - \underline{\mathbf{R}}', \omega) \cdot \underline{\mathbf{f}}(\underline{\mathbf{R}}', \omega) + \frac{1}{\mathrm{j}\omega\kappa} h(\underline{\mathbf{R}}', \omega) \Delta' G(\underline{\mathbf{R}} - \underline{\mathbf{R}}', \omega) \Big] \, \mathrm{d}^3\underline{\mathbf{R}}' + \frac{1}{\mathrm{j}\omega\kappa} h(\underline{\mathbf{R}}, \omega). \tag{5.81}$$

Equation 5.58 shows that the h-term outside the integral is canceled for $\underline{\mathbf{R}} \in V_Q$, it must not be accounted for by an "extra" δ-term; furthermore, we conclude with (5.67) utilizing (5.58) that $\nabla' \cdot \underline{\underline{\mathbf{G}}}_v = \nabla' G$ holds, finally realizing the development of (5.63) from (5.81) that is valid for $\underline{\mathbf{R}} \in V_Q$.

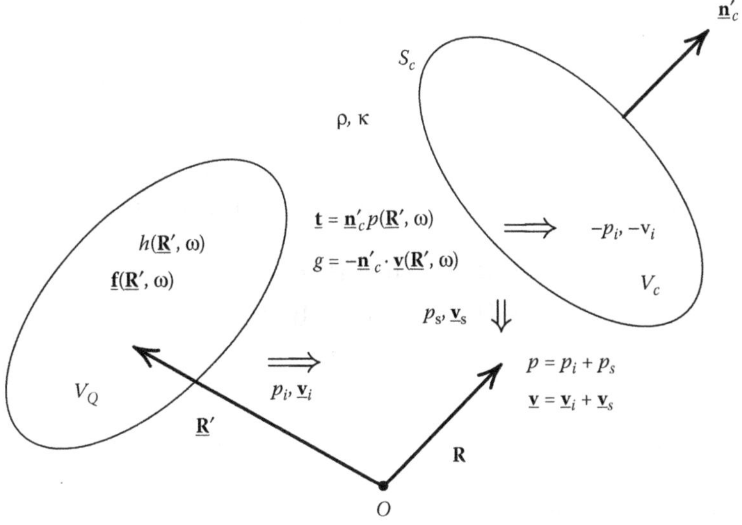

FIGURE 5.2
Acoustic scattering problem: surface sources for Huygens' principle.

5.6 Huygens' Principle for Acoustic Scattered Fields in Homogeneous Materials

5.6.1 Huygens' principle

In the presence of a scattering body with volume V_c and surface S_c—its material properties must not be specified at the moment—the source field, now called incident field $p_i(\underline{\mathbf{R}}, \omega)$, $\underline{\mathbf{v}}_i(\underline{\mathbf{R}}, \omega)$, has to be complemented by a scattered field $p_s(\underline{\mathbf{R}}, \omega)$, $\underline{\mathbf{v}}_s(\underline{\mathbf{R}}, \omega)$: The incident field "is not aware" of the scattering body and, therefore, cannot satisfy the necessary boundary or transition conditions on S_c; hence, these conditions enforce the existence of a scattered field in that way that they are fulfilled by the total field $p(\underline{\mathbf{R}}, \omega)$, $\underline{\mathbf{v}}(\underline{\mathbf{R}}, \omega)$ as the superposition of the incident and the scattered field[89] (Figure 5.2). Huygens' principle postulates a point source synthesis of the scattered field in terms of elementary spherical waves that are weighted with the field values on S_c; they superimpose to the scattered field in the exterior of V_c and they cancel the incident field in the interior of V_c, yielding a zero total field (extinction theorem). The exact mathematical derivation for scalar wave fields (Section 15.1.2)

[89]Per definitionem, the sources of the incident field are defined as prescribed sources without any feedback to the scattered field: V_Q does not represent a scattering body for the scattered field. Of course, this is an idealized model.

reveals[90] that besides spherical waves with direction-independent amplitude and phase dipole waves have to be included, and it renders more precisely the field weights of spherical and dipole waves in terms of the Helmholtz integral (representation theorem). Here, we want to present stringent but heuristic arguments going back to Larmor (1903) for electromagnetic waves: Postulating the extinction theorem, we start with a zero field volume V_c; then the transition conditions (5.11) and (5.12) define field-dependent surface sources

$$\underline{\mathbf{t}}(\underline{\mathbf{R}}', t) = \underline{\mathbf{n}}_c' \, p(\underline{\mathbf{R}}', t), \tag{5.82}$$

$$g(\underline{\mathbf{R}}', t) = -\underline{\mathbf{n}}_c' \cdot \underline{\mathbf{v}}(\underline{\mathbf{R}}', t) \tag{5.83}$$

for $\underline{\mathbf{R}}' \in S_c$, where $\underline{\mathbf{n}}_c'$ denotes the outward normal in the source point on S_c. Exactly, those surface sources are the sources of the scattered field (Figure 5.2)! Therefore they are called equivalent or secondary sources. Hence, we utilize the source field representation (5.63) reduced to[91] on S_c and insert the time harmonic surface sources (5.82) and (5.83):

$$p_s(\underline{\mathbf{R}}, \omega) = \iint_{S_c} \big[-\mathrm{j}\omega\rho \underline{\mathbf{n}}_c' \cdot \underline{\mathbf{v}}(\underline{\mathbf{R}}', \omega) G(\underline{\mathbf{R}} - \underline{\mathbf{R}}', \omega) + p(\underline{\mathbf{R}}', \omega) \underline{\mathbf{n}}_c' \cdot \nabla' G(\underline{\mathbf{R}} - \underline{\mathbf{R}}', \omega) \big] \, \mathrm{d}S'. \tag{5.84}$$

With (5.74), we can represent the first term in (5.84) by the normal derivative of the pressure:

$$p_s(\underline{\mathbf{R}}, \omega) = \iint_{S_c} \big[p(\underline{\mathbf{R}}', \omega) \nabla' G(\underline{\mathbf{R}} - \underline{\mathbf{R}}', \omega) - G(\underline{\mathbf{R}} - \underline{\mathbf{R}}', \omega) \nabla' p(\underline{\mathbf{R}}', \omega) \big] \cdot \underline{\mathbf{n}}_c' \, \mathrm{d}S'; \tag{5.85}$$

this is the exact version of Huygens' principle, i.e., the Helmholtz integral as a point source synthesis of the scattered field.[92] Note: The Helmholtz integral (5.85)—namely, the integral representation (5.84)—actually yields the extinction theorem $p(\underline{\mathbf{R}}, \omega) = p_i(\underline{\mathbf{R}}, \omega) + p_s(\underline{\mathbf{R}}, \omega) = 0$ for $\underline{\mathbf{R}} \in V_c$, hence $p_s(\underline{\mathbf{R}}, \omega) = -p_i(\underline{\mathbf{R}}, \omega)$ for $\underline{\mathbf{R}} \in V_c$; as a consequence, this point source synthesis is only meaningful in the exterior of V_c.

Excluding observation points on S_c in (5.85) and (5.84), respectively, due to the singularity of Green's function, we do not face mathematical problems applying (5.74) to (5.84) to obtain the formulation in terms of a point source

[90] It turns out that Huygens' principle is a mathematical consequence of the Helmholtz equation.

[91] We define volume sources multiplying the surface sources (5.82), (5.83) with the singular function $\gamma_c(\underline{\mathbf{R}})$ of S_c, thus transforming the volume integral of the source field representation into a surface integral (Section 2.4.5).

[92] We recognize the right-hand side of the second Green formula (2.202) that is used to derive the representation (5.85).

synthesis for the scattered particle velocity field:

$$\underline{\mathbf{v}}_s(\underline{\mathbf{R}}, \omega) = \iint_{S_c} \left[\underline{\mathbf{n}}_c' \cdot \underline{\mathbf{v}}(\underline{\mathbf{R}}', \omega) \nabla' G(\underline{\mathbf{R}} - \underline{\mathbf{R}}', \omega) \right.$$
$$\left. - \frac{1}{\mathrm{j}\omega\rho} p(\underline{\mathbf{R}}', \omega) \underline{\mathbf{n}}_c' \cdot \nabla'\nabla' G(\underline{\mathbf{R}} - \underline{\mathbf{R}}', \omega) \right] \mathrm{d}S'$$
$$= \iint_{S_c} \left[\underline{\mathbf{n}}_c' \cdot \underline{\mathbf{v}}(\underline{\mathbf{R}}', \omega) \nabla' G(\underline{\mathbf{R}} - \underline{\mathbf{R}}', \omega) \right.$$
$$\left. - \mathrm{j}\omega\kappa\, p(\underline{\mathbf{R}}', \omega) \underline{\mathbf{n}}_c' \cdot \underline{\underline{\mathbf{G}}}_v(\underline{\mathbf{R}} - \underline{\mathbf{R}}', \omega) \right] \mathrm{d}S'. \qquad (5.86)$$

Obviously, the same result is obtained if we use (5.82) and (5.83) in (5.73).

5.6.2 Acoustic fields scattered by inhomogeneities with soft and rigid boundaries, Kirchhoff approximation

Even though the point source synthesis is physically intuitive it is useless for actual scattering bodies: Two unknown quantities appear under the integral, i.e., the boundary values $\underline{\mathbf{n}}_c' \cdot \underline{\mathbf{v}}(\underline{\mathbf{R}}', \omega)$, $p(\underline{\mathbf{R}}', \omega)$ or $\underline{\mathbf{n}}_c' \cdot \nabla' p(\underline{\mathbf{R}}', \omega)$, $p(\underline{\mathbf{R}}', \omega)$, respectively, of the total field whose scattered contribution should be calculated beforehand with (5.84) and (5.85). How do we proceed to compute the boundary values? We execute the limit $\underline{\mathbf{R}} \longrightarrow S_c$ with the integral representations! Admittedly, this has to be performed with "great care" due to the singularity of Green's function; it turns out that the term with $\nabla' G(\underline{\mathbf{R}} - \underline{\mathbf{R}}', \omega)$ indeed causes problems that are nevertheless solvable. We obtain for instance with (5.85) (Colton and Kress 1983; Langenberg 2005):

$$p_s(\underline{\mathbf{R}}, \omega) = \frac{1}{2} p(\underline{\mathbf{R}}, \omega)$$
$$+ \iint_{S_c} \left[p(\underline{\mathbf{R}}', \omega) \nabla' G(\underline{\mathbf{R}} - \underline{\mathbf{R}}', \omega) - G(\underline{\mathbf{R}} - \underline{\mathbf{R}}', \omega) \nabla' p(\underline{\mathbf{R}}', \omega) \right]$$
$$\cdot \underline{\mathbf{n}}_c' \, \mathrm{d}S', \quad \underline{\mathbf{R}} \in S_c. \qquad (5.87)$$

Replacing p_s on the left-hand side by $p - p_i$, we find

$$\frac{1}{2} p(\underline{\mathbf{R}}, \omega) = p_i(\underline{\mathbf{R}}, \omega)$$
$$+ \iint_{S_c} \left[p(\underline{\mathbf{R}}', \omega) \nabla' G(\underline{\mathbf{R}} - \underline{\mathbf{R}}', \omega) - G(\underline{\mathbf{R}} - \underline{\mathbf{R}}', \omega) \nabla' p(\underline{\mathbf{R}}', \omega) \right]$$
$$\cdot \underline{\mathbf{n}}_c' \, \mathrm{d}S', \quad \underline{\mathbf{R}} \in S_c, \qquad (5.88)$$

as an integral equation relation between the two unknown quantities; this reveals that they are not independent upon each other, prescribing one of them makes the other one calculable. Meaningful, even though idealized assumptions now come into play through the actual physical nature of the scatterer: The volume V_c cannot only be kept "Huygens field-free" but physically field-free by prescribing the boundary condition for either a soft or a rigid scatterer.

The Dirichlet boundary condition (5.16) characterizes the acoustically soft scatterer; inserted into (5.88) results in an integral equation of the first kind (the unknown quantity appears only under the integral)

$$\int\!\!\int_{S_c} G(\underline{\mathbf{R}} - \underline{\mathbf{R}}', \omega)\nabla' p(\underline{\mathbf{R}}', \omega) \cdot \underline{\mathbf{n}}_c' \, \mathrm{d}S' = p_i(\underline{\mathbf{R}}, \omega), \quad \underline{\mathbf{R}} \in S_c, \tag{5.89}$$

for the normal derivative $\underline{\mathbf{n}}_c' \cdot \nabla' p(\underline{\mathbf{R}}', \omega)$ of the pressure or the weighted normal component $\mathrm{j}\omega\rho\underline{\mathbf{n}}_c' \cdot \underline{\mathbf{v}}(\underline{\mathbf{R}}', \omega)$ of the particle velocity, respectively. That way, the surface deformation (5.17) basically becomes calculable.

The Neumann boundary condition (5.18) characterizes the acoustically rigid scatterer; inserted into (5.88) results in an integral equation of the second kind (the unknown quantity also appears outside of the integral)

$$\frac{1}{2} p(\underline{\mathbf{R}}, \omega) - \int\!\!\int_{S_c} p(\underline{\mathbf{R}}', \omega)\underline{\mathbf{n}}_c' \cdot \nabla' G(\underline{\mathbf{R}} - \underline{\mathbf{R}}', \omega) \, \mathrm{d}S' = p_i(\underline{\mathbf{R}}, \omega), \quad \underline{\mathbf{R}} \in S_c, \tag{5.90}$$

for the pressure on S_c.

Only few scattering geometries allow for an analytic solution of the integral equations (5.89) and (5.90) (Bowman et al. 1987); in general, we have to rely on numerical methods that have been developed as the method(s) of moments to simulate electromagnetic fields (Harrington 1968; Wilton 2002); nowadays, also fast multipole methods are under concern (Chew et al. 2002; Michielssen et al. 2002).

The integral equation of the second kind is accessible to a very intuitive physical interpretation—and, hence, to a plausible approximation: Elementary dipole waves originate from each point $\underline{\mathbf{R}}'$ of the surface S_c that are "recorded" at each observation point $\underline{\mathbf{R}}$ on S_c, that is to say, the integral in (5.90) represents the radiation interaction of the surface points of the scatterer. If physical arguments can be found to neglect this interaction, an approximate solution of (5.90) turns out to be

$$p(\underline{\mathbf{R}}, \omega) \simeq 2p_i(\underline{\mathbf{R}}, \omega), \quad \underline{\mathbf{R}} \in S_c. \tag{5.91}$$

This is Kirchhoff's approximation of physical optics[93] (PO: Section 15.2.3). Are there any surfaces S_c for which (5.91) is exact? Yes, planar surfaces: Because $\nabla' G(\underline{\mathbf{R}} - \underline{\mathbf{R}}', \omega) \sim \underline{\mathbf{R}} - \underline{\mathbf{R}}'$ is a vector oriented within the planar surface the normal $\underline{\mathbf{n}}_c'$ is always perpendicular to it, the scalar product

[93] Even though light is an electromagnetic vector wave a scalar notation is often sufficient. The Helmholtz integral consequently describes light diffraction physically as a wave phenomenon, thus distinguishing it from geometrical optics. Today the notation "PO" explicitly stands for the approximation (5.91) of scalar wave fields.

$\underline{\mathbf{n}}_c' \cdot \nabla' G(\underline{\mathbf{R}} - \underline{\mathbf{R}}', \omega)$ is identically zero with the intuitive consequence: The Kirchhoff approximation might be a useful approximation if the surface of the scatterer is only weakly curved relative to the wavelength; the Kirchhoff approximation is a high frequency approximation! Therefore, convex scatterers with a closed surface exhibit, just like the infinitely large planar perforated screen as the originally Kirchhoff approximated problem, an illuminated and a shadow side complementing (5.91) with the requirement: $p(\underline{\mathbf{R}}, \omega) = 0$ on the shadow side of S_c. Inserting this Kirchhoff approximation into the integral representation (5.84), we obtain a PO-approximation of a point source synthesis for the scattered field that can immediately be evaluated because the incident field is supposed to be known. The generalization to elastodynamics reveals itself as one of the simulation methods of US-NDT (Section 15.5). Therefore, we discuss a comparison between exact and Kirchhoff-approximated scattered fields referring to Figures 5.3 and 5.4 to point out the differences for an NDT-relevant example. For simplicity, we choose a two-dimensional problem; the scatterer is supposed to be a planar "crack" of width $2a$ with a Neumann boundary condition, the incident field is supposed to be a plane impulse wave. Such a two-dimensional strip may be considered as the limiting case of a cylinder with elliptic cross-section. The eigenfunctions in elliptic cylindrical coordinates are Mathieu functions (Schäfke 1967); the coefficients of a series

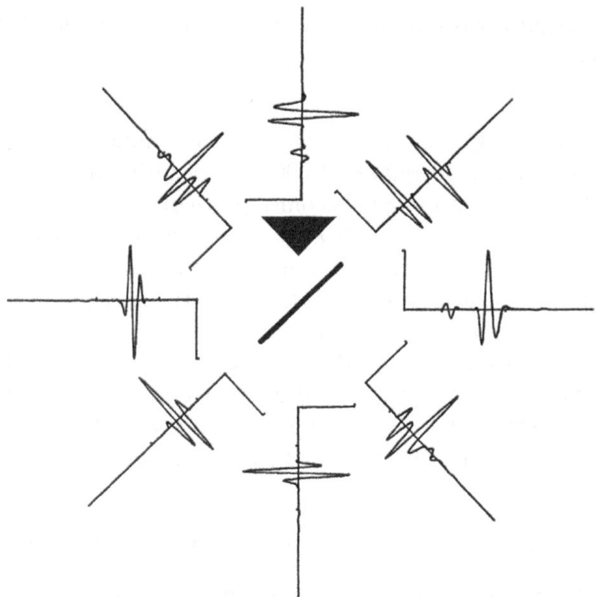

FIGURE 5.3
Acoustic scattered impulses in different far-field directions of an acoustically rigid two-dimensional strip illuminated by a plane wave under 45°: exact calculation.

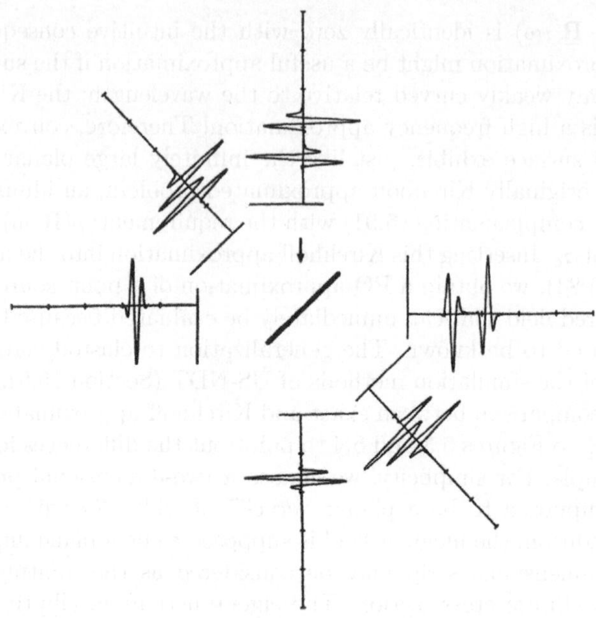

FIGURE 5.4
Acoustic scattered impulses in different far-field directions of an acoustically rigid two-dimensional strip illuminated by a plane wave under 45°: Kirchhoff approximation.

expansion of $p(\underline{\mathbf{R}}, \omega)$, $\underline{\mathbf{R}} \in S_c$, in terms of Mathieu functions can be exactly calculated with the integral equation (5.90); the resulting representation of the scattered field according to (5.84) becomes particularly simple if the far-field approximation for $R \gg a$ and $R \gg \lambda$—λ being the wavelength—is inserted for the normal derivative of the (two-dimensional) Green function (Section 13.1.3). Continuing as such for each frequency in the spectrum of the incident impulse $p_i(t)$, we can calculate the impulsive scattered field via Fourier inversion of (5.84). We have chosen an RC2(t)-pulse for $p_i(t)$ in both figures (Figure 2.20); we have displayed the impulsive scattered far-field $p_s(\underline{\mathbf{R}}, t - R/c)$ retarded with regard of the coordinate origin (in the middle of the strip) for different directions as function of time. Typical features of this scattered field are the following:

- In reflection direction, only a single scattered impulse is observed;
- In backscattering direction (for example), we nicely recognize the two crack tip impulses with opposite phase and unequal amplitude;
- In particular between reflection and backscattering direction, we observe small trailing impulses: They can be traced back to the radiation interaction of the crack tips (so-called resonances).

The results obtained with the Kirchhoff approximation—the evaluation of the integral (5.84) is trivial—as displayed in Figure 5.4 essentially differ with regard to the following:

- The crack tips always exhibit equal amplitudes and
- The radiation interaction impulses are missing (per definitionem).

However, the reflected impulse is exactly reproduced. This means in the time harmonic language: The main lobes in the scattered far-field are sufficiently accurate within the Kirchhoff approximation whereas this is not true for the side lobes. This general statement always holds.

Relying on the point source representation (5.86) for the particle velocity, we can derive an integral equation of the second kind for the Dirichlet problem:

$$\begin{aligned}&\frac{1}{2}\underline{\mathbf{n}}_c \cdot \underline{\mathbf{v}}(\underline{\mathbf{R}}, \omega) + \iint_{S_c} \underline{\mathbf{n}}_c' \cdot \underline{\mathbf{v}}(\underline{\mathbf{R}}', \omega)\underline{\mathbf{n}}_c \cdot \nabla G(\underline{\mathbf{R}} - \underline{\mathbf{R}}', \omega)\,\mathrm{d}S',\\ &\qquad \underline{\mathbf{R}} \in S_c = \underline{\mathbf{n}}_c \cdot \underline{\mathbf{v}}_i(\underline{\mathbf{R}}, \omega),\end{aligned} \tag{5.92}$$

that similarly allows for the justification of the Kirchhoff approximation for acoustically soft scatterers.

For the Neumann problem, Equation 5.86 yields the integral equation of the first kind:

$$\begin{aligned}&-\mathrm{j}\omega\rho\,\underline{\mathbf{n}}_c \cdot \underline{\mathbf{v}}_i(\underline{\mathbf{R}}, \omega)\\ &\quad = \mathrm{PV}_\epsilon \iint_{S_c} p(\underline{\mathbf{R}}', \omega)\underline{\mathbf{n}}_c\underline{\mathbf{n}}_c' : \nabla'\nabla G(\underline{\mathbf{R}} - \underline{\mathbf{R}}', \omega)\,\mathrm{d}S', \quad \underline{\mathbf{R}} \in S_c;\end{aligned} \tag{5.93}$$

here, the specially defined principal value PV_ϵ indicates (Langenberg 2005) that some thoughts have to be devoted to the double normal derivative of the singular Green function.

5.6.3 Acoustic fields scattered by penetrable inhomogeneities, Born approximation

Specifying the scatterer as penetrable for acoustic waves principally allows for a homogeneous or inhomogeneous material filling of V_c. For the first case, the homogeneous equation (5.40) has to be solved with the corresponding material parameters matching it to the solution in the exterior via the transition conditions. Conveniently, another Helmholtz integral representation is chosen for the interior solution[94] with the Green function of the homogeneous interior material (it differs from the exterior Green function with respect to the wave number); by the way, the extinction theorem for this representation tells us that it produces a null field in the exterior. On behalf of the transition conditions (5.13) and (5.14), we now need the respective integral representations

[94]Due to the extinction theorem, the exterior integral representation does not interfere with the solution in the interior.

(5.86) in the exterior and in the interior as solutions of (5.41). With the limits $\underline{\mathbf{R}} \underset{\text{ext}}{\longrightarrow} S_c$ (from the exterior) in the integral representations of the exterior field and $\underline{\mathbf{R}} \underset{\text{int}}{\longrightarrow} S_c$ (from the interior) in the integral representations of the interior field, we finally obtain a coupled system of two integral equations for the unknown Huygens surface sources.

An alternative procedure is applicable to homogeneous as well as inhomogeneous[95] material fillings of V_c; only the knowledge of the Green function of the homogeneous exterior material is required. The basis is the embedding of an inhomogeneous $\rho^{(i)}(\underline{\mathbf{R}})\kappa^{(i)}(\underline{\mathbf{R}})$-volume into the homogeneous $\rho\kappa$-material as sketched in Figure 7.2 (for the elastic case). With the characteristic function $\Gamma_c(\underline{\mathbf{R}})$ of V_c, we define rho- and kappa-contrast functions:

$$\chi_\rho(\underline{\mathbf{R}}) = \frac{1}{\rho}\left[\rho^{(i)}(\underline{\mathbf{R}}) - \rho\right]\Gamma_c(\underline{\mathbf{R}}), \tag{5.94}$$

$$\chi_\kappa(\underline{\mathbf{R}}) = \frac{1}{\kappa}\left[\kappa^{(i)}(\underline{\mathbf{R}}) - \kappa\right]\Gamma_c(\underline{\mathbf{R}}), \tag{5.95}$$

which are evidently zero outside V_c. Within the general differential equations (5.26) and (5.29) for $p(\underline{\mathbf{R}}, \omega)$ and $\underline{\mathbf{v}}(\underline{\mathbf{R}}, \omega)$ of the inhomogeneous $\rho(\underline{\mathbf{R}})\kappa(\underline{\mathbf{R}})$-material, we only have to put

$$\rho(\underline{\mathbf{R}}) = \rho\,[1 + \chi_\rho(\underline{\mathbf{R}})] = \begin{cases} \rho & \text{for } \underline{\mathbf{R}} \notin V_c \\ \rho^{(i)}(\underline{\mathbf{R}}) & \text{for } \underline{\mathbf{R}} \in V_c, \end{cases} \tag{5.96}$$

$$\kappa(\underline{\mathbf{R}}) = \kappa\,[1 + \chi_\kappa(\underline{\mathbf{R}})] = \begin{cases} \kappa & \text{for } \underline{\mathbf{R}} \notin V_c \\ \kappa^{(i)}(\underline{\mathbf{R}}) & \text{for } \underline{\mathbf{R}} \in V_c \end{cases} \tag{5.97}$$

and to arrange the resulting terms in (5.26) and (5.29) in a way[96] that only the differential operators (5.32) and (5.33) of the homogeneous embedding material remain on the left-hand side of the differential equations:

$$\begin{aligned} \Delta p(\underline{\mathbf{R}}, t) - \kappa\rho\,\frac{\partial^2 p(\underline{\mathbf{R}}, t)}{\partial t^2} &= \nabla \cdot \left[\underline{\mathbf{f}}(\underline{\mathbf{R}}, t) + \underline{\mathbf{f}}_\rho(\underline{\mathbf{R}}, t)\right] \\ &\quad + \rho\,\frac{\partial}{\partial t}\left[h(\underline{\mathbf{R}}, t) + h_\kappa(\underline{\mathbf{R}}, t)\right], \end{aligned} \tag{5.98}$$

$$\begin{aligned} \nabla\nabla \cdot \underline{\mathbf{v}}(\underline{\mathbf{R}}, t) - \kappa\rho\,\frac{\partial^2 \underline{\mathbf{v}}(\underline{\mathbf{R}}, t)}{\partial t^2} &= -\nabla\left[h(\underline{\mathbf{R}}, t) + h_\kappa(\underline{\mathbf{R}}, t)\right] \\ &\quad - \kappa\,\frac{\partial}{\partial t}\left[\underline{\mathbf{f}}(\underline{\mathbf{R}}, t) + \underline{\mathbf{f}}_\rho(\underline{\mathbf{R}}, t)\right]. \end{aligned} \tag{5.99}$$

The terms

$$\underline{\mathbf{f}}_\rho(\underline{\mathbf{R}}, t) = \Gamma_c(\underline{\mathbf{R}})\left[\rho - \rho^{(i)}(\underline{\mathbf{R}})\right]\frac{\partial \underline{\mathbf{v}}(\underline{\mathbf{R}}, t)}{\partial t}, \tag{5.100}$$

$$h_\kappa(\underline{\mathbf{R}}, t) = -\Gamma_c(\underline{\mathbf{R}})\left[\kappa - \kappa^{(i)}(\underline{\mathbf{R}})\right]\frac{\partial p(\underline{\mathbf{R}}, t)}{\partial t} \tag{5.101}$$

[95] If the Green functions are known, the above method also works.

[96] Note: We first formulate contrast function representations according to (5.96) and (5.97) for $1/\rho(\underline{\mathbf{R}})$ and $1/\kappa(\underline{\mathbf{R}})$ only to undo it subsequently.

appear as secondary volume sources representing the scatterer equivalently. Note: Just like Huygens surface sources, they are dependent on the total field. This becomes immediately clear if we formally consider the right-hand sides of (5.98) and (5.99) as inhomogeneities and apply—after the Fourier transform with respect to t—the point source synthesis method with the Green functions of the homogeneous ρκ-material:

$$\begin{aligned} p(\underline{\mathbf{R}}, \omega) = &\iiint_{V_Q} \big[\mathrm{j}\omega\rho h(\underline{\mathbf{R}}', \omega) G(\underline{\mathbf{R}} - \underline{\mathbf{R}}', \omega) \\ &\qquad + \underline{\mathbf{f}}(\underline{\mathbf{R}}', \omega) \cdot \nabla' G(\underline{\mathbf{R}} - \underline{\mathbf{R}}', \omega) \big] \mathrm{d}^3\underline{\mathbf{R}}' \\ &+ \iiint_{V_c} \big[\mathrm{j}\omega\rho h_\kappa(\underline{\mathbf{R}}', \omega) G(\underline{\mathbf{R}} - \underline{\mathbf{R}}', \omega) \\ &\qquad + \underline{\mathbf{f}}_\rho(\underline{\mathbf{R}}', \omega) \cdot \nabla' G(\underline{\mathbf{R}} - \underline{\mathbf{R}}', \omega) \big] \mathrm{d}^3\underline{\mathbf{R}}', \end{aligned} \tag{5.102}$$

$$\begin{aligned} \underline{\mathbf{v}}(\underline{\mathbf{R}}, \omega) = &\iiint_{V_Q} \big[-\mathrm{j}\omega\kappa \underline{\mathbf{f}}(\underline{\mathbf{R}}', \omega) \cdot \underline{\underline{\mathbf{G}}}_v(\underline{\mathbf{R}} - \underline{\mathbf{R}}', \omega) \\ &\qquad - h(\underline{\mathbf{R}}', \omega) \nabla' G(\underline{\mathbf{R}} - \underline{\mathbf{R}}', \omega) \big] \mathrm{d}^3\underline{\mathbf{R}}' \\ &+ \iiint_{V_c} \big[-\mathrm{j}\omega\kappa \underline{\mathbf{f}}_\rho(\underline{\mathbf{R}}', \omega) \cdot \underline{\underline{\mathbf{G}}}_v(\underline{\mathbf{R}} - \underline{\mathbf{R}}', \omega) \\ &\qquad - h_\kappa(\underline{\mathbf{R}}', \omega) \nabla' G(\underline{\mathbf{R}} - \underline{\mathbf{R}}', \omega) \big] \mathrm{d}^3\underline{\mathbf{R}}'. \end{aligned} \tag{5.103}$$

In each case, we obtain two volume integrals extending over $V_Q \not\subset V_c$ with the true sources—they are only nonvanishing in V_Q—and over V_c with the secondary sources—those are only nonvanishing in $V_c \not\subset V_Q$. Obviously, the integral representations (5.102) and (5.103) typify a separation of the total field $p(\underline{\mathbf{R}}, \omega)$, $\underline{\mathbf{v}}(\underline{\mathbf{R}}, \omega)$ into an incident field

$$\begin{aligned} p_i(\underline{\mathbf{R}}, \omega) = &\iiint_{V_Q} \big[\mathrm{j}\omega\rho h(\underline{\mathbf{R}}', \omega) G(\underline{\mathbf{R}} - \underline{\mathbf{R}}', \omega) \\ &\qquad + \underline{\mathbf{f}}(\underline{\mathbf{R}}', \omega) \cdot \nabla' G(\underline{\mathbf{R}} - \underline{\mathbf{R}}', \omega) \big] \mathrm{d}^3\underline{\mathbf{R}}', \end{aligned} \tag{5.104}$$

$$\begin{aligned} \underline{\mathbf{v}}_i(\underline{\mathbf{R}}, \omega) = &\iiint_{V_Q} \big[-\mathrm{j}\omega\kappa \underline{\mathbf{f}}(\underline{\mathbf{R}}', \omega) \cdot \underline{\underline{\mathbf{G}}}_v(\underline{\mathbf{R}} - \underline{\mathbf{R}}', \omega) \\ &\qquad - h(\underline{\mathbf{R}}', \omega) \nabla' G(\underline{\mathbf{R}} - \underline{\mathbf{R}}', \omega) \big] \mathrm{d}^3\underline{\mathbf{R}}' \end{aligned} \tag{5.105}$$

and a scattered field

$$\begin{aligned} p_s(\underline{\mathbf{R}}, \omega) = &\iiint_{V_c} \big[\mathrm{j}\omega\rho h_\kappa(\underline{\mathbf{R}}', \omega) G(\underline{\mathbf{R}} - \underline{\mathbf{R}}', \omega) \\ &\qquad + \underline{\mathbf{f}}_\rho(\underline{\mathbf{R}}', \omega) \cdot \nabla' G(\underline{\mathbf{R}} - \underline{\mathbf{R}}', \omega) \big] \mathrm{d}^3\underline{\mathbf{R}}', \end{aligned} \tag{5.106}$$

$$\begin{aligned} \underline{\mathbf{v}}_s(\underline{\mathbf{R}}, \omega) = &\iiint_{V_c} \big[-\mathrm{j}\omega\kappa \underline{\mathbf{f}}_\rho(\underline{\mathbf{R}}', \omega) \cdot \underline{\underline{\mathbf{G}}}_v(\underline{\mathbf{R}} - \underline{\mathbf{R}}', \omega) \\ &\qquad - h_\kappa(\underline{\mathbf{R}}', \omega) \nabla' G(\underline{\mathbf{R}} - \underline{\mathbf{R}}', \omega) \big] \mathrm{d}^3\underline{\mathbf{R}}'. \end{aligned} \tag{5.107}$$

It is for this reason that the secondary sources (5.100) and (5.101) as they enter the scattering integrals (5.106), (5.107) depend upon the total field $p(\underline{\mathbf{R}}, \omega)$, $\underline{\mathbf{v}}(\underline{\mathbf{R}}, \omega)$, that is to say, the volume point source synthesis for the field scattered by a penetrable body is, at that point, useless for its explicit calculation as it is true for the Huygens surface point source synthesis (5.84) and (5.86) for the scattered field of a perfectly rigid or soft body. Yet, the calculation of the total field in V_c equally relies on the solution of integral equations. Here we can immediately write down this system of so-called Lippmann–Schwinger integral equations, because we only have to realize that (5.102) holds in all space, especially in the interior of V_c:

$$p(\underline{\mathbf{R}}, \omega) = p_i(\underline{\mathbf{R}}, \omega) + \iiint_{V_c} \big[\mathrm{j}\omega\rho h_\kappa(\underline{\mathbf{R}}', \omega) G(\underline{\mathbf{R}} - \underline{\mathbf{R}}', \omega) + \underline{\mathbf{f}}_\rho(\underline{\mathbf{R}}', \omega) \cdot \nabla' G(\underline{\mathbf{R}} - \underline{\mathbf{R}}', \omega) \big] \mathrm{d}^3\underline{\mathbf{R}}', \quad \underline{\mathbf{R}} \in V_c. \tag{5.108}$$

Merely, some care has to be taken into account with (5.103) due to the singularity $\underline{\underline{\mathbf{G}}}_v$ as it was already discussed; yet, Equation 5.71 tells us the explicit source point behavior of $\underline{\underline{\mathbf{G}}}_v(\underline{\mathbf{R}} - \underline{\mathbf{R}}', \omega)$ for $\underline{\mathbf{R}} = \underline{\mathbf{R}}' \in V_c$ that we only must have in mind evaluating the integral equation

$$\underline{\mathbf{v}}(\underline{\mathbf{R}}, \omega) = \underline{\mathbf{v}}_i(\underline{\mathbf{R}}, \omega) + \iiint_{V_c} \big[-\mathrm{j}\omega\kappa \underline{\mathbf{f}}_\rho(\underline{\mathbf{R}}', \omega) \cdot \underline{\underline{\mathbf{G}}}_v(\underline{\mathbf{R}} - \underline{\mathbf{R}}', \omega) - h_\kappa(\underline{\mathbf{R}}', \omega) \nabla' G(\underline{\mathbf{R}} - \underline{\mathbf{R}}', \omega) \big] \mathrm{d}^3\underline{\mathbf{R}}', \quad \underline{\mathbf{R}} \in V_c. \tag{5.109}$$

Note: We obtain two coupled integral equations if a nonvanishing contrast of the scatterer with respect to the embedding material exists in the density and in the compressibility. Furthermore: A contrast only in the density results in a single vector integral equation—the Lippmann–Schwinger integral equation (5.109)—whereas a single contrast in the compressibility yields the single scalar Lippmann–Schwinger equation (5.108).

As it is true for the surface integral equation (5.90), both volume integrals in (5.108) and (5.109) stand for the radiation interaction in the interior of the scatterer; if this interaction is only weak, we may approximate

$$p(\underline{\mathbf{R}}, \omega) \simeq p_i(\underline{\mathbf{R}}, \omega), \tag{5.110}$$

$$\underline{\mathbf{v}}(\underline{\mathbf{R}}, \omega) \simeq \underline{\mathbf{v}}_i(\underline{\mathbf{R}}, \omega) \tag{5.111}$$

for $\underline{\mathbf{R}} \in V_c$ for insertion into (5.106) and (5.107). This is called Born's approximation. The validity of the Born approximation can only be globally expressed: It is a low frequency approximation for low contrast (Chew 1990).

The actual consequence of Born's and Kirchhoff's approximation can be discussed with the help of (5.108) for vanishing density contrast; in that case, the scalar Lippmann–Schwinger equation reads as

$$p(\underline{\mathbf{R}}, \omega) = p_i(\underline{\mathbf{R}}, \omega) + k^2 \iiint_{V_c} \chi_\kappa(\underline{\mathbf{R}}')p(\underline{\mathbf{R}}', \omega)G(\underline{\mathbf{R}} - \underline{\mathbf{R}}', \omega)\, \mathrm{d}^3\underline{\mathbf{R}}', \quad \underline{\mathbf{R}} \in V_c. \tag{5.112}$$

This integral equation can be nicely abbreviated according to

$$(\mathcal{I} - \mathcal{V}_c)\{p\}(\underline{\mathbf{R}}, \omega) = p_i(\underline{\mathbf{R}}, \omega) \tag{5.113}$$

if a volume scattering operator

$$\mathcal{V}_c\{p\}(\underline{\mathbf{R}}, \omega) = k^2 \iiint_{V_c} \chi_\kappa(\underline{\mathbf{R}}')p(\underline{\mathbf{R}}', \omega)G(\underline{\mathbf{R}} - \underline{\mathbf{R}}', \omega)\, \mathrm{d}^3\underline{\mathbf{R}}', \quad \underline{\mathbf{R}} \in V_c, \tag{5.114}$$

and an identity operator $\mathcal{I}$ with $\mathcal{I}\{p\}(\underline{\mathbf{R}}, \omega) = p(\underline{\mathbf{R}}, \omega)$ are introduced. The formal solution of (5.113) is obtained inverting the operator $\mathcal{I} - \mathcal{V}_c$ according to

$$p(\underline{\mathbf{R}}, \omega) = (\mathcal{I} - \mathcal{V}_c)^{-1}\{p_i\}(\underline{\mathbf{R}}, \omega), \quad \underline{\mathbf{R}} \in V_c. \tag{5.115}$$

It is now explicitly evident that the contrast χ_κ as present in $\mathcal{V}_c$ enters the interior total field nonlinearly and, therefore, also the exterior field:[97] This is the genuine difficulty solving a scattering problem even though the governing equations are linear. So, what is the essence of the Born approximation? It *linearizes* the scattering problem for penetrable scatterers because it simply deletes $\mathcal{V}_c$ in (5.115)!

In the same manner, the Kirchhoff approximation effectively linearizes scattering by a perfectly soft or rigid body. In addition, even the equally nonlinear inverse scattering problem, i.e., the problem to retrieve scatterers from the knowledge of the (measured) scattered field is linearized through Born's and Kirchhoff's approximations; that is why the imaging method termed synthetic aperture focusing technique (SAFT) exactly implies this linearization. (Langenberg 1987; Langenberg et al. 1993a; Langenberg et al. 1999a; Langenberg 2002). Not least, Kirchhoff's and Born's approximations are valued because the nonlinear inversion essentially causes trouble (e.g.: van den Berg 1999; Belkebir and Saillard 2001).

The linearizations according to Born and Kirchhoff, even though physically plausible, are physically absurd: Born's approximation violates energy conservation and Kirchhoff's approximation violates reciprocity (Langenberg 2002).

[97] If we add something to the geometry or to the material parameters of V_c, the field does not change additively.

6

Electromagnetism

6.1 Maxwell Equations, Poynting Vector, Lorentz Force

6.1.1 Maxwell equations

The basis of (macroscopic) electromagnetism are Maxwell's equations

$$\frac{\partial \underline{\mathbf{D}}(\underline{\mathbf{R}}, t)}{\partial t} = \nabla \times \underline{\mathbf{H}}(\underline{\mathbf{R}}, t) - \underline{\mathbf{J}}_\mathrm{e}(\underline{\mathbf{R}}, t), \tag{6.1}$$

$$\frac{\partial \underline{\mathbf{B}}(\underline{\mathbf{R}}, t)}{\partial t} = -\nabla \times \underline{\mathbf{E}}(\underline{\mathbf{R}}, t) - \underline{\mathbf{J}}_\mathrm{m}(\underline{\mathbf{R}}, t), \tag{6.2}$$

$$\nabla \cdot \underline{\mathbf{D}}(\underline{\mathbf{R}}, t) = \varrho_\mathrm{e}(\underline{\mathbf{R}}, t), \tag{6.3}$$

$$\nabla \cdot \underline{\mathbf{B}}(\underline{\mathbf{R}}, t) = \varrho_\mathrm{m}(\underline{\mathbf{R}}, t) \tag{6.4}$$

for the field quantities

- Electric field strength $\underline{\mathbf{E}}(\underline{\mathbf{R}}, t)$,
- Magnetic field strength $\underline{\mathbf{H}}(\underline{\mathbf{R}}, t)$,
- Electric flux density $\underline{\mathbf{D}}(\underline{\mathbf{R}}, t)$,
- Magnetic flux density $\underline{\mathbf{B}}(\underline{\mathbf{R}}, t)$

and the source quantities

- Electric current density $\underline{\mathbf{J}}_\mathrm{e}(\underline{\mathbf{R}}, t)$,
- Magnetic current density $\underline{\mathbf{J}}_\mathrm{m}(\underline{\mathbf{R}}, t)$,
- Electric charge density $\varrho_\mathrm{e}(\underline{\mathbf{R}}, t)$,
- Magnetic charge density $\varrho_\mathrm{m}(\underline{\mathbf{R}}, t)$.

A real physical meaning can only be devoted to the electric current density $\underline{\mathbf{J}}_\mathrm{e}(\underline{\mathbf{R}}, t)$ defined as transport of electric charge; attributing an electric charge q to a specific particle density $n(\underline{\mathbf{R}}, t)$ instead of a mass according to (3.25), we obtain the electric current density

$$\varrho_\mathrm{e}(\underline{\mathbf{R}}, t) = q\, n(\underline{\mathbf{R}}, t), \tag{6.5}$$

and $\underline{\mathbf{J}}_e(\underline{\mathbf{R}}, t)$ is analogously to the mechanical momentum density (Equation 3.26) defined as the corresponding transport quantity

$$\underline{\mathbf{J}}_e(\underline{\mathbf{R}}, t) = \varrho_e(\underline{\mathbf{R}}, t)\underline{\mathbf{v}}(\underline{\mathbf{R}}, t). \tag{6.6}$$

Accordingly, mass conservation (3.28) yields charge conservation

$$\nabla \cdot \underline{\mathbf{J}}_e(\underline{\mathbf{R}}, t) + \frac{\partial \varrho_e(\underline{\mathbf{R}}, t)}{\partial t} = 0 \tag{6.7}$$

in terms of a continuity equation (de Hoop 1995). If magnetic charges would physically exist, we could define the magnetic current density

$$\underline{\mathbf{J}}_m(\underline{\mathbf{R}}, t) = \varrho_m(\underline{\mathbf{R}}, t)\underline{\mathbf{v}}(\underline{\mathbf{R}}, t) \tag{6.8}$$

similarly to (6.6), and we would obtain the continuity equation

$$\nabla \cdot \underline{\mathbf{J}}_m(\underline{\mathbf{R}}, t) + \frac{\partial \varrho_m(\underline{\mathbf{R}}, t)}{\partial t} = 0. \tag{6.9}$$

As a matter of fact, magnetic charge and current densities are only auxiliary quantities primarily resulting from symmetry considerations for Maxwell's equations.

Nevertheless, we find that electric-physical phenomena—interatomic electric loop currents correspond to magnetic moments whose time derivatives are equivalent to magnetic current densities—can be interpreted as magnetic-physical phenomena; furthermore, the jump discontinuity of the tangential component of the electric field strength on an arbitrary closed surface defines a magnetic surface current density that plays an important role within Huygens' principle (Langenberg 2005).

The divergence relations (6.3) and (6.4) are referred to as compatibility relations, because they follow from the "proper" Maxwell equations (6.1) and (6.2) for causal fields with (6.7) and (6.9) (de Hoop 1995); nevertheless, they must be explicitly satisfied by any physical Maxwell field.

6.1.2 Vacuum Maxwell equations

In contrast to acoustic and elastic waves, electromagnetic waves even propagate in vacuum. In vacuum, field strengths and flux densities are related by the "constitutive equations"

$$\underline{\mathbf{D}}(\underline{\mathbf{R}}, t) = \epsilon_0 \underline{\mathbf{E}}(\underline{\mathbf{R}}, t), \tag{6.10}$$

$$\underline{\mathbf{B}}(\underline{\mathbf{R}}, t) = \mu_0 \underline{\mathbf{H}}(\underline{\mathbf{R}}, t), \tag{6.11}$$

where the magnetic field constant $\mu_0 = 4\pi \cdot 10^{-7}$ H/m and the electric field constant ϵ_0 appear; the latter is given by the definition of the speed of light (in vacuum) $c_0 = 299792458$ m/s through $\epsilon_0 = 1/(\mu_0 c_0^2) \simeq 8.8541878 \cdot 10^{-12}$ F/m. Maxwell's equations read in this case

$$\epsilon_0 \frac{\partial \underline{\mathbf{E}}(\underline{\mathbf{R}}, t)}{\partial t} = \nabla \times \underline{\mathbf{H}}(\underline{\mathbf{R}}, t), \tag{6.12}$$

$$\mu_0 \frac{\partial \underline{\mathbf{H}}(\underline{\mathbf{R}}, t)}{\partial t} = -\nabla \times \underline{\mathbf{E}}(\underline{\mathbf{R}}, t), \tag{6.13}$$

$$\nabla \cdot \underline{\mathbf{E}}(\underline{\mathbf{R}}, t) = 0, \tag{6.14}$$

$$\nabla \cdot \underline{\mathbf{H}}(\underline{\mathbf{R}}, t) = 0, \tag{6.15}$$

where the field strengths $\underline{\mathbf{E}}(\underline{\mathbf{R}}, t)$ and $\underline{\mathbf{H}}(\underline{\mathbf{R}}, t)$ (for a stationary observer relative to the $\underline{\mathbf{R}}$-coordinate system) are defined as forces on an infinitesimally small probing charge q moving with a velocity $\underline{\mathbf{v}}(\underline{\mathbf{R}}, t)$ that does not disturb the field. The total force is the sum of the Coulomb force applied by $\underline{\mathbf{E}}$ and the Lorentz force applied by $\underline{\mathbf{H}}$ (Section 6.1.4):

$$\underline{\mathbf{F}}(\underline{\mathbf{R}}, t) = q\underline{\mathbf{E}}(\underline{\mathbf{R}}, t) + q\mu_0 \underline{\mathbf{v}}(\underline{\mathbf{R}}, t) \times \underline{\mathbf{H}}(\underline{\mathbf{R}}, t). \tag{6.16}$$

Of course, sources must be the origin of electromagnetic fields; very often, they reside as prescribed (field-independent) charge and current densities in a spatially restricted source volume V_Q to be added to Maxwell's equations (6.12) through (6.15) in vacuum:

$$\epsilon_0 \frac{\partial \underline{\mathbf{E}}(\underline{\mathbf{R}}, t)}{\partial t} = \nabla \times \underline{\mathbf{H}}(\underline{\mathbf{R}}, t) - \underline{\mathbf{J}}_{\mathrm{e}}(\underline{\mathbf{R}}, t), \tag{6.17}$$

$$\mu_0 \frac{\partial \underline{\mathbf{H}}(\underline{\mathbf{R}}, t)}{\partial t} = -\nabla \times \underline{\mathbf{E}}(\underline{\mathbf{R}}, t) - \underline{\mathbf{J}}_{\mathrm{m}}(\underline{\mathbf{R}}, t), \tag{6.18}$$

$$\nabla \cdot \underline{\mathbf{E}}(\underline{\mathbf{R}}, t) = \frac{1}{\epsilon_0} \varrho_{\mathrm{e}}(\underline{\mathbf{R}}, t), \tag{6.19}$$

$$\nabla \cdot \underline{\mathbf{H}}(\underline{\mathbf{R}}, t) = \frac{1}{\mu_0} \varrho_{\mathrm{m}}(\underline{\mathbf{R}}, t). \tag{6.20}$$

In nonvacuum, the source volume is embedded in matter, and we have to rely on Maxwell equations (6.1) through (6.4) where the relation between $\underline{\mathbf{D}}$ and $\underline{\mathbf{E}}$, respectively, $\underline{\mathbf{B}}$ and $\underline{\mathbf{H}}$ has to be specified by constitutive equations.

6.1.3 Poynting's theorem

We define the vector of electromagnetic energy flux density (energy per time and area), the Poynting vector, as

$$\underline{\mathbf{S}}(\underline{\mathbf{R}}, t) = \underline{\mathbf{E}}(\underline{\mathbf{R}}, t) \times \underline{\mathbf{H}}(\underline{\mathbf{R}}, t). \tag{6.21}$$

Taking the divergence and insertion into Maxwell equations (6.1) through (6.4) results in Poynting's energy conservation law

$$
\begin{aligned}
&\underbrace{-\underline{\mathbf{H}}(\underline{\mathbf{R}},t)\cdot\underline{\mathbf{J}}_{\mathrm{m}}(\underline{\mathbf{R}},t)-\underline{\mathbf{E}}(\underline{\mathbf{R}},t)\cdot\underline{\mathbf{J}}_{\mathrm{e}}(\underline{\mathbf{R}},t)-\nabla\cdot\underline{\mathbf{S}}(\underline{\mathbf{R}},t)}_{=\frac{\partial w_Q(\underline{\mathbf{R}},t)}{\partial t}}\\
&\qquad=\underbrace{\underline{\mathbf{H}}(\underline{\mathbf{R}},t)\cdot\frac{\partial\underline{\mathbf{B}}(\underline{\mathbf{R}},t)}{\partial t}+\underline{\mathbf{E}}(\underline{\mathbf{R}},t)\cdot\frac{\partial\underline{\mathbf{D}}(\underline{\mathbf{R}},t)}{\partial t}}_{=\frac{\partial w_{\mathrm{em}}(\underline{\mathbf{R}},t)}{\partial t}},
\end{aligned}
\tag{6.22}
$$

where $\partial w_{\mathrm{em}}(\underline{\mathbf{R}},t)/\partial t$ defines the time variation of electromagnetic energy density that is also obtained in vacuum (and in linear nondissipative materials with symmetric $\underline{\underline{\epsilon}}_r$- and $\underline{\underline{\mu}}_r$-tensors: Section 6.3) if the energy density defined for electro/magneto-statics

$$
w_{\mathrm{em}}(\underline{\mathbf{R}},t)=\frac{1}{2}\underline{\mathbf{E}}(\underline{\mathbf{R}},t)\cdot\underline{\mathbf{D}}(\underline{\mathbf{R}},t)+\frac{1}{2}\underline{\mathbf{H}}(\underline{\mathbf{R}},t)\cdot\underline{\mathbf{B}}(\underline{\mathbf{R}},t)
\tag{6.23}
$$

is differentiated with regard to time. In (6.23), $\partial w_Q(\underline{\mathbf{R}},t)/\partial t$ denotes the time variation of the externally applied energy density; to ensure its positiveness, the prescribed current densities must be opposite to the fields.

For real valued time harmonic fields and after time averaging,

$$
\underline{\mathbf{S}}_{\mathrm{K}}(\underline{\mathbf{R}},\omega)=\frac{1}{2}\underline{\mathbf{E}}(\underline{\mathbf{R}},\omega)\times\underline{\mathbf{H}}^*(\underline{\mathbf{R}},\omega)
\tag{6.24}
$$

defines the complex Poynting vector, and the energy conservation law (6.22) takes the form

$$
\nabla\cdot\Re\{\underline{\mathbf{S}}_{\mathrm{K}}(\underline{\mathbf{R}},\omega)\}=-\frac{1}{2}\Re\{\underline{\mathbf{E}}(\underline{\mathbf{R}},\omega)\cdot\underline{\mathbf{J}}_{\mathrm{e}}^*(\underline{\mathbf{R}},\omega)+\underline{\mathbf{H}}(\underline{\mathbf{R}},\omega)\cdot\underline{\mathbf{J}}_{\mathrm{m}}^*(\underline{\mathbf{R}},\omega)\}
\tag{6.25}
$$

for linear nondissipative materials with symmetric $\underline{\underline{\epsilon}}_r$- and $\underline{\underline{\mu}}_r$-tensors.

6.1.4 Lorentz force

Electromagnetic fields bear forces on charges and currents (moving charges); due to these effects, they have actually been discovered (the electrically charged amber gave its Greek name to electricity). Forces appear in a conservation law for the momentum: Equation 3.78 is one of the governing equations of elastodynamics. To find an electromagnetic pendant to (3.78), we must first define an electromagnetic momentum density in a way that its time derivative yields a respective conservation law together with an electromagnetic stress tensor and an electromagnetic force density on the basis of Maxwell equations. We immediately verify that a product of $\underline{\mathbf{D}}(\underline{\mathbf{R}},t)$ and $\underline{\mathbf{B}}(\underline{\mathbf{R}},t)$ has the physical dimension of a momentum density formally defining a momentum density vector through a vector product $\underline{\mathbf{D}}(\underline{\mathbf{R}},t)\times\underline{\mathbf{B}}(\underline{\mathbf{R}},t)$. Investigating its time derivative in vacuum, we immediately obtain utilizing Maxwell equations—we exclusively consider electric charges and electric currents:

$$\epsilon_0\mu_0\frac{\partial}{\partial t}\left[\underline{\mathbf{E}}(\underline{\mathbf{R}},t)\times\underline{\mathbf{H}}(\underline{\mathbf{R}},t)\right] = -\mu_0\underline{\mathbf{H}}(\underline{\mathbf{R}},t)\times\left[\nabla\times\underline{\mathbf{H}}(\underline{\mathbf{R}},t)\right] - \epsilon_0\underline{\mathbf{E}}(\underline{\mathbf{R}},t) \times\left[\nabla\times\underline{\mathbf{E}}(\underline{\mathbf{R}},t)\right] - \mu_0\underline{\mathbf{J}}_e(\underline{\mathbf{R}},t). \tag{6.26}$$

Analogous to (3.78), we must try to create the divergence of a second rank tensor on the right-hand side of (6.26) via adequate conversions; we succeed if we introduce the electromagnetic vacuum stress tensor (Maxwell's stress tensor)

$$\underline{\underline{\mathbf{T}}}_{\mathrm{em}}(\underline{\mathbf{R}},t) = \epsilon_0\underline{\mathbf{E}}(\underline{\mathbf{R}},t)\underline{\mathbf{E}}(\underline{\mathbf{R}},t) + \mu_0\underline{\mathbf{H}}(\underline{\mathbf{R}},t)\underline{\mathbf{H}}(\underline{\mathbf{R}},t) - \left[\frac{\epsilon_0}{2}|\underline{\mathbf{E}}(\underline{\mathbf{R}},t)|^2 + \frac{\mu_0}{2}|\underline{\mathbf{H}}(\underline{\mathbf{R}},t)|^2\right]\underline{\underline{\mathbf{I}}}. \tag{6.27}$$

The result turns out to be the electromagnetic momentum conversation law (in vacuum)

$$\epsilon_0\mu_0\frac{\partial}{\partial t}\left[\underline{\mathbf{E}}(\underline{\mathbf{R}},t)\times\underline{\mathbf{H}}(\underline{\mathbf{R}},t)\right] = \nabla\cdot\underline{\underline{\mathbf{T}}}_{\mathrm{em}}(\underline{\mathbf{R}},t) - \underline{\mathbf{f}}_{\mathrm{em}}(\underline{\mathbf{R}},t) \tag{6.28}$$

with the force density

$$\underline{\mathbf{f}}_{\mathrm{em}}(\underline{\mathbf{R}},t) = \rho_e(\underline{\mathbf{R}},t)\underline{\mathbf{E}}(\underline{\mathbf{R}},t) + \mu_0\underline{\mathbf{J}}_e(\underline{\mathbf{R}},t)\times\underline{\mathbf{H}}(\underline{\mathbf{R}},t). \tag{6.29}$$

In (6.29), the first term represents the Coulomb force density and the second term the Lorentz force density, the latter being the essential basis for the construction of EMATs.

A unique separation of the right-hand side of (6.28) into $\nabla\cdot\underline{\underline{\mathbf{T}}}_{\mathrm{em}}$ and $\underline{\mathbf{f}}_{\mathrm{em}}$ is not possible for electrically and/or magnetically polarizable materials: Maxwell equations only define the sum of both terms. Nevertheless, with a certain arbitrariness explicit expressions for $\underline{\underline{\mathbf{T}}}_{\mathrm{em}}$ and $\underline{\mathbf{f}}_{\mathrm{em}}$ can also be obtained (Jackson 1975).

6.2 Transition and Boundary Conditions

From Maxwell equations (6.1) and (6.2), inhomogeneous transition conditions

$$\underline{\mathbf{n}}\times\left[\underline{\mathbf{H}}^{(1)}(\underline{\mathbf{R}}_S,t) - \underline{\mathbf{H}}^{(2)}(\underline{\mathbf{R}}_S,t)\right] = \underline{\mathbf{K}}_e(\underline{\mathbf{R}}_S,t), \tag{6.30}$$

$$\underline{\mathbf{n}}\times\left[\underline{\mathbf{E}}^{(1)}(\underline{\mathbf{R}}_S,t) - \underline{\mathbf{E}}^{(2)}(\underline{\mathbf{R}}_S,t)\right] = -\underline{\mathbf{K}}_m(\underline{\mathbf{R}}_S,t) \tag{6.31}$$

can immediately be deduced for a surface $\underline{\mathbf{R}} = \underline{\mathbf{R}}_S \in S$ separating two materials with a jump discontinuity of electromagnetic properties; $\underline{\mathbf{K}}_e(\underline{\mathbf{R}}_S,t)$ and

$\underline{\mathbf{K}}_{\mathrm{m}}(\underline{\mathbf{R}}_S, t)$ are electric and magnetic surface current densities that are related to the time variation of surface charge densities $\eta_{\mathrm{e,m}}(\underline{\mathbf{R}}_S, t)$ via the surface divergence operator $\boldsymbol{\nabla}_S$ according to

$$\boldsymbol{\nabla}_S \cdot \underline{\mathbf{K}}_{\mathrm{e,m}}(\underline{\mathbf{R}}_S, t) + \frac{\partial \eta_{\mathrm{e,m}}(\underline{\mathbf{R}}_S, t)}{\partial t} = 0, \tag{6.32}$$

provided the materials separated by S are nonconducting, because, in that case, the surface charges may drift into the material(s) as volume currents. Exactly, those surface charge densities appear in the compatibility transition conditions

$$\underline{\mathbf{n}} \cdot [\underline{\mathbf{D}}^{(1)}(\underline{\mathbf{R}}_S, t) - \underline{\mathbf{D}}^{(2)}(\underline{\mathbf{R}}_S, t)] = \eta_{\mathrm{e}}(\underline{\mathbf{R}}_S, t), \tag{6.33}$$

$$\underline{\mathbf{n}} \cdot [\underline{\mathbf{B}}^{(1)}(\underline{\mathbf{R}}_S, t) - \underline{\mathbf{B}}^{(2)}(\underline{\mathbf{R}}_S, t)] = \eta_{\mathrm{m}}(\underline{\mathbf{R}}_S, t) \tag{6.34}$$

that follow from the compatibility relations (6.3) and (6.4). If there are no prescribed surface currents, the homogeneous transition conditions

$$\underline{\mathbf{n}} \times [\underline{\mathbf{H}}^{(1)}(\underline{\mathbf{R}}_S, t) - \underline{\mathbf{H}}^{(2)}(\underline{\mathbf{R}}_S, t)] = \underline{\mathbf{0}}, \tag{6.35}$$

$$\underline{\mathbf{n}} \times [\underline{\mathbf{E}}^{(1)}(\underline{\mathbf{R}}_S, t) - \underline{\mathbf{E}}^{(2)}(\underline{\mathbf{R}}_S, t)] = \underline{\mathbf{0}}, \tag{6.36}$$

$$\underline{\mathbf{n}} \cdot [\underline{\mathbf{D}}^{(1)}(\underline{\mathbf{R}}_S, t) - \underline{\mathbf{D}}^{(2)}(\underline{\mathbf{R}}_S, t)] = 0, \tag{6.37}$$

$$\underline{\mathbf{n}} \cdot [\underline{\mathbf{B}}^{(1)}(\underline{\mathbf{R}}_S, t) - \underline{\mathbf{B}}^{(2)}(\underline{\mathbf{R}}_S, t)] = 0, \tag{6.38}$$

are obtained that involve the continuity of the tangential components of $\underline{\mathbf{E}}$ and $\underline{\mathbf{H}}$ and—as compatibility—the normal components of $\underline{\mathbf{D}}$ and $\underline{\mathbf{B}}$. Therefore, to maintain discontinuities of these field components surface currents and charges are indispensable: Postulating a field-free "material" (2) according to Huygens' principle yields the definition of tangential components of $\underline{\mathbf{E}}$ and $\underline{\mathbf{H}}$ and normal components of $\underline{\mathbf{D}}$ and $\underline{\mathbf{B}}$ in material (1)—we can omit the index—in terms of surface current and charge densities:

$$\underline{\mathbf{n}} \times \underline{\mathbf{H}}(\underline{\mathbf{R}}_S, t) = \underline{\mathbf{K}}_{\mathrm{e}}(\underline{\mathbf{R}}_S, t), \tag{6.39}$$

$$\underline{\mathbf{n}} \times \underline{\mathbf{E}}(\underline{\mathbf{R}}_S, t) = -\underline{\mathbf{K}}_{\mathrm{m}}(\underline{\mathbf{R}}_S, t), \tag{6.40}$$

$$\underline{\mathbf{n}} \cdot \underline{\mathbf{D}}(\underline{\mathbf{R}}_S, t) = \eta_{\mathrm{e}}(\underline{\mathbf{R}}_S, t), \tag{6.41}$$

$$\underline{\mathbf{n}} \cdot \underline{\mathbf{B}}(\underline{\mathbf{R}}_S, t) = \eta_{\mathrm{m}}(\underline{\mathbf{R}}_S, t), \tag{6.42}$$

that exactly maintain this discontinuity; the normal points away from the null-field. Idealized realizations of field-free materials may be materials with infinite electric or infinite magnetic conductivity; they either allow only for electric current and charge densities or magnetic current and charge densities resulting in the homogeneous boundary conditions

$$\underline{\mathbf{n}} \times \underline{\mathbf{E}}(\underline{\mathbf{R}}_S, t) = \underline{\mathbf{0}}, \tag{6.43}$$

$$\underline{\mathbf{n}} \cdot \underline{\mathbf{B}}(\underline{\mathbf{R}}_S, t) = 0 \tag{6.44}$$

for infinite electrically conducting surfaces S, which exactly define those surface current and charge densities:

$$\underline{\mathbf{n}} \times \underline{\mathbf{H}}(\underline{\mathbf{R}}_S, t) = \underline{\mathbf{K}}_e(\underline{\mathbf{R}}_S, t), \tag{6.45}$$

$$\underline{\mathbf{n}} \cdot \underline{\mathbf{D}}(\underline{\mathbf{R}}_S, t) = \eta_e(\underline{\mathbf{R}}_S, t). \tag{6.46}$$

Equivalently, infinite magnetically conducting surfaces are characterized by the boundary conditions

$$\underline{\mathbf{n}} \times \underline{\mathbf{H}}(\underline{\mathbf{R}}_S, t) = \underline{\mathbf{0}}, \tag{6.47}$$

$$\underline{\mathbf{n}} \cdot \underline{\mathbf{D}}(\underline{\mathbf{R}}_S, t) = 0 \tag{6.48}$$

that define magnetic surface current and charge densities according to

$$\underline{\mathbf{n}} \times \underline{\mathbf{E}}(\underline{\mathbf{R}}_S, t) = -\underline{\mathbf{K}}_m(\underline{\mathbf{R}}_S, t), \tag{6.49}$$

$$\underline{\mathbf{n}} \cdot \underline{\mathbf{B}}(\underline{\mathbf{R}}_S, t) = \eta_m(\underline{\mathbf{R}}_S, t). \tag{6.50}$$

6.3 Constitutive Equations: Permittivity and Permeability; Dissipation: Susceptibility Kernels and Conductivity

6.3.1 Permittivity and permeability

With

$$\underline{\mathbf{D}}(\underline{\mathbf{R}}, t) = \epsilon_0 \underline{\underline{\epsilon}}_r(\underline{\mathbf{R}}) \cdot \underline{\mathbf{E}}(\underline{\mathbf{R}}, t), \tag{6.51}$$

$$\underline{\mathbf{B}}(\underline{\mathbf{R}}, t) = \mu_0 \underline{\underline{\mu}}_r(\underline{\mathbf{R}}) \cdot \underline{\mathbf{H}}(\underline{\mathbf{R}}, t), \tag{6.52}$$

we postulate electromagnetic constitutive equations for a linear time invariant instantaneously and locally reacting inhomogeneous anisotropic material, thus defining the (dimensionless) permittivity tensor of second rank $\underline{\underline{\epsilon}}_r(\underline{\mathbf{R}})$ and the (dimension-less) permeability tensor of second rank $\underline{\underline{\mu}}_r(\underline{\mathbf{R}})$. Generalizations—$\underline{\mathbf{D}}$ also depends upon $\underline{\mathbf{B}}$ and $\underline{\mathbf{B}}$ upon $\underline{\mathbf{E}}$—are called bianisotropic materials (Karlsson and Kristensson 1992). Isotropic materials are characterized by two numbers, the scalar permittivity $\epsilon_r(\underline{\mathbf{R}})$ (dielectric constant) and the scalar permeability $\mu_r(\underline{\mathbf{R}})$, where $\underline{\underline{\epsilon}}_r(\underline{\mathbf{R}}) = \epsilon_r(\underline{\mathbf{R}})\,\underline{\underline{\mathbf{I}}}$, $\underline{\underline{\mu}}_r(\underline{\mathbf{R}}) = \mu_r(\underline{\mathbf{R}})\,\underline{\underline{\mathbf{I}}}$. The time derivative of the electromagnetic energy density (6.23) with the above constitutive equations is only consistent with the respective time derivative appearing in Poynting's theorem (6.22) if $\underline{\underline{\epsilon}}_r(\underline{\mathbf{R}})$ and $\underline{\underline{\mu}}_r(\underline{\mathbf{R}})$ are symmetric tensors.

With regard to permittivity anisotropy, we distinguish uniaxial materials according to

$$\underline{\underline{\epsilon}}_r = \epsilon_\perp \underline{\underline{\mathbf{I}}} + (\epsilon_\parallel - \epsilon_\perp)\,\underline{\hat{\mathbf{c}}}\,\underline{\hat{\mathbf{c}}} \tag{6.53}$$

as well as biaxial materials according to (Chen 1983)

$$\underline{\underline{\epsilon}}_r = \alpha \underline{\underline{I}} + \beta(\underline{\hat{n}}\,\underline{\hat{m}} + \underline{\hat{m}}\,\underline{\hat{n}}). \tag{6.54}$$

For corresponding inhomogeneous anisotropic materials, the permittivities $\epsilon_\perp, \epsilon_\parallel, \alpha, \beta$ and the preference directions $\underline{\hat{c}}$, $\underline{\hat{n}}$, and $\underline{\hat{m}}$ are spatially dependent.

6.3.2 Susceptibility kernels

Dissipation is introduced into linear constitutive equations via susceptibility kernels $\underline{\underline{\chi}}_{\mathrm{e,m}}(\underline{\mathbf{R}}, t)$ within relaxation terms (Karlsson and Kristensson 1992):

$$\underline{\mathbf{D}}(\underline{\mathbf{R}}, t) = \epsilon_0 \underline{\underline{\epsilon}}_r(\underline{\mathbf{R}}) \cdot \underline{\mathbf{E}}(\underline{\mathbf{R}}, t) + \epsilon_0 \int_0^t \underline{\underline{\chi}}_{\mathrm{e}}(\underline{\mathbf{R}}, t-\tau) \cdot \underline{\mathbf{E}}(\underline{\mathbf{R}}, \tau)\, \mathrm{d}\tau, \tag{6.55}$$

$$\underline{\mathbf{B}}(\underline{\mathbf{R}}, t) = \mu_0 \underline{\underline{\mu}}_r(\underline{\mathbf{R}}) \cdot \underline{\mathbf{H}}(\underline{\mathbf{R}}, t) + \mu_0 \int_0^t \underline{\underline{\chi}}_{\mathrm{m}}(\underline{\mathbf{R}}, t-\tau) \cdot \underline{\mathbf{H}}(\underline{\mathbf{R}}, \tau)\, \mathrm{d}\tau. \tag{6.56}$$

For causal fields and kernels the convolution integrals extend from 0 to t; the dissipative material characterized by (6.55) and (6.56) is still time invariant because the susceptibility kernels only depend upon $t - \tau$. Simple models for the susceptibility kernels are related to the names of Lorentz and Debye (Langenberg 2005).

The instantaneous reaction terms in (6.55) and (6.56) enter the electromagnetic (instantaneous) energy density

$$w_{\mathrm{em}}(\underline{\mathbf{R}}, t) = \frac{\epsilon_0}{2} \underline{\mathbf{E}}(\underline{\mathbf{R}}, t) \cdot \underline{\underline{\epsilon}}_r(\underline{\mathbf{R}}) \cdot \underline{\mathbf{E}}(\underline{\mathbf{R}}, t) + \frac{\mu_0}{2} \underline{\mathbf{H}}(\underline{\mathbf{R}}, t) \cdot \underline{\underline{\mu}}_r(\underline{\mathbf{R}}) \cdot \underline{\mathbf{H}}(\underline{\mathbf{R}}, t), \tag{6.57}$$

whereas the relaxation terms define the time derivative of the dissipation energy density

$$\begin{aligned} \frac{\partial w_{\mathrm{d}}(\underline{\mathbf{R}}, t)}{\partial t} = {} & \epsilon_0 \underline{\mathbf{E}}(\underline{\mathbf{R}}, t) \cdot \frac{\partial}{\partial t} \int_0^t \underline{\underline{\chi}}_{\mathrm{e}}(\underline{\mathbf{R}}, t-\tau) \cdot \underline{\mathbf{E}}(\underline{\mathbf{R}}, \tau)\, \mathrm{d}\tau + \mu_0 \underline{\mathbf{H}}(\underline{\mathbf{R}}, t) \\ & \cdot \frac{\partial}{\partial t} \int_0^t \underline{\underline{\chi}}_{\mathrm{m}}(\underline{\mathbf{R}}, t-\tau) \cdot \underline{\mathbf{H}}(\underline{\mathbf{R}}, \tau)\, \mathrm{d}\tau; \end{aligned} \tag{6.58}$$

that is why $\underline{\underline{\chi}}_{\mathrm{e,m}}$ must be nonnegative definite (but not symmetric). Poynting's energy theorem now reads as

$$\frac{\partial w_Q(\underline{\mathbf{R}}, t)}{\partial t} - \nabla \cdot \underline{\mathbf{S}}(\underline{\mathbf{R}}, t) = \frac{\partial w_{\mathrm{em}}(\underline{\mathbf{R}}, t)}{\partial t} + \frac{\partial w_{\mathrm{d}}(\underline{\mathbf{R}}, t)}{\partial t}, \tag{6.59}$$

where $w_{\mathrm{em}}(\underline{\mathbf{R}}, t) + w_{\mathrm{d}}(\underline{\mathbf{R}}, t)$ turns out to be the electromagnetic energy density in dissipative materials; $w_{\mathrm{d}}(\underline{\mathbf{R}}, t)$ emerges from causal integration of (6.58).

After time averaging, we obtain for time harmonic fields

$$
\begin{aligned}
&-\frac{1}{2}\Re\{\underline{\mathbf{H}}(\underline{\mathbf{R}},\omega)\cdot\underline{\mathbf{J}}_{\mathrm{m}}^{*}(\underline{\mathbf{R}},\omega)+\underline{\mathbf{E}}(\underline{\mathbf{R}},\omega)\cdot\underline{\mathbf{J}}_{\mathrm{e}}^{*}(\underline{\mathbf{R}},\omega)\}-\nabla\cdot\underline{\mathbf{S}}_{\mathrm{K}}(\underline{\mathbf{R}},\omega)\\
&\quad=\frac{\epsilon_0}{2}\Re\{\underline{\mathbf{E}}(\underline{\mathbf{R}},\omega)\cdot[-\mathrm{j}\omega\underline{\underline{\boldsymbol{\chi}}}_{\mathrm{e}}(\underline{\mathbf{R}},\omega)\cdot\underline{\mathbf{E}}(\underline{\mathbf{R}},\omega)]^{*}\}\\
&\qquad+\frac{\mu_0}{2}\Re\{\underline{\mathbf{H}}(\underline{\mathbf{R}},\omega)\cdot[-\mathrm{j}\omega\underline{\underline{\boldsymbol{\chi}}}_{\mathrm{m}}(\underline{\mathbf{R}},\omega)\cdot\underline{\mathbf{H}}(\underline{\mathbf{R}},\omega)]^{*}\}\\
&\quad=\frac{\omega\epsilon_0}{2}\Re\{\mathrm{j}\,\underline{\underline{\boldsymbol{\chi}}}_{\mathrm{e}}^{*}(\underline{\mathbf{R}},\omega):\underline{\mathbf{E}}(\underline{\mathbf{R}},\omega)\underline{\mathbf{E}}^{*}(\underline{\mathbf{R}},\omega)\}\\
&\qquad+\frac{\omega\mu_0}{2}\Re\{\mathrm{j}\,\underline{\underline{\boldsymbol{\chi}}}_{\mathrm{m}}^{*}(\underline{\mathbf{R}},\omega):\underline{\mathbf{H}}(\underline{\mathbf{R}},\omega)\underline{\mathbf{H}}^{*}(\underline{\mathbf{R}},\omega)\}\\
&\quad=\frac{\omega\epsilon_0}{4}\Big\{\underline{\mathbf{E}}(\underline{\mathbf{R}},\omega)\cdot\Im\{\underline{\underline{\boldsymbol{\chi}}}_{\mathrm{e}}(\underline{\mathbf{R}},\omega)\}\cdot\underline{\mathbf{E}}^{*}(\underline{\mathbf{R}},\omega)\\
&\qquad\qquad+\underline{\mathbf{E}}^{*}(\underline{\mathbf{R}},\omega)\cdot\Im\{\underline{\underline{\boldsymbol{\chi}}}_{\mathrm{e}}(\underline{\mathbf{R}},\omega)\}\cdot\underline{\mathbf{E}}(\underline{\mathbf{R}},\omega)\\
&\qquad\qquad+\mathrm{j}[\underline{\mathbf{E}}(\underline{\mathbf{R}},\omega)\cdot\Re\{\underline{\underline{\boldsymbol{\chi}}}_{\mathrm{e}}(\underline{\mathbf{R}},\omega)\}\cdot\underline{\mathbf{E}}^{*}(\underline{\mathbf{R}},\omega)\\
&\qquad\qquad-\underline{\mathbf{E}}^{*}(\underline{\mathbf{R}},\omega)\cdot\Re\{\underline{\underline{\boldsymbol{\chi}}}_{\mathrm{e}}(\underline{\mathbf{R}},\omega)\}\cdot\underline{\mathbf{E}}(\underline{\mathbf{R}},\omega)\Big\}\\
&\qquad+\frac{\omega\mu_0}{4}\Big\{\underline{\mathbf{H}}(\underline{\mathbf{R}},\omega)\cdot\Im\{\underline{\underline{\boldsymbol{\chi}}}_{\mathrm{m}}(\underline{\mathbf{R}},\omega)\}\cdot\underline{\mathbf{H}}^{*}(\underline{\mathbf{R}},\omega)\\
&\qquad\qquad+\underline{\mathbf{H}}^{*}(\underline{\mathbf{R}},\omega)\cdot\Im\{\underline{\underline{\boldsymbol{\chi}}}_{\mathrm{m}}(\underline{\mathbf{R}},\omega)\}\cdot\underline{\mathbf{H}}(\underline{\mathbf{R}},\omega)\\
&\qquad\qquad+\mathrm{j}[\underline{\mathbf{H}}(\underline{\mathbf{R}},\omega)\cdot\Re\{\underline{\underline{\boldsymbol{\chi}}}_{\mathrm{m}}(\underline{\mathbf{R}},\omega)\}\cdot\underline{\mathbf{H}}^{*}(\underline{\mathbf{R}},\omega)\\
&\qquad\qquad-\underline{\mathbf{H}}^{*}(\underline{\mathbf{R}},\omega)\cdot\Re\{\underline{\underline{\boldsymbol{\chi}}}_{\mathrm{m}}(\underline{\mathbf{R}},\omega)\}\cdot\underline{\mathbf{H}}(\underline{\mathbf{R}},\omega)\Big\}.
\end{aligned}
\tag{6.60}
$$

Note: The right-hand side of (6.60) is by no means a separation into real and imaginary parts; it is real valued. It becomes obvious from (6.60) that real and imaginary parts of susceptibility kernel spectra, being by the way mutual Hilbert transforms, must be responsible for dissipation. For isotropic kernels—$\underline{\underline{\boldsymbol{\chi}}}_{\mathrm{e,m}}(\underline{\mathbf{R}},t)=\chi_{\mathrm{e,m}}(\underline{\mathbf{R}},t)\,\underline{\underline{\mathbf{I}}}$—we have

$$
\begin{aligned}
&-\frac{1}{2}\Re\{\underline{\mathbf{H}}(\underline{\mathbf{R}},\omega)\cdot\underline{\mathbf{J}}_{\mathrm{m}}^{*}(\underline{\mathbf{R}},\omega)+\underline{\mathbf{E}}(\underline{\mathbf{R}},\omega)\cdot\underline{\mathbf{J}}_{\mathrm{e}}^{*}(\underline{\mathbf{R}},\omega)\}-\nabla\cdot\underline{\mathbf{S}}_{\mathrm{K}}(\underline{\mathbf{R}},\omega)\\
&\quad=\frac{\omega\epsilon_0}{2}\Im\{\chi_{\mathrm{e}}(\underline{\mathbf{R}},\omega)\}|\underline{\mathbf{E}}(\underline{\mathbf{R}},\omega)|^2+\frac{\omega\mu_0}{2}\Im\{\chi_{\mathrm{m}}(\underline{\mathbf{R}},\omega)\}|\underline{\mathbf{H}}(\underline{\mathbf{R}},\omega)|^2;
\end{aligned}
\tag{6.61}
$$

and similarly

$$
\begin{aligned}
&-\frac{1}{2}\Re\{\underline{\mathbf{H}}(\underline{\mathbf{R}},\omega)\cdot\underline{\mathbf{J}}_{\mathrm{m}}^{*}(\underline{\mathbf{R}},\omega)+\underline{\mathbf{E}}(\underline{\mathbf{R}},\omega)\cdot\underline{\mathbf{J}}_{\mathrm{e}}^{*}(\underline{\mathbf{R}},\omega)\}-\nabla\cdot\underline{\mathbf{S}}_{\mathrm{K}}(\underline{\mathbf{R}},\omega)\\
&\quad=\frac{\omega\epsilon_0}{2}\Im\{\underline{\underline{\boldsymbol{\chi}}}_{\mathrm{e}}(\underline{\mathbf{R}},\omega)\}:\underline{\mathbf{E}}^{*}(\underline{\mathbf{R}},\omega)\underline{\mathbf{E}}(\underline{\mathbf{R}},\omega)\\
&\qquad+\frac{\omega\mu_0}{2}\Im\{\underline{\underline{\boldsymbol{\chi}}}_{\mathrm{m}}(\underline{\mathbf{R}},\omega)\}:\underline{\mathbf{H}}^{*}(\underline{\mathbf{R}},\omega)\underline{\mathbf{H}}(\underline{\mathbf{R}},\omega);
\end{aligned}
\tag{6.62}
$$

for the symmetric kernels of a reciprocal material—$\underline{\underline{\boldsymbol{\chi}}}_{\mathrm{e,m}}(\underline{\mathbf{R}},t)=\underline{\underline{\boldsymbol{\chi}}}^{21}_{\mathrm{e,m}}(\underline{\mathbf{R}},t)$. In both cases, only the susceptibility kernels account for dissipation.

6.3.3 Conductivity

It is well known that the (finite) electric conductivity of a material complies with an ohmic resistor that transforms electromagnetic energy into thermal energy. In general, these losses are accounted for by a conduction current term

$$\underline{\mathbf{J}}_{\mathrm{e}}(\underline{\mathbf{R}}, t) \Longrightarrow \underline{\mathbf{J}}_{\mathrm{e}}(\underline{\mathbf{R}}, t) + \underline{\mathbf{J}}_{l}(\underline{\mathbf{R}}, t) \tag{6.63}$$

that complements the electric current density in Maxwell equation (6.1). Ohm's law

$$\underline{\mathbf{J}}_{l}(\underline{\mathbf{R}}, t) = \underline{\underline{\sigma}}_{\mathrm{e}}(\underline{\mathbf{R}}) \cdot \underline{\mathbf{E}}(\underline{\mathbf{R}}, t) \tag{6.64}$$

with the real valued second rank tensor of electric conductivity $\underline{\underline{\sigma}}_{\mathrm{e}}(\underline{\mathbf{R}})$ postulates a linear instantaneous reaction between electric field strength and conduction current density. The same result—complemented by the real valued second rank tensor $\underline{\underline{\sigma}}_{\mathrm{m}}(\underline{\mathbf{R}})$ of magnetic conductivity—is obtained with the constitutive equations

$$\frac{\partial \underline{\mathbf{D}}(\underline{\mathbf{R}}, t)}{\partial t} = \epsilon_0 \underline{\underline{\epsilon}}_r(\underline{\mathbf{R}}) \cdot \frac{\partial \underline{\mathbf{E}}(\underline{\mathbf{R}}, t)}{\partial t} + \underline{\underline{\sigma}}_{\mathrm{e}}(\underline{\mathbf{R}}) \cdot \underline{\mathbf{E}}(\underline{\mathbf{R}}, t), \tag{6.65}$$

$$\frac{\partial \underline{\mathbf{B}}(\underline{\mathbf{R}}, t)}{\partial t} = \mu_0 \underline{\underline{\mu}}_r(\underline{\mathbf{R}}) \cdot \frac{\partial \underline{\mathbf{H}}(\underline{\mathbf{R}}, t)}{\partial t} + \underline{\underline{\sigma}}_{\mathrm{m}}(\underline{\mathbf{R}}) \cdot \underline{\mathbf{H}}(\underline{\mathbf{R}}, t) \tag{6.66}$$

of the Maxwell model of electric–magnetic conductivity [compare the Maxwell model (4.77) and (4.78) of elastodynamic dissipation]. The time variation of electromagnetic energy density

$$\frac{\partial w(\underline{\mathbf{R}}, t)}{\partial t} = \underline{\mathbf{E}}(\underline{\mathbf{R}}, t) \cdot \frac{\partial \underline{\mathbf{D}}(\underline{\mathbf{R}}, t)}{\partial t} + \underline{\mathbf{H}}(\underline{\mathbf{R}}, t) \cdot \frac{\partial \underline{\mathbf{B}}(\underline{\mathbf{R}}, t)}{\partial t} \tag{6.67}$$

then contains the term $\partial w_{\mathrm{e,m}}(\underline{\mathbf{R}}, t)/\partial t$ as it results from (6.57) as well as the term of the time variation of the dissipation energy density:

$$\begin{aligned}\frac{\partial w(\underline{\mathbf{R}}, t)}{\partial t} = &\underbrace{\frac{\epsilon_0}{2} \underline{\underline{\epsilon}}_r(\underline{\mathbf{R}}) : \frac{\partial}{\partial t}[\underline{\mathbf{E}}(\underline{\mathbf{R}}, t)\underline{\mathbf{E}}(\underline{\mathbf{R}}, t)] + \frac{\mu_0}{2} \underline{\underline{\mu}}_r(\underline{\mathbf{R}}) : \frac{\partial}{\partial t}[\underline{\mathbf{H}}(\underline{\mathbf{R}}, t)\underline{\mathbf{H}}(\underline{\mathbf{R}}, t)]}_{= \frac{\partial w_{\mathrm{em}}(\underline{\mathbf{R}}, t)}{\partial t}} \\ &+ \underbrace{\underline{\underline{\sigma}}_{\mathrm{e}}(\underline{\mathbf{R}}) : \underline{\mathbf{E}}(\underline{\mathbf{R}}, t)\underline{\mathbf{E}}(\underline{\mathbf{R}}, t) + \underline{\underline{\sigma}}_{\mathrm{m}}(\underline{\mathbf{R}}) : \underline{\mathbf{H}}(\underline{\mathbf{R}}, t)\underline{\mathbf{H}}(\underline{\mathbf{R}}, t)}_{= \frac{\partial w_{\mathrm{d}}(\underline{\mathbf{R}}, t)}{\partial t}}.\end{aligned} \tag{6.68}$$

Therefore, the conductivity tensors must be nonnegative definite.

If we write (6.65) and (6.66) for Fourier spectra of fields

$$\underline{\mathbf{D}}(\underline{\mathbf{R}}, \omega) = \epsilon_0 \underbrace{\left[\underline{\underline{\epsilon}}_r(\underline{\mathbf{R}}) + \mathrm{j}\,\frac{\underline{\underline{\sigma}}_{\mathrm{e}}(\omega)}{\epsilon_0 \omega}\right]}_{= \underline{\underline{\epsilon}}_c(\underline{\mathbf{R}})} \cdot \underline{\mathbf{E}}(\underline{\mathbf{R}}, \omega), \tag{6.69}$$

$$\underline{\mathbf{B}}(\underline{\mathbf{R}}, \omega) = \mu_0 \underbrace{\left[\underline{\underline{\mu}}_r(\underline{\mathbf{R}}) + \mathrm{j} \frac{\underline{\underline{\sigma}}_{\mathrm{m}}(\omega)}{\mu_0 \omega} \right]}_{= \underline{\underline{\mu}}_c(\underline{\mathbf{R}})} \cdot \underline{\mathbf{H}}(\underline{\mathbf{R}}, \omega), \tag{6.70}$$

we define $\underline{\underline{\epsilon}}_c(\underline{\mathbf{R}})$, $\underline{\underline{\mu}}_c(\underline{\mathbf{R}})$ as complex material tensors with frequency-dependent imaginary parts; a consequence will be wave attenuation and dispersion in conducting materials. Note: Real and imaginary parts of (6.69) and (6.70) do not show up as Hilbert transform relations, not even if $\underline{\underline{\epsilon}}_r(\underline{\mathbf{R}}), \underline{\underline{\mu}}_r(\underline{\mathbf{R}})$ and $\underline{\underline{\sigma}}_{\mathrm{e,m}}(\underline{\mathbf{R}})$ are defined complex valued and frequency-dependent as required by Hilbert transforms (Langenberg 2005). Therefore, strictly speaking, the Maxwell model of conductivity is nonphysical; it even leads to discrepancies regarding dispersion of electromagnetic pulses (Langenberg 2005).

Utilizing (6.68), Poynting's theorem reads for real valued time harmonic fields after time averaging:

$$\begin{aligned} &-\frac{1}{2}\Re\{\underline{\mathbf{E}}(\underline{\mathbf{R}}, \omega) \cdot \underline{\mathbf{J}}_{\mathrm{e}}^*(\underline{\mathbf{R}}, \omega) + \underline{\mathbf{H}}(\underline{\mathbf{R}}, \omega) \cdot \underline{\mathbf{J}}_{\mathrm{m}}^*(\underline{\mathbf{R}}, \omega)\} - \nabla \cdot \underline{\mathbf{S}}_{\mathrm{K}}(\underline{\mathbf{R}}, \omega) \\ &= \frac{1}{2}\Re\{\underline{\underline{\sigma}}_{\mathrm{e}}(\underline{\mathbf{R}}) : \underline{\mathbf{E}}^*(\underline{\mathbf{R}}, \omega)\underline{\mathbf{E}}(\underline{\mathbf{R}}, \omega) + \underline{\underline{\sigma}}_{\mathrm{m}}(\underline{\mathbf{R}}) : \underline{\mathbf{H}}^*(\underline{\mathbf{R}}, \omega)\underline{\mathbf{H}}(\underline{\mathbf{R}}, \omega)\} \end{aligned} \tag{6.71}$$

and for scalar conductivities $\underline{\underline{\sigma}}_{\mathrm{e,m}}(\underline{\mathbf{R}}) = \sigma_{\mathrm{e,m}}(\underline{\mathbf{R}})\,\underline{\underline{\mathbf{I}}}$, respectively,

$$\begin{aligned} &-\frac{1}{2}\Re\{\underline{\mathbf{E}}(\underline{\mathbf{R}}, \omega) \cdot \underline{\mathbf{J}}_{\mathrm{e}}^*(\underline{\mathbf{R}}, \omega) + \underline{\mathbf{H}}(\underline{\mathbf{R}}, \omega) \cdot \underline{\mathbf{J}}_{\mathrm{m}}^*(\underline{\mathbf{R}}, \omega)\} - \nabla \cdot \underline{\mathbf{S}}_{\mathrm{K}}(\underline{\mathbf{R}}, \omega) \\ &= \frac{1}{2}\sigma_{\mathrm{e}}(\underline{\mathbf{R}})|\underline{\mathbf{E}}(\underline{\mathbf{R}}, \omega)|^2 + \frac{1}{2}\sigma_{\mathrm{m}}(\underline{\mathbf{R}})|\underline{\mathbf{H}}(\underline{\mathbf{R}}, \omega)|^2. \end{aligned} \tag{6.72}$$

6.4 Wave Equations in the Time and Frequency Domains

6.4.1 Wave equations in the time domain

From Maxwell's equations

$$\epsilon_0 \underline{\underline{\epsilon}}_r(\underline{\mathbf{R}}) \cdot \frac{\partial \underline{\mathbf{E}}(\underline{\mathbf{R}}, t)}{\partial t} = \nabla \times \underline{\mathbf{H}}(\underline{\mathbf{R}}, t) - \underline{\mathbf{J}}_{\mathrm{e}}(\underline{\mathbf{R}}, t), \tag{6.73}$$

$$\mu_0 \underline{\underline{\mu}}_r(\underline{\mathbf{R}}) \cdot \frac{\partial \underline{\mathbf{H}}(\underline{\mathbf{R}}, t)}{\partial t} = -\nabla \times \underline{\mathbf{E}}(\underline{\mathbf{R}}, t) - \underline{\mathbf{J}}_{\mathrm{m}}(\underline{\mathbf{R}}, t) \tag{6.74}$$

for linear time invariant instantaneously (nondissipative) and locally reacting inhomogeneous anisotropic materials, we deduce vector wave equations for the field strengths via mutual insertion:

$$\begin{aligned} &-\nabla \times \left[\underline{\underline{\mu}}_r^{-1}(\underline{\mathbf{R}}) \cdot \nabla \times \underline{\mathbf{E}}(\underline{\mathbf{R}}, t)\right] - \epsilon_0\mu_0 \underline{\underline{\epsilon}}_r(\underline{\mathbf{R}}) \cdot \frac{\partial^2 \underline{\mathbf{E}}(\underline{\mathbf{R}}, t)}{\partial t^2} \\ &= \mu_0 \frac{\partial \underline{\mathbf{J}}_{\mathrm{e}}(\underline{\mathbf{R}}, t)}{\partial t} + \nabla \times \left[\underline{\underline{\mu}}_r^{-1}(\underline{\mathbf{R}}) \cdot \underline{\mathbf{J}}_{\mathrm{m}}(\underline{\mathbf{R}}, t)\right], \end{aligned} \tag{6.75}$$

$$- \nabla \times \left[\underline{\underline{\epsilon}}_r^{-1}(\underline{\mathbf{R}}) \cdot \nabla \times \underline{\mathbf{H}}(\underline{\mathbf{R}}, t) \right] - \epsilon_0 \mu_0 \underline{\underline{\mu}}_r(\underline{\mathbf{R}}) \cdot \frac{\partial^2 \underline{\mathbf{H}}(\underline{\mathbf{R}}, t)}{\partial t^2}$$
$$= \epsilon_0 \frac{\partial \underline{\mathbf{J}}_{\mathrm{m}}(\underline{\mathbf{R}}, t)}{\partial t} - \nabla \times \left[\underline{\underline{\epsilon}}_r^{-1}(\underline{\mathbf{R}}) \cdot \underline{\mathbf{J}}_{\mathrm{e}}(\underline{\mathbf{R}}, t) \right]; \qquad (6.76)$$

That way, Maxwell equations are physically decoupled, yet not without constitutive equations.

For inhomogeneous anisotropic materials, Equations 6.75 and 6.76 do not really lead somewhere, why we assume "simple" homogeneous isotropic materials with ϵ_r and μ_r in the following:

$$\nabla \times \nabla \times \underline{\mathbf{E}}(\underline{\mathbf{R}}, t) + \frac{1}{c^2} \frac{\partial^2 \underline{\mathbf{E}}(\underline{\mathbf{R}}, t)}{\partial t^2} = -\mu_0 \mu_r \frac{\partial \underline{\mathbf{J}}_{\mathrm{e}}(\underline{\mathbf{R}}, t)}{\partial t} - \nabla \times \underline{\mathbf{J}}_{\mathrm{m}}(\underline{\mathbf{R}}, t), \qquad (6.77)$$

$$\nabla \times \nabla \times \underline{\mathbf{H}}(\underline{\mathbf{R}}, t) + \frac{1}{c^2} \frac{\partial^2 \underline{\mathbf{H}}(\underline{\mathbf{R}}, t)}{\partial t^2} = -\epsilon_0 \epsilon_r \frac{\partial \underline{\mathbf{J}}_{\mathrm{m}}(\underline{\mathbf{R}}, t)}{\partial t} + \nabla \times \underline{\mathbf{J}}_{\mathrm{e}}(\underline{\mathbf{R}}, t), \qquad (6.78)$$

where

$$c = \frac{1}{\sqrt{\epsilon_0 \epsilon_r \mu_0 \mu_r}} \qquad (6.79)$$

denotes the (phase) propagation velocity of electromagnetic waves within the $\epsilon_r \mu_r$-material. The double curl can be transformed according to $\nabla \times \nabla \times = \nabla \nabla \cdot - \Delta$ into the delta operator and the divergence of field strengths; additionally, utilizing the compatibility relations (6.3) and (6.4) for $\epsilon_r \mu_r$-materials equations (6.77) and (6.78) convert into d'Alembert vector wave equations:

$$\Delta \underline{\mathbf{E}}(\underline{\mathbf{R}}, t) - \frac{1}{c^2} \frac{\partial^2 \underline{\mathbf{E}}(\underline{\mathbf{R}}, t)}{\partial t^2}$$
$$= \mu_0 \mu_r \frac{\partial \underline{\mathbf{J}}_{\mathrm{e}}(\underline{\mathbf{R}}, t)}{\partial t} + \nabla \times \underline{\mathbf{J}}_{\mathrm{m}}(\underline{\mathbf{R}}, t) + \frac{1}{\epsilon_0 \epsilon_r} \nabla \varrho_{\mathrm{e}}(\underline{\mathbf{R}}, t), \qquad (6.80)$$

$$\Delta \underline{\mathbf{H}}(\underline{\mathbf{R}}, t) - \frac{1}{c^2} \frac{\partial^2 \underline{\mathbf{H}}(\underline{\mathbf{R}}, t)}{\partial t^2}$$
$$= \epsilon_0 \epsilon_r \frac{\partial \underline{\mathbf{J}}_{\mathrm{m}}(\underline{\mathbf{R}}, t)}{\partial t} - \nabla \times \underline{\mathbf{J}}_{\mathrm{e}}(\underline{\mathbf{R}}, t) + \frac{1}{\mu_0 \mu_r} \nabla \varrho_{\mathrm{m}}(\underline{\mathbf{R}}, t). \qquad (6.81)$$

We point out that both equations are symmetric; this is a consequence of the symmetry of Maxwell equations—in both equations the curl operator appears as spatial derivative—which is neither true in acoustics (Equations 5.32 and 5.33) nor in elastodynamics (Equations 7.21 and 13.211).

In that sense, Maxwell's equations are degenerate, and, as a consequence, the occurring Green functions do not exhibit explicit δ-terms. For electrically homogeneous isotropic conducting materials (we put $\sigma_{\mathrm{m}} = 0$), the Maxwell model (6.65) yields additional terms with the first time derivative of the field strengths, e.g., for (6.80) and (6.81):

$$\Delta \underline{\mathbf{E}}(\underline{\mathbf{R}}, t) - \frac{1}{c^2} \frac{\partial^2 \underline{\mathbf{E}}(\underline{\mathbf{R}}, t)}{\partial t^2} - \mu_0 \mu_r \sigma_{\mathrm{e}} \frac{\partial \underline{\mathbf{E}}(\underline{\mathbf{R}}, t)}{\partial t}$$
$$= \mu_0 \mu_r \frac{\partial \underline{\mathbf{J}}_{\mathrm{e}}(\underline{\mathbf{R}}, t)}{\partial t} + \nabla \times \underline{\mathbf{J}}_{\mathrm{m}}(\underline{\mathbf{R}}, t) + \frac{1}{\epsilon_0 \epsilon_r} \nabla \varrho_{\mathrm{e}}(\underline{\mathbf{R}}, t), \qquad (6.82)$$

$$\Delta\underline{\mathbf{H}}(\underline{\mathbf{R}},t) - \frac{1}{c^2}\frac{\partial^2\underline{\mathbf{H}}(\underline{\mathbf{R}},t)}{\partial t^2} - \mu_0\mu_r\sigma_e\frac{\partial\underline{\mathbf{H}}(\underline{\mathbf{R}},t)}{\partial t}$$
$$= \epsilon_0\epsilon_r\frac{\partial\underline{\mathbf{J}}_m(\underline{\mathbf{R}},t)}{\partial t} - \nabla\times\underline{\mathbf{J}}_e(\underline{\mathbf{R}},t) + \frac{1}{\mu_0\mu_r}\nabla\varrho_m(\underline{\mathbf{R}},t). \tag{6.83}$$

6.4.2 Wave equations in the frequency domain

Via Fourier transform with regard to t, we obtain vector Helmholtz equations from (6.80) and (6.81)

$$\Delta\underline{\mathbf{E}}(\underline{\mathbf{R}},\omega) + k^2\underline{\mathbf{E}}(\underline{\mathbf{R}},\omega)$$
$$= -\mathrm{j}\omega\mu_0\mu_r\underline{\mathbf{J}}_e(\underline{\mathbf{R}},\omega) + \nabla\times\underline{\mathbf{J}}_m(\underline{\mathbf{R}},\omega) + \frac{1}{\epsilon_0\epsilon_r}\nabla\varrho_e(\underline{\mathbf{R}},\omega), \tag{6.84}$$
$$\Delta\underline{\mathbf{H}}(\underline{\mathbf{R}},\omega) + k^2\underline{\mathbf{H}}(\underline{\mathbf{R}},\omega)$$
$$= -\mathrm{j}\omega\epsilon_0\epsilon_r\underline{\mathbf{J}}_m(\underline{\mathbf{R}},\omega) - \nabla\times\underline{\mathbf{J}}_e(\underline{\mathbf{R}},\omega) + \frac{1}{\mu_0\mu_r}\nabla\varrho_m(\underline{\mathbf{R}},\omega), \tag{6.85}$$

where

$$k = \frac{\omega}{c} = \omega\sqrt{\epsilon_0\epsilon_r\mu_0\mu_r} \tag{6.86}$$

denotes the wave number that is proportional to frequency.

As a homogeneous equation

$$\nabla\times\nabla\times\underline{\mathbf{E}}(\underline{\mathbf{R}},\omega) - k^2\underline{\mathbf{E}}(\underline{\mathbf{R}},\omega) = \underline{\mathbf{0}}, \tag{6.87}$$

the Fourier transformed version of (6.77) has advantages over

$$\Delta\underline{\mathbf{E}}(\underline{\mathbf{R}},\omega) + k^2\underline{\mathbf{E}}(\underline{\mathbf{R}},\omega) = \underline{\mathbf{0}}, \tag{6.88}$$

because the solutions of (6.87) are definitely divergence-free ($\operatorname{div}\operatorname{curl}\equiv 0$), whereas this physically necessary condition must be additionally stipulated for the solutions of (6.88). In addition: As a consequence being divergence-free plane electromagnetic waves are transversely polarized.

There are also advantages with (6.77) as an inhomogeneous equation because only the current densities appear; moreover, a dyadic differential operator results from the evaluation of the double curl according to[98]

$$\left[(\Delta + k^2)\underline{\underline{\mathbf{I}}} - \nabla\nabla\right]\cdot\underline{\mathbf{E}}(\underline{\mathbf{R}},\omega) = -\mathrm{j}\omega\mu_0\mu_r\underline{\mathbf{J}}_e(\underline{\mathbf{R}},\omega) + \nabla\times\underline{\mathbf{J}}_m(\underline{\mathbf{R}},\omega) \tag{6.89}$$

that has to be inverted with the method of Green's function to calculate source fields.

If dissipation occurs on behalf of a homogeneous isotropic electric conductivity, the square of the real valued wave number $k^2 = \omega^2\epsilon_0\epsilon_r\mu_0\mu_r$ in the Helmholtz equations (6.84) and (6.89) has to be replaced by the square of the complex wave number

$$k_c^2(\omega) = \omega^2\epsilon_0\epsilon_r\mu_0\mu_r + \mathrm{j}\omega\mu_0\mu_r\sigma_e. \tag{6.90}$$

[98]The divergence stays on the left-hand side.

As a result, neither $\Re k_c(\omega)$ nor $\Im k_c(\omega)$ is proportional to frequency, leading not only to dissipation based wave attenuation but also, due to $\Im k_c(\omega) \neq 0$, wave dispersion. For the limit of very large conductivity defined by $\sigma_e/\omega \gg \epsilon_0\epsilon_r$, the real part $k_c^2(\omega)$ can be neglected for comparatively low frequencies, thus defining eddy current fields; the differential equation (6.82)—the term with the second time derivative is missing—is no longer a wave equation but a diffusion equation. Therefore, we must be careful to use terms for wave propagation—e.g., far-field—for eddy current fields.

6.5 Solutions of Homogeneous Electromagnetic Wave Equations in Homogeneous Isotropic Materials: Plane Transverse Electromagnetic Waves

6.5.1 Nondissipative materials

Solutions of the homogeneous vector Helmholtz equation

$$\Delta\underline{\mathbf{E}}(\underline{\mathbf{R}}, \omega) + k^2\underline{\mathbf{E}}(\underline{\mathbf{R}}, \omega) = \underline{\mathbf{0}} \tag{6.91}$$

must satisfy the compatibility relation $\nabla \cdot \underline{\mathbf{E}}(\underline{\mathbf{R}}, \omega) = 0$; the special solution "plane wave"

$$\underline{\mathbf{E}}(\underline{\mathbf{R}}, \omega) = \underline{\mathbf{E}}(\omega)\, \mathrm{e}^{\pm \mathrm{j}\underline{\mathbf{k}}\cdot\underline{\mathbf{R}}} \tag{6.92}$$

therefore requires the dispersion relation

$$\underline{\mathbf{k}} \cdot \underline{\mathbf{k}} = \omega^2\epsilon_0\epsilon_r\mu_0\mu_r \tag{6.93}$$

for the wave vector and the orthogonality condition

$$\underline{\mathbf{E}}(\omega) \cdot \underline{\mathbf{k}} = 0 \tag{6.94}$$

for the vector amplitude $\underline{\mathbf{E}}(\omega)$. For example, we satisfy[99] (6.93) by

$$\underline{\mathbf{k}} = k\hat{\underline{\mathbf{k}}}. \tag{6.95}$$

According to (6.94), the electric field strength $\underline{\mathbf{E}}(\omega)$ is not allowed to have components in $\hat{\underline{\mathbf{k}}}$-direction; by choosing two orthogonal unit vectors $\hat{\underline{\mathbf{h}}}$ and $\hat{\underline{\mathbf{v}}}$

[99] The complex vector $\underline{\mathbf{k}} = \Re\underline{\mathbf{k}} + \mathrm{j}\Im\underline{\mathbf{k}}$ also satisfies (6.93) if

$$\Re\underline{\mathbf{k}} \cdot \Im\underline{\mathbf{k}} = 0,$$
$$(\Re\underline{\mathbf{k}})^2 - (\Im\underline{\mathbf{k}})^2 = k^2.$$

These are exactly the conditions for evanescent plane waves in nondissipative materials (Figure 9.7).

that are orthogonal to $\hat{\underline{\mathbf{k}}}$, we obtain a right-handed orthogonal trihedron $\hat{\underline{\mathbf{h}}}, \hat{\underline{\mathbf{v}}}, \hat{\underline{\mathbf{k}}}$ with regard to the propagation direction $+\hat{\underline{\mathbf{k}}}$ to consider

$$\underline{\mathbf{E}}(\omega) = E_h(\omega)\hat{\underline{\mathbf{h}}} + E_v(\omega)\hat{\underline{\mathbf{v}}} \tag{6.96}$$

as a two-component so-called Jones vector, where $E_h(\omega) = |E_h(\omega)|\mathrm{e}^{\mathrm{j}\phi_h(\omega)}$, $E_v(\omega) = |E_v(\omega)|\mathrm{e}^{\mathrm{j}\phi_v(\omega)}$ denote two arbitrary complex vector components in the polarization basis $\hat{\underline{\mathbf{h}}}, \hat{\underline{\mathbf{v}}}$.

From the time harmonic Maxwell equation (6.1), we obtain the magnetic field as it belongs to (6.92):

$$\begin{aligned} \underline{\mathbf{H}}(\underline{\mathbf{R}}, \omega) &= \frac{1}{\mathrm{j}\omega\mu_0\mu_r} \,\boldsymbol{\nabla} \times \underline{\mathbf{E}}(\underline{\mathbf{R}}, \omega) \\ &= \pm\frac{1}{Z} \underbrace{\hat{\underline{\mathbf{k}}} \times \underline{\mathbf{E}}(\omega)}_{=\,\underline{\mathbf{H}}(\omega)} \mathrm{e}^{\pm\mathrm{j}k\hat{\underline{\mathbf{k}}}\cdot\underline{\mathbf{R}}}, \end{aligned} \tag{6.97}$$

where

$$Z = \sqrt{\frac{\mu_0\mu_r}{\epsilon_0\epsilon_r}} \tag{6.98}$$

denotes the wave impedance of the $\epsilon_r\mu_r$-material. The vector amplitude of the magnetic field

$$\begin{aligned} \underline{\mathbf{H}}(\omega) &= \hat{\underline{\mathbf{k}}} \times \underline{\mathbf{E}}(\omega) \\ &= -E_v(\omega)\hat{\underline{\mathbf{h}}} + E_h(\omega)\hat{\underline{\mathbf{v}}} \end{aligned} \tag{6.99}$$

is orthogonal to $\hat{\underline{\mathbf{k}}}$ and also orthogonal to $\underline{\mathbf{E}}(\omega)$ due to $\underline{\mathbf{H}}(\omega) \cdot \underline{\mathbf{E}}(\omega) = 0$: A plane electromagnetic wave in an $\epsilon_r\mu_r$-material is electromagnetically transversely polarized. The energy transport occurs in $\pm\hat{\underline{\mathbf{k}}}$-direction because the complex Poynting vector is given by

$$\underline{\mathbf{S}}_{\mathrm{K}}(\underline{\mathbf{R}}, \omega) = \pm\frac{|E_h(\omega)|^2 + |E_v(\omega)|^2}{2Z}\,\hat{\underline{\mathbf{k}}}. \tag{6.100}$$

The notation

$$\underline{\mathbf{E}}(\underline{\mathbf{R}}, \omega) = E_h(\omega)\,\mathrm{e}^{\mathrm{j}k\hat{\underline{\mathbf{k}}}\cdot\underline{\mathbf{R}}} \left[\hat{\underline{\mathbf{h}}} + \frac{E_v(\omega)}{E_h(\omega)}\,\hat{\underline{\mathbf{v}}}\right] \tag{6.101}$$

of (6.92) with (6.96) defines as

$$\begin{aligned} A(\omega) &= \frac{E_v(\omega)}{E_h(\omega)} \\ &= |A(\omega)|\,\mathrm{e}^{\mathrm{j}\Delta\phi(\omega)} \end{aligned} \tag{6.102}$$

the complex polarization number, where $\Delta\phi(\omega) = \phi_v(\omega) - \phi_h(\omega)$ represents the phase difference of both orthogonal field strength components. For

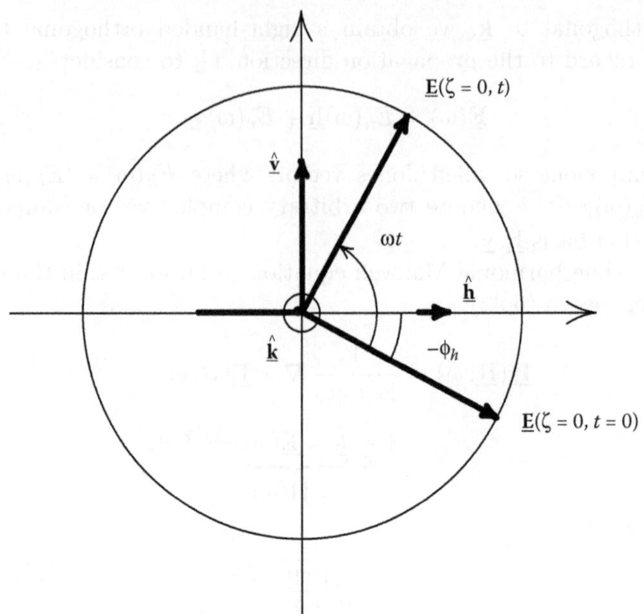

FIGURE 6.1
Right circular polarization.

$A = 0$, we obviously have linear $\hat{\underline{\mathbf{h}}}$orizontal polarization and for[100] $A = \infty$ ($E_h(\omega) = 0$), we have linear $\hat{\underline{\mathbf{v}}}$ertical polarization. For $A = \mathrm{j}$ ($|A| = 0$; $\Delta\phi = \pi/2$), the tip of the real valued time harmonic $\underline{\mathbf{E}}$-field vector

$$\begin{aligned}\underline{\mathbf{E}}(\underline{\mathbf{R}}, t) &= \Re\left\{E_h(\omega)\,\mathrm{e}^{\mathrm{j}k\hat{\underline{\mathbf{k}}}\cdot\underline{\mathbf{R}}}\,\mathrm{e}^{-\mathrm{j}\omega t}(\hat{\underline{\mathbf{h}}} + \mathrm{j}\hat{\underline{\mathbf{v}}})\right\}\\ &= |E_h|[\cos(\omega t - k\hat{\underline{\mathbf{k}}}\cdot\underline{\mathbf{R}} - \phi_h)\hat{\underline{\mathbf{h}}} + \underbrace{\cos(\omega t - k\hat{\underline{\mathbf{k}}}\cdot\underline{\mathbf{R}} - \phi_h - \frac{\pi}{2})}_{= \sin(\omega t - k\hat{\underline{\mathbf{k}}}\cdot\underline{\mathbf{R}} - \phi_h)}\hat{\underline{\mathbf{v}}}]\end{aligned} \tag{6.103}$$

moves on a circle with angular velocity ωt and the initial phase $-\phi_h$ as a function of time in a fixed plane $\hat{\underline{\mathbf{k}}}\cdot\underline{\mathbf{R}} = \zeta = \text{const}$ orthogonal to the propagation direction, for example, in the plane $\zeta = 0$: In electrical engineering, it is called right circular polarization because the movement is in the direction of the bent fingers of the right hand if the thumb points into propagation direction $\hat{\underline{\mathbf{k}}}$ (Figure 6.1). Watching the wave from behind $\underline{\mathbf{E}}(\zeta = 0, t)$ moves clockwise yielding the terminology cw-polarization for right circular. That way, each value of the polarization number $A(\omega)$ in a complex plane defines a characteristic polarization—in general, right or left elliptical polarization

[100]In the complex A-plane there is one point ∞ (Behnke and Sommer 1965).

with arbitrary ellipse orientation relative to the $\hat{\underline{\mathbf{h}}}\,\hat{\underline{\mathbf{v}}}$-basis—we call it the polarization diagram (Langenberg 2005). The change in polarization state of an electromagnetic wave through reflection, diffraction, or scattering is visible in the change of polarization number, i.e., in the transformation of the Jones vector of the incident (plane) wave into the Jones vector of the scattered wave[101]; this transformation is described by 2×2-matrices, the Jones and Sinclair matrices, where the difference is simply in the polarization state of the incident and the scattered waves. These matrices contain the total information about the scatterer, therefore, their algebraic analysis provides an excellent tool for nondestructive testing with microwaves (Cloude 2002).

First of all, the polarization of electromagnetic waves is a concept for time harmonic waves. For arbitrary time dependence, the Fourier inversion of (6.92) with (6.96) according to—we assume $E_{h,v}(-\omega) = E^*_{h,v}(\omega)$—

$$\underline{\mathbf{E}}(\underline{\mathbf{R}}, t) = E_h\left(t \mp \frac{\hat{\underline{\mathbf{k}}} \cdot \underline{\mathbf{R}}}{c}\right)\hat{\underline{\mathbf{h}}} + E_v\left(t \mp \frac{\hat{\underline{\mathbf{k}}} \cdot \underline{\mathbf{R}}}{c}\right)\hat{\underline{\mathbf{v}}} \tag{6.104}$$

allows for a simple interpretation only if we put $E_h(t) = E_v(t)$ assuming linear polarization. If this is not true, we have to switch to time averages (Langenberg 2005).

6.5.2 Dissipative materials

We imply an electrically homogeneous isotropic conducting $\epsilon_r\mu_r$-material with conductivity σ_e. With (6.90), the dispersion relation (6.93) reads as

$$\begin{aligned}\underline{\mathbf{k}} \cdot \underline{\mathbf{k}} &= k_c^2(\omega)\\ &= \omega^2\epsilon_0\epsilon_r\mu_0\mu_r + \mathrm{j}\omega\mu_0\mu_r\sigma_e,\end{aligned} \tag{6.105}$$

i.e., the wave vector is complex, it may[102] be chosen as

$$\underline{\mathbf{k}} = k_c(\omega)\hat{\underline{\mathbf{k}}}. \tag{6.106}$$

Calculation of the complex root yields (Equations 2.268 and 2.269)

$$\Re k_c(\omega) = k\sqrt{\frac{1}{2}\left[1 + \sqrt{1 + \left(\frac{\sigma_e}{\omega\epsilon_0\epsilon_r}\right)}\right]}, \tag{6.107}$$

[101] In the far-field of the scatterer, the scattered field is locally a plane wave.

[102] The general solution of (6.105) is given by: $\underline{\mathbf{k}} = \Re\underline{\mathbf{k}} + \mathrm{j}\Im\underline{\mathbf{k}}$ with

$$\begin{aligned}(\Re\underline{\mathbf{k}})^2 - (\Im\underline{\mathbf{k}})^2 &= \omega^2\epsilon_0\epsilon_r\mu_0\mu_r,\\ \Re\underline{\mathbf{k}} \cdot \Im\underline{\mathbf{k}} &= \frac{1}{2}\,\omega\mu_0\mu_r\sigma_e,\end{aligned}$$

where, in contrast to Footnote 99, we must have $\Re\underline{\mathbf{k}} \cdot \Im\underline{\mathbf{k}} \neq 0$; this case occurs for the transmitted wave if a plane electromagnetic wave impinges on a conducting material: The propagation direction of the phase satisfies Snell's law for $\Re\underline{\mathbf{k}}$, where $\Im\underline{\mathbf{k}}$ is always orthogonal to the surface (Langenberg 2005). For $\Re\underline{\mathbf{k}}$ parallel to $\Im\underline{\mathbf{k}}$, we meet the case as above.

$$\Im k_c(\omega) = k\sqrt{\frac{1}{2}\left[-1+\sqrt{1+\left(\frac{\sigma_e}{\omega\epsilon_0\epsilon_r}\right)}\right]}, \tag{6.108}$$

where this choice of the sign in the plane wave (6.92) according to—we have $\hat{\underline{\mathbf{k}}} \cdot \underline{\mathbf{R}} = \zeta$—

$$\underline{\mathbf{E}}(\zeta, \omega) = \underline{\mathbf{E}}(\omega)e^{j\zeta\Re k_c(\omega)}e^{-\zeta\Im k_c(\omega)} \tag{6.109}$$

ensures an exponential attenuation in $+\zeta$-direction in the half-space $\zeta > 0$. Because (6.107) and (6.108) are no longer proportional to frequency as in the nondissipative material, the impulsive wave corresponding to (6.109) experiences dispersion [a numerical example can be found in (Langenberg 2005)]. This fact must be recognized calculating travel times (for example, for electromagnetic waves propagating in humid masonry).

6.6 Electromagnetic Source Fields in Homogeneous Isotropic Materials, Electromagnetic Tensor Green Functions

The differential equation (6.89) with its dyadic differential operator immediately reveals that the definition of a (time harmonic) dyadic Green function through the differential equation

$$\left[(\Delta + k^2)\underline{\underline{\mathbf{I}}} - \nabla\nabla\right] \cdot \underline{\underline{\mathbf{G}}}_{\mathrm{e}}(\underline{\mathbf{R}}, \underline{\mathbf{R}}', \omega) = -\underline{\underline{\mathbf{I}}}\,\delta(\underline{\mathbf{R}} - \underline{\mathbf{R}}') \tag{6.110}$$

is meaningful; it is denoted as the electric Green dyadic $\underline{\underline{\mathbf{G}}}_{\mathrm{e}}(\underline{\mathbf{R}}, \underline{\mathbf{R}}', \omega)$, because it should finally relate the electric current density with the electric field strength in terms of a point source synthesis. With the same arguments as in Section 13.1.1, we conclude that $\underline{\underline{\mathbf{G}}}_{\mathrm{e}}(\underline{\mathbf{R}}, \underline{\mathbf{R}}', \omega) = \underline{\underline{\mathbf{G}}}_{\mathrm{e}}(\underline{\mathbf{R}} - \underline{\mathbf{R}}', \omega)$ holds. Before we further elaborate the idea of a dyadic Green function to utilize it successfully within the electromagnetic Huygens principle, we cite the usual procedure with electromagnetic potentials as it is applied in the theory of electromagnetic source fields (antenna fields).

6.6.1 Electric scalar potential and magnetic vector potential

We concentrate on electric current and charge densities being nonzero only in the source volume V_Q and "solve" the resulting Maxwell compatibility relation (6.4) with zero divergence of $\underline{\mathbf{B}}(\underline{\mathbf{R}}, t)$ through the *ansatz*

$$\underline{\mathbf{B}}(\underline{\mathbf{R}}, t) = \nabla \times \underline{\mathbf{A}}(\underline{\mathbf{R}}, t) \tag{6.111}$$

of a magnetic vector potential $\underline{\mathbf{A}}(\underline{\mathbf{R}}, t)$. Maxwell equation (6.2) subsequently proposes the representation

$$\underline{\mathbf{E}}(\underline{\mathbf{R}}, t) = -\boldsymbol{\nabla}\Phi(\underline{\mathbf{R}}, t) - \frac{\partial \underline{\mathbf{A}}(\underline{\mathbf{R}}, t)}{\partial t} \tag{6.112}$$

of the electric field strength by a scalar potential $\Phi(\underline{\mathbf{R}}, t)$, where the minus sign of $\boldsymbol{\nabla}\Phi$ comes from the voltage definition in electrostatics. The remaining equations (6.1) and (6.3) finally yield d'Alembert wave equations for $\epsilon_r\mu_r$-materials:

$$\Delta\Phi(\underline{\mathbf{R}}, t) - \frac{1}{c^2}\frac{\partial^2\Phi(\underline{\mathbf{R}}, t)}{\partial t^2} = -\frac{1}{\epsilon_0\epsilon_r}\,\varrho_{\mathrm{e}}(\underline{\mathbf{R}}, t), \tag{6.113}$$

$$\Delta\underline{\mathbf{A}}(\underline{\mathbf{R}}, t) - \frac{1}{c^2}\frac{\partial^2\underline{\mathbf{A}}(\underline{\mathbf{R}}, t)}{\partial t^2} = -\mu_0\mu_r\underline{\mathbf{J}}_{\mathrm{e}}(\underline{\mathbf{R}}, t) \tag{6.114}$$

if the potentials are related by the so-called Lorenz condition

$$\boldsymbol{\nabla}\cdot\underline{\mathbf{A}}(\underline{\mathbf{R}}, t) + \frac{1}{c^2}\frac{\partial\Phi(\underline{\mathbf{R}}, t)}{\partial t} = 0; \tag{6.115}$$

this is always possible in terms of a gauge transform of the potentials[103] (Langenberg 2005). Equations 6.113 and 6.114 reveal that the electric current and charge densities explicitly appear as sources of the potentials, and: We have reduced the vector wave equations (6.80) and (6.81) for the field strengths to *one vector* wave equation and *one scalar* wave equation. The solution

$$\Phi(\underline{\mathbf{R}}, t) = \frac{1}{4\pi\epsilon_0\epsilon_r}\iiint_{V_Q}\frac{\varrho_{\mathrm{e}}\left(\underline{\mathbf{R}}', t - \frac{|\underline{\mathbf{R}}-\underline{\mathbf{R}}'|}{c}\right)}{|\underline{\mathbf{R}}-\underline{\mathbf{R}}'|}\,\mathrm{d}^3\underline{\mathbf{R}}' \tag{6.116}$$

of (6.113) (Equation 13.59) is found with the scalar Green function in the time domain (Equation 13.25). To solve (6.114), we write down the three scalar components of this equation, solve each one separately with the scalar Green function, and combine the three solutions afterward to a vector[104]

$$\underline{\mathbf{A}}(\underline{\mathbf{R}}, t) = \frac{\mu_0\mu_r}{4\pi}\iiint_{V_Q}\frac{\underline{\mathbf{J}}_{\mathrm{e}}\left(\underline{\mathbf{R}}', t - \frac{|\underline{\mathbf{R}}-\underline{\mathbf{R}}'|}{c}\right)}{|\underline{\mathbf{R}}-\underline{\mathbf{R}}'|}\,\mathrm{d}^3\underline{\mathbf{R}}'. \tag{6.117}$$

The solutions (6.116) and (6.117) are called retarded potentials. Their Fourier spectra

$$\Phi(\underline{\mathbf{R}}, \omega) = \frac{1}{4\pi\epsilon_0\epsilon_r}\iiint_{V_Q}\varrho_{\mathrm{e}}(\underline{\mathbf{R}}', \omega)\,\frac{\mathrm{e}^{\mathrm{j}k|\underline{\mathbf{R}}-\underline{\mathbf{R}}'|}}{|\underline{\mathbf{R}}-\underline{\mathbf{R}}'|}\,\mathrm{d}^3\underline{\mathbf{R}}', \tag{6.118}$$

$$\underline{\mathbf{A}}(\underline{\mathbf{R}}, \omega) = \frac{\mu_0\mu_r}{4\pi}\iiint_{V_Q}\underline{\mathbf{J}}_{\mathrm{e}}(\underline{\mathbf{R}}', \omega)\,\frac{\mathrm{e}^{\mathrm{j}k|\underline{\mathbf{R}}-\underline{\mathbf{R}}'|}}{|\underline{\mathbf{R}}-\underline{\mathbf{R}}'|}\,\mathrm{d}^3\underline{\mathbf{R}}' \tag{6.119}$$

---

[103]Originally, the Lorenz condition is due to Ludvig Lorenz, yet very often it is associated with the name of Hendrik Antoon Lorentz (Sihvola 1991; Nevels and Shin 2001).

[104]The spatially independent cartesian trihedron can come out of the volume integral as well as enter it.

form the basis to calculate the Fourier spectra of the fields:

$$\underline{\mathbf{B}}(\underline{\mathbf{R}}, \omega) = \boldsymbol{\nabla} \times \underline{\mathbf{A}}(\underline{\mathbf{R}}, \omega), \tag{6.120}$$

$$\underline{\mathbf{E}}(\underline{\mathbf{R}}, \omega) = -\boldsymbol{\nabla}\Phi(\underline{\mathbf{R}}, \omega) + \mathrm{j}\omega\underline{\mathbf{A}}(\underline{\mathbf{R}}, \omega); \tag{6.121}$$

with regard to (6.121), there is the alternative[105] "Maxwell equation (6.1)":

$$\begin{aligned}
\underline{\mathbf{E}}(\underline{\mathbf{R}}, \omega) &= \frac{\mathrm{j}}{\omega\epsilon_0\epsilon_r\mu_0\mu_r} [\boldsymbol{\nabla} \times \underline{\mathbf{B}}(\underline{\mathbf{R}}, \omega) - \mu_0\mu_r\underline{\mathbf{J}}_\mathrm{e}(\underline{\mathbf{R}}, \omega)] \\
&= \frac{\mathrm{j}\omega}{k^2} \underbrace{[\boldsymbol{\nabla} \times \boldsymbol{\nabla} \times \underline{\mathbf{A}}(\underline{\mathbf{R}}, \omega) - \mu_0\mu_r\underline{\mathbf{J}}_\mathrm{e}(\underline{\mathbf{R}}, \omega)]}_{\begin{subarray}{l} = \boldsymbol{\nabla}\boldsymbol{\nabla}\cdot\underline{\mathbf{A}} - \Delta\underline{\mathbf{A}} - \mu_0\mu_r\underline{\mathbf{J}}_\mathrm{e}(\underline{\mathbf{R}}, \omega) \\ = \boldsymbol{\nabla}\boldsymbol{\nabla}\cdot\underline{\mathbf{A}} + k^2\underline{\mathbf{A}} \end{subarray}} \\
&= \mathrm{j}\omega\left(\underline{\underline{\mathbf{I}}} + \frac{1}{k^2}\boldsymbol{\nabla}\boldsymbol{\nabla}\right)\cdot\underline{\mathbf{A}}(\underline{\mathbf{R}}, \omega).
\end{aligned} \tag{6.122}$$

Due to the Lorenz convention, the single vector potential is evidently sufficient.

6.6.2 Electric second rank Green tensor

With (6.119), Equation 6.122 yields the source field representation:

$$\begin{aligned}
\underline{\mathbf{E}}(\underline{\mathbf{R}}, \omega) &= \mathrm{j}\omega\mu_0\mu_r\left(\underline{\underline{\mathbf{I}}} + \frac{1}{k^2}\boldsymbol{\nabla}\boldsymbol{\nabla}\right)\cdot\iiint_{V_Q} \underline{\mathbf{J}}_\mathrm{e}(\underline{\mathbf{R}}', \omega)\frac{\mathrm{e}^{\mathrm{j}k|\underline{\mathbf{R}}-\underline{\mathbf{R}}'|}}{4\pi|\underline{\mathbf{R}}-\underline{\mathbf{R}}'|}\,\mathrm{d}^3\underline{\mathbf{R}}' \\
&= \mathrm{j}\omega\mu_0\mu_r\iiint_{V_Q} \underline{\mathbf{J}}_\mathrm{e}(\underline{\mathbf{R}}', \omega)\cdot\left(\underline{\underline{\mathbf{I}}} + \frac{1}{k^2}\boldsymbol{\nabla}\boldsymbol{\nabla}\right)\frac{\mathrm{e}^{\mathrm{j}k|\underline{\mathbf{R}}-\underline{\mathbf{R}}'|}}{4\pi|\underline{\mathbf{R}}-\underline{\mathbf{R}}'|}\,\mathrm{d}^3\underline{\mathbf{R}}' \\
&= \mathrm{j}\omega\mu_0\mu_r\iiint_{V_Q} \underline{\mathbf{J}}_\mathrm{e}(\underline{\mathbf{R}}', \omega)\cdot\underline{\underline{\mathbf{G}}}_\mathrm{e}(\underline{\mathbf{R}}-\underline{\mathbf{R}}', \omega)\,\mathrm{d}^3\underline{\mathbf{R}}',
\end{aligned} \tag{6.123}$$

where

$$\begin{aligned}
\underline{\underline{\mathbf{G}}}_\mathrm{e}(\underline{\mathbf{R}}-\underline{\mathbf{R}}', \omega) &= \left(\underline{\underline{\mathbf{I}}} + \frac{1}{k^2}\boldsymbol{\nabla}\boldsymbol{\nabla}\right)\frac{\mathrm{e}^{\mathrm{j}k|\underline{\mathbf{R}}-\underline{\mathbf{R}}'|}}{4\pi|\underline{\mathbf{R}}-\underline{\mathbf{R}}'|} \\
&= \left(\underline{\underline{\mathbf{I}}} + \frac{1}{k^2}\boldsymbol{\nabla}'\boldsymbol{\nabla}'\right)\frac{\mathrm{e}^{\mathrm{j}k|\underline{\mathbf{R}}-\underline{\mathbf{R}}'|}}{4\pi|\underline{\mathbf{R}}-\underline{\mathbf{R}}'|}
\end{aligned} \tag{6.124}$$

denotes[106] the symmetric second rank electric tensor, the electric dyadic Green function, because the differentiation under the integral causes no problems for $\underline{\mathbf{R}} \notin V_Q$; $\underline{\underline{\mathbf{G}}}_\mathrm{e}(\underline{\mathbf{R}}-\underline{\mathbf{R}}', \omega)$ is a dyadic Green function because it describes time harmonic electromagnetic (spherical) elementary waves that originate from the source point $\underline{\mathbf{R}}' \in V_Q$, resulting in the electric field strength of the

[105]We can equally introduce the gradient of the time harmonic version of the Lorenz convention (6.125) into (6.121).

[106]Due to the symmetry of the differential operator, we can perform the contraction with $\underline{\mathbf{J}}_\mathrm{e}$ in (6.123) with regard to either the first or the second index.

$\underline{\mathbf{J}}_e$-source through the $\underline{\mathbf{J}}_e(\underline{\mathbf{R}}', \omega)$-weighted (point source) synthesis according to (6.123). Actually, the field strength

$$\begin{aligned}\underline{\mathbf{E}}^{\mathrm{PS_e}}(\underline{\mathbf{R}}, \omega) &= \mathrm{j}\omega\mu_0\mu_r \underline{\underline{\mathbf{G}}}_e(\underline{\mathbf{R}}, \omega) \cdot \hat{\underline{\mathbf{j}}}_e \\ &= \mathrm{j}\omega\mu_0\mu_r \hat{\underline{\mathbf{j}}}_e \cdot \underline{\underline{\mathbf{G}}}_e(\underline{\mathbf{R}}, \omega)\end{aligned} \tag{6.125}$$

of a $\underline{\mathbf{J}}_e$-unit point source

$$\underline{\mathbf{J}}_e(\underline{\mathbf{R}}, \omega) = \hat{\underline{\mathbf{j}}}_e \delta(\underline{\mathbf{R}}) \tag{6.126}$$

located in the coordinate origin yields Green's dyadic multiplied by $\mathrm{j}\omega\mu_0\mu_r$ and contracted by $\hat{\underline{\mathbf{j}}}_e$.

The physical meaning of the electric second rank Green tensor must be mathematically reflected in terms of a differential equation for $\underline{\underline{\mathbf{G}}}_e(\underline{\mathbf{R}} - \underline{\mathbf{R}}', \omega)$; apparently, Equation 6.110 is this differential equation because the application of the respective differential operator to (6.123) immediately reveals that (6.123) is indeed a solution of (6.89) for $\underline{\mathbf{J}}_m \equiv \underline{\mathbf{0}}$. Applying a three-dimensional spatial Fourier transform, we show (compare Section 13.2.1) that Green's dyadic (6.124) is a solution of (6.110). Insofar, the correct source point behavior of $\underline{\underline{\mathbf{G}}}_e$ is given by (6.124); however, we have to be careful applying the explicit $\nabla\nabla$-differentiation to the scalar Green function that is singular for $\underline{\mathbf{R}} = \underline{\mathbf{R}}'$. As already stated in Section 5.5, we obtain for a spherical exclusion volume (van Bladel 1961; van Bladel 1991; Chew 1990; Langenberg 2005)

$$\underline{\underline{\mathbf{G}}}_e(\underline{\mathbf{R}} - \underline{\mathbf{R}}', \omega) = \mathrm{PV}\, \underline{\underline{\mathbf{G}}}_e^{(0)}(\underline{\mathbf{R}} - \underline{\mathbf{R}}', \omega) - \frac{1}{3k^2} \underline{\underline{\mathbf{I}}}\, \delta(\underline{\mathbf{R}} - \underline{\mathbf{R}}') \tag{6.127}$$

with

$$\underline{\underline{\mathbf{G}}}_e^{(0)}(\underline{\mathbf{R}}, \omega) = \left[\underline{\underline{\mathbf{I}}} - \hat{\underline{\mathbf{R}}}\hat{\underline{\mathbf{R}}} + \frac{\mathrm{j}}{kR} (\underline{\underline{\mathbf{I}}} - 3\hat{\underline{\mathbf{R}}}\hat{\underline{\mathbf{R}}}) - \frac{1}{k^2R^2} (\underline{\underline{\mathbf{I}}} - 3\hat{\underline{\mathbf{R}}}\hat{\underline{\mathbf{R}}}) \right] \frac{\mathrm{e}^{\mathrm{j}kR}}{4\pi R}. \tag{6.128}$$

Here, PV is a well-defined Cauchy principal value to evaluate the integral (6.123) that is even—and especially—existent for the R^{-3}-term in (6.128). For $\underline{\mathbf{R}} \neq \underline{\mathbf{R}}'$, we evidently have

$$\underline{\underline{\mathbf{G}}}_e(\underline{\mathbf{R}} - \underline{\mathbf{R}}', \omega) = \underline{\underline{\mathbf{G}}}_e^{(0)}(\underline{\mathbf{R}} - \underline{\mathbf{R}}', \omega). \tag{6.129}$$

6.6.3 Far-field approximation

With the "substitution" of $\nabla \Longrightarrow \mathrm{j}k\hat{\underline{\mathbf{R}}}$ (Equation 13.47), we obtain from (6.124), and therefore from (6.123), the far-field approximation

$$\underline{\mathbf{E}}^{\mathrm{far}}(\underline{\mathbf{R}}, \omega) = \frac{\mathrm{e}^{\mathrm{j}kR}}{R} \underline{\mathbf{H}}_E^{\mathrm{e}}(\hat{\underline{\mathbf{R}}}, \omega) \tag{6.130}$$

with the vector radiation pattern

$$\underline{\mathbf{H}}_E^{\mathrm{e}}(\hat{\underline{\mathbf{R}}}, \omega) = \mathrm{j}\omega \frac{\mu_0\mu_r}{4\pi} (\underline{\underline{\mathbf{I}}} - \hat{\underline{\mathbf{R}}}\hat{\underline{\mathbf{R}}}) \cdot \iiint_{V_Q} \underline{\mathbf{J}}_e(\underline{\mathbf{R}}', \omega) \mathrm{e}^{-\mathrm{j}k\hat{\underline{\mathbf{R}}}\cdot\underline{\mathbf{R}}'}\, \mathrm{d}^3\underline{\mathbf{R}}' \tag{6.131}$$

of the electric field strength. Obviously, we have

$$\underline{\mathbf{E}}^{\text{far}}(\underline{\mathbf{R}}, \omega) \cdot \hat{\underline{\mathbf{R}}} = 0, \tag{6.132}$$

i.e., electromagnetic waves are transversely polarized with respect to the propagation direction $\hat{\underline{\mathbf{R}}}$ in the far-field of an *arbitrary* current distribution because the far-field magnetic field strength resulting from (6.2)

$$\underline{\mathbf{H}}^{\text{far}}(\underline{\mathbf{R}}, \omega) = \frac{\mathrm{e}^{\mathrm{j}kR}}{R} \underline{\mathbf{H}}_H^{\mathrm{e}}(\hat{\underline{\mathbf{R}}}, \omega) \tag{6.133}$$

with

$$\underline{\mathbf{H}}_H^{\mathrm{e}}(\underline{\mathbf{R}}, \omega) = \frac{\mathrm{j}k}{4\pi} \hat{\underline{\mathbf{R}}} \times \iiint_{V_Q} \underline{\mathbf{J}}_{\mathrm{e}}(\underline{\mathbf{R}}', \omega) \mathrm{e}^{-\mathrm{j}k\hat{\underline{\mathbf{R}}}\cdot\underline{\mathbf{R}}'} \,\mathrm{d}^3\underline{\mathbf{R}}' \tag{6.134}$$

equally satisfies the orthogonality

$$\underline{\mathbf{H}}^{\text{far}}(\underline{\mathbf{R}}, \omega) \cdot \hat{\underline{\mathbf{R}}} = 0; \tag{6.135}$$

in addition, the transformation $(\underline{\underline{\mathbf{I}}} - \hat{\underline{\mathbf{R}}}\hat{\underline{\mathbf{R}}}) \cdot \underline{\mathbf{J}}_{\mathrm{e}} = (\hat{\underline{\mathbf{R}}} \times \underline{\mathbf{J}}_{\mathrm{e}}) \times \hat{\underline{\mathbf{R}}}$ yields the orthogonality

$$\underline{\mathbf{E}}^{\text{far}}(\underline{\mathbf{R}}, \omega) = Z\underline{\mathbf{H}}^{\text{far}}(\underline{\mathbf{R}}, \omega) \times \hat{\underline{\mathbf{R}}}, \tag{6.136}$$

that is to say, $\hat{\underline{\mathbf{R}}}, \underline{\mathbf{E}}^{\text{far}}$, and $\underline{\mathbf{H}}^{\text{far}}$ form a right-handed orthogonal trihedron: The electromagnetic far-field of an arbitrary electric current distribution locally behaves as a plane wave. In the next section, we will see that this also holds for the source fields of magnetic current densities; as a consequence, scattered fields with induced current densities as sources must also have this property. Accordingly, we can define a polarization base for incident and scattered fields to describe the change in polarization state due to scattering by scattering matrices, Jones and Sinclair matrices, respectively (Ulaby and Elachi 1990; Langenberg 2005).

6.6.4 Hertzian dipole

In antenna theory, the point source (6.126) is especially named for historical reasons: Hertzian dipole. Due to (6.128), its electric field (6.125) exhibits near-, transition-, and far-fields. The latter one has the structure

$$\underline{\mathbf{E}}^{\mathrm{PS_e,far}}(\underline{\mathbf{R}}, \omega) = \mathrm{j}\omega \frac{\mu_0\mu_r}{4\pi} \frac{\mathrm{e}^{\mathrm{j}kR}}{R} (\underline{\underline{\mathbf{I}}} - \hat{\underline{\mathbf{R}}}\hat{\underline{\mathbf{R}}}) \cdot \hat{\underline{\mathbf{j}}}_{\mathrm{e}}; \tag{6.137}$$

with the choice $\hat{\underline{\mathbf{j}}}_{\mathrm{e}} = \underline{\mathbf{e}}_z$, the electric field component $E_\vartheta^{\mathrm{PS_e,far}}(\underline{\mathbf{R}}, \omega)$ is the only one being nonzero; it is proportional to $\sin\vartheta$, that is to say, the Hertzian dipole does not radiate in the direction of its axis.

6.6.5 Magnetic second-rank Green tensor

The differential equation (6.89) for the electric field strength shows not only the electric current density on the right-hand side but also the curl of the magnetic current density. Alternative to the last paragraph, we put $\underline{\mathbf{J}}_\mathrm{m} \equiv \underline{\mathbf{0}}$ to obtain the solution of (6.89)

$$\underline{\mathbf{E}}(\underline{\mathbf{R}}, \omega) = -\iiint_{V_Q} [\nabla' \times \underline{\mathbf{J}}_\mathrm{m}(\underline{\mathbf{R}}', \omega)] \cdot \underline{\underline{\mathbf{G}}}_\mathrm{e}(\underline{\mathbf{R}} - \underline{\mathbf{R}}', \omega)\, \mathrm{d}^3\underline{\mathbf{R}}', \tag{6.138}$$

utilizing the electric Green dyadic. With the identity

$$(\nabla' \times \underline{\mathbf{J}}_\mathrm{m}) \cdot \underline{\underline{\mathbf{G}}}_\mathrm{e} = \nabla' \cdot (\underline{\mathbf{J}}_\mathrm{m} \times \underline{\underline{\mathbf{G}}}_\mathrm{e}) + \underline{\mathbf{J}}_\mathrm{m} \cdot (\nabla' \times \underline{\underline{\mathbf{G}}}_\mathrm{e}) \tag{6.139}$$

and the same arguments as in Section 5.5, we can shift the curl operator in (6.138) to $\underline{\underline{\mathbf{G}}}_\mathrm{e}$:

$$\underline{\mathbf{E}}(\underline{\mathbf{R}}, \omega) = -\iiint_{V_Q} \underline{\mathbf{J}}_\mathrm{m}(\underline{\mathbf{R}}', \omega) \cdot \nabla' \times \underline{\underline{\mathbf{G}}}_\mathrm{e}(\underline{\mathbf{R}} - \underline{\mathbf{R}}', \omega)\, \mathrm{d}^3\underline{\mathbf{R}}', \tag{6.140}$$

thus producing the magnetic second rank Green tensor

$$\begin{aligned} \underline{\underline{\mathbf{G}}}_\mathrm{m}(\underline{\mathbf{R}} - \underline{\mathbf{R}}', \omega) &= -\nabla' \times \underline{\underline{\mathbf{G}}}_\mathrm{e}(\underline{\mathbf{R}} - \underline{\mathbf{R}}', \omega) \\ &= \nabla \times \underline{\underline{\mathbf{G}}}_\mathrm{e}(\underline{\mathbf{R}} - \underline{\mathbf{R}}', \omega) \\ &= \nabla \times [G(\underline{\mathbf{R}} - \underline{\mathbf{R}}', \omega)\, \underline{\underline{\mathbf{I}}}] \\ &= \nabla G(\underline{\mathbf{R}} - \underline{\mathbf{R}}', \omega) \times \underline{\underline{\mathbf{I}}} \\ &= -\nabla' G(\underline{\mathbf{R}} - \underline{\mathbf{R}}', \omega) \times \underline{\underline{\mathbf{I}}} \end{aligned} \tag{6.141}$$

with the properties

$$\underline{\underline{\mathbf{G}}}^{21}_\mathrm{m}(\underline{\mathbf{R}} - \underline{\mathbf{R}}', \omega) = -\underline{\underline{\mathbf{G}}}_\mathrm{m}(\underline{\mathbf{R}} - \underline{\mathbf{R}}', \omega), \tag{6.142}$$

$$\underline{\underline{\mathbf{G}}}_\mathrm{m}(\underline{\mathbf{R}} - \underline{\mathbf{R}}', \omega) = -\underline{\underline{\mathbf{G}}}_\mathrm{m}(\underline{\mathbf{R}}' - \underline{\mathbf{R}}, \omega). \tag{6.143}$$

The superposition of (6.123) and (6.140) finally results in the solution of (6.89) for $\underline{\mathbf{J}}_\mathrm{e} \neq \underline{\mathbf{0}}$ and $\underline{\mathbf{J}}_\mathrm{m} \neq \underline{\mathbf{0}}$:

$$\begin{aligned} \underline{\mathbf{E}}(\underline{\mathbf{R}}, \omega) = \iiint_{V_Q} & \big[\mathrm{j}\omega\mu_0\mu_r \underline{\mathbf{J}}_\mathrm{e}(\underline{\mathbf{R}}', \omega) \cdot \underline{\underline{\mathbf{G}}}_\mathrm{e}(\underline{\mathbf{R}} - \underline{\mathbf{R}}', \omega) \\ & + \underline{\mathbf{J}}_\mathrm{m}(\underline{\mathbf{R}}', \omega) \cdot \underline{\underline{\mathbf{G}}}_\mathrm{m}(\underline{\mathbf{R}} - \underline{\mathbf{R}}', \omega)\big]\, \mathrm{d}^3\underline{\mathbf{R}}'. \end{aligned} \tag{6.144}$$

That way, we can immediately write down the solution of the Fourier transformed wave equation (6.81) for the magnetic field strength:

$$\begin{aligned} \underline{\mathbf{H}}(\underline{\mathbf{R}}, \omega) = \iiint_{V_Q} & \big[\mathrm{j}\omega\epsilon_0\epsilon_r \underline{\mathbf{J}}_\mathrm{m}(\underline{\mathbf{R}}', \omega) \cdot \underline{\underline{\mathbf{G}}}_\mathrm{e}(\underline{\mathbf{R}} - \underline{\mathbf{R}}', \omega) \\ & - \underline{\mathbf{J}}_\mathrm{e}(\underline{\mathbf{R}}', \omega) \cdot \underline{\underline{\mathbf{G}}}_\mathrm{m}(\underline{\mathbf{R}} - \underline{\mathbf{R}}', \omega)\big]\, \mathrm{d}^3\underline{\mathbf{R}}'. \end{aligned} \tag{6.145}$$

In contrast to acoustics and elastodynamics, we need only two Green tensors (functions) in electromagnetics; this is a consequence of the symmetry of Maxwell equations (6.1) and (6.2). Figure 6.2 graphically displays this issue.

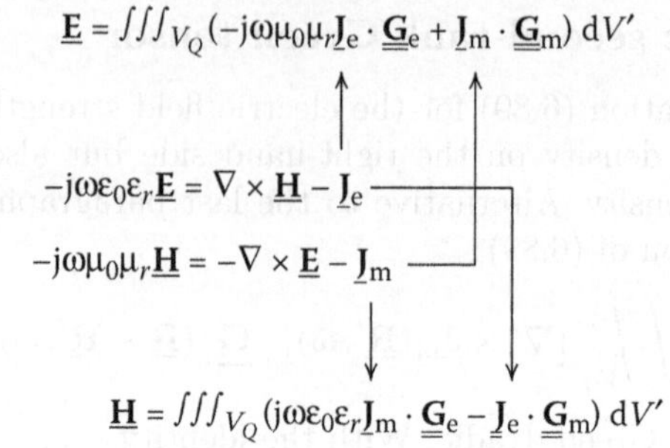

$$\underline{\mathbf{E}} = \iiint_{V_Q} (-j\omega\mu_0\mu_r \underline{\mathbf{J}}_e \cdot \underline{\underline{\mathbf{G}}}_e + \underline{\mathbf{J}}_m \cdot \underline{\underline{\mathbf{G}}}_m)\, dV'$$

$$-j\omega\varepsilon_0\varepsilon_r \underline{\mathbf{E}} = \nabla \times \underline{\mathbf{H}} - \underline{\mathbf{J}}_e$$

$$-j\omega\mu_0\mu_r \underline{\mathbf{H}} = -\nabla \times \underline{\mathbf{E}} - \underline{\mathbf{J}}_m$$

$$\underline{\mathbf{H}} = \iiint_{V_Q} (j\omega\varepsilon_0\varepsilon_r \underline{\mathbf{J}}_m \cdot \underline{\underline{\mathbf{G}}}_e - \underline{\mathbf{J}}_e \cdot \underline{\underline{\mathbf{G}}}_m)\, dV'$$

FIGURE 6.2
Assignment of Green functions in homogeneous isotropic electromagnetic materials to the source densities $\underline{\mathbf{J}}_e$ and $\underline{\mathbf{J}}_m$.

6.7 Electromagnetic Scattered Fields; Electromagnetic Formulation of Huygens' Principle

6.7.1 Electromagnetic formulation of Huygens' principle

The mathematical formulation of Huygens' principle comprises the representation theorem—the representation of a wave field on one side of a (mathematically virtual) closed surface S_g by the boundary values on S_g—and the extinction theorem—the extinction of the wave field on the respective other side of the surface (Sections 5.6 and 15.1.2): The representation of the field outside S_g enforces a null-field inside S_g and vice versa.

Going back to Larmor (1903), we can define the boundary values of the wave field with the transition conditions (6.30) and (6.31). For example, we keep the interior of S_g, characterized by the index (2), field-free; then

$$\underline{\mathbf{n}} \times \underline{\mathbf{H}}(\underline{\mathbf{R}}_S, t) = \underline{\mathbf{K}}_e(\underline{\mathbf{R}}_S, t), \quad \underline{\mathbf{R}}_S \in S_g, \tag{6.146}$$

$$\underline{\mathbf{n}} \times \underline{\mathbf{E}}(\underline{\mathbf{R}}_S, t) = -\underline{\mathbf{K}}_m(\underline{\mathbf{R}}_S, t), \quad \underline{\mathbf{R}}_S \in S_g, \tag{6.147}$$

explicitly define surface current densities maintaining the jump discontinuity from the interior null-field to the exterior $\underline{\mathbf{E}}(\underline{\mathbf{R}}, t), \underline{\mathbf{H}}(\underline{\mathbf{R}}, t)$-field; per definitionem, the normal points from (2) to (1), that is to say, into the exterior of S_g and consequently away from the null-field. Let S_g now be a physically existing surface S_c of a scattering body embedded in an $\epsilon_r\mu_r$-material that is illuminated by the incident field $\underline{\mathbf{E}}_i, \underline{\mathbf{H}}_i$ of a source distribution; that way, $\underline{\mathbf{K}}_e$ and $\underline{\mathbf{K}}_m$ become sources of the scattered field $\underline{\mathbf{E}}_s, \underline{\mathbf{H}}_s$ superimposing to the incident field in the exterior of S_c as the total field $\underline{\mathbf{E}} = \underline{\mathbf{E}}_i + \underline{\mathbf{E}}_s$, $\underline{\mathbf{H}} = \underline{\mathbf{H}}_i + \underline{\mathbf{H}}_s$ and compensating the incident field in the interior of S_c, thus leaving the total field there as a null-field (Figure 6.3). Therefore, the source

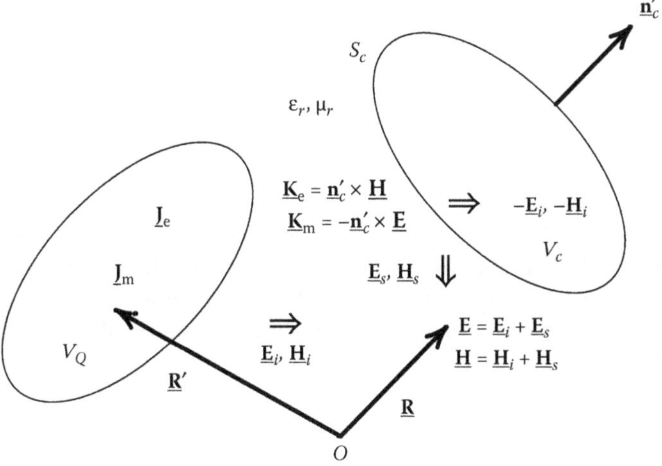

FIGURE 6.3
Electromagnetic scattering problem: Surface currents for Huygens' principle.

field representations (6.144) and (6.145) can be considered as the basis to calculate the time harmonic scattered field $\underline{\mathbf{E}}_s(\underline{\mathbf{R}}, \omega), \underline{\mathbf{H}}_s(\underline{\mathbf{R}}, \omega)$ if we insert the volume source densities

$$\underline{\mathbf{J}}_{\mathrm{e,m}}(\underline{\mathbf{R}}', \omega) = \underline{\mathbf{K}}_{\mathrm{e,m}}(\underline{\mathbf{R}}', \omega)\, \gamma_c(\underline{\mathbf{R}}'), \quad \underline{\mathbf{R}}' \in S_c, \tag{6.148}$$

which reduce the volume integrals (6.143) and (6.145) to surface integrals

$$\begin{aligned}
\underline{\mathbf{E}}_s(\underline{\mathbf{R}}, \omega) &= \iint_{S_c} \big[\mathrm{j}\omega\mu_0\mu_r \underline{\mathbf{K}}_\mathrm{e}(\underline{\mathbf{R}}', \omega) \cdot \underline{\underline{\mathbf{G}}}_\mathrm{e}(\underline{\mathbf{R}} - \underline{\mathbf{R}}', \omega) \\
&\qquad + \underline{\mathbf{K}}_\mathrm{m}(\underline{\mathbf{R}}', \omega) \cdot \underline{\underline{\mathbf{G}}}_\mathrm{m}(\underline{\mathbf{R}} - \underline{\mathbf{R}}', \omega)\big]\, \mathrm{d}S' \\
&= \iint_{S_c} \big[\mathrm{j}\omega\mu_0\mu_r \underline{\mathbf{n}}_c' \times \underline{\mathbf{H}}(\underline{\mathbf{R}}', \omega) \cdot \underline{\underline{\mathbf{G}}}_\mathrm{e}(\underline{\mathbf{R}} - \underline{\mathbf{R}}', \omega) \\
&\qquad - \underline{\mathbf{n}}_c' \times \underline{\mathbf{E}}(\underline{\mathbf{R}}', \omega) \cdot \underline{\underline{\mathbf{G}}}_\mathrm{m}(\underline{\mathbf{R}} - \underline{\mathbf{R}}', \omega)\big]\, \mathrm{d}S', \qquad (6.149) \\
\underline{\mathbf{H}}_s(\underline{\mathbf{R}}, \omega) &= \iint_{S_c} \big[\mathrm{j}\omega\epsilon_0\epsilon_r \underline{\mathbf{K}}_\mathrm{m}(\underline{\mathbf{R}}', \omega) \cdot \underline{\underline{\mathbf{G}}}_\mathrm{e}(\underline{\mathbf{R}} - \underline{\mathbf{R}}', \omega) \\
&\qquad - \underline{\mathbf{K}}_\mathrm{e}(\underline{\mathbf{R}}', \omega) \cdot \underline{\underline{\mathbf{G}}}_\mathrm{m}(\underline{\mathbf{R}} - \underline{\mathbf{R}}', \omega)\big]\, \mathrm{d}S' \\
&= -\iint_{S_c} \big[\mathrm{j}\omega\epsilon_0\epsilon_r \underline{\mathbf{n}}_c' \times \underline{\mathbf{E}}(\underline{\mathbf{R}}', \omega) \cdot \underline{\underline{\mathbf{G}}}_\mathrm{e}(\underline{\mathbf{R}} - \underline{\mathbf{R}}', \omega) \\
&\qquad + \underline{\mathbf{n}}_c' \times \underline{\mathbf{H}}(\underline{\mathbf{R}}', \omega) \cdot \underline{\underline{\mathbf{G}}}_\mathrm{m}(\underline{\mathbf{R}} - \underline{\mathbf{R}}', \omega)\big]\, \mathrm{d}S' \qquad (6.150)
\end{aligned}$$

due to the sifting property (2.382) of the singular function $\gamma_c(\underline{\mathbf{R}}')$ of S_c. We have found the mathematical formulation of Huygens' principle for electromagnetic waves! The particular mathematical version (6.149) and (6.150) comes "naturally" as a point source synthesis involving Green tensors and electric and magnetic current densities (that is the reason why we need *both* current densities from the beginning!), revealing that the wave field boundary values in (6.149) and (6.150) can only be tangential components of field strengths. The analytical derivation of (6.149) and (6.150) was given by Franz (Langenberg 2005), we speak of the Franz–Larmor version. Even though being physically compelling, there are numerical problems on behalf of the hyper singularity of $\underline{\underline{\mathbf{G}}}_{\mathrm{e}}$; therefore, transformations come into play—Stratton–Chu version—that only contain the scalar Green function (Langenberg 2005).

Incidentally, in the sense of an equivalence principle, the surface current densities $\underline{\mathbf{K}}_{\mathrm{e}}$ and $\underline{\mathbf{K}}_{\mathrm{m}}$, even if they flow on a mathematically virtual surface, are equivalent to a physically present scatterer; it could be removed without affecting the scattered field.

6.7.2 Electromagnetic fields scattered by perfect electrical conductors: EFIE and MFIE

The scattered field integrals (6.149) through (6.150) are insofar only a formal solution of the electromagnetic scattering problem as they contain unknown sources: $\underline{\mathbf{n}}_c' \times \underline{\mathbf{E}}$ and $\underline{\mathbf{n}}_c' \times \underline{\mathbf{H}}$ are tangential components of the total field[107] that also contain tangential components of the scattered field to be calculated. As in the scalar acoustic case (Section 5.6), integral equations are formulated via the limit $\underline{\mathbf{R}} \longrightarrow S_c$ for the observation point $\underline{\mathbf{R}}$ in the tangential components resulting from (6.149) to (6.150). Taking enough care regarding the hyper singularity of $\underline{\underline{\mathbf{G}}}_{\mathrm{e}}$—a special PV$_\varepsilon$-principle value has to be defined (Langenberg 2005)—we obtain for $\underline{\mathbf{R}}' \in S_c$

$$
\begin{aligned}
-\frac{1}{2}\underline{\mathbf{K}}_{\mathrm{m}}(\underline{\mathbf{R}},\omega) &= \underline{\mathbf{n}}_c \times \underline{\mathbf{E}}_i(\underline{\mathbf{R}},\omega) \\
&\quad + \mathrm{PV}_\varepsilon \underline{\mathbf{n}}_c \times \iint_{S_c} \big[\mathrm{j}\omega\mu_0\mu_r \underline{\mathbf{K}}_{\mathrm{e}}(\underline{\mathbf{R}}',\omega)\cdot \underline{\underline{\mathbf{G}}}_{\mathrm{e}}(\underline{\mathbf{R}}-\underline{\mathbf{R}}',\omega) \\
&\qquad + \underline{\mathbf{K}}_{\mathrm{m}}(\underline{\mathbf{R}}',\omega)\cdot \underline{\underline{\mathbf{G}}}_{\mathrm{m}}(\underline{\mathbf{R}}-\underline{\mathbf{R}}',\omega)\big]\,\mathrm{d}S'
\end{aligned}
\tag{6.151}
$$

from (6.149), respectively,

$$
\begin{aligned}
\frac{1}{2}\underline{\mathbf{K}}_{\mathrm{e}}(\underline{\mathbf{R}},\omega) &= \underline{\mathbf{n}}_c \times \underline{\mathbf{H}}_i(\underline{\mathbf{R}},\omega) \\
&\quad + \mathrm{PV}_\varepsilon \underline{\mathbf{n}}_c \times \iint_{S_c} \big[\mathrm{j}\omega\epsilon_0\epsilon_r \underline{\mathbf{K}}_{\mathrm{m}}(\underline{\mathbf{R}}',\omega)\cdot \underline{\underline{\mathbf{G}}}_{\mathrm{e}}(\underline{\mathbf{R}}-\underline{\mathbf{R}}',\omega) \\
&\qquad - \underline{\mathbf{K}}_{\mathrm{e}}(\underline{\mathbf{R}}',\omega)\cdot \underline{\underline{\mathbf{G}}}_{\mathrm{m}}(\underline{\mathbf{R}}-\underline{\mathbf{R}}',\omega)\big]\,\mathrm{d}S'
\end{aligned}
\tag{6.152}
$$

[107] The jump discontinuity from the null-field to the $\underline{\mathbf{E}}, \underline{\mathbf{H}}$-field has been postulated for the total field.

from (6.150). Both integral equation relations (6.151) and (6.152) reveal that $\underline{\mathbf{K}}_\mathrm{e}$ and $\underline{\mathbf{K}}_\mathrm{m}$ are dependent upon each other. Concentrating therefore on the practically important case—of course an idealization—of a perfectly electrically conducting scatterer, we have $\underline{\mathbf{K}}_\mathrm{m} \equiv \underline{\mathbf{0}}$; that way, Equation 6.151 is reduced to the electric field integral equation (of the first kind: EFIE):

$$\begin{aligned} &\mathrm{j}\omega\mu_0\mu_r \mathrm{PV}_\varepsilon \underline{\mathbf{n}}_c \times \iint_{S_c} \underline{\mathbf{K}}_\mathrm{e}(\underline{\mathbf{R}}', \omega) \cdot \underline{\underline{\mathbf{G}}}_\mathrm{e}(\underline{\mathbf{R}} - \underline{\mathbf{R}}', \omega)\, \mathrm{d}S' \\ &\qquad = -\underline{\mathbf{n}}_c \times \underline{\mathbf{E}}_i(\underline{\mathbf{R}}, \omega), \quad \underline{\mathbf{R}} \in S_c, \end{aligned} \tag{6.153}$$

and Equation 6.152 to the magnetic field integral equation (of the second kind: MFIE):

$$\begin{aligned} &\frac{1}{2}\underline{\mathbf{K}}_\mathrm{e}(\underline{\mathbf{R}}, \omega) + \underline{\mathbf{n}}_c \times \iint_{S_c} \underline{\mathbf{K}}_\mathrm{e}(\underline{\mathbf{R}}', \omega) \cdot \underline{\underline{\mathbf{G}}}_\mathrm{m}(\underline{\mathbf{R}} - \underline{\mathbf{R}}', \omega)\, \mathrm{d}S' \\ &\qquad = \underline{\mathbf{n}}_c \times \underline{\mathbf{H}}_i(\underline{\mathbf{R}}, \omega), \quad \underline{\mathbf{R}} \in S_c, \end{aligned} \tag{6.154}$$

each time for the unknown electric current density $\underline{\mathbf{K}}_\mathrm{e}$. In (6.154), we can refrain from the PV_ε-evaluation due to the missing dyadic Green function.

Only few geometries—among them the perfectly conducting sphere—allow for an analytic solution of the integral equations (6.153) and (6.154), and even in the case of the sphere, this solution shows up as an infinite series of spherical harmonics (Stratton 1941; Bowman et al. 1987) whose evaluation is by no means trivial. For a perfectly electrically conducting sphere of radius a located in vacuum and illuminated by a plane wave with linear polarization

$$\underline{\mathbf{E}}_i(\underline{\mathbf{R}}, \omega) = F(\omega)\, \mathrm{e}^{\mathrm{j}k_0 \hat{\underline{\mathbf{k}}}_i \cdot \underline{\mathbf{R}}}\, \hat{\underline{\mathbf{E}}}_0 \tag{6.155}$$

and a Gaussian spectrum

$$F(\omega) = \sqrt{\frac{\pi}{\alpha}}\, \mathrm{e}^{-\frac{\omega^2}{4\alpha}} \tag{6.156}$$

the ϑ- and φ-far-field components of the electric Gaussian impulsive scattered field are displayed in Figure 6.4; we have $\hat{\underline{\mathbf{k}}}_i = -\underline{\mathbf{e}}_z$, $\hat{\underline{\mathbf{E}}}_0 = \underline{\mathbf{e}}_x$, and $\alpha = 177.85\, c_0^2/a^2$. For angles ϑ in the vicinity of the backscattering direction, we nicely recognize the specularly reflected Gaussian impulse (in the ϑ-component with reversed sign) followed by the scattering contributions from the vicinity of the specular point; further pulses are identified as due to the current impulses that have circulated the sphere, they are called creeping waves[108] (Hönl et al. 1961).

If there is no analytical solution for either the EFIE or the MFIE at hand, we have to rely on numerical methods: the classical method of moments

[108] This terminology is due to W. Franz and it should not be utilized a second time—compare creeping wave transducer—for another physical phenomenon.

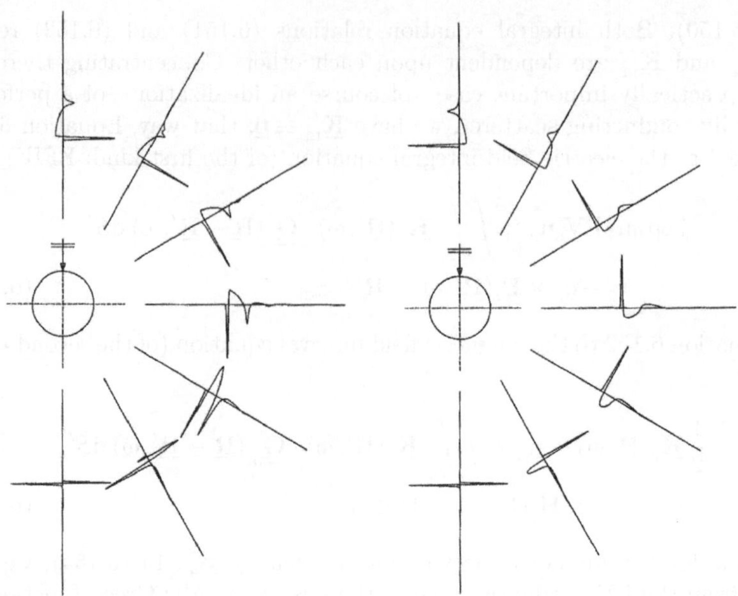

FIGURE 6.4
ϑ-Component of the electric scattered far-field in the time domain for different ϑ and $\varphi = 0$ (left); φ-component of the electric scattered far-field in the time domain for different ϑ and $\varphi = \pi/2$ (right).

(Harrington 1968; Poggio and Miller 1987; Wilton 2002) or the more recently developed fast multipole methods (Chew et al. 2002; Michielssen et al. 2002).

6.7.3 Kirchhoff approximation

As for the scalar acoustic case (Section 5.6), Kirchhoff's approximation of physical optics (compare Footnote 93) may be an equally useful approximation for the scattering of electromagnetic waves (high-frequency fields; convex scatterers). The integral equation (6.154) immediately suggests to neglect the radiation interaction integral—it is exactly zero for plane surfaces S_c—and to utilize

$$\underline{\mathbf{K}}_e^{\mathrm{PO}}(\underline{\mathbf{R}}, \omega) = 2\underline{\mathbf{n}}_c \times \underline{\mathbf{H}}_i(\underline{\mathbf{R}}, \omega) \tag{6.157}$$

as PO-approximated surface current density for $\underline{\mathbf{R}}$ on the illuminated side of S_c; for $\underline{\mathbf{R}}$ on the shadow side, we put $\underline{\mathbf{K}}_e^{\mathrm{PO}}(\underline{\mathbf{R}}, \omega) \equiv \underline{\mathbf{0}}$. In particular in the theory of electromagnetic wave inverse scattering, the Kirchhoff approximation plays a crucial role to linearize the problem (Langenberg et al. 1994; Langenberg et al. 1999b).

Calculating the Gaussian impulse scattered far-fields of a perfectly electrically conducting sphere within the Kirchhoff approximation reveals that this approximation, apart from small differences in the vicinity of specular reflection, does not provide any creeping waves.

6.7.4 Electromagnetic fields scattered by penetrable inhomogeneities: Lippmann–Schwinger integral equation

The scattering of electromagnetic waves by a penetrable scatterer can be formally solved defining equivalent volume sources analogously to acoustics (Section 5.6) (Langenberg 2005). We consider a homogeneous anisotropic $\underline{\underline{\epsilon}}_r^{(i)}(\underline{\mathbf{R}})\underline{\underline{\mu}}_r^{(i)}(\underline{\mathbf{R}})$-scatterer of volume V_c in an $\epsilon_r\mu_r$-embedding material that also contains the source volume V_Q with $\underline{\mathbf{J}}_{\mathrm{e,m}}(\underline{\mathbf{R}}, \omega) \neq \underline{\mathbf{0}}$. Similar to (5.94) and (5.95), we define contrast tensors[109]

$$\underline{\underline{\chi}}_{\mathrm{e}}(\underline{\mathbf{R}}) = \frac{1}{\epsilon_r}\left[\underline{\underline{\epsilon}}_r^{(i)}(\underline{\mathbf{R}}) - \epsilon_r \underline{\underline{\mathbf{I}}}\right] \Gamma_c(\underline{\mathbf{R}}), \tag{6.158}$$

$$\underline{\underline{\chi}}_{\mathrm{m}}(\underline{\mathbf{R}}) = \frac{1}{\mu_r}\left[\underline{\underline{\mu}}_r^{(i)}(\underline{\mathbf{R}}) - \mu_r \underline{\underline{\mathbf{I}}}\right] \Gamma_c(\underline{\mathbf{R}}) \tag{6.159}$$

that are equal to the null tensor outside V_c due to the characteristic function $\Gamma_c(\underline{\mathbf{R}})$ of V_c. In the differential equations (6.75) and (6.76) for $\underline{\mathbf{E}}(\underline{\mathbf{R}}, t)$ and $\underline{\mathbf{H}}(\underline{\mathbf{R}}, t)$ for the homogeneous anisotropic $\underline{\underline{\epsilon}}_r(\underline{\mathbf{R}})\underline{\underline{\mu}}_r(\underline{\mathbf{R}})$-material, we now insert[110]

$$\underline{\underline{\epsilon}}_r(\underline{\mathbf{R}}) = \epsilon_r\left[\underline{\underline{\mathbf{I}}} + \underline{\underline{\chi}}_{\mathrm{e}}(\underline{\mathbf{R}})\right] = \begin{cases} \epsilon_r & \text{for } \underline{\mathbf{R}} \notin V_c \\ \underline{\underline{\epsilon}}_r^{(i)}(\underline{\mathbf{R}}) & \text{for } \underline{\mathbf{R}} \in V_c \end{cases} \tag{6.160}$$

$$\underline{\underline{\mu}}_r(\underline{\mathbf{R}}) = \mu_r\left[\underline{\underline{\mathbf{I}}} + \underline{\underline{\chi}}_{\mathrm{m}}(\underline{\mathbf{R}})\right] = \begin{cases} \mu_r & \text{for } \underline{\mathbf{R}} \notin V_c \\ \underline{\underline{\mu}}_r^{(i)}(\underline{\mathbf{R}}) & \text{for } \underline{\mathbf{R}} \in V_c \end{cases} \tag{6.161}$$

---

[109]The $\underline{\underline{\chi}}_{\mathrm{e,m}}(\underline{\mathbf{R}})$-tensors are the susceptibility tensors of an $\underline{\underline{\epsilon}}_r^{(i)}(\underline{\mathbf{R}})\underline{\underline{\mu}}_r^{(i)}(\underline{\mathbf{R}})/\epsilon_r\mu_r$-material.

[110]It is convenient to represent the inverse material tensors

$$\underline{\underline{\epsilon}}_r^{-1}(\underline{\mathbf{R}}) = \frac{1}{\epsilon_r}\left[\underline{\underline{\mathbf{I}}} + \underline{\underline{\imath}}_{\mathrm{e}}(\underline{\mathbf{R}})\right],$$

$$\underline{\underline{\mu}}_r^{-1}(\underline{\mathbf{R}}) = \frac{1}{\mu_r}\left[\underline{\underline{\mathbf{I}}} + \underline{\underline{\imath}}_{\mathrm{m}}(\underline{\mathbf{R}})\right]$$

also by "inverse" contrast tensors (contrast functions of inverse material tensors)

$$\underline{\underline{\imath}}_{\mathrm{e}}(\underline{\mathbf{R}}) = \epsilon_r\left[\underline{\underline{\epsilon}}_r^{-1}(\underline{\mathbf{R}}) - \frac{1}{\epsilon_r}\underline{\underline{\mathbf{I}}}\right],$$

$$\underline{\underline{\imath}}_{\mathrm{m}}(\underline{\mathbf{R}}) = \mu_r\left[\underline{\underline{\mu}}_r^{-1}(\underline{\mathbf{R}}) - \frac{1}{\mu_r}\underline{\underline{\mathbf{I}}}\right].$$

and sort terms in a way that, on the left-hand side of the resulting differential equations, only the differential operators (6.77) and (6.78) of the homogeneous isotropic embedding material appear:

$$\nabla \times \nabla \times \underline{\mathbf{E}}(\underline{\mathbf{R}}, t) + \frac{1}{c^2} \frac{\partial^2 \underline{\mathbf{E}}(\underline{\mathbf{R}}, t)}{\partial t^2} = -\mu_0 \mu_r \frac{\partial}{\partial t} [\underline{\mathbf{J}}_{\mathrm{e}}(\underline{\mathbf{R}}, t) + \underline{\mathbf{J}}_{\mathrm{ec}}(\underline{\mathbf{R}}, t)] - \nabla \times [\underline{\mathbf{J}}_{\mathrm{m}}(\underline{\mathbf{R}}, t) + \underline{\mathbf{J}}_{\mathrm{mc}}(\underline{\mathbf{R}}, t)], \quad (6.162)$$

$$\nabla \times \nabla \times \underline{\mathbf{H}}(\underline{\mathbf{R}}, t) + \frac{1}{c^2} \frac{\partial^2 \underline{\mathbf{H}}(\underline{\mathbf{R}}, t)}{\partial t^2} = -\epsilon_0 \epsilon_r \frac{\partial}{\partial t} [\underline{\mathbf{J}}_{\mathrm{m}}(\underline{\mathbf{R}}, t) + \underline{\mathbf{J}}_{\mathrm{mc}}(\underline{\mathbf{R}}, t)] + \nabla \times [\underline{\mathbf{J}}_{\mathrm{e}}(\underline{\mathbf{R}}, t) + \underline{\mathbf{J}}_{\mathrm{ec}}(\underline{\mathbf{R}}, t)]; \quad (6.163)$$

the terms

$$\begin{aligned} \underline{\mathbf{J}}_{\mathrm{ec}}(\underline{\mathbf{R}}, t) &= -\epsilon_0 \epsilon_r \Gamma_c(\underline{\mathbf{R}}) \left[\underline{\underline{\mathbf{I}}} - \frac{\underline{\underline{\epsilon}}_r^{(i)}(\underline{\mathbf{R}})}{\epsilon_r} \right] \cdot \frac{\partial \underline{\mathbf{E}}(\underline{\mathbf{R}}, t)}{\partial t} \\ &= \epsilon_0 \epsilon_r \underline{\underline{\chi}}_{\mathrm{e}}(\underline{\mathbf{R}}) \cdot \frac{\partial \underline{\mathbf{E}}(\underline{\mathbf{R}}, t)}{\partial t}, \end{aligned} \quad (6.164)$$

$$\begin{aligned} \underline{\mathbf{J}}_{\mathrm{mc}}(\underline{\mathbf{R}}, t) &= -\mu_0 \mu_r \Gamma_c(\underline{\mathbf{R}}) \left[\underline{\underline{\mathbf{I}}} - \frac{\underline{\underline{\mu}}_r^{(i)}(\underline{\mathbf{R}})}{\mu_r} \right] \cdot \frac{\partial \underline{\mathbf{H}}(\underline{\mathbf{R}}, t)}{\partial t} \\ &= \mu_0 \mu_r \underline{\underline{\chi}}_{\mathrm{m}}(\underline{\mathbf{R}}) \cdot \frac{\partial \underline{\mathbf{H}}(\underline{\mathbf{R}}, t)}{\partial t} \end{aligned} \quad (6.165)$$

result as equivalent secondary volume sources representing the scatterer. They are equally dependent upon the total field as Huygens' surface sources. After Fourier transforming the differential equations (6.162) and (6.163) with respect to time, we find the integral representations for the incident field (Equations 6.144 through 6.145)

$$\begin{aligned} \underline{\mathbf{E}}_i(\underline{\mathbf{R}}, \omega) = \mathrm{j}\omega\mu_0\mu_r \iiint_{V_Q} \underline{\mathbf{J}}_{\mathrm{e}}(\underline{\mathbf{R}}', \omega) \cdot \underline{\underline{\mathbf{G}}}_{\mathrm{e}}(\underline{\mathbf{R}} - \underline{\mathbf{R}}', \omega) \, \mathrm{d}^3\underline{\mathbf{R}}' \\ + \iiint_{V_Q} \underline{\mathbf{J}}_{\mathrm{m}}(\underline{\mathbf{R}}', \omega) \cdot \underline{\underline{\mathbf{G}}}_{\mathrm{m}}(\underline{\mathbf{R}} - \underline{\mathbf{R}}', \omega) \, \mathrm{d}^3\underline{\mathbf{R}}', \end{aligned} \quad (6.166)$$

$$\begin{aligned} \underline{\mathbf{H}}_i(\underline{\mathbf{R}}, \omega) = \mathrm{j}\omega\epsilon_0\epsilon_r \iiint_{V_Q} \underline{\mathbf{J}}_{\mathrm{m}}(\underline{\mathbf{R}}', \omega) \cdot \underline{\underline{\mathbf{G}}}_{\mathrm{e}}(\underline{\mathbf{R}} - \underline{\mathbf{R}}', \omega) \, \mathrm{d}^3\underline{\mathbf{R}}' \\ - \iiint_{V_Q} \underline{\mathbf{J}}_{\mathrm{e}}(\underline{\mathbf{R}}', \omega) \cdot \underline{\underline{\mathbf{G}}}_{\mathrm{m}}(\underline{\mathbf{R}} - \underline{\mathbf{R}}', \omega) \, \mathrm{d}^3\underline{\mathbf{R}}', \quad \underline{\mathbf{R}} \in \mathbb{R}^3, \end{aligned} \quad (6.167)$$

with Green tensors $\underline{\underline{\mathbf{G}}}_{\mathrm{e}}$ and $\underline{\underline{\mathbf{G}}}_{\mathrm{m}}$ of the homogeneous isotropic embedding material. For $\underline{\mathbf{R}} \in V_Q$, we have to account for the distributional term in $\underline{\underline{\mathbf{G}}}_{\mathrm{e}}$, forcing us to define a suitable principle value of the integral; for the $\epsilon_r\mu_r$-material, it is given by Equation 6.127.

Completely analogously, we obtain for the electromagnetic scattered field for $\underline{\mathbf{R}} \in \mathbb{R}^3$:

$$\begin{aligned}\underline{\mathbf{E}}_s(\underline{\mathbf{R}}, \omega) = \mathrm{j}\omega\mu_0\mu_r \iiint_{V_c} \underline{\mathbf{J}}_{ec}(\underline{\mathbf{R}}', \omega) \cdot \underline{\underline{\mathbf{G}}}_{\mathrm{e}}(\underline{\mathbf{R}} - \underline{\mathbf{R}}', \omega)\, \mathrm{d}^3\underline{\mathbf{R}}' \\ + \iiint_{V_c} \underline{\mathbf{J}}_{\mathrm{mc}}(\underline{\mathbf{R}}', \omega) \cdot \underline{\underline{\mathbf{G}}}_{\mathrm{m}}(\underline{\mathbf{R}} - \underline{\mathbf{R}}', \omega)\, \mathrm{d}^3\underline{\mathbf{R}}', \end{aligned} \tag{6.168}$$

$$\begin{aligned}\underline{\mathbf{H}}_s(\underline{\mathbf{R}}, \omega) = \mathrm{j}\omega\epsilon_0\epsilon_r \iiint_{V_c} \underline{\mathbf{J}}_{\mathrm{mc}}(\underline{\mathbf{R}}', \omega) \cdot \underline{\underline{\mathbf{G}}}_{\mathrm{e}}(\underline{\mathbf{R}} - \underline{\mathbf{R}}', \omega)\, \mathrm{d}^3\underline{\mathbf{R}}' \\ - \iiint_{V_c} \underline{\mathbf{J}}_{ec}(\underline{\mathbf{R}}', \omega) \cdot \underline{\underline{\mathbf{G}}}_{\mathrm{m}}(\underline{\mathbf{R}} - \underline{\mathbf{R}}', \omega)\, \mathrm{d}^3\underline{\mathbf{R}}'; \end{aligned} \tag{6.169}$$

this time the hyper singularity of $\underline{\underline{\mathbf{G}}}_{\mathrm{e}}$ for $\underline{\mathbf{R}} \in V_c$ comes into play.

This formal solution immediately tells us how the present scattering problem can be generalized to an inhomogeneous anisotropic embedding material: Choose the Green tensors of the respective material in (6.166), (6.167) and (6.168), (6.169) (de Hoop 1995)!

For vanishing permeability contrast, we explicitly write down the Lippmann–Schwinger integral equation—similar to the scalar acoustic case in Section 5.6—for the total electric field in the interior of V_c (object equation[111]) by adding $\underline{\mathbf{E}}_i(\underline{\mathbf{R}}, \omega)$ on both sides of (6.168):

$$\begin{aligned}\underline{\mathbf{E}}(\underline{\mathbf{R}}, \omega) = \underline{\mathbf{E}}_i(\underline{\mathbf{R}}, \omega) + k^2 \iiint_{V_c} [\underline{\underline{\boldsymbol{\chi}}}_{\mathrm{e}}(\underline{\mathbf{R}}', \omega) \cdot \underline{\mathbf{E}}(\underline{\mathbf{R}}', \omega)] \cdot \underline{\underline{\mathbf{G}}}_{\mathrm{e}}(\underline{\mathbf{R}} - \underline{\mathbf{R}}', \omega)\, \mathrm{d}^3\underline{\mathbf{R}}' \\ \text{für } \underline{\mathbf{R}} \in V_c. \end{aligned} \tag{6.170}$$

Similar to (5.115), we can formally resolve (6.170) with regard to the scattered field:

$$\underline{\mathbf{E}}_s(\underline{\mathbf{R}}, \omega) = (\underline{\underline{\mathcal{I}}} - \underline{\underline{\mathcal{V}}}_c)^{-1} \cdot \underline{\underline{\mathcal{V}}}_c\{\underline{\mathbf{E}}_i\}(\underline{\mathbf{R}}, \omega), \quad \underline{\mathbf{R}} \in V_c. \tag{6.171}$$

We define the tensor integral operator $\underline{\underline{\mathcal{V}}}_c$ according to

$$\begin{aligned}\underline{\underline{\mathcal{V}}}_c\{\underline{\mathbf{E}}\}(\underline{\mathbf{R}}, \omega) = k^2 \iiint_{V_c} \underline{\underline{\mathbf{G}}}_{\mathrm{e}}(\underline{\mathbf{R}} - \underline{\mathbf{R}}', \omega) \\ \cdot \underline{\underline{\boldsymbol{\chi}}}_{\mathrm{e}}(\underline{\mathbf{R}}', \omega) \cdot \underline{\mathbf{E}}(\underline{\mathbf{R}}', \omega)\, \mathrm{d}^3\underline{\mathbf{R}}', \quad \underline{\mathbf{R}} \in V_c; \end{aligned} \tag{6.172}$$

its application to a vector yields another vector. Assuming $\underline{\mathbf{E}}_i(\underline{\mathbf{R}}, \omega)$ to be a plane wave with amplitude (Jones vector) $\underline{\mathbf{E}}_{0i}(\hat{\underline{\mathbf{k}}}_i, \omega)$—yielding $\underline{\mathbf{E}}_i(\underline{\mathbf{R}}, \omega, \hat{\underline{\mathbf{k}}}_i) = \underline{\mathbf{E}}_{0i}(\hat{\underline{\mathbf{k}}}_i, \omega)\, \mathrm{e}^{\mathrm{j}k\hat{\underline{\mathbf{k}}}_i \cdot \underline{\mathbf{R}}}$—Equation 6.171 shows according to

$$\begin{aligned}\underline{\mathbf{E}}_s(\underline{\mathbf{R}}, \omega, \hat{\underline{\mathbf{k}}}_i) &= (\underline{\underline{\mathcal{I}}} - \underline{\underline{\mathcal{V}}}_c)^{-1} \cdot \underline{\underline{\mathcal{V}}}_c\{\mathrm{e}^{\mathrm{j}k\hat{\underline{\mathbf{k}}}_i \cdot \underline{\mathbf{R}}'}\}(\underline{\mathbf{R}}, \omega) \cdot \underline{\mathbf{E}}_{0i}(\hat{\underline{\mathbf{k}}}_i, \omega) \\ &\stackrel{\text{def}}{=} \underline{\underline{\boldsymbol{\Sigma}}}_c(\underline{\mathbf{R}}, \omega, \hat{\underline{\mathbf{k}}}_i) \cdot \underline{\mathbf{E}}_{0i}(\hat{\underline{\mathbf{k}}}_i, \omega), \quad \underline{\mathbf{R}} \in V_c, \end{aligned} \tag{6.173}$$

[111]Note: In the object equation, we have $\underline{\mathbf{R}}, \underline{\mathbf{R}}' \in V_c$, i.e., both variables of the integral operator vary in the same domain.

that $\underline{\mathbf{E}}_s$ linearly depends upon this amplitude factor: Obviously, this is a consequence of the linearity of Maxwell equations!

In contrast to the scalar case, a numerical solution of (6.170) has to cope with the unfriendly hyper singularity of Green's dyadic for $\underline{\mathbf{R}} = \underline{\mathbf{R}}'$ requiring special care. For example, discretizing the volume V_c in terms of spherical voxels, we can utilize (6.127) resulting explicitly in:

$$\left[\underline{\underline{\mathbf{I}}} + \frac{1}{3}\underline{\underline{\chi}}_{\mathrm{e}}(\underline{\mathbf{R}}, \omega)\right] \cdot \underline{\mathbf{E}}(\underline{\mathbf{R}}, \omega) = \underline{\mathbf{E}}_i(\underline{\mathbf{R}}, \omega) + k^2\, \mathrm{PV} \iiint_{V_c} [\underline{\underline{\chi}}_{\mathrm{e}}(\underline{\mathbf{R}}', \omega) \cdot \underline{\mathbf{E}}(\underline{\mathbf{R}}', \omega)] \cdot \underline{\underline{\mathbf{G}}}_{\mathrm{e}}^{(0)}(\underline{\mathbf{R}} - \underline{\mathbf{R}}', \omega)\, \mathrm{d}^3\underline{\mathbf{R}}' \text{ for } \underline{\mathbf{R}} \in V_c. \tag{6.174}$$

For cubic voxels, the resulting integral equation identically looks like (6.174), yet the integral "principal value" has to be understood as pseudofunction (Langenberg and Fellinger 1995); for other voxel geometries, even the distributional term in $\underline{\underline{\mathbf{G}}}_{\mathrm{e}}$ looks different (van Bladel 1991; Chew 1990).

Having finally determined $\underline{\mathbf{E}}(\underline{\mathbf{R}}, \omega)$ for $\underline{\mathbf{R}} \in V_c$, the scattered field outside the scatterer can be comparatively easily calculated utilizing (6.168) for $\underline{\mathbf{R}} \in \mathbb{R}^3 \backslash \overline{V}_c$ in terms of the data equation[112]: The scattering problem has been solved!

In case if permittivity and permeability contrasts are nonzero, a coupled system of Lippmann–Schwinger equations has to be established and solved utilizing (6.166), (6.168) and (6.167), (6.169).

Lippmann–Schwinger integral equations have the advantage that, even for arbitrary inhomogeneous anisotropic scatterers, it is sufficient to know the Green tensors of the homogeneous isotropic embedding material; the disadvantage is that they are volume integral equations with a high discretization cost. Therefore, if the scatterer is equally homogeneous and isotropic, a different procedure is appropriate: We formulate surface current integral equations of the interior of V_c corresponding to (6.151) and (6.152), performing the limit $\underline{\mathbf{R}} \longrightarrow S_c$ from the interior in the respective Franz–Larmor integral representations of the electromagnetic Huygens principle; in these integrals, the Green tensors of the homogeneous isotropic scatterer material appear that differ only in the wave number from those of the exterior. The transition conditions require the continuity of the surface current densities for the exterior and the interior scattered fields, resulting in a coupled system of two surface integral equations (Langenberg 2005).

6.7.5 Born approximation

For large wavelengths as compared to the scatterer dimension and low contrast, the volume current densities

[112]Note: In the data equation, we have $\underline{\mathbf{R}}' \in V_c$ and $\underline{\mathbf{R}} \in \mathbb{R}^3 \backslash \overline{V}_c$, i.e., both variables of the integral operator vary in different domains.

$$\underline{\mathbf{J}}_{ec}^{\text{Born}}(\underline{\mathbf{R}}, \omega) = -\mathrm{j}\omega\epsilon_0\epsilon_r \underline{\underline{\chi}}_{\mathrm{e}}(\underline{\mathbf{R}}) \cdot \underline{\mathbf{E}}_i(\underline{\mathbf{R}}, \omega), \tag{6.175}$$

$$\underline{\mathbf{J}}_{mc}^{\text{Born}}(\underline{\mathbf{R}}, \omega) = -\mathrm{j}\omega\mu_0\mu_r \underline{\underline{\chi}}_{\mathrm{m}}(\underline{\mathbf{R}}) \cdot \underline{\mathbf{H}}_i(\underline{\mathbf{R}}, \omega) \tag{6.176}$$

represent an acceptable approximation. Then, the scattered field within the Born approximation can immediately be calculated with the integrals (6.168) and (6.169).

6.7.6 Scattering tensor

Remote sensing with electromagnetic waves is particularly interested in the polarization of the scattered wave as it depends upon the polarization of the incident wave (Ulaby and Elachi 1990; Cloude 2002). Utilizing the representation (6.168) of the scattered field, the representation (6.164)—for example—of an equivalent electric current density, as well as the result (6.173), we can directly derive the linear dependence

$$\underline{\mathbf{E}}_s^{\text{far}}(\underline{\mathbf{R}}, \omega, \hat{\underline{\mathbf{k}}}_i) = \frac{\mathrm{e}^{\mathrm{j}kR}}{R} \underbrace{\underline{\underline{\mathbf{\Sigma}}}(\hat{\underline{\mathbf{R}}}, \omega, \hat{\underline{\mathbf{k}}}_i) \cdot \underline{\mathbf{E}}_{0i}(\hat{\underline{\mathbf{k}}}_i, \omega)}_{= \underline{\mathbf{E}}_{0s}(\hat{\underline{\mathbf{R}}}, \omega, \hat{\underline{\mathbf{k}}}_i)} \tag{6.177}$$

of the vector amplitude $\underline{\mathbf{E}}_{0s}(\hat{\underline{\mathbf{R}}}, \omega, \hat{\underline{\mathbf{k}}}_i)$ of the scattered field from the Jones vector of the incident (plane) wave where the scattering tensor

$$\begin{aligned}\underline{\underline{\mathbf{\Sigma}}}(\hat{\underline{\mathbf{R}}}, \omega, \hat{\underline{\mathbf{k}}}_i) = \frac{k^2}{4\pi} (\underline{\underline{\mathbf{I}}} - \hat{\underline{\mathbf{R}}}\hat{\underline{\mathbf{R}}}) \cdot \iiint_{V_c} \underline{\underline{\chi}}_{\mathrm{e}}(\underline{\mathbf{R}}') \\ \cdot \left[\underline{\underline{\mathbf{\Sigma}}}_c(\underline{\mathbf{R}}', \omega, \hat{\underline{\mathbf{k}}}_i)\, \mathrm{e}^{-\mathrm{j}k\hat{\underline{\mathbf{k}}}_i \cdot \underline{\mathbf{R}}'} + \underline{\underline{\mathbf{I}}}\right] \mathrm{e}^{-\mathrm{j}k(\hat{\underline{\mathbf{R}}} - \hat{\underline{\mathbf{k}}}_i) \cdot \underline{\mathbf{R}}'}\, \mathrm{d}^3\underline{\mathbf{R}}'\end{aligned} \tag{6.178}$$

contains the complete information about the scatterer. The scattering tensor is indeed a second rank tensor with nine components in a well-defined coordinate system that can actually be reduced to a 2×2-scattering matrix if we recognize that $\underline{\mathbf{E}}_{0i}(\hat{\underline{\mathbf{k}}}_i, \omega)$ is orthogonal with regard to $\hat{\underline{\mathbf{k}}}_i$ and that $\underline{\mathbf{E}}_{0s}(\hat{\underline{\mathbf{R}}}, \omega, \hat{\underline{\mathbf{k}}}_i)$ is orthogonal with regard to $\hat{\underline{\mathbf{R}}}$; the Jones vectors of the incident and of the scattered waves in the far-field is given by two components in a suitable polarization basis. Depending on the polarization basis, 2×2-Jones and Sinclair scattering matrices arise (Langenberg 2005) whose measurement and evaluation comprise remote sensing. First ideas for a generalization to elastodynamics are presented in Section 15.4.1.

6.8 Two-Dimensional Electromagnetism: TM- and TE-Decoupling

To describe acoustic waves mathematically, we can rely on the pressure as a scalar field quantity, be it either in two or in three spatial dimensions.

A corresponding simplification for three-dimensional electromagnetism and three-dimensional elastodynamics is generally not possible; yet, in two-dimensional electromagnetism, a separation into scalar partial fields that do not depend upon each other may be possible[113]; both fields are mathematically completely equivalent to two-dimensional acoustics. In elastodynamics, not even *that* is possible; only two-dimensional SH-waves are comparable to scalar acoustics.

6.8.1 TM-field

We postulate two-dimensional Maxwell equations with $\partial/\partial y \equiv 0$ and claim that:

Assuming $\epsilon_r(x, z)$, $\mu_r = \text{const}$, zero field and current components according to

$$\begin{aligned} H_y(x, z, t) &= 0, \\ E_x(x, z, t) = E_z(x, z, t) &= 0; \end{aligned} \tag{6.179}$$

$$\begin{aligned} J_{\text{m}y}(x, z, t) &= 0, \\ J_{\text{e}x}(x, z, t) = J_{\text{e}z}(x, z, t) &= 0 \end{aligned} \tag{6.180}$$

result in a consistent system of equations for the remaining field components $H_x(x, z, t) \neq 0$, $H_z(x, z, t) \neq 0$ and $E_y(x, z, t) \neq 0$ if excited by the current components $J_{\text{m}x}(x, z, t) \neq 0$, $J_{\text{m}z}(x, z, t) \neq 0$, as well as $J_{\text{e}y}(x, z, t) \neq 0$, where $E_y(x, z, t)$ can be chosen as a scalar potential to calculate $H_x(x, z, t)$, $H_z(x, z, t)$, prescribing the current components and the inhomogeneity of the permittivity $\epsilon_r(x, z)$. This field is transversely magnetic with regard to the y-axis defining the two-dimensionality because the magnetic field does not have a component in this direction; Figure 6.5(a) illustrates the components of the TM-field.

For a proof, we write Maxwell equations (6.1) and (6.2) implying the above assumptions:

$$\frac{\partial D_y(x, z, t)}{\partial t} = \frac{\partial H_x(x, z, t)}{\partial z} - \frac{\partial H_z(x, z, t)}{\partial x} - J_{\text{e}y}(x, z, t), \tag{6.181}$$

$$\frac{\partial B_x(x, z, t)}{\partial t} = \frac{\partial E_y(x, z, t)}{\partial z} - J_{\text{m}x}(x, z, t), \tag{6.182}$$

$$\frac{\partial B_z(x, z, t)}{\partial t} = - \frac{\partial E_y(x, z, t)}{\partial x} - J_{\text{m}z}(x, z, t). \tag{6.183}$$

Differentiation of (6.182) with regard to z and (6.183) with regard to x, subtraction of the remaining equations, and insertion of (6.181) recognizing the constitutive equations

[113]Bromwich's theorem (Langenberg 2005) defines material inhomogeneities for certain coordinate systems that allow for a separation into two scalar TM- and TE-fields, respectively, even in three spatial dimensions.

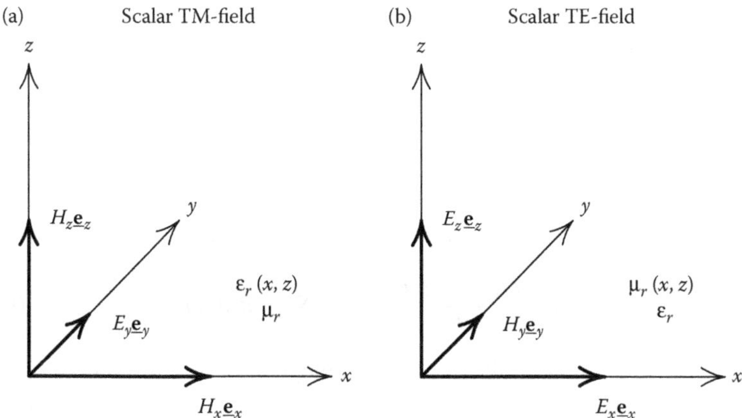

FIGURE 6.5
Two-dimensional electromagnetism ($\partial/\partial y \equiv 0$): a) TM-field, b) TE-field.

$$B_{x,z}(x,z,t) = \mu_0\mu_r H_{x,z}(x,z,t), \tag{6.184}$$
$$D_y(x,z,t) = \epsilon_0\epsilon_r(x,z)E_y(x,z,\omega) \tag{6.185}$$

results in a wave equation for $E_y(x,z,t)$:

$$\Delta_{xz}E_y(x,z,t) - \epsilon_0\epsilon_r(x,z)\mu_0\mu_r \frac{\partial^2 E_y(x,z,t)}{\partial t^2} = \mu_0\mu_r \frac{\partial J_{\mathrm{e}y}(x,z,t)}{\partial t} - \underline{\mathbf{e}}_y \cdot \boldsymbol{\nabla}_{xz} \times \underline{\mathbf{J}}_{\mathrm{m}}(x,z,t). \tag{6.186}$$

This proves the above statement.

If the material turns out to be inhomogeneous with regard to the permeability—no matter whether the permittivity is homogeneous or inhomogeneous—an additional term $\boldsymbol{\nabla} \ln \mu_r(x,z) \cdot \boldsymbol{\nabla} E_y(x,z,t)$ appears on the right-hand side of (6.186).

If the permittivity has jump discontinuities on two-dimensional "surfaces"—curves in an xz-plane—transition conditions (6.30) and (6.31) must be satisfied. Without prescribed surface current densities, it readily follows from (6.30) that E_y must be continuous. Furthermore, from (6.31), it follows the continuity of the normal derivative $\underline{\mathbf{n}} \cdot \boldsymbol{\nabla}_{xz}E_y$ of the scalar "potential" E_y; to show that, Equations 6.181 and 6.182 are used in the only remaining nonzero y-component of (6.31). The two-dimensional electromagnetic TM-case complies with (two-dimensional) scalar acoustics if the density is constant and the compressibility is discontinuous. If the curves are boundaries of a scatterer with perfect electric conductivity, the tangential component E_y as scalar "potential" must be equal to zero; $\underline{\mathbf{n}} \times \underline{\mathbf{H}}$ then defines the electric surface current density induced in y-direction. In this case, the two-dimensional electromagnetic TM-scattering problem corresponds to a scalar Dirichlet problem.

6.8.2 TE-field

As displayed in Figure 6.5(b), the only nonzero components $H_y \neq 0$, E_x, $E_z \neq 0$ lead to a transverse electric, a TE-field. Under the assumptions $\mu_r(x,z)$, $\epsilon_r = \text{const}$, and $J_{my} \neq 0$, $J_{ex}, J_{ez} \neq 0$, we obtain a scalar wave equation for the "potential" $H_y(x,z,t)$:

$$\Delta_{xz} H_y(x,z,t) - \epsilon_0 \epsilon_r \mu_0 \mu_r(x,z) \frac{\partial^2 H_y(x,z,t)}{\partial t^2} \\ = \epsilon_0 \epsilon_r \frac{\partial J_{\mathrm{m}y}(x,z,t)}{\partial t} - \underline{\mathbf{e}}_y \cdot \nabla_{xz} \times \underline{\mathbf{J}}_{\mathrm{e}}(x,z,t). \tag{6.187}$$

Again, an additional or alternative inhomogeneity of the "complementary" material parameter, in this case, the permittivity, leads to the additional term $\nabla \ln \epsilon_r(x,z) \cdot \nabla H_y(x,z,t)$; in that case, it is advisable to stay with a vector wave equation for the electric field strength.

Potential transition conditions require the continuity of H_y and the continuity of the normal derivative of H_y; for perfect electric conductivity, the condition $\underline{\mathbf{n}} \times \underline{\mathbf{E}} = \underline{\mathbf{0}}$ transforms into a Neumann boundary condition for H_y, and $\underline{\mathbf{n}} \times \underline{\mathbf{H}}$ then defines the induced $K_{\mathrm{e}x,z}$-current density components. In the TM-case, only axial currents are flowing, whereas in the TE-case, only circumferential currents are flowing.

7

Vector Wave Equations

The elastodynamic governing equations (4.33) and (4.34) represent a coupled system of partial differential equations of first order for the field quantities $\underline{\mathbf{v}}(\underline{\mathbf{R}}, t)$ and $\underline{\underline{\mathbf{T}}}(\underline{\mathbf{R}}, t)$ after introducing constitutive equations of linear time invariant instantaneously and locally reacting materials, that are always considered in this section. To neutralize this coupling in terms of a decoupling both equations are inserted into each other: We obtain partial differential equations of second order either for $\underline{\mathbf{v}}(\underline{\mathbf{R}}, t)$ or for $\underline{\underline{\mathbf{T}}}(\underline{\mathbf{R}}, t)$. Since both equations allow for waves as solutions, we generally call them "wave equations" even though they are more complicated than the simple d'Alembert wave equation (5.32), hence the terminology "Navier wave equations" is sometimes used in the literature. Another decoupling method, even though valid only for homogeneous isotropic materials, is dealt with in connection with the Helmholtz decomposition of a vector into potentials.

7.1 Wave Equations for Anisotropic and Isotropic Nondissipative Materials

We perform the insertion steps for the elastodynamic governing equations (4.33) and (4.34) subsequently for inhomogeneous anisotropic, homogeneous anisotropic, and homogeneous isotropic materials. That way, we learn something about the general structure of the wave equation and afterward we have two special cases for homogeneous materials at hand that are particularly important for US-NDT.

7.1.1 Inhomogeneous anisotropic materials

We perform a time derivative of Newton–Cauchy's equation of motion (4.33)

$$\rho(\underline{\mathbf{R}}) \frac{\partial^2 \underline{\mathbf{v}}(\underline{\mathbf{R}}, t)}{\partial t^2} = \nabla \cdot \frac{\partial \underline{\underline{\mathbf{T}}}(\underline{\mathbf{R}}, t)}{\partial t} + \frac{\partial \underline{\mathbf{f}}(\underline{\mathbf{R}}, t)}{\partial t} \tag{7.1}$$

and insert the deformation rate equation (4.34) in the form

$$\frac{\partial \underline{\underline{\mathbf{T}}}(\underline{\mathbf{R}}, t)}{\partial t} = \underline{\underline{\underline{\underline{\mathbf{c}}}}}(\underline{\mathbf{R}}) : \nabla \underline{\mathbf{v}}(\underline{\mathbf{R}}, t) + \underline{\underline{\underline{\underline{\mathbf{c}}}}} : \underline{\underline{\mathbf{h}}}(\underline{\mathbf{R}}, t); \tag{7.2}$$

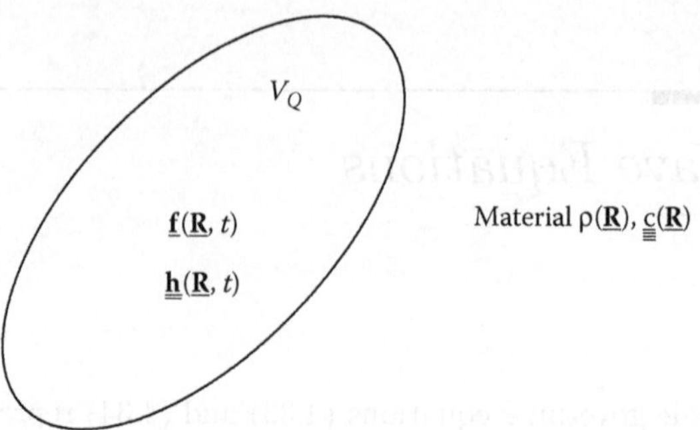

FIGURE 7.1
Source volume V_Q of elastic waves in an inhomogeneous anisotropic material.

the symmetry of the stiffness tensor with regard to the last two indices allows for the short-hand notation $\underline{\underline{\underline{\underline{\mathbf{c}}}}} : \nabla\underline{\mathbf{v}}$ for the double contraction of $\underline{\underline{\underline{\underline{\mathbf{c}}}}}$ with $\frac{1}{2}(\nabla\underline{\mathbf{v}} + \nabla\underline{\mathbf{v}}^{21})$ resulting in the Navier equation[114]

$$\nabla \cdot \left[\underline{\underline{\underline{\underline{\mathbf{c}}}}}(\underline{\mathbf{R}}) : \nabla\underline{\mathbf{v}}(\underline{\mathbf{R}}, t)\right] - \rho(\underline{\mathbf{R}}) \frac{\partial^2 \underline{\mathbf{v}}(\underline{\mathbf{R}}, t)}{\partial t^2} = -\frac{\partial \underline{\mathbf{f}}(\underline{\mathbf{R}}, t)}{\partial t} - \nabla \cdot \left[\underline{\underline{\underline{\underline{\mathbf{c}}}}}(\underline{\mathbf{R}}) : \underline{\underline{\mathbf{h}}}(\underline{\mathbf{R}}, t)\right] \tag{7.3}$$

for $\underline{\mathbf{v}}(\underline{\mathbf{R}}, t)$ [Equation 5.29 is the acoustic counterpart]. Note: For sources $\underline{\mathbf{f}}(\underline{\mathbf{R}}, t)$, $\underline{\underline{\mathbf{h}}}(\underline{\mathbf{R}}, t)$ confined to V_Q (Figure 7.1) embedded in the inhomogeneous anisotropic material, the $\underline{\underline{\underline{\underline{\mathbf{c}}}}}(\underline{\mathbf{R}})$-inhomogeneity of the material (inside V_Q) enters the inhomogeneity of the differential equation. Generally—and typically in US-NDT—we encounter the situation as sketched in Figure 7.2: The inhomogeneous anisotropic embedding material with the material parameters $\rho^{(e)}(\underline{\mathbf{R}})$, $\underline{\underline{\underline{\underline{\mathbf{c}}}}}^{(e)}(\underline{\mathbf{R}})$ contains a contrast volume V_c with the material parameters $\rho^{(i)}(\underline{\mathbf{R}})$, $\underline{\underline{\underline{\underline{\mathbf{c}}}}}^{(i)}(\underline{\mathbf{R}})$ with nonoverlapping V_Q and V_c; then, V_c can be considered as a defect for US-NDT that is illuminated by the sources in V_Q. We want to show how we can trickily "hide" the defect within an additional inhomogeneity term in the differential equation (7.3) (Snieder 2002). At first, we define the (dimensionless) contrast of the defect

[114]In elastodynamics, we are primarily interested in the particle velocity or the particle displacement as field quantities whereas acoustics is devoted to the pressure that would correspond to the stress tensor in the present context; therefore, the elastodynamic pendant to (5.26) would be the differential equation

$$\underline{\underline{\underline{\underline{\mathbf{I}}}}}^+ : \nabla \left[\frac{1}{\rho(\underline{\mathbf{R}})} \nabla \cdot \underline{\underline{\mathbf{T}}}(\underline{\mathbf{R}}, t)\right] - \underline{\underline{\underline{\underline{\mathbf{s}}}}}(\underline{\mathbf{R}}) : \frac{\partial^2 \underline{\underline{\mathbf{T}}}(\underline{\mathbf{R}}, t)}{\partial t^2} = -\underline{\underline{\underline{\underline{\mathbf{I}}}}}^+ : \nabla \left[\frac{1}{\rho(\underline{\mathbf{R}})} \underline{\mathbf{f}}(\underline{\mathbf{R}}, t)\right] - \frac{\partial \underline{\underline{\mathbf{h}}}(\underline{\mathbf{R}}, t)}{\partial t}.$$

For simplicity, we have used the symmetrization tensor $\underline{\underline{\underline{\underline{\mathbf{I}}}}}^+$ according to (2.109).

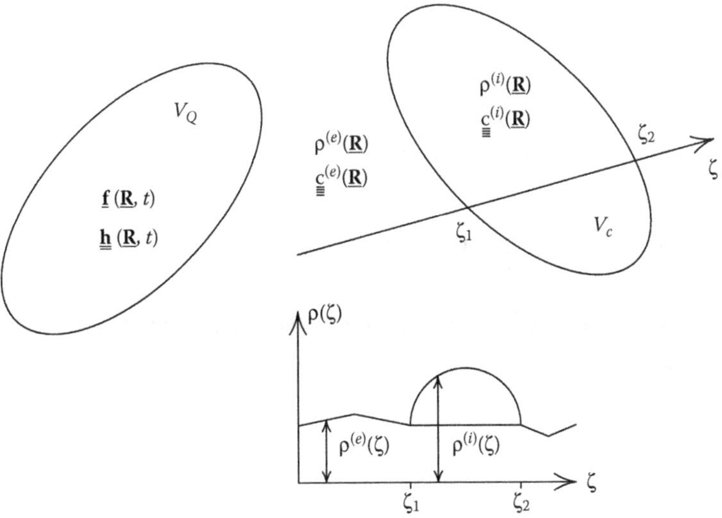

FIGURE 7.2
Source volume V_Q of elastic waves in an inhomogeneous anisotropic material.

$$\chi_\rho(\underline{\mathbf{R}}) = \frac{1}{\rho^{(e)}(\underline{\mathbf{R}})} \left[\rho^{(i)}(\underline{\mathbf{R}}) - \rho^{(e)}(\underline{\mathbf{R}})\right] \Gamma_c(\underline{\mathbf{R}}) \tag{7.4}$$

in the density and the (dimensionless) contrast

$$\underline{\underline{\underline{\underline{\boldsymbol{\chi}}}}}_{\mathbf{c}}(\underline{\mathbf{R}}) = \underline{\underline{\underline{\underline{\mathbf{s}}}}}^{(e)}(\underline{\mathbf{R}}) : \left[\underline{\underline{\underline{\underline{\mathbf{c}}}}}^{(i)}(\underline{\mathbf{R}}) - \underline{\underline{\underline{\underline{\mathbf{c}}}}}^{(e)}(\underline{\mathbf{R}})\right] \Gamma_c(\underline{\mathbf{R}}) \tag{7.5}$$

in the stiffness tensor, where $\underline{\underline{\underline{\underline{\mathbf{s}}}}}^{(e)}(\underline{\mathbf{R}})$ with

$$\begin{aligned} \underline{\underline{\underline{\underline{\mathbf{s}}}}}^{(e)}(\underline{\mathbf{R}}) : \underline{\underline{\underline{\underline{\mathbf{c}}}}}^{(e)}(\underline{\mathbf{R}}) &= \underline{\underline{\underline{\underline{\mathbf{c}}}}}^{(e)}(\underline{\mathbf{R}}) : \underline{\underline{\underline{\underline{\mathbf{s}}}}}^{(e)}(\underline{\mathbf{R}}) \\ &= \underline{\underline{\underline{\underline{\mathbf{I}}}}}^{+} \end{aligned} \tag{7.6}$$

is the compliance tensor of the embedding material; due to the characteristic function $\Gamma_c(\underline{\mathbf{R}})$, both $\chi_\rho(\underline{\mathbf{R}})$ and $\underline{\underline{\underline{\underline{\boldsymbol{\chi}}}}}_{\mathbf{c}}(\underline{\mathbf{R}})$ are actually equal to zero outside V_c. The material inhomogeneities in (7.3) can now be written as

$$\rho(\underline{\mathbf{R}}) = \rho^{(e)}(\underline{\mathbf{R}})\,[1 + \chi_\rho(\underline{\mathbf{R}})] = \begin{cases} \rho^{(e)}(\underline{\mathbf{R}}) & \text{for } \underline{\mathbf{R}} \notin V_c \\ \rho^{(i)}(\underline{\mathbf{R}}) & \text{for } \underline{\mathbf{R}} \in V_c, \end{cases} \tag{7.7}$$

$$\underline{\underline{\underline{\underline{\mathbf{c}}}}}(\underline{\mathbf{R}}) = \underline{\underline{\underline{\underline{\mathbf{c}}}}}^{(e)}(\underline{\mathbf{R}}) : \left[\underline{\underline{\underline{\underline{\mathbf{I}}}}}^{+} + \underline{\underline{\underline{\underline{\boldsymbol{\chi}}}}}_{\mathbf{c}}(\underline{\mathbf{R}})\right] = \begin{cases} \underline{\underline{\underline{\underline{\mathbf{c}}}}}^{(e)}(\underline{\mathbf{R}}) & \text{for } \underline{\mathbf{R}} \notin V_c \\ \underline{\underline{\underline{\underline{\mathbf{c}}}}}^{(i)}(\underline{\mathbf{R}}) & \text{for } \underline{\mathbf{R}} \in V_c \end{cases}; \tag{7.8}$$

in Figure 7.2, we have displayed $\rho(\underline{\mathbf{R}})$ in terms of a cross-section of V_c along the coordinate ζ.

Inserting (7.7) and (7.8) into (7.3), we obtain after rearranging terms:

$$\begin{aligned}
&\nabla \cdot \left[\underline{\underline{\underline{\underline{\mathbf{c}}}}}^{(e)}(\underline{\mathbf{R}}) : \nabla \underline{\mathbf{v}}(\underline{\mathbf{R}}, t)\right] - \rho^{(e)}(\underline{\mathbf{R}}) \frac{\partial^2 \underline{\mathbf{v}}(\underline{\mathbf{R}}, t)}{\partial t^2} \\
&= -\frac{\partial \underline{\mathbf{f}}(\underline{\mathbf{R}}, t)}{\partial t} - \nabla \cdot \left[\underline{\underline{\underline{\underline{\mathbf{c}}}}}^{(e)}(\underline{\mathbf{R}}) : \underline{\underline{\mathbf{h}}}(\underline{\mathbf{R}}, t)\right] + \rho^{(e)}(\underline{\mathbf{R}}) \chi_\rho(\underline{\mathbf{R}}) \frac{\partial^2 \underline{\mathbf{v}}(\underline{\mathbf{R}}, t)}{\partial t^2} \\
&\quad - \nabla \cdot \left[\underline{\underline{\underline{\underline{\mathbf{c}}}}}^{(e)}(\underline{\mathbf{R}}) : \underline{\underline{\underline{\underline{\chi}}}}_{\mathbf{c}}(\underline{\mathbf{R}}) : \nabla \underline{\mathbf{v}}(\underline{\mathbf{R}}, t)\right] ; \qquad (7.9)
\end{aligned}$$

in the term $\nabla \cdot \left[\underline{\underline{\underline{\underline{\mathbf{c}}}}}(\underline{\mathbf{R}}) : \underline{\underline{\mathbf{h}}}\right]$, we can replace $\underline{\underline{\underline{\underline{\mathbf{c}}}}}(\underline{\mathbf{R}})$ by $\underline{\underline{\underline{\underline{\mathbf{c}}}}}^{(e)}(\underline{\mathbf{R}})$ because V_Q with $\underline{\underline{\mathbf{h}}} \neq \underline{\underline{\mathbf{0}}}$ has been assumed to be in the exterior of V_c. The notation (7.9) suggests as if $\underline{\mathbf{v}}(\underline{\mathbf{R}}, t)$ would satisfy a Navier equation for the inhomogeneous anisotropic embedding material with an "extended" inhomogeneity that represents the contrast volume V_c. This is apparently not really true because the extended inhomogeneity contains the unknown field quantity; nevertheless, the form (7.9) has advantages for the approximate solution of scattering problems (keyword: Born approximation) consequently stimulating to explicitly express the two terms $\rho^{(e)}\chi_\rho \partial^2 \underline{\mathbf{v}}/\partial t^2$ and $\nabla \cdot \left[\underline{\underline{\underline{\underline{\mathbf{c}}}}}^{(e)} : \underline{\underline{\underline{\underline{\chi}}}}_{\mathbf{c}} : \nabla \underline{\mathbf{v}}\right]$ by (equivalent) sources $\underline{\mathbf{f}}_\rho$ and $\underline{\underline{\mathbf{h}}}_c$ where

$$\begin{aligned}
\underline{\mathbf{f}}_\rho(\underline{\mathbf{R}}, t) &= -\rho^{(e)}(\underline{\mathbf{R}}) \chi_\rho(\underline{\mathbf{R}}) \frac{\partial \underline{\mathbf{v}}(\underline{\mathbf{R}}, t)}{\partial t} \\
&= \Gamma_c(\underline{\mathbf{R}}) \left[\rho^{(e)}(\underline{\mathbf{R}}) - \rho^{(i)}(\underline{\mathbf{R}})\right] \frac{\partial \underline{\mathbf{v}}(\underline{\mathbf{R}}, t)}{\partial t}, \qquad (7.10) \\
\underline{\underline{\mathbf{h}}}_{\mathbf{c}}(\underline{\mathbf{R}}, t) &= \underline{\underline{\underline{\underline{\chi}}}}_{\mathbf{c}}(\underline{\mathbf{R}}) : \nabla \underline{\mathbf{v}}(\underline{\mathbf{R}}, t) \\
&= \underline{\underline{\underline{\underline{\chi}}}}_{\mathbf{c}}(\underline{\mathbf{R}}) : \underline{\underline{\underline{\underline{\mathbf{s}}}}}^{(i)}(\underline{\mathbf{R}}) : \frac{\partial \underline{\underline{\mathbf{T}}}(\underline{\mathbf{R}}, t)}{\partial t} \\
&= \Gamma_c(\underline{\mathbf{R}}) \left[\underline{\underline{\underline{\underline{\mathbf{s}}}}}^{(e)}(\underline{\mathbf{R}}) - \underline{\underline{\underline{\underline{\mathbf{s}}}}}^{(i)}(\underline{\mathbf{R}})\right] : \frac{\partial \underline{\underline{\mathbf{T}}}(\underline{\mathbf{R}}, t)}{\partial t}, \qquad (7.11)
\end{aligned}$$

because (7.9) then reads as

$$\begin{aligned}
\nabla \cdot \left[\underline{\underline{\underline{\underline{\mathbf{c}}}}}^{(e)}(\underline{\mathbf{R}}) : \nabla \underline{\mathbf{v}}(\underline{\mathbf{R}}, t)\right] - \rho^{(e)}(\underline{\mathbf{R}}) \frac{\partial^2 \underline{\mathbf{v}}(\underline{\mathbf{R}}, t)}{\partial t^2} &= -\frac{\partial}{\partial t} \left[\underline{\mathbf{f}}(\underline{\mathbf{R}}, t) + \underline{\mathbf{f}}_\rho(\underline{\mathbf{R}}, t)\right] \\
-\nabla \cdot \left\{\underline{\underline{\underline{\underline{\mathbf{c}}}}}^{(e)}(\underline{\mathbf{R}}) : \left[\underline{\underline{\mathbf{h}}}(\underline{\mathbf{R}}, t) + \underline{\underline{\mathbf{h}}}_{\mathbf{c}}(\underline{\mathbf{R}}, t)\right]\right\}&. \qquad (7.12)
\end{aligned}$$

To find the last line of (7.11), we have to use the symmetry of the material anisotropy tensor and the Newton–Cauchy equation (4.33) that is homogeneous in the contrast volume.[115]

[115]Equation 5.99 with 5.94 and 5.95 is the acoustic pendant to 7.12 (for a homogeneous embedding material).

In the Navier equation (7.12), the contrast volume V_c is equivalent to the sources $\underline{\mathbf{f}}_\rho(\underline{\mathbf{R}}, t)$, $\underline{\underline{\mathbf{h}}}_{\mathbf{c}}(\underline{\mathbf{R}}, t)$ why they are called equivalent sources. They are also called secondary (or induced) sources because they are created as kind of a feed-back by the incident field $\underline{\mathbf{v}}_i(\underline{\mathbf{R}}, t)$ with the sources $\underline{\mathbf{f}}(\underline{\mathbf{R}}, t)$, $\underline{\underline{\mathbf{h}}}(\underline{\mathbf{R}}, t)$ superimposed by their own scattered field $\underline{\mathbf{v}}_s(\underline{\mathbf{R}}, t)$ to the total field $\underline{\mathbf{v}}(\underline{\mathbf{R}}, t) = \underline{\mathbf{v}}_i(\underline{\mathbf{R}}, t) + \underline{\mathbf{v}}_s(\underline{\mathbf{R}}, t)$ in V_c. On one hand, this dependence of the equivalent sources upon the total field can be exploited to formulate a volume integral equation for their calculation (Section 15.3.1), and on the other hand, the linear superposition of the incident and the scattered field suggests to neglect the scattered field in V_c: This is the so-called Born approximation (Section 15.3.2; acoustics: Section 5.6).

The Navier equation (7.12) holds in all space ($\underline{\mathbf{R}} \in \mathbb{R}^3$). Nevertheless, we can specialize it to $\underline{\mathbf{R}} \notin V_c$ and $\underline{\mathbf{R}} \in V_c$ recognizing that the equivalent sources are restricted to:[116]

$$\begin{aligned} &\nabla \cdot \left[\underline{\underline{\underline{\underline{\mathbf{c}}}}}^{(e)}(\underline{\mathbf{R}}) : \nabla \underline{\mathbf{v}}(\underline{\mathbf{R}}, t)\right] - \rho^{(e)}(\underline{\mathbf{R}}) \frac{\partial^2 \underline{\mathbf{v}}(\underline{\mathbf{R}}, t)}{\partial t^2} \\ &\qquad = -\frac{\partial \underline{\mathbf{f}}(\underline{\mathbf{R}}, t)}{\partial t} - \nabla \cdot \left[\underline{\underline{\underline{\underline{\mathbf{c}}}}}^{(e)}(\underline{\mathbf{R}}) : \underline{\underline{\mathbf{h}}}(\underline{\mathbf{R}}, t)\right] \quad \text{for } \underline{\mathbf{R}} \notin V_c; \end{aligned} \tag{7.13}$$

in the interior of V_c apparently the homogeneous equation (7.3) with the material parameters of V_c holds

$$\nabla \cdot \left[\underline{\underline{\underline{\underline{\mathbf{c}}}}}^{(i)}(\underline{\mathbf{R}}) : \underline{\mathbf{v}}(\underline{\mathbf{R}}, t)\right] - \rho^{(i)}(\underline{\mathbf{R}}) \frac{\partial^2 \underline{\mathbf{v}}(\underline{\mathbf{R}}, t)}{\partial t^2} = \underline{\mathbf{0}} \text{ for } \underline{\mathbf{R}} \in V_c. \tag{7.14}$$

At this point, it should already not be unmentioned that even scattering by a void in an embedding material—our previous contrast volume has been an inclusion—can be reduced to equivalent (surface) sources (Section 15.1.3).

7.1.2 Homogeneous anisotropic materials

Specializing the wave equation (7.3) to homogeneous anisotropic materials $\rho(\underline{\mathbf{R}}) = \rho$, $\underline{\underline{\underline{\underline{\mathbf{c}}}}}(\underline{\mathbf{R}}) = \underline{\underline{\underline{\underline{\mathbf{c}}}}}$ is evident:

$$\nabla \cdot \underline{\underline{\underline{\underline{\mathbf{c}}}}} : \nabla \underline{\mathbf{v}}(\underline{\mathbf{R}}, t) - \rho \frac{\partial^2 \underline{\mathbf{v}}(\underline{\mathbf{R}}, t)}{\partial t^2} = -\frac{\partial \underline{\mathbf{f}}(\underline{\mathbf{R}}, t)}{\partial t} - \nabla \cdot \underline{\underline{\underline{\underline{\mathbf{c}}}}} : \underline{\underline{\mathbf{h}}}(\underline{\mathbf{R}}, t). \tag{7.15}$$

Note: To shift the $\underline{\mathbf{R}}$-independent $\underline{\underline{\underline{\underline{\mathbf{c}}}}}$-tensor in front of the divergence operator an interchange of indices is necessary, we obtain

$$\underline{\underline{\underline{\underline{\mathbf{c}}}}}^{2341} : \nabla \nabla \underline{\mathbf{v}}(\underline{\mathbf{R}}, t) - \rho \frac{\partial^2 \underline{\mathbf{v}}(\underline{\mathbf{R}}, t)}{\partial t^2} = -\frac{\partial \underline{\mathbf{f}}(\underline{\mathbf{R}}, t)}{\partial t} - \underline{\underline{\underline{\underline{\mathbf{c}}}}}^{2341} : \nabla \underline{\underline{\mathbf{h}}}(\underline{\mathbf{R}}, t). \tag{7.16}$$

[116]One says that: They are of compact support.

The specialization of (7.12) to a homogeneous anisotropic embedding material for inhomogeneous anisotropic contrast volumes equivalently reads as

$$\begin{aligned}\nabla \cdot \underline{\underline{\underline{\underline{c}}}} : \nabla \underline{\mathbf{v}}(\underline{\mathbf{R}}, t) - \rho \frac{\partial^2 \underline{\mathbf{v}}(\underline{\mathbf{R}}, t)}{\partial t^2} &= -\frac{\partial}{\partial t}\left[\underline{\mathbf{f}}(\underline{\mathbf{R}}, t) + \underline{\mathbf{f}}_\rho(\underline{\mathbf{R}}, t)\right] \\ &\quad - \nabla \cdot \underline{\underline{\underline{\underline{c}}}} : \left[\underline{\underline{\mathbf{h}}}(\underline{\mathbf{R}}, t) + \underline{\underline{\mathbf{h}}}_{\mathbf{c}}(\underline{\mathbf{R}}, t)\right], \end{aligned} \tag{7.17}$$

where $\rho^{(e)} \Longrightarrow \rho$ has to be replaced in $\underline{\mathbf{f}}_\rho$ and $\underline{\underline{\underline{\underline{c}}}}^{(e)} \Longrightarrow \underline{\underline{\underline{\underline{c}}}}$ in $\underline{\underline{\mathbf{h}}}_{\mathbf{c}}$.

7.1.3 Homogeneous isotropic materials

For a homogeneous isotropic material, we have to insert the stiffness tensor

$$\underline{\underline{\underline{\underline{c}}}} = \lambda \underline{\underline{\underline{\underline{I}}}}^{\delta} + 2\mu \underline{\underline{\underline{\underline{I}}}}^{+} \tag{7.18}$$

with Lamé constants λ and μ in (7.15) and to pay attention to (2.115) and (2.116):

$$\begin{aligned}&\lambda \nabla \cdot \underline{\underline{\mathbf{I}}} \underbrace{\operatorname{trace}\left[\nabla \underline{\mathbf{v}}(\underline{\mathbf{R}}, t)\right]}_{= \nabla \cdot \underline{\mathbf{v}}(\underline{\mathbf{R}}, t)} + \underbrace{2\mu \nabla \cdot \left[\nabla \underline{\mathbf{v}}(\underline{\mathbf{R}}, t)\right]_{\mathrm{s}}}_{= \mu[\nabla \cdot \nabla \underline{\mathbf{v}}(\underline{\mathbf{R}}, t) + \nabla \nabla \cdot \underline{\mathbf{v}}(\underline{\mathbf{R}}, t)]} - \rho \frac{\partial^2 \underline{\mathbf{v}}(\underline{\mathbf{R}}, t)}{\partial t^2} \\ &= -\frac{\partial \underline{\mathbf{f}}(\underline{\mathbf{R}}, t)}{\partial t} - \lambda \nabla [\operatorname{trace} \underline{\underline{\mathbf{h}}}(\underline{\mathbf{R}}, t)] - 2\mu \nabla \cdot \underline{\underline{\mathbf{h}}}(\underline{\mathbf{R}}, t), \end{aligned} \tag{7.19}$$

where we exploited the symmetry of $\underline{\underline{\mathbf{h}}}$ in the $\underline{\underline{\mathbf{h}}}$-term. Explicitly written Equation 7.19 finally reads as

$$\begin{aligned}&\mu \Delta \underline{\mathbf{v}}(\underline{\mathbf{R}}, t) + (\lambda + \mu) \nabla \nabla \cdot \underline{\mathbf{v}}(\underline{\mathbf{R}}, t) - \rho \frac{\partial^2 \underline{\mathbf{v}}(\underline{\mathbf{R}}, t)}{\partial t^2} \\ &\quad = -\frac{\partial \underline{\mathbf{f}}(\underline{\mathbf{R}}, t)}{\partial t} - \lambda \nabla [\operatorname{trace} \underline{\underline{\mathbf{h}}}(\underline{\mathbf{R}}, t)] - 2\mu \nabla \cdot \underline{\underline{\mathbf{h}}}(\underline{\mathbf{R}}, t), \end{aligned} \tag{7.20}$$

and another version emerges if we express the Δ-Operator according to (2.187) by $\nabla \times \nabla \times$ and $\nabla\nabla$:

$$\begin{aligned}&(\lambda + 2\mu) \nabla \nabla \cdot \underline{\mathbf{v}}(\underline{\mathbf{R}}, t) - \mu \nabla \times \nabla \times \underline{\mathbf{v}}(\underline{\mathbf{R}}, t) - \rho \frac{\partial^2 \underline{\mathbf{v}}(\underline{\mathbf{R}}, t)}{\partial t^2} \\ &\quad = -\frac{\partial \underline{\mathbf{f}}(\underline{\mathbf{R}}, t)}{\partial t} - \lambda \nabla [\operatorname{trace} \underline{\underline{\mathbf{h}}}(\underline{\mathbf{R}}, t)] - 2\mu \nabla \cdot \underline{\underline{\mathbf{h}}}(\underline{\mathbf{R}}, t). \end{aligned} \tag{7.21}$$

The notation (7.21) nicely reveals[117] the pressure wave term $(\lambda + 2\mu)\nabla\nabla \cdot \underline{\mathbf{v}}$ and the shear wave term $\mu \nabla \times \nabla \times \underline{\mathbf{v}}$ (Section 8.1.2).

[117] From the homogeneous equation in this writing, we immediately deduce that respective solutions—for example, plane waves—may either be polarized longitudinally or transversely with different velocities. For instance, we take the divergence resulting in a vanishing double curl operator and leaving us with only the differential operator for acoustic longitudinal pressure waves with $\rho/(\lambda + 2\mu)$ as inverse square of the pressure wave velocity. However, if we take the curl the first term vanishes and ρ/μ appears to be the inverse square of the velocity of transverse waves: The remaining differential operator is equivalent to the one for transverse electromagnetic waves.

In the homogeneous isotropic material outside the source volume Equation 7.2 specially reads as

$$\frac{\partial \underline{\underline{\mathbf{T}}}(\underline{\mathbf{R}}, t)}{\partial t} = \lambda \underline{\underline{\mathbf{I}}} \nabla \cdot \underline{\mathbf{v}}(\underline{\mathbf{R}}, t) + \mu \left\{ \nabla \underline{\mathbf{v}}(\underline{\mathbf{R}}, t) + [\nabla \underline{\mathbf{v}}(\underline{\mathbf{R}}, t)]^{21} \right\}; \tag{7.22}$$

for causal fields (Section 3.3) time integration leads to

$$\underline{\underline{\mathbf{T}}}(\underline{\mathbf{R}}, t) = \lambda \underline{\underline{\mathbf{I}}} \nabla \cdot \underline{\mathbf{u}}(\underline{\mathbf{R}}, t) + \mu \left\{ \nabla \underline{\mathbf{u}}(\underline{\mathbf{R}}, t) + [\nabla \underline{\mathbf{u}}(\underline{\mathbf{R}}, t)]^{21} \right\}, \tag{7.23}$$

where $\underline{\mathbf{u}}(\underline{\mathbf{R}}, t)$ according to (3.94) denotes the particle displacement. If $\underline{\underline{\mathbf{h}}}$ is the null tensor, we equally write (7.20) and (7.21) often for the particle displacement instead the particle velocity:

$$\mu \Delta \underline{\mathbf{u}}(\underline{\mathbf{R}}, t) + (\lambda + \mu) \nabla \nabla \cdot \underline{\mathbf{u}}(\underline{\mathbf{R}}, t) - \rho \frac{\partial^2 \underline{\mathbf{u}}(\underline{\mathbf{R}}, t)}{\partial t^2} = -\underline{\mathbf{f}}(\underline{\mathbf{R}}, t), \tag{7.24}$$

$$(\lambda + 2\mu) \nabla \nabla \cdot \underline{\mathbf{u}}(\underline{\mathbf{R}}, t) - \mu \nabla \times \nabla \times \underline{\mathbf{u}}(\underline{\mathbf{R}}, t) - \rho \frac{\partial^2 \underline{\mathbf{u}}(\underline{\mathbf{R}}, t)}{\partial t^2} = -\underline{\mathbf{f}}(\underline{\mathbf{R}}, t). \tag{7.25}$$

7.1.4 Inhomogeneous isotropic materials

For inhomogeneous isotropic materials, the isotropic stiffness tensor (7.18)

$$\underline{\underline{\underline{\underline{\mathbf{c}}}}}(\underline{\mathbf{R}}) = \lambda(\underline{\mathbf{R}}) \underline{\underline{\underline{\underline{\mathbf{I}}}}}^{\delta} + 2\mu(\underline{\mathbf{R}}) \underline{\underline{\underline{\underline{\mathbf{I}}}}}^{+} \tag{7.26}$$

and the mass density $\rho(\underline{\mathbf{R}})$ are spatially dependent; calculating $\nabla \cdot \left[\underline{\underline{\underline{\underline{\mathbf{c}}}}}(\underline{\mathbf{R}}) : \nabla \underline{\mathbf{v}}(\underline{\mathbf{R}}, t) \right]$ in (7.3) further terms appear in addition to the terms $\mu(\underline{\mathbf{R}}) \Delta \underline{\mathbf{v}}(\underline{\mathbf{R}}, t)$ and $[\lambda(\underline{\mathbf{R}}) + \mu(\underline{\mathbf{R}})] \nabla \nabla \cdot \underline{\mathbf{v}}(\underline{\mathbf{R}}, t)$ that result from the differentiation of $\underline{\underline{\underline{\underline{\mathbf{c}}}}}(\underline{\mathbf{R}})$ (for clearness, we omit the evaluation of this differentiation on the right-hand side):

$$\begin{aligned}
&[\lambda(\underline{\mathbf{R}}) + \mu(\underline{\mathbf{R}})] \nabla \nabla \cdot \underline{\mathbf{v}}(\underline{\mathbf{R}}, t) + \mu(\underline{\mathbf{R}}) \Delta \underline{\mathbf{v}}(\underline{\mathbf{R}}, t) - \rho(\underline{\mathbf{R}}) \frac{\partial^2 \underline{\mathbf{v}}(\underline{\mathbf{R}}, t)}{\partial t^2} \\
&\quad + [\nabla \lambda(\underline{\mathbf{R}})] [\nabla \cdot \underline{\mathbf{v}}(\underline{\mathbf{R}}, t)] + [\nabla \mu(\underline{\mathbf{R}})] \cdot [\nabla \underline{\mathbf{v}}(\underline{\mathbf{R}}, t)] + [\nabla \underline{\mathbf{v}}(\underline{\mathbf{R}}, t)] \cdot [\nabla \mu(\underline{\mathbf{R}})] \\
&\quad = -\frac{\partial \underline{\mathbf{f}}(\underline{\mathbf{R}}, t)}{\partial t} - \nabla \cdot \underline{\underline{\underline{\underline{\mathbf{c}}}}}(\underline{\mathbf{R}}) : \underline{\underline{\mathbf{h}}}(\underline{\mathbf{R}}, t).
\end{aligned} \tag{7.27}$$

This differential equation serves to calculate the ray propagation in inhomogeneous isotropic materials (Section 12.3.2).

7.2 Helmholtz Decomposition for Homogeneous Isotropic Materials: Pressure and Shear Waves

The Helmholtz decomposition of a vector field, for example, the particle velocity $\underline{\mathbf{u}}(\underline{\mathbf{R}}, t)$, into a scalar potential $\Phi(\underline{\mathbf{R}}, t)$ and a vector potential $\underline{\boldsymbol{\Psi}}(\underline{\mathbf{R}}, t)$ according to

$$\underline{\mathbf{u}}(\underline{\mathbf{R}}, t) = \boldsymbol{\nabla}\Phi(\underline{\mathbf{R}}, t) + \boldsymbol{\nabla} \times \underline{\boldsymbol{\Psi}}(\underline{\mathbf{R}}, t) \tag{7.28}$$

with the gauge[118]

$$\boldsymbol{\nabla} \cdot \underline{\boldsymbol{\Psi}}(\underline{\mathbf{R}}, t) = 0 \tag{7.29}$$

is always possible because Φ and $\underline{\boldsymbol{\Psi}}$ can be calculated knowing $\underline{\mathbf{u}}$ (Achenbach 1973; Achenbach et al. 1982). Obviously, Equation 7.28 is the decomposition of $\underline{\mathbf{u}}(\underline{\mathbf{R}}, t)$ into a divergence-free shear term $\boldsymbol{\nabla} \times \underline{\boldsymbol{\Psi}}(\underline{\mathbf{R}}, t)$ and a curl-free pressure term $\boldsymbol{\nabla}\Phi(\underline{\mathbf{R}}, t)$. Turning to the Helmholtz decomposition of the particle velocity

$$\underline{\mathbf{v}}(\underline{\mathbf{R}}, t) = \boldsymbol{\nabla}\frac{\partial\Phi(\underline{\mathbf{R}}, t)}{\partial t} + \boldsymbol{\nabla} \times \frac{\partial\underline{\boldsymbol{\Psi}}(\underline{\mathbf{R}}, t)}{\partial t}, \tag{7.30}$$

it becomes obvious that the Navier equation (7.21) for homogeneous isotropic materials decouples in two independent equations for the potentials due to the inherent divergence and curl. Insertion of (7.28) into (7.21) directly yields

$$\begin{aligned} &\boldsymbol{\nabla}\frac{\partial}{\partial t}\left[(\lambda + 2\mu)\Delta\Phi(\underline{\mathbf{R}}, t) - \rho\frac{\partial^2\Phi(\underline{\mathbf{R}}, t)}{\partial t^2}\right] \\ &\quad + \boldsymbol{\nabla} \times \frac{\partial}{\partial t}\left[\underbrace{-\mu\boldsymbol{\nabla} \times \boldsymbol{\nabla} \times \underline{\boldsymbol{\Psi}}(\underline{\mathbf{R}}, t)}_{\overset{(7.29)}{=}\ \mu\Delta\underline{\boldsymbol{\Psi}}(\underline{\mathbf{R}}, t)} - \rho\frac{\partial^2\underline{\boldsymbol{\Psi}}(\underline{\mathbf{R}}, t)}{\partial t^2}\right] \end{aligned} \tag{7.31}$$

for the left-hand side of (7.21). Similarly, by decomposing the right-hand side of (7.21) into a (curl-free) gradient and a (divergence-free) curl term, we can equalize the respective curl-free and divergence-free terms as Helmholtz decompositions of the left- and right-hand sides of (7.21). The force density vector $\underline{\mathbf{f}}(\underline{\mathbf{R}}, t)$ is directly assigned to the Helmholtz potentials $(\lambda + 2\mu)\Phi_f(\underline{\mathbf{R}}, t)$ and $\mu\underline{\boldsymbol{\Psi}}_f(\underline{\mathbf{R}}, t)$ according to

$$\underline{\mathbf{f}}(\underline{\mathbf{R}}, t) = (\lambda + 2\mu)\boldsymbol{\nabla}\Phi_f(\underline{\mathbf{R}}, t) + \mu\boldsymbol{\nabla} \times \underline{\boldsymbol{\Psi}}_f(\underline{\mathbf{R}}, t), \quad \boldsymbol{\nabla} \cdot \underline{\boldsymbol{\Psi}}_f(\underline{\mathbf{R}}, t) = 0. \tag{7.32}$$

For the symmetric tensor of the injected deformation rate $\underline{\underline{\mathbf{h}}}(\underline{\mathbf{R}}, t)$, we establish a Helmholtz decomposition similar to (7.32) for each fixed value of the second index only to combine the three equations to a tensor equation (of second rank) afterward; nevertheless, we have to enforce the symmetry:

$$\begin{aligned} \underline{\underline{\mathbf{h}}}(\underline{\mathbf{R}}, t) = &\frac{1}{2}\left\{\boldsymbol{\nabla}\underline{\boldsymbol{\Theta}}_h(\underline{\mathbf{R}}, t) + [\boldsymbol{\nabla}\underline{\boldsymbol{\Theta}}_h(\underline{\mathbf{R}}, t)]^{21}\right\} \\ &+ \frac{1}{2}\left\{\boldsymbol{\nabla} \times \underline{\underline{\boldsymbol{\Xi}}}_h(\underline{\mathbf{R}}, t) + [\boldsymbol{\nabla} \times \underline{\underline{\boldsymbol{\Xi}}}_h(\underline{\mathbf{R}}, t)]^{21}\right\}, \end{aligned} \tag{7.33}$$

[118] A vector—here: $\underline{\boldsymbol{\Psi}}$—is only uniquely determined if its curl and its divergence are assessed.

where

$$\nabla \cdot \underline{\underline{\Xi}}(\underline{\mathbf{R}}, t) = \underline{\mathbf{0}} \tag{7.34}$$

is a gauge requirement. For the right-hand side complementing (7.31), we have to calculate

$$\begin{aligned}
&\lambda \nabla \operatorname{trace} \underline{\underline{\mathbf{h}}}(\underline{\mathbf{R}}, t) + 2\mu \nabla \cdot \underline{\underline{\mathbf{h}}}(\underline{\mathbf{R}}, t) \\
&\quad = \lambda \nabla\nabla \cdot \underline{\mathbf{\Theta}}_h(\underline{\mathbf{R}}, t) + \mu \nabla \cdot \nabla \underline{\mathbf{\Theta}}_h(\underline{\mathbf{R}}, t) + \mu \nabla\nabla \cdot \underline{\mathbf{\Theta}}_h(\underline{\mathbf{R}}, t) \\
&\qquad + \lambda \nabla \operatorname{trace} [\nabla \times \underline{\underline{\Xi}}(\underline{\mathbf{R}}, t)] + \frac{1}{2} \mu \nabla \times \nabla \cdot \underline{\underline{\Xi}}^{21}(\underline{\mathbf{R}}, t);
\end{aligned}$$

also decomposing $\nabla \cdot \nabla \underline{\mathbf{\Theta}}_h$ into $\nabla\nabla \cdot \underline{\mathbf{\Theta}}_h - \nabla \times \nabla \times \underline{\mathbf{\Theta}}_h$, the equalization of the respective divergence-free and curl-free terms of (7.31) with the corresponding terms of the inhomogeneity results in the decoupled equations for $\Phi(\underline{\mathbf{R}}, t)$ and $\underline{\mathbf{\Psi}}(\underline{\mathbf{R}}, t)$:

$$\begin{aligned}
&\Delta \Phi(\underline{\mathbf{R}}, t) - \frac{\rho}{\lambda + 2\mu} \frac{\partial^2 \Phi(\underline{\mathbf{R}}, t)}{\partial t^2} \\
&\quad = -\Phi_f(\underline{\mathbf{R}}, t) - \nabla \cdot \underline{\mathbf{\Theta}}_h(\underline{\mathbf{R}}, t) - \frac{\lambda}{\lambda + 2\mu} \operatorname{trace} \left[\nabla \times \int_0^t \underline{\underline{\Xi}}_h(\underline{\mathbf{R}}, \tau)\, d\tau\right],
\end{aligned} \tag{7.35}$$

$$\begin{aligned}
&\Delta \underline{\mathbf{\Psi}}(\underline{\mathbf{R}}, t) - \frac{\rho}{\mu} \frac{\partial^2 \underline{\mathbf{\Psi}}(\underline{\mathbf{R}}, t)}{\partial t^2} \\
&\quad = -\underline{\mathbf{\Psi}}_f(\underline{\mathbf{R}}, t) + \nabla \times \int_0^t \underline{\mathbf{\Theta}}_h(\underline{\mathbf{R}}, \tau)\, d\tau - \frac{1}{2} \nabla \cdot \int_0^t \underline{\underline{\Xi}}_h^{21}(\underline{\mathbf{R}}, \tau)\, d\tau.
\end{aligned} \tag{7.36}$$

Since the inhomogeneities of the differential equations for the potentials are related to the "true" physical sources $\underline{\mathbf{f}}(\underline{\mathbf{R}}, t)$ and $\underline{\underline{\mathbf{h}}}(\underline{\mathbf{R}}, t)$ in a rather complicated way only the homogeneous versions of (7.35) and (7.36) are generally used (e.g., Section 8.1.2; Schmerr 1998); however, we then have d'Alembert wave equations in contrast to the homogeneous Navier equation (7.21), and one of them is even scalar. Simultaneously, the mathematical decoupling into two d'Alembert equations is also a physical decoupling into pressure and shear waves resulting, at least for pressure waves in a "scalarization." However, the pressure–shear coupling due to boundary and transition conditions destroys the above decoupling hence it is advisable to stay with the Navier equation in particular because there is no alternative for inhomogeneous and/or anisotropic materials. Even plane waves do not decouple into pressure and shear waves in infinite homogeneous anisotropic materials, they are even no longer longitudinally and transversely polarized (Section 8.3.1). In inhomogeneous isotropic materials, a respective separation into longitudinal pressure and transverse shear waves only exists if the material properties are slowly varying within the framework of a high frequency approximation (Červený 2001; Section 12.3).

7.3 Decoupling of Scalar SH-Waves for Inhomogeneous Isotropic Two-Dimensional Materials

There is a possibility for a complete "scalarization" of elastic wave propagation—more precisely, elastic shear waves: We simply have to postulate a single nonvanishing cartesian component of $\underline{\mathbf{v}}(\underline{\mathbf{R}}, t)$ and require the independence of the wave propagation from this coordinate, i.e., we confine ourselves to a two-dimensional problem with—for example—$\partial/\partial y \equiv 0$ for $\underline{\mathbf{v}}(\underline{\mathbf{R}}, t) = v_y(x, z, t)\underline{\mathbf{e}}_y$. This two-dimensional SH-propagation problem is mathematically completely equivalent to the two-dimensional electromagnetic TE-problem.

We rely on a homogeneous isotropic material, hence on the Navier equation (7.13) with

$$\underline{\underline{\underline{\underline{\mathbf{c}}}}}^{(e)}(\underline{\mathbf{R}}) = \lambda(\underline{\mathbf{R}})\,\underline{\underline{\underline{\underline{\mathbf{I}}}}}^{\delta} + 2\mu(\underline{\mathbf{R}})\,\underline{\underline{\underline{\underline{\mathbf{I}}}}}^{+} \tag{7.37}$$

for $\underline{\mathbf{R}} \in \mathrm{I\!R}^3$. Under the assumptions

$$\frac{\partial}{\partial y} \equiv 0, \tag{7.38}$$

$$\underline{\mathbf{v}}(\underline{\mathbf{R}}, t) = v_y(x, z, t)\underline{\mathbf{e}}_y, \tag{7.39}$$

its left-hand side is reduced to the only remaining y-component—we have $\nabla \cdot \underline{\mathbf{v}} \equiv 0$ and $\nabla\mu \cdot \underline{\mathbf{v}} \equiv 0$—

$$\begin{aligned} &\mu(x, z)\left(\frac{\partial^2 v_y(x, z, t)}{\partial x^2} + \frac{\partial^2 v_y(x, z, t)}{\partial z^2}\right) \\ &\quad + \nabla\mu(x, z) \cdot \nabla v_y(x, z, t) - \rho(x, z)\,\frac{\partial^2 v_y(x, z, t)}{\partial t^2}. \end{aligned} \tag{7.40}$$

We must require

$$\underline{\mathbf{f}}(\underline{\mathbf{R}}, t) = f_y(x, z, t)\underline{\mathbf{e}}_y, \tag{7.41}$$

$$\underline{\mathbf{e}}_x \cdot \nabla[\lambda(x, z)\,\mathrm{trace}\,\underline{\underline{\mathbf{h}}}(x, z, t)] = 0, \tag{7.42}$$

$$\underline{\mathbf{e}}_z \cdot \nabla[\lambda(x, z)\,\mathrm{trace}\,\underline{\underline{\mathbf{h}}}(x, z, t)] = 0, \tag{7.43}$$

$$\nabla \cdot [\mu(x, z)\underline{\underline{\mathbf{h}}}(x, z, t)] \cdot \underline{\mathbf{e}}_x = 0, \tag{7.44}$$

$$\nabla \cdot [\mu(x, z)\underline{\underline{\mathbf{h}}}(x, z, t)] \cdot \underline{\mathbf{e}}_z = 0 \tag{7.45}$$

to get only a y-component also on the right-hand side. With the assumption (7.38), we then obtain

$$\begin{aligned} &\mu(x, z)\Delta v_y(x, z, t) + \nabla\mu(x, z) \cdot \nabla v_y(x, z, t) - \rho(x, z)\,\frac{\partial^2 v_y(x, z, t)}{\partial t^2} \\ &= -\frac{\partial f_y(x, z, t)}{\partial t} - 2\frac{\partial}{\partial x}[\mu(x, z)h_{xy}(x, z, t)] - 2\frac{\partial}{\partial z}[\mu(x, z)h_{zy}(x, z, t)] \end{aligned} \tag{7.46}$$

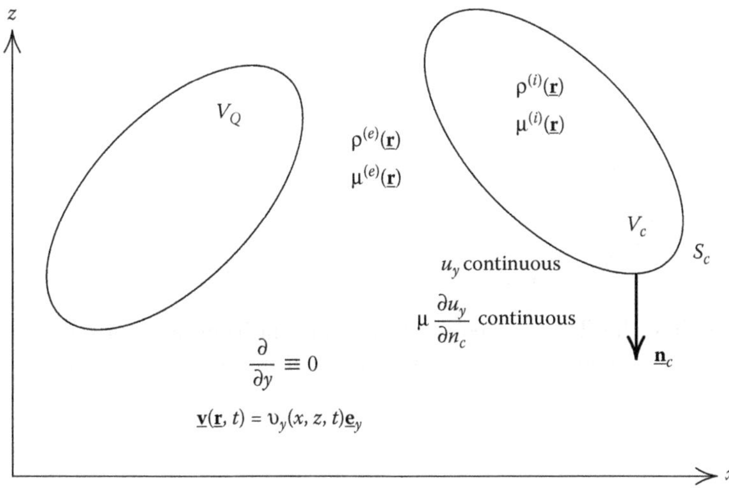

FIGURE 7.3
Two-dimensional scalar SH-wave scattering problem ($\underline{\mathbf{r}} = x\underline{\mathbf{e}}_x + z\underline{\mathbf{e}}_z$).

as a two-dimensional scalar wave equation for shear waves with a polarization parallel to the independency axis y. In a homogeneous isotropic material, (7.46) reduces to the d'Alembert wave equation

$$\Delta v_y(x,z,t) - \frac{\rho}{\mu}\frac{\partial^2 v_y(x,z,t)}{\partial t^2} = -\frac{1}{\mu}\frac{\partial f_y(x,z,t)}{\partial t} - 2\frac{\partial h_{xy}(x,z,t)}{\partial x} - 2\frac{\partial h_{zy}(x,z,t)}{\partial z}. \tag{7.47}$$

In the presence of (two-dimensional[119]) inclusion- or defect "volumes" V_c ("surface" S_c with outer normal $\underline{\mathbf{n}}_c$: Figure 7.3) with parameters $\rho^{(i)}(\underline{\mathbf{r}})$, $\mu^{(i)}(\underline{\mathbf{r}})$ of an inhomogeneous isotropic material embedded in a $\rho^{(e)}(\underline{\mathbf{r}})\mu^{(e)}(\underline{\mathbf{r}})$-material the scalar two-dimensionality also remains intact. Stress tensor and particle velocity have to satisfy the transition conditions (3.88) and (3.97); with (7.23), we calculate the stress tensor for the present two-dimensional case as

$$\underline{\underline{\mathbf{T}}}(x,z,t) = \mu(x,z)[\nabla u_y(x,z,t)\underline{\mathbf{e}}_y + \underline{\mathbf{e}}_y \nabla u_y(x,z,t)], \tag{7.48}$$

thus reducing (3.88)—we have $\underline{\mathbf{n}}_c \cdot \underline{\mathbf{e}}_y = 0$—to

$$\underline{\mathbf{n}}_c \cdot \underline{\underline{\mathbf{T}}}^{(e)}(\underline{\mathbf{r}}_S, t) - \underline{\mathbf{n}}_c \cdot \underline{\underline{\mathbf{T}}}^{(i)}(\underline{\mathbf{r}}_S, t) = \mu^{(e)}(\underline{\mathbf{r}}_S)\underline{\mathbf{n}}_c \cdot \nabla u_y^{(e)}(\underline{\mathbf{r}}, t)|_{\underline{\mathbf{r}}=\underline{\mathbf{r}}_S}\underline{\mathbf{e}}_y - \mu^{(i)}(\underline{\mathbf{r}}_S)\underline{\mathbf{n}}_c \cdot \nabla u_y^{(i)}(\underline{\mathbf{r}}, t)|_{\underline{\mathbf{r}}=\underline{\mathbf{r}}_S}\underline{\mathbf{e}}_y = \underline{\mathbf{0}}, \quad \underline{\mathbf{r}}_S \in S_c, \tag{7.49}$$

[119]Two-dimensional "volumes" are domains in $\mathbb{R}^2$ whose "surfaces" are curves in $\mathbb{R}^2$. Instead of using the mathematical notations Ω and $\partial\Omega$ for domains and their boundaries, we stay with the more intuitive notations V and S.

where

$$\underline{\mathbf{r}} = x\underline{\mathbf{e}}_x + z\underline{\mathbf{e}}_z. \tag{7.50}$$

The transition condition equation (7.49) only has a y-component—we refer to the notation (2.146) of the normal derivative—

$$\mu^{(e)}(\underline{\mathbf{r}}_S) \left. \frac{\partial u_y^{(e)}(\underline{\mathbf{r}}, t)}{\partial n_c} \right|_{\underline{\mathbf{r}}=\underline{\mathbf{r}}_S} = \mu^{(i)}(\underline{\mathbf{r}}_S) \left. \frac{\partial u_y^{(i)}(\underline{\mathbf{r}}, t)}{\partial n_c} \right|_{\underline{\mathbf{r}}=\underline{\mathbf{r}}_S}, \quad \underline{\mathbf{r}}_S \in S_c, \tag{7.51}$$

finally yielding the (homogeneous) transition conditions as continuity requirements of the particle velocity component according to (Equation 3.97)

$$u_y^{(e)}(\underline{\mathbf{r}}_S, t) = u_y^{(i)}(\underline{\mathbf{r}}_S, t), \quad \underline{\mathbf{r}}_S \in S_c, \tag{7.52}$$

and its μ-multiplied normal derivative according to (7.51).

For perfectly soft "inclusions," (7.51) degenerates to a Neumann boundary condition

$$\left. \frac{\partial u_y(\underline{\mathbf{r}}, t)}{\partial n_c} \right|_{\underline{\mathbf{r}}=\underline{\mathbf{r}}_S} = 0, \quad \underline{\mathbf{r}}_S \in S_c, \tag{7.53}$$

and for perfectly rigid inclusions, (7.52) degenerates to a Dirichlet boundary condition

$$u_y(\underline{\mathbf{r}}_S, t) = 0, \quad \underline{\mathbf{r}}_S \in S_c, \tag{7.54}$$

for the particle displacement component. Scattering of elastic shear waves polarized parallel to the independency axis of a two-dimensional soft scatterer—a two-dimensional void with a stress-free boundary—(SH-waves for shear-horizontal) consequently is a scalar Neumann boundary value problem.[120] Note: The respective scattering of a pressure wave in acoustics is a scalar Dirichlet problem (for the pressure). Furthermore, two-dimensional P-SV-wave scattering with a polarization vector parallel to the independency axis is not a scalar problem.

7.4 Frequency Domain Wave Equations for Nondissipative and Dissipative Materials

With a Fourier transform with regard to time a real valued time and space-dependent field quantity—for example, $\underline{\mathbf{v}}(\underline{\mathbf{R}}, t)$—changes into a complex valued frequency and space dependent Fourier spectrum $\underline{\mathbf{v}}(\underline{\mathbf{R}}, \omega)$ with the

[120] It complies with the two-dimensional electromagnetic TE-problem (Langenberg 2005).

property $\underline{\mathbf{v}}(\underline{\mathbf{R}}, -\omega) = \underline{\mathbf{v}}^*(\underline{\mathbf{R}}, \omega)$. The differentiation theorem of the Fourier transform (2.292) changes a time derivative of a field quantity $\underline{\mathbf{v}}(\underline{\mathbf{R}}, t)$ into a multiplication of the spectrum with $-\mathrm{j}\omega$:

$$\frac{\partial}{\partial t}\underline{\mathbf{v}}(\underline{\mathbf{R}}, t) \circ\!\!-\!\!\bullet\ -\mathrm{j}\omega\, \underline{\mathbf{v}}(\underline{\mathbf{R}}, \omega), \tag{7.55}$$

$$\frac{\partial^2}{\partial t^2}\underline{\mathbf{v}}(\underline{\mathbf{R}}, t) \circ\!\!-\!\!\bullet\ -\omega^2\, \underline{\mathbf{v}}(\underline{\mathbf{R}}, \omega) \tag{7.56}$$

transforming the d'Alembert operator of a scalar hyperbolic wave equation (for example: Equation 5.32)

$$\Delta - \frac{1}{c^2}\frac{\partial^2}{\partial t^2} \tag{7.57}$$

into the Helmholtz operator

$$\Delta + k^2 \text{ mit } k = \frac{\omega}{c} \tag{7.58}$$

of a scalar elliptic—the so-called reduced—wave equation. According to (3.21), the Fourier spectrum for a fixed (circular) frequency ω can be related to the phasor of the real valued time harmonic field quantity $\underline{\mathbf{v}}(\underline{\mathbf{R}}, t, \omega_0)$ that consequently oscillates at each spatial point with different amplitudes $\underline{\mathbf{v}}(\underline{\mathbf{R}}, \omega_0)$ but always with the same (circular) frequency ω_0.

7.4.1 Frequency domain wave equations for nondissipative materials

Due to (7.55) and (7.56), we can directly cite all reduced wave equations corresponding to the time domain wave equations as derived in Section 7.1; being actually trivial, we will refrain from a "derivation" only to utilize them upon request. Yet one should note that: This elementary Fourier transform of the wave equations is strictly related to the assumptions of instantaneously reacting materials yielding only frequency-independent material parameters in the respective reduced wave equations. Nevertheless, we can always insert those material parameters into an ω_0-equation that are given for this particular frequency; with that we basically have introduced constitutive equations for dissipative materials, i.e., for noninstantaneously reacting materials. This is the topic of the next section.

7.4.2 Frequency domain wave equations for dissipative materials

For dissipative (linear) materials, we refer to the physically consistent relaxation models (4.109), (4.110) of dissipation, where the relaxation kernels $\underline{\underline{\mu}}(\underline{\mathbf{R}}, t)$ and $\underline{\underline{\underline{\underline{\chi}}}}(\underline{\mathbf{R}}, t)$ should be—physically stringent—causal square integrable time functions. As a consequence, the fields are causal and possess finite energy (Karlsson and Kristensson 1992). As a further consequence, the real and imaginary parts of the necessarily complex Fourier spectra $\underline{\underline{\mu}}(\underline{\mathbf{R}}, \omega)$ and $\underline{\underline{\underline{\underline{\chi}}}}(\underline{\mathbf{R}}, \omega)$ (all components of these tensors) are mutual Hilbert transforms; they are not independent upon each other (Section 2.3.4).

Inserting the constitutive equations (4.109) and (4.110) into the governing equations (3.1) and (3.2) leads to the system of equations[121]

$$\underline{\underline{\rho}}(\underline{\mathbf{R}}) \cdot \frac{\partial \underline{\mathbf{v}}(\underline{\mathbf{R}}, t)}{\partial t} + \int_0^t \underline{\underline{\mu}}(\underline{\mathbf{R}}, t-\tau) \cdot \frac{\partial \underline{\mathbf{v}}(\underline{\mathbf{R}}, \tau)}{\partial \tau} \, \mathrm{d}\tau = \nabla \cdot \underline{\underline{\mathbf{T}}}(\underline{\mathbf{R}}, t) + \underline{\mathbf{f}}(\underline{\mathbf{R}}, t), \tag{7.59}$$

$$\underline{\underline{\underline{\underline{\mathbf{s}}}}}(\underline{\mathbf{R}}) : \frac{\partial \underline{\underline{\mathbf{T}}}(\underline{\mathbf{R}}, t)}{\partial t} + \int_0^t \underline{\underline{\underline{\underline{\chi}}}}(\underline{\mathbf{R}}, t-\tau) : \frac{\partial \underline{\underline{\mathbf{T}}}(\underline{\mathbf{R}}, \tau)}{\partial \tau} \, \mathrm{d}\tau = \underline{\underline{\underline{\underline{\mathbf{I}}}}}^{+} : \nabla \underline{\mathbf{v}}(\underline{\mathbf{R}}, t) + \underline{\underline{\mathbf{h}}}(\underline{\mathbf{R}}, t); \tag{7.60}$$

due to the convolution rule (2.295), its Fourier transform results in

$$-\mathrm{j}\omega \left[\underline{\underline{\rho}}(\underline{\mathbf{R}}) + \underline{\underline{\mu}}(\underline{\mathbf{R}}, \omega) \right] \cdot \underline{\mathbf{v}}(\underline{\mathbf{R}}, \omega) = \nabla \cdot \underline{\underline{\mathbf{T}}}(\underline{\mathbf{R}}, \omega) + \underline{\mathbf{f}}(\underline{\mathbf{R}}, \omega), \tag{7.61}$$

$$-\mathrm{j}\omega \left[\underline{\underline{\underline{\underline{\mathbf{s}}}}}(\underline{\mathbf{R}}) + \underline{\underline{\underline{\underline{\chi}}}}(\underline{\mathbf{R}}, \omega) \right] : \underline{\underline{\mathbf{T}}}(\underline{\mathbf{R}}, \omega) = \underline{\underline{\underline{\underline{\mathbf{I}}}}}^{+} : \nabla \underline{\mathbf{v}}(\underline{\mathbf{R}}, \omega) + \underline{\underline{\mathbf{h}}}(\underline{\mathbf{R}}, \omega). \tag{7.62}$$

Eliminating $\underline{\underline{\mathbf{T}}}(\underline{\mathbf{R}}, \omega)$ yields the reduced wave equation

$$\begin{aligned} \nabla \cdot \underline{\underline{\underline{\underline{\mathbf{c}}}}}_c(\underline{\mathbf{R}}, \omega) : \nabla \underline{\mathbf{v}}(\underline{\mathbf{R}}, \omega) + \omega^2 \underline{\underline{\rho}}_c(\underline{\mathbf{R}}, \omega) \cdot \underline{\mathbf{v}}(\underline{\mathbf{R}}, \omega) \\ = \mathrm{j}\omega \underline{\mathbf{f}}(\underline{\mathbf{R}}, \omega) - \underline{\underline{\underline{\underline{\mathbf{c}}}}}_c(\underline{\mathbf{R}}, \omega) : \underline{\underline{\mathbf{h}}}(\underline{\mathbf{R}}, \omega), \end{aligned} \tag{7.63}$$

where

$$\underline{\underline{\underline{\underline{\mathbf{s}}}}}_c(\underline{\mathbf{R}}, \omega) = \underline{\underline{\underline{\underline{\mathbf{s}}}}}(\underline{\mathbf{R}}) + \underline{\underline{\underline{\underline{\chi}}}}(\underline{\mathbf{R}}, \omega), \tag{7.64}$$

$$\underline{\underline{\underline{\underline{\mathbf{c}}}}}_c(\underline{\mathbf{R}}, \omega) = \underline{\underline{\underline{\underline{\mathbf{s}}}}}_c^{-1}(\underline{\mathbf{R}}, \omega), \tag{7.65}$$

$$\underline{\underline{\rho}}_c(\underline{\mathbf{R}}, \omega) = \underline{\underline{\rho}}(\underline{\mathbf{R}}) + \underline{\underline{\mu}}(\underline{\mathbf{R}}, \omega) \tag{7.66}$$

denote complex valued frequency-dependent material tensors that replace the real valued instantaneously reacting and therefore nondissipative materials: The reduced wave equation (7.63) is complementary to (7.3) for relaxation dissipative materials. Should the Maxwell model (4.77), (4.78) of dissipation be sufficient the respective reduced wave equation emerges through Fourier transform of (4.81) and should a Rayleigh–Kelvin–Voigt model be appropriate, we find the respective reduced wave equation under (4.108). Yet, we once more emphasize that only the relaxation models (4.109) and (4.110) are physically consistent, all other models can only have a restricted validity.

---

[121] For a shorter notation, we use the symmetrization operator

$$\underline{\underline{\underline{\underline{\mathbf{I}}}}}^{+} : \nabla \underline{\mathbf{v}} = \frac{1}{2} \left[\nabla \underline{\mathbf{v}} + (\nabla \underline{\mathbf{v}})^{21} \right].$$

8

Elastic Plane Waves in Homogeneous Materials

8.1 Homogeneous Plane Waves in Isotropic Nondissipative Materials

Plane waves emerge as very special solutions of homogeneous wave equations[122] for homogeneous materials, that is to say, we look for solutions of the homogeneous equation (7.20)

$$\mu\,\Delta\underline{\mathbf{v}}(\underline{\mathbf{R}},t)+(\lambda+\mu)\nabla\nabla\cdot\underline{\mathbf{v}}(\underline{\mathbf{R}},t)-\rho\,\frac{\partial^2\underline{\mathbf{v}}(\underline{\mathbf{R}},t)}{\partial t^2}=\underline{\mathbf{0}} \tag{8.1}$$

for isotropic nondissipative materials; applying one time integration, we can equally write this equation in terms of the particle displacement (Equation 7.24):

$$\mu\,\Delta\underline{\mathbf{u}}(\underline{\mathbf{R}},t)+(\lambda+\mu)\nabla\nabla\cdot\underline{\mathbf{u}}(\underline{\mathbf{R}},t)-\rho\,\frac{\partial^2\underline{\mathbf{u}}(\underline{\mathbf{R}},t)}{\partial t^2}=\underline{\mathbf{0}}. \tag{8.2}$$

8.1.1 One-dimensional plane waves: Primary longitudinal and secondary transverse waves

"One-dimensional" means that all field quantities should only depend upon one (Cartesian) coordinate. We choose the z-coordinate, that is to say, we postulate independence of x and y putting all derivatives with regard to x and y to zero:

$$\frac{\partial}{\partial x}\equiv 0,\quad \frac{\partial}{\partial y}\equiv 0. \tag{8.3}$$

With (2.182), (2.180), and (2.186), respectively, the requirements (8.3) yield as

$$\mu\frac{\partial^2\underline{\mathbf{u}}(z,t)}{\partial z^2}+(\lambda+\mu)\frac{\partial^2 u_z(z,t)}{\partial z^2}\,\underline{\mathbf{e}}_z-\rho\frac{\partial\underline{\mathbf{u}}(z,t)}{\partial t^2}=\underline{\mathbf{0}}. \tag{8.4}$$

[122]Gaussian wave packets or plane wave spectra represent other special solutions of the homogeneous wave equation (Chapter 12).

We take the three Cartesian components of this one-dimensional vector wave equation:

$$\mu \frac{\partial^2 u_x(z,t)}{\partial z^2} - \rho \frac{\partial^2 u_x(z,t)}{\partial t^2} = 0, \tag{8.5}$$

$$\mu \frac{\partial^2 u_y(z,t)}{\partial z^2} - \rho \frac{\partial^2 u_y(z,t)}{\partial t^2} = 0, \tag{8.6}$$

$$(\lambda + 2\mu) \frac{\partial^2 u_z(z,t)}{\partial z^2} - \rho \frac{\partial^2 u_z(z,t)}{\partial t^2} = 0. \tag{8.7}$$

We obtain three mutual independent (decoupled) equations for the respective components of $\underline{\mathbf{u}}(\underline{\mathbf{R}}, t)$ with a similar mathematical structure that can be solved independently. For example, from the outset, we can choose trivial solutions for two equations, e.g., $u_x(z,t) = u_y(z,t) \equiv 0$ or $u_x(z,t) = u_z(z,t) \equiv 0$ or $u_y(z,t) = u_z(z,t) \equiv 0$.

Pulsed primary longitudinal waves: We choose trivial solutions for (8.5) and (8.6) and investigate the Fourier transformed equation (8.7), that is to say, the reduced wave equation corresponding to:

$$(\lambda + 2\mu) \frac{\partial^2 u_z(z,\omega)}{\partial z^2} + \omega^2 \rho u_z(z,\omega) = 0. \tag{8.8}$$

Because now only one differential equation variable remains, we can write (8.8) as ordinary differential equation of second order

$$(\lambda + 2\mu) \frac{\mathrm{d}^2 u_z(z,\omega)}{\mathrm{d}z^2} + \omega^2 \rho u_z(z,\omega) = 0 \tag{8.9}$$

with constant coefficients; writing it as

$$\frac{\mathrm{d}^2 u_z(z,\omega)}{\mathrm{d}z^2} + k_\mathrm{P}^2 u_z(z,\omega) = 0 \tag{8.10}$$

with the short-hand notations

$$\begin{aligned} k_\mathrm{P} &= \omega \sqrt{\frac{\rho}{\lambda + 2\mu}} \\ &= \frac{\omega}{c_\mathrm{P}} \end{aligned} \tag{8.11}$$

it definitely has $\sin k_\mathrm{P} z$ and/or $\cos k_\mathrm{P} z$ as solutions. With (2.259), we combine both possibilities to the complex valued solutions[123]

$$u_z(z,\omega) = u(\omega)\, \mathrm{e}^{\pm \mathrm{j} k_\mathrm{P} z}, \tag{8.12}$$

where, due to the homogeneity of the differential equation (8.10), $u(\omega)$ turns out to be an arbitrary amplitude eventually depending upon the parameter ω

[123]The physical meaning of the sine/cosine solutions will be discussed later on.

that is "hidden" in k_P; we understand $u(\omega)$ as function of ω, as eventually complex spectrum of the equally arbitrary time function $u(t)$ where

$$u(t) \circ\!\!-\!\!\bullet\, u(\omega). \tag{8.13}$$

Note: $u(t)$ has the unit m whereas $u(\omega)$ has the unit ms.

To answer the question which sign can or must be chosen in the exponent of the exponential function (8.12), we transform the solutions (8.12) into the time domain applying the translation rule (2.290) of the Fourier transform:

$$\begin{aligned} u_z(z,t) &= \mathcal{F}^{-1}\{u_z(z,\omega)\} \\ &= \mathcal{F}^{-1}\{u(\omega)\,\mathrm{e}^{\pm \mathrm{j} k_P z}\} \\ &= \mathcal{F}^{-1}\{u(\omega)\,\mathrm{e}^{\pm \mathrm{j}\frac{z}{c_P}\omega}\} \\ &= u\left(t \mp \frac{z}{c_P}\right). \end{aligned} \tag{8.14}$$

Equation 8.14 represents the plane wave solution of the wave equation (8.7); why is it a wave? At first, we investigate the upper sign according to

$$u_z(z,t) = u\left(t - \frac{z}{c_P}\right) \tag{8.15}$$

and choose a particular location, e.g., $z = 0$: An observer of the plane wave "measures" there the time function—the impulse—$u_z(0,t) = u(t)$ [Figure 8.1(b); $u(t)$ is an RC2(t)-pulse displaced by half its duration to the right side of the t-axis: Section 2.3.2]. Another observer at $z_0 > 0$ measures with $u_z(z_0,t) = u(t - z_0/c_P)$ the same pulse, yet delayed by the time $t = z_0/c_P$ (Figure 8.1a): The impulse—the pulsed wave—has propagated into the direction of positive z-values during the time z_0/c_P with the velocity c_P. At the location $z = -z_0$, $z_0 > 0$, the respective observer has measured the impulse already at time $t = -z_0/c_P$ (Figure 8.1c), that is to say, the one-dimensional plane wave (8.15) has the z-axis as propagation direction, it comes from negative infinity and propagates in the direction of positive z-values into positive infinity. Similarly, the one-dimensional plane wave

$$u_z(z,t) = u\left(t + \frac{z}{c_P}\right) \tag{8.16}$$

propagates with velocity c_P into negative z-direction. Both signs in (8.14) are physically meaningful and definitely mathematically possible,[124] they characterize the propagation direction.

Since $u_z(z,t)$ is a function of two variables, we can also display, alternatively to Figure 8.1 that discusses the dependence of t for fixed values of z, the dependence of z for fixed values of t (Figure 8.2). At time $t = 0$ [Figure 8.2(b)],

[124]The derivation of the Green function reveals that mathematically possible signs must not at all be physically meaningful (Section 13.1).

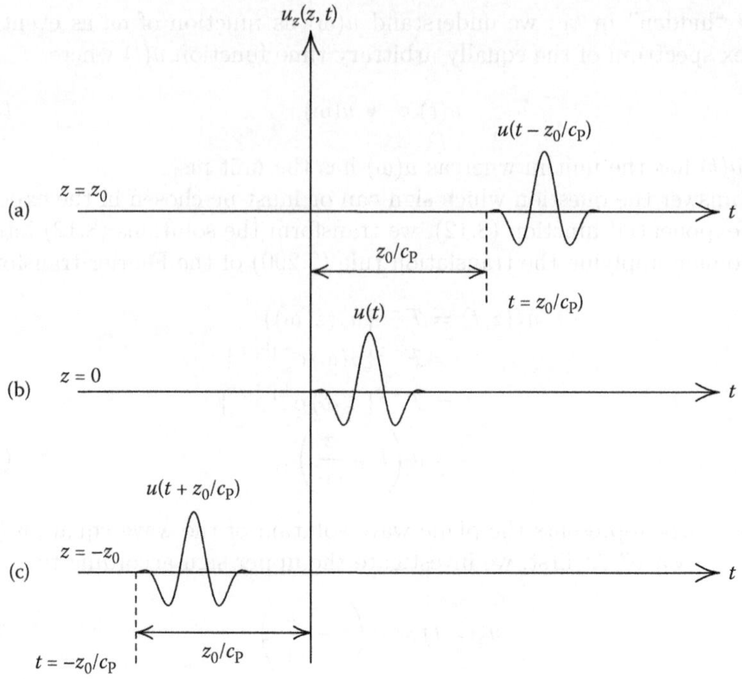

FIGURE 8.1
Propagation of a one-dimensional pulsed plane wave with the velocity c_P; time dependence for various locations.

the wave amplitude distribution is given by $u(-z/c_P)$, that is to say, by the time impulse mirrored at the origin with a differently normalized argument: If the time impulse $u(t)$ has duration T (in seconds), then the "spatial impulse" is Tc_P meters long.[125] For our chosen symmetric time impulse, we exactly "see" the same impulse in the spatial domain that propagates—compare Figure 8.2(a)—during the time t_0 from $z = 0$ to $z = t_0 c_P$. In the spatial domain, we also sketch the particle velocity vector $u_z \underline{\mathbf{e}}_z$.

The so-called phases

$$\phi(z,t) = t \mp \frac{z}{c_P} \tag{8.17}$$

of the plane ($\mp$)-waves are constant on all planes perpendicular to the propagation coordinate: That is the reason why they are called plane waves.

[125]Consequently, a longitudinal wave pulse in steel of duration 1 μs is 5900 μm long. In Figures 8.1 and 8.2, we have geometrically sketched the time and the spatial pulses with equal length; this means that we either have normalized the velocity to 1—kind of a brute force—or we have agreed to a dimensionless axis: t in Figure 8.1 may be normalized to the duration T and z in Figure 8.2 to the spatial length Tc_P. That way, travel times are geometrically equal to travel distances.

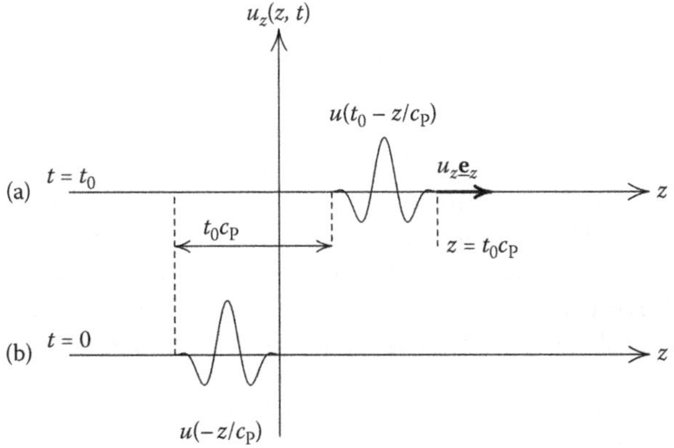

FIGURE 8.2
Propagation of a one-dimensional pulsed plane wave with velocity c_P: spatial dependence for times $t = 0$ (b) and $t = t_0$ (a).

Simultaneously, the amplitudes are equally constant on all these planes, the planes of constant phase and constant amplitude coincide: These are homogeneous plane waves. The velocity

$$c_P = \sqrt{\frac{\lambda + 2\mu}{\rho}} \tag{8.18}$$

of one-dimensional homogeneous plane waves evidently is the velocity of phase propagation, hence it is called phase velocity.[126] Let us consider two z-values z_1, z_2 and two times t_1, t_2 with equal respective phases; then, we obtain

$$\begin{aligned} \phi(z_1, t_1) - \phi(z_2, t_2) &= 0 \\ &= t_1 - t_2 \mp \frac{z_1 - z_2}{c_P} \end{aligned} \tag{8.19}$$

or

$$\Delta t = \pm \frac{\Delta z}{c_P}; \tag{8.20}$$

switching to differential time and space intervals, we have

$$\frac{\mathrm{d}z}{\mathrm{d}t} = \pm c_P \tag{8.21}$$

as spatial change of phase with time, i.e., as phase velocity.

The particle displacement of the one-dimensional wave (8.15) only has a z-component pointing into propagation direction (Figure 8.2), it is a longitudinal (homogeneous plane) wave with phase velocity c_P.

[126] In Section 8.1.2, we define the velocity of energy propagation, and in Section 8.3, we find that it may differ from the phase velocity both in magnitude and direction.

Solving the wave equations (8.5) and (8.6) in the next paragraph, we will find that the respective phase velocity is always smaller than (8.18), revealing that the presently discussed wave arrives always first at a particular observation point, it is the primary wave; this explains the index P that we attach to the phase velocity (8.18). In Section 8.1.2, we will find that P may also indicate pressure, i.e., the primary wave is a pressure wave from a physical point of view. Yet, the primary or pressure wave is only strictly longitudinal if it is a homogeneous plane wave; hence, we do not use the notation c_L (e.g.: Kutzner, 1983; Krautkrämer and Krautkrämer, 1986); and similarly not c_T for the secondary or shear wave that is transversely polarized as a homogeneous plane wave.

We have implicitly assumed (Figures 8.1 and 8.2) that the arbitrary[127] time function $u(t)$ is an impulse with finite duration; this must not necessarily be so, we can also consider a time harmonic function of infinite duration like $\sin \omega_0 t$ or $\cos \omega_0 t$, $-\infty < t < \infty$.

Time harmonic longitudinal waves: At first, we choose the complex valued time harmonic function[128]

$$u(t, \omega_0) = f(\omega_0)\, \mathrm{e}^{-\mathrm{j}\omega_0 t} \tag{8.22}$$

as complex valued combination of $\cos \omega_0 t$ and $\sin \omega_0 t$ with the circular frequency $\omega_0 = 2\pi f_0$—the frequency f_0 has the unit Hz, the circular frequency ω_0 the unit s^{-1}—and the eventually complex valued amplitude $f(\omega_0)$. The *ansatz* (8.22) involves a single spectral line at exactly this circular frequency with amplitude $2\pi f(\omega_0)$:

$$u(\omega, \omega_0) = 2\pi f(\omega_0)\, \delta(\omega - \omega_0). \tag{8.23}$$

Hence, our time harmonic one-dimensional homogeneous plane waves are characterized by

$$\begin{aligned} u_z(z, t, \omega_0) &= f(\omega_0)\, \mathrm{e}^{-\mathrm{j}\omega_0 \left(t \mp \frac{z}{c_P}\right)} \\ &= f(\omega_0)\, \mathrm{e}^{\pm \mathrm{j} k_P z} \mathrm{e}^{-\mathrm{j}\omega_0 t} \end{aligned} \tag{8.24}$$

now exhibiting the so-called wave number

$$k_P = \frac{\omega_0}{c_P} \tag{8.25}$$

corresponding to ω_0 in the time function $u_z(z, t, \omega_0)$ of the primary wave: With (8.11), we defined it as short-hand notation in the spectra (8.12). The

[127] If not arbitrary, there would be no US-NDT.

[128] The sign in the exponent matches the sign in the kernel of the inverse Fourier transform (2.270). Using the *ansatz*

$$u(t, \omega_0) = f(\omega_0)\, \mathrm{e}^{\mathrm{j}\omega_0 t},$$

we have to switch to the complex conjugate in all complex valued formulas in the elaboration.

complex exponential function $e^{j\varphi}$ is 2π-periodic with regard to φ, hence the time function $e^{-j\omega_0 t}$ is 2π-periodic with regard to $\omega_0 t$; the periodicity interval in the time domain

$$T_0 = \frac{2\pi}{\omega_0} \tag{8.26}$$

is called the period duration of time harmonic waves. Correspondingly, the spatial function $e^{\pm jk_P z}$ is 2π-periodic in $k_P z$; the periodicity interval in the spatial domain

$$\lambda_P = \frac{2\pi}{k_P} \tag{8.27}$$

is called wavelength of the time harmonic longitudinal plane wave. Time harmonic longitudinal plane waves are periodic in space and time with periodicity intervals depending upon each other due to (8.25):

$$\begin{aligned}\lambda_P &= c_P T_0 \\ &= \frac{c_P}{f_0}.\end{aligned} \tag{8.28}$$

This fundamental relation among frequency, wavelength, and phase velocity is an immediate consequence of the wave equation, written as

$$\omega = c_P k_P; \tag{8.29}$$

it is called dispersion relation of the underlying material even though this is not obvious at this point.[129] We will multiply come back to that.

We explicitly point out that the term "wavelength" is originally related to time harmonic waves. As far as pulses are concerned, we have to consider a particular spectral component, for example, the carrier frequency of the RCN(t)-pulse (2.275).

Sometimes the time harmonic plane wave (8.24) is written with suppressed time dependence:

$$\begin{aligned}u_z(z, \omega_0) &\stackrel{\text{def}}{=} u_z(z, t, \omega_0)\, e^{j\omega_0 t} \\ &= f(\omega_0)\, e^{\pm jk_P z}.\end{aligned} \tag{8.30}$$

In (8.24), it is immediately evident that the sign combination $+jk_P z$ with $-j\omega_0 t$ stands for a harmonic plane wave propagation into positive z-direction in contrast to the sign combination $-jk_P z$ with $-j\omega_0 t$ standing for propagation into negative z-direction. Yet, in order to guess the propagation direction from (8.30), we must know the underlying time dependence; it could easily have been $e^{+j\omega_0 t}$ yielding a change of the signs that determine the propagation

[129]The dispersion relation (8.29) of the homogeneous isotropic nondissipative material is a linear relation between frequency and wave number; a material with such a dispersion relation does not at all exhibit dispersion of a propagating wave impulse (Figures 8.1 and 8.2).

direction: The wave $\mathrm{e}^{-\mathrm{j}k_{\mathrm{P}}z}\,\mathrm{e}^{\mathrm{j}\omega t}$ propagates into positive z-direction and the wave $\mathrm{e}^{\mathrm{j}k_{\mathrm{P}}z}\,\mathrm{e}^{\mathrm{j}\omega t}$ into negative z-direction.

It is clear that time harmonic waves are idealizations, yet, this is already true for plane waves: They "require" an infinite propagation space. Nevertheless, both constructs are extremely useful to compose less idealized wave fields like transducer radiation fields. We only have to switch from the single spectral line to a complete Fourier spectrum and from a single propagation direction to a spatial spectrum of directions!

If anybody is irritated by complex valued time harmonic longitudinal plane waves (8.24)—US-NDT is a real valued "application space"!—he can choose two spectral lines at $\omega = \omega_0$ and $\omega = -\omega_0$ with respective one half of the previous amplitude (Equation 3.20) instead of the single spectral line at $\omega = \omega_0$ (Equation 8.23):

$$u(\omega) = \pi[f(\omega_0)\,\delta(\omega - \omega_0) + f^*(\omega_0)\,\delta(\omega + \omega_0)]. \tag{8.31}$$

If we additionally assume the amplitude at $\omega = -\omega_0$ to be conjugate complex as compared to the amplitude at $\omega = \omega_0$, we obtain

$$u(t, \omega_0) = \Re\{f(\omega_0)\,\mathrm{e}^{-\mathrm{j}\omega_0 t}\} \tag{8.32}$$

via inverse Fourier transform of (8.31) and consequently

$$u_z(z, t, \omega_0) = \Re\{f(\omega_0)\,\mathrm{e}^{\pm\mathrm{j}k_{\mathrm{P}}z}\mathrm{e}^{-\mathrm{j}\omega_0 t}\}. \tag{8.33}$$

The expression $f(\omega_0)\,\mathrm{e}^{\pm\mathrm{j}\frac{\omega_0}{c_{\mathrm{P}}}z}$ is called phasor of the time harmonic wave (8.33) (Section 3.2.6).

One remark concerning the $\sin k_{\mathrm{P}}z$- and/or $\cos k_{\mathrm{P}}z$-solutions of (8.10): We obtain them from (8.12) if we superimpose back and forth traveling waves with equal and opposite amplitudes according to

$$\begin{aligned} u_z(z, t) &= f(\omega_0)\left(\mathrm{e}^{\mathrm{j}k_{\mathrm{P}}z} \pm \mathrm{e}^{-\mathrm{j}k_{\mathrm{P}}z}\right)\mathrm{e}^{-\mathrm{j}\omega_0 t} \\ &= f(\omega_0)\,\mathrm{e}^{-\mathrm{j}\omega_0 t}\begin{cases} 2\cos k_{\mathrm{P}}z & \text{for the positive sign} \\ 2\mathrm{j}\sin k_{\mathrm{P}}z & \text{for the negative sign} \end{cases}. \end{aligned} \tag{8.34}$$

Obviously, these are standing "waves" oscillating with ω_0 at a certain location z with the respective sine or cosine amplitudes, they represent time-dependent elastic oscillations similar to those of a violin string. Such oscillations can be effectively used to compose modes in an elastic wave guide (Rose 1999).

Pulsed secondary transverse waves: We now turn to the remaining differential equations (8.5) and (8.6), yet we keep this shorter because we have already learned the essential facts about waves. We select (8.5) arbitrarily putting $u_y(z, t) \equiv 0$. Again, we switch to a reduced wave equation for the frequency spectrum applying a Fourier transform with regard to time t:

$$\mu\frac{\partial^2 u_x(z, \omega)}{\partial z^2} + \omega^2\rho u_x(z, \omega) = 0. \tag{8.35}$$

Introducing the wave number for the secondary wave as well as the velocity—from a physical view point, the index S equally stands for *shear* wave (Section 8.1.2)—

$$\begin{aligned} k_\mathrm{S} &= \omega\sqrt{\frac{\rho}{\mu}} \\ &= \frac{\omega}{c_\mathrm{S}} \end{aligned} \tag{8.36}$$

we find the solutions according to (8.12)

$$u_x(z,\omega) = u(\omega)\,\mathrm{e}^{\pm \mathrm{j} k_\mathrm{S} z} \tag{8.37}$$

of

$$\frac{\mathrm{d}^2 u_x(z,\omega)}{\mathrm{d}z^2} + k_\mathrm{S}^2 u_x(z,\omega) = 0. \tag{8.38}$$

The inverse Fourier transform of (8.37) leads us to the one-dimensional pulsed secondary plane waves

$$u_x(z,t) = u\left(t \mp \frac{z}{c_\mathrm{S}}\right) \tag{8.39}$$

that propagate with the phase velocity c_S in $\pm z$-direction. Indeed, Equation 8.39 refers to secondary waves because with

$$c_\mathrm{S} = \sqrt{\frac{\mu}{\rho}}, \tag{8.40}$$

we always have $c_\mathrm{S} < c_\mathrm{P}$. The time domain representation of the pulsed secondary wave

$$u_x(z,t) = u\left(t - \frac{z}{c_\mathrm{S}}\right) \tag{8.41}$$

for different locations principally looks identical to the one for the primary wave in Figure 8.1, we simply have to replace c_P by c_S. Yet, in a representation scaled to Figure 8.1, we have to choose the travel time z_0/c_S of the secondary wave at z_0 larger than the travel time z_0/c_P of the primary wave at the same location z_0. It is for this reason that primary and secondary waves separate with increasing time, a fact that we have displayed in Figure 8.3. Note: Both waves must not be necessarily identically pulsed as $u(t)$. In US-NDT, this travel time separation is used to identify either primary or secondary waves through time gating.

It is quite clear that the picture in Figure 8.3 is equally valid for $u_y(z,t)$ because we obtain the solutions of (8.6) as

$$u_y(z,t) = u\left(t \mp \frac{z}{c_\mathrm{S}}\right). \tag{8.42}$$

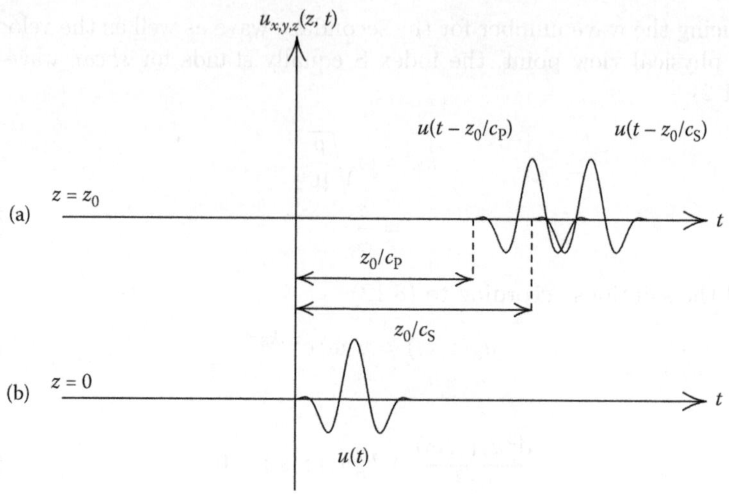

FIGURE 8.3
Propagation of one-dimensional pulsed P- and S-waves.

The spatial picture of pulsed secondary waves as compared to Figure 8.2 is depicted in Figure 8.4; two facts are worth being noted:

- The impulse $u(-z/c_S)$ mirrored for $t = 0$ exhibits a different normalization of the z-coordinate; since $c_P < c_S$, it appears compressed as compared to[130] $u(-z/c_P)$: We have normalized the primary wave velocity c_P to 1; this corresponds to the normalized axis scaling z/Tc_P in Figure 8.4; due to the relation $\lambda_S = c_S/f_0$ corresponding to (8.28), the wavelength λ_S of the carrier frequency of the RC2(t)-pulse is smaller than λ_P.

- During the same time interval t_0 that we considered in Figure 8.2, the pulsed secondary wave has only reached the location $z = t_0 c_S < t_0 c_P$, it arrives as pulsed secondary wave at this point.

The particle displacement vector (8.39) of the secondary wave only has an x-component; hence, it is oriented *perpendicular* to the propagation direction: We encounter *transverse* (homogeneous plane) waves. In Figure 8.4, this is indicated by an arrow.

The sketches 8.2 and 8.4 are even getting more intuitive if we animate the one-dimensional pulsed homogeneous plane P- and S-waves in a two-dimensional xz-space (or yz-space)—a two-dimensional xz-plane (or yz-plane)—as a movie; for this purpose, the wave amplitudes $u(t_i - z/c_{P,S})$, $i = 1, 2, 3, \ldots, I$, are displayed either color or gray coded in an xz-plane for a dense sequence of times t_i, $i = 1, 2, 3, \ldots, I$. In Figure 8.5, two times $t = t_1$

[130]Compare Footnote 125.

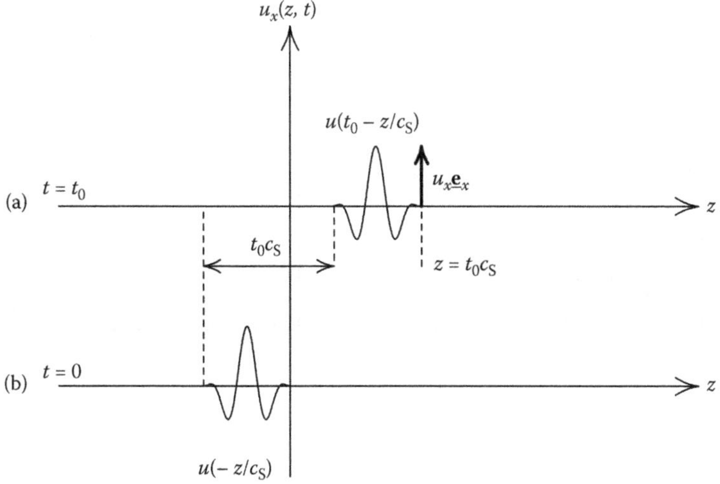

FIGURE 8.4
Spatial propagation picture of a one-dimensional pulsed S-wave with velocity c_S for $t = 0$ (b) and $t = t_0$ (a).

and $t = t_2 > t_1$ are selected: *Wavefronts* of pulsed plane P- and S-waves become very demonstrative that way; in particular, we nicely recognize the orthogonality of planes of constant phase and amplitude with respect to the propagation direction, identifying the waves as pulsed homogeneous plane waves.

We additionally emphasize that plane P- and S-waves are completely independent of each other in a homogeneous material of infinite extent. The so-called mode conversion, P $\Longrightarrow$ S and S $\Longrightarrow$ P, only appears in inhomogeneous materials, for example, at the plane boundary between two half-spaces of infinite extent, and even then only for nonnormal incidence.

Just as the transverse secondary waves (8.39), the secondary waves (8.42) are also transversely polarized, and they have only a y-component. Should both components $u_x(z,t)$ and $u_y(z,t)$ be equal to zero, we can combine them—identical pulse structure $u(t)$ anticipated[131]—with different amplitudes u_x and u_y to the transverse vector

$$\begin{aligned}\underline{\mathbf{u}}_S(z,t) &= u\left(t \mp \frac{z}{c_S}\right)(u_x\underline{\mathbf{e}}_x + u_y\underline{\mathbf{e}}_y)\\ &= u\left(t \mp \frac{z}{c_S}\right)\underline{\mathbf{u}}_S. \end{aligned} \tag{8.43}$$

[131]The general case of nonidentical pulse structure is discussed in connection with time harmonic waves.

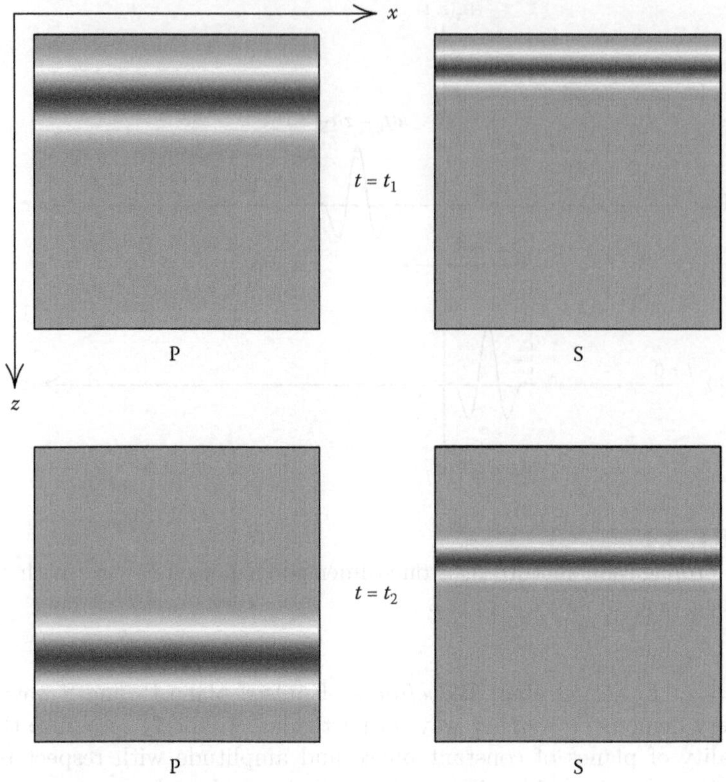

FIGURE 8.5
Plane P- and S-wavefronts: two-dimensional spatial display—a window to infinite space—of one-dimensional pulsed homogeneous plane P- and S-waves for two times t_1 and $t_2 > t_1$ (RC2-pulses).

Without loosing generality, we can normalize $\underline{\mathbf{u}}_S$ as unit vector $\underline{\hat{\mathbf{u}}}_S$ [we "hide" the magnitude of $\underline{\mathbf{u}}_S$ in $u(t)$]:

$$\underline{\mathbf{u}}_S(z,t) = u\left(t \mp \frac{z}{c_S}\right)\underline{\hat{\mathbf{u}}}_S, \tag{8.44}$$

where $\underline{\hat{\mathbf{u}}}_S \cdot \underline{\mathbf{e}}_z = 0$; we have introduced transverse waves with linear polarization $\underline{\hat{\mathbf{u}}}_S$. The homogeneous infinitely extended material, as considered here, does not exhibit any preference direction that we can arrange for a cartesian coordinate system with the z-axis as propagation direction and—for example—the x-axis pointing into $\underline{\hat{\mathbf{u}}}_S$-direction. This is no longer possible if any preference directions exist; then, we must ascertain the propagation direction of plane waves in three dimensions (Section 8.1.2), where the linear transverse polarization separates into two well distinguishable transverse waves (e.g.: SH and SV).

Time harmonic transverse waves: As in the paragraph on time harmonic longitudinal waves, we can "excite" time harmonic transverse waves with circular frequency ω_0. In contrast to (8.24), we then obtain

$$u_{x,y}(z,t,\omega_0) = f(\omega_0)\,\mathrm{e}^{\pm \mathrm{j}k_\mathrm{S}z}\mathrm{e}^{-\mathrm{j}\omega_0 t} \tag{8.45}$$

with the secondary wave number

$$k_\mathrm{S} = \frac{\omega}{c_\mathrm{S}} \tag{8.46}$$

and the secondary wavelength

$$\lambda_\mathrm{S} = \frac{c_\mathrm{S}}{f_0}. \tag{8.47}$$

The wavelengths of time harmonic plane P- and S-waves of equal frequency similarly compare to the respective phase velocities.

Superimposing x- and y-components of time harmonic plane waves propagating into $+z$-direction according to

$$\begin{aligned} \underline{\mathbf{u}}_\mathrm{S}(z,t,\omega_0) &= \mathrm{e}^{\mathrm{j}k_\mathrm{S}z-\mathrm{j}\omega_0 t}\left[u_x(\omega_0)\,\underline{\mathbf{e}}_x + u_y(\omega_0)\,\underline{\mathbf{e}}_y\right], \\ &= \mathrm{e}^{\mathrm{j}k_\mathrm{S}z-\mathrm{j}\omega_0 t}\underline{\mathbf{u}}_\mathrm{S}(\omega_0) \end{aligned} \tag{8.48}$$

we note that the choice of equal amplitudes of both components is by no means mandatory: We can choose $u_x(\omega_0)$ and $u_y(\omega_0)$ differently complex valued for each circular frequency[132] ω_0. The resulting complex ratio

$$A(\omega_0) = \frac{u_y(\omega_0)}{u_x(\omega_0)} \tag{8.49}$$

is called polarization number in the theory of electromagnetic waves; $A(\omega_0)$ in a complex A-plane uniquely determines the curve of the vector tip $\underline{\mathbf{u}}_\mathrm{S}(z_0,t,\omega_0)$ in the xy-plane for a fixed location z_0 as function of time. For example, $A(\omega_0) = \mathrm{j}$ means right-circular polarization (RC) of the time harmonic plane wave: If the thumb of the right hand points into the $+z$-propagation direction of the wave, the tip of $\underline{\mathbf{u}}_\mathrm{S}(z_0,t,\omega_0)$ moves on a circle following the bent fingers of the right hand and consequently in clockwise direction (CW) if we observe the wave from behind; for this observation, mode RC is identical to CW and left-circular (LC) to counter-clockwise (CCW).[133] If we would have

[132]This is the reason why the concept of wave polarization is at first only applicable to time harmonic waves. For impulses, one has to consider time averages (Langenberg 2005).

[133]Note: This is the definition of electrical engineering; in the physics/optics literature (e.g., Born and Wolf 1975), the thumb is held opposite to the propagation direction. To relate this definition to the engineering sense of rotation for right-circular CW polarized waves the left hand is needed—the same wave is optically LC-polarized—and looking toward the propagating wave this again corresponds to CCW. In addition, the assignment of the polarization state to the point $A(\omega_0)$ in the complex plane depends on the chosen time function $\mathrm{e}^{-\mathrm{j}\omega_0 t}$ or $\mathrm{e}^{\mathrm{j}\omega_0 t}$: For $\mathrm{e}^{\mathrm{j}\omega_0 t}$, the value $A(\omega_0) = \mathrm{j}$ corresponds to left-circular polarization in electrical engineering.

to use the left hand for the same thumb orientation to describe the sense of rotation *opposite* to the clock while watching the wave from behind, it would be left-circular polarized. Due to missing respective transducers to generate arbitrarily elliptically polarized transverse waves, the above terminology is not widely known in US-NDT. Yet, in the theory of electromagnetic waves, an essential part of communication and radar technology is based upon the concept of polarization (Cloude 2002; Langenberg 2005).

8.1.2 Three-dimensional plane waves: Primary longitudinal pressure and secondary transverse shear waves

Mathematical representation of homogeneous plane waves in three dimensions: As we already stated: In an infinitely extended homogeneous isotropic material—in an elastic full-space—the previous discussion of one-dimensional plane waves is sufficient because we can always rotate a cartesian coordinate system such that—for example—the z-axis coincides with the propagation direction. Yet, the simplest case of an inhomogeneous material, the one-sided infinitely extended homogeneous isotropic half-space, its planar surface defines a preference plane whose embedding into a cartesian coordinate system is, even not strictly necessary, rather advisable to calculate the reflection and mode conversion of elastic waves. By choosing the xy-plane as surface, we would be left only with the z-axis as one-dimensional propagation direction limiting our investigation to normal incidence, angular transducers would not exist. As a consequence, we need the mathematical representation of plane waves propagating three-dimensionally in a fixed cartesian coordinate system in an arbitrary direction given, for instance, by the unit vector $\hat{\underline{\mathbf{k}}}$. In Figure 8.6, we have sketched this situation: In the direction $\hat{\underline{\mathbf{k}}}$, we define a coordinate axis ζ for one-dimensional plane wave propagation in three-dimensional xyz-space. Hence, we postulate pulsed elastic plane waves with phase velocity c and the linear—longitudinal or transverse—polarization $\hat{\underline{\mathbf{u}}}$ according to

$$\underline{\mathbf{u}}(\zeta, t) = u\left(t \mp \frac{\zeta}{c}\right)\hat{\underline{\mathbf{u}}}. \tag{8.50}$$

We know that the phases and amplitudes of homogeneous plane waves are per definitionem constant in planes perpendicular to the propagation direction. Therefore, we must find planes orthogonal to $\hat{\underline{\mathbf{k}}}$ as the geometric location of all vectors of position $\underline{\mathbf{R}}$ for which

$$\begin{aligned}\phi(\zeta, t) &= \phi[\zeta(\underline{\mathbf{R}}), t]\\ &= t \mp \frac{\zeta(\underline{\mathbf{R}})}{c}\end{aligned} \tag{8.51}$$

is constant for fixed times. Figures 8.6 and 2.9 illustrate that such planes are described by[134] $\hat{\underline{\mathbf{k}}} \cdot \underline{\mathbf{R}} = \zeta(\underline{\mathbf{R}}) = \text{const}$ leading to the mathematical

[134] For $\hat{\underline{\mathbf{k}}} = \underline{\mathbf{e}}_z$, we obtain $\zeta = z$ as before: planes perpendicular to the z-axis.

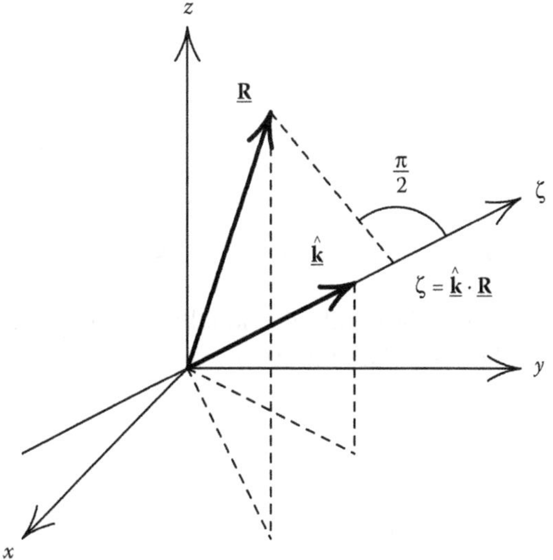

FIGURE 8.6
Propagation coordinate ζ of a one-dimensional plane wave in three-dimensional xyz-space.

representation of pulsed homogeneous elastic plane waves in three-dimensional $\underline{\mathbf{R}}$-space—we affiliate the propagation vector $\underline{\hat{\mathbf{k}}}$ in the list of arguments of the particle displacement—

$$\underline{\mathbf{u}}(\underline{\mathbf{R}}, t, \underline{\hat{\mathbf{k}}}) = u\left(t \mp \frac{\underline{\hat{\mathbf{k}}} \cdot \underline{\mathbf{R}}}{c}\right) \underline{\hat{\mathbf{u}}}. \tag{8.52}$$

We actually know that we have to distinguish between the phase velocities $c = c_{\mathrm{P}}$ and $c = c_{\mathrm{S}}$ of primary and secondary waves with longitudinal ($\underline{\hat{\mathbf{u}}} = \underline{\hat{\mathbf{u}}}_{\mathrm{P}} \parallel \underline{\hat{\mathbf{k}}}$) and transverse ($\underline{\hat{\mathbf{u}}} = \underline{\hat{\mathbf{u}}}_{\mathrm{S}} \perp \underline{\hat{\mathbf{k}}}$) polarization. But exactly this fact should emerge from a formal mathematical procedure because discussion of wave propagation in homogeneous anisotropic materials leaves no other choice. Therefore it is beneficial to equally formulate and solve the present problem of plane elastic wave propagation in a homogeneous isotropic material as a so-called eigenvalue problem even though we already know the solution.

Phase velocities and polarizations of plane waves in three dimensions: Solution of an eigenvalue problem: Phase velocities and polarizations of plane elastic waves necessarily resulted from the wave equations (8.5) through (8.7) decoupled with regard to the components of $\underline{\mathbf{u}}(\underline{\mathbf{R}}, t)$. This decoupling was a consequence of the assumption $\partial/\partial y \equiv 0$ that is here not possible, why we have to work with the vector wave equation (8.2). The Fourier transform with regard to time leads us to the homogeneous vectorial reduced wave equation

$$\mu\,\Delta\underline{\mathbf{u}}(\underline{\mathbf{R}},\omega) + (\lambda+\mu)\nabla\nabla\cdot\underline{\mathbf{u}}(\underline{\mathbf{R}},\omega) + \omega^2\rho\,\underline{\mathbf{u}}(\underline{\mathbf{R}},\omega) = \underline{\mathbf{0}} \tag{8.53}$$

for the frequency spectrum $\underline{\mathbf{u}}(\underline{\mathbf{R}},\omega)$ of the particle velocity. To solve this differential equation, we make the *ansatz* of homogeneous[135] plane waves

$$\underline{\mathbf{u}}(\underline{\mathbf{R}},\omega) = \underline{\mathbf{u}}(\underline{\mathbf{R}},\omega,\hat{\underline{\mathbf{k}}}) = \underline{\mathbf{u}}(\omega,\hat{\underline{\mathbf{k}}})\,\mathrm{e}^{\pm \mathrm{j}\frac{\hat{\underline{\mathbf{k}}}\cdot\underline{\mathbf{R}}}{c(\hat{\underline{\mathbf{k}}})}\omega} \tag{8.54}$$

by Fourier transforming (8.50) and additionally admitting that the polarization vector $\hat{\underline{\mathbf{u}}}(\hat{\underline{\mathbf{k}}})$ of the vectorial amplitude

$$\underline{\mathbf{u}}(\omega,\hat{\underline{\mathbf{k}}}) = \hat{\underline{\mathbf{u}}}(\hat{\underline{\mathbf{k}}})u(\omega) \tag{8.55}$$

and the phase velocity $c(\hat{\underline{\mathbf{k}}})$ depend upon the parameter $\hat{\underline{\mathbf{k}}}$ characterizing the propagation direction.[136] If we now insert $\underline{\mathbf{u}}(\underline{\mathbf{R}},\omega)$ according to (8.54) into (8.53), we must calculate $\Delta\underline{\mathbf{u}}(\underline{\mathbf{R}},\omega,\hat{\underline{\mathbf{k}}}) = \nabla\cdot\nabla\underline{\mathbf{u}}(\underline{\mathbf{R}},\omega,\hat{\underline{\mathbf{k}}})$ and $\nabla\nabla\cdot\underline{\mathbf{u}}(\underline{\mathbf{R}},\omega,\hat{\underline{\mathbf{k}}})$; for this purpose, we use product and chain rules as given in Section 2.2.2 as well as the result (2.175) of the calculation[137] of $\nabla[\mathrm{j}k\hat{\underline{\mathbf{k}}}\cdot\underline{\mathbf{R}}]$:

$$\begin{aligned}
\nabla\underline{\mathbf{u}}(\underline{\mathbf{R}},\omega,\hat{\underline{\mathbf{k}}}) &\stackrel{(2.174)}{=} \underbrace{\nabla\left[\pm\mathrm{j}\frac{\omega}{c(\hat{\underline{\mathbf{k}}})}\hat{\underline{\mathbf{k}}}\cdot\underline{\mathbf{R}}\right]}_{\stackrel{(2.175)}{=}\pm\mathrm{j}\frac{\omega}{c(\hat{\underline{\mathbf{k}}})}\hat{\underline{\mathbf{k}}}}\underline{\mathbf{u}}(\underline{\mathbf{R}},\omega,\hat{\underline{\mathbf{k}}}) \\
&= \pm\mathrm{j}\frac{\omega}{c(\hat{\underline{\mathbf{k}}})}\hat{\underline{\mathbf{k}}}\,\underline{\mathbf{u}}(\underline{\mathbf{R}},\omega,\hat{\underline{\mathbf{k}}}),
\end{aligned} \tag{8.56}$$

$$\begin{aligned}
\nabla\cdot\nabla\underline{\mathbf{u}}(\underline{\mathbf{R}},\omega,\hat{\underline{\mathbf{k}}}) &\stackrel{(2.167)}{=} \pm\mathrm{j}\frac{\omega}{c(\hat{\underline{\mathbf{k}}})}\hat{\underline{\mathbf{k}}}\cdot\nabla\underline{\mathbf{u}}(\underline{\mathbf{R}},\omega,\hat{\underline{\mathbf{k}}}) \\
&= -\frac{\omega^2}{c^2(\hat{\underline{\mathbf{k}}})}\underline{\mathbf{u}}(\underline{\mathbf{R}},\omega,\hat{\underline{\mathbf{k}}});
\end{aligned} \tag{8.57}$$

---

[135]It is an *ansatz* of homogeneous plane waves because $\hat{\underline{\mathbf{k}}}$ is assumed to be real valued (Section 8.2).

[136]In principle, also the impulse spectrum $u(\omega)$ could depend upon $\hat{\underline{\mathbf{k}}}$; yet we will see that we can ignore this idea even for the anisotropic case. Only the concept of the spatial spectrum of plane waves is built on it.

However, for plane waves in elastic half-spaces, we are forced to accept a dependence of the amplitude upon the propagation direction in a once again factorized version:

$$\begin{aligned}
\underline{\mathbf{u}}(\omega,\hat{\underline{\mathbf{k}}}) &= u(\omega,\hat{\underline{\mathbf{k}}})\hat{\underline{\mathbf{u}}}(\hat{\underline{\mathbf{k}}}) \\
&= u(\omega)u(\hat{\underline{\mathbf{k}}})\hat{\underline{\mathbf{u}}}(\hat{\underline{\mathbf{k}}}),
\end{aligned}$$

where $u(\hat{\underline{\mathbf{k}}})$ will turn out to be a reflection, transmission, or mode conversion factor (Chapter 9).

[137]Note: If we assume (2.175), we do not need coordinates to perform the $(\nabla\cdot\nabla)$- and $(\nabla\nabla\cdot)$-differentiations.

$$\nabla \cdot \underline{\mathbf{u}}(\underline{\mathbf{R}}, \omega, \hat{\underline{\mathbf{k}}}) \overset{(2.172)}{=} \underline{\mathbf{u}}(\underline{\mathbf{R}}, \omega, \hat{\underline{\mathbf{k}}}) \cdot \nabla \left[\pm \mathrm{j} \frac{\omega}{c(\hat{\underline{\mathbf{k}}})} \hat{\underline{\mathbf{k}}} \cdot \underline{\mathbf{R}} \right]$$

$$= \pm \mathrm{j} \frac{\omega}{c(\hat{\underline{\mathbf{k}}})} \underline{\mathbf{u}}(\underline{\mathbf{R}}, \omega, \hat{\underline{\mathbf{k}}}) \cdot \hat{\underline{\mathbf{k}}}, \tag{8.58}$$

$$\nabla\nabla \cdot \underline{\mathbf{u}}(\underline{\mathbf{R}}, \omega, \hat{\underline{\mathbf{k}}}) \overset{(2.166)}{=} \pm \mathrm{j} \frac{\omega}{c(\hat{\underline{\mathbf{k}}})} \nabla \underline{\mathbf{u}}(\underline{\mathbf{R}}, \omega, \hat{\underline{\mathbf{k}}}) \cdot \hat{\underline{\mathbf{k}}}$$

$$= -\frac{\omega^2}{c^2(\hat{\underline{\mathbf{k}}})} \hat{\underline{\mathbf{k}}}\hat{\underline{\mathbf{k}}} \cdot \underline{\mathbf{u}}(\underline{\mathbf{R}}, \omega, \hat{\underline{\mathbf{k}}}). \tag{8.59}$$

It follows:

$$-\frac{\omega^2}{c^2(\hat{\underline{\mathbf{k}}})} \mu\, \underline{\mathbf{u}}(\underline{\mathbf{R}}, \omega, \hat{\underline{\mathbf{k}}}) - \frac{\omega^2}{c^2(\hat{\underline{\mathbf{k}}})} (\lambda + \mu)\, \hat{\underline{\mathbf{k}}}\hat{\underline{\mathbf{k}}} \cdot \underline{\mathbf{u}}(\underline{\mathbf{R}}, \omega, \hat{\underline{\mathbf{k}}}) + \omega^2 \rho\, \underline{\mathbf{u}}(\underline{\mathbf{R}}, \omega, \hat{\underline{\mathbf{k}}}) = \underline{\mathbf{0}}; \tag{8.60}$$

we multiply with $c^2(\hat{\underline{\mathbf{k}}})$, divide by ρ, and factor out $-\omega^2 \underline{\mathbf{u}}(\underline{\mathbf{R}}, \omega, \hat{\underline{\mathbf{k}}})$ recognizing (2.96):

$$\left[\frac{\mu}{\rho} \underline{\underline{\mathbf{I}}} + \frac{\lambda + \mu}{\rho} \hat{\underline{\mathbf{k}}}\hat{\underline{\mathbf{k}}} - c^2(\hat{\underline{\mathbf{k}}}) \underline{\underline{\mathbf{I}}} \right] \cdot \omega^2 \underline{\mathbf{u}}(\underline{\mathbf{R}}, \omega, \hat{\underline{\mathbf{k}}}) = \underline{\mathbf{0}}. \tag{8.61}$$

The exponential function $\mathrm{e}^{\pm \mathrm{j} \frac{\omega}{c(\hat{\underline{\mathbf{k}}})} \hat{\underline{\mathbf{k}}} \cdot \underline{\mathbf{R}}}\, \underline{\mathbf{u}}(\underline{\mathbf{R}}, \omega, \hat{\underline{\mathbf{k}}})$ is always nonzero; in the time domain, $-\omega^2 u(\omega)$ is nothing else than the second derivative of the arbitrary impulse $u(t)$ finally resulting in the notation of (8.61) as an eigenvalue problem (Equation 2.128)

$$\left(\frac{\mu}{\rho} \underline{\underline{\mathbf{I}}} + \frac{\lambda + \mu}{\rho} \hat{\underline{\mathbf{k}}}\hat{\underline{\mathbf{k}}} \right) \cdot \hat{\underline{\mathbf{u}}}(\hat{\underline{\mathbf{k}}}) = c^2(\hat{\underline{\mathbf{k}}}) \hat{\underline{\mathbf{u}}}(\hat{\underline{\mathbf{k}}}) \tag{8.62}$$

of the real valued symmetric tensor

$$\underline{\underline{\mathbf{D}}}(\hat{\underline{\mathbf{k}}}) = \frac{\mu}{\rho} \underline{\underline{\mathbf{I}}} + \frac{\lambda + \mu}{\rho} \hat{\underline{\mathbf{k}}}\hat{\underline{\mathbf{k}}}. \tag{8.63}$$

Eigenvalues of $\underline{\underline{\mathbf{D}}}(\hat{\underline{\mathbf{k}}})$ are the squares of phase velocities $c^2(\hat{\underline{\mathbf{k}}})$ and eigenvectors are the polarization vectors $\hat{\underline{\mathbf{u}}}(\hat{\underline{\mathbf{k}}})$; because $\underline{\underline{\mathbf{D}}}(\hat{\underline{\mathbf{k}}})$ is given by the prescribed material properties and the plane wave *ansatz*, this is consequently also true for the possible phase velocities and the corresponding polarizations.

In Section 2.1.4, we alluded to real valued eigenvalues of real valued symmetric tensors; hence, the squares of the phase velocities are real. In addition, the tensor $\underline{\underline{\mathbf{D}}}(\hat{\underline{\mathbf{k}}})$ is positive-definite because

$$\underline{\mathbf{R}} \cdot \underline{\underline{\mathbf{D}}}(\hat{\underline{\mathbf{k}}}) \cdot \underline{\mathbf{R}} = \frac{\mu}{\rho} R^2 + \frac{\lambda + \mu}{\rho} (\hat{\underline{\mathbf{k}}} \cdot \underline{\mathbf{R}})^2 \tag{8.64}$$

is always greater than zero for $R > 0$ and equal to zero only for $R = 0$. Therefore, its eigenvalues—the squares of the phase velocities—are greater than zero, that is to say, the phase velocities themselves are real valued and positive, as it has to be. We will see that explicitly below.

To calculate the eigenvalues $c^2(\hat{\underline{\mathbf{k}}})$, the determinant of the homogeneous system of equations

$$\left[\frac{\mu}{\rho}\underline{\underline{\mathbf{I}}} + \frac{\lambda+\mu}{\rho}\hat{\underline{\mathbf{k}}}\hat{\underline{\mathbf{k}}} - c^2(\hat{\underline{\mathbf{k}}})\underline{\underline{\mathbf{I}}}\right] \cdot \hat{\underline{\mathbf{u}}}(\hat{\underline{\mathbf{k}}}) = \underline{\mathbf{0}} \tag{8.65}$$

must be required to be zero (Equation 2.131). We write (8.65) in Chen's standard form (1983)

$$\underbrace{\left[\frac{\mu - \rho c^2(\hat{\underline{\mathbf{k}}})}{\lambda+\mu}\underline{\underline{\mathbf{I}}} + \hat{\underline{\mathbf{k}}}\hat{\underline{\mathbf{k}}}\right]}_{=\underline{\underline{\mathbf{W}}}(\hat{\underline{\mathbf{k}}}, c^2)} \cdot \hat{\underline{\mathbf{u}}}(\hat{\underline{\mathbf{k}}}) = \underline{\mathbf{0}} \tag{8.66}$$

with the wave tensor $\underline{\underline{\mathbf{W}}}(\hat{\underline{\mathbf{k}}}, c^2)$ and utilize one of Chen's identities:

$$\det(\beta\,\underline{\underline{\mathbf{I}}} + \underline{\mathbf{C}}\,\underline{\mathbf{D}}) = \beta^2(\beta + \underline{\mathbf{C}} \cdot \underline{\mathbf{D}}). \tag{8.67}$$

It follows:

$$\det \underline{\underline{\mathbf{W}}}(\hat{\underline{\mathbf{k}}}, c^2) = \left[\frac{\mu - \rho c^2(\hat{\underline{\mathbf{k}}})}{\lambda+\mu}\right]^2 \left[\frac{\mu - \rho c^2(\hat{\underline{\mathbf{k}}})}{\lambda+\mu} + \hat{\underline{\mathbf{k}}} \cdot \hat{\underline{\mathbf{k}}}\right], \tag{8.68}$$

and this determinant is definitely equal to zero if either

$$\left[\frac{\mu - \rho c^2(\hat{\underline{\mathbf{k}}})}{\lambda+\mu}\right]^2 = 0 \tag{8.69}$$

or

$$\frac{\mu - \rho c^2(\hat{\underline{\mathbf{k}}})}{\lambda+\mu} + 1 = 0 \tag{8.70}$$

holds. This third degree polynomial (8.68) in $c^2(\hat{\underline{\mathbf{k}}})$ factorizes into a polynomial of second and into a polynomial of first degree in $c^2(\hat{\underline{\mathbf{k}}})$; the coefficients of both polynomials do not depend on $\hat{\underline{\mathbf{k}}}$, hence, the eigenvalues $c^2(\hat{\underline{\mathbf{k}}})$ are not functions of $\hat{\underline{\mathbf{k}}}$. Actually, the material is isotropic because the phase velocities c of plane waves do not depend upon the propagation direction, and vice versa, $\hat{\underline{\mathbf{k}}}$-independent phase velocities define the material as isotropic. This is a consequence of the stiffness tensor (7.18).

The quadratic equation (8.69) possesses two equal solutions

$$c^2 = \frac{\mu}{\rho}, \tag{8.71}$$

and the linear equation (8.70) has the solution

$$c^2 = \frac{\lambda + 2\mu}{\rho}. \tag{8.72}$$

The double eigenvalue (8.71) and the single eigenvalue (8.72) are real and larger than zero, a consequence of the real valued symmetry and the positive definiteness of the tensor (8.63). From the eigenvalues, we obtain the phase velocities

$$c_\mathrm{S} = \sqrt{\frac{\mu}{\rho}} \tag{8.73}$$

of secondary

$$c_\mathrm{P} = \sqrt{\frac{\lambda + 2\mu}{\rho}} \tag{8.74}$$

and primary waves. In the strict sense, phase velocities characterize plane elastic waves in first place;[138] the transverse polarization of secondary and the longitudinal polarization of primary homogeneous waves result from the second step of the eigenvalue problem solution via the calculation of the eigenvectors. The fact that, from a physical view point, secondary waves are shear waves and primary waves are pressure waves (in homogeneous materials) follows from the Helmholtz decomposition of the particle velocity vector (even for nonplane waves).[139]

In Section 2.1.4, we claimed that eigenvectors of real symmetric tensors are orthogonal to each other if they belong to different eigenvalues. Both eigenvectors belonging to the twofold eigenvalue c_S^2 must therefore be orthogonal to the eigenvector belonging to the eigenvalue c_P^2: We expect that the polarization vectors of primary and secondary plane waves are orthogonal to each other; the additional fact that primary plane waves are longitudinally and secondary plane waves are transversely polarized will be explicitly shown in the following. We simply calculate $\underline{\underline{\mathbf{W}}}(\hat{\underline{\mathbf{k}}}, c_\mathrm{P}^2)$ and $\underline{\underline{\mathbf{W}}}(\hat{\underline{\mathbf{k}}}, c_\mathrm{S}^2)$ and search for vectors $\hat{\underline{\mathbf{u}}}(\hat{\underline{\mathbf{k}}})$ satisfying (8.66).

We find

$$\underline{\underline{\mathbf{W}}}(\hat{\underline{\mathbf{k}}}, c_\mathrm{P}^2) = -\underline{\underline{\mathbf{I}}} + \hat{\underline{\mathbf{k}}}\,\hat{\underline{\mathbf{k}}}; \tag{8.75}$$

therefore, each vector $\hat{\underline{\mathbf{u}}}_\mathrm{P}(\hat{\underline{\mathbf{k}}})$ is eigenvector to the (single) c_P^2-eigenvalue, for which

$$(\underline{\underline{\mathbf{I}}} - \hat{\underline{\mathbf{k}}}\,\hat{\underline{\mathbf{k}}}) \cdot \hat{\underline{\mathbf{u}}}_\mathrm{P}(\hat{\underline{\mathbf{k}}}) = \underline{\mathbf{0}} \tag{8.76}$$

[138] This is especially true in homogeneous and/or anisotropic materials because even plane waves in such materials are generally no pure longitudinal pressure or shear waves (Sections 8.3.1 and 12.3)

[139] This equally follows from the divergence and curl of the particle velocity vectors (8.82) through (8.84).

and, respectively,

$$\begin{aligned}\underline{\hat{\mathbf{u}}}_\mathrm{P}(\underline{\hat{\mathbf{k}}}) &= \underline{\hat{\mathbf{k}}}\,\underline{\hat{\mathbf{k}}} \cdot \underline{\hat{\mathbf{u}}}_\mathrm{P}(\underline{\hat{\mathbf{k}}}) \\ &= [\underline{\hat{\mathbf{k}}} \cdot \underline{\hat{\mathbf{u}}}_\mathrm{P}(\underline{\hat{\mathbf{k}}})]\,\underline{\hat{\mathbf{k}}}\end{aligned} \tag{8.77}$$

holds. It follows that: Depending upon the sign of $\underline{\hat{\mathbf{k}}} \cdot \underline{\hat{\mathbf{u}}}_\mathrm{P}(\underline{\hat{\mathbf{k}}})$, the vector $\underline{\hat{\mathbf{u}}}_\mathrm{P}(\underline{\hat{\mathbf{k}}})$ points into the direction $\pm\underline{\hat{\mathbf{k}}}$—P-waves are longitudinally polarized!—resulting per definition in $\underline{\hat{\mathbf{k}}} \cdot \underline{\hat{\mathbf{u}}}_\mathrm{P}(\underline{\hat{\mathbf{k}}}) = \pm 1$ and[140]

$$\underline{\hat{\mathbf{u}}}_\mathrm{P}(\underline{\hat{\mathbf{k}}}) = \pm\underline{\hat{\mathbf{k}}}. \tag{8.78}$$

According to (2.97), the expression $(\underline{\underline{\mathbf{I}}} - \underline{\hat{\mathbf{k}}}\,\underline{\hat{\mathbf{k}}}) \cdot \underline{\hat{\mathbf{u}}}_\mathrm{P}(\underline{\hat{\mathbf{k}}})$ is the vectorial component $\underline{\hat{\mathbf{u}}}_{\mathrm{P}t}(\underline{\hat{\mathbf{k}}})$ of $\underline{\hat{\mathbf{u}}}_\mathrm{P}(\underline{\hat{\mathbf{k}}})$ orthogonal to $\underline{\hat{\mathbf{k}}}$; if it is requested to be zero, $\underline{\hat{\mathbf{u}}}_\mathrm{P}(\underline{\hat{\mathbf{k}}})$ must point into the direction $\pm\underline{\hat{\mathbf{k}}}$, that is to say, $\underline{\hat{\mathbf{u}}}_\mathrm{P}(\underline{\hat{\mathbf{k}}})$ is uniquely determined apart from the sign: Since plane waves are solutions of the homogenenous wave equation, the sign can be arbitrarily chosen such that $\underline{\hat{\mathbf{u}}}_\mathrm{P}(\underline{\hat{\mathbf{k}}})$ always points into propagation direction, namely into $(+\underline{\hat{\mathbf{k}}})$-direction for propagation in $(+\underline{\hat{\mathbf{k}}})$-direction (Figure 8.2) and into $(-\underline{\hat{\mathbf{k}}})$-direction for propagation in $(-\underline{\hat{\mathbf{k}}})$-direction.

Now, we calculate

$$\underline{\underline{\mathbf{W}}}(\underline{\hat{\mathbf{k}}}, c_\mathrm{S}^2) = \underline{\hat{\mathbf{k}}}\,\underline{\hat{\mathbf{k}}}; \tag{8.79}$$

consequently, each vector $\underline{\hat{\mathbf{u}}}_\mathrm{S}(\underline{\hat{\mathbf{k}}})$ is eigenvector to the (twofold) c_S^2-eigenvalue, for which

$$\underline{\hat{\mathbf{k}}}\,\underline{\hat{\mathbf{k}}} \cdot \underline{\hat{\mathbf{u}}}_\mathrm{S}(\underline{\hat{\mathbf{k}}}) = \underline{\mathbf{0}} \tag{8.80}$$

and, respectively,

$$\underline{\hat{\mathbf{k}}} \cdot \underline{\hat{\mathbf{u}}}_\mathrm{S}(\underline{\hat{\mathbf{k}}}) = 0 \tag{8.81}$$

holds. This is true for each vector orthogonal to $\underline{\hat{\mathbf{k}}}$, i.e., the nonnormalized eigenvectors $\underline{\mathbf{u}}_\mathrm{S}(\underline{\hat{\mathbf{k}}})$ are located in a plane orthogonal to $\underline{\hat{\mathbf{k}}}$, where the arbitrariness is a consequence of the eigenvalue c_S^2 to be twofold. If we choose any unit vector $\underline{\hat{\mathbf{u}}}_{\mathrm{S}1}(\underline{\hat{\mathbf{k}}})$ with $\underline{\hat{\mathbf{u}}}_{\mathrm{S}1}(\underline{\hat{\mathbf{k}}}) \cdot \underline{\hat{\mathbf{k}}} = 0$ in this plane as (normalized) eigenvector, then $\underline{\hat{\mathbf{u}}}_{\mathrm{S}2}(\underline{\hat{\mathbf{k}}}) = \pm\underline{\hat{\mathbf{k}}} \times \underline{\hat{\mathbf{u}}}_{\mathrm{S}1}(\underline{\hat{\mathbf{k}}})$ with $\underline{\hat{\mathbf{u}}}_{\mathrm{S}2}(\underline{\hat{\mathbf{k}}}) \cdot \underline{\hat{\mathbf{k}}} = 0$ is another (normalized) eigenvector orthogonal to $\underline{\hat{\mathbf{u}}}_{\mathrm{S}1}(\underline{\hat{\mathbf{k}}})$; together with $\underline{\hat{\mathbf{u}}}_\mathrm{P}(\underline{\hat{\mathbf{k}}})$, we now have an orthonormal trihedron $\underline{\hat{\mathbf{u}}}_\mathrm{P}(\underline{\hat{\mathbf{k}}}), \underline{\hat{\mathbf{u}}}_{\mathrm{S}1}(\underline{\hat{\mathbf{k}}}), \underline{\hat{\mathbf{u}}}_{\mathrm{S}2}(\underline{\hat{\mathbf{k}}})$ of eigenvectors and polarizations, respectively, and applying the already practiced sign choice for $\underline{\hat{\mathbf{u}}}_\mathrm{P}(\underline{\hat{\mathbf{k}}})$ to $\underline{\hat{\mathbf{u}}}_{\mathrm{S}2}(\underline{\hat{\mathbf{k}}})$, this trihedron is right handed. Note: Such a trihedron consists of linearly independent vectors (Equation 2.59), i.e., the two transversely polarized secondary waves and the longitudinally polarized primary wave are mutually independent upon each other.

---

[140]By the way: In contrast to the eigenvalues, the eigenvectors depend upon $\underline{\hat{\mathbf{k}}}$; this is depicted in Figure 8.8.

With the solution of the preceding eigenvalue problem, we have ascertained the phase velocities and polarizations in the *ansatz* (8.54) of homogeneous elastic plane waves:

$$\underline{\mathbf{u}}_{\mathrm{P}}(\underline{\mathbf{R}}, \omega, \hat{\underline{\mathbf{k}}}) = \pm\, u_{\mathrm{P}}(\omega)\, \mathrm{e}^{\pm \mathrm{j} k_{\mathrm{P}} \hat{\underline{\mathbf{k}}}\cdot\underline{\mathbf{R}}}\, \hat{\underline{\mathbf{k}}}, \tag{8.82}$$

$$\underline{\mathbf{u}}_{\mathrm{S1}}(\underline{\mathbf{R}}, \omega, \hat{\underline{\mathbf{k}}}) = \pm\, u_{\mathrm{S1}}(\omega)\, \mathrm{e}^{\pm \mathrm{j} k_{\mathrm{S}} \hat{\underline{\mathbf{k}}}\cdot\underline{\mathbf{R}}}\, \hat{\underline{\mathbf{u}}}_{\mathrm{S1}}(\hat{\underline{\mathbf{k}}}) \quad \text{with} \quad \hat{\underline{\mathbf{u}}}_{\mathrm{S1}}(\hat{\underline{\mathbf{k}}}) \cdot \hat{\underline{\mathbf{k}}} = 0, \tag{8.83}$$

$$\underline{\mathbf{u}}_{\mathrm{S2}}(\underline{\mathbf{R}}, \omega, \hat{\underline{\mathbf{k}}}) = \pm\, u_{\mathrm{S2}}(\omega)\, \mathrm{e}^{\pm \mathrm{j} k_{\mathrm{S}} \hat{\underline{\mathbf{k}}}\cdot\underline{\mathbf{R}}}\, \hat{\underline{\mathbf{k}}} \times \hat{\underline{\mathbf{u}}}_{\mathrm{S1}}(\hat{\underline{\mathbf{k}}}); \tag{8.84}$$

For the notation of these Fourier spectra, we again fell back on wave numbers

$$k_{\mathrm{P,S}} = \frac{\omega}{c_{\mathrm{P,S}}}; \tag{8.85}$$

due to the independence of the three wave modes (8.82) through (8.84), we can choose different amplitude spectra. In the time domain, we finally obtain

$$\underline{\mathbf{u}}_{\mathrm{P}}(\underline{\mathbf{R}}, t, \hat{\underline{\mathbf{k}}}) = \pm u_{\mathrm{P}} \left(t \mp \frac{\hat{\underline{\mathbf{k}}} \cdot \underline{\mathbf{R}}}{c_{\mathrm{P}}} \right) \hat{\underline{\mathbf{k}}}, \tag{8.86}$$

$$\underline{\mathbf{u}}_{\mathrm{S1}}(\underline{\mathbf{R}}, t, \hat{\underline{\mathbf{k}}}) = \pm u_{\mathrm{S1}} \left(t \mp \frac{\hat{\underline{\mathbf{k}}} \cdot \underline{\mathbf{R}}}{c_{\mathrm{S}}} \right) \hat{\underline{\mathbf{u}}}_{\mathrm{S1}}(\hat{\underline{\mathbf{k}}}) \quad \text{mit} \quad \hat{\underline{\mathbf{u}}}_{\mathrm{S1}}(\hat{\underline{\mathbf{k}}}) \cdot \hat{\underline{\mathbf{k}}} = 0, \tag{8.87}$$

$$\underline{\mathbf{u}}_{\mathrm{S2}}(\underline{\mathbf{R}}, t, \hat{\underline{\mathbf{k}}}) = \pm u_{\mathrm{S2}} \left(t \mp \frac{\hat{\underline{\mathbf{k}}} \cdot \underline{\mathbf{R}}}{c_{\mathrm{S}}} \right) \hat{\underline{\mathbf{k}}} \times \hat{\underline{\mathbf{u}}}_{\mathrm{S1}}(\hat{\underline{\mathbf{k}}}). \tag{8.88}$$

Besides the amplitude spectra, we could also choose different propagation directions of the P,S_1,S_2-waves, because the wave modes are independent upon each other in the elastic full-space.

In Figure 8.7, the propagation of a P-RC2(t)-pulse is illustrated in a way that is comparable to Figure 8.5.

The coordinate-free representations (8.82) through (8.84) and (8.86) through (8.88), respectively, of plane elastic waves may be embedded—often it must be done!—in a cartesian coordinate system. In Figure 8.8(a), the vectors $\hat{\underline{\mathbf{k}}}, \hat{\underline{\mathbf{u}}}_{\mathrm{S1}}(\hat{\underline{\mathbf{k}}}), \hat{\underline{\mathbf{u}}}_{\mathrm{S2}}(\hat{\underline{\mathbf{k}}}) = \hat{\underline{\mathbf{k}}} \times \hat{\underline{\mathbf{u}}}_{\mathrm{S1}}(\hat{\underline{\mathbf{k}}})$ have, for example, the components (2.225) of the orthonormal trihedron of spherical coordinates with regard to the polar angle ϑ_k and the azimuth angle φ_k (compare Figure 2.16):

$$\hat{\underline{\mathbf{k}}} = \sin\vartheta_k \cos\varphi_k\, \underline{\mathbf{e}}_x + \sin\vartheta_k \sin\varphi_k\, \underline{\mathbf{e}}_y + \cos\vartheta_k\, \underline{\mathbf{e}}_z, \tag{8.89}$$

$$\hat{\underline{\mathbf{u}}}_{\mathrm{S1}}(\hat{\underline{\mathbf{k}}}) = \cos\vartheta_k \cos\varphi_k\, \underline{\mathbf{e}}_x + \cos\vartheta_k \sin\varphi_k\, \underline{\mathbf{e}}_y - \sin\vartheta_k\, \underline{\mathbf{e}}_z, \tag{8.90}$$

$$\hat{\underline{\mathbf{u}}}_{\mathrm{S2}}(\hat{\underline{\mathbf{k}}}) = -\sin\varphi_k\, \underline{\mathbf{e}}_x + \cos\varphi_k\, \underline{\mathbf{e}}_y. \tag{8.91}$$

The scalar product $\hat{\underline{\mathbf{k}}} \cdot \underline{\mathbf{R}}$ reads as follows

$$\hat{\underline{\mathbf{k}}} \cdot \underline{\mathbf{R}} = \sin\vartheta_k \cos\varphi_k\, x + \sin\vartheta_k \sin\varphi_k\, y + \cos\vartheta_k\, z. \tag{8.92}$$

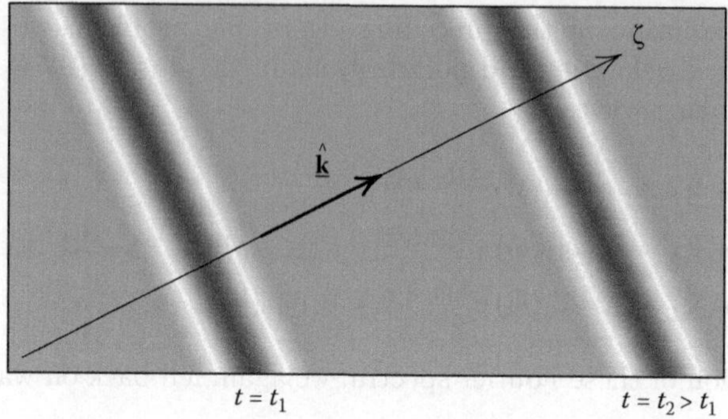

FIGURE 8.7
Two-dimensional spatial representation of pulsed wavefronts of a plane elastic P-wave for two different times $t = t_1$ and $t = t_2 > t_1$ propagating into $+\hat{\underline{\mathbf{k}}}$-direction; note: $\hat{\underline{\mathbf{k}}}$ is orthogonal to the wavefronts.

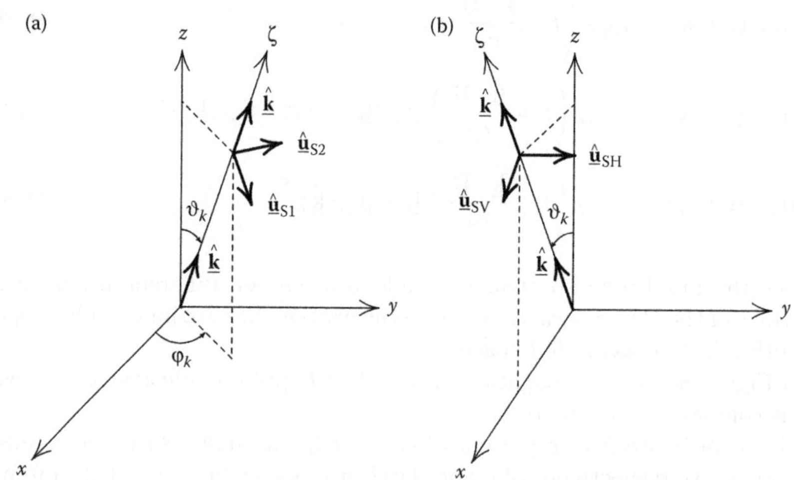

FIGURE 8.8
Orthogonal polarization of secondary waves: S1- and S2-waves (a) as well as SH- and SV-waves (b).

Specially choosing $\varphi_k = 0$, we confine the propagation direction $\hat{\underline{\mathbf{k}}}$ to vectors in the xz-plane [Figure 8.8(b)]:

$$\hat{\underline{\mathbf{k}}} = \sin\vartheta_k\,\underline{\mathbf{e}}_x + \cos\vartheta_k\,\underline{\mathbf{e}}_z, \tag{8.93}$$

$$\begin{aligned} \hat{\underline{\mathbf{u}}}_{\mathrm{S1}}(\hat{\underline{\mathbf{k}}}) &= \cos\vartheta_k\,\underline{\mathbf{e}}_x - \sin\vartheta_k\,\underline{\mathbf{e}}_z \\ &\stackrel{\mathrm{def}}{=} \hat{\underline{\mathbf{u}}}_{\mathrm{SV}}(\hat{\underline{\mathbf{k}}}), \end{aligned} \tag{8.94}$$

$$\hat{\underline{\mathbf{u}}}_{S2}(\hat{\underline{\mathbf{k}}}) = \underline{\mathbf{e}}_y$$
$$\stackrel{\text{def}}{=} \hat{\underline{\mathbf{u}}}_{SH}(\hat{\underline{\mathbf{k}}}); \tag{8.95}$$
$$\hat{\underline{\mathbf{k}}} \cdot \underline{\mathbf{R}} = \sin\vartheta_k\, x + \cos\vartheta_k\, z. \tag{8.96}$$

If the xy-plane accidentally constitutes a reference plane—boundary between two materials, specimen surface—$\hat{\underline{\mathbf{u}}}_{S2}(\hat{\underline{\mathbf{k}}}) = \hat{\underline{\mathbf{u}}}_{SH}(\hat{\underline{\mathbf{k}}}) = \underline{\mathbf{e}}_y$ becomes a horizontal polarization of the transverse secondary wave, i.e., an SH-wave, and $\hat{\underline{\mathbf{u}}}_{S1}(\hat{\underline{\mathbf{k}}}) = \hat{\underline{\mathbf{u}}}_{SV}(\hat{\underline{\mathbf{k}}})$ becomes the polarization of the SV-wave that generally exhibits a nonzero vertical component with respect to the xy-plane; the notations[141] SH and SV always refer to a reference plane (or to the axis of independence of a two-dimensional problem; Section 7.3)! For $\hat{\underline{\mathbf{k}}} = \underline{\mathbf{e}}_z$ (Figures 8.2 and 8.4), we have $\hat{\underline{\mathbf{u}}}_{SV} = \underline{\mathbf{e}}_x$, and with $\hat{\underline{\mathbf{u}}}_{SH} = \underline{\mathbf{e}}_y$ both transversely polarized wave modes are then horizontal and, hence, physically undistinguishable.

With (8.75), (8.76) and (8.79), (8.80), respectively, we have found the polarization vectors for plane elastic waves by a "close look"; yet, in Section 2.1.4, we referred to a formal evaluation being based on the knowledge of the adjoint $\text{adj}\,\underline{\underline{\mathbf{W}}}(\hat{\underline{\mathbf{k}}}, c^2)$ of the wave tensor. With Chen's formula (Chen 1983)

$$\text{adj}\,(\beta\,\underline{\underline{\mathbf{I}}} + \underline{\mathbf{C}}\,\underline{\mathbf{D}}) = \beta\,[(\beta + \underline{\mathbf{C}} \cdot \underline{\mathbf{D}})\,\underline{\underline{\mathbf{I}}} - \underline{\mathbf{C}}\,\underline{\mathbf{D}}], \tag{8.97}$$

we calculate

$$\text{adj}\,\underline{\underline{\mathbf{W}}}(\hat{\underline{\mathbf{k}}}, c^2) = \frac{\mu - \rho c^2}{\lambda + \mu}\left(\frac{\lambda + 2\mu - \rho c^2}{\lambda + \mu}\,\underline{\underline{\mathbf{I}}} - \hat{\underline{\mathbf{k}}}\,\hat{\underline{\mathbf{k}}}\right) \tag{8.98}$$

and consequently

$$\text{adj}\,\underline{\underline{\mathbf{W}}}(\hat{\underline{\mathbf{k}}}, c_P^2) = \hat{\underline{\mathbf{k}}}\,\hat{\underline{\mathbf{k}}}, \tag{8.99}$$
$$\text{adj}\,\underline{\underline{\mathbf{W}}}(\hat{\underline{\mathbf{k}}}, c_S^2) = \underline{\underline{\mathbf{0}}}; \tag{8.100}$$

$\underline{\underline{\mathbf{0}}}$ denotes the null tensor. With (8.75) and (8.79), we state that: A second rank tensor (here: $\underline{\underline{\mathbf{W}}}$), whose determinant is identically zero is either, as one says (Chen 1983), planar (roughly speaking, it consists of two terms: Equation 8.75), then the adjoint tensor (8.99) is linear (dyadic product of two vectors, a dyadic), or the tensor is linear (Equation 8.79), then the adjoint tensor is the null tensor. In the first case, the column vector (left vector) of the adjoint tensor (here: $\hat{\underline{\mathbf{k}}}$) is an eigenvector, in the second case, any vector orthogonal to the row vector (right vector) of the tensor itself (here: $\hat{\underline{\mathbf{k}}}$) is an eigenvector. The fact that the tensor is linear and the adjoint tensor is the null

[141]Figure 8.8(a) immediately tells us that $\hat{\underline{\mathbf{u}}}_{S2}$ and $\hat{\underline{\mathbf{u}}}_{S1}$ are obviously also SH and SV with regard to the xy-plane; nevertheless, Figure 8.8(b) stands for the "proper" SH- and SV-definition, namely, in the SH-case for an orientation parallel to an independency axis—here: y—of a two-dimensional problem. Furthermore: If the xz-plane is declared to be an incidence plane (Chapter 9), SH is a polarization orthogonal and SV a polarization parallel to the incidence plane.

tensor typically occurs for a twofold eigenvalue. This so-called degeneracy—the two respective eigenvectors are indeed orthogonal to the third eigenvector, yet otherwise arbitrary—is characteristic for isotropic materials; in anisotropic materials, it is nullified.

Primary longitudinal pressure and secondary transverse shear waves: The longitudinal polarization of primary homogeneous plane waves and the transverse polarization of secondary homogeneous plane waves in isotropic materials similarly result—and actually a little bit less formal than in the preceding paragraph—with the help of the Helmholtz decomposition (Section 7.2); this reveals that primary waves are pressure and secondary waves are shear waves: The letters P and S then stand for pressure and shear. Yet, we emphasize once again: In anisotropic materials, this procedure does not lead us to the destination, why in that case we can only expect plane quasipressure and plane quasishear waves.

Using the gauge (7.29), the Helmholtz decomposition (7.28) of the particle displacement vector in homogeneous isotropic materials resulted in decoupled d'Alembert wave equations (7.35) and (7.36) for the scalar potential $\Phi(\underline{\mathbf{R}}, t)$ and the vector potential $\underline{\mathbf{\Psi}}(\underline{\mathbf{R}}, t)$ whose homogeneous and Fourier transformed versions

$$\Delta\Phi(\underline{\mathbf{R}}, \omega) + \omega^2 \frac{\rho}{\lambda + 2\mu} \Phi(\underline{\mathbf{R}}, \omega) = 0, \tag{8.101}$$

$$\Delta\underline{\mathbf{\Psi}}(\underline{\mathbf{R}}, \omega) + \omega^2 \frac{\rho}{\mu} \underline{\mathbf{\Psi}}(\underline{\mathbf{R}}, \omega) = \underline{\mathbf{0}} \tag{8.102}$$

are now investigated. With the *ansatz* of plane waves

$$\Phi(\underline{\mathbf{R}}, \omega, \hat{\underline{\mathbf{k}}}) = \Phi(\omega, \hat{\underline{\mathbf{k}}})\, \mathrm{e}^{\pm \mathrm{j} \frac{\hat{\underline{\mathbf{k}}} \cdot \underline{\mathbf{R}}}{c(\hat{\underline{\mathbf{k}}})} \omega}, \tag{8.103}$$

$$\underline{\mathbf{\Psi}}(\underline{\mathbf{R}}, \omega, \hat{\underline{\mathbf{k}}}) = \underline{\mathbf{\Psi}}(\omega, \hat{\underline{\mathbf{k}}})\, \mathrm{e}^{\pm \mathrm{j} \frac{\hat{\underline{\mathbf{k}}} \cdot \underline{\mathbf{R}}}{c(\hat{\underline{\mathbf{k}}})} \omega} \tag{8.104}$$

corresponding to (8.54), we obtain

$$-\frac{\omega^2}{c^2(\hat{\underline{\mathbf{k}}})} \Phi(\underline{\mathbf{R}}, \omega, \hat{\underline{\mathbf{k}}}) + \omega^2 \frac{\rho}{\lambda + 2\mu} \Phi(\underline{\mathbf{R}}, \omega, \hat{\underline{\mathbf{k}}}) = 0, \tag{8.105}$$

$$-\frac{\omega^2}{c^2(\hat{\underline{\mathbf{k}}})} \underline{\mathbf{\Psi}}(\underline{\mathbf{R}}, \omega, \hat{\underline{\mathbf{k}}}) + \omega^2 \frac{\rho}{\mu} \underline{\mathbf{\Psi}}(\underline{\mathbf{R}}, \omega, \hat{\underline{\mathbf{k}}}) = \underline{\mathbf{0}} \tag{8.106}$$

if we utilize (2.175), (2.172), (2.174), and (2.167).[142] Because the exponential functions in (8.103) and (8.104) are always nonzero, the equations (8.105) and (8.106) are equivalent to

$$\left(\frac{1}{c^2(\hat{\underline{\mathbf{k}}})} - \frac{\rho}{\lambda + 2\mu} \right) \omega^2 \Phi(\omega, \hat{\underline{\mathbf{k}}}) = 0, \tag{8.107}$$

[142] That way, we calculate $\Delta\left[\underline{\mathbf{\Psi}}(\omega, \hat{\underline{\mathbf{k}}})\, \mathrm{e}^{\pm \mathrm{j} \frac{\hat{\underline{\mathbf{k}}} \cdot \underline{\mathbf{R}}}{c(\hat{\underline{\mathbf{k}}})} \omega}\right]$ coordinate free without too much paperwork!

$$\left(\frac{1}{c^2(\hat{\underline{\mathbf{k}}})} - \frac{\rho}{\mu}\right)\omega^2 \underline{\mathbf{\Psi}}(\omega, \hat{\underline{\mathbf{k}}}) = \underline{\mathbf{0}}. \tag{8.108}$$

Since $\Phi(\omega, \hat{\underline{\mathbf{k}}})$ and $\underline{\mathbf{\Psi}}(\omega, \hat{\underline{\mathbf{k}}})$ are arbitrary amplitudes, it follows that the brackets in (8.107) and (8.108) must be equal to zero,[143] hence $c(\hat{\underline{\mathbf{k}}})$ adopts the values

$$c_P = \sqrt{\frac{\lambda + 2\mu}{\rho}}, \tag{8.109}$$

$$c_S = \sqrt{\frac{\mu}{\rho}}. \tag{8.110}$$

As usual we have characterized both possible "bracket solutions" for the phase velocities by indices P and S, and evidently, both do not depend on $\hat{\underline{\mathbf{k}}}$ in isotropic materials. Instead of $\omega/c_{P,S}$, we may also use the wave numbers $k_{P,S}$ in (8.103) and (8.104).

The potential $\Phi(\underline{\mathbf{R}}, \omega)$ is a scalar field quantity; hence, we must not check its polarization. This is different for the vector potential $\underline{\mathbf{\Psi}}(\underline{\mathbf{R}}, \omega, \hat{\underline{\mathbf{k}}})$: Those who are familiar with the theory of electromagnetic waves know that the divergence condition (7.29) implies $\underline{\mathbf{\Psi}}(\omega, \hat{\underline{\mathbf{k}}})$ to be transverse. We calculate

$$\begin{aligned}\nabla \cdot \left[\underline{\mathbf{\Psi}}(\omega, \hat{\underline{\mathbf{k}}})\, e^{\pm jk_S \hat{\underline{\mathbf{k}}}\cdot\underline{\mathbf{R}}}\right] &= \underline{\mathbf{\Psi}}(\omega, \hat{\underline{\mathbf{k}}})\, e^{\pm jk_S \hat{\underline{\mathbf{k}}}\cdot\underline{\mathbf{R}}} \cdot [\pm jk_S \hat{\underline{\mathbf{k}}}] \\ &= \pm\, jk_S\, e^{\pm jk_S \hat{\underline{\mathbf{k}}}\cdot\underline{\mathbf{R}}}\, \underline{\mathbf{\Psi}}(\omega, \hat{\underline{\mathbf{k}}}) \cdot \hat{\underline{\mathbf{k}}};\end{aligned} \tag{8.111}$$

according to (7.29), the result of this calculation should always be zero, and on behalf of the nonzero exponential function,

$$\hat{\underline{\mathbf{k}}} \cdot \underline{\mathbf{\Psi}}(\omega, \hat{\underline{\mathbf{k}}}) = 0 \tag{8.112}$$

must hold (for $\omega > 0$), i.e., $\underline{\mathbf{\Psi}}(\omega, \hat{\underline{\mathbf{k}}})$ must be transverse to the propagation direction. This is a consequence of the arbitrary[144] gauge (7.29) for the vector potential of secondary waves.

With (7.28), we obtain—we utilize (2.172), (2.173), and (2.175)—

$$\begin{aligned}\underline{\mathbf{u}}(\underline{\mathbf{R}}, \omega, \hat{\underline{\mathbf{k}}}) &= \nabla\Phi(\underline{\mathbf{R}}, \omega, \hat{\underline{\mathbf{k}}}) + \nabla \times \underline{\mathbf{\Psi}}(\underline{\mathbf{R}}, \omega, \hat{\underline{\mathbf{k}}}) \\ &= \nabla\left[\Phi(\omega, \hat{\underline{\mathbf{k}}})\, e^{\pm jk_P \hat{\underline{\mathbf{k}}}\cdot\underline{\mathbf{R}}}\right] + \nabla \times \left[\underline{\mathbf{\Psi}}(\omega, \hat{\underline{\mathbf{k}}})\, e^{\pm jk_S \hat{\underline{\mathbf{k}}}\cdot\underline{\mathbf{R}}}\right] \\ &= \pm\, jk_P \Phi(\omega, \hat{\underline{\mathbf{k}}})\, e^{\pm jk_P \hat{\underline{\mathbf{k}}}\cdot\underline{\mathbf{R}}}\, \hat{\underline{\mathbf{k}}} \pm jk_S\, e^{\pm jk_S \hat{\underline{\mathbf{k}}}\cdot\underline{\mathbf{R}}}\, \hat{\underline{\mathbf{k}}} \times \underline{\mathbf{\Psi}}(\omega, \hat{\underline{\mathbf{k}}}).\end{aligned} \tag{8.113}$$

We immediately realize that: The primary part of the particle displacement of plane elastic waves is longitudinally and the secondary part is transversely

[143] Naturally, we are primarily interested in the case $\omega \neq 0$.

[144] For electromagnetic plane waves, the zero divergence of the electric flux density is a physical law formulated by Maxwell equations.

polarized (independent upon the arbitrary gauge!). In addition, we know due to equations (2.188) and (2.189) that the primary part is curl free and the secondary part is divergence free [it can be explicitly calculated with (8.113): Footnote 139]. Specially choosing (compare Footnote 136 and Equation 8.55)

$$\Phi(\omega, \hat{\underline{\mathbf{k}}}) = \Phi(\omega), \tag{8.114}$$

$$\underline{\mathbf{\Psi}}(\omega, \hat{\underline{\mathbf{k}}}) = \hat{\underline{\mathbf{\Psi}}}(\hat{\underline{\mathbf{k}}})\Psi(\omega) \tag{8.115}$$

and by comparing Equations 8.82 and 8.83, we obtain the following relations between the pulse spectra $\Phi(\omega)$, $\Psi(\omega)$ of the potentials and the pulse spectra $u_{\mathrm{P}}(\omega)$, $u_{\mathrm{S}}(\omega)$ of the particle displacements:

$$u_{\mathrm{P}}(\omega) = \mathrm{j}k_{\mathrm{P}}\Phi(\omega), \tag{8.116}$$

$$u_{\mathrm{S}}(\omega) = \mathrm{j}k_{\mathrm{S}}\Psi(\omega); \tag{8.117}$$

for the time functions, this means

$$u_{\mathrm{P}}(t) = -\frac{1}{c_{\mathrm{P}}}\frac{\mathrm{d}\Phi(t)}{\mathrm{d}t}, \tag{8.118}$$

$$u_{\mathrm{S}}(t) = -\frac{1}{c_{\mathrm{S}}}\frac{\mathrm{d}\Psi(t)}{\mathrm{d}t}. \tag{8.119}$$

Regarding the relation of the polarization vectors $\hat{\underline{\mathbf{u}}}_{\mathrm{S1}}(\hat{\underline{\mathbf{k}}})$, $\hat{\underline{\mathbf{u}}}_{\mathrm{S2}}(\hat{\underline{\mathbf{k}}})$, and $\hat{\underline{\mathbf{\Psi}}}(\hat{\underline{\mathbf{k}}})$, we have the choice: Either we can choose $\hat{\underline{\mathbf{\Psi}}}(\hat{\underline{\mathbf{k}}})$ "SV-oriented" according to $\hat{\underline{\mathbf{\Psi}}}(\hat{\underline{\mathbf{k}}}) = \hat{\underline{\mathbf{u}}}_{\mathrm{S1}}(\hat{\underline{\mathbf{k}}})$ or "SH-oriented" according to $\hat{\underline{\mathbf{\Psi}}}(\hat{\underline{\mathbf{k}}}) = \hat{\underline{\mathbf{u}}}_{\mathrm{S2}}(\hat{\underline{\mathbf{k}}})$; in the first case, we have $\hat{\underline{\mathbf{u}}}_{\mathrm{S2}}(\hat{\underline{\mathbf{k}}}) = \hat{\underline{\mathbf{k}}} \times \hat{\underline{\mathbf{\Psi}}}(\hat{\underline{\mathbf{k}}})$, and in the second case we have $\hat{\underline{\mathbf{u}}}_{\mathrm{S1}}(\hat{\underline{\mathbf{k}}}) = \hat{\underline{\mathbf{\Psi}}}(\hat{\underline{\mathbf{k}}}) \times \hat{\underline{\mathbf{k}}}$. In (cartesian) coordinates, this means (compare Figure 8.8): With $\hat{\underline{\mathbf{k}}}$ in the xz-plane, $\hat{\underline{\mathbf{\Psi}}} = \hat{\Psi}_x\underline{\mathbf{e}}_x + \hat{\Psi}_z\underline{\mathbf{e}}_z$ yields an SH-wave with $\hat{\underline{\mathbf{u}}}_{\mathrm{SH}} = \hat{\underline{\mathbf{k}}} \times \hat{\underline{\mathbf{\Psi}}} = \underline{\mathbf{e}}_y$, and $\hat{\underline{\mathbf{\Psi}}} = \underline{\mathbf{e}}_y$ yields an SV-wave with $\hat{\underline{\mathbf{u}}}_{\mathrm{SV}} = \hat{\underline{\mathbf{\Psi}}} \times \hat{\underline{\mathbf{k}}} = \hat{u}_{\mathrm{SV}x}\underline{\mathbf{e}}_x + \hat{u}_{\mathrm{SV}z}\underline{\mathbf{e}}_z$.

Sound pressure of plane elastic waves: Generally, the field quantity "sound pressure" in the sense of acoustics only exists iff (if and only if) the stress tensor is equal to the isotropic pressure tensor:

$$\underline{\underline{\mathbf{T}}}(\underline{\mathbf{R}}, t) = \underline{\underline{\mathbf{P}}}(\underline{\mathbf{R}}, t) = -p(\underline{\mathbf{R}}, t)\,\underline{\underline{\mathbf{I}}}. \tag{8.120}$$

With (7.23), the stress tensor for homogeneous isotropic materials, we calculate for plane waves utilizing (8.58) and (8.56):

$$\underline{\underline{\mathbf{T}}}_{\mathrm{P}}(\underline{\mathbf{R}}, \omega, \hat{\underline{\mathbf{k}}}) = \mathrm{j}k_{\mathrm{P}}\,u_{\mathrm{P}}(\omega)\,\mathrm{e}^{\mathrm{j}k_{\mathrm{P}}\hat{\underline{\mathbf{k}}}\cdot\underline{\mathbf{R}}}\,(\lambda\,\underline{\underline{\mathbf{I}}} + 2\mu\,\hat{\underline{\mathbf{k}}}\hat{\underline{\mathbf{k}}}), \tag{8.121}$$

$$\underline{\underline{\mathbf{T}}}_{\mathrm{S}}(\underline{\mathbf{R}}, \omega, \hat{\underline{\mathbf{k}}}) = \mathrm{j}k_{\mathrm{S}}\mu\,u_{\mathrm{S}}(\omega)\,\mathrm{e}^{\mathrm{j}k_{\mathrm{S}}\hat{\underline{\mathbf{k}}}\cdot\underline{\mathbf{R}}}\,[\hat{\underline{\mathbf{k}}}\,\hat{\underline{\mathbf{u}}}_{\mathrm{S}}(\hat{\underline{\mathbf{k}}}) + \hat{\underline{\mathbf{u}}}_{\mathrm{S}}(\hat{\underline{\mathbf{k}}})\,\hat{\underline{\mathbf{k}}}]. \tag{8.122}$$

Due to the zero divergence of the shear wave, the λ-term is missing in $\underline{\underline{\mathbf{T}}}_{\mathrm{S}}(\underline{\mathbf{R}}, \omega, \hat{\underline{\mathbf{k}}})$. Apparently, the tensors (8.121) and (8.122) are not proportional to the unit tensor,[145] resulting in the nonexistence of $p(\underline{\mathbf{R}}, t)$ or $p(\underline{\mathbf{R}}, \omega)$,

[145]Only if $\mu = 0$, but then its acoustics.

respectively, in elastic materials in the sense of (8.120). Nevertheless, we can deduce a scalar sound pressure from the stress tensor for plane elastic waves in analogy to (5.54) if we project (8.121) and (8.122) first to the propagation direction $\hat{\underline{\mathbf{k}}}$ and then to the respective polarization $\hat{\underline{\mathbf{k}}}$ or $\hat{\underline{\mathbf{u}}}_\mathrm{S}(\hat{\underline{\mathbf{k}}})$ of the plane elastic wave:[146]

$$\begin{aligned} p_\mathrm{P}(\underline{\mathbf{R}}, \omega, \hat{\underline{\mathbf{k}}}) &\stackrel{\text{def}}{=} -\underline{\underline{\mathbf{T}}}_\mathrm{P}(\underline{\mathbf{R}}, \omega, \hat{\underline{\mathbf{k}}}) : \hat{\underline{\mathbf{k}}}\,\hat{\underline{\mathbf{k}}} = -\mathrm{j}\omega Z_\mathrm{P} \underbrace{u_\mathrm{P}(\omega)\,\mathrm{e}^{\mathrm{j}k_\mathrm{P}\hat{\underline{\mathbf{k}}}\cdot\underline{\mathbf{R}}}}_{= u_\mathrm{P}(\underline{\mathbf{R}}, \omega, \hat{\underline{\mathbf{k}}})} \\ &= p_\mathrm{P}(\omega)\,\mathrm{e}^{\mathrm{j}k_\mathrm{P}\hat{\underline{\mathbf{k}}}\cdot\underline{\mathbf{R}}}, \end{aligned} \tag{8.123}$$

$$\begin{aligned} p_\mathrm{S}(\underline{\mathbf{R}}, \omega, \hat{\underline{\mathbf{k}}}) &\stackrel{\text{def}}{=} -\underline{\underline{\mathbf{T}}}_\mathrm{S}(\underline{\mathbf{R}}, \omega, \hat{\underline{\mathbf{k}}}) : \hat{\underline{\mathbf{k}}}\,\hat{\underline{\mathbf{u}}}_\mathrm{S}(\hat{\underline{\mathbf{k}}}) = -\mathrm{j}\omega Z_\mathrm{S} \underbrace{u_\mathrm{S}(\omega)\,\mathrm{e}^{\mathrm{j}k_\mathrm{S}\hat{\underline{\mathbf{k}}}\cdot\underline{\mathbf{R}}}}_{= u_\mathrm{S}(\underline{\mathbf{R}}, \omega, \hat{\underline{\mathbf{k}}})} \\ &= p_\mathrm{S}(\omega)\,\mathrm{e}^{\mathrm{j}k_\mathrm{S}\hat{\underline{\mathbf{k}}}\cdot\underline{\mathbf{R}}}; \end{aligned} \tag{8.124}$$

here, we have

$$Z_{\mathrm{P,S}} = \rho c_{\mathrm{P,S}} \tag{8.125}$$

as acoustic wave impedances according to (5.52). As $u_\mathrm{P}(\underline{\mathbf{R}}, \omega, \hat{\underline{\mathbf{k}}})$ and $u_\mathrm{S}(\underline{\mathbf{R}}, \omega, \hat{\underline{\mathbf{k}}})$, we understand scalar particle displacements that are consequently proportional to the respective sound pressure. Note: Krautkrämer and Krautkrämer (1986) do not consider the factor $-\mathrm{j}$; yet, it is important if we transform (8.123) and (8.124) into the time domain:

$$p_\mathrm{P}(\underline{\mathbf{R}}, t, \hat{\underline{\mathbf{k}}}) = Z_\mathrm{P}\,\frac{\partial u_\mathrm{P}(\underline{\mathbf{R}}, t, \hat{\underline{\mathbf{k}}})}{\partial t}, \tag{8.126}$$

$$p_\mathrm{S}(\underline{\mathbf{R}}, t, \hat{\underline{\mathbf{k}}}) = Z_\mathrm{S}\,\frac{\partial u_\mathrm{S}(\underline{\mathbf{R}}, t, \hat{\underline{\mathbf{k}}})}{\partial t}. \tag{8.127}$$

It is quite clear that we can simply write down particle displacement pressure relations of the kind (5.54) according to (8.123) and (8.124), yet a physical meaning for the pressure written as such is only obtained through the respectively defined double contractions of the stress tensor if the required projections are subsequently interpreted (Figure 8.9). For the plane P-wave, the projection of $\underline{\underline{\mathbf{T}}}_\mathrm{P}$ onto the propagation direction $\hat{\underline{\mathbf{k}}}$ yields a (traction) vector $\underline{\mathbf{t}}_\mathrm{P} \sim \hat{\underline{\mathbf{k}}}$ with the dimension of a force density = force/area pointing into the direction of the polarization vector $\hat{\underline{\mathbf{k}}}$; if we now define a unit area S_P with the normal $\underline{\mathbf{n}}_\mathrm{P} = \hat{\underline{\mathbf{k}}}$ in polarization direction, we actually obtain $-\underline{\mathbf{t}}_\mathrm{P} \cdot \underline{\mathbf{n}}_\mathrm{P} = -\underline{\mathbf{t}}_\mathrm{P} \cdot \hat{\underline{\mathbf{k}}} = -\underline{\underline{\mathbf{T}}}_\mathrm{P} : \hat{\underline{\mathbf{k}}}\,\hat{\underline{\mathbf{k}}} = p_\mathrm{P}$ as P-sound pressure p_P on S_P (Equation 8.123). The right side of Figure 8.9 is relevant for the S-wave: The projection of $\underline{\underline{\mathbf{T}}}_\mathrm{S}$ onto the propagation direction $\hat{\underline{\mathbf{k}}}$ yields the force density vector $\underline{\mathbf{t}}_\mathrm{S} \sim \hat{\underline{\mathbf{u}}}_\mathrm{S}$ pointing into the polarization direction $\hat{\underline{\mathbf{u}}}_\mathrm{S}$; the projection of $\underline{\mathbf{t}}_\mathrm{S}$ onto the normal $\underline{\mathbf{n}}_\mathrm{S} = \hat{\underline{\mathbf{u}}}_\mathrm{S}$ of the unit area S_S yields $-\underline{\mathbf{t}}_\mathrm{S} \cdot \underline{\mathbf{n}}_\mathrm{S} = -\underline{\mathbf{t}}_\mathrm{S} \cdot \hat{\underline{\mathbf{u}}}_\mathrm{S} = -\underline{\underline{\mathbf{T}}}_\mathrm{S} : \hat{\underline{\mathbf{k}}}\,\hat{\underline{\mathbf{u}}}_\mathrm{S} = p_\mathrm{S}$

[146] This becomes only obvious if we write the stress tensors according to (8.121) and (8.122) coordinate free!

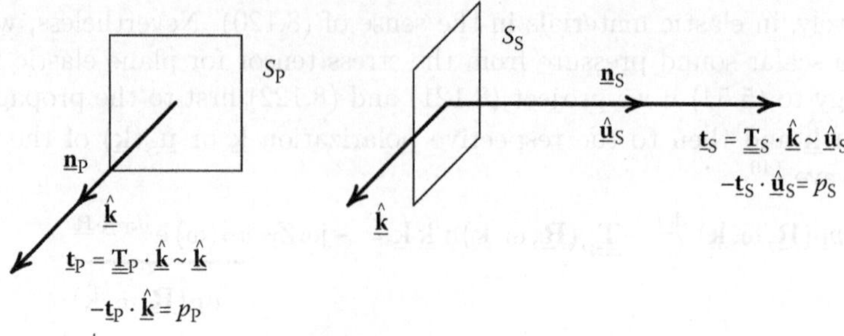

FIGURE 8.9
Sound pressure of plane P- and S-waves.

(Equation 8.124). Note: Rotating the polarization $\hat{\underline{\mathbf{u}}}_S \Longrightarrow -\hat{\underline{\mathbf{u}}}_S$ results in the same expression (8.124).

Energy velocities of homogeneous plane elastic waves: We will investigate to what extent a homogeneous plane elastic wave transports energy.[147] According to Section 4.3, we must calculate the elastodynamic Poynting vector; we advantageously concentrate on time harmonic plane elastic waves of circular frequency ω_0, thus calculating time averages of the energy flux density in terms of the real part of the complex elastodynamic Poynting vector

$$\begin{aligned}\underline{\mathbf{S}}_K(\underline{\mathbf{R}}, \omega_0) &= -\frac{1}{2}\,\underline{\mathbf{v}}(\underline{\mathbf{R}}, \omega_0) \cdot \underline{\underline{\mathbf{T}}}^*(\underline{\mathbf{R}}, \omega_0) \\ &= \mathrm{j}\frac{\omega_0}{2}\,\underline{\mathbf{u}}(\underline{\mathbf{R}}, \omega_0) \cdot \underline{\underline{\mathbf{T}}}^*(\underline{\mathbf{R}}, \omega_0) \end{aligned} \tag{8.128}$$

utilizing the complex valued phasors of time harmonic homogeneous plane P- and S-waves (Equations 8.82, 8.83, and 8.84)

$$\underline{\mathbf{u}}_P(\underline{\mathbf{R}}, \omega_0, \hat{\underline{\mathbf{k}}}) = u_P(\omega_0)\,\mathrm{e}^{\mathrm{j}\frac{\hat{\underline{\mathbf{k}}}\cdot\underline{\mathbf{R}}}{c_P}\omega_0}\,\hat{\underline{\mathbf{k}}}, \tag{8.129}$$

$$\underline{\mathbf{u}}_S(\underline{\mathbf{R}}, \omega_0, \hat{\underline{\mathbf{k}}}) = u_S(\omega_0)\,\mathrm{e}^{\mathrm{j}\frac{\hat{\underline{\mathbf{k}}}\cdot\underline{\mathbf{R}}}{c_S}\omega_0}\,\hat{\underline{\mathbf{u}}}_S(\hat{\underline{\mathbf{k}}}), \quad \hat{\underline{\mathbf{u}}}_S(\hat{\underline{\mathbf{k}}}) \cdot \hat{\underline{\mathbf{k}}} = 0. \tag{8.130}$$

With (8.121) and (8.122), we immediately have[148]

$$\begin{aligned}\underline{\mathbf{S}}_{KP}(\underline{\mathbf{R}}, \omega_0, \hat{\underline{\mathbf{k}}}) &= \frac{\omega_0^2\rho}{2}\,c_P|u_P(\omega_0)|^2\,\hat{\underline{\mathbf{k}}} \\ &= \frac{|p_P(\omega_0)|^2}{2Z_P}\,\hat{\underline{\mathbf{k}}}, \end{aligned} \tag{8.131}$$

[147]The energy transport and the energy density of inhomogeneous plane waves with orthogonal phase and amplitude planes due to a complex valued $\hat{\underline{\mathbf{k}}}$ is calculated in Section 8.2; their physical realization is discussed in Section 9.1.2.

[148]For time harmonic inhomogeneous plane waves (in nondissipative materials), different expressions are obtained: Section 8.2.

$$\underline{\mathbf{S}}_{\mathrm{KS}}(\underline{\mathbf{R}}, \omega_0, \hat{\underline{\mathbf{k}}}) = \frac{\omega_0^2 \rho}{2}\, c_{\mathrm{S}} |u_{\mathrm{S}}(\omega_0)|^2\, \hat{\underline{\mathbf{k}}}$$
$$= \frac{|p_{\mathrm{S}}(\omega_0)|^2}{2Z_{\mathrm{S}}}\, \hat{\underline{\mathbf{k}}}. \tag{8.132}$$

Both Poynting vectors are real valued.

As result of our investigation, we find that: Homogeneous plane elastic waves in isotropic materials transport energy for $\omega_0 > 0$ into the direction $\hat{\underline{\mathbf{k}}}$, i.e., the propagation direction of the phase coincides with the propagation direction of energy. This is a degeneration of the isotropic material because this is generally not true in anisotropic materials.

Regarding dimension, the Poynting vector stands for energy per time per area, thus dividing by an energy per volume—an energy density—we obtain the dimension of a velocity. It is reasonable to choose the energy density as it is stored in the time average of a time harmonic elastic wave (Section 4.3):

$$\langle w(\underline{\mathbf{R}}, t, \omega_0)\rangle = \frac{\rho}{4}\, \underline{\mathbf{v}}(\underline{\mathbf{R}}, \omega_0) \cdot \underline{\mathbf{v}}^*(\underline{\mathbf{R}}, \omega_0) + \frac{1}{4}\, \underline{\underline{\mathbf{S}}}(\underline{\mathbf{R}}, \omega_0) : \underline{\underline{\underline{\underline{\mathbf{c}}}}} : \underline{\underline{\mathbf{S}}}^*(\underline{\mathbf{R}}, \omega_0) \tag{8.133}$$
$$= \frac{\omega_0^2 \rho}{4}\, \underline{\mathbf{u}}(\underline{\mathbf{R}}, \omega_0) \cdot \underline{\mathbf{u}}^*(\underline{\mathbf{R}}, \omega_0) + \frac{1}{4}\, \underline{\underline{\mathbf{S}}}(\underline{\mathbf{R}}, \omega_0) : \underline{\underline{\underline{\underline{\mathbf{c}}}}} : \underline{\underline{\mathbf{S}}}^*(\underline{\mathbf{R}}, \omega_0). \tag{8.134}$$

With the definition (3.2) of the deformation tensor for source-free materials, we obtain for the particle displacements (8.129) and (8.130)

$$\underline{\underline{\mathbf{S}}}_{\mathrm{P}}(\underline{\mathbf{R}}, \omega_0, \hat{\underline{\mathbf{k}}}) = \mathrm{j}k_{\mathrm{P}}\, u_{\mathrm{P}}(\omega_0)\, \mathrm{e}^{\mathrm{j}k_{\mathrm{P}} \hat{\underline{\mathbf{k}}} \cdot \underline{\mathbf{R}}}\, \hat{\underline{\mathbf{k}}}\, \hat{\underline{\mathbf{k}}}, \tag{8.135}$$
$$\underline{\underline{\mathbf{S}}}_{\mathrm{S}}(\underline{\mathbf{R}}, \omega_0, \hat{\underline{\mathbf{k}}}) = \frac{1}{2}\, \mathrm{j}k_{\mathrm{S}}\, u_{\mathrm{S}}(\omega_0)\, \mathrm{e}^{\mathrm{j}k_{\mathrm{S}} \hat{\underline{\mathbf{k}}} \cdot \underline{\mathbf{R}}}\, [\hat{\underline{\mathbf{k}}}\, \hat{\underline{\mathbf{u}}}_{\mathrm{S}}(\hat{\underline{\mathbf{k}}}) + \hat{\underline{\mathbf{u}}}_{\mathrm{S}}(\hat{\underline{\mathbf{k}}})\, \hat{\underline{\mathbf{k}}}] \tag{8.136}$$

and consequently for the present homogeneous isotropic nondissipative material[149]

$$\langle w_{\mathrm{P}}(\underline{\mathbf{R}}, t, \omega_0, \hat{\underline{\mathbf{k}}})\rangle = \frac{\omega_0^2 \rho}{2}\, |u_{\mathrm{P}}(\omega_0)|^2, \tag{8.137}$$
$$\langle w_{\mathrm{S}}(\underline{\mathbf{R}}, t, \omega_0, \hat{\underline{\mathbf{k}}})\rangle = \frac{\omega_0^2 \rho}{2}\, |u_{\mathrm{S}}(\omega_0)|^2. \tag{8.138}$$

Note: The energy densities of homogeneous plane waves are spatially independent, i.e., continuously distributed over infinite space, resulting in an infinite total energy of such waves; they are nonrealizable.

With

$$\underline{\mathbf{c}}_{\mathrm{EP}}(\hat{\underline{\mathbf{k}}}) \stackrel{\mathrm{def}}{=} \frac{\Re\{\underline{\mathbf{S}}_{\mathrm{KP}}(\underline{\mathbf{R}}, \omega_0, \hat{\underline{\mathbf{k}}})\}}{\langle w_{\mathrm{P}}(\underline{\mathbf{R}}, t, \omega_0, \hat{\underline{\mathbf{k}}})\rangle}, \tag{8.139}$$
$$\underline{\mathbf{c}}_{\mathrm{ES}}(\hat{\underline{\mathbf{k}}}) \stackrel{\mathrm{def}}{=} \frac{\Re\{\underline{\mathbf{S}}_{\mathrm{KS}}(\underline{\mathbf{R}}, \omega_0, \hat{\underline{\mathbf{k}}})\}}{\langle w_{\mathrm{S}}(\underline{\mathbf{R}}, t, \omega_0, \hat{\underline{\mathbf{k}}})\rangle}, \tag{8.140}$$

[149] Footnote 148 similarly holds.

we now define energy velocity vectors and calculate them for homogeneous plane elastic waves in homogeneous isotropic nondissipative materials as follows:

$$\underline{\mathbf{c}}_{\mathrm{EP}}(\hat{\underline{\mathbf{k}}}) = c_{\mathrm{P}}\,\hat{\underline{\mathbf{k}}}, \tag{8.141}$$

$$\underline{\mathbf{c}}_{\mathrm{ES}}(\hat{\underline{\mathbf{k}}}) = c_{\mathrm{S}}\,\hat{\underline{\mathbf{k}}}. \tag{8.142}$$

These energy velocity vectors have direction of phase propagation and their magnitudes are equal to the phase velocities. With the definition of phase velocity vectors (of homogeneous plane waves in nondissipative materials)

$$\underline{\mathbf{c}}_{\mathrm{P,S}}(\hat{\underline{\mathbf{k}}}) \stackrel{\mathrm{def}}{=} c_{\mathrm{P,S}}\,\hat{\underline{\mathbf{k}}}, \tag{8.143}$$

it follows

$$\underline{\mathbf{c}}_{\mathrm{EP,S}}(\hat{\underline{\mathbf{k}}}) = \underline{\mathbf{c}}_{\mathrm{P,S}}(\hat{\underline{\mathbf{k}}}). \tag{8.144}$$

Due to the isotropy of the material surfaces of constant, phase velocity vectors are spherical surfaces with radii c_{P} and c_{S}. Similarly, the surfaces of constant so-called slowness vectors

$$\begin{aligned}\underline{\mathbf{s}}_{\mathrm{P,S}}(\hat{\underline{\mathbf{k}}}) &= \frac{1}{c_{\mathrm{P,S}}}\,\hat{\underline{\mathbf{k}}}\\ &= s_{\mathrm{P,S}}\,\hat{\underline{\mathbf{k}}}\end{aligned} \tag{8.145}$$

are spherical surfaces with radii $s_{\mathrm{P,S}} = 1/c_{\mathrm{P,S}}$; in Figure 8.10(a), cross-sections through slowness surfaces and slowness vectors $\underline{\mathbf{s}}_{\mathrm{P}}(\hat{\underline{\mathbf{k}}}_{\mathrm{P}})$, $\underline{\mathbf{s}}_{\mathrm{S}}(\hat{\underline{\mathbf{k}}}_{\mathrm{S}})$ are depicted for two given phase vectors $\hat{\underline{\mathbf{k}}}_{\mathrm{P}}$, $\hat{\underline{\mathbf{k}}}_{\mathrm{S}}$; with (8.141) and (8.142), we know that the energy velocity vectors belonging to the directions $\hat{\underline{\mathbf{k}}}_{\mathrm{P}}$, $\hat{\underline{\mathbf{k}}}_{\mathrm{S}}$ are orthogonal to

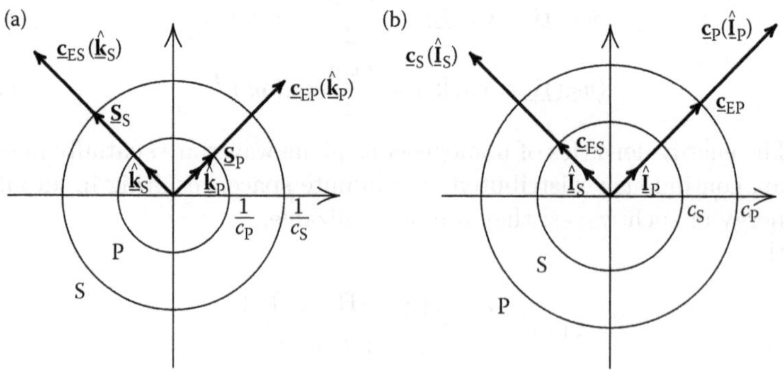

FIGURE 8.10
Cross-sections through slowness (a) and energy velocity surfaces (b) of the isotropic material.

the slowness surfaces at points $\underline{\mathbf{s}}_\mathrm{P}(\hat{\underline{\mathbf{k}}}_\mathrm{P})$, $\underline{\mathbf{s}}_\mathrm{S}(\hat{\underline{\mathbf{k}}}_\mathrm{S})$. If we otherwise prescribe unit ray vectors $\hat{\underline{\mathbf{l}}}_\mathrm{P}, \hat{\underline{\mathbf{l}}}_\mathrm{S}$ of energy propagation, the resulting surfaces of constant energy velocity are spherical surfaces with radii $c_\mathrm{EP}(\hat{\underline{\mathbf{l}}}_\mathrm{P}) = c_\mathrm{P}$, $c_\mathrm{ES}(\hat{\underline{\mathbf{l}}}_\mathrm{S}) = c_\mathrm{S}$, and due to (8.145) and (8.144), we know that the slowness vectors belonging to the directions $\hat{\underline{\mathbf{l}}}_\mathrm{P}, \hat{\underline{\mathbf{l}}}_\mathrm{S}$ are orthogonal to the energy velocity surfaces at points $\underline{\mathbf{c}}_\mathrm{EP}(\hat{\underline{\mathbf{l}}}_\mathrm{P})$, $\underline{\mathbf{c}}_\mathrm{ES}(\hat{\underline{\mathbf{l}}}_\mathrm{S})$ [Figure 8.10(b)]. Having displayed these trivialities of the isotropic material in a separate figure finds a reason when considering anisotropic materials: The respective surfaces are no longer spherical but the above orthogonalities are kept.

8.2 Inhomogeneous Plane Waves in Isotropic Nondissipative Materials

Complex wave number vectors: We generalize the *ansatz* (8.54) for the Fourier spectra of plane waves to complex wave number vectors $\underline{\mathbf{k}}$ (in Section 9.3.1, we repeat the following arguments for complex slowness vectors in cartesian coordinates):

$$\underline{\mathbf{u}}(\mathbf{R}, \omega, \underline{\mathbf{k}}) = u(\omega)\,\mathrm{e}^{\pm \mathrm{j}\underline{\mathbf{k}}\cdot\mathbf{R}}\,\hat{\underline{\mathbf{u}}}(\underline{\mathbf{k}}). \tag{8.146}$$

That way, we obtain

$$\underline{\underline{\mathbf{W}}}(\underline{\mathbf{k}}, \omega) = \frac{\mu}{\rho}\,\underline{\mathbf{k}}\cdot\underline{\mathbf{k}}\,\underline{\underline{\mathbf{I}}} + \frac{\lambda+\mu}{\rho}\,\underline{\mathbf{k}}\,\underline{\mathbf{k}} - \omega^2\,\underline{\underline{\mathbf{I}}} \tag{8.147}$$

instead of the wave tensor (8.66). Equating the determinant

$$\det \underline{\underline{\mathbf{W}}}(\underline{\mathbf{k}}, \omega) = \left(\frac{\mu\underline{\mathbf{k}}\cdot\underline{\mathbf{k}} - \rho\omega^2}{\lambda+\mu}\right)^2 \left(\frac{\mu\underline{\mathbf{k}}\cdot\underline{\mathbf{k}} - \rho\omega^2}{\lambda+\mu} + \underline{\mathbf{k}}\cdot\underline{\mathbf{k}}\right) \tag{8.148}$$

to zero, we obtain the dispersion relations

$$\underline{\mathbf{k}}\cdot\underline{\mathbf{k}} = \frac{\rho}{\mu}\,\omega^2 = \frac{\omega^2}{c_\mathrm{S}^2} = k_\mathrm{S}^2, \tag{8.149}$$

$$\underline{\mathbf{k}}\cdot\underline{\mathbf{k}} = \frac{\rho}{\lambda+2\mu}\,\omega^2 = \frac{\omega^2}{c_\mathrm{P}^2} = k_\mathrm{P}^2 \tag{8.150}$$

that are understood as the dependence of the wave number vector $\underline{\mathbf{k}}(\omega)$ upon frequency (and the material parameters) and, respectively, the dependence of the frequency $\omega(\underline{\mathbf{k}})$ upon the wave number vector. On the left-hand side, the structure of these dispersion relations is a consequence of the isotropy of the material, and on the right-hand side, it is a consequence of its lack of dissipation (vanishing dissipation) because (here) k_P and k_S are real valued.

The homogeneity of the material is already reflected by Equation 8.146 of the plane wave.[150] If we "solve" (8.149) and (8.150) with the *ansatz*

$$\underline{\mathbf{k}}_{\mathrm{P,S}} = k_{\mathrm{P,S}}\hat{\underline{\mathbf{k}}}, \tag{8.151}$$

we obtain homogeneous plane P- and S-waves whose planes of constant phase and amplitude are parallel to another; they have been discussed in the previous section.

Yet, we do not have to satisfy the dispersion relations (8.149) or (8.150) through the *ansatz* (8.151), by all means we can accept complex wave number vectors

$$\underline{\mathbf{k}} = \Re\underline{\mathbf{k}} + \mathrm{j}\Im\underline{\mathbf{k}} \tag{8.152}$$

if we only require

$$\Re\underline{\mathbf{k}} \cdot \Im\underline{\mathbf{k}} = 0. \tag{8.153}$$

This constraint ensures that

$$\underline{\mathbf{k}} \cdot \underline{\mathbf{k}} = |\Re\underline{\mathbf{k}}|^2 - |\Im\underline{\mathbf{k}}|^2 + 2\mathrm{j}\underbrace{\Re\underline{\mathbf{k}} \cdot \Im\underline{\mathbf{k}}}_{=0} \tag{8.154}$$

is real as claimed by the dispersion relations for nondissipative (isotropic) materials: In a nondissipative material, $\Re\underline{\mathbf{k}}$ and $\Im\underline{\mathbf{k}}$ must be orthogonal to each other if we allow for $\Im\underline{\mathbf{k}} \neq \underline{\mathbf{0}}$. As a matter of fact, the *ansatz* (8.152) according to

$$\underline{\mathbf{u}}(\underline{\mathbf{R}}, \omega, \underline{\mathbf{k}}) = u(\omega)\, \mathrm{e}^{\pm\mathrm{j}\Re\underline{\mathbf{k}}\cdot\underline{\mathbf{R}}}\, \mathrm{e}^{\mp\Im\underline{\mathbf{k}}\cdot\underline{\mathbf{R}}}\, \hat{\underline{\mathbf{u}}}(\underline{\mathbf{k}}) \tag{8.155}$$

with (8.153) generates an evanescent inhomogeneous plane wave that is attenuated perpendicularly to $\Re\underline{\mathbf{k}}$ if we point with the attenuation vector $\Im\underline{\mathbf{k}}$ into the "correct" half-space, depending upon the propagation direction. It is illustrated in Figure 8.11: For propagation in $(+\Re\underline{\mathbf{k}})$-direction [Figure 8.11(a)], $\mathrm{e}^{-\Im\underline{\mathbf{k}}\cdot\underline{\mathbf{R}}}$ is an exponential attenuation for that half-space into which $\Im\underline{\mathbf{k}}$ points because then we have $\Im\underline{\mathbf{k}} \cdot \underline{\mathbf{R}} > 0$; for propagation in $(-\Re\underline{\mathbf{k}})$-direction [Figure 8.11(b)], $\mathrm{e}^{\Im\underline{\mathbf{k}}\cdot\underline{\mathbf{R}}}$ is only an exponential attenuation for that half-space into which $\Im\underline{\mathbf{k}}$ does not point because then we have $\Im\underline{\mathbf{k}} \cdot \underline{\mathbf{R}} < 0$. The complex wave number vector $\underline{\mathbf{k}}$ has the phase propagation vector $\Re\underline{\mathbf{k}}$ as real part and the attenuation vector $\Im\underline{\mathbf{K}}$ as imaginary part, whose direction must be determined to assess an attenuation. Such inhomogeneous plane waves in nondissipative materials will be met for the first time while discussing the total reflection of a plane SV-wave at the plane boundary of an elastic half-space (Section 9.1.2).

[150]The generalization to inhomogeneous materials in terms of eikonal equations is discussed in Section 12.3.

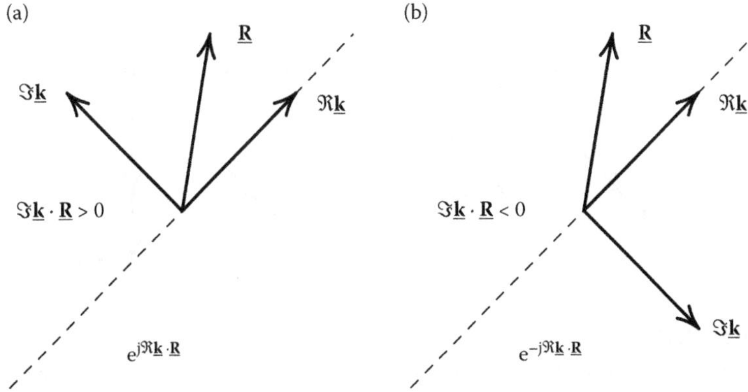

FIGURE 8.11
Evanescent inhomogeneous plane waves (elastic full-space); (a) propagation direction $+\Re\underline{\mathbf{k}}$, (b) propagation direction $-\Re\underline{\mathbf{k}}$.

Complex polarization vectors: As pressure waves, primary waves in isotropic materials are curl-free, and as shear waves, secondary waves are divergence-free; therefore, we determine the polarization (unit) vectors $\underline{\hat{\mathbf{u}}}_{\mathrm{P,S}}(\underline{\mathbf{k}})$ via the requirements:

$$\underline{\mathbf{k}}_{\mathrm{P}} \times \underline{\hat{\mathbf{u}}}_{\mathrm{P}}(\underline{\mathbf{k}}_{\mathrm{P}}) = \underline{\mathbf{0}}, \tag{8.156}$$

$$\underline{\mathbf{k}}_{\mathrm{S}} \cdot \underline{\hat{\mathbf{u}}}_{\mathrm{S}}(\underline{\mathbf{k}}_{\mathrm{S}}) = 0. \tag{8.157}$$

Obviously, (8.156) is satisfied by

$$\underline{\hat{\mathbf{u}}}_{\mathrm{P}}(\underline{\mathbf{k}}_{\mathrm{P}}) \sim \underline{\mathbf{k}}_{\mathrm{P}} \tag{8.158}$$

because, even for complex wave number vectors $\underline{\mathbf{k}}_{\mathrm{P}} = \Re\underline{\mathbf{k}}_{\mathrm{P}} + \mathrm{j}\Im\underline{\mathbf{k}}_{\mathrm{P}}$, we always have $\underline{\mathbf{k}}_{\mathrm{P}} \times \underline{\mathbf{k}}_{\mathrm{P}} \equiv \underline{\mathbf{0}}$. With (8.158), $\underline{\hat{\mathbf{u}}}_{\mathrm{P}}(\underline{\mathbf{k}}_{\mathrm{P}})$ is also complex defining the P-polarization unit vector $\underline{\hat{\mathbf{u}}}_{\mathrm{P}}(\underline{\mathbf{k}}_{\mathrm{P}})$ in the Hermitian sense by

$$\underline{\hat{\mathbf{u}}}_{\mathrm{P}}(\underline{\mathbf{k}}_{\mathrm{P}}) = \frac{\underline{\mathbf{k}}_{\mathrm{P}}}{\sqrt{\underline{\mathbf{k}}_{\mathrm{P}} \cdot \underline{\mathbf{k}}_{\mathrm{P}}^*}} \tag{8.159}$$

because then we have $\underline{\hat{\mathbf{u}}}_{\mathrm{P}}(\underline{\mathbf{k}}_{\mathrm{P}}) \cdot \underline{\hat{\mathbf{u}}}_{\mathrm{P}}^*(\underline{\mathbf{k}}_{\mathrm{P}}) = 1$. With (8.153) and (8.159), the polarization picture of an inhomogeneous plane pressure wave in a nondissipative material results as sketched in Figure 8.12. For propagation into positive $\Re\underline{\mathbf{k}}$-direction, the exponential attenuation results for the half-space indicated by $\underline{\mathbf{R}}$.

The requirement (8.157) is satisfied for each complex vector $\underline{\hat{\mathbf{u}}}_{\mathrm{S}}(\underline{\mathbf{k}}_{\mathrm{S}})$ for which

$$\Re\underline{\mathbf{k}}_{\mathrm{S}} \cdot \Re\underline{\hat{\mathbf{u}}}_{\mathrm{S}}(\underline{\mathbf{k}}_{\mathrm{S}}) - \Im\underline{\mathbf{k}}_{\mathrm{S}} \cdot \Im\underline{\hat{\mathbf{u}}}_{\mathrm{S}}(\underline{\mathbf{k}}_{\mathrm{S}}) = 0, \tag{8.160}$$

$$\Im\underline{\mathbf{k}}_{\mathrm{S}} \cdot \Re\underline{\hat{\mathbf{u}}}_{\mathrm{S}}(\underline{\mathbf{k}}_{\mathrm{S}}) + \Re\underline{\mathbf{k}}_{\mathrm{S}} \cdot \Im\underline{\hat{\mathbf{u}}}_{\mathrm{S}}(\underline{\mathbf{k}}_{\mathrm{S}}) = 0 \tag{8.161}$$

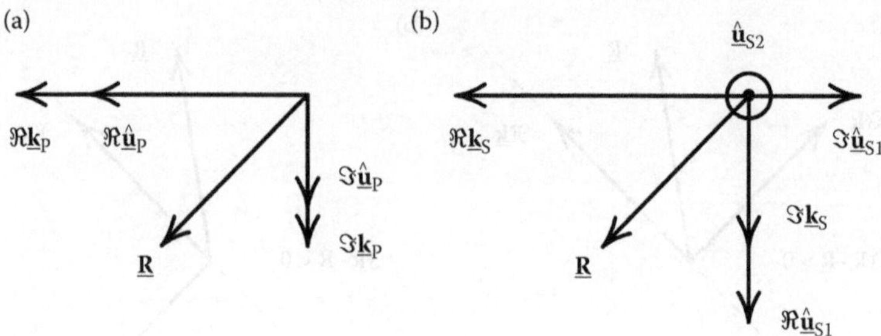

FIGURE 8.12
(a) Polarization of an inhomogeneous plane pressure wave in a nondissipative material; (b) S1/S2-polarization of an inhomogeneous plane shear wave in a nondissipative material.

holds, (8.157) does not imply orthogonality of $\underline{\mathbf{k}}_S$, $\hat{\underline{\mathbf{u}}}_S(\underline{\mathbf{k}}_S)$ in a geometric sense. As in the previous section for the case of real valued wave numbers, we choose $\hat{\underline{\mathbf{u}}}_{S2}$ real valued, hence $\Im\hat{\underline{\mathbf{u}}}_{S2} = \underline{\mathbf{0}}$,

$$\Re\underline{\mathbf{k}}_S \cdot \hat{\underline{\mathbf{u}}}_{S2} \stackrel{(8.160)}{=} 0, \tag{8.162}$$

$$\Im\underline{\mathbf{k}}_S \cdot \hat{\underline{\mathbf{u}}}_{S2} \stackrel{(8.161)}{=} 0 \tag{8.163}$$

ensuring that, independent of $\underline{\mathbf{k}}_S$, $\hat{\underline{\mathbf{u}}}_{S2}$ is always orthogonal to the plane spanned by $\Re\underline{\mathbf{k}}_S$ and $\Im\underline{\mathbf{k}}_S$. As allusion to Figure 8.8(a), we now define

$$\begin{aligned}\hat{\underline{\mathbf{u}}}_{S1}(\underline{\mathbf{k}}_S) &= \frac{\hat{\underline{\mathbf{u}}}_{S2} \times \underline{\mathbf{k}}_S}{\sqrt{(\hat{\underline{\mathbf{u}}}_{S2} \times \underline{\mathbf{k}}_S) \cdot (\hat{\underline{\mathbf{u}}}_{S2}^* \times \underline{\mathbf{k}}_S^*)}} \\ &= \frac{\hat{\underline{\mathbf{u}}}_{S2} \times \underline{\mathbf{k}}_S}{\sqrt{\underline{\mathbf{k}}_S \cdot \underline{\mathbf{k}}_S^*}},\end{aligned} \tag{8.164}$$

where, due to $\hat{\underline{\mathbf{u}}}_{S2} \cdot \underline{\mathbf{k}}_S = 0$, $\hat{\underline{\mathbf{u}}}_{S2} \cdot \underline{\mathbf{k}}_S^* = 0$, $\hat{\underline{\mathbf{u}}}_{S2} \cdot \hat{\underline{\mathbf{u}}}_{S2} = 1$, the same normalization results than for $\hat{\underline{\mathbf{u}}}_P(\underline{\mathbf{k}}_P)$. The definition (8.164) assures

$$\hat{\underline{\mathbf{u}}}_{S1}(\underline{\mathbf{k}}_S) \cdot \underline{\mathbf{k}}_S = 0, \tag{8.165}$$

$$\hat{\underline{\mathbf{u}}}_{S1}(\underline{\mathbf{k}}_S) \cdot \hat{\underline{\mathbf{u}}}_{S2} = 0; \tag{8.166}$$

due to[151]

$$\Re\hat{\underline{\mathbf{u}}}_{S1}(\underline{\mathbf{k}}_S) = \frac{\hat{\underline{\mathbf{u}}}_{S2} \times \Re\underline{\mathbf{k}}_S}{\sqrt{\underline{\mathbf{k}}_S \cdot \underline{\mathbf{k}}_S^*}}, \tag{8.167}$$

$$\Im\hat{\underline{\mathbf{u}}}_{S1}(\underline{\mathbf{k}}_S) = \frac{\hat{\underline{\mathbf{u}}}_{S2} \times \Im\underline{\mathbf{k}}_S}{\sqrt{\underline{\mathbf{k}}_S \cdot \underline{\mathbf{k}}_S^*}}, \tag{8.168}$$

[151] Note: $\Re\hat{\underline{\mathbf{u}}}_{S1}, \Re\hat{\underline{\mathbf{u}}}_{S2}, \Re\underline{\mathbf{k}}_S$ always constitute our familiar right-handed orthogonal trihedron of shear polarizations; yet, $\Im\hat{\underline{\mathbf{u}}}_{S1}$ is connected to $\Im\underline{\mathbf{k}}_S$ that must point into the "evanescence space" $(+\Re\underline{\mathbf{k}}_S)$-direction.

it is a special solution of (8.160); the relations (8.167) and (8.168) yield the situation as sketched in Figure 8.12(b), together with (8.153) and (8.161).

According to Figure 8.8(b), we also can certainly postulate in a cartesian coordinate system $\underline{\mathbf{k}}_{\mathrm{P,S}} \cdot \underline{\mathbf{e}}_y = 0$; it follows

$$\hat{\underline{\mathbf{u}}}_{\mathrm{P}}(\underline{\mathbf{k}}_{\mathrm{P}}) \cdot \underline{\mathbf{e}}_y = 0, \tag{8.169}$$

$$\hat{\underline{\mathbf{u}}}_{\mathrm{S2}} \Longrightarrow \hat{\underline{\mathbf{u}}}_{\mathrm{SH}} = \underline{\mathbf{e}}_y, \tag{8.170}$$

$$\hat{\underline{\mathbf{u}}}_{\mathrm{S1}}(\underline{\mathbf{k}}_{\mathrm{S}}) \Longrightarrow \hat{\underline{\mathbf{u}}}_{\mathrm{SV}}(\underline{\mathbf{k}}_{\mathrm{S}}) = \frac{\underline{\mathbf{e}}_y \times \underline{\mathbf{k}}_{\mathrm{S}}}{\sqrt{\underline{\mathbf{k}}_{\mathrm{S}} \cdot \underline{\mathbf{k}}_{\mathrm{S}}^*}}. \tag{8.171}$$

Phase velocity: According to (8.54), the respective *ansatz* of homogeneous plane waves,

$$c_{\mathrm{P,S}}(\underline{\mathbf{k}}_{\mathrm{P,S}}) = \frac{\omega}{|\Re \underline{\mathbf{k}}_{\mathrm{P,S}}|} \tag{8.172}$$

defines the phase velocity of inhomogeneous plane waves in (8.155) because the phase can be written as

$$\Re \underline{\mathbf{k}}_{\mathrm{P,S}} \cdot \underline{\mathbf{R}} = \frac{|\Re \underline{\mathbf{k}}_{\mathrm{P,S}}|}{\omega} \widehat{\Re \underline{\mathbf{k}}}_{\mathrm{P,S}} \cdot \underline{\mathbf{R}}\, \omega. \tag{8.173}$$

Thus we obtain as vectorial phase velocity similarly to (8.144)

$$\underline{\mathbf{c}}_{\mathrm{P,S}}(\underline{\mathbf{k}}_{\mathrm{P,S}}) = \frac{\omega}{|\Re \underline{\mathbf{k}}_{\mathrm{P,S}}|} \widehat{\Re \underline{\mathbf{k}}}_{\mathrm{P,S}}. \tag{8.174}$$

Due to $\underline{\mathbf{k}}_{\mathrm{P,S}} + \underline{\mathbf{k}}_{\mathrm{P,S}}^* = 2\Re \underline{\mathbf{k}}_{\mathrm{P,S}}$, we can explicitly write $c_{\mathrm{P,S}}(\underline{\mathbf{k}}_{\mathrm{P,S}})$ showing the dependence of the complex vectors $\underline{\mathbf{k}}_{\mathrm{P,S}}$:

$$c_{\mathrm{P,S}}(\underline{\mathbf{k}}_{\mathrm{P,S}}) = \frac{2}{\sqrt{\underline{\mathbf{k}}_{\mathrm{P,S}} \cdot \underline{\mathbf{k}}_{\mathrm{P,S}} + \underline{\mathbf{k}}_{\mathrm{P,S}}^* \cdot \underline{\mathbf{k}}_{\mathrm{P,S}}^* + 2\underline{\mathbf{k}}_{\mathrm{P,S}} \cdot \underline{\mathbf{k}}_{\mathrm{P,S}}^*}}\, \omega. \tag{8.175}$$

Due to the dispersion relations (8.149) and (8.150) of nondissipative materials, it follows $\underline{\mathbf{k}}_{\mathrm{P,S}} \cdot \underline{\mathbf{k}}_{\mathrm{P,S}} = k_{\mathrm{P,S}}^2$ and $\underline{\mathbf{k}}_{\mathrm{P,S}}^* \cdot \underline{\mathbf{k}}_{\mathrm{P,S}}^* = k_{\mathrm{P,S}}^2$, hence

$$c_{\mathrm{P,S}}(\underline{\mathbf{k}}_{\mathrm{P,S}}) = \sqrt{\frac{2}{k_{\mathrm{P,S}}^2 + \underline{\mathbf{k}}_{\mathrm{P,S}} \cdot \underline{\mathbf{k}}_{\mathrm{P,S}}^*}}\, \omega. \tag{8.176}$$

Via $\underline{\mathbf{k}}_{\mathrm{P}} \cdot \underline{\mathbf{k}}_{\mathrm{P}}^*$, we can only say further things about *homogeneous* plane waves (Sections 8.1.2 and 8.4.1). With

$$\hat{\underline{\mathbf{k}}}_{\mathrm{P,S}} = \frac{\underline{\mathbf{k}}_{\mathrm{P,S}}}{k_{\mathrm{P,S}}}, \tag{8.177}$$

we admittedly define in the present case of nondissipative materials non-Hermitian complex unit vectors[152] $\hat{\underline{\mathbf{k}}}_{\mathrm{P,S}}$ because the dispersion relations then read

$$\hat{\underline{\mathbf{k}}}_{\mathrm{P,S}} \cdot \hat{\underline{\mathbf{k}}}_{\mathrm{P,S}} = 1. \tag{8.178}$$

[152] With (8.178), we have $\hat{\underline{\mathbf{k}}}_{\mathrm{P,S}} \cdot \hat{\underline{\mathbf{k}}}_{\mathrm{P,S}}^* \geq 1$, and the equality sign only holds for $\Im \hat{\underline{\mathbf{k}}}_{\mathrm{P,S}} = \underline{\mathbf{0}}$.

That way, we finally obtain

$$c_{P,S}(\hat{\underline{\mathbf{k}}}_{P,S}) = \sqrt{\frac{2}{1 + \hat{\underline{\mathbf{k}}}_{P,S} \cdot \hat{\underline{\mathbf{k}}}^*_{P,S}}}\, c_{P,S}. \tag{8.179}$$

The phase velocities of homogeneous and inhomogeneous plane waves are different! In Section 9.1.2, we will give $c_P(\hat{\underline{\mathbf{k}}}_P)$ for an actual wave number vector $\underline{\mathbf{k}}_P$. Since $\hat{\underline{\mathbf{k}}}_{P,S} \cdot \hat{\underline{\mathbf{k}}}_{P,S} = |\Re\hat{\underline{\mathbf{k}}}_{P,S}|^2 - |\Im\hat{\underline{\mathbf{k}}}_{P,S}|^2 = 1$, we have $|\Re\hat{\underline{\mathbf{k}}}_{P,S}| \geq 1$, and due to $\hat{\underline{\mathbf{k}}}_{P,S} \cdot \hat{\underline{\mathbf{k}}}^*_{P,S} = |\Re\hat{\underline{\mathbf{k}}}_{P,S}|^2 + |\Im\hat{\underline{\mathbf{k}}}_{P,S}|^2$, it follows $\hat{\underline{\mathbf{k}}}_{P,S} \cdot \hat{\underline{\mathbf{k}}}^*_{P,S} \geq 1$ (Footnote 152) with the result $c_{P,S}(\hat{\underline{\mathbf{k}}}_{P,S}) \leq c_{P,S}$.

Energy velocities for complex pressure and shear wave number vectors: To calculate energy velocities of inhomogeneous plane waves, we must first compute the complex Poynting vector and the time averaged energy density of time harmonic waves of circular frequency ω_0 for complex wave number vectors.

We resort to (8.128):

$$\begin{aligned} &\Re\underline{\mathbf{S}}_K(\underline{\mathbf{R}}, \omega_0, \underline{\mathbf{k}}) \\ &\quad = \mathrm{j}\,\frac{\omega_0}{4}\left[\underline{\mathbf{u}}(\underline{\mathbf{R}}, \omega_0, \underline{\mathbf{k}}) \cdot \underline{\underline{\mathbf{T}}}^*(\underline{\mathbf{R}}, \omega_0, \underline{\mathbf{k}}) + \underline{\mathbf{u}}^*(\underline{\mathbf{R}}, \omega_0, \underline{\mathbf{k}}) \cdot \underline{\underline{\mathbf{T}}}(\underline{\mathbf{R}}, \omega_0, \underline{\mathbf{k}})\right]; \end{aligned} \tag{8.180}$$

with

$$\begin{aligned} \underline{\underline{\mathbf{T}}}(\underline{\mathbf{R}}, \omega_0, \underline{\mathbf{k}}) &= \underline{\underline{\underline{\mathbf{c}}}} : \underline{\underline{\mathbf{S}}}(\underline{\mathbf{R}}, \omega_0, \underline{\mathbf{k}}) \\ &= \frac{\mathrm{j}}{2}\,\underline{\underline{\underline{\mathbf{c}}}} : [\underline{\mathbf{k}}\,\underline{\mathbf{u}}(\underline{\mathbf{R}}, \omega_0, \underline{\mathbf{k}}) + \underline{\mathbf{u}}(\underline{\mathbf{R}}, \omega_0, \underline{\mathbf{k}})\underline{\mathbf{k}}], \end{aligned} \tag{8.181}$$

the *ansatz* (8.146) with (8.152) yields

$$\begin{aligned} &\Re\underline{\mathbf{S}}_K(\underline{\mathbf{R}}, \omega_0, \underline{\mathbf{k}}) \\ &\quad = \frac{\omega_0}{4}\,|u(\omega_0)|^2 \mathrm{e}^{-2\Im\underline{\mathbf{k}}\cdot\underline{\mathbf{R}}}\big\{\lambda\left[\underline{\mathbf{k}}^* \cdot \underline{\hat{\mathbf{u}}}^*(\underline{\mathbf{k}})\underline{\hat{\mathbf{u}}}(\underline{\mathbf{k}}) + \underline{\mathbf{k}} \cdot \underline{\hat{\mathbf{u}}}(\underline{\mathbf{k}})\underline{\hat{\mathbf{u}}}^*(\underline{\mathbf{k}})\right] + \mu(\underline{\mathbf{k}} + \underline{\mathbf{k}}^*) \\ &\qquad + \mu\left[\underline{\mathbf{k}}^* \cdot \underline{\hat{\mathbf{u}}}(\underline{\mathbf{k}})\underline{\hat{\mathbf{u}}}^*(\underline{\mathbf{k}}) + \underline{\mathbf{k}} \cdot \underline{\hat{\mathbf{u}}}^*(\underline{\mathbf{k}})\underline{\hat{\mathbf{u}}}(\underline{\mathbf{k}})\right]\big\} \end{aligned} \tag{8.182}$$

for isotropic materials; we have incorporated $\underline{\hat{\mathbf{u}}}(\underline{\mathbf{k}}) \cdot \underline{\hat{\mathbf{u}}}^*(\underline{\mathbf{k}}) = 1$.

With (8.134), we similarly obtain for the time averaged energy density:

$$\begin{aligned} &\langle w(\underline{\mathbf{R}}, t, \omega_0, \underline{\mathbf{k}})\rangle \\ &\quad = \frac{1}{4}\,|u(\omega_0)|^2 \mathrm{e}^{-2\Im\underline{\mathbf{k}}\cdot\underline{\mathbf{R}}}\big[\varrho\omega_0^2 + \lambda\underline{\hat{\mathbf{u}}}(\underline{\mathbf{k}})\underline{\hat{\mathbf{u}}}^*(\underline{\mathbf{k}}) : \underline{\mathbf{k}}^*\underline{\mathbf{k}} + \mu\underline{\mathbf{k}} \cdot \underline{\mathbf{k}}^* \\ &\qquad + \mu\underline{\hat{\mathbf{u}}}^*(\underline{\mathbf{k}})\underline{\hat{\mathbf{u}}}(\underline{\mathbf{k}}) : \underline{\mathbf{k}}^*\underline{\mathbf{k}}\big]. \end{aligned} \tag{8.183}$$

Obviously, the (time averaged) potential and kinetic energy densities are no longer equal for complex wave number vectors, be it for inhomogeneous plane waves in nondissipative or for homogeneous/inhomogeneous plane waves in dissipative materials.[153]

[153] This is also true for electromagnetic waves (Chen 1983).

Specialization of (8.182) and (8.183) to pressure waves results through insertion of (8.159):

$$
\begin{aligned}
&\Re \underline{\mathbf{S}}_{\mathrm{KP}}(\underline{\mathbf{R}}, \omega_0, \underline{\mathbf{k}}_\mathrm{P}) \\
&\quad = \frac{\omega_0}{4} |u(\omega_0)|^2 \mathrm{e}^{-2\Im \underline{\mathbf{k}}_\mathrm{P} \cdot \underline{\mathbf{R}}} \\
&\qquad \times \frac{\lambda \big(\underline{\mathbf{k}}_\mathrm{P}^* \cdot \underline{\mathbf{k}}_\mathrm{P}^* \underline{\mathbf{k}}_\mathrm{P} + \underline{\mathbf{k}}_\mathrm{P} \cdot \underline{\mathbf{k}}_\mathrm{P} \underline{\mathbf{k}}_\mathrm{P}^*\big) + 2\mu \underline{\mathbf{k}}_\mathrm{P} \cdot \underline{\mathbf{k}}_\mathrm{P}^* \big(\underline{\mathbf{k}}_\mathrm{P} + \underline{\mathbf{k}}_\mathrm{P}^*\big)}{\underline{\mathbf{k}}_\mathrm{P} \cdot \underline{\mathbf{k}}_\mathrm{P}^*},
\end{aligned}
\tag{8.184}
$$

$$
\begin{aligned}
&\langle w_\mathrm{P}(\underline{\mathbf{R}}, t, \omega_0, \underline{\mathbf{k}}_\mathrm{P}) \rangle \\
&\quad = \frac{1}{4} |u(\omega_0)|^2 \mathrm{e}^{-2\Im \underline{\mathbf{k}}_\mathrm{P} \cdot \underline{\mathbf{R}}} \\
&\qquad \times \frac{\lambda \big(k_\mathrm{P}^2 \underline{\mathbf{k}}_\mathrm{P} \cdot \underline{\mathbf{k}}_\mathrm{P}^* + \underline{\mathbf{k}}_\mathrm{P} \underline{\mathbf{k}}_\mathrm{P}^* : \underline{\mathbf{k}}_\mathrm{P}^* \underline{\mathbf{k}}_\mathrm{P}\big) + 2\mu \underline{\mathbf{k}}_\mathrm{P} \cdot \underline{\mathbf{k}}_\mathrm{P}^* \big(k_\mathrm{P}^2 + \underline{\mathbf{k}}_\mathrm{P} \cdot \underline{\mathbf{k}}_\mathrm{P}^*\big)}{\underline{\mathbf{k}}_\mathrm{P} \cdot \underline{\mathbf{k}}_\mathrm{P}^*}.
\end{aligned}
\tag{8.185}
$$

Up to now, we have not yet used dispersion relations; yet, for inhomogeneous plane pressure waves in nondissipative materials, we search complex wave number vectors $\underline{\mathbf{k}}_\mathrm{P}$, whose scalar product with themselves is real valued; hence, we have $\underline{\mathbf{k}}_\mathrm{P} \cdot \underline{\mathbf{k}}_\mathrm{P} = k_\mathrm{P}^2$ as well as $\underline{\mathbf{k}}_\mathrm{P}^* \cdot \underline{\mathbf{k}}_\mathrm{P}^* = k_\mathrm{P}^2$. Complementing (8.131) and (8.137), we obtain

$$
\Re \underline{\mathbf{S}}_{\mathrm{KP}}(\underline{\mathbf{R}}, \omega_0, \hat{\underline{\mathbf{k}}}_\mathrm{P}) = \frac{\omega_0^2}{2c_\mathrm{P}} |u(\omega_0)|^2 \mathrm{e}^{-2k_\mathrm{P} \Im \hat{\underline{\mathbf{k}}}_\mathrm{P} \cdot \underline{\mathbf{R}}} \, \frac{\lambda + 2\mu \hat{\underline{\mathbf{k}}}_\mathrm{P} \cdot \hat{\underline{\mathbf{k}}}_\mathrm{P}^*}{\hat{\underline{\mathbf{k}}}_\mathrm{P} \cdot \hat{\underline{\mathbf{k}}}_\mathrm{P}^*} \, \Re \hat{\underline{\mathbf{k}}}_\mathrm{P},
\tag{8.186}
$$

$$
\langle w_\mathrm{P}(\underline{\mathbf{R}}, t, \omega_0, \hat{\underline{\mathbf{k}}}_\mathrm{P}) \rangle = \frac{\omega_0^2}{4c_\mathrm{P}^2} |u(\omega_0)|^2 \mathrm{e}^{-2k_\mathrm{P} \Im \hat{\underline{\mathbf{k}}}_\mathrm{P} \cdot \underline{\mathbf{R}}} \, \frac{\lambda + 2\mu \hat{\underline{\mathbf{k}}}_\mathrm{P} \cdot \hat{\underline{\mathbf{k}}}_\mathrm{P}^*}{\hat{\underline{\mathbf{k}}}_\mathrm{P} \cdot \hat{\underline{\mathbf{k}}}_\mathrm{P}^*} \left(1 + \hat{\underline{\mathbf{k}}}_\mathrm{P} \cdot \hat{\underline{\mathbf{k}}}_\mathrm{P}^*\right)
\tag{8.187}
$$

and, therefore,

$$
\underline{\mathbf{c}}_{\mathrm{EP}}(\hat{\underline{\mathbf{k}}}_\mathrm{P}) = \frac{2c_\mathrm{P}}{1 + \hat{\underline{\mathbf{k}}}_\mathrm{P} \cdot \hat{\underline{\mathbf{k}}}_\mathrm{P}^*} \, \Re \hat{\underline{\mathbf{k}}}_\mathrm{P};
\tag{8.188}
$$

the energy flux is perpendicular to the phase surfaces as it is true for homogeneous plane waves. But not only the directions of phase and energy velocity are equal, they have also equal magnitudes because we find $|\underline{\mathbf{c}}_{\mathrm{EP}}(\hat{\underline{\mathbf{k}}}_\mathrm{P})| = c_\mathrm{P}(\hat{\underline{\mathbf{k}}}_\mathrm{P})$ with (8.172), (8.177), and (8.176) by taking the magnitude of (8.188).

Because of the occurrence of complex and conjugate complex unit vectors in (8.182) and (8.183), we can anticipate that we will obtain different expressions for S2/SH- and S1/SV-polarizations. For the real valued $\hat{\underline{\mathbf{u}}}_{\mathrm{S2}}$-vector and for the complex valued $\hat{\underline{\mathbf{u}}}_{\mathrm{S1}}(\underline{\mathbf{k}}_\mathrm{S})$-vector, we have $\underline{\mathbf{k}}_\mathrm{S} \cdot \hat{\underline{\mathbf{u}}}_{\mathrm{S2/S1}} = 0$ and $\underline{\mathbf{k}}_\mathrm{S}^* \cdot \hat{\underline{\mathbf{u}}}_{\mathrm{S2/S1}}^* = 0$; for $\hat{\underline{\mathbf{u}}}_{\mathrm{S2}}$, we additionally have $\underline{\mathbf{k}}_\mathrm{S}^* \cdot \hat{\underline{\mathbf{u}}}_{\mathrm{S2}} = 0$, reducing (8.182) and (8.183) to

$$
\Re \underline{\mathbf{S}}_{\mathrm{KS2}}(\underline{\mathbf{R}}, \omega_0, \underline{\mathbf{k}}_\mathrm{S}) = \frac{\omega_0}{2} \, \mu \, |u(\omega_0)|^2 \mathrm{e}^{-2\Im \underline{\mathbf{k}}_\mathrm{S} \cdot \underline{\mathbf{R}}} \, \Re \underline{\mathbf{k}}_\mathrm{S},
\tag{8.189}
$$

$$
\begin{aligned}
\langle w_{\mathrm{S2}}(\underline{\mathbf{R}}, t, \omega_0, \underline{\mathbf{k}}_\mathrm{S}) \rangle &= \frac{1}{4} |u(\omega_0)|^2 \mathrm{e}^{-2\Im \underline{\mathbf{k}}_\mathrm{S} \cdot \underline{\mathbf{R}}} \left(\varrho \omega_0^2 + \mu \underline{\mathbf{k}}_\mathrm{S} \cdot \underline{\mathbf{k}}_\mathrm{S}^*\right) \\
&= \frac{1}{4} \, \mu \, |u(\omega_0)|^2 \mathrm{e}^{-2\Im \underline{\mathbf{k}}_\mathrm{S} \cdot \underline{\mathbf{R}}} \left(k_\mathrm{S}^2 + \underline{\mathbf{k}}_\mathrm{S} \cdot \underline{\mathbf{k}}_\mathrm{S}^*\right).
\end{aligned}
\tag{8.190}
$$

With a definition similar to (8.177)

$$\hat{\underline{\mathbf{k}}}_S = \frac{\underline{\mathbf{k}}_S}{k_S} \tag{8.191}$$

it follows:

$$\Re\underline{\mathbf{S}}_{KS2}(\underline{\mathbf{R}}, \omega_0, \underline{\mathbf{k}}_S) = \frac{\varrho\omega_0^2}{2}\, c_S\, |u(\omega_0)|^2 e^{-2k_S \Im \hat{\underline{\mathbf{k}}}_S \underline{\mathbf{R}}}\, \Re\hat{\underline{\mathbf{k}}}_S, \tag{8.192}$$

$$\langle w_{S2}(\underline{\mathbf{R}}, t, \omega_0, \underline{\mathbf{k}}_S)\rangle = \frac{\varrho\omega_0^2}{4}\, |u(\omega_0)|^2 e^{-2k_S \Im \hat{\underline{\mathbf{k}}}_S \cdot \underline{\mathbf{R}}} \left(1 + \hat{\underline{\mathbf{k}}}_S \cdot \hat{\underline{\mathbf{k}}}_S^*\right). \tag{8.193}$$

We obtain

$$\underline{\mathbf{c}}_{ES2}(\hat{\underline{\mathbf{k}}}_S) = \frac{2c_S}{1 + \hat{\underline{\mathbf{k}}}_S \cdot \hat{\underline{\mathbf{k}}}_S^*}\, \Re\hat{\underline{\mathbf{k}}}_S \tag{8.194}$$

with—as above—$|\underline{\mathbf{c}}_{ES2}(\hat{\underline{\mathbf{k}}}_S)| = c_{S2}(\hat{\underline{\mathbf{k}}}_S)$ as energy velocity. The expressions (8.189), (8.190) and (8.192), (8.193), respectively, as well as (8.194) hold for each real valued S2-polarization vector, therefore also for $\hat{\underline{\mathbf{u}}}_{S2} = \hat{\underline{\mathbf{u}}}_{SH} = \underline{\mathbf{e}}_y$.

With the identity

$$\begin{aligned}(\underline{\mathbf{A}} \times \underline{\mathbf{B}})(\underline{\mathbf{C}} \times \underline{\mathbf{D}}) = (\underline{\mathbf{A}} \times \underline{\mathbf{B}}) \cdot (\underline{\mathbf{C}} \times \underline{\mathbf{D}})\, \underline{\underline{\mathbf{I}}} + (\underline{\mathbf{A}} \cdot \underline{\mathbf{D}})\underline{\mathbf{C}}\,\underline{\mathbf{B}} + (\underline{\mathbf{B}} \cdot \underline{\mathbf{C}})\underline{\mathbf{D}}\,\underline{\mathbf{A}}\\ - (\underline{\mathbf{A}} \cdot \underline{\mathbf{C}})\underline{\mathbf{D}}\,\underline{\mathbf{B}} - (\underline{\mathbf{B}} \cdot \underline{\mathbf{D}})\underline{\mathbf{C}}\,\underline{\mathbf{A}},\end{aligned} \tag{8.195}$$

we compute

$$\begin{aligned}\Re\underline{\mathbf{S}}_{KS1}(\underline{\mathbf{R}}, \omega_0, \underline{\mathbf{k}}_S) &= \frac{1}{4}\, \mu\, |u(\omega_0)|^2 e^{-2\Im \underline{\mathbf{k}}_S \cdot \underline{\mathbf{R}}}\\ &\quad\times \frac{2\underline{\mathbf{k}}_S \cdot \underline{\mathbf{k}}_S^*\, (\underline{\mathbf{k}}_S + \underline{\mathbf{k}}_S^*) - (\underline{\mathbf{k}}_S^* \cdot \underline{\mathbf{k}}_S^* \underline{\mathbf{k}}_S + \underline{\mathbf{k}}_S \cdot \underline{\mathbf{k}}_S \underline{\mathbf{k}}_S^*)}{\underline{\mathbf{k}}_S \cdot \underline{\mathbf{k}}_S^*},\end{aligned} \tag{8.196}$$

$$\begin{aligned}\langle w_{S1}(\underline{\mathbf{R}}, t, \omega_0, \underline{\mathbf{k}}_S)\rangle &= \frac{1}{4}\, \mu\, |u(\omega_0)|^2 e^{-\Im \underline{\mathbf{k}}_S \cdot \underline{\mathbf{R}}}\\ &\quad\times \frac{\underline{\mathbf{k}}_S \cdot \underline{\mathbf{k}}_S^*\, (k_S^2 + \underline{\mathbf{k}}_S \cdot \underline{\mathbf{k}}_S^*) + (\underline{\mathbf{k}}_S \cdot \underline{\mathbf{k}}_S^*)^2 - \underline{\mathbf{k}}_S \underline{\mathbf{k}}_S^* : \underline{\mathbf{k}}_S^* \underline{\mathbf{k}}_S}{\underline{\mathbf{k}}_S \cdot \underline{\mathbf{k}}_S^*}\end{aligned} \tag{8.197}$$

for the S1-polarization. With the dispersion relation of nondissipative materials and the definition (8.191), we further have

$$\Re\underline{\mathbf{S}}_{KS1}(\underline{\mathbf{R}}, \omega_0, \underline{\mathbf{k}}_S) = \frac{\varrho\omega_0^2}{2}\, c_S\, |u(\omega_0)|^2 e^{-2k_S \Im \hat{\underline{\mathbf{k}}}_S \cdot \underline{\mathbf{R}}}\, \frac{2\hat{\underline{\mathbf{k}}}_S \cdot \hat{\underline{\mathbf{k}}}_S^* - 1}{\hat{\underline{\mathbf{k}}}_S \cdot \hat{\underline{\mathbf{k}}}_S^*}\, \Re\hat{\underline{\mathbf{k}}}_S, \tag{8.198}$$

$$\langle w_{S1}(\underline{\mathbf{R}}, t, \omega_0, \underline{\mathbf{k}}_S)\rangle = \frac{\varrho\omega_0^2}{4}\, |u(\omega_0)|^2 e^{-2k_S \Im \hat{\underline{\mathbf{k}}}_S \cdot \underline{\mathbf{R}}}\, \frac{(1 + \hat{\underline{\mathbf{k}}}_S \cdot \hat{\underline{\mathbf{k}}}_S^*)(2\hat{\underline{\mathbf{k}}}_S \cdot \hat{\underline{\mathbf{k}}}_S^* - 1)}{\hat{\underline{\mathbf{k}}}_S \cdot \hat{\underline{\mathbf{k}}}_S^*}. \tag{8.199}$$

Therefore, energy density and energy flux density are different for real and complex S-polarization vectors. Yet, in the expressions

$$\underline{\mathbf{c}}_{\mathrm{ES1}}(\hat{\underline{\mathbf{k}}}_{\mathrm{S}}) = \frac{2c_{\mathrm{S}}}{1 + \hat{\underline{\mathbf{k}}}_{\mathrm{S}} \cdot \hat{\underline{\mathbf{k}}}_{\mathrm{S}}^{*}} \Re \hat{\underline{\mathbf{k}}}_{\mathrm{S}} \tag{8.200}$$

for the energy density, these differences cancel; we certainly have $|\underline{\mathbf{c}}_{\mathrm{ES1}}(\hat{\underline{\mathbf{k}}}_{\mathrm{S}})| = c_{\mathrm{S1}}(\hat{\underline{\mathbf{k}}}_{\mathrm{S}})$.

Specializing $\hat{\underline{\mathbf{u}}}_{\mathrm{S1}} = \hat{\underline{\mathbf{u}}}_{\mathrm{SV}}$ with $\hat{\underline{\mathbf{u}}}_{\mathrm{SV}} \cdot \underline{\mathbf{e}}_y = 0$ does not change the results.

8.3 Plane Waves in Anisotropic Nondissipative Materials

In this section, we first treat the important case of homogeneous plane waves; inhomogeneous plane waves (in nondissipative) materials are dealt with when they physically appear with the reflection/transmission at the plane boundary between isotropic and anisotropic (transversely isotropic) half-spaces (Section 9.3).

8.3.1 Plane waves in anisotropic materials

Wave tensor: The spectral particle displacement $\underline{\mathbf{u}}(\underline{\mathbf{R}}, \omega)$ satisfies the Fourier transformed homogeneous wave equation (7.15) for homogeneous anisotropic materials:

$$\nabla \cdot \underline{\underline{\underline{\underline{\mathbf{c}}}}} : \nabla \underline{\mathbf{u}}(\underline{\mathbf{R}}, \omega) + \rho\omega^2 \underline{\mathbf{u}}(\underline{\mathbf{R}}, \omega) = \underline{\mathbf{0}}. \tag{8.201}$$

Similar to (8.54), we make the solution *ansatz* of homogeneous plane waves

$$\underline{\mathbf{u}}(\underline{\mathbf{R}}, \omega) \Longrightarrow \underline{\mathbf{u}}(\underline{\mathbf{R}}, \omega, \hat{\underline{\mathbf{k}}}) = \underline{\mathbf{u}}(\omega, \hat{\underline{\mathbf{k}}})\, \mathrm{e}^{\pm \mathrm{j} \frac{\hat{\underline{\mathbf{k}}} \cdot \underline{\mathbf{R}}}{c(\hat{\underline{\mathbf{k}}})} \omega}, \tag{8.202}$$

where $\hat{\underline{\mathbf{k}}}$ is the (real) unit vector of the (phase) propagation direction; it will turn out that the phase velocity $c(\hat{\underline{\mathbf{k}}})$ explicitly depends on that vector as it is true for the polarization vector $\underline{\mathbf{u}}(\omega, \hat{\underline{\mathbf{k}}})$; we factorize as in (8.55):

$$\underline{\mathbf{u}}(\omega, \hat{\underline{\mathbf{k}}}) = \hat{\underline{\mathbf{u}}}(\hat{\underline{\mathbf{k}}}) u(\omega). \tag{8.203}$$

With the *ansatz* (8.202), we obtain the counterpart to (8.61) for homogeneous anisotropic materials from (8.201):

$$\left[\frac{1}{\rho} \hat{\underline{\mathbf{k}}} \cdot \underline{\underline{\underline{\underline{\mathbf{c}}}}} \cdot \hat{\underline{\mathbf{k}}} - c^2(\hat{\underline{\mathbf{k}}}) \underline{\underline{\mathbf{I}}} \right] \cdot \omega^2 \underline{\mathbf{u}}(\underline{\mathbf{R}}, \omega, \hat{\underline{\mathbf{k}}}) = \underline{\mathbf{0}}. \tag{8.204}$$

Consequently, the calculation of the phase velocity and the polarization of homogeneous plane waves in homogeneous anisotropic materials turns out to be an eigenvalue problem

$$\underline{\underline{\mathbf{D}}}^{\text{aniso}}(\hat{\underline{\mathbf{k}}}) \cdot \hat{\underline{\mathbf{u}}}(\hat{\underline{\mathbf{k}}}) = c^2(\hat{\underline{\mathbf{k}}})\hat{\underline{\mathbf{u}}}(\hat{\underline{\mathbf{k}}}), \tag{8.205}$$

where

$$\underline{\underline{\mathbf{D}}}^{\text{aniso}}(\hat{\underline{\mathbf{k}}}) = \frac{1}{\rho}\hat{\underline{\mathbf{k}}} \cdot \underline{\underline{\underline{\underline{\mathbf{c}}}}} \cdot \hat{\underline{\mathbf{k}}} \tag{8.206}$$

is the real symmetric Kelvin–Christoffel tensor due to the symmetries (4.15) and (4.17).

The eigenvalues $c^2(\hat{\underline{\mathbf{k}}})$ must be positive to ensure real phase velocities $c(\hat{\underline{\mathbf{k}}})$; therefore, $\underline{\underline{\mathbf{D}}}^{\text{aniso}}(\hat{\underline{\mathbf{k}}})$ must be positive definite. Hence, $\underline{\mathbf{R}} \cdot \underline{\underline{\mathbf{D}}}^{\text{aniso}}(\hat{\underline{\mathbf{k}}}) \cdot \underline{\mathbf{R}}$ must be larger than zero and equal to zero only if $\underline{\mathbf{R}} = \underline{\mathbf{0}}$. This requirement is carried over to the stiffness tensor due to the definition (8.206) of $\underline{\underline{\mathbf{D}}}^{\text{aniso}}(\hat{\underline{\mathbf{k}}})$: $\underline{\mathbf{R}}\hat{\underline{\mathbf{k}}} : \underline{\underline{\underline{\underline{\mathbf{c}}}}} : \hat{\underline{\mathbf{k}}}\underline{\mathbf{R}} = \underline{\mathbf{R}}\hat{\underline{\mathbf{k}}} : \underline{\underline{\underline{\underline{\mathbf{c}}}}} : \underline{\mathbf{R}}\hat{\underline{\mathbf{k}}}$ must be larger than zero and equal to zero only if $\underline{\mathbf{R}} = \underline{\mathbf{0}}$, that is to say, $\underline{\mathbf{R}}\hat{\underline{\mathbf{k}}} = \underline{\underline{\mathbf{0}}}$. Since $\frac{1}{2}\underline{\underline{\mathbf{S}}}(\underline{\mathbf{R}}, t) : \underline{\underline{\underline{\underline{\mathbf{c}}}}} : \underline{\underline{\mathbf{S}}}(\underline{\mathbf{R}}, t)$ has the meaning of the elastodynamic potential deformation energy density for arbitrary time-dependent fields due to (4.36), this requirement is guaranteed. For homogeneous isotropic materials, we have explicitly confirmed the positive definiteness $\underline{\underline{\mathbf{D}}}(\hat{\underline{\mathbf{k}}})$ with (8.64).

To calculate the eigenvalues $c^2(\hat{\underline{\mathbf{k}}})$, the determinant of the wave tensor

$$\underline{\underline{\mathbf{W}}}^{\text{aniso}}(\hat{\underline{\mathbf{k}}}, c^2) = \hat{\underline{\mathbf{k}}} \cdot \underline{\underline{\underline{\underline{\mathbf{c}}}}} \cdot \hat{\underline{\mathbf{k}}} - \rho\, c^2(\hat{\underline{\mathbf{k}}})\,\underline{\underline{\mathbf{I}}} \tag{8.207}$$

must be equal to zero because the eigenvectors $\hat{\underline{\mathbf{u}}}(\hat{\underline{\mathbf{k}}})$ are solutions of the homogeneous system of equations (of the Kelvin–Christoffel equation)

$$\left[\underline{\underline{\mathbf{D}}}^{\text{aniso}}(\hat{\underline{\mathbf{k}}}) - c^2(\hat{\underline{\mathbf{k}}})\,\underline{\underline{\mathbf{I}}}\right] \cdot \hat{\underline{\mathbf{u}}}(\hat{\underline{\mathbf{k}}}) = \underline{\mathbf{0}} \iff \underbrace{\left[\hat{\underline{\mathbf{k}}} \cdot \underline{\underline{\underline{\underline{\mathbf{c}}}}} \cdot \hat{\underline{\mathbf{k}}} - \rho\, c^2(\hat{\underline{\mathbf{k}}})\,\underline{\underline{\mathbf{I}}}\right]}_{= \underline{\underline{\mathbf{W}}}^{\text{aniso}}(\hat{\underline{\mathbf{k}}}, c^2)} \cdot \hat{\underline{\mathbf{u}}}(\hat{\underline{\mathbf{k}}}) = \underline{\mathbf{0}}. \tag{8.208}$$

To calculate the eigenvectors, we need the adjoint of $\underline{\underline{\mathbf{W}}}^{\text{aniso}}(\hat{\underline{\mathbf{k}}}, c^2)$; the calculation of the determinant and the adjoint requires specification of the anisotropy through $\underline{\underline{\underline{\underline{\mathbf{c}}}}}$.

Nevertheless, there are some general results even for the arbitrary $\underline{\underline{\underline{\underline{\mathbf{c}}}}}$-case that will be discussed in the following (Auld 1973; Helbig 1994; Royer and Dieulesaint 2000; Snieder 2002).

Energy velocities: With (8.139) and (8.140), respectively, we defined energy velocity vectors of plane elastic wave modes; we cite

$$\underline{\mathbf{c}}_{\text{E}}(\hat{\underline{\mathbf{k}}}) = \frac{\Re\{\underline{\mathbf{S}}_{\text{K}}(\underline{\mathbf{R}}, \omega, \hat{\underline{\mathbf{k}}})\}}{\langle w_{\text{el}}(\underline{\mathbf{R}}, t, \hat{\underline{\mathbf{k}}})\rangle}. \tag{8.209}$$

Here, $\Re\{\underline{\mathbf{S}}_{\mathrm{K}}(\underline{\mathbf{R}}, \omega, \hat{\underline{\mathbf{k}}})\}$ is the real part of the complex elastodynamic Poynting vector and $\langle w_{\mathrm{el}}(\underline{\mathbf{R}}, t, \hat{\underline{\mathbf{k}}})\rangle$ is the time averaged elastodynamic energy density of time harmonic plane waves with circular frequency ω, hence according to (8.128) and (8.134), respectively:

$$\underline{\mathbf{S}}_{\mathrm{K}}(\underline{\mathbf{R}}, \omega, \hat{\underline{\mathbf{k}}}) = \frac{\mathrm{j}\omega}{2}\, \underline{\mathbf{u}}(\underline{\mathbf{R}}, \omega, \hat{\underline{\mathbf{k}}}) \cdot \underline{\underline{\mathbf{T}}}^*(\underline{\mathbf{R}}, \omega, \hat{\underline{\mathbf{k}}}), \tag{8.210}$$

$$\langle w_{\mathrm{el}}(\underline{\mathbf{R}}, t, \hat{\underline{\mathbf{k}}})\rangle = \frac{\rho\omega^2}{4}\, \underline{\mathbf{u}}(\underline{\mathbf{R}}, \omega, \hat{\underline{\mathbf{k}}}) \cdot \underline{\mathbf{u}}^*(\underline{\mathbf{R}}, \omega, \hat{\underline{\mathbf{k}}}) + \frac{1}{4}\, \underline{\underline{\mathbf{S}}}(\underline{\mathbf{R}}, \omega, \hat{\underline{\mathbf{k}}}) : \underline{\underline{\underline{\underline{\mathbf{c}}}}} : \underline{\underline{\mathbf{S}}}^*(\underline{\mathbf{R}}, \omega, \hat{\underline{\mathbf{k}}}). \tag{8.211}$$

The *ansatz* (8.202) of homogeneous plane waves with (8.203) leads to

$$\underline{\underline{\mathbf{S}}}(\underline{\mathbf{R}}, \omega, \hat{\underline{\mathbf{k}}}) = \frac{\mathrm{j}\omega}{c(\hat{\underline{\mathbf{k}}})}\, u(\omega)\, \mathrm{e}^{\mathrm{j}\frac{\omega}{c(\hat{\underline{\mathbf{k}}})}\hat{\underline{\mathbf{k}}}\cdot\underline{\mathbf{R}}}\, \frac{1}{2}\left[\hat{\underline{\mathbf{k}}}\,\hat{\underline{\mathbf{u}}}(\hat{\underline{\mathbf{k}}}) + \hat{\underline{\mathbf{u}}}(\hat{\underline{\mathbf{k}}})\hat{\underline{\mathbf{k}}}\right], \tag{8.212}$$

and consequently, we have on behalf of the $\underline{\underline{\underline{\underline{\mathbf{c}}}}}$-symmetries:

$$\begin{aligned}
\underline{\mathbf{S}}_{\mathrm{K}}(\underline{\mathbf{R}}, \omega, \hat{\underline{\mathbf{k}}}) &= \frac{\omega^2}{2c(\hat{\underline{\mathbf{k}}})}\, |u(\omega)|^2\, \hat{\underline{\mathbf{u}}}(\hat{\underline{\mathbf{k}}}) \cdot \underline{\underline{\underline{\underline{\mathbf{c}}}}} : \hat{\underline{\mathbf{k}}}\,\hat{\underline{\mathbf{u}}}(\hat{\underline{\mathbf{k}}}) \\
&= \frac{\omega^2}{2c(\hat{\underline{\mathbf{k}}})}\, |u(\omega)|^2\, \underline{\underline{\underline{\underline{\mathbf{c}}}}} \,\vdots\, \hat{\underline{\mathbf{k}}}\,\hat{\underline{\mathbf{u}}}(\hat{\underline{\mathbf{k}}})\hat{\underline{\mathbf{u}}}(\hat{\underline{\mathbf{k}}}),
\end{aligned} \tag{8.213}$$

$$\langle w_{\mathrm{el}}(\underline{\mathbf{R}}, t, \hat{\underline{\mathbf{k}}})\rangle = \frac{\rho\omega^2}{4}\, |u(\omega)|^2 + \frac{\omega^2}{4c^2(\hat{\underline{\mathbf{k}}})}\, |u(\omega)|^2\, \hat{\underline{\mathbf{u}}}(\hat{\underline{\mathbf{k}}})\hat{\underline{\mathbf{k}}} : \underline{\underline{\underline{\underline{\mathbf{c}}}}} : \hat{\underline{\mathbf{k}}}\,\hat{\underline{\mathbf{u}}}(\hat{\underline{\mathbf{k}}}). \tag{8.214}$$

The double contraction of (8.205) with $\hat{\underline{\mathbf{u}}}(\hat{\underline{\mathbf{k}}})$ and recognizing (8.206) leads to

$$\hat{\underline{\mathbf{u}}}(\hat{\underline{\mathbf{k}}})\hat{\underline{\mathbf{k}}} : \underline{\underline{\underline{\underline{\mathbf{c}}}}} : \hat{\underline{\mathbf{k}}}\,\hat{\underline{\mathbf{u}}}(\hat{\underline{\mathbf{k}}}) = \rho c^2(\hat{\underline{\mathbf{k}}}), \tag{8.215}$$

thus simplifying the expression (8.214):

$$\langle w_{\mathrm{el}}(\underline{\mathbf{R}}, t, \hat{\underline{\mathbf{k}}})\rangle = \frac{\rho\omega^2}{2}\, |u(\omega)|^2. \tag{8.216}$$

For homogeneous plane waves, the time averaged kinetic energy density is equal to the time averaged deformation energy density[154] (Royer and Dieulesaint 2000). As a consequence, we obtain the calculation instruction[155]

$$\underline{\mathbf{c}}_{\mathrm{E}}(\hat{\underline{\mathbf{k}}}) = \frac{1}{\rho c(\hat{\underline{\mathbf{k}}})}\, \underline{\underline{\underline{\underline{\mathbf{c}}}}} \,\vdots\, \hat{\underline{\mathbf{k}}}\,\hat{\underline{\mathbf{u}}}(\hat{\underline{\mathbf{k}}})\hat{\underline{\mathbf{u}}}(\hat{\underline{\mathbf{k}}}) \tag{8.217}$$

[154]This is not true for inhomogeneous plane waves (compare inhomogeneous electromagnetic waves: Chen 1983) in either nondissipative or dissipative materials: compare Equations 9.156 and 9.159, respectively, for isotropic nondissipative and (9.346) and (9.349), respectively, for anisotropic nondissipative materials as well as Section 8.4 for isotropic dissipative materials.

[155]For energetically evanescent waves, we have (9.349).

for the energy velocity vectors of respective wave modes if the corresponding polarization vectors are known.

The combination of (8.217) and (8.215) yields the result

$$\underline{\mathrm{c}}_{\mathrm{E}}(\hat{\underline{\mathbf{k}}}) \cdot \hat{\underline{\mathbf{k}}} = c(\hat{\underline{\mathbf{k}}}), \tag{8.218}$$

that is to say, the magnitude of the energy velocity vector is always greater than or equal to the phase velocity of the wave mode under concern: $c_{\mathrm{E}}(\hat{\underline{\mathbf{k}}}) \geq c(\hat{\underline{\mathbf{k}}})$. The equality sign exactly holds if energy and phase velocity vectors $\underline{\mathbf{c}}(\hat{\underline{\mathbf{k}}}) = c(\hat{\underline{\mathbf{k}}})\hat{\underline{\mathbf{k}}}$ are parallel as it is always the case in isotropic materials according to (8.144) but only exceptionally in anisotropic materials.

Defining a ray vector according to

$$\underline{\mathbf{l}}(\hat{\underline{\mathbf{k}}}) = \frac{\underline{\mathbf{c}}_{\mathrm{E}}(\hat{\underline{\mathbf{k}}})}{\omega} \tag{8.219}$$

—the ultrasonic ray points into the direction of energy flux!—and contracting it with the phase vector

$$\underline{\mathbf{k}} = \frac{\omega}{c(\hat{\underline{\mathbf{k}}})}\,\hat{\underline{\mathbf{k}}} \tag{8.220}$$

results in

$$\underline{\mathbf{l}}(\hat{\underline{\mathbf{k}}}) \cdot \underline{\mathbf{k}} = 1. \tag{8.221}$$

The dependence of the phase velocity upon $\hat{\underline{\mathbf{k}}}$ is the reason that the ray vector $\underline{\mathbf{l}}(\hat{\underline{\mathbf{k}}})$ does not point into the direction of $\hat{\underline{\mathbf{k}}}$; namely, taking the $\nabla_{\hat{\underline{\mathbf{k}}}}$-gradient of the eigenvalue equation[156] (8.208)—we have $\nabla_{\hat{\underline{\mathbf{k}}}}\hat{\underline{\mathbf{k}}} = \underline{\underline{\mathbf{I}}}$ and $\nabla_{\hat{\underline{\mathbf{k}}}}(\hat{\underline{\mathbf{k}}} \cdot \underline{\underline{\underline{\underline{\mathbf{c}}}}} \cdot \hat{\underline{\mathbf{k}}}) = 2\,\underline{\underline{\underline{\underline{\mathbf{c}}}}} \cdot \hat{\underline{\mathbf{k}}}$—

$$\begin{aligned}
\nabla_{\hat{\underline{\mathbf{k}}}}\left\{\left[\hat{\underline{\mathbf{k}}} \cdot \underline{\underline{\underline{\underline{\mathbf{c}}}}} \cdot \hat{\underline{\mathbf{k}}} - \rho\, c^2(\hat{\underline{\mathbf{k}}})\,\underline{\underline{\mathbf{I}}}\right] \cdot \hat{\underline{\mathbf{u}}}(\hat{\underline{\mathbf{k}}})\right\} &= \left\{\nabla_{\hat{\underline{\mathbf{k}}}}\left[\hat{\underline{\mathbf{k}}} \cdot \underline{\underline{\underline{\underline{\mathbf{c}}}}} \cdot \hat{\underline{\mathbf{k}}} - \rho\, c^2(\hat{\underline{\mathbf{k}}})\,\underline{\underline{\mathbf{I}}}\right]\right\} \cdot \hat{\underline{\mathbf{u}}}(\hat{\underline{\mathbf{k}}}) \\
&\quad + \left[\nabla_{\hat{\underline{\mathbf{k}}}}\hat{\underline{\mathbf{u}}}(\hat{\underline{\mathbf{k}}})\right] \cdot \left[\hat{\underline{\mathbf{k}}} \cdot \underline{\underline{\underline{\underline{\mathbf{c}}}}} \cdot \hat{\underline{\mathbf{k}}} - \rho\, c^2(\hat{\underline{\mathbf{k}}})\,\underline{\underline{\mathbf{I}}}\right]^{21} \\
&= 2\left[\underline{\underline{\underline{\underline{\mathbf{c}}}}} \cdot \hat{\underline{\mathbf{k}}} - \rho\, c(\hat{\underline{\mathbf{k}}})\,\nabla_{\hat{\underline{\mathbf{k}}}} c(\hat{\underline{\mathbf{k}}})\,\underline{\underline{\mathbf{I}}}\right] \cdot \hat{\underline{\mathbf{u}}}(\hat{\underline{\mathbf{k}}}) \\
&\quad + \left[\nabla_{\hat{\underline{\mathbf{k}}}}\hat{\underline{\mathbf{u}}}(\hat{\underline{\mathbf{k}}})\right] \cdot \left[\hat{\underline{\mathbf{k}}} \cdot \underline{\underline{\underline{\underline{\mathbf{c}}}}} \cdot \hat{\underline{\mathbf{k}}} - \rho\, c^2(\hat{\underline{\mathbf{k}}})\,\underline{\underline{\mathbf{I}}}\right] \\
&= \underline{\underline{\mathbf{0}}}
\end{aligned} \tag{8.222}$$

yields after right-contraction with $\hat{\underline{\mathbf{u}}}(\hat{\underline{\mathbf{k}}})$ and another utilization of the eigenvalue equation (8.208)

$$\underline{\underline{\underline{\underline{\mathbf{c}}}}} \vdots \hat{\underline{\mathbf{k}}}\,\hat{\underline{\mathbf{u}}}(\hat{\underline{\mathbf{k}}})\hat{\underline{\mathbf{u}}}(\hat{\underline{\mathbf{k}}}) = \rho\, c(\hat{\underline{\mathbf{k}}})\,\nabla_{\hat{\underline{\mathbf{k}}}} c(\hat{\underline{\mathbf{k}}}) \tag{8.223}$$

[156]This makes sense only if $\nabla_{\hat{\underline{\mathbf{k}}}} c(\hat{\underline{\mathbf{k}}}) \neq \underline{\mathbf{0}}$, that is to say, if $c(\hat{\underline{\mathbf{k}}})$ is actually a function of $\hat{\underline{\mathbf{k}}}$. In addition: The result of the calculation of Royer and Dieulesaint (2000) is correct but not complete; the authors ignore the dependence of the polarization vector $\hat{\underline{\mathbf{u}}}(\hat{\underline{\mathbf{k}}})$ upon $\hat{\underline{\mathbf{k}}}$.

with the consequence

$$\underline{\mathbf{c}}_{\mathrm{E}}(\hat{\underline{\mathbf{k}}}) = \nabla_{\hat{\underline{\mathbf{k}}}} c(\hat{\underline{\mathbf{k}}}) \tag{8.224}$$

due to (8.217). Result: Since we excluded the isotropic case with $c(\hat{\underline{\mathbf{k}}}) = c$, that is to say, $\underline{\mathbf{c}}_{\mathrm{E}} = c\hat{\underline{\mathbf{k}}}$ (Equation 8.144), we find that $\underline{\mathbf{c}}_{\mathrm{E}}(\hat{\underline{\mathbf{k}}})$ does not point into the direction of $\hat{\underline{\mathbf{k}}}$. Yet, there is a simple geometric way to find this direction. We take the $\nabla_{\hat{\underline{\mathbf{k}}}}$-gradient of the slowness vector

$$\begin{aligned}\underline{\mathbf{s}}(\hat{\underline{\mathbf{k}}}) &= s(\hat{\underline{\mathbf{k}}})\hat{\underline{\mathbf{k}}} \\ &= \frac{1}{c(\hat{\underline{\mathbf{k}}})}\,\hat{\underline{\mathbf{k}}}\end{aligned} \tag{8.225}$$

according to

$$\nabla_{\hat{\underline{\mathbf{k}}}}\underline{\mathbf{s}}(\hat{\underline{\mathbf{k}}}) = -\frac{1}{c^2(\hat{\underline{\mathbf{k}}})}\nabla_{\hat{\underline{\mathbf{k}}}} c(\hat{\underline{\mathbf{k}}})\,\hat{\underline{\mathbf{k}}} + \frac{1}{c(\hat{\underline{\mathbf{k}}})}\,\underline{\underline{\mathbf{I}}} \tag{8.226}$$

and find with (8.224) and (8.218)

$$\nabla_{\hat{\underline{\mathbf{k}}}}\underline{\mathbf{s}}(\hat{\underline{\mathbf{k}}}) \cdot \underline{\mathbf{c}}_{\mathrm{E}}(\hat{\underline{\mathbf{k}}}) = \underline{\mathbf{0}}. \tag{8.227}$$

On the other hand, the differential slowness vector

$$\underline{\mathbf{ds}}(\hat{\underline{\mathbf{k}}}) = \underline{\mathbf{d}\hat{\mathbf{k}}} \cdot \nabla_{\hat{\underline{\mathbf{k}}}}\underline{\mathbf{s}}(\hat{\underline{\mathbf{k}}}) \tag{8.228}$$

is tangential to the slowness surface according to Figure 8.13 $\underline{\mathbf{s}}(\hat{\underline{\mathbf{k}}})$ so that (8.227) implies

$$\underline{\mathbf{ds}}(\hat{\underline{\mathbf{k}}}) \cdot \underline{\mathbf{c}}_{\mathrm{E}}(\hat{\underline{\mathbf{k}}}) = 0. \tag{8.229}$$

The energy velocity vector $\underline{\mathbf{c}}_{\mathrm{E}}(\hat{\underline{\mathbf{k}}})$ is perpendicular to the slowness surface in each point characterized by $\hat{\underline{\mathbf{k}}}$. This result is extraordinarily important for US-NDT because an ultrasonic angle transducer designed for isotropic steel in terms of its aperture phase distribution may radiate into a completely different direction in anisotropic (austenitic) steel. Yet, to understand that intuitively, we must consider a finite aperture size of the transducer because, for infinitely extended wavefronts of plane waves, the different directions of $\hat{\underline{\mathbf{k}}}$ and $\underline{\mathbf{l}}(\hat{\underline{\mathbf{k}}})$ are not visible: Figure 8.14(a) shows plane wavefronts for two different times t_1 and $t_2 > t_1$; they are orthogonal to the phase propagation vector $\hat{\underline{\mathbf{k}}}$ and hence to the phase velocity vector $\underline{\mathbf{c}}(\hat{\underline{\mathbf{k}}}) = c(\hat{\underline{\mathbf{k}}})\hat{\underline{\mathbf{k}}}$. Per definition, the amplitude of a (homogeneous) plane wave is constant on its wavefront merely exhibiting a constant shift of the energy coming from infinity and drifting to infinity due to the direction of the energy velocity vector $\underline{\mathbf{c}}_{\mathrm{E}}(\hat{\underline{\mathbf{k}}})$ differing from the direction $\hat{\underline{\mathbf{k}}}$. Yet, turning to spatially (and timely) constrained wave packets

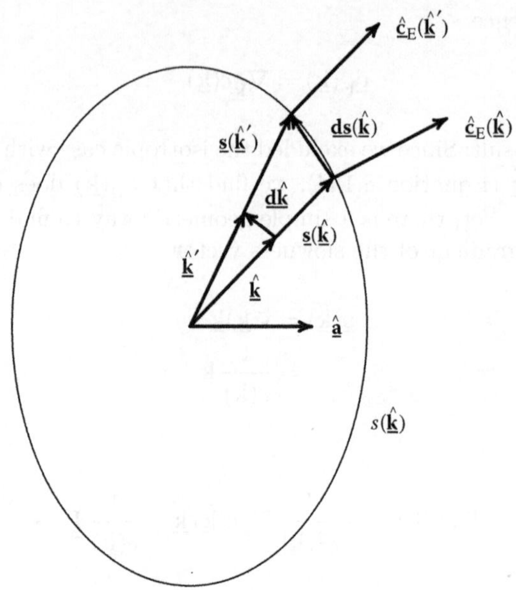

FIGURE 8.13
Orientation of the (unit) vector energy velocity relative to the slowness surface $s(\hat{\underline{\mathbf{k}}})$.

[Figure 8.14(b)], this energy shift becomes immediately visible. Hence, the propagation direction of an impulse is given by the energy velocity, i.e., the formal Fourier inversion of (8.202) makes no physical sense; instead, we must write

$$\underline{\mathbf{u}}(\underline{\mathbf{R}}, t, \hat{\underline{\mathbf{l}}}) = u\left(t \pm \frac{\hat{\underline{\mathbf{l}}} \cdot \underline{\mathbf{R}}}{c_{\mathrm{E}}(\hat{\underline{\mathbf{l}}})}\right) \hat{\underline{\mathbf{u}}}(\hat{\underline{\mathbf{l}}}). \tag{8.230}$$

The angle between $\underline{\mathbf{c}}_{\mathrm{E}}(\hat{\underline{\mathbf{k}}})$—or $\hat{\underline{\mathbf{l}}}(\hat{\underline{\mathbf{k}}})$—and $\hat{\underline{\mathbf{k}}}$ is called skewing angle.

By the way: The phase propagation vector $\hat{\underline{\mathbf{k}}}$ corresponding to a given ray vector $\underline{\mathbf{l}}(\hat{\underline{\mathbf{k}}})$ is perpendicular to the so-called wave surface originating from the end points of all ray vectors. For that purpose, we take the $\nabla_{\hat{\underline{\mathbf{l}}}}$-gradient of $\underline{\mathbf{c}}_{\mathrm{E}}(\hat{\underline{\mathbf{l}}})$, i.e., of the energy velocity vector with respect to the ray $\hat{\underline{\mathbf{l}}}$ corresponding to a given phase propagation direction $\hat{\underline{\mathbf{k}}}$. Due to (8.219) and (8.221), we have

$$\underline{\mathbf{c}}_{\mathrm{E}}(\hat{\underline{\mathbf{l}}}) \cdot \underline{\mathbf{k}} = \omega \tag{8.231}$$

and, therefore,

$$\nabla_{\hat{\underline{\mathbf{l}}}}\, \underline{\mathbf{c}}_{\mathrm{E}}(\hat{\underline{\mathbf{l}}}) \cdot \hat{\underline{\mathbf{k}}} = \underline{\mathbf{0}}. \tag{8.232}$$

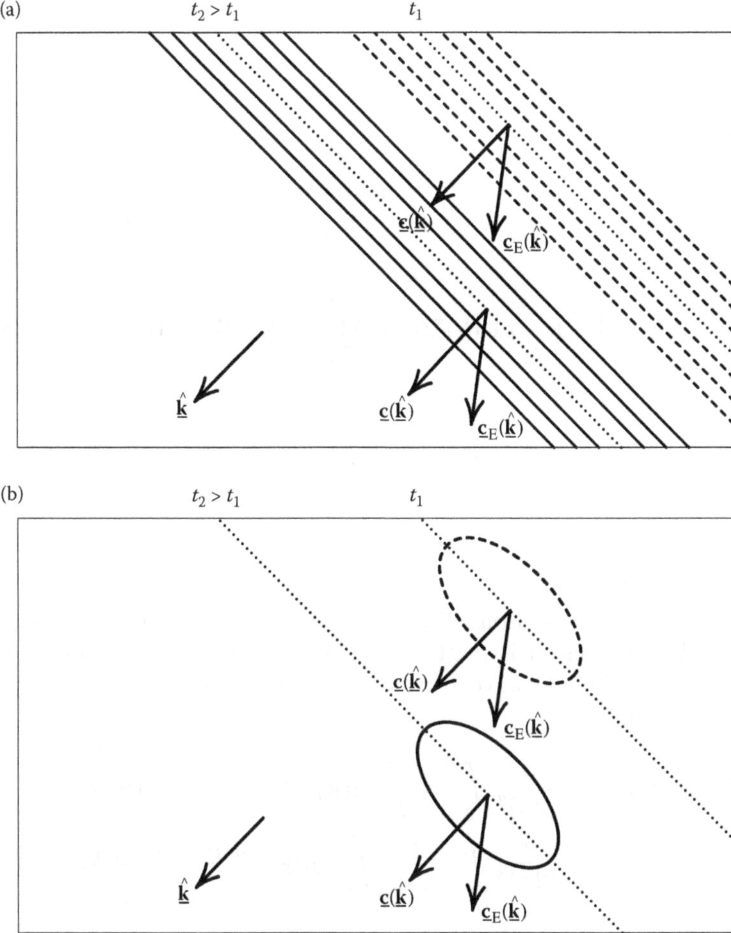

FIGURE 8.14
Consequence of different phase and energy velocity vectors for plane waves (a) and plane wave packets (b).

On the other hand, the differential vector

$$\underline{\mathbf{dc}}_{\mathrm{E}}(\hat{\underline{\mathbf{l}}}) = \underline{\mathbf{d}\hat{\mathbf{l}}} \cdot \nabla_{\hat{\underline{\mathbf{l}}}}\, \underline{\mathbf{c}}_{\mathrm{E}}(\hat{\underline{\mathbf{l}}}) \tag{8.233}$$

is tangential to the wave surface resulting in

$$\underline{\mathbf{dc}}_{\mathrm{E}}(\hat{\underline{\mathbf{l}}}) \cdot \hat{\underline{\mathbf{k}}} = 0 \tag{8.234}$$

with (8.232).

Group velocities: There is a different formula to calculate the energy velocity of homogeneous plane waves in nondissipative materials—besides (8.217) and (8.224)—relating it to the group velocity that may be defined under certain assumptions. According to

$$\left[\underline{\mathbf{k}} \cdot \underline{\underline{\underline{\mathbf{c}}}} \cdot \underline{\mathbf{k}} - \rho\,\omega^2(\underline{\mathbf{k}})\,\underline{\underline{\mathbf{I}}}\right] \cdot \underline{\hat{\mathbf{u}}}(\underline{\mathbf{k}}) = \underline{\mathbf{0}}, \tag{8.235}$$

we insert the phase vector (8.220) in the eigenvalue equation (8.208) and define $\omega^2(\underline{\mathbf{k}})$ as eigenvalue. Taking the $\nabla_{\underline{\mathbf{k}}}$-gradient of (8.235) yields similar to (8.222):

$$\left[\underline{\underline{\underline{\mathbf{c}}}} \cdot \underline{\mathbf{k}} - \rho\,\omega(\underline{\mathbf{k}})\nabla_{\underline{\mathbf{k}}}\omega(\underline{\mathbf{k}})\,\underline{\underline{\mathbf{I}}}\right] : \underline{\hat{\mathbf{u}}}(\underline{\mathbf{k}})\underline{\hat{\mathbf{u}}}(\underline{\mathbf{k}}) = \underline{\mathbf{0}}, \tag{8.236}$$

thus resulting in

$$\begin{aligned}\underline{\mathbf{c}}_{\mathrm{E}}(\underline{\mathbf{k}}) &= \nabla_{\underline{\mathbf{k}}}\omega(\underline{\mathbf{k}}) \\ &\stackrel{\mathrm{def}}{=} \frac{\partial\omega(\underline{\mathbf{k}})}{\partial\underline{\mathbf{k}}}\end{aligned} \tag{8.237}$$

with (8.217) and (8.220).

Under certain assumptions, $\nabla_{\underline{\mathbf{k}}}\omega(\underline{\mathbf{k}})$ is the group velocity vector $\underline{\mathbf{c}}_{\mathrm{gr}}(\underline{\mathbf{k}})$ of a wave packet. The latter is synthesized via integration of a given $\underline{\mathbf{k}}$-spectrum $\underline{\mathbf{u}}(\underline{\mathbf{k}}) = \underline{\mathbf{u}}(\omega, \hat{\underline{\mathbf{k}}})$ of plane waves with the volume element $\mathrm{d}^3\underline{\mathbf{k}} = k^2\mathrm{d}^2\hat{\underline{\mathbf{k}}}\,\mathrm{d}k$ of $\underline{\mathbf{k}}$-space according to

$$\begin{aligned}\underline{\mathbf{u}}(\underline{\mathbf{R}}, t) &= \frac{1}{(2\pi)^3}\int_0^\infty \iint_{S^2} \underline{\mathbf{u}}(\underline{\mathbf{k}})\,\mathrm{e}^{\mathrm{j}\underline{\mathbf{k}}\cdot\underline{\mathbf{R}}-\mathrm{j}\omega(\underline{\mathbf{k}})t}\,k^2\mathrm{d}^2\hat{\underline{\mathbf{k}}}\,\mathrm{d}k \\ &= \frac{1}{(2\pi)^3}\int_{-\infty}^\infty\int_{-\infty}^\infty\int_{-\infty}^\infty \underline{\mathbf{u}}(\underline{\mathbf{k}})\,\mathrm{e}^{\mathrm{j}\underline{\mathbf{k}}\cdot\underline{\mathbf{R}}-\mathrm{j}\omega(\underline{\mathbf{k}})t}\,\mathrm{d}^3\underline{\mathbf{k}},\end{aligned} \tag{8.238}$$

where we assume that $\underline{\mathbf{u}}(\underline{\mathbf{k}})$ is concentrated around the spectral center of gravity $\underline{\mathbf{k}}_0 = \hat{\underline{\mathbf{k}}}_0\omega_0/c(\hat{\underline{\mathbf{k}}}_0)$; S^2 is the unit sphere that hosts all directions $\hat{\underline{\mathbf{k}}}$. If $\omega(\underline{\mathbf{k}})$ is slowly varying in the volume occupied by $\underline{\mathbf{u}}(\underline{\mathbf{k}})$—this is essentially the already mentioned assumption—we can truncate a Taylor expansion after the linear term:

$$\omega(\underline{\mathbf{k}}) = \omega(\underline{\mathbf{k}}_0) + (\underline{\mathbf{k}} - \underline{\mathbf{k}}_0) \cdot \nabla_{\underline{\mathbf{k}}}\omega(\underline{\mathbf{k}})\Big|_{\underline{\mathbf{k}}=\underline{\mathbf{k}}_0}. \tag{8.239}$$

With the short-hand notation

$$\underline{\mathbf{c}}_{\mathrm{gr}}(\underline{\mathbf{k}}_0) = \nabla_{\underline{\mathbf{k}}}\omega(\underline{\mathbf{k}})\Big|_{\underline{\mathbf{k}}=\underline{\mathbf{k}}_0}, \tag{8.240}$$

we evidently define a velocity vector determining the retardation of the envelope $\underline{\mathbf{A}}_{\underline{\mathbf{k}}_0}(\underline{\mathbf{R}})$ of the wave packet after inserting (8.239) into (8.238):

$$\underline{\mathbf{u}}(\underline{\mathbf{R}}, t) \simeq \underline{\mathbf{A}}_{\underline{\mathbf{k}}_0}\left[\underline{\mathbf{R}} - \underline{\mathbf{c}}_{\mathrm{gr}}(\underline{\mathbf{k}}_0)t\right]\,\mathrm{e}^{\mathrm{j}\underline{\mathbf{k}}_0\cdot\underline{\mathbf{R}}-\;ij\omega(\underline{\mathbf{k}}_0)t}, \tag{8.241}$$

where

$$\underline{\mathbf{A}}_{\underline{\mathbf{k}}_0}\left[\underline{\mathbf{R}} - \underline{\mathbf{c}}_{\mathrm{gr}}(\underline{\mathbf{k}}_0)t\right] = \frac{1}{(2\pi)^3}\int_{-\infty}^{\infty}\int_{-\infty}^{\infty}\int_{-\infty}^{\infty} \underline{\mathbf{u}}(\underline{\mathbf{k}})\,\mathrm{e}^{\mathrm{j}(\underline{\mathbf{k}}-\underline{\mathbf{k}}_0)\cdot[\underline{\mathbf{R}}-\underline{\mathbf{c}}_{\mathrm{gr}}(\underline{\mathbf{k}}_0)t]}\,\mathrm{d}^3\underline{\mathbf{k}}. \tag{8.242}$$

Therefore, the velocity vector of the phase of the wave packet is given by

$$\underline{\mathbf{c}}(\underline{\mathbf{k}}_0) = \omega(\underline{\mathbf{k}}_0)\,\underline{\mathbf{k}}_0, \tag{8.243}$$

and the group velocity vector is given by (8.240). Consulting Figure 8.14, illustrates the consequence of the different directions of $\underline{\mathbf{c}}(\underline{\mathbf{k}}_0)$ and $\underline{\mathbf{c}}_{\mathrm{gr}}(\underline{\mathbf{k}}_0)$ as far as the propagation of wave packets is concerned. These wave packets emerge as pulsed radiation of a transducer; mathematically, its radiation field is synthesized with point sources; hence, physical intuition suggests that the pulsed radiation field of a point source reflects the direction dependence of the group (energy) velocity vector (of a plane wave) in terms of time domain wavefronts (wave surfaces)[157] (Helbig 1994; Snieder 2002; Langenberg et al. 2002b). For transversely isotropic materials, this is confirmed by the EFIT-simulation of Figure 8.26.

Sound pressure of plane waves: Equation (8.124) equally defines the sound pressure of a plane wave mode in anisotropic materials:

$$p(\underline{\mathbf{R}}, \omega, \hat{\underline{\mathbf{k}}}) = -\underline{\underline{\mathbf{T}}}(\underline{\mathbf{R}}, \omega, \hat{\underline{\mathbf{k}}}) : \hat{\underline{\mathbf{k}}}\,\hat{\underline{\mathbf{u}}}(\hat{\underline{\mathbf{k}}}); \tag{8.244}$$

with

$$\underline{\underline{\mathbf{T}}}(\underline{\mathbf{R}}, \omega, \hat{\underline{\mathbf{k}}}) = \frac{\mathrm{j}\omega}{c(\hat{\underline{\mathbf{k}}})}\,u(\omega)\,\mathrm{e}^{\mathrm{j}k\hat{\underline{\mathbf{k}}}\cdot\underline{\mathbf{R}}}\,\underline{\underline{\underline{\underline{\mathbf{c}}}}} : \hat{\underline{\mathbf{k}}}\,\hat{\underline{\mathbf{u}}}(\hat{\underline{\mathbf{k}}}), \tag{8.245}$$

the $\underline{\underline{\underline{\underline{\mathbf{c}}}}}$-symmetries and Equation 8.215, we obtain

$$\begin{aligned} p(\underline{\mathbf{R}}, \omega, \hat{\underline{\mathbf{k}}}) &= -\mathrm{j}\omega u(\omega)\,\mathrm{e}^{\mathrm{j}k\hat{\underline{\mathbf{k}}}\cdot\underline{\mathbf{R}}}\frac{1}{c(\hat{\underline{\mathbf{k}}})}\underbrace{\hat{\underline{\mathbf{k}}}\,\hat{\underline{\mathbf{u}}}(\hat{\underline{\mathbf{k}}}) : \underline{\underline{\underline{\underline{\mathbf{c}}}}} : \hat{\underline{\mathbf{k}}}\,\hat{\underline{\mathbf{u}}}(\hat{\underline{\mathbf{k}}})}_{=\rho c^2(\hat{\underline{\mathbf{k}}})} \\ &= -\mathrm{j}\omega\underbrace{\rho\, c(\hat{\underline{\mathbf{k}}})}_{=Z(\hat{\underline{\mathbf{k}}})}\,u(\omega)\,\mathrm{e}^{\mathrm{j}k\hat{\underline{\mathbf{k}}}\cdot\underline{\mathbf{R}}}. \end{aligned} \tag{8.246}$$

8.3.2 Plane waves in transversely isotropic materials

Wave tensor: We specify $\underline{\underline{\underline{\underline{\mathbf{c}}}}}$ according to (4.24) considering a material with transverse isotropy, where the preference direction is given by $\hat{\underline{\mathbf{a}}}$, i.e., in planes

[157]Therefore, the following quotation (Born and Wolf 1975) is only relevant for plane waves: It may be noted that the ray velocity, being derived from the Poynting vector shares with it a certain degree of arbitrariness. It is nevertheless a useful concept although, like the phase velocity, it has no directly verifiable physical significance.

orthogonal to $\underline{\hat{\mathbf{a}}}$, the material is isotropic. Through calculation, we obtain the respective wave tensor:

$$\underline{\underline{\mathbf{W}}}^{\text{triso}}(\underline{\hat{\mathbf{k}}}, c^2) = \gamma_1 \underline{\underline{\mathbf{I}}} + \gamma_2 \underline{\hat{\mathbf{k}}}\,\underline{\hat{\mathbf{k}}} + \gamma_3 \underline{\hat{\mathbf{a}}}\,\underline{\hat{\mathbf{a}}} + \gamma_4 (\underline{\hat{\mathbf{k}}}\,\underline{\hat{\mathbf{a}}} + \underline{\hat{\mathbf{a}}}\,\underline{\hat{\mathbf{k}}}), \tag{8.247}$$

where

$$\gamma_1 = \mu_\perp + (\mu_\parallel - \mu_\perp)(\underline{\hat{\mathbf{k}}} \cdot \underline{\hat{\mathbf{a}}})^2 - \rho c^2(\underline{\hat{\mathbf{k}}}), \tag{8.248}$$

$$\gamma_2 = \lambda_\perp + \mu_\perp, \tag{8.249}$$

$$\gamma_3 = (\lambda_\perp + 2\mu_\perp + \lambda_\parallel - 2\mu_\parallel - 2\nu)(\underline{\hat{\mathbf{k}}} \cdot \underline{\hat{\mathbf{a}}})^2 + \mu_\parallel - \mu_\perp, \tag{8.250}$$

$$\gamma_4 = (\mu_\parallel - \mu_\perp + \nu - \lambda_\perp)\underline{\hat{\mathbf{k}}} \cdot \underline{\hat{\mathbf{a}}}. \tag{8.251}$$

Phase velocities: To calculate the eigenvalues $c^2(\underline{\hat{\mathbf{k}}})$ of $\underline{\underline{\mathbf{D}}}^{\text{triso}}(\underline{\hat{\mathbf{k}}})$, we write the wave tensor in the form

$$\underline{\underline{\mathbf{W}}}^{\text{triso}}(\underline{\hat{\mathbf{k}}}, c^2) = \gamma_1 \underline{\underline{\mathbf{I}}} + \underline{\hat{\mathbf{k}}}\,(\gamma_2 \underline{\hat{\mathbf{k}}} + \gamma_4 \underline{\hat{\mathbf{a}}}) + \underline{\hat{\mathbf{a}}}\,(\gamma_3 \underline{\hat{\mathbf{a}}} + \gamma_4 \underline{\hat{\mathbf{k}}}) \tag{8.252}$$

to be able to utilize Chen's formula (Chen 1983)

$$\begin{aligned} &\det(\alpha \underline{\underline{\mathbf{I}}} + \underline{\mathbf{A}}_1 \underline{\mathbf{C}}_1 + \underline{\mathbf{A}}_2 \underline{\mathbf{C}}_2) \\ &\quad = \alpha \left[\alpha^2 + \alpha(\underline{\mathbf{A}}_1 \cdot \underline{\mathbf{C}}_1 + \underline{\mathbf{A}}_2 \cdot \underline{\mathbf{C}}_2) + (\underline{\mathbf{C}}_1 \times \underline{\mathbf{C}}_2) \cdot (\underline{\mathbf{A}}_1 \times \underline{\mathbf{A}}_2)\right] \end{aligned} \tag{8.253}$$

for the determinant:

$$\begin{aligned} &\det \underline{\underline{\mathbf{W}}}^{\text{triso}}(\underline{\hat{\mathbf{k}}}, c^2) \\ &\quad = \gamma_1 \left\{\gamma_1^2 + \gamma_1(\gamma_2 + \gamma_3 + 2\gamma_4 \underline{\hat{\mathbf{k}}} \cdot \underline{\hat{\mathbf{a}}}) + (\gamma_2\gamma_3 - \gamma_4^2)\left[1 - (\underline{\hat{\mathbf{k}}} \cdot \underline{\hat{\mathbf{a}}})^2\right]\right\}. \end{aligned} \tag{8.254}$$

Note that $c^2(\underline{\hat{\mathbf{k}}})$ is hidden in γ_1 identifying the first eigenvalue via

$$\gamma_1 = 0 \tag{8.255}$$

according to

$$c_{\text{SH}}^2(\underline{\hat{\mathbf{k}}}) = \frac{\mu_\perp + (\mu_\parallel - \mu_\perp)(\underline{\hat{\mathbf{k}}} \cdot \underline{\hat{\mathbf{a}}})^2}{\rho}. \tag{8.256}$$

Anticipating the orientation of the corresponding eigenvector, we have appended the index SH. We state that: The phase velocity $c_{\text{SH}}(\underline{\hat{\mathbf{k}}})$ actually depends explicitly upon the phase propagation direction $\underline{\hat{\mathbf{k}}}$ for $\mu_\parallel \neq \mu_\perp$ as it is "allowed" for an anisotropic material! However, we have transverse isotropy with respect to $\underline{\hat{\mathbf{a}}}$, hence we obtain the $\underline{\hat{\mathbf{k}}}$-independent (isotropic) value

$$c_{\text{SH}}^2(\underline{\hat{\mathbf{k}}} \perp \underline{\hat{\mathbf{a}}}) = \frac{\mu_\perp}{\rho} \tag{8.257}$$

for $\underline{\hat{\mathbf{k}}} \cdot \underline{\hat{\mathbf{a}}} = 0$; that way, the notation for Lamé's parameter $\mu_\perp$ is enlightened. Furthermore, following from transverse isotropy, the phase velocity diagram $c_{\text{SH}}(\underline{\hat{\mathbf{k}}})$ in a fixed $\underline{\hat{\mathbf{k}}}\,\underline{\hat{\mathbf{a}}}$-plane is rotationally symmetric around $\underline{\hat{\mathbf{a}}}$.

The second and third eigenvalues of $\underline{\underline{\mathbf{D}}}^{\text{triso}}(\hat{\underline{\mathbf{k}}})$ are obtained putting the cambered bracket of (8.254) to zero

$$\gamma_1 = -\frac{1}{2}\left(\gamma_2 + \gamma_3 + 2\gamma_4 \hat{\underline{\mathbf{k}}} \cdot \hat{\underline{\mathbf{a}}}\right) \mp \frac{1}{2}\sqrt{(\gamma_2 + \gamma_3 + 2\gamma_4 \hat{\underline{\mathbf{k}}} \cdot \hat{\underline{\mathbf{a}}})^2 - 4(\gamma_2\gamma_3 - \gamma_4^2)[1 - (\hat{\underline{\mathbf{k}}} \cdot \hat{\underline{\mathbf{a}}})^2]} \tag{8.258}$$

and solving the quadratic equation[158] in γ_1

$$\gamma_1^2 + \gamma_1(\gamma_2 + \gamma_3 + 2\gamma_4 \hat{\underline{\mathbf{k}}} \cdot \hat{\underline{\mathbf{a}}}) + (\gamma_2\gamma_3 - \gamma_4^2)[1 - (\hat{\underline{\mathbf{k}}} \cdot \hat{\underline{\mathbf{a}}})^2] = 0 \tag{8.259}$$

while considering the short-hand notation γ_1 and Equation 8.256:

$$c_{\text{qP,qSV}}^2(\hat{\underline{\mathbf{k}}}) = c_{\text{SH}}^2(\hat{\underline{\mathbf{k}}}) - \frac{\gamma_1^{\text{qP,qSV}}}{\rho} \tag{8.260}$$

$$= c_{\text{SH}}^2(\hat{\underline{\mathbf{k}}}) + \frac{\frac{1}{2}\left(\gamma_2 + \gamma_3 + 2\gamma_4 \hat{\underline{\mathbf{k}}} \cdot \hat{\underline{\mathbf{a}}}\right)}{\rho} \pm \frac{\frac{1}{2}\sqrt{(\gamma_2 + \gamma_3 + 2\gamma_4 \hat{\underline{\mathbf{k}}} \cdot \hat{\underline{\mathbf{a}}})^2 - 4(\gamma_2\gamma_3 - \gamma_4^2)[1 - (\hat{\underline{\mathbf{k}}} \cdot \hat{\underline{\mathbf{a}}})^2]}}{\rho}. \tag{8.261}$$

As a consequence of transverse isotropy, the diagrams of the phase velocities $c_{\text{qP,qSV}}(\hat{\underline{\mathbf{k}}})$ are rotationally symmetric around $\hat{\underline{\mathbf{a}}}$ in a fixed $\hat{\underline{\mathbf{k}}}\hat{\underline{\mathbf{a}}}$-plane. The indices stand for quasi-P and quasi-SV that become comprehensible when we calculate the corresponding eigenvectors, i.e., the polarization vectors. The assignment qP $\Longrightarrow +\sqrt{\ }$ and qSV $\Longrightarrow -\sqrt{\ }$ in (8.261) becomes immediately clear if we investigate once more the $\hat{\underline{\mathbf{k}}}$-specialization to the isotropy plane: For $\hat{\underline{\mathbf{k}}} \cdot \hat{\underline{\mathbf{a}}} = 0$, we obtain

$$c_{\text{qP}}^2(\hat{\underline{\mathbf{k}}} \perp \hat{\underline{\mathbf{a}}}) = \frac{\lambda_\perp + 2\mu_\perp}{\rho} \quad \text{for } +\sqrt{\ }, \tag{8.262}$$

$$c_{\text{qSV}}^2(\hat{\underline{\mathbf{k}}} \perp \hat{\underline{\mathbf{a}}}) = \frac{\mu_\parallel}{\rho} \quad \text{for } -\sqrt{\ }, \tag{8.263}$$

that is to say, for $+\sqrt{\ }$, we actually find the primary wave velocity for the $\perp$-Lamé parameters, and for $-\sqrt{\ }$, the secondary wave velocity for the $\|$-Lamé parameters. With Figure 8.15, we anticipate the respective orientation of the eigenvectors, i.e., polarization vectors; since all eigenvalues are different for $\hat{\underline{\mathbf{k}}} \cdot \hat{\underline{\mathbf{a}}} = 0$, the eigenvectors must constitute an orthogonal trihedron that apparently has no "quasi" properties: Apart from the fact that both secondary

[158] The minus sign in front of the square root yields the polarization vectors as displayed in Figures 8.18 and 8.19, after evaluating (8.278); with (8.278), the positive sign in fact results in a qSV-polarization perpendicular to the SH- and qP-polarizations, yet opposite to qSV-polarization vectors as depicted in the above figures. With the choice of (8.270) for SH, no right-handed orthonormal trihedron $\underline{\hat{\mathbf{u}}}_{\text{qP}}, \underline{\hat{\mathbf{u}}}_{\text{qSV}}, \underline{\hat{\mathbf{u}}}_{\text{SH}}$ would result; that is why we choose the negative sign in (8.278).

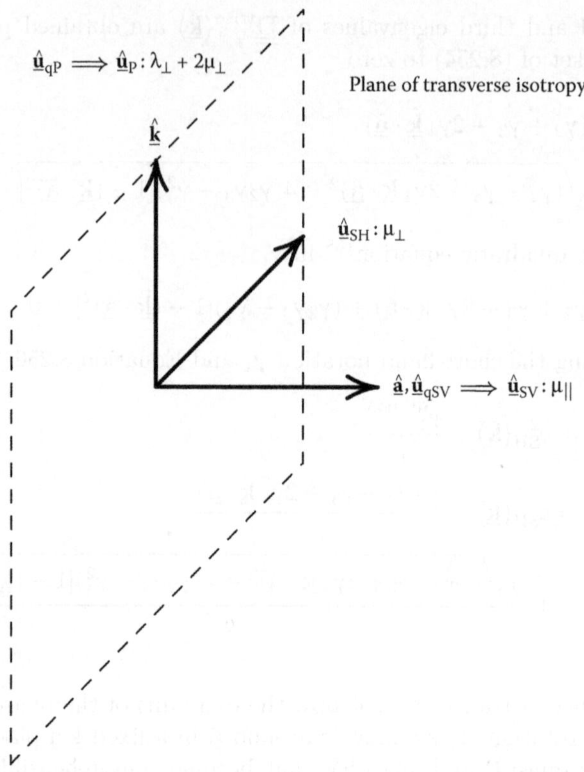

FIGURE 8.15
Polarization vectors and phase velocities of wave modes for phase propagation in the plane of transverse isotropy perpendicular to the preference direction.

wave polarization vectors must be orthogonal to each other due to the different eigenvalues, we find the usual polarization vectors of a (homogeneous) plane wave in an isotropic material (compare Figure 8.8) yet with the speciality that the bipod $\hat{\underline{u}}_{SH}$, $\hat{\underline{u}}_{qSV}$ cannot be rotated around the propagation direction $\hat{\underline{k}}$ as in the isotropic case, $\hat{\underline{u}}_{qSV}$ must point into $\hat{\underline{a}}$-direction. Therefore, an arbitrarily given ($\hat{\underline{u}}_S \perp \hat{\underline{k}}$)-polarization splits into two orthogonal $\hat{\underline{u}}_{SH}$- and $\hat{\underline{u}}_{qSV}$-polarizations that propagate with different velocities; after a finite travel distance, a phase difference results, the originally linearly polarized shear wave is elliptically polarized.[159]

Phase propagation along the preference direction, namely $\hat{\underline{k}} \cdot \hat{\underline{a}} = 1$, leads to another simple case; we find

[159]The same effect is observed for electromagnetic waves in biaxial materials along the main axis; it is utilized in photoelasticity (Born and Wolf 1975; Wolf 1976).

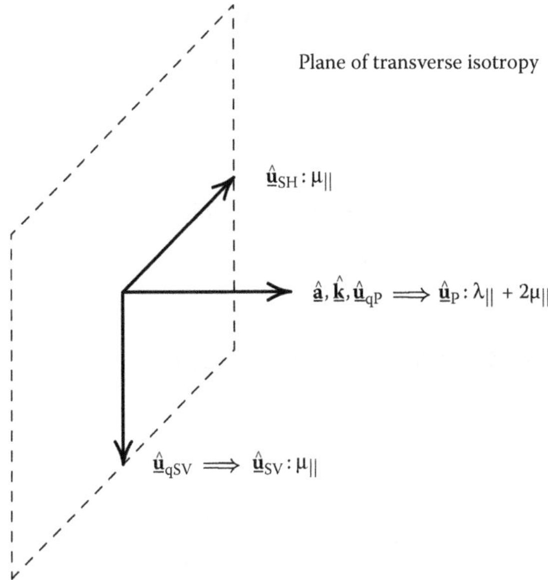

FIGURE 8.16
Polarization vectors and phase velocities of wave modes for phase propagation perpendicular to the plane of transverse isotropy parallel to the preference direction.

$$c_{\text{SH}}^2(\hat{\underline{\mathbf{k}}} \parallel \hat{\underline{\mathbf{a}}}) = \frac{\mu_{\|}}{\rho}, \tag{8.264}$$

$$c_{\text{qP}}^2(\hat{\underline{\mathbf{k}}} \parallel \hat{\underline{\mathbf{a}}}) = \frac{\lambda_{\|} + 2\mu_{\|}}{\rho}, \tag{8.265}$$

$$c_{\text{qSV}}^2(\hat{\underline{\mathbf{k}}} \parallel \hat{\underline{\mathbf{a}}}) = \frac{\mu_{\|}}{\rho}. \tag{8.266}$$

The corresponding polarization vectors are depicted in Figure 8.16; since in that case only $\|$-Lamé parameters are involved and the eigenvalues (8.264) and (8.266) are equal, we encounter a standard (homogeneous) plane wave (Figure 8.8): Both shear wave polarizations may be chosen orthogonal to each other, but it is not mandatory because each vector perpendicular to $\hat{\underline{\mathbf{k}}}$ and $\hat{\underline{\mathbf{a}}}$, respectively, is an eigenvector to the eigenvalue $\mu_{\|}/\rho$; this time, we can even rotate the $(\hat{\underline{\mathbf{u}}}_{\text{SH}}, \hat{\underline{\mathbf{u}}}_{\text{qSV}})$-bipod around the $\hat{\underline{\mathbf{a}}}$-axis.

Polarization vectors: For the eigenvalue $c_{\text{SH}}^2(\hat{\underline{\mathbf{k}}})$, we have $\gamma_1 = 0$ (Equation 8.255), and therefore we immediately see that, according to (8.252), $\underline{\underline{\mathbf{W}}}^{\text{triso}}(\hat{\underline{\mathbf{k}}}, c_{\text{SH}}^2)$ is a planar tensor (sum of two dyadic products). Consequently, $\text{adj}\,\underline{\underline{\mathbf{W}}}^{\text{triso}}(\hat{\underline{\mathbf{k}}}, c_{\text{SH}}^2)$ must be a linear tensor (a single dyadic product; Chen 1983). With Chen's formula

$$\operatorname{adj}(\alpha\,\underline{\underline{\mathbf{I}}} + \underline{\mathbf{A}}_1\underline{\mathbf{C}}_1 + \underline{\mathbf{A}}_2\underline{\mathbf{C}}_2) = \alpha[(\alpha + \underline{\mathbf{A}}_1\cdot\underline{\mathbf{C}}_1 + \underline{\mathbf{A}}_2\cdot\underline{\mathbf{C}}_2)\underline{\underline{\mathbf{I}}} - \underline{\mathbf{A}}_1\underline{\mathbf{C}}_1 - \underline{\mathbf{A}}_2\underline{\mathbf{C}}_2] + (\underline{\mathbf{C}}_1\times\underline{\mathbf{C}}_2)(\underline{\mathbf{A}}_1\times\underline{\mathbf{A}}_2), \tag{8.267}$$

we calculate

$$\begin{aligned}&\operatorname{adj}\underline{\underline{\mathbf{W}}}^{\text{triso}}(\hat{\underline{\mathbf{k}}}, c^2)\\ &= \gamma_1\left[(\gamma_1+\gamma_2+\gamma_3+2\gamma_4\,\hat{\underline{\mathbf{k}}}\cdot\hat{\underline{\mathbf{a}}})\,\underline{\underline{\mathbf{I}}} - \gamma_2\,\hat{\underline{\mathbf{k}}}\hat{\underline{\mathbf{k}}} - \gamma_3\,\hat{\underline{\mathbf{a}}}\hat{\underline{\mathbf{a}}} - \gamma_4(\hat{\underline{\mathbf{a}}}\hat{\underline{\mathbf{k}}} + \hat{\underline{\mathbf{k}}}\hat{\underline{\mathbf{a}}})\right]\\ &\quad + (\gamma_2\gamma_3 - \gamma_4^2)(\hat{\underline{\mathbf{k}}}\times\hat{\underline{\mathbf{a}}})(\hat{\underline{\mathbf{k}}}\times\hat{\underline{\mathbf{a}}}),\end{aligned} \tag{8.268}$$

and obviously,

$$\operatorname{adj}\underline{\underline{\mathbf{W}}}^{\text{triso}}(\hat{\underline{\mathbf{k}}}, c_{\text{SH}}^2) = (\gamma_2\gamma_3 - \gamma_4^2)(\hat{\underline{\mathbf{k}}}\times\hat{\underline{\mathbf{a}}})(\hat{\underline{\mathbf{k}}}\times\hat{\underline{\mathbf{a}}}) \tag{8.269}$$

is linear! Consequence: The eigenvector $\hat{\underline{\mathbf{u}}}_{\text{SH}}(\hat{\underline{\mathbf{k}}})$ is proportional to the left vector $\hat{\underline{\mathbf{k}}}\times\hat{\underline{\mathbf{a}}}$ in the dyadic product (8.269), i.e., normalization to $|\hat{\underline{\mathbf{k}}}\times\hat{\underline{\mathbf{a}}}|$ results in:[160]

$$\hat{\underline{\mathbf{u}}}_{\text{SH}}(\hat{\underline{\mathbf{k}}}) = \frac{\hat{\underline{\mathbf{k}}}\times\hat{\underline{\mathbf{a}}}}{\sqrt{1-(\hat{\underline{\mathbf{k}}}\cdot\hat{\underline{\mathbf{a}}})^2}}. \tag{8.270}$$

This eigenvector that gives the SH-polarization direction is orthogonal to the plane spanned by $\hat{\underline{\mathbf{k}}}$ and $\hat{\underline{\mathbf{a}}}$ (Figure 8.17); it does not explain the notation SH for shear-horizontal. Usually, this notation only becomes meaningful if a reference plane, e.g., the xy-plane as a plane surface of a specimen, is present and if $\hat{\underline{\mathbf{k}}}$ and $\hat{\underline{\mathbf{a}}}$ span a plane perpendicular to it [Figure 8.17(b)]; then, $\hat{\underline{\mathbf{u}}}_{\text{SH}}$ (by the way, independent upon $\hat{\underline{\mathbf{k}}}$) is horizontal with regard to that plane. Incidentally, $\hat{\underline{\mathbf{u}}}_{\text{SH}}$ defines a "real" shear wave because it is divergence-free.

The special case $\hat{\underline{\mathbf{k}}}\perp\hat{\underline{\mathbf{a}}}$ is contained in (8.269) and (8.270) but not the special case $\hat{\underline{\mathbf{k}}}\parallel\hat{\underline{\mathbf{a}}}$ because then we have $\operatorname{adj}\underline{\underline{\mathbf{W}}}(\hat{\underline{\mathbf{k}}}\parallel\hat{\underline{\mathbf{a}}}, c_{\text{SH}}^2)$ as null tensor. Therefore, $\underline{\underline{\mathbf{W}}}(\hat{\underline{\mathbf{k}}}\parallel\hat{\underline{\mathbf{a}}}, c_{\text{SH}}^2)$ must be linear:

$$\underline{\underline{\mathbf{W}}}^{\text{triso}}(\hat{\underline{\mathbf{k}}}\parallel\hat{\underline{\mathbf{a}}}, c_{\text{SH}}^2) = (\gamma_2+\gamma_3+2\gamma_4)\,\hat{\underline{\mathbf{a}}}\hat{\underline{\mathbf{a}}}. \tag{8.271}$$

Hence, the eigenvector is proportional to any vector orthogonal to the right factor $\hat{\underline{\mathbf{a}}}$ of the dyadic (8.271); this is exactly so as we plotted it as an example in Figure 8.16.

With the knowledge of *all* eigenvalues and one eigenvector as well as with the fact of the real symmetry of $\underline{\underline{\mathbf{D}}}^{\text{triso}}(\hat{\underline{\mathbf{k}}})$ before our eyes, we conclude that the two remaining eigenvectors $\hat{\underline{\mathbf{u}}}_{\text{qP}}(\hat{\underline{\mathbf{k}}})$ and $\hat{\underline{\mathbf{u}}}_{\text{qSV}}(\hat{\underline{\mathbf{k}}})$ reside in a plane spanned by $\hat{\underline{\mathbf{k}}}$ and $\hat{\underline{\mathbf{a}}}$ because they must be orthogonal to $\hat{\underline{\mathbf{u}}}_{\text{SH}}(\hat{\underline{\mathbf{k}}})$; in addition, they must be orthogonal to each other. Therefore, we make the *ansatz*[161]

$$\hat{\underline{\mathbf{u}}}_{\text{qP,qSV}}(\hat{\underline{\mathbf{k}}}) \sim \alpha_{\text{qP,qSV}}\,\hat{\underline{\mathbf{k}}} + \beta_{\text{qP,qSV}}\,\hat{\underline{\mathbf{a}}} \tag{8.272}$$

---

[160] Evidently, we can also choose $-\hat{\underline{\mathbf{k}}}\times\hat{\underline{\mathbf{a}}} = \hat{\underline{\mathbf{a}}}\times\hat{\underline{\mathbf{k}}}$ (Spies 1992), yet we stay with the convention as given in Figure 8.8.

[161] The formal calculation using the adjoint of $\underline{\underline{\mathbf{W}}}$, respectively, Chen's formulas (1983) for (electromagnetic) biaxial materials is far more strenuous.

(a)

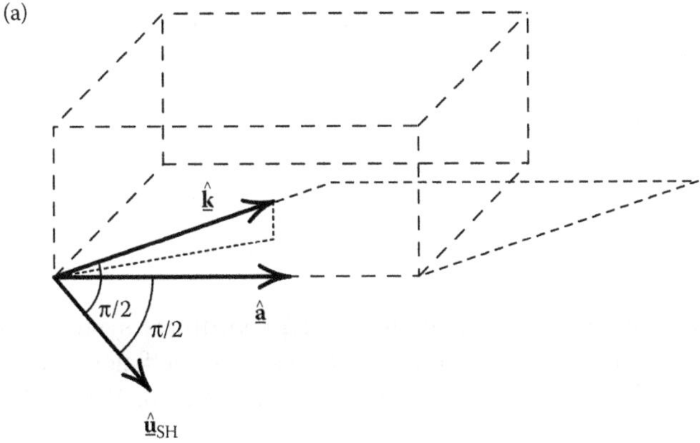

(b)

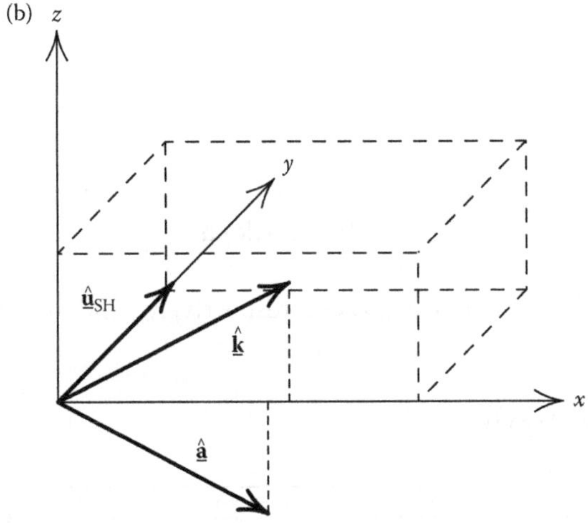

FIGURE 8.17
Orientation of SH-polarization in a transversely isotropic material without (a) and with (b) reference plane.

and determine $\beta_{\mathrm{qP,qSV}}/\alpha_{\mathrm{qP,qSV}}$ from the equation

$$\underline{\underline{\mathbf{W}}}^{\mathrm{triso}}(\hat{\underline{\mathbf{k}}}, c^2_{\mathrm{qP,qSV}}) \cdot (\alpha_{\mathrm{qP,qSV}}\,\hat{\underline{\mathbf{k}}} + \beta_{\mathrm{qP,qSV}}\,\hat{\underline{\mathbf{a}}}) \equiv \underline{\mathbf{0}} \tag{8.273}$$

of the eigenvalue problem. With (8.252), we obtain from (8.273)

$$\begin{aligned}&\Big[\alpha_{\mathrm{qP,qSV}}(\gamma_1^{\mathrm{qP,qSV}} + \gamma_2 + \gamma_4\,\hat{\underline{\mathbf{k}}}\cdot\hat{\underline{\mathbf{a}}}) + \beta_{\mathrm{qP,qSV}}(\gamma_2\,\hat{\underline{\mathbf{k}}}\cdot\hat{\underline{\mathbf{a}}} + \gamma_4)\Big]\,\hat{\underline{\mathbf{k}}}\\&\quad + \Big[\alpha_{\mathrm{qP,qSV}}(\gamma_3\,\hat{\underline{\mathbf{k}}}\cdot\hat{\underline{\mathbf{a}}} + \gamma_4) + \beta_{\mathrm{qP,qSV}}(\gamma_1^{\mathrm{qP,qSV}} + \gamma_3 + \gamma_4\,\hat{\underline{\mathbf{k}}}\cdot\hat{\underline{\mathbf{a}}})\Big]\,\hat{\underline{\mathbf{a}}} = \underline{\mathbf{0}};\end{aligned} \tag{8.274}$$

Both solutions of (8.259) have been written according to (8.258) as $\gamma_1^{\mathrm{qP,qSV}}$. Because $\hat{\underline{\mathbf{k}}}$ and $\hat{\underline{\mathbf{a}}}$ can be considered as linearly independent—the special case $\hat{\underline{\mathbf{k}}} \parallel \hat{\underline{\mathbf{a}}}$ is separately treated—both brackets in (8.274) must equally be zero; the result is the homogeneous system of equations

$$\alpha_{\mathrm{qP,qSV}}(\gamma_1^{\mathrm{qP,qSV}} + \gamma_2 + \gamma_4 \hat{\underline{\mathbf{k}}} \cdot \hat{\underline{\mathbf{a}}}) + \beta_{\mathrm{qP,qSV}}(\gamma_2 \hat{\underline{\mathbf{k}}} \cdot \hat{\underline{\mathbf{a}}} + \gamma_4) = 0, \tag{8.275}$$

$$\alpha_{\mathrm{qP,qSV}}(\gamma_3 \hat{\underline{\mathbf{k}}} \cdot \hat{\underline{\mathbf{a}}} + \gamma_4) + \beta_{\mathrm{qP,qSV}}(\gamma_1^{\mathrm{qP,qSV}} + \gamma_3 + \gamma_4 \hat{\underline{\mathbf{k}}} \cdot \hat{\underline{\mathbf{a}}}) = 0 \tag{8.276}$$

for the components of the eigenvalues in a $\hat{\underline{\mathbf{k}}}\,\hat{\underline{\mathbf{a}}}$-coordinate system. This system of equations has a nontrivial solution because its coefficient determinant is equal to $\det \underline{\underline{\mathbf{W}}}^{\mathrm{triso}}(\hat{\underline{\mathbf{k}}}, c_{\mathrm{qP,qSV}}^2)$ and, hence, equal to zero. We find

$$\begin{aligned}
\frac{\beta_{\mathrm{qP,qSV}}}{\alpha_{\mathrm{qP,qSV}}} &\stackrel{\mathrm{def}}{=} \gamma_{\mathrm{qP,qSV}} \\
&= -\frac{\gamma_1^{\mathrm{qP,qSV}} + \gamma_2 + \gamma_4 \hat{\underline{\mathbf{k}}} \cdot \hat{\underline{\mathbf{a}}}}{\gamma_4 + \gamma_2 \hat{\underline{\mathbf{k}}} \cdot \hat{\underline{\mathbf{a}}}} \\
&= -\frac{\rho\, c_{\mathrm{SH}}^2 - \rho\, c_{\mathrm{qP,qSV}}^2 + \gamma_2 + \gamma_4 \hat{\underline{\mathbf{k}}} \cdot \hat{\underline{\mathbf{a}}}}{(\nu + \mu_\parallel)\hat{\underline{\mathbf{k}}} \cdot \hat{\underline{\mathbf{a}}}};
\end{aligned} \tag{8.277}$$

after insertion of $\gamma_1^{\mathrm{qP,qSV}}$ and $c_{\mathrm{SH,qP,qSV}}$, respectively, and normalization, we get[162]

$$\begin{aligned}
\hat{\underline{\mathbf{u}}}_{\mathrm{qP,qSV}}(\hat{\underline{\mathbf{k}}}) &= \frac{\hat{\underline{\mathbf{k}}} + \gamma_{\mathrm{qP,qSV}} \hat{\underline{\mathbf{a}}}}{U_{\mathrm{qP,qSV}}} \\
&= \frac{\hat{\underline{\mathbf{k}}} - \frac{\frac{1}{2}(\gamma_2 - \gamma_3) \mp \frac{1}{2}\sqrt{(\gamma_2 + \gamma_3 + 2\gamma_4 \hat{\underline{\mathbf{k}}} \cdot \hat{\underline{\mathbf{a}}})^2 - 4(\gamma_2\gamma_3 - \gamma_4^2)[1 - (\hat{\underline{\mathbf{k}}} \cdot \hat{\underline{\mathbf{a}}})^2]}}{\gamma_4 + \gamma_2 \hat{\underline{\mathbf{k}}} \cdot \hat{\underline{\mathbf{a}}}} \hat{\underline{\mathbf{a}}}}{U_{\mathrm{qP,qSV}}},
\end{aligned} \tag{8.278}$$

where

$$\begin{aligned}
U_{\mathrm{qP,qSV}} &= |\hat{\underline{\mathbf{k}}} + \gamma_{\mathrm{qP,qSV}} \hat{\underline{\mathbf{a}}}| \\
&= \sqrt{1 + \gamma_{\mathrm{qP,qSV}}^2 + 2\gamma_{\mathrm{qP,qSV}} \hat{\underline{\mathbf{k}}} \cdot \hat{\underline{\mathbf{a}}}}.
\end{aligned} \tag{8.279}$$

A short calculation reveals that in fact

$$\hat{\underline{\mathbf{u}}}_{\mathrm{qP}}(\hat{\underline{\mathbf{k}}}) \cdot \hat{\underline{\mathbf{u}}}_{\mathrm{qSV}}(\hat{\underline{\mathbf{k}}}) = 0 \tag{8.280}$$

holds.

For the special case $\hat{\underline{\mathbf{k}}} \cdot \hat{\underline{\mathbf{a}}} = 0$, we have $\gamma_4 = 0$; therefore, we cannot readily use (8.278). A remedy consists in a transformation of $\operatorname{adj} \underline{\underline{\mathbf{W}}}^{\mathrm{triso}}(\hat{\underline{\mathbf{k}}}, c^2)$:

[162] The result looks much simpler than the one given and published by Spies (1992, 1994).

$$\begin{aligned}
\operatorname{adj} \underline{\underline{\mathbf{W}}}^{\text{triso}}(\hat{\underline{\mathbf{k}}}, c^2) = & \underbrace{\left\{ \gamma_1(\gamma_1 + \gamma_2 + \gamma_3 + 2\gamma_4 \hat{\underline{\mathbf{k}}} \cdot \hat{\underline{\mathbf{a}}}) + (\gamma_2\gamma_3 - \gamma_4^2)[1 - (\hat{\underline{\mathbf{k}}} \cdot \hat{\underline{\mathbf{a}}})^2] \right\}}_{= 0 \text{ für } \gamma_1 = \gamma_1^{\text{qP,qSV}}} \underline{\underline{\mathbf{I}}} \\
& - (\gamma_1\gamma_2 + \gamma_2\gamma_3 - \gamma_4^2)\, \hat{\underline{\mathbf{k}}}\hat{\underline{\mathbf{k}}} - (\gamma_1\gamma_3 + \gamma_2\gamma_3 - \gamma_4^2)\, \hat{\underline{\mathbf{a}}}\hat{\underline{\mathbf{a}}} \\
& - \left[\gamma_1\gamma_4 - (\gamma_2\gamma_3 - \gamma_4^2)\hat{\underline{\mathbf{k}}} \cdot \hat{\underline{\mathbf{a}}}\right](\hat{\underline{\mathbf{k}}}\hat{\underline{\mathbf{a}}} + \hat{\underline{\mathbf{a}}}\hat{\underline{\mathbf{k}}});
\end{aligned} \tag{8.281}$$

we obtain

$$\operatorname{adj} \underline{\underline{\mathbf{W}}}^{\text{triso}}(\hat{\underline{\mathbf{k}}} \perp \hat{\underline{\mathbf{a}}}, c_{\text{qP}}^2) = \gamma_2(\gamma_2 - \gamma_3)\, \hat{\underline{\mathbf{k}}}\hat{\underline{\mathbf{k}}} \tag{8.282}$$

and, respectively,

$$\operatorname{adj} \underline{\underline{\mathbf{W}}}^{\text{triso}}(\hat{\underline{\mathbf{k}}} \perp \hat{\underline{\mathbf{a}}}, c_{\text{qSV}}^2) = \gamma_3(\gamma_3 - \gamma_2)\, \hat{\underline{\mathbf{a}}}\hat{\underline{\mathbf{a}}}. \tag{8.283}$$

Consequently, $\hat{\underline{\mathbf{u}}}_{\text{qP}}(\hat{\underline{\mathbf{k}}} \perp \hat{\underline{\mathbf{a}}})$ is parallel to $\hat{\underline{\mathbf{k}}}$ and $\hat{\underline{\mathbf{u}}}_{\text{qSV}}(\hat{\underline{\mathbf{k}}} \perp \hat{\underline{\mathbf{a}}})$ is parallel to $\hat{\underline{\mathbf{a}}}$ as it is sketched in Figure 8.15.

For $\hat{\underline{\mathbf{k}}} \parallel \hat{\underline{\mathbf{a}}}$, the two vectors $\hat{\underline{\mathbf{k}}}$ and $\hat{\underline{\mathbf{a}}}$ are not linearly independent making the derivation of (8.277) not applicable. Yet, calculating $\operatorname{adj} \underline{\underline{\mathbf{W}}}^{\text{triso}}(\hat{\underline{\mathbf{k}}} \parallel \hat{\underline{\mathbf{a}}}, c_{\text{SH,qSV}}^2)$ results in the null tensor; hence,

$$\underline{\underline{\mathbf{W}}}^{\text{triso}}(\hat{\underline{\mathbf{k}}} \parallel \hat{\underline{\mathbf{a}}}, c_{\text{SH,qSV}}^2) = (\gamma_2 + \gamma_3 + 2\gamma_4)\, \hat{\underline{\mathbf{a}}}\hat{\underline{\mathbf{a}}} \tag{8.284}$$

is linear and $\hat{\underline{\mathbf{u}}}_{\text{SH,qSV}}(\hat{\underline{\mathbf{k}}} \parallel \hat{\underline{\mathbf{a}}})$ orthogonal to the respective right factor $\hat{\underline{\mathbf{a}}}$: We choose $\hat{\underline{\mathbf{u}}}_{\text{SH}}$ and $\hat{\underline{\mathbf{u}}}_{\text{qSV}}$ orthogonal to each other. This is the same situation as for a standard plane wave in an isotropic material because both eigenvalues $c_{\text{SH}}^2(\hat{\underline{\mathbf{k}}} \parallel \hat{\underline{\mathbf{a}}})$ and $c_{\text{qSV}}^2(\hat{\underline{\mathbf{k}}} \parallel \hat{\underline{\mathbf{a}}})$ are equal. Obviously, even $\hat{\underline{\mathbf{u}}}_{\text{qP}}(\hat{\underline{\mathbf{k}}} \parallel \hat{\underline{\mathbf{a}}})$ fits into this picture of a standard plane wave because we immediately realize $\operatorname{adj} \underline{\underline{\mathbf{W}}}^{\text{triso}}(\hat{\underline{\mathbf{k}}} \parallel \hat{\underline{\mathbf{a}}}, c_{\text{qP}}^2) \sim \hat{\underline{\mathbf{a}}}\hat{\underline{\mathbf{a}}}$ so that $\hat{\underline{\mathbf{u}}}_{\text{qP}}(\hat{\underline{\mathbf{k}}} \parallel \hat{\underline{\mathbf{a}}})$ is parallel to the left factor $\hat{\underline{\mathbf{a}}}$. With Figure 8.16, we have anticipated this fact.

In Figures 8.18 and 8.19, we have depicted the $\hat{\underline{\mathbf{k}}}$-dependence of the phase velocities $c_{\text{SH}}(\hat{\underline{\mathbf{k}}})$, $c_{\text{qSV}}(\hat{\underline{\mathbf{k}}})$, $c_{\text{qP}}(\hat{\underline{\mathbf{k}}})$ for two typical materials with transverse isotropy: fiber-reinforced composite and austenitic steel 308; hence, $\hat{\underline{\mathbf{a}}}$ is given by the direction of the carbon fibers and, respectively, by the crystal orientation. The velocity diagrams are displayed in the plane spanned by $\hat{\underline{\mathbf{k}}}$ and $\hat{\underline{\mathbf{a}}}$, and they are rotationally symmetric with respect to the $\hat{\underline{\mathbf{a}}}$-direction.[163] Generally, we have[164] $c_{\text{qP}}(\hat{\underline{\mathbf{k}}}) > c_{\text{SH,qSV}}(\hat{\underline{\mathbf{k}}})$ obviously relating the addendum quasi-P and quasi-SV not to the characterization of wave modes according to pressure and shear waves—distinguished by velocities—but to the polarization (and to

[163]Spies (1992) also discusses the dependence upon $\hat{\underline{\mathbf{a}}}$ if the diagrams are displayed in a fixed plane (the plane of incidence) perpendicular to a reference plane; if $\underline{\mathbf{n}}$ denotes the normal to the reference plane, the plane of incidence is spanned by $\hat{\underline{\mathbf{k}}}$ and $\underline{\mathbf{n}}$.

[164]There are exceptions (Royer and Dieulesaint 2000): TeO_2.

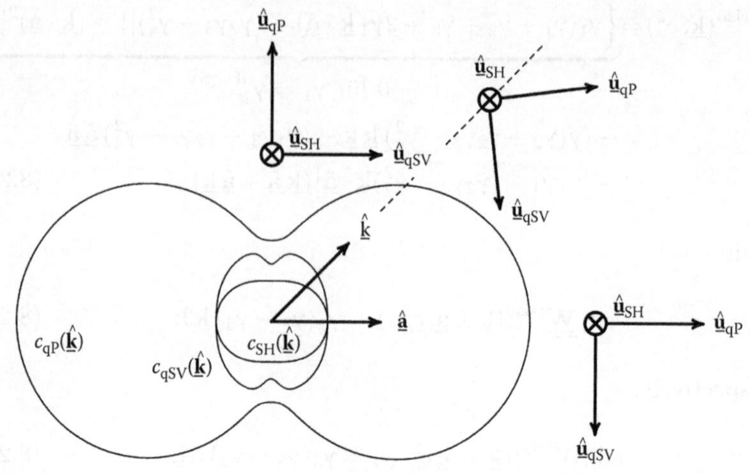

FIGURE 8.18
Phase velocities $c_{\mathrm{SH}}(\hat{\underline{\mathbf{k}}})$, $c_{\mathrm{qSV}}(\hat{\underline{\mathbf{k}}})$ and $c_{\mathrm{qP}}(\hat{\underline{\mathbf{k}}})$ and polarization vectors for fiber reinforced composite: $\lambda_{\perp} + 2\mu_{\perp} = 13.5$, $\mu_{\perp} = 3.4$, $\lambda_{\|} + 2\mu_{\|} = 145.8$, $\mu_{\|} = 6.8$, $\nu = 10.2$ [GPa]; $\rho = 1.6$ g/cm^3.

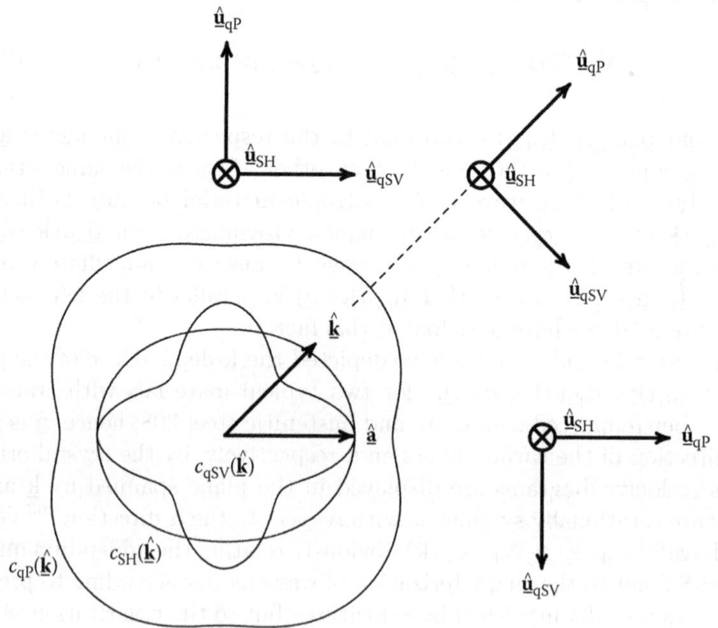

FIGURE 8.19
Phase velocities $c_{\mathrm{qSV}}(\hat{\underline{\mathbf{k}}})$, $c_{\mathrm{SH}}(\hat{\underline{\mathbf{k}}})$ and $c_{\mathrm{qP}}(\hat{\underline{\mathbf{k}}})$ and polarization vectors for austenitic steel 308: $\lambda_{\perp} + 2\mu_{\perp} = 262.75$, $\mu_{\perp} = 82.25$, $\lambda_{\|} + 2\mu_{\|} = 216$, $\mu_{\|} = 129$, $\nu = 145$ [GPa]; $\rho = 8$ g/cm^3.

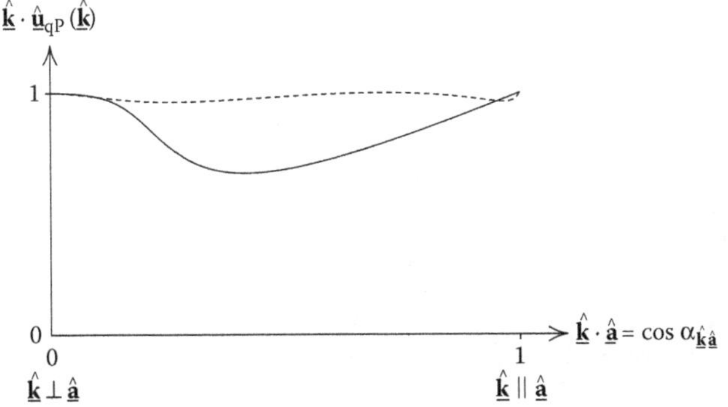

FIGURE 8.20
Longitudinal deviation of the qP-polarization vector for fiber-reinforced composite (—) and austenitic steel 308 (- - -).

nonvanishing curl and divergence); we learn from Figure 8.18 that the longitudinal deviation of the qP-mode and the transverse deviation of the qSV-mode may be significant, i.e., the primary wave is quasilongitudinally and the secondary wave is quasitransversely polarized, whereas the polarization vector of the SH-mode is always perpendicular to the $\hat{\underline{\mathbf{k}}}\,\hat{\underline{\mathbf{a}}}$-plane and, hence, to the propagation direction. This is the reason why a qP-wave is only a quasipressure wave due to its nonvanishing curl, and a qSV-wave is only a quasishear wave due to its nonvanishing divergence: We cannot separate the particle velocity according to the Helmholtz decomposition (7.28). In Figure 8.20, we have displayed the longitudinal deviation of the qP-polarization vector for both materials as a function of $\hat{\underline{\mathbf{k}}} \cdot \hat{\underline{\mathbf{a}}} = \cos\alpha_{\hat{\underline{\mathbf{k}}}\hat{\underline{\mathbf{a}}}}$, that is to say, as a function of the phase propagation angle relative to[165] $\hat{\underline{\mathbf{a}}}$: Evidently, the longitudinal deviation of the qP-mode and, hence, also for the transverse deviation of the qSV-mode is only marginal for austenitic steel.

To derive reflection, transmission, and mode conversion laws of plane waves, we use phase matching for the (plane) boundary of two materials, and to geometrically construct the respective phase propagation directions, we use the slowness diagrams (Chapter 9). For this reason, we have also plotted the corresponding slowness diagrams in Figure 8.21 corresponding to Figures 8.18 and 8.19. As we have seen in Section 8.3.1, the planes of constant slowness $s(\hat{\underline{\mathbf{k}}}) = 1/c(\hat{\underline{\mathbf{k}}})$ in anisotropic materials are important for another reason: The energy velocity vector $\underline{\mathbf{c}}_E(\hat{\underline{\mathbf{k}}})$ is orthogonal to the respective slowness surface, i.e., with the knowledge of the slowness diagram $s(\hat{\underline{\mathbf{k}}})$, we can immediately

[165]Additionally, calculating the sign of $(\hat{\underline{\mathbf{k}}} \times \hat{\underline{\mathbf{u}}}_{SH}) \cdot \hat{\underline{\mathbf{u}}}_{qP}$, we actually obtain the orientation of $\hat{\underline{\mathbf{u}}}_{qP}$ "between $\hat{\underline{\mathbf{k}}}$ and $\hat{\underline{\mathbf{a}}}$" as displayed in Figure 8.18.

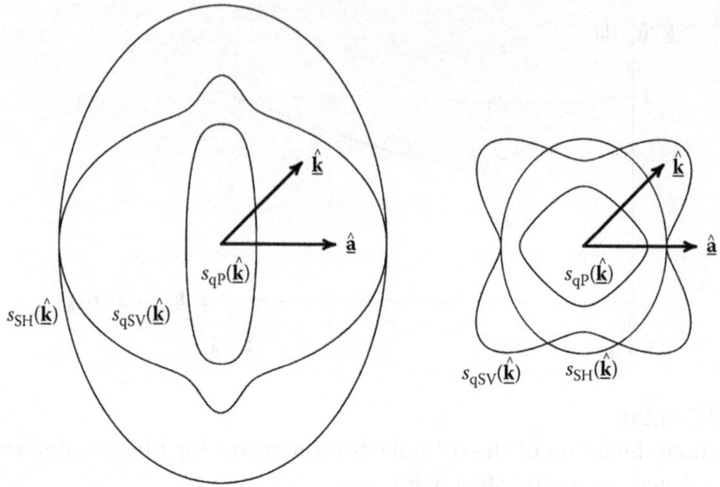

FIGURE 8.21
Slownesses $s_{\mathrm{SH}}(\hat{\underline{\mathbf{k}}})$, $s_{\mathrm{qSV}}(\hat{\underline{\mathbf{k}}})$ and $s_{\mathrm{qP}}(\hat{\underline{\mathbf{k}}})$ for fiber-reinforced composite (left) and austenitic steel 308 (right).

construct the ray direction $\hat{\underline{\mathbf{l}}}(\hat{\underline{\mathbf{k}}})$ that only coincides with the phase propagation direction $\hat{\underline{\mathbf{k}}}$ in isotropic materials (Section 8.1.2).

As we will see further below, the concave depressions of the qSV-slowness diagrams in Figure 8.21 lead to nonunique phase propagation vectors in the direction of certain ray vectors.

Energy velocities: Knowing explicitly the phase velocities and polarization vectors for the transversely isotropic material, we utilize Equation 8.217 to calculate the energy velocities.[166] With (4.24), according to

$$\begin{aligned}\underline{\underline{\underline{\underline{\mathbf{c}}}}}^{\mathrm{triso}} &= \lambda_\perp \underline{\underline{\underline{\underline{\mathbf{I}}}}}^{\delta} + 2\mu_\perp \underline{\underline{\underline{\underline{\mathbf{I}}}}}^{+} + \\ &\quad + \alpha_1 \hat{\underline{\mathbf{a}}}\hat{\underline{\mathbf{a}}}\hat{\underline{\mathbf{a}}}\hat{\underline{\mathbf{a}}} + \alpha_2(\underline{\underline{\mathbf{I}}}\,\hat{\underline{\mathbf{a}}}\hat{\underline{\mathbf{a}}} + \hat{\underline{\mathbf{a}}}\hat{\underline{\mathbf{a}}}\,\underline{\underline{\mathbf{I}}}) + \\ &\quad + \alpha_3(\underline{\underline{\mathbf{I}}}\,\hat{\underline{\mathbf{a}}}\hat{\underline{\mathbf{a}}} + \hat{\underline{\mathbf{a}}}\hat{\underline{\mathbf{a}}}\,\underline{\underline{\mathbf{I}}})^{1324} + \alpha_3(\underline{\underline{\mathbf{I}}}\,\hat{\underline{\mathbf{a}}}\hat{\underline{\mathbf{a}}} + \hat{\underline{\mathbf{a}}}\hat{\underline{\mathbf{a}}}\,\underline{\underline{\mathbf{I}}})^{1342}, \end{aligned} \tag{8.285}$$

where

$$\alpha_1 = \lambda_\perp + 2\mu_\perp + \lambda_\parallel + 2\mu_\parallel - 2(\nu + 2\mu_\parallel), \tag{8.286}$$

$$\alpha_2 = \nu - \lambda_\perp, \tag{8.287}$$

$$\alpha_3 = \mu_\parallel - \mu_\perp, \tag{8.288}$$

[166] Spies (1992, 1994) utilizes Equation 8.237 that leads to much more complicated expressions.

we immediately calculate[167]

$$\underline{\underline{\underline{\underline{\mathbf{c}}}}}^{\text{triso}} : \hat{\underline{\mathbf{k}}}\,\hat{\underline{\mathbf{u}}} = \lambda_\perp \hat{\underline{\mathbf{k}}}\cdot\hat{\underline{\mathbf{u}}}\,\underline{\underline{\mathbf{I}}} + \mu_\perp(\hat{\underline{\mathbf{k}}}\,\hat{\underline{\mathbf{u}}} + \hat{\underline{\mathbf{u}}}\,\hat{\underline{\mathbf{k}}})$$
$$+ \alpha_1 \hat{\underline{\mathbf{k}}}\cdot\hat{\underline{\mathbf{a}}}\,\hat{\underline{\mathbf{a}}}\cdot\hat{\underline{\mathbf{u}}}\,\hat{\underline{\mathbf{a}}}\,\hat{\underline{\mathbf{a}}} + \alpha_2(\hat{\underline{\mathbf{k}}}\cdot\hat{\underline{\mathbf{a}}}\,\hat{\underline{\mathbf{a}}}\cdot\hat{\underline{\mathbf{u}}}\,\underline{\underline{\mathbf{I}}} + \hat{\underline{\mathbf{k}}}\cdot\hat{\underline{\mathbf{u}}}\,\hat{\underline{\mathbf{a}}}\,\hat{\underline{\mathbf{a}}})$$
$$+ \alpha_3(\hat{\underline{\mathbf{k}}}\cdot\hat{\underline{\mathbf{a}}}\,\hat{\underline{\mathbf{u}}}\,\hat{\underline{\mathbf{a}}} + \hat{\underline{\mathbf{u}}}\cdot\hat{\underline{\mathbf{a}}}\,\hat{\underline{\mathbf{a}}}\,\hat{\underline{\mathbf{k}}} + \hat{\underline{\mathbf{u}}}\cdot\hat{\underline{\mathbf{a}}}\,\hat{\underline{\mathbf{k}}}\,\hat{\underline{\mathbf{a}}} + \hat{\underline{\mathbf{k}}}\cdot\hat{\underline{\mathbf{a}}}\,\hat{\underline{\mathbf{a}}}\,\hat{\underline{\mathbf{u}}}) \tag{8.289}$$

and subsequently

$$\underline{\mathbf{c}}_{\mathrm{E}}(\hat{\underline{\mathbf{k}}}) = \frac{1}{\rho\, c(\hat{\underline{\mathbf{k}}})}\Big[\lambda_\perp \hat{\underline{\mathbf{k}}}\cdot\hat{\underline{\mathbf{u}}}\,\hat{\underline{\mathbf{u}}} + \mu_\perp \hat{\underline{\mathbf{k}}} + \mu_\perp \hat{\underline{\mathbf{k}}}\cdot\hat{\underline{\mathbf{u}}}\,\hat{\underline{\mathbf{u}}} + \alpha_1 \hat{\underline{\mathbf{k}}}\cdot\hat{\underline{\mathbf{a}}}(\hat{\underline{\mathbf{u}}}\cdot\hat{\underline{\mathbf{a}}})^2\hat{\underline{\mathbf{a}}}$$
$$+ \alpha_2(\hat{\underline{\mathbf{k}}}\cdot\hat{\underline{\mathbf{a}}}\,\hat{\underline{\mathbf{a}}}\cdot\hat{\underline{\mathbf{u}}}\,\hat{\underline{\mathbf{u}}} + \hat{\underline{\mathbf{k}}}\cdot\hat{\underline{\mathbf{u}}}\,\hat{\underline{\mathbf{u}}}\cdot\hat{\underline{\mathbf{a}}}\,\hat{\underline{\mathbf{a}}}) + \alpha_3(\hat{\underline{\mathbf{k}}}\cdot\hat{\underline{\mathbf{a}}}\,\hat{\underline{\mathbf{a}}}\cdot\hat{\underline{\mathbf{u}}}\,\hat{\underline{\mathbf{u}}}$$
$$+ \hat{\underline{\mathbf{a}}}\cdot\hat{\underline{\mathbf{u}}}\,\hat{\underline{\mathbf{u}}}\cdot\hat{\underline{\mathbf{k}}}\,\hat{\underline{\mathbf{a}}} + \hat{\underline{\mathbf{u}}}\cdot\hat{\underline{\mathbf{a}}}\,\hat{\underline{\mathbf{a}}}\cdot\hat{\underline{\mathbf{u}}}\,\hat{\underline{\mathbf{k}}} + \hat{\underline{\mathbf{k}}}\cdot\hat{\underline{\mathbf{a}}}\,\hat{\underline{\mathbf{a}}})\Big]. \tag{8.290}$$

With $\hat{\underline{\mathbf{k}}}\cdot\hat{\underline{\mathbf{u}}}_{\mathrm{SH}}(\hat{\underline{\mathbf{k}}}) = 0$ and $\hat{\underline{\mathbf{u}}}_{\mathrm{SH}}(\hat{\underline{\mathbf{k}}})\cdot\hat{\underline{\mathbf{a}}} = 0$, we directly obtain $\underline{\mathbf{c}}_{\mathrm{ESH}}(\hat{\underline{\mathbf{k}}})$ as:

$$\underline{\mathbf{c}}_{\mathrm{ESH}}(\hat{\underline{\mathbf{k}}}) = \frac{\mu_\perp \hat{\underline{\mathbf{k}}} + (\mu_\parallel - \mu_\perp)\hat{\underline{\mathbf{k}}}\cdot\hat{\underline{\mathbf{a}}}\,\hat{\underline{\mathbf{a}}}}{\rho\, c_{\mathrm{SH}}(\hat{\underline{\mathbf{k}}})}. \tag{8.291}$$

Disregarding the special cases

$$\underline{\mathbf{c}}_{\mathrm{ESH}}(\hat{\underline{\mathbf{k}}}\perp\hat{\underline{\mathbf{a}}}) = c_{\mathrm{SH}}(\hat{\underline{\mathbf{k}}}\perp\hat{\underline{\mathbf{a}}})\,\hat{\underline{\mathbf{k}}}, \tag{8.292}$$
$$\underline{\mathbf{c}}_{\mathrm{ESH}}(\hat{\underline{\mathbf{k}}}\parallel\hat{\underline{\mathbf{a}}}) = c_{\mathrm{SH}}(\hat{\underline{\mathbf{k}}}\parallel\hat{\underline{\mathbf{a}}})\,\hat{\underline{\mathbf{k}}} \tag{8.293}$$

the energy velocity vector $\underline{\mathbf{c}}_{\mathrm{ESH}}(\hat{\underline{\mathbf{k}}})$ actually does not have the direction of the phase velocity vector for $\mu_\parallel \neq \mu_\perp$. For $\mu_\parallel = \mu_\perp$, the propagation of the SH-mode is isotropic.

To calculate $\underline{\mathbf{c}}_{\mathrm{EqP,qSV}}(\hat{\underline{\mathbf{k}}})$ numerically, it is advisable to insert the respective polarization vectors

$$\hat{\underline{\mathbf{u}}}_{\mathrm{qP,qSV}}(\hat{\underline{\mathbf{k}}}) = \frac{\hat{\underline{\mathbf{k}}} + \gamma_{\mathrm{qP,qSV}}\hat{\underline{\mathbf{a}}}}{U_{\mathrm{qP,qSV}}} \tag{8.294}$$

[167]We have, for example:

$$\hat{\underline{\mathbf{a}}}\,\hat{\underline{\mathbf{a}}}\,\underline{\underline{\mathbf{I}}}^{1342} : \hat{\underline{\mathbf{k}}}\,\hat{\underline{\mathbf{u}}} = \hat{a}_i\hat{a}_j\delta_{kl}\underline{\mathbf{e}}_{x_i}\underline{\mathbf{e}}_{x_k}\underline{\mathbf{e}}_{x_l}\underline{\mathbf{e}}_{x_j} : \hat{k}_n\hat{u}_m\underline{\mathbf{e}}_{x_n}\underline{\mathbf{e}}_{x_m}$$
$$= \hat{a}_i\hat{a}_j\delta_{kl}\delta_{jn}\delta_{lm}\hat{k}_n\hat{u}_m\underline{\mathbf{e}}_{x_i}\underline{\mathbf{e}}_{x_k}$$
$$= \hat{a}_i\hat{a}_n\hat{k}_n\hat{u}_m\underline{\mathbf{e}}_{x_i}\underline{\mathbf{e}}_{x_m}$$
$$= (\hat{\underline{\mathbf{k}}}\cdot\hat{\underline{\mathbf{a}}})\hat{\underline{\mathbf{a}}}\,\hat{\underline{\mathbf{u}}}.$$

in a way that a $\hat{\underline{\mathbf{k}}}\,\hat{\underline{\mathbf{a}}}$-component decomposition of the velocity vectors results:

$$\begin{aligned}
&\rho\, c_{\mathrm{qP,qSV}}(\hat{\underline{\mathbf{k}}})\underline{\mathbf{c}}_{\mathrm{EqP,qSV}}(\hat{\underline{\mathbf{k}}}) \\
&\quad = \Big\{\mu_\perp + \alpha_3 \hat{\underline{\mathbf{u}}}_{\mathrm{qP,qSV}} \cdot \hat{\underline{\mathbf{a}}}\,\hat{\underline{\mathbf{a}}} \cdot \hat{\underline{\mathbf{u}}}_{\mathrm{qP,qSV}} \\
&\qquad + \frac{1}{U_{\mathrm{qP,qSV}}}\Big[(\lambda_\perp + \mu_\perp)\hat{\underline{\mathbf{k}}} \cdot \hat{\underline{\mathbf{u}}}_{\mathrm{qP,qSV}} + (\alpha_2 + \alpha_3)\hat{\underline{\mathbf{k}}} \cdot \hat{\underline{\mathbf{a}}}\,\hat{\underline{\mathbf{a}}} \cdot \hat{\underline{\mathbf{u}}}_{\mathrm{qP,qSV}}\Big]\Big\}\,\hat{\underline{\mathbf{k}}} \\
&\quad + \Big\{(\alpha_1 \hat{\underline{\mathbf{k}}} \cdot \hat{\underline{\mathbf{a}}}\,\hat{\underline{\mathbf{a}}} \cdot \hat{\underline{\mathbf{u}}}_{\mathrm{qP,qSV}} + \alpha_2 \hat{\underline{\mathbf{k}}} \cdot \hat{\underline{\mathbf{u}}}_{\mathrm{qP,qSV}})\hat{\underline{\mathbf{u}}}_{\mathrm{qP,qSV}} \cdot \hat{\underline{\mathbf{a}}} \\
&\qquad + \alpha_3(\hat{\underline{\mathbf{k}}} \cdot \hat{\underline{\mathbf{a}}} + \hat{\underline{\mathbf{k}}} \cdot \hat{\underline{\mathbf{u}}}_{\mathrm{qP,qSV}} \hat{\underline{\mathbf{u}}}_{\mathrm{qP,qSV}} \cdot \hat{\underline{\mathbf{a}}}) \\
&\qquad + \frac{\gamma_{\mathrm{qP,qSV}}}{U_{\mathrm{qP,qSV}}}\Big[(\lambda_\perp + \mu_\perp)\hat{\underline{\mathbf{k}}} \cdot \hat{\underline{\mathbf{u}}}_{\mathrm{qP,qSV}} + (\alpha_2 + \alpha_3)\hat{\underline{\mathbf{k}}} \cdot \hat{\underline{\mathbf{a}}}\,\hat{\underline{\mathbf{a}}} \cdot \hat{\underline{\mathbf{u}}}_{\mathrm{qP,qSV}}\Big]\Big\}\,\hat{\underline{\mathbf{a}}}.
\end{aligned} \tag{8.295}$$

With Figure 8.13, we directly conclude from the slowness diagrams of Figure 8.21 that all energy velocity vectors $\underline{\mathbf{c}}_{\mathrm{ESH,qP,qSV}}(\hat{\underline{\mathbf{k}}})$ have the direction of $\hat{\underline{\mathbf{k}}}$ for $\hat{\underline{\mathbf{k}}} \perp \hat{\underline{\mathbf{a}}}$ and $\hat{\underline{\mathbf{k}}} \parallel \hat{\underline{\mathbf{a}}}$; as a matter of fact, we even have

$$\underline{\mathbf{c}}_{\mathrm{EqP,qSV}}(\hat{\underline{\mathbf{k}}} \perp \hat{\underline{\mathbf{a}}}) = c_{\mathrm{qP,qSV}}(\hat{\underline{\mathbf{k}}} \perp \hat{\underline{\mathbf{a}}})\,\hat{\underline{\mathbf{k}}}, \tag{8.296}$$

$$\underline{\mathbf{c}}_{\mathrm{EqP,qSV}}(\hat{\underline{\mathbf{k}}} \parallel \hat{\underline{\mathbf{a}}}) = c_{\mathrm{qP,qSV}}(\hat{\underline{\mathbf{k}}} \parallel \hat{\underline{\mathbf{a}}})\,\hat{\underline{\mathbf{k}}} \tag{8.297}$$

as completion of (8.292) and (8.293). Numerical evaluation of (8.291) and (8.295), respectively, that is to say, production of energy velocity diagrams goes as such: We prescribe $\hat{\underline{\mathbf{k}}}$-vectors and plot $|\underline{\mathbf{c}}_{\mathrm{ESH,qSV,qP}}(\hat{\underline{\mathbf{k}}})| = c_{\mathrm{ESH,qSV,qP}}(\hat{\underline{\mathbf{k}}})$ in the direction of the unit vectors $\hat{\underline{\mathbf{c}}}_{\mathrm{ESH,qSV,qP}}(\hat{\underline{\mathbf{k}}})$, i.e., in the direction of the (unit) ray vectors $\hat{\underline{\mathbf{l}}}(\hat{\underline{\mathbf{k}}})$, where the functions $\hat{\underline{\mathbf{l}}}(\hat{\underline{\mathbf{k}}})$ for the three wave modes are generally nonlinear. For our model of the fiber-reinforced composite, the diagrams of Figure 8.22 and for our model austenitic steel 308 those of Figure 8.23 are obtained. Strikingly appealing are the cusps of qSV-diagrams, whose origin is the nonuniqueness of the respective

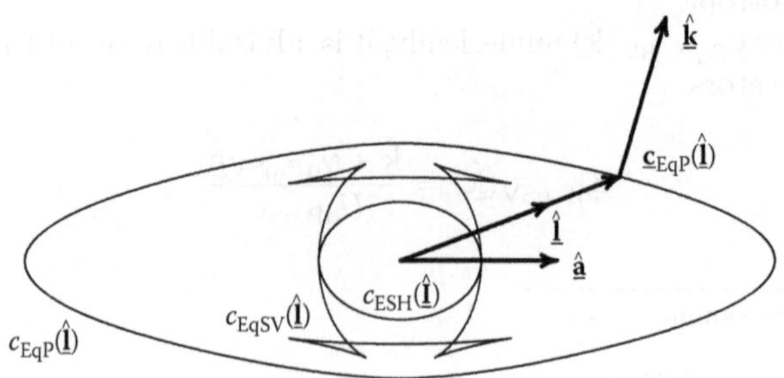

FIGURE 8.22
Energy velocities $c_{\mathrm{ESH}}(\hat{\underline{\mathbf{k}}})$, $c_{\mathrm{EqSV}}(\hat{\underline{\mathbf{k}}})$ and $c_{\mathrm{EqP}}(\hat{\underline{\mathbf{k}}})$ for fiber reinforced composite: $\lambda_\perp + 2\mu_\perp = 13.5$, $\mu_\perp = 3.4$, $\lambda_\parallel + 2\mu_\parallel = 145.8$, $\mu_\parallel = 6.8$, $\nu = 10.2$ [GPa]; $\rho = 1.6$ g/cm^3.

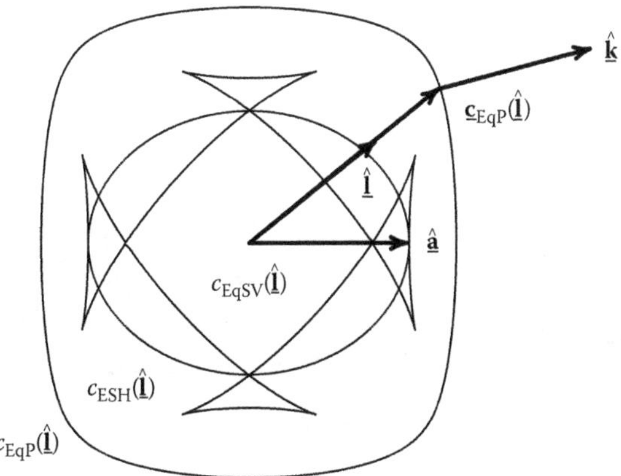

FIGURE 8.23
Energy velocities $c_{\mathrm{ESH}}(\hat{\underline{\mathbf{k}}})$, $c_{\mathrm{EqSV}}(\hat{\underline{\mathbf{k}}})$ and $c_{\mathrm{EqP}}(\hat{\underline{\mathbf{k}}})$ for austenitic steel 308: $\lambda_\perp + 2\mu_\perp = 262.75$, $\mu_\perp = 82.25$, $\lambda_\parallel + 2\mu_\parallel = 216$, $\mu_\parallel = 129$, $\nu = 145$ [GPa]; $\rho = 8$ g/cm^3.

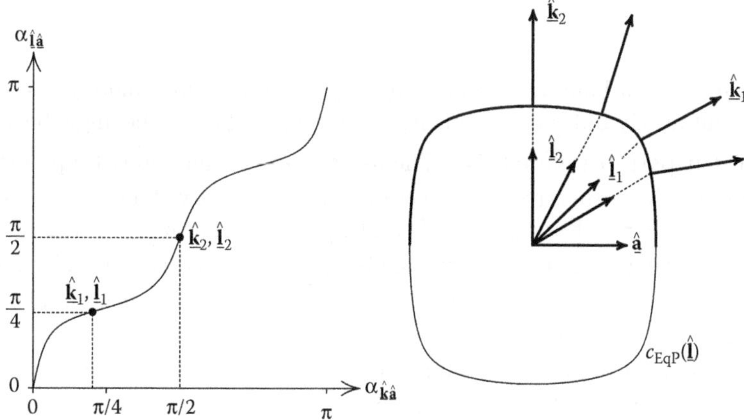

FIGURE 8.24
Illustration of the nonlinear relation between the phase propagation vector $\hat{\underline{\mathbf{k}}}$ and the ray vector $\hat{\underline{\mathbf{l}}}$ for the qP-energy velocity diagram of austenitic steel 308.

$\hat{\underline{\mathbf{k}}}(\hat{\underline{\mathbf{l}}})$-function. Before we discuss that in detail with the help of Figure 8.25, we first consider the simpler (and nonambiguous) case of a qP-diagram (Figure 8.24). We know from Section 8.3.1 (Equation 8.229) that, for a given ray vector $\hat{\underline{\mathbf{l}}}(\hat{\underline{\mathbf{k}}})$, the phase propagation vector $\hat{\underline{\mathbf{k}}}$ is always perpendicular to the surface of the energy velocity diagram—the wave surface. This is illustrated in Figure 8.24 for several ray vectors for the qP-diagram of austenitic steel 308.

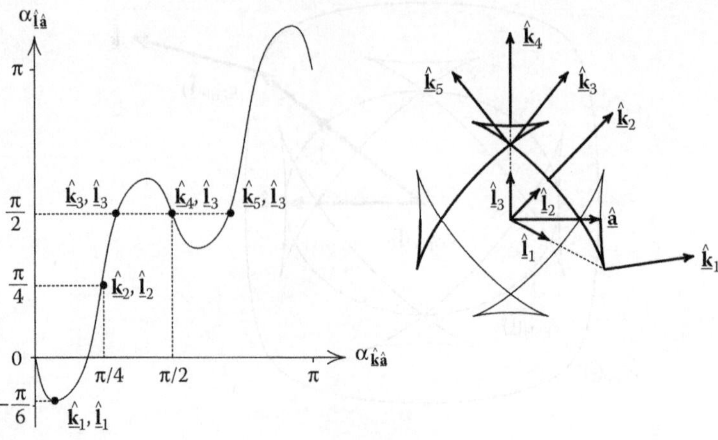

FIGURE 8.25
Illustration of the nonlinear relation of the phase propagation vector $\hat{\underline{k}}$ and the ray vector $\hat{\underline{l}}$ for the qSV-energy velocity diagram of austenitic steel 308 (for clarity, the $\hat{\underline{l}}$-(unit)vectors have only half the length as compared to the $\hat{\underline{k}}$-(unit)vectors).

The angle between $\hat{\underline{l}}$ and $\hat{\underline{a}}$ is denoted by $\alpha_{\hat{\underline{l}}\hat{\underline{a}}}$, obviously this angle grows, even though nonlinearly but monotonously, as a function of $\alpha_{\hat{\underline{k}}\hat{\underline{a}}}$, the angle between $\hat{\underline{k}}$ and $\hat{\underline{a}}$ (left part of Figure 8.24). The assignment of characteristic points on this curve to the respective $\hat{\underline{l}}, \hat{\underline{k}}$-directions of the velocity diagram is readily obtained: $\alpha_{\hat{\underline{l}}_1\hat{\underline{a}}} = \pi/4$ has $\alpha_{\hat{\underline{k}}_1\hat{\underline{a}}} = \pi/4$ and $\alpha_{\hat{\underline{l}}_2\hat{\underline{a}}} = \pi/2$ has $\alpha_{\hat{\underline{k}}_2\hat{\underline{a}}} = \pi/2$ as consequence; starting from $\alpha_{\hat{\underline{l}}\hat{\underline{a}}} = 0$, the angle $\alpha_{\hat{\underline{k}}\hat{\underline{a}}}$ slowly and then rapidly grows until the direction $\alpha_{\hat{\underline{l}}_1\hat{\underline{a}}} = \alpha_{\hat{\underline{k}}_1\hat{\underline{a}}} = \pi/4$ is reached for both vectors, and then, for $\pi/4 < \alpha_{\hat{\underline{l}}\hat{\underline{a}}} < \pi/2$, we observe the reverse behavior, and for $\pi/2 < \alpha_{\hat{\underline{l}}\hat{\underline{a}}} < \pi$, everything starts from the beginning. The boldface part of the energy velocity diagram refers to the interval $0 \leq \alpha_{\hat{\underline{k}}\hat{\underline{a}}} \leq \pi$.

With Figure 8.25, we turn to the already mentioned cusps of the energy velocity diagrams of austenitic steel 308. At first, the $\alpha_{\hat{\underline{l}}\hat{\underline{a}}}(\alpha_{\hat{\underline{k}}\hat{\underline{a}}})$-curve on the left side of the figure reveals its nonlinearity and additionally being nonmonotonous: A specific $\hat{\underline{l}}$-value—for example: $\hat{\underline{l}}_3$—may belong to three $\hat{\underline{k}}$-values, here: $\hat{\underline{k}}_3, \hat{\underline{k}}_4, \hat{\underline{k}}_5$. Again, we discuss the mapping of the $(0 \leq \alpha_{\hat{\underline{k}}\hat{\underline{a}}} \leq \pi)$-interval into a respective $\alpha_{\hat{\underline{l}}\hat{\underline{a}}}$-interval: Obviously—we directly notice it from the velocity diagram—$\alpha_{\hat{\underline{k}}\hat{\underline{a}}} = 0$ belongs to $\alpha_{\hat{\underline{l}}\hat{\underline{a}}} = 0$. Starting from this origin $\alpha_{\hat{\underline{k}}\hat{\underline{a}}}$ increases only if $\alpha_{\hat{\underline{l}}\hat{\underline{a}}}$ becomes negative until it reaches the lower tip of the boldface part of the right-hand cusp; afterward, $\alpha_{\hat{\underline{k}}\hat{\underline{a}}}$ still increases whereas $\alpha_{\hat{\underline{l}}\hat{\underline{a}}}$ heads for a second zero. The value $\alpha_{\hat{\underline{l}}\hat{\underline{a}}} = \pi/4$ is related—as for the qP-diagram—to the value $\alpha_{\hat{\underline{k}}\hat{\underline{a}}} = \pi/4$. In the following, $\alpha_{\hat{\underline{l}}\hat{\underline{a}}}$ passes through the complete upper cusp that manifests itself in the $\alpha_{\hat{\underline{l}}\hat{\underline{a}}}(\alpha_{\hat{\underline{k}}\hat{\underline{a}}})$-curve as consecutiveness of a local maximum and a local minimum. To the $\hat{\underline{\mathbf{l}}}_3$-direction—$\alpha_{\hat{\underline{\mathbf{l}}}_3\hat{\underline{\mathbf{a}}}} = \pi/2$—belongs a $\hat{\underline{\mathbf{k}}}_3$-direction with $\pi/2 < \alpha_{\hat{\underline{\mathbf{k}}}_3\hat{\underline{\mathbf{a}}}} \underset{\sim}{\to} > \pi/4$, the left-hand cusp tip corresponds to the subsequent maximum of the $\alpha_{\hat{\underline{\mathbf{l}}}\hat{\underline{\mathbf{a}}}}(\alpha_{\hat{\underline{\mathbf{k}}}\hat{\underline{\mathbf{a}}}})$-curve, and finally, we find the value $\alpha_{\hat{\underline{\mathbf{k}}}_4\hat{\underline{\mathbf{a}}}} = \pi/2$ for $\alpha_{\hat{\underline{\mathbf{l}}}_3\hat{\underline{\mathbf{a}}}} = \pi/2$. The subsequent minimum corresponds to the right-hand tip of the upper cusp and the ensuing double point of the velocity diagram (double point in the $\hat{\underline{\mathbf{k}}}\,\hat{\underline{\mathbf{a}}}$-plane) now belongs to the phase propagation vector $\hat{\underline{\mathbf{k}}}_5$: One ray direction $\hat{\underline{\mathbf{l}}}_3$ with $\alpha_{\hat{\underline{\mathbf{l}}}_3\hat{\underline{\mathbf{a}}}} = \pi/2$ relates to three differently oriented phase surfaces where two of them propagate with the same energy velocity and one with a larger energy velocity, that is to say, $\hat{\underline{\mathbf{k}}}$ is no longer orthogonal to the wavefronts as in Figure 8.7. It becomes exceptionally clear in Figure 8.26: Here, we have

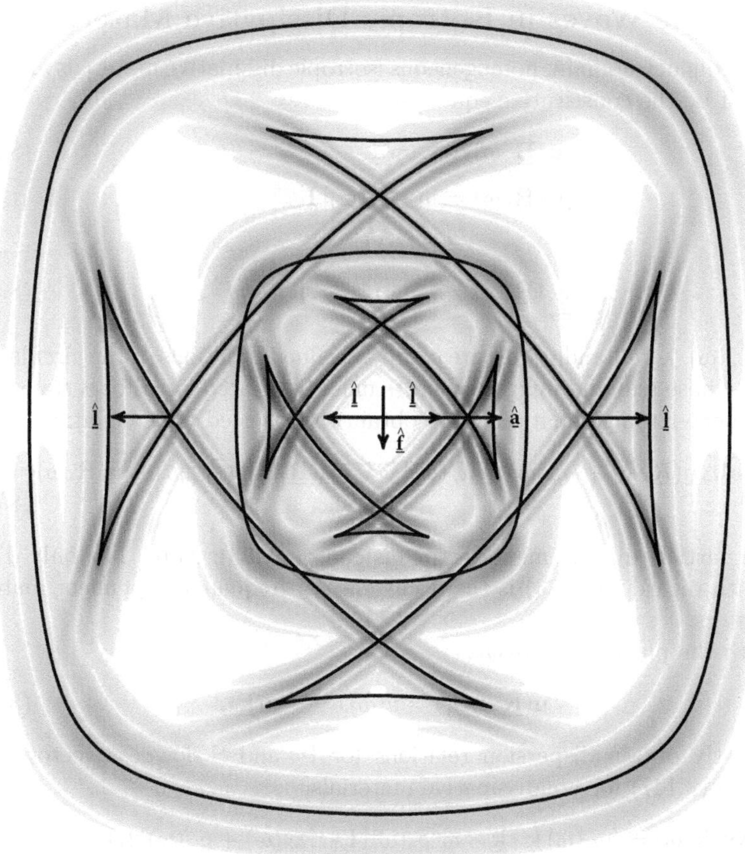

FIGURE 8.26
2D-EFIT-qP/qSV-RC2(t)-wavefronts of a line force density $\hat{\underline{\mathbf{f}}}$ in austenitic steel 308 for two different times.

calculated EFIT-RC2(t)-wavefronts for two different times that originate as qP- and qSV-wavefronts from a line force density in austenitic steel 308. In Section 8.3.1, we already mentioned that time domain wave fronts of point and line sources (Green functions) must coincide with the energy velocity diagrams of plane waves; this is nicely highlighted by the respective velocity diagrams, where we observe that indeed three phase surfaces exist in the directions of the cusp double points. They do not propagate into the directions of the three phase vectors but into the direction of the one ray vector (Equation 8.230). Two of them even possess different energy velocities that may lead to confusion interpreting seismic bore hole signals (Wang 2002).

8.4 Plane Waves in Isotropic Dissipative Materials

In the frequency domain, homogeneous isotropic dissipative materials are characterized by the constitutive equations

$$\underline{\underline{\mu}}(\underline{\mathbf{R}}, \omega) = \varepsilon_c(\omega)\underline{\underline{\mathbf{I}}}, \tag{8.298}$$

$$\begin{aligned} \underline{\underline{\rho}}_c(\underline{\mathbf{R}}, \omega) &= [\rho + \varepsilon_c(\omega)]\,\underline{\underline{\mathbf{I}}} \\ &= \rho_c(\omega)\,\underline{\underline{\mathbf{I}}}, \end{aligned} \tag{8.299}$$

$$\underline{\underline{\underline{\underline{\mathbf{c}}}}}_c(\underline{\mathbf{R}}, \omega) = \lambda_c(\omega)\,\underline{\underline{\underline{\underline{\mathbf{I}}}}}^{\delta} + 2\mu_c(\omega)\,\underline{\underline{\underline{\underline{\mathbf{I}}}}}^{+}, \tag{8.300}$$

where $\varepsilon_c(\omega)$, $\lambda_c(\omega)$ and $\mu_c(\omega)$ are complex-valued functions of the circular frequency whose real and imaginary parts must be related by Hilbert transforms as Kramers–Kronig equations. We obtain the reduced wave equation

$$\mu_c(\omega)\Delta\underline{\mathbf{u}}(\underline{\mathbf{R}}, \omega) + [\lambda_c(\omega) + \mu_c(\omega)]\nabla\nabla \cdot \underline{\mathbf{u}}(\underline{\mathbf{R}}, \omega) + \omega^2\rho_c(\omega)\underline{\mathbf{u}}(\underline{\mathbf{R}}, \omega) = \underline{\mathbf{0}} \tag{8.301}$$

for the particle displacement vector instead of (8.53). Note: The only difference with regard to (8.53) is the occurrence of frequency-dependent material parameters.

With the plane wave *ansatz*

$$\underline{\mathbf{u}}(\underline{\mathbf{R}}, \omega, \underline{\mathbf{k}}) = u(\omega)\,\mathrm{e}^{\mathrm{j}\underline{\mathbf{k}}\cdot\underline{\mathbf{R}}}\,\hat{\underline{\mathbf{u}}}(\underline{\mathbf{k}}), \tag{8.302}$$

we directly obtain dispersion relations for P- and S-slowness vectors from [compare (8.147) for nondissipative materials]

$$\underline{\underline{\mathbf{W}}}(\underline{\mathbf{k}}, \omega) = [\mu_c(\omega)\,\underline{\mathbf{k}} \cdot \underline{\mathbf{k}} - \omega^2\rho_c(\omega)]\,\underline{\underline{\mathbf{I}}} + [\lambda_c(\omega) + \mu_c(\omega)]\,\underline{\mathbf{k}}\,\underline{\mathbf{k}} \tag{8.303}$$

and from the requirement $\det\underline{\underline{\mathbf{W}}}(\underline{\mathbf{k}}, \omega) = 0$:

$$\underline{\mathbf{k}}_{\mathrm{P}} \cdot \underline{\mathbf{k}}_{\mathrm{P}} = \omega^2\,\frac{\rho_c(\omega)}{\lambda_c(\omega) + 2\mu_c(\omega)} \stackrel{\mathrm{def}}{=} k_{\mathrm{P}c}^2(\omega), \tag{8.304}$$

$$\underline{\mathbf{k}}_S \cdot \underline{\mathbf{k}}_S = \omega^2 \frac{\rho_c(\omega)}{\mu_c(\omega)} \stackrel{\text{def}}{=} k_{Sc}^2(\omega); \tag{8.305}$$

the respective right-hand sides identify them as actual dispersion equations. The second sign of equality in (8.304) and (8.305) defines complex valued non-linear frequency-dependent[168] wave numbers $k_{P,Sc}(\omega)$ in each case. Evidently, the possible solutions $\underline{\mathbf{k}}_{P,S}$ of the dispersion relations are also complex valued vectors:

$$\underline{\mathbf{k}}_{P,S} = \Re \underline{\mathbf{k}}_{P,S} + j\Im \underline{\mathbf{k}}_{P,S}. \tag{8.306}$$

The P,S-indices already suggest that we are searching for plane pressure and shear waves as solutions of the reduced wave equation (8.301). Therefore, we again determine the polarization vectors —similar to Section 8.2—from the requirement

$$\nabla \times \underline{\mathbf{u}}_P(\underline{\mathbf{R}}, \omega, \underline{\mathbf{k}}_P) = \underline{\mathbf{0}}, \tag{8.307}$$

$$\nabla \cdot \underline{\mathbf{u}}_S(\underline{\mathbf{R}}, \omega, \underline{\mathbf{k}}_S) = 0 \tag{8.308}$$

and, respectively,

$$\underline{\mathbf{k}}_P \times \underline{\hat{\mathbf{u}}}_P(\underline{\mathbf{k}}_P) = \underline{\mathbf{0}}, \tag{8.309}$$

$$\underline{\mathbf{k}}_S \cdot \underline{\hat{\mathbf{u}}}_S(\underline{\mathbf{k}}_S) = 0; \tag{8.310}$$

together with the Hermitian normalization condition

$$\underline{\hat{\mathbf{u}}}_{P,S} \cdot \underline{\hat{\mathbf{u}}}_{P,S}^* = 1, \tag{8.311}$$

we obtain polarization unit vectors as in Section 8.2:

$$\underline{\hat{\mathbf{u}}}_P(\underline{\mathbf{k}}_P) = \frac{\underline{\mathbf{k}}_P}{\sqrt{\underline{\mathbf{k}}_P \cdot \underline{\mathbf{k}}_P^*}}, \tag{8.312}$$

$$\underline{\hat{\mathbf{u}}}_{S1}(\underline{\mathbf{k}}_S) = \frac{\underline{\hat{\mathbf{u}}}_{S2} \times \underline{\mathbf{k}}_S}{\sqrt{\underline{\mathbf{k}}_S \cdot \underline{\mathbf{k}}_S^*}}, \tag{8.313}$$

if we assume $\underline{\hat{\mathbf{u}}}_{S2}$ to be real valued with $\underline{\hat{\mathbf{u}}}_{S2} \cdot \underline{\mathbf{k}}_S = 0$.

After embedding into a cartesian coordinate system, we can choose $\underline{\hat{\mathbf{u}}}_{S2} = \underline{\hat{\mathbf{u}}}_{SH} = \underline{\mathbf{e}}_y$ and $\underline{\hat{\mathbf{u}}}_{S1} = \underline{\hat{\mathbf{u}}}_{SV}$ with $\underline{\hat{\mathbf{u}}}_{SV} \cdot \underline{\mathbf{e}}_y = 0$ alluding to Figure 8.8.

8.4.1 Homogeneous plane waves

The dispersion relations (8.304) and (8.305) for dissipative materials—as in nondissipative materials—allow for homogeneous and inhomogeneous plane P,S-wave solutions.

[168]With $k_{P,Sc}(\omega)$, we adumbrate the nonproportionality of wave numbers of complex valued materials to the circular frequency.

In the *ansatz* (8.306) for the complex wave number vectors, we identify according to

$$\underline{\mathbf{u}}_{\mathrm{P,S1/2}}(\underline{\mathbf{R}}, \omega, \underline{\mathbf{k}}_{\mathrm{P,S}}) = u(\omega)\, \mathrm{e}^{\mathrm{j}\Re\underline{\mathbf{k}}_{\mathrm{P,S}}\cdot\underline{\mathbf{R}}}\, \mathrm{e}^{-\Im\underline{\mathbf{k}}_{\mathrm{P,S}}\cdot\underline{\mathbf{R}}}\, \underline{\hat{\mathbf{u}}}_{\mathrm{P,S1/2}}(\underline{\mathbf{k}}_{\mathrm{P,S}}) \tag{8.314}$$

the real part $\Re\underline{\mathbf{k}}_{\mathrm{P,S}}$ as phase (propagation) vector and the imaginary part as attenuation vector $\Im\underline{\mathbf{k}}_{\mathrm{P,S}}$. In nondissipative (isotropic) materials, we have exactly two possibilities for $\Im\underline{\mathbf{k}}_{\mathrm{P,S}}$: either equal to the null vector—representing homogeneous plane waves—or perpendicular to $\Re\underline{\mathbf{k}}_{\mathrm{P,S}}$—representing inhomogeneous plane waves—with the orientation as given in Figure 8.11 (propagation into $+\Re\underline{\mathbf{k}}_{\mathrm{P,S}}$-direction only results in an attenuation in the half-space $\Im\underline{\mathbf{k}}_{\mathrm{P,S}} \cdot \underline{\mathbf{R}} > 0$). The resulting plane waves are either nonattenuated or evanescent with regard to the propagation direction. Exactly these two cases are excluded in dissipative materials! Namely inserting (8.306) into the dispersion equation, we obtain the separation into real and imaginary parts:

$$\Re\underline{\mathbf{k}}_{\mathrm{P,S}} \cdot \Re\underline{\mathbf{k}}_{\mathrm{P,S}} - \Im\underline{\mathbf{k}}_{\mathrm{P,S}} \cdot \Im\underline{\mathbf{k}}_{\mathrm{P,S}} = \Re k^2_{\mathrm{P,S}}(\omega), \tag{8.315}$$

$$\Re\underline{\mathbf{k}}_{\mathrm{P,S}} \cdot \Im\underline{\mathbf{k}}_{\mathrm{P,S}} = \frac{1}{2}\, \Im k^2_{\mathrm{P,S}}(\omega), \tag{8.316}$$

and on behalf of (8.316), the imaginary part $\Im\underline{\mathbf{k}}_{\mathrm{P,S}}$ can neither be zero nor can it be perpendicular to $\Re\underline{\mathbf{k}}_{\mathrm{P,S}}$ for $\Im\underline{\mathbf{k}}_{\mathrm{P,S}} \neq 0$. Two alternatives remain: $\Im\underline{\mathbf{k}}_{\mathrm{P,S}}$ parallel to $\Re\underline{\mathbf{k}}_{\mathrm{P,S}}$—resulting in homogeneous attenuated plane waves because phase and amplitude surfaces coincide—or arbitrary nonorthogonal orientation relative to $\Re\underline{\mathbf{k}}_{\mathrm{P,S}}$—resulting in inhomogeneous plane waves, whose phase and amplitude surfaces are neither parallel nor perpendicular to each other. The angle $0 \leq \sphericalangle(\Re\underline{\mathbf{k}}_{\mathrm{P,S}}, \Im\underline{\mathbf{k}}_{\mathrm{P,S}}) < \pi/2$ is not determined by the elastodynamic governing equations, it is an arbitrary parameter.

We obtain homogeneous plane waves that are attenuated in propagation direction $\underline{\mathbf{e}}_\zeta$ for $\zeta > 0$ as solutions

$$\underline{\mathbf{k}}_{\mathrm{P,S}} = k_{\mathrm{P,Sc}}(\omega)\, \underline{\mathbf{e}}_\zeta \tag{8.317}$$

of the dispersion equations yielding

$$\Re\underline{\mathbf{k}}_{\mathrm{P,S}} = \Re k_{\mathrm{P,Sc}}(\omega)\, \underline{\mathbf{e}}_\zeta, \tag{8.318}$$

$$\Im\underline{\mathbf{k}}_{\mathrm{P,S}} = \Im k_{\mathrm{P,Sc}}(\omega)\, \underline{\mathbf{e}}_\zeta \tag{8.319}$$

(compare Section 4.4.1). To give these phase and attenuation vectors explicitly, we must calculate the real and imaginary part of $k_{\mathrm{P,Sc}}(\omega) = \Re k_{\mathrm{P,Sc}}(\omega) + \mathrm{j}\Im k_{\mathrm{P,Sc}}(\omega)$ for prescribed real and imaginary part of $k^2_{\mathrm{P,Sc}}(\omega) = \Re k^2_{\mathrm{P,Sc}}(\omega) + \mathrm{j}\Im k^2_{\mathrm{P,Sc}}(\omega)$; we obtain (Equations 2.268 and 2.269 with the adequate choice of sign):

$$\begin{aligned}\Re k_{\mathrm{P,Sc}}(\omega) &= \frac{1}{\sqrt{2}}\sqrt{\Re k^2_{\mathrm{P,Sc}}(\omega) + \sqrt{\left[\Re k^2_{\mathrm{P,Sc}}(\omega)\right]^2 + \left[\Im k^2_{\mathrm{P,Sc}}(\omega)\right]^2}} \\ &= \frac{1}{\sqrt{2}}\sqrt{|k^2_{\mathrm{P,Sc}}(\omega)| + \Re k^2_{\mathrm{P,Sc}}(\omega)},\end{aligned} \tag{8.320}$$

$$
\begin{aligned}
\Im k_{\mathrm{P,Sc}}(\omega) &= \frac{1}{\sqrt{2}}\sqrt{-\Re k_{\mathrm{P,Sc}}^2(\omega) + \sqrt{\left[\Re k_{\mathrm{P,Sc}}^2(\omega)\right]^2 + \left[\Im k_{\mathrm{P,Sc}}^2(\omega)\right]^2}} \\
&= \frac{1}{\sqrt{2}}\sqrt{|k_{\mathrm{P,Sc}}^2(\omega)| - \Re k_{\mathrm{P,Sc}}^2(\omega)},
\end{aligned} \tag{8.321}
$$

as solution of the system of equations

$$
[\Re k_{\mathrm{P,Sc}}(\omega)]^2 - [\Im k_{\mathrm{P,Sc}}(\omega)]^2 = \Re k_{\mathrm{P,Sc}}^2(\omega), \tag{8.322}
$$

$$
\Re k_{\mathrm{P,Sc}}(\omega)\Im k_{\mathrm{P,Sc}}(\omega) = \frac{1}{2}\,\Im k_{\mathrm{P,Sc}}^2(\omega) \tag{8.323}
$$

that follows from the separation into real and imaginary part of

$$
\begin{aligned}
k_{\mathrm{P,Sc}}^2(\omega) &= \Re k_{\mathrm{P,Sc}}^2(\omega) + \mathrm{j}\Im k_{\mathrm{P,Sc}}^2(\omega) \\
&= [\Re k_{\mathrm{P,Sc}}(\omega) + \mathrm{j}\Im k_{\mathrm{P,Sc}}(\omega)]^2
\end{aligned} \tag{8.324}
$$

(Equations 8.304, 8.305, with 8.317). Hence, we obtain

$$
c_{\mathrm{P,S}}(\omega) = \frac{\omega}{\Re k_{\mathrm{P,Sc}}(\omega)} \tag{8.325}
$$

as frequency-dependent phase velocity of homogeneous dispersive P- and S-waves in homogeneous isotropic dissipative materials. Of course, this expression is also obtained through specialization of the general formula (8.175).

The energy velocities of homogeneous plane waves in dissipative materials are directly found inserting (8.317) into (8.184), (8.185), (8.189), (8.190), (8.196), and (8.197) and subsequent division:

$$
\begin{aligned}
&\Re \underline{\mathbf{S}}_{\mathrm{KP,S1/S2}}(\underline{\mathbf{R}}, \omega, \underline{\mathbf{k}}_{\mathrm{P,S}}) \\
&\quad = \frac{\varrho\omega}{2}\, c_{\mathrm{P,S}}^2\, |u(\omega)|^2 \mathrm{e}^{-2\Im k_{\mathrm{P,Sc}}(\omega)\underline{\mathbf{e}}_\zeta\cdot\underline{\mathbf{R}}}\, \Re k_{\mathrm{P,Sc}}(\omega)\, \underline{\mathbf{e}}_\zeta,
\end{aligned} \tag{8.326}
$$

$$
\begin{aligned}
&\langle w_{\mathrm{P,S1/2}}(\underline{\mathbf{R}}, t, \omega, \underline{\mathbf{k}}_{\mathrm{P,S}}\rangle \\
&\quad = \frac{\varrho}{4}\, c_{\mathrm{P,S}}^2\, |u(\omega)|^2 \mathrm{e}^{-2\Im k_{\mathrm{P,Sc}}(\omega)\underline{\mathbf{e}}_\zeta\cdot\underline{\mathbf{R}}}\, \left[k_{\mathrm{P,S}}^2 + |k_{\mathrm{P,Sc}}(\omega)|^2\right],
\end{aligned} \tag{8.327}
$$

$$
\begin{aligned}
\underline{\mathbf{c}}_{\mathrm{EP,S1/2}}(\omega) &= \frac{2\omega\Re k_{\mathrm{P,Sc}}(\omega)}{k_{\mathrm{P,S}}^2 + |k_{\mathrm{P,Sc}}(\omega)|^2}\,\underline{\mathbf{e}}_\zeta \\
&\overset{(8.320)}{=} c_{\mathrm{P,S}}(\omega)\,\frac{\Re k_{\mathrm{P,Sc}}^2(\omega) + |k_{\mathrm{P,Sc}}(\omega)|^2}{k_{\mathrm{P,S}}^2 + |k_{\mathrm{P,Sc}}(\omega)|^2}\,\underline{\mathbf{e}}_\zeta.
\end{aligned} \tag{8.328}
$$

Four remarks are appropriate:

- In (8.326) and (8.327), we switched to the circular frequency variable ω of Fourier spectra starting from the arbitrary but fixed circular frequency ω_0 of time harmonic fields.
- The formulas (8.326) and (8.327) emerge through specialization of the general expressions (8.182) and (8.183) to P,S1/2-polarizations and dispersion

relations of dissipative materials, yet they contain with ϱ, λ, and μ as isotropic specialization of the $\underline{\underline{\underline{\mathbf{c}}}}$-tensor in (8.181), that is to say, of the $\underline{\underline{\underline{\mathbf{c}}}}$-tensor without losses, the (frequency-independent) material parameters of the instantaneous reaction of the material according to (4.109), (4.110), and (7.64) through (7.66), respectively; the wave numbers $k_{\mathrm{P,S}}$ and the velocities $c_{\mathrm{P,S}}$ are assigned to this instantaneous reaction, i.e., the material parameters[169] ϱ, λ, μ.

- The complex frequency-dependent wave numbers $k_{\mathrm{P,Sc}}(\omega)$ according to (8.304) and (8.305) define the frequency-dependent phase velocity (8.325); under certain assumptions (Langenberg 2005), we can also define a frequency-dependent group velocity of pulsed plane waves

$$c_{\mathrm{grP,S}}(\omega) = \frac{1}{\frac{\mathrm{d}k_{\mathrm{P,Sc}}(\omega)}{\mathrm{d}\omega}}. \tag{8.329}$$

Yet, it is not equal to the magnitude of the energy velocity (8.328). The equality of energy and group velocity has been shown in Section 8.3.1 for anisotropic nondissipative materials.

Apropos pulsed waves: Even though homogeneous plane waves in dissipative materials appear simple in the frequency domain, it is becoming much more complicated in the time domain; to calculate an ultrasonic pulse based on (8.314), we have to evaluate an inverse Fourier transform:

$$\begin{aligned}&\underline{\mathbf{u}}_{\mathrm{P,S1/2}}(\underline{\mathbf{R}}, t, \underline{\mathbf{k}}_{\mathrm{P,S}})\\ &\quad = \mathcal{F}^{-1}\{u(\omega)\mathrm{e}^{\mathrm{j}\Re k_{\mathrm{P,Sc}}(\omega)\underline{\mathbf{e}}_{\zeta}\cdot\underline{\mathbf{R}}}\,\mathrm{e}^{-\Im k_{\mathrm{P,Sc}}(\omega)\underline{\mathbf{e}}_{\zeta}\cdot\underline{\mathbf{R}}}\,\underline{\hat{\mathbf{u}}}_{\mathrm{P,S1/2}}(\underline{\mathbf{k}}_{\mathrm{P,S}})\},\end{aligned} \tag{8.330}$$

and this might not be possible analytically due to the general nonlinear frequency dependence of $k_{\mathrm{P,Sc}}(\omega)$; an approximate evaluation leads to the above concept of the group velocity: The envelope of a bandlimited impulse propagates nondispersively whereas the phase "slides" through the envelope. Yet, one fundamental remark must always be considered: A causal pulse always remains causal while propagating in a dissipative material, even so-called precursors do not arrive earlier at the point of observation as it is allowed by the phase velocity related to the material parameters ϱ and $\underline{\underline{\underline{\mathbf{c}}}}$ of the instantaneous reaction of the material (Sommerfeld 1914; Brillouin 1914; Kristensson et al. 2000).

- For the special Maxwell model of dissipation (only ϱ_c is complex: Equation 4.85) with $\Re k_{\mathrm{P,Sc}}(\omega) = k_{\mathrm{P,S}}$, we have $|\underline{\mathbf{c}}_{\mathrm{EP,S1/2}}(\omega)| = c_{\mathrm{P,S}}(\omega)$.

Figures 8.27(a) and (b) depict the phase propagation, attenuation, and polarization vectors.

[169] Note: In general, we have $\varrho, \lambda, \mu \neq \Re\varrho_c, \Re\lambda_c, \Re\mu_c$.

8.4.2 Inhomogeneous plane waves

The dispersion equations (8.315) and (8.316) for dissipative materials separated into real and imaginary parts allow for homogeneous as well as inhomogeneous plane waves as solutions. The first ones are characterized by parallel $\Re\underline{\mathbf{k}}_{P,S}$ and $\Im\underline{\mathbf{k}}_{P,S}$ and the second ones contain arbitrary parameter, namely, the angle $\langle(\Re\underline{\mathbf{k}}_{P,S}, \Im\underline{\mathbf{k}}_{P,S})$ between phase and attenuation vector with $0 < \langle(\Re\underline{\mathbf{k}}_{P,S}, \Im\underline{\mathbf{k}}_{P,S}) < \pi/2$. Based on (8.162), (8.163) and (8.167), (8.168), we obtain the graphical display of phase, attenuation, and polarization vectors in Figure 8.28(a) and (b). Note: The special case of homogeneous plane waves in dissipative materials [Figure 8.27(a) and (b)] is contained in Figure 8.28(a) and (b) but not the case of Figure 8.12 because the orthogonality $\Re\underline{\mathbf{k}}_{P,S}$ and $\Im\underline{\mathbf{k}}_{P,S}$ is only possible in nondissipative materials.

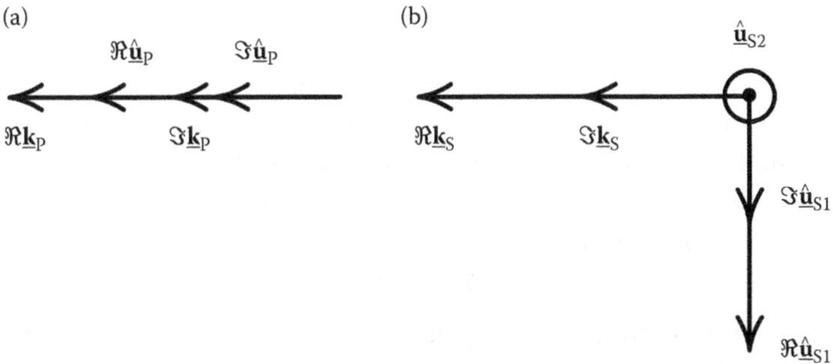

FIGURE 8.27
(a) Polarization of a homogeneous plane pressure wave in a dissipative material; (b) S1/S2-polarization of a homogeneous plane shear wave in a dissipative material.

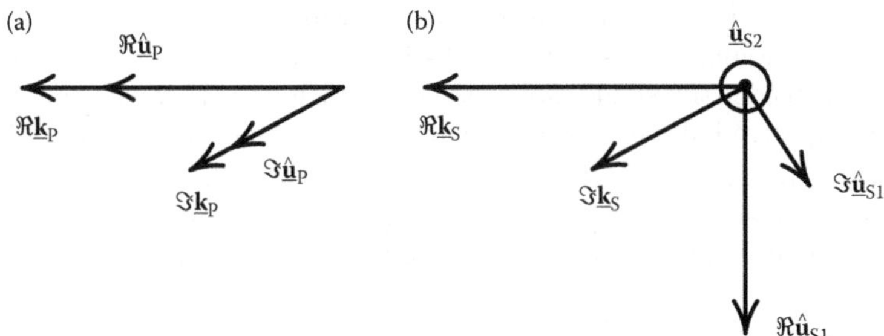

FIGURE 8.28
(a) Polarization of an inhomogeneous plane pressure wave in a dissipative material; (b) S1/S2-polarization of an inhomogeneous plane shear wave in a dissipative material.

In dissipative materials, we can only say something about $\underline{\mathbf{k}}_{\mathrm{P,S}} \cdot \underline{\mathbf{k}}^*_{\mathrm{P,S}}$ for homogeneous plane waves, for inhomogeneous waves, we must generally stay with this term in Equations 8.184, 8.192, and 8.196 for the real part of the complex Poynting vector and in Equations 8.185, 8.193, and 8.197 for the time averaged elastodynamic energy density; only $\underline{\mathbf{k}}_{\mathrm{P,S}} \cdot \underline{\mathbf{k}}_{\mathrm{P,S}}$ can be replaced by $k^2_{\mathrm{P,S}c}$ and $\underline{\mathbf{k}}^*_{\mathrm{P,S}} \cdot \underline{\mathbf{k}}^*_{\mathrm{P,S}}$ by $k^{*2}_{\mathrm{P,S}c}$. The respective quotient ($\omega_0 \Longrightarrow \omega$)

$$\underline{\mathbf{c}}_{\mathrm{EP}}(\underline{\mathbf{k}}_{\mathrm{P}}) = \frac{\lambda\left[k^{*2}_{\mathrm{P}c}(\omega)\underline{\mathbf{k}}_{\mathrm{P}} + k^2_{\mathrm{P}c}(\omega)\underline{\mathbf{k}}^*_{\mathrm{P}}\right] + 2\mu\underline{\mathbf{k}}_{\mathrm{P}}\cdot\underline{\mathbf{k}}^*_{\mathrm{P}}\left(\underline{\mathbf{k}}_{\mathrm{P}} + \underline{\mathbf{k}}^*_{\mathrm{P}}\right)}{\lambda\left[k^2_{\mathrm{P}}\underline{\mathbf{k}}_{\mathrm{P}}\cdot\underline{\mathbf{k}}^*_{\mathrm{P}} + |k^2_{\mathrm{P}c}(\omega)|^2\right] + 2\mu\underline{\mathbf{k}}_{\mathrm{P}}\cdot\underline{\mathbf{k}}^*_{\mathrm{P}}\left(k^2_{\mathrm{P}} + \underline{\mathbf{k}}_{\mathrm{P}}\cdot\underline{\mathbf{k}}^*_{\mathrm{P}}\right)}\,\omega, \tag{8.331}$$

$$\underline{\mathbf{c}}_{\mathrm{ES2}}(\underline{\mathbf{k}}_{\mathrm{S}}) = \frac{2\Re\underline{\mathbf{k}}_{\mathrm{S}}}{k^2_{\mathrm{S}} + \underline{\mathbf{k}}_{\mathrm{S}}\cdot\underline{\mathbf{k}}^*_{\mathrm{S}}}\,\omega, \tag{8.332}$$

$$\underline{\mathbf{c}}_{\mathrm{ES1}}(\underline{\mathbf{k}}_{\mathrm{S}}) = \frac{2\underline{\mathbf{k}}_{\mathrm{S}}\cdot\underline{\mathbf{k}}^*_{\mathrm{S}}\left(\underline{\mathbf{k}}_{\mathrm{S}} + \underline{\mathbf{k}}^*_{\mathrm{S}}\right) - \left[k^{*2}_{\mathrm{S}c}(\omega)\underline{\mathbf{k}}_{\mathrm{S}} + k^2_{\mathrm{S}c}(\omega)\underline{\mathbf{k}}^*_{\mathrm{S}}\right]}{\underline{\mathbf{k}}_{\mathrm{S}}\cdot\underline{\mathbf{k}}^*_{\mathrm{S}}\left(k^2_{\mathrm{S}} + \underline{\mathbf{k}}_{\mathrm{S}}\cdot\underline{\mathbf{k}}^*_{\mathrm{S}}\right) + \left(\underline{\mathbf{k}}_{\mathrm{S}}\cdot\underline{\mathbf{k}}^*_{\mathrm{S}}\right)^2 - |k^2_{\mathrm{P}c}(\omega)|^2}\,\omega, \tag{8.333}$$

therefore, yields the energy velocity of P-, S1(SH)-, and S2(SV)-waves. For all previous cases of complex phase vectors, we always found expressions with comparable structures for the energy velocities of P-, S1-, and S2-wave modes but this is no longer true for the general case of inhomogeneous waves in dissipative materials. Only the phase velocities according to (8.175)

$$c_{\mathrm{P,S1/2}}(\underline{\mathbf{k}}_{\mathrm{P,S}}) = \sqrt{\frac{2}{\Re k^2_{\mathrm{P,S}c} + \underline{\mathbf{k}}_{\mathrm{P,S}}\cdot\underline{\mathbf{k}}^*_{\mathrm{P,S}}}}\,\omega \tag{8.334}$$

are of the same structure. Note: $\underline{\mathbf{c}}_{\mathrm{EP}}(\underline{\mathbf{k}}_{\mathrm{P}})$ and $\underline{\mathbf{c}}_{\mathrm{ES1}}(\underline{\mathbf{k}}_{\mathrm{S}})$ do not have the direction of $\Re\underline{\mathbf{k}}_{\mathrm{P}}$ and $\Re\underline{\mathbf{k}}_{\mathrm{S}}$, respectively, i.e., the energy flux is not orthogonal to the phase surfaces.

In nondissipative materials inhomogeneous plane waves basically exist as mathematical solutions of a homogeneous reduced wave equation in an infinite elastic full-space, yet their physical realization is tied to the boundary of a half-space (Section 9.2.1). That is the reason why we also discuss inhomogeneous plane waves in dissipative materials with respect to a reference plane, namely, the xy-boundary between a homogeneous isotropic nondissipative half-space ($z > 0$) and a homogeneous isotropic dissipative half-space ($z < 0$). For simplicity, we consider the single mode case of an incident SH-wave coming from the nondissipative half-space with the angle of incidence $\vartheta_{i\mathrm{S}}$ [Figure 9.12 mit $\lambda^{(2)} \Longrightarrow \lambda^{(2)}_c(\omega)$, $\mu^{(2)} \Longrightarrow \mu^{(2)}_c(\omega)$, $\rho^{(2)} \Longrightarrow \rho^{(2)}_c(\omega)$]; the already mentioned arbitrary parameter—the angle between $\Re\underline{\mathbf{k}}_{\mathrm{S}}$ and $\Im\underline{\mathbf{k}}_{\mathrm{S}}$—is appointed by the angle of incidence. The phase matching for the phase vectors (Equation 9.19)

$$\begin{aligned}\underline{\mathbf{n}} \times \underline{\mathbf{k}}_{i\mathrm{S}} &= \underline{\mathbf{n}} \times \underline{\mathbf{k}}_{t\mathrm{S}} \\ &= \underline{\mathbf{n}} \times \Re\underline{\mathbf{k}}_{t\mathrm{S}} + \mathrm{j}\,\underline{\mathbf{n}} \times \Im\underline{\mathbf{k}}_{t\mathrm{S}}\end{aligned} \tag{8.335}$$

with $\underline{\mathbf{n}} = \underline{\mathbf{e}}_z$ requires

$$\underline{\mathbf{n}} \times \Re\underline{\mathbf{k}}_{t\mathrm{S}} = \underline{\mathbf{n}} \times \underline{\mathbf{k}}_{i\mathrm{S}}, \tag{8.336}$$

$$\underline{\mathbf{n}} \times \Im\underline{\mathbf{k}}_{t\mathrm{S}} = \underline{\mathbf{0}} \tag{8.337}$$

after separation into real and imaginary parts. Obviously, $\Im\underline{\mathbf{k}}_{tS}$ must be orthogonal to the boundary and hence parallel to the z-axis; note: For each phase propagation direction given by $\Re\underline{\mathbf{k}}_{tS}$, the attenuation is always orthogonal to the boundary, i.e., the transmitted SH-wave is generally an inhomogeneous plane wave (except for vertical incidence). To calculate this nonzero z-component of $\Im\underline{\mathbf{k}}_{tS}$ as well as the still open z-component of $\Re\underline{\mathbf{k}}_{tS}$—the phase matching (8.337) only assigns the x-component—we use the dispersion relation separated into real and imaginary parts, namely, Equations 8.315 and 8.316. With

$$\underline{\mathbf{k}}_{tS} = k_{tSx}\underline{\mathbf{e}}_x + k_{tSz}\underline{\mathbf{e}}_z \tag{8.338}$$

and

$$k_{tSx} = -\, k_{S}^{(1)} \sin\vartheta_{iS}, \tag{8.339}$$

$$k_{tSz} = \Re k_{tSz} + \mathrm{j}\,\Im k_{tSz}, \tag{8.340}$$

hence

$$\Re\underline{\mathbf{k}}_{tS} = k_{tSx}\underline{\mathbf{e}}_x + \Re k_{tSz}\underline{\mathbf{e}}_z, \tag{8.341}$$

$$\Im\underline{\mathbf{k}}_{tS} = \Im k_{tSz}\underline{\mathbf{e}}_z, \tag{8.342}$$

we obtain

$$k_{tSx}^2 + \left(\Re k_{tSz}\right)^2 - \left(\Im k_{tSz}\right)^2 = \Re k_{Sc}^{(2)^2}(\omega), \tag{8.343}$$

$$\Re k_{tSz}\Im k_{tSz} = \frac{1}{2}\,\Im k_{Sc}^{(2)^2}(\omega); \tag{8.344}$$

if we write (8.343) according to

$$\left(\Re k_{tSz}\right)^2 - \left(\Im k_{tSz}\right)^2 = \Re k_{Sc}^{(2)^2}(\omega) - k_{S}^{(1)^2}\sin^2\vartheta_{iS}, \tag{8.345}$$

it becomes obvious that (8.345) and (8.344) are analogous to (8.322) and (8.323), whose solutions can be readily given by (8.320) and (8.321) just spending some thoughts to the correct signs:

$$\Re k_{tSz}(\omega) = -\frac{1}{\sqrt{2}}\sqrt{\Re k_{Sc}^{(2)^2}(\omega) - k_{S}^{(1)^2}\sin^2\vartheta_{iS} + \sqrt{\left[\Re k_{Sc}^{(2)^2}(\omega) - k_{S}^{(1)^2}\sin^2\vartheta_{iS}\right]^2 + \left[\Im k_{Sc}^{(2)^2}(\omega)\right]^2}}, \tag{8.346}$$

$$\Im k_{tSz}(\omega) = -\frac{1}{\sqrt{2}}\sqrt{-\Re k_{Sc}^{(2)^2}(\omega) + k_{S}^{(1)^2}\sin^2\vartheta_{iS} + \sqrt{\left[\Re k_{Sc}^{(2)^2}(\omega) - k_{S}^{(1)^2}\sin^2\vartheta_{iS}\right]^2 + \left[\Im k_{Sc}^{(2)^2}(\omega)\right]^2}}. \tag{8.347}$$

Once again this illustrates explicitly that phase and attenuation vector in a dissipative material depend upon the frequency distribution of dissipation.

The phase velocity of the transmitted wave is calculated according to (8.172) with (8.341):

$$c_{t\mathrm{S}}(\omega) = \frac{\omega}{\sqrt{k_{t\mathrm{S}x}^2 + (\Re k_{t\mathrm{S}z})^2}}$$

$$= \sqrt{\frac{2}{\Re k_{\mathrm{S}c}^{(2)^2}(\omega) + k_{\mathrm{S}}^{(1)^2} \sin^2 \vartheta_{i\mathrm{S}} + \sqrt{[\Re k_{\mathrm{S}c}^{(2)^2}(\omega) - k_{\mathrm{S}}^{(1)^2} \sin^2 \vartheta_{i\mathrm{S}}]^2 + [\Im k_{\mathrm{S}c}^{(2)^2}(\omega)]^2}}}\ \omega; \tag{8.348}$$

due to (8.325), it depends nonlinearly upon the frequency via $k_{\mathrm{S}c}^{(2)}(\omega)$, similar to $c_{\mathrm{S}}(\omega)$ according to (8.325), why a dissipative material is always dispersive. Furthermore, the phase velocity depends upon the propagation direction: The dissipative half-space appears to be anisotropic (Langenberg 2005). This is also true for the energy velocity because with (8.346) and (8.347), we obtain an expression

$$\underline{\mathbf{k}}_{t\mathrm{S}} \cdot \underline{\mathbf{k}}_{t\mathrm{S}}^* = k_{\mathrm{S}}^{(1)^2} \sin^2 \vartheta_{i\mathrm{S}} + \sqrt{\left[\Re k_{\mathrm{S}c}^{(2)^2}(\omega) - k_{\mathrm{S}}^{(1)^2} \sin^2 \vartheta_{i\mathrm{S}}\right]^2 + \left[\Im k_{\mathrm{S}c}^{(2)^2}(\omega)\right]^2} \tag{8.349}$$

that can be inserted into (8.332) with $k_{\mathrm{S}} = k_{\mathrm{S}}^{(2)}$, where, as usual, $k_{\mathrm{S}}^{(2)}$ denotes the wave number of the instantaneous reaction of the dissipative half-space.

For the special case of vertical incidence (and only for this case), the inhomogeneous transmitted wave is generally reduced to a homogeneous plane wave attenuated in propagation direction $-\underline{\mathbf{e}}_z$.

The above thoughts and results can be immediately transferred to inhomogeneous SV- and P-waves in dissipative materials. To calculate the energy velocities, the formulas (8.331) and (8.333), respectively, must then be used.

9

Reflection, Transmission, and Mode Conversion of Elastic Plane Waves at Planar Boundaries between Homogeneous Nondissipative Materials

9.1 Stress-Free Planar Boundary of a Homogeneous Isotropic Nondissipative Elastic Half-Space

At first, we investigate reflection and—eventually—mode conversion of elastic plane waves for the planar boundary of a homogeneous isotropic nondissipative elastic half-space, assuming the "material" of the complementary half-space to be vacuum. Because vacuum with mass density zero does not support elastic waves, the complementary half-space is field-free so that the field in the elastic half-space must satisfy the (homogeneous[170]) boundary condition

$$\underline{\underline{\mathbf{T}}}(\underline{\mathbf{R}}, t) \cdot \underline{\mathbf{n}} = \underline{\mathbf{0}}, \quad \underline{\mathbf{R}} \in S, \tag{9.1}$$

according to Section 3.3: The boundary S with normal $\underline{\mathbf{n}}$ is stress-free (traction-free). This standard problem of elastic wave propagation has often been treated in the literature (e.g., Achenbach 1973; Auld 1973; Ben-Menahem and Singh 1981; Harker 1988; Langenberg 1983), and even US-NDT is well aware if it (Krautkrämer and Krautkrämer 1986; Kutzner 1983; Schmerr 1998). Nevertheless, we discuss it here: First for the sake of completeness and second to appreciate, a (nearly) coordinate-free calculation[171] that essentially shortens the procedure and makes it clearer. In addition, we want to depict the result with reflection and mode conversion of pulsed waves because US-NDT is in fact pulsed testing; if at all, we only find illustrations for time harmonic waves in the literature.

9.1.1 Primary longitudinal pressure wave incidence

Reflected primary longitudinal P- and mode converted secondary transverse SV-waves: The boundary of the half-space is a physically

[170]We disregard prescribed tractions in the boundary.

[171]A *completely* coordinate-free treatise of Fresnel's reflection of electromagnetic waves can be found in Chen (1983).

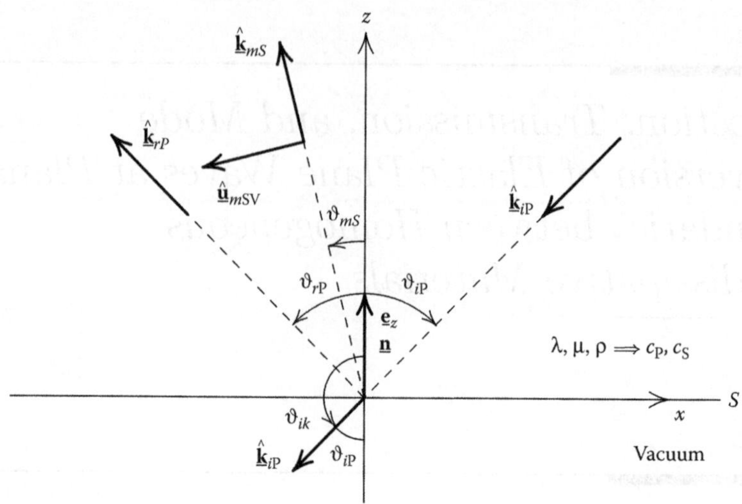

FIGURE 9.1
Reflection and mode conversion of a P-wave at the boundary S of a homogeneous isotropic nondissipative elastic half-space with the material parameters λ, μ, ρ (the y-axis points into the page).

existing reference plane that we conveniently identify as a coordinate plane of a cartesian coordinate system: We choose the xy-plane and count positive z into direction $\underline{\mathbf{n}}$, where $\underline{\mathbf{n}}$ denotes the outer normal of the vacuum half-space (Figure 9.1). We are able to rotate the coordinate system around the z-axis until the given phase propagation vector $\hat{\underline{\mathbf{k}}}_{iP}$ of the incident P-wave lies in the xz-plane; then the plane of incidence spanned by $\underline{\mathbf{e}}_z = \underline{\mathbf{n}}$ and $\hat{\underline{\mathbf{k}}}_{iP}$ coincides with the xz-plane. According to Figure 8.8, ϑ_{ik} is the polar angle of $\hat{\underline{\mathbf{k}}}_{iP}$—$\varphi_{ik} = \pi$ is the azimuth angle—yet as angle of incidence, we denote the angle ϑ_{iP} counted from the z-axis with

$$\vartheta_{iP} = \pi - \vartheta_{ik}. \tag{9.2}$$

For $\pi/2 \leq \vartheta_{ik} \leq \pi$, we have $\pi/2 \geq \vartheta_{iP} \geq 0$. According to (8.89), $\hat{\underline{\mathbf{k}}}_{iP}$ then has the component representation

$$\begin{aligned} \hat{\underline{\mathbf{k}}}_{iP} &= -\sin\vartheta_{ik}\,\underline{\mathbf{e}}_x + \cos\vartheta_{ik}\,\underline{\mathbf{e}}_z \\ &= -\sin\vartheta_{iP}\,\underline{\mathbf{e}}_x - \cos\vartheta_{iP}\,\underline{\mathbf{e}}_z; \end{aligned} \tag{9.3}$$

it can also be read from Figure 9.1.

The incident pulsed longitudinal plane P-wave is given according to (8.86) as

$$\underline{\mathbf{u}}_{iP}(\underline{\mathbf{R}}, t, \hat{\underline{\mathbf{k}}}_{iP}) = u_{iP}\left(t - \frac{\hat{\underline{\mathbf{k}}}_{iP} \cdot \underline{\mathbf{R}}}{c_P}\right)\hat{\underline{\mathbf{k}}}_{iP}, \tag{9.4}$$

yet we immediately turn to the Fourier spectrum

$$\underline{\mathbf{u}}_{i\mathrm{P}}(\underline{\mathbf{R}}, \omega, \hat{\underline{\mathbf{k}}}_{i\mathrm{P}}) = u_{i\mathrm{P}}(\omega)\, \mathrm{e}^{\mathrm{j}k_\mathrm{P}\hat{\underline{\mathbf{k}}}_{i\mathrm{P}}\cdot\underline{\mathbf{R}}}\, \hat{\underline{\mathbf{k}}}_{i\mathrm{P}}. \tag{9.5}$$

Equation 8.121 tells that we cannot satisfy the boundary condition (9.1) with (9.5) alone: The physically existing boundary condition enforces the existence of reflected P- and, as we will see, mode converted SV-waves, whose amplitudes depend upon the propagation direction (compare Footnote 136). We use the *ansatz* for both waves as follows:

$$\underline{\mathbf{u}}_{r\mathrm{P}}(\underline{\mathbf{R}}, \omega, \hat{\underline{\mathbf{k}}}_{r\mathrm{P}}) = u_{r\mathrm{P}}(\omega, \hat{\underline{\mathbf{k}}}_{r\mathrm{P}})\, \mathrm{e}^{\mathrm{j}k_\mathrm{P}\hat{\underline{\mathbf{k}}}_{r\mathrm{P}}\cdot\underline{\mathbf{R}}}\, \hat{\underline{\mathbf{k}}}_{r\mathrm{P}}, \tag{9.6}$$

$$\begin{aligned}\underline{\mathbf{u}}_{m\mathrm{SV}}(\underline{\mathbf{R}}, \omega, \hat{\underline{\mathbf{k}}}_{m\mathrm{S}}) &= u_{m\mathrm{S}}(\omega, \hat{\underline{\mathbf{k}}}_{m\mathrm{S}})\, \mathrm{e}^{\mathrm{j}k_\mathrm{S}\hat{\underline{\mathbf{k}}}_{m\mathrm{S}}\cdot\underline{\mathbf{R}}}\, \hat{\underline{\mathbf{u}}}_{m\mathrm{SV}}(\hat{\underline{\mathbf{k}}}_{m\mathrm{S}}),\\ \hat{\underline{\mathbf{u}}}_{m\mathrm{SV}}(\hat{\underline{\mathbf{k}}}_{m\mathrm{S}}) \cdot \hat{\underline{\mathbf{k}}}_{m\mathrm{S}} &= 0,\end{aligned} \tag{9.7}$$

where we denote the propagation vectors of these waves as $\hat{\underline{\mathbf{k}}}_{r\mathrm{P}} \cdot \underline{\mathbf{e}}_y = \hat{\underline{\mathbf{k}}}_{m\mathrm{S}} \cdot \underline{\mathbf{e}}_y = 0$, indicating that they also lie in the xz-plane (we will see that this is a must). With

$$\hat{\underline{\mathbf{u}}}_{m\mathrm{SV}}(\hat{\underline{\mathbf{k}}}_{m\mathrm{S}}) = \hat{\underline{\mathbf{k}}}_{m\mathrm{S}} \times \underline{\mathbf{e}}_y, \tag{9.8}$$

the polarization of the shear wave (9.7) follows indeed the convention for an SV-wave according to Figure 8.8(b) being SV with regard to the boundary; $\hat{\underline{\mathbf{u}}}_{m\mathrm{SV}}(\hat{\underline{\mathbf{k}}}_{m\mathrm{S}})$ lies also in the plane of incidence. By using (8.89) and (8.90), we calculate the following components:

$$\hat{\underline{\mathbf{k}}}_{r\mathrm{P}} = -\sin\vartheta_{r\mathrm{P}}\, \underline{\mathbf{e}}_x + \cos\vartheta_{r\mathrm{P}}\, \underline{\mathbf{e}}_z, \tag{9.9}$$

$$\hat{\underline{\mathbf{k}}}_{m\mathrm{S}} = -\sin\vartheta_{m\mathrm{S}}\, \underline{\mathbf{e}}_x + \cos\vartheta_{m\mathrm{S}}\, \underline{\mathbf{e}}_z, \tag{9.10}$$

$$\hat{\underline{\mathbf{u}}}_{m\mathrm{SV}}(\hat{\underline{\mathbf{k}}}_{m\mathrm{S}}) = -\cos\vartheta_{m\mathrm{S}}\, \underline{\mathbf{e}}_x - \sin\vartheta_{m\mathrm{S}}\, \underline{\mathbf{e}}_z. \tag{9.11}$$

Reflection and mode conversion laws: With (9.5) through (9.7) and (8.121) and (8.122), we must now try to satisfy the boundary condition

$$\begin{aligned}&\underline{\underline{\mathbf{T}}}(\underline{\mathbf{R}}_S, \omega) \cdot \underline{\mathbf{e}}_z\\ &\quad = \left[\underline{\underline{\mathbf{T}}}_{i\mathrm{P}}(\underline{\mathbf{R}}_S, \omega, \hat{\underline{\mathbf{k}}}_{i\mathrm{P}}) + \underline{\underline{\mathbf{T}}}_{r\mathrm{P}}(\underline{\mathbf{R}}_S, \omega, \hat{\underline{\mathbf{k}}}_{r\mathrm{P}}) + \underline{\underline{\mathbf{T}}}_{m\mathrm{SV}}(\underline{\mathbf{R}}_S, \omega, \hat{\underline{\mathbf{k}}}_{m\mathrm{S}})\right] \cdot \underline{\mathbf{e}}_z\\ &\quad = \underline{\mathbf{0}}\end{aligned} \tag{9.12}$$

for *each* vector of position $\underline{\mathbf{R}}_S = x\underline{\mathbf{e}}_x + y\underline{\mathbf{e}}_y$ arbitrarily located in the xy-plane. To "try" means that: The components of the vector equation (9.12) must be sufficient to determine the unknowns $\hat{\underline{\mathbf{k}}}_{r\mathrm{P}}, \hat{\underline{\mathbf{k}}}_{m\mathrm{S}}, u_{r\mathrm{P}}(\omega), u_{m\mathrm{S}}(\omega)$; if not, the *ansatz* (9.6) and (9.7) would have failed. Since we have

$$\underline{\underline{\mathbf{T}}}_{i\mathrm{P}}(\underline{\mathbf{R}}, \omega, \hat{\underline{\mathbf{k}}}_{i\mathrm{P}}) = \mathrm{j}k_\mathrm{P}\, u_{i\mathrm{P}}(\omega)\, \mathrm{e}^{\mathrm{j}k_\mathrm{P}\hat{\underline{\mathbf{k}}}_{i\mathrm{P}}\cdot\underline{\mathbf{R}}}\, (\lambda\, \underline{\underline{\mathbf{I}}} + 2\mu\, \hat{\underline{\mathbf{k}}}_{i\mathrm{P}}\, \hat{\underline{\mathbf{k}}}_{i\mathrm{P}}), \tag{9.13}$$

$$\underline{\underline{\mathbf{T}}}_{r\mathrm{P}}(\underline{\mathbf{R}}, \omega, \hat{\underline{\mathbf{k}}}_{r\mathrm{P}}) = \mathrm{j}k_\mathrm{P}\, u_{r\mathrm{P}}(\omega, \hat{\underline{\mathbf{k}}}_{r\mathrm{P}})\, \mathrm{e}^{\mathrm{j}k_\mathrm{P}\hat{\underline{\mathbf{k}}}_{r\mathrm{P}}\cdot\underline{\mathbf{R}}}\, (\lambda\, \underline{\underline{\mathbf{I}}} + 2\mu\, \hat{\underline{\mathbf{k}}}_{r\mathrm{P}}\, \hat{\underline{\mathbf{k}}}_{r\mathrm{P}}), \tag{9.14}$$

$$\begin{aligned}\underline{\underline{\mathbf{T}}}_{m\mathrm{SV}}(\underline{\mathbf{R}}, \omega, \hat{\underline{\mathbf{k}}}_{m\mathrm{S}}) &= \mathrm{j}k_\mathrm{S}\mu\, u_{m\mathrm{S}}(\omega, \hat{\underline{\mathbf{k}}}_{m\mathrm{S}})\, \mathrm{e}^{\mathrm{j}k_\mathrm{S}\hat{\underline{\mathbf{k}}}_{m\mathrm{S}}\cdot\underline{\mathbf{R}}}\\ &\quad \times [\hat{\underline{\mathbf{k}}}_{m\mathrm{S}}\, \hat{\underline{\mathbf{u}}}_{m\mathrm{SV}}(\hat{\underline{\mathbf{k}}}_{m\mathrm{S}}) + \hat{\underline{\mathbf{u}}}_{m\mathrm{SV}}(\hat{\underline{\mathbf{k}}}_{m\mathrm{S}})\, \hat{\underline{\mathbf{k}}}_{m\mathrm{S}}],\end{aligned} \tag{9.15}$$

we directly see that (9.12) has no y-component: If we would have eventually considered a mode converted SH-wave in (9.12) with $\underline{\hat{\mathbf{u}}}_{m\mathrm{SH}} = -\underline{\mathbf{e}}_y$, it would totally drop while considering the remaining x- and z-components of (9.12); it is obviously not required in our *ansatz*, it is decoupled from the P-SV-waves (Section 7.3). If we now insert (9.13) through (9.15) into (9.12), the resulting vector equation must be satisfied for all points $\underline{\mathbf{R}}_S$, and this is only possible if the arguments of the exponential functions in (9.13) through (9.15) are equal for[172] $\underline{\mathbf{R}} = \underline{\mathbf{R}}_S$:

$$k_\mathrm{P}\, \underline{\hat{\mathbf{k}}}_{i\mathrm{P}} \cdot \underline{\mathbf{R}}_S = k_\mathrm{P}\, \underline{\hat{\mathbf{k}}}_{r\mathrm{P}} \cdot \underline{\mathbf{R}}_S = k_\mathrm{S}\, \underline{\hat{\mathbf{k}}}_{m\mathrm{S}} \cdot \underline{\mathbf{R}}_S. \tag{9.16}$$

An arbitrary point $\underline{\mathbf{R}}_S \in S$ can be represented according to

$$\underline{\mathbf{R}}_S = \underline{\mathbf{n}} \times \underline{\mathbf{R}} \tag{9.17}$$

by an arbitrary point $\underline{\mathbf{R}} \in \mathbb{R}^3$; therefore, we can write (9.16) according to

$$k_\mathrm{P}(\underline{\hat{\mathbf{k}}}_{i\mathrm{P}} \times \underline{\mathbf{n}}) \cdot \underline{\mathbf{R}} = k_\mathrm{P}(\underline{\hat{\mathbf{k}}}_{r\mathrm{P}} \times \underline{\mathbf{n}}) \cdot \underline{\mathbf{R}} = k_\mathrm{S}(\underline{\hat{\mathbf{k}}}_{m\mathrm{S}} \times \underline{\mathbf{n}}) \cdot \underline{\mathbf{R}}, \tag{9.18}$$

and because $\underline{\mathbf{R}}$ is a completely arbitrary vector, not only the $(\underline{\hat{\mathbf{k}}} \times \underline{\mathbf{n}})$-projections in (9.18) but also the $(\underline{\hat{\mathbf{k}}} \times \underline{\mathbf{n}})$-vectors themselves must be equal:

$$k_\mathrm{P}\, \underline{\hat{\mathbf{k}}}_{i\mathrm{P}} \times \underline{\mathbf{n}} = k_\mathrm{P}\, \underline{\hat{\mathbf{k}}}_{r\mathrm{P}} \times \underline{\mathbf{n}} = k_\mathrm{S}\, \underline{\hat{\mathbf{k}}}_{m\mathrm{S}} \times \underline{\mathbf{n}}. \tag{9.19}$$

As sketched in Figure 9.1, we have assumed that $\underline{\hat{\mathbf{k}}}_{i\mathrm{P}}$ lies in the plane of incidence, yielding $\underline{\hat{\mathbf{k}}}_{i\mathrm{P}} \times \underline{\mathbf{n}} \sim \underline{\mathbf{e}}_y$; Equation 9.19 says that this must also be valid for $\underline{\hat{\mathbf{k}}}_{r\mathrm{P}}$ and $\underline{\hat{\mathbf{k}}}_{m\mathrm{S}}$ so that in fact $\underline{\hat{\mathbf{k}}}_{r\mathrm{P}}$ and $\underline{\hat{\mathbf{k}}}_{m\mathrm{S}}$ lie in the plane of incidence[173] (check Figure 9.1). If we take the component representations (9.3), (9.9), and (9.10) into consideration, only the y-component of the vector equations (9.19) remains and the double equality sign yields the two equations

$$\sin\vartheta_{r\mathrm{P}} = \sin\vartheta_{i\mathrm{P}}, \tag{9.20}$$

$$k_\mathrm{S}\sin\vartheta_{m\mathrm{S}} = k_\mathrm{P}\sin\vartheta_{i\mathrm{P}}. \tag{9.21}$$

The reflection law (9.20) delivers the equality

$$\vartheta_{r\mathrm{P}} = \vartheta_{i\mathrm{P}} \tag{9.22}$$

of reflection and incidence angles and the mode conversion law (9.21) allows for the calculation of the propagation direction of the mode converted SV-wave for a given angle of incidence and given wave number and phase velocity ratio, respectively. Note: Since $k_\mathrm{P} < k_\mathrm{S}$, we always have $\sin\vartheta_{m\mathrm{S}} < \sin\vartheta_{i\mathrm{P}} < 1$

[172] Exponential functions with different arguments x_1, x_2, x_3 are linearly independent; the equation

$$\alpha_1\,\mathrm{e}^{x_1} + \alpha_2\,\mathrm{e}^{x_2} + \alpha_3\,\mathrm{e}^{x_3} = 0$$

would yield $\alpha_1 = \alpha_2 = \alpha_3 = 0$.

[173] Therefore, the present boundary value problem of elastic waves is a two-dimensional problem: We have $\partial/\partial y \equiv 0$.

and hence $\vartheta_{mS} < \vartheta_{iP}$. The laws of reflection and mode conversion follow from the boundary condition simply through the phase matching condition.

Reflection and mode conversion coefficients for the vector particle displacement: We take the z- and x-components of (9.12) with (9.13) through (9.15):

$$\begin{aligned} &k_P\, u_{iP}(\omega)[\lambda + 2\mu(\hat{\underline{\mathbf{k}}}_{iP} \cdot \underline{\mathbf{e}}_z)^2] + k_P\, u_{rP}(\omega, \hat{\underline{\mathbf{k}}}_{rP})[\lambda + 2\mu(\hat{\underline{\mathbf{k}}}_{rP} \cdot \underline{\mathbf{e}}_z)^2] \\ &\qquad + 2k_S\mu\, u_{mS}(\omega, \hat{\underline{\mathbf{k}}}_{mS})(\hat{\underline{\mathbf{k}}}_{mS} \cdot \underline{\mathbf{e}}_z)(\hat{\underline{\mathbf{k}}}_{mS} \cdot \underline{\mathbf{e}}_x) = 0, \end{aligned} \tag{9.23}$$

$$\begin{aligned} &2k_P\, u_{iP}(\omega)(\hat{\underline{\mathbf{k}}}_{iP} \cdot \underline{\mathbf{e}}_x)(\hat{\underline{\mathbf{k}}}_{iP} \cdot \underline{\mathbf{e}}_z) + 2k_P\, u_{rP}(\omega, \hat{\underline{\mathbf{k}}}_{rP})(\hat{\underline{\mathbf{k}}}_{rP} \cdot \underline{\mathbf{e}}_x)(\hat{\underline{\mathbf{k}}}_{rP} \cdot \underline{\mathbf{e}}_z) \\ &\qquad + k_S\, u_{mS}(\omega, \hat{\underline{\mathbf{k}}}_{mS})[(\hat{\underline{\mathbf{k}}}_{mS} \cdot \underline{\mathbf{e}}_x)^2 - (\hat{\underline{\mathbf{k}}}_{mS} \cdot \underline{\mathbf{e}}_z)^2] = 0, \end{aligned} \tag{9.24}$$

where we consider

$$\hat{\underline{\mathbf{u}}}_{mSV}(\hat{\underline{\mathbf{k}}}_{mS}) \cdot \underline{\mathbf{e}}_z = \hat{\underline{\mathbf{k}}}_{mS} \cdot \underline{\mathbf{e}}_x, \tag{9.25}$$

$$\hat{\underline{\mathbf{u}}}_{mSV}(\hat{\underline{\mathbf{k}}}_{mS}) \cdot \underline{\mathbf{e}}_x = -\hat{\underline{\mathbf{k}}}_{mS} \cdot \underline{\mathbf{e}}_z \tag{9.26}$$

due to (9.8). With (9.23) and (9.24), we have found two equations[174] for both still unknown amplitudes $u_{rP}(\omega, \hat{\underline{\mathbf{k}}}_{rP})$ and $u_{mS}(\omega, \hat{\underline{\mathbf{k}}}_{mS})$. We now define reflection and mode conversion coefficients—we will see that they depend upon the angle of incidence but not on frequency—

$$R_P(\vartheta_{iP}) = \frac{u_{rP}(\omega, \hat{\underline{\mathbf{k}}}_{rP})}{u_{iP}(\omega)}, \tag{9.27}$$

$$M_S(\vartheta_{iP}) = \frac{u_{mS}(\omega, \hat{\underline{\mathbf{k}}}_{mS})}{u_{iP}(\omega)}, \tag{9.28}$$

and introduce angles just now via (9.3), (9.9), and (9.10); we consider (9.21) converting $\hat{\underline{\mathbf{k}}}_{iP} \cdot \underline{\mathbf{e}}_z$ as follows:

$$\begin{aligned} \lambda + 2\mu(\hat{\underline{\mathbf{k}}}_{iP} \cdot \underline{\mathbf{e}}_z)^2 &= \lambda + 2\mu\cos^2\vartheta_{iP} \\ &= \lambda + 2\mu\left(1 - \frac{\lambda + 2\mu}{\mu}\sin^2\vartheta_{mS}\right) \\ &= (\lambda + 2\mu)(1 - 2\sin^2\vartheta_{mS}) \\ &= (\lambda + 2\mu)(\cos^2\vartheta_{mS} - \sin^2\vartheta_{mS}) \end{aligned} \tag{9.29}$$

to arrive at the following system of equations:

$$R_P(\vartheta_{iP}) - \frac{\mu}{\lambda + 2\mu}\frac{k_S}{k_P}\frac{2\cos\vartheta_{mS}\sin\vartheta_{mS}}{\cos^2\vartheta_{mS} - \sin^2\vartheta_{mS}} M_S(\vartheta_{iP}) = -1, \tag{9.30}$$

$$-R_P(\vartheta_{iP}) - \frac{k_S}{k_P}\frac{\cos^2\vartheta_{mS} - \sin^2\vartheta_{mS}}{2\sin\vartheta_{iP}\cos\vartheta_{iP}} M_S(\vartheta_{iP}) = -1. \tag{9.31}$$

---

[174]Working without the *ansatz* of mode converted waves putting $u_{mS}(\omega, \hat{\underline{\mathbf{k}}}_{mS})$ equal to zero, the two equations for $u_{rP}(\omega, \hat{\underline{\mathbf{k}}}_{rP})$ would contradict each other.

Its solution is readily obtained:

$$R_{\mathrm{P}}(\vartheta_{i\mathrm{P}}) = \frac{\sin 2\vartheta_{i\mathrm{P}} \sin 2\vartheta_{m\mathrm{S}} - \kappa^2 \cos^2 2\vartheta_{m\mathrm{S}}}{\sin 2\vartheta_{i\mathrm{P}} \sin 2\vartheta_{m\mathrm{S}} + \kappa^2 \cos^2 2\vartheta_{m\mathrm{S}}}, \tag{9.32}$$

$$M_{\mathrm{S}}(\vartheta_{i\mathrm{P}}) = \kappa \frac{2 \sin 2\vartheta_{i\mathrm{P}} \cos 2\vartheta_{m\mathrm{S}}}{\sin 2\vartheta_{i\mathrm{P}} \sin 2\vartheta_{m\mathrm{S}} + \kappa^2 \cos^2 2\vartheta_{m\mathrm{S}}}, \tag{9.33}$$

where we have used trigonometric formulas for the double angles and the usual notation

$$\kappa = \frac{k_{\mathrm{S}}}{k_{\mathrm{P}}} = \frac{c_{\mathrm{P}}}{c_{\mathrm{S}}} > 1. \tag{9.34}$$

If $M_{\mathrm{S}}(\vartheta_{i\mathrm{P}})$ is found in the literature with a different sign (e.g.: Schmerr 1998; Ben-Menahem and Singh 1981; Langenberg 1983), the respective authors have chosen the opposite direction of $\underline{\hat{\mathbf{u}}}_{m\mathrm{SV}}(\hat{\underline{\mathbf{k}}}_{m\mathrm{S}})$. Sometimes (Krautkrämer and Krautkrämer 1986; Schmerr 1998; Harker 1988), reflection and mode conversion coefficients are given for the Helmholtz potentials and not for the particle displacement; then, due to (8.113), the factor $\kappa = c_{\mathrm{P}}/c_{\mathrm{S}}$ is missing in (9.33).

Another item: $R_{\mathrm{P}}(\vartheta_{i\mathrm{P}})$ and $M_{\mathrm{S}}(\vartheta_{i\mathrm{P}})$ are amplitude factors of vectorial particle displacements (9.6) and (9.7), that is to say, we have explicitly

$$\underline{\mathbf{u}}_{r\mathrm{P}}(\underline{\mathbf{R}}, \omega, \hat{\underline{\mathbf{k}}}_{r\mathrm{P}}) = R_{\mathrm{P}}(\vartheta_{i\mathrm{P}})\, u_{i\mathrm{P}}(\omega)\, \mathrm{e}^{\mathrm{j}k_{\mathrm{P}}\hat{\underline{\mathbf{k}}}_{r\mathrm{P}}\cdot\underline{\mathbf{R}}}\, \hat{\underline{\mathbf{k}}}_{r\mathrm{P}}, \tag{9.35}$$

$$\underline{\mathbf{u}}_{m\mathrm{SV}}(\underline{\mathbf{R}}, \omega, \hat{\underline{\mathbf{k}}}_{m\mathrm{S}}) = M_{\mathrm{S}}(\vartheta_{i\mathrm{P}})\, u_{i\mathrm{P}}(\omega)\, \mathrm{e}^{\mathrm{j}k_{\mathrm{S}}\hat{\underline{\mathbf{k}}}_{m\mathrm{S}}\cdot\underline{\mathbf{R}}}\, \hat{\underline{\mathbf{k}}}_{m\mathrm{S}} \times \underline{\mathbf{e}}_y \tag{9.36}$$

together with (9.5). Hence, switching to (scalar) cartesian components, additional angle functions appear:

$$u_{i\mathrm{P}x}(x, z, \omega, \vartheta_{i\mathrm{P}}) = -\sin\vartheta_{i\mathrm{P}}\, u_{i\mathrm{P}}(\omega)\, \mathrm{e}^{-\mathrm{j}k_{\mathrm{P}}(\sin\vartheta_{i\mathrm{P}}\, x + \cos\vartheta_{i\mathrm{P}}\, z)}, \tag{9.37}$$

$$u_{i\mathrm{P}z}(x, z, \omega, \vartheta_{i\mathrm{P}}) = -\cos\vartheta_{i\mathrm{P}}\, u_{i\mathrm{P}}(\omega)\, \mathrm{e}^{-\mathrm{j}k_{\mathrm{P}}(\sin\vartheta_{i\mathrm{P}}\, x + \cos\vartheta_{i\mathrm{P}}\, z)}; \tag{9.38}$$

$$u_{r\mathrm{P}x}(x, z, \omega, \vartheta_{i\mathrm{P}}) = -\sin\vartheta_{i\mathrm{P}}\, R_{\mathrm{P}}(\vartheta_{i\mathrm{P}})\, u_{i\mathrm{P}}(\omega)\, \mathrm{e}^{-\mathrm{j}k_{\mathrm{P}}(\sin\vartheta_{i\mathrm{P}}\, x - \cos\vartheta_{i\mathrm{P}}\, z)}, \tag{9.39}$$

$$u_{r\mathrm{P}z}(x, z, \omega, \vartheta_{i\mathrm{P}}) = \cos\vartheta_{i\mathrm{P}}\, R_{\mathrm{P}}(\vartheta_{i\mathrm{P}})\, u_{i\mathrm{P}}(\omega)\, \mathrm{e}^{-\mathrm{j}k_{\mathrm{P}}(\sin\vartheta_{i\mathrm{P}}\, x - \cos\vartheta_{i\mathrm{P}}\, z)}; \tag{9.40}$$

$$\begin{aligned} u_{m\mathrm{S}x}(x, z, \omega, \vartheta_{i\mathrm{P}}) = &-\frac{1}{\kappa}\sqrt{\kappa^2 - \sin^2\vartheta_{i\mathrm{P}}}\, M_{\mathrm{S}}(\vartheta_{i\mathrm{P}})\, u_{i\mathrm{P}}(\omega) \\ &\times \mathrm{e}^{-\mathrm{j}k_{\mathrm{P}}(\sin\vartheta_{i\mathrm{P}}\, x - \sqrt{\kappa^2 - \sin^2\vartheta_{i\mathrm{P}}}\, z)}, \end{aligned} \tag{9.41}$$

$$u_{m\mathrm{S}z}(x, z, \omega, \vartheta_{i\mathrm{P}}) = -\frac{1}{\kappa}\sin\vartheta_{i\mathrm{P}}\, M_{\mathrm{S}}(\vartheta_{i\mathrm{P}})\, u_{i\mathrm{P}}(\omega)\, \mathrm{e}^{-\mathrm{j}k_{\mathrm{P}}(\sin\vartheta_{i\mathrm{P}}\, x - \sqrt{\kappa^2 - \sin^2\vartheta_{i\mathrm{P}}}\, z)}. \tag{9.42}$$

For vertical incidence—$\vartheta_{i\mathrm{P}} = 0$: $R_{\mathrm{P}}(0) = -1$, $M_{\mathrm{S}}(0) = 0$—only the two components $u_{i\mathrm{P}z}$ and $u_{r\mathrm{P}z}$ remain:

$$u_{i\mathrm{P}z}(x, z, \omega, 0) = -\, u_{i\mathrm{P}}(\omega)\, \mathrm{e}^{-\mathrm{j}k_{\mathrm{P}}z}, \tag{9.43}$$

$$u_{r\mathrm{P}z}(x, z, \omega, 0) = -\, u_{i\mathrm{P}}(\omega)\, \mathrm{e}^{\mathrm{j}k_{\mathrm{P}}z}, \tag{9.44}$$

and their ratio for $z = 0$

$$\frac{u_{r\mathrm{P}z}(x,0,\omega,0)}{u_{i\mathrm{P}z}(x,0,\omega,0)} = 1 \tag{9.45}$$

is +1: The particle velocity of the incident wave has a positive $u_{i\mathrm{P}}(\omega)$-amplitude in $(+\hat{\underline{\mathbf{k}}}_{i\mathrm{P}} = -\underline{\mathbf{e}}_z)$-direction, and the particle velocity of the reflected wave has a negative $u_{i\mathrm{P}}(\omega)$-amplitude in $(+\hat{\underline{\mathbf{k}}}_{r\mathrm{P}} = \underline{\mathbf{e}}_z)$-direction due to $R_\mathrm{P}(0) = -1$. Yet, the respective vector components exhibit equal signs.

The stress tensors (9.13) through (9.15) related to the particle velocities (9.4), (9.35), and (9.36) have the following cartesian components:

$$\begin{aligned}\underline{\underline{\mathbf{T}}}_{i\mathrm{P}}(x,z,\omega,\vartheta_{i\mathrm{P}}) &= \mathrm{j}k_\mathrm{P} u_{i\mathrm{P}}(\omega)\,\mathrm{e}^{-\mathrm{j}k_\mathrm{P}(\sin\vartheta_{i\mathrm{P}}\,x+\cos\vartheta_{i\mathrm{P}}\,z)}\\ &\times\Big\{\lambda\,\underline{\underline{\mathbf{I}}} + \mu[2\sin^2\vartheta_{i\mathrm{P}}\,\underline{\mathbf{e}}_x\underline{\mathbf{e}}_x + \sin 2\vartheta_{i\mathrm{P}}\,(\underline{\mathbf{e}}_x\underline{\mathbf{e}}_z + \underline{\mathbf{e}}_z\underline{\mathbf{e}}_x)\\ &\quad + 2\cos^2\vartheta_{i\mathrm{P}}\,\underline{\mathbf{e}}_z\underline{\mathbf{e}}_z]\Big\};\end{aligned} \tag{9.46}$$

$$\begin{aligned}\underline{\underline{\mathbf{T}}}_{r\mathrm{P}}(x,z,\omega,\vartheta_{i\mathrm{P}}) &= \mathrm{j}k_\mathrm{P}\,R_\mathrm{P}(\vartheta_{i\mathrm{P}})\,u_{i\mathrm{P}}(\omega)\,\mathrm{e}^{-\mathrm{j}k_\mathrm{P}(\sin\vartheta_{i\mathrm{P}}\,x-\cos\vartheta_{i\mathrm{P}}\,z)}\\ &\times\Big\{\lambda\,\underline{\underline{\mathbf{I}}} + \mu[2\sin^2\vartheta_{i\mathrm{P}}\,\underline{\mathbf{e}}_x\underline{\mathbf{e}}_x - \sin 2\vartheta_{i\mathrm{P}}\,(\underline{\mathbf{e}}_x\underline{\mathbf{e}}_z + \underline{\mathbf{e}}_z\underline{\mathbf{e}}_x)\\ &\quad + 2\cos^2\vartheta_{i\mathrm{P}}\,\underline{\mathbf{e}}_z\underline{\mathbf{e}}_z]\Big\};\end{aligned} \tag{9.47}$$

$$\begin{aligned}\underline{\underline{\mathbf{T}}}_{m\mathrm{SV}}(x,z,\omega,\vartheta_{i\mathrm{P}}) &= \mathrm{j}\frac{k_\mathrm{P}}{\kappa}\,M_\mathrm{S}(\vartheta_{i\mathrm{P}})\,u_{i\mathrm{P}}(\omega)\,\mathrm{e}^{-\mathrm{j}k_\mathrm{P}(\sin\vartheta_{i\mathrm{P}}\,x-\sqrt{\kappa^2-\sin^2\vartheta_{i\mathrm{P}}}\,z)}\\ &\times\Big[2\mu\sin\vartheta_{i\mathrm{P}}\sqrt{\kappa^2-\sin^2\vartheta_{i\mathrm{P}}}\,\underline{\mathbf{e}}_x\underline{\mathbf{e}}_x\\ &\quad - (\lambda+2\mu\cos^2\vartheta_{i\mathrm{P}})(\underline{\mathbf{e}}_x\underline{\mathbf{e}}_z + \underline{\mathbf{e}}_z\underline{\mathbf{e}}_x)\\ &\quad - 2\mu\sin\vartheta_{i\mathrm{P}}\sqrt{\kappa^2-\sin^2\vartheta_{i\mathrm{P}}}\,\underline{\mathbf{e}}_z\underline{\mathbf{e}}_z\Big].\end{aligned} \tag{9.48}$$

For the special case $\vartheta_{i\mathrm{P}} = 0$, we realize that:

$$\underline{\underline{\mathbf{T}}}_{i\mathrm{P}}(x,z,\omega,0) = \mathrm{j}k_\mathrm{P} u_{i\mathrm{P}}(\omega)\,\mathrm{e}^{-\mathrm{j}k_\mathrm{P}z}\,(\lambda\,\underline{\underline{\mathbf{I}}} + 2\mu\,\underline{\mathbf{e}}_z\underline{\mathbf{e}}_z); \tag{9.49}$$

$$\underline{\underline{\mathbf{T}}}_{r\mathrm{P}}(x,z,\omega,0) = -\,\mathrm{j}k_\mathrm{P} u_{i\mathrm{P}}(\omega)\,\mathrm{e}^{\mathrm{j}k_\mathrm{P}z}\,(\lambda\,\underline{\underline{\mathbf{I}}} + 2\mu\,\underline{\mathbf{e}}_z\underline{\mathbf{e}}_z); \tag{9.50}$$

$$\underline{\underline{\mathbf{T}}}_{m\mathrm{SV}}(x,z,\omega,0) = \underline{\underline{\mathbf{0}}}. \tag{9.51}$$

Not even for vertical incidence can we define an isotropic pressure tensor for $\mu \neq 0$—and hence a scalar pressure—because the matrix representations of both tensors (9.49) and (9.50),

$$\begin{pmatrix}\lambda & 0 & 0\\ 0 & \lambda & 0\\ 0 & 0 & \lambda+2\mu\end{pmatrix},$$

are not proportional to the unit matrix even though they are diagonal. Yet, with (8.123) and (8.124), we have assigned a scalar pressure as a physically meaningful quantity to plane pressure and shear waves through two double contractions of the stress tensor; hence, we obtain here:

$$\begin{aligned} p_{i\mathrm{P}}(\underline{\mathbf{R}}, \omega, \vartheta_{i\mathrm{P}}) &= -\,\underline{\underline{\mathbf{T}}}_{i\mathrm{P}}(\underline{\mathbf{R}}, \omega, \hat{\underline{\mathbf{k}}}_{i\mathrm{P}}) : \hat{\underline{\mathbf{k}}}_{i\mathrm{P}}\hat{\underline{\mathbf{k}}}_{i\mathrm{P}} \\ &= -\,\mathrm{j}\omega Z_\mathrm{P}\, u_{i\mathrm{P}}(\omega)\, \mathrm{e}^{\mathrm{j}k_\mathrm{P}\hat{\underline{\mathbf{k}}}_{i\mathrm{P}}\cdot\underline{\mathbf{R}}}, \end{aligned} \tag{9.52}$$

$$\begin{aligned} p_{r\mathrm{P}}(\underline{\mathbf{R}}, \omega, \vartheta_{i\mathrm{P}}) &= -\,\underline{\underline{\mathbf{T}}}_{r\mathrm{P}}(\underline{\mathbf{R}}, \omega, \hat{\underline{\mathbf{k}}}_{r\mathrm{P}}) : \hat{\underline{\mathbf{k}}}_{r\mathrm{P}}\hat{\underline{\mathbf{k}}}_{r\mathrm{P}} \\ &= -\,\mathrm{j}\omega Z_\mathrm{P}\, R_\mathrm{P}(\vartheta_{i\mathrm{P}})\, u_{i\mathrm{P}}(\omega)\, \mathrm{e}^{\mathrm{j}k_\mathrm{P}\hat{\underline{\mathbf{k}}}_{r\mathrm{P}}\cdot\underline{\mathbf{R}}}, \end{aligned} \tag{9.53}$$

$$\begin{aligned} p_{m\mathrm{SV}}(\underline{\mathbf{R}}, \omega, \vartheta_{i\mathrm{P}}) &= -\,\underline{\underline{\mathbf{T}}}_{m\mathrm{SV}}(\underline{\mathbf{R}}, \omega, \hat{\underline{\mathbf{k}}}_{m\mathrm{S}}) : \hat{\underline{\mathbf{k}}}_{m\mathrm{S}}(\hat{\underline{\mathbf{k}}}_{m\mathrm{S}} \times \underline{\mathbf{e}}_y) \\ &= -\,\mathrm{j}\omega Z_\mathrm{S}\, M_\mathrm{S}(\vartheta_{i\mathrm{P}})\, u_{i\mathrm{P}}(\omega)\, \mathrm{e}^{\mathrm{j}k_\mathrm{S}\hat{\underline{\mathbf{k}}}_{m\mathrm{S}}\cdot\underline{\mathbf{R}}} \\ &= -\,\mathrm{j}\omega Z_\mathrm{P}\, \underbrace{\frac{Z_\mathrm{S} M_\mathrm{S}(\vartheta_{i\mathrm{P}})}{Z_\mathrm{P}}}_{=\,M_{p\mathrm{S}}(\vartheta_{i\mathrm{P}})}\, u_{i\mathrm{P}}(\omega)\, \mathrm{e}^{\mathrm{j}k_\mathrm{S}\hat{\underline{\mathbf{k}}}_{m\mathrm{S}}\cdot\underline{\mathbf{R}}}, \end{aligned} \tag{9.54}$$

identifying $R_\mathrm{P}(\vartheta_{i\mathrm{P}})$ and $M_{p\mathrm{S}}(\vartheta_{i\mathrm{P}})$ as reflection and mode conversion coefficients of the respectively defined sound pressure. Yet we note that: This scalar sound pressure does not satisfy, for example, the condition of a "pressure-free" boundary for $\underline{\mathbf{R}} = \underline{\mathbf{R}}_S$ because we have $p_{i\mathrm{P}}(\underline{\mathbf{R}}_S, \omega, \vartheta_{i\mathrm{P}}) + p_{r\mathrm{P}}(\underline{\mathbf{R}}_S, \omega, \vartheta_{i\mathrm{P}}) + p_{m\mathrm{SV}}(\underline{\mathbf{R}}_S, \omega, \vartheta_{i\mathrm{P}}) \neq 0$; for an illustration, we refer to Figure 8.9: The preceding equation relates "pressure surfaces" parallel and perpendicular to the boundary. Only for vertical incidence, that is not the standard case in US-NDT, we have $p_{i\mathrm{P}}(\underline{\mathbf{R}}_S, \omega, 0) + p_{r\mathrm{P}}(\underline{\mathbf{R}}_S, \omega, 0) = 0$, and this is an immediate consequence of the continuity of the T_{zz}-stress tensor component (for vertical incidence); for an illustration, consult Figure 8.9(a): The vertically incident pressure wave "presses" into the opposite direction on the boundary than the vertically reflected pressure wave, and "shear pressure surfaces" are not present.

Obviously, both factors $R_\mathrm{P}(\vartheta_{i\mathrm{P}})$ and $M_\mathrm{S}(\vartheta_{i\mathrm{P}})$ depend upon the angle of incidence but not upon frequency. This means that the reflected and the mode converted pulse have the same pulse spectrum, a consequence of the nondissipative half-space. With (9.4) and the inversely Fourier transformed representations (9.35) and (9.36), we therefore obtain the pulsed waves:

$$\underline{\mathbf{u}}_{i\mathrm{P}}(\underline{\mathbf{R}}, t, \hat{\underline{\mathbf{k}}}_{i\mathrm{P}}) = u_{i\mathrm{P}}\left(t - \frac{\hat{\underline{\mathbf{k}}}_{i\mathrm{P}}\cdot\underline{\mathbf{R}}}{c_\mathrm{P}}\right)\hat{\underline{\mathbf{k}}}_{i\mathrm{P}}, \tag{9.55}$$

$$\underline{\mathbf{u}}_{r\mathrm{P}}(\underline{\mathbf{R}}, t, \hat{\underline{\mathbf{k}}}_{r\mathrm{P}}) = R_\mathrm{P}(\vartheta_{i\mathrm{P}})\, u_{i\mathrm{P}}\left(t - \frac{\hat{\underline{\mathbf{k}}}_{r\mathrm{P}}\cdot\underline{\mathbf{R}}}{c_\mathrm{P}}\right)\hat{\underline{\mathbf{k}}}_{r\mathrm{P}}, \tag{9.56}$$

$$\underline{\mathbf{u}}_{m\mathrm{SV}}(\underline{\mathbf{R}}, t, \hat{\underline{\mathbf{k}}}_{m\mathrm{S}}) = M_\mathrm{S}(\vartheta_{i\mathrm{P}})\, u_{i\mathrm{P}}\left(t - \frac{\hat{\underline{\mathbf{k}}}_{m\mathrm{S}}\cdot\underline{\mathbf{R}}}{c_\mathrm{P}}\right)\hat{\underline{\mathbf{k}}}_{m\mathrm{S}} \times \underline{\mathbf{e}}_y. \tag{9.57}$$

The superposition of (9.55) through (9.57) results in the total particle displacement $\underline{\mathbf{u}}^\mathrm{P}(x, z, t, \vartheta_{i\mathrm{P}})$ for P-wave incidence for $z \geq 0$, and specially, we obtain for $z = 0$:

$$\underline{\mathbf{u}}^\mathrm{P}(x, 0, t, \vartheta_{i\mathrm{P}}) = u_{i\mathrm{P}}\left(t + \frac{\sin\vartheta_{i\mathrm{P}}}{c_\mathrm{P}}\,x\right)[\hat{\underline{\mathbf{k}}}_{i\mathrm{P}} + R_\mathrm{P}(\vartheta_{i\mathrm{P}})\,\hat{\underline{\mathbf{k}}}_{r\mathrm{P}} + M_\mathrm{S}(\vartheta_{i\mathrm{P}})\,\hat{\underline{\mathbf{k}}}_{m\mathrm{S}} \times \underline{\mathbf{e}}_y]. \tag{9.58}$$

In contrast to the boundary condition $\underline{\underline{\mathbf{T}}}^{\mathrm{P}}(x,0,t,\vartheta_{i\mathrm{P}}) \cdot \underline{\mathbf{e}}_z = \underline{\mathbf{0}}$ for the total stress tensor that we used as starting point, the particle velocity itself is nonzero on the boundary: According to Section 3.3 (Equation 3.110), it defines an induced boundary deformation rate tensor[175]

$$\underline{\underline{\mathbf{g}}}^{\mathrm{P}}(x,t,\vartheta_{i\mathrm{P}}) = -\frac{1}{2}\frac{\mathrm{d}}{\mathrm{d}t}u_{i\mathrm{P}}\left(t+\frac{\sin\vartheta_{i\mathrm{P}}}{c_{\mathrm{P}}}x\right)[\underline{\mathbf{e}}_z\hat{\underline{\mathbf{k}}}_{i\mathrm{P}} + \hat{\underline{\mathbf{k}}}_{i\mathrm{P}}\underline{\mathbf{e}}_z + R_{\mathrm{P}}(\vartheta_{i\mathrm{P}})(\underline{\mathbf{e}}_z\hat{\underline{\mathbf{k}}}_{r\mathrm{P}} + \hat{\underline{\mathbf{k}}}_{r\mathrm{P}}\underline{\mathbf{e}}_z) + M_{\mathrm{S}}(\vartheta_{i\mathrm{P}})(\underline{\mathbf{e}}_z\hat{\underline{\mathbf{u}}}_{m\mathrm{SV}} + \hat{\underline{\mathbf{u}}}_{m\mathrm{SV}}\underline{\mathbf{e}}_z)] \tag{9.59}$$

for P-wave incidence that, as we will see in Section 15.1.3, plays the role of an xy-plane localized equivalent source of the particle velocity in Huygens' principle of elastodynamics, i.e., the physically existing boundary can be replaced by "Huygens integration" of (9.59) (Sections 15.1 and 15.2), $\underline{\underline{\mathbf{g}}}^{\mathrm{P}}(x,t,\hat{\underline{\mathbf{k}}}_{i\mathrm{P}})$ is the source of the reflected and mode converted impulses (9.56) and (9.57). As a matter of fact,

$$u_{i\mathrm{P}}\left(t+\frac{\sin\vartheta_{i\mathrm{P}}}{c_{\mathrm{P}}}x\right)$$

is a surface impulse propagating with the surface (phase) velocity

$$c_{i\mathrm{P}S}(\vartheta_{i\mathrm{P}}) = \frac{c_{\mathrm{P}}}{\sin\vartheta_{i\mathrm{P}}} \tag{9.60}$$

into negative x-direction. For $\pi/2 \geq \vartheta_{i\mathrm{P}} > 0$, we have $c_{\mathrm{P}} \leq c_{i\mathrm{P}S}(\vartheta_{i\mathrm{P}}) < \infty$. A wavefront representation similar to Figure 8.7 of the superposition of (9.55) through (9.57) especially reveals very nicely as a movie how the surface impulse "hauls" the wavefronts of the incident, reflected, and mode converted pulses, and this is only possible if the surface velocities $c_{i\mathrm{P}S}, c_{r\mathrm{P}S}, c_{m\mathrm{S}S}$ of all three pulses are equal; but this is exactly the consequence of phase matching coming along in the law of reflection (9.20) and in the mode conversion law (9.21). As an extraction of a movie, Figure 9.2 shows wavefronts for four different times, where the magnitude of the particle velocity vector is displayed, i.e., possible different pulse signs are not visible.

For the sake of completeness: The Fourier transforms of (9.58) and (9.59) are given by

$$\underline{\mathbf{u}}^{\mathrm{P}}(x,0,\omega,\vartheta_{i\mathrm{P}}) = u_{i\mathrm{P}}(\omega)\,\mathrm{e}^{-\mathrm{j}k_{\mathrm{P}}\sin\vartheta_{i\mathrm{P}}\,x}\,[\hat{\underline{\mathbf{k}}}_{i\mathrm{P}} + R_{\mathrm{P}}(\vartheta_{i\mathrm{P}})\,\hat{\underline{\mathbf{k}}}_{r\mathrm{P}} + M_{\mathrm{S}}(\vartheta_{i\mathrm{P}})\,\hat{\underline{\mathbf{k}}}_{m\mathrm{S}} \times \underline{\mathbf{e}}_y], \tag{9.61}$$

$$\underline{\underline{\mathbf{g}}}^{\mathrm{P}}(x,\omega,\vartheta_{i\mathrm{P}}) = \frac{\mathrm{j}\omega}{2}\,u_{i\mathrm{P}}(\omega)\,\mathrm{e}^{-\mathrm{j}k_{\mathrm{P}}\sin\vartheta_{i\mathrm{P}}\,x}\,[\underline{\mathbf{e}}_z\hat{\underline{\mathbf{k}}}_{i\mathrm{P}} + \hat{\underline{\mathbf{k}}}_{i\mathrm{P}}\underline{\mathbf{e}}_z + R_{\mathrm{P}}(\vartheta_{i\mathrm{P}})(\underline{\mathbf{e}}_z\hat{\underline{\mathbf{k}}}_{r\mathrm{P}} + \hat{\underline{\mathbf{k}}}_{r\mathrm{P}}\underline{\mathbf{e}}_z) + M_{\mathrm{S}}(\vartheta_{i\mathrm{P}})(\underline{\mathbf{e}}_z\hat{\underline{\mathbf{u}}}_{m\mathrm{SV}} + \hat{\underline{\mathbf{u}}}_{m\mathrm{SV}}\underline{\mathbf{e}}_z)], \tag{9.62}$$

and (9.62) may be inserted into a time harmonic elastic Huygens integral as equivalent source.

[175] As claimed in Section 3.3, it may only be equal to zero for a vanishing incident field.

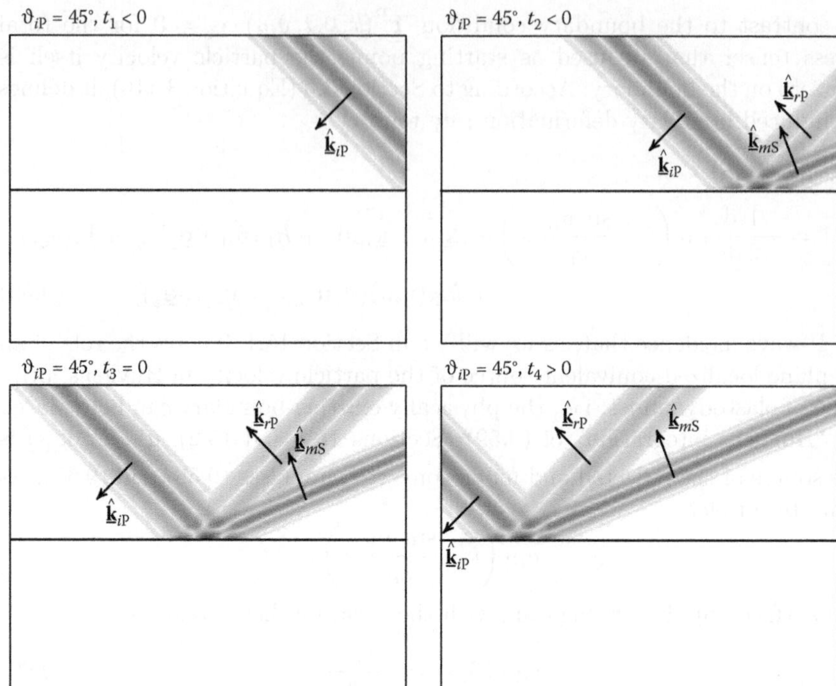

FIGURE 9.2
Wavefronts of incident, reflected, and mode converted RC2(t)-pulses for a stress-free boundary of a steel half-space with $c_P = 5900$ m/s and $c_S = 3200$ m/s; $\kappa = 1.84$.

Dispersion relations and slowness-diagrams: We want to render the concept of slowness surface or slowness diagram (Figure 8.10) more precisely to utilize it for a geometric construction of the mode conversion angle for given angle of incidence.

With the definition of the slowness vector

$$\underline{\mathbf{s}} = \frac{\underline{\mathbf{k}}}{\omega} \tag{9.63}$$

with the dimension of a reciprocal velocity—whence the name comes from—the dispersion relations (8.149) and (8.150) for P- and S-waves in isotropic nondissipative materials, respectively, can be written as:

$$\underline{\mathbf{s}}_{P,S} \cdot \underline{\mathbf{s}}_{P,S} = \frac{1}{c_{P,S}^2} = s_{P,S}^2 \tag{9.64}$$

that we satisfy with

$$\underline{\mathbf{s}}_{P,S} = s_{P,S}\hat{\underline{\mathbf{k}}} \tag{9.65}$$

in allusion to (8.151).

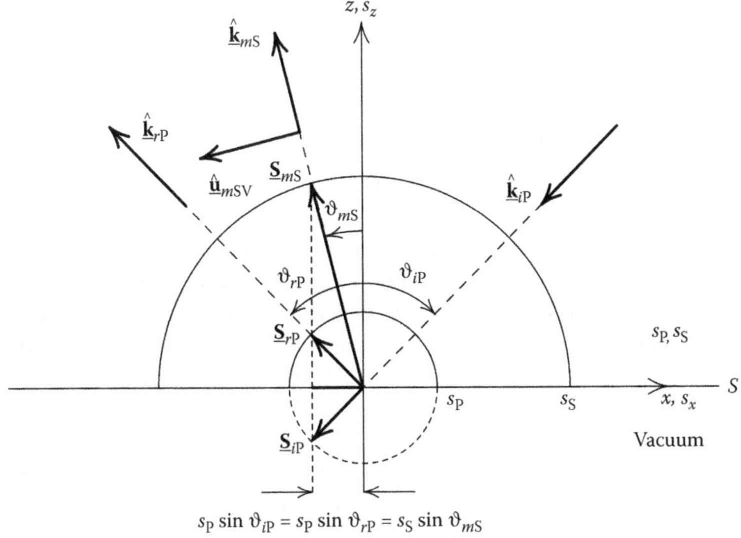

FIGURE 9.3
Slowness diagrams for reflection and mode conversion of a plane P-wave at a stress-free boundary.

Therefore, according to Figure 9.1, we can introduce the following slowness vectors:

$$\underline{\mathbf{s}}_{i\mathrm{P}} = s_\mathrm{P}\hat{\underline{\mathbf{k}}}_{i\mathrm{P}}, \tag{9.66}$$

$$\underline{\mathbf{s}}_{r\mathrm{P}} = s_\mathrm{P}\hat{\underline{\mathbf{k}}}_{r\mathrm{P}}, \tag{9.67}$$

$$\underline{\mathbf{s}}_{m\mathrm{S}} = s_\mathrm{S}\hat{\underline{\mathbf{k}}}_{m\mathrm{S}} \tag{9.68}$$

for the present problem of reflection and mode conversion of an incident P-wave. The respective dispersion relations

$$\underline{\mathbf{s}}_{i\mathrm{P}} \cdot \underline{\mathbf{s}}_{i\mathrm{P}} = |\underline{\mathbf{s}}_{i\mathrm{P}}|^2 = s_{i\mathrm{P}x}^2 + s_{i\mathrm{P}z}^2 = s_\mathrm{P}^2, \tag{9.69}$$

$$\underline{\mathbf{s}}_{r\mathrm{P}} \cdot \underline{\mathbf{s}}_{r\mathrm{P}} = |\underline{\mathbf{s}}_{r\mathrm{P}}|^2 = s_{r\mathrm{P}x}^2 + s_{r\mathrm{P}z}^2 = s_\mathrm{P}^2, \tag{9.70}$$

$$\underline{\mathbf{s}}_{m\mathrm{S}} \cdot \underline{\mathbf{s}}_{m\mathrm{S}} = |\underline{\mathbf{s}}_{m\mathrm{S}}|^2 = s_{m\mathrm{S}x}^2 + s_{m\mathrm{S}z}^2 = s_\mathrm{S}^2 \tag{9.71}$$

according to (9.64) are identically fulfilled by (9.66) through (9.68). In an $s_x s_z$-coordinate system (Figure 9.3), the right-hand equations of (9.69) through (9.71) are equations describing circles, i.e., the endpoints of the slowness vectors $\underline{\mathbf{s}}_{i\mathrm{P}}, \underline{\mathbf{s}}_{r\mathrm{P}}$ lie on a circle with radius s_P, and the endpoint of the slowness vector $\underline{\mathbf{s}}_{m\mathrm{S}}$ is located on a circle with radius s_S: Figure 9.3 depicts these slowness diagrams[176] for the relevant half-space $z \geq 0$. Due to phase

[176]For arbitrary directions of incidence in the fixed cartesian coordinate system with eventually nonzero $\hat{\underline{\mathbf{k}}}_{i\mathrm{P}} \cdot \underline{\mathbf{e}}_y$, we would obtain spherical surfaces as three-dimensional slowness diagrams.

matching at the boundary according to (9.16), the x-components of the slowness vectors must be equal:

$$\underline{\mathbf{s}}_{i\mathrm{P}} \cdot \underline{\mathbf{e}}_x = \underline{\mathbf{s}}_{r\mathrm{P}} \cdot \underline{\mathbf{e}}_x = \underline{\mathbf{s}}_{m\mathrm{S}} \cdot \underline{\mathbf{e}}_x$$
$$\Longrightarrow s_\mathrm{P} \sin \vartheta_{i\mathrm{P}} = s_\mathrm{P} \sin \vartheta_{r\mathrm{P}} = s_\mathrm{S} \sin \vartheta_{m\mathrm{S}}, \tag{9.72}$$

that is to say, we "find" the laws of reflection and mode conversion in terms of a slowness notation. The x-components also determine the z-components via the circles (9.69) through (9.71)—the endpoints of the slowness vectors are located on their respective slowness circle—enforcing the geometric construction of the slowness vectors and, hence, the phase propagation vectors $\hat{\underline{\mathbf{k}}}_{r\mathrm{P}}, \hat{\underline{\mathbf{k}}}_{m\mathrm{S}}$ as sketched in Figure 9.3: The projections of $\underline{\mathbf{s}}_{i\mathrm{P}}, \underline{\mathbf{s}}_{r\mathrm{P}}, \underline{\mathbf{s}}_{m\mathrm{S}}$ onto the x-axis have the same magnitude as the boldface line.

Wavefronts of reflected P- and mode converted SV-waves: The expressions (9.32) and (9.33) for reflection and mode conversion coefficients are discussed in dependence of the angle of incidence and the phase velocity ratio: neither explicitly appear Lamé constants nor the density. For $\kappa = 1.84$—$c_\mathrm{P} = 5900$ m/s, $c_\mathrm{S} = 3200$ m/s: steel—the angular dependence of R_P and M_S is displayed in Figure 9.4. Since we will encounter complex valued reflection and transmission coefficients in the sections to follow, we generally display $|R_\mathrm{P}(\vartheta_{i\mathrm{P}})|$, $|M_\mathrm{S}(\vartheta_{i\mathrm{P}})|$. Yet the formulas (9.32) and (9.33) directly reveal that $M_\mathrm{S}(\vartheta_{i\mathrm{P}}) \geq 0$, $0 \leq \vartheta_{i\mathrm{P}} \leq \pi/2$ and $R_\mathrm{P}(0) = -1$, $R_\mathrm{P}(\pi/2) = -1$, that is to say, at the endpoints of the angle of incidence interval, the reflection coefficient is negative. For the κ-value chosen in Figure 9.4, this is true for the total

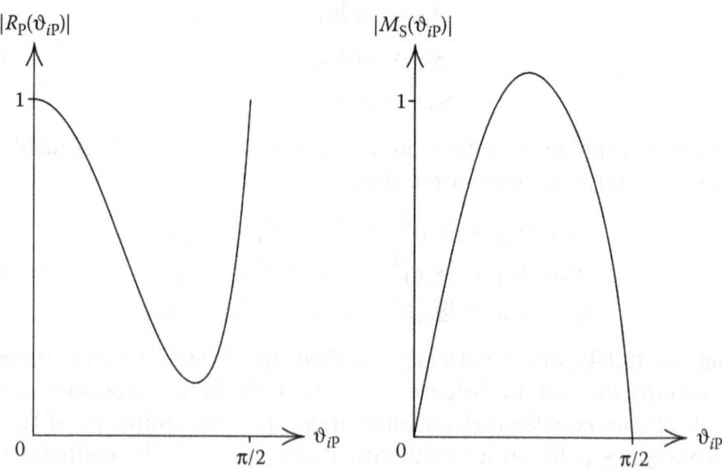

FIGURE 9.4
Magnitudes of reflection and mode conversion coefficients for P-wave incidence as a function of the angle of incidence (steel: $c_\mathrm{P} = 5900$ m/s, $c_\mathrm{S} = 3200$ m/s; $\kappa = 1.84$).

interval, yet for smaller κ-values $R_P(\vartheta_{iP})$, $0 < \vartheta_{iP} < \pi/2$ may also become positive yielding two zeroes.

The angle-dependent magnitudes of reflection and mode conversion coefficients also manifest themselves in the amplitudes of pulsed wavefronts; Figure 9.2 gives an example. Here, i.e., in Figure 9.5, we display respective

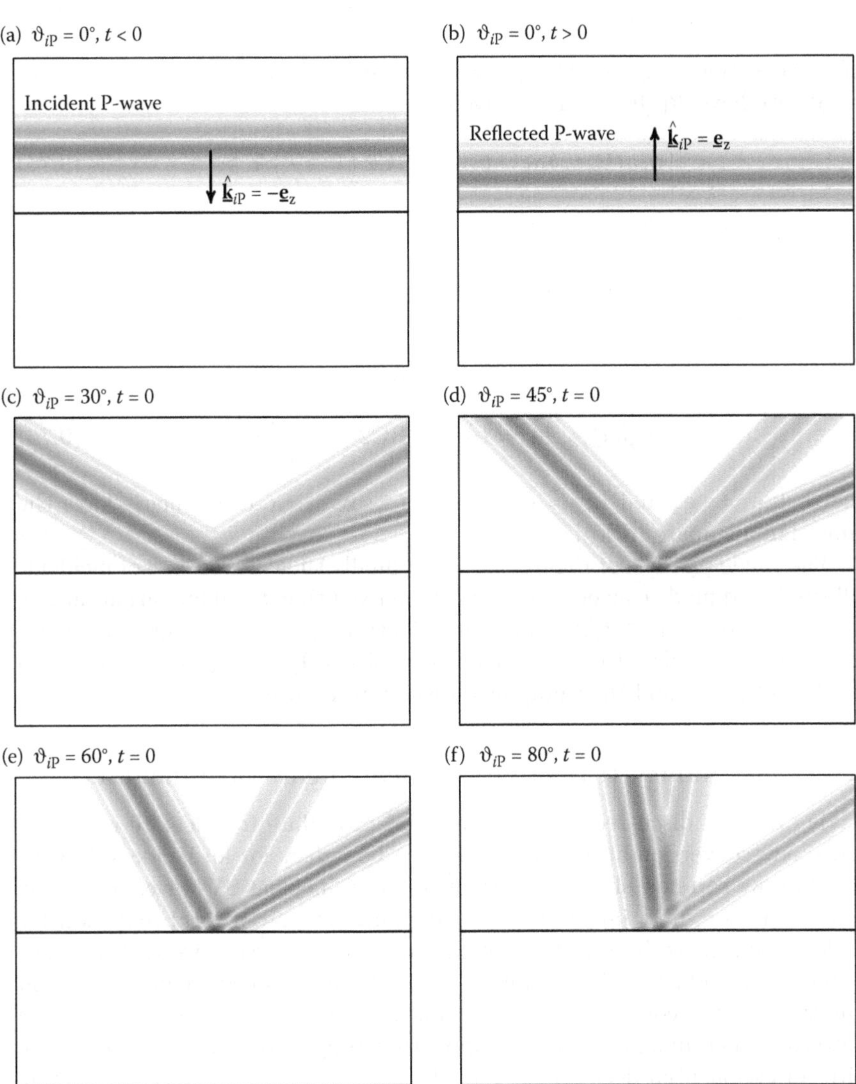

FIGURE 9.5
Wavefronts of incident and reflected P- as well as mode converted SV-waves for different angles of incidence (a)–(f) (material: steel with $c_P = 5900$ m/s, $c_S = 3200$ m/s; $\kappa = 1.84$).

wavefronts for various angles of incidence, where, as in Figure 9.2, the gray scale represents the magnitude of the particle velocity not allowing for a sign recognition. Yet, for vertical incidence, we are able to conclude the following: An RC2(t)-pulse is impressed as time function $u_{i\mathrm{P}}(t)$ to the vertically incident P-wave [Figure 9.5(a)] yielding for vertical incidence:

$$\underline{\mathbf{u}}_{i\mathrm{P}}(\underline{\mathbf{R}}, t, \vartheta_{i\mathrm{P}} = 0) = \mathrm{RC2}\left(t + \frac{z}{c}\right)\hat{\underline{\mathbf{k}}}_{i\mathrm{P}}. \tag{9.73}$$

After reflection—Figure 9.5(b)—the only existing reflected wavefront for $z > 0$—we have $R_\mathrm{P}(0) = -1$—is given by

$$\underline{\mathbf{u}}_{r\mathrm{P}}(\underline{\mathbf{R}}, t, \vartheta_{i\mathrm{P}} = 0) = -\mathrm{RC2}\left(t - \frac{z}{c}\right)\hat{\underline{\mathbf{k}}}_{r\mathrm{P}}; \tag{9.74}$$

this wavefront has only a negative longitudinal component with regard to the propagation direction $\hat{\underline{\mathbf{k}}}_{r\mathrm{P}}$, a consequence of $R_\mathrm{P}(0) = -1$. Yet considering the vector components of the particle velocity in the xz-coordinate system instead of the longitudinal components, we have

$$\underline{\mathbf{u}}_{i\mathrm{P}}(\underline{\mathbf{R}}, t, \vartheta_{i\mathrm{P}} = 0) = -\,\mathrm{RC2}\left(t + \frac{z}{c}\right)\underline{\mathbf{e}}_z, \tag{9.75}$$

$$\underline{\mathbf{u}}_{r\mathrm{P}}(\underline{\mathbf{R}}, t, \vartheta_{i\mathrm{P}} = 0) = -\,\mathrm{RC2}\left(t - \frac{z}{c}\right)\underline{\mathbf{e}}_z \tag{9.76}$$

due to $\hat{\underline{\mathbf{k}}}_{i\mathrm{P}} = -\underline{\mathbf{e}}_z$ and $\hat{\underline{\mathbf{k}}}_{r\mathrm{P}} = \underline{\mathbf{e}}_z$, that is to say, these components have the same sign [compare (9.45)].

The single pictures of Figure 9.5(c)–(f) finally show wavefronts of incident, reflected, and mode converted waves for the fixed time $t = 0$ for various angles of incidence $0 < \vartheta_{i\mathrm{P}} < \pi/2$ (Equations 9.55 through 9.57). The gray scales for the amplitudes reflect the respective values of $R_\mathrm{P}(\vartheta_{i\mathrm{P}})$, $M_\mathrm{S}(\vartheta_{i\mathrm{P}})$ according to (9.32) and (9.33) and their graphical display in Figure 9.4.

9.1.2 Secondary transverse vertical shear wave incidence

Having already discussed the mathematical procedure to calculate reflection and mode conversion at a plane stress-free boundary for an incident P-wave in detail, we can now shorten the treatise for an incident SV-wave; yet on the other hand, we must investigate the new phenomenon of total reflection for angles of incidence beyond the critical angle: In that case, we realize phase matching introducing the definition of a complex valued mode conversion "angle" with the consequence of a complex wave number vector of the mode converted, then inhomogeneous, P-wave that propagates along the boundary with an exponential decay for $z > 0$; the extent of the attenuation is determined by the imaginary part of the mode conversion "angle" identifying it as an attenuation constant.

Reflected secondary transverse SV-waves and mode converted primary longitudinal P-waves below the critical angle: To illustrate the

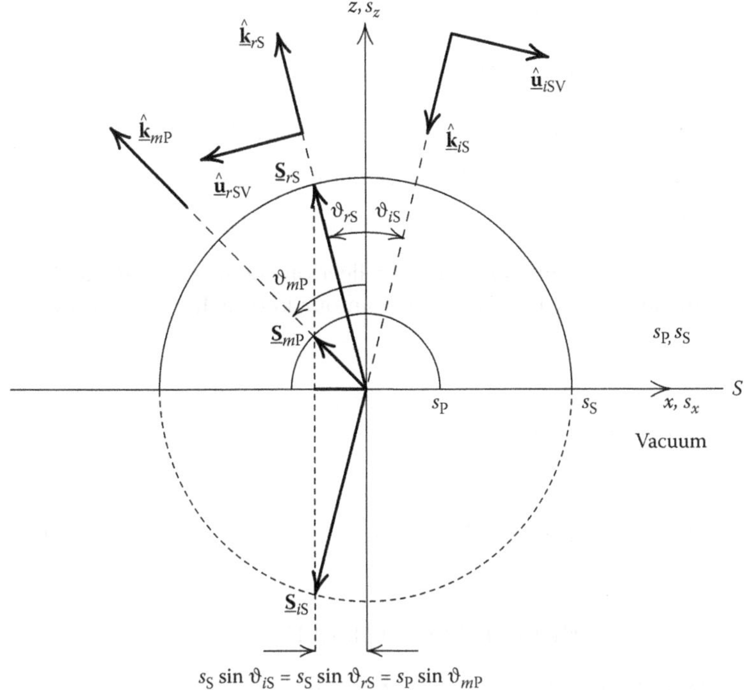

FIGURE 9.6
Slowness diagrams for reflection and mode conversion of an SV-wave at the stress-free boundary for angles of incidence below the critical angle.

propagation directions, reflection and mode conversion angles that have been visualized with two figures in the previous section—Figures 9.1 and 9.3—we only plot the slowness diagrams of the single Figure 9.6, yet pointing out that this figure is only relevant for angles of incidence below the critical angle; for larger angles of incidence, we have to consult Figure 8.11. We immediately come back to the definition and calculation of the critical angle when discussing phase matching.

From Figure 9.6, we directly take the component representations of the respective phase propagation vectors:

$$\hat{\underline{\mathbf{k}}}_{i\mathrm{S}} = -\sin\vartheta_{i\mathrm{S}}\,\underline{\mathbf{e}}_x - \cos\vartheta_{i\mathrm{S}}\,\underline{\mathbf{e}}_z, \tag{9.77}$$

$$\hat{\underline{\mathbf{k}}}_{r\mathrm{S}} = -\sin\vartheta_{r\mathrm{S}}\,\underline{\mathbf{e}}_x + \cos\vartheta_{r\mathrm{S}}\,\underline{\mathbf{e}}_z, \tag{9.78}$$

$$\hat{\underline{\mathbf{k}}}_{m\mathrm{P}} = -\sin\vartheta_{m\mathrm{P}}\,\underline{\mathbf{e}}_x + \cos\vartheta_{m\mathrm{P}}\,\underline{\mathbf{e}}_z. \tag{9.79}$$

With

$$\hat{\underline{\mathbf{u}}}_{i\mathrm{SV}}(\hat{\underline{\mathbf{k}}}_{i\mathrm{S}}) = \hat{\underline{\mathbf{k}}}_{i\mathrm{S}} \times \underline{\mathbf{e}}_y, \tag{9.80}$$

$$\hat{\underline{\mathbf{u}}}_{r\mathrm{SV}}(\hat{\underline{\mathbf{k}}}_{r\mathrm{S}}) = \hat{\underline{\mathbf{k}}}_{r\mathrm{S}} \times \underline{\mathbf{e}}_y \tag{9.81}$$

we define the polarization vectors of incident and reflected SV-waves in the usual way. The Fourier spectra of the partial waves that are needed due to the boundary condition (9.1) for the total particle displacement field look like that:

$$\underline{\mathbf{u}}_{i\mathrm{SV}}(\underline{\mathbf{R}}, \omega, \hat{\underline{\mathbf{k}}}_{i\mathrm{S}}) = u_{i\mathrm{S}}(\omega)\, \mathrm{e}^{\mathrm{j}k_\mathrm{S}\hat{\underline{\mathbf{k}}}_{i\mathrm{S}}\cdot\underline{\mathbf{R}}}\, \hat{\underline{\mathbf{u}}}_{i\mathrm{SV}}(\hat{\underline{\mathbf{k}}}_{i\mathrm{S}}), \tag{9.82}$$

$$\underline{\mathbf{u}}_{r\mathrm{SV}}(\underline{\mathbf{R}}, \omega, \hat{\underline{\mathbf{k}}}_{r\mathrm{S}}) = R_\mathrm{SV}(\vartheta_{i\mathrm{S}})\, u_{i\mathrm{S}}(\omega)\, \mathrm{e}^{\mathrm{j}k_\mathrm{S}\hat{\underline{\mathbf{k}}}_{r\mathrm{S}}\cdot\underline{\mathbf{R}}}\, \hat{\underline{\mathbf{u}}}_{r\mathrm{SV}}(\hat{\underline{\mathbf{k}}}_{r\mathrm{S}}), \tag{9.83}$$

$$\underline{\mathbf{u}}_{m\mathrm{P}}(\underline{\mathbf{R}}, \omega, \hat{\underline{\mathbf{k}}}_{m\mathrm{P}}) = M_\mathrm{P}(\vartheta_{i\mathrm{S}})\, u_{i\mathrm{S}}(\omega)\, \mathrm{e}^{\mathrm{j}k_\mathrm{P}\hat{\underline{\mathbf{k}}}_{m\mathrm{P}}\cdot\underline{\mathbf{R}}}\, \hat{\underline{\mathbf{k}}}_{m\mathrm{P}}, \tag{9.84}$$

where we have already introduced the reflection coefficient and the mode conversion coefficient $R_\mathrm{SV}(\vartheta_{i\mathrm{S}})\, M_\mathrm{P}(\vartheta_{i\mathrm{S}})$ from our knowledge of the previous subsection.

Reflection and mode conversion law: critical angle: The condition (9.1) for a stress-free boundary reads explicitly:

$$\begin{aligned}
&\underline{\underline{\mathbf{T}}}(\underline{\mathbf{R}}_S, \omega) \cdot \underline{\mathbf{e}}_z \\
&= \left[\underline{\underline{\mathbf{T}}}_{i\mathrm{SV}}(\underline{\mathbf{R}}_S, \omega, \hat{\underline{\mathbf{k}}}_{i\mathrm{S}}) + \underline{\underline{\mathbf{T}}}_{r\mathrm{SV}}(\underline{\mathbf{R}}_S, \omega, \hat{\underline{\mathbf{k}}}_{r\mathrm{S}}) + \underline{\underline{\mathbf{T}}}_{m\mathrm{P}}(\underline{\mathbf{R}}_S, \omega, \hat{\underline{\mathbf{k}}}_{m\mathrm{P}})\right] \cdot \underline{\mathbf{e}}_z \\
&= \underline{\mathbf{0}},
\end{aligned} \tag{9.85}$$

where we have similarly to (9.13) through (9.15)

$$\underline{\underline{\mathbf{T}}}_{i\mathrm{SV}}(\underline{\mathbf{R}}, \omega, \hat{\underline{\mathbf{k}}}_{i\mathrm{S}}) = \mathrm{j}k_\mathrm{S}\mu\, u_{i\mathrm{S}}(\omega)\, \mathrm{e}^{\mathrm{j}k_\mathrm{S}\hat{\underline{\mathbf{k}}}_{i\mathrm{S}}\cdot\underline{\mathbf{R}}}\, [\hat{\underline{\mathbf{k}}}_{i\mathrm{S}}\, \hat{\underline{\mathbf{u}}}_{i\mathrm{SV}}(\hat{\underline{\mathbf{k}}}_{i\mathrm{S}}) + \hat{\underline{\mathbf{u}}}_{i\mathrm{SV}}(\hat{\underline{\mathbf{k}}}_{i\mathrm{S}})\, \hat{\underline{\mathbf{k}}}_{i\mathrm{S}}], \tag{9.86}$$

$$\begin{aligned}
\underline{\underline{\mathbf{T}}}_{r\mathrm{SV}}(\underline{\mathbf{R}}, \omega, \hat{\underline{\mathbf{k}}}_{r\mathrm{S}}) &= \mathrm{j}k_\mathrm{S}\mu\, R_\mathrm{SV}(\vartheta_{i\mathrm{S}})\, u_{i\mathrm{S}}(\omega) \\
&\quad \times \mathrm{e}^{\mathrm{j}k_\mathrm{S}\hat{\underline{\mathbf{k}}}_{r\mathrm{S}}\cdot\underline{\mathbf{R}}}\, [\hat{\underline{\mathbf{k}}}_{r\mathrm{S}}\, \hat{\underline{\mathbf{u}}}_{r\mathrm{SV}}(\hat{\underline{\mathbf{k}}}_{r\mathrm{S}}) + \hat{\underline{\mathbf{u}}}_{r\mathrm{SV}}(\hat{\underline{\mathbf{k}}}_{r\mathrm{S}})\, \hat{\underline{\mathbf{k}}}_{r\mathrm{S}}],
\end{aligned} \tag{9.87}$$

$$\underline{\underline{\mathbf{T}}}_{m\mathrm{P}}(\underline{\mathbf{R}}, \omega, \hat{\underline{\mathbf{k}}}_{m\mathrm{P}}) = \mathrm{j}k_\mathrm{P}\, M_\mathrm{P}(\vartheta_{i\mathrm{S}})\, u_{i\mathrm{S}}(\omega)\, \mathrm{e}^{\mathrm{j}k_\mathrm{P}\hat{\underline{\mathbf{k}}}_{m\mathrm{P}}\cdot\underline{\mathbf{R}}}\, (\lambda\, \underline{\underline{\mathbf{I}}} + 2\mu\, \hat{\underline{\mathbf{k}}}_{m\mathrm{P}}\, \hat{\underline{\mathbf{k}}}_{m\mathrm{P}}). \tag{9.88}$$

Again, it is obvious that an eventually considered SH-wave in the *ansatz* (9.82) through (9.84) stands isolated when taking the components of (9.85); it is always decoupled from the two-dimensional P,SV-reflection problem (Section 7.3).

With the stress tensors (9.86) through (9.88) of plane waves the boundary condition (9.85) enforces phase matching

$$k_\mathrm{S}\, \hat{\underline{\mathbf{k}}}_{i\mathrm{S}} \cdot \underline{\mathbf{R}}_S = k_\mathrm{S}\, \hat{\underline{\mathbf{k}}}_{r\mathrm{S}} \cdot \underline{\mathbf{R}}_S = k_\mathrm{P}\, \hat{\underline{\mathbf{k}}}_{m\mathrm{P}} \cdot \underline{\mathbf{R}}_S \tag{9.89}$$

that must hold for all boundary points $\underline{\mathbf{R}}_S = x\underline{\mathbf{e}}_x + y\underline{\mathbf{e}}_y$. With (9.77) through (9.79), we once again obtain the reflection law

$$\vartheta_{r\mathrm{S}} = \vartheta_{i\mathrm{S}} \tag{9.90}$$

as well as the mode conversion law

$$\sin\vartheta_{m\mathrm{P}} = \frac{k_\mathrm{S}}{k_\mathrm{P}}\, \sin\vartheta_{i\mathrm{S}}. \tag{9.91}$$

As compared to P-incidence—Equation 9.21—to calculate the mode conversion angle $\vartheta_{m\mathrm{P}}$ only the factor in front of the sine of the angle of incidence has reversed: Since $k_\mathrm{S}/k_\mathrm{P} = \kappa > 1$, we can expect the sine of the mode conversion angle $\vartheta_{m\mathrm{P}}$ to become larger than 1 depending on the angle of incidence $\vartheta_{i\mathrm{S}}$; Obviously, this is "critical" for a sine function; hence, we denote

$$\vartheta_{cm\mathrm{P}} = \arcsin \frac{k_\mathrm{P}}{k_\mathrm{S}} \tag{9.92}$$

as critical angle (of incidence) for the mode conversion of an incident SV-wave into a P-wave as a consequence of $\sin \vartheta_{m\mathrm{P}} = 1$. For $\vartheta_{i\mathrm{S}} > \vartheta_{cm\mathrm{P}}$, we meet the phenomenon of a "critical sine."

We do not face any problems for angles of incidence below the critical angle to construct the phase propagation directions in Figure 9.6 based on the geometry of slowness diagrams; even the $\underline{\hat{\mathbf{u}}}_{i\mathrm{SV}}, \underline{\hat{\mathbf{u}}}_{r\mathrm{SV}}$-polarizations can be depicted without difficulty.[177]

But now we face the following question: How do we "create" a sine bigger than 1 since angles of incidence beyond the critical angle are physically permitted? The answer comes with the theory of complex valued functions of a complex variable: As it is true for the equation $x^2 + 1 = 0$ to have solutions only for complex numbers, a sine with complex argument according to (2.255) may well have a real part bigger than 1 because of the emerging hyperbolic cosine that is larger than 1; we simply must allow for complex values of the "angle" $\vartheta_{m\mathrm{P}}$ in[178] $\sin \vartheta_{m\mathrm{P}}$. Those complex values are defined by the phase matching (9.91); yet, $\sin \vartheta_{m\mathrm{P}}$ should be larger than 1 but *real* valued requesting a vanishing imaginary part; this is achieved attaching the fixed real part $\pi/2$ to the complex "angle" $\vartheta_{m\mathrm{P}}$

$$\vartheta_{m\mathrm{P}} = \frac{\pi}{2} + \mathrm{j}\,\Im\vartheta_{m\mathrm{P}}. \tag{9.93}$$

Concerning the sign of $\Im\vartheta_{m\mathrm{P}}$, we make a decision below. There are intuitive arguments for (9.93): Since $\vartheta_{m\mathrm{P}}$ according to (9.91) is always larger than $\vartheta_{i\mathrm{S}}$, the mode conversion eventually reaches the value $\pi/2$ before we have $\vartheta_{i\mathrm{S}} = \pi/2$; a further increase for the angle $\vartheta_{m\mathrm{P}}$ for increasing angle of incidence is not possible yielding the "escape" of $\vartheta_{m\mathrm{P}}$ to complex values with the real part remaining constant according to $\Re\vartheta_{m\mathrm{P}} = \pi/2$; afterward, the physical meaning of the imaginary part has to be clarified. With (9.93), (2.255), and (9.91), we have for $\vartheta_{i\mathrm{S}} > \vartheta_{cm\mathrm{P}}$

$$\sin \vartheta_{m\mathrm{P}} = \cosh \Im\vartheta_{m\mathrm{P}} = \frac{k_\mathrm{S}}{k_\mathrm{P}} \sin \vartheta_{i\mathrm{S}} > 1. \tag{9.94}$$

We also have to clarify the consequences of complex mode conversion "angles" for the field structure of $\underline{\mathbf{u}}_{m\mathrm{P}}(\underline{\mathbf{R}}, \omega, \hat{\underline{\mathbf{k}}}_{m\mathrm{P}})$: Via $\hat{\underline{\mathbf{k}}}_{m\mathrm{P}}$ according to (9.79),

[177]Note: Per definition, both polarization vectors point into the direction of increasing polar angle; this has consequences regarding the sign of the reflection coefficient.

[178]Of course, a complex "angle" $\vartheta_{m\mathrm{P}}$ is no longer visible as an angle in a slowness diagram.

$\sin\vartheta_{mP}$ and $\cos\vartheta_{mP}$ appear; due to phase matching, $\sin\vartheta_{mP}$ has always to be replaced by $\kappa\sin\vartheta_{iS}$, leaving us only with $\cos\vartheta_{mP} = \cos(\pi/2 + \mathrm{j}\,\Im\vartheta_{mP})$. According to (2.256), we have

$$\cos\vartheta_{mP} = -\mathrm{j}\sinh\Im\vartheta_{mP}, \tag{9.95}$$

i.e., the cosine of the complex mode conversion angle (9.93) is purely imaginary! Inserting (9.94) and (9.95) into (9.79), we find a complex wave number (unit) vector

$$\hat{\underline{\mathbf{k}}}_{mP} = -\cosh\Im\vartheta_{mP}\,\underline{\mathbf{e}}_x - \mathrm{j}\sinh\Im\vartheta_{mP}\,\underline{\mathbf{e}}_z; \tag{9.96}$$

even if its Hermitian scalar product $\hat{\underline{\mathbf{k}}}_{mP}\cdot\hat{\underline{\mathbf{k}}}^*_{mP}$ is not equal[179] to 1, its "normal" scalar product $\hat{\underline{\mathbf{k}}}_{mP}\cdot\hat{\underline{\mathbf{k}}}_{mP}$ is equal to 1 due to the dispersion relation, retaining the characterization as unit vector. Insertion into (9.84) yields

$$\underline{\mathbf{u}}_{mP}(\underline{\mathbf{R}},\omega,\hat{\underline{\mathbf{k}}}_{mP}) = M_P(\vartheta_{iS})\,u_{iS}(\omega)\,\mathrm{e}^{-\mathrm{j}k_P\cosh\Im\vartheta_{mP}\,x}\,\mathrm{e}^{k_P\sinh\Im\vartheta_{mP}\,z}\,\hat{\underline{\mathbf{k}}}_{mP}. \tag{9.97}$$

First: The exponential function $\mathrm{e}^{\mathrm{j}k_P\hat{\underline{\mathbf{k}}}_{mP}\cdot\underline{\mathbf{R}}}$ being complex for $\vartheta_{iS} < \vartheta_{cmP}$ separates into a complex exponential function $\mathrm{e}^{-\mathrm{j}k_P\cosh\Im\vartheta_{mP}\,x}$ and a real valued exponential function $\mathrm{e}^{k_P\sinh\Im\vartheta_{mP}\,z}$ for $\vartheta_{iS} > \vartheta_{cmP}$; the latter one should not tend to infinity for $z \longrightarrow \infty$, instead it should decrease requesting the choice

$$\Im\vartheta_{mP} \le 0. \tag{9.98}$$

Further: This choice actually gives a physical meaning to $\Im\vartheta_{mP}$ because $\sinh\Im\vartheta_{mP}$ obviously is a (negative) attenuation constant! With (9.97), we obtain an inhomogeneous plane wave propagating into $(-x)$-direction (Section 8.2): The surfaces of constant phase are planes perpendicular to the x-axis, yet the surfaces of constant amplitude are planes perpendicular to the z-axis, both planes are orthogonal to each other. It is an evanescent inhomogeneous plane wave.

In (9.97), the imaginary part of the mode conversion "angle" still explicitly appears; but with (9.94), we can replace $k_P\cosh\Im\vartheta_{mP}$ by $k_S\sin\vartheta_{iS}$, and for $\sinh\Im\vartheta_{mP}$, this is possible via the relation

$$\cosh^2\Im\vartheta_{mP} - \sinh^2\Im\vartheta_{mP} = 1, \tag{9.99}$$

where we only have to take care that $\Im\vartheta_{mP} \le 0$, and hence $\sinh\Im\vartheta_{mP} \le 0$, holds after resolving for $\sinh\Im\vartheta_{mP}$:

$$\sinh\Im\vartheta_{mP} = -\sqrt{\kappa^2\sin^2\vartheta_{iS} - 1}. \tag{9.100}$$

---

[179]We calculate

$$\hat{\underline{\mathbf{k}}}_{mP}\cdot\hat{\underline{\mathbf{k}}}^*_{mP} = \cosh^2\Im\vartheta_{mP} + \sinh^2\Im\vartheta_{mP} \ne 1$$

and

$$\hat{\underline{\mathbf{k}}}_{mP}\cdot\hat{\underline{\mathbf{k}}}_{mP} = \cosh^2\Im\vartheta_{mP} - \sinh^2\Im\vartheta_{mP} = 1.$$

We finally obtain

$$\underline{\mathbf{u}}_{m\mathrm{P}}(\underline{\mathbf{R}}, \omega, \vartheta_{i\mathrm{S}}) = -\, M_{\mathrm{P}}(\vartheta_{i\mathrm{S}})\, u_{i\mathrm{S}}(\omega)\, \mathrm{e}^{-\mathrm{j}k_{\mathrm{S}} \sin\vartheta_{i\mathrm{S}}\, x}\, \mathrm{e}^{-\sqrt{k_{\mathrm{S}}^2 \sin^2\vartheta_{i\mathrm{S}} - k_{\mathrm{P}}^2}\, z} \times \left(\kappa \sin\vartheta_{i\mathrm{S}}\, \underline{\mathbf{e}}_x - \mathrm{j}\sqrt{\kappa^2 \sin^2\vartheta_{i\mathrm{S}} - 1}\, \underline{\mathbf{e}}_z \right). \tag{9.101}$$

This representation of the mode converted wave, that is (a) a solution of the homogeneous wave equation and (b) satisfies the physically required boundary condition together with $\underline{\mathbf{u}}_{i\mathrm{P}}$ and $\underline{\mathbf{u}}_{r\mathrm{P}}$, does no longer exhibit any trace of a complex "angle": Only the given quantities $c_{\mathrm{P}}, c_{\mathrm{S}}$, and $\vartheta_{i\mathrm{S}}$ appear.

We postpone the discussion of differently complex components of $\underline{\mathbf{u}}_{m\mathrm{P}}(\underline{\mathbf{R}}, \omega, \vartheta_{i\mathrm{S}})$ until we have calculated $M_{\mathrm{P}}(\vartheta_{i\mathrm{S}})$ [and $R_{\mathrm{SV}}(\vartheta_{i\mathrm{S}})$]. An illustration can be found in Figure 8.12.

Dispersion relations and slowness-diagrams; evanescent inhomogeneous plane waves: The dispersion relations for the participating slowness vectors are completely analogous to (9.69) through (9.71) for all angles of incidence $0 \leq \vartheta_{i\mathrm{S}} \leq \pi/2$ because these relations are based on (8.149) and (8.150) as a consequence of the wave equation:

$$\underline{\mathbf{s}}_{i\mathrm{S}} \cdot \underline{\mathbf{s}}_{i\mathrm{S}} = |\underline{\mathbf{s}}_{i\mathrm{S}}|^2 = s_{i\mathrm{S}x}^2 + s_{i\mathrm{S}z}^2 = s_{\mathrm{S}}^2, \tag{9.102}$$

$$\underline{\mathbf{s}}_{r\mathrm{S}} \cdot \underline{\mathbf{s}}_{r\mathrm{S}} = |\underline{\mathbf{s}}_{r\mathrm{S}}|^2 = s_{r\mathrm{S}x}^2 + s_{r\mathrm{S}z}^2 = s_{\mathrm{S}}^2, \tag{9.103}$$

$$\underline{\mathbf{s}}_{m\mathrm{P}} \cdot \underline{\mathbf{s}}_{m\mathrm{P}} = |\underline{\mathbf{s}}_{m\mathrm{P}}|^2 = s_{m\mathrm{P}x}^2 + s_{m\mathrm{P}z}^2 = s_{\mathrm{P}}^2. \tag{9.104}$$

For angles of incidence below the critical angle—$\vartheta_{i\mathrm{S}} < \vartheta_{cm\mathrm{P}}$—the phase propagation vectors $\hat{\underline{\mathbf{k}}}_{i\mathrm{S}}, \hat{\underline{\mathbf{k}}}_{r\mathrm{S}}, \hat{\underline{\mathbf{k}}}_{m\mathrm{P}}$—Equations 9.77 through 9.79—are real valued; and that also holds for the slowness vectors

$$\underline{\mathbf{s}}_{i\mathrm{S}} = s_{\mathrm{S}} \hat{\underline{\mathbf{k}}}_{i\mathrm{S}}, \tag{9.105}$$

$$\underline{\mathbf{s}}_{r\mathrm{S}} = s_{\mathrm{S}} \hat{\underline{\mathbf{k}}}_{r\mathrm{S}}, \tag{9.106}$$

$$\underline{\mathbf{s}}_{m\mathrm{P}} = s_{\mathrm{P}} \hat{\underline{\mathbf{k}}}_{m\mathrm{P}} \tag{9.107}$$

due to (9.65). Due to phase matching, their x-components

$$\underline{\mathbf{s}}_{i\mathrm{S}} \cdot \underline{\mathbf{e}}_x = \underline{\mathbf{s}}_{r\mathrm{S}} \cdot \underline{\mathbf{e}}_x = \underline{\mathbf{s}}_{m\mathrm{P}} \cdot \underline{\mathbf{e}}_x \Longrightarrow s_{\mathrm{S}} \sin\vartheta_{i\mathrm{S}} = s_{\mathrm{S}} \sin\vartheta_{r\mathrm{S}} = s_{\mathrm{P}} \sin\vartheta_{m\mathrm{P}} \tag{9.108}$$

must be equal, yielding Figure 9.6 in complete analogy to Figure 9.3 as construction recipe for the phase propagation vectors.

For $\vartheta_{i\mathrm{S}} > \vartheta_{cm\mathrm{P}}$, we have with

$$\underline{\mathbf{k}}_{m\mathrm{P}} = -k_{\mathrm{P}} \cosh \Im\vartheta_{m\mathrm{P}}\, \underline{\mathbf{e}}_x - \mathrm{j}\, k_{\mathrm{P}} \sinh \Im\vartheta_{m\mathrm{P}}\, \underline{\mathbf{e}}_z = \underbrace{-k_{\mathrm{S}} \sin\vartheta_{i\mathrm{S}}\, \underline{\mathbf{e}}_x}_{=\Re\underline{\mathbf{k}}_{m\mathrm{P}}} + \mathrm{j} \underbrace{\sqrt{k_{\mathrm{S}}^2 \sin^2\vartheta_{i\mathrm{S}} - k_{\mathrm{P}}^2}\, \underline{\mathbf{e}}_z}_{=\Im\underline{\mathbf{k}}_{m\mathrm{P}}} \tag{9.109}$$

exactly the case (8.152) of a complex valued solution of the dispersion relation $\underline{\mathbf{k}} \cdot \underline{\mathbf{k}} = k_\mathrm{P}^2$, where

$$\Re\underline{\mathbf{k}}_{m\mathrm{P}} = -k_\mathrm{S} \sin \vartheta_{i\mathrm{S}}\, \underline{\mathbf{e}}_x \tag{9.110}$$

is the phase propagation vector and

$$\Im\underline{\mathbf{k}}_{m\mathrm{P}} = \sqrt{k_\mathrm{S}^2 \sin^2 \vartheta_{i\mathrm{S}} - k_\mathrm{P}^2}\, \underline{\mathbf{e}}_z \tag{9.111}$$

is the amplitude attenuation vector of an evanescent inhomogeneous plane wave. With (9.84), the (phase) propagates into $(+\Re\underline{\mathbf{k}}_{m\mathrm{P}})$-direction, i.e., in $(-x)$-direction, and adequately (Figure 8.11) $\Im\underline{\mathbf{k}}_{m\mathrm{P}}$ points into that half-space $z \geq 0$, where the inhomogeneous wave is evanescent. According to (8.177), the unit vector

$$\hat{\underline{\mathbf{k}}}_{m\mathrm{P}} = \frac{\underline{\mathbf{k}}_{m\mathrm{P}}}{k_\mathrm{P}} \tag{9.112}$$

belongs to (9.109), and the slowness vector related to (9.109)

$$\underline{\mathbf{s}}_{m\mathrm{P}} = \underbrace{-s_\mathrm{S} \sin \vartheta_{i\mathrm{S}}\, \underline{\mathbf{e}}_x}_{= \Re\underline{\mathbf{s}}_{m\mathrm{P}}} + \mathrm{j} \underbrace{\sqrt{s_\mathrm{S}^2 \sin^2 \vartheta_{i\mathrm{S}} - s_\mathrm{P}^2}\, \underline{\mathbf{e}}_z}_{= \Im\underline{\mathbf{s}}_{m\mathrm{P}}} \tag{9.113}$$

separates into the "original" slowness vector $\Re\underline{\mathbf{s}}_{m\mathrm{P}}$ determining the (phase) propagation direction and the slowness attenuation vector $\Im\underline{\mathbf{s}}_{m\mathrm{P}}$; $\Re\underline{\mathbf{s}}_{m\mathrm{P}}$ only has an x-component

$$\Re s_{m\mathrm{P}x} = -s_\mathrm{S} \sin \vartheta_{i\mathrm{S}} \tag{9.114}$$

and $\Im\underline{\mathbf{k}}_{m\mathrm{P}}$ only has a z-component

$$\Im s_{m\mathrm{P}z} = \sqrt{s_\mathrm{S}^2 \sin^2 \vartheta_{i\mathrm{S}} - s_\mathrm{P}^2}, \tag{9.115}$$

and due to the slowness dispersion relations

$$\begin{aligned} \underline{\mathbf{s}}_{m\mathrm{P}} \cdot \underline{\mathbf{s}}_{m\mathrm{P}} &= |\Re\underline{\mathbf{s}}_{m\mathrm{P}}|^2 - |\Im\underline{\mathbf{s}}_{m\mathrm{P}}|^2 = s_\mathrm{P}^2 \\ &= (\Re s_{m\mathrm{P}x})^2 - (\Im s_{m\mathrm{P}z})^2 = s_\mathrm{P}^2, \end{aligned} \tag{9.116}$$

both are related by the hyperbola equation (9.116) in a $\Re s_x \Im s_z$-coordinate system. Figure 9.7 shows $\Re\underline{\mathbf{s}}_{m\mathrm{P}}$ and $\Im\underline{\mathbf{s}}_{m\mathrm{P}}$ as well as the hyperbola construction of the attenuation constant $\Im s_{m\mathrm{P}z}$ via phase matching.

Note: With (9.97), we did not choose the polarization vector $\hat{\underline{\mathbf{u}}}_{m\mathrm{P}} = \hat{\underline{\mathbf{k}}}_{m\mathrm{P}}$ as a Hermitian unit vector in contrast to the definition (8.159) that has to be considered deriving general expressions like (8.186) and (8.187) for the real part of the complex Poynting vector and for the time averaged energy density. Yet, the energy velocity according to (8.188) is not affected because the Hermitian normalization cancels.

Reflection and mode conversion coefficients for the particle displacement vector; Hilbert transformed wavefronts beyond the critical angle: We insert the stress tensors (9.86) through (9.88) into the boundary

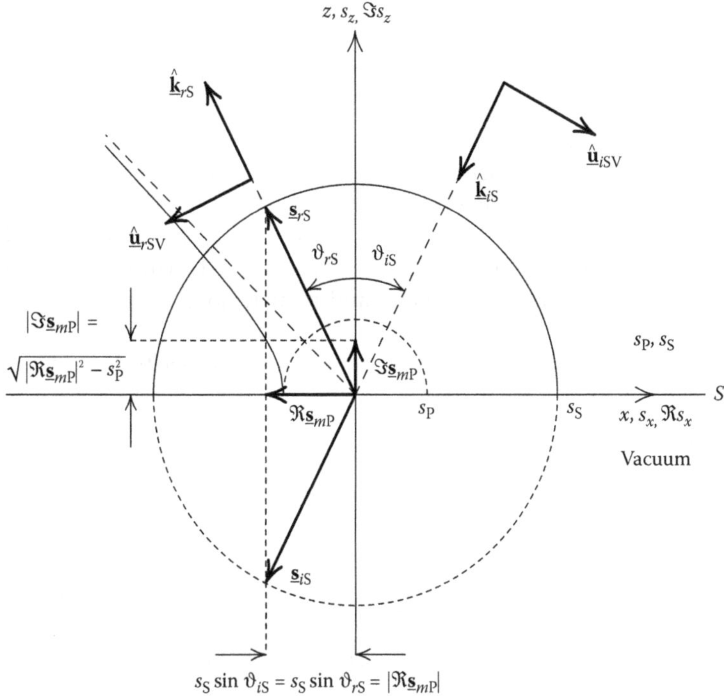

FIGURE 9.7
Slowness diagram for reflection and mode conversion of a plane SV-wave at a stress-free boundary for angles of incidence beyond the angle of incidence.

condition (9.85) and calculate z- and x-components considering (9.80) and (9.81):

$$2k_S\mu\,(\hat{\underline{\mathbf{k}}}_{iS}\cdot\underline{\mathbf{e}}_z)(\hat{\underline{\mathbf{k}}}_{iS}\cdot\underline{\mathbf{e}}_x) + 2k_S\mu\,R_{SV}(\vartheta_{iS})(\hat{\underline{\mathbf{k}}}_{rS}\cdot\underline{\mathbf{e}}_z)(\hat{\underline{\mathbf{k}}}_{rS}\cdot\underline{\mathbf{e}}_x)$$
$$+ k_P M_P(\vartheta_{iS})[\lambda + 2\mu(\hat{\underline{\mathbf{k}}}_{mP}\cdot\underline{\mathbf{e}}_z)^2] = 0, \qquad (9.117)$$

$$k_S[(\hat{\underline{\mathbf{k}}}_{iS}\cdot\underline{\mathbf{e}}_x)^2 - (\hat{\underline{\mathbf{k}}}_{iS}\cdot\underline{\mathbf{e}}_z)^2] + k_S\,R_{SV}(\vartheta_{iS})[(\hat{\underline{\mathbf{k}}}_{rS}\cdot\underline{\mathbf{e}}_x)^2 - (\hat{\underline{\mathbf{k}}}_{rS}\cdot\underline{\mathbf{e}}_z)^2]$$
$$+ 2k_P\,M_P(\vartheta_{iS})(\hat{\underline{\mathbf{k}}}_{mP}\cdot\underline{\mathbf{e}}_x)(\hat{\underline{\mathbf{k}}}_{mP}\cdot\underline{\mathbf{e}}_z) = 0. \qquad (9.118)$$

Introduction of angles and the conversion

$$\lambda + 2\mu\cos^2\vartheta_{mP} = (\lambda + 2\mu)(\cos^2\vartheta_{iS} - \sin^2\vartheta_{iS}) \qquad (9.119)$$

yields as solution of this system of equations

$$R_{SV}(\vartheta_{iS}) = \frac{\sin 2\vartheta_{iS}\sin 2\vartheta_{mP} - \kappa^2\cos^2 2\vartheta_{iS}}{\sin 2\vartheta_{iS}\sin 2\vartheta_{mP} + \kappa^2\cos^2 2\vartheta_{iS}}, \qquad (9.120)$$

$$M_P(\vartheta_{iS}) = -\kappa\,\frac{\sin 4\vartheta_{iS}}{\sin 2\vartheta_{iS}\sin 2\vartheta_{mP} + \kappa^2\cos^2 2\vartheta_{iS}}. \qquad (9.121)$$

Again we state that: Based on the arbitrary choice of SV-polarization directions, we may find $R_{\rm SV}(\vartheta_{i\rm S})$ as well as $M_{\rm P}(\vartheta_{i\rm S})$ with different signs in the literature (Auld 1973; Langenberg 1983; Ben-Menahem and Singh 1981). If the calculation is performed for the scalar and vector potentials as field quantities, a factor κ^2 appears in (9.121) (Krautkrämer and Krautkrämer 1986; Schmerr 1998; Harker 1988). And further: $R_{\rm SV}(\vartheta_{i\rm S})$ and $M_{\rm P}(\vartheta_{i\rm S})$ are amplitude factors of the vector displacement according to (9.83) and (9.84); switching to components, additional angle functions appear in (9.37) through (9.42) as factors. For vertical incidence—$\vartheta_{i\rm S} = 0 : R_{\rm SV}(0) = -1$, $M_{\rm P}(0) = 0$—only the $u_{i{\rm SV}x}$- and $u_{r{\rm SV}x}$-components remain, and their ratio is $+1$ as a consequence of $R_{\rm SV}(0) = -1$.

Analogous to (9.46) through (9.48), we calculate the cartesian stress tensor components:

$$\begin{aligned}\underline{\underline{\mathbf{T}}}_{i\rm SV}(x, z, \omega, \vartheta_{i\rm S}) = {} & {\rm j}k_{\rm S}\mu\, u_{i\rm S}(\omega)\, {\rm e}^{-k_{\rm S}(\sin\vartheta_{i\rm S}\, x + \cos\vartheta_{i\rm S}\, z)} \\ & \times \Big[-\sin 2\vartheta_{i\rm S}\, \underline{\mathbf{e}}_x\underline{\mathbf{e}}_x - \cos 2\vartheta_{i\rm S}(\underline{\mathbf{e}}_x\underline{\mathbf{e}}_z + \underline{\mathbf{e}}_z\underline{\mathbf{e}}_x) \\ & + \sin 2\vartheta_{i\rm S}\, \underline{\mathbf{e}}_z\underline{\mathbf{e}}_z\Big]; \end{aligned} \tag{9.122}$$

$$\begin{aligned}\underline{\underline{\mathbf{T}}}_{r\rm SV}(x, z, \omega, \vartheta_{i\rm S}) = {} & {\rm j}k_{\rm S}\mu\, R_{\rm SV}(\vartheta_{i\rm S})\, u_{i\rm S}(\omega)\, {\rm e}^{-k_{\rm S}(\sin\vartheta_{i\rm S}\, x - \cos\vartheta_{i\rm S}\, z)} \\ & \times \Big[\sin 2\vartheta_{i\rm S}\, \underline{\mathbf{e}}_x\underline{\mathbf{e}}_x - \cos 2\vartheta_{i\rm S}(\underline{\mathbf{e}}_x\underline{\mathbf{e}}_z + \underline{\mathbf{e}}_z\underline{\mathbf{e}}_x) \\ & - \sin 2\vartheta_{i\rm S}\, \underline{\mathbf{e}}_z\underline{\mathbf{e}}_z\Big]; \end{aligned} \tag{9.123}$$

$$\begin{aligned}\underline{\underline{\mathbf{T}}}_{m\rm P}(x, z, \omega, \vartheta_{i\rm S} \le \vartheta_{cm\rm P}) = {} & {\rm j}k_{\rm P}\, M_{\rm P}(\vartheta_{i\rm S})\, u_{i\rm S}(\omega)\, {\rm e}^{-{\rm j}k_{\rm S}(\sin\vartheta_{i\rm S}\, x - \sqrt{\kappa^{-2} - \sin^2\vartheta_{i\rm S}}\, z)} \\ & \times \Big[\lambda\, \underline{\underline{\mathbf{I}}} + 2\mu\kappa^2 \sin^2\vartheta_{i\rm S}\, \underline{\mathbf{e}}_x\underline{\mathbf{e}}_x \\ & - 2\mu\kappa \sin\vartheta_{i\rm S}\sqrt{1 - \kappa^2 \sin\vartheta_{i\rm S}}\, (\underline{\mathbf{e}}_x\underline{\mathbf{e}}_z + \underline{\mathbf{e}}_z\underline{\mathbf{e}}_x) \\ & + 2\mu(1 - \kappa^2 \sin^2\vartheta_{i\rm S})\, \underline{\mathbf{e}}_z\underline{\mathbf{e}}_z\Big]. \end{aligned} \tag{9.124}$$

With coordinate-free representations (9.86) through (9.88) of P- and SV-wave stress tensors, we once more define the sound pressure of the respective plane wave modes [compare (9.52) through (9.54)]:

$$\begin{aligned} p_{i\rm SV}(\underline{\mathbf{R}}, \omega, \vartheta_{i\rm S}) &= -\underline{\underline{\mathbf{T}}}_{i\rm SV}(\underline{\mathbf{R}}, \omega, \underline{\hat{\mathbf{k}}}_{i\rm S}) : \underline{\hat{\mathbf{k}}}_{i\rm S}\underline{\hat{\mathbf{u}}}_{i\rm SV}(\underline{\hat{\mathbf{k}}}_{i\rm S}) \\ &= -{\rm j}\omega Z_{\rm S}\, u_{i\rm S}(\omega)\, {\rm e}^{{\rm j}k_{\rm S}\underline{\hat{\mathbf{k}}}_{i\rm S}\cdot\underline{\mathbf{R}}}, \end{aligned} \tag{9.125}$$

$$\begin{aligned} p_{r\rm SV}(\underline{\mathbf{R}}, \omega, \vartheta_{i\rm S}) &= -\underline{\underline{\mathbf{T}}}_{r\rm SV}(\underline{\mathbf{R}}, \omega, \underline{\hat{\mathbf{k}}}_{r\rm S}) : \underline{\hat{\mathbf{k}}}_{r\rm S}\underline{\hat{\mathbf{u}}}_{r\rm SV}(\underline{\hat{\mathbf{k}}}_{r\rm S}) \\ &= -{\rm j}\omega Z_{\rm S}\, R_{\rm SV}(\vartheta_{i\rm S})\, u_{i\rm S}(\omega)\, {\rm e}^{{\rm j}k_{\rm S}\underline{\hat{\mathbf{k}}}_{r\rm S}\cdot\underline{\mathbf{R}}}, \end{aligned} \tag{9.126}$$

$$\begin{aligned} p_{m\rm P}(\underline{\mathbf{R}}, \omega, \vartheta_{i\rm S}) &= -\underline{\underline{\mathbf{T}}}_{m\rm P}(\underline{\mathbf{R}}, \omega, \underline{\hat{\mathbf{k}}}_{m\rm P}) : \underline{\hat{\mathbf{k}}}_{m\rm P}\underline{\hat{\mathbf{k}}}_{m\rm P} \\ &= -{\rm j}\omega Z_{\rm P}\, M_{\rm P}(\vartheta_{i\rm S})\, u_{i\rm S}(\omega)\, {\rm e}^{{\rm j}k_{\rm P}\underline{\hat{\mathbf{k}}}_{m\rm P}\cdot\underline{\mathbf{R}}} \\ &= -{\rm j}\omega Z_{\rm S}\, \underbrace{\frac{Z_{\rm P} M_{\rm P}(\vartheta_{i\rm S})}{Z_{\rm S}}}_{= M_{p\rm P}(\vartheta_{i\rm S})}\, u_{i\rm S}(\omega)\, {\rm e}^{{\rm j}k_{\rm P}\underline{\hat{\mathbf{k}}}_{m\rm P}\cdot\underline{\mathbf{R}}}. \end{aligned} \tag{9.127}$$

Once again we state that: Except for $\vartheta_{i\mathrm{S}} = 0$, the sound pressure does not satisfy a Dirichlet boundary condition at the boundary $\underline{\mathbf{R}} = \underline{\mathbf{R}}_S$ (compare Figure 8.9).

Before utilizing the frequency independence of $R_{\mathrm{SV}}(\vartheta_{i\mathrm{S}})$ and $M_{\mathrm{P}}(\vartheta_{i\mathrm{S}})$ to obtain pulsed wave representations for the particle displacements, we have to clarify how to deal with complex reflection and mode conversion coefficients for $\vartheta_{i\mathrm{S}} > \vartheta_{cm\mathrm{P}}$ because pulsed waves should be real valued. First, we investigate $R_{\mathrm{SV}}(\vartheta_{i\mathrm{S}})$ and $M_{\mathrm{P}}(\vartheta_{i\mathrm{S}})$ for angles of incidence beyond the critical angle. We have

$$\begin{aligned}\sin 2\vartheta_{m\mathrm{P}} &= 2\sin\vartheta_{m\mathrm{P}}\cos\vartheta_{m\mathrm{P}}\\ &= \begin{cases} 2\kappa\sin\vartheta_{i\mathrm{S}}\sqrt{1-\kappa^2\sin^2\vartheta_{i\mathrm{S}}} & \text{for } 0 \le \vartheta_{i\mathrm{S}} \le \vartheta_{cm\mathrm{P}}\\ 2\mathrm{j}\kappa\sin\vartheta_{i\mathrm{S}}\sqrt{\kappa^2\sin^2\vartheta_{i\mathrm{S}}-1} & \text{for } \vartheta_{cm\mathrm{P}} < \vartheta_{i\mathrm{S}} \le \pi/2. \end{cases}\end{aligned} \tag{9.128}$$

It turns out that $R_{\mathrm{SV}}(\vartheta_{i\mathrm{S}})$ is the quotient of two respective complex conjugate numbers[180] for $\vartheta_{i\mathrm{S}} > \vartheta_{cm\mathrm{P}}$; hence, in this case, we obtain

$$\begin{aligned}R_{\mathrm{SV}}(\vartheta_{i\mathrm{S}}) &= \exp\left(-2\mathrm{j}\arctan\frac{2\sin 2\vartheta_{i\mathrm{S}}\sin\vartheta_{i\mathrm{S}}\sqrt{\kappa^2\sin^2\vartheta_{i\mathrm{S}}-1}}{\kappa\cos^2 2\vartheta_{i\mathrm{S}}} + \mathrm{j}\pi\right)\\ &= \mathrm{e}^{\mathrm{j}\phi_{R_{\mathrm{SV}}}(\vartheta_{i\mathrm{S}})}\end{aligned} \tag{9.129}$$

with $|R_{\mathrm{SV}}(\vartheta_{i\mathrm{S}})| = 1$.

A real valued pulsed wave $\underline{\mathbf{u}}_{r\mathrm{SV}}(\underline{\mathbf{R}}, t, \hat{\underline{\mathbf{k}}}_{r\mathrm{S}})$ is obtained through Fourier inversion of (9.83) if $\underline{\mathbf{u}}_{r\mathrm{SV}}(\underline{\mathbf{R}}, -\omega, \hat{\underline{\mathbf{k}}}_{r\mathrm{S}}) = \underline{\mathbf{u}}^*_{r\mathrm{SV}}(\underline{\mathbf{R}}, \omega, \hat{\underline{\mathbf{k}}}_{r\mathrm{S}})$ holds; for $u_{i\mathrm{S}}(\omega)$, this is true according to our assumption, for $\mathrm{e}^{\mathrm{j}k_{\mathrm{S}}\hat{\underline{\mathbf{k}}}_{r\mathrm{S}}\cdot\underline{\mathbf{R}}}$, this is true due to $k_{\mathrm{S}} = \omega/c_{\mathrm{S}}$; only $R_{\mathrm{SV}}(\vartheta_{i\mathrm{S}})$ has to be *complemented* for negative frequencies according to[181]

$$R_{\mathrm{SV}}(\vartheta_{i\mathrm{S}}, \omega) \stackrel{\mathrm{def}}{=} \mathrm{e}^{\mathrm{j}\,\mathrm{sign}(\omega)\,\phi_{R_{\mathrm{SV}}}(\vartheta_{i\mathrm{S}})} \tag{9.130}$$

yielding a frequency-dependent reflection coefficient. Now we can apply (2.328) to calculate the inverse Fourier transform of (9.83). Before doing this explicitly, we look at $\underline{\mathbf{u}}_{m\mathrm{P}}(\underline{\mathbf{R}}, \omega, \hat{\underline{\mathbf{k}}}_{m\mathrm{P}})$ whether we can perform the inverse Fourier transform of this partial wave in the same way for $\vartheta_{i\mathrm{S}} > \vartheta_{cm\mathrm{P}}$. Certainly, $M_{\mathrm{P}}(\vartheta_{i\mathrm{S}})$ is also complex in that case so that we will equally proceed as for (9.130):

$$\begin{aligned}M_{\mathrm{P}}(\vartheta_{i\mathrm{S}}) &= -\frac{\sin 4\vartheta_{i\mathrm{S}}}{2\mathrm{j}\sin 2\vartheta_{i\mathrm{S}}\sin\vartheta_{i\mathrm{S}}\sqrt{\kappa^2\sin^2\vartheta_{i\mathrm{S}}-1}+\kappa\cos^2 2\vartheta_{i\mathrm{S}}}\\ &= |M_{\mathrm{P}}(\vartheta_{i\mathrm{S}})|\,\mathrm{e}^{\mathrm{j}\,\mathrm{sign}(\omega)\,\phi_{M_{\mathrm{P}}}(\vartheta_{i\mathrm{S}})};\end{aligned} \tag{9.131}$$

[180]For example, $R_{\mathrm{SV}}(\vartheta_{i\mathrm{S}}) = -z^*/z$ with

$$\begin{aligned}z &= \kappa^2\cos^2 2\vartheta_{i\mathrm{S}} + 2\mathrm{j}\kappa\sin 2\vartheta_{i\mathrm{S}}\sin\vartheta_{i\mathrm{S}}\sqrt{\kappa^2\sin^2\vartheta_{i\mathrm{S}}-1}\\ &= x + \mathrm{j}y;\end{aligned}$$

then we have $R_{\mathrm{SV}}(\vartheta_{i\mathrm{S}}) = -\mathrm{e}^{-2\mathrm{j}\varphi} = \mathrm{e}^{-2\mathrm{j}\varphi+\mathrm{j}\pi}$ with $\varphi = \mathrm{PV}\arctan y/x$ (compare Footnote 40) and consequently $\phi_{R_{\mathrm{SV}}}(\vartheta_{i\mathrm{S}}) = -2\varphi + \pi$.

[181]It is also mentioned by Schmerr (1998).

since $|M_{\rm P}(\vartheta_{i\rm S})| \neq 1$, there is a slight difference[182] as compared to (9.130). It is much more essential that, in contrast to $\underline{\mathbf{u}}_{r\rm SV}(\underline{\mathbf{R}}, \omega, \hat{\underline{\mathbf{k}}}_{r\rm S})$, the converted wave mode $\underline{\mathbf{u}}_{m\rm P}(\underline{\mathbf{R}}, \omega, \hat{\underline{\mathbf{k}}}_{m\rm P})$ has differently complex valued components for $\vartheta_{i\rm S} > \vartheta_{cm\rm P}$ and, additionally, containing the attenuation factor

$$
\mathrm{e}^{-\sqrt{k_{\rm S}^2 \sin^2\vartheta_{i\rm S} - k_{\rm P}^2}\, z}.
$$

Therefore, the Fourier inversion must be separately performed for the respective components while considering the frequency-dependent factor

$$
\mathrm{e}^{-\frac{|\omega|}{c_{\rm P}}\sqrt{\kappa^2 \sin^2\vartheta_{i\rm S} - 1}\, z}. \tag{9.132}
$$

Before performing this for an arbitrary pulse spectrum $u_{i\rm S}(\omega)$, we first consider two complex conjugated spectral lines according to (8.31) with circular frequencies $\pm\omega_0$ to obtain the real valued ω_0-time harmonic particle displacement

$$
\begin{aligned}
&\underline{\mathbf{u}}_{m\rm P}(x, z, t, \omega_0, \vartheta_{i\rm S}) \\
&\quad = \Re\Bigg\{ - M_{\rm P}(\vartheta_{i\rm S})\, u_{i\rm S}(\omega_0) \left(\kappa \sin\vartheta_{i\rm S}\, \underline{\mathbf{e}}_x - \mathrm{j}\sqrt{\kappa^2 \sin^2\vartheta_{i\rm S} - 1}\, \underline{\mathbf{e}}_z \right) \\
&\qquad \times \mathrm{e}^{-\mathrm{j}k_{\rm S} \sin\vartheta_{i\rm S}\, x}\, \mathrm{e}^{-\sqrt{k_{\rm S}^2 \sin^2\vartheta_{i\rm S} - k_{\rm P}^2}\, z}\, \mathrm{e}^{-\mathrm{j}\omega_0 t} \Bigg\}.
\end{aligned} \tag{9.133}
$$

We calculate the curve for the tip of $\underline{\mathbf{u}}_{m\rm P}(x, z, t, \omega_0, \vartheta_{i\rm S})$ as a function of time for a fixed location, for example, $x = 0, z = 0$; for a real phase propagation vector $\hat{\underline{\mathbf{k}}}_{m\rm P}$ $(\vartheta_{i\rm S} \leq \vartheta_{cm\rm P})$, the vector $\underline{\mathbf{u}}_{m\rm P}(0, 0, t, \omega_0, \vartheta_{i\rm S})$ oscillates according to longitudinal polarization along a line extending in $\hat{\underline{\mathbf{k}}}_{m\rm P}$-direction. With the notation

$$
\begin{aligned}
u_{0m\rm P}(\vartheta_{i\rm S}, \omega_0) &= - M_{\rm P}(\vartheta_{i\rm S})\, u_{i\rm S}(\omega_0) \\
&= |u_{0m\rm P}(\vartheta_{i\rm S}, \omega_0)|\, \mathrm{e}^{\mathrm{j}\phi_{0m\rm P}(\vartheta_{i\rm S}, \omega_0)},
\end{aligned} \tag{9.134}
$$

where $\phi_{0m\rm P}(\vartheta_{i\rm S}, \omega_0) = \pi + \phi_{M_{\rm P}}(\vartheta_{i\rm S}) + \phi_{u_{i\rm S}}(\omega_0)$ and $|u_{0m\rm P}(\vartheta_{i\rm S}, \omega_0)| = |M_{\rm P}(\vartheta_{i\rm S})\, u_{i\rm S}(\omega_0)|$, we can write (9.133) for $x = 0, z = 0$ as follows:

$$
\begin{aligned}
\underline{\mathbf{u}}_{m\rm P}(0, 0, t, \omega_0, \vartheta_{i\rm S}) = |u_{0m\rm P}| \Big[&\kappa \sin\vartheta_{i\rm S} \cos(\omega_0 t - \phi_{0m\rm P})\, \underline{\mathbf{e}}_x \\
&- \sqrt{\kappa^2 \sin^2\vartheta_{i\rm S} - 1}\, \sin(\omega_0 t - \phi_{0m\rm P})\, \underline{\mathbf{e}}_z \Big].
\end{aligned} \tag{9.135}
$$

Now, it becomes obvious that the tip of the vector

$$
\begin{aligned}
\frac{\underline{\mathbf{u}}_{m\rm P}(0, 0, t, \omega_0, \vartheta_{i\rm S})}{|u_{0m\rm P}|} &= \kappa \sin\vartheta_{i\rm S} \cos(\omega_0 t - \phi_{0m\rm P})\, \underline{\mathbf{e}}_x \\
&\quad - \sqrt{\kappa^2 \sin^2\vartheta_{i\rm S} - 1}\, \sin(\omega_0 t - \phi_{0m\rm P})\, \underline{\mathbf{e}}_z
\end{aligned} \tag{9.136}
$$

[182] Under the keyword "total reflection," we explain the nonzero mode conversion coefficient even though the magnitude of the reflection coefficient is equal to one.

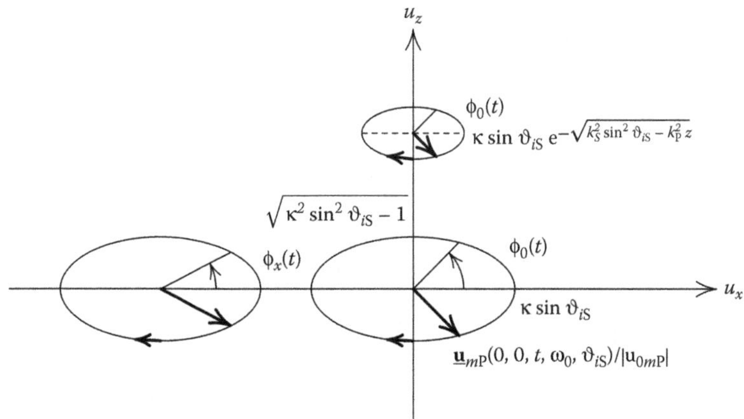

FIGURE 9.8
Elliptical curve as location of the real-valued particle displacement vector of the mode converted P-wave for $\vartheta_{iS} > \vartheta_{cmP}$: $\underline{\mathbf{u}}_{mP}$ rotates clockwise with increasing angle $\phi_0(t) = \omega_0 t - \phi_{0mP}$.

rotates clockwise in the xz-plane with increasing phase angle

$$\phi_0(t) = \omega_0 t - \phi_{0mP} \tag{9.137}$$

on a canonical[183] ellipse with the main axis $\kappa \sin \vartheta_{iS}$ and the smaller axis $\sqrt{\kappa^2 \sin^2 \vartheta_{iS} - 1}$ (Figure 9.8), because (9.136) is nothing else than the parameter representation of such an ellipse. For $x \neq 0$—e.g., $x < 0$—and $z = 0$ only the initial phase changes for $t = 0$ because the phase angle is then given by $\phi_x(t) = \omega_0 t - \phi_{0mP} + k_S \sin \vartheta_{iS}\, x$ (Figure 9.8); for $z > 0$, the length of both axes decreases exponentially (Figure 9.8).

At this point, we come back to the Fourier inversion of (9.82) through (9.84) for arbitrary pulse spectra $u_{iS}(\omega)$, where we may also consider the case $\vartheta_{iS} > \vartheta_{cmP}$ with (9.130), (9.131), and (9.132); the resulting (for a real positive x-component) $(-\mathrm{j})$-imaginary z-component of $\underline{\mathbf{u}}_{mP}(\underline{\mathbf{R}}, \omega, \vartheta_{iS})$ (Equation 9.101) is included in the phase of (9.131) in terms of an additional $\mathrm{e}^{-\mathrm{j}\pi/2}$-phase of the z-component as compared to the x-component. With this agreement, we use the relation (2.328) for $\vartheta_{iS} > \vartheta_{cmP}$ to complement (9.55) through (9.57):

$$\underline{\mathbf{u}}_{iSV}(\underline{\mathbf{R}}, t, \vartheta_{iS}) = u_{iS}\left(t - \frac{\hat{\underline{\mathbf{k}}}_{iS} \cdot \underline{\mathbf{R}}}{c_S}\right) \hat{\underline{\mathbf{k}}}_{iS} \times \underline{\mathbf{e}}_y, \quad 0 \leq \vartheta_{iS} \leq \pi/2, \tag{9.138}$$

[183] Both axes of the canonical ellipse point into the directions of the coordinate axes, and the ellipse is not rotated in the xz-coordinate system.

$$
\underline{\mathbf{u}}_{r\mathrm{SV}}(\underline{\mathbf{R}}, t, \vartheta_{i\mathrm{S}}) = \begin{cases} R_{\mathrm{SV}}(\vartheta_{i\mathrm{S}})\, u_{i\mathrm{S}}\left(t - \dfrac{\hat{\underline{\mathbf{k}}}_{r\mathrm{S}} \cdot \underline{\mathbf{R}}}{c_{\mathrm{S}}}\right) \hat{\underline{\mathbf{k}}}_{r\mathrm{S}} \times \underline{\mathbf{e}}_y, \quad 0 \le \vartheta_{i\mathrm{S}} \le \vartheta_{cm\mathrm{P}} \\ \Bigg[\cos\phi_{R_{\mathrm{SV}}}(\vartheta_{i\mathrm{S}})\, u_{i\mathrm{S}}\left(t - \dfrac{\hat{\underline{\mathbf{k}}}_{r\mathrm{S}} \cdot \underline{\mathbf{R}}}{c_{\mathrm{S}}}\right) \\ \quad - \sin\phi_{R_{\mathrm{SV}}}(\vartheta_{i\mathrm{S}})\, \mathcal{H}\left\{ u_{i\mathrm{S}}\left(\tau - \dfrac{\hat{\underline{\mathbf{k}}}_{r\mathrm{S}} \cdot \underline{\mathbf{R}}}{c_{\mathrm{S}}}\right)\right\}\Bigg] \hat{\underline{\mathbf{k}}}_{r\mathrm{S}} \times \underline{\mathbf{e}}_y, \\ \qquad \vartheta_{cm\mathrm{P}} < \vartheta_{i\mathrm{S}} \le \pi/2, \end{cases} \tag{9.139}
$$

$$
\underline{\mathbf{u}}_{m\mathrm{P}}(\underline{\mathbf{R}}, t, \vartheta_{i\mathrm{S}}) = \begin{cases} M_{\mathrm{P}}(\vartheta_{i\mathrm{S}})\, u_{i\mathrm{S}}\left(t - \dfrac{\hat{\underline{\mathbf{k}}}_{m\mathrm{P}} \cdot \underline{\mathbf{R}}}{c_{\mathrm{P}}}\right) \hat{\underline{\mathbf{k}}}_{m\mathrm{P}}, \quad 0 \le \vartheta_{i\mathrm{S}} \le \vartheta_{cm\mathrm{P}} \\ -|M_{\mathrm{P}}(\vartheta_{i\mathrm{S}})|\, \alpha(z, t, \vartheta_{i\mathrm{S}}) \\ \quad * \Bigg\{ \kappa \sin\vartheta_{i\mathrm{S}} \Bigg[\cos\phi_{M_{\mathrm{P}}}(\vartheta_{i\mathrm{S}})\, u_{i\mathrm{S}}\left(t + \dfrac{\sin\vartheta_{i\mathrm{S}}\, x}{c_{\mathrm{S}}}\right) \\ \qquad - \sin\phi_{M_{\mathrm{P}}}(\vartheta_{i\mathrm{S}}) \mathcal{H}\left\{ u_{i\mathrm{S}}\left(\tau + \dfrac{\sin\vartheta_{i\mathrm{S}}\, x}{c_{\mathrm{S}}}\right)\right\}\Bigg] \underline{\mathbf{e}}_x \\ \quad + \sqrt{\kappa^2 \sin^2\vartheta_{i\mathrm{S}} - 1}\, \Bigg[\sin\phi_{M_{\mathrm{P}}}(\vartheta_{i\mathrm{S}})\, u_{i\mathrm{S}}\left(t + \dfrac{\sin\vartheta_{i\mathrm{S}}\, x}{c_{\mathrm{S}}}\right) \\ \qquad + \cos\phi_{M_{\mathrm{P}}}(\vartheta_{i\mathrm{S}}) \mathcal{H}\left\{ u_{i\mathrm{S}}\left(\tau + \dfrac{\sin\vartheta_{i\mathrm{S}}\, x}{c_{\mathrm{S}}}\right)\right\}\Bigg] \underline{\mathbf{e}}_z \Bigg\}, \\ \qquad \vartheta_{cm\mathrm{P}} < \vartheta_{i\mathrm{S}} \le \pi/2, \end{cases} \tag{9.140}
$$

where

$$
\begin{aligned} \alpha(z, t, \vartheta_{i\mathrm{S}}) &= \mathcal{F}^{-1}\left\{ \mathrm{e}^{-\frac{|\omega|}{c_{\mathrm{P}}} \sqrt{\kappa^2 \sin^2\vartheta_{i\mathrm{S}} - 1}\, z} \right\} \\ &\overset{(2.278)}{=} \frac{c_{\mathrm{P}}}{\pi} \frac{\sqrt{\kappa^2 \sin^2\vartheta_{i\mathrm{S}} - 1}\, z}{(\kappa^2 \sin^2\vartheta_{i\mathrm{S}} - 1)\, z^2 + c_{\mathrm{P}}^2 t^2} \end{aligned} \tag{9.141}
$$

denotes the Fourier transform of the attenuation function to be used in (9.140) for convolution of the impulse $u_{i\mathrm{S}}(t)$ and its Hilbert transform. Due to its maximum amplitude

$$
\alpha(z, 0, \vartheta_{i\mathrm{S}}) = \frac{c_{\mathrm{P}}}{\pi} \frac{1}{\sqrt{\kappa^2 \sin^2\vartheta_{i\mathrm{S}} - 1}\, z} \tag{9.142}
$$

being proportional to z^{-1}, we observe that this convolution again results in the evanescence of $\underline{\mathbf{u}}_{m\mathrm{P}}(\underline{\mathbf{R}}, t, \vartheta_{i\mathrm{S}})$ for $z > 0$. As a consequence of the convolution, the $u_{i\mathrm{S}}(t)$-impulse is z-dependent dispersive (Figure 9.11). At least some intuition from (9.140) is obtained for $\vartheta_{i\mathrm{S}} > \vartheta_{cm\mathrm{P}}$ putting (9.135) $x = 0, z = 0$

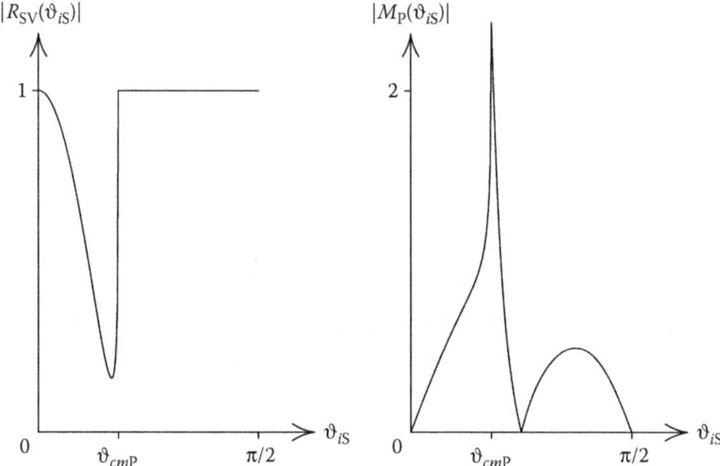

FIGURE 9.9
Magnitudes of reflection and mode conversion coefficients for SV-incidence as function of the angle of incidence (material: steel with $c_P = 5900$ m/s, $c_S = 3200$ m/s; $\kappa = 1.84$).

as in the time harmonic case, and choosing an RC$N(t)$-pulse for $u_{iS}(t)$, and observing[184] (2.306):

$$\underline{\mathbf{u}}_{mP}^{\mathrm{RC}N}(0,0,t,\vartheta_{iS} > \vartheta_{cmP}) \simeq - |M_P(\vartheta_{iS})| \, eN(t) \Big[\kappa \sin\vartheta_{iS} \cos(\omega_0 t - \phi_{M_P}) \, \underline{\mathbf{e}}_x - \sqrt{\kappa^2 \sin^2\vartheta_{iS} - 1} \sin(\omega_0 t - \phi_{M_P}) \, \underline{\mathbf{e}}_z \Big]; \quad (9.143)$$

the similarity with (9.136) and, hence, with Figure 9.8 may not be overlooked:[185] The tip of the vector $\underline{\mathbf{u}}_{mP}^{\mathrm{RC}N}(0,0,t,\vartheta_{iS})$ moves as a function of time on the same ellipse as in the time harmonic case, only the axes of the ellipse change with the RC$N(t)$-envelope $eN(t)$.

With (9.59) and based on (9.58), we presented the surface deformation rate tensor in the time domain for P-wave incidence that can be considered as source density of the reflected total field; here, due to the complexity of the

[184]For the time domain attenuation function $\alpha(z,t,\vartheta_{iS})$, we have

$$\lim_{z\to 0} \alpha(z,t,\vartheta_{iS}) = \delta(t)$$

canceling the convolution for $z = 0$.

[185]In (9.136), the angle π that is contained in ϕ_{0mP} is changed into a negative sign; the additional phase angle $\phi_{u_{iS}}$ is contained in the envelope $eN(t)$ of the RC$N(t)$-pulse.

expressions (9.138) through (9.140), we only present the Fourier spectrum of the respective tensor for SV-incidence, an equation that corresponds to (9.62):

$$\begin{aligned}
&\underline{\underline{\mathbf{g}}}^{\mathrm{SV}}(x, \omega, \vartheta_{i\mathrm{S}}) \\
&\quad = \frac{\mathrm{j}\omega}{2}\, u_{i\mathrm{S}}(\omega)\, \mathrm{e}^{-\mathrm{j}k_{\mathrm{S}} \sin\vartheta_{i\mathrm{S}}\, x} \Big[\underline{\mathbf{e}}_z \underline{\hat{\mathbf{k}}}_{i\mathrm{S}} \times \underline{\mathbf{e}}_y + \underline{\hat{\mathbf{k}}}_{i\mathrm{S}} \times \underline{\mathbf{e}}_y \underline{\mathbf{e}}_z \\
&\qquad + R_{\mathrm{SV}}(\vartheta_{i\mathrm{S}})(\underline{\mathbf{e}}_z \underline{\hat{\mathbf{k}}}_{r\mathrm{S}} \times \underline{\mathbf{e}}_y + \underline{\hat{\mathbf{k}}}_{r\mathrm{S}} \times \underline{\mathbf{e}}_y \underline{\mathbf{e}}_z) \\
&\qquad + M_{\mathrm{P}}(\vartheta_{i\mathrm{S}})(\underline{\mathbf{e}}_z \underline{\hat{\mathbf{k}}}_{m\mathrm{P}} + \underline{\hat{\mathbf{k}}}_{m\mathrm{P}} \underline{\mathbf{e}}_z)\Big]. \qquad (9.144)
\end{aligned}$$

Note: (9.144) holds for $0 \leq \vartheta_{i\mathrm{S}} \leq \pi/2$, hence also for $\vartheta_{i\mathrm{S}} > \vartheta_{cm\mathrm{P}}$! Yet, actually inserting this source density into a Huygens integral extending over the xy-plane, the radiation pattern of the mode converted P-"ray" disappears for $\vartheta_{i\mathrm{S}} > \vartheta_{c\mathrm{P}}$ (Figure 15.5).

Wavefronts of reflected SV- and mode converted P-waves: As with Figure 9.4, we discuss reflection and mode conversion coefficients (9.120) and (9.121) for fixed κ as function of the angle of incidence. Since both factors are complex for $\vartheta_{i\mathrm{S}} > \vartheta_{cm\mathrm{P}}$, only the display of the magnitude (and eventually the phase) makes sense. For $\kappa = 1.84$ (steel), the angular dependence of $|R_{\mathrm{SV}}(\vartheta_{i\mathrm{S}})|$ and $|M_{\mathrm{P}}(\vartheta_{i\mathrm{S}})|$ is depicted in Figure 9.9; obviously, we have $|R_{\mathrm{SV}}(\vartheta_{i\mathrm{S}} > \vartheta_{cm\mathrm{P}})| = 1$, and from Equation 9.120, we take $R_{\mathrm{SV}}(0) = -1$, $R_{\mathrm{SV}}(\vartheta_{m\mathrm{P}}) = -1$, and $|R_{\mathrm{SV}}(\pi/2)| = 1$ with $\phi_{R_{\mathrm{SV}}}(\pi/2) = \pi$.

In Figure 9.10(a) and (b), we show—similarly to Figure 9.5(a) and (b)—wavefronts of vertically incident and reflected waves for two different times, and, as before, the magnitude of the vector particle velocity is displayed. Again, the time function of the incident SV-wave is given as an RC2(t)-pulse—note the smaller wavelength of the shear wave—resulting in

$$\underline{\mathbf{u}}_{i\mathrm{SV}}(\underline{\mathbf{R}}, t, \vartheta_{i\mathrm{S}} = 0) = \mathrm{RC2}\left(t + \frac{z}{c_{\mathrm{S}}}\right) \underline{\mathbf{e}}_x \qquad (9.145)$$

for vertical incidence based on (9.138). After reflection—Figure 9.10(b)—the only existing reflected wavefront for $z > 0$—we have $R_{\mathrm{SV}}(0) = -1$ and $\underline{\hat{\mathbf{u}}}_{r\mathrm{SV}}(\vartheta_{i\mathrm{S}} = 0) = -\underline{\mathbf{e}}_x$ (Figure 9.7)—

$$\underline{\mathbf{u}}_{r\mathrm{SV}}(\underline{\mathbf{R}}, t, \vartheta_{i\mathrm{S}} = 0) = \mathrm{RC2}\left(t - \frac{z}{c_{\mathrm{S}}}\right) \underline{\mathbf{e}}_x \qquad (9.146)$$

is displayed; it is evident that we obtain an RC2-pulse with the same sign as (9.145): The negative reflection coefficient turns the negative x-component of $\underline{\hat{\mathbf{u}}}_{r\mathrm{SV}}(\vartheta_{i\mathrm{S}} = 0)$ into the positive x-direction.

The other pictures of Figure 9.10 display wavefronts of incident, reflected, and mode converted waves for $\vartheta_{i\mathrm{S}} > 0$ for fixed time $t = 0$ (Equations 9.138 through 9.140). Particularly, Figure 9.10(e) is interesting: Here, the angle $\vartheta_{i\mathrm{S}}$ is "a little bit" larger than the critical angle $\vartheta_{cm\mathrm{P}}$ for the mode converted P-wave, and apparently, the evanescence of this wavefront is visible.

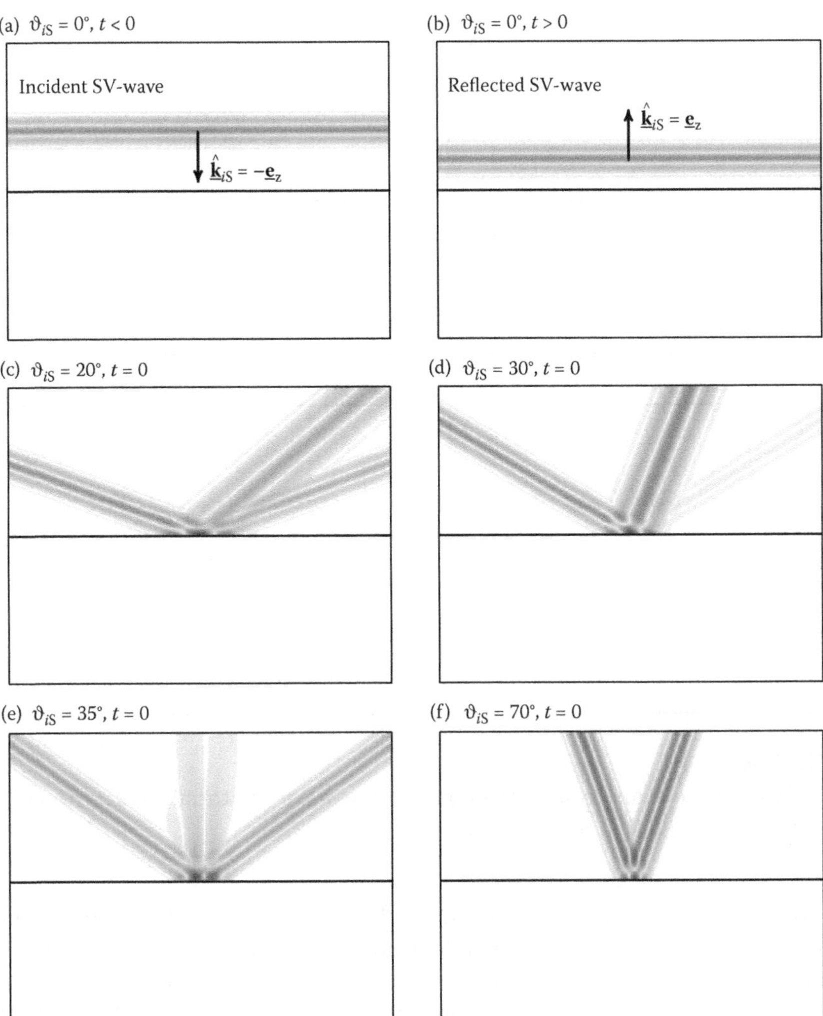

FIGURE 9.10
Wavefronts of incident and reflected SV- as well as mode converted P-waves for various angles of incidence (a)–(f) (material: steel mit $c_P = 5900$ m/s, $c_S = 3200$ m/s; $\kappa = 1.84$).

Displaying it alone (Figure 9.11), we nicely notice the z-dependent dispersion caused by the convolution in (9.140); the nonsymmetry with regard to $x = 0$ results from the superposition of the symmetric RC2(t)-pulse with its antisymmetric Hilbert transform. Figure 9.10(f) intuitively illustrates total reflection to be immediately discussed, where the phase shift of the reflected SV-wave is due to the superposition with the Hilbert transformed RC2(t)-pulse (Equation 9.139).

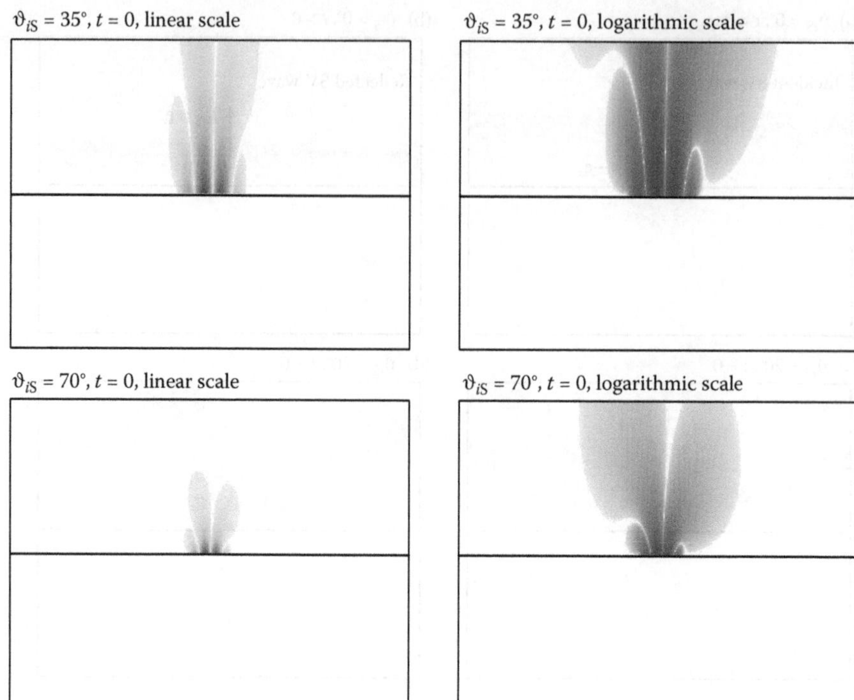

FIGURE 9.11
Wavefront of the evanescent mode converted p-wave (material: steel with $c_P =$ 5900 m/s, $c_S = 3200$ m/s; $\kappa = 1.84$).

For "running" times, all wavefronts in Figure 9.10(c)–(f) move simultaneously with the constant (trace) velocity $c_S/\sin\vartheta_{iS}$ (Equations 9.138 through 9.140) of the boundary phase center from right to left (in negative x-direction) through their respective observation windows (compare Figure 9.2). Due to (9.92), this trace velocity becomes equal to c_P for $\vartheta_{iS} = \vartheta_{cmP}$: The SV-wavefront being reflected under the angle ϑ_{cmP} manifests itself as the "head wave" (Section 14.2) of a phase center moving along the boundary with this velocity.

Energy balance for reflection and mode conversion: Total reflection: In both diagrams in Figure 9.9, two things are worth being noted:

- Even though the magnitude of the reflection coefficient $R_{SV}(\vartheta_{iS})$ for the incident SV-wave is equal to 1 for $\vartheta_{iS} > \vartheta_{cmP}$, the amplitude factor $M_P(\vartheta_{iS})$ of the mode converted P-wave is not at all equal to 0;
- the amplitude factor of the mode converted P-wave normalized to the amplitude of the incident wave may be larger than 1.

Yet, both items do not contradict an energy balance because it must hold for the energy flux density and not for amplitudes. So, let us calculate the complex elastodynamic Poynting vector! For the incident and reflected SV-waves, we have access to (8.132):

$$\underline{\mathbf{S}}_{\mathrm{K}i\mathrm{SV}}(\underline{\mathbf{R}}, \omega_0, \hat{\underline{\mathbf{k}}}_{i\mathrm{S}}) = \frac{\omega_0^2 \rho}{2} c_{\mathrm{S}} |u_{i\mathrm{S}}(\omega_0)|^2 \, \hat{\underline{\mathbf{k}}}_{i\mathrm{S}}, \tag{9.147}$$

$$\underline{\mathbf{S}}_{\mathrm{K}r\mathrm{SV}}(\underline{\mathbf{R}}, \omega_0, \hat{\underline{\mathbf{k}}}_{r\mathrm{S}}) = \frac{\omega_0^2 \rho}{2} c_{\mathrm{S}} |R_{\mathrm{SV}}(\vartheta_{i\mathrm{S}})|^2 |u_{i\mathrm{S}}(\omega_0)|^2 \, \hat{\underline{\mathbf{k}}}_{r\mathrm{S}}. \tag{9.148}$$

Both Poynting vectors are real valued directly standing for the time averaged energy transported through a unit area.

Yet, for the mode converted P-wave, we have to admit complex values of the wave number vector $\hat{\underline{\mathbf{k}}}_{m\mathrm{P}}$. Generalizing (8.131), we obtain[186]

$$\begin{aligned}\underline{\mathbf{S}}_{\mathrm{K}m\mathrm{P}}(\underline{\mathbf{R}}, \omega_0, \hat{\underline{\mathbf{k}}}_{m\mathrm{P}}) = {} & \frac{\omega_0}{2} k_{\mathrm{P}} |M_{\mathrm{P}}(\vartheta_{i\mathrm{S}})|^2 |u_{i\mathrm{S}}(\omega_0)|^2 \, \mathrm{e}^{-2k_{\mathrm{P}} \Im \hat{\underline{\mathbf{k}}}_{m\mathrm{P}} \cdot \underline{\mathbf{R}}} \\ & \times \left(\lambda \hat{\underline{\mathbf{k}}}_{m\mathrm{P}} + 2\mu \hat{\underline{\mathbf{k}}}_{m\mathrm{P}} \cdot \hat{\underline{\mathbf{k}}}^*_{m\mathrm{P}} \hat{\underline{\mathbf{k}}}^*_{m\mathrm{P}} \right),\end{aligned} \tag{9.149}$$

resulting in (8.131) of a homogeneous plane wave for $\Im \hat{\underline{\mathbf{k}}}_{m\mathrm{P}} = \underline{\mathbf{0}}$, namely for $\vartheta_{i\mathrm{S}} \le \vartheta_{cm\mathrm{P}}$. Even though $\hat{\underline{\mathbf{k}}}_{m\mathrm{P}} \cdot \hat{\underline{\mathbf{k}}}^*_{m\mathrm{P}}$ is always real, $\underline{\mathbf{S}}_{\mathrm{K}m\mathrm{P}}$ is eventually complex due to the eventually complex wave number vector that explicitly appears; we calculate the real part

$$\begin{aligned}\Re \left\{ \underline{\mathbf{S}}_{\mathrm{K}m\mathrm{P}}(\underline{\mathbf{R}}, \omega_0, \hat{\underline{\mathbf{k}}}_{m\mathrm{P}}) \right\} = {} & \frac{\omega_0^2}{2c_{\mathrm{P}}} |M_{\mathrm{P}}(\vartheta_{i\mathrm{S}})|^2 |u_{i\mathrm{S}}(\omega_0)|^2 \, \mathrm{e}^{-2k_{\mathrm{P}} \Im \hat{\underline{\mathbf{k}}}_{m\mathrm{P}} \cdot \underline{\mathbf{R}}} \\ & \times \left(\lambda + 2\mu \hat{\underline{\mathbf{k}}}_{m\mathrm{P}} \cdot \hat{\underline{\mathbf{k}}}^*_{m\mathrm{P}} \right) \Re \hat{\underline{\mathbf{k}}}^*_{m\mathrm{P}}.\end{aligned} \tag{9.150}$$

Now the "boundary energy balance" looks like that:

- The sum of the real parts of the normal components of all three Poynting vectors must be equal to zero in the boundary, else we would observe a (positive) energy accumulation:

$$\begin{aligned}& \underline{\mathbf{n}} \cdot \Re \left\{ \underline{\mathbf{S}}_{\mathrm{K}i\mathrm{SV}}(x, 0, \omega_0, \hat{\underline{\mathbf{k}}}_{i\mathrm{S}}) \right\} + \underline{\mathbf{n}} \cdot \Re \left\{ \underline{\mathbf{S}}_{\mathrm{K}r\mathrm{SV}}(x, 0, \omega_0, \hat{\underline{\mathbf{k}}}_{r\mathrm{S}}) \right\} \\ & \quad + \underline{\mathbf{n}} \cdot \Re \left\{ \underline{\mathbf{S}}_{\mathrm{K}m\mathrm{P}}(x, 0, \omega_0, \hat{\underline{\mathbf{k}}}_{m\mathrm{P}}) \right\} = 0.\end{aligned} \tag{9.151}$$

  This balance equation is nothing more than the energy conservation law (4.64) for nondissipative materials specialized to the boundary (utilizing the method to derive transition and boundary conditions in Section 3.3).

[186] We can directly come back to (8.186), yet we must observe that we have defined $\hat{\underline{\mathbf{u}}}_{m\mathrm{P}}$ according to (8.159) in Section 8.2; for this definition, a mode conversion coefficient $M'_{\mathrm{P}} = M_{\mathrm{P}} \sqrt{\hat{\underline{\mathbf{k}}}_{m\mathrm{P}} \cdot \hat{\underline{\mathbf{k}}}^*_{m\mathrm{P}}}$ results that has to be inserted into (8.186): We obtain (9.149).

- The real parts of the tangential components of the three respective Poynting vectors may be arbitrary in the boundary because of the infinite total energy of plane waves the corresponding energy flux prevents balancing. In other words: The energy conservation law (4.64) provides nothing regarding the tangential components in the boundary.

With (9.147), (9.148), and (9.150), we calculate (9.151) for $\vartheta_{i\mathrm{S}} \leq \vartheta_{c m\mathrm{P}}$ as—we have $\underline{\mathbf{n}} = \underline{\mathbf{e}}_z$—

$$\cos\vartheta_{i\mathrm{S}} - \cos\vartheta_{i\mathrm{S}}|R_{\mathrm{SV}}(\vartheta_{i\mathrm{S}})|^2 - \kappa|M_{\mathrm{P}}(\vartheta_{i\mathrm{S}})|^2\sqrt{1-\kappa^2\sin^2\vartheta_{i\mathrm{S}}} = 0; \tag{9.152}$$

the expression

$$|M_{\mathrm{P}}(\vartheta_{i\mathrm{S}})|^2 = \frac{\cos\vartheta_{i\mathrm{S}}[1-|R_{\mathrm{SV}}(\vartheta_{i\mathrm{S}})|^2]}{\kappa\sqrt{1-\kappa^2\sin^2\vartheta_{i\mathrm{S}}}} \tag{9.153}$$

emerges that evidently permits values bigger than 1 for $|M_{\mathrm{P}}(\vartheta_{i\mathrm{S}} \leq \vartheta_{c m\mathrm{P}})|^2$. Hence, this is not a contradiction to the energy conservation law.

For $\vartheta_{i\mathrm{S}} > \vartheta_{c m\mathrm{P}}$, we have $\underline{\mathbf{e}}_z \cdot \Re\{\underline{\mathbf{S}}_{\mathrm{K}m\mathrm{P}}(\underline{\mathbf{R}}, \omega_0, \hat{\underline{\mathbf{k}}}_{m\mathrm{P}})\} = 0$ due to $\underline{\mathbf{e}}_z \cdot \Re\hat{\underline{\mathbf{k}}}_{m\mathrm{P}} = 0$, i.e., (9.151) expresses the fact that the evanescent wave does not take part in the energy exchange at the boundary resulting in

$$|R_{\mathrm{SV}}(\vartheta_{i\mathrm{S}} > \vartheta_{c m\mathrm{P}})|^2 = 1 \tag{9.154}$$

from (9.151) independently upon M_{P}. Therefore, we speak of the total reflection of the incident wave.

With (9.150), we calculate the nonvanishing component of $\Re\{\underline{\mathbf{S}}_{\mathrm{K}m\mathrm{P}}\}$ as

$$\begin{aligned}\Re\left\{\underline{\mathbf{S}}_{\mathrm{K}m\mathrm{P}}(\underline{\mathbf{R}}, \omega_0, \hat{\underline{\mathbf{k}}}_{m\mathrm{P}})\right\} &= -\frac{\omega_0^2}{2c_{\mathrm{S}}}\,|M_{\mathrm{P}}(\vartheta_{i\mathrm{S}})|^2|u_{i\mathrm{S}}(\omega_0)|^2\,\mathrm{e}^{-2k_{\mathrm{P}}\Im\hat{\underline{\mathbf{k}}}_{m\mathrm{P}}\cdot\underline{\mathbf{R}}}\\ &\quad\times[\lambda + 2\mu(2\kappa^2\sin^2\vartheta_{i\mathrm{S}} - 1)]\sin\vartheta_{i\mathrm{S}}\,\underline{\mathbf{e}}_x\end{aligned} \tag{9.155}$$

and note that the inhomogeneous—evanescent—plane wave propagating into $(-x)$-direction transports energy into propagation direction with exponentially decaying energy flux density in z-direction. The origin of this energy for an evanescent wave being energetically not coupled to the incident SV-wave may not be clarified with the model of plane waves due to their infinite total energy; we have to switch to spatially constrained rays.

For the sake of completeness, we also compute the time averaged energy density of time harmonic inhomogeneous (evanescent) plane waves. Generalizing (8.137) yields with some calculus [Footnote 186 similarly holds if we apply (8.187)]:

$$\begin{aligned}\langle w_{m\mathrm{P}}(\underline{\mathbf{R}}, t, \hat{\underline{\mathbf{k}}}_{m\mathrm{P}})\rangle &= \frac{\omega_0^2}{4}\frac{1}{c_{\mathrm{P}}^2}\,|M_{\mathrm{P}}(\vartheta_{i\mathrm{S}})|^2|u_{i\mathrm{S}}(\omega_0)|^2\,\mathrm{e}^{-2k_{\mathrm{P}}\Im\hat{\underline{\mathbf{k}}}_{m\mathrm{P}}\cdot\underline{\mathbf{R}}}\\ &\quad\times(1 + \hat{\underline{\mathbf{k}}}_{m\mathrm{P}}\cdot\hat{\underline{\mathbf{k}}}_{m\mathrm{P}}^*)(\lambda + 2\mu\hat{\underline{\mathbf{k}}}_{m\mathrm{P}}\cdot\hat{\underline{\mathbf{k}}}_{m\mathrm{P}}^*);\end{aligned} \tag{9.156}$$

for $\vartheta_{i\mathrm{S}} \leq \vartheta_{cm\mathrm{P}}$, hence $\Im\underline{\hat{\mathbf{k}}}_{m\mathrm{P}} = \underline{\mathbf{0}}$, this corresponds to the expression (8.137) for the homogeneous plane wave.

With (8.172) and (9.109), we calculate the phase velocity of the evanescent wave as (compare Equation 9.140):

$$c_{m\mathrm{P}}(\vartheta_{i\mathrm{S}}) = \frac{\omega_0}{|\Re\underline{\mathbf{k}}_{m\mathrm{P}}|} \tag{9.157}$$

$$= \frac{c_\mathrm{S}}{\sin\vartheta_{i\mathrm{S}}}, \qquad \vartheta_{i\mathrm{S}} > \vartheta_{cm\mathrm{P}}. \tag{9.158}$$

With (8.188) and (9.109), we calculate the energy velocity of the evanescent wave as:

$$\underline{\mathbf{c}}_{\mathrm{E}m\mathrm{P}}(\vartheta_{i\mathrm{S}} > \vartheta_{cm\mathrm{P}}) = \frac{2c_\mathrm{P}}{1 + \underline{\hat{\mathbf{k}}}_{m\mathrm{P}} \cdot \underline{\hat{\mathbf{k}}}^*_{m\mathrm{P}}} \Re\underline{\hat{\mathbf{k}}}_{m\mathrm{P}} \tag{9.159}$$

$$= -\frac{c_\mathrm{S}}{\sin\vartheta_{i\mathrm{S}}}\,\underline{\mathbf{e}}_x$$

$$= -\,c_{m\mathrm{P}}(\vartheta_{i\mathrm{S}})\,\underline{\mathbf{e}}_x. \tag{9.160}$$

The magnitudes of the phase and energy velocities of the evanescent wave are equal, and we have

$$c_\mathrm{P} > |\underline{\mathbf{c}}_{\mathrm{E}m\mathrm{P}}(\vartheta_{i\mathrm{S}} > \vartheta_{cm\mathrm{P}})| \geq c_\mathrm{S}, \tag{9.161}$$

because $\sin\vartheta_{i\mathrm{S}} > c_\mathrm{S}/c_\mathrm{P}$ for $\vartheta_{i\mathrm{S}} > \vartheta_{cm\mathrm{P}}$.

9.1.3 Secondary transverse horizontal shear wave incidence

The simplest case of shear wave reflection at a stress-free boundary is still missing: SH-wave reflection. Instead of (9.82), we assume

$$\underline{\mathbf{u}}_{i\mathrm{SH}}(\underline{\mathbf{R}}, \omega, \underline{\hat{\mathbf{k}}}_{i\mathrm{S}}) = -u_{i\mathrm{S}}(\omega)\,\mathrm{e}^{\mathrm{j}k_\mathrm{S}\underline{\hat{\mathbf{k}}}_{i\mathrm{S}}\cdot\underline{\mathbf{R}}}\,\underline{\mathbf{e}}_y, \tag{9.162}$$

where $\underline{\hat{\mathbf{k}}}_{i\mathrm{S}}$ is given by (9.77). The particle velocity (9.162) yields the stress tensor

$$\underline{\underline{\mathbf{T}}}_{i\mathrm{SH}}(\underline{\mathbf{R}}, \omega, \underline{\hat{\mathbf{k}}}_{i\mathrm{S}}) = -\mathrm{j}k_\mathrm{S}\mu\,u_{i\mathrm{S}}(\omega)\,\mathrm{e}^{\mathrm{j}k_\mathrm{S}\underline{\hat{\mathbf{k}}}_{i\mathrm{S}}\cdot\underline{\mathbf{R}}}\left(\underline{\hat{\mathbf{k}}}_{i\mathrm{S}}\underline{\mathbf{e}}_y + \underline{\mathbf{e}}_y\underline{\hat{\mathbf{k}}}_{i\mathrm{S}}\right) \tag{9.163}$$

in a homogeneous isotropic nondissipative material. Consequently, the projection

$$\underline{\underline{\mathbf{T}}}_{i\mathrm{SH}}(\underline{\mathbf{R}}, \omega, \underline{\hat{\mathbf{k}}}_{i\mathrm{S}}) \cdot \underline{\mathbf{e}}_z = -\mathrm{j}k_\mathrm{S}\mu\,u_{i\mathrm{S}}(\omega)\,\mathrm{e}^{\mathrm{j}k_\mathrm{S}\underline{\hat{\mathbf{k}}}_{i\mathrm{S}}\cdot\underline{\mathbf{R}}}\,\underline{\hat{\mathbf{k}}}_{i\mathrm{S}} \cdot \underline{\mathbf{e}}_z\,\underline{\mathbf{e}}_y \tag{9.164}$$

that appears in the requirement for the boundary to be stress-free has only a y-component. Since the respective projections for P- and SV-waves—Equations 9.12 and 9.85—exhibit only x- and z-components, SH-wave reflection is obviously decoupled from P-SV-reflection, and we might be successful with the *ansatz* of the single reflected SH-wave

$$\underline{\mathbf{u}}_{r\mathrm{SH}}(\underline{\mathbf{R}},\omega,\hat{\underline{\mathbf{k}}}_{r\mathrm{S}}) = -R_{\mathrm{SH}}(\vartheta_{i\mathrm{S}})u_{i\mathrm{S}}(\omega)\,\mathrm{e}^{\mathrm{j}k_{\mathrm{S}}\hat{\underline{\mathbf{k}}}_{r\mathrm{S}}\cdot\underline{\mathbf{R}}}\,\underline{\mathbf{e}}_y, \tag{9.165}$$

where $\hat{\underline{\mathbf{k}}}_{r\mathrm{S}}$ is given by (9.78). With its stress tensor

$$\underline{\underline{\mathbf{T}}}_{r\mathrm{SH}}(\underline{\mathbf{R}},\omega,\hat{\underline{\mathbf{k}}}_{r\mathrm{S}})\cdot\underline{\mathbf{e}}_z = -\mathrm{j}k_{\mathrm{S}}\mu R_{\mathrm{SH}}(\vartheta_{i\mathrm{S}})u_{i\mathrm{S}}(\omega)\,\mathrm{e}^{\mathrm{j}k_{\mathrm{S}}\hat{\underline{\mathbf{k}}}_{r\mathrm{S}}\cdot\underline{\mathbf{R}}}\left(\hat{\underline{\mathbf{k}}}_{r\mathrm{S}}\underline{\mathbf{e}}_y+\underline{\mathbf{e}}_y\hat{\underline{\mathbf{k}}}_{r\mathrm{S}}\right), \tag{9.166}$$

we satisfy the boundary condition

$$\left[\underline{\underline{\mathbf{T}}}_{i\mathrm{SH}}(\underline{\mathbf{R}}_S,\omega,\hat{\underline{\mathbf{k}}}_{r\mathrm{S}})+\underline{\underline{\mathbf{T}}}_{r\mathrm{SH}}(\underline{\mathbf{R}}_S,\omega,\hat{\underline{\mathbf{k}}}_{r\mathrm{S}})\right]\cdot\underline{\mathbf{e}}_z = \underline{\mathbf{0}} \tag{9.167}$$

via superposition with (9.163); it follows

$$-u_{i\mathrm{S}}(\omega)\,\mathrm{e}^{-\mathrm{j}k_{\mathrm{S}}\sin\vartheta_{i\mathrm{S}}\,x}\cos\vartheta_{i\mathrm{S}}+R_{\mathrm{SH}}(\vartheta_{i\mathrm{S}})u_{i\mathrm{S}}(\omega)\,\mathrm{e}^{-\mathrm{j}k_{\mathrm{S}}\sin\vartheta_{r\mathrm{S}}\,x}\cos\vartheta_{r\mathrm{S}} = 0. \tag{9.168}$$

Due to phase matching, we immediately obtain the reflection law

$$\vartheta_{r\mathrm{S}} = \vartheta_{i\mathrm{S}}, \tag{9.169}$$

and subsequently

$$R_{\mathrm{SH}}(\vartheta_{i\mathrm{S}}) = R_{\mathrm{SH}} = 1. \tag{9.170}$$

SH-waves are totally reflected at a stress-free boundary independent upon the angle of incidence!

For vertical incidence $\vartheta_{i\mathrm{S}} = 0$, SH-polarization is physically indistinguishable from SV-polarization; so, how do we explain the different signs of the reflection coefficients $R_{\mathrm{SV}}(0) = -1$ and $R_{\mathrm{SH}} = 1$? It is explicated with the *ansatz* for polarization directions of incident and reflected SV- and SH-waves, respectively: For vertical incidence, the pertinent SV-polarizations are opposite to each other (Figure 9.7), whereas the SH-polarizations are parallel (not only for vertical incidence) (Figure 9.12). This has consequences for the sound pressure: For this (accidental) choice of signs of polarizations, $p_{i\mathrm{SV}}(\underline{\mathbf{R}}_S,\omega,0)+p_{r\mathrm{SV}}(\underline{\mathbf{R}}_S,\omega,0)$ satisfies a Dirichlet boundary condition but not $p_{i\mathrm{SH}}(\underline{\mathbf{R}}_S,\omega,0)+p_{r\mathrm{SH}}(\underline{\mathbf{R}}_S,\omega,0)$. A repair is easy: Change SH-polarization directions accordingly! This illuminates that: The relative polarization directions are physically determined via the pertinent reflection coefficients but not the sound pressure "orientations."

With (9.170), we explicitly present—for the sake of completeness—the cartesian stress tensor components of the incident and reflected field:

$$\begin{aligned}\underline{\underline{\mathbf{T}}}_{i\mathrm{SH}}(x,z,\omega,\vartheta_{i\mathrm{S}}) &= \mathrm{j}k_{\mathrm{S}}\mu\,u_{i\mathrm{S}}(\omega)\,\mathrm{e}^{-\mathrm{j}k_{\mathrm{S}}(\sin\vartheta_{i\mathrm{S}}\,x+\mathrm{j}\cos\vartheta_{i\mathrm{S}}\,z)}\\ &\quad\times\left[\sin\vartheta_{i\mathrm{S}}(\underline{\mathbf{e}}_x\underline{\mathbf{e}}_y+\underline{\mathbf{e}}_y\underline{\mathbf{e}}_x)+\cos\vartheta_{i\mathrm{S}}(\underline{\mathbf{e}}_y\underline{\mathbf{e}}_z+\underline{\mathbf{e}}_z\underline{\mathbf{e}}_y)\right],\end{aligned} \tag{9.171}$$

$$\underline{\underline{\mathbf{T}}}_{r\mathrm{SH}}(x, z, \omega, \vartheta_{i\mathrm{S}}) = \mathrm{j}k_{\mathrm{S}}\mu\, u_{i\mathrm{S}}(\omega)\, \mathrm{e}^{-\mathrm{j}k_{\mathrm{S}}(\sin\vartheta_{i\mathrm{S}}\, x - \mathrm{j}\cos\vartheta_{i\mathrm{S}}\, z)} \times \left[\sin\vartheta_{i\mathrm{S}}(\underline{\mathbf{e}}_x\underline{\mathbf{e}}_y + \underline{\mathbf{e}}_y\underline{\mathbf{e}}_x) - \cos\vartheta_{i\mathrm{S}}(\underline{\mathbf{e}}_y\underline{\mathbf{e}}_z + \underline{\mathbf{e}}_z\underline{\mathbf{e}}_y)\right]; \tag{9.172}$$

furthermore, the sound pressure equations of plane SH-waves read

$$p_{i\mathrm{SH}}(\underline{\mathbf{R}}, \omega, \vartheta_{i\mathrm{S}}) = \underline{\underline{\mathbf{T}}}_{i\mathrm{SH}}(\underline{\mathbf{R}}, \omega, \hat{\underline{\mathbf{k}}}_{i\mathrm{S}}) : \hat{\underline{\mathbf{k}}}_{i\mathrm{S}}\underline{\mathbf{e}}_y = -\,\mathrm{j}\omega Z_{\mathrm{S}}\, u_{i\mathrm{S}}(\omega)\, \mathrm{e}^{\mathrm{j}k_{\mathrm{S}}\hat{\underline{\mathbf{k}}}_{i\mathrm{S}}\cdot\underline{\mathbf{R}}}, \tag{9.173}$$

$$p_{r\mathrm{SH}}(\underline{\mathbf{R}}, \omega, \vartheta_{i\mathrm{S}}) = \underline{\underline{\mathbf{T}}}_{r\mathrm{SH}}(\underline{\mathbf{R}}, \omega, \hat{\underline{\mathbf{k}}}_{r\mathrm{S}}) : \hat{\underline{\mathbf{k}}}_{r\mathrm{S}}\underline{\mathbf{e}}_y = -\,\mathrm{j}\omega Z_{\mathrm{S}}\, u_{i\mathrm{S}}(\omega)\, \mathrm{e}^{\mathrm{j}k_{\mathrm{S}}\hat{\underline{\mathbf{k}}}_{r\mathrm{S}}\cdot\underline{\mathbf{R}}}. \tag{9.174}$$

Due to $R_{\mathrm{SH}} = 1$, the surface deformation rate tensor as equivalent source of the reflected field is proportional to the incident field itself:

$$\underline{\underline{\mathbf{g}}}^{\mathrm{SH}}(x, \omega, \vartheta_{i\mathrm{S}}) = \mathrm{j}\omega\left[\underline{\mathbf{e}}_z\hat{\underline{\mathbf{u}}}_{i\mathrm{SH}}(x, 0, \omega, \vartheta_{i\mathrm{S}}) + \hat{\underline{\mathbf{u}}}_{i\mathrm{SH}}(x, 0, \omega, \vartheta_{i\mathrm{S}})\underline{\mathbf{e}}_z\right] = -\,\mathrm{j}\omega\, u_{i\mathrm{S}}(\omega)\, \mathrm{e}^{-\mathrm{j}k_{\mathrm{S}}\sin\vartheta_{i\mathrm{S}}\, x}(\underline{\mathbf{e}}_z\underline{\mathbf{e}}_y + \underline{\mathbf{e}}_y\underline{\mathbf{e}}_z). \tag{9.175}$$

We explicitly state the following for this special case (general remarks have already been made in Section 7.3):

- Since both the field quantity $\underline{\mathbf{u}}_{\mathrm{SH}}$ as well as the equation for vanishing stress only possess a y-component, and due to $\partial/\partial y \equiv 0$, SH-reflection turns out to be a two-dimensional scalar boundary value problem for the scalar field quantity $\underline{\mathbf{u}}_{\mathrm{SH}} \cdot \underline{\mathbf{e}}_y$.

- The boundary condition (9.168) corresponds to a (homogeneous) Neumann boundary condition (Equation 5.21)

$$\underline{\mathbf{n}} \cdot \nabla\left[\underline{\mathbf{u}}_{\mathrm{SH}}(\underline{\mathbf{R}}, \omega, \hat{\underline{\mathbf{k}}}_{\mathrm{S}}) \cdot \underline{\mathbf{e}}_y\right]_{\underline{\mathbf{R}}=\underline{\mathbf{R}}_S} = \left.\frac{\partial \underline{\mathbf{u}}_{\mathrm{SH}}(\underline{\mathbf{R}}, \omega, \hat{\underline{\mathbf{k}}}_{\mathrm{S}}) \cdot \underline{\mathbf{e}}_y}{\partial z}\right|_{z=0} = 0 \tag{9.176}$$

  just for this scalar field quantity. This is surprising because the respective boundary value problem of scalar acoustics—the soft boundary condition for a boundary to a vacuum half-space—is a Dirichlet boundary value problem; yet, it is a Dirichlet problem for the scalar pressure $p(\underline{\mathbf{R}}, \omega)$ as it would correspond to the isotropic stress tensor $\underline{\underline{\mathbf{T}}}(\underline{\mathbf{R}}, \omega) = -p(\underline{\mathbf{R}}, \omega)\underline{\underline{\mathbf{I}}}$; yet, according to (9.171) and (9.172), it may not be defined. We may define the sound pressure (9.173) and (9.174) for plane SH-waves, but it does not satisfy a Dirichlet boundary condition for the underlying polarization orientations:[187] The two-dimensional scalar SH-wave boundary value problem is a scalar Neumann problem (Section 7.3).

[187] The sound pressure of plane waves has nothing to do with the scalar field quantity "pressure."

9.2 Planar Boundary between Homogeneous Isotropic Nondissipative Elastic Half-Spaces

This standard problem of elastic waves is particularly important for US-NDT because the physically relevant and NDT-relevant wavefronts appearing in the ultrasonic field of piezoelectric transducers—vertical, angle, and "creeping wave" transducers—may be intuitively understood with the dispersion relations of homogeneous and inhomogeneous plane waves through phase matching at the boundary.

We first discuss the case of a "welded" connection of both half-spaces, and afterward, we investigate fluid coupling; as a special case, we obtain the combination "elastic half-space – fluid half-space."

Even though the underlying full-spaces constitute an inhomogeneous—more precisely: piecewise homogeneous—material, we succeed—interestingly enough—with the solution "plane wave" of the wave equation for homogeneous (isotropic nondissipative) materials (compare the flow chart 1.1). The reason is that: The plane wave is a so-called separation solution of the wave equation in cartesian coordinates (Langenberg 2005), and the planar boundary fits into such a coordinate system. This may be generalized: If the material discontinuities of piecewise homogeneous materials coincide with respective coordinate surfaces that allow for a separation of the wave equation for homogeneous materials, we get along with this wave equation. Unfortunately, only three coordinate systems are available: cartesian, circular cylindrical, and spherical coordinates. These so-called canonical scattering problems are treated in Sections 15.4.2 and 15.4.4.

9.2.1 SH-wave incidence

Reflection and transmission of elastic plane waves at the planar boundary of two homogeneous isotropic nondissipative half-spaces is first investigated for the simplest wave mode: the SH-wave mode $\underline{\hat{\mathbf{u}}}_{i\mathrm{SH}}(\underline{\mathbf{R}}, t, \underline{\hat{\mathbf{k}}}_{i\mathrm{S}})$; Section 9.1.3 revealed that the corresponding boundary condition equation is decoupled from the respective P-SV-equation, and this does not change if we switch to transition conditions, it is still a two-dimensional scalar problem (Section 7.3).

Reflection and transmission of SH-waves: At first, we consider the geometry (Figure 9.12). Generalizing Figure 9.1, we replace the half-space $z < 0$ by an elastic half-space with material parameters $\lambda^{(2)}, \mu^{(2)}, \rho^{(2)}$ and distinguish those from the material parameters $\lambda^{(1)}, \mu^{(1)}, \rho^{(1)}$ of the upper half-space through a corresponding index; similarly, this holds for the wave (phase) velocities $c_{\mathrm{P}}^{(1,2)}, c_{\mathrm{S}}^{(1,2)}$ and the slownesses $s_{\mathrm{P}}^{(1,2)}, s_{\mathrm{S}}^{(1,2)}$.

As incident wave, we postulate a plane wave with transverse horizontal polarization that is also assumed for the reflected and transmitted waves:

$$\underline{\mathbf{u}}_{i\mathrm{SH}}(\underline{\mathbf{R}}, t, \underline{\hat{\mathbf{k}}}_{i\mathrm{S}}) = -u_{i\mathrm{S}}\left(t - \frac{\underline{\hat{\mathbf{k}}}_{i\mathrm{S}} \cdot \underline{\mathbf{R}}}{c_{\mathrm{S}}^{(1)}}\right)\underline{\mathbf{e}}_y, \quad z \geq 0, \tag{9.177}$$

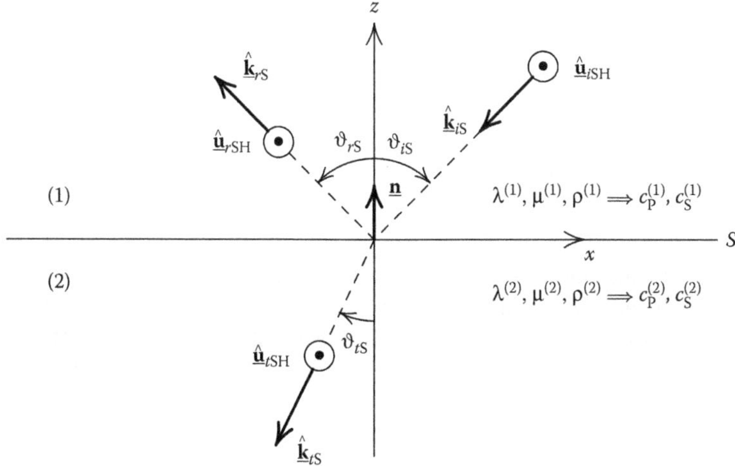

FIGURE 9.12
Reflection and transmission of a plane SH-wave at the boundary of two homogeneous isotropic nondissipative elastic half-spaces with material parameters $\lambda^{(1,2)}, \mu^{(1,2)}, \rho^{(1,2)}$ (the y-axis points into the page).

$$\underline{\mathbf{u}}_{rSH}(\underline{\mathbf{R}}, t, \hat{\underline{\mathbf{k}}}_{rS}) = -R_{SH}(\vartheta_{iS})\, u_{iS}\left(t - \frac{\hat{\underline{\mathbf{k}}}_{rS}\cdot\underline{\mathbf{R}}}{c_S^{(1)}}\right)\underline{\mathbf{e}}_y, \quad z \geq 0, \tag{9.178}$$

$$\underline{\mathbf{u}}_{tSH}(\underline{\mathbf{R}}, t, \hat{\underline{\mathbf{k}}}_{tS}) = -T_{SH}(\vartheta_{iS})\, u_{iS}\left(t - \frac{\hat{\underline{\mathbf{k}}}_{tS}\cdot\underline{\mathbf{R}}}{c_S^{(2)}}\right)\underline{\mathbf{e}}_y, \quad z \leq 0. \tag{9.179}$$

The amplitudes of $\underline{\mathbf{u}}_{rSH}$ and $\underline{\mathbf{u}}_{tSH}$ define angle-dependent reflection and transmission coefficients $R_{SH}(\vartheta_{iS})$ and $T_{SH}(\vartheta_{iS})$, respectively. The components of the phase propagation vectors are taken from Figure 9.12:

$$\hat{\underline{\mathbf{k}}}_{iS} = -\sin\vartheta_{iS}\,\underline{\mathbf{e}}_x - \cos\vartheta_{iS}\,\underline{\mathbf{e}}_z, \tag{9.180}$$

$$\hat{\underline{\mathbf{k}}}_{rS} = -\sin\vartheta_{rS}\,\underline{\mathbf{e}}_x + \cos\vartheta_{rS}\,\underline{\mathbf{e}}_z, \tag{9.181}$$

$$\hat{\underline{\mathbf{k}}}_{tS} = -\sin\vartheta_{tS}\,\underline{\mathbf{e}}_x - \cos\vartheta_{tS}\,\underline{\mathbf{e}}_z, \tag{9.182}$$

where we already note that the transmission angle ϑ_{tS} may be larger or smaller than ϑ_{iS} depending upon the ratio of the shear wave velocities $c_S^{(1)}/c_S^{(2)}$; if it is larger than ϑ_{iS}, we will encounter a critical angle ϑ_{ctS}, and ϑ_{tS} will then be complex. Regarding the SH-polarization, we refer to Figure 8.8 yielding $\hat{\underline{\mathbf{u}}}_{SH} = -\underline{\mathbf{e}}_y$; in Figure 9.12, we look at the tip of the polarization vector.

For the stress tensors—again we switch to Fourier spectra—

$$\underline{\underline{\mathbf{T}}}_{iSH}(\underline{\mathbf{R}}, \omega, \hat{\underline{\mathbf{k}}}_{iS}) = -\mathrm{j}k_S^{(1)}\mu^{(1)}\, u_{iS}(\omega)\,\mathrm{e}^{\mathrm{j}k_S^{(1)}\hat{\underline{\mathbf{k}}}_{iS}\cdot\underline{\mathbf{R}}}\left(\hat{\underline{\mathbf{k}}}_{iS}\underline{\mathbf{e}}_y + \underline{\mathbf{e}}_y\hat{\underline{\mathbf{k}}}_{iS}\right), \tag{9.183}$$

$$\underline{\underline{\mathbf{T}}}_{rSH}(\underline{\mathbf{R}}, \omega, \hat{\underline{\mathbf{k}}}_{rS}) = -\mathrm{j}k_S^{(1)}\mu^{(1)}R_{SH}(\vartheta_{iS})\, u_{iS}(\omega)\,\mathrm{e}^{\mathrm{j}k_S^{(1)}\hat{\underline{\mathbf{k}}}_{rS}\cdot\underline{\mathbf{R}}}\left(\hat{\underline{\mathbf{k}}}_{rS}\underline{\mathbf{e}}_y + \underline{\mathbf{e}}_y\hat{\underline{\mathbf{k}}}_{rS}\right), \tag{9.184}$$

$$\underline{\underline{\mathbf{T}}}_{t\mathrm{SH}}(\underline{\mathbf{R}}, \omega, \hat{\underline{\mathbf{k}}}_{t\mathrm{S}}) = -\mathrm{j}k_{\mathrm{S}}^{(2)}\mu^{(2)}T_{\mathrm{SH}}(\vartheta_{i\mathrm{S}})\, u_{i\mathrm{S}}(\omega)\, \mathrm{e}^{\mathrm{j}k_{\mathrm{S}}^{(2)}\hat{\underline{\mathbf{k}}}_{t\mathrm{S}}\cdot\underline{\mathbf{R}}} \left(\hat{\underline{\mathbf{k}}}_{t\mathrm{S}}\underline{\mathbf{e}}_y + \underline{\mathbf{e}}_y\hat{\underline{\mathbf{k}}}_{t\mathrm{S}}\right), \tag{9.185}$$

the transition condition (Equation 3.98)

$$\left[\underline{\underline{\mathbf{T}}}_{i\mathrm{SH}}(\underline{\mathbf{R}}_S, \omega, \hat{\underline{\mathbf{k}}}_{i\mathrm{S}}) + \underline{\underline{\mathbf{T}}}_{r\mathrm{SH}}(\underline{\mathbf{R}}_S, \omega, \hat{\underline{\mathbf{k}}}_{r\mathrm{S}})\right] \cdot \underline{\mathbf{e}}_z = \underline{\underline{\mathbf{T}}}_{t\mathrm{SH}}(\underline{\mathbf{R}}_S, \omega, \hat{\underline{\mathbf{k}}}_{t\mathrm{S}}) \cdot \underline{\mathbf{e}}_z \tag{9.186}$$

holds instead of the boundary condition (9.167), where $\underline{\mathbf{R}}_S \in S$. Since (9.186) has only a y-component, we are left with only one equation to determine both the coefficients R_{SH} and T_{SH}. Yet, a second equation emerges from the continuity requirement of the particle displacement vector for $\underline{\mathbf{R}} = \underline{\mathbf{R}}_S$, which is also given in terms of Fourier spectra [transition condition (3.99)]:

$$\underline{\mathbf{u}}_{i\mathrm{SH}}(\underline{\mathbf{R}}_S, \omega, \hat{\underline{\mathbf{k}}}_{i\mathrm{S}}) + \underline{\mathbf{u}}_{r\mathrm{SH}}(\underline{\mathbf{R}}_S, \omega, \hat{\underline{\mathbf{k}}}_{r\mathrm{S}}) = \underline{\mathbf{u}}_{t\mathrm{SH}}(\underline{\mathbf{R}}_S, \omega, \hat{\underline{\mathbf{k}}}_{t\mathrm{S}}). \tag{9.187}$$

This vector equation also has only a y-component finally providing two equations for two unknowns as soon as we have eliminated $\underline{\mathbf{R}}_S$ from (9.186) and (9.187).

Reflection and transmission law: critical angle: As already mentioned, the first step concerns the elimination of the arbitrary boundary vector $\underline{\mathbf{R}}_S$ from Equations 9.186 and 9.187 via phase matching

$$k_{\mathrm{S}}^{(1)}\hat{\underline{\mathbf{k}}}_{i\mathrm{S}} \cdot \underline{\mathbf{R}}_S = k_{\mathrm{S}}^{(1)}\hat{\underline{\mathbf{k}}}_{r\mathrm{S}} \cdot \underline{\mathbf{R}}_S = k_{\mathrm{S}}^{(2)}\hat{\underline{\mathbf{k}}}_{t\mathrm{S}} \cdot \underline{\mathbf{R}}_S \tag{9.188}$$

that on one hand—similar to the transition from (9.16) to (9.19)—enforces coplanar vectors $\hat{\underline{\mathbf{k}}}_{i\mathrm{S}}, \hat{\underline{\mathbf{k}}}_{r\mathrm{S}}, \hat{\underline{\mathbf{k}}}_{t\mathrm{S}}$ in the way as displayed in Figure 9.12; on the other hand, Equation 9.188 provides the reflection law

$$\vartheta_{r\mathrm{S}} = \vartheta_{i\mathrm{S}} \tag{9.189}$$

and the transmission law

$$\sin\vartheta_{t\mathrm{S}} = \frac{k_{\mathrm{S}}^{(1)}}{k_{\mathrm{S}}^{(2)}} \sin\vartheta_{i\mathrm{S}}. \tag{9.190}$$

The latter one again permits, together with the dispersion relations,

$$\underline{\mathbf{s}}_{i\mathrm{S}} \cdot \underline{\mathbf{s}}_{i\mathrm{S}} = s_{\mathrm{S}}^{(1)^2}, \tag{9.191}$$

$$\underline{\mathbf{s}}_{r\mathrm{S}} \cdot \underline{\mathbf{s}}_{r\mathrm{S}} = s_{\mathrm{S}}^{(1)^2}, \tag{9.192}$$

$$\underline{\mathbf{s}}_{t\mathrm{S}} \cdot \underline{\mathbf{s}}_{t\mathrm{S}} = s_{\mathrm{S}}^{(2)^2} \tag{9.193}$$

for the slowness vectors

$$\underline{\mathbf{s}}_{i\mathrm{S}} = s_{\mathrm{S}}^{(1)}\hat{\underline{\mathbf{k}}}_{i\mathrm{S}}, \tag{9.194}$$

$$\underline{\mathbf{s}}_{r\mathrm{S}} = s_{\mathrm{S}}^{(1)}\hat{\underline{\mathbf{k}}}_{r\mathrm{S}}, \tag{9.195}$$

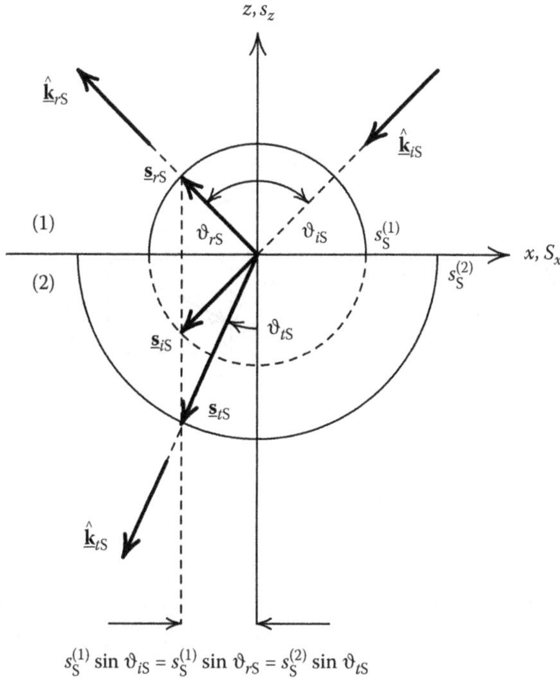

FIGURE 9.13
Slowness diagram for reflection and transmission of an SH-wave at the boundary of two homogeneous isotropic nondissipative elastic half-spaces: $c_S^{(1)} > c_S^{(2)}$.

$$\underline{\mathbf{s}}_{tS} = s_S^{(2)} \hat{\underline{\mathbf{k}}}_{tS}, \tag{9.196}$$

the geometric construction of the transmission angle ϑ_{tS} with the help of a slowness diagram; Figure 9.13 illustrates this construction for $c_S^{(1)} \geq c_S^{(2)}$ always resulting in $\vartheta_{tS} \leq \vartheta_{iS}$.

For $c_S^{(1)} < c_S^{(2)}$, the transmission law (9.190) defines the critical angle for $\sin \vartheta_{tS} = 1$:

$$\vartheta_{iS} = \vartheta_{ctS} = \arcsin \frac{s_S^{(2)}}{s_S^{(1)}}. \tag{9.197}$$

In the slowness diagram, we then have to distinguish whether $\vartheta_{iS} \leq \vartheta_{ctS}$ or $\vartheta_{iS} > \vartheta_{ctS}$ holds: Figure 9.14(a) gives an example for $\vartheta_{iS} < \vartheta_{ctS}$; apart from the fact that now we have $\vartheta_{tS} > \vartheta_{iS}$ there are no other peculiarities. Even for angles of incidence beyond the critical angle ϑ_{ctS}, we find no new peculiarities: The transmission angle ϑ_{tS} has to be chosen complex valued:

$$\vartheta_{tS} = \frac{\pi}{2} + \mathrm{j}\Im\vartheta_{tS}, \tag{9.198}$$

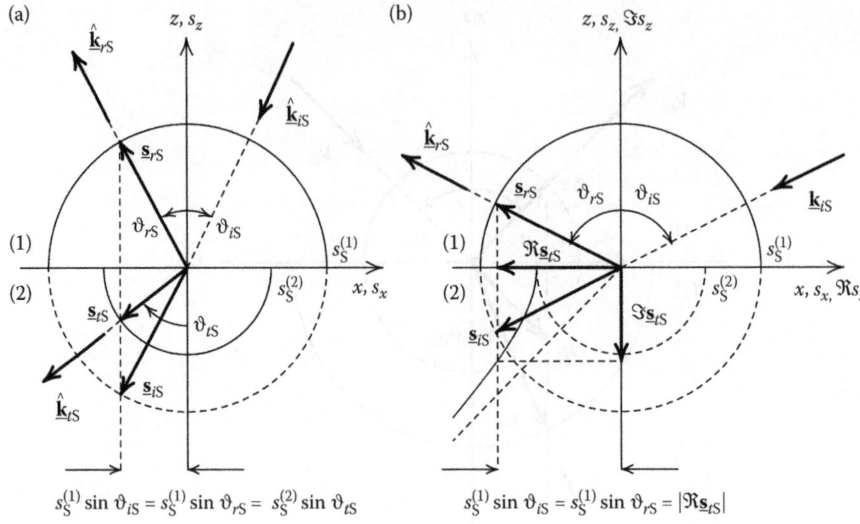

FIGURE 9.14
Slowness diagram for reflection and transmission of an SH-wave at the boundary of two homogeneous isotropic nondissipative elastic half-spaces: $c_S^{(1)} < c_S^{(2)}$; (a) $\vartheta_{iS} \leq \vartheta_{ctS}$ and (b) $\vartheta_{iS} > \vartheta_{ctS}$.

because it opens the door for the sine of ϑ_{tS} becoming bigger than 1 due to (9.198):

$$\sin \vartheta_{tS} = \cosh \Im\vartheta_{tS} = \frac{k_S^{(1)}}{k_S^{(2)}} \sin \vartheta_{iS}. \tag{9.199}$$

Consequently, the phase propagation vector (9.182) becomes a complex wave number (unit) vector

$$\begin{aligned} \hat{\underline{\mathbf{k}}}_{tS} &= - \sin \vartheta_{tS}\, \underline{\mathbf{e}}_x - \cos \vartheta_{tS}\, \underline{\mathbf{e}}_z \\ &= - \cosh \Im\vartheta_{tS}\, \underline{\mathbf{e}}_x + \mathrm{j} \sinh \Im\vartheta_{tS}\, \underline{\mathbf{e}}_z, \end{aligned} \tag{9.200}$$

where $\sinh \Im\vartheta_{tS}$ can be calculated as

$$\sinh \Im\vartheta_{tS} = - \sqrt{\frac{k_S^{(1)2}}{k_S^{(2)2}} \sin^2 \vartheta_{iS} - 1} \tag{9.201}$$

due to (9.99). As in (9.100), the negative sign in (9.201) is needed—therefore, $\Im\vartheta_{tS}$ in (9.198) must be negative—to ensure the exponential decay of the Fourier spectrum of the transmitted wave

$$\underline{\mathbf{u}}_{tSH}(\underline{\mathbf{R}}, \omega, \vartheta_{iS}) = -T_{SH}(\vartheta_{iS}) u_{iS}(\omega)\, \mathrm{e}^{-\mathrm{j}k_S^{(1)} \sin \vartheta_{iS}\, x}\, \mathrm{e}^{\sqrt{k_S^{(1)2} \sin^2 \vartheta_{iS} - k_S^{(2)2}}\, z}\, \underline{\mathbf{e}}_y \tag{9.202}$$

for $\vartheta_{i\text{S}} > \vartheta_{ct\text{S}}$ in the half-space $z < 0$; then the wave is evanescent because the imaginary part of the complex wave number vector

$$\begin{aligned}\underline{\mathbf{k}}_{t\text{S}} &= k_{\text{S}}^{(2)}\hat{\underline{\mathbf{k}}}_{t\text{S}} \\ &= \underbrace{-k_{\text{S}}^{(1)}\sin\vartheta_{i\text{S}}\,\underline{\mathbf{e}}_x}_{=\Re\underline{\mathbf{k}}_{t\text{S}}} + \mathrm{j}\underbrace{\left(-\sqrt{k_{\text{S}}^{(1)^2}\sin^2\vartheta_{i\text{S}} - k_{\text{S}}^{(2)^2}}\right)\underline{\mathbf{e}}_z}_{=\Im\underline{\mathbf{k}}_{t\text{S}}} \end{aligned} \tag{9.203}$$

points into $(-\underline{\mathbf{e}}_z)$-direction for wave propagation into $(+\Re\underline{\mathbf{k}}_{t\text{S}} \sim -\underline{\mathbf{e}}_x)$-direction, hence vertically into that half-space for which we require exponential attenuation [Figure 8.11(a)]. The dispersion relation (9.193) for the complex slowness vector [Figure 9.14(b)]

$$\begin{aligned}\underline{\mathbf{s}}_{t\text{S}} &= -\,s_{\text{S}}^{(1)}\sin\vartheta_{i\text{S}}\,\underline{\mathbf{e}}_x - \mathrm{j}\sqrt{s_{\text{S}}^{(1)^2}\sin^2\vartheta_{i\text{S}} - s_{\text{S}}^{(2)^2}}\,\underline{\mathbf{e}}_z \\ &= \Re\underline{\mathbf{s}}_{t\text{S}} + \mathrm{j}\Im\underline{\mathbf{s}}_{t\text{S}} \end{aligned} \tag{9.204}$$

yields the hyperbola equation

$$(\Re s_{t\text{S}x})^2 - (\Im s_{t\text{S}z})^2 = s_{\text{S}}^{(2)^2} \tag{9.205}$$

in a $\Re s_x\Im s_z$-coordinate system allowing for the construction of $\Im s_{t\text{S}z}$ via phase matching as sketched in Figure 9.14(b).

Note: For SV-wave incidence on a stress-free boundary, we always observe a critical angle; here, it only exists if the shear wave ratio of both half-spaces permits.

Reflection and transmission coefficients of the particle displacement vector; Hilbert transformed wavefronts beyond the critical angle: With (9.177) through (9.179) and (9.183) through (9.185), we obtain from the transition conditions (9.186) and (9.187) after some short calculus:

$$R_{\text{SH}}(\vartheta_{i\text{S}}) = \frac{Z_{\text{S}}^{(1)}\cos\vartheta_{i\text{S}} - Z_{\text{S}}^{(2)}\cos\vartheta_{t\text{S}}}{Z_{\text{S}}^{(1)}\cos\vartheta_{i\text{S}} + Z_{\text{S}}^{(2)}\cos\vartheta_{t\text{S}}}, \tag{9.206}$$

$$T_{\text{SH}}(\vartheta_{i\text{S}}) = \frac{2Z_{\text{S}}^{(1)}\cos\vartheta_{i\text{S}}}{Z_{\text{S}}^{(1)}\cos\vartheta_{i\text{S}} + Z_{\text{S}}^{(2)}\cos\vartheta_{t\text{S}}}, \tag{9.207}$$

where

$$Z_{\text{S}}^{(j)} = \sqrt{\rho^{(j)}\mu^{(j)}}, \tag{9.208}$$

$$= \rho^{(j)}c_{\text{S}}^{(j)}, \quad j = 1, 2, \tag{9.209}$$

as usual define acoustic (shear wave) impedances of the respective materials. For $\vartheta_{i\text{S}} = 0$, we obtain

$$R_{\text{SH}}(0) = -\frac{Z_{\text{S}}^{(2)} - Z_{\text{S}}^{(1)}}{Z_{\text{S}}^{(2)} + Z_{\text{S}}^{(1)}}, \tag{9.210}$$

$$T_{\mathrm{SH}}(0) = \frac{2Z_{\mathrm{S}}^{(1)}}{Z_{\mathrm{S}}^{(2)} + Z_{\mathrm{S}}^{(1)}}, \tag{9.211}$$

equations that one usually knows by heart for a rapid estimation of reflection and transmission close to vertical incidence. Again we state that: The resulting figures refer to the vector particle displacement (here, of course, to the only nonvanishing component).

As always, we present the cartesian components of the stress tensors; we adopt (9.171) and (9.172) with respective (1)-indexing

$$\begin{aligned} \underline{\underline{\mathbf{T}}}_{i\mathrm{SH}}(x, z, \omega, \vartheta_{i\mathrm{S}}) = {} & \mathrm{j}\omega Z_{\mathrm{S}}^{(1)}\, u_{i\mathrm{S}}(\omega)\, \mathrm{e}^{-\mathrm{j}k_{\mathrm{S}}^{(1)}(\sin\vartheta_{i\mathrm{S}}\, x + \cos\vartheta_{i\mathrm{S}}\, z)} \\ & \times [\sin\vartheta_{i\mathrm{S}}(\underline{\mathbf{e}}_x\underline{\mathbf{e}}_y + \underline{\mathbf{e}}_y\underline{\mathbf{e}}_x) + \cos\vartheta_{i\mathrm{S}}(\underline{\mathbf{e}}_y\underline{\mathbf{e}}_z + \underline{\mathbf{e}}_z\underline{\mathbf{e}}_y)], \end{aligned} \tag{9.212}$$

$$\begin{aligned} \underline{\underline{\mathbf{T}}}_{r\mathrm{SH}}(x, z, \omega, \vartheta_{i\mathrm{S}}) = {} & \mathrm{j}\omega Z_{\mathrm{S}}^{(1)}\, R_{\mathrm{SH}}(\vartheta_{i\mathrm{S}})\, u_{i\mathrm{S}}(\omega)\, \mathrm{e}^{-\mathrm{j}k_{\mathrm{S}}^{(1)}(\sin\vartheta_{i\mathrm{S}}\, x - \cos\vartheta_{i\mathrm{S}}\, z)} \\ & \times [\sin\vartheta_{i\mathrm{S}}(\underline{\mathbf{e}}_x\underline{\mathbf{e}}_y + \underline{\mathbf{e}}_y\underline{\mathbf{e}}_x) - \cos\vartheta_{i\mathrm{S}}(\underline{\mathbf{e}}_y\underline{\mathbf{e}}_z + \underline{\mathbf{e}}_z\underline{\mathbf{e}}_y)] \end{aligned} \tag{9.213}$$

and add $\underline{\underline{\mathbf{T}}}_{t\mathrm{SH}}(x, z, \omega, \vartheta_{i\mathrm{S}})$:

$$\begin{aligned} \underline{\underline{\mathbf{T}}}_{t\mathrm{SH}}(x, z, \omega, \vartheta_{i\mathrm{S}}) = {} & \mathrm{j}\omega Z_{\mathrm{S}}^{(2)}\, T_{\mathrm{SH}}(\vartheta_{i\mathrm{S}})\, u_{i\mathrm{S}}(\omega)\, \mathrm{e}^{-\mathrm{j}(k_{\mathrm{S}}^{(1)}\sin\vartheta_{i\mathrm{S}}\, x + k_{\mathrm{S}}^{(2)}\cos\vartheta_{t\mathrm{S}}\, z)} \\ & \times [\sin\vartheta_{t\mathrm{S}}(\underline{\mathbf{e}}_x\underline{\mathbf{e}}_y + \underline{\mathbf{e}}_y\underline{\mathbf{e}}_x) + \cos\vartheta_{t\mathrm{S}}(\underline{\mathbf{e}}_y\underline{\mathbf{e}}_z + \underline{\mathbf{e}}_z\underline{\mathbf{e}}_y)]. \end{aligned} \tag{9.214}$$

The sound pressure equations of plane waves read

$$\begin{aligned} p_{i\mathrm{SH}}(\underline{\mathbf{R}}, \omega, \vartheta_{i\mathrm{S}}) &= \underline{\underline{\mathbf{T}}}_{i\mathrm{SH}}(\underline{\mathbf{R}}, \omega, \hat{\underline{\mathbf{k}}}_{i\mathrm{S}}) : \hat{\underline{\mathbf{k}}}_{i\mathrm{S}}\underline{\mathbf{e}}_y \\ &= -\mathrm{j}\omega Z_{\mathrm{S}}^{(1)}\, u_{i\mathrm{S}}(\omega)\, \mathrm{e}^{\mathrm{j}k_{\mathrm{S}}^{(1)}\hat{\underline{\mathbf{k}}}_{i\mathrm{S}}\cdot\underline{\mathbf{R}}}, \end{aligned} \tag{9.215}$$

$$\begin{aligned} p_{r\mathrm{SH}}(\underline{\mathbf{R}}, \omega, \vartheta_{i\mathrm{S}}) &= \underline{\underline{\mathbf{T}}}_{r\mathrm{SH}}(\underline{\mathbf{R}}, \omega, \hat{\underline{\mathbf{k}}}_{r\mathrm{S}}) : \hat{\underline{\mathbf{k}}}_{r\mathrm{S}}\underline{\mathbf{e}}_y \\ &= -\mathrm{j}\omega Z_{\mathrm{S}}^{(1)}\, R_{\mathrm{SH}}(\vartheta_{i\mathrm{S}})\, u_{i\mathrm{S}}(\omega)\, \mathrm{e}^{\mathrm{j}k_{\mathrm{S}}^{(1)}\hat{\underline{\mathbf{k}}}_{r\mathrm{S}}\cdot\underline{\mathbf{R}}}, \end{aligned} \tag{9.216}$$

$$\begin{aligned} p_{t\mathrm{SH}}(\underline{\mathbf{R}}, \omega, \vartheta_{i\mathrm{S}}) &= \underline{\underline{\mathbf{T}}}_{t\mathrm{SH}}(\underline{\mathbf{R}}, \omega, \hat{\underline{\mathbf{k}}}_{t\mathrm{S}}) : \hat{\underline{\mathbf{k}}}_{t\mathrm{S}}\underline{\mathbf{e}}_y \\ &= -\mathrm{j}\omega Z_{\mathrm{S}}^{(2)}\, T_{\mathrm{SH}}(\vartheta_{i\mathrm{S}})\, u_{i\mathrm{S}}(\omega)\, \mathrm{e}^{\mathrm{j}k_{\mathrm{S}}^{(1)}\hat{\underline{\mathbf{k}}}_{t\mathrm{S}}\cdot\underline{\mathbf{R}}}; \end{aligned} \tag{9.217}$$

evidently, $R_{\mathrm{SH}}(\vartheta_{i\mathrm{S}})$ is also the reflection coefficient for the sound pressure; yet, as transmission coefficient, we obtain

$$\begin{aligned} T_{p\mathrm{SH}}(\vartheta_{i\mathrm{S}}) &= \frac{Z_{\mathrm{S}}^{(2)}}{Z_{\mathrm{S}}^{(1)}}\, T_{\mathrm{SH}}(\vartheta_{i\mathrm{S}}) \\ &= \frac{2Z_{\mathrm{S}}^{(2)}\cos\vartheta_{i\mathrm{S}}}{Z_{\mathrm{S}}^{(1)}\cos\vartheta_{i\mathrm{S}} + Z_{\mathrm{S}}^{(2)}\cos\vartheta_{t\mathrm{S}}}, \end{aligned} \tag{9.218}$$

because

$$p_{t\mathrm{SH}}(\underline{\mathbf{R}}, \omega, \vartheta_{i\mathrm{S}}) = -\mathrm{j}\omega Z_{\mathrm{S}}^{(1)}\, T_{p\mathrm{SH}}(\vartheta_{i\mathrm{S}})\, u_{i\mathrm{S}}(\omega)\, \mathrm{e}^{\mathrm{j}k_{\mathrm{S}}^{(2)}\hat{\underline{\mathbf{k}}}_{t\mathrm{S}}\cdot\underline{\mathbf{R}}} \tag{9.219}$$

only receives the same prefactor as p_{iSH} with the definition (9.218). Therefore, when giving numbers, we have to clearly identify the physical quantities that have been used to define a transmission coefficient. Furthermore: The resulting discontinuity of the sound pressure even for $\vartheta_{iS} = 0$—it is always discontinuous for $\vartheta_{iS} \neq 0$—is once more a consequence of the actual choice of SH-polarization directions of incident, reflected, and transmitted waves.

For $\vartheta_{iS} \leq \pi/2$, assuming a nonexistent critical angle, and for $\vartheta_{iS} \leq \vartheta_{ctS}$, assuming a critical angle, the pulsed SH-wavefront are given by (9.177) through (9.179). Yet, allowing for a critical angle and choosing $\vartheta_{iS} > \vartheta_{ctS}$, Fourier inversion of the frequency spectra of reflected and transmitted SH-waves has to account for the complex valued $R_{SH}(\vartheta_{iS})$ and $T_{SH}(\vartheta_{iS})$ due to (9.198), especially according to

$$R_{SH}(\vartheta_{iS} > \vartheta_{ctS}) = e^{j\,\mathrm{sign}(\omega)\phi_{R_{SH}}(\vartheta_{iS})}, \tag{9.220}$$

$$T_{SH}(\vartheta_{iS} > \vartheta_{ctS}) = |T_{SH}(\vartheta_{iS})|\, e^{j\,\mathrm{sign}(\omega)\phi_{T_{SH}}(\vartheta_{iS})}, \tag{9.221}$$

where the sign(ω)-functions must be inserted, because the pulsed wavefronts must be real valued for real pulse functions $u_{iS}(t)$; consequently, we obtain for $c_S^{(2)} > c_S^{(1)}$

$$\begin{aligned}
&\underline{\mathbf{u}}_{rSH}(\underline{\mathbf{R}}, t, \vartheta_{iS})\\
&\quad = -\Bigg[\cos\phi_{R_{SH}}(\vartheta_{iS})\, u_{iS}\left(t - \frac{\hat{\underline{\mathbf{k}}}_{rS}\cdot\underline{\mathbf{R}}}{c_S^{(1)}}\right)\\
&\qquad - \sin\phi_{R_{SH}}(\vartheta_{iS})\mathcal{H}\left\{u_{iS}\left(\tau - \frac{\hat{\underline{\mathbf{k}}}_{rS}\cdot\underline{\mathbf{R}}}{c_S^{(1)}}\right)\right\}\Bigg]\underline{\mathbf{e}}_y, \quad \vartheta_{ctS} < \vartheta_{iS} \leq \pi/2,
\end{aligned} \tag{9.222}$$

$$\begin{aligned}
&\underline{\mathbf{u}}_{tSH}(\underline{\mathbf{R}}, t, \vartheta_{iS})\\
&\quad = -|T_{SH}(\vartheta_{iS})|\alpha^{(2)}(z, t, \vartheta_{iS})\\
&\qquad * \Bigg[\cos\phi_{T_{SH}}(\vartheta_{iS})\, u_{iS}\left(t + \frac{\sin\vartheta_{iS}x}{c_S^{(1)}}\right)\\
&\qquad - \sin\phi_{T_{SH}}(\vartheta_{iS})\mathcal{H}\left\{u_{iS}\left(\tau + \frac{\sin\vartheta_{iS}x}{c_S^{(1)}}\right)\right\}\Bigg]\underline{\mathbf{e}}_y, \quad \vartheta_{ctS} < \vartheta_{iS} \leq \pi/2.
\end{aligned} \tag{9.223}$$

The evanescence follows form the convolution with (Equation 9.141)

$$\begin{aligned}
\alpha^{(2)}(z, t, \vartheta_{iS}) &= \mathcal{F}^{-1}\left\{e^{-\frac{|\omega|}{c_S^{(2)}}\sqrt{\kappa_S^2\sin^2\vartheta_{iS}-1}|z|}\right\}\\
&= \frac{c_S^{(2)}}{\pi}\frac{\sqrt{\kappa_S^2\sin^2\vartheta_{iS}-1}|z|}{\left(\kappa_S^2\sin^2\vartheta_{iS}-1\right)z^2 + c_S^{(2)^2}t^2},
\end{aligned} \tag{9.224}$$

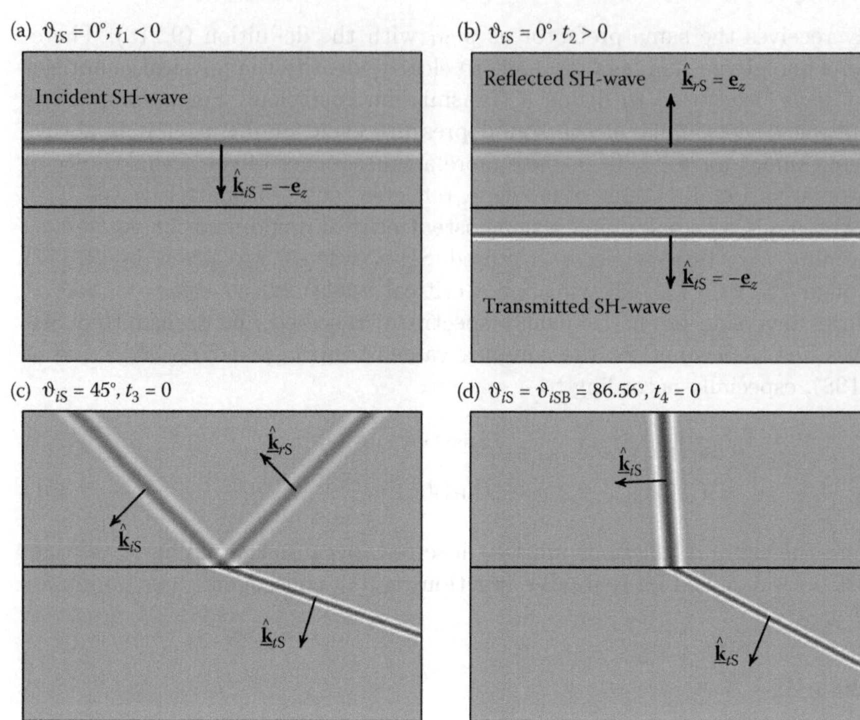

FIGURE 9.15
Wavefronts of incident, reflected, and transmitted SH-waves for various angles of incidence (a)–(d) (materials: steel(1)-plexiglas(2): compare the dashed curves in Figure 9.16).

where

$$\kappa_S = \frac{c_S^{(2)}}{c_S^{(1)}} > 1. \tag{9.225}$$

Wavefronts of reflected and transmitted SH-waves: As before, we display wavefronts of reflected and transmitted waves as spatial amplitude distribution for fixed times. Again, we choose an RC2(t)-pulse as prescribed time function that also manifests itself in space according to Figure 8.4. Figure 9.15(a) displays the incident wave for vertical incidence and $t = t_1 < 0$; the RC2-pulse appears as a positive component in the $(-\underline{\mathbf{e}}_y)$-SH-polarization:

$$\underline{\mathbf{u}}_{i\mathrm{SH}}(\underline{\mathbf{R}}, t, \vartheta_{i\mathrm{S}} = 0) = -\mathrm{RC2}\left(t + \frac{z}{c_S^{(1)}}\right)\underline{\mathbf{e}}_y. \tag{9.226}$$

For $Z_S^{(2)} < Z_S^{(1)}$ (here: steel(1)-plexiglas(2)), the reflected impulse [Figure 9.15(b)]

$$\underline{\mathbf{u}}_{r\mathrm{SH}}(\underline{\mathbf{R}}, t, \vartheta_{i\mathrm{S}} = 0) = -\, R_{\mathrm{SH}}(0)\,\mathrm{RC2}\left(t - \frac{z}{c_{\mathrm{S}}^{(1)}}\right)\underline{\mathbf{e}}_y$$
$$= \frac{Z_{\mathrm{S}}^{(2)} - Z_{\mathrm{S}}^{(1)}}{Z_{\mathrm{S}}^{(2)} + Z_{\mathrm{S}}^{(1)}}\,\mathrm{RC2}\left(t - \frac{z}{c_{\mathrm{S}}^{(1)}}\right)\underline{\mathbf{e}}_y \qquad (9.227)$$

has the same RC2-sign in the $(-\underline{\mathbf{e}}_y)$-SH-polarization; the same holds for the transmitted pulse having a smaller wavelength due to $c_{\mathrm{S}}^{(2)} < c_{\mathrm{S}}^{(1)}$. Since Figure 9.15 depicts scalar components of particle displacement vectors, the sign of the RC2(t)-pulse is also visible. For $\vartheta_{i\mathrm{S}} > 0$, Figure 9.15(c) and (d) finally transform the information contained in Figure 9.16 (dashed curves)—e.g., the zero[188] of the reflection coefficient for $\vartheta_{i\mathrm{S}} = \vartheta_{i\mathrm{SB}}$—into gray scales.

Exchanging materials—plexiglas(1)-steel(2)—yields a sign change of the reflected impulse due to $Z_{\mathrm{S}}^{(2)} > Z_{\mathrm{S}}^{(1)}$ [Figure 9.17(b)] and an increase in wavelength of the transmitted pulse. Figure 9.17(c), displays the reflection and transmission for $\vartheta_{i\mathrm{S}} < \vartheta_{ct\mathrm{S}}$: The reflection coefficient is still real valued negative. Once again, the zero of the reflection coefficient below the critical angle defines a Brewster angle,[188] and a little beyond but still below the critical

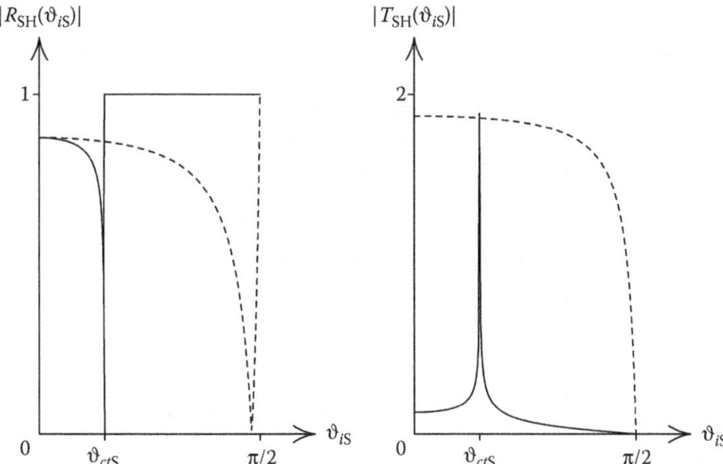

FIGURE 9.16
$|R_{\mathrm{SH}}|$ and $|T_{\mathrm{SH}}|$ without (steel(1)-plexiglas(2): - - -) and with (plexiglas(1)-steel(2):—) critical angle (steel: $c_{\mathrm{S}} = 3200$ m/s, $\rho = 7.7 \cdot 10^3$ kg/m^3; plexiglas: $c_{\mathrm{S}} = 1430$ m/s, $\rho = 1.18 \cdot 10^3$ kg/m^3).

[188]Note: As in the electromagnetic case for a parallel polarization of the electric field with regard to the plane of incidence (TM case for equal permeability of both materials), the elastodynamic SH-reflection exhibits a Brewster angle with zero reflection coefficient; it is always less than the critical angle.

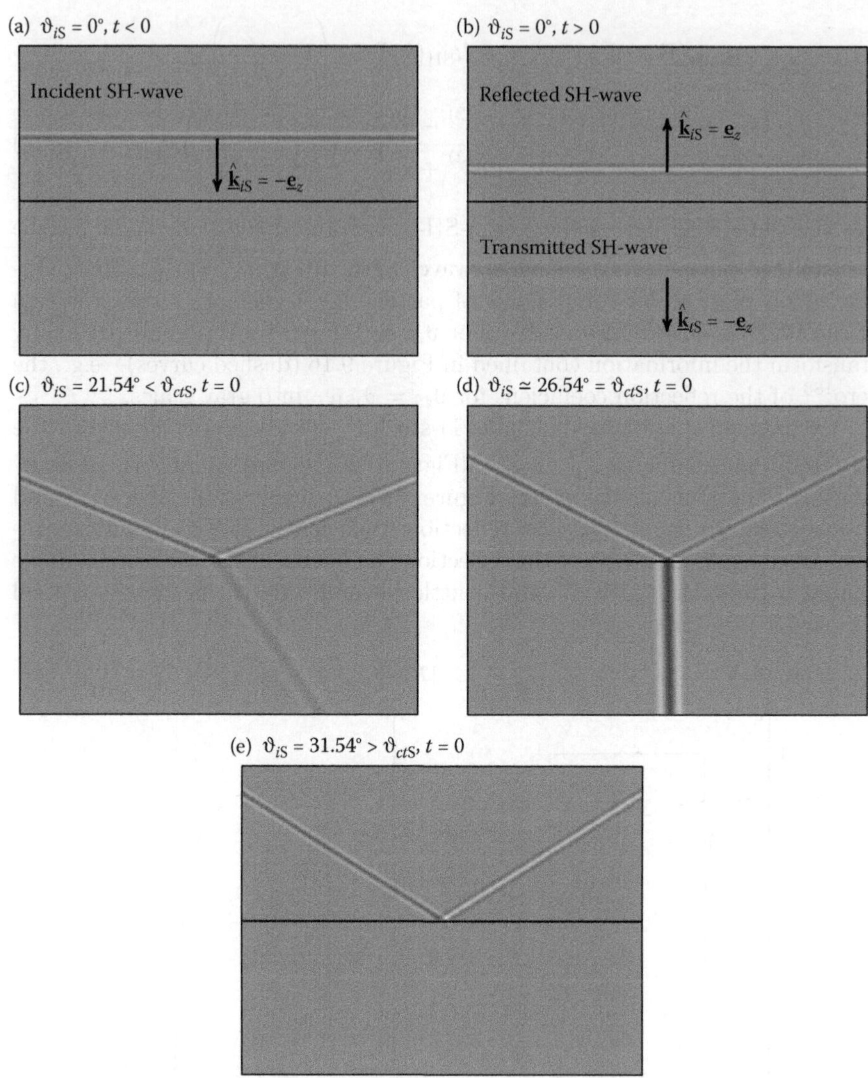

FIGURE 9.17
Wavefronts of incident, reflected, and transmitted SH-waves for various angles of incidence (a)–(e) (materials: plexiglas(1)-steel(2): compare the solid curves in Figure 9.16).

angle, the reflection coefficient becomes real valued positive and tends toward +1 [Figure 9.17(d): The transmitted wave is about to become evanescent]. For $\vartheta_{iS} > \vartheta_{tS}$, the phase $\phi_{R_{\rm SH}}$ tends very rapidly toward $-\pi$—the magnitude of $R_{\rm SH}$ is equal to 1—resulting in a negative sign of the reflected pulse in Figure 9.17(e); due to its attenuation, the transmitted wave is not visible.

Energy balance for reflection and transmission: Total reflection: As for the case of SV-wave incidence on a stress-free surface, we investigate the energy balance, in particular for $\vartheta_{iS} > \vartheta_{ctS}$, provided a critical angle exists. The energy conservation law (4.64) for nondissipative materials requires the continuity of the normal component of the real part of the complex Poynting vector for $z = 0$:

$$\begin{aligned} &\underline{\mathbf{n}} \cdot \Re\left\{\underline{\mathbf{S}}_{\mathrm{K}i\mathrm{SH}}(x, 0, \omega_0, \hat{\underline{\mathbf{k}}}_{i\mathrm{S}})\right\} + \underline{\mathbf{n}} \cdot \Re\left\{\underline{\mathbf{S}}_{\mathrm{K}r\mathrm{SH}}(x, 0, \omega_0, \hat{\underline{\mathbf{k}}}_{r\mathrm{S}})\right\} \\ &\quad = \underline{\mathbf{n}} \cdot \Re\left\{\underline{\mathbf{S}}_{\mathrm{K}t\mathrm{SH}}(x, 0, \omega_0, \hat{\underline{\mathbf{k}}}_{t\mathrm{S}})\right\}; \end{aligned} \tag{9.228}$$

regarding $\Re\underline{\mathbf{S}}_{\mathrm{K}i\mathrm{SH}}$ and $\Re\underline{\mathbf{S}}_{\mathrm{K}r\mathrm{SH}}$, we may go back to (8.132):

$$\Re\left\{\underline{\mathbf{S}}_{\mathrm{K}i\mathrm{SH}}(\underline{\mathbf{R}}, \omega_0, \hat{\underline{\mathbf{k}}}_{i\mathrm{S}})\right\} = \frac{\omega_0^2 \rho^{(1)}}{2} c_{\mathrm{S}}^{(1)} |u_{i\mathrm{S}}(\omega_0)|^2 \, \hat{\underline{\mathbf{k}}}_{i\mathrm{S}}, \tag{9.229}$$

$$\Re\left\{\underline{\mathbf{S}}_{\mathrm{K}r\mathrm{SH}}(\underline{\mathbf{R}}, \omega_0, \hat{\underline{\mathbf{k}}}_{r\mathrm{S}})\right\} = \frac{\omega_0^2 \rho^{(1)}}{2} c_{\mathrm{S}}^{(1)} |R_{\mathrm{SH}}(\vartheta_{i\mathrm{S}})|^2 |u_{i\mathrm{S}}(\omega_0)|^2 \, \hat{\underline{\mathbf{k}}}_{r\mathrm{S}}, \tag{9.230}$$

and $\Re\underline{\mathbf{S}}_{\mathrm{K}t\mathrm{SH}}$ is taken from (8.189):

$$\Re\left\{\underline{\mathbf{S}}_{\mathrm{K}t\mathrm{SH}}(\underline{\mathbf{R}}, \omega_0, \hat{\underline{\mathbf{k}}}_{t\mathrm{S}})\right\} = \frac{\omega_0^2 \rho^{(2)}}{2} c_{\mathrm{S}}^{(2)} |T_{\mathrm{SH}}(\vartheta_{i\mathrm{S}})|^2 |u_{i\mathrm{S}}(\omega_0)|^2 \, \mathrm{e}^{-2k_{\mathrm{S}}^{(2)} \Im\hat{\underline{\mathbf{k}}}_{t\mathrm{S}} \cdot \underline{\mathbf{R}}} \, \Re\hat{\underline{\mathbf{k}}}_{t\mathrm{S}}. \tag{9.231}$$

The only nonvanishing component of $\Re\underline{\mathbf{S}}_{\mathrm{K}t\mathrm{SH}}$ for $\vartheta_{i\mathrm{S}} > \vartheta_{ct\mathrm{S}}$ then reads

$$\begin{aligned} &\Re\left\{\underline{\mathbf{S}}_{\mathrm{K}t\mathrm{SH}}(\underline{\mathbf{R}}, \omega_0, \hat{\underline{\mathbf{k}}}_{t\mathrm{S}})\right\} \\ &\quad = -\frac{\omega_0^2}{2} \frac{\mu^{(2)}}{c_{\mathrm{S}}^{(1)}} |T_{\mathrm{SH}}(\vartheta_{i\mathrm{S}})|^2 |u_{i\mathrm{S}}(\omega_0)|^2 \, \mathrm{e}^{-2k_{\mathrm{S}}^{(2)} \Im\hat{\underline{\mathbf{k}}}_{t\mathrm{S}} \cdot \underline{\mathbf{R}}} \sin\vartheta_{i\mathrm{S}} \, \underline{\mathbf{e}}_x, \end{aligned} \tag{9.232}$$

preventing the appearance of $\Re\underline{\mathbf{S}}_{\mathrm{K}t\mathrm{SH}}$ in the balance equation (9.228) in that case; the remaining terms in (9.228) then yield the requirement $|R_{\mathrm{SH}}(\vartheta_{i\mathrm{S}} > \vartheta_{ct\mathrm{S}})| = 1$ for total reflection.

For the case of an existing critical angle and $\vartheta_{i\mathrm{S}} > \vartheta_{ct\mathrm{S}}$, we go back to (8.190) to calculate the time averaged energy density of the transmitted field:

$$\langle w_{t\mathrm{SH}}(\underline{\mathbf{R}}, t, \hat{\underline{\mathbf{k}}}_{t\mathrm{S}}) \rangle = \frac{\omega_0^2}{4} \rho^{(2)} |T_{\mathrm{SH}}(\vartheta_{i\mathrm{S}})|^2 |u_{i\mathrm{S}}(\omega_0)|^2 \, \mathrm{e}^{-2k_{\mathrm{S}}^{(2)} \Im\hat{\underline{\mathbf{k}}}_{t\mathrm{S}} \cdot \underline{\mathbf{R}}} (1 + \hat{\underline{\mathbf{k}}}_{t\mathrm{S}} \cdot \hat{\underline{\mathbf{k}}}_{t\mathrm{S}}^*). \tag{9.233}$$

According to (9.157), we obtain the phase velocity of the evanescent transmitted wave as

$$c_{t\mathrm{SH}}(\vartheta_{i\mathrm{S}}) = \frac{c_{\mathrm{S}}^{(1)}}{\sin\vartheta_{i\mathrm{S}}}, \quad \vartheta_{i\mathrm{S}} > \vartheta_{ct\mathrm{S}}, \tag{9.234}$$

and for its energy velocity vector, we obtain (9.160)

$$\underline{\mathbf{c}}_{\mathrm{E}t\mathrm{SH}}(\vartheta_{i\mathrm{S}}) = -\frac{c_{\mathrm{S}}^{(1)}}{\sin\vartheta_{i\mathrm{S}}} \, \underline{\mathbf{e}}_x, \quad \vartheta_{i\mathrm{S}} > \vartheta_{ct\mathrm{S}}, \tag{9.235}$$

using (9.160). It follows

$$c_S^{(2)} > |\underline{c}_{EtSH}(\vartheta_{iS})| > c_S^{(1)}. \tag{9.236}$$

9.2.2 P- and SV-waves incidence

Slowness diagrams and critical angles: If two (homogeneous isotropic nondissipative) elastic materials are under concern, the respective slowness diagrams for P- or SV-wave incidence are determined by four slownesses $s_P^{(1)}$, $s_S^{(1)}$, $s_P^{(2)}$, $s_S^{(2)}$, and it depends upon the incident wave mode and the slowness ratios whether any and if, how many critical angles exist. The Figures 9.18 and 9.19 summarize all possibilities that we may think of.[189] For the case of P-wave incidence, we may expect reflected and transmitted P-waves ($\underline{\mathbf{s}}_{rP} \Longrightarrow \hat{\underline{\mathbf{k}}}_{rP}$, $\underline{\mathbf{s}}_{tP} \Longrightarrow \hat{\underline{\mathbf{k}}}_{tP}$) as well as mode converted SV-waves in reflection ($\underline{\mathbf{s}}_{m_rS} \Longrightarrow \hat{\underline{\mathbf{k}}}_{m_rS}$) and transmission ($\underline{\mathbf{s}}_{m_tS} \Longrightarrow \hat{\underline{\mathbf{k}}}_{m_tS}$); for the case of SV-wave incidence, we obtain reflected and transmitted SV-waves ($\underline{\mathbf{s}}_{rS} \Longrightarrow \hat{\underline{\mathbf{k}}}_{rS}$, $\underline{\mathbf{s}}_{tS} \Longrightarrow \hat{\underline{\mathbf{k}}}_{tS}$) as well as mode converted P-waves in reflection ($\underline{\mathbf{s}}_{m_rP} \Longrightarrow \hat{\underline{\mathbf{k}}}_{m_rP}$) and transmission ($\underline{\mathbf{s}}_{m_tP} \Longrightarrow \hat{\underline{\mathbf{k}}}_{m_tP}$). The potential respective critical angles are distinguished by respective indices, where ϑ_{cm_rP} always exists for SV-wave incidence, whereas the existence of $\vartheta_{ctP}, \vartheta_{cm_tS}, \vartheta_{cm_tP}, \vartheta_{ctS}$ depends upon the ratio of the (phase) velocities of both materials. For the case of P-wave incidence, we face at most two critical angles $\vartheta_{ctP}, \vartheta_{cm_tS}$, and for the case of SV-wave incidence, at most three critical angles $\vartheta_{ctS}, \vartheta_{cm_rP}, \vartheta_{cm_tP}$ can exist. The number of possible critical angles may be immediately deduced from Figures 9.18 and 9.19: We have plotted boldface slowness diagrams for the respective incident wave modes making it directly obvious if there are other slowness diagrams in the interior of these boldface diagrams; their number is equal to the number of critical angles. By the way: Only the diagrams (Pa) and (SVa) in Figure 9.18 are displayed with all details, whereas (Pb), (SVb), (Pc), and (SVc) are only rudimentarily sketched to enhance the facility of inspection. The same is true for all diagrams of Figure 9.19.

Incident P-wave: System of equations for reflection, transmission, and mode conversion coefficients: In the present case, the transition conditions (3.99) and (3.98) must be satisfied with the following partial waves:

$$\underline{\mathbf{u}}_{iP}(\underline{\mathbf{R}}, \omega, \hat{\underline{\mathbf{k}}}_{iP}) = u_{iP}(\omega)\, e^{jk_P^{(1)}\hat{\underline{\mathbf{k}}}_{iP}\cdot\underline{\mathbf{R}}}\, \hat{\underline{\mathbf{k}}}_{iP}, \tag{9.237}$$

$$\underline{\mathbf{u}}_{rP}(\underline{\mathbf{R}}, \omega, \hat{\underline{\mathbf{k}}}_{rP}) = R_P(\vartheta_{iP})\, u_{iP}(\omega)\, e^{jk_P^{(1)}\hat{\underline{\mathbf{k}}}_{rP}\cdot\underline{\mathbf{R}}}\, \hat{\underline{\mathbf{k}}}_{rP}, \tag{9.238}$$

$$\underline{\mathbf{u}}_{m_rSV}(\underline{\mathbf{R}}, \omega, \hat{\underline{\mathbf{k}}}_{m_rS}) = M_{rS}(\vartheta_{iP})\, u_{iP}(\omega)\, e^{jk_S^{(1)}\hat{\underline{\mathbf{k}}}_{m_rS}\cdot\underline{\mathbf{R}}}\, \hat{\underline{\mathbf{k}}}_{m_rS} \times \underline{\mathbf{e}}_y \tag{9.239}$$

[189] All these possibilities can already be realized with the materials given in the tables by Krautkrämer and Krautkrämer (1986); (Pa) corresponds to the case (1)=steel, (2)=plexiglas and (Pa′) to (1)=plexiglas and (2)=steel.

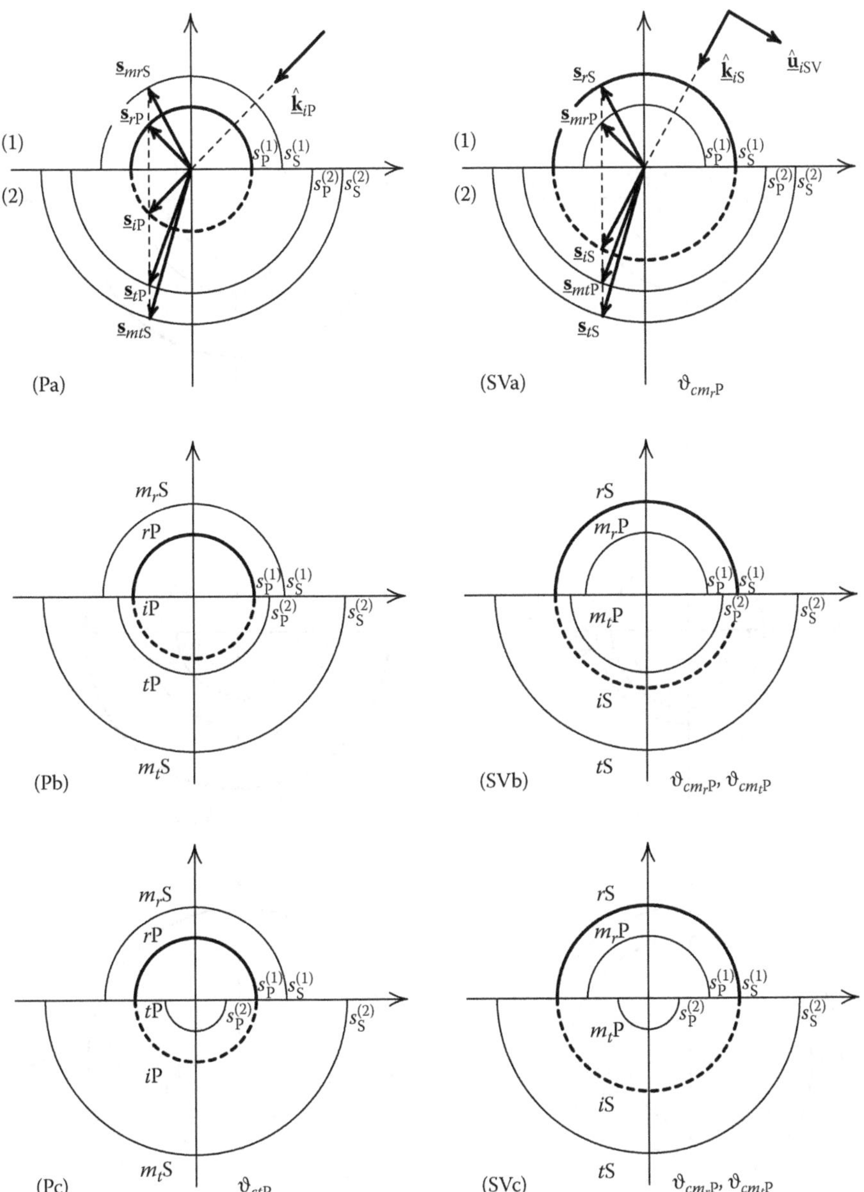

FIGURE 9.18
Boundary between two homogeneous isotropic nondissipative elastic half-spaces: slowness diagrams for reflection, transmission, and mode conversion of incident P- (left: Pa, Pb, Pc) and incident SV-waves (right: SVa, SVb, SVc).

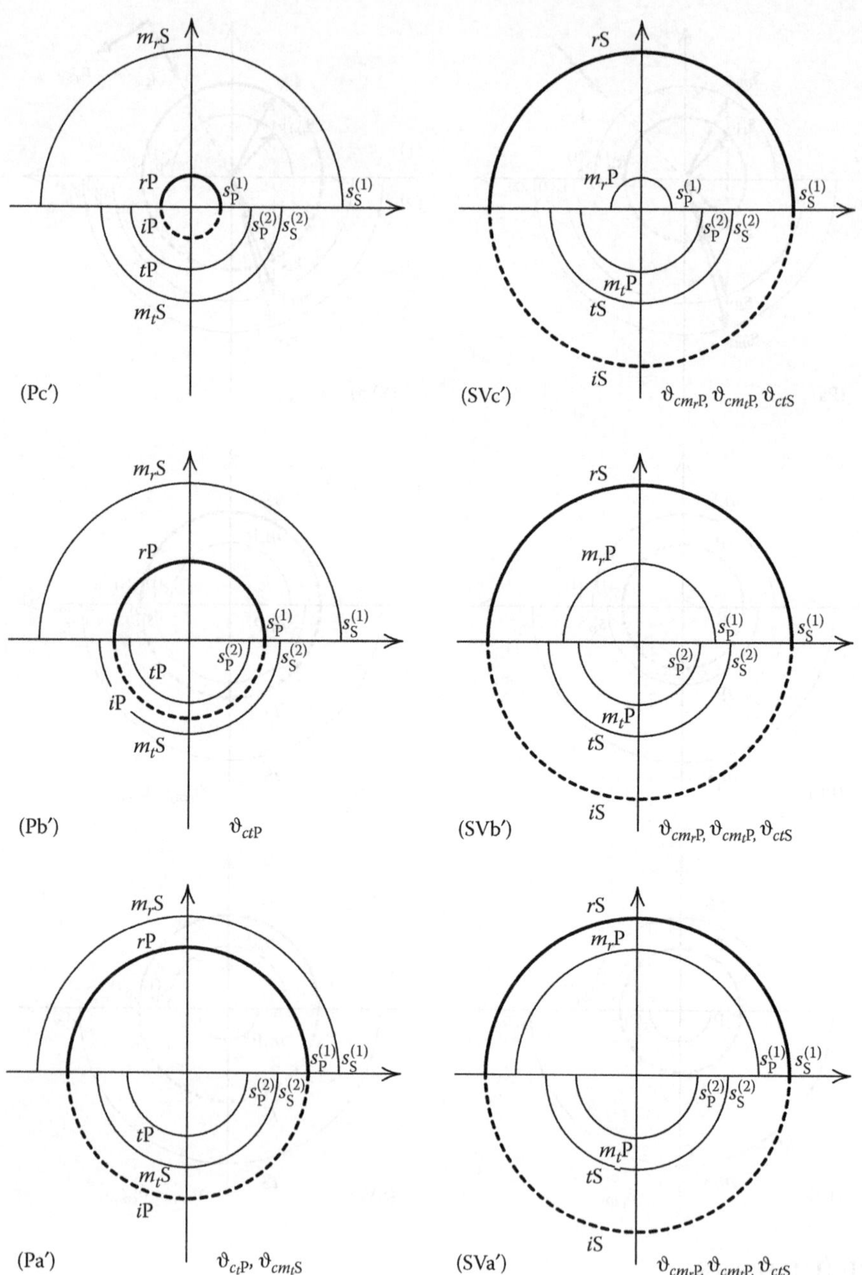

FIGURE 9.19
Continuation of Figure 9.18.

for $z \geq 0$ and

$$\underline{\mathbf{u}}_{t\mathrm{P}}(\underline{\mathbf{R}}, \omega, \hat{\underline{\mathbf{k}}}_{t\mathrm{P}}) = T_{\mathrm{P}}(\vartheta_{i\mathrm{P}})\, u_{i\mathrm{P}}(\omega)\, \mathrm{e}^{\mathrm{j}k_{\mathrm{P}}^{(2)}\hat{\underline{\mathbf{k}}}_{t\mathrm{P}}\cdot\underline{\mathbf{R}}}\, \hat{\underline{\mathbf{k}}}_{t\mathrm{P}}, \tag{9.240}$$

$$\underline{\mathbf{u}}_{m_t\mathrm{SV}}(\underline{\mathbf{R}}, \omega, \hat{\underline{\mathbf{k}}}_{m_t\mathrm{S}}) = M_{t\mathrm{S}}(\vartheta_{i\mathrm{P}})\, u_{i\mathrm{P}}(\omega)\, \mathrm{e}^{\mathrm{j}k_{\mathrm{S}}^{(2)}\hat{\underline{\mathbf{k}}}_{m_t\mathrm{S}}\cdot\underline{\mathbf{R}}}\, \hat{\underline{\mathbf{k}}}_{m_t\mathrm{S}} \times \underline{\mathbf{e}}_y \tag{9.241}$$

for $z \leq 0$, where

$$\hat{\underline{\mathbf{k}}}_{i\mathrm{P}} = -\sin\vartheta_{i\mathrm{P}}\, \underline{\mathbf{e}}_x - \cos\vartheta_{i\mathrm{P}}\, \underline{\mathbf{e}}_z, \tag{9.242}$$

$$\hat{\underline{\mathbf{k}}}_{r\mathrm{P}} = -\sin\vartheta_{r\mathrm{P}}\, \underline{\mathbf{e}}_x + \cos\vartheta_{r\mathrm{P}}\, \underline{\mathbf{e}}_z, \tag{9.243}$$

$$\hat{\underline{\mathbf{k}}}_{m_r\mathrm{S}} = -\sin\vartheta_{m_r\mathrm{S}}\, \underline{\mathbf{e}}_x + \cos\vartheta_{m_r\mathrm{S}}\, \underline{\mathbf{e}}_z, \tag{9.244}$$

$$\hat{\underline{\mathbf{k}}}_{t\mathrm{P}} = -\sin\vartheta_{t\mathrm{P}}\, \underline{\mathbf{e}}_x - \cos\vartheta_{t\mathrm{P}}\, \underline{\mathbf{e}}_z, \tag{9.245}$$

$$\hat{\underline{\mathbf{k}}}_{m_t\mathrm{S}} = -\sin\vartheta_{m_t\mathrm{S}}\, \underline{\mathbf{e}}_x - \cos\vartheta_{m_t\mathrm{S}}\, \underline{\mathbf{e}}_z. \tag{9.246}$$

The angles $\vartheta_{t\mathrm{P}}$ and $\vartheta_{m_t\mathrm{S}}$ must eventually be complex with $\pi/2$-real part; hence, their cosines must eventually be purely imaginary; to ensure the attenuation of $\underline{\mathbf{u}}_{t\mathrm{P}}$ and $\underline{\mathbf{u}}_{m_t\mathrm{SV}}$ in the half-space $z < 0$, the imaginary parts of the cosines must be positive.

The phase matching requirement yields

$$\begin{aligned} k_{\mathrm{P}}^{(1)} \sin\vartheta_{i\mathrm{P}} = k_{\mathrm{P}}^{(1)} \sin\vartheta_{r\mathrm{P}} = k_{\mathrm{S}}^{(1)} \sin\vartheta_{m_r\mathrm{S}} = k_{\mathrm{P}}^{(2)} \sin\vartheta_{t\mathrm{P}} \\ = k_{\mathrm{S}}^{(2)} \sin\vartheta_{m_t\mathrm{S}} \end{aligned} \tag{9.247}$$

leading to the reflection law

$$\vartheta_{r\mathrm{P}} = \vartheta_{i\mathrm{P}}, \tag{9.248}$$

the mode conversion law in reflection

$$\sin\vartheta_{m_r\mathrm{S}} = \underbrace{\frac{k_{\mathrm{P}}^{(1)}}{k_{\mathrm{S}}^{(1)}}}_{<1} \sin\vartheta_{i\mathrm{P}}, \tag{9.249}$$

the transmission law

$$\sin\vartheta_{t\mathrm{P}} = \underbrace{\frac{k_{\mathrm{P}}^{(1)}}{k_{\mathrm{P}}^{(2)}}}_{><1} \sin\vartheta_{i\mathrm{P}}, \tag{9.250}$$

as well as the mode conversion law in transmission

$$\sin\vartheta_{m_t\mathrm{S}} = \underbrace{\frac{k_{\mathrm{P}}^{(1)}}{k_{\mathrm{S}}^{(2)}}}_{><1} \sin\vartheta_{i\mathrm{P}}. \tag{9.251}$$

In Figure 9.18(Pa), the geometric construction of the angles $\vartheta_{r\mathrm{P}}, \vartheta_{m_r\mathrm{S}}, \vartheta_{t\mathrm{P}}$, $\vartheta_{m_t\mathrm{S}}$ is explicitly performed making it evident in the further pictures of Figures

9.18 and 9.19. Equations 9.250 and 9.251 reveal the existence of at most two critical angles.

The partial waves (9.237) through (9.241) define reflection and transmission coefficients $R_{\rm P}(\vartheta_{i{\rm P}})$, $T_{\rm P}(\vartheta_{i{\rm P}})$, and mode conversion coefficients $M_{r{\rm S}}(\vartheta_{i{\rm P}})$, $M_{t{\rm S}}(\vartheta_{i{\rm P}})$ in reflection and transmission. Their calculation requires four equations that have to be provided by the transition conditions (3.99) and (3.98). The transition condition (3.99) for the particle displacement vector has, according to (9.237) through (9.241) and (9.242) through (9.246), an x- and a z-component:

$$\begin{aligned} &-\sin\vartheta_{i{\rm P}} - R_{\rm P}(\vartheta_{i{\rm P}})\sin\vartheta_{i{\rm P}} - M_{r{\rm S}}(\vartheta_{i{\rm P}})\cos\vartheta_{m_r{\rm S}} \\ &\quad = -T_{\rm P}(\vartheta_{i{\rm P}})\sin\vartheta_{t{\rm P}} + M_{t{\rm S}}(\vartheta_{i{\rm P}})\cos\vartheta_{m_t{\rm S}}, \qquad (9.252) \\ &-\cos\vartheta_{i{\rm P}} + R_{\rm P}(\vartheta_{i{\rm P}})\cos\vartheta_{i{\rm P}} - M_{r{\rm S}}(\vartheta_{i{\rm P}})\sin\vartheta_{m_r{\rm S}} \\ &\quad = -T_{\rm P}(\vartheta_{i{\rm P}})\cos\vartheta_{t{\rm P}} - M_{t{\rm S}}(\vartheta_{i{\rm P}})\sin\vartheta_{m_t{\rm S}}. \qquad (9.253) \end{aligned}$$

With (8.121), (8.122), (9.8), and (9.237) through (9.246), we use (3.98) to calculate the $\underline{\mathbf{e}}_z$-projections of the stress tensors[190]

$$\begin{aligned} &\underline{\underline{\mathbf{T}}}_{i{\rm P}}(\underline{\mathbf{R}},\omega,\vartheta_{i{\rm P}})\cdot\underline{\mathbf{e}}_z \\ &\quad = {\rm j}k_{\rm P}^{(1)}u_{i{\rm P}}(\omega)\,{\rm e}^{{\rm j}k_{\rm P}^{(1)}\hat{\underline{\mathbf{k}}}_{i{\rm P}}\cdot\underline{\mathbf{R}}}\,(\lambda^{(1)}\underline{\mathbf{e}}_z - 2\mu^{(1)}\hat{\underline{\mathbf{k}}}_{i{\rm P}}\cos\vartheta_{i{\rm P}}), \qquad (9.254) \\ &\underline{\underline{\mathbf{T}}}_{r{\rm P}}(\underline{\mathbf{R}},\omega,\vartheta_{i{\rm P}})\cdot\underline{\mathbf{e}}_z \\ &\quad = {\rm j}k_{\rm P}^{(1)}R_{\rm P}(\vartheta_{i{\rm P}})u_{i{\rm P}}(\omega)\,{\rm e}^{{\rm j}k_{\rm P}^{(1)}\hat{\underline{\mathbf{k}}}_{r{\rm P}}\cdot\underline{\mathbf{R}}}\,(\lambda^{(1)}\underline{\mathbf{e}}_z + 2\mu^{(1)}\hat{\underline{\mathbf{k}}}_{r{\rm P}}\cos\vartheta_{i{\rm P}}), \\ &\qquad (9.255) \\ &\underline{\underline{\mathbf{T}}}_{m_r{\rm SV}}(\underline{\mathbf{R}},\omega,\vartheta_{i{\rm P}})\cdot\underline{\mathbf{e}}_z \\ &\quad = {\rm j}k_{\rm S}^{(1)}\mu^{(1)}M_{r{\rm S}}(\vartheta_{i{\rm P}})u_{i{\rm P}}(\omega)\,{\rm e}^{{\rm j}k_{\rm S}^{(2)}\hat{\underline{\mathbf{k}}}_{m_r{\rm S}}\cdot\underline{\mathbf{R}}}(-\hat{\underline{\mathbf{k}}}_{m_r{\rm S}}\sin\vartheta_{m_r{\rm S}} + \hat{\underline{\mathbf{u}}}_{m_r{\rm SV}}\cos\vartheta_{m_r{\rm S}}), \\ &\qquad (9.256) \\ &\underline{\underline{\mathbf{T}}}_{t{\rm P}}(\underline{\mathbf{R}},\omega,\vartheta_{i{\rm P}})\cdot\underline{\mathbf{e}}_z \\ &\quad = {\rm j}k_{\rm P}^{(2)}T_{\rm P}(\vartheta_{i{\rm P}})u_{i{\rm P}}(\omega)\,{\rm e}^{{\rm j}k_{\rm P}^{(2)}\hat{\underline{\mathbf{k}}}_{t{\rm P}}\cdot\underline{\mathbf{R}}}\,(\lambda^{(2)}\underline{\mathbf{e}}_z - 2\mu^{(2)}\hat{\underline{\mathbf{k}}}_{t{\rm P}}\cos\vartheta_{t{\rm P}}), \qquad (9.257) \\ &\underline{\underline{\mathbf{T}}}_{m_t{\rm SV}}(\underline{\mathbf{R}},\omega,\vartheta_{i{\rm P}})\cdot\underline{\mathbf{e}}_z \\ &\quad = {\rm j}k_{\rm S}^{(2)}\mu^{(2)}M_{t{\rm S}}(\vartheta_{i{\rm P}})u_{i{\rm P}}(\omega)\,{\rm e}^{{\rm j}k_{\rm S}^{(2)}\hat{\underline{\mathbf{k}}}_{m_t{\rm S}}\cdot\underline{\mathbf{R}}}(-\hat{\underline{\mathbf{k}}}_{m_t{\rm S}}\sin\vartheta_{m_t{\rm S}} - \hat{\underline{\mathbf{u}}}_{m_t{\rm SV}}\cos\vartheta_{m_t{\rm S}}) \\ &\qquad (9.258) \end{aligned}$$

and state that (9.254) through (9.258) also have an x- and a z-component leading to the explicit version of (3.98):

$$\begin{aligned} &k_{\rm P}^{(1)}\mu^{(1)}\sin 2\vartheta_{i{\rm P}} - k_{\rm P}^{(1)}\mu^{(1)}R_{\rm P}(\vartheta_{i{\rm P}})\sin 2\vartheta_{i{\rm P}} - k_{\rm S}^{(1)}\mu^{(1)}M_{r{\rm S}}(\vartheta_{i{\rm P}})\cos 2\vartheta_{m_r{\rm S}} \\ &\quad = k_{\rm P}^{(2)}\mu^{(2)}T_{\rm P}(\vartheta_{i{\rm P}})\sin 2\vartheta_{t{\rm P}} - k_{\rm S}^{(2)}\mu^{(2)}M_{t{\rm S}}(\vartheta_{i{\rm P}})\cos 2\vartheta_{m_t{\rm S}}, \qquad (9.259) \end{aligned}$$

[190]The polarization vectors $\hat{\underline{\mathbf{u}}}_{mr,t{\rm SV}}$ are always unit vectors in the sense of $\hat{\underline{\mathbf{u}}}_{mr,t{\rm SV}}\cdot\hat{\underline{\mathbf{u}}}_{mr,t{\rm SV}} = 1$.

$$
\begin{aligned}
& k_{\mathrm{P}}^{(1)}(\lambda^{(1)} + 2\mu^{(1)}) \cos 2\vartheta_{m_r\mathrm{S}} + k_{\mathrm{P}}^{(1)}(\lambda^{(1)} + 2\mu^{(1)}) R_{\mathrm{P}}(\vartheta_{i\mathrm{P}}) \cos 2\vartheta_{m_r\mathrm{S}} \\
& \quad - k_{\mathrm{S}}^{(1)} \mu^{(1)} M_{r\mathrm{S}}(\vartheta_{i\mathrm{P}}) \sin 2\vartheta_{m_r\mathrm{S}} \\
& \qquad = k_{\mathrm{P}}^{(2)}(\lambda^{(2)} + 2\mu^{(2)}) T_{\mathrm{P}}(\vartheta_{i\mathrm{P}}) \cos 2\vartheta_{m_t\mathrm{S}} + k_{\mathrm{S}}^{(2)} \mu^{(2)} M_{t\mathrm{S}}(\vartheta_{i\mathrm{P}}) \sin 2\vartheta_{m_t\mathrm{S}};
\end{aligned}
\tag{9.260}
$$

we have utilized the conversions

$$
\lambda^{(1)} + 2\mu^{(1)} \cos^2 \vartheta_{i\mathrm{P}} = (\lambda^{(1)} + 2\mu^{(1)}) \cos 2\vartheta_{m_r\mathrm{S}}, \tag{9.261}
$$

$$
\lambda^{(2)} + 2\mu^{(2)} \cos^2 \vartheta_{t\mathrm{P}} = (\lambda^{(2)} + 2\mu^{(2)}) \cos 2\vartheta_{m_t\mathrm{S}}. \tag{9.262}
$$

Finally, we obtain the following system of equations

$$
\underline{\underline{\boldsymbol{K}}}_{\mathrm{P}}(\vartheta_{i\mathrm{P}})\, \underline{\boldsymbol{f}}_{\mathrm{P}}(\vartheta_{i\mathrm{P}}) = \underline{\boldsymbol{i}}_{\mathrm{P}}(\vartheta_{i\mathrm{P}}) \tag{9.263}
$$

for the four-component solution (column) matrix (for the four-component column vector)

$$
\underline{\boldsymbol{f}}_{\mathrm{P}}(\vartheta_{i\mathrm{P}}) = \begin{pmatrix} R_{\mathrm{P}}(\vartheta_{i\mathrm{P}}) \\ M_{r\mathrm{S}}(\vartheta_{i\mathrm{P}}) \\ T_{\mathrm{P}}(\vartheta_{i\mathrm{P}}) \\ M_{t\mathrm{S}}(\vartheta_{i\mathrm{P}}) \end{pmatrix} \tag{9.264}
$$

for given inhomogeneity matrix (given inhomogeneity column vector)

$$
\underline{\boldsymbol{i}}_{\mathrm{P}}(\vartheta_{i\mathrm{P}}) = \begin{pmatrix} \sin \vartheta_{i\mathrm{P}} \\ \cos \vartheta_{i\mathrm{P}} \\ \sin 2\vartheta_{i\mathrm{P}} \\ \cos 2\vartheta_{m_r\mathrm{S}} \end{pmatrix}, \tag{9.265}
$$

and given (4×4)-coefficient matrix

$$
\underline{\underline{\boldsymbol{K}}}_{\mathrm{P}}(\vartheta_{i\mathrm{P}})
= \begin{pmatrix}
-\sin \vartheta_{i\mathrm{P}} & -\cos \vartheta_{m_r\mathrm{S}} & \sin \vartheta_{t\mathrm{P}} & -\cos \vartheta_{m_t\mathrm{S}} \\
\cos \vartheta_{i\mathrm{P}} & -\sin \vartheta_{m_r\mathrm{S}} & \cos \vartheta_{t\mathrm{P}} & \sin \vartheta_{m_t\mathrm{S}} \\
\sin 2\vartheta_{i\mathrm{P}} & \kappa^{(1)} \cos 2\vartheta_{m_r\mathrm{S}} & \frac{\kappa^{(1)}}{\kappa^{(2)}} \frac{Z_{\mathrm{S}}^{(2)}}{Z_{\mathrm{S}}^{(1)}} \sin 2\vartheta_{t\mathrm{P}} & -\kappa^{(1)} \frac{Z_{\mathrm{S}}^{(2)}}{Z_{\mathrm{S}}^{(1)}} \cos 2\vartheta_{m_t\mathrm{S}} \\
-\cos 2\vartheta_{m_r\mathrm{S}} & \frac{1}{\kappa^{(1)}} \sin 2\vartheta_{m_r\mathrm{S}} & \frac{Z_{\mathrm{P}}^{(2)}}{Z_{\mathrm{P}}^{(1)}} \cos 2\vartheta_{m_t\mathrm{S}} & \frac{Z_{\mathrm{S}}^{(2)}}{Z_{\mathrm{P}}^{(1)}} \sin 2\vartheta_{m_t\mathrm{S}}
\end{pmatrix} \tag{9.266}
$$

with the short-hand notations

$$
\kappa^{(j)} = \frac{c_{\mathrm{P}}^{(j)}}{c_{\mathrm{S}}^{(j)}}, \quad j = 1, 2, \tag{9.267}
$$

for the velocity ratios, and

$$
Z_{\mathrm{P,S}}^{(j)} = \rho^{(j)} c_{\mathrm{P,S}}^{(j)}, \quad j = 1, 2, \tag{9.268}
$$

for the acoustic impedances.

Even though the matrix (9.266) may be analytically inverted, the resulting expressions are rather intricate—even if the Ewing method (Schmerr 1988) is applied—providing no advantage over a numerical method of the system of Equations 9.263 to obtain numbers for $\underline{\boldsymbol{f}}_{\mathrm{P}}(\vartheta_{i\mathrm{P}})$. But in case one goes back to Schmerr (1998), one has to note that the respective coefficients are related to the amplitudes of potentials. Yet, Ben-Menahem and Singh (1981), Achenbach (1973), and Auld (1973) also present the matrix (9.266) for the amplitudes of the particle displacement, but care has to be taken regarding the signs of the polarization directions of SV-waves.

For vertical incidence—$\vartheta_{i\mathrm{P}} = 0$—we obtain a simple special case: The (4×4)-system of equations (9.263) separates into two (2×2)-systems of equations with only trivial solutions for $M_{r\mathrm{S}}(0)$ and $M_{t\mathrm{S}}(0)$, whereas we obtain for the P-reflection and transmission coefficients:

$$R_{\mathrm{P}}(0) = \frac{Z_{\mathrm{P}}^{(2)} - Z_{\mathrm{P}}^{(1)}}{Z_{\mathrm{P}}^{(2)} + Z_{\mathrm{P}}^{(1)}}, \tag{9.269}$$

$$T_{\mathrm{P}}(0) = \frac{2Z_{\mathrm{P}}^{(1)}}{Z_{\mathrm{P}}^{(2)} + Z_{\mathrm{P}}^{(1)}}. \tag{9.270}$$

Therefore, the sound pressure equations for arbitrary incidence angles

$$p_{i\mathrm{P}}(\underline{\mathbf{R}}, \omega, \vartheta_{i\mathrm{P}}) = -\mathrm{j}\omega Z_{\mathrm{P}}^{(1)}\, u_{i\mathrm{P}}(\omega)\, \mathrm{e}^{\mathrm{j}k_{\mathrm{P}}^{(1)}\hat{\underline{\mathbf{k}}}_{i\mathrm{P}}\cdot\underline{\mathbf{R}}}, \tag{9.271}$$

$$p_{r\mathrm{P}}(\underline{\mathbf{R}}, \omega, \vartheta_{i\mathrm{P}}) = -\mathrm{j}\omega Z_{\mathrm{P}}^{(1)}\, R_{\mathrm{P}}(\vartheta_{i\mathrm{P}})\, u_{i\mathrm{P}}(\omega)\, \mathrm{e}^{\mathrm{j}k_{\mathrm{P}}^{(1)}\hat{\underline{\mathbf{k}}}_{r\mathrm{P}}\cdot\underline{\mathbf{R}}}, \tag{9.272}$$

$$p_{m_r\mathrm{SV}}(\underline{\mathbf{R}}, \omega, \vartheta_{i\mathrm{P}}) = -\mathrm{j}\omega Z_{\mathrm{S}}^{(1)}\, M_{r\mathrm{S}}(\vartheta_{i\mathrm{P}})\, u_{i\mathrm{P}}(\omega)\, \mathrm{e}^{\mathrm{j}k_{\mathrm{S}}^{(1)}\hat{\underline{\mathbf{k}}}_{m_r\mathrm{S}}\cdot\underline{\mathbf{R}}}, \tag{9.273}$$

$$p_{t\mathrm{P}}(\underline{\mathbf{R}}, \omega, \vartheta_{i\mathrm{P}}) = -\mathrm{j}\omega Z_{\mathrm{P}}^{(2)}\, T_{\mathrm{P}}(\vartheta_{i\mathrm{P}})\, u_{i\mathrm{P}}(\omega)\, \mathrm{e}^{\mathrm{j}k_{\mathrm{P}}^{(2)}\hat{\underline{\mathbf{k}}}_{t\mathrm{P}}\cdot\underline{\mathbf{R}}}, \tag{9.274}$$

$$p_{m_t\mathrm{SV}}(\underline{\mathbf{R}}, \omega, \vartheta_{i\mathrm{P}}) = -\mathrm{j}\omega Z_{\mathrm{S}}^{(2)}\, M_{t\mathrm{S}}(\vartheta_{i\mathrm{P}})\, u_{i\mathrm{P}}(\omega)\, \mathrm{e}^{\mathrm{j}k_{\mathrm{S}}^{(2)}\hat{\underline{\mathbf{k}}}_{t\mathrm{S}}\cdot\underline{\mathbf{R}}} \tag{9.275}$$

reduce to the following ones for vertical incidence:

$$p_{i\mathrm{P}}(\underline{\mathbf{R}}, \omega, 0) = -\mathrm{j}\omega Z_{\mathrm{P}}^{(1)}\, u_{i\mathrm{P}}(\omega)\, \mathrm{e}^{-\mathrm{j}k_{\mathrm{P}}^{(1)}\cos\vartheta_{i\mathrm{P}}\, z}, \tag{9.276}$$

$$p_{r\mathrm{P}}(\underline{\mathbf{R}}, \omega, 0) = -\mathrm{j}\omega Z_{\mathrm{P}}^{(1)}\, R_{\mathrm{P}}(0)\, u_{i\mathrm{P}}(\omega)\, \mathrm{e}^{\mathrm{j}k_{\mathrm{P}}^{(1)}\cos\vartheta_{i\mathrm{P}}\, z}, \tag{9.277}$$

$$p_{t\mathrm{P}}(\underline{\mathbf{R}}, \omega, 0) = -\mathrm{j}\omega Z_{\mathrm{P}}^{(1)}\, T_{pt\mathrm{P}}(0)\, u_{i\mathrm{P}}(\omega)\, \mathrm{e}^{-\mathrm{j}k_{\mathrm{P}}^{(2)}\cos\vartheta_{t\mathrm{P}}\, z}, \tag{9.278}$$

where $T_{pt\mathrm{P}}(0)$ according to

$$T_{pt\mathrm{P}}(0) = \frac{2Z_{\mathrm{P}}^{(2)}}{Z_{\mathrm{P}}^{(2)} + Z_{\mathrm{P}}^{(1)}} \tag{9.279}$$

structurally coincides with (9.218). The polarization directions of incident, reflected, and transmitted P-waves are chosen such that the resulting sign of the reflection coefficient of the vector particle velocity entails the continuity of the sound pressure for $\vartheta_{i\mathrm{P}} = 0$ according to (9.276) through (9.278).

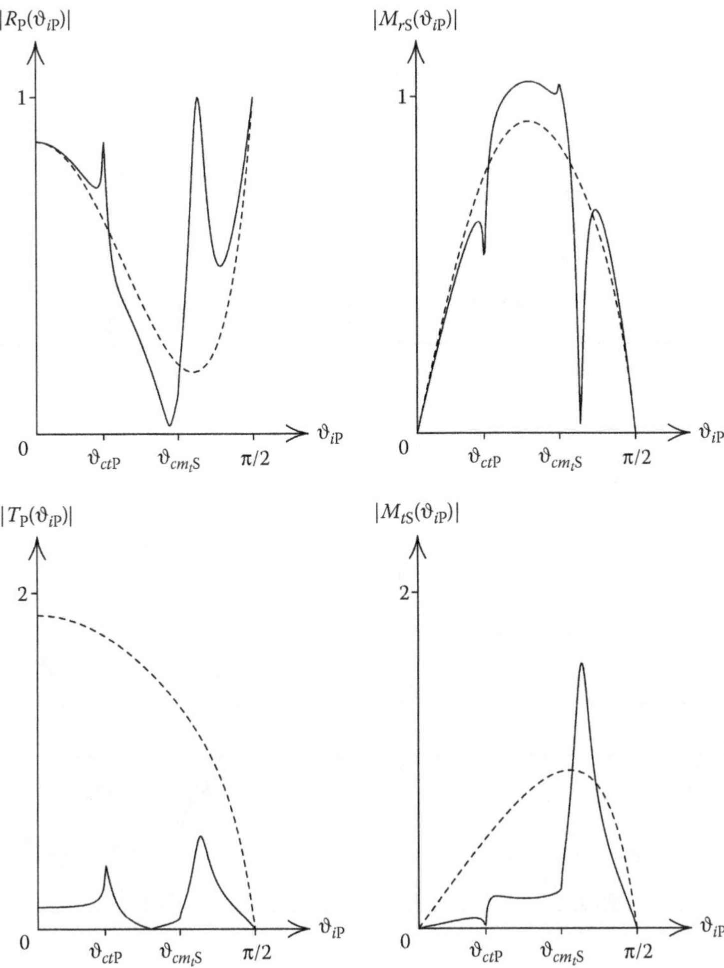

FIGURE 9.20
Magnitudes of reflection, transmission, and mode conversion coefficients for P-wave incidence as function of the angle of incidence: steel(1)-plexiglas(2): - - -; plexiglas(1)-steel(2):—(steel: $c_P = 5900$ m/s, $c_S = 3200$ m/s, $\rho = 7.7 \cdot 10^3$ kg/m^3; plexiglas: $c_P = 2730$ m/s, $c_S = 1430$ m/s, $\rho = 1.18 \cdot 10^3$ kg/m^3).

Wavefronts for P-wave incidence: Figure 9.21 displays wavefronts of incident, reflected and transmitted P-waves as well as wavefronts of mode converted reflected and mode converted transmitted SV-waves for the existence of two critical angles [Figure 9.19(Pa′)]; the display is logarithmic enhancing the “side lobes” of the RC2(t)-pulse relative to the “main lobe.”

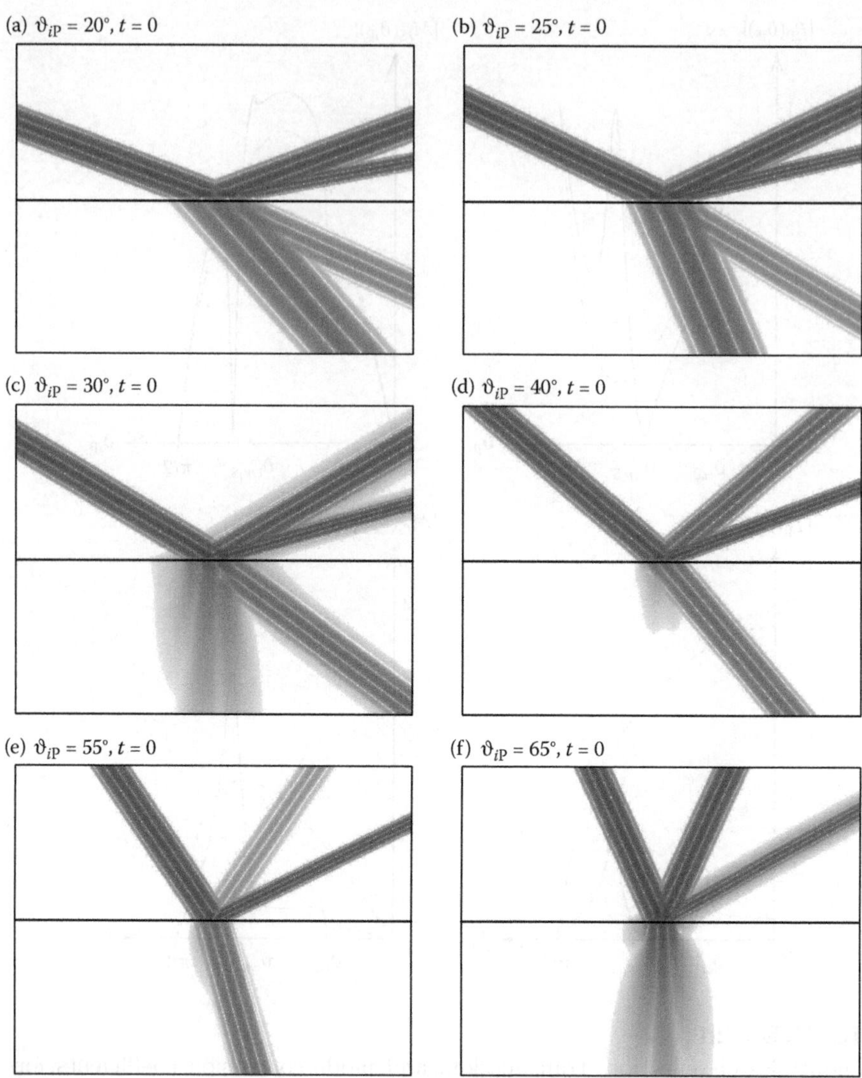

FIGURE 9.21
Wavefronts for P-wave incidence (a)–(f): plexiglas(1)-steel(2) (logarithmic magnitude of the vector particle velocity).

Incident SV-wave: System of equations for reflection, transmission, and mode conversion coefficients: We postulate the following plane waves

$$\underline{\mathbf{u}}_{i\mathrm{SV}}(\underline{\mathbf{R}}, \omega, \hat{\underline{\mathbf{k}}}_{i\mathrm{S}}) = u_{i\mathrm{S}}(\omega)\, \mathrm{e}^{\mathrm{j}k_{\mathrm{S}}^{(1)}\hat{\underline{\mathbf{k}}}_{i\mathrm{S}}\cdot\underline{\mathbf{R}}}\, \hat{\underline{\mathbf{k}}}_{i\mathrm{S}} \times \underline{\mathbf{e}}_y, \tag{9.280}$$

$$\underline{\mathbf{u}}_{r\mathrm{SV}}(\underline{\mathbf{R}}, \omega, \hat{\underline{\mathbf{k}}}_{r\mathrm{S}}) = R_{\mathrm{SV}}(\vartheta_{i\mathrm{S}})\, u_{i\mathrm{S}}(\omega)\, \mathrm{e}^{\mathrm{j}k_{\mathrm{S}}^{(1)}\hat{\underline{\mathbf{k}}}_{r\mathrm{S}}\cdot\underline{\mathbf{R}}}\, \hat{\underline{\mathbf{k}}}_{r\mathrm{S}} \times \underline{\mathbf{e}}_y, \tag{9.281}$$

$$\underline{\mathbf{u}}_{m_r\mathrm{P}}(\underline{\mathbf{R}}, \omega, \hat{\underline{\mathbf{k}}}_{m_r\mathrm{P}}) = M_{r\mathrm{P}}(\vartheta_{i\mathrm{S}})\, u_{i\mathrm{S}}(\omega)\, \mathrm{e}^{\mathrm{j}k_\mathrm{S}^{(1)}\hat{\underline{\mathbf{k}}}_{m_r\mathrm{P}}\cdot\underline{\mathbf{R}}}\,\hat{\underline{\mathbf{k}}}_{m_r\mathrm{P}} \tag{9.282}$$

for $z \geq 0$ and

$$\underline{\mathbf{u}}_{t\mathrm{SV}}(\underline{\mathbf{R}}, \omega, \hat{\underline{\mathbf{k}}}_{t\mathrm{S}}) = T_{\mathrm{SV}}(\vartheta_{i\mathrm{S}})\, u_{i\mathrm{S}}(\omega)\, \mathrm{e}^{\mathrm{j}k_\mathrm{S}^{(2)}\hat{\underline{\mathbf{k}}}_{t\mathrm{S}}\cdot\underline{\mathbf{R}}}\,\hat{\underline{\mathbf{k}}}_{t\mathrm{S}} \times \underline{\mathbf{e}}_y, \tag{9.283}$$

$$\underline{\mathbf{u}}_{m_t\mathrm{P}}(\underline{\mathbf{R}}, \omega, \hat{\underline{\mathbf{k}}}_{m_t\mathrm{P}}) = M_{t\mathrm{P}}(\vartheta_{i\mathrm{S}})\, u_{i\mathrm{S}}(\omega)\, \mathrm{e}^{\mathrm{j}k_\mathrm{S}^{(2)}\hat{\underline{\mathbf{k}}}_{m_t\mathrm{P}}\cdot\underline{\mathbf{R}}}\,\hat{\underline{\mathbf{k}}}_{m_t\mathrm{P}} \tag{9.284}$$

for $z \leq 0$ an; the respective polarization vectors read

$$\hat{\underline{\mathbf{k}}}_{i\mathrm{S}} = -\sin\vartheta_{i\mathrm{S}}\,\underline{\mathbf{e}}_x - \cos\vartheta_{i\mathrm{S}}\,\underline{\mathbf{e}}_z, \tag{9.285}$$

$$\hat{\underline{\mathbf{k}}}_{r\mathrm{S}} = -\sin\vartheta_{r\mathrm{S}}\,\underline{\mathbf{e}}_x + \cos\vartheta_{r\mathrm{S}}\,\underline{\mathbf{e}}_z, \tag{9.286}$$

$$\hat{\underline{\mathbf{k}}}_{m_r\mathrm{P}} = -\sin\vartheta_{m_r\mathrm{P}}\,\underline{\mathbf{e}}_x + \cos\vartheta_{m_r\mathrm{P}}\,\underline{\mathbf{e}}_z, \tag{9.287}$$

$$\hat{\underline{\mathbf{k}}}_{t\mathrm{S}} = -\sin\vartheta_{t\mathrm{S}}\,\underline{\mathbf{e}}_x - \cos\vartheta_{t\mathrm{S}}\,\underline{\mathbf{e}}_z, \tag{9.288}$$

$$\hat{\underline{\mathbf{k}}}_{m_t\mathrm{P}} = -\sin\vartheta_{m_t\mathrm{P}}\,\underline{\mathbf{e}}_x - \cos\vartheta_{m_t\mathrm{P}}\,\underline{\mathbf{e}}_z. \tag{9.289}$$

The angles $\vartheta_{m_r\mathrm{P}}$, $\vartheta_{t\mathrm{S}}$, and $\vartheta_{m_t\mathrm{P}}$ must eventually be complex with $\pi/2$-real parts; hence, their cosines must be purely imaginary with positive imaginary parts to ensure evanescent waves.

The phase matching requirement leads to

$$k_\mathrm{S}^{(1)}\sin\vartheta_{i\mathrm{S}} = k_\mathrm{S}^{(1)}\sin\vartheta_{r\mathrm{S}} = k_\mathrm{P}^{(1)}\sin\vartheta_{m_r\mathrm{P}} = k_\mathrm{S}^{(2)}\sin\vartheta_{t\mathrm{S}} = k_\mathrm{P}^{(2)}\sin\vartheta_{m_t\mathrm{P}} \tag{9.290}$$

resulting in the reflection law

$$\vartheta_{r\mathrm{S}} = \vartheta_{i\mathrm{S}}, \tag{9.291}$$

the mode conversion law for reflection

$$\sin\vartheta_{m_r\mathrm{P}} = \underbrace{\frac{k_\mathrm{S}^{(1)}}{k_\mathrm{P}^{(1)}}}_{>1}\sin\vartheta_{i\mathrm{S}}, \tag{9.292}$$

the transmission law

$$\sin\vartheta_{t\mathrm{S}} = \underbrace{\frac{k_\mathrm{S}^{(1)}}{k_\mathrm{S}^{(2)}}}_{><1}\sin\vartheta_{i\mathrm{S}}, \tag{9.293}$$

as well as the mode conversion law for transmission

$$\sin\vartheta_{m_t\mathrm{P}} = \underbrace{\frac{k_\mathrm{S}^{(1)}}{k_\mathrm{P}^{(2)}}}_{><1}\sin\vartheta_{i\mathrm{S}}. \tag{9.294}$$

The geometric construction of the angles $\vartheta_{r\mathrm{S}}, \vartheta_{m_r\mathrm{P}}, \vartheta_{t\mathrm{S}}, \vartheta_{m_t\mathrm{P}}$ is explicitly performed in Figure 9.18(SVa), thus becoming obvious for the subsequent

SV-pictures of Figures 9.18 and 9.19. Equations 9.292, 9.293, and 9.294 reveal the mandatory existence of at least one critical angle and at most three. The latter case leads to total reflection of the incident SV-wave.

The partial waves (9.280) through (9.284) define reflection and transmission coefficients $R_{\mathrm{SV}}(\vartheta_{i\mathrm{S}})$, $T_{\mathrm{SV}}(\vartheta_{i\mathrm{S}})$ as well as mode conversion coefficients $M_{r\mathrm{P}}(\vartheta_{i\mathrm{S}}), M_{t\mathrm{P}}(\vartheta_{i\mathrm{S}})$ in reflection and transmission. Again, we need four equations to calculate them; still missing are the respective stress tensor projections onto the boundary normal $\underline{\mathbf{n}} = \underline{\mathbf{e}}_z$ related to (9.280) through (9.284):

$$\begin{aligned}
&\underline{\underline{\mathbf{T}}}_{i\mathrm{SV}}(\underline{\mathbf{R}}, \omega, \vartheta_{i\mathrm{S}}) \cdot \underline{\mathbf{e}}_z \\
&\quad = -\mathrm{j}k_\mathrm{S}^{(1)} \mu^{(1)} u_{i\mathrm{S}}(\omega)\, \mathrm{e}^{\mathrm{j}k_\mathrm{S}^{(1)} \hat{\underline{\mathbf{k}}}_{i\mathrm{S}} \cdot \underline{\mathbf{R}}}\, (\hat{\underline{\mathbf{k}}}_{i\mathrm{S}} \sin\vartheta_{i\mathrm{S}} + \hat{\underline{\mathbf{u}}}_{i\mathrm{SV}} \cos\vartheta_{i\mathrm{S}}),
\end{aligned} \tag{9.295}$$

$$\begin{aligned}
&\underline{\underline{\mathbf{T}}}_{r\mathrm{SV}}(\underline{\mathbf{R}}, \omega, \vartheta_{i\mathrm{S}}) \cdot \underline{\mathbf{e}}_z \\
&\quad = \mathrm{j}k_\mathrm{S}^{(1)} \mu^{(1)} R_{\mathrm{SV}}(\vartheta_{i\mathrm{S}}) u_{i\mathrm{S}}(\omega)\, \mathrm{e}^{\mathrm{j}k_\mathrm{S}^{(1)} \hat{\underline{\mathbf{k}}}_{r\mathrm{S}} \cdot \underline{\mathbf{R}}}\, (-\hat{\underline{\mathbf{k}}}_{r\mathrm{S}} \sin\vartheta_{i\mathrm{S}} + \hat{\underline{\mathbf{u}}}_{r\mathrm{SV}} \cos\vartheta_{i\mathrm{S}}),
\end{aligned} \tag{9.296}$$

$$\begin{aligned}
&\underline{\underline{\mathbf{T}}}_{m_r\mathrm{P}}(\underline{\mathbf{R}}, \omega, \vartheta_{i\mathrm{S}}) \cdot \underline{\mathbf{e}}_z \\
&\quad = \mathrm{j}k_\mathrm{P}^{(1)} M_{r\mathrm{P}}(\vartheta_{i\mathrm{S}}) u_{i\mathrm{S}}(\omega)\, \mathrm{e}^{\mathrm{j}k_\mathrm{P}^{(1)} \hat{\underline{\mathbf{k}}}_{m_r\mathrm{P}} \cdot \underline{\mathbf{R}}}\, (\lambda^{(1)} \underline{\mathbf{e}}_z + 2\mu^{(1)} \hat{\underline{\mathbf{k}}}_{m_r\mathrm{P}} \cos\vartheta_{m_r\mathrm{P}}),
\end{aligned} \tag{9.297}$$

$$\begin{aligned}
&\underline{\underline{\mathbf{T}}}_{t\mathrm{SV}}(\underline{\mathbf{R}}, \omega, \vartheta_{i\mathrm{S}}) \cdot \underline{\mathbf{e}}_z \\
&\quad = \mathrm{j}k_\mathrm{S}^{(2)} \mu^{(2)} T_{\mathrm{SV}}(\vartheta_{i\mathrm{S}}) u_{i\mathrm{S}}(\omega)\, \mathrm{e}^{\mathrm{j}k_\mathrm{S}^{(2)} \hat{\underline{\mathbf{k}}}_{t\mathrm{S}} \cdot \underline{\mathbf{R}}}\, (-\hat{\underline{\mathbf{k}}}_{t\mathrm{S}} \sin\vartheta_{t\mathrm{S}} - \hat{\underline{\mathbf{u}}}_{t\mathrm{SV}} \cos\vartheta_{t\mathrm{S}}),
\end{aligned} \tag{9.298}$$

$$\begin{aligned}
&\underline{\underline{\mathbf{T}}}_{m_t\mathrm{P}}(\underline{\mathbf{R}}, \omega, \vartheta_{i\mathrm{S}}) \cdot \underline{\mathbf{e}}_z \\
&\quad = \mathrm{j}k_\mathrm{P}^{(2)} M_{t\mathrm{P}}(\vartheta_{i\mathrm{S}}) u_{i\mathrm{S}}(\omega)\, \mathrm{e}^{\mathrm{j}k_\mathrm{P}^{(2)} \hat{\underline{\mathbf{k}}}_{m_t\mathrm{P}} \cdot \underline{\mathbf{R}}}\, (\lambda^{(2)} \underline{\mathbf{e}}_z - 2\mu^{(2)} \hat{\underline{\mathbf{k}}}_{m_t\mathrm{P}} \cos\vartheta_{m_t\mathrm{P}}).
\end{aligned} \tag{9.299}$$

With the transition conditions (3.99) and (3.98), we finally find the system of equations

$$\underline{\underline{\boldsymbol{K}}}_{\mathrm{SV}}(\vartheta_{i\mathrm{S}})\, \underline{\boldsymbol{f}}_{\mathrm{SV}}(\vartheta_{i\mathrm{S}}) = \underline{\boldsymbol{i}}_{\mathrm{SV}}(\vartheta_{i\mathrm{S}}) \tag{9.300}$$

for the four-component solution (column) matrix

$$\underline{\boldsymbol{f}}_{\mathrm{SV}}(\vartheta_{i\mathrm{S}}) = \begin{pmatrix} R_{\mathrm{SV}}(\vartheta_{i\mathrm{S}}) \\ M_{r\mathrm{P}}(\vartheta_{i\mathrm{S}}) \\ T_{\mathrm{SV}}(\vartheta_{i\mathrm{S}}) \\ M_{t\mathrm{P}}(\vartheta_{i\mathrm{S}}) \end{pmatrix} \tag{9.301}$$

with given inhomogeneity matrix

$$\underline{\boldsymbol{i}}_{\mathrm{SV}}(\vartheta_{i\mathrm{S}}) = \begin{pmatrix} \cos\vartheta_{i\mathrm{S}} \\ \sin\vartheta_{i\mathrm{S}} \\ \cos 2\vartheta_{i\mathrm{S}} \\ \sin 2\vartheta_{i\mathrm{S}} \end{pmatrix} \tag{9.302}$$

and given (4×4)-coefficient matrix

$$\underline{\underline{\boldsymbol{K}}}_{\mathrm{SV}}(\vartheta_{i\mathrm{S}}) = \begin{pmatrix} \cos\vartheta_{i\mathrm{S}} & \sin\vartheta_{m_r\mathrm{P}} & \cos\vartheta_{t\mathrm{S}} & -\sin\vartheta_{m_t\mathrm{P}} \\ -\sin\vartheta_{i\mathrm{S}} & \cos\vartheta_{m_r\mathrm{P}} & \sin\vartheta_{t\mathrm{S}} & \cos\vartheta_{m_t\mathrm{P}} \\ -\cos 2\vartheta_{i\mathrm{S}} & -\frac{1}{\kappa^{(1)}}\sin 2\vartheta_{m_r\mathrm{P}} & \frac{Z_\mathrm{S}^{(2)}}{Z_\mathrm{S}^{(1)}}\cos 2\vartheta_{t\mathrm{S}} & -\frac{1}{\kappa^{(2)}}\frac{Z_\mathrm{S}^{(2)}}{Z_\mathrm{S}^{(1)}}\sin 2\vartheta_{m_t\mathrm{P}} \\ \sin 2\vartheta_{i\mathrm{S}} & -\kappa^{(1)}\cos 2\vartheta_{i\mathrm{S}} & \frac{Z_\mathrm{S}^{(2)}}{Z_\mathrm{S}^{(1)}}\sin 2\vartheta_{t\mathrm{S}} & \frac{Z_\mathrm{P}^{(2)}}{Z_\mathrm{S}^{(1)}}\cos 2\vartheta_{t\mathrm{S}} \end{pmatrix}. \tag{9.303}$$

As before we state that: The system of equations (9.300) is best solved numerically! Yet for vertical incidence, we immediately obtain $M_{r\mathrm{P}}(0) = M_{t\mathrm{P}}(0) = 0$ and

$$R_\mathrm{SV}(0) = \frac{Z_\mathrm{S}^{(2)} - Z_\mathrm{S}^{(1)}}{Z_\mathrm{S}^{(2)} + Z_\mathrm{S}^{(1)}}, \tag{9.304}$$

$$T_\mathrm{SV}(0) = \frac{2Z_\mathrm{S}^{(1)}}{Z_\mathrm{S}^{(2)} + Z_\mathrm{S}^{(1)}}. \tag{9.305}$$

For vertical incidence, the SV-polarization may not be distinguished from the SH-polarization; the still different signs of $R_\mathrm{SV}(0)$ and $R_\mathrm{SH}(0)$ (Equation 9.210) are explained by the different directions of incident and reflected polarizations in both cases. This is the reason why here the sound pressure is continuous for $\vartheta_{i\mathrm{S}} = 0$ in contrast to the SH-case. In the limit $\rho^{(2)} \longrightarrow 0$, hence $Z_\mathrm{S}^{(2)} \longrightarrow 0$, this remark corresponds to the one in connection with the SH- and SV-reflection at a boundary to vacuum.

Obviously, the content of Figure 9.22 may be visualized with wavefronts; this is especially intuitive for an animation in dependence of the angle of incidence.

9.3 Planar Boundary between a Homogeneous Isotropic Nondissipative and a Homogeneous Transversely Isotropic Nondissipative Half-Space

This selected special case of a planar boundary between two homogeneous anisotropic nondissipative half-spaces will not be investigated as detailed as "Fresnel's" reflection at the boundary between two isotropic half-spaces; the reason is its complexity with generally nonavailable analytical expressions. We will rather emphasize some peculiarities that do not appear in the isotropic case.

- Inhomogeneous plane waves with nonorthogonal real and imaginary parts of the phase propagation vector even though there is no dissipation: This

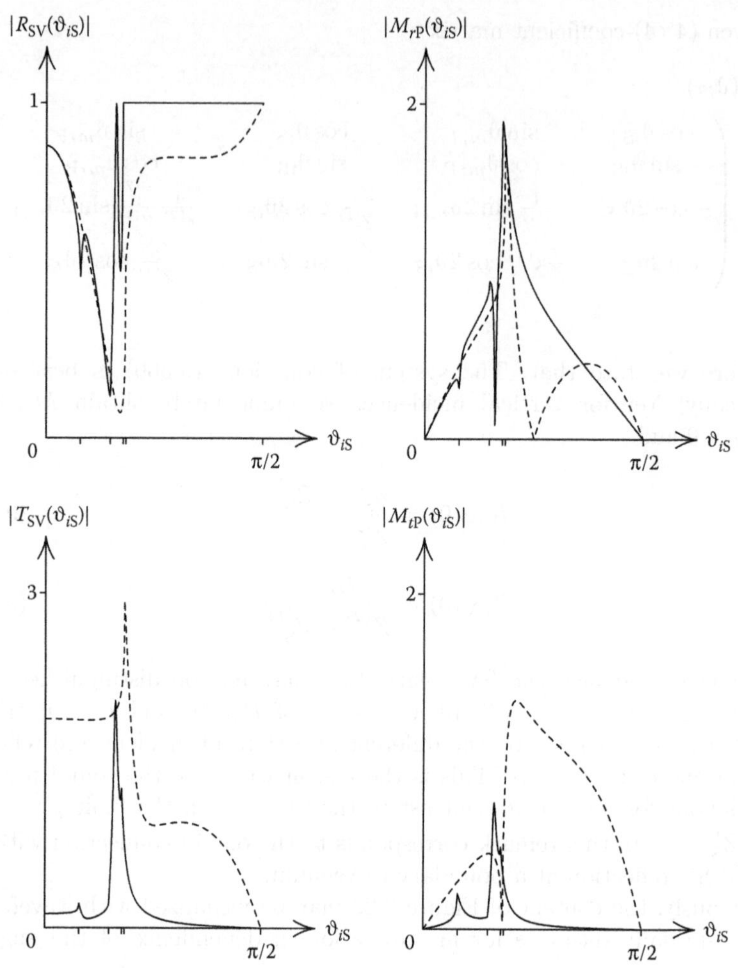

FIGURE 9.22
Magnitudes of reflection, transmission, and mode conversion coefficients for SV-wave incidence as function of the angle of incidence: steel(1)-plexiglas(2): - - -; plexiglas(1)-steel(2):—(Steel: $c_P = 5900$ m/s, $c_S = 3200$ m/s, $\rho = 7.7 \cdot 10^3$ kg/m^3; plexiglas: $c_P = 2730$ m/s, $c_S = 1430$ m/s, $\rho = 1.18 \cdot 10^3$ kg/m^3); the markers on the abscissae refer to critical angles in the following order: ϑ_{cm_tP}, ϑ_{ctS}, ϑ_{cm_rP} for—, ϑ_{cm_rP} for - - -.

is a generalization (annihilation of isotropic degeneracy) of evanescent (evanescent with regard to the phase propagation) of inhomogeneous plane waves as they occur at the boundary of an isotropic nondissipative half-space and at the boundary of two isotropic nondissipative half-spaces. Evanescence has basically to be understood with respect of the energy propagation (Langenberg and Marklein 2005), and only for coinciding phase and energy propagation evanescence refers to the phase surfaces.

- Simultaneous excitation of two plane SV-waves with different phase and energy velocities: We will discuss a relevant example for US-NDT of austenitic parts (Langenberg and Marklein 2005).

9.3.1 Inhomogeneous elastic plane waves in isotropic materials

As introduction, we will again explicitly prove (compare Section 8.2) that isotropic nondissipative materials exclusively support inhomogeneous plane waves that are evanescent with respect to the phase and energy propagation. Slowness diagrams constitute the physically intuitive geometric method to understand "Fresnel's" reflection of plane waves; therefore, we write the wave tensor (8.147) for isotropic materials in terms of slowness vectors $\underline{\mathbf{s}}$ instead of phase vectors $\underline{\mathbf{k}}$ as in Section 8.2:

$$\underline{\underline{\mathbf{W}}}(\underline{\mathbf{s}}) = \mu\, \underline{\mathbf{s}} \cdot \underline{\mathbf{s}}\, \underline{\underline{\mathbf{I}}} + (\lambda + \mu)\, \underline{\mathbf{s}}\, \underline{\mathbf{s}} - \rho\, \underline{\underline{\mathbf{I}}}; \tag{9.306}$$

with $\underline{\mathbf{s}} = s\,\hat{\underline{\mathbf{s}}}$—$s$ is per definition independent upon $\underline{\mathbf{s}}$ in isotropic materials—and a real valued unit vector $\hat{\underline{\mathbf{s}}} = \hat{\underline{\mathbf{k}}}$, we obtain the respective eigenvalue equation with (9.306)

$$\left[\mu\, \underline{\underline{\mathbf{I}}} + (\lambda + \mu)\, \hat{\underline{\mathbf{s}}}\, \hat{\underline{\mathbf{s}}} - \frac{\rho}{s^2}\, \underline{\underline{\mathbf{I}}}\right] \cdot \hat{\underline{\mathbf{u}}}(\underline{\mathbf{s}}) = \underline{\mathbf{0}} \tag{9.307}$$

as compared to (8.61) for the real eigenvalue ρ/s^2. Calculation of the determinant according to

$$\det \underline{\underline{\mathbf{W}}}(\hat{\underline{\mathbf{s}}}, s^2) = \left(\mu - \frac{\rho}{s^2}\right)^2 \left[\mu - \frac{\rho}{s^2} + (\lambda + \mu)\, \hat{\underline{\mathbf{s}}} \cdot \hat{\underline{\mathbf{s}}}\right] \tag{9.308}$$

yields

$$\underline{\mathbf{s}} \cdot \underline{\mathbf{s}} = s_{\mathrm{S}}^2 = \frac{\rho}{\mu} \tag{9.309}$$

on one hand and, on the other hand,

$$\underline{\mathbf{s}} \cdot \underline{\mathbf{s}} = s_{\mathrm{P}}^2 = \frac{\rho}{\lambda + 2\mu} \tag{9.310}$$

resulting in

$$s_{\mathrm{P}x}^2 + s_{\mathrm{P}z}^2 = s_{\mathrm{P}}^2 \tag{9.311}$$

or

$$s_{\mathrm{S}x}^2 + s_{\mathrm{S}z}^2 = s_{\mathrm{S}}^2 \tag{9.312}$$

for the real components s_x, s_z of

$$\underline{\mathbf{s}} = s_x\underline{\mathbf{e}}_x + s_z\underline{\mathbf{e}}_z. \tag{9.313}$$

If, for example, $s_{\mathrm{P,S}x}$-components are prescribed via phase matching the pertinent $s_{\mathrm{P,S}z}$-components are obtained as solutions

$$s_{\mathrm{P,S}z} = \pm\sqrt{s_{\mathrm{P,S}}^2 - s_{\mathrm{P,S}x}^2} \tag{9.314}$$

of the quadratic equations (9.311) and (9.312). Note: For $s_{\mathrm{P,S}x} < s_{\mathrm{P,S}}$, we find two real solutions $s_{\mathrm{P,S}z}$ characterizing homogeneous plane waves each; selection of one of the zeroes is tied to the physical requirement of a "meaningful" propagation of the respective homogeneous plane wave, i.e., "away" from a boundary for reflected/transmitted waves.

As we stated for the first time in Section 9.1.2 investigating the mode conversion SV$\Longrightarrow$P, the given x-component of the slowness vector may be too large to result in a real z-component; in that case, we obtain purely imaginary $s_{\mathrm{P,S}z}$-components

$$s_{\mathrm{P,S}z} = \pm\mathrm{j}\sqrt{s_{\mathrm{P,S}x}^2 - s_{\mathrm{P,S}}^2} \tag{9.315}$$

describing inhomogeneous evanescent plane waves. Again, sign selection is done with physical arguments, here: The inhomogeneous plane wave must exponentially decay in the z-half-space under concern (Section 9.1.2). It follows: A solution of the eigenvalue problem for elastic plane waves in isotropic nondissipative materials is possible for real slowness vector (9.313) and for complex slowness vector

$$\underline{\mathbf{s}} = \Re\underline{\mathbf{s}} + \mathrm{j}\Im\underline{\mathbf{s}} \tag{9.316}$$

with

$$\Re\underline{\mathbf{s}} = s_{\mathrm{P,S}x}\underline{\mathbf{e}}_x, \tag{9.317}$$

$$\Im\underline{\mathbf{s}} = \pm\sqrt{s_{\mathrm{P,S}x}^2 - s_{\mathrm{P,S}}^2}\,\underline{\mathbf{e}}_z, \tag{9.318}$$

the solutions $s_{\mathrm{P,S}z}$ of the quadratic equations (with real coefficients) (9.311) and (9.312) are either real or conjugate purely imaginary; the real part (9.317) and the imaginary part (9.318) of the slowness vector are orthogonal to each other. Now, we ask ourselves whether the eigenvalue equations (9.309) and (9.310) allow for complex solutions with nonorthogonal $\Re\underline{\mathbf{s}}$ and $\Im\underline{\mathbf{s}}$. In that case, we would have to assume $s_z = s_{\mathrm{R}z} + \mathrm{j}s_{\mathrm{I}z}$ with real part and, as a solution of a quadratic equation with real coefficients, with complex conjugate imaginary part; as a consequence, the representation $\Re\underline{\mathbf{s}} = s_x\underline{\mathbf{e}}_x + s_{\mathrm{R}z}\underline{\mathbf{e}}_z$ and $\Im\underline{\mathbf{s}} = s_{\mathrm{I}z}\underline{\mathbf{e}}_z$ of the slowness vector with nonorthogonal real and imaginary vectors would result, i.e., the emerging inhomogeneous plane wave would not be evanescent with regard to the phase propagation direction $\widehat{\Re\underline{\mathbf{s}}}$. We will immediately see that, as already shown in Section 8.2, this is not possible in isotropic nondissipative materials.

We calculate the determinant of (9.306) according to

$$\det \underline{\underline{\mathbf{W}}}(\underline{\mathbf{s}}) = (\mu\, \underline{\mathbf{s}} \cdot \underline{\mathbf{s}} - \rho)^2[(\lambda + 2\mu)\, \underline{\mathbf{s}} \cdot \underline{\mathbf{s}} - \rho] \tag{9.319}$$

and assume complex slowness vectors (9.316) to satisfy the equation

$$\det \underline{\underline{\mathbf{W}}}(\underline{\mathbf{s}}) = 0; \tag{9.320}$$

since both factors in (9.319) are similarly structured, we concentrate on the first one:

$$\mu(\Re\underline{\mathbf{s}} + \mathrm{j}\Im\underline{\mathbf{s}}) \cdot (\Re\underline{\mathbf{s}} + \mathrm{j}\Im\underline{\mathbf{s}}) - \rho = 0, \tag{9.321}$$

hence

$$\Re\underline{\mathbf{s}} \cdot \Re\underline{\mathbf{s}} - \Im\underline{\mathbf{s}} \cdot \Im\underline{\mathbf{s}} + 2\mathrm{j}\Re\underline{\mathbf{s}} \cdot \Im\underline{\mathbf{s}} = s_{\mathrm{S}}^2, \tag{9.322}$$

where s_{S} is real due to the material being nondissipative resulting in the real and imaginary part separation of (9.321)

$$\Re\underline{\mathbf{s}} \cdot \Re\underline{\mathbf{s}} - \Im\underline{\mathbf{s}} \cdot \Im\underline{\mathbf{s}} = s_{\mathrm{S}}^2, \tag{9.323}$$

$$\Re\underline{\mathbf{s}} \cdot \Im\underline{\mathbf{s}} = 0; \tag{9.324}$$

$\Im\underline{\mathbf{s}}$ ($\Re\underline{\mathbf{s}}$) must be orthogonal to $\Re\underline{\mathbf{s}}$ ($\Im\underline{\mathbf{s}}$) in isotropic nondissipative materials (Figure 8.11), inhomogeneous plane waves must be evanescent perpendicular to the phase propagation direction and the coinciding energy propagation direction in those materials, and this is exactly expressed with Equations (9.317) and (9.318): For given $\Re\underline{\mathbf{s}} = s_x\underline{\mathbf{e}}_x$, the imaginary part $\Im\underline{\mathbf{s}}$ may only have a z-component.

9.3.2 Inhomogeneous plane SH-waves in transversely isotropic materials

The comparatively simple example of plane SH-waves will show us that the requirement (9.324) emerges from the degeneracy of isotropic nondissipative materials concerning phase and energy velocities having the same direction; plane waves in anisotropic nondissipative materials are also evanescent, yet evanescent with regard to the energy velocity (Červený 2001) leading to eventually nonorthogonal phase $\Re\underline{\mathbf{s}}$ and attenuation vector $\Im\underline{\mathbf{s}}$. For SH-waves in transversely isotropic nondissipative materials, we can derive simple analytical expressions for $\det \underline{\underline{\mathbf{W}}}^{\mathrm{triso}}(\underline{\mathbf{s}})$ and $\underline{\mathbf{c}}_{\mathrm{ESH}}(\hat{\underline{\mathbf{s}}})$, allowing for an immediate verification of the preceding assertion; a general coordinate-free proof for arbitrary wave modes and arbitrary anisotropic materials is given by Langenberg and Marklein (2005).

Based on

$$\underline{\underline{\mathbf{W}}}^{\mathrm{triso}}(\underline{\mathbf{s}}) = \beta_1\, \underline{\underline{\mathbf{I}}} + \beta_2\, \underline{\mathbf{s}}\,\underline{\mathbf{s}} + \beta_3\, \hat{\underline{\mathbf{a}}}\,\hat{\underline{\mathbf{a}}} + \beta_4\, (\underline{\mathbf{s}}\,\hat{\underline{\mathbf{a}}} + \hat{\underline{\mathbf{a}}}\,\underline{\mathbf{s}}) \tag{9.325}$$

with—the α-expressions are defined by (8.286) through (8.288)—

$$\beta_1 = \mu_\perp \underline{\mathbf{s}} \cdot \underline{\mathbf{s}} + \alpha_3 (\underline{\mathbf{s}} \cdot \hat{\underline{\mathbf{a}}})^2 - \rho, \tag{9.326}$$

$$\beta_2 = \lambda_\perp + \mu_\perp, \tag{9.327}$$

$$\beta_3 = \alpha_1 (\underline{\mathbf{s}} \cdot \hat{\underline{\mathbf{a}}})^2 + \alpha_3 \underline{\mathbf{s}} \cdot \underline{\mathbf{s}}, \tag{9.328}$$

$$\beta_4 = (\alpha_2 + \alpha_3) \underline{\mathbf{s}} \cdot \hat{\underline{\mathbf{a}}}, \tag{9.329}$$

we calculate

$$\det \underline{\underline{\mathbf{W}}}^{\text{triso}}(\underline{\mathbf{s}}) = \beta_1 \Big[\beta_1^2 + \beta_1 (\beta_3 + \beta_2 \underline{\mathbf{s}} \cdot \underline{\mathbf{s}} + 2\beta_4 \underline{\mathbf{s}} \cdot \hat{\underline{\mathbf{a}}}) + (\beta_2 \beta_3 - \beta_4^2)(\underline{\underline{\mathbf{I}}} - \hat{\underline{\mathbf{a}}}\hat{\underline{\mathbf{a}}}) : \underline{\mathbf{s}}\underline{\mathbf{s}} \Big] \tag{9.330}$$

similar to (8.254). As mentioned before, we focus on the SH-case and investigate the equation $\beta_1 = 0$ with the *ansatz*

$$\underline{\mathbf{s}} = \Re \underline{\mathbf{s}} + \mathrm{j} \Im \underline{\mathbf{s}} \tag{9.331}$$

according to

$$\mu_\perp (\Re \underline{\mathbf{s}} \cdot \Re \underline{\mathbf{s}} - \Im \underline{\mathbf{s}} \cdot \Im \underline{\mathbf{s}} + 2\mathrm{j}\, \Re \underline{\mathbf{s}} \cdot \Im \underline{\mathbf{s}}) + \alpha_3 \left[(\Re \underline{\mathbf{s}} \cdot \hat{\underline{\mathbf{a}}})^2 - (\Im \underline{\mathbf{s}} \cdot \hat{\underline{\mathbf{a}}})^2 + 2\mathrm{j} (\Re \underline{\mathbf{s}} \cdot \hat{\underline{\mathbf{a}}})(\Im \underline{\mathbf{s}} \cdot \hat{\underline{\mathbf{a}}}) \right] - \rho = 0 \tag{9.332}$$

after its separation into real and imaginary parts:

$$\mu_\perp (\Re \underline{\mathbf{s}} \cdot \Re \underline{\mathbf{s}} - \Im \underline{\mathbf{s}} \cdot \Im \underline{\mathbf{s}}) + (\mu_\parallel - \mu_\perp) \left[(\Re \underline{\mathbf{s}} \cdot \hat{\underline{\mathbf{a}}})^2 - (\Im \underline{\mathbf{s}} \cdot \hat{\underline{\mathbf{a}}})^2 \right] - \rho = 0, \tag{9.333}$$

$$\mu_\perp \Re \underline{\mathbf{s}} \cdot \Im \underline{\mathbf{s}} + (\mu_\parallel - \mu_\perp) (\Re \underline{\mathbf{s}} \cdot \hat{\underline{\mathbf{a}}})(\Im \underline{\mathbf{s}} \cdot \hat{\underline{\mathbf{a}}}) = 0. \tag{9.334}$$

The isotropic case $\mu_\parallel = \mu_\perp$ again yields (9.323) and (9.324); the same is true for the phase propagation in the isotropy plane according to $\Re \underline{\mathbf{s}} \cdot \hat{\underline{\mathbf{a}}} = 0$.

The planar boundary—presently still virtual—between two elastic half-spaces is, as usual, identified with the xy-plane of a cartesian coordinate system. As $\underline{\mathbf{e}}_z \underline{\mathbf{s}}$-plane of incidence we choose the xz-plane; in the following, we confine ourselves to those $\hat{\underline{\mathbf{a}}}$-directions lying in the plane of incidence. With that definition of the xz-plane, we would generally obtain the component representation $\underline{\mathbf{s}} = (s_{\mathrm{R}x} + \mathrm{j} s_{\mathrm{I}x}) \underline{\mathbf{e}}_x + (s_{\mathrm{R}z} + \mathrm{j} s_{\mathrm{I}z}) \underline{\mathbf{e}}_z$ of the complex slowness vector according to the real and imaginary part separation $\Re \underline{\mathbf{s}} = s_{\mathrm{R}x} \underline{\mathbf{e}}_x + s_{\mathrm{R}z} \underline{\mathbf{e}}_z$, $\Im \underline{\mathbf{s}} = s_{\mathrm{I}x} \underline{\mathbf{e}}_x + s_{\mathrm{I}z} \underline{\mathbf{e}}_z$. Via phase matching in the xy-plane, the x-component of the slowness vector is prescribed real valued, thus coercively yielding the component separation

$$\begin{aligned} \underline{\mathbf{s}} &= s_x \underline{\mathbf{e}}_x + s_z \underline{\mathbf{e}}_z \\ &= s_{\mathrm{R}x} \underline{\mathbf{e}}_x + (s_{\mathrm{R}z} + \mathrm{j} s_{\mathrm{I}z}) \underline{\mathbf{e}}_z && (9.335) \\ &= \underbrace{(s_{\mathrm{R}x} \underline{\mathbf{e}}_x + s_{\mathrm{R}z} \underline{\mathbf{e}}_z)}_{= \Re \underline{\mathbf{s}}} + \mathrm{j} \underbrace{s_{\mathrm{I}z} \underline{\mathbf{e}}_z}_{= \Im \underline{\mathbf{s}}} && (9.336) \end{aligned}$$

of the slowness vector, i.e., $\Im\underline{\mathbf{s}}$ may only have a z-component. But this does not mean "evanescence with regard to phase propagation" because $\Re\underline{\mathbf{s}}$ may possess an x and a z-component if this is compatible with the imaginary part (9.334) of the SH-eigenvalue equation. Similar to the geometric interpretation of (9.324) (Figure 8.11), we can use (9.334) to construct $\Re\underline{\mathbf{s}}$ geometrically. We write (9.334) as

$$\Re\underline{\mathbf{s}} \cdot \left[\mu_\perp(\underline{\underline{\mathbf{I}}} - \hat{\underline{\mathbf{a}}}\,\hat{\underline{\mathbf{a}}}) + \mu_\parallel\,\hat{\underline{\mathbf{a}}}\,\hat{\underline{\mathbf{a}}}\right] \cdot \Im\underline{\mathbf{s}} = 0 \tag{9.337}$$

and insert $\Im\underline{\mathbf{s}} = s_{\mathrm{I}z}\underline{\mathbf{e}}_z \neq \underline{\mathbf{0}}$:

$$\Re\underline{\mathbf{s}} \cdot \left[\mu_\perp(\underline{\underline{\mathbf{I}}} - \hat{\underline{\mathbf{a}}}\,\hat{\underline{\mathbf{a}}}) + \mu_\parallel\,\hat{\underline{\mathbf{a}}}\,\hat{\underline{\mathbf{a}}}\right] \cdot \underline{\mathbf{e}}_z = 0. \tag{9.338}$$

For $\hat{\underline{\mathbf{a}}} = \underline{\mathbf{e}}_x$, we have $\Re\underline{\mathbf{s}} \cdot \underline{\mathbf{e}}_z = 0$ to obtain, as for the isotropic case, inhomogeneous plane SH-wave evanescence with regard to the phase propagation for such a preference direction. Figure 9.23 translates Equation 9.338 accordingly: With regard to the given preference direction $\hat{\underline{\mathbf{a}}}$ lying in the plane of incidence, we calculate the projection vector $\mu_\parallel\,\hat{\underline{\mathbf{a}}}\,\hat{\underline{\mathbf{a}}} \cdot \underline{\mathbf{e}}_z$ of $\underline{\mathbf{e}}_z$ in the direction of $\hat{\underline{\mathbf{a}}}$ and the projection vector $\mu_\perp(\underline{\underline{\mathbf{I}}} - \hat{\underline{\mathbf{a}}}\,\hat{\underline{\mathbf{a}}}) \cdot \underline{\mathbf{e}}_z$ orthogonal to $\hat{\underline{\mathbf{a}}}$; according to (9.338), $\Re\underline{\mathbf{s}}$ is orthogonal to the sum of these two projection vectors, where the (real) vectorial $s_{\mathrm{R}x}\underline{\mathbf{e}}_x$-component of $\Re\underline{\mathbf{s}}$ is prescribed per definition: If $\hat{\underline{\mathbf{a}}} \neq \underline{\mathbf{e}}_x$, the imaginary part (9.334) of the SH-eigenvalue equation for a particular transversely isotropic material allows for (slowness) phase vectors with an arbitrary material specific orientation with regard to the attenuation vector. This

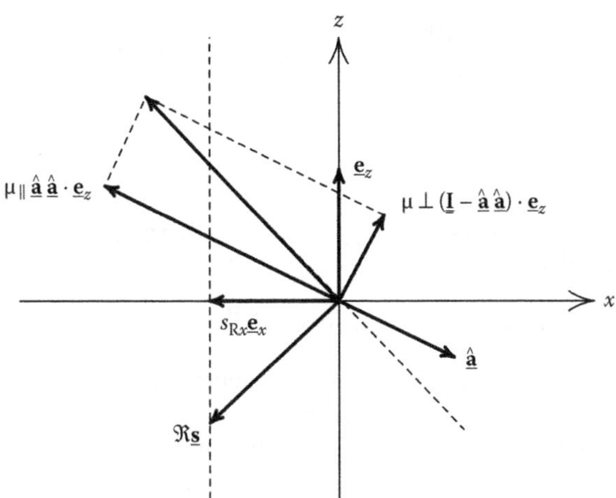

FIGURE 9.23
Geometric construction of the real part of the slowness vector for the SH-wave mode in a transversely isotropic nondissipative material with $\mu_\parallel > \mu_\perp$.

obviously complies for given real valued s_x with the possible complex valued solutions s_z of the quadratic eigenvalue equation

$$\mu_\perp(s_x^2 + s_z^2) + (\mu_\parallel - \mu_\perp)(s_x\hat{a}_x + s_z\hat{a}_z)^2 - \rho = 0 \tag{9.339}$$

with, due to the real coefficients, conjugate complex imaginary parts according to (9.335). The exclusive evanescence of inhomogeneous plane waves in the previous sense, namely with regard to the phase or slowness vector, evidently turns out to be a degeneracy of the isotropic material; we will see that in the general case evanescence has to be understood with regard to the energy velocity vector.

Using (9.334) according to

$$(s_{Rx}\underline{\mathbf{e}}_x + s_{Rz}\underline{\mathbf{e}}_z) \cdot \left[\mu_\perp\,(\underline{\underline{\mathbf{I}}} - \underline{\hat{\mathbf{a}}}\,\underline{\hat{\mathbf{a}}}) \cdot \underline{\mathbf{e}}_z + \mu_\parallel\,\underline{\hat{\mathbf{a}}}\,\underline{\hat{\mathbf{a}}} \cdot \underline{\mathbf{e}}_z\right] = 0, \tag{9.340}$$

we calculate the ratio of the z- and x-components of $\Re\underline{\mathbf{s}}$:

$$s_{Rz} = \underbrace{-\frac{(\mu_\parallel - \mu_\perp)\underline{\mathbf{e}}_x \cdot \underline{\hat{\mathbf{a}}}\,\underline{\hat{\mathbf{a}}} \cdot \underline{\mathbf{e}}_z}{\rho c_{SH}^2(\underline{\mathbf{e}}_z)}}_{=\sigma_R}\, s_{Rx}; \tag{9.341}$$

we state that σ_R is constant for a specific material, the direction of $\Re\underline{\mathbf{s}}$ remains constant with increasing s_{Rx} (as in the isotropic case). By the way: For $\mu_\parallel > \mu_\perp$, $\underline{\hat{\mathbf{a}}} \cdot \underline{\mathbf{e}}_z < 0$, $\underline{\hat{\mathbf{a}}} \cdot \underline{\mathbf{e}}_x > 0$ and $s_{Rx} < 0$, i.e., for the parameters of Figure 9.23, we have $s_{Rz} < 0$ as it corresponds to the figure. Accordingly, we find $s_{Rz} > 0$ if $\mu_\parallel$ is chosen smaller than $\mu_\perp$ for unmodified $\underline{\hat{\mathbf{a}}}$ and s_{Rx}; then $\Re\underline{\mathbf{s}}$ points into the half-space $z > 0$.

With (9.341), we find s_{Iz} from (9.332):

$$s_{Iz} = \pm\frac{s_{SH}(\underline{\mathbf{e}}_z)}{s_{SH}(\widehat{\Re\underline{\mathbf{s}}})}\sqrt{1+\sigma_R^2}\sqrt{s_{Rx}^2 - \frac{s_{SH}^2(\widehat{\Re\underline{\mathbf{s}}})}{1+\sigma_R^2}}; \tag{9.342}$$

the configuration of Figure 9.23 requests the choice of the negative sign. Via explicit insertion of $\widehat{\Re\underline{\mathbf{s}}}$ according to (9.369), Equation 9.342 adopts the form (9.400).

With $\Im\underline{\mathbf{s}} = s_{Iz}\underline{\mathbf{e}}_z$ and $\Re\underline{\mathbf{s}} = s_{Rx}\underline{\mathbf{e}}_x + s_{Rz}\underline{\mathbf{e}}_z$, where $s_{Rz} = \sigma_R s_{Rx}$, the inhomogeneous plane SH-wave

$$\underline{\mathbf{u}}(\underline{\mathbf{R}}, \omega, \underline{\mathbf{s}}) = -u(\omega)\, \mathrm{e}^{\mathrm{j}\omega\Re\underline{\mathbf{s}}\cdot\underline{\mathbf{R}}}\, \mathrm{e}^{\pm\omega\Im\underline{\mathbf{s}}\cdot\underline{\mathbf{R}}}\, \underline{\mathbf{e}}_y \tag{9.343}$$

is actually evanescent with regard to the energy propagation direction if we have chosen the physically meaningful sign for s_{Iz}. Namely, generalizing (8.332), we calculate for transversely isotropic materials

$$\begin{aligned}\underline{\mathbf{c}}_{ESH}(\underline{\mathbf{s}}) &= \frac{2\left[\mu_\perp\,\Re\underline{\mathbf{s}} + (\mu_\parallel - \mu_\perp)\,\Re\underline{\mathbf{s}} \cdot \underline{\hat{\mathbf{a}}}\,\underline{\hat{\mathbf{a}}}\right]}{\rho + \mu_\perp\underline{\mathbf{s}} \cdot \underline{\mathbf{s}}^* + (\mu_\parallel - \mu_\perp)\underline{\mathbf{s}}\,\underline{\mathbf{s}}^* : \underline{\hat{\mathbf{a}}}\,\underline{\hat{\mathbf{a}}}} \\ &= \frac{2\Re\underline{\mathbf{s}} \cdot \left[\mu_\perp(\underline{\underline{\mathbf{I}}} - \underline{\hat{\mathbf{a}}}\,\underline{\hat{\mathbf{a}}}) + \mu_\parallel\,\underline{\hat{\mathbf{a}}}\,\underline{\hat{\mathbf{a}}}\right]}{\rho + \mu_\perp\underline{\mathbf{s}} \cdot \underline{\mathbf{s}}^* + (\mu_\parallel - \mu_\perp)\underline{\mathbf{s}}\,\underline{\mathbf{s}}^* : \underline{\hat{\mathbf{a}}}\,\underline{\hat{\mathbf{a}}}}\end{aligned} \tag{9.344}$$

because we obtain the anisotropic generalization[191]

$$\underline{\mathbf{S}}_{\mathrm{K}}(\underline{\mathbf{R}}, \omega, \underline{\mathbf{s}}) = \frac{\omega^2}{2} |u(\omega)|^2 \,\mathrm{e}^{-2\omega\Im\underline{\mathbf{s}}\cdot\underline{\mathbf{R}}}\, \underline{\hat{\mathbf{u}}}(\underline{\mathbf{s}}) \cdot \underline{\underline{\underline{\underline{\mathbf{c}}}}} : \underline{\mathbf{s}}^*\underline{\hat{\mathbf{u}}}^*(\underline{\mathbf{s}}), \tag{9.345}$$

$$\langle w(\underline{\mathbf{R}}, t, \underline{\mathbf{s}})\rangle = \frac{\omega^2\rho}{4} |u(\omega)|^2 \,\mathrm{e}^{-2\omega\Im\underline{\mathbf{s}}\cdot\underline{\mathbf{R}}} + \frac{\omega^2}{4} |u(\omega)|^2 \,\mathrm{e}^{-2\omega\Im\underline{\mathbf{s}}\cdot\underline{\mathbf{R}}}\, \underline{\mathbf{s}}\,\underline{\hat{\mathbf{u}}}(\underline{\mathbf{s}}) : \underline{\underline{\underline{\underline{\mathbf{c}}}}} : \underline{\mathbf{s}}^*\underline{\hat{\mathbf{u}}}^*(\underline{\mathbf{s}}) \tag{9.346}$$

of (8.182) and (8.183) for complex slowness vectors with

$$\underline{\mathbf{u}}(\underline{\mathbf{R}}, \omega, \underline{\mathbf{s}}) = u(\omega)\,\mathrm{e}^{\mathrm{j}\omega\underline{\mathbf{s}}\cdot\underline{\mathbf{R}}}\,\underline{\hat{\mathbf{u}}}(\underline{\mathbf{s}}) \tag{9.347}$$

and

$$\underline{\hat{\mathbf{u}}}(\underline{\mathbf{s}}) \cdot \underline{\hat{\mathbf{u}}}^*(\underline{\mathbf{s}}) = 1; \tag{9.348}$$

it follows

$$\underline{\mathbf{c}}_{\mathrm{E}}(\underline{\mathbf{s}}) = \frac{2\Re\left\{\underline{\underline{\underline{\underline{\mathbf{c}}}}} \,\vdots\, \underline{\mathbf{s}}^*\underline{\hat{\mathbf{u}}}^*(\underline{\mathbf{s}})\underline{\hat{\mathbf{u}}}(\underline{\mathbf{s}})\right\}}{\rho + \underline{\mathbf{s}}\,\underline{\hat{\mathbf{u}}}(\underline{\mathbf{s}}) : \underline{\underline{\underline{\underline{\mathbf{c}}}}} : \underline{\mathbf{s}}^*\underline{\hat{\mathbf{u}}}^*(\underline{\mathbf{s}})}. \tag{9.349}$$

With (4.24) and recognizing $\underline{\mathbf{s}}^* \cdot \underline{\hat{\mathbf{u}}}_{\mathrm{SH}} = 0$, $\hat{\underline{\mathbf{a}}} \cdot \underline{\hat{\mathbf{u}}}_{\mathrm{SH}} = 0$, $\underline{\hat{\mathbf{u}}}_{\mathrm{SH}} \cdot \underline{\hat{\mathbf{u}}}^*_{\mathrm{SH}} = 1$ (Footnote 196), we find

$$\underline{\underline{\underline{\underline{\mathbf{c}}}}}^{\mathrm{triso}} \,\vdots\, \underline{\mathbf{s}}^*\underline{\hat{\mathbf{u}}}^*_{\mathrm{SH}}\underline{\hat{\mathbf{u}}}_{\mathrm{SH}} = \underline{\mathbf{s}}^* \cdot \left[\mu_\perp \underline{\underline{\mathbf{I}}} + (\mu_\parallel - \mu_\perp)\hat{\underline{\mathbf{a}}}\,\hat{\underline{\mathbf{a}}}\right] \tag{9.350}$$

and (9.344), respectively, and due to (9.340), we apparently have

$$\underline{\mathbf{c}}_{\mathrm{ESH}}(\underline{\mathbf{s}}) \cdot \underline{\mathbf{e}}_z = 0. \tag{9.351}$$

This fact is not noticeable for the isotropic nondissipative material because phase and energy velocity vectors have the same direction.

9.3.3 Reflection and transmission of plane SH-Waves at the planar boundary between homogeneous isotropic and homogeneous transversely isotropic nondissipative materials

We have already multiply used the idea of a given x-component of a real or complex slowness vector via phase matching at material discontinuity

---

[191]The second term in (9.346) may no longer be converted with the Kelvin–Christoffel equation

$$(\underline{\mathbf{s}} \cdot \underline{\underline{\underline{\underline{\mathbf{c}}}}} \cdot \underline{\mathbf{s}} - \rho\,\underline{\underline{\mathbf{I}}}) \cdot \underline{\hat{\mathbf{u}}}(\underline{\mathbf{s}}) = \underline{\mathbf{0}},$$

thus yielding the inequality of the time averaged kinetic energy density and the time averaged potential energy density for inhomogeneous plane waves as it is true for the isotropic nondissipative material.

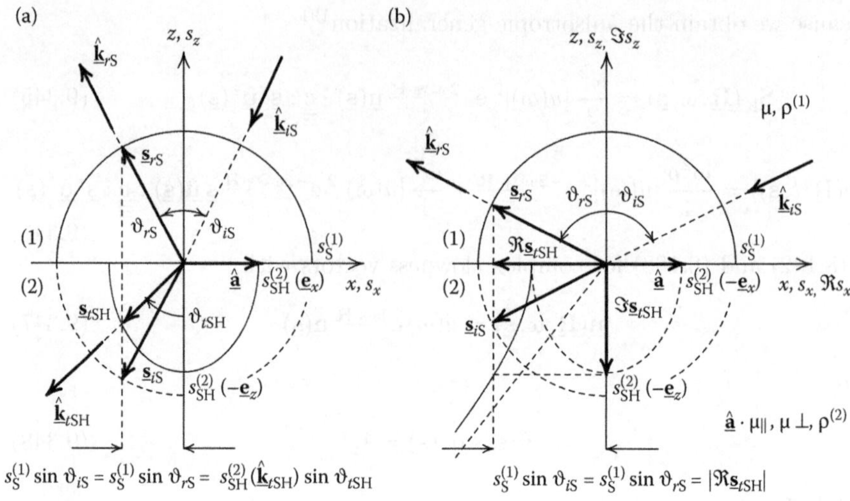

FIGURE 9.24
Slowness diagram for reflection and transmission of an SH-wave at the boundary of a homogeneous isotropic nondissipative and a homogeneous transversely isotropic nondissipative elastic half-space: $\underline{\hat{\mathbf{a}}} = \underline{\mathbf{e}}_x$, $\mu_{\|} > \mu_{\perp}$, $c_{\mathrm{S}}^{(1)} < c_{\mathrm{SH}}^{(2)}(\underline{\mathbf{e}}_x)$; (a) $\vartheta_{i\mathrm{S}} \leq \vartheta_{ct\mathrm{SH}}$ and (b) $\vartheta_{i\mathrm{S}} > \vartheta_{ct\mathrm{SH}}$.

boundaries. Consequently, we should illustrate the conclusions of the previous subsection with the help of slowness diagrams. First, we confine ourselves to the simplest case of a boundary between a homogeneous isotropic nondissipative ($z > 0: \ \mu, \rho^{(1)}$) and a homogeneous transversely isotropic nondissipative half-space ($z < 0: \ \underline{\hat{\mathbf{a}}}, \mu_{\|}, \mu_{\perp}, \rho^{(2)}$) for SH-wave incidence from the isotropic half-space.

The assumption "$\underline{\hat{\mathbf{a}}}$ lies in the plane of incidence" yields $-\underline{\mathbf{e}}_y$ as polarization vector of SH-waves in both half-spaces conveying physical intuition to the vanishing mode conversion into qP- and/or qSV-waves: In this case, SH-waves are decoupled from qP- and qSV-waves.

Preference direction $\underline{\hat{\mathbf{a}}}$ in the boundary: With Figure 9.24(a) and (b); we consign Figure 9.14(a) and (b) to the case as defined above: We have attributed the SH-slowness diagram for fiber reinforced composite with $\underline{\hat{\mathbf{a}}} = \underline{\mathbf{e}}_x$ (Figure 8.21) to the transversely isotropic half-space $z < 0$. Having directly selected the more interesting case $s^{(1)} > s^{(2)}(\underline{\mathbf{e}}_x)$ yielding the existence of a critical angle $\vartheta_{ct\mathrm{SH}}$, we distinguish between $\vartheta_{i\mathrm{S}} \leq \vartheta_{ct\mathrm{SH}}$ [Figure 9.24(a)] and $\vartheta_{i\mathrm{S}} > \vartheta_{ct\mathrm{SH}}$ [Figure 9.24(b)].

We start with $\vartheta_{i\mathrm{S}} \leq \vartheta_{t\mathrm{SH}}$. Phase matching of incident, reflected, and mode converted plane SH-waves [compare (9.177) through (9.179)]

$$\underline{\mathbf{u}}_{i\mathrm{S}}(\underline{\mathbf{R}}, \omega, \hat{\underline{\mathbf{k}}}_{i\mathrm{S}}) = -u_{i\mathrm{S}}(\omega)\, \mathrm{e}^{\mathrm{j}\omega s_{\mathrm{S}}^{(1)} \hat{\underline{\mathbf{k}}}_{i\mathrm{S}} \cdot \underline{\mathbf{R}}}\, \underline{\mathbf{e}}_y, \quad z \geq 0, \tag{9.352}$$

$$\underline{\mathbf{u}}_{rS}(\underline{\mathbf{R}}, \omega, \hat{\underline{\mathbf{k}}}_{rS}) = -R_{\mathrm{SH}}^{\mathrm{triso}}(\vartheta_{iS})\, u_{iS}(\omega)\, \mathrm{e}^{\mathrm{j}\omega s_{\mathrm{S}}^{(1)} \hat{\underline{\mathbf{k}}}_{rS} \cdot \underline{\mathbf{R}}}\, \underline{\mathbf{e}}_y, \quad z \geq 0, \tag{9.353}$$

$$\underline{\mathbf{u}}_{t\mathrm{SH}}(\underline{\mathbf{R}}, \omega, \hat{\underline{\mathbf{k}}}_{t\mathrm{SH}}) = -T_{\mathrm{SH}}^{\mathrm{triso}}(\vartheta_{iS})\, u_{iS}(\omega)\, \mathrm{e}^{\mathrm{j}\omega s_{\mathrm{SH}}^{(2)}(\hat{\underline{\mathbf{k}}}_{t\mathrm{SH}}) \hat{\underline{\mathbf{k}}}_{t\mathrm{SH}} \cdot \underline{\mathbf{R}}}\, \underline{\mathbf{e}}_y, \quad z \leq 0, \tag{9.354}$$

for all x, y in the boundary $z = 0$ results in the reflection and transmission laws

$$\vartheta_{rS} = \vartheta_{iS}, \tag{9.355}$$

$$s_{\mathrm{SH}}^{(2)}(\hat{\underline{\mathbf{k}}}_{t\mathrm{SH}}) \sin\vartheta_{t\mathrm{SH}} = s_{\mathrm{S}}^{(1)} \sin\vartheta_{iS}, \tag{9.356}$$

respectively, via the chain of equations

$$s_{\mathrm{S}}^{(1)} \sin\vartheta_{iS} = s_{\mathrm{S}}^{(1)} \sin\vartheta_{rS} = s_{\mathrm{SH}}^{(2)}(\hat{\underline{\mathbf{k}}}_{tS}) \sin\vartheta_{t\mathrm{SH}}. \tag{9.357}$$

Note: The transmission law (9.356) comes as an implicit equation for the transmission angle because

$$\hat{\underline{\mathbf{k}}}_{t\mathrm{SH}} = -\sin\vartheta_{t\mathrm{SH}}\, \underline{\mathbf{e}}_x - \cos\vartheta_{t\mathrm{SH}}\, \underline{\mathbf{e}}_z \tag{9.358}$$

tells us that the slowness $s_{\mathrm{SH}}^{(2)}(\hat{\underline{\mathbf{k}}}_{t\mathrm{SH}})$ also depends upon $\vartheta_{t\mathrm{SH}}$; we must visualize the explicit dependence of ϑ_{iS} utilizing the eigenvalue equation; this is often successful—for the SH-wave mode even for $\hat{\underline{\mathbf{a}}} \neq \underline{\mathbf{e}}_x$, and for qP- and qSV-wave modes at least for $\hat{\underline{\mathbf{a}}} \cdot \underline{\mathbf{e}}_z = 0$—because in these cases, only quadratic equations have to be solved; nevertheless, the paper work can be considerable. Here, for the SH-case, we will explain the procedure in detail.

Figure 9.24(a) reveals that a critical angle $\vartheta_{ct\mathrm{SH}}$ exists for $s_{\mathrm{S}}^{(1)} > s_{\mathrm{SH}}^{(2)}(-\underline{\mathbf{e}}_x)$; if $\vartheta_{iS} = \vartheta_{ct\mathrm{SH}}$, we obviously have $\vartheta_{t\mathrm{SH}} = \pi/2$, and therefore $\hat{\underline{\mathbf{k}}}_{t\mathrm{SH}} = -\underline{\mathbf{e}}_x$; we obtain

$$\begin{aligned} \sin\vartheta_{ct\mathrm{SH}} &= \frac{s_{\mathrm{SH}}^{(2)}(-\underline{\mathbf{e}}_x)}{s_{\mathrm{S}}^{(1)}} \\ &= \frac{s_{\mathrm{SH}}^{(2)}(\hat{\underline{\mathbf{a}}})}{s_{\mathrm{S}}^{(1)}} \\ &= \sqrt{\frac{\rho^{(2)}\, \mu}{\rho^{(1)}\, \mu_{\|}}}. \end{aligned} \tag{9.359}$$

For $\vartheta_{iS} < \vartheta_{ct\mathrm{SH}}$, the SH-eigenvalue equation

$$\mu_{\perp}\, s^2 + (\mu_{\|} - \mu_{\perp})\, (\underline{\mathbf{s}} \cdot \hat{\underline{\mathbf{a}}})^2 - \rho^{(2)} = 0 \tag{9.360}$$

with real slowness vector

$$\underline{\mathbf{s}} \Longrightarrow \underline{\mathbf{s}}_{t\mathrm{SH}} = s_{t\mathrm{SH}x}\, \underline{\mathbf{e}}_x + s_{t\mathrm{SH}z}\, \underline{\mathbf{e}}_z \tag{9.361}$$

and given

$$s_{t\mathrm{SH}x} = s_{\mathrm{S}}^{(1)} \sin\vartheta_{iS} \tag{9.362}$$

possesses two real zeroes s_{tSHz} that can immediately be found[192] from the eigenvalue equation (9.360) due to $\underline{\mathbf{s}}_{tSH} \cdot \hat{\underline{\mathbf{a}}} = s_{tSHx}$—we have $\hat{\underline{\mathbf{a}}} = \underline{\mathbf{e}}_x$—and $s_{tSHx}^2 + s_{tSHz}^2 = s_{\mathrm{SH}}^{(2)^2}$:

$$s_{tSHz} = \pm\sqrt{\frac{\mu_\parallel}{\mu_\perp}}\sqrt{\frac{\rho^{(2)}}{\mu_\parallel} - s_{\mathrm{S}}^{(1)^2}\sin^2\vartheta_{i\mathrm{S}}}. \tag{9.363}$$

The selection of the "correct" z-component of the slowness vector is performed with physical arguments: For transmitted waves in the half-space $z < 0$, only the negative sign is useful. Yet, we would only have to interchange the two half-spaces including the wave incidence for a selection of the other sign. With both components (9.362) and (9.363)—or, in the case $\hat{\underline{\mathbf{a}}} = \underline{\mathbf{e}}_x$, directly with (9.360) and (9.362)—$s_{\mathrm{SH}}^{(2)}(\hat{\underline{\mathbf{k}}}_{tSH})$ is computable from the transmission law as a function of the angle of incidence according to

$$s_{\mathrm{SH}}^{(2)}(\hat{\underline{\mathbf{k}}}_{tSH}) = \sqrt{\frac{\rho^{(2)} - (\mu_\parallel - \mu_\perp)\, s_{\mathrm{S}}^{(1)^2}\sin^2\vartheta_{i\mathrm{S}}}{\mu_\perp}}, \tag{9.364}$$

and hence

$$\sin\vartheta_{tSH} = s_{\mathrm{S}}^{(1)}\sin\vartheta_{i\mathrm{S}}\sqrt{\frac{\mu_\perp}{\rho^{(2)} - (\mu_\parallel - \mu_\perp)s_{\mathrm{S}}^{(1)^2}\sin^2\vartheta_{i\mathrm{S}}}}. \tag{9.365}$$

For $\vartheta_{i\mathrm{S}} > \vartheta_{tSH}$, we expect transmitted inhomogeneous plane waves as solutions of the SH-eigenvalue equation (9.360) with

$$\Re\underline{\mathbf{s}}_{tSH} = s_{\mathrm{R}tSHx}\,\underline{\mathbf{e}}_x + s_{\mathrm{R}tSHz}\,\underline{\mathbf{e}}_z, \tag{9.366}$$

$$\Im\underline{\mathbf{s}}_{tSH} = s_{\mathrm{I}tSHz}\,\underline{\mathbf{e}}_z, \tag{9.367}$$

where according to (9.338), we immediately obtain $s_{\mathrm{R}tSHz} = 0$ ($\sigma_{\mathrm{R}} = 0$ in (9.341)) for $\hat{\underline{\mathbf{a}}} = \underline{\mathbf{e}}_x$; with (9.362), we derive

$$\begin{aligned} s_{\mathrm{I}tSHz} &= \pm\frac{s_{\mathrm{SH}}^{(2)}(\underline{\mathbf{e}}_z)}{s_{\mathrm{SH}}^{(2)}(-\underline{\mathbf{e}}_x)}\sqrt{s_{\mathrm{R}tSHx}^2 - s_{\mathrm{SH}}^{(2)^2}(-\underline{\mathbf{e}}_x)} \\ &= \pm\sqrt{\frac{\mu_\parallel}{\mu_\perp}}\sqrt{s_{\mathrm{S}}^{(1)^2}\sin^2\vartheta_{i\mathrm{S}} - \frac{\rho^{(2)}}{\mu_\parallel}} \end{aligned} \tag{9.368}$$

from Equation 9.342. As in Figure 9.14, this is a hyperbola as function of $s_{\mathrm{S}}^{(1)}\sin\vartheta_{i\mathrm{S}}$, yet with an asymptote given by the ratio $\sqrt{\mu_\parallel/\mu_\perp}$ [Figure 9.24(b)]; it is understood that the negative sign is relevant for the half-space $z < 0$ (and the time dependence $\mathrm{e}^{-\mathrm{j}\omega t}$).

[192]For $\hat{\underline{\mathbf{a}}} = \underline{\mathbf{e}}_x$ or, even more general, $\hat{\underline{\mathbf{a}}}$ in the xy-plane, we always find s_z, i.e., even for qP- and qSV-wave modes, in terms of this simple explicit form (Spies 1994) because the qP- and qSV-eigenvalue equations also only contain $\underline{\mathbf{s}} \cdot \hat{\underline{\mathbf{a}}}$ [compare (9.330)]; yet we have to solve a quadratic equation in s^2.

Preference direction $\underline{\hat{\mathbf{a}}}$ inclined to the boundary: With Figures 9.25 and 9.26, we turn to the case of a preference direction of the transversely isotropic half-space inclined to the boundary. We will see that, despite the "harmless" ellipticity of the SH-slowness diagram, a rich variety of possibilities exists. Yet, we continue to assume $\underline{\hat{\mathbf{a}}} \cdot \underline{\mathbf{e}}_y = 0$ to assure decoupling of the SH-wave modes from the qP- and qSV-wave modes.

Based on Figure 9.25—the pertinent equations are discussed in the next paragraph—we first discuss the orientation of $\underline{\hat{\mathbf{a}}}$ "toward the bottom," i.e., we assume $\underline{\hat{\mathbf{a}}} \cdot \underline{\mathbf{e}}_z < 0$, $0 < \underline{\hat{\mathbf{a}}} \cdot \underline{\mathbf{e}}_x < 1$ resulting in a rotation of the slowness diagram with respect to the boundary with the consequence of an asymmetry with regard to the wave incidence either from the "left" or from the "right"; with Figure 9.25, we particularly assume a negative x-component of $\underline{\mathbf{s}}_{i\mathrm{S}}$ as it is true for Figure 9.24. The dashed phase matching lines in Figure 9.25(a) separate the SH-slowness diagram for $s_x \leq 0$ into three regions: region I containing one real zero $s_z < 0$ of the SH-eigenvalue equation, region II (in that case very narrow) containing two real zeroes $s_z < 0$, and region III containing two complex conjugate s_z-zeroes. With Figures 9.25(b) and (c), we select angles of incidence with $s_{i\mathrm{S}x}$ in region I: Obviously, for either one, we obtain one intersection of the respective phase matching lines with the slowness diagram in the lower half-space leading to a negative z-component of the slowness vector $\underline{\mathbf{s}}_{t\mathrm{SH}}$ of the transmitted SH-wave in both cases, i.e., phase propagation is away from the boundary as we are used to. Yet, this is not the mathematical criterion for a physically meaningful solution $s_{t\mathrm{SH}z}$ of the eigenvalue equation: It depends on whether the energy flux given by the energy velocity vector $\underline{\mathbf{c}}_{\mathrm{E}t\mathrm{SH}}(\hat{\underline{\mathbf{s}}}_{t\mathrm{SH}})$ is directed away from the boundary. With a negative z-component, this is true for both cases—in fact, the energy velocity is perpendicular to the slowness surface—nevertheless, we face a surprise: Even though the incident SH-wave coming from the isotropic half-space is inclined to the boundary the energy flux in the transversely isotropic half-space is perpendicular to the boundary [Figure 9.25(c)] or even "backwards" [Figure 9.25(b)]. This opens the door for wrong assessments of ultrasonic signals not taking into account the potential anisotropy of a material. Yet from a physical point of view, such a behavior can be easily understood: For example, vertical incidence yields a bending of energy propagation toward $\underline{\hat{\mathbf{a}}}$, because for $\mu_{\parallel} > \mu_{\perp}$, we observe the largest energy velocity in the inclined $\underline{\hat{\mathbf{a}}}$-direction (Figure 8.22), the vertically incident waves "feel" an appropriate component perpendicular to the boundary and "except this offer."

With Figure 9.25(d), we turn to an example for region II: An appropriately chosen $s_{i\mathrm{S}x}$-component obviously leads to two different negative real $s_{t\mathrm{SH}z}$-solutions of the eigenvalue equation (9.339), yet the one with smaller magnitude must be excluded for a physical reason: For such a slowness vector, the energy velocity has a positive z-component; physical arguments yield uniqueness of the mathematically nonunique wave propagation![193] With increasing

[193]Therefore, the inverted direction of the energy velocity is the real reason to exclude ($s_z > 0$)-solutions in region I.

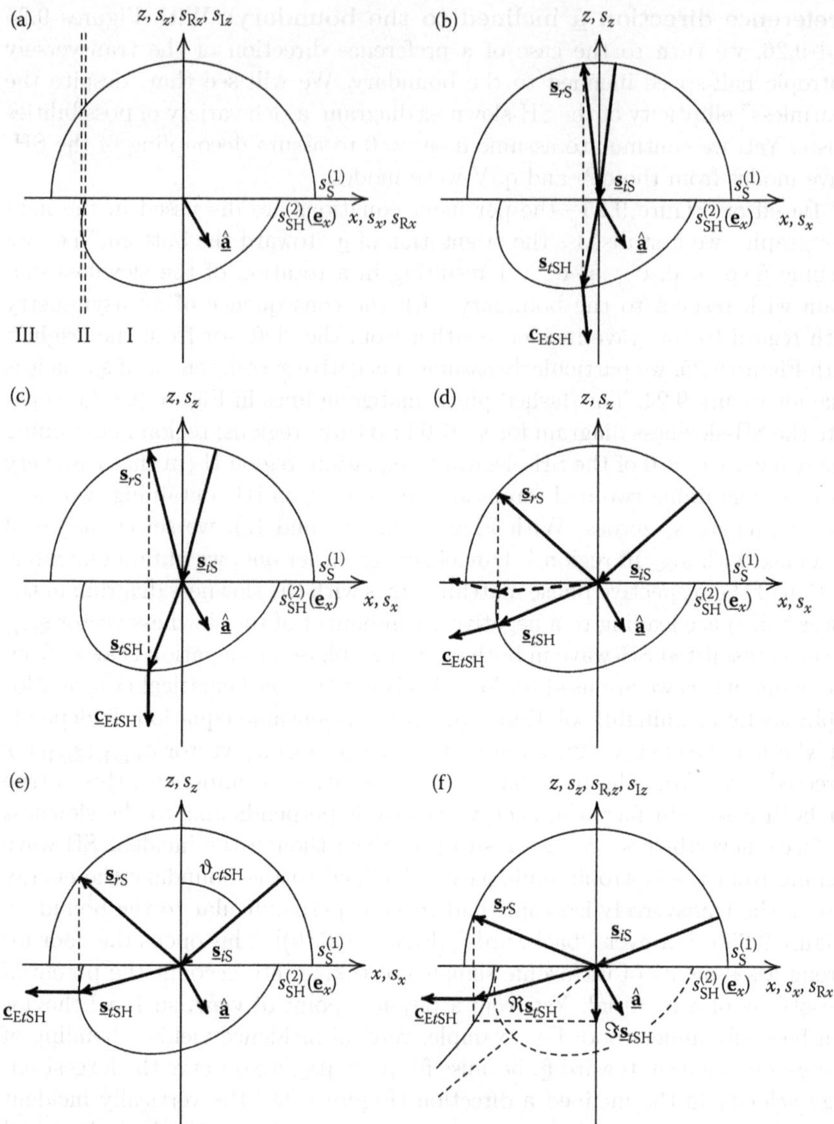

FIGURE 9.25
Slowness diagrams for reflection and transmission of an SH-wave at the boundary of a homogeneous isotropic nondissipative and a homogeneous transversely isotropic nondissipative elastic half-space for increasing angle of incidence (b)–(f); (a) defines different regions for slowness components through phase matching lines: $0 < \underline{\hat{\mathbf{a}}} \cdot \underline{\mathbf{e}}_x < 1$, $\underline{\hat{\mathbf{a}}} \cdot \underline{\mathbf{e}}_z < 0$, $\mu_{\|} > \mu_{\perp}$, $c_{\mathrm{S}}^{(1)} < c_{\mathrm{SH}}^{(2)}(-\underline{\mathbf{e}}_x)$.

magnitude of the s_{iSx}-component, the two possible slowness vectors in region II tend to coincide to one with an energy velocity having a (negative) x-component, i.e., the energy flux is parallel to the boundary [Figure 9.25(e)]. The corresponding angle of incidence is the critical angle ϑ_{ctSH}, and we expect evanescence of this energy flux for angles of incidence larger than ϑ_{ctSH} due to the complex conjugate s_{tSHz}-zeroes of the SH-eigenvalue equation having chosen the appropriate sign of the imaginary part. As a matter of fact, in Figure 9.25(f), we construct the imaginary part $\Im \underline{\mathbf{s}}_{tSH} = s_{ItSHz}\,\underline{\mathbf{e}}_z$ of the slowness vector as (energy) attenuation vector for s_{iSx} located in region III with the negative branch of the hyperbola; for the real part vector, we obtain

$$\Re \underline{\mathbf{s}}_{tSH} = -s_{\mathrm{S}}^{(1)} \sin\vartheta_{i\mathrm{S}}(\underline{\mathbf{e}}_x + \sigma_{\mathrm{R}}\,\underline{\mathbf{e}}_z) \tag{9.369}$$

with (9.341); its direction remains unchanged as compared to the "critical slowness vector" of Figure 9.25, and this is also true for the energy velocity due to (9.351).

With Figure 9.26, we discuss the orientation of the preference direction $\hat{\underline{\mathbf{a}}}$ "to the top," i.e., we assume $\hat{\underline{\mathbf{a}}} \cdot \underline{\mathbf{e}}_z > 0$, and still $0 < \hat{\underline{\mathbf{a}}} \cdot \underline{\mathbf{e}}_x < 1$. Hence, the slowness diagram is rotated into the opposite direction as compared to Figure 9.25, and to be on the safer side, we have plotted the whole 360°-diagram; the same phase matching lines as in Figure 9.25(a) equally define regions I, II, and III for different characteristic values of $-s_{\mathrm{S}}^{(1)} \sin\vartheta_{i\mathrm{S}}$. The single pictures of Figure 9.26 corresponding to the pictures 9.25(b)–(d) have been omitted because the respective results can be easily imagined: For example, vertical incidence leads to a negative x-component of the energy velocity of the

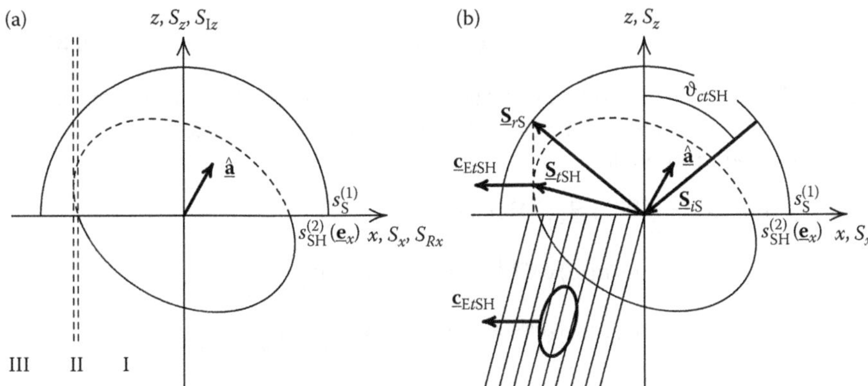

FIGURE 9.26
Slowness diagrams for reflection and transmission of an SH-wave at the boundary between a homogeneous isotopic nondissipative and a homogeneous transversely isotropic nondissipative elastic half-space; (a) defines different regions for slowness components through phase matching lines; (b) energy evanescent transmitted waves: $0 < \hat{\underline{\mathbf{a}}} \cdot \underline{\mathbf{e}}_x < 1$, $\hat{\underline{\mathbf{a}}} \cdot \underline{\mathbf{e}}_z > 0$, $\mu_{\|} > \mu_{\perp}$, $c_{\mathrm{S}}^{(1)} < c_{\mathrm{SH}}^{(2)}(-\underline{\mathbf{e}}_x)$.

transmitted wave, and the same is true for all incidence angles $\vartheta_{i\mathrm{S}} > 0$ with $-s_\mathrm{S}^{(1)} \sin\vartheta_{i\mathrm{S}}$ located in region I. Entering region II with the phase x-component of the incident wave, we again obtain two real s_z-solutions of the eigenvalue equation, yet in contrast to region I, both are located on the dashed part of the slowness diagram for $z > 0$ [compare Figure 9.26(a)]. Consequently, it seems that both solutions are irrelevant or transmitted waves because the pertinent slowness vectors point away from the boundary; yet, this holds for the phase propagation, and the energy propagation corresponding to the solution with the smaller magnitude is directed into the lower half-space $z < 0$, representing a physically meaningful transmitted wave. The coincidence of both solutions marks the boundary of region III; even though the resulting slowness vector also points into the half-space $z > 0$, we encounter an energy velocity vector parallel to the boundary indicating the transition to the energy evanescent transmitted waves of region III [Figure 9.26(b)]. This wave mode is sketched as a wave packet in Figure 9.26(b) (compare Figure 8.14 and Equation 8.230): Hence, imagining a transmitted wave with phase propagation into the upper half-space and energy propagation into the lower half-space or parallel to the boundary does not cause any problems.

Reflection and transmission coefficients: We write the deformation tensor (8.212) of a plane elastic wave in terms of an explicit visibility of the slowness vector:

$$\underline{\underline{\mathbf{S}}}(\underline{\mathbf{R}}, \omega, \underline{\mathbf{s}}) = \mathrm{j}\omega\, u(\omega)\, \mathrm{e}^{\mathrm{j}\omega \underline{\mathbf{s}}(\hat{\underline{\mathbf{k}}})\cdot\underline{\mathbf{R}}}\, \frac{1}{2}\left[\underline{\mathbf{s}}(\hat{\underline{\mathbf{k}}})\,\hat{\underline{\mathbf{u}}}(\hat{\underline{\mathbf{k}}}) + \hat{\underline{\mathbf{u}}}(\hat{\underline{\mathbf{k}}})\,\underline{\mathbf{s}}(\hat{\underline{\mathbf{k}}})\right]; \tag{9.370}$$

here, the phase propagation (unit) vector $\hat{\underline{\mathbf{k}}}$ according to

$$\hat{\underline{\mathbf{k}}} = \widehat{\Re\underline{\mathbf{s}}} \tag{9.371}$$

is given by the unit vector $\widehat{\Re\underline{\mathbf{s}}}$ of the real part vector $\Re\underline{\mathbf{s}}$ of the eventually complex slowness vector $\underline{\mathbf{s}}$. For the SH-wave mode, we have $\hat{\underline{\mathbf{u}}}(\hat{\underline{\mathbf{k}}}) = -\underline{\mathbf{e}}_y$, hence (8.289) immediately yields

$$\begin{aligned}
\underline{\underline{\mathbf{T}}}_\mathrm{SH}(\underline{\mathbf{R}}, \omega, \underline{\mathbf{s}}) &= \underline{\underline{\underline{\underline{\mathbf{c}}}}}^\mathrm{triso} : \underline{\underline{\mathbf{S}}}_\mathrm{SH}(\underline{\mathbf{R}}, \omega, \underline{\mathbf{s}}) \\
&= -\mathrm{j}\omega\, u(\omega)\, \mathrm{e}^{\mathrm{j}\omega \underline{\mathbf{s}}(\hat{\underline{\mathbf{k}}})\cdot\underline{\mathbf{R}}}\, \underline{\underline{\underline{\underline{\mathbf{c}}}}}^\mathrm{triso} : \underline{\mathbf{s}}(\hat{\underline{\mathbf{k}}})\,\underline{\mathbf{e}}_y \\
&= -\mathrm{j}\omega\, u(\omega)\, \mathrm{e}^{\mathrm{j}\omega \underline{\mathbf{s}}(\hat{\underline{\mathbf{k}}})\cdot\underline{\mathbf{R}}} \Big\{ \mu_\perp \left[\underline{\mathbf{s}}(\hat{\underline{\mathbf{k}}})\,\underline{\mathbf{e}}_y + \underline{\mathbf{e}}_y\,\underline{\mathbf{s}}(\hat{\underline{\mathbf{k}}})\right] \\
&\qquad + (\mu_\| - \mu_\perp)\,\underline{\mathbf{s}}(\hat{\underline{\mathbf{k}}})\cdot\hat{\underline{\mathbf{a}}}\,(\underline{\mathbf{e}}_y\,\hat{\underline{\mathbf{a}}} + \hat{\underline{\mathbf{a}}}\,\underline{\mathbf{e}}_y)\Big\}
\end{aligned} \tag{9.372}$$

for the stress tensor of a plane SH-wave in a transversely isotropic material. The particle displacements (9.352) through (9.354), written for arbitrary real or complex valued slowness vectors,

$$\underline{\mathbf{u}}_{i\mathrm{SH}}(\underline{\mathbf{R}}, \omega, \underline{\mathbf{s}}_{i\mathrm{S}}) = -u_{i\mathrm{S}}(\omega)\, \mathrm{e}^{\mathrm{j}\omega \underline{\mathbf{s}}_{i\mathrm{S}}(\hat{\underline{\mathbf{k}}}_{i\mathrm{S}})\cdot\underline{\mathbf{R}}}\, \underline{\mathbf{e}}_y, \quad z \geq 0, \tag{9.373}$$

$$\underline{\mathbf{u}}_{r\mathrm{SH}}(\underline{\mathbf{R}},\omega,\underline{\mathbf{s}}_{r\mathrm{S}}) = -R_{\mathrm{SH}}^{\mathrm{triso}}(\vartheta_{i\mathrm{S}})\,u_{i\mathrm{S}}(\omega)\,\mathrm{e}^{\mathrm{j}\omega\underline{\mathbf{s}}_{r\mathrm{S}}(\hat{\underline{\mathbf{k}}}_{r\mathrm{S}})\cdot\underline{\mathbf{R}}}\,\underline{\mathbf{e}}_y, \quad z\geq 0, \tag{9.374}$$

$$\underline{\mathbf{u}}_{t\mathrm{SH}}(\underline{\mathbf{R}},\omega,\underline{\mathbf{s}}_{t\mathrm{SH}}) = -T_{\mathrm{SH}}^{\mathrm{triso}}(\vartheta_{i\mathrm{S}})\,u_{i\mathrm{S}}(\omega)\,\mathrm{e}^{\mathrm{j}\omega\underline{\mathbf{s}}_{t\mathrm{SH}}(\hat{\underline{\mathbf{k}}}_{t\mathrm{SH}})\cdot\underline{\mathbf{R}}}\,\underline{\mathbf{e}}_y, \quad z\leq 0, \tag{9.375}$$

therefore correspond to the stress tensors

$$\begin{aligned}&\underline{\underline{\mathbf{T}}}_{i\mathrm{SH}}(\underline{\mathbf{R}},\omega,\underline{\mathbf{s}}_{i\mathrm{S}})\\ &\quad= -\mathrm{j}\omega\mu\,u_{i\mathrm{S}}(\omega)\,\mathrm{e}^{\mathrm{j}\omega\underline{\mathbf{s}}_{i\mathrm{S}}(\hat{\underline{\mathbf{k}}}_{i\mathrm{S}})\cdot\underline{\mathbf{R}}}\left[\underline{\mathbf{s}}_{i\mathrm{S}}(\hat{\underline{\mathbf{k}}}_{i\mathrm{S}})\,\underline{\mathbf{e}}_y + \underline{\mathbf{e}}_y\,\underline{\mathbf{s}}_{i\mathrm{S}}(\hat{\underline{\mathbf{k}}}_{i\mathrm{S}})\right],\end{aligned} \tag{9.376}$$

$$\begin{aligned}&\underline{\underline{\mathbf{T}}}_{r\mathrm{SH}}(\underline{\mathbf{R}},\omega,\underline{\mathbf{s}}_{r\mathrm{S}})\\ &\quad= -\mathrm{j}\omega\mu\,R_{\mathrm{SH}}^{\mathrm{triso}}(\vartheta_{i\mathrm{S}})\,u_{i\mathrm{S}}(\omega)\,\mathrm{e}^{\mathrm{j}\omega\underline{\mathbf{s}}_{r\mathrm{S}}(\hat{\underline{\mathbf{k}}}_{r\mathrm{S}})\cdot\underline{\mathbf{R}}}\left[\underline{\mathbf{s}}_{r\mathrm{S}}(\hat{\underline{\mathbf{k}}}_{r\mathrm{S}})\,\underline{\mathbf{e}}_y + \underline{\mathbf{e}}_y\,\underline{\mathbf{s}}_{r\mathrm{S}}(\hat{\underline{\mathbf{k}}}_{r\mathrm{S}})\right],\end{aligned} \tag{9.377}$$

$$\begin{aligned}&\underline{\underline{\mathbf{T}}}_{t\mathrm{SH}}(\underline{\mathbf{R}},\omega,\underline{\mathbf{s}}_{t\mathrm{SH}})\\ &\quad= -\mathrm{j}\omega\,T_{\mathrm{SH}}^{\mathrm{triso}}(\vartheta_{i\mathrm{S}})\,u_{i\mathrm{S}}(\omega)\,\mathrm{e}^{\mathrm{j}\omega\underline{\mathbf{s}}_{t\mathrm{SH}}(\hat{\underline{\mathbf{k}}}_{t\mathrm{SH}})\cdot\underline{\mathbf{R}}}\Big\{\mu_\perp\left[\underline{\mathbf{s}}_{t\mathrm{SH}}(\hat{\underline{\mathbf{k}}}_{t\mathrm{SH}})\,\underline{\mathbf{e}}_y + \underline{\mathbf{e}}_y\,\underline{\mathbf{s}}_{t\mathrm{SH}}(\hat{\underline{\mathbf{k}}}_{t\mathrm{SH}})\right]\\ &\qquad\qquad + (\mu_\| - \mu_\perp)\,\underline{\mathbf{s}}_{t\mathrm{SH}}(\hat{\underline{\mathbf{k}}}_{t\mathrm{SH}})\cdot\hat{\underline{\mathbf{a}}}\,(\underline{\mathbf{e}}_y\,\hat{\underline{\mathbf{a}}} + \hat{\underline{\mathbf{a}}}\,\underline{\mathbf{e}}_y)\Big\}.\end{aligned} \tag{9.378}$$

To satisfy the transition conditions

$$\begin{aligned}&\underline{\mathbf{u}}_{i\mathrm{SH}}(x,y,z=0,\omega,\underline{\mathbf{s}}_{i\mathrm{S}}) + \underline{\mathbf{u}}_{r\mathrm{SH}}(x,y,z=0,\omega,\underline{\mathbf{s}}_{r\mathrm{S}})\\ &\quad= \underline{\mathbf{u}}_{t\mathrm{SH}}(x,y,z=0,\omega,\underline{\mathbf{s}}_{t\mathrm{SH}}),\end{aligned} \tag{9.379}$$

$$\begin{aligned}&\underline{\underline{\mathbf{T}}}_{i\mathrm{SH}}(x,y,z=0,\omega,\underline{\mathbf{s}}_{i\mathrm{S}})\cdot\underline{\mathbf{e}}_z + \underline{\underline{\mathbf{T}}}_{r\mathrm{SH}}(x,y,z=0,\omega,\underline{\mathbf{s}}_{r\mathrm{S}})\cdot\underline{\mathbf{e}}_z\\ &\quad= \underline{\underline{\mathbf{T}}}_{t\mathrm{SH}}(x,y,z=0,\omega,\underline{\mathbf{s}}_{t\mathrm{SH}})\cdot\underline{\mathbf{e}}_z\end{aligned} \tag{9.380}$$

for all x and y, we have to match the phases of all contributing wave modes in the boundary according to

$$\underline{\mathbf{s}}_{i\mathrm{S}}\cdot\underline{\mathbf{e}}_x = \underline{\mathbf{s}}_{r\mathrm{S}}\cdot\underline{\mathbf{e}}_x = \underline{\mathbf{s}}_{t\mathrm{SH}}\cdot\underline{\mathbf{e}}_x, \tag{9.381}$$

thus assessing the x-components of the slowness vectors:

$$s_{r\mathrm{S}x} = s_{t\mathrm{SH}x} = s_{i\mathrm{S}x} = -s^{(1)}\sin\vartheta_{i\mathrm{S}}. \tag{9.382}$$

To calculate the still open z-components, we utilize the dispersion relations (eigenvalue equations) of the respective wave modes:

$$\underline{\mathbf{s}}_{i\mathrm{S}}\cdot\underline{\mathbf{s}}_{i\mathrm{S}} = \frac{\rho^{(1)}}{\mu} = s_{\mathrm{S}}^{(1)^2}, \tag{9.383}$$

$$\underline{\mathbf{s}}_{r\mathrm{S}}\cdot\underline{\mathbf{s}}_{r\mathrm{S}} = \frac{\rho^{(1)}}{\mu} = s_{\mathrm{S}}^{(1)^2}, \tag{9.384}$$

$$\underline{\mathbf{s}}_{t\mathrm{SH}}\cdot\underline{\mathbf{s}}_{t\mathrm{SH}} + \frac{\mu_\| - \mu_\perp}{\mu_\perp}\,(\underline{\mathbf{s}}_{t\mathrm{SH}}\cdot\hat{\underline{\mathbf{a}}})^2 = \frac{\rho^{(2)}}{\mu_\perp}; \tag{9.385}$$

we nicely recognize the disturbance of the isotropy of the half-space $z<0$ due to the $(\mu_\| - \mu_\perp)/\mu_\perp$-anisotropy ratio. With

$$s_{i,r\mathrm{S}}^2 = s_{i,r\mathrm{S}x}^2 + s_{i,r\mathrm{S}z}^2 \tag{9.386}$$

and (9.382), we obtain

$$s_{i\mathrm{S}z} = \pm\sqrt{s_\mathrm{S}^{(1)2} - s_{i\mathrm{S}x}^2} = \pm\, s_\mathrm{S}^{(1)} \cos\vartheta_{i\mathrm{S}}; \tag{9.387}$$

$$s_{r\mathrm{S}z} = \pm\sqrt{s_\mathrm{S}^{(1)2} - s_{r\mathrm{S}x}^2} = \pm\, s_\mathrm{S}^{(1)} \cos\vartheta_{i\mathrm{S}} \tag{9.388}$$

from (9.383) and (9.384), respectively, i.e., the eigenvalue equations (9.383) and (9.384) that are quadratic in $s_{i,r\mathrm{S}z}$ each have two *real* solutions for the z-components of the slowness vectors due to the phase matching (9.382) for the x-components from which we may choose the physically meaningful ones prescribing the phase propagation directions $\underline{\hat{\mathbf{k}}}_{i\mathrm{S}}$ and $\underline{\hat{\mathbf{k}}}_{r\mathrm{S}}$:

$$s_{i\mathrm{S}z} = -\, s_\mathrm{S}^{(1)} \cos\vartheta_{i\mathrm{S}}, \tag{9.389}$$

$$s_{r\mathrm{S}z} = s_\mathrm{S}^{(1)} \cos\vartheta_{i\mathrm{S}}. \tag{9.390}$$

Therefore, the reflection law expresses the equality of the x- and the opposite equality of the z-components of $\underline{\mathbf{s}}_{i\mathrm{S}}$ and $\underline{\mathbf{s}}_{r\mathrm{S}}$, respectively, it is a consequence of phase matching and eigenvalue equations. Obviously, the same is true for the transmission law between the isotropic and the transversely isotropic half-space, and we may equally use the above formalism—that is why, we demonstrated it once more explicitly—to arrive at the respective result.

With

$$\underline{\mathbf{s}}_{t\mathrm{SH}} = s_{t\mathrm{SH}x}\,\underline{\mathbf{e}}_x + s_{t\mathrm{SH}z}\,\underline{\mathbf{e}}_z, \tag{9.391}$$

hence[194]

$$\underline{\mathbf{s}}_{t\mathrm{SH}} \cdot \underline{\mathbf{s}}_{t\mathrm{SH}} = s_{t\mathrm{SH}}^2 = s_{t\mathrm{SH}x}^2 + s_{t\mathrm{SH}z}^2, \tag{9.392}$$

the eigenvalue equation (9.385) together with the given x-component of $\underline{\mathbf{s}}_{t\mathrm{SH}}$ via phase matching (9.382) is a quadratic equation for the z-component:

$$\mu_\perp\,(s_{t\mathrm{SH}x}^2 + s_{t\mathrm{SH}z}^2) + (\mu_\parallel - \mu_\perp)\,(s_{t\mathrm{SH}x}\hat{a}_x + s_{t\mathrm{SH}z}\hat{a}_z)^2 = \rho^{(2)}; \tag{9.393}$$

after some short calculation, its solution can be written according to

$$s_{t\mathrm{SH}z} = \sigma_\mathrm{R}\, s_{t\mathrm{SH}x} \pm s_\mathrm{SH}^2(\underline{\mathbf{e}}_z)\sqrt{\frac{\mu_\perp}{\rho^{(2)}}}\sqrt{\frac{\mu_\parallel}{\rho^{(2)}}}\sqrt{\frac{\rho^{(2)2}}{\mu_\perp\mu_\parallel s_\mathrm{SH}^2(\underline{\mathbf{e}}_z)} - s_{t\mathrm{SH}x}^2} \tag{9.394}$$

utilizing (Equation 9.341)

$$\sigma_\mathrm{R} = -\frac{(\mu_\parallel - \mu_\perp)\underline{\mathbf{e}}_x \cdot \underline{\hat{\mathbf{a}}}\,\underline{\hat{\mathbf{a}}} \cdot \underline{\mathbf{e}}_z}{\rho^{(2)} c_\mathrm{SH}^2(\underline{\mathbf{e}}_z)}. \tag{9.395}$$

For $\underline{\hat{\mathbf{a}}} \cdot \underline{\mathbf{e}}_z = 0$, Equation 9.394 reduces to (9.363).

[194]Note: $s_{t\mathrm{SH}}^2$ is the eventually complex non-Hermitian magnitude of the slowness vector $\underline{\mathbf{s}}_{t\mathrm{SH}}$ and not the square of the slowness $s_\mathrm{SH}^{(2)}$.

With (9.394), we have found two real valued $s_{t\mathrm{SH}z}$-zeroes of the SH-eigenvalue equation for

$$\sin\vartheta_{i\mathrm{S}} \leq \frac{\rho^{(2)}}{s_{\mathrm{S}}^{(1)} s_{\mathrm{SH}}^{(2)}(\underline{\mathbf{e}}_z)\sqrt{\mu_\perp \mu_\parallel}} \tag{9.396}$$

that obviously define regions I and II of Figure 9.25(a), namely region I for zeroes with opposite signs and region II for two negative zeroes. For all other cases, the choice of the physically meaningful zero has to be made consulting the direction of the pertinent energy velocity, not only for region II [Figure 9.25(d)], but strictly speaking also for region I; based on Figure 9.26, we have encountered an example for a phase propagation direction belonging to the positive $s_{t\mathrm{SH}z}$-zero if the pertinent energy velocity is physically possible. With the above example together with the example of Figure 9.25(e), we illustrate the case of one (double) real zero that is given according to (9.394) with the equality sign in (9.396). Therefore, this equality sign defines the critical angle $\vartheta_{ct\mathrm{SH}}$ as

$$\sin\vartheta_{ct\mathrm{SH}} = \frac{\rho^{(2)}}{s_{\mathrm{S}}^{(1)} s_{\mathrm{SH}}^{(2)}(\underline{\mathbf{e}}_z)\sqrt{\mu_\perp \mu_\parallel}}; \tag{9.397}$$

for $\vartheta_{i\mathrm{S}} \leq \vartheta_{ct\mathrm{SH}}$, and with

$$\begin{aligned} \tan\vartheta_{t\mathrm{SH}} &= \frac{s_{t\mathrm{SH}x}}{s_{t\mathrm{SH}z}} \\ &= -\frac{s_{\mathrm{S}}^{(1)}\sin\vartheta_{i\mathrm{S}}}{s_{t\mathrm{SH}z}}, \end{aligned} \tag{9.398}$$

we obtain the (real) transmission angle for the phase propagation vector, but only the pertinent direction of the energy velocity vector tells us what actually happens in the transversely isotropic half-space.

For $\vartheta_{i\mathrm{S}} > \vartheta_{ct\mathrm{SH}}$, we obtain from (9.394) two conjugate complex zeroes; together with $s_{t\mathrm{SH}x}$, their real parts compose the real part vector

$$\Re\underline{\mathbf{s}}_{t\mathrm{SH}} = -s_{\mathrm{S}}^{(1)}\sin\vartheta_{i\mathrm{S}}(\underline{\mathbf{e}}_x + \sigma_{\mathrm{R}}\,\underline{\mathbf{e}}_z) \tag{9.399}$$

and their imaginary parts the imaginary part vector

$$\Im\underline{\mathbf{s}}_{t\mathrm{SH}} = \pm s_{\mathrm{SH}}^2(\underline{\mathbf{e}}_z)\sqrt{\frac{\mu_\perp}{\rho^{(2)}}}\sqrt{\frac{\mu_\parallel}{\rho^{(2)}}}\sqrt{s_{\mathrm{S}}^{(1)^2}\sin^2\vartheta_{i\mathrm{S}} - \frac{\rho^{(2)^2}}{\mu_\perp\mu_\parallel s_{\mathrm{SH}}^2(\underline{\mathbf{e}}_z)}}\,\underline{\mathbf{e}}_z \tag{9.400}$$

of the transmission slowness vector [Equation 9.369 as well as Equation 9.342 that may actually be transformed to (9.400) with (9.399) utilizing the eigenvalue equation]. In case, the half-space $z < 0$ is isotropic or transversely isotropic with $\hat{\underline{\mathbf{a}}}\cdot\underline{\mathbf{e}}_z = 0$, the (phase) transmission angle $\vartheta_{t\mathrm{SH}}$ becomes complex for $\vartheta_{i\mathrm{S}} > \vartheta_{ct\mathrm{SH}}$, its real part $\Re\vartheta_{t\mathrm{SH}} = \pi/2$ then giving the direction of the phase as well as the energy propagation [Figure 9.24(b)]. Yet, if we have

$\underline{\hat{\mathbf{a}}} \cdot \underline{\mathbf{e}}_z \neq 0$ for the transversely isotropic half-space, it is completely inappropriate to describe the phase and energy propagations of the transmitted wave by a complex angle for $\vartheta_{i\mathrm{S}} > \vartheta_{ct\mathrm{SH}}$; however, we obtain the direction of the phase propagation vector independently of the angle of incidence from (9.399) as

$$\tan \vartheta_{t\mathrm{SH}} = \frac{1}{\sigma_\mathrm{R}}, \tag{9.401}$$

and we know that the pertinent energy velocity vector is parallel to the boundary. The evanescence of this energy flux is evident through the physical meaningful selection of the imaginary part of the complex valued—not purely imaginary—$s_{t\mathrm{SH}z}$-zero of the eigenvalue equation.

The vector equations (9.379) and (9.380) of the transition conditions are indeed scalar equations with only y-components; therefore, we immediately calculate

$$R_\mathrm{SH}^\mathrm{triso}(\vartheta_{i\mathrm{S}}) = \frac{\mu\, s_\mathrm{S}^{(1)} \cos \vartheta_{i\mathrm{S}} + \left[\mu_\perp\, \underline{\mathbf{s}}_{t\mathrm{SH}} \cdot \underline{\mathbf{e}}_z + (\mu_\parallel - \mu_\perp)\, \underline{\mathbf{s}}_{t\mathrm{SH}} \cdot \underline{\hat{\mathbf{a}}}\,\underline{\hat{\mathbf{a}}} \cdot \underline{\mathbf{e}}_z\right]}{\mu\, s_\mathrm{S}^{(1)} \cos \vartheta_{i\mathrm{S}} - \left[\mu_\perp\, \underline{\mathbf{s}}_{t\mathrm{SH}} \cdot \underline{\mathbf{e}}_z + (\mu_\parallel - \mu_\perp)\, \underline{\mathbf{s}}_{t\mathrm{SH}} \cdot \underline{\hat{\mathbf{a}}}\,\underline{\hat{\mathbf{a}}} \cdot \underline{\mathbf{e}}_z\right]}, \tag{9.402}$$

$$T_\mathrm{SH}^\mathrm{triso}(\vartheta_{i\mathrm{S}}) = \frac{2\mu\, s_\mathrm{S}^{(1)} \cos \vartheta_{i\mathrm{S}}}{\mu\, s_\mathrm{S}^{(1)} \cos \vartheta_{i\mathrm{S}} - \left[\mu_\perp\, \underline{\mathbf{s}}_{t\mathrm{SH}} \cdot \underline{\mathbf{e}}_z + (\mu_\parallel - \mu_\perp)\, \underline{\mathbf{s}}_{t\mathrm{SH}} \cdot \underline{\hat{\mathbf{a}}}\,\underline{\hat{\mathbf{a}}} \cdot \underline{\mathbf{e}}_z\right]}. \tag{9.403}$$

For the special case $\underline{\hat{\mathbf{a}}} \cdot \underline{\mathbf{e}}_z = 0$ and $\vartheta_{i\mathrm{S}} \leq \vartheta_{ct\mathrm{SH}}$, these two formulas reduce to

$$R_\mathrm{SH}^\mathrm{triso}(\vartheta_{i\mathrm{S}} = \frac{\mu\, s_\mathrm{S}^{(1)} \cos \vartheta_{i\mathrm{S}} - \mu_\perp\, s_\mathrm{SH}^{(2)}(\vartheta_{t\mathrm{SH}}) \cos \vartheta_{t\mathrm{SH}}}{\mu\, s_\mathrm{S}^{(1)} \cos \vartheta_{i\mathrm{S}} + \mu_\perp\, s_\mathrm{SH}^{(2)}(\vartheta_{t\mathrm{SH}}) \cos \vartheta_{t\mathrm{SH}}}, \tag{9.404}$$

$$T_\mathrm{SH}^\mathrm{triso}(\vartheta_{i\mathrm{S}} = \frac{2\mu\, s_\mathrm{S}^{(1)} \cos \vartheta_{t\mathrm{S}}}{\mu\, s_\mathrm{S}^{(1)} \cos \vartheta_{i\mathrm{S}} + \mu_\perp\, s_\mathrm{SH}^{(2)}(\vartheta_{t\mathrm{SH}}) \cos \vartheta_{t\mathrm{SH}}}, \tag{9.405}$$

and for $\vartheta_{i\mathrm{S}} > \vartheta_{ct\mathrm{SH}}$, we obtain

$$R_\mathrm{SH}^\mathrm{triso}(\vartheta_{i\mathrm{S}}) = \frac{\mu\, s_\mathrm{S}^{(1)} \cos \vartheta_{i\mathrm{S}} + \mathrm{j}\mu_\perp\, s_{\mathrm{I}t\mathrm{SH}z}}{\mu\, s_\mathrm{S}^{(1)} \cos \vartheta_{i\mathrm{S}} - \mathrm{j}\mu_\perp\, s_{\mathrm{I}t\mathrm{SH}z}}, \tag{9.406}$$

$$T_\mathrm{SH}^\mathrm{triso}(\vartheta_{i\mathrm{S}}) = \frac{2\mu\, s_\mathrm{S}^{(1)} \cos \vartheta_{i\mathrm{S}}}{\mu\, s_\mathrm{S}^{(1)} \cos \vartheta_{i\mathrm{S}} - \mathrm{j}\mu_\perp\, s_{\mathrm{I}t\mathrm{SH}z}}, \tag{9.407}$$

where $s_{\mathrm{I}t\mathrm{SH}z}$ is given with the negative sign through (9.368). With (9.404) and (9.405), we are "very close" to (9.206) and (9.207); furthermore: Since nominator and denominator in (9.406) are complex conjugate to each other, we observe total reflection for $\vartheta_{i\mathrm{S}} > \vartheta_{ct\mathrm{SH}}$ in the case $\underline{\hat{\mathbf{a}}} \cdot \underline{\mathbf{e}}_z = 0$. For the more general case $\underline{\hat{\mathbf{a}}} \cdot \underline{\mathbf{e}}_z \neq 0$ (Equations 9.402 and 9.403), we do not observe total reflection because $s_{t\mathrm{SH}z}$ is not purely imaginary.

9.3.4 Reflection, transmission, and mode conversion of plane SV-waves at the planar boundary between homogeneous isotropic and homogeneous transversely isotropic nondissipative materials

The case of reflection, transmission, and mode conversion of P/SV-qP/qSV-wave modes at the boundary of an isotropic and a transversely isotropic half-space is exemplarily discussed in terms of an SV-wave coming from an isotropic steel half-space and impinging onto an austenitic steel 308 half-space. Apart from the practical relevance for US-NDT, this case serves again for the finding of some new basic results.

Preference direction $\underline{\hat{\mathbf{a}}}$ in the boundary: We start with the assumption $\underline{\hat{\mathbf{a}}} \cdot \underline{\mathbf{e}}_z = 0$ and turn to the slowness diagram of Figure 9.27: For $z > 0$, the circular (in three dimensions: spherical) slowness diagrams for isotropic steel are plotted in the xz-plane; yet, we know already [Figure 9.25(b)] that the slowness diagrams of the transversely isotropic austenitic steel half-space (austenitic steel 308) for $z < 0$ may even play a role for $z > 0$, hence we have continued the respective diagrams of Figure 8.21 as dashed lines for $z > 0$. First: The P-diagram for steel does not essentially differ from the qP-diagram for austenitic steel 308, whereas the qSV-diagram of austenitic steel 308 is considerably distinct from the steel S-diagram. Since we assume $\underline{\hat{\mathbf{a}}}$ to be

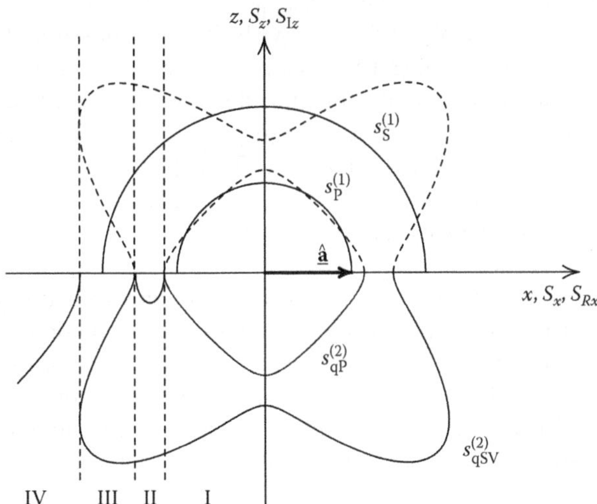

FIGURE 9.27
Slowness diagram of reflection, transmission, and mode conversion of an SV-wave at the boundary of a homogeneous isotropic nondissipative (steel) and a homogeneous transversely isotropic nondissipative half-space (austenitic steel): $\underline{\hat{\mathbf{a}}} \cdot \underline{\mathbf{e}}_z = 0$.

parallel to the plane of incidence, SH-wave modes are decoupled from P/SV-qP/qSV-wave modes. According to these slowness diagrams, an incident SV-wave from the half-space $z > 0$ should result in a reflected SV-wave (reflection law!) and a reflected mode converted P-wave including a critical angle for exactly those waves [Figure 9.18(SVa)]. Yet, the performance in the half-space $z < 0$ is much more interesting: For "small" incidence angles $\vartheta_{i\mathrm{S}}$ characterizing region I a transmitted qSV- and a transmitted mode converted qP-wave are excited according to two (real) physically meaningful $s_{t\mathrm{qP},\mathrm{qSV}z}$-zeroes of the qP/qSV-slowness eigenvalue equation [equating the square bracket in (9.330) to zero]; the latter exhibits a critical angle $\vartheta_{cm_t\mathrm{qP}} > \vartheta_{cm_r\mathrm{P}}$, i.e., we reach region II of Figure 9.27 for $\vartheta_{i\mathrm{S}} > \vartheta_{cm_t\mathrm{qP}}$: The transmitted mode converted qP-wave becomes inhomogeneous, it is evanescent with regard to phase and energy because phase and energy velocity vector point into (negative) x-direction; the magnitude of the attenuation results from the purely imaginary $s_{t\mathrm{qP}z}$-zeroes in region II (with different signs). The imaginary part of the negative imaginary zero varies on the quasielliptical curve plotted for region II with increasing $\vartheta_{i\mathrm{S}}$; it is not a hyperbola because two additional real $s_{t\mathrm{qSV}z}$-zeroes (with different signs) exist, the negative one belonging to the (energetically) propagating transmitted qSV-wave. Region III of Figure 9.27 is characterized by the existence of four real $s_{t\mathrm{qSV}z}$-zeroes with pairwise different signs, the previous purely imaginary $s_{t\mathrm{qP}z}$-zeroes can no longer exist, the inhomogeneous qP-wave "disappears" beyond the critical angle $\vartheta_{cm_t\mathrm{qP}_{\mathrm{inh}}}$ that defines region III as incidence angle. Consequently, there is no transmitted quasipressure wave. For $\vartheta_{i\mathrm{S}} > \vartheta_{cm_t\mathrm{qP}_{\mathrm{inh}}}$, we observe two energetically propagating qSV-waves qSV_1 and qSV_2 with different phase and energy velocity: qSV_1 belongs to the (negative) $s_{t\mathrm{qSV}z}$-zero with larger magnitude and qSV_2 to the one with the smaller magnitude. For the underlying slowness diagrams of Figure 9.27, we have $s_{\mathrm{S}}^{(1)} < \max_{\hat{\underline{\mathbf{k}}}} s_{\mathrm{qSV}x}(\hat{\underline{\mathbf{k}}})$ making region IV inaccessible; its boundary is characterized by the coincidence of the two negative with the two positive $s_{t\mathrm{qSV}z}$-zeroes yielding a double zero in each case with pertinent energy velocity parallel to the boundary. That way, critical angles $\vartheta_{ct\mathrm{qSV}_1} = \vartheta_{cm_t\mathrm{qSV}_2} = \vartheta_{ct\mathrm{qSV}}$ are defined; for $\vartheta_{i\mathrm{S}} > \vartheta_{ct\mathrm{qSV}}$, we observe energy evanescence of both $\mathrm{qSV}_{1,2}$-waves due to complex valued $s_{t\mathrm{qSV}z}$-zeroes, where the magnitude of the respective attenuation must be determined via a slowness hyperbola as variation curve of the imaginary part of the physically meaningful zero. Note: Even though the energy velocity of both waves has the same direction for $\vartheta_{i\mathrm{S}} > \vartheta_{ct\mathrm{qSV}}$, their phase velocities are differently oriented making them distinguishable as energetically evanescent waves.

Preference direction $\hat{\underline{\mathbf{a}}}$ inclined to the boundary: As for the SH-case (Figure 9.26), we now turn our attention to inclined slowness diagrams confining ourselves to SV/qSV-diagrams (Figure 9.29) because we aim at a particular application: Figure 9.28 shows the model of a steel test specimen with an austenitic weld that stood for extensive parameter variations for EFIT-simulations (Hannemann 2001); the simultaneous existence of measurements

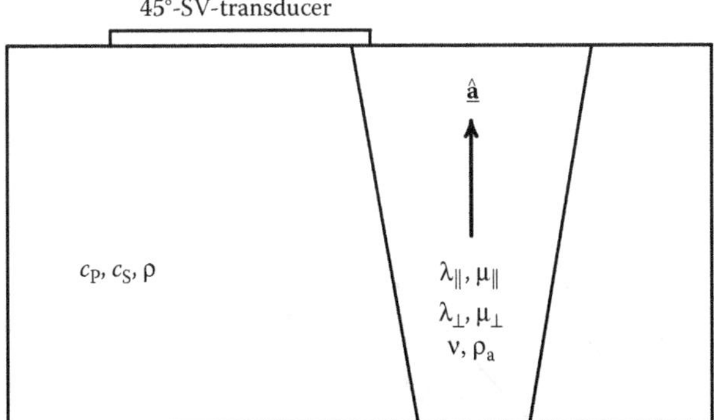

FIGURE 9.28
Model of a test specimen made of ferritic steel with a weld made of austenitic steel: $c_P = 5900$ m/s, $c_S = 3200$ m/s, $\rho = 7.7 \cdot 10^3$ kg/m^3; $\lambda_\| + 2\mu_\| = 216$, $\lambda_\perp + 2\mu_\perp = 262.75$, $\mu_\perp = 82.5$, $\mu_\| = 129$, $\nu = 145$ GPa, $\rho_a = 7.7999 \cdot 10^3$ kg/m^3.

(Langenberg et al. 2000) suggests the choice of material parameters for austenitic steel 308SS90; they differ slightly from those of austenitic steel 308 (Figure 9.27) yielding slightly different slowness diagrams. This weld is to be tested with a 45°-SV-transducer (dimensioned for ferritic steel) as shown in Figure 9.28; the weld boundary makes an angle of 10° relative to the surface of the specimen leading to an incidence angle of 55° for the SV-wave; $\hat{\underline{\mathbf{a}}}$ is also inclined by 10° with regard to the weld boundary. Hence, we face the configuration of Figure 9.29 with the respective direction of $\underline{\mathbf{s}}_{i\mathrm{SV}}$. Figure 9.27 reveals the evanescence of the reflected mode converted P-wave for this particular direction of $\underline{\mathbf{s}}_{i\mathrm{SV}}$ as well as the non-existence of the transmitted mode converted qP-wave; we are actually only confronted with the qSV-diagram with respect to transmitted waves. The phase matching line to the reflected SV-wave (located at a distance $\underline{\mathbf{s}}_{r\mathrm{SV}} \cdot \underline{\mathbf{e}}_x$ parallel to the z-axis) creates four intersections with the qSV-diagram, where $\underline{\mathbf{s}}_{tq\mathrm{SV}_1}$ and $\underline{\mathbf{s}}_{tq\mathrm{SV}_2}$ yield energy velocity directions pointing into the austenitic half-space: The incident SV-wave excites two transmitted qSV-waves with different directions and magnitudes of phase and energy velocities. We realign the coordinate system of Figure 9.29 as in Figure 9.30 with the x, s_x-axis coinciding with the weld boundary; that way, we should have a basis to interpret the (two-dimensional) EFIT-simulation of the test problem under concern in Figure 9.28; this is confirmed through the wavefronts of Figure 9.31.

First: From Figure 9.29, we conclude facts about reflection, transmission, and mode conversion of plane waves, yet, regarding the incident wave, the test problem of Figure 9.28 is a source field problem of a transducer of finite extent on a stress-free surface (Chapter 14), and due to the presence of the weld, a scattering problem has to be solved for this incident field (Section 15.1.3).

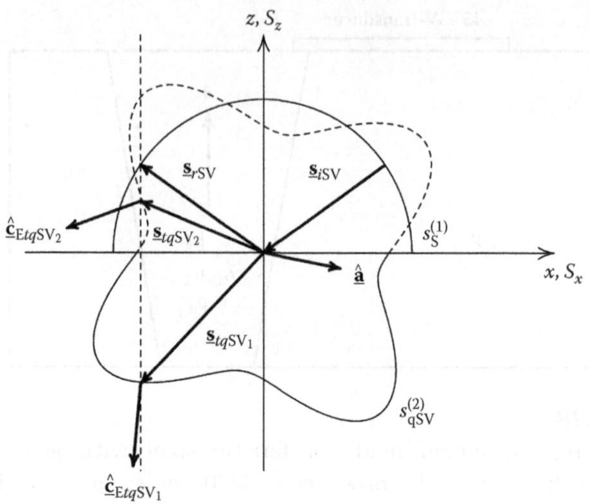

FIGURE 9.29
Slowness diagram construction of two transmitted qSV-waves at the boundary of a homogeneous isotropic nondissipative (steel) and a homogeneous transversely isotropic nondissipative (austenitic) half-space.

The formal solution via point source synthesis utilizes Green functions, not plane waves. Yet, we may represent Green functions in terms of spatial spectra of plane waves (Section 14.2.3), thus approaching our desired goal at least verbally. In addition, the near-field evaluation of respective integral representations (Section 11.1) reveals the presence of quasiplane waves in the source field of an aperture transducer that may be understood as wave packets in the sense of Figure 8.14; moreover, this conception is backed by EFIT-simulations (Figure 14.19). Therefore, we may conversely interpret the EFIT-simulations of the transducer problem of Figure 9.28 with the help of the slowness diagrams of plane waves according to Figure 9.30: It is achieved in Figure 9.31. There, we have displayed wavefronts of a two-dimensional EFIT-simulation for three different times; these simulations are closely related to measurements[195] (Langenberg et al. 2000), hence the transducer aperture is amplitude tapered with a measured excitation distribution and not just with a rectangular function: Due to this reason, the wave packet of the incident SV-RC5(t)-pulse does not emerge from the center of the aperture; yet it propagates at least under 45° in the direction $\hat{\underline{\mathbf{s}}}_{i\mathrm{SV}}$ relative to the surface normal. Reaching the weld, a reflected SV-RC5(t)-wave packet is created propagating in the direction $\hat{\underline{\mathbf{s}}}_{r\mathrm{SV}}$. Regarding the phase propagation directions—the normals to the phase fronts—the two transmitted wave packets coincide with the directions

[195]These measurements have been performed by B. Köhler of the Fraunhofer Institute for Non-Destructive Testing in Dresden, Germany.

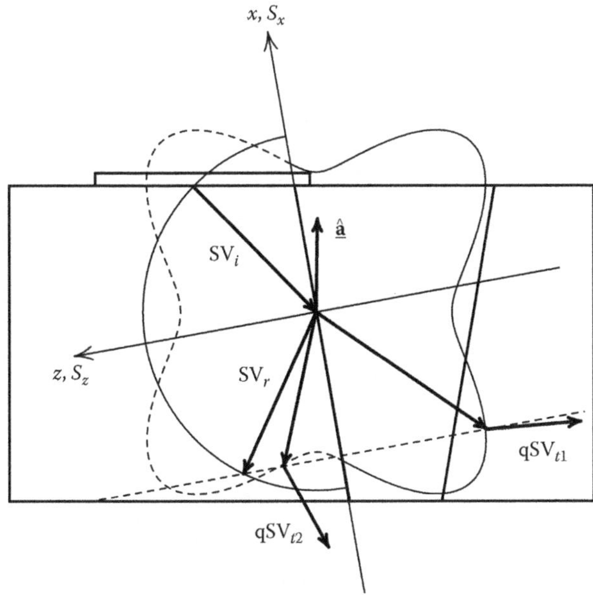

FIGURE 9.30
Excitation of two qSV-waves in an austenitic weld with a 45°-SV-transducer.

$\hat{\underline{\mathbf{s}}}_{tq\mathrm{SV}_1}$ and $\hat{\underline{\mathbf{s}}}_{tq\mathrm{SV}_2}$ of Figure 9.30. The pertinent different directions of the energy velocities $\hat{\underline{\mathbf{c}}}_{\mathrm{E}tq\mathrm{SV}_{1,2}}$ become noticeable in the last wavefront picture of Figure 9.31 through the drift of the wave packets in the respective directions according to Figure 8.14, keeping the original phase surface orientations.

Reflection and transmission coefficients: We are not at all capable to give equally "nice" analytical expressions for the reflection and transmission coefficients for the qP/qSV-case as we could for the SH-case with (9.402) and (9.403), yet we can systematize the approach:

- At first, we have to calculate 360°-slowness diagrams for the material parameters $\hat{\underline{\mathbf{a}}}$, $\lambda_{\|}$, $\mu_{\|}$, $\lambda_{\perp}$, $\mu_{\perp}$, ν, ρ of the transversely isotropic half-space; together with the diagrams for the isotropic half-space, their graphical display illustrates the occurrence of respective wave modes as function of the incidence angle: rP, m_rSV, and tqP, m_tqSV as well as $m_t\mathrm{qSV}_{1,2}$ for P-incidence, and rSV, m_rP as well as m_tqP, tqSV, and $t\mathrm{qSV}_{1,2}$ for SV-incidence.

- The given angle of incidence defines the phase matching line through $\underline{\mathbf{s}}_{i\mathrm{P,SV}} \cdot \underline{\mathbf{e}}_x$, namely the (real) x-components of all existing slowness vectors, hence, it prescribes any possible evanescence, i.e., whether the z-component of a particular wave mode slowness vector becomes complex.

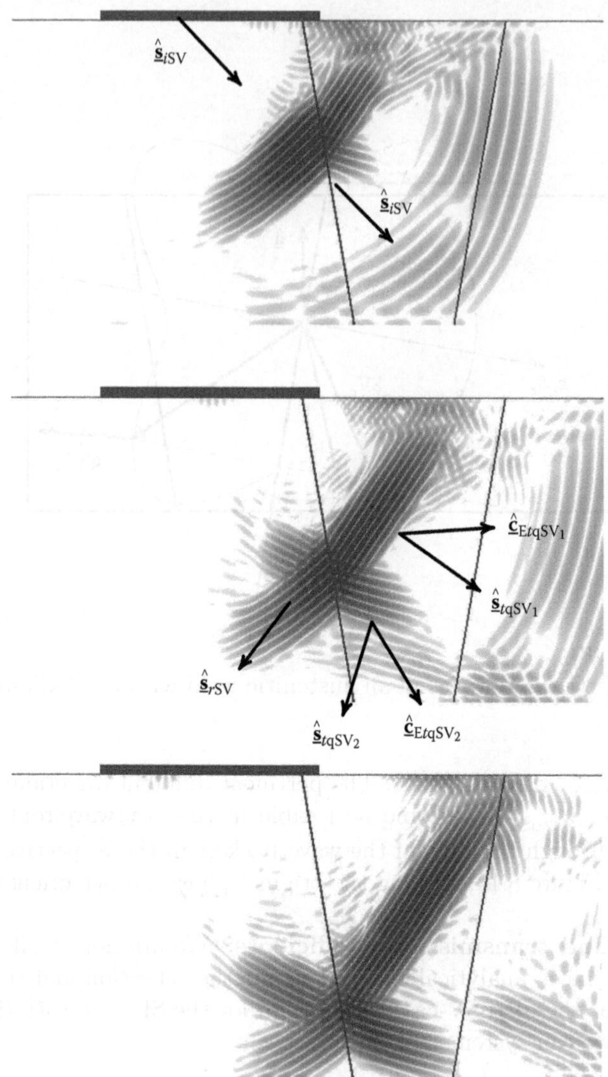

FIGURE 9.31
2D-EFIT-simulation of US-NDT of an austenitic weld: excitation of two transmitted waves.

- For the given real x-component of $\underline{\mathbf{s}}$, we have to find the four solutions of the fourth order polynomial

$$\beta_1^2(\underline{\mathbf{s}}) + \beta_1(\underline{\mathbf{s}})\left[\beta_2\underline{\mathbf{s}}\cdot\underline{\mathbf{s}} + \beta_3(\underline{\mathbf{s}}) + 2\beta_4(\underline{\mathbf{s}})\underline{\mathbf{s}}\cdot\hat{\underline{\mathbf{a}}}\right] + \left[\beta_2\beta_3(\underline{\mathbf{s}}) - \beta_4^2(\underline{\mathbf{s}})\right]\left(\underline{\underline{\mathbf{I}}} - \hat{\underline{\mathbf{a}}}\,\hat{\underline{\mathbf{a}}}\right) : \underline{\mathbf{s}}\,\underline{\mathbf{s}} = 0 \tag{9.408}$$

for the z-component of $\underline{\mathbf{s}}$ (for $\hat{\underline{\mathbf{a}}} \cdot \underline{\mathbf{e}}_z \neq 0$) to determine the transmitted wave modes, in general numerically, where $\beta_1(\underline{\mathbf{s}})$, β_2, $\beta_3(\underline{\mathbf{s}})$, $\beta_4(\underline{\mathbf{s}})$ are given by (9.326) through (9.329). With the slowness diagrams, the real and/or complex zeroes can be assigned to distinct wave modes.

- The physically meaningful selection of wave modes belonging to the four s_z-zeroes has to be based upon the direction of the energy velocity. To utilize (9.349), we need the polarization vectors as Hermitian unit vectors; even for complex slowness vectors, their calculation is performed according to Section 8.3.2, and we explicitly obtain:[196]

$$\hat{\underline{\mathbf{u}}}(\underline{\mathbf{s}}) = \frac{\underline{\mathbf{s}} + \gamma(\underline{\mathbf{s}})\hat{\underline{\mathbf{a}}}}{\sqrt{\underline{\mathbf{s}} \cdot \underline{\mathbf{s}}^* + \gamma(\underline{\mathbf{s}})\underline{\mathbf{s}}^* \cdot \hat{\underline{\mathbf{a}}} + \gamma^*(\underline{\mathbf{s}})\underline{\mathbf{s}} \cdot \hat{\underline{\mathbf{a}}} + |\gamma(\underline{\mathbf{s}})|^2}}, \tag{9.409}$$

where

$$\gamma(\underline{\mathbf{s}}) = -\frac{\beta_1(\underline{\mathbf{s}}) + \beta_2 \underline{\mathbf{s}} \cdot \underline{\mathbf{s}} + \beta_4(\underline{\mathbf{s}})\underline{\mathbf{s}} \cdot \hat{\underline{\mathbf{a}}}}{\beta_4(\underline{\mathbf{s}}) + \beta_2 \underline{\mathbf{s}} \cdot \hat{\underline{\mathbf{a}}}}. \tag{9.410}$$

- Now we can calculate the energy velocity according to (9.349) for any real or complex slowness vector; the selection criterion for physically meaningful slowness vectors $\underline{\mathbf{s}}_{t\mathrm{M}}$, and hence for those transmitted wave modes indexed tM that actually exist, and to be accounted for in the transition conditions, reads $\underline{\mathbf{c}}_{\mathrm{E}}(\underline{\mathbf{s}}_{t\mathrm{M}}) \cdot \underline{\mathbf{e}}_z \leq 0$ (and not $\Re \underline{\mathbf{s}}_{t\mathrm{M}} \leq 0$).

- From the transition conditions

$$\begin{aligned}
&\underline{\mathbf{u}}_{i\mathrm{P,SV}}(\underline{\mathbf{R}}, \omega, \underline{\mathbf{s}}_{i\mathrm{P,S}}) + \underline{\mathbf{u}}_{r\mathrm{P,SV}}(\underline{\mathbf{R}}, \omega, \underline{\mathbf{s}}_{r\mathrm{P,S}}) + \underline{\mathbf{u}}_{m_r\mathrm{SV,P}}(\underline{\mathbf{R}}, \omega, \underline{\mathbf{s}}_{m_r\mathrm{S,P}}) \\
&\quad = \sum_{t\mathrm{M}} \underline{\mathbf{u}}_{t\mathrm{M}}(\underline{\mathbf{R}}, \omega, \underline{\mathbf{s}}_{t\mathrm{M}}),
\end{aligned} \tag{9.411}$$

$$\begin{aligned}
&\underline{\mathbf{e}}_z \cdot \underline{\underline{\mathbf{T}}}_{i\mathrm{P,SV}}(\underline{\mathbf{R}}, \omega, \underline{\mathbf{s}}_{i\mathrm{P,S}}) + \underline{\mathbf{e}}_z \cdot \underline{\underline{\mathbf{T}}}_{r\mathrm{P,SV}}(\underline{\mathbf{R}}, \omega, \underline{\mathbf{s}}_{r\mathrm{P,S}}) \\
&\quad + \underline{\mathbf{e}}_z \cdot \underline{\underline{\mathbf{T}}}_{m_r\mathrm{SV,P}}(\underline{\mathbf{R}}, \omega, \underline{\mathbf{s}}_{m_r\mathrm{S,P}}) \\
&\qquad = \sum_{t\mathrm{M}} \underline{\mathbf{e}}_z \cdot \underline{\underline{\mathbf{T}}}_{t\mathrm{M}}(\underline{\mathbf{R}}, \omega, \underline{\mathbf{s}}_{t\mathrm{M}})
\end{aligned} \tag{9.412}$$

for $z = 0$, we then obtain reflection, transmission, and mode conversion coefficients. Of course, eventually evanescent wave modes have to be accounted for in (9.411) and (9.412).

Marklein (1997), among others, discusses the generalization of this procedure to arbitrary anisotropic half-spaces; even numerical examples are given. It

[196] By the way, for complex slowness vectors, the SH-polarization unit vector reads in generalization of (8.270):

$$\hat{\underline{\mathbf{u}}}_{\mathrm{SH}}(\underline{\mathbf{s}}) = \frac{\underline{\mathbf{s}} \times \hat{\underline{\mathbf{a}}}}{\sqrt{\underline{\mathbf{s}} \cdot \underline{\mathbf{s}}^* - \underline{\mathbf{s}}\,\underline{\mathbf{s}}^* : \hat{\underline{\mathbf{a}}}\hat{\underline{\mathbf{a}}}}},$$

because we then have $\hat{\underline{\mathbf{u}}}_{\mathrm{SH}}(\underline{\mathbf{s}}) \cdot \hat{\underline{\mathbf{u}}}^*_{\mathrm{SH}}(\underline{\mathbf{s}}) = 1$.

should be pointed out that the respective work of Marklein consequently continues (to a large extent) with the coordinate-free notation as introduced into elastodynamics by Fellinger (1991) and utilized in this elaboration; once used to it, it provides a much clearer lay out than the index notation in the standard literature (Auld 1973; Helbig 1994; de Hoop 1995; Nayfeh 1995; Royer and Dieulesaint 2000).

10

Rayleigh Surface Waves

10.1 Planar Surfaces

Inhomogeneous plane waves as solutions of the (Fourier transformed) wave equation for isotropic nondissipative materials (Section 8.2) have found physical realizations concerning reflection and transmission of plane waves at the planar boundary between such materials or at their planar (stress-free) surface, respectively (for electromagnetic waves, this topic is called Fresnel's reflection): The existence of up to three critical angles yields an equal number of evanescent plane waves (Sections 9.1.2, 9.2.1, and 9.2.2). Plane waves on one selected side of the boundary exhibit at most two critical angles [Figures 9.19 (SVc)–(SVa′)], one for the pressure and one for the shear wave, where the amplitudes of the resulting evanescent waves are given by complex transmission coefficients. However, a stress-free surface only allows for one evanescent pressure wave with an amplitude given by a complex mode conversion coefficient. The question leading to Rayleigh surface waves is the following: Does the homogeneous wave equation allow for a solution in terms of the superposition of two evanescent plane waves along the stress-free surface of an isotropic nondissipative material, namely in terms of the superposition of an inhomogeneous pressure and an inhomogeneous shear wave? As for the case of a boundary separating two materials, both waves should propagate with the same phase velocity along the surface, however with different attenuation constants. What we already know: Neither an incident plane pressure nor an incident plane shear wave can excite such a "free" surface wave!

Hence, we make the following *ansatz* for a time harmonic inhomogeneous plane wave:

$$\underline{\mathbf{u}}(\underline{\mathbf{R}}, \omega, \underline{\mathbf{k}}) = u_{\mathrm{P}}(\omega)\,\mathrm{e}^{\mathrm{j}\underline{\mathbf{k}}_{\mathrm{P}}\cdot\underline{\mathbf{R}}}\,\hat{\underline{\mathbf{u}}}_{\mathrm{P}}(\underline{\mathbf{k}}_{\mathrm{P}}) + u_{\mathrm{S}}(\omega)\,\mathrm{e}^{\mathrm{j}\underline{\mathbf{k}}_{\mathrm{S}}\cdot\underline{\mathbf{R}}}\,\hat{\underline{\mathbf{u}}}_{\mathrm{SV}}(\underline{\mathbf{k}}_{\mathrm{S}}), \tag{10.1}$$

where the amplitude ratio $u_{\mathrm{P}}(\omega)/u_{\mathrm{S}}(\omega)$ must be determined from the stress-free boundary condition. We assume the shear partial wave in (10.1) to be SV-polarized since a horizontally polarized shear wave is decoupled from either an SV- or a P-wave. Note: The superposition of these partial waves resulting in a Rayleigh wave is neither a pressure nor a shear wave because (10.1) is neither divergence- nor curl-free. The evanescence of both partial waves is expressed by the complex phase vectors

$$\underline{\mathbf{k}}_{\mathrm{P,S}} = \Re\underline{\mathbf{k}} + \mathrm{j}\Im\underline{\mathbf{k}}_{\mathrm{P,S}}, \tag{10.2}$$

where we have

$$\Re\underline{\mathbf{k}} \cdot \Im\underline{\mathbf{k}}_{\mathrm{P,S}} = 0 \tag{10.3}$$

as orthogonality of the phase propagation and the attenuation direction due to the lossless material (Equation 8.153). According to our assumption, the phases of both partial waves of (10.1) should propagate into the same direction with the same magnitude (with the same phase velocity), hence $\underline{\mathbf{k}}_{\mathrm{P}}$ and $\underline{\mathbf{k}}_{\mathrm{S}}$ should have the same real part vector $\Re\underline{\mathbf{k}}$. Due to the dispersion relations for the isotropic nondissipative material

$$\underline{\mathbf{k}}_{\mathrm{P,S}} \cdot \underline{\mathbf{k}}_{\mathrm{P,S}} = k_{\mathrm{P,S}}^2, \tag{10.4}$$

we have

$$\begin{aligned} |\Im\underline{\mathbf{k}}_{\mathrm{P,S}}|^2 &= |\Re\underline{\mathbf{k}}|^2 - k_{\mathrm{P,S}}^2 \\ &= k_{\mathrm{R}}^2 - k_{\mathrm{P,S}}^2 \end{aligned} \tag{10.5}$$

for the magnitudes of the imaginary part vectors under the assumption (10.3); we have used the short-hand notation k_{R}—R for Rayleigh—for the wave number $|\Re\underline{\mathbf{k}}|$ of the surface wave we are searching for. To ensure a real valued $|\Im\underline{\mathbf{k}}_{\mathrm{P,S}}|$, we must have $k_{\mathrm{R}} > k_{\mathrm{P,S}}$, i.e., the phase velocity of the Rayleigh wave will turn out to be smaller than the velocity of pressure and shear waves.

As already mentioned, the partial P-wave of (10.1) should be a curl-free inhomogeneous pressure and the partial SV-wave should be a divergence-free inhomogeneous shear wave. Accordingly, Equations 8.159 and 8.164 apply for the orientation of real and imaginary part vectors of the complex polarization vectors $\underline{\hat{\mathbf{u}}}_{\mathrm{P,SV}}(\underline{\mathbf{k}}_{\mathrm{P,S}})$ with $\underline{\hat{\mathbf{u}}}_{\mathrm{SV}} = \underline{\hat{\mathbf{u}}}_{\mathrm{S1}}$:

$$\underline{\hat{\mathbf{u}}}_{\mathrm{P}}(\underline{\mathbf{k}}_{\mathrm{P}}) = \frac{\underline{\mathbf{k}}_{\mathrm{P}}}{\sqrt{\underline{\mathbf{k}}_{\mathrm{P}} \cdot \underline{\mathbf{k}}_{\mathrm{P}}^*}}, \tag{10.6}$$

$$\underline{\hat{\mathbf{u}}}_{\mathrm{SV}}(\underline{\mathbf{k}}_{\mathrm{S}}) = \frac{\underline{\hat{\mathbf{u}}}_{\mathrm{S2}} \times \underline{\mathbf{k}}_{\mathrm{S}}}{\sqrt{\underline{\mathbf{k}}_{\mathrm{S}} \cdot \underline{\mathbf{k}}_{\mathrm{S}}^*}}; \tag{10.7}$$

here, $\underline{\hat{\mathbf{u}}}_{\mathrm{S2}}$ is a real valued unit vector with $\underline{\hat{\mathbf{u}}}_{\mathrm{S2}} \cdot \underline{\mathbf{n}} = 0$ and $\underline{\hat{\mathbf{u}}}_{\mathrm{S2}} \cdot \Re\underline{\mathbf{k}} = 0$, where $\underline{\mathbf{n}}$ denotes the normal to the surface. Even though, as already stated, a Rayleigh wave may not be excited by an incident plane wave, we stay with the "reflection related" choice of a cartesian coordinate system and configure the phase propagation vector $\Re\underline{\mathbf{k}}$ as well as the unit vector $\underline{\hat{\mathbf{u}}}_{\mathrm{S2}}$ as follows (Figures 9.1 and 9.7):

$$\Re\underline{\mathbf{k}} = -k_{\mathrm{R}}\,\underline{\mathbf{e}}_x \tag{10.8}$$

$$\underline{\hat{\mathbf{u}}}_{\mathrm{S2}} = -\underline{\mathbf{e}}_y; \tag{10.9}$$

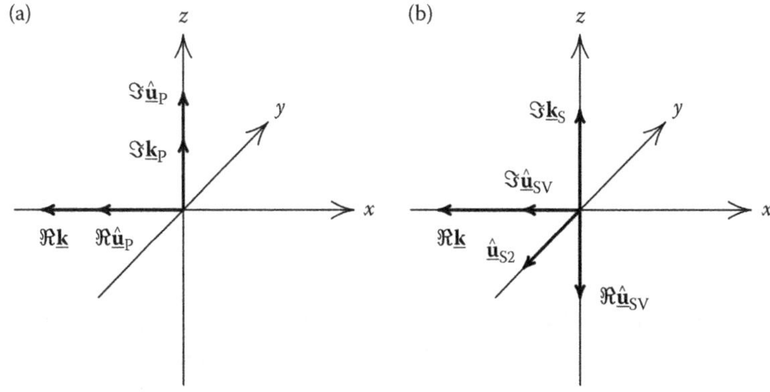

FIGURE 10.1
Complex polarization vectors of (a) the partial pressure and (b) the partial shear constituents of a Rayleigh surface wave.

since the attenuation vectors $\Im\underline{\mathbf{k}}_{\mathrm{P,S}}$ must point into the material half-space $z > 0$, we assess

$$\Im\underline{\mathbf{k}}_{\mathrm{P,S}} = \gamma_{\mathrm{P,S}}\underline{\mathbf{e}}_z \tag{10.10}$$

with

$$\gamma_{\mathrm{P,S}} = \sqrt{k_{\mathrm{R}}^2 - k_{\mathrm{P,S}}^2} > 0. \tag{10.11}$$

With (10.6) and (10.7), we obtain polarization vectors as sketched in Figure 10.1 (note Figure 8.12 while considering Footnote 151). The particle velocity vector of the potential Rayleigh wave then reads as

$$\begin{aligned}
&\underline{\mathbf{u}}_{\mathrm{R}}(\underline{\mathbf{R}}, \omega, -k_{\mathrm{R}}\underline{\mathbf{e}}_x) \\
&\quad = \underbrace{u_{\mathrm{P}}(\omega)\,\mathrm{e}^{-\gamma_{\mathrm{P}} z}\,\frac{-k_{\mathrm{R}}\underline{\mathbf{e}}_x + \mathrm{j}\gamma_{\mathrm{P}}\underline{\mathbf{e}}_z}{\sqrt{2k_{\mathrm{R}}^2 - k_{\mathrm{P}}^2}}\,\mathrm{e}^{-\mathrm{j}k_{\mathrm{R}}x}}_{=\,\underline{\mathbf{u}}_{\mathrm{P}}(x, z, \omega, -k_{\mathrm{R}}\underline{\mathbf{e}}_x)} \underbrace{-u_{\mathrm{S}}(\omega)\,\mathrm{e}^{-\gamma_{\mathrm{S}} z}\,\frac{\mathrm{j}\gamma_{\mathrm{S}}\underline{\mathbf{e}}_x + k_{\mathrm{R}}\underline{\mathbf{e}}_z}{\sqrt{2k_{\mathrm{R}}^2 - k_{\mathrm{S}}^2}}\,\mathrm{e}^{-\mathrm{j}k_{\mathrm{R}}x}}_{=\,\underline{\mathbf{u}}_{\mathrm{SV}}(x, z, \omega, -k_{\mathrm{R}}\underline{\mathbf{e}}_x)},
\end{aligned} \tag{10.12}$$

where we can rely on the stress-free boundary condition $\underline{\mathbf{e}}_z \cdot \underline{\underline{\mathbf{T}}}_{\mathrm{R}}(\underline{\mathbf{R}}_S, \omega, -k_{\mathrm{R}}\underline{\mathbf{e}}_x) = \underline{\mathbf{0}}$, $\underline{\mathbf{R}}_S \cdot \underline{\mathbf{e}}_z = 0$, to determine the phase velocity $c_{\mathrm{R}} = \omega/k_{\mathrm{R}}$ and the amplitude ratio $u_{\mathrm{P}}(\omega)/u_{\mathrm{S}}(\omega)$. With (8.121) and (8.122), we find

$$\begin{aligned}
&\underline{\mathbf{e}}_z \cdot \underline{\underline{\mathbf{T}}}_{\mathrm{R}}(\underline{\mathbf{R}}_S, \omega, -k_{\mathrm{R}}\underline{\mathbf{e}}_x) \cdot \underline{\mathbf{e}}_z \\
&\quad = \mathrm{j}\lambda\left[\underline{\mathbf{k}}_{\mathrm{P}} \cdot \underline{\mathbf{u}}_{\mathrm{P}}(x, 0, \omega, -k_{\mathrm{R}}\underline{\mathbf{e}}_x) + \underline{\mathbf{k}}_{\mathrm{S}} \cdot \underline{\mathbf{u}}_{\mathrm{SV}}(x, 0, \omega, -k_{\mathrm{R}}\underline{\mathbf{e}}_x)\right] \\
&\qquad + 2\mathrm{j}\mu\left[\underline{\mathbf{k}}_{\mathrm{P}} \cdot \underline{\mathbf{e}}_z\underline{\mathbf{u}}_{\mathrm{P}}(x, 0, \omega, -k_{\mathrm{R}}\underline{\mathbf{e}}_x) \cdot \underline{\mathbf{e}}_z + \underline{\mathbf{k}}_{\mathrm{S}} \cdot \underline{\mathbf{e}}_z\underline{\mathbf{u}}_{\mathrm{SV}}(x, 0, \omega, -k_{\mathrm{R}}\underline{\mathbf{e}}_x) \cdot \underline{\mathbf{e}}_z\right]
\end{aligned} \tag{10.13}$$

for the normal component $\underline{\mathbf{e}}_z \cdot \underline{\underline{\mathbf{T}}}_{\mathrm{R}}(\underline{\mathbf{R}}_S, \omega, -k_{\mathrm{R}}\underline{\mathbf{e}}_x) \cdot \underline{\mathbf{e}}_z$ and

$$\begin{aligned}
&\underline{\mathbf{e}}_z \cdot \underline{\underline{\mathbf{T}}}_{\mathrm{R}}(\underline{\mathbf{R}}_S, \omega, -k_{\mathrm{R}}\underline{\mathbf{e}}_x) \times \underline{\mathbf{e}}_z \\
&\quad = \mathrm{j}\mu \Big[\underline{\mathbf{k}}_{\mathrm{P}} \cdot \underline{\mathbf{e}}_z \underline{\mathbf{u}}_{\mathrm{P}}(x, 0, \omega, -k_{\mathrm{R}}\underline{\mathbf{e}}_x) \times \underline{\mathbf{e}}_z + \underline{\mathbf{u}}_{\mathrm{P}}(x, 0, \omega, -k_{\mathrm{R}}\underline{\mathbf{e}}_x) \cdot \underline{\mathbf{e}}_z \underline{\mathbf{k}}_{\mathrm{P}} \times \underline{\mathbf{e}}_z \\
&\qquad + \underline{\mathbf{k}}_{\mathrm{S}} \cdot \underline{\mathbf{e}}_z \underline{\mathbf{u}}_{\mathrm{SV}}(x, 0, \omega, -k_{\mathrm{R}}\underline{\mathbf{e}}_x) \times \underline{\mathbf{e}}_z + \underline{\mathbf{u}}_{\mathrm{SV}}(x, 0, \omega, -k_{\mathrm{R}}\underline{\mathbf{e}}_x) \cdot \underline{\mathbf{e}}_z \underline{\mathbf{k}}_{\mathrm{S}} \times \underline{\mathbf{e}}_z\Big]
\end{aligned} \tag{10.14}$$

for the tangential (vector) component $\underline{\mathbf{e}}_z \cdot \underline{\underline{\mathbf{T}}}_{\mathrm{R}}(\underline{\mathbf{R}}_S, \omega, -k_{\mathrm{R}}\underline{\mathbf{e}}_x) \times \underline{\mathbf{e}}_z$. Since phase matching is already assured through the joint factor $\mathrm{e}^{-\mathrm{j}k_{\mathrm{R}}x}$ of $\underline{\mathbf{u}}_{\mathrm{P}}$ and $\underline{\mathbf{u}}_{\mathrm{SV}}$, we obtain a homogeneous system of equations equating (10.13) and (10.14) to zero[197] for the amplitudes $u_{\mathrm{P}}(\omega)$ and $u_{\mathrm{S}}(\omega)$; this system has a nontrivial solution only if its coefficient determinant is vanishing, hence a conditional equation for the Rayleigh wave number k_{R} and the Rayleigh slowness $s_{\mathrm{R}} = 1/c_{\mathrm{R}} = k_{\mathrm{R}}/\omega$ emerges after some calculation:

$$\left(2s_{\mathrm{R}}^2 - s_{\mathrm{S}}^2\right)^2 - 4s_{\mathrm{R}}^2\sqrt{s_{\mathrm{R}}^2 - s_{\mathrm{S}}^2}\sqrt{s_{\mathrm{R}}^2 - s_{\mathrm{P}}^2} = 0. \tag{10.15}$$

As a matter of fact, this equation has a real valued solution $s_{\mathrm{R}} > s_{\mathrm{P,S}}$—it is shown by Achenbach (1973) using complex function analysis—and it is explicitly found applying Cardani's formulas (Vinh and Ogden 2004).

Utilizing the Rayleigh equation we immediately obtain the (frequency-dependent) amplitude ratio from (10.14)

$$\frac{u_{\mathrm{P}}(\omega)}{u_{\mathrm{S}}(\omega)} = \frac{1}{2\mathrm{j}k_{\mathrm{R}}\gamma_{\mathrm{P}}}\sqrt{2k_{\mathrm{R}}^2 - k_{\mathrm{S}}^2}\sqrt{2k_{\mathrm{R}}^2 - k_{\mathrm{P}}^2}, \tag{10.16}$$

and hence

$$u_{\mathrm{R}x}(x, z, \omega, -k_{\mathrm{R}}\underline{\mathbf{e}}_x) = \frac{\mathrm{j}\gamma_{\mathrm{S}}}{\sqrt{2k_{\mathrm{R}}^2 - k_{\mathrm{S}}^2}}\, u_{\mathrm{S}}(\omega)\left(\frac{2k_{\mathrm{R}}^2 - k_{\mathrm{S}}^2}{2\gamma_{\mathrm{P}}\gamma_{\mathrm{S}}}\,\mathrm{e}^{-\gamma_{\mathrm{P}}z} - \mathrm{e}^{-\gamma_{\mathrm{S}}z}\right)\mathrm{e}^{-\mathrm{j}k_{\mathrm{R}}x}, \tag{10.17}$$

$$u_{\mathrm{R}z}(x, z, \omega, -k_{\mathrm{R}}\underline{\mathbf{e}}_x) = \frac{k_{\mathrm{R}}}{\sqrt{2k_{\mathrm{R}}^2 - k_{\mathrm{S}}^2}}\, u_{\mathrm{S}}(\omega)\left(\frac{2k_{\mathrm{R}}^2 - k_{\mathrm{S}}^2}{2k_{\mathrm{R}}^2}\,\mathrm{e}^{-\gamma_{\mathrm{P}}z} - \mathrm{e}^{-\gamma_{\mathrm{S}}z}\right)\mathrm{e}^{-\mathrm{j}k_{\mathrm{R}}z} \tag{10.18}$$

for the x- and z-components of the Rayleigh wave particle velocity. Both components are displayed in Figure 10.2 for steel as function of z. Since $u_{\mathrm{R}x}$ is purely imaginary due to the j-factor, and since the magnitudes of the components are different, the Rayleigh wave is elliptically polarized in the xz-plane (not in a plane perpendicular to the propagation direction); a similar behavior has already been observed for "Fresnel's" evanescent waves (e.g., Figure 9.8).

[197]The excitation by an incident wave is missing.

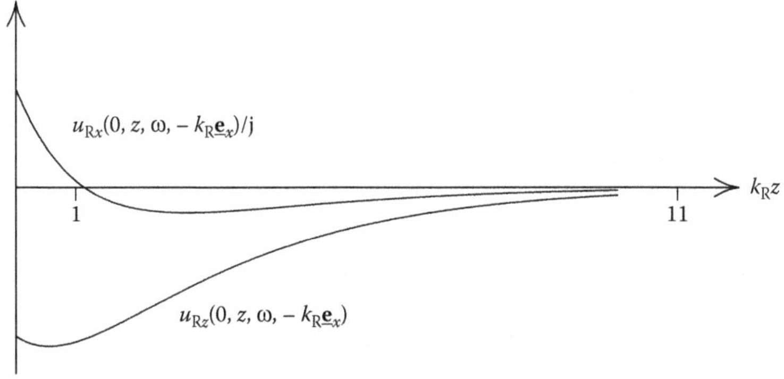

FIGURE 10.2
Particle displacement components of a Rayleigh surface wave ($c_P = 5900$ m/s, $c_S = 3200$ m/s, $c_R = 2963$ m/s).

10.2 Lightly Curved Surfaces

For lightly curved stress-free surfaces, we can equally prove the existence of Rayleigh surface waves with the same phase velocity c_R as for planar surfaces utilizing the formalism of the preceding section. For the sake of simplicity, we postulate a circular cylindrical surface with radius $a \gg \lambda_{P,S}$ (Figure 10.3) and introduce the orthogonal trihedron $\underline{\mathbf{e}}_\rho, \underline{\mathbf{e}}_s, \underline{\mathbf{e}}_z$ of the so-called Dupin coordinates (orthogonal curvilinear surface coordinates combined with the normal): ρ is

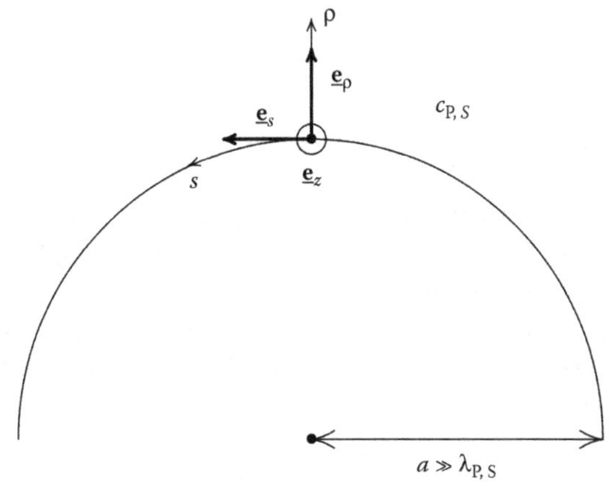

FIGURE 10.3
Dupin coordinates $\underline{\mathbf{e}}_\rho, \underline{\mathbf{e}}_s, \underline{\mathbf{e}}_z$ of a circular cylindrical surface.

the radial variable counted from the boundary at $\rho = 0$ into the (nondissipative homogeneous isotropic) $c_{P,S}$-material, s is an arc length variable, and $\underline{\mathbf{e}}_z$ characterizes the cylinder axis.

Again, a Rayleigh wave is found by superimposing inhomogeneous P- and SV-waves with wave number vectors

$$\underline{\mathbf{k}}_{P,S} = k_R \underline{\mathbf{e}}_s + j\gamma_{P,S} \underline{\mathbf{e}}_\rho \tag{10.19}$$

that obviously satisfy the necessary constraint

$$\Re \underline{\mathbf{k}}_{P,S} \cdot \Im \underline{\mathbf{k}}_{P,S} = 0. \tag{10.20}$$

The dispersion relation corresponding to (10.4) must be found from the reduced wave equation in Dupin coordinates. Therefore, we come back to two-dimensional Helmholtz potentials $\Phi(\rho, s, \omega, k_R)$ and $\underline{\Psi}(\rho, s, \omega, k_R) = \Psi(\rho, s, \omega, k_R)\underline{\mathbf{e}}_z$ with

$$\Delta\Phi(\rho, s, \omega, k_R) + k_P^2 \Phi(\rho, s, \omega, k_R) = 0, \tag{10.21}$$

$$\Delta\Psi(\rho, s, \omega, k_R) + k_S^2 \Psi(\rho, s, \omega, k_R) = 0, \tag{10.22}$$

adopting the Δ-operator according to

$$\Delta = \frac{\partial^2}{\partial \rho^2} + \frac{1}{\rho + a}\frac{\partial}{\partial \rho} + \frac{a^2}{(\rho + a)^2}\frac{\partial^2}{\partial s^2} \tag{10.23}$$

from circular cylindrical coordinates: Obviously, we have $s = a\varphi$ and $\rho = r - a$. With the *ansatz*

$$\Phi(\rho, s, \omega, k_R) = \Phi_0(\omega)\, e^{jk_R s - \gamma_P \rho}, \tag{10.24}$$

$$\Psi(\rho, s, \omega, k_R) = \Psi_0(\omega)\, e^{jk_R s - \gamma_S \rho} \tag{10.25}$$

for locally plane inhomogeneous P- and SV-waves traveling along the surface with the wave number vectors (10.19), we find

$$\gamma_{P,S}^2 - \frac{1}{\rho + a}\gamma_{P,S} - \frac{a^2}{(\rho + a)^2} k_R^2 + k_{P,S}^2 = 0. \tag{10.26}$$

The *ansatz* (10.24) and (10.25) would only be a solution of (10.21) and (10.22) if the Rayleigh wave number k_R would be ρ-dependent—logical: the larger ρ the faster the wave has to be in order not to be delayed—in that case, $k_R(\rho)$ would also have to be differentiated while being inserted into the differential equations. Hence, we confine ourselves to $\rho \simeq 0$—the resulting Rayleigh wave should be a surface wave—and obtain

$$\gamma_{P,S}^2 - \frac{1}{a}\gamma_{P,S} - k_R^2 + k_{P,S}^2 = 0. \tag{10.27}$$

Figure 10.2 additionally exhibits that $\gamma_{P,S} \simeq 1/\lambda_{P,S}$, i.e., $\gamma_{P,S}^2 - \gamma_{P,S}/a$ is approximately given by $(1 - \lambda_{P,S}/a)/\lambda_{P,S}^2$, where we have neglected the term

$\lambda_{\mathrm{P,S}}/a$ for $a \gg \lambda_{\mathrm{P,S}}$. Therefore, we find the same dispersion relation for lightly curved surfaces

$$k_{\mathrm{R}}^2 = k_{\mathrm{P,S}}^2 + \gamma_{\mathrm{P,S}}^2 \tag{10.28}$$

as for planar surfaces. If we assume $\rho \simeq 0$ after having performed the derivatives $\nabla\Phi$ and $\nabla \times \underline{\Psi}$ with

$$\nabla = \underline{\mathbf{e}}_\rho \frac{\partial}{\partial\rho} + \frac{a}{\rho + a} \underline{\mathbf{e}}_s \frac{\partial}{\partial s}, \tag{10.29}$$

we obtain

$$\underline{\mathbf{u}}_{\mathrm{R}}(\rho, s, \omega, k_{\mathrm{R}}) = u_{\mathrm{P}}(\omega)\, \mathrm{e}^{\mathrm{j}k_{\mathrm{R}}s - \gamma_{\mathrm{P}}\rho}\, \underline{\hat{\mathbf{u}}}_{\mathrm{P}} + u_{\mathrm{S}}(\omega)\, \mathrm{e}^{\mathrm{j}k_{\mathrm{R}}s - \gamma_{\mathrm{S}}\rho}\, \underline{\hat{\mathbf{u}}}_{\mathrm{SV}}, \tag{10.30}$$

where $\underline{\hat{\mathbf{u}}}_{\mathrm{P}}$ and $\underline{\hat{\mathbf{u}}}_{\mathrm{SV}}$ are given by

$$\underline{\hat{\mathbf{u}}}_{\mathrm{P}} = \frac{\mathrm{j}\gamma_{\mathrm{P}}\underline{\mathbf{e}}_\rho + k_{\mathrm{R}}\underline{\mathbf{e}}_s}{\sqrt{2k_{\mathrm{R}}^2 - k_{\mathrm{P}}^2}}, \tag{10.31}$$

$$\underline{\hat{\mathbf{u}}}_{\mathrm{SV}} = \frac{-k_{\mathrm{R}}\underline{\mathbf{e}}_\rho + \mathrm{j}\gamma_{\mathrm{S}}\underline{\mathbf{e}}_s}{\sqrt{2k_{\mathrm{R}}^2 - k_{\mathrm{S}}^2}}. \tag{10.32}$$

Comparison with (10.12) reveals that we obtain an expression for $\underline{\mathbf{u}}_{\mathrm{R}}$ in the $\underline{\mathbf{e}}_\rho\underline{\mathbf{e}}_s$-coordinate system of the lightly curved surface that is equally structured as the one for planar surfaces. Once more assuming $\rho \simeq 0$ after the calculation of the stress tensor components

$$\underline{\underline{\mathbf{T}}}_{\mathrm{R}} = \lambda \underline{\underline{\mathbf{I}}}\, \nabla \cdot \underline{\mathbf{u}}_{\mathrm{R}} + \mu \left[\nabla\underline{\mathbf{u}}_{\mathrm{R}} + (\nabla\underline{\mathbf{u}}_{\mathrm{R}})^{21}\right], \tag{10.33}$$

and—as in the dispersion equation—neglecting terms of the order[198] $\gamma_{\mathrm{P,S}}/a$ and k_{R}/a with respect to quadratic or product terms of k_{R} and $\gamma_{\mathrm{P,S}}$ the boundary condition $\underline{\mathbf{e}}_\rho \cdot \underline{\underline{\mathbf{T}}}_{\mathrm{R}} = \underline{\mathbf{0}}$ for $\rho = 0$ yields a homogeneous system of equations for the components $\underline{\mathbf{e}}_\rho \cdot \underline{\underline{\mathbf{T}}}_{\mathrm{R}} \cdot \underline{\mathbf{e}}_\rho$ and $\underline{\mathbf{e}}_\rho \cdot \underline{\underline{\mathbf{T}}}_{\mathrm{R}} \cdot \underline{\mathbf{e}}_s$, leading once more to the particle displacement components (10.7) and (10.8) and the Rayleigh equation via equating the determinant to zero.

[198]These terms result from divergence and gradient calculation of $\underline{\hat{\mathbf{u}}}_{\mathrm{P,SV}}$ in $\underline{\mathbf{e}}_\rho\underline{\mathbf{e}}_s$-coordinates. It is only within this approximation that the first term of (10.30) represents a pressure wave, and the second one a shear wave.

11

Plane Wave Spatial Spectrum

Due to their infinite energy, plane waves are not immediately appropriate for applications in US-NDT. Yet, as solutions of homogeneous wave equations, they may be superimposed with different amplitudes into arbitrary propagation directions: We obtain a spatial spectrum of plane waves that is literally a spatial spectrum because the propagation directions of the respective plane waves are given by the Fourier vector $\underline{\mathbf{K}}$ as conjugate vector to $\underline{\mathbf{R}}$! Transducer radiation fields and Gaussian beams may be adequately described by these spectra (Chapters 14 and 12). Formally, we find the spatial plane wave spectrum by Fourier transforming the reduced wave equations with respect to two Cartesian coordinates; consequently, the resulting two-dimensional inverse Fourier integral has to be interpreted accordingly. To avoid unnecessary algebraic ballast, we will first deal with scalar acoustic wave spectra.

11.1 Acoustic Plane Wave Spatial Spectrum

11.1.1 Plane wave spatial spectrum

We start with the homogeneous acoustic Helmholtz equation (5.40)

$$\Delta p(x, y, z, \omega) + k^2 p(x, y, z, \omega) = 0 \tag{11.1}$$

for the frequency spectrum of the acoustic pressure $p(x, y, z, \omega)$ in cartesian coordinates and take its Fourier transform with regard x and y:

$$\frac{\partial^2}{\partial z^2}\hat{p}(K_x, K_y, z, \omega) + (k^2 - K_x^2 - K_y^2)\hat{p}(K_x, K_y, z, \omega) = 0. \tag{11.2}$$

As solution of this one-dimensional differential equation, we can immediately write down a plane wave propagating in $+z$-direction

$$\hat{p}(K_x, K_y, z, \omega) = \hat{p}_0(K_x, K_y, \omega)\, \mathrm{e}^{\mathrm{j}K_z(K_x, K_y, \omega)z} \tag{11.3}$$

with the wave number

$$K_z(K_x, K_y, \omega) = \sqrt{k^2 - K_x^2 - K_y^2}, \tag{11.4}$$

and the arbitrary amplitude $\hat{p}_0(K_x, K_y, \omega)$ for $z = 0$. The two-dimensional inverse Fourier integral

$$p(x,y,z,\omega) = \frac{1}{(2\pi)^2}\int_{-\infty}^{\infty}\int_{-\infty}^{\infty}\hat{p}_0(K_x,K_y,\omega)\mathrm{e}^{\mathrm{j}K_z(K_x,K_y,\omega)z}\mathrm{e}^{\mathrm{j}K_x x+\mathrm{j}K_y y}\,\mathrm{d}K_x\mathrm{d}K_y \tag{11.5}$$

may immediately be written as superposition of plane waves according to

$$p(x,y,z,\omega) = \frac{1}{(2\pi)^2}\int_{-\infty}^{\infty}\int_{-\infty}^{\infty}\hat{p}_0(K_x,K_y,\omega)\,\mathrm{e}^{\mathrm{j}\underline{\mathbf{K}}\cdot\underline{\mathbf{R}}}\,\mathrm{d}K_x\mathrm{d}K_y \tag{11.6}$$

if a phase vector

$$\begin{aligned}\underline{\mathbf{K}} &= K_x\underline{\mathbf{e}}_x + K_y\underline{\mathbf{e}}_y + K_z\underline{\mathbf{e}}_z\\ &= K_x\underline{\mathbf{e}}_x + K_y\underline{\mathbf{e}}_y + \sqrt{k^2 - K_x^2 - K_y^2}\,\underline{\mathbf{e}}_z\end{aligned} \tag{11.7}$$

is defined that satisfies the dispersion relation

$$\underline{\mathbf{K}}\cdot\underline{\mathbf{K}} = k^2. \tag{11.8}$$

Since $-\infty < K_x < \infty$, $-\infty < K_y < \infty$, and due to

$$K_z = \begin{cases}\sqrt{k^2 - K_x^2 - K_y^2} & \text{for } K_x^2 + K_y^2 \le k^2\\ \pm\mathrm{j}\sqrt{K_x^2 + K_y^2 - k^2} & \text{for } K_x^2 + K_y^2 > k^2,\end{cases} \tag{11.9}$$

$\underline{\mathbf{K}}$ may become a complex vector

$$\underline{\mathbf{K}} = \Re\underline{\mathbf{K}} + \mathrm{j}\Im\underline{\mathbf{K}} \tag{11.10}$$

for $K_x^2 + K_y^2 > k^2$; its imaginary part

$$\Im\underline{\mathbf{K}} = \sqrt{K_x^2 + K_y^2 - k^2}\,\underline{\mathbf{e}}_z \tag{11.11}$$

has to be chosen with a positive sign to ensure an exponential attenuation

$$\mathrm{e}^{\mathrm{j}\Re\underline{\mathbf{K}}\cdot\underline{\mathbf{R}} - \Im\underline{\mathbf{K}}\cdot\underline{\mathbf{R}}} \tag{11.12}$$

in the half-space $z > 0$ (Figure 8.11). The dispersion relation (11.8) allows us to calculate K_z according to

$$K_x^2 + K_y^2 + K_z^2 = k^2 \tag{11.13}$$

for K_x, K_y-values in the interior of the circle $K_x^2 + K_y^2 \le k^2$—resulting in propagating spectral components of the spatial plane wave spectrum (11.6)—and K_z according to

$$K_x^2 + K_y^2 - K_z^2 = k^2 \tag{11.14}$$

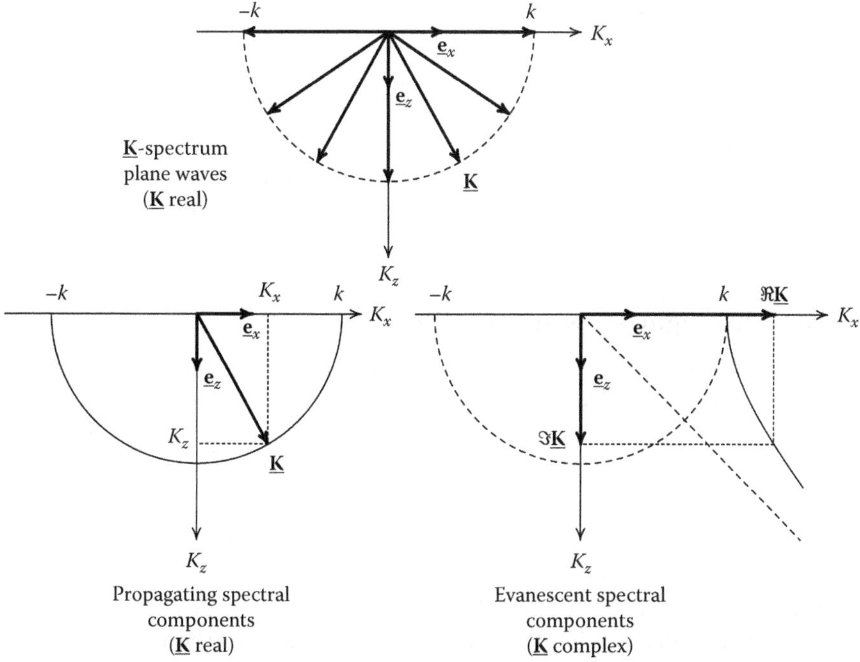

FIGURE 11.1
Plane wave spatial spectrum.

for K_x, K_y-values in the exterior of the circle $K_x^2 + K_y^2 \leq k^2$—resulting in evanescent spectral components. Figure 11.1 illustrates these facts for $K_y = 0$; we have encountered them numerous times in Chapter 9 for a single plane wave with phase vector $\underline{\mathbf{k}}$. Here, we integrate over all phase vectors $\underline{\mathbf{K}}$ according to (11.7), where each spatial spectral component, i.e., each single propagating or evanescent plane wave, is given an arbitrary amplitude $\hat{p}_0(K_x, K_y, \omega)$. According to (11.5), this amplitude is the two-dimensional Fourier transform of the pressure distribution $p(x, y, z = 0, \omega) = p_0(x, y, \omega)$ in the xy-plane for $z = 0$:

$$\hat{p}_0(K_x, K_y, \omega) = \int_{-\infty}^{\infty}\int_{-\infty}^{\infty} p_0(x, y, \omega)\,\mathrm{e}^{-\mathrm{j}K_x x - \mathrm{j}K_y y}\,\mathrm{d}x\mathrm{d}y. \tag{11.15}$$

Therefore, we can finally interpret (11.6) like so: A given field distribution in the xy-plane is "transported" into the half-space $z > 0$ by means of a spatial spectrum of plane waves; or: Equation 11.6 is the mathematical plane wave spatial spectrum representation of the radiation field in the half-space $z > 0$ excited by an "aperture distribution" prescribed in the xy-plane for $z = 0$. This illustrates the relevance of such a representation for the calculation of transducer radiation fields; we only have to consider a physically existing aperture placed on a physically existing half-space with a planar surface

(Chapter 14). Note: Even though the contribution of the evanescent spectral components appears to be marginal, it is absolutely necessary to ensure the completeness of the spectrum (Tygel and Hubral 1987).

By the way: Equation 11.6 becomes applicable for the half-space $z < 0$ if $\underline{\mathbf{K}}$ is given a negative z-component for the propagating spectral waves and if $\Im\underline{\mathbf{K}}$ is equally given a negative z-component for the evanescent waves, i.e., in (11.3), we choose the negative z-direction as propagation direction matching the attenuation appropriately.

11.1.2 Propagator as spatial filter

The role of the evanescent waves becomes particularly apparent with another interpretation of the integral representation (11.5), namely with the interpretation of the propagator

$$\hat{P}(K_x, K_y, z, \omega) = \mathrm{e}^{\mathrm{j}z\sqrt{k^2-K_x^2-K_y^2}} \tag{11.16}$$

as spatial filter. Simultaneously, this interpretation implies an appropriate calculation of sound fields.

Regarding the Fourier variables K_x and K_y, the propagator $\hat{P}(K_x, K_y, z, \omega)$ represents a spatially varying frequency-dependent filter of the Fourier transformed initial field distribution $\hat{p}_0(K_x, K_y, \omega)$. The influence of the filter parameters z and ω are discussed with the help of Figure 11.2 clearly exhibiting that

$$|\hat{P}(K_x, K_y, z, \omega)| = \begin{cases} 1 & \text{for } k \geq \sqrt{K_x^2 + K_y^2} \\ \mathrm{e}^{-z\sqrt{K_x^2+K_y^2-k^2}} & \text{for } k < \sqrt{K_x^2 + K_y^2} \end{cases} \tag{11.17}$$

exponentially suppresses the spatial frequencies contained in $\hat{p}_0(K_x, K_y, \omega)$ for $k < \sqrt{K_x^2 + K_y^2}$ and large values of z.

For a simplified graphical display of these facts in Figure 11.2, we choose a two-dimensional problem, i.e., we assume $p_0(x, y, \omega) = p_0(x, \omega)$ being independent of y with the consequence $\hat{p}_0(K_x, K_y, \omega) = \hat{p}_0(K_x, \omega)\delta(K_y)$. As "aperture distribution" for $z = 0$, a rectangular function $p_0(x, \omega) = q_a(x)$ should be fine, it is depicted in Figure 11.2 together with its spatial spectrum $\hat{p}_0(K_x, \omega) = 2 \sin aK_x/K_x$; the aperture width $2a$ appears in the zero locations of the sinc function with the first one at $K_x = \pm\pi/a$. Due to (11.17), the transmission region of the spatial filter (11.16) for $z > 0$ is essentially determined by the wave number k: The boundaries of the spatial low-pass filter are located at $K_x = \pm k$. Due to our actual example $ka = 10$, we have $3\pi/a < ka < 4\pi/a$ allowing for a respective number of oscillations of the sinc function to pass the filter; we nicely recognize the significant exponential attenuation of the spatial filter beyond $K_x = \pm k$ already for $z = 0.5a$, and for $z = a$, we can almost speak of a perfect low-pass filter. According to the spectral superposition of plane waves, this obviously indicates that evanescent

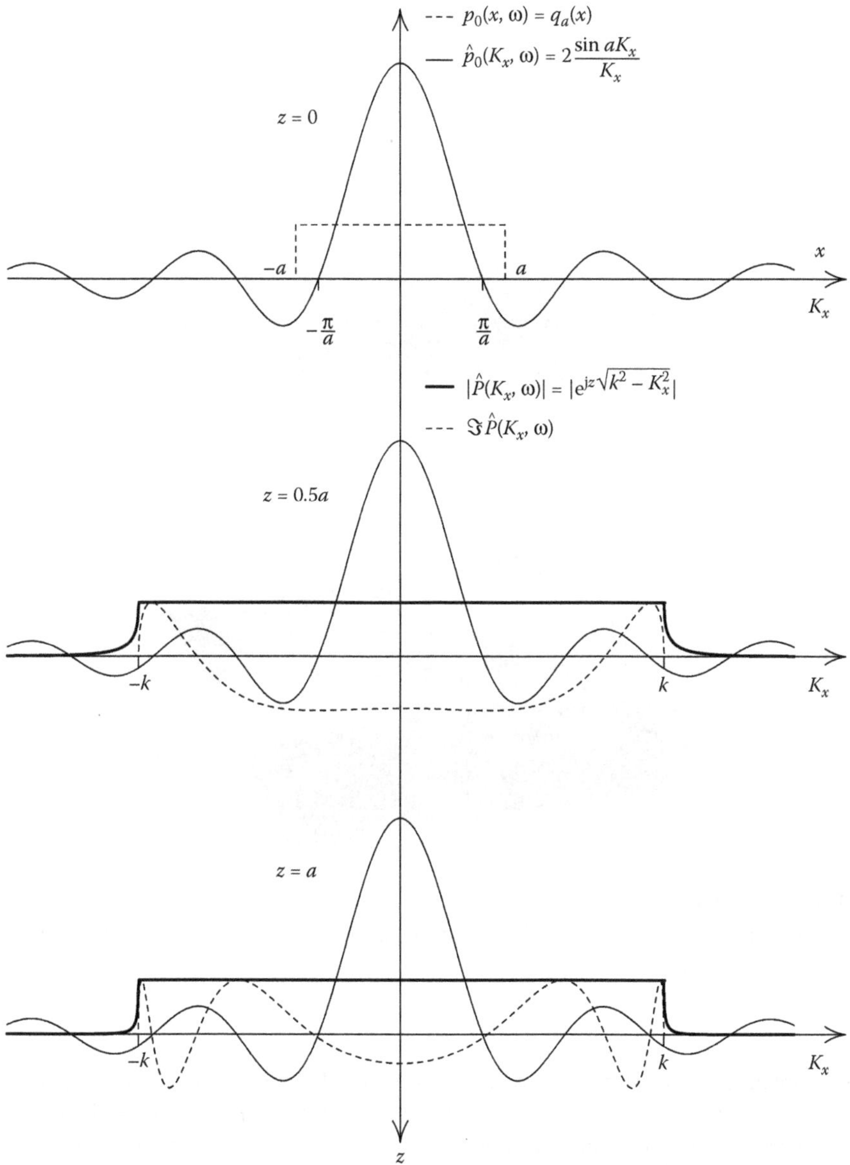

FIGURE 11.2
Interpretation of the spatial plane wave spectrum as spatial filter ($ka = 10$).

waves have already decayed. Figure 11.3 displays the result of the spatial filtering of the sinc function for $ka = 10$ and $ka = \pi$ as function of kz in terms of a three-dimensional plot; it is apparent that the interaction of aperture width and filter bandwidth determines the structure of $\hat{p}(K_x, z, \omega)$ as function of z

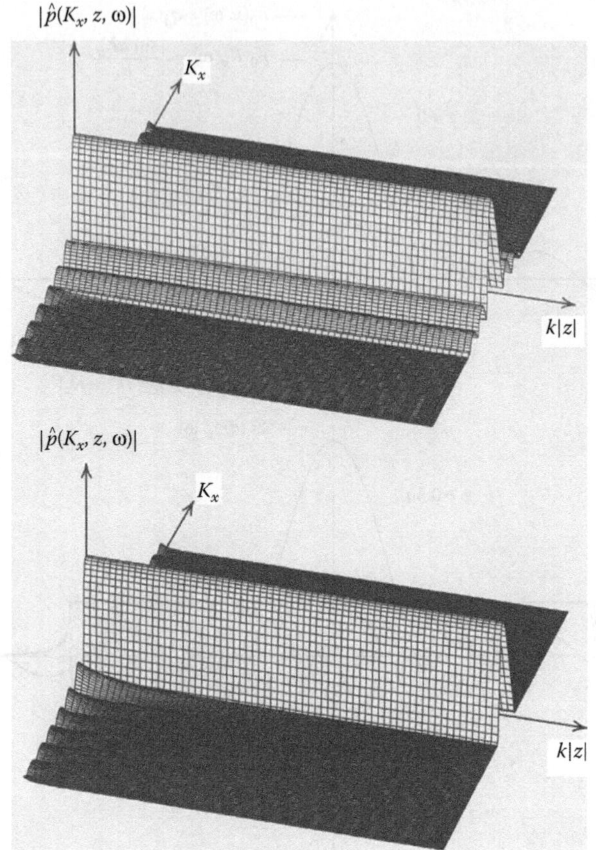

FIGURE 11.3
Spatial spectrum of a strip-like radiating aperture; $0 \leq k|z| \leq 10ka$; top: $ka = 10$, bottom: $ka = \pi$.

and, naturally, the structure of $p(x, z, \omega)$ via the inverse Fourier transform with regard to K_x: In Figure 11.4, the magnitude of $p(x, z, \omega)$ is displayed as function of x and z for the two cases $ka = 10$, and $ka = \pi$. For $ka = 10$, we recognize a well-structured near and transition region finally ending up—for $x = 0$—in the $1/\sqrt{z}$-decay of the amplitude of the far-field (square root of z because Figures 11.3 and 11.4 stand for a two-dimensional problem). For $ka = \pi$, the magnitude of the far-field parallel to the x-axis looks like a sinc function, i.e., like the Fourier transform of the initial aperture distribution for $z = 0$; we will later on indeed prove this evaluating (11.5) with the stationary phase method, and to illustrate it properly, Figure 11.2 additionally shows the imaginary part of the propagator (dashed curve): Obviously, it exhibits an increasing oscillatory behavior at the filter boundaries with increasing z,

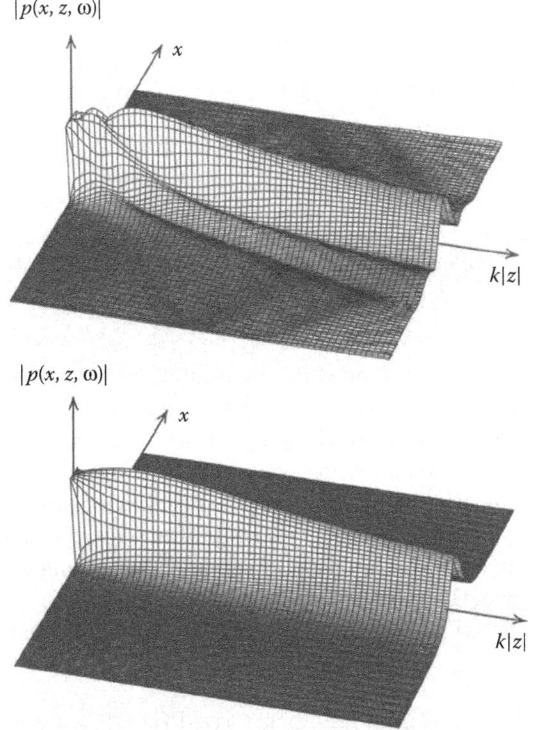

FIGURE 11.4
Time harmonic radiation field of a strip-like aperture; $0 \leq k|z| \leq 10ka$; top: $ka = 10$, bottom: $ka = \pi$.

while it remains stationary in the vicinity of $K_x = 0$. For $ka = \pi$, Figure 11.4 teaches us—the width $2a$ of the initial rectangular distribution is just as big as one wavelength—that the far-field is nothing but the Green function of a quasipoint-(here: line-)source (Sections 13.4.1 and 14.1.2).

11.1.3 Approximate evaluation with the stationary phase method

One-dimensional inverse Fourier integrals: We emphasized the convenience of the spatial plane wave spectrum to represent transducer radiation fields mathematically. Since a transducer has a finite lateral extension, we could assume an aperture distribution $p_0(x, y, \omega)$ with a compact support, i.e., being zero outside the rectangle $q_a(x)q_b(y)$. Then, we can choose a field observation point $\underline{\mathbf{R}}$ with $R \gg a, b$, and simultaneously $kR \gg 1$, implying the relevant assumptions for a far-field approximation (compare Section 13.1.3). Figure 11.4 (top) suggests that such an approximate evaluation of the inverse

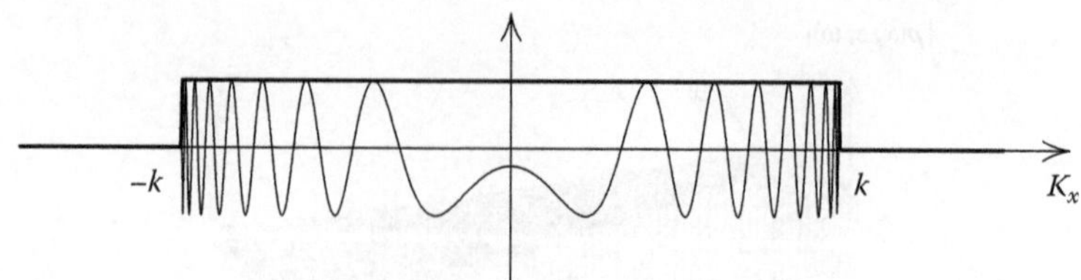

FIGURE 11.5
Magnitude (bold face) and imaginary part of the propagator for $kz = 50$ ($z = 5a$).

Fourier integral (11.5) would produce something like the Fourier transform of the aperture distribution. The method of stationary phase is based upon the strong oscillatory behavior of the propagator for $kz \gg 1$ close to the filter boundaries, whereas it remains comparatively stationary for K_x, K_y in the vicinity of the origin. Figure 11.2 indicates this behavior already for small kz, and Figure 11.5 is convincing for $kz = 50$; simultaneously, this figure confirms that evanescent waves have already decayed and do no longer contribute. Insofar, we neglect exactly these spatial components in (11.5)—more precisely: in the two-dimensional version of (11.5), i.e., in the one-dimensional inverse Fourier integral

$$p(x, z, \omega) = \frac{1}{2\pi} \int_{-\infty}^{\infty} \hat{p}_0(K_x, \omega)\, \mathrm{e}^{\mathrm{j}K_x x + \mathrm{j}K_z z}\, \mathrm{d}K_x; \tag{11.18}$$

hence:

$$p(x, z, \omega) \simeq \frac{1}{2\pi} \int_{-k}^{k} \hat{p}_0(K_x, \omega)\, \mathrm{e}^{\mathrm{j}K_x x + \mathrm{j}K_z z}\, \mathrm{d}K_x; \tag{11.19}$$

K_z is real valued in the interval $-k \leq K_x \leq k$. In the far-field, the observation direction plays the essential role suggesting the introduction of polar coordinates

$$x = \rho \sin \theta, \tag{11.20}$$

$$z = \rho \cos \theta \tag{11.21}$$

in the xz-plane (Figure 11.6: $\boldsymbol{\varrho} = \rho \sin \theta \underline{\mathbf{e}}_x + \rho \cos \theta \underline{\mathbf{e}}_z$). Our cartesian $\underline{\mathbf{e}}_x\underline{\mathbf{e}}_z$-bipod is equally used for cartesian $\underline{\mathbf{K}}$-space coordinates (Figure 11.6), i.e., we put

$$K_x = k \sin \theta_K, \tag{11.22}$$

$$K_z = k \cos \theta_K. \tag{11.23}$$

Consequently, the integral (11.19) is written as:

$$p(x, z, \omega) \simeq \frac{k}{2\pi} \int_{-\pi/2}^{\pi/2} \hat{p}_0(K_x, \omega)\, \mathrm{e}^{\mathrm{j}k\rho \cos(\theta - \theta_K)}\, \cos \theta_K\, \mathrm{d}\theta_K. \tag{11.24}$$

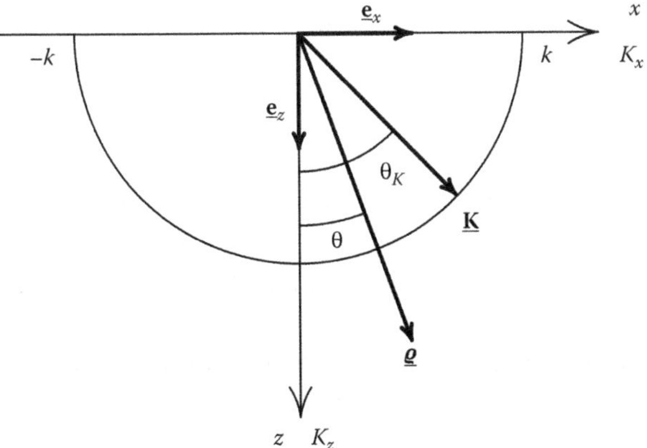

FIGURE 11.6
Polar coordinates in the spatial and spectral domain.

For $k\rho \gg 1$, the essential contributions arise from the neighborhood of the stationary phase point defined by

$$\frac{\mathrm{d}}{\mathrm{d}\theta_K}\cos(\theta - \theta_K) = \sin(\theta - \theta_K) \tag{11.25}$$

being zero; it follows $\theta_K = \theta$, i.e., the phase is stationary for a spatial spectral component if we observe exactly in this direction, or, differently expressed, that plane wave within the spatial spectrum propagating into the observation directions mostly contributes to the integral. Therefore, we truncate the Taylor series

$$\cos(\theta - \theta_K) = 1 - \frac{1}{2}(\theta - \theta_K)^2 \tag{11.26}$$

after the quadratic term and calculate

$$p^{\mathrm{far}}(\rho, \theta, \omega) = \frac{k}{2\pi}\cos\theta\, \mathrm{e}^{\mathrm{j}k\rho}\hat{p}_0(k\sin\theta, \omega)\underbrace{\int_{-\pi/2}^{\pi/2} \mathrm{e}^{-\mathrm{j}\frac{k\rho}{2}(\theta-\theta_K)^2}\,\mathrm{d}\theta_K}_{=\int_{\theta-\pi/2}^{\theta+\pi/2}\mathrm{e}^{-\mathrm{j}\frac{k\rho}{2}\theta_K^2}\,\mathrm{d}\theta_K} . \tag{11.27}$$

Now we wish to extend the integration boundaries in (11.27) to infinity in order to make the correspondence (2.280) applicable; this should be approximately feasible because the strongly oscillating integrand for $k\rho \gg 1$ would make sure to cancel the contributions for $\theta_K \longrightarrow \pm\infty$; yet the resulting integral does not converge in the sense of classical analysis. Therefore, we "smuggle" a convergence generating factor in calculating

$$\lim_{\alpha' \to 0} \int_{-\infty}^{\infty} e^{-\frac{1}{4\alpha}\omega^2} e^{-j\omega t} \, d\omega \Big|_{t=0} = \lim_{\alpha' \to 0} 2\pi \sqrt{\frac{\alpha}{\pi}}$$
$$= 2\pi \sqrt{\frac{-j}{\pi\alpha''}} \tag{11.28}$$

with a complex valued $1/\alpha = \alpha' + j\alpha''$. Application to (11.27) yields

$$p^{\text{far}}(\rho, \theta, \omega) = k \cos\theta \, \hat{p}_0(k \sin\theta, \omega) \, \frac{e^{jk\rho - j\frac{\pi}{4}}}{\sqrt{2\pi k\rho}}. \tag{11.29}$$

In fact, in the far-field, we "see" the Fourier transform of the aperture distribution according to Figure 11.4 (top). The term $e^{jk\rho}/\sqrt{\rho}$ represents a cylindrical wave with an amplitude proportional to $\cos\theta \, \hat{p}_0(k \sin\theta, \omega)$ impressed in the direction θ. The respective representation of source fields with the help of Green functions is given in Sections 13.1.4 and 14.1.2.

Two-dimensional inverse Fourier integrals: Schmerr (1998) explicitly performs the evaluation of two-dimensional inverse Fourier integrals[199]

$$p(x, y, z, \omega)$$
$$= \frac{1}{(2\pi)^2} \int_{-\infty}^{\infty} \int_{-\infty}^{\infty} \hat{p}_0(K_x, K_y, \omega) \, e^{jz\sqrt{k^2 - K_x^2 - K_y^2}} e^{jK_x x + jK_y y} \, dK_x dK_y \tag{11.30}$$

with the stationary phase method; spherical coordinates are introduced in the spatial domain, and the resulting integral

$$p(R, \vartheta, \varphi, \omega)$$
$$= \frac{1}{(2\pi)^2} \int_{-\infty}^{\infty} \int_{-\infty}^{\infty} \hat{p}_0(K_x, K_y, \omega)$$
$$\times e^{jR\cos\vartheta\sqrt{k^2 - K_x^2 - K_y^2} + jK_x R \sin\vartheta \cos\varphi + jK_y R \sin\vartheta \sin\varphi} \, dK_x dK_y \tag{11.31}$$

is asymptotically evaluated for $kR \gg 1$. We cite the result:

$$p^{\text{far}}(R, \vartheta, \varphi, \omega) = -\frac{j}{2\pi} \, k \, \hat{p}_0(k \sin\vartheta \cos\varphi, k \sin\vartheta \sin\varphi, \omega) \, \cos\vartheta \, \frac{e^{jkR}}{R}. \tag{11.32}$$

11.2 Elastic Plane Wave Spatial Spectrum

The spatial spectrum of elastic plane waves is readily obtained utilizing the Helmholtz decomposition (7.28) of the particle displacement; the homogeneous

[199] Yet, in the last line of his Equation E.21, a minus sign is missing.

scalar Helmholtz equation for $\Phi(x,y,z,\omega)$ is solved analogous to (11.6):

$$\Phi(x,y,z,\omega) = \frac{1}{(2\pi)^2}\int_{-\infty}^{\infty}\int_{-\infty}^{\infty}\hat{\Phi}_0(K_x,K_y,\omega)\,\mathrm{e}^{\mathrm{j}\underline{\mathbf{K}}_\mathrm{P}\cdot\underline{\mathbf{R}}}\,\mathrm{d}K_x\mathrm{d}K_y, \tag{11.33}$$

where

$$\underline{\mathbf{K}}_\mathrm{P} = K_x\underline{\mathbf{e}}_x + K_y\underline{\mathbf{e}}_y + \underbrace{\sqrt{k_\mathrm{P}^2 - K_x^2 - K_y^2}\,\underline{\mathbf{e}}_z}_{=K_{\mathrm{P}z}}; \tag{11.34}$$

in Cartesian coordinates, we find the solution of the homogeneous vector Helmholtz equation for $\underline{\mathbf{\Psi}}(x,y,z,\omega)$ component wise as (11.33):

$$\underline{\mathbf{\Psi}}(x,y,z,\omega) = \frac{1}{(2\pi)^2}\int_{-\infty}^{\infty}\int_{-\infty}^{\infty}\hat{\underline{\mathbf{\Psi}}}_0(K_x,K_y,\omega)\,\mathrm{e}^{\mathrm{j}\underline{\mathbf{K}}_\mathrm{S}\cdot\underline{\mathbf{R}}}\,\mathrm{d}K_x\mathrm{d}K_y, \tag{11.35}$$

where

$$\underline{\mathbf{K}}_\mathrm{S} = K_x\underline{\mathbf{e}}_x + K_y\underline{\mathbf{e}}_y + \underbrace{\sqrt{k_\mathrm{S}^2 - K_x^2 - K_y^2}\,\underline{\mathbf{e}}_z}_{=K_{\mathrm{S}z}}. \tag{11.36}$$

The spatial spectrum $\hat{\Phi}_0(K_x,K_y,\omega)$ is the two-dimensional Fourier transform of the arbitrarily given aperture distribution $\Phi_0(x,y,\omega) = \Phi(x,y,z=0,\omega)$ of the pressure wave spectrum; yet, the spatial shear wave spectrum (11.35) has to satisfy the constraint $\nabla\cdot\underline{\mathbf{\Psi}}(x,y,z,\omega) = 0$ (Equation 7.29), and this is carried over for $z=0$ to (the hat does not characterize a unit vector)

$$\underline{\mathbf{K}}_\mathrm{S}\cdot\hat{\underline{\mathbf{\Psi}}}_0(K_x,K_y,\omega) = 0 \tag{11.37}$$

in the spectral domain, i.e., only two (Cartesian) components of $\underline{\mathbf{\Psi}}_0(x,y,\omega) = \underline{\mathbf{\Psi}}(x,y,z=0,\omega)$ can be arbitrarily prescribed. The evaluation of the differentiations (7.28) on (11.33) and (11.35) finally yields the spatial spectrum of elastic plane pressure and shear waves:

$$\begin{aligned}\underline{\mathbf{u}}&(x,y,z,\omega)\\ &= \frac{1}{(2\pi)^2}\int_{-\infty}^{\infty}\int_{-\infty}^{\infty}\\ &\quad\times\left[\mathrm{j}\underline{\mathbf{K}}_\mathrm{P}\hat{\Phi}_0(K_x,K_y,\omega)\,\mathrm{e}^{\mathrm{j}\underline{\mathbf{K}}_\mathrm{P}\cdot\underline{\mathbf{R}}} + \mathrm{j}\underline{\mathbf{K}}_\mathrm{S}\times\hat{\underline{\mathbf{\Psi}}}_0(K_x,K_y,\omega)\,\mathrm{e}^{\mathrm{j}\underline{\mathbf{K}}_\mathrm{S}\cdot\underline{\mathbf{R}}}\right]\mathrm{d}K_x\mathrm{d}K_y.\end{aligned} \tag{11.38}$$

In Section 14.2.2, we will utilize the representation (11.38) to calculate the radiation field of a given strip-like normal as well as tangential force density on the stress-free surface of an elastic half-space. The respective far-field evaluation with the stationary phase method gives us the Miller–Pursey point directivities (Miller and Pursey 1954).

12

Ultrasonic Beams and Wave Packets

US-NDT language often uses the term "ultrasonic beam" instead of "ultrasonic wave"; it is advantageous that we may distinguish between "beams" and "rays" for a correct mathematical definition of respective approximate solutions of homogeneous wave equations: Basically, rays come as generalization of time harmonic plane waves to time harmonic locally plane waves in inhomogeneous materials—they are geometric trajectories of a wave normal—whereas time harmonic beams are paraxial solutions of homogeneous reduced wave equations propagating along rays and keeping their functional, for example, Gaussian, structure; as wave packets or pulsed beams, they represent respective impulse solutions physically visible in the time domain. Beams have a lot of conceptual, mathematical, and practical benefits as compared to rays.

12.1 Gaussian Beams as Paraxial Approximation of a Spatial Plane Wave Spectrum

In homogeneous isotropic materials, the plane wave spatial spectrum as solution of a homogeneous Helmholtz equation (Section 11.1)

$$\Delta p(x, y, z, \omega) + k^2 p(x, y, z, \omega) = 0 \tag{12.1}$$

offers a simple access to scalar acoustic Gaussian beams accomplishing the paraxial approximation in the solution. Another access is offered by the immediate parabolic approximation of the hyperbolic wave equation with the subsequent exact solution; we refer to it later on.

A two-dimensional Fourier transform of (12.1) with regard to x and y leads to

$$\frac{\partial^2}{\partial z^2}\hat{p}(K_x, K_y, z, \omega) + (k^2 - K_x^2 - K_y^2)\hat{p}(K_x, K_y, z, \omega) = 0; \tag{12.2}$$

the solution is

$$\hat{p}(K_x, K_y, z, \omega) = \hat{p}_0(K_x, K_y, \omega)\, \mathrm{e}^{\mathrm{j}K_z z} \tag{12.3}$$

with

$$K_z = \sqrt{k^2 - K_x^2 - K_y^2} \tag{12.4}$$

and arbitrarily chosen amplitude $\hat{p}_0(K_x, K_y, \omega)$ is a plane wave propagating into positive z-direction with wave number K_z (Section 11.1). The two-dimensional inverse Fourier transform of (12.3) according to

$$p(x, y, z, \omega) = \frac{1}{(2\pi)^2} \int_{-\infty}^{\infty} \int_{-\infty}^{\infty} \hat{p}_0(K_x, K_y, \omega)\, e^{jK_z z}\, e^{jK_x x + jK_y y}\, dK_x dK_y \tag{12.5}$$

reveals this integral representation as solution "spatial plane wave spectrum" of (12.1) propagating from the Fourier transformed[200] "aperture" distribution $\mathcal{F}_{xy}\{p(x, y, 0, \omega)\} = \hat{p}_0(K_x, K_y, \omega)$ in the plane $z = 0$ with the phase vectors $\underline{\mathbf{K}} = K_x \underline{\mathbf{e}}_x + K_y \underline{\mathbf{e}}_y + K_z \underline{\mathbf{e}}_z$ into the half-space $z > 0$ (Figure 11.1). We arrive at the paraxial approximation of (12.5) assuming that the maximum geometric extension of the "aperture" $p(x, y, 0, \omega)$—denoting it by $2a$—is large with regard to the wavelength, i.e., that the high-frequency approximation $ka \gg 1$ holds; due to the uncertainty relation of the Fourier transform, the Fourier variables K_x and K_y being conjugate to the "aperture" variables x and y satisfy the constraint $k \gg \sqrt{K_x^2 + K_y^2}$, i.e., summation only affects those propagating spectral components whose K_z is only slightly different from k, that is to say, whose $\underline{\mathbf{K}}$-vector almost points into $\underline{\mathbf{e}}_z$-direction being paraxial in this sense (Figure 11.1). Hence, the approximation

$$K_z \simeq k - \frac{K_x^2 + K_y^2}{2k} \tag{12.6}$$

is suggested that has

$$p^{\text{pax}}(x, y, z, \omega) = \frac{1}{(2\pi)^2}\, e^{jkz} \int_{-\infty}^{\infty} \int_{-\infty}^{\infty} \hat{p}_0(K_x, K_y, \omega) \times e^{-j\frac{z}{2k}(K_x^2 + K_y^2)}\, e^{jK_x x + jK_y y}\, dK_x dK_y \tag{12.7}$$

as consequence; we may stay with the infinite integration boundaries because presetting of $\hat{p}_0(K_x, K_y, \omega)$ arranges for a respective limitation of the integration interval. For a further analytical evaluation of (12.7), we have to prescribe $\hat{p}_0(K_x, K_y, \omega)$, and consequently $p^{\text{pax}}(x, y, 0, \omega)$ explicitly: We choose the Gaussian Fourier spectrum

$$\hat{p}_0(K_x, K_y, \omega) = p_0(\omega)\, \frac{\pi}{\alpha}\, e^{-\frac{1}{4\alpha(\omega)}(K_x^2 + K_y^2)} \tag{12.8}$$

because, in that case, Equation 12.7 is a two-dimensional inverse Fourier integral of a two-dimensional Gauss function that is explicitly known[201] with (2.280). The Gaussian "aperture" distribution

$$p^{\text{pax}}(x, y, 0, \omega) = p_0(x, y, \omega) = p_0(\omega)\, e^{-\alpha(\omega)(x^2 + y^2)} \tag{12.9}$$

[200] We put "aperture" in quotes to indicate that it is not a physically existing field distribution within an aperture but the cross-section profile of a Gaussian spectrum of plane waves in the xy-plane.

[201] Being a two-dimensional transform, the factor $\sqrt{\pi/\alpha}$ appears twice.

belongs to (12.8), where $\alpha(\omega)$ is a frequency-dependent parameter that can be played with; hence, we speak of time harmonic Gaussian beams (Ishimaru 1991; Heyman 1994; Heyman and Felsen 2001).

Based on the rotational symmetry in the xy-plane,[202] we may also write the Gaussian distribution (12.9) as

$$p_0(x, y, \omega) = p_0(\omega)\, \mathrm{e}^{-\alpha(\omega) r^2}. \tag{12.10}$$

Assuming $\alpha(\omega)$ to be a complex valued parameter

$$\alpha(\omega) = \alpha' + \mathrm{j}\omega\alpha'', \quad \alpha' > 0, \tag{12.11}$$

the time function

$$p_0(x, y, t) = \mathrm{e}^{-\alpha' r^2}\, p_0(t + \alpha'' r^2) \tag{12.12}$$

belongs to (12.10). With α', we control the geometric extension of the "aperture," and with α'', an r-dependent precipitate radiation of the $p_0(t)$-pulse with regard to $t = 0$. We now want to adjust this timely lead in a way that each impulse $p_0(t + \alpha'' r^2)$ arrives at the same time at the axial observation point $\underline{\mathbf{R}} = R_\mathrm{f} \underline{\mathbf{e}}_z$, the so-called focus point; to achieve this, the travel time difference $\sqrt{R_\mathrm{f}^2 + r^2}/c - R_\mathrm{f}/c$ for "aperture" points $r \neq 0$ must be compensated by α'' with respect to $r = 0$. For $r \ll R_\mathrm{f}$, we may approximate this travel time difference according to

$$\frac{\sqrt{R_\mathrm{f}^2 - r^2}}{c} - \frac{R_\mathrm{f}}{c} \simeq \frac{1}{2cR_\mathrm{f}}\, r^2 \tag{12.13}$$

resulting in

$$\alpha'' = \frac{1}{2cR_\mathrm{f}}. \tag{12.14}$$

With $\alpha' = 1/a^2 \ll k^2$, we ensure that the extension of the "aperture" is approximately $2a$ with $ka \gg 1$ finally yielding

$$\alpha(\omega) = \frac{1}{a^2} + \mathrm{j}\frac{1}{2R_\mathrm{f}}\, k. \tag{12.15}$$

With $R_\mathrm{f} \gg a$, we satisfy the necessary requirement that (12.13) holds; due to the paraxial approximation, the real part of $\alpha(\omega)$ is in fact also frequency dependent because $ka \gg 1$.

With (12.15), we can now calculate

$$p^{\mathrm{pax}}(x, y, z, \omega) = p_0(\omega)\, \frac{\pi}{\alpha(\omega)}\, \mathrm{e}^{\mathrm{j}kz}\, \frac{1}{(2\pi)^2} \int_{-\infty}^{\infty}\int_{-\infty}^{\infty} \mathrm{e}^{-\left(\frac{1}{4\alpha(\omega)} + \mathrm{j}\frac{z}{2k}\right)(K_x^2 + K_y^2)} \times \mathrm{e}^{\mathrm{j}K_x x + \mathrm{j}K_y y}\, \mathrm{d}K_x \mathrm{d}K_y \tag{12.16}$$

[202]The rotational symmetry characterizes this Gaussian beam as stigmatic; we meet astigmatic Gaussian beams in the next section.

as a two-dimensional inverse Fourier integral; we find

$$p^{\mathrm{pax}}(x,y,z,\omega) = p_0(\omega)\,\frac{\beta(z,\omega)}{\beta(0,\omega)}\,\mathrm{e}^{-\beta(z,\omega)r^2}\,\mathrm{e}^{\mathrm{j}kz} \tag{12.17}$$

with

$$\beta(z,\omega) = \frac{\alpha(\omega)}{1+\mathrm{j}\frac{2\alpha(\omega)}{k}\,z}. \tag{12.18}$$

Equation 12.17 is the (approximate) solution "paraxial Gaussian beam" of the Helmholtz equation (12.1).

With the separation of $\beta(z,\omega)$ into real and imaginary parts,

$$\Re\beta(z,\omega) = \frac{1}{a^2}\,\frac{1}{\left(1-\frac{z}{R_{\mathrm{f}}}\right)^2+\frac{z^2}{R_{\mathrm{f}}^2}\frac{1}{\gamma_{\mathrm{f}}^2(\omega)}}, \tag{12.19}$$

$$\Im\beta(z,\omega) = \frac{\gamma_{\mathrm{f}}(\omega)}{a^2}\,\frac{1-\frac{z}{R_{\mathrm{f}}}\left(1+\frac{1}{\gamma_{\mathrm{f}}^2(\omega)}\right)}{\left(1-\frac{z}{R_{\mathrm{f}}}\right)^2+\frac{z^2}{R_{\mathrm{f}}^2}\frac{1}{\gamma_{\mathrm{f}}^2(\omega)}} \tag{12.20}$$

with

$$\gamma_{\mathrm{f}}(\omega) = \frac{ka^2}{2R_{\mathrm{f}}}, \tag{12.21}$$

we recognize according to

$$p^{\mathrm{pax}}(x,y,z,\omega) = p_0(\omega)\,\frac{\beta(z,\omega)}{\beta(0,\omega)}\,\mathrm{e}^{-\Re\beta(z,\omega)r^2}\,\mathrm{e}^{\mathrm{j}kz-\mathrm{j}\Im\beta(z,\omega)r^2} \tag{12.22}$$

that $\Re\beta(z,\omega)$ determines the z-dependent beam geometry, namely the exponential decay in r-direction for fixed z; we define that r-value for which the amplitude $p_0(\omega)\beta(z,\omega)/\beta(0,\omega)$ for $r=0$ has decreased by the factor e^{-1} as half the beam width $w(z,\omega)$:

$$w(z,\omega) = \frac{1}{\sqrt{\Re\beta(z,\omega)}}, \tag{12.23}$$

hence,

$$\frac{w(z,\omega)}{a} = \sqrt{\left(1-\frac{z}{R_{\mathrm{f}}}\right)^2+\frac{z^2}{R_{\mathrm{f}}^2}\frac{1}{\gamma_{\mathrm{f}}^2(\omega)}}. \tag{12.24}$$

For $z=0$, we have $w(0,\omega)=a$, i.e., we recover the given (circular) "aperture" distribution in the xy-plane. According to (12.22), $\Im\beta(z,\omega)$ defines the z-dependent deviation of the phase surfaces

$$kz-\Im\beta(z,\omega)r^2 = \mathrm{const} \tag{12.25}$$

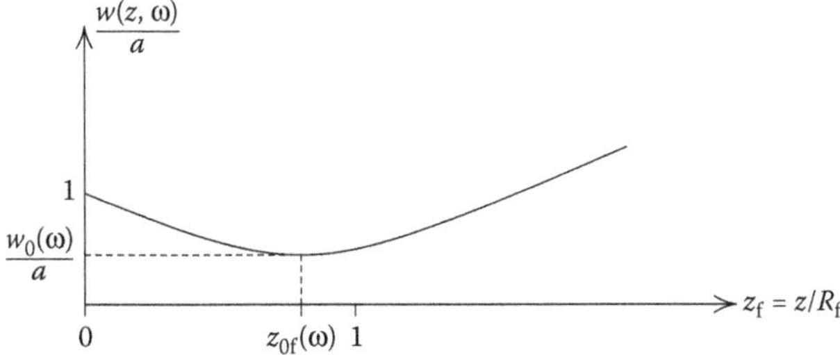

FIGURE 12.1
Waist profile of a Gaussian beam for $\gamma_\mathrm{f} = 2$.

of a Gaussian beam with regard to a plane wave propagating into z-direction. Figure 12.1 illustrates the waist profile $w(z,\omega)/a$ of a Gaussian beam as a function of the normalized axis coordinate $z_\mathrm{f} = z/R_\mathrm{f}$. At

$$\frac{z_0(\omega)}{R_\mathrm{f}} = z_{0\mathrm{f}}(\omega) = \frac{\gamma_\mathrm{f}^2(\omega)}{1+\gamma_\mathrm{f}^2(\omega)} < 1, \tag{12.26}$$

we find a minimum of $w(z,\omega)$, and its value

$$\begin{aligned} \frac{w_0(\omega)}{a} &= \frac{1}{\sqrt{1+\gamma_\mathrm{f}^2(\omega)}} \\ &= \frac{\sqrt{z_{0\mathrm{f}}(\omega)}}{\gamma_\mathrm{f}(\omega)} \end{aligned} \tag{12.27}$$

gives us the half waist width; by the way: Only for $\gamma_\mathrm{f} \gg 1$, we have $z_0 \simeq R_\mathrm{f}$ for the location of the minimum. For $z \longrightarrow \infty$, we find

$$\begin{aligned} \frac{w(z,\omega)}{a} &\stackrel{z\to\infty}{\longrightarrow} \frac{\sqrt{1+\gamma_\mathrm{f}^2(\omega)}}{\gamma_\mathrm{f}(\omega)}\, z_\mathrm{f} \\ &= \frac{z_\mathrm{f}}{\sqrt{z_{0\mathrm{f}}(\omega)}}, \end{aligned} \tag{12.28}$$

i.e., the Gaussian beam diverges linearly with z. The location and the width of the waist as well as its divergence are only determined by the frequency-dependent dimension-less "play parameter" $\gamma_\mathrm{f}(\omega)$ as function of the normalized axis. For $\gamma_\mathrm{f} \gg 1$, we have a considerably waisted Gaussian beam, $\gamma_\mathrm{f} = 1$ yields the collimated Gaussian beam, and $\gamma_\mathrm{f} \ll 1$ results in a strongly diverging Gaussian beam.[203]

The phase surfaces of a plane wave propagating into z-direction are given by planes perpendicular to the z-axis, the phase surfaces (12.25) of a Gaussian

[203] Due to the strong divergence $\sim z_\mathrm{f}/\gamma_\mathrm{f}$, the waist of the beam is now close to $z_\mathrm{f} = 0$; even though a little bit misleading, this beam is called narrow-waisted Gaussian beam (Galdi et al. 2001, 2003).

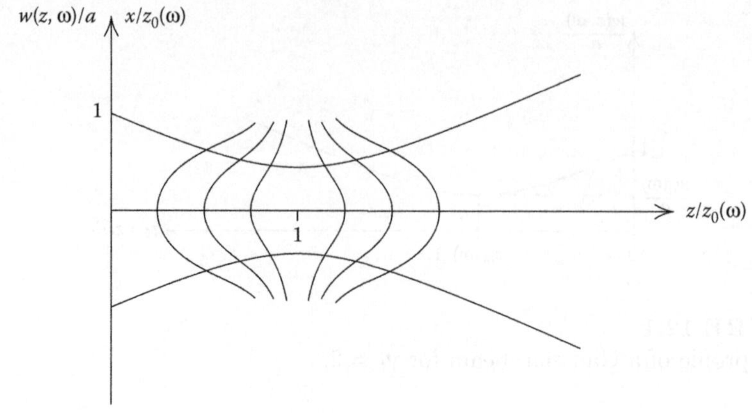

FIGURE 12.2
Phase surfaces of a Gaussian beam for $\gamma_{\mathrm{f}} = 2$.

beam are only locally planar in the proximity of the axis $r \simeq 0$; Figure 12.2 illustrates their curvature in the xz-plane for different values of the constants on the right-hand side of (12.25): Obviously, the sign of the curvature radius changes while passing the waist plane $z = z_0$.

Instead of parameterizing a Gaussian beam through $\gamma_{\mathrm{f}}(\omega)$ as function of an R_{f}-normalized axis, a parametrization through $z_0(\omega)$ is essentially offered (Siegman 1986; Heyman and Felsen 2001):

$$\Re\beta(z,\omega) = \frac{k}{2}\,\frac{z_{\mathrm{R}}(\omega)}{[z - z_0(\omega)]^2 + z_{\mathrm{R}}^2(\omega)}, \tag{12.29}$$

$$\Im\beta(z,\omega) = -\,\frac{k}{2}\,\frac{z - z_0(\omega)}{[z - z_0(\omega)]^2 + z_{\mathrm{R}}^2(\omega)}; \tag{12.30}$$

that way, a new parameter

$$z_{\mathrm{R}}(\omega) = \frac{z_0(\omega)}{\gamma_{\mathrm{f}}(\omega)} \tag{12.31}$$

immediately appears that has the dimension of a length being called Rayleigh distance: Because of

$$w(z_0 + z_{\mathrm{R}}) = \sqrt{2}\,w_0(\omega), \tag{12.32}$$

the beam remains collimated within the Rayleigh distance. Combining real and imaginary parts within the z_0z_{R}-parametrization to a complex valued function $\beta(z,\omega)$, it is recognized that $\beta(z,\omega)$ now has the simple representation

$$\beta(z,\omega) = \frac{k}{2}\,\frac{1}{z_{\mathrm{R}}(\omega) + \mathrm{j}[z - z_0(\omega)]}. \tag{12.33}$$

This conversion produces a factor k suggesting to write (12.17) in the following way:

$$
\begin{aligned}
p^{\mathrm{pax}}(x,y,z,\omega) &= p_0(\omega)\,\frac{\beta(z,\omega)}{\beta(0,\omega)}\,\mathrm{e}^{\mathrm{j}k\left[z+\mathrm{j}\,\frac{1}{k}\,\beta(z,\omega)r^2\right]}\\
&= p_0(\omega)\,\frac{q(0,\omega)}{q(z,\omega)}\,\mathrm{e}^{\mathrm{j}k\left[z+\frac{1}{2}\,\frac{r^2}{q(z,\omega)}\right]},
\end{aligned}
\qquad (12.34)
$$

where

$$
\begin{aligned}
q(z,\omega) &= -\,\mathrm{j}\,\frac{k}{2}\,\frac{1}{\beta(z,\omega)}\\
&= z - z_0(\omega) - \mathrm{j}z_{\mathrm{R}}(\omega).
\end{aligned}
\qquad (12.35)
$$

With (12.29) and (12.30), we obviously have

$$
\Re\frac{1}{q(z,\omega)} = \frac{z - z_0(\omega)}{[z - z_0(\omega)]^2 + z_{\mathrm{R}}^2(\omega)}, \qquad (12.36)
$$

$$
\Im\frac{1}{q(z,\omega)} = \frac{z_{\mathrm{R}}(\omega)}{[z - z_0(\omega)]^2 + z_{\mathrm{R}}^2(\omega)}. \qquad (12.37)
$$

As we will show in the next section, (12.34) with (12.35) is a natural parametrization of a Gaussian beam if it is derived as an exact solution of a parabolically approximated wave equation. Using this parametrization, we obtain

$$
w(z,\omega) = \sqrt{\frac{2}{kz_{\mathrm{R}}(\omega)}}\sqrt{[z - z_0(\omega)]^2 + z_{\mathrm{R}}^2(\omega)} \qquad (12.38)
$$

for the half beam width with

$$
w_0(\omega) = \sqrt{\frac{2z_{\mathrm{R}}(\omega)}{k}}. \qquad (12.39)
$$

Due to (12.39), we have

$$
z_{\mathrm{R}}(\omega) = \frac{\pi w_0^2(\omega)}{\lambda}, \qquad (12.40)
$$

i.e., z_{R} is kind of a near-field length of the waist "aperture" again revealing the above characterization as a collimation length.

To arrive at wave packets, namely pulsed beams that are localized in space and time, we have to calculate

$$
p^{\mathrm{pax}}(x,y,z,t) = \mathcal{F}_{\omega}^{-1}\{p^{\mathrm{pax}}(x,y,z,\omega)\}; \qquad (12.41)
$$

yet the frequency dependence of $\beta(z,\omega)$ and $q(z,\omega)$, respectively, barely offers a chance to perform the Fourier inversion analytically. Even postulating frequency independence does not lead any further: Within the physical

$R_\mathrm{f}\gamma_\mathrm{f}$-parametrization, we should then have $R_\mathrm{f} \sim \omega$ due to (12.21), and consequently, $\beta(z, \omega)$ would still be frequency dependent; within the mathematical $z_0 z_\mathrm{R}$-parametrization, the frequency independence of these parameters yields the frequency independence of $q(z, \omega)$; however, with (12.37), a term

$$\mathrm{e}^{-\frac{\omega}{2c}\Im\frac{1}{q(z)}r^2} \tag{12.42}$$

results whose Fourier inversion does not converge. To yield convergence, the negative frequencies must be omitted: The resulting impulses without negative frequencies represent analytical signals (Section 2.3.4), and with that, a further evaluation is possible (Heyman 1994); we will follow these ideas in the next section.

12.2 Pulsed Beams as Exact Solutions of an Approximate Wave Equation

The wave equation belonging to (12.1) reads

$$\Delta p(x, y, z, t) - \frac{1}{c^2}\frac{\partial^2}{\partial t^2}p(x, y, z, t) = 0. \tag{12.43}$$

The phase term $\mathrm{e}^{\mathrm{j}kz}$ in (12.34) suggests to represent $p(x, y, z, t)$ in a moving coordinate system

$$p(x, y, z, t) = P(x, y, z, \tau) \tag{12.44}$$

according to the substitution

$$\tau = t - \frac{z}{c}. \tag{12.45}$$

The conversion of the z- and t-differentiations in (12.43) to $P(x, y, z, \tau)$ yields

$$\begin{aligned}
&\frac{\partial^2}{\partial z^2}p(x, y, z, t) - \frac{1}{c^2}\frac{\partial^2}{\partial t^2}p(x, y, z, t)\\
&\quad = \frac{\partial}{\partial z}\left[\frac{\partial}{\partial z}P(x, y, z, \tau) - \frac{2}{c}\frac{\partial}{\partial \tau}P(x, y, z, \tau)\right],
\end{aligned} \tag{12.46}$$

because $\partial\tau(t, z)/\partial z = -1/c$ and $\partial\tau(t, z)/\partial t = 1$. The constraint to high-frequency Gaussian beams allows for the approximation

$$\left|\frac{\partial}{\partial z}P(x, y, z, \tau)\right| \ll \left|\frac{1}{c}\frac{\partial}{\partial \tau}P(x, y, z, \tau)\right| \tag{12.47}$$

leading us to the "wave packet equation"

$$\frac{\partial^2}{\partial x^2}P(x, y, z, \tau) + \frac{\partial^2}{\partial y^2}P(x, y, z, \tau) - \frac{2}{c}\frac{\partial^2}{\partial z \partial \tau}P(x, y, z, \tau) = 0. \tag{12.48}$$

The concluding remarks of the last section suggest to seek solutions of this equation in terms of analytical signals

$$P_+(x, y, z, \tau) = P(x, y, z, \tau) + j\mathcal{H}_{\tau'}\{P(x, y, z, \tau')\}, \tag{12.49}$$

whose imaginary part is the Hilbert transform of the real valued function $P(x, y, z, \tau)$ for real values of τ. We are especially looking for solutions with the special xyz-structure of an astigmatic Gaussian beam

$$P_+(x, y, z, \tau) = A(z) f_+\left(\tau - \frac{1}{2}\underline{\boldsymbol{r}}^{\mathrm{T}}\underline{\underline{\boldsymbol{B}}}(z)\underline{\boldsymbol{r}}\right), \tag{12.50}$$

where $\underline{\underline{\boldsymbol{B}}}(z)$ is a complex symmetric 2×2-matrix

$$\underline{\underline{\boldsymbol{B}}}(z) = \begin{pmatrix} B_{xx}(z) & B_{xy}(z) \\ B_{xy}(z) & B_{yy}(z) \end{pmatrix} \tag{12.51}$$

with a positive-definite imaginary part matrix, while $\underline{\boldsymbol{r}}$ is the 2×1-matrix

$$\underline{\boldsymbol{r}} = \begin{pmatrix} x \\ y \end{pmatrix} \tag{12.52}$$

corresponding to the observation point vector $\underline{\mathbf{r}}$ in the xy-plane, and $f_+(t)$ denotes a high-frequency analytic signal. The term

$$\underline{\boldsymbol{r}}^{\mathrm{T}}\underline{\underline{\boldsymbol{B}}}(z)\underline{\boldsymbol{r}} = B_{xx}(z)x^2 + 2B_{xy}(z)xy + B_{yy}(z)y^2 \tag{12.53}$$

is a quadratic form, where the positive definiteness $\underline{\boldsymbol{r}}^{\mathrm{T}}\Im\underline{\underline{\boldsymbol{B}}}(z)\underline{\boldsymbol{r}} > 0$ of the imaginary part matrix ensures the exponential decay of the Gaussian beam perpendicular to the z-axis (compare Equation 2.320). In the solution *ansatz* (12.50), we now have to calculate $\underline{\underline{\boldsymbol{B}}}(z)$ and $A(z)$ via insertion into the differential equation (12.48). We obtain that:

$$\begin{aligned} A(z)\underline{\boldsymbol{r}}^{\mathrm{T}} &\left[c\underline{\underline{\boldsymbol{B}}}(z)\underline{\underline{\boldsymbol{B}}}(z) + \frac{\mathrm{d}\underline{\underline{\boldsymbol{B}}}(z)}{\mathrm{d}z}\right]\underline{\boldsymbol{r}}\,\frac{\mathrm{d}^2 f_+(t)}{\mathrm{d}t^2} \\ &- c\left[A(z)\,\mathrm{trace}\,\underline{\underline{\boldsymbol{B}}}(z) + \frac{2}{c}\frac{\mathrm{d}A(z)}{\mathrm{d}z}\right]\frac{\mathrm{d}f_+(t)}{\mathrm{d}t} = 0. \end{aligned} \tag{12.54}$$

This equation should hold for any arbitrary impulse $f_+(t)$; therefore, the factors[204] of these derivatives must be separately equal to zero:

$$c\underline{\underline{\boldsymbol{B}}}(z)\underline{\underline{\boldsymbol{B}}}(z) + \frac{\mathrm{d}\underline{\underline{\boldsymbol{B}}}(z)}{\mathrm{d}z} = \underline{\underline{\boldsymbol{O}}}, \tag{12.55}$$

$$A(z)\,\mathrm{trace}\,\underline{\underline{\boldsymbol{B}}}(z) + \frac{2}{c}\frac{\mathrm{d}A(z)}{\mathrm{d}z} = 0; \tag{12.56}$$

[204]By the way, the validity of (12.55) requires the symmetry of $\underline{\underline{\boldsymbol{B}}}(z)$.

$\underline{\underline{O}}$ denotes the 2×2-zero matrix. The conversion from $\beta(z,\omega)$ to $q(z,\omega)$ according to (12.35) equally suggests to introduce the inverse of $\underline{\underline{B}}(z)$:

$$\underline{\underline{B}}(z) = \underline{\underline{Q}}^{-1}(z). \tag{12.57}$$

With

$$\begin{aligned}\frac{\mathrm{d}\underline{\underline{B}}(z)}{\mathrm{d}z} &= -\left[\underline{\underline{Q}}(z)\underline{\underline{Q}}(z)\right]^{-1}\frac{\mathrm{d}\underline{\underline{Q}}(z)}{\mathrm{d}z}\\ &= -\underline{\underline{Q}}^{-1}(z)\underline{\underline{Q}}^{-1}(z)\frac{\mathrm{d}\underline{\underline{Q}}(z)}{\mathrm{d}z},\end{aligned} \tag{12.58}$$

we obtain

$$\left[c\underline{\underline{I}} - \frac{\mathrm{d}\underline{\underline{Q}}(z)}{\mathrm{d}z}\right]\underline{\underline{Q}}^{-1}(z)\underline{\underline{Q}}^{-1}(z) = \underline{\underline{O}} \tag{12.59}$$

instead of (12.55) yielding

$$\frac{\mathrm{d}\underline{\underline{Q}}(z)}{\mathrm{d}z} = c\underline{\underline{I}}, \tag{12.60}$$

and finally

$$\underline{\underline{Q}}(z) = \underline{\underline{Q}}(0) + cz\,\underline{\underline{I}}; \tag{12.61}$$

$\underline{\underline{I}}$ is the 2×2-unit matrix. The proximity to (12.35) is apparent!

The solution of (12.56) is immediately obtained:

$$A(z) = A_0\,\mathrm{e}^{-\frac{c}{2}\int \mathrm{trace}\,\underline{\underline{B}}(z)\,\mathrm{d}z}; \tag{12.62}$$

it would be desirable to be able to calculate at least the integral in the exponent of (12.62); as a matter of fact, even more can be achieved! We write

$$c\,\underline{\underline{B}}(z) = c\,\underline{\underline{I}}\,\underline{\underline{B}}(z) = \frac{\mathrm{d}\underline{\underline{Q}}(z)}{\mathrm{d}z}\underline{\underline{Q}}^{-1}(z). \tag{12.63}$$

The explicit calculation of the trace in the last term of (12.63) miraculously yields the following identity for each 2×2-matrix[205] $\underline{\underline{Q}}(z)$:

$$\mathrm{trace}\left[\frac{\mathrm{d}\underline{\underline{Q}}(z)}{\mathrm{d}z}\underline{\underline{Q}}^{-1}(z)\right] = \frac{\mathrm{d}}{\mathrm{d}z}\ln\det\underline{\underline{Q}}(z). \tag{12.64}$$

[205] This identity may even be proved for each rectangular matrix starting, for instance, from an eigenvalue decomposition of the matrix (Shlivinski 2004).

With $A_0 = \sqrt{\det \underline{\underline{\boldsymbol{Q}}}(0)}$, we consequently obtain

$$A(z) = \sqrt{\frac{\det \underline{\underline{\boldsymbol{Q}}}(0)}{\det \underline{\underline{\boldsymbol{Q}}}(z)}}, \tag{12.65}$$

and hence

$$P_+(x, y, z, \tau) = \sqrt{\frac{\det \underline{\underline{\boldsymbol{Q}}}(0)}{\det \underline{\underline{\boldsymbol{Q}}}(z)}} f_+ \left[\tau - \frac{1}{2} \underline{\boldsymbol{r}}^{\mathrm{T}} \underline{\underline{\boldsymbol{Q}}}^{-1}(z) \underline{\boldsymbol{r}} \right], \tag{12.66}$$

as exact solution of the approximately valid wave packet equation (12.48). As the equation of a complex valued pulsed beam, we obtain

$$p_+(x, y, z, t) = \sqrt{\frac{\det \underline{\underline{\boldsymbol{Q}}}(0)}{\det \underline{\underline{\boldsymbol{Q}}}(z)}} f_+ \left[t - \frac{z}{c} - \frac{1}{2} \underline{\boldsymbol{r}}^{\mathrm{T}} \underline{\underline{\boldsymbol{Q}}}^{-1}(z) \underline{\mathbf{r}} \right]. \tag{12.67}$$

This beam becomes rotationally symmetric in the xy-plane if we assume the matrix $\underline{\underline{\boldsymbol{Q}}}(0)$ to be proportional to a unit matrix according to

$$\underline{\underline{\boldsymbol{Q}}}(0) = cq(0) \underline{\underline{\boldsymbol{I}}}. \tag{12.68}$$

We then have

$$\underline{\underline{\boldsymbol{Q}}}(z) = c \underbrace{[q(0) + z)]}_{= q(z)} \underline{\underline{\boldsymbol{I}}}, \tag{12.69}$$

hence

$$\underline{\underline{\boldsymbol{Q}}}^{-1}(z) = \frac{1}{cq(z)} \underline{\underline{\boldsymbol{I}}}. \tag{12.70}$$

Now we only need to calculate

$$\det \underline{\underline{\boldsymbol{Q}}}(z) = c^2 q^2(z) \tag{12.71}$$

and to choose $q(0) = -z_0 - \mathrm{j} z_{\mathrm{R}}$ (Equation 12.35) to rediscover the stigmatic time harmonic Gaussian beam with frequency-independent parameters z_0 and z_{R} choosing the analytic signal

$$f_+(t) = p_0(\omega) \mathrm{e}^{-\mathrm{j}\omega t}. \tag{12.72}$$

Hence, the paraxially approximated solution of the Helmholtz equation is an exact solution of the wave packet equation as a parabolic approximation of the exact wave equation. On the other hand, we are now able, for the case of frequency-independent parameters z_0 and z_{R} in the representation (12.34) of a time harmonic Gaussian beam, to produce a pulsed beam applying a $\mathcal{F}_+^{-1}$-Fourier inversion (Equation 2.319)

$$p_+(x, y, z, t) = \frac{q(0)}{q(z)} f_+ \left[t - \frac{z}{c} - \frac{1}{2c} \frac{r^2}{q(z)} \right], \tag{12.73}$$

if we assume $p_0(t) \Longrightarrow f_+(t)$ as analytic signal; therefore, we may well speak of a Gaussian beam even though there is no Gauss function in (12.73). With the choice of an analytic signal of finite duration, the pulsed Gaussian beam is consequently localized in space and time.

Evidently, the choice of (12.68) yields a rotationally symmetric pulsed beam for all z and all times. With the obvious generalization

$$\underline{\underline{\boldsymbol{Q}}}(0) = c \begin{pmatrix} q_1(0) & 0 \\ 0 & q_2(0) \end{pmatrix} \tag{12.74}$$

with $q_{1,2}(0) = -z_{0_{1,2}} - \mathrm{j} z_{\mathrm{R}_{1,2}}$, we can generate a canonical elliptic "aperture" distribution in the $(z = 0)$-plane—the main axes of the ellipse coincide with the coordinate axes—that also remains the beam cross-section for all z and all times. With a full (symmetric) $\underline{\underline{\boldsymbol{Q}}}(0)$-matrix, we can finally generate astigmatic wave packets with z-dependent noncoinciding main axes of the wavefronts (phase surfaces) and the amplitude distributions perpendicular to the propagation direction (Heyman 1994).

Real valued pulsed (stigmatic) beams are obtained if we separate

$$\begin{aligned} p_+(r, z, t) &= \frac{q(0)}{q(z)} f_+ \left[t - \frac{z}{c} - \frac{1}{2c} \frac{r^2}{q(z)} \right] \\ &= a(z)\, f_+ \left[t - \overline{\tau}(r, z) - \mathrm{j}\gamma(r, z) \right] \end{aligned} \tag{12.75}$$

with

$$a(z) = \frac{q(0)}{q(z)}, \tag{12.76}$$

$$\overline{\tau}(r, z) = \frac{r^2}{2c} \Re \frac{1}{q(z)}, \tag{12.77}$$

$$\gamma(r, z) = \frac{r^2}{2c} \Im \frac{1}{q(z)} \tag{12.78}$$

into a real and an imaginary part. To achieve this, we use (2.321):

$$f_+[t - \overline{\tau}(r, z) - \mathrm{j}\gamma(r, z)] = f_\gamma[t - \overline{\tau}(r, z)] + \mathrm{j}\mathcal{H}\{f_\gamma(\tau)\}_{t - \overline{\tau}(r, z)}, \tag{12.79}$$

where

$$f_\gamma(t) = \Re \left\{ \frac{1}{\pi} \int_0^\infty F(\omega) \mathrm{e}^{-\gamma(r,z)\omega} \mathrm{e}^{-\mathrm{j}\omega t} \, \mathrm{d}\omega \right\}, \tag{12.80}$$

$$\mathcal{H}\{f_\gamma(t)\} = \Im \left\{ \frac{1}{\pi} \int_0^\infty F(\omega) \mathrm{e}^{-\gamma(r,z)\omega} \mathrm{e}^{-\mathrm{j}\omega t} \, \mathrm{d}\omega \right\}. \tag{12.81}$$

The evanescence parameter $\gamma(r, z)$ of the beam is related to the spectrum $F(\omega)$ of the given real valued pulse $f(t) = \mathcal{F}^{-1}\{F(\omega)\}$ via the integrands of (12.80) and (12.81), a fact that reflects the frequency-dependent beam width according to (12.38) even for frequency-independent beam parameters.

We finally obtain

$$\Re p_{+}(r, z, t) = f_{\gamma}[t - \overline{\tau}(r, z)]\Re a(z) - \mathcal{H}\{f_{\gamma}(\tau)\}_{t-\overline{\tau}(r,z)}\Im a(z), \quad (12.82)$$

$$\Im p_{+}(r, z, t) = f_{\gamma}[t - \overline{\tau}(r, z)]\Im a(z) + \mathcal{H}\{f_{\gamma}(\tau)\}_{t-\overline{\tau}(r,z)}\Re a(z), \quad (12.83)$$

where

$$\Re a(z) = \frac{z_0(z_0 - z) + z_{\mathrm{R}}^2}{(z - z_0)^2 + z_{\mathrm{R}}^2}, \quad (12.84)$$

$$\Im a(z) = -\frac{z_{\mathrm{R}} z}{(z - z_0)^2 + z_{\mathrm{R}}^2}. \quad (12.85)$$

Now we can choose $f(t)$, for example, as an RC$N(t)$-pulse and insert $F(\omega) = \mathrm{RCN}(\omega)$ according to (2.276) in (12.80) and (12.81): We obtain RC$N(t)$-wave packets (12.82) and (12.83) being traced as localized pulses through space and time. The balance between $\Re a(z)$ and $\Im a(z)$ determines whether we see the RCN(t)-pulse itself sor its Hilbert transform. Figure 12.3 exhibits time snap-shots matching the parameters chosen for Figure 12.2 according to the numerical evaluation[206] of (12.82) for an RC2(t)-pulse with $k_0 z_0 = 18$. We nicely recognize how the beam starts as RC2-pulse converting into its Hilbert transform while passing the focus point [we apply (2.306)] and finally propagating as negative RC2-pulse: The pulsed Gaussian beam changes its sign passing the focus point; this special dissipative behavior is induced by a Hilbert transform.

It is interesting to note that the explicit appearance of k in front of the brackets in the exponential of (12.34) also yields different pulse structures of the "aperture" distributions of $p^{\mathrm{pax}}(r, z, t)$ according to (12.41) and $p_{+}(r, z, t)$ according to (12.75) for $z = 0$:

$$p^{\mathrm{pax}}(r, 0, t) \overset{(12.12)}{=} \mathrm{e}^{-\frac{r^2}{a^2}}\, p_0\left(t + \frac{r^2}{2cR_{\mathrm{f}}}\right); \quad (12.86)$$

$$p_{+}(r, 0, t) = f_{\gamma}\left[t - \overline{\tau}(r, 0)\right], \quad (12.87)$$

where

$$\overline{\tau}(r, 0) = -\frac{r^2}{2c}\frac{z_0}{z_0^2 + z_{\mathrm{R}}^2} = -\frac{r^2}{2cR_{\mathrm{f}}}, \quad (12.88)$$

$$f_{\gamma}(t) = \Re\left\{\frac{1}{\pi}\int_0^{\infty} F(\omega)\,\mathrm{e}^{-\gamma(r,0)\omega}\mathrm{e}^{-\mathrm{j}\omega t}\,\mathrm{d}\omega\right\}, \quad (12.89)$$

[206]This evaluation is by no means trivial; it is not simply based on programming Equation 12.82: We have to thank Shlivinski.

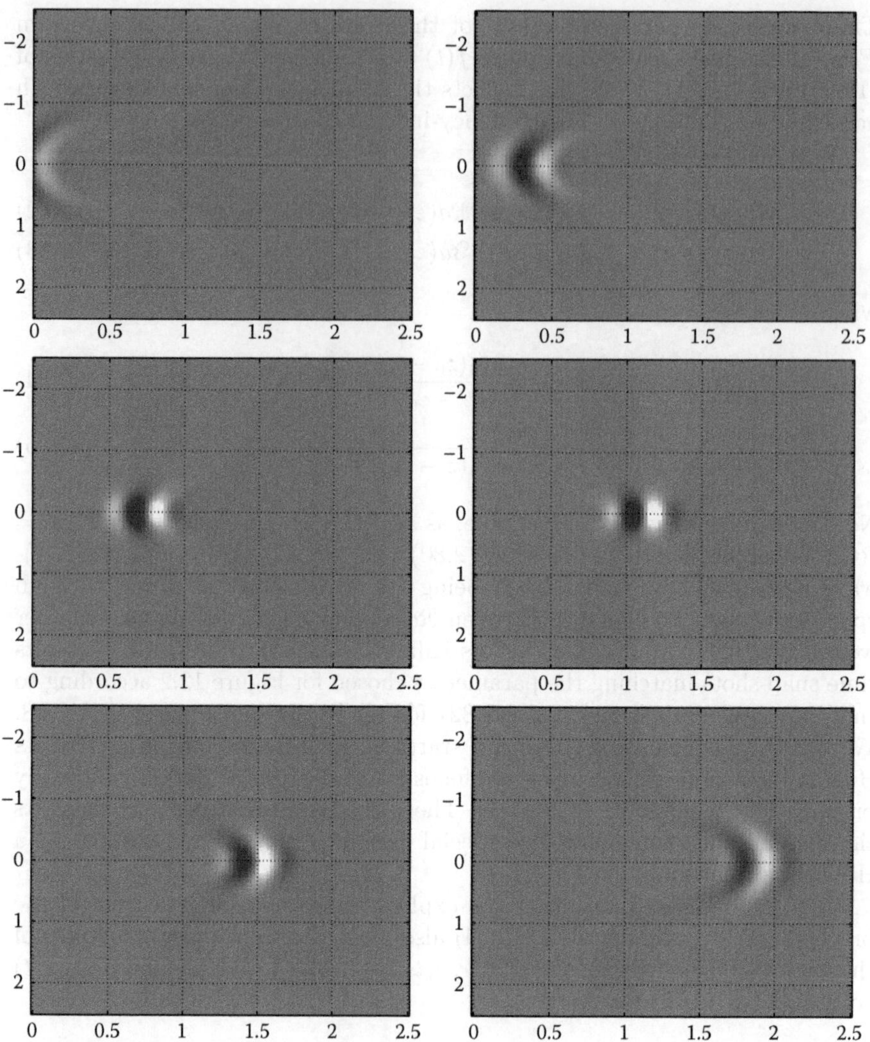

FIGURE 12.3
Time snap-shots of an RC2-pulsed Gaussian beam (black: maximum positive values; white: maximum negative values).

$$\gamma(r,0) = \frac{r^2}{2c} \frac{z_\mathrm{R}}{z_0^2 + z_\mathrm{R}^2}. \tag{12.90}$$

The pulse structure of $f_\gamma(t)$ is r-dependent! Yet, if we express z_R and z_0 in $\gamma(r,0)$ according to (12.90) through γ_f and R_f explicitly inserting γ_f, we have $\gamma(r,0) = r^2/(\omega a^2)$ allowing for the attenuation exponential function to come out of the integral (12.89).

12.3 Pulsed Beams as Approximate Solutions of Eikonal and Transport Equations

Rays are solutions of eikonal and transport equations resulting from a high-frequency approximation of the wave equation, recognizing that this only makes sense for inhomogeneous materials because a ray in homogeneous material is initially nothing else but a slowness vector. Nevertheless, this offers another approach to Gaussian, and generally, pulsed beams if the eikonal equation—even in homogeneous materials—is approximately solved by a truncated Taylor series: The result is a Gaussian beam in the frequency domain and a wave packet (pulsed Gaussian beam) in the time domain as it is given by (12.67). Since rays, as already mentioned, are particularly useful for ray tracing in inhomogeneous materials, we now have a possibility for generalization: We identify the ray in an inhomogeneous material as the axis of a Gaussian beam!

We will again explain this third approach to Gaussian and pulsed beams for the scalar acoustic wave equation for homogeneous materials after having defined the respective ray eikonal and transport equations; for inhomogeneous materials, we refer to the literature. Afterwards, we define P- and S-rays in inhomogeneous isotropic elastic materials; concerning the generalization to P- and S-beams in such materials, we again refer to the literature, and this also holds for the derivation of Gaussian beams via complex source point coordinates.

12.3.1 Eikonal and transport equations for acoustic beams

We are looking for ray solutions of the acoustic wave equation

$$\Delta p(\underline{\mathbf{R}}, t) - \frac{1}{c^2(\underline{\mathbf{R}})} \frac{\partial^2 p(\underline{\mathbf{R}}, t)}{\partial t^2} = 0 \tag{12.91}$$

in inhomogeneous materials implying a spatially dependent compressibility $\kappa(\underline{\mathbf{R}})$ according to (5.27), yet a constant density. The time harmonic *ansatz*

$$p(\underline{\mathbf{R}}, t) = P(\underline{\mathbf{R}})\, \mathrm{e}^{\mathrm{j}\omega_0 T(\underline{\mathbf{R}})} \mathrm{e}^{-\mathrm{j}\omega_0 t} \tag{12.92}$$

with slowly varying functions $\kappa(\underline{\mathbf{R}})$, $P(\underline{\mathbf{R}})$, and $T(\underline{\mathbf{R}})$ as compared to the wavelength emerges as a plane wave for a constant phase velocity $c(\underline{\mathbf{R}}) = c$ with $T(\underline{\mathbf{R}}) = \hat{\underline{\mathbf{k}}} \cdot \underline{\mathbf{R}}/c$; consequently, we interpret (12.92) as a local plane wave in particular requiring now its $\underline{\mathbf{R}}$-dependent propagation direction. Insertion of (12.92) into (12.91) yields terms with ω_0^2- and ω_0^1-factors that are predominantly important for large ω_0. Equating them separately to zero results in the eikonal equation

$$\nabla T(\underline{\mathbf{R}}) \cdot \nabla T(\underline{\mathbf{R}}) = \frac{1}{c^2(\underline{\mathbf{R}})} \tag{12.93}$$

and in the transport equation

$$2\nabla P(\underline{\mathbf{R}}) \cdot \nabla T(\underline{\mathbf{R}}) + P(\underline{\mathbf{R}})\Delta T(\underline{\mathbf{R}}) = 0. \tag{12.94}$$

On the other hand, the ω_0^0-term $\Delta P(\underline{\mathbf{R}}) = 0$ is neglected (else we would have three equations for two unknowns); Červený (2001) cites possibilities to incorporate it.

With the phase $T(\underline{\mathbf{R}}) = \hat{\underline{\mathbf{k}}} \cdot \underline{\mathbf{R}}/c$ of a plane wave in homogeneous materials, we have $\nabla T(\underline{\mathbf{R}}) = \hat{\underline{\mathbf{k}}}/c$, and consequently, the respective eikonal equation reads $\nabla T(\underline{\mathbf{R}}) \cdot \nabla T(\underline{\mathbf{R}}) = \hat{\underline{\mathbf{k}}} \cdot \hat{\underline{\mathbf{k}}}/c^2 = 1/c^2$ or $\underline{\mathbf{s}} \cdot \underline{\mathbf{s}} = 1/c^2$ if $\underline{\mathbf{s}} = \hat{\underline{\mathbf{k}}}/c$ denotes the slowness vector. Therefore, defining a spatially dependent slowness vector even in inhomogeneous materials according to

$$\underline{\mathbf{s}}(\underline{\mathbf{R}}) = \nabla T(\underline{\mathbf{R}}), \tag{12.95}$$

the eikonal equation

$$\underline{\mathbf{s}}(\underline{\mathbf{R}}) \cdot \underline{\mathbf{s}}(\underline{\mathbf{R}}) = \frac{1}{c^2(\underline{\mathbf{R}})} = \rho\kappa(\underline{\mathbf{R}}) \tag{12.96}$$

appears as generalization of the dispersion relation (5.45) for the $\kappa(\underline{\mathbf{R}})$-inhomogeneous material; hence, the unit vector $\hat{\underline{\mathbf{s}}}(\underline{\mathbf{R}})$ defines a ray trajectory. As soon as this trajectory has been calculated as so-called characteristic of the eikonal equation, a ray coordinate s can be assigned and

$$\nabla T(\underline{\mathbf{R}}) \cdot \hat{\underline{\mathbf{s}}}(\underline{\mathbf{R}}) \stackrel{\text{def}}{=} \frac{\partial T(\underline{\mathbf{R}})}{\partial s} \tag{12.97}$$

can be defined as wave(phase)front travel time along the ray. With (12.95) and (12.96), we then have

$$T(\underline{\mathbf{R}}) = \int_{L(\underline{\mathbf{R}}_0,\underline{\mathbf{R}})} \frac{1}{c(s)}\,\mathrm{d}s \tag{12.98}$$

as travel time of the wavefront from the starting point $\underline{\mathbf{R}}_0$ to the actual observation point $\underline{\mathbf{R}}$. To calculate characteristics—and also to solve transport equations—we have a handful of possibilities that are discussed in detail by Červený (2001). With Figure 12.4, we illustrate a graphical "solution" for a one-dimensionally layered inhomogeneous material exploiting the law of refraction. The ray starting from $\underline{\mathbf{R}}_0$ follows a piecewise straight line trajectory, and we nicely recognize why the material may only be slowly varying (within the extent of several wavelengths) for such a ray solution of the wave equation: Internal (multiple) reflections are disregarded, similar to the Born approximation for the field scattered by a penetrable body (Section 5.6.3).

Červený (2001) also shows that an inhomogeneous density $\rho(\underline{\mathbf{R}})$ does not really complicate the ray equations: The eikonal equation remains unchanged, and similarly the transport equation, if $P(\underline{\mathbf{R}})$ is replaced by $\tilde{P}(\underline{\mathbf{R}}) = P(\underline{\mathbf{R}})/\sqrt{\rho(\underline{\mathbf{R}})}$.

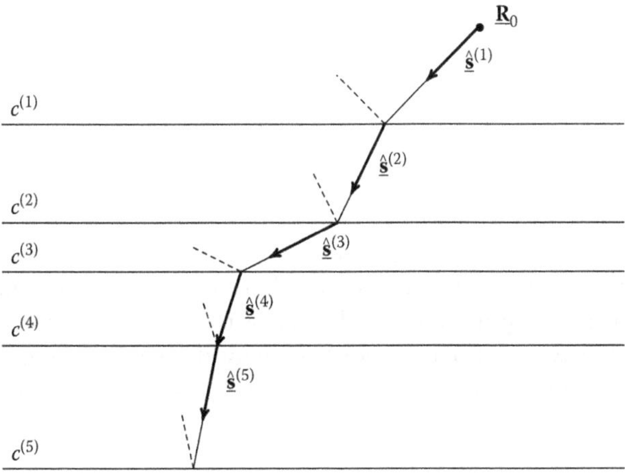

FIGURE 12.4
Ray trajectory in a one-dimensionally layered inhomogeneous material (ray tracing).

Likewise, the transition to pulsed rays is straightforward:

$$p(\underline{\mathbf{R}}, t) = P(\underline{\mathbf{R}}) f_+ \left[t - T(\underline{\mathbf{R}})\right], \tag{12.99}$$

where $f_+(t)$ is again an analytic signal.

The formal similarity of (12.99) with (12.67) suggests, at least for a homogeneous material, to transform the ray into a beam testing whether

$$T(\underline{\mathbf{R}}) = \frac{z}{c} + \frac{1}{2}\underline{\boldsymbol{r}}^{\mathrm{T}} \underline{\underline{\boldsymbol{Q}}}^{-1}(z)\underline{\mathbf{r}}, \tag{12.100}$$

$$P(\underline{\mathbf{R}}) = \sqrt{\frac{\det \underline{\underline{\boldsymbol{Q}}}(0)}{\det \underline{\underline{\boldsymbol{Q}}}(z)}} \tag{12.101}$$

are solutions of the eikonal and transport equations for $c(\underline{\mathbf{R}}) = c$; $\hat{\underline{\mathbf{s}}} = \underline{\mathbf{e}}_z$, i.e., the z-axis as ray trajectory would then be the axis of the beam. Precisely checking (12.100), we realize that it obviously represents a truncated Taylor series with regard to the coordinates perpendicular to the ray axis—here: $\underline{\mathbf{r}}$, respectively, x and y:

$$T(\underline{\mathbf{r}}, z) = T(\underline{\mathbf{r}}, z)\Big|_{\underline{\mathbf{r}}=\underline{\mathbf{0}}} + \nabla_{\underline{\mathbf{r}}} T(\underline{\mathbf{r}}, z)\Big|_{\underline{\mathbf{r}}=\underline{\mathbf{0}}} \cdot \underline{\mathbf{r}} + \frac{1}{2}\nabla_{\underline{\mathbf{r}}}\nabla_{\underline{\mathbf{r}}} T(\underline{\mathbf{r}}, z)\Big|_{\underline{\mathbf{r}}=\underline{\mathbf{0}}} : \underline{\mathbf{r}}\,\underline{\mathbf{r}}, \tag{12.102}$$

if we put $T(\underline{\mathbf{r}}, z)\big|_{\underline{\mathbf{r}}=\underline{\mathbf{0}}} = z/c$ and observe

$$\begin{aligned} &\nabla_{\underline{\mathbf{r}}} T(\underline{\mathbf{r}}, z)\Big|_{\underline{\mathbf{r}}=\underline{\mathbf{0}}} \cdot \underline{\mathbf{r}} \\ &= \left[\underline{\mathbf{s}}(\underline{\mathbf{r}}, z) - \frac{\partial s_z(\underline{\mathbf{r}}, z)}{\partial z}\underline{\mathbf{e}}_z\right]_{\underline{\mathbf{r}}=\underline{\mathbf{0}}} \cdot \underline{\mathbf{r}} = \underline{\mathbf{s}}(\underline{\mathbf{r}} = \underline{\mathbf{0}}, z) \cdot \underline{\mathbf{r}} = \frac{1}{c}\underline{\mathbf{e}}_z \cdot \underline{\mathbf{r}} = 0. \end{aligned} \tag{12.103}$$

We may now either insert (12.102) into the eikonal equation to find (12.100) with (12.61), and then (12.101) as solution of the transport equation, or we directly check whether (12.100) and (12.101) are solutions of the eikonal and transport equations while considering (12.55) and (12.56); in both cases, we obtain the correct result utilizing the paraxial approximation, namely: In a homogeneous material, the *ansatz* (12.92) with the Taylor expansion (12.102) of the phase applying eikonal and transport equations leads to a pulsed Gaussian beam with (12.100) and (12.101) as paraxially approximated solution of the wave equation. We already know that the pulsed beam in a homogeneous material is an approximate solution of the wave equation (or an exact solution of the wave packet equation), so why do we again emphasize it for different approaches? It is because the above procedure can be generalized to (weakly) inhomogeneous materials (Červený 2001; Klimeš 1989; Popov 1982, 2004)! Then, we obtain Gaussian or pulsed beams in inhomogeneous materials having the ray trajectories as a beam axes. Interesting enough, this result is also obtained as exact solution of an approximately valid wave packet equation for inhomogeneous materials, i.e., according to the approach discussed in Section 12.2 (Červený et al. 1982; Heyman 1994).

The conceptual advantage of beams as compared to rays is in their nature having no singularities at focus points or caustics due to det $\underline{\underline{Q}}(l) \neq 0$, where l is a coordinate along the ray trajectory. In addition: A physically existing aperture distribution may be quantitatively separated into "aperture" distributions of beams to be launched along rays into well-defined directions (superposition of Gaussian or pulsed beams) to calculate acoustic radiation fields in inhomogeneous materials (Maciel and Felsen 1990; Shlivinski et al. 2004, 2005). Beam rays may also formally be derived choosing a complex source point in Green functions for source fields (Heyman 2002); yet we do not want to get deeper into that approach even though it has already been applied to calculate transducer radiation fields (Zeroug et al. 1996).

12.3.2 Eikonal and transport equations for elastic beams

Inhomogeneous isotropic materials: Up to now, we have only dealt with acoustic (scalar) ultrasonic beams, yet US-NDT requires elastic longitudinal and transverse beams. To define respective rays, we need the pertinent eikonal and transport equations. Starting point is the homogeneous Navier equation (7.27)—we write it in terms of the particle velocity—

$$\begin{aligned}
&[\lambda(\underline{\mathbf{R}}) + \mu(\underline{\mathbf{R}})]\, \nabla\nabla \cdot \underline{\mathbf{u}}(\underline{\mathbf{R}}, t) + \mu(\underline{\mathbf{R}})\Delta\underline{\mathbf{u}}(\underline{\mathbf{R}}, t) - \rho(\underline{\mathbf{R}})\, \frac{\partial^2 \underline{\mathbf{u}}(\underline{\mathbf{R}}, t)}{\partial t^2} \\
&\quad + [\nabla\lambda(\underline{\mathbf{R}})]\, [\nabla \cdot \underline{\mathbf{u}}(\underline{\mathbf{R}}, t)] + [\nabla\mu(\underline{\mathbf{R}})] \cdot [\nabla\underline{\mathbf{u}}(\underline{\mathbf{R}}, t)] \\
&\quad + [\nabla\underline{\mathbf{u}}(\underline{\mathbf{R}}, t)] \cdot [\nabla\mu(\underline{\mathbf{R}})] = \underline{\mathbf{0}}
\end{aligned} \tag{12.104}$$

that may be approximately solved with the time harmonic ray *ansatz*

$$\underline{\mathbf{u}}(\underline{\mathbf{R}}, t) = \underline{\mathbf{U}}(\underline{\mathbf{R}})\, \mathrm{e}^{\mathrm{j}\omega_0 T(\underline{\mathbf{R}})}\, \mathrm{e}^{-\mathrm{j}\omega_0 t}. \tag{12.105}$$

Equating the quadratic term in ω_0 to zero results in

$$\left[\frac{\lambda(\underline{\mathbf{R}})+\mu(\underline{\mathbf{R}})}{\rho(\underline{\mathbf{R}})}\nabla T(\underline{\mathbf{R}})\nabla T(\underline{\mathbf{R}})+\frac{\mu(\underline{\mathbf{R}})}{\rho(\underline{\mathbf{R}})}\nabla T(\underline{\mathbf{R}})\cdot\nabla T(\underline{\mathbf{R}})\,\underline{\underline{\mathbf{I}}}-\underline{\underline{\mathbf{I}}}\right]\cdot\underline{\mathbf{U}}(\underline{\mathbf{R}})=\underline{\mathbf{0}} \tag{12.106}$$

after some calculus; with the definition of the slowness vector according to (12.95), we find the alternative form

$$\left[\frac{\mu(\underline{\mathbf{R}})\underline{\mathbf{s}}(\underline{\mathbf{R}})\cdot\underline{\mathbf{s}}(\underline{\mathbf{R}})-\rho(\underline{\mathbf{R}})}{\rho(\underline{\mathbf{R}})}\underline{\underline{\mathbf{I}}}+\frac{\lambda(\underline{\mathbf{R}})+\mu(\underline{\mathbf{R}})}{\rho(\underline{\mathbf{R}})}\underline{\mathbf{s}}(\underline{\mathbf{R}})\underline{\mathbf{s}}(\underline{\mathbf{R}})\right]\cdot\underline{\mathbf{U}}(\underline{\mathbf{R}})=\underline{\mathbf{0}}. \tag{12.107}$$

With

$$\underline{\mathbf{s}}(\underline{\mathbf{R}})=\hat{\underline{\mathbf{s}}}(\underline{\mathbf{R}})/c(\underline{\mathbf{R}}), \tag{12.108}$$

this condition is structurally identical to the eigenvalue problem (8.62) of plane waves in homogeneous materials. Equating to zero the determinant of the wave tensor $\underline{\underline{\mathbf{W}}}[\underline{\mathbf{s}}(\underline{\mathbf{R}})]$ in the square brackets, we immediately find the dispersion relation

$$\begin{aligned}\underline{\mathbf{s}}_{\mathrm{S}}(\underline{\mathbf{R}})\cdot\underline{\mathbf{s}}_{\mathrm{S}}(\underline{\mathbf{R}})&=\frac{\rho(\underline{\mathbf{R}})}{\mu(\underline{\mathbf{R}})}\\&=\frac{1}{c_{\mathrm{S}}^2(\underline{\mathbf{R}})},\end{aligned} \tag{12.109}$$

$$\begin{aligned}\underline{\mathbf{s}}_{\mathrm{P}}(\underline{\mathbf{R}})\cdot\underline{\mathbf{s}}_{\mathrm{P}}(\underline{\mathbf{R}})&=\frac{\rho(\underline{\mathbf{R}})}{\lambda(\underline{\mathbf{R}})+2\mu(\underline{\mathbf{R}})}\\&=\frac{1}{c_{\mathrm{P}}^2(\underline{\mathbf{R}})}\end{aligned} \tag{12.110}$$

with (8.67), and consequently, the eikonal equation

$$\nabla T_{\mathrm{P}}(\underline{\mathbf{R}})\cdot\nabla T_{\mathrm{P}}(\underline{\mathbf{R}})=\frac{1}{c_{\mathrm{P}}^2(\underline{\mathbf{R}})} \tag{12.111}$$

for a P-ray as well as the eikonal equation

$$\nabla T_{\mathrm{S}}(\underline{\mathbf{R}})\cdot\nabla T_{\mathrm{S}}(\underline{\mathbf{R}})=\frac{1}{c_{\mathrm{S}}^2(\underline{\mathbf{R}})} \tag{12.112}$$

for an S-ray; however, "P" only stands for "primary" and "S" only for "secondary" because a decoupling into pressure and shear waves is no longer possible in inhomogeneous materials.[207]

[207] The divergence and the curl of (12.105) contain $\nabla\cdot\underline{\mathbf{U}}(\underline{\mathbf{R}})$ and $\nabla\times\underline{\mathbf{U}}(\underline{\mathbf{R}})$, and both expressions are generally nonzero.

Concerning polarizations of P- and S-rays, we argue exactly in the same way as for plane waves in homogeneous materials (Section 8.1.2). We calculate

$$\underline{\underline{\mathbf{W}}}[\underline{\mathbf{s}}_{\mathrm{P}}(\underline{\mathbf{R}})] = \frac{\lambda(\underline{\mathbf{R}}) + \mu(\underline{\mathbf{R}})}{\rho(\underline{\mathbf{R}})} \left[\underline{\mathbf{s}}_{\mathrm{P}}(\underline{\mathbf{R}})\underline{\mathbf{s}}_{\mathrm{P}}(\underline{\mathbf{R}}) - \frac{1}{c_{\mathrm{P}}^2(\underline{\mathbf{R}})} \underline{\underline{\mathbf{I}}} \right] \tag{12.113}$$

and conclude

$$\underline{\mathbf{s}}_{\mathrm{P}}(\underline{\mathbf{R}})\underline{\mathbf{s}}_{\mathrm{P}}(\underline{\mathbf{R}}) \cdot \underline{\mathbf{U}}_{\mathrm{P}}(\underline{\mathbf{R}}) = \frac{1}{c_{\mathrm{P}}^2(\underline{\mathbf{R}})} \underline{\mathbf{U}}_{\mathrm{P}}(\underline{\mathbf{R}}) \tag{12.114}$$

from the eigenvalue equation $\underline{\underline{\mathbf{W}}}[\underline{\mathbf{s}}_{\mathrm{P}}(\underline{\mathbf{R}})] \cdot \underline{\mathbf{U}}_{\mathrm{P}}(\underline{\mathbf{R}}) \equiv \underline{\mathbf{0}}$. Equation 12.114 tells us that the amplitude vector $\underline{\mathbf{U}}_{\mathrm{P}}(\underline{\mathbf{R}})$ of the P-ray has the direction of the slowness vector $\underline{\mathbf{s}}_{\mathrm{P}}(\underline{\mathbf{R}})$, it is therefore perpendicular to the phase front $T_{\mathrm{P}}(\underline{\mathbf{R}}) = \mathrm{const}$: The P-ray is longitudinally polarized with respect to the ray trajectory just as a plane wave. This is a consequence of the high-frequency approximation because the eigenvalue problem (12.107) exclusively follows from equating the quadratic ω_0-term to zero. From (12.114), we obtain the representation

$$\begin{aligned} \underline{\mathbf{U}}_{\mathrm{P}}(\underline{\mathbf{R}}) &= c_{\mathrm{P}}(\underline{\mathbf{R}}) \underbrace{c_{\mathrm{P}}(\underline{\mathbf{R}})\underline{\mathbf{s}}_{\mathrm{P}}(\underline{\mathbf{R}}) \cdot \underline{\mathbf{U}}_{\mathrm{P}}(\underline{\mathbf{R}})}_{\stackrel{\mathrm{def}}{=} U_{\mathrm{P}}(\underline{\mathbf{R}})} \underline{\mathbf{s}}_{\mathrm{P}}(\underline{\mathbf{R}}) \\ &= U_{\mathrm{P}}(\underline{\mathbf{R}})\underline{\hat{\mathbf{s}}}_{\mathrm{P}}(\underline{\mathbf{R}}). \end{aligned} \tag{12.115}$$

We have $\underline{\mathbf{U}}_{\mathrm{P}}(\underline{\mathbf{R}}) \cdot \underline{\mathbf{U}}_{\mathrm{P}}^*(\underline{\mathbf{R}}) = |U_{\mathrm{P}}(\underline{\mathbf{R}})|^2$, where $|U_{\mathrm{P}}(\underline{\mathbf{R}})|$ is the magnitude of the complex scalar amplitude $U_{\mathrm{P}}(\underline{\mathbf{R}})$.

With

$$\underline{\underline{\mathbf{W}}}[\underline{\mathbf{s}}_{\mathrm{S}}(\underline{\mathbf{R}})] = \frac{\lambda(\underline{\mathbf{R}}) + \mu(\underline{\mathbf{R}})}{\rho(\underline{\mathbf{R}})} \underline{\mathbf{s}}_{\mathrm{S}}(\underline{\mathbf{R}})\underline{\mathbf{s}}_{\mathrm{S}}(\underline{\mathbf{R}}), \tag{12.116}$$

we find the condition

$$\underline{\mathbf{s}}_{\mathrm{S}}(\underline{\mathbf{R}})\underline{\mathbf{s}}_{\mathrm{S}}(\underline{\mathbf{R}}) \cdot \underline{\mathbf{U}}_{\mathrm{S}}(\underline{\mathbf{R}}) = \underline{\mathbf{0}} \tag{12.117}$$

for the polarization of the S-ray from the eigenvalue equation $\underline{\underline{\mathbf{W}}}[\underline{\mathbf{s}}_{\mathrm{S}}(\underline{\mathbf{R}})] \cdot \underline{\mathbf{U}}_{\mathrm{S}}(\underline{\mathbf{R}}) = \underline{\mathbf{0}}$. It follows

$$\underline{\mathbf{s}}_{\mathrm{S}}(\underline{\mathbf{R}}) \cdot \underline{\mathbf{U}}_{\mathrm{S}}(\underline{\mathbf{R}}) = 0, \tag{12.118}$$

i.e., the S-ray polarization vector is perpendicular to the pertinent ray trajectory, the S-ray is transversely polarized with respect to the ray trajectory. Therefore, we introduce two orthogonal unit vectors $\underline{\mathbf{e}}_{\mathrm{S1}}(\underline{\mathbf{R}})$ and $\underline{\mathbf{e}}_{\mathrm{S2}}(\underline{\mathbf{R}})$ tangentially to the phase front $T_{\mathrm{S}}(\underline{\mathbf{R}}) = \mathrm{const}$ satisfying $\underline{\mathbf{e}}_{\mathrm{S1,S2}}(\underline{\mathbf{R}}) \cdot \underline{\hat{\mathbf{s}}}_{\mathrm{S}}(\underline{\mathbf{R}}) = 0$ besides $\underline{\mathbf{e}}_{\mathrm{S1}}(\underline{\mathbf{R}}) \cdot \underline{\mathbf{e}}_{S2}(\underline{\mathbf{R}}) = 0$; these two vectors are used as polarization basis for the S-ray:

$$\begin{aligned} \underline{\mathbf{U}}_{\mathrm{S}}(\underline{\mathbf{R}}) &= U_{\mathrm{S1}}(\underline{\mathbf{R}})\underline{\mathbf{e}}_{\mathrm{S1}}(\underline{\mathbf{R}}) + U_{\mathrm{S2}}(\underline{\mathbf{R}})\underline{\mathbf{e}}_{\mathrm{S2}}(\underline{\mathbf{R}}) \\ &= \underline{\mathbf{U}}_{\mathrm{S1}}(\underline{\mathbf{R}}) + \underline{\mathbf{U}}_{\mathrm{S2}}(\underline{\mathbf{R}}), \end{aligned} \tag{12.119}$$

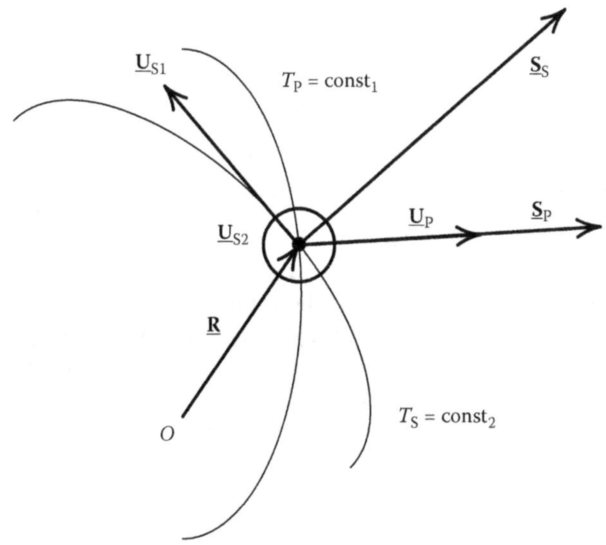

FIGURE 12.5
Polarization of P- and S-rays.

where $U_{S1,S2}(\underline{\mathbf{R}})$ are complex valued scalar amplitude factors: We have $\underline{\mathbf{U}}_S(\underline{\mathbf{R}}) \cdot \underline{\mathbf{U}}_S^*(\underline{\mathbf{R}}) = |U_{S1}(\underline{\mathbf{R}})|^2 + |U_{S2}(\underline{\mathbf{R}})|^2$. As for the case of plane waves (Figure 8.8), $\underline{\mathbf{s}}_S(\underline{\mathbf{R}})$, $\underline{\mathbf{U}}_{S1}(\underline{\mathbf{R}})$, and $\underline{\mathbf{U}}_{S2}(\underline{\mathbf{R}})$ should establish a (local) right-handed system, why we look at the tip of the vector $\underline{\mathbf{U}}_{S2}(\underline{\mathbf{R}})$ in Figure 12.5. Because of the different and spatially dependent phase velocities $c_P(\underline{\mathbf{R}})$ and $c_S(\underline{\mathbf{R}})$, the phase fronts $T_P(\underline{\mathbf{R}}) = \text{const}$ and $T_S(\underline{\mathbf{R}}) = \text{const}$ are nonplanar, and they do not coincide at the same time; therefore, they are displayed for two different times in Figure 12.5. Per definition, the trajectories defined by $\underline{\mathbf{s}}_P(\underline{\mathbf{R}})$ and $\underline{\mathbf{s}}_S(\underline{\mathbf{R}})$ are orthogonal to the respective phase fronts at the point $\underline{\mathbf{R}}$; hence, they generally do not have the same direction. As indicated above, the polarization vector $\underline{\mathbf{U}}_P(\underline{\mathbf{R}})$ is also orthogonal to the phase front $T_P(\underline{\mathbf{R}}) = \text{const}$, while the polarization vectors $\underline{\mathbf{U}}_{S1,S2}(\underline{\mathbf{R}})$ are tangentially oriented with respect to $T_S(\underline{\mathbf{R}}) = \text{const}$.[208]

For a complete description of P- and S-rays, we must also know the transport equations for both rays according to (12.94). To derive them, we

[208]The three polarization vectors $\underline{\mathbf{U}}_P(\underline{\mathbf{R}})$, $\underline{\mathbf{U}}_{S1,S2}(\underline{\mathbf{R}})$ are generally nonorthogonal to each other, even though the wave tensor is real-symmetric, and should consequently have real orthogonal eigenvectors! For plane waves in homogeneous materials, be they isotropic or anisotropic, the wave tensor contains the parameter $\hat{\underline{\mathbf{k}}}$ as phase unit vector; it may be chosen equal for P,S1,S2-, respectively, qP,qS1,qS2-wave modes yielding the same real-symmetric tensor for all wave modes with, therefore, real orthogonal polarization vectors regarding the direction $\hat{\underline{\mathbf{k}}}$. Yet, for rays in inhomogeneous materials, the wave tensor is a function of the phase fronts $T_{P,S1,S2}(\underline{\mathbf{R}}) = \text{const}$, the solutions $\underline{\mathbf{s}}_{P,S1,S2}(\underline{\mathbf{R}}) = \nabla T_{P,S1,S2}(\underline{\mathbf{R}})$ of the respective eikonal equations determine the directions $\hat{\underline{\mathbf{s}}}_{P,S1,S2}(\underline{\mathbf{R}})$ yielding a different wave tensor for each wave mode.

concentrate on the linear ω_0-term—we denote it by $\underline{\mathbf{F}}(\underline{\mathbf{R}})$—after we have inserted (12.105) into (12.104). We find

$$\begin{aligned}\underline{\mathbf{F}}(\underline{\mathbf{R}}) = {} & [\lambda(\underline{\mathbf{R}}) + \mu(\underline{\mathbf{R}})][\boldsymbol{\nabla\nabla} T(\underline{\mathbf{R}}) \cdot \underline{\mathbf{U}}(\underline{\mathbf{R}}) + \boldsymbol{\nabla} T(\underline{\mathbf{R}})\boldsymbol{\nabla} \cdot \underline{\mathbf{U}}(\underline{\mathbf{R}}) \\ & + \boldsymbol{\nabla}\underline{\mathbf{U}}(\underline{\mathbf{R}}) \cdot \boldsymbol{\nabla} T(\underline{\mathbf{R}})] + \mu(\underline{\mathbf{R}})[2\boldsymbol{\nabla} T(\underline{\mathbf{R}}) \cdot \boldsymbol{\nabla}\underline{\mathbf{U}}(\underline{\mathbf{R}}) + \Delta T(\underline{\mathbf{R}})\underline{\mathbf{U}}(\underline{\mathbf{R}})] \\ & + [\boldsymbol{\nabla}\lambda(\underline{\mathbf{R}})\boldsymbol{\nabla} T(\underline{\mathbf{R}}) \cdot \underline{\mathbf{U}}(\underline{\mathbf{R}}) + \boldsymbol{\nabla} T(\underline{\mathbf{R}})\boldsymbol{\nabla}\mu(\underline{\mathbf{R}}) \cdot \underline{\mathbf{U}}(\underline{\mathbf{R}}) \\ & + \boldsymbol{\nabla}\mu(\underline{\mathbf{R}}) \cdot \boldsymbol{\nabla} T(\underline{\mathbf{R}})\underline{\mathbf{U}}(\underline{\mathbf{R}}); \end{aligned} \tag{12.120}$$

the $\boldsymbol{\nabla}$-differentiations always refer to the immediate adjacent term. Now we must insert either the P- or the S-ray for $T(\underline{\mathbf{R}})$, and $\underline{\mathbf{U}}(\underline{\mathbf{R}})$, respectively; we start with the P-ray and carry out the relevant differentiations on $\underline{\mathbf{U}}_{\mathrm{P}}(\underline{\mathbf{R}}) = U_{\mathrm{P}}(\underline{\mathbf{R}})c_{\mathrm{P}}(\underline{\mathbf{R}})\underline{\mathbf{s}}_{\mathrm{P}}(\underline{\mathbf{R}})$ with $\underline{\mathbf{s}}_{\mathrm{P}}(\underline{\mathbf{R}}) = \boldsymbol{\nabla} T_{\mathrm{P}}(\underline{\mathbf{R}})$; that way, terms with $\boldsymbol{\nabla} T_{\mathrm{P}}(\underline{\mathbf{R}})$ appear in the gradient dyadic $\boldsymbol{\nabla}\underline{\mathbf{U}}_{\mathrm{P}}(\underline{\mathbf{R}})$ suggesting to investigate the vector equation $\underline{\mathbf{F}}_{\mathrm{P}}(\underline{\mathbf{R}}) = \underline{\mathbf{0}}$ projected onto the P-trajectory $\underline{\mathbf{s}}_{\mathrm{P}}(\underline{\mathbf{R}}) = \boldsymbol{\nabla} T_{\mathrm{P}}(\underline{\mathbf{R}})$ (Červený 2001), thus allowing for the multiple use of the eikonal equation (12.111). We obtain after some calculus[209]

$$\begin{aligned}& \underline{\mathbf{F}}_{\mathrm{P}}(\underline{\mathbf{R}}) \cdot \boldsymbol{\nabla} T_{\mathrm{P}}(\underline{\mathbf{R}}) \\ & \quad = \boldsymbol{\nabla} T_{\mathrm{P}}(\underline{\mathbf{R}}) \cdot \left\{2c_{\mathrm{P}}^2(\underline{\mathbf{R}})\rho(\underline{\mathbf{R}})\boldsymbol{\nabla} U_{\mathrm{P}}(\underline{\mathbf{R}}) + U_{\mathrm{P}}(\underline{\mathbf{R}})\boldsymbol{\nabla}\left[c_{\mathrm{P}}^2(\underline{\mathbf{R}})\rho(\underline{\mathbf{R}})\right]\right\} \\ & \qquad + c_{\mathrm{P}}^2(\underline{\mathbf{R}})\rho(\underline{\mathbf{R}})U_{\mathrm{P}}(\underline{\mathbf{R}})\Delta T_{\mathrm{P}}(\underline{\mathbf{R}}), \end{aligned} \tag{12.121}$$

resulting in the following form of the transport equation

$$2\boldsymbol{\nabla}\tilde{U}_{\mathrm{P}}(\underline{\mathbf{R}}) \cdot \boldsymbol{\nabla} T_{\mathrm{P}}(\underline{\mathbf{R}}) + \tilde{U}_{\mathrm{P}}(\underline{\mathbf{R}})\Delta T_{\mathrm{P}}(\underline{\mathbf{R}}) = 0 \tag{12.122}$$

with the substitution $\tilde{U}_{\mathrm{P}}(\underline{\mathbf{R}}) = \sqrt{c_{\mathrm{P}}^2(\underline{\mathbf{R}})\rho(\underline{\mathbf{R}})}U_{\mathrm{P}}(\underline{\mathbf{R}})$. It is structurally identical to the transport equation (12.94) for acoustic waves.

For the S-ray, we investigate projections of $\underline{\mathbf{F}}_{\mathrm{S}}(\underline{\mathbf{R}})$ onto the polarization vectors[210] $\underline{\mathbf{F}}_{\mathrm{S}}(\underline{\mathbf{R}}) \cdot \underline{\mathbf{e}}_{\alpha}(\underline{\mathbf{R}})$ with $\alpha = \mathrm{S1}, \mathrm{S2}$ because three terms immediately disappear due to the orthogonality of $\boldsymbol{\nabla} T_{\mathrm{S}}(\underline{\mathbf{R}}) \cdot \underline{\mathbf{e}}_{\alpha} = 0$:

$$\begin{aligned}& \underline{\mathbf{F}}_{\mathrm{S}}(\underline{\mathbf{R}}) \cdot \underline{\mathbf{e}}_{\alpha}(\underline{\mathbf{R}}) \\ & \quad = [\lambda(\underline{\mathbf{R}}) + \mu(\underline{\mathbf{R}})][\underline{\mathbf{e}}_{\alpha}(\underline{\mathbf{R}}) \cdot \boldsymbol{\nabla\nabla} T_{\mathrm{S}}(\underline{\mathbf{R}}) \cdot \underline{\mathbf{U}}_{\mathrm{S}}(\underline{\mathbf{R}}) + \underline{\mathbf{e}}_{\alpha}(\underline{\mathbf{R}}) \cdot \boldsymbol{\nabla}\underline{\mathbf{U}}_{\mathrm{S}}(\underline{\mathbf{R}}) \cdot \boldsymbol{\nabla} T_{\mathrm{S}}(\underline{\mathbf{R}})] \\ & \qquad + \mu(\underline{\mathbf{R}})[2\boldsymbol{\nabla} T_{\mathrm{S}}(\underline{\mathbf{R}}) \cdot \boldsymbol{\nabla}\underline{\mathbf{U}}_{\mathrm{S}}(\underline{\mathbf{R}}) \cdot \underline{\mathbf{e}}_{\alpha}(\underline{\mathbf{R}}) + \Delta T_{\mathrm{S}}(\underline{\mathbf{R}})\underline{\mathbf{U}}_{\mathrm{S}}(\underline{\mathbf{R}}) \cdot \underline{\mathbf{e}}_{\alpha}(\underline{\mathbf{R}})] \\ & \qquad + \boldsymbol{\nabla}\mu(\underline{\mathbf{R}}) \cdot \boldsymbol{\nabla} T_{\mathrm{S}}(\underline{\mathbf{R}})\underline{\mathbf{U}}_{\mathrm{S}}(\underline{\mathbf{R}}) \cdot \underline{\mathbf{e}}_{\alpha}(\underline{\mathbf{R}}); \end{aligned} \tag{12.123}$$

since $\boldsymbol{\nabla}[\boldsymbol{\nabla} T_{\mathrm{S}}(\underline{\mathbf{R}}) \cdot \underline{\mathbf{e}}_{\alpha}(\underline{\mathbf{R}})] = 0 = \boldsymbol{\nabla\nabla} T_{\mathrm{S}}(\underline{\mathbf{R}}) \cdot \underline{\mathbf{e}}_{\alpha}(\underline{\mathbf{R}}) + \boldsymbol{\nabla}\underline{\mathbf{e}}_{\alpha}(\underline{\mathbf{R}}) \cdot \boldsymbol{\nabla} T_{\mathrm{S}}(\underline{\mathbf{R}})$, the terms in the $[\lambda(\underline{\mathbf{R}}) + \mu(\underline{\mathbf{R}})]$-factor cancel as soon as we have inserted $\underline{\mathbf{U}}_{\mathrm{S}}(\underline{\mathbf{R}})$ according to (12.119), and the resulting expression $\boldsymbol{\nabla}\underline{\mathbf{U}}_{\mathrm{S}}(\underline{\mathbf{R}})$. It remains

---

[209] From $\boldsymbol{\nabla} T \cdot \boldsymbol{\nabla\nabla} T \cdot \underline{\mathbf{U}}$, a term $\boldsymbol{\nabla} T \cdot \boldsymbol{\nabla\nabla} T \cdot \boldsymbol{\nabla} T$ results that may be brought into the form $\frac{1}{2}\boldsymbol{\nabla}(\boldsymbol{\nabla} T \cdot \boldsymbol{\nabla} T) \cdot \boldsymbol{\nabla} T$ with the formula $\boldsymbol{\nabla}(\underline{\mathbf{A}} \cdot \underline{\mathbf{B}}) = (\boldsymbol{\nabla}\underline{\mathbf{A}}) \cdot \underline{\mathbf{B}} + (\boldsymbol{\nabla}\underline{\mathbf{B}}) \cdot \underline{\mathbf{A}}$ allowing for another application of the eikonal equation.

[210] We did it similarly for the P-ray.

$$\begin{aligned}
&\underline{\mathbf{F}}_{\mathrm{S}}(\underline{\mathbf{R}}) \cdot \underline{\mathbf{e}}_\alpha(\underline{\mathbf{R}}) \\
&\quad = \mu(\underline{\mathbf{R}})\{2\nabla T_{\mathrm{S}}(\underline{\mathbf{R}}) \cdot \nabla U_\alpha(\underline{\mathbf{R}}) + 2\nabla T_{\mathrm{S}}(\underline{\mathbf{R}}) \\
&\qquad \cdot [U_{\mathrm{S1}}(\underline{\mathbf{R}})\nabla \underline{\mathbf{e}}_{\mathrm{S1}}(\underline{\mathbf{R}}) + U_{\mathrm{S2}}(\underline{\mathbf{R}}) \cdot \nabla \underline{\mathbf{e}}_{\mathrm{S2}}(\underline{\mathbf{R}})] \cdot \underline{\mathbf{e}}_\alpha(\underline{\mathbf{R}}) \\
&\qquad + U_\alpha(\underline{\mathbf{R}})\Delta T_{\mathrm{S}}(\underline{\mathbf{R}})\} + U_\alpha(\underline{\mathbf{R}})\nabla\mu(\underline{\mathbf{R}}) \cdot \nabla T_{\mathrm{S}}(\underline{\mathbf{R}}),
\end{aligned} \tag{12.124}$$

and finally—we write (12.124) separately for $\alpha = \mathrm{S1}$ and $\alpha = \mathrm{S2}$—

$$\begin{aligned}
&2\nabla\tilde{U}_{\mathrm{S1}}(\underline{\mathbf{R}}) \cdot \nabla T_{\mathrm{S}}(\underline{\mathbf{R}}) + \tilde{U}_{\mathrm{S1}}(\underline{\mathbf{R}})\Delta T_{\mathrm{S}}(\underline{\mathbf{R}}) \\
&\quad + 2\tilde{U}_{\mathrm{S2}}(\underline{\mathbf{R}})\nabla T_{\mathrm{S}}(\underline{\mathbf{R}}) \cdot \nabla \underline{\mathbf{e}}_{\mathrm{S2}}(\underline{\mathbf{R}}) \cdot \underline{\mathbf{e}}_{\mathrm{S1}}(\underline{\mathbf{R}}) = 0,
\end{aligned} \tag{12.125}$$

$$\begin{aligned}
&2\nabla\tilde{U}_{\mathrm{S2}}(\underline{\mathbf{R}}) \cdot \nabla T_{\mathrm{S}}(\underline{\mathbf{R}}) + \tilde{U}_{\mathrm{S2}}(\underline{\mathbf{R}})\Delta T_{\mathrm{S}}(\underline{\mathbf{R}}) \\
&\quad + 2\tilde{U}_{\mathrm{S1}}(\underline{\mathbf{R}})\nabla T_{\mathrm{S}}(\underline{\mathbf{R}}) \cdot \nabla \underline{\mathbf{e}}_{\mathrm{S1}}(\underline{\mathbf{R}}) \cdot \underline{\mathbf{e}}_{\mathrm{S2}}(\underline{\mathbf{R}}) = 0
\end{aligned} \tag{12.126}$$

if we once more put $\tilde{U}_\alpha(\underline{\mathbf{R}}) = \sqrt{c_{\mathrm{S}}^2(\underline{\mathbf{R}})\rho(\underline{\mathbf{R}})}U_\alpha(\underline{\mathbf{R}})$ and observe that $\nabla[\underline{\mathbf{e}}_\alpha(\underline{\mathbf{R}}) \cdot \underline{\mathbf{e}}_\alpha(\underline{\mathbf{R}})] = 0 = 2\nabla\underline{\mathbf{e}}_\alpha(\underline{\mathbf{R}}) \cdot \underline{\mathbf{e}}_\alpha(\underline{\mathbf{R}})$ holds. Hence, the transport equations for both polarizations of the S-ray are coupled for an arbitrary choice of the polarization basis $\underline{\mathbf{e}}_\alpha(\underline{\mathbf{R}})$, $\alpha = \mathrm{S1}, \mathrm{S2}$, the polarizations are not—as for plane waves in homogeneous materials—*a priori* independent upon each other. However, deciding on a specific basis

$$\nabla T_{\mathrm{S}}(\underline{\mathbf{R}}) \cdot \nabla \underline{\mathbf{e}}_\alpha(\underline{\mathbf{R}}) = \beta_\alpha(\underline{\mathbf{R}})\nabla T_{\mathrm{S}}(\underline{\mathbf{R}}) \tag{12.127}$$

with normalization constants $\beta_\alpha(\underline{\mathbf{R}})$, the coupling disappears; Červený (2001) discusses algorithms for a practical calculation of such a basis.

As for the case of acoustic rays, we may now convert elastic rays into beams inserting respective Taylor expansions of $T_{\mathrm{P,S}}(\underline{\mathbf{R}})$ into the eikonal equations (Červený 2001; Norris 1988); Spies (1994) has applied this formalism to the calculation of transducer fields in homogeneous materials.

Inhomogeneous anisotropic materials: The formalism to define and to calculate P- and S-rays in inhomogeneous isotropic materials suggests to proceed similarly for the case of inhomogeneous anisotropic material because the *ansatz* (12.105) immediately yields

$$\left[\nabla T(\underline{\mathbf{R}}) \cdot \underline{\underline{\underline{\underline{\mathbf{c}}}}}(\underline{\mathbf{R}}) \cdot \nabla T(\underline{\mathbf{R}}) - \rho(\underline{\mathbf{R}})\,\underline{\underline{\mathbf{I}}}\right] \cdot \underline{\mathbf{U}}(\underline{\mathbf{R}}) = \underline{\mathbf{0}} \tag{12.128}$$

after insertion of the homogeneous Navier equation (7.3), and equating the ω_0^2-factor to zero. As usual, the determinant of the wave tensor in the squared brackets must vanish:

$$\det\left[\nabla T(\underline{\mathbf{R}}) \cdot \underline{\underline{\underline{\underline{\mathbf{c}}}}}(\underline{\mathbf{R}}) \cdot \nabla T(\underline{\mathbf{R}}) - \rho(\underline{\mathbf{R}})\,\underline{\underline{\mathbf{I}}}\right] = 0, \tag{12.129}$$

in order to ensure nontrivial solutions for $\underline{\mathbf{U}}(\underline{\mathbf{R}})$; due to $\underline{\mathbf{s}}(\underline{\mathbf{R}}) = \nabla T(\underline{\mathbf{R}})$, the eigenvalue problem known from Section 8.3 results, i.e., the dispersion relation turns into the eikonal equation with the solutions $\underline{\mathbf{s}}_{\mathrm{qP}}(\underline{\mathbf{R}})$, $\underline{\mathbf{s}}_{\mathrm{qS1}}(\underline{\mathbf{R}})$, and

$\underline{\mathbf{s}}_{\mathrm{qS2}}(\underline{\mathbf{R}})$ for qP-, qS1-, and qS2-slowness vectors, respectively, whose magnitudes are equal to the pertinent phase velocities, and their directions being orthogonal to the phase fronts $T_{\mathrm{qP,qS1,qS2}}(\underline{\mathbf{R}}) = \mathrm{const.}$

Transport equations result equating the ω_0-factor to zero:

$$\nabla T(\underline{\mathbf{R}}) \cdot \underline{\underline{\underline{\underline{\mathbf{c}}}}}(\underline{\mathbf{R}}) : \nabla \underline{\mathbf{U}}(\underline{\mathbf{R}}) + \nabla \cdot \left[\underline{\underline{\underline{\underline{\mathbf{c}}}}}(\underline{\mathbf{R}}) : \nabla T(\underline{\mathbf{R}}) \underline{\mathbf{U}}(\underline{\mathbf{R}}) \right] = \underline{\mathbf{0}}. \tag{12.130}$$

In the second term, the divergence operates on all factors in the squared brackets; we do not explicitly evaluate it because later on we want to bring the energy velocity vector according to (8.217) into play; this gets apparent if we insert the slowness vector into (12.130) as well as the representation

$$\underline{\mathbf{U}}(\underline{\mathbf{R}}) = U(\underline{\mathbf{R}}) \underline{\hat{\mathbf{u}}}(\underline{\mathbf{R}}) \tag{12.131}$$

and contract the resulting vector equation with $\underline{\hat{\mathbf{u}}}(\underline{\mathbf{R}})$, i.e., project it onto the polarization unit vector (we have similarly done that for the isotropic case):

$$\begin{aligned} &\underline{\hat{\mathbf{u}}}(\underline{\mathbf{R}}) \underline{\mathbf{s}}(\underline{\mathbf{R}}) : \underline{\underline{\underline{\underline{\mathbf{c}}}}}(\underline{\mathbf{R}}) : \nabla U(\underline{\mathbf{R}}) \underline{\hat{\mathbf{u}}}(\underline{\mathbf{R}}) + \underline{\hat{\mathbf{u}}}(\underline{\mathbf{R}}) \underline{\mathbf{s}}(\underline{\mathbf{R}}) : \underline{\underline{\underline{\underline{\mathbf{c}}}}}(\underline{\mathbf{R}}) : \nabla \underline{\hat{\mathbf{u}}}(\underline{\mathbf{R}}) U(\underline{\mathbf{R}}) \\ &\quad + \underline{\hat{\mathbf{u}}}(\underline{\mathbf{R}}) \nabla : \left[\underline{\underline{\underline{\underline{\mathbf{c}}}}}(\underline{\mathbf{R}}) : \underline{\mathbf{s}}(\underline{\mathbf{R}}) \underline{\hat{\mathbf{u}}}(\underline{\mathbf{R}}) U(\underline{\mathbf{R}}) \right] = 0; \end{aligned} \tag{12.132}$$

in addition, we have calculated the $\nabla \underline{\mathbf{U}}(\underline{\mathbf{R}})$-term in (12.130) with (12.131). Pursuing our above formulated goal, it would be nice if $\underline{\hat{\mathbf{u}}}(\underline{\mathbf{R}})$ would appear in the last term of (12.132) within the squared brackets. Hence, we tentatively calculate

$$\begin{aligned} &\nabla \cdot \left[U(\underline{\mathbf{R}}) \underline{\underline{\underline{\underline{\mathbf{c}}}}}(\underline{\mathbf{R}}) \cdot \underline{\mathbf{s}}(\underline{\mathbf{R}}) \cdot \underline{\hat{\mathbf{u}}}(\underline{\mathbf{R}}) \cdot \underline{\hat{\mathbf{u}}}(\underline{\mathbf{R}}) \right] \\ &\quad = \nabla \cdot \left[U(\underline{\mathbf{R}}) \underline{\underline{\underline{\underline{\mathbf{c}}}}}(\underline{\mathbf{R}}) \cdot \underline{\mathbf{s}}(\underline{\mathbf{R}}) \cdot \underline{\hat{\mathbf{u}}}(\underline{\mathbf{R}}) \right] \cdot \underline{\hat{\mathbf{u}}}(\underline{\mathbf{R}}) \\ &\qquad + \left[U(\underline{\mathbf{R}}) \underline{\underline{\underline{\underline{\mathbf{c}}}}}(\underline{\mathbf{R}}) \cdot \underline{\mathbf{s}}(\underline{\mathbf{R}}) \cdot \underline{\hat{\mathbf{u}}}(\underline{\mathbf{R}}) \right] : \nabla \underline{\hat{\mathbf{u}}}(\underline{\mathbf{R}}), \end{aligned} \tag{12.133}$$

where we have exploited the symmetry of $\underline{\underline{\underline{\underline{\mathbf{c}}}}}(\underline{\mathbf{R}})$ with respect to the first two indices in the second term on the right-hand side. The first term on the right-hand side is equal to the third one in (12.132), it may replace it; due to the symmetry of $c_{ijkl}(\underline{\mathbf{R}}) = c_{klij}(\underline{\mathbf{R}})$, the second term on the right-hand side of (12.133) compensates the second term in (12.132), and it remains

$$\underline{\hat{\mathbf{u}}}(\underline{\mathbf{R}}) \underline{\mathbf{s}}(\underline{\mathbf{R}}) : \underline{\underline{\underline{\underline{\mathbf{c}}}}}(\underline{\mathbf{R}}) : \nabla U(\underline{\mathbf{R}}) \underline{\hat{\mathbf{u}}}(\underline{\mathbf{R}}) + \nabla \cdot \left[U(\underline{\mathbf{R}}) \underline{\underline{\underline{\underline{\mathbf{c}}}}}(\underline{\mathbf{R}}) \cdot \underline{\mathbf{s}}(\underline{\mathbf{R}}) \cdot \underline{\hat{\mathbf{u}}}(\underline{\mathbf{R}}) \cdot \underline{\hat{\mathbf{u}}}(\underline{\mathbf{R}}) \right] = 0. \tag{12.134}$$

The repeated exploitation of the $\underline{\underline{\underline{\underline{\mathbf{c}}}}}(\underline{\mathbf{R}})$-symmetries allows us to apply (8.217), and we obtain

$$\rho(\underline{\mathbf{R}}) \underline{\mathbf{c}}_{\mathrm{E}}(\underline{\mathbf{R}}) \cdot \nabla U(\underline{\mathbf{R}}) + \nabla \cdot [\rho(\underline{\mathbf{R}}) \underline{\mathbf{c}}_{\mathrm{E}}(\underline{\mathbf{R}}) U(\underline{\mathbf{R}})] = 0, \tag{12.135}$$

respectively,

$$2\rho(\underline{\mathbf{R}})\underline{\mathbf{c}}_{\mathrm{E}}(\underline{\mathbf{R}}) \cdot \boldsymbol{\nabla} U(\underline{\mathbf{R}}) + \boldsymbol{\nabla} \cdot [\rho(\underline{\mathbf{R}})\underline{\mathbf{c}}_{\mathrm{E}}(\underline{\mathbf{R}})]\, U(\underline{\mathbf{R}}) = 0. \tag{12.136}$$

The definition of $\tilde{U}(\underline{\mathbf{R}}) = \sqrt{\rho(\underline{\mathbf{R}})}U(\underline{\mathbf{R}})$ yields the final form

$$2\underline{\mathbf{c}}_{\mathrm{E}}(\underline{\mathbf{R}}) \cdot \boldsymbol{\nabla}\tilde{U}(\underline{\mathbf{R}}) + \tilde{U}(\underline{\mathbf{R}})\boldsymbol{\nabla} \cdot \underline{\mathbf{c}}_{\mathrm{E}}(\underline{\mathbf{R}}) = 0 \tag{12.137}$$

of the transport equation. For the isotropic case, it reduces to the transport equation (12.122) for P-rays, and (12.125), respectively, (12.126) for S-rays under the constraint (12.127). With the definition of the ray vector according to (8.219), we obtain

$$2\underline{\mathbf{l}}(\underline{\mathbf{R}}) \cdot \boldsymbol{\nabla}\tilde{U}(\underline{\mathbf{R}}) + \tilde{U}(\underline{\mathbf{R}})\boldsymbol{\nabla} \cdot \underline{\mathbf{l}}(\underline{\mathbf{R}}) = 0, \tag{12.138}$$

making it obvious that the (energy) transport is in the direction of the ray vector, namely the energy velocity vector that replaces the slowness vector to calculate the ray trajectories in anisotropic materials.

For plane waves in homogeneous materials, we had derived the relation (8.220) between ray vector and phase vector utilizing the eigenvalue equation (8.205) and the formula (8.217) for the energy velocity; in inhomogeneous materials, the respective eigenvalue equation (12.128) only holds within the high-frequency approximation; however, it appropriately follows

$$\underline{\mathbf{c}}_{\mathrm{E}}(\underline{\mathbf{R}}) \cdot \underline{\mathbf{s}}(\underline{\mathbf{R}}) = 1. \tag{12.139}$$

Instead of calculating the ray travel time (12.98) with the phase velocity, we have to use

$$T(\underline{\mathbf{R}}) = \int_{L(\underline{\mathbf{R}}_0,\underline{\mathbf{R}})} \frac{1}{c_{\mathrm{E}}(l)}\,\mathrm{d}l \tag{12.140}$$

from the respectively converted equation (12.139):

$$\hat{\underline{\mathbf{l}}}(\underline{\mathbf{R}}) \cdot \boldsymbol{\nabla} T(\underline{\mathbf{R}}) = \frac{1}{c_{\mathrm{E}}(\underline{\mathbf{R}})}. \tag{12.141}$$

Typical NDT-relevant ray tracing results have been published, for instance, by Ogilvy (1986) and Harker et al. (1990); beyond that, the travel time equation (12.141) may be applied to formulate and implement an ultrasonic imaging technique for inhomogeneous anisotropic welds (InASAFT for Inhomogeneous Anisotropic Synthetic Aperture Focusing Technique: Shlivinski et al. 2004b).

The above cited theory of ray trajectories in anisotropic materials may again be applied to pulsed and Gaussian beams propagating along the ray vector trajectories (Červený 2001; Norris 1987); Spies (1998, 2000a) has applied this formalism to the calculation of transducer fields in inhomogeneous anisotropic materials.

13

Point Sources in Homogeneous Isotropic Infinite Space, Elastodynamic Source Fields

13.1 Homogeneous Infinite Space Scalar Green Function

In Section 5.5, we had utilized the scalar Green function $G(\underline{\mathbf{R}}, \underline{\mathbf{R}}', \omega)$ for homogeneous infinite space to represent acoustic source fields and, in Section 5.6, to represent acoustic scattered fields in terms of a point source synthesis. There, we emphasized the appropriateness to define additional Green functions via gradient operations—$\nabla G(\underline{\mathbf{R}}, \underline{\mathbf{R}}', \omega)$, $\nabla\nabla G(\underline{\mathbf{R}}, \underline{\mathbf{R}}', \omega)$—because they explicitly appear in the integral representations of acoustic fields. For electromagnetism, the constructions $(\underline{\underline{\mathbf{I}}} + k^{-2}\nabla\nabla)G(\underline{\mathbf{R}}, \underline{\mathbf{R}}', \omega)$ and $\nabla G(\underline{\mathbf{R}}, \underline{\mathbf{R}}', \omega) \times \underline{\underline{\mathbf{I}}}$ played a fundamental role, and we will see in Section 13.2.3 that elastodynamics enforces additional derivatives. Despite the mathematical complexity of Green functions differential equations and their respective solutions, the physical concept of Green functions as elementary waves superimposing to wave fields in terms of the point source synthesis is very intuitive, why we start again from the beginning with $G(\underline{\mathbf{R}}, \underline{\mathbf{R}}', \omega)$ to clarify some aspects that we ignored in Section 5.5.

13.1.1 Time harmonic Green function

The differential equation (Equation 5.58)

$$\Delta G(\underline{\mathbf{R}}, \underline{\mathbf{R}}', \omega) + k^2 G(\underline{\mathbf{R}}, \underline{\mathbf{R}}', \omega) = -\delta(\underline{\mathbf{R}} - \underline{\mathbf{R}}') \tag{13.1}$$

has been written down as defining equation for the three-dimensional scalar Green function of homogeneous infinite space with the (initially real) wave number k. The wave number may be representative for acoustics—$k = \omega\sqrt{\rho\kappa}$—electromagnetism in an $\epsilon_r\mu_r$-material—$k = \omega\sqrt{\epsilon_0\epsilon_r\mu_0\mu_r}$—or for elastic pressure or shear waves: $k = k_{\mathrm{P}} = \omega\sqrt{\rho/(\lambda + 2\mu)}$; $k = k_{\mathrm{S}} = \omega\sqrt{\rho/\mu}$; in any case, we write $k = \omega/c$. The inhomogeneity of (13.1) is a unit point source at $\underline{\mathbf{R}}'$ according to the property

$$\iiint_V \delta(\underline{\mathbf{R}} - \underline{\mathbf{R}}')\, \mathrm{d}^3\underline{\mathbf{R}}' = \begin{cases} 1 & \text{for } \underline{\mathbf{R}} \in V \\ 0 & \text{for } \underline{\mathbf{R}} \notin V \end{cases} \tag{13.2}$$

of the δ-distribution. It is due to this distributional inhomogeneity that (13.1) must strictly be mathematically solved in the sense of distributions. Hence, the intuitive approaches to solve (13.1) mostly resemble patchwork. However, the result is a distribution defined by a function (regular distribution) rendering it legitimate to stay on the bottom of (distributionally reinforced) classical analysis.

In the following, we make two proposals to solve (13.1): a more intuitive and a formal one that is particularly practical for vector and tensor Green functions.

Solution of the differential equation for the scalar Green function: Physical reasoning: Per definition, the Green function should represent a time harmonic elementary wave that emerges form the point source at $\underline{\mathbf{R}}'$. The surrounding homogeneous infinite space will not disturb the propagation of this elementary wave suggesting, based on symmetry arguments, the phase and amplitude surfaces of this elementary wave to be spherical surfaces with midpoint $\underline{\mathbf{R}}'$ (Figure 13.1). For an arbitrary observation point $\underline{\mathbf{R}}$, the propagation of the elementary spherical wave with wave number k into the direction $\underline{\mathbf{R}} - \underline{\mathbf{R}}'$ should be outbound with regard to $\underline{\mathbf{R}} - \underline{\mathbf{R}}'$. With time dependence $\mathrm{e}^{-\mathrm{j}\omega t}$, we therefore have

$$G(\underline{\mathbf{R}}, \underline{\mathbf{R}}', \omega) \sim \mathrm{e}^{-\mathrm{j}k|\underline{\mathbf{R}} - \underline{\mathbf{R}}'|}. \tag{13.3}$$

With increasing distance $|\underline{\mathbf{R}} - \underline{\mathbf{R}}'|$, the spherical surface increases proportionally to $|\underline{\mathbf{R}} - \underline{\mathbf{R}}'|^2$; it is reasonable to claim that the power density transported by the elementary wave decreases proportionally to $|\underline{\mathbf{R}} - \underline{\mathbf{R}}|^{-2}$ to ensure a

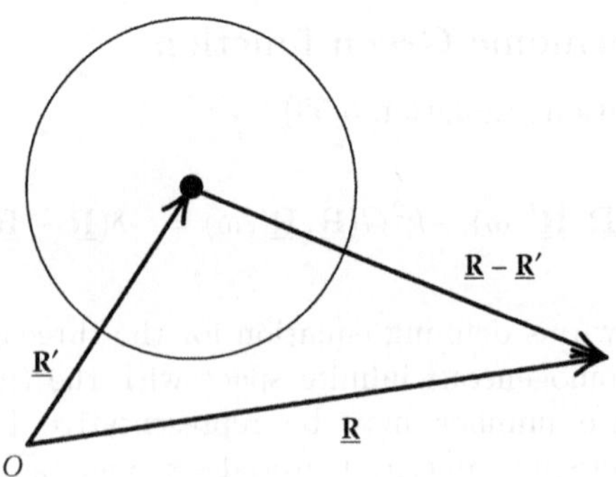

FIGURE 13.1
Phase and amplitude surfaces of a time harmonic elementary spherical wave emanating from the source point $\underline{\mathbf{R}}'$.

constant total radiated power. Hence, (13.3) has to be rendered more precisely according to[211]

$$G(\underline{\mathbf{R}}, \underline{\mathbf{R}}', \omega) \sim \frac{e^{jk|\underline{\mathbf{R}}-\underline{\mathbf{R}}'|}}{|\underline{\mathbf{R}}-\underline{\mathbf{R}}'|}. \tag{13.4}$$

Now, we rapidly show via differentiation (advantageously for $\underline{\mathbf{R}}' = \underline{\mathbf{0}}$, and in spherical coordinates) that (13.4) is in fact a solution of the homogeneous equation (13.1) for $\underline{\mathbf{R}} \neq \underline{\mathbf{R}}'$! It now depends to introduce a factor in (13.4) to yield the correct source point behavior of the Green function for $\underline{\mathbf{R}} = \underline{\mathbf{R}}'$.

Solution of the differential equation for the scalar Green function: The factor $1/4\pi$: Instead of proving for the "initial guess" function (13.4) to be a solution of the differential equation (13.1), we may even derive it analytically. For simplicity, we choose $\underline{\mathbf{R}}' = \underline{\mathbf{0}}$ and write (13.1) in spherical coordinates R, ϑ, φ immediately exploiting the request for spherical symmetry, i.e., independence of ϑ and φ:

$$\underbrace{\frac{1}{R^2}\frac{\partial}{\partial R}\left[R^2\frac{\partial G(R,\omega)}{\partial R}\right]}_{= \frac{\partial^2 G(R,\omega)}{\partial R^2} + \frac{2}{R}\frac{\partial G(R,\omega)}{\partial R}} + k^2 G(R,\omega) = -\delta(\underline{\mathbf{R}}), \tag{13.5}$$

where $G(R, \omega)$ denotes the rotationally symmetric Green function with source point $\underline{\mathbf{R}}' = \underline{\mathbf{0}}$. With the substitution $G(R, \omega) = U(R, \omega)/R$, the homogeneous equation (13.5)—$\underline{\mathbf{R}} \neq \underline{\mathbf{0}}$—turns into

$$\frac{d^2 U(R,\omega)}{dR^2} + k^2 U(R,\omega) = 0 \tag{13.6}$$

whose solutions "time harmonic plane waves propagating into positive R-direction" is given by

$$U(R,\omega) = U(\omega)\, e^{jkR}. \tag{13.7}$$

It follows

$$G(R,\omega) = U(\omega)\,\frac{e^{jkR}}{R}. \tag{13.8}$$

The amplitude factor may not be arbitrary because (13.8) should become solution of the inhomogeneous equation (13.1). For its determination, there is no way around distributional arguments. Starting from (13.2), we integrate

[211]At a slower rate than $1/R$ is physically meaningless because the total power would increase with increasing distance. On the other hand, a famous lemma of F. Rellich (Colton and Kress 1983) tells us that a solution of the inhomogeneous Helmholtz equation approaching zero faster than $1/R$ is equal to the trivial solution.

(13.5) over a spherical volume V_K with radius R_0 and surface S_K containing the origin and observe that the first term in (13.5) has emerged from $\Delta = \nabla \cdot \nabla$ and apply Gauss' theorem[212]—$\hat{\underline{\mathbf{R}}}$ is the outer normal on S_K—

$$\int\!\!\int_{S_K} \underbrace{\hat{\underline{\mathbf{R}}} \cdot \nabla G(R,\omega)\Big|_{R=R_0}}_{= \left.\frac{\partial G(R,\omega)}{\partial R}\right|_{R=R_0}} \mathrm{d}S + k^2 \int\!\!\int\!\!\int_{V_K} G(R,\omega)\,\mathrm{d}V = -1. \tag{13.9}$$

The right-hand side of this equation is R_0-independent; hence, this must also hold for the left-hand side; therefore, we investigate it for $R_0 \longrightarrow 0$. In the volume integral, we have $\mathrm{d}V = R^2 \sin\vartheta \mathrm{d}R\mathrm{d}\vartheta\mathrm{d}\varphi$ making it obvious that it tends to zero for $R_0 \longrightarrow 0$ if we insert (13.8). With $\mathrm{d}S = R_0^2 \sin\vartheta\mathrm{d}\vartheta\mathrm{d}\varphi$, we obtain the respective limit in the surface integral

$$\begin{aligned}
&\lim_{R_0\to 0} \int_0^{2\pi}\int_0^{\pi} \left.\frac{\partial G(R,\omega)}{\partial R}\right|_{R=R_0} R_0^2 \sin\vartheta\, \mathrm{d}\vartheta\mathrm{d}\varphi \\
&= \lim_{R_0\to 0} \Bigg[\underbrace{R_0^2 \left.\frac{\partial G(R,\omega)}{\partial R}\right|_{R=R_0}}_{= U(\omega)\,\mathrm{e}^{\mathrm{j}kR_0}(\mathrm{j}kR_0 - 1)} \underbrace{\int_0^{2\pi}\int_0^{\pi} \sin\vartheta\, \mathrm{d}\vartheta\mathrm{d}\varphi}_{= 4\pi} \Bigg] \\
&= -4\pi U(\omega);
\end{aligned} \tag{13.10}$$

it follows $U(\omega) = 1/4\pi$, and hence[213]

$$\begin{aligned}
G(\underline{\mathbf{R}}, \underline{\mathbf{R}}', \omega) &= G(\underline{\mathbf{R}} - \underline{\mathbf{R}}', \omega) \\
&= \frac{\mathrm{e}^{\mathrm{j}k|\underline{\mathbf{R}} - \underline{\mathbf{R}}'|}}{4\pi|\underline{\mathbf{R}} - \underline{\mathbf{R}}'|}.
\end{aligned} \tag{13.11}$$

From an engineering point of view, we can be quite comfortable with this derivation of the Green function for the scalar Helmholtz equation (5.32). Yet to derive the Green function for (5.33)—defined by (5.64) and given by (5.67)—we already had to utilize the three-dimensional spatial Fourier transform tool requiring the knowledge of the inverse transform of $1/(K^2 - k^2)$. A couple of remarks might be adequate in this connection.

Solution of the differential equation for the scalar Green function: Three-dimensional spatial Fourier transform: Let

$$\tilde{G}(\underline{\mathbf{K}}, \underline{\mathbf{R}}', \omega) = \int_{-\infty}^{\infty}\int_{-\infty}^{\infty}\int_{-\infty}^{\infty} G(\underline{\mathbf{R}}, \underline{\mathbf{R}}', \omega)\,\mathrm{e}^{-\mathrm{j}\underline{\mathbf{K}}\cdot\underline{\mathbf{R}}}\,\mathrm{d}^3\underline{\mathbf{R}} \tag{13.12}$$

---

[212]Mathematically, this is not truly correct because $G(R,\omega)$ in V_K is not continuously differentiable; we should apply Gauss' theorem in a distributional sense with ∇ as distributional gradient choosing respective test functions.

[213]In the physics literature, the factor 4π is often found in front of the δ-function in the differential equation (13.1); consequently, it is missing in the denominator of (13.11). Yet, this normalization does not provide a unit source.

be the three-dimensional spatial Fourier transform of $G(\underline{\mathbf{R}}, \underline{\mathbf{R}}', \omega)$ with regard to $\underline{\mathbf{R}}$; due to the double position vector argument of $G(\underline{\mathbf{R}}, \underline{\mathbf{R}}', \omega)$, we distinguish the three-dimensional Fourier transform with regard to $\underline{\mathbf{R}}$ by a tilde. Hence, the transform of (13.1) yields

$$(\mathrm{j}\underline{\mathbf{K}}) \cdot (\mathrm{j}\underline{\mathbf{K}})\, \tilde{G}(\underline{\mathbf{K}}, \underline{\mathbf{R}}', \omega) + k^2 \tilde{G}(\underline{\mathbf{K}}, \underline{\mathbf{R}}', \omega) = -\mathrm{e}^{-\mathrm{j}\underline{\mathbf{K}} \cdot \underline{\mathbf{R}}'}; \tag{13.13}$$

due to $\Delta = \boldsymbol{\nabla} \cdot \boldsymbol{\nabla}$, we have used the differentiation rule (2.292) twice, and the shifting rule (2.339) as well as (2.379). It follows with $\underline{\mathbf{K}} \cdot \underline{\mathbf{K}} = K^2$

$$\tilde{G}(\underline{\mathbf{K}}, \underline{\mathbf{R}}', \omega) = \frac{1}{K^2 - k^2}\, \mathrm{e}^{-\mathrm{j}\underline{\mathbf{K}} \cdot \underline{\mathbf{R}}'} \tag{13.14}$$

as well as the inversion integral

$$G(\underline{\mathbf{R}}, \underline{\mathbf{R}}', \omega) = \frac{1}{(2\pi)^3} \int_{-\infty}^{\infty} \int_{-\infty}^{\infty} \int_{-\infty}^{\infty} \frac{1}{K^2 - k^2}\, \mathrm{e}^{\mathrm{j}\underline{\mathbf{K}} \cdot (\underline{\mathbf{R}} - \underline{\mathbf{R}}')}\, \mathrm{d}^3\underline{\mathbf{K}}. \tag{13.15}$$

Obviously, $G(\underline{\mathbf{R}}, \underline{\mathbf{R}}', \omega)$ is only a function of $\underline{\mathbf{R}} - \underline{\mathbf{R}}'$ that is denoted by $G(\underline{\mathbf{R}} - \underline{\mathbf{R}}', \omega)$, where

$$G(\underline{\mathbf{R}}, \omega) = \frac{1}{(2\pi)^3} \int_{-\infty}^{\infty} \int_{-\infty}^{\infty} \int_{-\infty}^{\infty} \frac{1}{K^2 - k^2}\, \mathrm{e}^{\mathrm{j}\underline{\mathbf{K}} \cdot \underline{\mathbf{R}}}\, \mathrm{d}^3\underline{\mathbf{K}} \tag{13.16}$$

has now to be calculated. To achieve this, the $K_x K_y K_z$-coordinate system is adjusted to the fixed point of observation $\underline{\mathbf{R}}$ (Figure 13.2). With $\underline{\mathbf{K}} \cdot \underline{\mathbf{R}} = KR\cos\vartheta_K$, we write (13.16) in $\underline{\mathbf{K}}$-space spherical coordinates

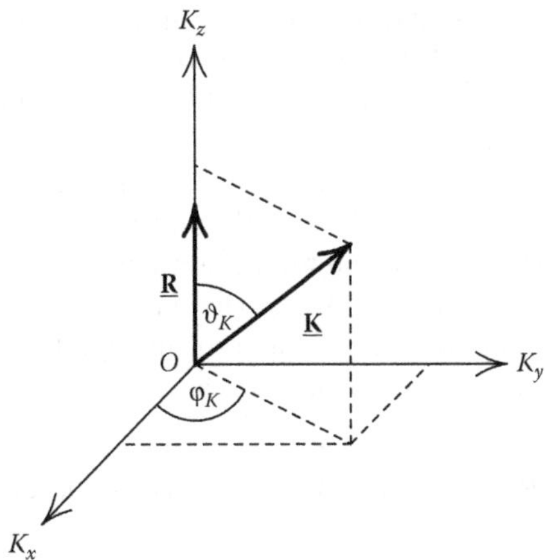

FIGURE 13.2
$\underline{\mathbf{K}}$-space cartesian coordinate system K_x, K_y, K_z and spherical coordinates $K, \vartheta_K, \varphi_K$.

$$G(\underline{\mathbf{R}}, \omega) = \frac{1}{(2\pi)^3} \int_0^{2\pi} \int_0^{\pi} \int_0^{\infty} \frac{K^2}{K^2 - k^2} \, \mathrm{e}^{\mathrm{j}KR\cos\vartheta_K} \sin\vartheta_K \, \mathrm{d}K \mathrm{d}\vartheta_K \mathrm{d}\varphi_K, \tag{13.17}$$

where the φ_K-integration immediately yields 2π. With the substitution $\cos\vartheta_K = \eta$, elementary evaluation of the η-integration, and combination of the two resulting integrals, we obtain

$$G(\underline{\mathbf{R}}, \omega) = \frac{1}{\mathrm{j}R} \frac{1}{(2\pi)^2} \int_{-\infty}^{\infty} \frac{K}{K^2 - k^2} \, \mathrm{e}^{\mathrm{j}KR} \, \mathrm{d}K. \tag{13.18}$$

Now we have to cope with the singularity of the integrand on the so-called Ewald sphere $K = k$; we have the choice between complex valued functions analysis (Langenberg 2005; Becker 1974; de Hoop 1995; King and Harrison 1969) or distributional calculus (DeSanto 1992). Using complex functions analysis, we would like to apply the residue theorem; this requires an appropriate closure of the $(-\infty, \infty)$-integration path in a complex K-plane, and this is performed in a way that physically an outbound traveling spherical wave is obtained. The mathematically possible solution of (13.1) as an inbound traveling spherical wave must be excluded due this “radiation condition.” Therefore, we uniquely obtain

$$\begin{aligned} G(\underline{\mathbf{R}}, \omega) &= G(R, \omega) \\ &= \frac{\mathrm{e}^{\mathrm{j}kR}}{4\pi R} \end{aligned} \tag{13.19}$$

as Green function. Using distributional calculus, the Fourier integrals resulting from a partial fraction decomposition of the $K/(K^2 - k^2)$-integrand are simply calculated:[214] We immediately obtain—however, physically meaningless—$G(R, \omega) = \cos kR/4\pi R$ as standing wave. To make it a physically meaningful outbound traveling wave, (13.14) has to be complemented by an appropriate solution of the homogeneous equation[215] (13.1), and a fitting solution is[216] $\mathrm{j}\sin kR/4\pi R$ again resulting in (13.19).

Hence, we see—and this is the reason why we presented some details—that the physically intuitive concept of an outbound time harmonic (elementary) spherical wave is not that easy to be found mathematically. Yet, once we have adopted (13.11) as the solution of (13.1), there are no limits for further interpretations; this becomes particularly apparent interpreting Green functions in the time domain as it is true for plane waves.

[214] The correspondence (2.373) is required.

[215] The function $\sin kR/R$ is nonsingular for $R = 0$; therefore, it satisfies the homogeneous Helmholtz equation.

[216] According to

$$\tilde{G}(\underline{\mathbf{K}}, \omega) = \frac{1}{K^2 - k^2} + \mathrm{j}\Im\{\tilde{G}(\underline{\mathbf{K}}, \omega)\},$$

we have to complement with an imaginary part

$$\Im\{\tilde{G}(\underline{\mathbf{K}}, \omega)\} = \frac{\pi}{2k} \, \delta(K - |k|),$$

whose inversion yields $\sin kR/4\pi R$.

Very often, for the sake of simplicity, simulations and model calculations for US-NDT are carried out in two spatial dimensions. This requires a completely different scalar Green function again suggesting to present the respective result at first as a time harmonic cylindrical wave.

Two-dimensional time harmonic Green function: We postulate two-dimensionality with $\partial/\partial y \equiv 0$ and introduce polar coordinates r, θ in the xz-plane (counting θ from the z-axis, then $\underline{\mathbf{e}}_r, \underline{\mathbf{e}}_\theta, \underline{\mathbf{e}}_y$ is a right-handed trihedron). Due to the expected rotational symmetry of the Green function $G(\underline{\mathbf{r}}, \omega) = G(r, \omega)$, the two-dimensional pendant to (13.5) reads

$$\underbrace{\frac{1}{r}\frac{\partial}{\partial r}\left[r\,\frac{\partial G(r,\omega)}{\partial r}\right]}_{=\,\frac{\partial^2 G(r,\omega)}{\partial r^2}+\frac{1}{r}\frac{\partial G(r,\omega)}{\partial r}} + k^2 G(r,\omega) = -\delta(\underline{\mathbf{r}}). \tag{13.20}$$

Even though the difference as compared to (13.5) seems to be marginal—$1/r$ instead $2/R$ as variable dependent coefficient function—it has considerable consequences: The *homogeneous* differential equation (13.20) defines cylindrical functions[217] $\mathrm{J}_0(kr)$, $\mathrm{N}_0(kr)$, $\mathrm{H}_0^{(1)}(kr)$, $\mathrm{H}_0^{(2)}(kr)$ (Bessel function, Neumann function, and Hankel functions of first and second kind, all of order zero) as solutions. Which one to choose out of the four? With (13.7), we immediately selected that solution representing an outbound time harmonic wave with respect to R. Here, $G(r,\omega) \sim \mathrm{e}^{\mathrm{j}kr}$ would be appropriate; as amplitude decrease with increasing r, the function $1/\sqrt{r}$ should do because the phase surfaces $\mathrm{e}^{\mathrm{j}kr}$ are circles, whose circumference increases proportional to r accounting for the power density associated with

$$G(r,\omega) \sim \frac{\mathrm{e}^{\mathrm{j}kr}}{\sqrt{r}} \tag{13.21}$$

as a quadratic quantity to decay in fact equally fast as the "surface" increases yielding a constant total power independent of r. The power radiated by the unit line source is transported to infinity as an outbound radiation. The solution structure (13.21) now helps us to select[218] the correct cylindrical function because only $\mathrm{H}_0^{(1)}(kr)$ exhibits exactly this behavior for large kr (Abramowitz and Stegun 1965):

$$\mathrm{H}_0^{(1)}(kr) \simeq \mathrm{e}^{-\mathrm{j}\frac{\pi}{4}}\sqrt{\frac{2}{\pi}}\,\frac{\mathrm{e}^{\mathrm{j}kr}}{\sqrt{kr}} \quad \text{für } kr \gg 1. \tag{13.22}$$

[217]Comparable to the solutions $\sin kz$, $\cos kz$, $\mathrm{e}^{\mathrm{j}kz}$, $\mathrm{e}^{-\mathrm{j}kz}$ of

$$\frac{\partial^2 \phi(z,\omega)}{\partial z^2} + k^2 \phi(z,\omega) = 0$$

that have been identified in Chapter 8 as time harmonic plane waves together with the time dependence $\mathrm{e}^{-\mathrm{j}\omega t}$.

[218]This selection is related to the time dependence $\mathrm{e}^{-\mathrm{j}\omega t}$; $\mathrm{e}^{\mathrm{j}\omega t}$ would yield the choice of $\mathrm{H}_0^{(2)}(kr)$.

The correct amplitude factor is found analogous to (13.9) and (13.10) (Langenberg 2005), finally yielding the two-dimensional scalar Green function

$$G(\underline{\mathbf{r}} - \underline{\mathbf{r}}', \omega) = \frac{\mathrm{j}}{4}\,\mathrm{H}_0^{(1)}(k|\underline{\mathbf{r}} - \underline{\mathbf{r}}'|) \tag{13.23}$$

if we again displace the source point from the origin to $\underline{\mathbf{r}}'$.

In connection with the point source synthesis of source and scattered fields, it is useful to know that the integration of the three-dimensional Green function along the independence axis of the two-dimensional problem leads to the two-dimensional Green function; namely, the Hankel function $\mathrm{H}_0^{(1)}(kr)$ has the integral representation (Abramowitz and Stegun 1965)

$$\frac{\mathrm{j}}{4}\,\mathrm{H}_0^{(1)}(kr) = \int_{-\infty}^{\infty} \frac{\mathrm{e}^{\mathrm{j}k\sqrt{r^2+y^2}}}{4\pi\sqrt{r^2+y^2}}\,\mathrm{d}y. \tag{13.24}$$

13.1.2 Time domain Green function

Evidently, the two- and three-dimensional Green functions (13.23) and (13.11) are functions of the circular frequency ω via $k = \omega/c$; therefore, they may be considered to represent Fourier spectra of time domain Green functions $G(\underline{\mathbf{r}} - \underline{\mathbf{r}}', t)$, respectively, $G(\underline{\mathbf{R}} - \underline{\mathbf{R}}', t)$.

Three-dimensional Green function in the time domain: With (2.366) and the shifting rule (2.290), $G(\underline{\mathbf{R}} - \underline{\mathbf{R}}', t)$ may be rapidly found:

$$G(\underline{\mathbf{R}} - \underline{\mathbf{R}}', t) = \frac{1}{4\pi|\underline{\mathbf{R}} - \underline{\mathbf{R}}'|}\,\delta\left(t - \frac{|\underline{\mathbf{R}} - \underline{\mathbf{R}}'|}{c}\right). \tag{13.25}$$

The differential equation defining (13.25) is found via Fourier inversion of (13.1):

$$\Delta G(\underline{\mathbf{R}} - \underline{\mathbf{R}}', t) - \frac{1}{c^2}\,\frac{\partial^2 G(\underline{\mathbf{R}} - \underline{\mathbf{R}}', t)}{\partial t^2} = -\delta(\underline{\mathbf{R}} - \underline{\mathbf{R}}')\,\delta(t). \tag{13.26}$$

Apparently, the right-hand side of (13.26) is now a pulsed unit point source flashing "briefly" at the source point $\underline{\mathbf{R}}'$ at time $t = 0$. Its field is a pulsed elementary spherical wave according to (13.25), whose time dependence reproduces the source time function: The propagation of elementary waves in three-dimensional space (filled with homogeneous nondissipative material) is dispersion-free!

A slight generalization of (13.26) and (13.25) introduces a nonzero switch-on time at $t = t'$:

$$\Delta G(\underline{\mathbf{R}} - \underline{\mathbf{R}}', t, t') - \frac{1}{c^2}\,\frac{\partial^2 G(\underline{\mathbf{R}} - \underline{\mathbf{R}}', t, t')}{\partial t^2} = -\delta(\underline{\mathbf{R}} - \underline{\mathbf{R}}')\,\delta(t - t'); \tag{13.27}$$

then the Fourier transform with respect to t leads to

$$\Delta G(\underline{\mathbf{R}} - \underline{\mathbf{R}}', \omega, t') + k^2 G(\underline{\mathbf{R}} - \underline{\mathbf{R}}', \omega, t') = -\delta(\underline{\mathbf{R}} - \underline{\mathbf{R}}')\, \mathrm{e}^{\mathrm{j}t'\omega} \qquad (13.28)$$

revealing

$$G(\underline{\mathbf{R}} - \underline{\mathbf{R}}', \omega, t') = G(\underline{\mathbf{R}} - \underline{\mathbf{R}}', \omega)\, \mathrm{e}^{\mathrm{j}t'\omega} \qquad (13.29)$$

to hold. Therefore, the Fourier inversion of (13.29) yields together with (13.25):

$$\begin{aligned} G(\underline{\mathbf{R}} - \underline{\mathbf{R}}', t, t') &= \frac{1}{4\pi|\underline{\mathbf{R}} - \underline{\mathbf{R}}'|}\, \delta\left(t - t' - \frac{|\underline{\mathbf{R}} - \underline{\mathbf{R}}'|}{c}\right) \\ &= G(\underline{\mathbf{R}} - \underline{\mathbf{R}}', t - t'). \end{aligned} \qquad (13.30)$$

Figure 13.3 interprets the Green function (13.30) as elementary δ-impulse spherical wave: Similar to a plane wave, we may either plot A-"scans" at fixed spatial points (compare Figure 8.1), or we may plot the amplitude distribution for fixed times (compare Figure 8.5); in Figure 13.3, both representations have been combined. The two circles represent δ-impulse wavefronts at two fixed times t_1 and t_2 with $t_2 > t_1$, i.e., they join those points $\underline{\mathbf{R}}$ for which

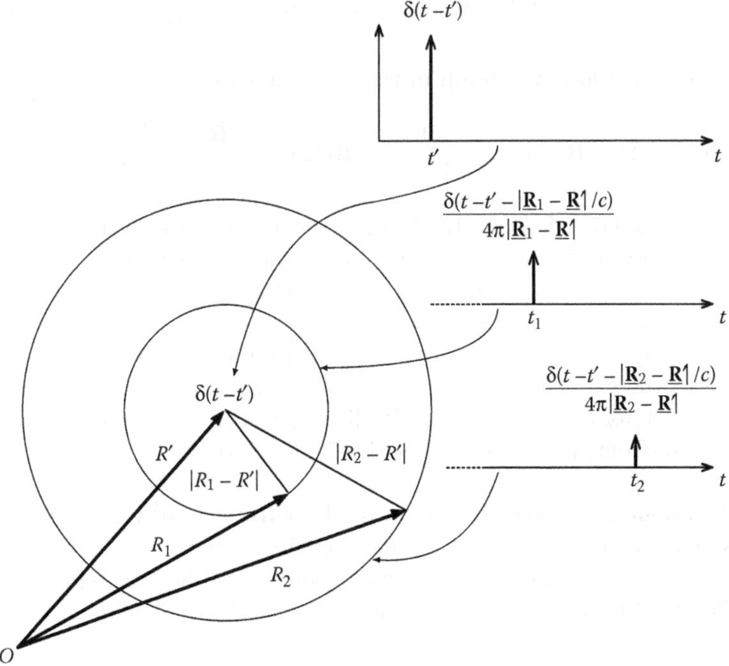

FIGURE 13.3
Pulsed δ-wavefronts of the Green function $G(\underline{\mathbf{R}} - \underline{\mathbf{R}}', t - t')$ for two different times and A-"scans" for two different spatial points.

$$t_1 - t' - \frac{|\underline{\mathbf{R}} - \underline{\mathbf{R}}'|}{c} = 0 \tag{13.31}$$

holds or, respectively,

$$t_2 - t' - \frac{|\underline{\mathbf{R}} - \underline{\mathbf{R}}'|}{c} = 0. \tag{13.32}$$

A-"scans" are depicted for two selected points $\underline{\mathbf{R}}_1$ and $\underline{\mathbf{R}}_2$; these points are located on the respective wavefronts resulting in a δ-impulse at time t_1, respectively, t_2 in the A-"scan," as symbolized by an arrow in Figure 2.22, where the "amplitude" of the δ-pulse decays due to the $|\underline{\mathbf{R}} - \underline{\mathbf{R}}'|^{-1}$-factor in (13.30).

For Green functions, the δ-pulse time dependence is mandatorily dictated, however not for plane waves allowing to choose immediately RC2(t)-pulses for NDT-relevant reasons (Section 8). Similarly here: We define an RC2(ω)-bandlimited[219] "Green function" $G^{\mathrm{RC2}}(\underline{\mathbf{R}} - \underline{\mathbf{R}}', \omega)$ according to

$$\Delta G^{\mathrm{RC2}}(\underline{\mathbf{R}} - \underline{\mathbf{R}}', \omega) + k^2 G^{\mathrm{RC2}}(\underline{\mathbf{R}} - \underline{\mathbf{R}}', \omega) = -\delta(\underline{\mathbf{R}} - \underline{\mathbf{R}}')\,\mathrm{RC2}(\omega) \tag{13.33}$$

that may immediately be given comparing (13.33) with (13.1):

$$G^{\mathrm{RC2}}(\underline{\mathbf{R}} - \underline{\mathbf{R}}', \omega) = \mathrm{RC2}(\omega)\, \frac{\mathrm{e}^{\mathrm{j}k|\underline{\mathbf{R}} - \underline{\mathbf{R}}'|}}{4\pi|\underline{\mathbf{R}} - \underline{\mathbf{R}}'|}. \tag{13.34}$$

Via Fourier-inversion, we obtain in the time domain:

$$G^{\mathrm{RC2}}(\underline{\mathbf{R}} - \underline{\mathbf{R}}', t) = \frac{1}{4\pi|\underline{\mathbf{R}} - \underline{\mathbf{R}}'|}\,\mathrm{RC2}\left(t - \frac{|\underline{\mathbf{R}} - \underline{\mathbf{R}}'|}{c}\right). \tag{13.35}$$

Figure 13.4 similarly depicts (13.35) in terms of three wavefronts (as usual in gray-coded amplitude) and three A-"scans" for different distances from the source point. The display of this bandlimited elementary spherical wave is self-guiding: However, we want to point out that the RC2-pulse in the spatial wavefronts appears spatially mirrored as in Figure 8.2, which is not visible here due to the symmetry to the origin. That way, a causal excitation pulse $f(t)$—$f(t) \equiv 0$ for $t < 0$—manifests itself as spatially causal wavefront: Before a causal wavefront, the wave field is identically zero.

Two-dimensional Green function in the time domain: To find the two-dimensional Green function in the time domain, we have to subject (13.23) to an inverse Fourier transform; assistance comes from integral representations of the Hankel function (Abramowitz and Stegun 1965). We find

$$G(\underline{\mathbf{r}} - \underline{\mathbf{r}}', t) = \frac{c}{2\pi}\, \frac{1}{\sqrt{c^2t^2 - |\underline{\mathbf{r}} - \underline{\mathbf{r}}'|^2}}\, u(ct - |\underline{\mathbf{r}} - \underline{\mathbf{r}}'|), \tag{13.36}$$

[219]The spectrum (13.29) of the Green function (13.30) always has the same magnitude for $-\infty < \omega < \infty$.

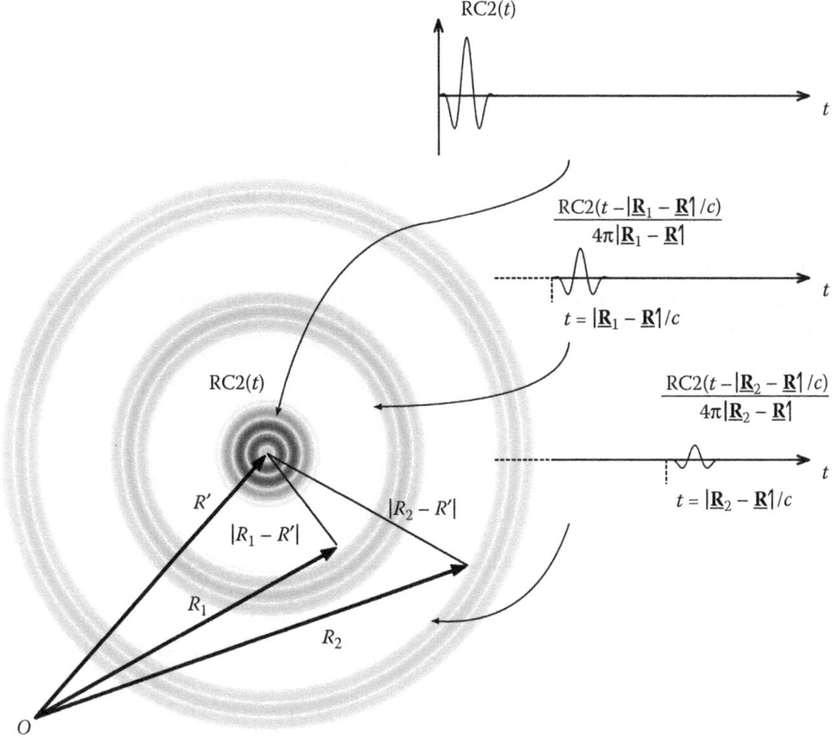

FIGURE 13.4
RC2(t)-impulse wavefronts of the bandlimited Green function $G^{\mathrm{RC2}}(\underline{\mathbf{R}} - \underline{\mathbf{R}}',$ $t)$ for three different times, and A-"scans" for three different locations (as in Figure 8.1, the RC2(t)-impulse from Figure 2.20 is displaced by half a width to the right to ensure its causality for $t < 0$).

where $u(ct - |\underline{\mathbf{r}} - \underline{\mathbf{r}}'|)$ as unit step-function ensures the causality of $G(\underline{\mathbf{r}} - \underline{\mathbf{r}}', t)$, i.e., $G(\underline{\mathbf{r}} - \underline{\mathbf{r}}', t) \equiv 0$ for $t < |\underline{\mathbf{r}} - \underline{\mathbf{r}}'|/c$. Two typical A-"scans" for (13.36) are displayed in Figure 13.5 for two different observation points $\underline{\mathbf{r}}_1$ and $\underline{\mathbf{r}}_2$ with $|\underline{\mathbf{r}}_2 - \underline{\mathbf{r}}'| > |\underline{\mathbf{r}}_1 - \underline{\mathbf{r}}'|$. Two facts are immediately apparent:

- Even though the line source radiates a δ-impulse, the time variation of the radiated field is *not* δ-like.[220]
- The square root singularity $t = |\underline{\mathbf{r}} - \underline{\mathbf{r}}'|/c$ is the same for each distance from the line source; the amplitude decay is hidden in the decreasing area under the square root function with increasing distance.

[220]This result is also obtained if all point sources stringed along the line source are superimposed with the three-dimensional Green function (Langenberg 2005): For a selected observation point, the contributions from more distant point sources arrive later with lower amplitudes.

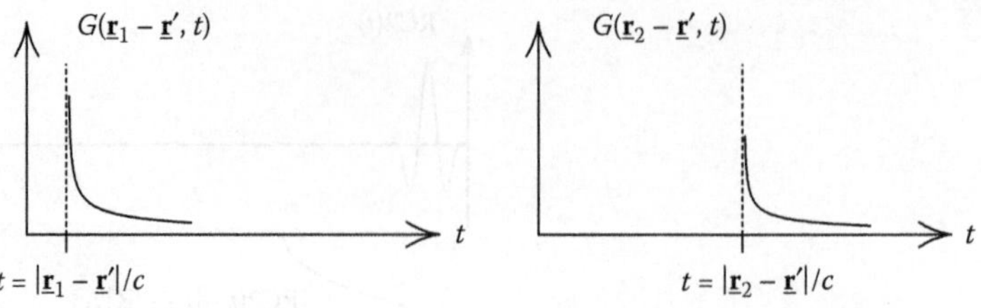

FIGURE 13.5
Two-dimensional Green function as function of time for two different observation points $\underline{\mathbf{r}}_1$ and $\underline{\mathbf{r}}_2$ mit $|\underline{\mathbf{r}}_2 - \underline{\mathbf{r}}'| > |\underline{\mathbf{r}}_1 - \underline{\mathbf{r}}'|$.

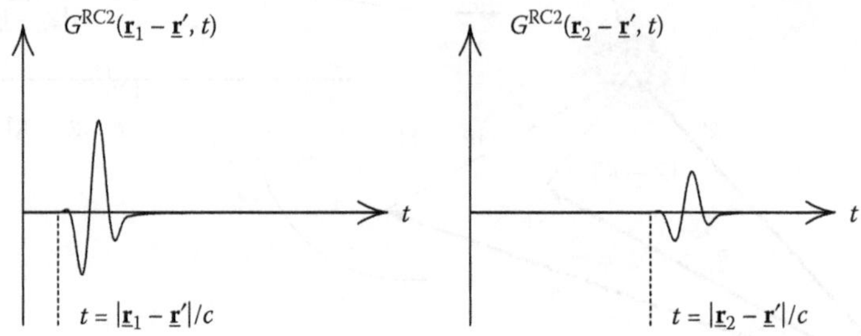

FIGURE 13.6
Two-dimensional bandlimited Green function as function of time for two different observation points $\underline{\mathbf{r}}_1$ and $\underline{\mathbf{r}}_2$ mit $|\underline{\mathbf{r}}_2 - \underline{\mathbf{r}}'| > |\underline{\mathbf{r}}_1 - \underline{\mathbf{r}}'|$.

To make the amplitude decay visible with increasing distance, we calculate analogous to (13.33)

$$G^{\mathrm{RC2}}(\underline{\mathbf{r}} - \underline{\mathbf{r}}', \omega) = \mathrm{RC2}(\omega)\,\frac{\mathrm{j}}{4}\,\mathrm{H}_0^{(1)}\left(\frac{|\underline{\mathbf{r}} - \underline{\mathbf{r}}'|}{c}\,\omega\right); \tag{13.37}$$

consequently, the expression

$$G^{\mathrm{RC2}}(\underline{\mathbf{r}} - \underline{\mathbf{r}}', t) = \mathrm{RC2}(t) * \frac{c}{2\pi}\,\frac{1}{\sqrt{c^2t^2 - |\underline{\mathbf{r}} - \underline{\mathbf{r}}'|^2}}\,u(ct - |\underline{\mathbf{r}} - \underline{\mathbf{r}}'|) \tag{13.38}$$

may be evaluated as a convolution integral. The decreasing area under the square root function then yields the amplitude decay and the dispersion of the RC2(t)-pulse (Figure 13.6). We simply have to insert the asymptotic (13.22) of the Hankel function into (13.37) to recognize a $1/\sqrt{\omega}$-multiplication of the RC2(ω)-spectrum:

$$G^{\mathrm{RC2}}(\underline{\mathbf{r}} - \underline{\mathbf{r}}', \omega) \simeq \frac{1}{4}\,\mathrm{e}^{\mathrm{j}\frac{\pi}{4}}\sqrt{\frac{2c}{\pi}}\,\frac{\mathrm{RC2}(\omega)}{\sqrt{\omega}}\,\frac{\mathrm{e}^{\mathrm{j}k|\underline{\mathbf{r}} - \underline{\mathbf{r}}'|}}{\sqrt{|\underline{\mathbf{r}} - \underline{\mathbf{r}}'|}}, \tag{13.39}$$

where the constraint $kr \gg 1$ must hold for all frequencies contained in the RC2(ω)-spectrum.

These differences in two- and three-dimensional time domain Green functions definitely effect respective sound field calculations, and restrain the quantitative comparison of two-dimensional simulations with measurements that are always three-dimensional.

13.1.3 Far-field approximation

A convenient location for only one source point is the coordinate origin; yet in general—in particular in US-NDT—we have to deal with extended sources necessitating a superposition of the contributions from "many" (continuously distributed) point sources. In that case, the coordinate origin might be conveniently located close to the source volume or even right in the "middle." If additionally the observation point distance R is large with regard to the maximum linear dimension of the source volume (and large with regard to the wavelength), the so-called far-field approximation may be introduced simplifying the field calculation considerably.

Figure 13.7 supports the following argument: If $R \gg R'$ (for all $\underline{\mathbf{R}}'$ in the interior of an eventual source volume) holds, then $\underline{\mathbf{R}} - \underline{\mathbf{R}}'$ is "nearly" parallel to $\underline{\mathbf{R}}$ allowing for the approximation

$$|\underline{\mathbf{R}} - \underline{\mathbf{R}}'| \simeq R - \underline{\mathbf{R}}' \cdot \hat{\underline{\mathbf{R}}}. \tag{13.40}$$

Yet, in the Green function (13.11), the expression $|\underline{\mathbf{R}} - \underline{\mathbf{R}}'|$ appears twice; due to the major sensitivity of the phase with regard to approximations, we will use (13.40) in the exponential function yet $|\underline{\mathbf{R}} - \underline{\mathbf{R}}'|^{-1} \simeq R^{-1}$ for the amplitude. This results in the far-field approximation

$$G^{\text{far}}(\underline{\mathbf{R}}, \underline{\mathbf{R}}', \omega) = \frac{\mathrm{e}^{\mathrm{j}kR}}{4\pi R}\, \mathrm{e}^{-\mathrm{j}k\underline{\mathbf{R}}' \cdot \hat{\underline{\mathbf{R}}}} \tag{13.41}$$

of the Green function. A precise calculation and estimate reveals (Langenberg 2005) that this approximation is practical for $R \gg R'$ and $kR \gg 1$.

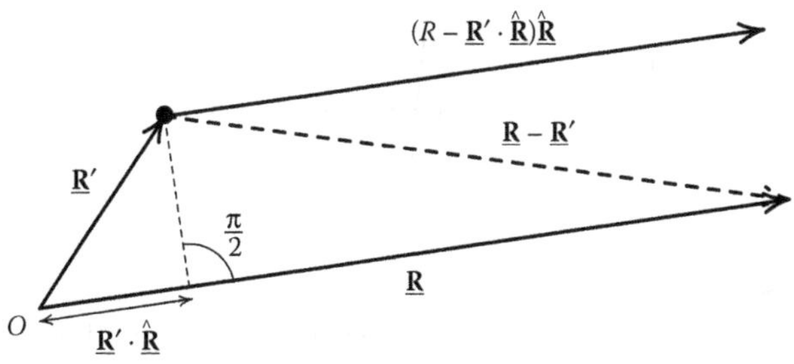

FIGURE 13.7
Geometry of the far-field approximation.

The far-field approximated Green function (13.41) exhibits a characteristic structure: It is an elementary spherical wave that emerges from the origin instead as from the actual source point[221] $\underline{\mathbf{R}}'$, hence it must be direction dependent phase corrected through multiplication with the phase directivity characteristic

$$H(\hat{\underline{\mathbf{R}}}, \underline{\mathbf{R}}', \omega) = \mathrm{e}^{-\mathrm{j}k\underline{\mathbf{R}}'\cdot\hat{\underline{\mathbf{R}}}}. \tag{13.42}$$

To present a time domain far-field A-“scan” of the Green function, we must be somehow careful because (13.41) only holds for $kR \gg 1$; a “plain” Fourier inversion is not permitted because $\delta(t)$ contains all frequencies with the same amplitude [the spectrum $\mathcal{F}\{\delta(t)\} = 1$ appears as a factor in (13.11)]. Fortunately, we have the bandlimited Green function (13.34) at hand: For the frequencies contained in RC2(ω), we surely may globally require $R \gg c/\omega$ resulting in

$$G^{\mathrm{RC2,far}}(\underline{\mathbf{R}}, \underline{\mathbf{R}}', t) = \frac{1}{4\pi R}\,\mathrm{RC2}\left(t - \frac{R}{c} + \frac{\underline{\mathbf{R}}'\cdot\hat{\underline{\mathbf{R}}}}{c}\right) \tag{13.43}$$

due to the convolution of the Fourier inversion of (13.41) with RC2(t). Figure 13.8 displays a respective A-“scan.” Obviously, the phase correction manifests itself as a travel time correction for the coordinate origin source.

The far-field approximation is particularly useful to give Green functions $\nabla' G(\underline{\mathbf{R}} - \underline{\mathbf{R}}', \omega)$, $\nabla'\nabla' G(\underline{\mathbf{R}} - \underline{\mathbf{R}}', \omega)$ that have been required in Section 5.5 for acoustic source fields a comparatively simple mathematical structure. According to (2.177), the gradient $\nabla' G(\underline{\mathbf{R}} - \underline{\mathbf{R}}', \omega)$ always has the direction of $|\underline{\mathbf{R}} - \underline{\mathbf{R}}'|$, hence it generally has three components in spherical coordinates. We may now either neglect the term with $|\underline{\mathbf{R}} - \underline{\mathbf{R}}'|^{-2}$ in (2.177) as compared to

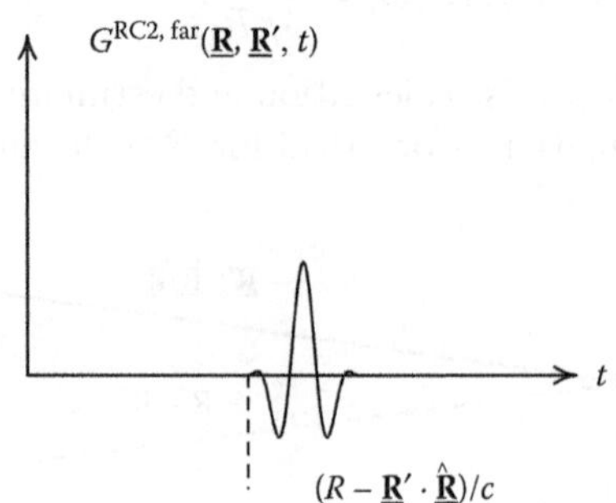

FIGURE 13.8
A-“scan” of the RC2(ω)-bandlimited Green function in the far-field.

[221]Within a point source synthesis, this term is common to all point sources; hence, it can be ignored in the synthesis volume integral.

the $|\underline{\mathbf{R}} - \underline{\mathbf{R}}'|^{-1}$-term to approximate $\nabla|\underline{\mathbf{R}} - \underline{\mathbf{R}}'| \simeq \hat{\underline{\mathbf{R}}}$ or we can immediately calculate

$$\nabla' G^{\text{far}}(\underline{\mathbf{R}}, \underline{\mathbf{R}}', \omega) = -\mathrm{j}k\hat{\underline{\mathbf{R}}}\,\frac{\mathrm{e}^{\mathrm{j}kR}}{4\pi R}\,\mathrm{e}^{-\mathrm{j}k\underline{\mathbf{R}}'\cdot\hat{\underline{\mathbf{R}}}}. \tag{13.44}$$

Due to $\nabla G(\underline{\mathbf{R}} - \underline{\mathbf{R}}', \omega) = -\nabla' G(\underline{\mathbf{R}} - \underline{\mathbf{R}}', \omega)$, we have

$$\nabla G^{\text{far}}(\underline{\mathbf{R}}, \underline{\mathbf{R}}', \omega) = \mathrm{j}k\hat{\underline{\mathbf{R}}}\,\frac{\mathrm{e}^{\mathrm{j}kR}}{4\pi R}\,\mathrm{e}^{-\mathrm{j}k\underline{\mathbf{R}}'\cdot\hat{\underline{\mathbf{R}}}}. \tag{13.45}$$

We conclude that: In the far-field ∇'-, respectively, ∇-differential operations, on the Green function may be approximated by algebraic $(-\mathrm{j}k\hat{\underline{\mathbf{R}}})$-, respectively, $(\mathrm{j}k\hat{\underline{\mathbf{R}}})$-multiplications:

$$\nabla' \overset{\text{far}}{\Longrightarrow} -\mathrm{j}k\hat{\underline{\mathbf{R}}}, \tag{13.46}$$

$$\nabla \overset{\text{far}}{\Longrightarrow} \mathrm{j}k\hat{\underline{\mathbf{R}}}. \tag{13.47}$$

That way, $\nabla' G^{\text{far}}(\underline{\mathbf{R}}, \underline{\mathbf{R}}', \omega)$ has only one $\underline{\mathbf{e}}_R$-component in spherical coordinates!

The del-operation simultaneously produces a factor ω, yielding in the time domain after—for example—RC2(ω)-band limitation

$$\nabla' G^{\text{RC2,far}}(\underline{\mathbf{R}}, \underline{\mathbf{R}}', t) = \frac{1}{c}\,\frac{\partial G^{\text{RC2,far}}(\underline{\mathbf{R}}, \underline{\mathbf{R}}', t)}{\partial t}\,\hat{\underline{\mathbf{R}}}; \tag{13.48}$$

as compared to the A-"scan" in Figure 13.8, we observe the derivative of an RC2(t)-impulse (in three dimensions)!

With the help of (13.46), respectively (13.47), we may immediately present the far-field approximation of the acoustic dyadic Green function [Equation 5.67, respectively (5.68), for $\underline{\mathbf{R}} \neq \underline{\mathbf{R}}'$]

$$\begin{aligned} \underline{\underline{\mathbf{G}}}_v(\underline{\mathbf{R}} - \underline{\mathbf{R}}', \omega) &= -\frac{1}{k^2}\,\nabla'\nabla' G(\underline{\mathbf{R}} - \underline{\mathbf{R}}', \omega) \\ &= -\frac{1}{k^2}\,\nabla\nabla G(\underline{\mathbf{R}} - \underline{\mathbf{R}}', \omega), \end{aligned} \tag{13.49}$$

namely

$$\underline{\underline{\mathbf{G}}}_v^{\text{far}}(\underline{\mathbf{R}}, \underline{\mathbf{R}}', \omega) = \frac{\mathrm{e}^{\mathrm{j}kR}}{4\pi R}\,\mathrm{e}^{-\mathrm{j}k\underline{\mathbf{R}}'\cdot\hat{\underline{\mathbf{R}}}}\,\hat{\underline{\mathbf{R}}}\,\hat{\underline{\mathbf{R}}}; \tag{13.50}$$

after band limitation with RC2(ω), we obtain in the time domain

$$\underline{\underline{\mathbf{G}}}_v^{\text{RC2,far}}(\underline{\mathbf{R}}, \underline{\mathbf{R}}', t) = \frac{1}{4\pi R}\,\text{RC2}\left(t - \frac{R}{c} + \frac{\underline{\mathbf{R}}'\cdot\hat{\underline{\mathbf{R}}}}{c}\right)\hat{\underline{\mathbf{R}}}\,\hat{\underline{\mathbf{R}}}, \tag{13.51}$$

that is to say, another RC2(t)-pulse. The single components of (13.51) exhibit differently direction dependent weighted amplitudes.

With (13.22) and the above arguments, we obtain the far-field approximation of the two-dimensional Green function (13.23):

$$G^{\mathrm{far}}(\underline{\mathbf{r}},\underline{\mathbf{r}}',\omega) = \frac{1}{4}\,\mathrm{e}^{\mathrm{j}\frac{\pi}{4}}\sqrt{\frac{2}{\pi}}\,\frac{\mathrm{e}^{\mathrm{j}kr}}{\sqrt{kr}}\,\mathrm{e}^{-\mathrm{j}k\underline{\mathbf{r}}'\cdot\hat{\underline{\mathbf{r}}}}, \quad kr \gg 1,\; r \gg r'. \tag{13.52}$$

As before, the transform into the time domain is only meaningful for a band-limited "Green" function. Analogous to (13.37), respectively (13.38), we multiply with an RC2(ω)-spectrum and utilize the correspondence[222] (Doetsch 1967)

$$\frac{1}{\sqrt{t}}\,u(t) \circ\!\!-\!\!\bullet \sqrt{\pi}\frac{1}{\sqrt{|\omega|}}\,\mathrm{e}^{\mathrm{j}\frac{\pi}{4}\,\mathrm{sign}(\omega)} \tag{13.53}$$

as well as the convolution and shifting rules:

$$G^{\mathrm{RC2,far}}(\underline{\mathbf{r}},\underline{\mathbf{r}}',t) = \frac{c}{2\sqrt{2}\pi}\,\frac{1}{\sqrt{r}}\,\mathrm{RC2}(t) * \frac{1}{\sqrt{ct - r + \underline{\mathbf{r}}'\cdot\hat{\underline{\mathbf{r}}}}}\,u(ct - r + \underline{\mathbf{r}}'\cdot\hat{\underline{\mathbf{r}}}). \tag{13.54}$$

In contrast to (13.38), the square root function does not experience an area change while shifted on the t-axis leading to constant dispersion of the RC2(t)-pulse (due to the convolution) and to the expected $1/\sqrt{r}$-dependence.

13.1.4 Point source synthesis of scalar source fields with the scalar Green function

As a matter of fact, elastodynamics in a three-dimensional homogeneous isotropic material offers with (Equation 7.35)

$$\Delta\Phi(\underline{\mathbf{R}},t) - \frac{1}{c_{\mathrm{P}}^2}\,\frac{\partial^2\Phi(\underline{\mathbf{R}},t)}{\partial t^2} = -q(\underline{\mathbf{R}},t) \tag{13.55}$$

one scalar equation for the scalar Helmholtz potential $\Phi(\underline{\mathbf{R}},t)$; for simplicity, we have abbreviated the right-hand side of (7.35) with $q(\underline{\mathbf{R}},t)$, where $q(\underline{\mathbf{R}},t) \neq 0$ only for $\underline{\mathbf{R}} \in V_Q$.

Even though, we already addressed the point source synthesis of acoustic (scalar) wave fields in Section 5.5, we want to demonstrate one more time the

[222]We have to continue the ($\omega > 0$)-spectrum (13.52) as complex conjugate for $\omega < 0$! Alternatively, instead of using (13.53), we may approximate

$$\frac{1}{\sqrt{c^2t^2 - |\underline{\mathbf{r}} - \underline{\mathbf{r}}'|^2}} = \frac{1}{\sqrt{ct - |\underline{\mathbf{r}} - \underline{\mathbf{r}}'|}}\,\underbrace{\frac{1}{\sqrt{ct + |\underline{\mathbf{r}} - \underline{\mathbf{r}}'|}}}_{\simeq \frac{1}{\sqrt{2}}}$$

in (13.36).

usefulness of the scalar Green function to solve the scalar equation (13.55). We claim that: If the solution of (13.55) for a point source at $\underline{\mathbf{R}}'$ radiating a δ-impulse for t', namely the scalar Green function $G(\underline{\mathbf{R}} - \underline{\mathbf{R}}', t - t')$ of the three-dimensional homogeneous isotropic material is known, we obtain the solution of (13.55) for arbitrary sources $q(\underline{\mathbf{R}}, t)$ superimposing all elementary spherical waves given by $G(\underline{\mathbf{R}} - \underline{\mathbf{R}}', t - t')\ q(\underline{\mathbf{R}}', t')$-weighted for all source points $\underline{\mathbf{R}}' \in V_Q$ and all switch-on times $-\infty < t' < \infty$, i.e., we should have

$$\Phi(\underline{\mathbf{R}}, t) = \iiint_{V_Q} \int_{-\infty}^{\infty} q(\underline{\mathbf{R}}', t') G(\underline{\mathbf{R}} - \underline{\mathbf{R}}', t - t')\, \mathrm{d}t' \mathrm{d}^3\underline{\mathbf{R}}'. \tag{13.56}$$

To prove it, we insert (13.56) into (13.55):

$$\begin{aligned} &\left(\Delta - \frac{1}{c_{\mathrm{P}}^2} \frac{\partial^2}{\partial t^2}\right) \Phi(\underline{\mathbf{R}}, t) \\ &\quad = \iiint_{V_Q} \int_{-\infty}^{\infty} q(\underline{\mathbf{R}}', t') \underbrace{\left(\Delta - \frac{1}{c_{\mathrm{P}}^2} \frac{\partial^2}{\partial t^2}\right) G(\underline{\mathbf{R}} - \underline{\mathbf{R}}', t - t')}_{\stackrel{\text{def}}{=}\ -\delta(\underline{\mathbf{R}} - \underline{\mathbf{R}}')\,\delta(t - t')} \mathrm{d}t' \mathrm{d}^3\underline{\mathbf{R}}', \end{aligned} \tag{13.57}$$

where we could shift the differential operator acting on $\underline{\mathbf{R}}$ and t to $G(\underline{\mathbf{R}} - \underline{\mathbf{R}}', t - t')$. Due to the sifting property of the δ-functions, we finally have

$$\left(\Delta - \frac{1}{c_{\mathrm{P}}^2} \frac{\partial^2}{\partial t^2}\right) \Phi(\underline{\mathbf{R}}, t) = \begin{cases} -q(\underline{\mathbf{R}}, t) & \text{for } \underline{\mathbf{R}} \in V_Q \\ 0 & \text{for } \underline{\mathbf{R}} \notin V_Q, \end{cases} \tag{13.58}$$

i.e., (13.56) is a solution of (13.55). In the point source synthesis integral (13.56), the Green function appears—in cartesian coordinates—as kernel of a four-dimensional convolution integral.[223] With the explicit representation (13.29) of the Green function ($c \Longrightarrow c_{\mathrm{P}}$), we can immediately calculate the t'-integral due to (2.362):

$$\Phi(\underline{\mathbf{R}}, t) = \frac{1}{4\pi} \iiint_{V_Q} \frac{q\left(\underline{\mathbf{R}}', t - \frac{|\underline{\mathbf{R}} - \underline{\mathbf{R}}'|}{c_{\mathrm{P}}}\right)}{|\underline{\mathbf{R}} - \underline{\mathbf{R}}'|}\, \mathrm{d}^3\underline{\mathbf{R}}', \tag{13.59}$$

where the remaining volume integral is now no longer a (three-dimensional) convolution integral. The representation (13.59) of a scalar potential is called a retarded potential: The potential at $\underline{\mathbf{R}}$ for time t is composed by all those point sources with their respective source point amplitude at $\underline{\mathbf{R}}' \in V_Q$ for a

[223] The theory of linear time invariant systems knows the impulse response of the system as reaction to a $\delta(t)$-pulse input. An arbitrary input signal then yields the convolution with the impulse response as output signal. The Green function is nothing but the impulse response in time and space.

time t' that is as far in the past with regard to the actual observation time as it corresponds to the travel time of the pertinent spherical waves from the source point $\underline{\mathbf{R}}'$ to the observation point $\underline{\mathbf{R}}$. The Fourier transform (13.59) according to

$$\Phi(\underline{\mathbf{R}}, \omega) = \frac{1}{4\pi} \iiint_{V_Q} q(\underline{\mathbf{R}}', \omega) \frac{e^{jk_P|\underline{\mathbf{R}}-\underline{\mathbf{R}}'|}}{|\underline{\mathbf{R}}-\underline{\mathbf{R}}'|} \, d^3\underline{\mathbf{R}}' \tag{13.60}$$

is recognized as superposition of time harmonic spherical waves:[224] The time retardation of the spherical waves is now obvious from their phase.

Note: Even for the frequent case of so-called synchronous sources

$$q(\underline{\mathbf{R}}, t) = f(t) q(\underline{\mathbf{R}}) \circ\!\!-\!\!\bullet\, q(\underline{\mathbf{R}}, \omega) = F(\omega) q(\underline{\mathbf{R}}), \tag{13.61}$$

whose single point sources all radiate with the same time dependence $f(t)$ the potential $\Phi(\underline{\mathbf{R}}, t)$ generally does not exhibit the time variation $f(t)$ because in the pertinent spectrum

$$\Phi(\underline{\mathbf{R}}, \omega) = F(\omega) \underbrace{\frac{1}{4\pi} \iiint_{V_Q} q(\underline{\mathbf{R}}') \frac{e^{jk_P|\underline{\mathbf{R}}-\underline{\mathbf{R}}'|}}{|\underline{\mathbf{R}}-\underline{\mathbf{R}}'|} \, d^3\underline{\mathbf{R}}'}_{= \phi(\underline{\mathbf{R}}, \omega)}, \tag{13.62}$$

$F(\omega)$ is weighted with the frequency-dependent function $\phi(\underline{\mathbf{R}}, \omega)$, i.e., $f(t)$ is convolved with $\phi(\underline{\mathbf{R}}, t)$.

In the far-field (13.60) adopts a particularly simple form (compare Footnote 221):

$$\Phi^{\mathrm{far}}(\underline{\mathbf{R}}, \omega) = \frac{e^{jk_P R}}{R} \underbrace{\frac{1}{4\pi} \iiint_{V_Q} q(\underline{\mathbf{R}}', \omega) \, e^{-jk_P \hat{\underline{\mathbf{R}}} \cdot \underline{\mathbf{R}}'} \, d^3\underline{\mathbf{R}}'}_{= H^q(\hat{\underline{\mathbf{R}}}, \omega)}. \tag{13.63}$$

A spherical wave emanating from the coordinate origin is weighted by the direction- and frequency-dependent radiation characteristic $H^q(\hat{\underline{\mathbf{R}}}, \omega)$ of the source, where $H^q(\hat{\underline{\mathbf{R}}}, \omega)$ generally accounts for an amplitude and phase weighting. Note: Not even in the far-field, the assumption of a synchronous $f(t)$-source leads to an $f(t)$-dependence of $\Phi^{\mathrm{far}}(\underline{\mathbf{R}}, t)$.

Once again, we concentrate on the point source superposition of the acoustic pressure as scalar quantity (Equation 5.63)

$$p(\underline{\mathbf{R}}, \omega) = \iiint_{V_Q} \left[j\omega\rho \, h(\underline{\mathbf{R}}', \omega) G(\underline{\mathbf{R}} - \underline{\mathbf{R}}', \omega) + \underline{\mathbf{f}}(\underline{\mathbf{R}}', \omega) \cdot \nabla' G(\underline{\mathbf{R}} - \underline{\mathbf{R}}', \omega) \right] d^3\underline{\mathbf{R}}' \tag{13.64}$$

[224] In Cartesian coordinates, this is once again a three-dimensional convolution integral.

to write down the far-field approximation of both terms:

$$
\begin{aligned}
p^{\text{far}}(\underline{\mathbf{R}}, \omega) &= \frac{\mathrm{e}^{\mathrm{j}kR}}{R}\,\mathrm{j}\omega \frac{Z}{4\pi c} \iiint_{V_Q} h(\underline{\mathbf{R}}', \omega)\, \mathrm{e}^{-\mathrm{j}k\hat{\underline{\mathbf{R}}}\cdot\underline{\mathbf{R}}'}\, \mathrm{d}^3\underline{\mathbf{R}}' \\
&\quad - \frac{\mathrm{e}^{\mathrm{j}kR}}{R}\,\mathrm{j}\omega \frac{1}{4\pi c} \iiint_{V_Q} \hat{\underline{\mathbf{R}}}\cdot \underline{\mathbf{f}}(\underline{\mathbf{R}}', \omega)\, \mathrm{e}^{-\mathrm{j}k\hat{\underline{\mathbf{R}}}\cdot\underline{\mathbf{R}}'}\, \mathrm{d}^3\underline{\mathbf{R}}' \\
&= \frac{\mathrm{e}^{\mathrm{j}kR}}{R} \left[H^h(\hat{\underline{\mathbf{R}}}, \omega) + H^f(\hat{\underline{\mathbf{R}}}, \omega) \right]. \qquad (13.65)
\end{aligned}
$$

For $H^h(\hat{\underline{\mathbf{R}}}, \omega)$, the relative far-field phases of elementary spherical waves are superimposed with direction-dependent amplitudes, whereas for $H^f(\hat{\underline{\mathbf{R}}}, \omega)$, an additional amplitude weighting is observed due to the direction dependence of the spherical waves: The reason for it is the $\boldsymbol{\nabla}'$-operation on the Green function. If—for instance—$\underline{\mathbf{f}}$ represents a z-directed force density at the origin according to

$$
\underline{\mathbf{f}}(\underline{\mathbf{R}}, \omega) = F(\omega)\, \delta(\underline{\mathbf{R}})\, \underline{\mathbf{e}}_z, \qquad (13.66)
$$

we have $H^f(\hat{\underline{\mathbf{R}}}, \omega) \sim \hat{\underline{\mathbf{R}}} \cdot \underline{\mathbf{e}}_z = \cos\vartheta$: The pertinent elementary spherical wave is a dipole wave (Langenberg 2005).

With the special choice of the direction of $\underline{\mathbf{f}}(\underline{\mathbf{R}}, \omega)$ according to

$$
\begin{aligned}
\underline{\mathbf{f}}(\underline{\mathbf{R}}, \omega) &= f(\underline{\mathbf{R}}, \omega)\, \underline{\mathbf{e}}_R \\
&= f(\underline{\mathbf{R}}, \omega)\, \hat{\underline{\mathbf{R}}} \qquad (13.67)
\end{aligned}
$$

(so-called breathing sphere), we can eliminate the dipole waves from the acoustic radiation field.

The point source synthesis (13.64) of the acoustic pressure suggests another illustration of the Green function. We put $\underline{\mathbf{f}}(\underline{\mathbf{R}}, \omega) \equiv \underline{\mathbf{0}}$ and $h(\underline{\mathbf{R}}, \omega) = \delta(\underline{\mathbf{R}})$ postulating a point-like (unit) injected dilatation rate at the origin to find its pressure radiation field

$$
p^{\mathrm{PS}_h}(\underline{\mathbf{R}}, \omega) = \mathrm{j}\omega\rho\, G(R, \omega) \qquad (13.68)
$$

with (13.64). The $\mathrm{j}\omega\rho$-multiplied scalar Green function is the time harmonic pressure radiation field of a (scalar) point source. In the time domain, this multiplication accounts for a differentiation:

$$
p^{\mathrm{PS}_h}(\underline{\mathbf{R}}, t) = -\rho\, \frac{\partial G(R, t)}{\partial t}, \qquad (13.69)
$$

i.e., a $\delta(t)$-h-source yields a $\delta'(t)$-pressure wavefront. However, if we put $h(\underline{\mathbf{R}}, \omega) \equiv 0$ and $\underline{\mathbf{f}}(\underline{\mathbf{R}}, \omega) = \delta(\underline{\mathbf{R}})\hat{\underline{\mathbf{f}}}$ assuming a point-like (unit) force density with the given arbitrary direction $\hat{\underline{\mathbf{f}}}$, we can also observe the dipole wave in the pressure wave (projected onto $\hat{\underline{\mathbf{f}}}$):

$$p^{\mathrm{PS}_f}(\underline{\mathbf{R}}, \omega) = -\hat{\underline{\mathbf{f}}} \cdot \nabla G(R, \omega). \tag{13.70}$$

In the far-field, the gradient operation again accounts for a differentiation in the time domain.

Another hint: If we insert the far-field approximation

$$\underline{\mathbf{v}}^{\mathrm{far}}(\underline{\mathbf{R}}, \omega) = \frac{1}{Z} p^{\mathrm{far}}(\underline{\mathbf{R}}, \omega)\, \hat{\underline{\mathbf{R}}} \tag{13.71}$$

into the time harmonic acoustic governing equation

$$\mathrm{j}\omega\rho\, \underline{\mathbf{v}}(\underline{\mathbf{R}}, \omega) = \nabla p(\underline{\mathbf{R}}, \omega), \tag{13.72}$$

we also obtain the expression (5.53) valid for plane waves for acoustic source fields:

$$\begin{aligned} \underline{\mathbf{v}}^{\mathrm{far}}(\underline{\mathbf{R}}, \omega) \cdot \hat{\underline{\mathbf{R}}} &= \frac{1}{Z} p^{\mathrm{far}}(\underline{\mathbf{R}}, \omega) \\ &= \frac{1}{Z} [H^h(\hat{\underline{\mathbf{R}}}, \omega) + H^f(\hat{\underline{\mathbf{R}}}, \omega)] \frac{\mathrm{e}^{\mathrm{j}kR}}{R}. \end{aligned} \tag{13.73}$$

13.2 Homogeneous Isotropic Infinite Space Green Tensors of Elastodynamics

13.2.1 Second-rank Green tensor

In acoustics, the differential equations (5.32) and (5.33) for the field quantities $p(\underline{\mathbf{R}}, t)$ and $\underline{\mathbf{v}}(\underline{\mathbf{R}}, t)$ result from the acoustic governing equations for a homogeneous infinite (nondissipative) space. After a Fourier transform with respect to t, we gave their solutions in terms of the point source synthesis according to (5.63) and (5.73), where Green functions played the role of superimposing time harmonic elementary spherical waves. All necessary Green functions could be lead back to ∇-differentiations of the scalar Green function $G(\underline{\mathbf{R}} - \underline{\mathbf{R}}', \omega)$. This is also true in elastodynamics (and electromagnetics).

We refrain ourselves to homogeneous isotropic materials because only these materials allow for explicit mathematical expressions for Green functions (tensors); hence, we assume

$$\underline{\underline{\underline{\underline{\mathbf{c}}}}} = \lambda\, \underline{\underline{\underline{\underline{\mathbf{I}}}}}^{\delta} + 2\mu\, \underline{\underline{\underline{\underline{\mathbf{I}}}}}^{+}, \tag{13.74}$$

respectively

$$\underline{\underline{\underline{\underline{\mathbf{s}}}}} = \Lambda\, \underline{\underline{\underline{\underline{\mathbf{I}}}}}^{\delta} + 2M\, \underline{\underline{\underline{\underline{\mathbf{I}}}}}^{+}, \tag{13.75}$$

where Λ and M are related to λ and μ via (4.30) and (4.31). The basis to calculate elastodynamic source fields are the respective governing Equations (4.33) and (4.34) resulting in the differential equations for the field quantities $\underline{\underline{\mathbf{T}}}(\underline{\mathbf{R}}, t)$ and $\underline{\mathbf{v}}(\underline{\mathbf{R}}, t)$:

$$\underline{\underline{\underline{\underline{\mathbf{I}}}}}^+ : \nabla\left[\nabla \cdot \underline{\underline{\mathbf{T}}}(\underline{\mathbf{R}}, t)\right] - \rho\, \underline{\underline{\underline{\underline{\mathbf{s}}}}} : \frac{\partial^2 \underline{\underline{\mathbf{T}}}(\underline{\mathbf{R}}, t)}{\partial t^2} = -\underline{\underline{\underline{\underline{\mathbf{I}}}}}^+ : \nabla \underline{\mathbf{f}}(\underline{\mathbf{R}}, t) - \rho \frac{\partial \underline{\underline{\mathbf{h}}}(\underline{\mathbf{R}}, t)}{\partial t}, \tag{13.76}$$

$$\begin{aligned} &\mu\, \Delta \underline{\mathbf{v}}(\underline{\mathbf{R}}, t) + (\lambda + \mu)\, \nabla\nabla \cdot \underline{\mathbf{v}}(\underline{\mathbf{R}}, t) - \rho \frac{\partial^2 \underline{\mathbf{v}}(\underline{\mathbf{R}}, t)}{\partial t^2} \\ &\quad = -\frac{\partial \underline{\mathbf{f}}(\underline{\mathbf{R}}, t)}{\partial t} - \nabla \cdot \underline{\underline{\underline{\underline{\mathbf{c}}}}} : \underline{\underline{\mathbf{h}}}(\underline{\mathbf{R}}, t). \end{aligned} \tag{13.77}$$

The differential equation for $\underline{\mathbf{v}}(\underline{\mathbf{R}}, t)$ has already been given with (7.20), here, for simplicity, we keep $\underline{\underline{\underline{\underline{\mathbf{c}}}}}$ on the right-hand side, even though we have (13.74) in mind. The differential equation for $\underline{\underline{\mathbf{T}}}(\underline{\mathbf{R}}, t)$ also results from inserting the elastodynamic governing equations into each other, where the utilization of the symmetrization tensor $\underline{\underline{\underline{\underline{\mathbf{I}}}}}^+$ is useful for a short-hand notation; despite the specification (13.75), we explicitly keep the compliance tensor $\underline{\underline{\underline{\underline{\mathbf{s}}}}}$.

De Hoop (1995) simultaneously elaborates the theory of Green functions for both field quantities $\underline{\mathbf{v}}(\underline{\mathbf{R}}, t)$ and $\underline{\underline{\mathbf{T}}}(\underline{\mathbf{R}}, t)$, establishing a reciprocity theorem with the elastodynamic governing equations; here, we start with the Navier equation (13.77) for $\underline{\mathbf{v}}(\underline{\mathbf{R}}, t)$ and calculate $\underline{\underline{\mathbf{T}}}(\underline{\mathbf{R}}, t)$ from its solution via the deformation rate equation, as we have done it already for plane waves. Equation 13.77 is essentially distinguished form the respective acoustic Equation 5.33 by the additional $\Delta \underline{\mathbf{v}}(\underline{\mathbf{R}}, t)$-term. From the specialization (8.5) through (8.7) to the homogeneous one-dimensional form, we conclude that this term stands for the additionally occurring shear waves. Hence, we will expect from the respective Green function for Equation 13.77 that it accounts for pressure and shear elementary waves.

After the Fourier transform of (13.77) with regard to t according to

$$(\mu\, \Delta + \rho\omega^2) \underline{\mathbf{v}}(\underline{\mathbf{R}}, \omega) + (\lambda + \mu) \nabla\nabla \cdot \underline{\mathbf{v}}(\underline{\mathbf{R}}, \omega) = -\underline{\mathbf{Q}}(\underline{\mathbf{R}}, \omega), \tag{13.78}$$

where

$$\underline{\mathbf{Q}}(\underline{\mathbf{R}}, \omega) = -\mathrm{j}\omega\, \underline{\mathbf{f}}(\underline{\mathbf{R}}, \omega) + \nabla \cdot \underline{\underline{\underline{\underline{\mathbf{c}}}}} : \underline{\underline{\mathbf{h}}}(\underline{\mathbf{R}}, \omega), \tag{13.79}$$

we define a time harmonic (second rank) Green tensor $\underline{\underline{\mathbf{G}}}(\underline{\mathbf{R}}, \underline{\mathbf{R}}', \omega)$ through

$$\left[(\mu\, \Delta + \rho\omega^2)\, \underline{\underline{\mathbf{I}}} + (\lambda + \mu)\, \nabla\nabla\right] \cdot \underline{\underline{\mathbf{G}}}(\underline{\mathbf{R}}, \underline{\mathbf{R}}', \omega) = -\delta(\underline{\mathbf{R}} - \underline{\mathbf{R}}')\, \underline{\underline{\mathbf{I}}}. \tag{13.80}$$

To calculate $\underline{\underline{\mathbf{G}}}(\underline{\mathbf{R}}, \underline{\mathbf{R}}', \omega)$, the dyadic differential operator $(\mu\, \Delta + \rho\omega^2)\, \underline{\underline{\mathbf{I}}} + (\lambda + \mu) \nabla\nabla$ must on one hand be inverted as a dyadic operator and, on the other hand, as a differential operator. We formally apply a three-dimensional Fourier transform with respect to $\underline{\mathbf{R}}$ [compare (13.12)]:

$$\tilde{\underline{\underline{\mathbf{G}}}}(\underline{\mathbf{K}}, \underline{\mathbf{R}}', \omega) = \int_{-\infty}^{\infty}\int_{-\infty}^{\infty}\int_{-\infty}^{\infty} \underline{\underline{\mathbf{G}}}(\underline{\mathbf{R}}, \underline{\mathbf{R}}', \omega)\, e^{-j\underline{\mathbf{K}}\cdot\underline{\mathbf{R}}}\, d^3\underline{\mathbf{R}}'. \tag{13.81}$$

That way the differential equation (13.80) turns into the algebraic equation

$$\underbrace{\left(\frac{\mu K^2 - \rho\omega^2}{\lambda + \mu} \underline{\underline{\mathbf{I}}} + \underline{\mathbf{K}}\,\underline{\mathbf{K}}\right)}_{=\tilde{\underline{\underline{\mathbf{W}}}}(\underline{\mathbf{K}})} \cdot \tilde{\underline{\underline{\mathbf{G}}}}(\underline{\mathbf{K}}, \underline{\mathbf{R}}', \omega) = \frac{1}{\lambda + \mu}\, e^{-j\underline{\mathbf{R}}'\cdot\underline{\mathbf{K}}}\, \underline{\underline{\mathbf{I}}} \tag{13.82}$$

allowing for the inversion of the wave tensor $\tilde{\underline{\underline{\mathbf{W}}}}(\underline{\mathbf{K}})$ with algebraic methods according to the instruction

$$\tilde{\underline{\underline{\mathbf{W}}}}^{-1}(\underline{\mathbf{K}}) = \frac{\operatorname{adj} \tilde{\underline{\underline{\mathbf{W}}}}(\underline{\mathbf{K}})}{\det \tilde{\underline{\underline{\mathbf{W}}}}(\underline{\mathbf{K}})}. \tag{13.83}$$

For $\underline{\mathbf{K}} \Longrightarrow \omega\hat{\underline{\mathbf{k}}}/c$, we already know the wave tensor with Equation (8.66): With (8.67), we calculated its determinant and equalized it to zero to determine the phase velocities $c = c_{\mathrm{P}}$, $c = c_{\mathrm{S}}$. Here, $|\underline{\mathbf{K}}| = K$ is the magnitude of the Fourier vector $\underline{\mathbf{K}}$ with $0 \leq K < \infty$ allowing at least for the expectation $\det \tilde{\underline{\underline{\mathbf{W}}}}(\underline{\mathbf{K}}) \neq 0$ for $K \neq \omega/c_{\mathrm{P}}$ and $K \neq \omega/c_{\mathrm{S}}$. Analogous to (8.68), we calculate

$$\det \tilde{\underline{\underline{\mathbf{W}}}}(\underline{\mathbf{K}}) = \left(\frac{\mu K^2 - \rho\omega^2}{\lambda + \mu}\right)^2 \left[\frac{(\lambda + 2\mu)K^2 - \rho\omega^2}{\lambda + \mu}\right] \tag{13.84}$$

and note that $\det \tilde{\underline{\underline{\mathbf{W}}}}(K = \omega/c_{\mathrm{P,S}}, \hat{\underline{\mathbf{K}}})$ is in fact equal to zero, but nonzero otherwise. With Chen's formula (Chen 1983)

$$\operatorname{adj}(\beta\,\underline{\underline{\mathbf{I}}} + \underline{\mathbf{C}}\,\underline{\mathbf{D}}) = \beta[(\beta + \underline{\mathbf{C}} \cdot \underline{\mathbf{D}})\,\underline{\underline{\mathbf{I}}} - \underline{\mathbf{C}}\,\underline{\mathbf{D}}], \tag{13.85}$$

we calculate the adjoint of $\tilde{\underline{\underline{\mathbf{W}}}}(\underline{\mathbf{K}})$ according to:

$$\operatorname{adj} \tilde{\underline{\underline{\mathbf{W}}}}(\underline{\mathbf{K}}) = \frac{\mu K^2 - \rho\omega^2}{\lambda + \mu} \left\{\left[\frac{(\lambda + 2\mu)K^2 - \rho\omega^2}{\lambda + \mu}\right] \underline{\underline{\mathbf{I}}} - \underline{\mathbf{K}}\,\underline{\mathbf{K}}\right\}; \tag{13.86}$$

consequently, we obtain for the inverse

$$\begin{aligned}\tilde{\underline{\underline{\mathbf{W}}}}^{-1}(\underline{\mathbf{K}}) &= \frac{\lambda + \mu}{\mu K^2 - \rho\omega^2} \left[\underline{\underline{\mathbf{I}}} - \frac{\lambda + \mu}{(\lambda + 2\mu)K^2 - \rho\omega^2} \underline{\mathbf{K}}\,\underline{\mathbf{K}}\right] \\ &= \frac{\lambda + \mu}{\mu} \left[\frac{1}{K^2 - k_{\mathrm{S}}^2} \underline{\underline{\mathbf{I}}} - \frac{\lambda + \mu}{\lambda + 2\mu} \frac{1}{(K^2 - k_{\mathrm{S}}^2)(K^2 - k_{\mathrm{P}}^2)} \underline{\mathbf{K}}\,\underline{\mathbf{K}}\right];\end{aligned} \tag{13.87}$$

as expected $\underline{\underline{\tilde{\mathbf{W}}}}^{-1}(\underline{\mathbf{K}})$, and hence $\underline{\underline{\tilde{\mathbf{G}}}}(\underline{\mathbf{K}}, \underline{\mathbf{R}}', \omega)$, is singular on the two so-called Ewald spheres $K = k_P$ and $K = k_S$ in $\underline{\mathbf{K}}$-space; however, such a singularity already appeared in the Fourier spectrum (13.14) of the scalar Green function! Due to this analogy, it appears meaningful to separate the second term in the square brackets of (13.87) via partial fraction decomposition

$$\frac{\lambda+\mu}{\lambda+2\mu} \frac{1}{(K^2-k_S^2)(K^2-k_P^2)} = \frac{1}{k_S^2}\left(\frac{1}{K^2-k_P^2} - \frac{1}{K^2-k_S^2}\right) \tag{13.88}$$

into the two terms $1/(K^2-k_P^2)$ and $1/(K^2-k_S^2)$. It follows

$$\begin{aligned}\underline{\underline{\tilde{\mathbf{G}}}}(\underline{\mathbf{K}}, \underline{\mathbf{R}}', \omega) &= \frac{1}{\lambda+\mu} \mathrm{e}^{-\mathrm{j}\underline{\mathbf{R}}'\cdot\underline{\mathbf{K}}} \underline{\underline{\tilde{\mathbf{W}}}}^{-1}(\underline{\mathbf{K}}) \cdot \underline{\underline{\mathbf{I}}} \\ &= \frac{1}{\mu}\left[\left(\underline{\underline{\mathbf{I}}} - \frac{1}{k_S^2}\underline{\mathbf{K}}\,\underline{\mathbf{K}}\right)\frac{1}{K^2-k_S^2} + \frac{1}{k_S^2}\frac{1}{K^2-k_P^2}\underline{\mathbf{K}}\,\underline{\mathbf{K}}\right]\mathrm{e}^{-\mathrm{j}\underline{\mathbf{R}}'\cdot\underline{\mathbf{K}}}.\end{aligned} \tag{13.89}$$

The inverse Fourier integral suggests [compare (13.15)]

$$\underline{\underline{\mathbf{G}}}(\underline{\mathbf{R}}, \underline{\mathbf{R}}', \omega) = \underline{\underline{\mathbf{G}}}(\underline{\mathbf{R}} - \underline{\mathbf{R}}', \omega). \tag{13.90}$$

Furthermore, $\underline{\underline{\tilde{\mathbf{G}}}}(\underline{\mathbf{K}}, \underline{\mathbf{R}}', \omega)$, and hence $\underline{\underline{\mathbf{G}}}(\underline{\mathbf{R}} - \underline{\mathbf{R}}', \omega)$, is a symmetric second rank tensor. We find that:

$$\begin{aligned}&\underline{\underline{\mathbf{G}}}(\underline{\mathbf{R}} - \underline{\mathbf{R}}', \omega) \\ &= \frac{1}{\mu}\left[\left(\underline{\underline{\mathbf{I}}} + \frac{1}{k_S^2}\nabla'\nabla'\right)\frac{\mathrm{e}^{\mathrm{j}k_S|\underline{\mathbf{R}}-\underline{\mathbf{R}}'|}}{4\pi|\underline{\mathbf{R}}-\underline{\mathbf{R}}'|} - \frac{1}{k_S^2}\nabla'\nabla'\frac{\mathrm{e}^{\mathrm{j}k_P|\underline{\mathbf{R}}-\underline{\mathbf{R}}'|}}{4\pi|\underline{\mathbf{R}}-\underline{\mathbf{R}}'|}\right] \\ &= \frac{1}{\mu}\left[\left(\underline{\underline{\mathbf{I}}} + \frac{1}{k_S^2}\nabla\nabla\right)\frac{\mathrm{e}^{\mathrm{j}k_S|\underline{\mathbf{R}}-\underline{\mathbf{R}}'|}}{4\pi|\underline{\mathbf{R}}-\underline{\mathbf{R}}'|} - \frac{1}{k_S^2}\nabla\nabla\frac{\mathrm{e}^{\mathrm{j}k_P|\underline{\mathbf{R}}-\underline{\mathbf{R}}'|}}{4\pi|\underline{\mathbf{R}}-\underline{\mathbf{R}}'|}\right] \\ &= \frac{1}{\mu}\left[\left(\underline{\underline{\mathbf{I}}} + \frac{1}{k_S^2}\nabla\nabla\right)G_S(\underline{\mathbf{R}}-\underline{\mathbf{R}}', \omega) - \frac{1}{k_S^2}\nabla\nabla G_P(\underline{\mathbf{R}}-\underline{\mathbf{R}}', \omega)\right] \\ &= \frac{1}{\mu}\left\{\underline{\underline{\mathbf{I}}}\,G_S(\underline{\mathbf{R}}-\underline{\mathbf{R}}', \omega) + \frac{1}{k_S^2}\nabla\nabla\left[G_S(\underline{\mathbf{R}}-\underline{\mathbf{R}}', \omega) - G_P(\underline{\mathbf{R}}-\underline{\mathbf{R}}', \omega)\right]\right\},\end{aligned} \tag{13.91}$$

where

$$G_{P,S}(\underline{\mathbf{R}}-\underline{\mathbf{R}}', \omega) = \frac{\mathrm{e}^{\mathrm{j}k_{P,S}|\underline{\mathbf{R}}-\underline{\mathbf{R}}'|}}{4\pi|\underline{\mathbf{R}}-\underline{\mathbf{R}}'|}. \tag{13.92}$$

A further remark concerning the mathematical structure: As it was already obvious with (5.71), the $\nabla\nabla$-operation on the scalar Green function contains a distributional term for $\underline{\mathbf{R}} = \underline{\mathbf{R}}'$ that is, by the way, also present in elastostatics in the limit $k_{S,P} \longrightarrow 0$ because it is not due to the exponential function but to the $1/|\underline{\mathbf{R}} - \underline{\mathbf{R}}'|$-term. Yet, this behavior is exhibited by both $\nabla\nabla$-terms in (13.91); due to the different signs, the distributional terms cancel. In addition, the necessary principal value calculation in (5.71) for $\underline{\mathbf{R}} \in V_Q$ becomes obsolete allowing for a "tranquilized" calculation of the differentiations in (13.91) with a subsequent integration. Only if we evaluate both terms separately, we must take care.[225]

From a physical point of view, the two terms

$$\underline{\underline{\mathbf{G}}}_{\mathrm{P}}(\underline{\mathbf{R}}, \omega) = -\frac{1}{k_{\mathrm{P}}^2} \nabla\nabla \frac{\mathrm{e}^{\mathrm{j}k_{\mathrm{P}}R}}{4\pi R}, \tag{13.93}$$

$$\underline{\underline{\mathbf{G}}}_{\mathrm{S}}(\underline{\mathbf{R}}, \omega) = \left(\underline{\underline{\mathbf{I}}} + \frac{1}{k_{\mathrm{S}}^2} \nabla\nabla\right) \frac{\mathrm{e}^{\mathrm{j}k_{\mathrm{S}}R}}{4\pi R}, \tag{13.94}$$

composing $\underline{\underline{\mathbf{G}}}(\underline{\mathbf{R}}, \omega)$ according to

$$\underline{\underline{\mathbf{G}}}(\underline{\mathbf{R}}, \omega) = \frac{1}{\lambda + 2\mu} \underline{\underline{\mathbf{G}}}_{\mathrm{P}}(\underline{\mathbf{R}}, \omega) + \frac{1}{\mu} \underline{\underline{\mathbf{G}}}_{\mathrm{S}}(\underline{\mathbf{R}}, \omega) \tag{13.95}$$

exactly satisfy our expectations: $\underline{\underline{\mathbf{G}}}_{\mathrm{P}}(\underline{\mathbf{R}}, \omega)$ represents primary and $\underline{\underline{\mathbf{G}}}_{\mathrm{S}}(\underline{\mathbf{R}}, \omega)$ secondary waves. We immediately show that for $\underline{\mathbf{R}} \neq \underline{\mathbf{0}}$, we have

$$\nabla \times \underline{\underline{\mathbf{G}}}_{\mathrm{P}}(\underline{\mathbf{R}}, \omega) = \underline{\underline{\mathbf{0}}}, \tag{13.96}$$

$$\nabla \cdot \underline{\underline{\mathbf{G}}}_{\mathrm{S}}(\underline{\mathbf{R}}, \omega) = \underline{\mathbf{0}}, \tag{13.97}$$

i.e., $\underline{\underline{\mathbf{G}}}_{\mathrm{P}}(\underline{\mathbf{R}}, \omega)$ equally stands for pressure and $\underline{\underline{\mathbf{G}}}_{\mathrm{S}}(\underline{\mathbf{R}}, \omega)$ for shear waves. Insofar, $\underline{\underline{\mathbf{G}}}_{\mathrm{P}}(\underline{\mathbf{R}}, \omega)$ is identical to the acoustic tensor[226] $\underline{\underline{\mathbf{G}}}_{v}(\underline{\mathbf{R}}, \omega)$ according to (5.67) for $\underline{\mathbf{R}} \neq \underline{\mathbf{0}}$[227] (exterior to the source point); yet even $\underline{\underline{\mathbf{G}}}_{\mathrm{S}}(\underline{\mathbf{R}}, \omega)$ has a counterpart, namely the tensor $\underline{\underline{\mathbf{G}}}_{\mathrm{e}}(\underline{\mathbf{R}}, \omega)$ of electromagnetics (Section 6.6): An electric current density radiates elementary waves (of the electric field strength) of the same mathematical and physical structure as an elastodynamic force density (regarding the particle velocity) if one solely concentrates on the secondary shear waves. Including pressure waves ($\underline{\mathbf{f}} \Longrightarrow \underline{\mathbf{v}}$)-elastodynamics structurally appears as a "superposition" of ($\underline{\mathbf{f}} \Longrightarrow \underline{\mathbf{v}}$)-acoustics and ($\underline{\mathbf{J}}_{\mathrm{e}} \Longrightarrow \underline{\mathbf{E}}$)-electromagnetics!

For $\underline{\mathbf{R}} \neq \underline{\mathbf{0}}$, the evaluation of the $\nabla\nabla$-differentiation in the separately appearing terms (13.93) and (13.94) does not cause any problems:

$$\underline{\underline{\mathbf{G}}}_{\mathrm{P}}(\underline{\mathbf{R}}, \omega) = \left[\underline{\hat{\mathbf{R}}}\,\underline{\hat{\mathbf{R}}} - \frac{\mathrm{j}}{k_{\mathrm{P}}R}(\underline{\underline{\mathbf{I}}} - 3\underline{\hat{\mathbf{R}}}\,\underline{\hat{\mathbf{R}}}) + \frac{1}{k_{\mathrm{P}}^2 R^2}(\underline{\underline{\mathbf{I}}} - 3\underline{\hat{\mathbf{R}}}\,\underline{\hat{\mathbf{R}}})\right] \frac{\mathrm{e}^{\mathrm{j}k_{\mathrm{P}}R}}{4\pi R}, \tag{13.98}$$

$$\underline{\underline{\mathbf{G}}}_{\mathrm{S}}(\underline{\mathbf{R}}, \omega) = \left[\underline{\underline{\mathbf{I}}} - \underline{\hat{\mathbf{R}}}\,\underline{\hat{\mathbf{R}}} + \frac{\mathrm{j}}{k_{\mathrm{S}}R}(\underline{\underline{\mathbf{I}}} - 3\underline{\hat{\mathbf{R}}}\,\underline{\hat{\mathbf{R}}}) - \frac{1}{k_{\mathrm{S}}^2 R^2}(\underline{\underline{\mathbf{I}}} - 3\underline{\hat{\mathbf{R}}}\,\underline{\hat{\mathbf{R}}})\right] \frac{\mathrm{e}^{\mathrm{j}k_{\mathrm{S}}R}}{4\pi R}. \tag{13.99}$$

---

[225] Regarding a separate evaluation, Footnote 232 is relevant.

[226] Hence, all subsequent discussions also hold for acoustic pressure waves.

[227] Note that the limit $\mu \longrightarrow 0$ must be performed in the first line of (13.87).

The following is attracting attention:

- Even if the source point is located in the coordinate origin, the elementary elastodynamic pressure and shear waves possess direction-dependent amplitudes that are functions of R and $\hat{\underline{\mathbf{R}}}$.

- The amplitudes of elementary elastodynamic pressure and shear waves each contain terms with characteristic R-dependencies: $1/R$, $1/R^2$, and $1/R^3$. Obviously, we have

$$\begin{aligned}\underline{\underline{\mathbf{G}}}_{\mathrm{P}}^{\mathrm{far}}(\underline{\mathbf{R}},\omega) &= \frac{\mathrm{e}^{\mathrm{j}k_{\mathrm{P}}R}}{4\pi R}\,\hat{\underline{\mathbf{R}}}\,\hat{\underline{\mathbf{R}}} \\ &= \frac{\mathrm{e}^{\mathrm{j}k_{\mathrm{P}}R}}{R}\,\underline{\underline{\mathbf{g}}}_{\mathrm{P}}(\hat{\underline{\mathbf{R}}}), \end{aligned} \tag{13.100}$$

$$\begin{aligned}\underline{\underline{\mathbf{G}}}_{\mathrm{S}}^{\mathrm{far}}(\underline{\mathbf{R}},\omega) &= \frac{\mathrm{e}^{\mathrm{j}k_{\mathrm{S}}R}}{4\pi R}\,(\underline{\underline{\mathbf{I}}}-\hat{\underline{\mathbf{R}}}\,\hat{\underline{\mathbf{R}}}) \\ &= \frac{\mathrm{e}^{\mathrm{j}k_{\mathrm{S}}R}}{R}\,\underline{\underline{\mathbf{g}}}_{\mathrm{S}}(\hat{\underline{\mathbf{R}}}) \end{aligned} \tag{13.101}$$

  as far-field terms with the (frequency-independent) tensorial radiations characteristics

$$\underline{\underline{\mathbf{g}}}_{\mathrm{P}}(\hat{\underline{\mathbf{R}}}) = \frac{1}{4\pi}\,\hat{\underline{\mathbf{R}}}\,\hat{\underline{\mathbf{R}}}, \tag{13.102}$$

$$\underline{\underline{\mathbf{g}}}_{\mathrm{S}}(\hat{\underline{\mathbf{R}}}) = \frac{1}{4\pi}\,(\underline{\underline{\mathbf{I}}}-\hat{\underline{\mathbf{R}}}\,\hat{\underline{\mathbf{R}}}). \tag{13.103}$$

  Accordingly, the $1/R^3$-terms represent the near-field and the $1/R^2$-terms a transition field.

- Near-field, transition field, and far-field are differently frequency dependent; only in the far-field, the $\delta(t)$-impulse of the source in the time domain version of (13.80) appears as $\delta(t)$-$\underline{\underline{\mathbf{G}}}_{\mathrm{P,S}}$-elementary wavefront.

- With respect to a sphere (in the far-field),

$$\underline{\underline{\mathbf{G}}}_{\mathrm{P}}^{\mathrm{far}}(\underline{\mathbf{R}},\omega) = \frac{\mathrm{e}^{\mathrm{j}k_{\mathrm{P}}R}}{4\pi R}\,\underline{\mathbf{e}}_R\,\underline{\mathbf{e}}_R \tag{13.104}$$

  has only a radial (tensor) component and

$$\underline{\underline{\mathbf{G}}}_{\mathrm{S}}^{\mathrm{far}}(\underline{\mathbf{R}},\omega) = \frac{\mathrm{e}^{\mathrm{j}k_{\mathrm{S}}R}}{4\pi R}\,(\underline{\mathbf{e}}_\vartheta\,\underline{\mathbf{e}}_\vartheta+\underline{\mathbf{e}}_\varphi\,\underline{\mathbf{e}}_\varphi) \tag{13.105}$$

  only tangential (tensor) components, because we have $\underline{\underline{\mathbf{I}}} = \underline{\mathbf{e}}_R\,\underline{\mathbf{e}}_R + \underline{\mathbf{e}}_\vartheta\,\underline{\mathbf{e}}_\vartheta + \underline{\mathbf{e}}_\varphi\,\underline{\mathbf{e}}_\varphi$ in spherical coordinates.

The discussion of $\underline{\underline{\mathbf{G}}}(\underline{\mathbf{R}}-\underline{\mathbf{R}}',\omega)$, respectively $\underline{\underline{\mathbf{G}}}(\underline{\mathbf{R}}-\underline{\mathbf{R}}',t)$, in the sense of Section 13.1.2 for the scalar Green function is a little more cumbersome due to the tensorial character of $\underline{\underline{\mathbf{G}}}$ and due to the different frequency dependence of near-, transition-, and far-fields. Therefore, we refer to (13.68) through (13.70) considering immediately the physical field quantity, in the present case, the particle velocity. Beforehand, we give the tensor components of $\underline{\underline{\mathbf{G}}}_{\mathrm{P}}(\underline{\mathbf{R}}, \omega)$ and $\underline{\underline{\mathbf{G}}}_{\mathrm{S}}(\underline{\mathbf{R}}, \omega)$ in spherical coordinates:

$$\underline{\underline{\mathbf{G}}}_{\mathrm{P}}(\underline{\mathbf{R}}, \omega) : \underline{\mathbf{e}}_R\underline{\mathbf{e}}_R = \left(1 + \mathrm{j}\frac{2}{k_{\mathrm{P}}R} - \frac{2}{k_{\mathrm{P}}^2R^2}\right)\frac{\mathrm{e}^{\mathrm{j}k_{\mathrm{P}}R}}{4\pi R}, \tag{13.106}$$

$$\underline{\underline{\mathbf{G}}}_{\mathrm{P}}(\underline{\mathbf{R}}, \omega) : \underline{\mathbf{e}}_\vartheta\underline{\mathbf{e}}_\vartheta = \left(-\mathrm{j}\frac{1}{k_{\mathrm{P}}R} + \frac{1}{k_{\mathrm{P}}^2R^2}\right)\frac{\mathrm{e}^{\mathrm{j}k_{\mathrm{P}}R}}{4\pi R}, \tag{13.107}$$

$$\underline{\underline{\mathbf{G}}}_{\mathrm{P}}(\underline{\mathbf{R}}, \omega) : \underline{\mathbf{e}}_\varphi\underline{\mathbf{e}}_\varphi = \left(-\mathrm{j}\frac{1}{k_{\mathrm{P}}R} + \frac{1}{k_{\mathrm{P}}^2R^2}\right)\frac{\mathrm{e}^{\mathrm{j}k_{\mathrm{P}}R}}{4\pi R}; \tag{13.108}$$

$$\underline{\underline{\mathbf{G}}}_{\mathrm{S}}(\underline{\mathbf{R}}, \omega) : \underline{\mathbf{e}}_R\underline{\mathbf{e}}_R = \left(-\mathrm{j}\frac{2}{k_{\mathrm{S}}R} + \frac{2}{k_{\mathrm{S}}^2R^2}\right)\frac{\mathrm{e}^{\mathrm{j}k_{\mathrm{S}}R}}{4\pi R}, \tag{13.109}$$

$$\underline{\underline{\mathbf{G}}}_{\mathrm{S}}(\underline{\mathbf{R}}, \omega) : \underline{\mathbf{e}}_\vartheta\underline{\mathbf{e}}_\vartheta = \left(1 + \mathrm{j}\frac{1}{k_{\mathrm{S}}R} - \frac{1}{k_{\mathrm{S}}^2R^2}\right)\frac{\mathrm{e}^{\mathrm{j}k_{\mathrm{S}}R}}{4\pi R}, \tag{13.110}$$

$$\underline{\underline{\mathbf{G}}}_{\mathrm{S}}(\underline{\mathbf{R}}, \omega) : \underline{\mathbf{e}}_\varphi\underline{\mathbf{e}}_\varphi = \left(1 + \mathrm{j}\frac{1}{k_{\mathrm{S}}R} - \frac{1}{k_{\mathrm{S}}^2R^2}\right)\frac{\mathrm{e}^{\mathrm{j}k_{\mathrm{S}}R}}{4\pi R}. \tag{13.111}$$

In spherical coordinates $\underline{\underline{\mathbf{G}}}(\underline{\mathbf{R}}, \omega)$ is diagonal, and apparently the far-field approximation (13.104) and (13.105) is found to be related to the 1-terms in the brackets of (13.106), (13.110), and (13.111).

13.2.2 Particle displacement of a point source force density, point radiation characteristic

Analogous to (13.56), we claim that

$$\underline{\mathbf{v}}(\underline{\mathbf{R}}, \omega) = \iiint_{V_Q} \underline{\underline{\mathbf{G}}}(\underline{\mathbf{R}} - \underline{\mathbf{R}}', \omega) \cdot \underline{\mathbf{Q}}(\underline{\mathbf{R}}', \omega)\, \mathrm{d}^3\underline{\mathbf{R}}' \tag{13.112}$$

is a solution of the differential equation (13.78) if $\underline{\underline{\mathbf{G}}}(\underline{\mathbf{R}} - \underline{\mathbf{R}}', \omega)$ satisfies (13.80). For a proof insert (13.112) into (13.78), use (13.80), and argue with the sifting property of the δ-function. Equation 13.112 is the point source synthesis of a time harmonic elastodynamic particle velocity field that is radiated by the source density $\underline{\mathbf{Q}}(\underline{\mathbf{R}}, \omega)$ being nonzero in V_Q: The tensorial elementary waves are composed of pressure and shear waves with spherical amplitude and phase surfaces. As for plane waves, to "observe" the elementary waves, we switch to the particle velocity[228] and initially consider $\underline{\underline{\mathbf{h}}}(\underline{\mathbf{R}}, \omega) \equiv \underline{\underline{\mathbf{0}}}$:

[228] De Hoop (1995) denotes the Green tensor $-\mathrm{j}\omega\underline{\underline{\mathbf{G}}}$ as $\underline{\underline{\mathbf{G}}}^{vf}$:

$$\underline{\mathbf{v}}(\underline{\mathbf{R}}, \omega) = \iiint_{V_Q} \underline{\underline{\mathbf{G}}}^{vf}(\underline{\mathbf{R}} - \underline{\mathbf{R}}', \omega) \cdot \underline{\mathbf{f}}(\underline{\mathbf{R}}', \omega)\, \mathrm{d}^3\underline{\mathbf{R}}'.$$

$$\underline{\mathbf{u}}(\underline{\mathbf{R}}, \omega) = \iiint_{V_Q} \underline{\underline{\mathbf{G}}}(\underline{\mathbf{R}} - \underline{\mathbf{R}}', \omega) \cdot \underline{\mathbf{f}}(\underline{\mathbf{R}}', \omega) \, \mathrm{d}^3\underline{\mathbf{R}}' \tag{13.113}$$

$$= \iiint_{V_Q} \underline{\mathbf{f}}(\underline{\mathbf{R}}', \omega) \cdot \underline{\underline{\mathbf{G}}}^{21}(\underline{\mathbf{R}} - \underline{\mathbf{R}}', \omega) \, \mathrm{d}^3\underline{\mathbf{R}}'$$

$$= \iiint_{V_Q} \underline{\mathbf{f}}(\underline{\mathbf{R}}', \omega) \cdot \underline{\underline{\mathbf{G}}}(\underline{\mathbf{R}} - \underline{\mathbf{R}}', \omega) \, \mathrm{d}^3\underline{\mathbf{R}}'; \tag{13.114}$$

we have exploited the symmetry of $\underline{\underline{\mathbf{G}}}(\underline{\mathbf{R}} - \underline{\mathbf{R}}', \omega)$. The vector particle displacement field of a point-like (unit) force density[229]

$$\underline{\mathbf{f}}(\underline{\mathbf{R}}, \omega) = \delta(\underline{\mathbf{R}}) \, \hat{\underline{\mathbf{f}}} \tag{13.115}$$

at the origin is the tensor Green function projected onto the direction $\hat{\underline{\mathbf{f}}}$ of the force density [compare (13.70)]

$$\underline{\mathbf{u}}^{\mathrm{PS}_f}(\underline{\mathbf{R}}, \omega) = \underline{\underline{\mathbf{G}}}(\underline{\mathbf{R}}, \omega) \cdot \hat{\underline{\mathbf{f}}} \tag{13.116}$$

$$= \hat{\underline{\mathbf{f}}} \cdot \underline{\underline{\mathbf{G}}}(\underline{\mathbf{R}}, \omega). \tag{13.117}$$

This explains the tensor character of the elementary waves: The vector force density $\hat{\underline{\mathbf{f}}}$ is rotated into the vector particle displacement, both do not have the same direction. The vector (13.116) may be separated into its spherical coordinate components:

$$u_{R,\vartheta,\varphi}^{\mathrm{PS}_f}(\underline{\mathbf{R}}, \omega) = \underline{\mathbf{u}}^{\mathrm{PS}_f}(\underline{\mathbf{R}}, \omega) \cdot \underline{\mathbf{e}}_{R,\vartheta,\varphi}$$

$$= \underline{\underline{\mathbf{G}}}(\underline{\mathbf{R}}, \omega) : \hat{\underline{\mathbf{f}}} \, \underline{\mathbf{e}}_{R,\vartheta,\varphi}; \tag{13.118}$$

with (13.106) through (13.111), we find

$$u_R^{\mathrm{PS}_f}(\underline{\mathbf{R}}, \omega) = \left[\frac{1}{\lambda + 2\mu} \left(1 + \mathrm{j} \frac{2}{k_{\mathrm{P}} R} - \frac{2}{k_{\mathrm{P}}^2 R^2} \right) \frac{\mathrm{e}^{\mathrm{j} k_{\mathrm{P}} R}}{4\pi R} + \frac{1}{\mu} \left(-\mathrm{j} \frac{2}{k_{\mathrm{S}} R} + \frac{2}{k_{\mathrm{S}}^2 R^2} \right) \frac{\mathrm{e}^{\mathrm{j} k_{\mathrm{S}} R}}{4\pi R} \right] \hat{\underline{\mathbf{f}}} \cdot \underline{\mathbf{e}}_R, \tag{13.119}$$

$$u_\vartheta^{\mathrm{PS}_f}(\underline{\mathbf{R}}, \omega) = \left[\frac{1}{\lambda + 2\mu} \left(-\mathrm{j} \frac{1}{k_{\mathrm{P}} R} + \frac{1}{k_{\mathrm{P}}^2 R^2} \right) \frac{\mathrm{e}^{\mathrm{j} k_{\mathrm{P}} R}}{4\pi R} + \frac{1}{\mu} \left(1 + \mathrm{j} \frac{1}{k_{\mathrm{S}} R} - \frac{1}{k_{\mathrm{S}}^2 R^2} \right) \frac{\mathrm{e}^{\mathrm{j} k_{\mathrm{S}} R}}{4\pi R} \right] \hat{\underline{\mathbf{f}}} \cdot \underline{\mathbf{e}}_\vartheta, \tag{13.120}$$

$$u_\varphi^{\mathrm{PS}_f}(\underline{\mathbf{R}}, \omega) = \left[\frac{1}{\lambda + 2\mu} \left(-\mathrm{j} \frac{1}{k_{\mathrm{P}} R} + \frac{1}{k_{\mathrm{P}}^2 R^2} \right) \frac{\mathrm{e}^{\mathrm{j} k_{\mathrm{P}} R}}{4\pi R} + \frac{1}{\mu} \left(1 + \mathrm{j} \frac{1}{k_{\mathrm{S}} R} - \frac{1}{k_{\mathrm{S}}^2 R^2} \right) \frac{\mathrm{e}^{\mathrm{j} k_{\mathrm{S}} R}}{4\pi R} \right] \hat{\underline{\mathbf{f}}} \cdot \underline{\mathbf{e}}_\varphi. \tag{13.121}$$

[229]The $\delta(\underline{\mathbf{R}})$-function has the unit m^{-3}, a force $f_0 = 1$ has the unit N.

At first, we keep in mind: Each elastodynamic (force density) point source radiates pressure and shear waves. In addition: Generally, the primary pressure wave is not longitudinal, and the secondary shear wave is not transverse since both waves appear in each component. The terminology "long and shear waves" common to US-NDT is strictly only valid for plane waves; pressure as well as shear elementary waves have both longitudinal $\underline{\mathbf{e}}_R$- as well as transverse $\underline{\mathbf{e}}_\vartheta, \underline{\mathbf{e}}_\varphi$-particle displacement components.[230]

Only in the far-field approximation

$$u_R^{\mathrm{PS}_f,\mathrm{far}}(\underline{\mathbf{R}}, \omega) = \frac{1}{\lambda + 2\mu} \frac{\mathrm{e}^{\mathrm{j}k_\mathrm{P}R}}{4\pi R} \hat{\underline{\mathbf{f}}} \cdot \underline{\mathbf{e}}_R, \tag{13.122}$$

$$u_\vartheta^{\mathrm{PS}_f,\mathrm{far}}(\underline{\mathbf{R}}, \omega) = \frac{1}{\mu} \frac{\mathrm{e}^{\mathrm{j}k_\mathrm{S}R}}{4\pi R} \hat{\underline{\mathbf{f}}} \cdot \underline{\mathbf{e}}_\vartheta, \tag{13.123}$$

$$u_\varphi^{\mathrm{PS}_f,\mathrm{far}}(\underline{\mathbf{R}}, \omega) = \frac{1}{\mu} \frac{\mathrm{e}^{\mathrm{j}k_\mathrm{S}R}}{4\pi R} \hat{\underline{\mathbf{f}}} \cdot \underline{\mathbf{e}}_\varphi, \tag{13.124}$$

the familiar terminology holds: Primary far-field pressure waves are longitudinally polarized, secondary far-field shear waves are transversely polarized. Otherwise spoken: In the far-field, primary waves only appear in the R-component and shear waves only in the ϑ, φ-components of the particle velocity. Specially choosing $\hat{\underline{\mathbf{f}}} = \hat{\underline{\mathbf{f}}}(\hat{\underline{\mathbf{R}}}) = \underline{\mathbf{e}}_R$ (so-called breathing sphere), tangential components basically do not appear in the particle displacement radiation field, and the far-field of this special source only consists of a pressure wave.

For US-NDT, the special choice $\hat{\underline{\mathbf{f}}} = \underline{\mathbf{e}}_z$ is particularly important (the simplest model of a piezoelectric transducer is a normal point force on an elastic half-space with a surface parallel to the xy-plane); that way, the φ-component of the shear wave disappears, and the R-component of the pressure wave exhibits the directivity $\hat{\underline{\mathbf{f}}} \cdot \underline{\mathbf{e}}_R = \cos\vartheta$, i.e., a zero perpendicular to the force orientation ($\vartheta = \pi/2$), while the ϑ-component of the shear wave has the directivity $\hat{\underline{\mathbf{f}}} \cdot \underline{\mathbf{e}}_\vartheta = -\sin\vartheta$, i.e., zeroes in the direction of the force orientation ($\vartheta = 0$, $\vartheta = \pi$). RC2(ω)-bandlimited time domain far-field wavefronts of the particle velocity—multiply (13.122) through (13.124) by RC2(ω)—are displayed as such in Figure 13.9; note: The pressure and shear wavefronts appear in different components of the particle velocity. This figure carries the wavefront representation of plane pressure and shear waves in Figure 8.5 over to respective elementary waves from point sources; in both cases, the shear wave is recognizable by its smaller wavelength, hence both wavefront types are longitudinally, respectively transversely, polarized in both cases, yet, here, the far-field approximation is assumed, and the point source parameter $\hat{\underline{\mathbf{f}}}$ comes into play. For example, the breathing sphere $\hat{\underline{\mathbf{f}}} = \hat{\underline{\mathbf{f}}}(\hat{\underline{\mathbf{R}}}) = \underline{\mathbf{e}}_R$ yields the wavefront picture of Figure 13.10. An A-"scan" complementing Figure 13.9 (Figure 13.11) once

[230]Therefore, we necessarily find shear waves on the acoustic axis in the near-field of a normal point force on the surface of a half-space (the zeroes of the point directivity—Figure 14.12—only appear in the far-field) yielding applications in medical diagnostics (Fink et al. 2002).

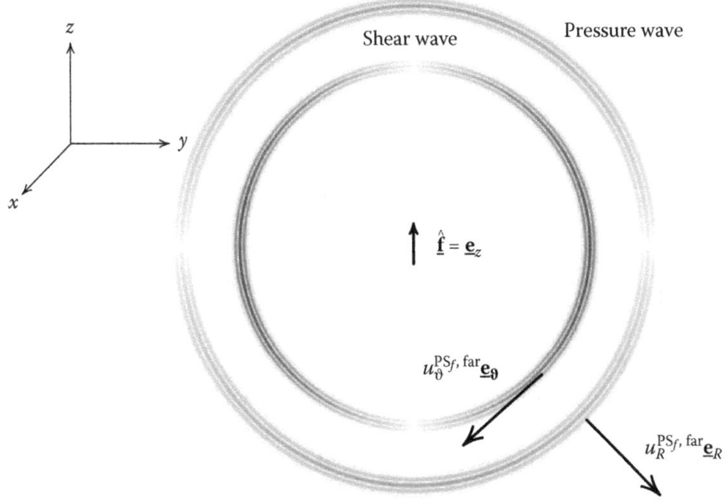

FIGURE 13.9
RC2(ω)-bandlimited far-field wavefronts of the particle velocity with force density $\hat{\underline{\mathbf{f}}} = \underline{\mathbf{e}}_z$ [with regard to the pressure wave, the shear wave basically exhibits the amplitude factor $(\lambda + 2\mu)/\mu = \kappa^2$; a second factor κ is inherent because at fixed time t the wavefront of the shear wave has only reached $R = c_\mathrm{S}t$, while the pressure wave is already observed at $R = c_\mathrm{P}t$].

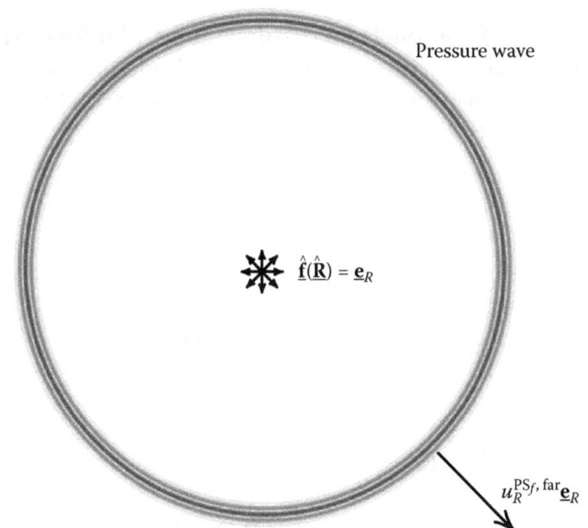

FIGURE 13.10
RC2(ω)-bandlimited far-field wavefronts of the particle velocity for given force density $\hat{\underline{\mathbf{f}}} = \underline{\mathbf{e}}_R$.

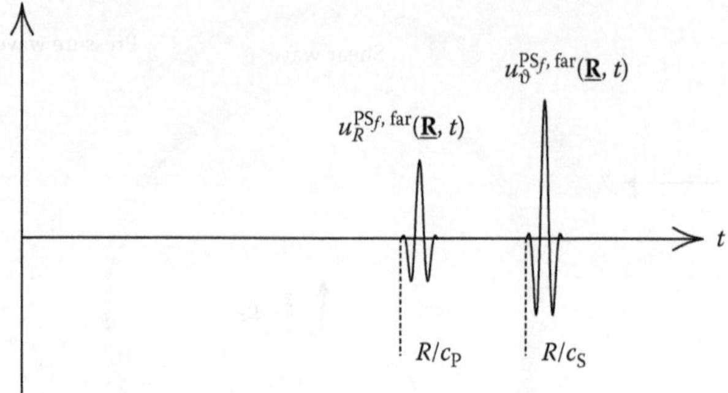

FIGURE 13.11
A-"scan" to Figure 13.9 for a fixed location $\underline{\mathbf{R}}$ (with regard to the pressure wave, the shear wave exhibits a larger amplitude by a factor of $(\lambda + 2\mu)/\mu = \kappa^2$ as compared to the pressure wave at a fixed location).

again explicitly illustrates that an RC2(t)-pulse in the force density manifests itself again as RC2(t)-pulse in the far-field of the particle velocity, in fact in the pressure wave pulse as well as in the shear wave pulse. For plane waves in infinite space, we could independently prescribe the time dependence of pressure and shear waves; for elementary waves from point sources, the time dependence of the shear wave is coupled to the time dependence of the pressure wave via the time dependence of the source.

Point directivities of the particle displacement of a force density: As already mentioned, the direction-dependent amplitudes of the far-field wavefronts in Figure 13.9 according to (13.118), and (13.100), (13.101) are given by the (non-normalized) $\hat{\underline{\mathbf{f}}}$-directivities

$$\begin{aligned} H_\mathrm{P}^{\mathrm{PS}_f}(\hat{\underline{\mathbf{R}}}) &= \frac{1}{\lambda + 2\mu}\,\underline{\underline{\mathbf{g}}}_\mathrm{P}(\hat{\underline{\mathbf{R}}}) : \hat{\underline{\mathbf{f}}}\,\underline{\mathbf{e}}_R \\ &= \frac{1}{4\pi(\lambda + 2\mu)}\,\hat{\underline{\mathbf{R}}}\,\hat{\underline{\mathbf{R}}} : \hat{\underline{\mathbf{f}}}\,\underline{\mathbf{e}}_R \\ &= \frac{1}{4\pi(\lambda + 2\mu)}\,\hat{\underline{\mathbf{f}}} \cdot \hat{\underline{\mathbf{R}}} \\ &= R\,\mathrm{e}^{-\mathrm{j}k_\mathrm{P}R}\,u_R^{\mathrm{PS}_f,\mathrm{far}}(\underline{\mathbf{R}}, \omega), \end{aligned} \tag{13.125}$$

$$\begin{aligned} H_{\mathrm{S}\vartheta}^{\mathrm{PS}_f}(\hat{\underline{\mathbf{R}}}) &= \frac{1}{\mu}\,\underline{\underline{\mathbf{g}}}_\mathrm{S}(\hat{\underline{\mathbf{R}}}) : \hat{\underline{\mathbf{f}}}\,\underline{\mathbf{e}}_\vartheta \\ &= \frac{1}{4\pi\mu}\,(\underline{\underline{\mathbf{I}}} - \hat{\underline{\mathbf{R}}}\,\hat{\underline{\mathbf{R}}}) : \hat{\underline{\mathbf{f}}}\,\underline{\mathbf{e}}_\vartheta \\ &= \frac{1}{4\pi\mu}\,\hat{\underline{\mathbf{f}}} \cdot \underline{\mathbf{e}}_\vartheta \\ &= R\,\mathrm{e}^{-\mathrm{j}k_\mathrm{S}R}\,u_\vartheta^{\mathrm{PS}_f,\mathrm{far}}(\underline{\mathbf{R}}, \omega), \end{aligned} \tag{13.126}$$

$$\begin{aligned} H_{\mathrm{S}\varphi}^{\mathrm{PS}_f}(\hat{\underline{\mathbf{R}}}) &= \frac{1}{\mu}\,\underline{\underline{\mathbf{g}}}_{\mathrm{S}}(\hat{\underline{\mathbf{R}}}) : \hat{\underline{\mathbf{f}}}\,\underline{\mathbf{e}}_{\varphi} \\ &= \frac{1}{4\pi\mu}\,(\underline{\underline{\mathbf{I}}} - \hat{\underline{\mathbf{R}}}\,\hat{\underline{\mathbf{R}}}) : \hat{\underline{\mathbf{f}}}\,\underline{\mathbf{e}}_{\varphi} \\ &= \frac{1}{4\pi\mu}\,\hat{\underline{\mathbf{f}}}\cdot\underline{\mathbf{e}}_{\varphi} \\ &= R\,\mathrm{e}^{-\mathrm{j}k_{\mathrm{S}}R}\,u_{\varphi}^{\mathrm{PS}_f,\mathrm{far}}(\underline{\mathbf{R}},\omega), \end{aligned} \tag{13.127}$$

where we have assumed $\hat{\underline{\mathbf{f}}} = \underline{\mathbf{e}}_z$ in the alluded figure. These point directivities are frequency independent. Figure 13.12 exhibits graphical displays of the point directivities $|H_{\mathrm{P}}^{\mathrm{PS}_f}(\hat{\underline{\mathbf{R}}})|$ and $|H_{\mathrm{S}\vartheta}^{\mathrm{PS}_f}(\hat{\underline{\mathbf{R}}})|$ (radiation patterns) for $\hat{\underline{\mathbf{f}}} = \underline{\mathbf{e}}_z$ and $\hat{\underline{\mathbf{f}}} = \underline{\mathbf{e}}_x$ in the xz-plane. The single pictures 13.12(a) and (b) refer to the amplitude weighting of the wavefronts in Figure 13.9.

Note: The ($\hat{\underline{\mathbf{f}}} = \underline{\mathbf{e}}_z$)-radiation pattern of the transverse shear wave corresponds to the one of an accordingly oriented Hertzian dipole for electromagnetic waves (Section 6.6). This is mandatory because the Green tensors $\underline{\underline{\mathbf{G}}}_{\mathrm{e}}$ and $\underline{\underline{\mathbf{G}}}_{\mathrm{S}}$ agree with each other, and the force density in the $\underline{\mathbf{v}}$-differential

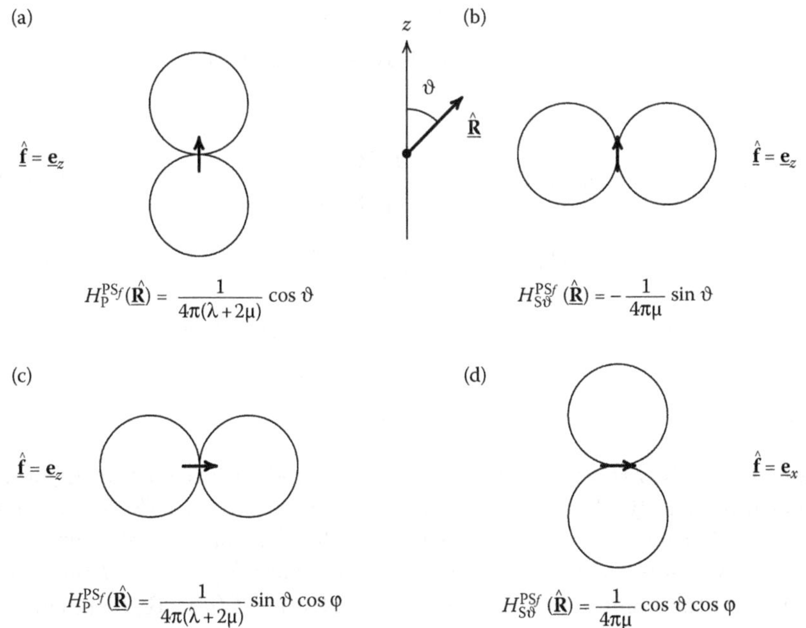

FIGURE 13.12
Radiation patterns of the particle displacement for two orthogonal [(a), (b) vs. (c), (d)] point-like force density sources ($\hat{\underline{\mathbf{f}}}$-point directivities): in the direction $\hat{\underline{\mathbf{R}}}$, the magnitude normalized to 1 of the point directivities is displayed for $\varphi = 0$ as function of ϑ.

equation has a similar significance as the electric current density in the $\underline{\mathbf{E}}$-differential equation.

Locally plane waves: By the way, with (13.116) and (13.95) as well as with (13.100) and (13.101), we may write the far-field approximation (13.122) through (13.124) coordinate-free:

$$\begin{aligned}\underline{\mathbf{u}}_{\mathrm{P}}^{\mathrm{PS}_f,\mathrm{far}}(\underline{\mathbf{R}}, \omega) &= \frac{1}{\lambda + 2\mu}\,\frac{\mathrm{e}^{\mathrm{j}k_{\mathrm{P}}R}}{4\pi R}\,\hat{\underline{\mathbf{R}}}\hat{\underline{\mathbf{R}}} \cdot \hat{\underline{\mathbf{f}}} \\ &= u_{\mathrm{P}}(\hat{\underline{\mathbf{f}}}, \hat{\underline{\mathbf{R}}})\,\frac{\mathrm{e}^{\mathrm{j}k_{\mathrm{P}}R}}{R}\,\hat{\underline{\mathbf{u}}}_{\mathrm{P}}(\hat{\underline{\mathbf{f}}}, \hat{\underline{\mathbf{R}}}) \qquad (13.128)\end{aligned}$$

with

$$\hat{\underline{\mathbf{u}}}_{\mathrm{P}}(\hat{\underline{\mathbf{f}}}, \hat{\underline{\mathbf{R}}}) = \hat{\underline{\mathbf{R}}}, \qquad (13.129)$$

$$u_{\mathrm{P}}(\hat{\underline{\mathbf{f}}}, \hat{\underline{\mathbf{R}}}) = \frac{1}{4\pi(\lambda + 2\mu)}\,\hat{\underline{\mathbf{R}}} \cdot \hat{\underline{\mathbf{f}}} \qquad (13.130)$$

as well as

$$\begin{aligned}\underline{\mathbf{u}}_{\mathrm{S}}^{\mathrm{PS}_f,\mathrm{far}}(\underline{\mathbf{R}}, \omega) &= \frac{1}{\mu}\,\frac{\mathrm{e}^{\mathrm{j}k_{\mathrm{S}}R}}{4\pi R}\,(\underline{\underline{\mathbf{I}}} - \hat{\underline{\mathbf{R}}}\hat{\underline{\mathbf{R}}}) \cdot \hat{\underline{\mathbf{f}}} \\ &= u_{\mathrm{S}}(\hat{\underline{\mathbf{f}}}, \hat{\underline{\mathbf{R}}})\,\frac{\mathrm{e}^{\mathrm{j}k_{\mathrm{S}}R}}{R}\,\hat{\underline{\mathbf{u}}}_{\mathrm{S}}(\hat{\underline{\mathbf{f}}}, \hat{\underline{\mathbf{R}}}) \qquad (13.131)\end{aligned}$$

with

$$\hat{\underline{\mathbf{u}}}_{\mathrm{S}}(\hat{\underline{\mathbf{f}}}, \hat{\underline{\mathbf{R}}}) = \frac{(\underline{\underline{\mathbf{I}}} - \hat{\underline{\mathbf{R}}}\hat{\underline{\mathbf{R}}}) \cdot \hat{\underline{\mathbf{f}}}}{|(\underline{\underline{\mathbf{I}}} - \hat{\underline{\mathbf{R}}}\hat{\underline{\mathbf{R}}}) \cdot \hat{\underline{\mathbf{f}}}|}, \qquad (13.132)$$

$$u_{\mathrm{S}}(\hat{\underline{\mathbf{f}}}, \hat{\underline{\mathbf{R}}}) = \frac{1}{4\pi\mu}\,|(\underline{\underline{\mathbf{I}}} - \hat{\underline{\mathbf{R}}}\hat{\underline{\mathbf{R}}}) \cdot \hat{\underline{\mathbf{f}}}|, \qquad (13.133)$$

where the amplitude (unit) vectors $\hat{\underline{\mathbf{u}}}_{\mathrm{P}}(\hat{\underline{\mathbf{f}}}, \hat{\underline{\mathbf{R}}})$ and $\hat{\underline{\mathbf{u}}}_{\mathrm{S}}(\hat{\underline{\mathbf{f}}}, \hat{\underline{\mathbf{R}}})$ according to

$$(\underline{\underline{\mathbf{I}}} - \hat{\underline{\mathbf{R}}}\hat{\underline{\mathbf{R}}}) \cdot \hat{\underline{\mathbf{u}}}_{\mathrm{P}}(\hat{\underline{\mathbf{f}}}, \hat{\underline{\mathbf{R}}}) = \underline{\mathbf{0}}, \qquad (13.134)$$

$$\hat{\underline{\mathbf{R}}} \cdot \hat{\underline{\mathbf{u}}}_{\mathrm{S}}(\hat{\underline{\mathbf{f}}}, \hat{\underline{\mathbf{R}}}) = 0 \qquad (13.135)$$

are longitudinally, respectively transversely, oriented with regard to the propagation direction $\hat{\underline{\mathbf{R}}}$. The comparison of this notation with (8.82) through (8.84) reveals that the particle displacement far-field of a point source behaves locally as a plane wave. One essential difference: A potential $F(\omega)$-band limitation of the point source according to $\underline{\mathbf{f}}(\underline{\mathbf{R}}, \omega) = F(\omega)\delta(\underline{\mathbf{R}})\,\hat{\underline{\mathbf{f}}}$ appears as a factor in both amplitudes $u_{\mathrm{P,S}}(\hat{\underline{\mathbf{f}}}, \hat{\underline{\mathbf{R}}}) \Longrightarrow u_{\mathrm{P,S}}(\hat{\underline{\mathbf{f}}}, \hat{\underline{\mathbf{R}}}, \omega) = F(\omega)u_{\mathrm{P,S}}(\hat{\underline{\mathbf{f}}}, \hat{\underline{\mathbf{R}}})$.

Obviously, this analogy—Equations 8.121 and 8.122—also holds for the stress tensor. With (13.74) (Equation 7.23), we calculate

$$\begin{aligned}\underline{\underline{\mathbf{T}}}(\underline{\mathbf{R}}, \omega) &= \underline{\underline{\underline{\underline{\mathbf{c}}}}} : \nabla\underline{\mathbf{u}}(\underline{\mathbf{R}}, \omega) \\ &= \lambda\,\underline{\underline{\mathbf{I}}}\,\nabla \cdot \underline{\mathbf{u}}(\underline{\mathbf{R}}, \omega) + \mu\left\{\nabla\underline{\mathbf{u}}(\underline{\mathbf{R}}, \omega) + [\nabla\underline{\mathbf{u}}(\underline{\mathbf{R}}, \omega)]^{21}\right\}, \qquad (13.136)\end{aligned}$$

i.e., in the far-field of the $\underline{\hat{\mathbf{f}}}$-point source

$$\underline{\underline{\mathbf{T}}}^{\mathrm{PS}_f,\mathrm{far}}(\underline{\mathbf{R}},\omega) = \underline{\underline{\mathbf{T}}}_{\mathrm{P}}^{\mathrm{PS}_f,\mathrm{far}}(\underline{\mathbf{R}},\omega) + \underline{\underline{\mathbf{T}}}_{\mathrm{S}}^{\mathrm{PS}_f,\mathrm{far}}(\underline{\mathbf{R}},\omega) \tag{13.137}$$

with—note (13.135)—

$$\underline{\underline{\mathbf{T}}}_{\mathrm{P}}^{\mathrm{PS}_f,\mathrm{far}}(\underline{\mathbf{R}},\omega) = \mathrm{j}k_{\mathrm{P}}\, u_{\mathrm{P}}(\underline{\hat{\mathbf{f}}},\underline{\hat{\mathbf{R}}})\, \frac{\mathrm{e}^{\mathrm{j}k_{\mathrm{P}}R}}{R}\, (\lambda\,\underline{\underline{\mathbf{I}}} + 2\mu\,\underline{\hat{\mathbf{R}}}\,\underline{\hat{\mathbf{R}}}), \tag{13.138}$$

$$\underline{\underline{\mathbf{T}}}_{\mathrm{S}}^{\mathrm{PS}_f,\mathrm{far}}(\underline{\mathbf{R}},\omega) = \mathrm{j}k_{\mathrm{S}}\mu\, u_{\mathrm{S}}(\underline{\hat{\mathbf{f}}},\underline{\hat{\mathbf{R}}})\, \frac{\mathrm{e}^{\mathrm{j}k_{\mathrm{S}}R}}{R}\, [\underline{\hat{\mathbf{R}}}\,\underline{\hat{\mathbf{u}}}_{\mathrm{S}}(\underline{\hat{\mathbf{f}}},\underline{\hat{\mathbf{R}}}) + \underline{\hat{\mathbf{u}}}_{\mathrm{S}}(\underline{\hat{\mathbf{f}}},\underline{\hat{\mathbf{R}}})\underline{\hat{\mathbf{R}}}]. \tag{13.139}$$

We will see in Chapter 14 that the local analogy between spherical and plane waves is even true for the far-field of arbitrarily finite sources.

The separation (13.128) and (13.131) into pressure and shear elementary waves also holds, recognizing (13.95) and (13.116), for arbitrary distances:

$$\underline{\mathbf{u}}^{\mathrm{PS}_f}(\underline{\mathbf{R}},\omega) = \underbrace{\frac{1}{\lambda+2\mu}\,\underline{\underline{\mathbf{G}}}_{\mathrm{P}}(\underline{\mathbf{R}},\omega)\cdot\underline{\hat{\mathbf{f}}}}_{=\,\underline{\mathbf{u}}_{\mathrm{P}}^{\mathrm{PS}_f}(\underline{\mathbf{R}},\omega)} + \underbrace{\frac{1}{\mu}\,\underline{\underline{\mathbf{G}}}_{\mathrm{S}}(\underline{\mathbf{R}},\omega)\cdot\underline{\hat{\mathbf{f}}}}_{=\,\underline{\mathbf{u}}_{\mathrm{S}}^{\mathrm{PS}_f}(\underline{\mathbf{R}},\omega)}; \tag{13.140}$$

yet for arbitrary distances, this is not a separation into longitudinal and transverse waves. Hence, we have for the stress tensor

$$\begin{aligned}\underline{\underline{\mathbf{T}}}^{\mathrm{PS}_f}(\underline{\mathbf{R}},\omega) &= \frac{1}{\lambda+2\mu}\,\underline{\underline{\underline{\underline{\mathbf{c}}}}} : \nabla\underline{\underline{\mathbf{G}}}_{\mathrm{P}}(\underline{\mathbf{R}},\omega)\cdot\underline{\hat{\mathbf{f}}} + \frac{1}{\mu}\,\underline{\underline{\underline{\underline{\mathbf{c}}}}} : \nabla\underline{\underline{\mathbf{G}}}_{\mathrm{S}}(\underline{\mathbf{R}},\omega)\cdot\underline{\hat{\mathbf{f}}}\\ &= \underline{\underline{\underline{\boldsymbol{\Sigma}}}}(\underline{\mathbf{R}},\omega)\cdot\underline{\hat{\mathbf{f}}},\end{aligned} \tag{13.141}$$

where

$$\begin{aligned}\underline{\underline{\underline{\boldsymbol{\Sigma}}}}(\underline{\mathbf{R}},\omega) &= \frac{1}{\lambda+2\mu}\,\underline{\underline{\underline{\underline{\mathbf{c}}}}} : \nabla\underline{\underline{\mathbf{G}}}_{\mathrm{P}}(\underline{\mathbf{R}},\omega) + \frac{1}{\mu}\,\underline{\underline{\underline{\underline{\mathbf{c}}}}} : \nabla\underline{\underline{\mathbf{G}}}_{\mathrm{S}}(\underline{\mathbf{R}},\omega)\\ &= \underline{\underline{\underline{\underline{\mathbf{c}}}}} : \nabla\underline{\underline{\mathbf{G}}}(\underline{\mathbf{R}},\omega) \qquad (13.142)\\ &= \left(\lambda\,\underline{\underline{\mathbf{I}}}\nabla + \mu\nabla\,\underline{\underline{\mathbf{I}}} + \mu\nabla\,\underline{\underline{\mathbf{I}}}^{213}\right)\cdot\underline{\underline{\mathbf{G}}}(\underline{\mathbf{R}},\omega) \qquad (13.143)\end{aligned}$$

obviously—compare (13.141) with (13.116)—plays the role of a Green tensor characterizing the elementary stress waves emerging from a point force density source.

Impulse responses of the particle velocity components: With Figures 13.9 and 13.11, we have discussed bandlimited far-field wavefronts and A-"scans" of the particle displacement, where typically different components of the particle displacement vector can be displayed in one respective figure because they appear separately in space and time in case the duration of the excitation bandlimited pulse is short enough. This might not be true for the near- and transition fields, yet we could principally excite with a δ-pulse with infinite bandwidth. The only difficulty: We must observe the different frequency dependence of the near-, transition-, and far-field terms evaluating the Fourier inversion of (13.119) through (13.121). So, let us write, for example, Equation 13.119 with the view of the frequency dependence:

$$u_R^{\mathrm{PS}_f}(\underline{\mathbf{R}}, \omega) = \frac{\hat{\underline{\mathbf{f}}} \cdot \underline{\mathbf{e}}_R}{4\pi\rho R} \left[\left(\frac{1}{c_{\mathrm{P}}^2} + \mathrm{j} \frac{2}{c_{\mathrm{P}} R} \frac{1}{\omega} - \frac{2}{R^2} \frac{1}{\omega^2} \right) \mathrm{e}^{\mathrm{j}\frac{R}{c_{\mathrm{P}}}\omega} + \left(-\mathrm{j} \frac{2}{c_{\mathrm{S}}} \frac{1}{\omega} + \frac{2}{R^2} \frac{1}{\omega^2} \right) \mathrm{e}^{\mathrm{j}\frac{R}{c_{\mathrm{S}}}\omega} \right]. \tag{13.144}$$

With the correspondences (2.373), the Fourier inversion yields[231] (Doetsch 1967)

$$|t| \circ\!\!-\!\!\bullet -\mathrm{pf}\frac{2}{\omega^2} \tag{13.145}$$

the impulse response of the R-component of the particle displacement because we may consider the factor 1 in (13.115) as spectrum of a $\delta(t)$-excitation:

$$u_R^{\mathrm{PS}_f}(\underline{\mathbf{R}}, t) = \frac{\hat{\underline{\mathbf{f}}} \cdot \underline{\mathbf{e}}_R}{4\pi(\lambda + 2\mu)} \left[\frac{\delta\left(t - \frac{R}{c_{\mathrm{P}}}\right)}{R} + c_{\mathrm{P}} \frac{\mathrm{sign}\left(t - \frac{R}{c_{\mathrm{P}}}\right)}{R^2} + c_{\mathrm{P}}^2 \frac{\left|t - \frac{R}{c_{\mathrm{P}}}\right|}{R^3} - \kappa c_{\mathrm{P}} \frac{\mathrm{sign}\left(t - \frac{R}{c_{\mathrm{S}}}\right)}{R^2} - c_{\mathrm{P}}^2 \frac{\left|t - \frac{R}{c_{\mathrm{S}}}\right|}{R^3} \right]. \tag{13.146}$$

The following is striking: Taken as such, neither the primary wave terms—apart from the δ-pulse—nor the secondary wave terms are causal, equal to zero for $t < R/c_{\mathrm{P}}$, respectively $t < R/c_{\mathrm{S}}$. However, in the sum[232] the physically mandatory causality for $t < R/c_{\mathrm{P}}$ is realized! It is illustrated as graphical display of (13.146) in Figure 13.13. We recognize the far-field-δ-pulse with R^{-1}-dependence, and—in the impulse response!—for $R/c_{\mathrm{P}} < t < R/c_{\mathrm{S}}$, a (linear) ramp function with R^{-2}-amplitude dependence thus disappearing in the far-field. This ramp function is discontinuous at $t = R/c_{\mathrm{P}}$ and $t = R/c_{\mathrm{S}}$; hence, exciting with a bandlimited RC2(t)-pulse, the convolution of (13.146) with such an impulse results in an integration of the RC2(t)-pulse at the two discontinuous jumps with an amplitude factor given by the jump heights (once with a positive sign and once with a negative sign); in between, due to the very slow rise of the ramp, the convolution nearly yields zero. The final conclusions read as follows: In the near- and transition-regions, the primary pressure RC2(t)-pulse is marginally distorted by a small integrated RC2(t)-impulse, and a second small integrated RC2(t)-impulse arrives with the secondary wave velocity (Schleichert et al. 1989). Hence, the secondary wave is also visible in the longitudinal R-component of the particle displacement, yet it disappears in this component in the far-field. The pendant—a primary wave pulse in the

[231] The spectrum $1/\omega^2$ has to be considered as a pseudofunction in the distributional sense.

[232] For the pressure wave of acoustics, this "sum compensation" does not exist; in that case, causality must be enforced with an appropriate solution of the homogeneous equation.

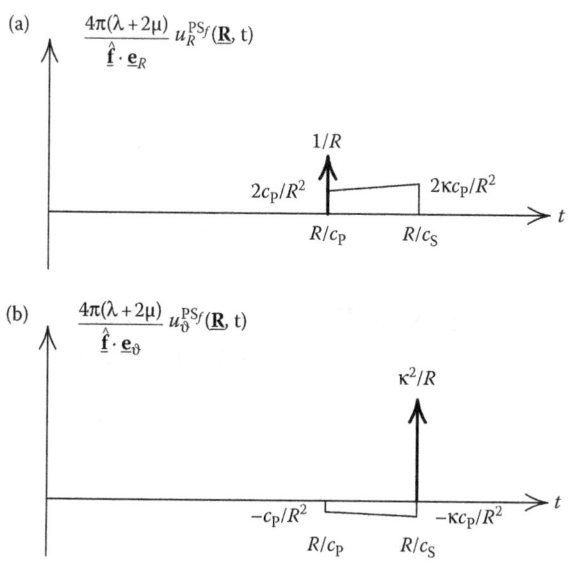

FIGURE 13.13
Impulse response of the particle displacement of a point-like force density source: (a) R-component, (b) ϑ-component.

transverse component—is observed in the ϑ- and φ-components of the particle velocity. For example, we calculate

$$u_\vartheta^{\mathrm{PS}_f}(\underline{\mathbf{R}}, t) = \frac{\hat{\underline{\mathbf{f}}}\cdot\underline{\mathbf{e}}_\vartheta}{4\pi(\lambda+2\mu)}\left[-\frac{c_P}{2}\frac{\mathrm{sign}\left(t-\frac{R}{c_P}\right)}{R^2} - \frac{c_P^2}{2}\frac{\left|t-\frac{R}{c_P}\right|}{R^3}\right.$$
$$\left. +\kappa^2\frac{\delta\left(t-\frac{R}{c_S}\right)}{R} + \kappa\frac{c_P}{2}\frac{\mathrm{sign}\left(t-\frac{R}{c_S}\right)}{R^2} + \frac{c_P^2}{2}\frac{\left|t-\frac{R}{c_S}\right|}{R^3}\right]. \tag{13.147}$$

The graphical display of (13.147) can be found in Figure 13.13(b): Compared to the R-component, attention is attracted by the amplitude of the ramp function: It is by a factor of 0.5 smaller, i.e., the primary impulse in the transverse ϑ-component is on one hand smaller than the secondary impulse in the longitudinal component by this factor and, on the other hand, once more by the factor κ^{-1}.

13.2.3 Third-rank Green tensor

We illustrated the second rank Green tensor with point-like force densities $\underline{\mathbf{f}}$ based on (13.113), respectively (13.114), the injected deformation rate tensor $\underline{\underline{\mathbf{h}}}$ was set equal to the null tensor. So, let us return to (13.112) utilizing (13.79):

$$\underline{\mathbf{v}}(\underline{\mathbf{R}},\omega) = \iiint_{V_Q} \underline{\underline{\mathbf{G}}}(\underline{\mathbf{R}}-\underline{\mathbf{R}}',\omega)\cdot\left[-\mathrm{j}\omega\underline{\mathbf{f}}(\underline{\mathbf{R}}',\omega) + \boldsymbol{\nabla}'\cdot\underline{\underline{\underline{\underline{\mathbf{c}}}}} : \underline{\underline{\mathbf{h}}}(\underline{\mathbf{R}}',\omega)\right]\,\mathrm{d}^3\underline{\mathbf{R}}', \tag{13.148}$$

and let us concentrate on the second term in (13.148). Similar to the transition from (5.61) to (5.63), we want to shift the operation $\boldsymbol{\nabla}'\cdot\underline{\underline{\underline{\underline{\mathbf{c}}}}}:$ on $\underline{\underline{\mathbf{h}}}(\underline{\mathbf{R}}',\omega)$ to $\underline{\underline{\mathbf{G}}}(\underline{\mathbf{R}}-\underline{\mathbf{R}}',\omega)$. For that purpose, we need the pendant to (5.62) (Langenberg 2005):

$$\begin{aligned}\boldsymbol{\nabla}\cdot(\underline{\underline{\mathbf{D}}}\cdot\underline{\underline{\mathbf{A}}}) &= (\boldsymbol{\nabla}\cdot\underline{\underline{\mathbf{D}}})\cdot\underline{\underline{\mathbf{A}}} + \underline{\underline{\mathbf{D}}}^{21}:\boldsymbol{\nabla}\underline{\underline{\mathbf{A}}}\\ &= \underline{\underline{\mathbf{A}}}^{21}\cdot(\boldsymbol{\nabla}\cdot\underline{\underline{\mathbf{D}}}) + \underline{\underline{\mathbf{D}}}^{21}:\boldsymbol{\nabla}\underline{\underline{\mathbf{A}}}.\end{aligned} \tag{13.149}$$

If we identify $\underline{\underline{\mathbf{A}}}^{21}$ with $\underline{\underline{\mathbf{G}}}$ and $\underline{\underline{\mathbf{D}}}$ with $\underline{\underline{\underline{\underline{\mathbf{c}}}}}:\underline{\underline{\mathbf{h}}}$, we have a suitable formula at hand:

$$\begin{aligned}\boldsymbol{\nabla}'\cdot[\underline{\underline{\underline{\underline{\mathbf{c}}}}}:\underline{\underline{\mathbf{h}}}(\underline{\mathbf{R}}',\omega)\cdot\underline{\underline{\mathbf{G}}}^{21}(\underline{\mathbf{R}}-\underline{\mathbf{R}}',\omega)] &= \underline{\underline{\mathbf{G}}}(\underline{\mathbf{R}}-\underline{\mathbf{R}}',\omega)\cdot\boldsymbol{\nabla}'\cdot[\underline{\underline{\underline{\underline{\mathbf{c}}}}}:\underline{\underline{\mathbf{h}}}(\underline{\mathbf{R}}',\omega)]\\ &\quad + [\underline{\underline{\underline{\underline{\mathbf{c}}}}}:\underline{\underline{\mathbf{h}}}(\underline{\mathbf{R}}',\omega)]^{21}:\boldsymbol{\nabla}'\underline{\underline{\mathbf{G}}}^{21}(\underline{\mathbf{R}}-\underline{\mathbf{R}}',\omega).\end{aligned} \tag{13.150}$$

Due to the symmetries $\underline{\underline{\underline{\underline{\mathbf{c}}}}} = \underline{\underline{\underline{\underline{\mathbf{c}}}}}^{2134} = \underline{\underline{\underline{\underline{\mathbf{c}}}}}^{3421} = \underline{\underline{\underline{\underline{\mathbf{c}}}}}^{3412}$, and $\underline{\underline{\mathbf{h}}} = \underline{\underline{\mathbf{h}}}^{21}$, we can convert: $(\underline{\underline{\underline{\underline{\mathbf{c}}}}}:\underline{\underline{\mathbf{h}}})^{21} = \underline{\underline{\mathbf{h}}}:\underline{\underline{\underline{\underline{\mathbf{c}}}}}$; since $\underline{\underline{\mathbf{G}}}(\underline{\mathbf{R}}-\underline{\mathbf{R}}',\omega)$ is symmetric, the expression

$$\begin{aligned}-\underline{\underline{\underline{\underline{\mathbf{c}}}}}:\boldsymbol{\nabla}'\underline{\underline{\mathbf{G}}}^{21}(\underline{\mathbf{R}}-\underline{\mathbf{R}}',\omega) &= -\,\underline{\underline{\underline{\underline{\mathbf{c}}}}}:\boldsymbol{\nabla}'\underline{\underline{\mathbf{G}}}(\underline{\mathbf{R}}-\underline{\mathbf{R}}',\omega)\\ &= \underline{\underline{\underline{\underline{\mathbf{c}}}}}:\boldsymbol{\nabla}\underline{\underline{\mathbf{G}}}(\underline{\mathbf{R}}-\underline{\mathbf{R}}',\omega)\\ &= \underline{\underline{\underline{\boldsymbol{\Sigma}}}}(\underline{\mathbf{R}}-\underline{\mathbf{R}}',\omega)\end{aligned} \tag{13.151}$$

obviously defines a third rank tensor [compare (13.142)] that plays the role of a ("right-sided") third rank Green tensor[233] for the injected deformation rate

[233] With the conversion

$$\begin{aligned}-(\underline{\underline{\underline{\underline{\mathbf{c}}}}}:\underline{\underline{\mathbf{h}}})^{21}:\boldsymbol{\nabla}'\underline{\underline{\mathbf{G}}}^{21} &= -\,(\boldsymbol{\nabla}'\underline{\underline{\mathbf{G}}})^{231}:\underline{\underline{\underline{\underline{\mathbf{c}}}}}:\underline{\underline{\mathbf{h}}}\\ &= (\boldsymbol{\nabla}\underline{\underline{\mathbf{G}}})^{231}:\underline{\underline{\underline{\underline{\mathbf{c}}}}}:\underline{\underline{\mathbf{h}}}\\ &= \underbrace{(\boldsymbol{\nabla}\underline{\underline{\mathbf{G}}})^{321}:\underline{\underline{\underline{\underline{\mathbf{c}}}}}}_{=\bar{\underline{\underline{\underline{\boldsymbol{\Sigma}}}}}}:\underline{\underline{\mathbf{h}}},\end{aligned}$$

we may also define a "left-sided" third rank Green tensor $\bar{\underline{\underline{\underline{\boldsymbol{\Sigma}}}}}$. In fact, based on the $\underline{\underline{\underline{\underline{\mathbf{c}}}}}$-symmetries, we have $\bar{\underline{\underline{\underline{\boldsymbol{\Sigma}}}}}^{231} = \underline{\underline{\underline{\boldsymbol{\Sigma}}}}$ resulting in

$$\begin{aligned}\bar{\underline{\underline{\underline{\boldsymbol{\Sigma}}}}}:\underline{\underline{\mathbf{h}}} &= \underline{\underline{\mathbf{h}}}:\bar{\underline{\underline{\underline{\boldsymbol{\Sigma}}}}}^{231}\\ &= \underline{\underline{\mathbf{h}}}:\underline{\underline{\underline{\boldsymbol{\Sigma}}}}.\end{aligned}$$

$$\underline{\mathbf{v}}(\underline{\mathbf{R}}, \omega) = \iiint_{V_Q} \Big[- \mathrm{j}\omega \underline{\mathbf{f}}(\underline{\mathbf{R}}', \omega) \cdot \underline{\underline{\mathbf{G}}}(\underline{\mathbf{R}} - \underline{\mathbf{R}}', \omega) + \underline{\underline{\mathbf{h}}}(\underline{\mathbf{R}}', \omega) : \underline{\underline{\underline{\boldsymbol{\Sigma}}}}(\underline{\mathbf{R}} - \underline{\mathbf{R}}', \omega) \Big] \, \mathrm{d}^3\underline{\mathbf{R}}' \tag{13.152}$$

that coincides with the third rank Green tensor in (13.141) due to the symmetry of $\underline{\underline{\mathbf{G}}}(\underline{\mathbf{R}} - \underline{\mathbf{R}}', \omega)$ [the volume integral on the left-hand side of (13.150) disappears as can be shown with the usual Gauss' theorem argument or with Footnote 83]. The first two indices of $\underline{\underline{\underline{\boldsymbol{\Sigma}}}}$ are those of $\underline{\underline{\underline{\underline{\mathbf{c}}}}}$ yielding the symmetry

$$\underline{\underline{\underline{\boldsymbol{\Sigma}}}}(\underline{\mathbf{R}} - \underline{\mathbf{R}}', \omega) = \underline{\underline{\underline{\boldsymbol{\Sigma}}}}^{213}(\underline{\mathbf{R}} - \underline{\mathbf{R}}', \omega); \tag{13.153}$$

the third index of $\underline{\underline{\underline{\boldsymbol{\Sigma}}}}$ is the second one of $\underline{\underline{\mathbf{G}}}$. The integral representation (13.152) holds for $\underline{\mathbf{R}} \in \mathbb{R}^3$, i.e., even for observation points in the interior of the source volume; the difference $G_\mathrm{S} - G_\mathrm{P}$ in (13.91) accounts for these nonexisting convergence problems.

We may further specialize (13.151) with (13.74) since we assume a homogeneous isotropic material:

$$\underline{\underline{\underline{\boldsymbol{\Sigma}}}}(\underline{\mathbf{R}} - \underline{\mathbf{R}}', \omega) = -\Big\{ \lambda \underline{\underline{\mathbf{I}}} \boldsymbol{\nabla}' \cdot \underline{\underline{\mathbf{G}}}(\underline{\mathbf{R}} - \underline{\mathbf{R}}', \omega) + \mu \boldsymbol{\nabla}' \underline{\underline{\mathbf{G}}}(\underline{\mathbf{R}} - \underline{\mathbf{R}}', \omega) + \mu \left[\boldsymbol{\nabla}' \underline{\underline{\mathbf{G}}}(\underline{\mathbf{R}} - \underline{\mathbf{R}}', \omega) \right]^{213} \Big\}, \tag{13.154}$$

respectively for $\underline{\mathbf{R}}' = \underline{\mathbf{0}}$

$$\underline{\underline{\underline{\boldsymbol{\Sigma}}}}(\underline{\mathbf{R}}, \omega) = \lambda \underline{\underline{\mathbf{I}}} \boldsymbol{\nabla} \cdot \underline{\underline{\mathbf{G}}}(\underline{\mathbf{R}}, \omega) + \mu \boldsymbol{\nabla} \underline{\underline{\mathbf{G}}}(\underline{\mathbf{R}}, \omega) + \mu \left[\boldsymbol{\nabla} \underline{\underline{\mathbf{G}}}(\underline{\mathbf{R}}, \omega) \right]^{213} \tag{13.155}$$

$$= \frac{\lambda}{\lambda + 2\mu} \underline{\underline{\mathbf{I}}} \boldsymbol{\nabla} G_\mathrm{P}(\underline{\mathbf{R}}, \omega) + \frac{2}{k_\mathrm{S}^2} \boldsymbol{\nabla}\boldsymbol{\nabla}\boldsymbol{\nabla} [G_\mathrm{S}(\underline{\mathbf{R}}, \omega) - G_\mathrm{P}(\underline{\mathbf{R}}, \omega)] + (\boldsymbol{\nabla} \underline{\underline{\mathbf{I}}} + \boldsymbol{\nabla} \underline{\underline{\mathbf{I}}}^{213}) G_\mathrm{S}(\underline{\mathbf{R}}, \omega) \, . \tag{13.156}$$

The representation (13.156) of $\underline{\underline{\underline{\boldsymbol{\Sigma}}}}$ nicely reveals the appearance of the difference $G_\mathrm{S} - G_\mathrm{P}$ in $\underline{\underline{\underline{\boldsymbol{\Sigma}}}}$, thus, as mentioned above, excluding problems with (13.152) for $\underline{\mathbf{R}} \in V_Q$.

Due to the separation (13.95), $\underline{\underline{\underline{\boldsymbol{\Sigma}}}}$ is also composed of a primary pressure and a secondary shear elementary wave term:[234]

$$\underline{\underline{\underline{\boldsymbol{\Sigma}}}}(\underline{\mathbf{R}}, \omega) = \frac{1}{\lambda + 2\mu} \underline{\underline{\underline{\boldsymbol{\Sigma}}}}_\mathrm{P}(\underline{\mathbf{R}}, \omega) + \frac{1}{\mu} \underline{\underline{\underline{\boldsymbol{\Sigma}}}}_\mathrm{S}(\underline{\mathbf{R}}, \omega), \tag{13.157}$$

[234]To show that $\underline{\underline{\underline{\boldsymbol{\Sigma}}}}_\mathrm{P}$ in (13.152) in fact creates a pressure wave term with $\boldsymbol{\nabla} \times \underline{\mathbf{v}} = \underline{\mathbf{0}}$ and $\underline{\underline{\underline{\boldsymbol{\Sigma}}}}_\mathrm{S}$ a shear wave term with $\boldsymbol{\nabla} \cdot \underline{\mathbf{v}} = 0$, we should use the left-sided $\bar{\underline{\underline{\underline{\boldsymbol{\Sigma}}}}}$-tensor according to Footnote 233.

where for $\underline{\mathbf{R}} \neq \underline{\mathbf{0}}$

$$\underline{\underline{\mathbf{\Sigma}}}_{\mathrm{P}}(\underline{\mathbf{R}}, \omega) = \lambda \underline{\underline{\mathbf{I}}} \nabla \cdot \underline{\underline{\mathbf{G}}}_{\mathrm{P}}(\underline{\mathbf{R}}, \omega) + \mu \left\{ \nabla \underline{\underline{\mathbf{G}}}_{\mathrm{P}}(\underline{\mathbf{R}}, \omega) + \left[\nabla \underline{\underline{\mathbf{G}}}_{\mathrm{P}}(\underline{\mathbf{R}}, \omega) \right]^{213} \right\}, \tag{13.158}$$

$$\underline{\underline{\mathbf{\Sigma}}}_{\mathrm{S}}(\underline{\mathbf{R}}, \omega) = \mu \left\{ \nabla \underline{\underline{\mathbf{G}}}_{\mathrm{S}}(\underline{\mathbf{R}}, \omega) + \left[\nabla \underline{\underline{\mathbf{G}}}_{\mathrm{S}}(\underline{\mathbf{R}}, \omega) \right]^{213} \right\} \tag{13.159}$$

holds because, for $\underline{\mathbf{R}} \neq \underline{\mathbf{0}}$, the term $\nabla \cdot \underline{\underline{\mathbf{G}}}_{\mathrm{S}}$ disappears on behalf of (13.97). For $\underline{\mathbf{R}} = \underline{\mathbf{0}}$, a term $\nabla \cdot \underline{\underline{\mathbf{G}}}_{\mathrm{S}} = -\nabla \delta(\underline{\mathbf{R}})/\rho\omega^2$ remains in (13.159) that is finally compensated in the sum (13.157), respectively (13.156), by a corresponding term from $\nabla \cdot \underline{\underline{\mathbf{G}}}_{\mathrm{P}}$.

To arrive at the differentiated representations for $\underline{\mathbf{R}} \neq \underline{\mathbf{0}}$ similar to (13.98) and (13.99), we actually must go through some calculus;[235] the result reads as

$$\begin{aligned}
\underline{\underline{\mathbf{\Sigma}}}_{\mathrm{P}}(\underline{\mathbf{R}}, \omega) = \Big[& \mathrm{j}k_{\mathrm{P}} \Big(\lambda \underline{\underline{\mathbf{I}}} \hat{\underline{\mathbf{R}}} + 2\mu \hat{\underline{\mathbf{R}}} \hat{\underline{\mathbf{R}}} \hat{\underline{\mathbf{R}}} \Big) \\
& + \frac{1}{R} \Big(-\lambda \underline{\underline{\mathbf{I}}} \hat{\underline{\mathbf{R}}} + 2\mu \hat{\underline{\mathbf{R}}} \underline{\underline{\mathbf{I}}} + 2\mu \hat{\underline{\mathbf{R}}} \underline{\underline{\mathbf{I}}}^{213} + 2\mu \underline{\underline{\mathbf{I}}} \hat{\underline{\mathbf{R}}} - 12\mu \hat{\underline{\mathbf{R}}} \hat{\underline{\mathbf{R}}} \hat{\underline{\mathbf{R}}} \Big) \\
& + \mathrm{j} \frac{6\mu}{k_{\mathrm{P}} R^2} \Big(\hat{\underline{\mathbf{R}}} \underline{\underline{\mathbf{I}}} + \hat{\underline{\mathbf{R}}} \underline{\underline{\mathbf{I}}}^{213} + \underline{\underline{\mathbf{I}}} \hat{\underline{\mathbf{R}}} - 5 \hat{\underline{\mathbf{R}}} \hat{\underline{\mathbf{R}}} \hat{\underline{\mathbf{R}}} \Big) \\
& - \frac{6\mu}{k_{\mathrm{P}}^2 R^3} \Big(\hat{\underline{\mathbf{R}}} \underline{\underline{\mathbf{I}}} + \hat{\underline{\mathbf{R}}} \underline{\underline{\mathbf{I}}}^{213} + \underline{\underline{\mathbf{I}}} \hat{\underline{\mathbf{R}}} - 5 \hat{\underline{\mathbf{R}}} \hat{\underline{\mathbf{R}}} \hat{\underline{\mathbf{R}}} \Big) \Big] \frac{\mathrm{e}^{\mathrm{j}k_{\mathrm{P}} R}}{4\pi R};
\end{aligned} \tag{13.160}$$

$$\begin{aligned}
\underline{\underline{\mathbf{\Sigma}}}_{\mathrm{S}}(\underline{\mathbf{R}}, \omega) = \mu \Big[& \mathrm{j}k_{\mathrm{S}} \Big(\hat{\underline{\mathbf{R}}} \underline{\underline{\mathbf{I}}} + \hat{\underline{\mathbf{R}}} \underline{\underline{\mathbf{I}}}^{213} - 2 \hat{\underline{\mathbf{R}}} \hat{\underline{\mathbf{R}}} \hat{\underline{\mathbf{R}}} \Big) \\
& - \frac{1}{R} \Big(3 \hat{\underline{\mathbf{R}}} \underline{\underline{\mathbf{I}}} + 3 \hat{\underline{\mathbf{R}}} \underline{\underline{\mathbf{I}}}^{213} + 2 \underline{\underline{\mathbf{I}}} \hat{\underline{\mathbf{R}}} - 12 \hat{\underline{\mathbf{R}}} \hat{\underline{\mathbf{R}}} \hat{\underline{\mathbf{R}}} \Big) \\
& - \mathrm{j} \frac{6}{k_{\mathrm{S}} R^2} \Big(\hat{\underline{\mathbf{R}}} \underline{\underline{\mathbf{I}}} + \hat{\underline{\mathbf{R}}} \underline{\underline{\mathbf{I}}}^{213} + \underline{\underline{\mathbf{I}}} \hat{\underline{\mathbf{R}}} - 5 \hat{\underline{\mathbf{R}}} \hat{\underline{\mathbf{R}}} \hat{\underline{\mathbf{R}}} \Big) \\
& + \frac{6}{k_{\mathrm{S}}^2 R^3} \Big(\hat{\underline{\mathbf{R}}} \underline{\underline{\mathbf{I}}} + \hat{\underline{\mathbf{R}}} \underline{\underline{\mathbf{I}}}^{213} + \underline{\underline{\mathbf{I}}} \hat{\underline{\mathbf{R}}} - 5 \hat{\underline{\mathbf{R}}} \hat{\underline{\mathbf{R}}} \hat{\underline{\mathbf{R}}} \Big) \Big] \frac{\mathrm{e}^{\mathrm{j}k_{\mathrm{S}} R}}{4\pi R};
\end{aligned} \tag{13.161}$$

The far-field approximations of $\underline{\underline{\mathbf{\Sigma}}}_{\mathrm{P}}$ and $\underline{\underline{\mathbf{\Sigma}}}_{\mathrm{S}}$ are not obtained from (13.160) and (13.161) but immediately from (13.158) and (13.159) applying (13.47):

---

[235] The following formulas are required:

$$\begin{aligned}
\nabla R &= \hat{\underline{\mathbf{R}}}, \\
\nabla \hat{\underline{\mathbf{R}}} &= \frac{1}{R} (\underline{\underline{\mathbf{I}}} - \hat{\underline{\mathbf{R}}} \hat{\underline{\mathbf{R}}}), \\
\nabla (\Phi \underline{\underline{\mathbf{D}}}) &= (\nabla \Phi) \underline{\underline{\mathbf{D}}} + \Phi \nabla \underline{\underline{\mathbf{D}}}, \\
\nabla (\underline{\mathbf{A}}\, \underline{\mathbf{B}}) &= (\nabla \underline{\mathbf{A}}) \underline{\mathbf{B}} + (\underline{\mathbf{A}} \nabla \underline{\mathbf{B}})^{213}.
\end{aligned}$$

$$\underline{\underline{\underline{\boldsymbol{\Sigma}}}}_{\mathrm{P}}^{\mathrm{far}}(\underline{\mathbf{R}}, \omega) = \mathrm{j}k_{\mathrm{P}} \frac{\mathrm{e}^{\mathrm{j}k_{\mathrm{P}}R}}{4\pi R} \left(\lambda \underline{\underline{\mathbf{I}}}\,\hat{\underline{\mathbf{R}}} + 2\mu \hat{\underline{\mathbf{R}}}\,\hat{\underline{\mathbf{R}}}\,\hat{\underline{\mathbf{R}}}\right)$$
$$= \mathrm{j}k_{\mathrm{P}} \frac{\mathrm{e}^{\mathrm{j}k_{\mathrm{P}}R}}{4\pi R} (\lambda \underline{\underline{\mathbf{I}}} + 2\mu \hat{\underline{\mathbf{R}}}\,\hat{\underline{\mathbf{R}}})\hat{\underline{\mathbf{R}}}$$
$$= -\mathrm{j}\omega \frac{\mathrm{e}^{\mathrm{j}k_{\mathrm{P}}R}}{R} \underline{\underline{\underline{\boldsymbol{\sigma}}}}_{\mathrm{P}}(\hat{\underline{\mathbf{R}}}), \tag{13.162}$$

$$\underline{\underline{\underline{\boldsymbol{\Sigma}}}}_{\mathrm{S}}^{\mathrm{far}}(\underline{\mathbf{R}}, \omega) = \mathrm{j}k_{\mathrm{S}}\mu \frac{\mathrm{e}^{\mathrm{j}k_{\mathrm{S}}R}}{4\pi R} \left(\hat{\underline{\mathbf{R}}}\,\underline{\underline{\mathbf{I}}} + \hat{\underline{\mathbf{R}}}\,\underline{\underline{\mathbf{I}}}^{213} - 2\hat{\underline{\mathbf{R}}}\,\hat{\underline{\mathbf{R}}}\,\hat{\underline{\mathbf{R}}}\right)$$
$$= -\mathrm{j}\omega \frac{\mathrm{e}^{\mathrm{j}k_{\mathrm{S}}R}}{R} \underline{\underline{\underline{\boldsymbol{\sigma}}}}_{\mathrm{S}}(\hat{\underline{\mathbf{R}}}). \tag{13.163}$$

We have

$$\underline{\underline{\underline{\boldsymbol{\Sigma}}}}_{\mathrm{P}}^{\mathrm{far}}(\underline{\mathbf{R}}, \omega) \cdot (\underline{\underline{\mathbf{I}}} - \hat{\underline{\mathbf{R}}}\,\hat{\underline{\mathbf{R}}}) = \underline{\underline{\underline{\mathbf{0}}}}, \tag{13.164}$$

$$\underline{\underline{\underline{\boldsymbol{\Sigma}}}}_{\mathrm{S}}^{\mathrm{far}}(\underline{\mathbf{R}}, \omega) \cdot \hat{\underline{\mathbf{R}}} = \underline{\underline{\mathbf{0}}}. \tag{13.165}$$

The radiation characteristics $\underline{\underline{\underline{\boldsymbol{\sigma}}}}_{\mathrm{P}}(\hat{\underline{\mathbf{R}}})$ and $\underline{\underline{\underline{\boldsymbol{\sigma}}}}_{\mathrm{S}}(\hat{\underline{\mathbf{R}}})$ of the $\underline{\underline{\underline{\boldsymbol{\Sigma}}}}_{\mathrm{P,S}}$-tensors are given by

$$\underline{\underline{\underline{\boldsymbol{\sigma}}}}_{\mathrm{P}}(\hat{\underline{\mathbf{R}}}) = -\frac{1}{4\pi c_{\mathrm{P}}} (\lambda \underline{\underline{\mathbf{I}}} + 2\mu \hat{\underline{\mathbf{R}}}\,\hat{\underline{\mathbf{R}}})\hat{\underline{\mathbf{R}}} \tag{13.166}$$

$$\underline{\underline{\underline{\boldsymbol{\sigma}}}}_{\mathrm{S}}(\hat{\underline{\mathbf{R}}}) = -\frac{\mu}{4\pi c_{\mathrm{S}}} (\hat{\underline{\mathbf{R}}}\,\underline{\underline{\mathbf{I}}} + \hat{\underline{\mathbf{R}}}\,\underline{\underline{\mathbf{I}}}^{213} - 2\hat{\underline{\mathbf{R}}}\,\hat{\underline{\mathbf{R}}}\,\hat{\underline{\mathbf{R}}}); \tag{13.167}$$

we define them independent of frequency similar to the radiation characteristics $\underline{\underline{\mathbf{g}}}_{\mathrm{P}}$, $\underline{\underline{\mathbf{g}}}_{\mathrm{S}}$ of the $\underline{\underline{\mathbf{G}}}_{\mathrm{P,S}}$-tensors.

13.2.4 Particle displacement of a point source deformation rate, point radiation characteristic

According to (13.152), the second rank tensor for the particle velocity is related to a force density source and the third rank Green tensor for the particle velocity to a deformation rate source. Accordingly, we now investigate the case $\underline{\mathbf{f}} = \underline{\mathbf{0}}$ and $\underline{\underline{\mathbf{h}}} \neq \underline{\underline{\mathbf{0}}}$ in some more detail:

$$\underline{\mathbf{v}}(\underline{\mathbf{R}}, \omega) = \iiint_{V_Q} \underline{\underline{\mathbf{h}}}(\underline{\mathbf{R}}', \omega) : \underline{\underline{\underline{\boldsymbol{\Sigma}}}}(\underline{\mathbf{R}} - \underline{\mathbf{R}}', \omega)\, \mathrm{d}^3\underline{\mathbf{R}}'. \tag{13.168}$$

With (Equation 2.127)

$$\hat{\underline{\underline{\mathbf{h}}}} = \frac{\underline{\underline{\mathbf{h}}}}{\sqrt{\underline{\underline{\mathbf{h}}} : \underline{\underline{\mathbf{h}}}^{+}}} \tag{13.169}$$

and

$$\underline{\underline{\mathbf{h}}}(\underline{\mathbf{R}}, \omega) = \delta(\underline{\mathbf{R}})\, \hat{\underline{\underline{\mathbf{h}}}}, \tag{13.170}$$

we postulate a point-like deformation rate unit source in the origin and find[236]

$$\underline{\mathbf{v}}^{\mathrm{PS}_h}(\underline{\mathbf{R}}, \omega) = \underline{\underline{\hat{\mathbf{h}}}} : \underline{\underline{\underline{\boldsymbol{\Sigma}}}}(\underline{\mathbf{R}}, \omega) \\ = \underline{\underline{\underline{\boldsymbol{\Sigma}}}}^{312}(\underline{\mathbf{R}}, \omega) : \underline{\underline{\hat{\mathbf{h}}}} \tag{13.171}$$

as the counterpart to (13.116). The calculation of the components of $\underline{\mathbf{v}}^{\mathrm{PS}_h}$ in spherical coordinates similar to (13.119) through (13.121) would result in a lot of paper work; therefore, we directly concentrate on the far-field similar to (13.128) and (13.131) utilizing (13.162) and (13.163):

$$\underline{\mathbf{v}}^{\mathrm{PS}_h,\mathrm{far}}(\underline{\mathbf{R}}, \omega) = \underline{\underline{\hat{\mathbf{h}}}} : \left[\frac{1}{\lambda + 2\mu} \underline{\underline{\underline{\boldsymbol{\Sigma}}}}_{\mathrm{P}}^{\mathrm{far}}(\underline{\mathbf{R}}, \omega) + \frac{1}{\mu} \underline{\underline{\underline{\boldsymbol{\Sigma}}}}_{\mathrm{S}}^{\mathrm{far}}(\underline{\mathbf{R}}, \omega) \right] \\ = \underline{\mathbf{v}}_{\mathrm{P}}^{\mathrm{PS}_h,\mathrm{far}}(\underline{\mathbf{R}}, \omega) + \underline{\mathbf{v}}_{\mathrm{S}}^{\mathrm{PS}_h,\mathrm{far}}(\underline{\mathbf{R}}, \omega), \tag{13.172}$$

where

$$\underline{\mathbf{v}}_{\mathrm{P}}^{\mathrm{PS}_h,\mathrm{far}}(\underline{\mathbf{R}}, \omega) = \mathrm{j} \frac{k_{\mathrm{P}}}{\lambda + 2\mu} \frac{\mathrm{e}^{\mathrm{j}k_{\mathrm{P}}R}}{4\pi R} \underline{\underline{\hat{\mathbf{h}}}} : (\lambda \underline{\underline{\mathbf{I}}} + 2\mu \underline{\hat{\mathbf{R}}}\,\underline{\hat{\mathbf{R}}})\underline{\hat{\mathbf{R}}} \\ = \underline{\mathbf{v}}_{\mathrm{P}}(\underline{\underline{\hat{\mathbf{h}}}}, \underline{\hat{\mathbf{R}}}) \frac{\mathrm{e}^{\mathrm{j}k_{\mathrm{P}}R}}{R} \tag{13.173}$$

with

$$\underline{\mathbf{v}}_{\mathrm{P}}(\underline{\underline{\hat{\mathbf{h}}}}, \underline{\hat{\mathbf{R}}}) = v_{\mathrm{P}}(\underline{\underline{\hat{\mathbf{h}}}}, \underline{\hat{\mathbf{R}}})\underline{\hat{\mathbf{R}}} \\ = \mathrm{j} \frac{k_{\mathrm{P}}}{4\pi(\lambda + 2\mu)} \underline{\underline{\hat{\mathbf{h}}}} : (\lambda \underline{\underline{\mathbf{I}}} + 2\mu \underline{\hat{\mathbf{R}}}\,\underline{\hat{\mathbf{R}}})\underline{\hat{\mathbf{R}}} \tag{13.174}$$

as well as

$$\underline{\mathbf{v}}_{\mathrm{S}}^{\mathrm{PS}_h,\mathrm{far}}(\underline{\mathbf{R}}, \omega) = 2\mathrm{j}k_{\mathrm{S}} \frac{\mathrm{e}^{\mathrm{j}k_{\mathrm{S}}R}}{4\pi R} \underline{\underline{\hat{\mathbf{h}}}} : \underline{\hat{\mathbf{R}}}(\underline{\underline{\mathbf{I}}} - \underline{\hat{\mathbf{R}}}\,\underline{\hat{\mathbf{R}}}) \\ = \underline{\mathbf{v}}_{\mathrm{S}}(\underline{\underline{\hat{\mathbf{h}}}}, \underline{\hat{\mathbf{R}}}) \frac{\mathrm{e}^{\mathrm{j}k_{\mathrm{S}}R}}{R} \tag{13.175}$$

with

$$\underline{\mathbf{v}}_{\mathrm{S}}(\underline{\underline{\hat{\mathbf{h}}}}, \underline{\hat{\mathbf{R}}}) = v_{\mathrm{S}}(\underline{\underline{\hat{\mathbf{h}}}}, \underline{\hat{\mathbf{R}}})\underline{\hat{\mathbf{v}}}_{\mathrm{S}}(\underline{\underline{\hat{\mathbf{h}}}}, \underline{\hat{\mathbf{R}}}) \\ = \mathrm{j} \frac{k_{\mathrm{S}}}{2\pi} \underline{\underline{\hat{\mathbf{h}}}} : \underline{\hat{\mathbf{R}}}(\underline{\underline{\mathbf{I}}} - \underline{\hat{\mathbf{R}}}\,\underline{\hat{\mathbf{R}}}). \tag{13.176}$$

With

$$\underline{\mathbf{u}}_{\mathrm{P}}^{\mathrm{PS}_h\mathrm{far}}(\underline{\mathbf{R}}, \omega) = -\frac{1}{4\pi c_{\mathrm{P}}(\lambda + 2\mu)} \frac{\mathrm{e}^{\mathrm{j}k_{\mathrm{P}}R}}{R} \underline{\underline{\hat{\mathbf{h}}}} : (\lambda \underline{\underline{\mathbf{I}}} + 2\mu \underline{\hat{\mathbf{R}}}\,\underline{\hat{\mathbf{R}}})\underline{\hat{\mathbf{R}}} \\ = u_{\mathrm{P}}(\underline{\underline{\hat{\mathbf{h}}}}, \underline{\hat{\mathbf{R}}}) \frac{\mathrm{e}^{\mathrm{j}k_{\mathrm{P}}R}}{R} \underline{\hat{\mathbf{u}}}_{\mathrm{P}}(\underline{\underline{\hat{\mathbf{h}}}}, \underline{\hat{\mathbf{R}}}); \tag{13.177}$$

[236]We repeatedly apply the symmetries $\underline{\underline{\mathbf{h}}} = \underline{\underline{\mathbf{h}}}^{21}$ and $\underline{\underline{\underline{\boldsymbol{\Sigma}}}} = \underline{\underline{\underline{\boldsymbol{\Sigma}}}}^{213}$.

$$\underline{\mathbf{u}}_{\mathrm{S}}^{\mathrm{PS}_h,\mathrm{far}}(\underline{\mathbf{R}},\omega) = -\frac{1}{2\pi c_{\mathrm{S}}}\frac{\mathrm{e}^{\mathrm{j}k_{\mathrm{S}}R}}{R}\,\hat{\underline{\underline{\mathbf{h}}}} : \hat{\underline{\mathbf{R}}}(\underline{\underline{\mathbf{I}}} - \hat{\underline{\mathbf{R}}}\,\hat{\underline{\mathbf{R}}})$$

$$= u_{\mathrm{S}}(\hat{\underline{\underline{\mathbf{h}}}},\hat{\underline{\mathbf{R}}})\,\frac{\mathrm{e}^{\mathrm{j}k_{\mathrm{S}}R}}{R}\,\hat{\underline{\mathbf{u}}}_{\mathrm{S}}(\hat{\underline{\underline{\mathbf{h}}}},\hat{\underline{\mathbf{R}}}), \tag{13.178}$$

where

$$\hat{\underline{\mathbf{u}}}_{\mathrm{P}}(\hat{\underline{\underline{\mathbf{h}}}},\hat{\underline{\mathbf{R}}}) = \hat{\underline{\mathbf{R}}}, \tag{13.179}$$

$$u_{\mathrm{P}}(\hat{\underline{\underline{\mathbf{h}}}},\hat{\underline{\mathbf{R}}}) = -\frac{\hat{\underline{\underline{\mathbf{h}}}} : (\lambda\underline{\underline{\mathbf{I}}} + 2\mu\hat{\underline{\mathbf{R}}}\,\hat{\underline{\mathbf{R}}})}{4\pi c_{\mathrm{P}}(\lambda+2\mu)}; \tag{13.180}$$

$$\hat{\underline{\mathbf{u}}}_{\mathrm{S}}(\hat{\underline{\underline{\mathbf{h}}}},\hat{\underline{\mathbf{R}}}) = \frac{\hat{\underline{\underline{\mathbf{h}}}} : \hat{\underline{\mathbf{R}}}(\underline{\underline{\mathbf{I}}} - \hat{\underline{\mathbf{R}}}\,\hat{\underline{\mathbf{R}}})}{|\hat{\underline{\underline{\mathbf{h}}}} : \hat{\underline{\mathbf{R}}}(\underline{\underline{\mathbf{I}}} - \hat{\underline{\mathbf{R}}}\,\hat{\underline{\mathbf{R}}})|}, \tag{13.181}$$

$$u_{\mathrm{S}}(\hat{\underline{\underline{\mathbf{h}}}},\hat{\underline{\mathbf{R}}}) = -\frac{|\hat{\underline{\underline{\mathbf{h}}}} : \hat{\underline{\mathbf{R}}}(\underline{\underline{\mathbf{I}}} - \hat{\underline{\mathbf{R}}}\,\hat{\underline{\mathbf{R}}})|}{2\pi c_{\mathrm{S}}}, \tag{13.182}$$

we switch to the particle displacement. Evidently, the polarization equations (13.134) and (13.135) also hold for the $\hat{\underline{\underline{\mathbf{h}}}}$-source; the stress tensor equations (13.138) and (13.139) may also be kept if (13.179) through (13.182) are inserted: Even the particle displacement far-field of a point-like deformation rate source behaves locally as a plane wave. In addition, we emphasize that the time dependence of a point-like $\underline{\mathbf{f}}$- as well as a point-like $\underline{\underline{\mathbf{h}}}$-source reproduces in the far-field A-"scan" of the particle displacement (compare Figures 13.9 and 13.11).

Based on Figure 13.10, we demonstrated that the special choice $\hat{\underline{\mathbf{f}}}(\hat{\underline{\mathbf{R}}}) = \hat{\underline{\mathbf{R}}}$ of the force density cuts off the pertinent shear wave in the far-field; with the special choice $\hat{\underline{\underline{\mathbf{h}}}} = \underline{\underline{\mathbf{I}}}/\sqrt{3}$ of a deformation rate unit source, we may completely cut off the pertinent shear wave, i.e., in the near- and far-fields because we have $\underline{\underline{\mathbf{I}}} : \underline{\underline{\mathbf{\Sigma}}}_{\mathrm{S}} \equiv \underline{\mathbf{0}}$ with (13.161). Deceptions of that kind will become useful for the physical interpretation of side echoes.

Point directivities for the particle displacement of a deformation rate source: Analogous to (13.125) through (13.127), we define frequency-independent[237] (non-normalized) far-field point directivities for the particle velocity of a deformation rate source with (13.166) and (13.167):

$$H_{\mathrm{P}}^{\mathrm{PS}_h}(\hat{\underline{\mathbf{R}}}) = \frac{1}{\lambda+2\mu}\,\hat{\underline{\underline{\mathbf{h}}}} : \underline{\underline{\boldsymbol{\sigma}}}_{\mathrm{P}}(\hat{\underline{\mathbf{R}}})\cdot\underline{\mathbf{e}}_R$$

$$= -\frac{1}{4\pi c_{\mathrm{P}}^2 Z_{\mathrm{P}}}\,\hat{\underline{\underline{\mathbf{h}}}} : (\lambda\underline{\underline{\mathbf{I}}} + 2\mu\hat{\underline{\mathbf{R}}}\,\hat{\underline{\mathbf{R}}})\hat{\underline{\mathbf{R}}}\cdot\underline{\mathbf{e}}_R$$

$$= -\frac{1}{4\pi c_{\mathrm{P}}^2 Z_{\mathrm{P}}}\,\hat{\underline{\underline{\mathbf{h}}}} : (\lambda\underline{\underline{\mathbf{I}}} + 2\mu\hat{\underline{\mathbf{R}}}\,\hat{\underline{\mathbf{R}}})$$

$$= R\,\mathrm{e}^{-\mathrm{j}k_{\mathrm{P}}R}u_R^{\mathrm{PS}_h,\mathrm{far}}(\underline{\mathbf{R}},\omega), \tag{13.183}$$

[237]Therefore, we defined $\underline{\underline{\boldsymbol{\sigma}}}_{\mathrm{P}}$ and $\underline{\underline{\boldsymbol{\sigma}}}_{\mathrm{S}}$ according to (13.162) and (13.163) without the factor $-\mathrm{j}\omega$.

$$
\begin{aligned}
H_{\mathrm{S}\vartheta}^{\mathrm{PS}_h}(\hat{\underline{\mathbf{R}}}) &= \frac{1}{\mu}\,\hat{\underline{\underline{\mathbf{h}}}} : \underline{\underline{\underline{\sigma}}}_{\mathrm{S}}(\hat{\underline{\mathbf{R}}}) \cdot \underline{\mathbf{e}}_\vartheta \\
&= -\frac{1}{4\pi c_{\mathrm{S}}}\,\hat{\underline{\underline{\mathbf{h}}}} : (\hat{\underline{\mathbf{R}}}\underline{\underline{\mathbf{I}}} + \hat{\underline{\mathbf{R}}}\underline{\underline{\mathbf{I}}}^{213} - 2\hat{\underline{\mathbf{R}}}\hat{\underline{\mathbf{R}}}\hat{\underline{\mathbf{R}}}) \cdot \underline{\mathbf{e}}_\vartheta \\
&= -\frac{1}{2\pi c_{\mathrm{S}}}\,\hat{\underline{\underline{\mathbf{h}}}} : \hat{\underline{\mathbf{R}}}(\underline{\underline{\mathbf{I}}} - \hat{\underline{\mathbf{R}}}\hat{\underline{\mathbf{R}}}) \cdot \underline{\mathbf{e}}_\vartheta \\
&= -\frac{1}{4\pi c_{\mathrm{S}}}\,\hat{\underline{\underline{\mathbf{h}}}} : \left(\hat{\underline{\mathbf{R}}}\underline{\mathbf{e}}_\vartheta + \underline{\mathbf{e}}_\vartheta\hat{\underline{\mathbf{R}}}\right) \\
&= R\,\mathrm{e}^{-\mathrm{j}k_{\mathrm{S}}R} u_\vartheta^{\mathrm{PS}_h,\mathrm{far}}(\underline{\mathbf{R}}, \omega),
\end{aligned}
\tag{13.184}
$$

$$
\begin{aligned}
H_{\mathrm{S}\varphi}^{\mathrm{PS}_h}(\hat{\underline{\mathbf{R}}}) &= \frac{1}{\mu}\,\hat{\underline{\underline{\mathbf{h}}}} : \underline{\underline{\underline{\sigma}}}_{\mathrm{S}}(\hat{\underline{\mathbf{R}}}) \cdot \underline{\mathbf{e}}_\varphi \\
&= -\frac{1}{4\pi c_{\mathrm{S}}}\,\hat{\underline{\underline{\mathbf{h}}}} : \left(\hat{\underline{\mathbf{R}}}\underline{\mathbf{e}}_\varphi + \underline{\mathbf{e}}_\varphi\hat{\underline{\mathbf{R}}}\right) \\
&= R\,\mathrm{e}^{-\mathrm{j}k_{\mathrm{S}}R} u_\varphi^{\mathrm{PS}_h,\mathrm{far}}(\underline{\mathbf{R}}, \omega).
\end{aligned}
\tag{13.185}
$$

Choosing subsequently single components of the $\hat{\underline{\underline{\mathbf{h}}}}$-tensor, we obtain, for example, the radiation patterns in Figure 13.14 (we constrain to $h_{xx}, h_{yy}, h_{zz}, h_{xz}$; note: In the S-contribution, the $\underline{\mathbf{e}}_x\underline{\mathbf{e}}_x$-, $\underline{\mathbf{e}}_y\underline{\mathbf{e}}_y$-, and $\underline{\mathbf{e}}_z\underline{\mathbf{e}}_z$-terms compensate), where we adhere to the fact that the $H_{\mathrm{P}}^{\mathrm{PS}_h}$-diagrams—even without the prefactors—may be structurally material dependent. Together with Figure 13.12, a quick overview concerning the directivity of point sources is available; the formalism of elastodynamic Green tensors is finally reduced to sine and cosine! With this perception, we will even approach the next section.

13.2.5 Fourth-rank Green tensor: Stress tensor of a point source deformation rate

Four Green tensors of elastodynamics: Why another tensor, even of rank four? With (13.152), the elastodynamic source field of the particle velocity for a given force and deformation rate source is already known! In view of the acoustic scheme in Figure 5.1, it becomes clear what we mean: With the source field representation for p containing the Green functions G and ∇G, everything is said; yet the gradient operation ∇p to calculate $\underline{\mathbf{v}}$ leads to a second rank Green tensor $\underline{\underline{\mathbf{G}}}_v$ in the resulting source field representation through ∇G! On the other hand, first, solving the differential equation for $\underline{\mathbf{v}}$ the tensor $\underline{\underline{\mathbf{G}}}_v$ comes immediately into play.

Here, the situation is quite similar: The source field representation of $\underline{\mathbf{v}}$ requires $\underline{\underline{\mathbf{G}}}$ and $\underline{\underline{\underline{\mathbf{\Sigma}}}}$; calculating the stress tensor from $\underline{\mathbf{v}}$ via “Hooke's differentiation” $\underline{\underline{\underline{\underline{\mathbf{c}}}}} : \nabla$ a forth rank Green tensor $\underline{\underline{\underline{\underline{\mathbf{\Pi}}}}}$ is emerging from $\underline{\underline{\underline{\mathbf{\Sigma}}}}$ that afterward relates given deformation rates and resulting stresses through double contraction with $\underline{\underline{\mathbf{h}}}$. From a physical point of view, $\underline{\underline{\underline{\underline{\mathbf{\Pi}}}}}$ represents (Huygens) stress elementary waves emanating form $\underline{\underline{\mathbf{h}}}$. Anticipating the following calculation, we display in Figure 13.15 a scheme similar to Figure 5.1, and before we

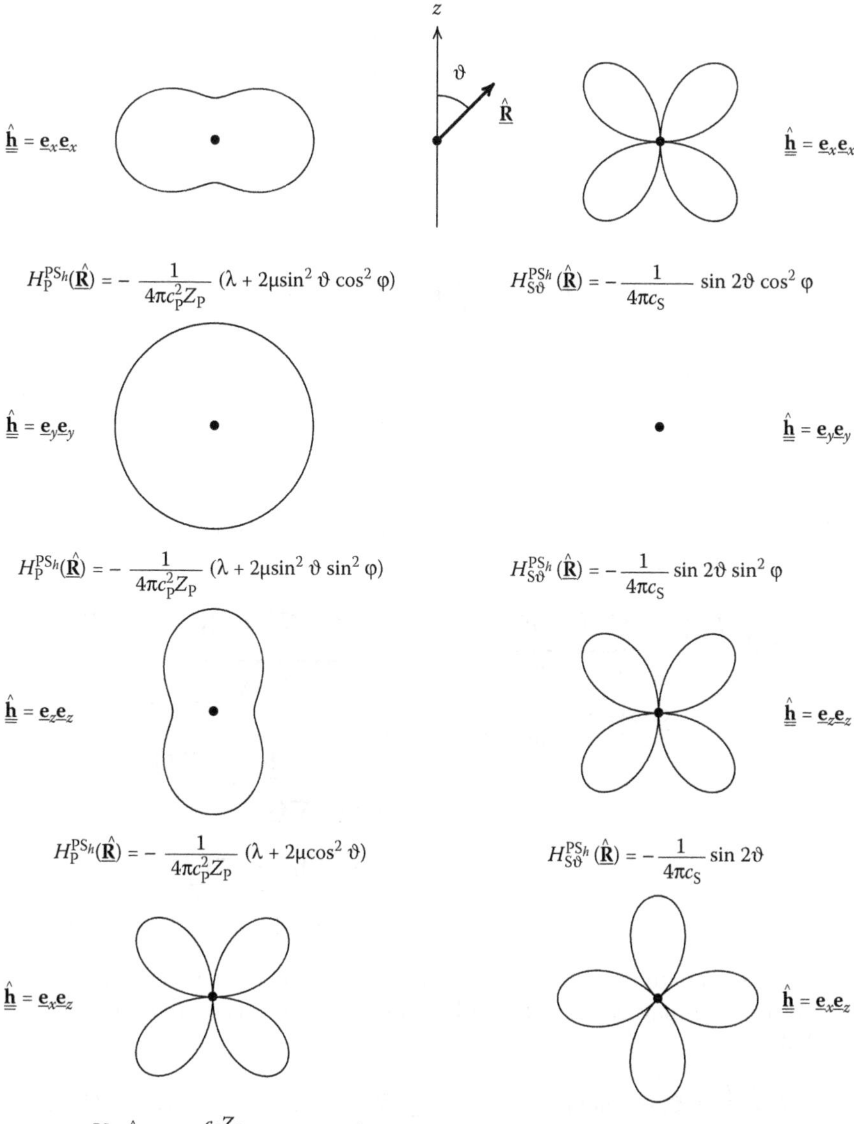

FIGURE 13.14

Radiation patterns of the particle velocity for point-like deformation rate sources ($\underline{\underline{\hat{\mathbf{h}}}}$-point directivities): In the direction $\underline{\hat{\mathbf{R}}}$, the normalized magnitude of the point-directivities is displayed as function of ϑ for $y = 0$ (material: steel with c_{P}=5900 m/s, c_{S}=3200 m/s, $\rho = 7.7 \cdot 10^3$ kg/m^3).

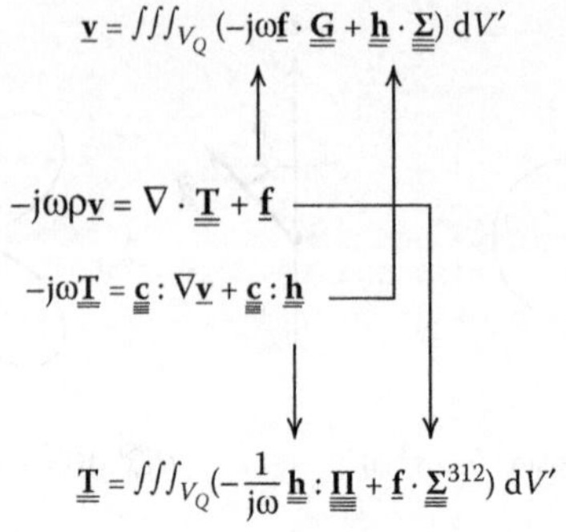

$$\underline{\mathbf{v}} = \iiint_{V_Q} (-j\omega\underline{\mathbf{f}} \cdot \underline{\underline{\mathbf{G}}} + \underline{\underline{\mathbf{h}}} \cdot \underline{\underline{\underline{\mathbf{\Sigma}}}})\, dV'$$

$$-j\omega\rho\underline{\mathbf{v}} = \nabla \cdot \underline{\underline{\mathbf{T}}} + \underline{\mathbf{f}}$$

$$-j\omega\underline{\underline{\mathbf{T}}} = \underline{\underline{\underline{\underline{\mathbf{c}}}}} : \nabla\underline{\mathbf{v}} + \underline{\underline{\underline{\underline{\mathbf{c}}}}} : \underline{\underline{\mathbf{h}}}$$

$$\underline{\underline{\mathbf{T}}} = \iiint_{V_Q} (-\frac{1}{j\omega}\underline{\underline{\mathbf{h}}} : \underline{\underline{\underline{\underline{\mathbf{\Pi}}}}} + \underline{\mathbf{f}} \cdot \underline{\underline{\underline{\mathbf{\Sigma}}}}^{312})\, dV'$$

FIGURE 13.15
Relation of Green functions for homogeneous isotropic elastic materials with the source densities $\underline{\mathbf{f}}$ and $\underline{\underline{\mathbf{h}}}$ (we have $\underline{\underline{\underline{\underline{\mathbf{c}}}}} = \lambda\underline{\underline{\mathbf{I}}}^{\delta} + 2\mu\underline{\underline{\underline{\underline{\mathbf{I}}}}}^{+}$).

explicitly perform this calculation, we summarize the formal operations that bestowed us (13.152):

$$\underline{\mathbf{v}} = \iiint_{V_Q} [\quad \underbrace{\underline{\underline{\mathbf{G}}} \cdot (-j\omega\underline{\mathbf{f}})} \quad + \quad \underbrace{\underline{\underline{\mathbf{G}}} \cdot \nabla' \cdot \underline{\underline{\underline{\underline{\mathbf{c}}}}} : \underline{\underline{\mathbf{h}}}} \quad]\, dV'.$$

$$\begin{aligned}
&= -j\omega\underline{\mathbf{f}} \cdot \underline{\underline{\mathbf{G}}}^{21} \\
&\overset{\underline{\underline{\mathbf{G}}}-\text{Symm.}}{=} -j\omega\underline{\mathbf{f}} \cdot \underline{\underline{\mathbf{G}}}
\end{aligned}$$

$$\begin{aligned}
&= -(\underline{\underline{\underline{\underline{\mathbf{c}}}}} : \underline{\underline{\mathbf{h}}})^{21} : \nabla'\underline{\underline{\mathbf{G}}}^{21} \\
&\overset{\underline{\underline{\underline{\underline{\mathbf{c}}}}},\underline{\underline{\mathbf{h}}}-\text{Symm.}}{=} -\underline{\underline{\mathbf{h}}} : \underline{\underline{\underline{\underline{\mathbf{c}}}}} : \nabla'\underline{\underline{\mathbf{G}}}^{21} \\
&= \underline{\underline{\mathbf{h}}} : \underline{\underline{\underline{\underline{\mathbf{c}}}}} : \nabla\underline{\underline{\mathbf{G}}}^{21} \\
&\overset{\underline{\underline{\mathbf{G}}}-\text{Symm.}}{=} \underline{\underline{\mathbf{h}}} : \underline{\underline{\underline{\underline{\mathbf{c}}}}} : \nabla\underline{\underline{\mathbf{G}}} \\
&\overset{(13.142)}{=} \underline{\underline{\mathbf{h}}} : \underline{\underline{\underline{\mathbf{\Sigma}}}}
\end{aligned} \tag{13.186}$$

That way, Green tensors result that appear on the right-hand side of the sources. Yet, we may equally define left-sided Green tensors (compare Footnote 233):

$$\underline{\mathbf{v}} = \iiint_{V_Q} [\, \underline{\underline{\mathbf{G}}} \cdot (-j\omega\underline{\mathbf{f}}) + \quad \underbrace{\underline{\underline{\mathbf{G}}} \cdot \nabla' \cdot \underline{\underline{\underline{\underline{\mathbf{c}}}}} : \underline{\underline{\mathbf{h}}}} \quad]\, dV'. \tag{13.187}$$

$$\begin{aligned}
&\overset{\underline{\underline{\underline{\underline{\mathbf{c}}}}},\underline{\underline{\mathbf{h}}}-\text{Symm.}}{=} -(\nabla'\underline{\underline{\mathbf{G}}})^{231} : \underline{\underline{\underline{\underline{\mathbf{c}}}}} : \underline{\underline{\mathbf{h}}} \\
&= (\nabla\underline{\underline{\mathbf{G}}})^{231} : \underline{\underline{\underline{\underline{\mathbf{c}}}}} : \underline{\underline{\mathbf{h}}} \\
&\overset{\text{def}}{=} \overline{\underline{\underline{\underline{\mathbf{\Sigma}}}}} : \underline{\underline{\mathbf{h}}} \\
&= \underline{\underline{\mathbf{h}}} : \overline{\underline{\underline{\underline{\mathbf{\Sigma}}}}}^{231} \\
&\overset{\underline{\underline{\underline{\underline{\mathbf{c}}}}}-\text{Symm.}}{=} \underline{\underline{\mathbf{h}}} : \underline{\underline{\underline{\mathbf{\Sigma}}}}
\end{aligned}$$

Stress elementary wave from a point-like deformation rate source: With the stress tensor $\underline{\underline{\mathbf{T}}}(\underline{\mathbf{R}}, \omega) = \underline{\underline{\underline{\underline{\mathbf{c}}}}} : \nabla \underline{\mathbf{u}}(\underline{\mathbf{R}}, \omega)$—to be on the safer side, we initially assume $\underline{\mathbf{R}} \notin V_Q$—and $-\mathrm{j}\omega \underline{\mathbf{u}}(\underline{\mathbf{R}}, \omega) = \underline{\mathbf{v}}(\underline{\mathbf{R}}, \omega)$, we immediately calculate with the help of the third equality sign of the horizontal bracket of (13.187):

$$
\begin{aligned}
\underline{\underline{\mathbf{T}}}&(\underline{\mathbf{R}}, \omega) \\
&= \iiint_{V_Q} [\underbrace{\underline{\underline{\underline{\underline{\mathbf{c}}}}} : \nabla \underline{\underline{\mathbf{G}}}(\underline{\mathbf{R}} - \underline{\mathbf{R}}', \omega) \cdot \underline{\mathbf{f}}(\underline{\mathbf{R}}', \omega)}_{\substack{= \underline{\underline{\mathbf{\Sigma}}}(\underline{\mathbf{R}} - \underline{\mathbf{R}}', \omega) \cdot \underline{\mathbf{f}}(\underline{\mathbf{R}}', \omega) \\ = \underline{\mathbf{f}}(\underline{\mathbf{R}}', \omega) \cdot \underline{\underline{\mathbf{\Sigma}}}^{312}(\underline{\mathbf{R}}', \omega)}} - \frac{1}{\mathrm{j}\omega} \\
&\times \underbrace{\underline{\underline{\underline{\underline{\mathbf{c}}}}} : \nabla \bar{\underline{\underline{\mathbf{\Sigma}}}}(\underline{\mathbf{R}} - \underline{\mathbf{R}}', \omega) : \underline{\underline{\mathbf{h}}}(\underline{\mathbf{R}}', \omega)}_{\substack{\stackrel{\text{def}}{=} \bar{\underline{\underline{\underline{\underline{\mathbf{\Pi}}}}}}(\underline{\mathbf{R}} - \underline{\mathbf{R}}', \omega) : \underline{\underline{\mathbf{h}}}(\underline{\mathbf{R}}', \omega) \\ = \underline{\underline{\mathbf{h}}}(\underline{\mathbf{R}}', \omega) : \bar{\underline{\underline{\underline{\underline{\mathbf{\Pi}}}}}}^{3412}(\underline{\mathbf{R}} - \underline{\mathbf{R}}', \omega) \\ \stackrel{\text{def}}{=} \underline{\underline{\mathbf{h}}}(\underline{\mathbf{R}}', \omega) : \underline{\underline{\underline{\underline{\mathbf{\Pi}}}}}(\underline{\mathbf{R}} - \underline{\mathbf{R}}', \omega) \\ \stackrel{\underline{\underline{\underline{\underline{\mathbf{c}}}}}-\text{Symm.}}{=} \underline{\underline{\mathbf{h}}}(\underline{\mathbf{R}}', \omega) : [\nabla \underline{\underline{\mathbf{\Sigma}}}(\underline{\mathbf{R}} - \underline{\mathbf{R}}', \omega)]^{2314} : \underline{\underline{\underline{\underline{\mathbf{c}}}}}}}]\, \mathrm{d}^3\underline{\mathbf{R}}'
\end{aligned}
\tag{13.188}
$$

$$
= \iiint_{V_Q} \left[\underline{\underline{\mathbf{\Sigma}}}(\underline{\mathbf{R}} - \underline{\mathbf{R}}', \omega) \cdot \underline{\mathbf{f}}(\underline{\mathbf{R}}', \omega) - \frac{1}{\mathrm{j}\omega} \bar{\underline{\underline{\underline{\underline{\mathbf{\Pi}}}}}}(\underline{\mathbf{R}} - \underline{\mathbf{R}}', \omega) : \underline{\underline{\mathbf{h}}}(\underline{\mathbf{R}}', \omega) \right] \mathrm{d}^3\underline{\mathbf{R}}'
\tag{13.189}
$$

$$
= \iiint_{V_Q} \left[\underline{\mathbf{f}}(\underline{\mathbf{R}}', \omega) \cdot \underline{\underline{\mathbf{\Sigma}}}^{312}(\underline{\mathbf{R}} - \underline{\mathbf{R}}', \omega) - \frac{1}{\mathrm{j}\omega} \underline{\underline{\mathbf{h}}}(\underline{\mathbf{R}}', \omega) : \underline{\underline{\underline{\underline{\mathbf{\Pi}}}}}(\underline{\mathbf{R}} - \underline{\mathbf{R}}', \omega) \right] \mathrm{d}^3\underline{\mathbf{R}}'.
\tag{13.190}
$$

The last equality sign of the horizontal bracket of the second term in (13.188) results from conversion of

$$
\begin{aligned}
\underline{\underline{\underline{\underline{\mathbf{\Pi}}}}}(\underline{\mathbf{R}}, \omega) &= \bar{\underline{\underline{\underline{\underline{\mathbf{\Pi}}}}}}^{3412}(\underline{\mathbf{R}}, \omega) \\
&= \left[\underline{\underline{\underline{\underline{\mathbf{c}}}}} : \nabla \bar{\underline{\underline{\mathbf{\Sigma}}}}(\underline{\mathbf{R}}, \omega) \right]^{3412} \\
&= \left[\underline{\underline{\underline{\underline{\mathbf{c}}}}} : \nabla \underline{\underline{\mathbf{\Sigma}}}^{312}(\underline{\mathbf{R}}, \omega) \right]^{3412}
\end{aligned}
\tag{13.191}
$$

in cartesian coordinates (most rapidly using the summation convention) consequently exploiting the definition of the index notation (commutation of the unit vectors in the fourfold dyadic product, not the indices!). Hence, we find

$$
\underline{\underline{\underline{\underline{\mathbf{\Pi}}}}}(\underline{\mathbf{R}}, \omega) = \left[\nabla \underline{\underline{\mathbf{\Sigma}}}(\underline{\mathbf{R}}, \omega) \right]^{2314} : \underline{\underline{\underline{\underline{\mathbf{c}}}}} \ , \quad \underline{\mathbf{R}} \neq \underline{\mathbf{0}},
\tag{13.192}
$$

respectively

$$\underline{\underline{\underline{\underline{\boldsymbol{\Pi}}}}}(\underline{\mathbf{R}}-\underline{\mathbf{R}}',\omega)=\left[\nabla\underline{\underline{\underline{\boldsymbol{\Sigma}}}}(\underline{\mathbf{R}}-\underline{\mathbf{R}}',\omega)\right]^{2314}:\underline{\underline{\underline{\underline{\mathbf{c}}}}}\;,\quad \underline{\mathbf{R}}\neq\underline{\mathbf{R}}'. \tag{13.193}$$

We can prove that $\underline{\underline{\underline{\underline{\boldsymbol{\Pi}}}}}$ (in a homogeneous material) exhibits the same symmetries as the $\underline{\underline{\underline{\underline{\mathbf{c}}}}}$-tensor (de Hoop 1995):

$$\begin{aligned}\underline{\underline{\underline{\underline{\boldsymbol{\Pi}}}}}(\underline{\mathbf{R}},\omega)&=\underline{\underline{\underline{\underline{\boldsymbol{\Pi}}}}}^{2134}(\underline{\mathbf{R}},\omega)\\&=\underline{\underline{\underline{\underline{\boldsymbol{\Pi}}}}}^{1243}(\underline{\mathbf{R}},\omega)\\&=\underline{\underline{\underline{\underline{\boldsymbol{\Pi}}}}}^{3412}(\underline{\mathbf{R}},\omega).\end{aligned} \tag{13.194}$$

Therefore, we have

$$\underline{\underline{\underline{\underline{\boldsymbol{\Pi}}}}}(\underline{\mathbf{R}},\omega)=\underline{\underline{\underline{\underline{\bar{\boldsymbol{\Pi}}}}}}(\underline{\mathbf{R}},\omega). \tag{13.195}$$

With the Green tensors $\underline{\underline{\underline{\boldsymbol{\Sigma}}}}$ and $\underline{\underline{\underline{\underline{\boldsymbol{\Pi}}}}}$, we are now able to give the stress elementary waves from point-like $\underline{\hat{\mathbf{f}}}$- and $\underline{\underline{\hat{\mathbf{h}}}}$-sources:

$$\underline{\underline{\mathbf{T}}}^{\mathrm{PS}_{f,h}}(\underline{\mathbf{R}},\omega)=\underline{\hat{\mathbf{f}}}\cdot\underline{\underline{\underline{\boldsymbol{\Sigma}}}}^{312}(\underline{\mathbf{R}},\omega)+\underline{\underline{\hat{\mathbf{h}}}}:\underline{\underline{\underline{\underline{\boldsymbol{\Pi}}}}}(\underline{\mathbf{R}},\omega). \tag{13.196}$$

Assuming a homogeneous isotropic material, we may insert (7.18) into (13.192):

$$\begin{aligned}\underline{\underline{\underline{\underline{\boldsymbol{\Pi}}}}}(\underline{\mathbf{R}},\omega)&=\lambda\nabla\cdot\underline{\underline{\underline{\boldsymbol{\Sigma}}}}^{312}(\underline{\mathbf{R}},\omega)\,\underline{\underline{\mathbf{I}}}\\&\quad+\mu\left\{\left[\nabla\underline{\underline{\underline{\boldsymbol{\Sigma}}}}(\underline{\mathbf{R}},\omega)\right]^{2314}+\left[\nabla\underline{\underline{\underline{\boldsymbol{\Sigma}}}}(\underline{\mathbf{R}},\omega)\right]^{2341}\right\},\quad \underline{\mathbf{R}}\neq\underline{\mathbf{0}}.\end{aligned} \tag{13.197}$$

Due to (13.157), this also leads to the separation

$$\underline{\underline{\underline{\underline{\boldsymbol{\Pi}}}}}(\underline{\mathbf{R}},\omega)=\frac{1}{\lambda+2\mu}\underline{\underline{\underline{\underline{\boldsymbol{\Pi}}}}}_{\mathrm{P}}(\underline{\mathbf{R}},\omega)+\frac{1}{\mu}\underline{\underline{\underline{\underline{\boldsymbol{\Pi}}}}}_{\mathrm{S}}(\underline{\mathbf{R}},\omega) \tag{13.198}$$

with

$$\underline{\underline{\underline{\underline{\boldsymbol{\Pi}}}}}_{\mathrm{P}}(\underline{\mathbf{R}},\omega)=\lambda\nabla\cdot\underline{\underline{\underline{\boldsymbol{\Sigma}}}}_{\mathrm{P}}^{312}\,\underline{\underline{\mathbf{I}}}+\mu\left\{\left[\nabla\underline{\underline{\underline{\boldsymbol{\Sigma}}}}_{\mathrm{P}}(\underline{\mathbf{R}},\omega)\right]^{2314}+\left[\nabla\underline{\underline{\underline{\boldsymbol{\Sigma}}}}_{\mathrm{P}}(\underline{\mathbf{R}},\omega)\right]^{2341}\right\}, \tag{13.199}$$

$$\underline{\underline{\underline{\underline{\boldsymbol{\Pi}}}}}_{\mathrm{S}}(\underline{\mathbf{R}},\omega)=\mu\left\{\left[\nabla\underline{\underline{\underline{\boldsymbol{\Sigma}}}}_{\mathrm{S}}(\underline{\mathbf{R}},\omega)\right]^{2314}+\left[\nabla\underline{\underline{\underline{\boldsymbol{\Sigma}}}}_{\mathrm{S}}(\underline{\mathbf{R}},\omega)\right]^{2341}\right\}; \tag{13.200}$$

on behalf of (13.159) and $\nabla\cdot\underline{\underline{\mathbf{G}}}_{\mathrm{S}}(\underline{\mathbf{R}},\omega)\equiv\underline{\mathbf{0}}$ (Equation 13.97), we have

$$\begin{aligned}\nabla\cdot\left[\nabla\underline{\underline{\mathbf{G}}}_{\mathrm{S}}(\underline{\mathbf{R}},\omega)\right]^{312}&=\nabla\nabla\cdot\underline{\underline{\mathbf{G}}}_{\mathrm{S}}^{21}(\underline{\mathbf{R}},\omega)\\&=\nabla\left[\nabla\cdot\underline{\underline{\mathbf{G}}}_{\mathrm{S}}(\underline{\mathbf{R}},\omega)\right]\\&=\underline{\underline{\mathbf{0}}}.\end{aligned} \tag{13.201}$$

The explicit representation of (13.199) and (13.200) similar to (13.160) and (13.161) requires (for $\underline{\mathbf{R}} \neq \underline{\mathbf{0}}$) the calculation of the fourfold gradient dyadic of the scalar Green function; we avoid that and refer to de Hoop (1995). However, we give the far-field representation utilizing (13.162) and (13.163):

$$\underline{\underline{\underline{\underline{\mathbf{\Pi}}}}}_{\mathrm{P}}^{\mathrm{far}}(\underline{\mathbf{R}}, \omega) = -k_{\mathrm{P}}^2 \frac{\mathrm{e}^{\mathrm{j}k_{\mathrm{P}}R}}{4\pi R} (\lambda \underline{\underline{\mathbf{I}}} + 2\mu \hat{\underline{\mathbf{R}}}\hat{\underline{\mathbf{R}}})(\lambda \underline{\underline{\mathbf{I}}} + 2\mu \hat{\underline{\mathbf{R}}}\hat{\underline{\mathbf{R}}}), \tag{13.202}$$

$$\underline{\underline{\underline{\underline{\mathbf{\Pi}}}}}_{\mathrm{S}}^{\mathrm{far}}(\underline{\mathbf{R}}, \omega) = -k_{\mathrm{S}}^2 \mu^2 \frac{\mathrm{e}^{\mathrm{j}k_{\mathrm{S}}R}}{4\pi R} \Big(\hat{\underline{\mathbf{R}}}\hat{\underline{\mathbf{R}}} \underline{\underline{\mathbf{I}}}^{2314} + \hat{\underline{\mathbf{R}}}\hat{\underline{\mathbf{R}}} \underline{\underline{\mathbf{I}}}^{2341} + \hat{\underline{\mathbf{R}}}\hat{\underline{\mathbf{R}}} \underline{\underline{\mathbf{I}}}^{3214} + \hat{\underline{\mathbf{R}}}\hat{\underline{\mathbf{R}}} \underline{\underline{\mathbf{I}}}^{3241} - 4 \hat{\underline{\mathbf{R}}}\hat{\underline{\mathbf{R}}}\hat{\underline{\mathbf{R}}}\hat{\underline{\mathbf{R}}} \Big). \tag{13.203}$$

The symmetries (13.194) are obvious.

The distributional term of the forth rank Green tensor: In the $\underline{\underline{\underline{\underline{\mathbf{\Pi}}}}}$-representation (13.192), we assumed $\underline{\mathbf{R}} \neq \underline{\mathbf{0}}$; what happens for $\underline{\mathbf{R}} = \underline{\mathbf{0}}$?

The second rank tensor $\underline{\underline{\mathbf{G}}}_v(\underline{\mathbf{R}} - \underline{\mathbf{R}}', \omega)$ contains an explicit distributional term for $\underline{\mathbf{R}} = \underline{\mathbf{R}}'$ according to (5.67) that was found solving the differential equation (5.64); with (5.75) through (5.79), we argued for its necessity. Using similar arguments, we will give the distributional term in $\underline{\underline{\underline{\underline{\mathbf{\Pi}}}}}(\underline{\mathbf{R}} - \underline{\mathbf{R}}', \omega)$.

For $\underline{\mathbf{R}} \in V_Q$,

$$\underline{\underline{\mathbf{T}}}(\underline{\mathbf{R}}, \omega) = -\frac{1}{\mathrm{j}\omega} \underline{\underline{\underline{\underline{\mathbf{c}}}}} : \nabla \underline{\mathbf{v}}(\underline{\mathbf{R}}, \omega) - \frac{1}{\mathrm{j}\omega} \underline{\underline{\underline{\underline{\mathbf{c}}}}} : \underline{\underline{\mathbf{h}}}(\underline{\mathbf{R}}, \omega) \tag{13.204}$$

must hold. With the source field representation (13.152) for the particle velocity that is also valid for $\underline{\mathbf{R}} \in V_Q$, we obtain for $\underline{\underline{\mathbf{T}}}(\underline{\mathbf{R}}, \omega)$ according to (13.204):

$$\underline{\underline{\mathbf{T}}}(\underline{\mathbf{R}}, \omega) = \underline{\underline{\underline{\underline{\mathbf{c}}}}} : \nabla \iiint_{V_Q} \Big[\underline{\underline{\mathbf{G}}}(\underline{\mathbf{R}} - \underline{\mathbf{R}}', \omega) \cdot \underline{\mathbf{f}}(\underline{\mathbf{R}}', \omega) - \frac{1}{\mathrm{j}\omega} \underline{\underline{\underline{\bar{\mathbf{\Sigma}}}}}(\underline{\mathbf{R}} - \underline{\mathbf{R}}', \omega) : \underline{\underline{\mathbf{h}}}(\underline{\mathbf{R}}', \omega) \Big] \, \mathrm{d}^3\underline{\mathbf{R}}' - \frac{1}{\mathrm{j}\omega} \underline{\underline{\underline{\underline{\mathbf{c}}}}} : \underline{\underline{\mathbf{h}}}(\underline{\mathbf{R}}, \omega). \tag{13.205}$$

Knowing that $\underline{\underline{\mathbf{G}}}(\underline{\mathbf{R}} - \underline{\mathbf{R}}', \omega)$, and hence $\underline{\underline{\underline{\bar{\mathbf{\Sigma}}}}}(\underline{\mathbf{R}} - \underline{\mathbf{R}}', \omega)$, does not exhibit hyper singularities for $\underline{\mathbf{R}} = \underline{\mathbf{R}}'$ (Equations 13.91 and 13.156), we shift $\underline{\underline{\underline{\underline{\mathbf{c}}}}} : \nabla$ under the integral:

$$\underline{\underline{\mathbf{T}}}(\underline{\mathbf{R}}, \omega) = \iiint_{V_Q} \Big[\underline{\underline{\underline{\underline{\mathbf{c}}}}} : \nabla \underline{\underline{\mathbf{G}}}(\underline{\mathbf{R}} - \underline{\mathbf{R}}', \omega) \cdot \underline{\mathbf{f}}(\underline{\mathbf{R}}', \omega) - \frac{1}{\mathrm{j}\omega} \underline{\underline{\underline{\underline{\mathbf{c}}}}} : \nabla \underline{\underline{\underline{\bar{\mathbf{\Sigma}}}}}(\underline{\mathbf{R}} - \underline{\mathbf{R}}', \omega) : \underline{\underline{\mathbf{h}}}(\underline{\mathbf{R}}', \omega) \Big] \, \mathrm{d}^3\underline{\mathbf{R}}' - \frac{1}{\mathrm{j}\omega} \underline{\underline{\underline{\underline{\mathbf{c}}}}} : \underline{\underline{\mathbf{h}}}(\underline{\mathbf{R}}, \omega). \tag{13.206}$$

For $\underline{\mathbf{R}} \in V_Q$, we may add the $\underline{\underline{\underline{\underline{\mathbf{c}}}}} : \underline{\underline{\mathbf{h}}}$-term in (13.206) outside the integral according to

$$\underline{\underline{\mathbf{T}}}(\underline{\mathbf{R}}, \omega) = \iiint_{V_Q} \Bigg\{ \underbrace{\underline{\underline{\underline{\underline{\mathbf{c}}}}} : \nabla \underline{\underline{\mathbf{G}}}(\underline{\mathbf{R}} - \underline{\mathbf{R}}', \omega)}_{= \underline{\underline{\underline{\mathbf{\Sigma}}}}(\underline{\mathbf{R}} - \underline{\mathbf{R}}', \omega)} \cdot \underline{\mathbf{f}}(\underline{\mathbf{R}}', \omega) - \frac{1}{j\omega} \Big[\underline{\underline{\underline{\underline{\mathbf{c}}}}} : \nabla \underline{\underline{\underline{\mathbf{\Sigma}}}}(\underline{\mathbf{R}} - \underline{\mathbf{R}}', \omega) + \delta(\underline{\mathbf{R}} - \underline{\mathbf{R}}') \underline{\underline{\underline{\underline{\mathbf{c}}}}} \Big] : \underline{\underline{\mathbf{h}}}(\underline{\mathbf{R}}', \omega) \Bigg\} \, d^3\underline{\mathbf{R}}' \tag{13.207}$$

to a tensor

$$\begin{aligned} \underline{\underline{\underline{\underline{\bar{\mathbf{\Pi}}}}}}(\underline{\mathbf{R}} - \underline{\mathbf{R}}', \omega) &= \underline{\underline{\underline{\underline{\mathbf{\Pi}}}}}(\underline{\mathbf{R}} - \underline{\mathbf{R}}', \omega) \\ &= \underline{\underline{\underline{\underline{\mathbf{c}}}}} : \nabla \underline{\underline{\underline{\bar{\mathbf{\Sigma}}}}}(\underline{\mathbf{R}} - \underline{\mathbf{R}}', \omega) + \underline{\underline{\underline{\underline{\mathbf{c}}}}}\, \delta(\underline{\mathbf{R}} - \underline{\mathbf{R}}') \\ &= \Big[\nabla \underline{\underline{\underline{\mathbf{\Sigma}}}}(\underline{\mathbf{R}} - \underline{\mathbf{R}}', \omega) \Big]^{2314} : \underline{\underline{\underline{\underline{\mathbf{c}}}}} + \underline{\underline{\underline{\underline{\mathbf{c}}}}}\, \delta(\underline{\mathbf{R}} - \underline{\mathbf{R}}') \end{aligned} \tag{13.208}$$

that replaces (13.193) if $\underline{\mathbf{R}} = \underline{\mathbf{R}}'$.

On the other hand, if we start from—similar to (5.80) through (5.81)—

$$\underline{\mathbf{v}}(\underline{\mathbf{R}}, \omega) = -\frac{1}{j\omega\rho} \nabla \cdot \underline{\underline{\mathbf{T}}}(\underline{\mathbf{R}}, \omega) - \frac{1}{j\omega\rho} \underline{\mathbf{f}}(\underline{\mathbf{R}}, \omega) \tag{13.209}$$

and insert (13.189) with (13.208) for $\underline{\mathbf{R}} \in V_Q$, the δ-term in (13.210) compensates[238] the $\underline{\mathbf{f}}$-term in (13.209) due to

$$\nabla \cdot \underline{\underline{\underline{\mathbf{\Sigma}}}}(\underline{\mathbf{R}} - \underline{\mathbf{R}}', \omega) + \rho\omega^2 \underline{\underline{\mathbf{G}}}(\underline{\mathbf{R}} - \underline{\mathbf{R}}', \omega) = -\delta(\underline{\mathbf{R}} - \underline{\mathbf{R}}')\, \underline{\underline{\mathbf{I}}}; \tag{13.210}$$

furthermore, $\nabla \cdot \underline{\underline{\underline{\underline{\bar{\mathbf{\Pi}}}}}}(\underline{\mathbf{R}} - \underline{\mathbf{R}}', \omega)$ leads to a distributional term due to (13.208) that is equally compensated via insertion of $\underline{\underline{\underline{\bar{\mathbf{\Sigma}}}}}(\underline{\mathbf{R}} - \underline{\mathbf{R}}', \omega)$, respectively $\underline{\underline{\mathbf{G}}}(\underline{\mathbf{R}} - \underline{\mathbf{R}}', \omega)$, and recognizing the differential equation for $\underline{\underline{\mathbf{G}}}(\underline{\mathbf{R}} - \underline{\mathbf{R}}', \omega)$.[238] Consequently, the representation (13.152) for $\underline{\mathbf{R}} \in V_Q$ emerges from (13.209) with (13.189) and (13.208).

Differential equation for the fourth rank Green tensor: From the governing equations of acoustics for homogeneous materials, we derive the reduced wave equations (5.37) and (5.39) for the Fourier transformed field quantities

[238] This is nothing but the differential equation (13.80), where $\underline{\underline{\underline{\underline{\mathbf{c}}}}} = \lambda \underline{\underline{\mathbf{I}}}^{\delta} + 2\mu \underline{\underline{\underline{\underline{\mathbf{I}}}}}^{+}$ has not yet been inserted:

$$\nabla \cdot \underline{\underline{\underline{\underline{\mathbf{c}}}}} : \nabla \underline{\underline{\mathbf{G}}}(\underline{\mathbf{R}} - \underline{\mathbf{R}}', \omega) + \rho\omega^2 \underline{\underline{\mathbf{G}}}(\underline{\mathbf{R}} - \underline{\mathbf{R}}', \omega) = -\delta(\underline{\mathbf{R}} - \underline{\mathbf{R}}')\, \underline{\underline{\mathbf{I}}}.$$

$p(\underline{\mathbf{R}}, \omega)$ and $\underline{\mathbf{v}}(\underline{\mathbf{R}}, \omega)$; with (5.58) and (5.64), each of these differential equations is assigned a differential equation for Green functions $G(\underline{\mathbf{R}} - \underline{\mathbf{R}}', \omega)$ and $\underline{\underline{\mathbf{G}}}_v(\underline{\mathbf{R}} - \underline{\mathbf{R}}', \omega)$ (also compare Figure 5.1). Here, in elastodynamics, the field quantities $\underline{\mathbf{v}}(\underline{\mathbf{R}}, \omega)$ and $\underline{\underline{\mathbf{T}}}(\underline{\mathbf{R}}, \omega)$ are under concern; the differential equation (13.78) for $\underline{\mathbf{v}}(\underline{\mathbf{R}}, \omega)$ is assigned the differential equation (13.80) for the Green function $\underline{\underline{\mathbf{G}}}(\underline{\mathbf{R}} - \underline{\mathbf{R}}', \omega)$, and Figure 13.15 illustrates that in fact $\underline{\underline{\underline{\bar{\mathbf{\Pi}}}}}(\underline{\mathbf{R}} - \underline{\mathbf{R}}', \omega)$, respectively $\underline{\underline{\underline{\mathbf{\Pi}}}}(\underline{\mathbf{R}} - \underline{\mathbf{R}}', \omega)$, should satisfy a differential equation assigned to the Fourier transformed equation (13.76). We double contract (13.76) with $\underline{\underline{\underline{\underline{\mathbf{c}}}}}$ according to

$$\underline{\underline{\underline{\underline{\mathbf{c}}}}} : \boldsymbol{\nabla}\left[\boldsymbol{\nabla} \cdot \underline{\underline{\mathbf{T}}}(\underline{\mathbf{R}}, \omega)\right] + \rho\omega^2 \underline{\underline{\mathbf{T}}}(\underline{\mathbf{R}}, \omega) = -\underline{\underline{\underline{\underline{\mathbf{c}}}}} : \boldsymbol{\nabla}\underline{\mathbf{f}}(\underline{\mathbf{R}}, \omega) + \mathrm{j}\omega\rho\, \underline{\underline{\underline{\underline{\mathbf{c}}}}} : \underline{\underline{\mathbf{h}}}(\underline{\mathbf{R}}, \omega) \tag{13.211}$$

and claim that $\underline{\underline{\underline{\mathbf{\Pi}}}}(\underline{\mathbf{R}} - \underline{\mathbf{R}}', \omega)$ [respectively $\underline{\underline{\underline{\bar{\mathbf{\Pi}}}}}(\underline{\mathbf{R}} - \underline{\mathbf{R}}', \omega)$] according to (13.208) is a solution of the differential equation

$$\underline{\underline{\underline{\underline{\mathbf{c}}}}} : \boldsymbol{\nabla}\left[\boldsymbol{\nabla} \cdot \underline{\underline{\underline{\mathbf{\Pi}}}}(\underline{\mathbf{R}} - \underline{\mathbf{R}}', \omega)\right] + \rho\omega^2 \underline{\underline{\underline{\mathbf{\Pi}}}}(\underline{\mathbf{R}} - \underline{\mathbf{R}}', \omega) = \rho\omega^2 \underline{\underline{\underline{\underline{\mathbf{c}}}}}\, \delta(\underline{\mathbf{R}} - \underline{\mathbf{R}}'). \tag{13.212}$$

This is comparatively easy to prove inserting and utilizing the three-dimensional Fourier transform ($\boldsymbol{\nabla} \Longrightarrow \mathrm{j}\underline{\mathbf{K}}$).

13.3 Two- and Three-Dimensional Elastodynamic Source Fields

13.3.1 Elastodynamic point source synthesis

With the integral representations (13.152) and (13.190), we formulate the point source synthesis for elastodynamic source fields in homogeneous isotropic materials:

$$\begin{aligned}\underline{\mathbf{v}}(\underline{\mathbf{R}}, \omega) = \iiint_{V_Q} \Big[&- \mathrm{j}\omega \underline{\mathbf{f}}(\underline{\mathbf{R}}', \omega) \cdot \underline{\underline{\mathbf{G}}}(\underline{\mathbf{R}} - \underline{\mathbf{R}}', \omega) \\ &+ \underline{\underline{\mathbf{h}}}(\underline{\mathbf{R}}', \omega) : \underline{\underline{\underline{\mathbf{\Sigma}}}}(\underline{\mathbf{R}} - \underline{\mathbf{R}}', \omega)\Big]\, \mathrm{d}^3\underline{\mathbf{R}}',\end{aligned} \tag{13.213}$$

$$\begin{aligned}\underline{\underline{\mathbf{T}}}(\underline{\mathbf{R}}, \omega) = \iiint_{V_Q} \Big[&\underline{\mathbf{f}}(\underline{\mathbf{R}}', \omega) \cdot \underline{\underline{\underline{\mathbf{\Sigma}}}}^{312}(\underline{\mathbf{R}}', \omega) \\ &- \frac{1}{\mathrm{j}\omega} \underline{\underline{\mathbf{h}}}(\underline{\mathbf{R}}', \omega) : \underline{\underline{\underline{\underline{\mathbf{\Pi}}}}}(\underline{\mathbf{R}} - \underline{\mathbf{R}}', \omega)\Big]\, \mathrm{d}^3\underline{\mathbf{R}}'.\end{aligned} \tag{13.214}$$

The amplitude and phase distributions of force density sources $\underline{\mathbf{f}}(\underline{\mathbf{R}}', \omega)$ and deformation rate sources $\underline{\underline{\mathbf{h}}}(\underline{\mathbf{R}}', \omega)$ in the interior of a source volume V_Q tune the elastodynamic elementary waves emanating from each source point $\underline{\mathbf{R}}'$ as given mathematically by the Green tensors $\underline{\underline{\mathbf{G}}}(\underline{\mathbf{R}} - \underline{\mathbf{R}}', \omega)$, $\underline{\underline{\underline{\mathbf{\Sigma}}}}(\underline{\mathbf{R}} - \underline{\mathbf{R}}', \omega)$ and $\underline{\underline{\underline{\underline{\mathbf{\Pi}}}}}(\underline{\mathbf{R}} - \underline{\mathbf{R}}', \omega)$; particle velocities and stresses of given sources become calculable for $\underline{\mathbf{R}} \in \mathbb{R}^3$, i.e., in the interior as well as the exterior of V_Q using (13.213) and (13.214).

13.3.2 Far-field approximations of three-dimensional elastodynamic source fields

Pressure and shear waves radiation characteristics of force density and deformation rate sources: Be a the maximum linear dimension of V_Q; then we obtain the far-field approximations of the source fields for $R \gg a$ and $k_{\mathrm{P,S}}R \gg 1$ with Equation 13.95 and the far-field approximations (13.100) and (13.101), with Equation 13.157 and the far-field approximations (13.162) and (13.163), as well as Equation 13.198 and the far-field approximations (13.202) and (13.203) of the Green tensors; these may be separated into primary and secondary terms. At first, we discuss the primary terms[239]

$$\underline{\mathbf{v}}_{\mathrm{P}}^{\mathrm{far}}(\underline{\mathbf{R}}, \omega) = -\mathrm{j}\omega \frac{1}{\lambda + 2\mu} \iiint_{V_Q} \underline{\mathbf{f}}(\underline{\mathbf{R}}', \omega) \cdot \underline{\underline{\mathbf{G}}}_{\mathrm{P}}^{\mathrm{far}}(\underline{\mathbf{R}}, \underline{\mathbf{R}}', \omega)\, \mathrm{d}^3\underline{\mathbf{R}}'$$

$$+ \frac{1}{\lambda + 2\mu} \iiint_{V_Q} \underline{\underline{\mathbf{h}}}(\underline{\mathbf{R}}', \omega) : \underline{\underline{\underline{\mathbf{\Sigma}}}}_{\mathrm{P}}^{\mathrm{far}}(\underline{\mathbf{R}}, \underline{\mathbf{R}}', \omega)\, \mathrm{d}^3\underline{\mathbf{R}}', \qquad (13.215)$$

$$\underline{\underline{\mathbf{T}}}_{\mathrm{P}}^{\mathrm{far}}(\underline{\mathbf{R}}, \omega) = \frac{1}{\lambda + 2\mu} \iiint_{V_Q} \underline{\mathbf{f}}(\underline{\mathbf{R}}', \omega) \cdot \underline{\underline{\underline{\mathbf{\Sigma}}}}_{\mathrm{P}}^{312\,\mathrm{far}}(\underline{\mathbf{R}}, \underline{\mathbf{R}}', \omega)\, \mathrm{d}^3\underline{\mathbf{R}}'$$

$$- \frac{1}{\mathrm{j}\omega} \frac{1}{\lambda + 2\mu} \iiint_{V_Q} \underline{\underline{\mathbf{h}}}(\underline{\mathbf{R}}', \omega) : \underline{\underline{\underline{\underline{\mathbf{\Pi}}}}}_{\mathrm{P}}^{\mathrm{far}}(\underline{\mathbf{R}}, \underline{\mathbf{R}}', \omega)\, \mathrm{d}^3\underline{\mathbf{R}}' \quad (13.216)$$

and write (13.215) in the particle displacement

$$\underline{\mathbf{u}}_{\mathrm{P}}^{\mathrm{far}}(\underline{\mathbf{R}}, \omega) = \frac{\mathrm{e}^{\mathrm{j}k_{\mathrm{P}}R}}{R} \left[\underline{\mathbf{H}}_{\mathrm{P}}^{f}(\hat{\underline{\mathbf{R}}}, \omega) + \underline{\mathbf{H}}_{\mathrm{P}}^{h}(\hat{\underline{\mathbf{R}}}, \omega) \right]$$

$$= \frac{\mathrm{e}^{\mathrm{j}k_{\mathrm{P}}R}}{R} \left[H_{\mathrm{P}}^{f}(\hat{\underline{\mathbf{R}}}, \omega) + H_{\mathrm{P}}^{h}(\hat{\underline{\mathbf{R}}}, \omega) \right] \hat{\underline{\mathbf{R}}} \qquad (13.217)$$

with the short-hand notations

$$\underline{\mathbf{H}}_{\mathrm{P}}^{f}(\hat{\underline{\mathbf{R}}}, \omega) = \underbrace{\frac{1}{4\pi c_{\mathrm{P}} Z_{\mathrm{P}}} \iiint_{V_Q} \underline{\mathbf{f}}(\underline{\mathbf{R}}', \omega) \cdot \hat{\underline{\mathbf{R}}}\, \mathrm{e}^{-\mathrm{j}k_{\mathrm{P}}\hat{\underline{\mathbf{R}}}\cdot\underline{\mathbf{R}}'}\, \mathrm{d}^3\underline{\mathbf{R}}'}_{= H_{\mathrm{P}}^{f}(\hat{\underline{\mathbf{R}}}, \omega)}\ \hat{\underline{\mathbf{R}}}, \qquad (13.218)$$

[239]Due to the argument $\underline{\mathbf{R}} - \underline{\mathbf{R}}'$ of the Green tensors in (13.213) and (13.214), and last but not least the insertion of (13.41), we have to write "$\underline{\mathbf{R}}$ comma $\underline{\mathbf{R}}'$" as argument in the far-field approximation.

$$\underline{\mathbf{H}}_{\mathrm{P}}^{h}(\hat{\underline{\mathbf{R}}},\omega) = \underbrace{-\frac{1}{4\pi c_{\mathrm{P}}^2 Z_{\mathrm{P}}} \iiint_{V_Q} \underline{\underline{\mathbf{h}}}(\underline{\mathbf{R}}',\omega) : (\lambda \underline{\underline{\mathbf{I}}} + 2\mu \hat{\underline{\mathbf{R}}}\hat{\underline{\mathbf{R}}})\, \mathrm{e}^{-\mathrm{j}k_{\mathrm{P}}\hat{\underline{\mathbf{R}}}\cdot\underline{\mathbf{R}}'}\, \mathrm{d}^3\underline{\mathbf{R}}'}_{= H_{\mathrm{P}}^{h}(\hat{\underline{\mathbf{R}}},\omega)}\ \hat{\underline{\mathbf{R}}}. \tag{13.219}$$

The functions $H_{\mathrm{P}}^{f}(\hat{\underline{\mathbf{R}}},\omega)$ and $H_{\mathrm{P}}^{h}(\hat{\underline{\mathbf{R}}},\omega)$ are scalar radiation characteristics of the source distributions that completely describe the particle velocity field of the primary wave (not only in the far-field: Colton and Kress 1983). With

$$\underline{\mathbf{H}}_{\mathrm{P}}^{f,h}(\hat{\underline{\mathbf{R}}},\omega) \cdot (\underline{\underline{\mathbf{I}}} - \hat{\underline{\mathbf{R}}}\hat{\underline{\mathbf{R}}}) = \underline{\mathbf{0}}, \tag{13.220}$$

the primary wave identifies itself as a longitudinal pressure wave. Nota bene: The primary displacement source field is longitudinally polarized only in the far-field!

For the stress tensor (13.216), we obtain the explicit representation

$$\underline{\underline{\mathbf{T}}}_{\mathrm{P}}^{\mathrm{far}}(\underline{\mathbf{R}},\omega) = \mathrm{j}k_{\mathrm{P}} \left[H_{\mathrm{P}}^{f}(\hat{\underline{\mathbf{R}}},\omega) + H_{\mathrm{P}}^{h}(\hat{\underline{\mathbf{R}}},\omega)\right] \frac{\mathrm{e}^{\mathrm{j}k_{\mathrm{P}}R}}{R} (\lambda \underline{\underline{\mathbf{I}}} + 2\mu \hat{\underline{\mathbf{R}}}\hat{\underline{\mathbf{R}}}). \tag{13.221}$$

The comparison of (13.217) with (8.82) and (13.221) with (8.121) reveals the local plane wave behavior of the primary far-field of elastodynamic source fields. We have already stated that for the special case of point-like $\underline{\mathbf{f}}$- and $\underline{\underline{\mathbf{h}}}$-sources.

With satisfaction, we note that: The necessarily prescribed path via the formalism of Green functions (we admit: not quite unlabored) has provided us with relatively simple expressions for the radiation far-field of elastodynamic source fields that are on one hand "close" to plane waves and, on the other hand "close" to acoustic source fields (Equation 13.65). In contrast to the purely scalar h-source of the acoustic pressure in (13.65), the $\underline{\mathbf{f}}$-source appears in the P-displacement according to (13.218) with a point directivity (compare Figure 13.12), by the way, as already observed for the $\underline{\mathbf{f}}$-source of the acoustic pressure in (13.65). The corresponding point directivity for the $\underline{\underline{\mathbf{h}}}$-source of the P-displacement in (13.219) was already discussed in connection with Figure 13.14.

Now we turn to the S-term of the source field; the counterparts to (13.215) and (13.216) read

$$\begin{aligned}\underline{\mathbf{v}}_{\mathrm{S}}^{\mathrm{far}}(\underline{\mathbf{R}},\omega) = {} & -\mathrm{j}\omega\frac{1}{\mu}\iiint_{V_Q} \underline{\mathbf{f}}(\underline{\mathbf{R}}',\omega)\cdot\underline{\underline{\mathbf{G}}}_{\mathrm{S}}^{\mathrm{far}}(\underline{\mathbf{R}},\underline{\mathbf{R}}',\omega)\,\mathrm{d}^3\underline{\mathbf{R}}' \\ & + \frac{1}{\mu}\iiint_{V_Q} \underline{\underline{\mathbf{h}}}(\underline{\mathbf{R}}',\omega) : \underline{\underline{\boldsymbol{\Sigma}}}_{\mathrm{S}}^{\mathrm{far}}(\underline{\mathbf{R}},\underline{\mathbf{R}}',\omega)\,\mathrm{d}^3\underline{\mathbf{R}}',\end{aligned} \tag{13.222}$$

$$\begin{aligned}\underline{\underline{\mathbf{T}}}_{\mathrm{S}}^{\mathrm{far}}(\underline{\mathbf{R}},\omega) = {} & \frac{1}{\mu}\iiint_{V_Q} \underline{\mathbf{f}}(\underline{\mathbf{R}}',\omega)\cdot\underline{\underline{\underline{\boldsymbol{\Sigma}}}}_{\mathrm{S}}^{312\,\mathrm{far}}(\underline{\mathbf{R}},\underline{\mathbf{R}}',\omega)\,\mathrm{d}^3\underline{\mathbf{R}}' \\ & - \frac{1}{\mathrm{j}\omega}\frac{1}{\mu}\iiint_{V_Q} \underline{\underline{\mathbf{h}}}(\underline{\mathbf{R}}',\omega) : \underline{\underline{\underline{\underline{\boldsymbol{\Pi}}}}}_{\mathrm{S}}^{\mathrm{far}}(\underline{\mathbf{R}},\underline{\mathbf{R}}',\omega)\,\mathrm{d}^3\underline{\mathbf{R}}'.\end{aligned} \tag{13.223}$$

With (13.101) and (13.163), we obtain for the particle velocity similar to (13.217):

$$\begin{aligned}\underline{\mathbf{u}}_\mathrm{S}^\mathrm{far}(\underline{\mathbf{R}},\omega) &= \frac{\mathrm{e}^{\mathrm{j}k_\mathrm{S}R}}{R}\left[\underline{\mathbf{H}}_\mathrm{S}^f(\hat{\underline{\mathbf{R}}},\omega)+\underline{\mathbf{H}}_\mathrm{S}^h(\hat{\underline{\mathbf{R}}},\omega)\right] \\ &= \underbrace{\frac{\mathrm{e}^{\mathrm{j}k_\mathrm{S}R}}{R}\left[H_{\mathrm{S}\vartheta}^f(\hat{\underline{\mathbf{R}}},\omega)+H_{\mathrm{S}\vartheta}^h(\hat{\underline{\mathbf{R}}},\omega)\right]}_{=u_{\mathrm{S}\vartheta}^\mathrm{far}(\underline{\mathbf{R}},\omega)}\underline{\mathbf{e}}_\vartheta \\ &\quad + \underbrace{\frac{\mathrm{e}^{\mathrm{j}k_\mathrm{S}R}}{R}\left[H_{\mathrm{S}\varphi}^f(\hat{\underline{\mathbf{R}}},\omega)+H_{\mathrm{S}\varphi}^h(\hat{\underline{\mathbf{R}}},\omega)\right]}_{=u_{\mathrm{S}\varphi}^\mathrm{far}(\underline{\mathbf{R}},\omega)}\underline{\mathbf{e}}_\varphi\end{aligned} \tag{13.224}$$

with the short-hand notations

$$\begin{aligned}\underline{\mathbf{H}}_\mathrm{S}^f(\hat{\underline{\mathbf{R}}},\omega) &= \frac{1}{4\pi c_\mathrm{S} Z_\mathrm{S}}\iiint_{V_Q}\underline{\mathbf{f}}(\underline{\mathbf{R}}',\omega)\,\mathrm{e}^{-\mathrm{j}k_\mathrm{S}\hat{\underline{\mathbf{R}}}\cdot\underline{\mathbf{R}}'}\,\mathrm{d}^3\underline{\mathbf{R}}'\cdot(\underline{\underline{\mathbf{I}}}-\hat{\underline{\mathbf{R}}}\hat{\underline{\mathbf{R}}}) \\ &= \frac{1}{4\pi c_\mathrm{S} Z_\mathrm{S}}\iiint_{V_Q}\underline{\mathbf{f}}(\underline{\mathbf{R}}',\omega)\,\mathrm{e}^{-\mathrm{j}k_\mathrm{S}\hat{\underline{\mathbf{R}}}\cdot\underline{\mathbf{R}}'}\,\mathrm{d}^3\underline{\mathbf{R}}'\cdot(\underline{\mathbf{e}}_\vartheta\underline{\mathbf{e}}_\vartheta+\underline{\mathbf{e}}_\varphi\underline{\mathbf{e}}_\varphi) \\ &= \underbrace{\frac{1}{4\pi c_\mathrm{S} Z_\mathrm{S}}\iiint_{V_Q}\underline{\mathbf{f}}(\underline{\mathbf{R}}',\omega)\cdot\underline{\mathbf{e}}_\vartheta\,\mathrm{e}^{-\mathrm{j}k_\mathrm{S}\hat{\underline{\mathbf{R}}}\cdot\underline{\mathbf{R}}'}\,\mathrm{d}^3\underline{\mathbf{R}}'}_{=H_{\mathrm{S}\vartheta}^f(\hat{\underline{\mathbf{R}}},\omega)}\,\underline{\mathbf{e}}_\vartheta \\ &\quad + \underbrace{\frac{1}{4\pi c_\mathrm{S} Z_\mathrm{S}}\iiint_{V_Q}\underline{\mathbf{f}}(\underline{\mathbf{R}}',\omega)\cdot\underline{\mathbf{e}}_\varphi\,\mathrm{e}^{-\mathrm{j}k_\mathrm{S}\hat{\underline{\mathbf{R}}}\cdot\underline{\mathbf{R}}'}\,\mathrm{d}^3\underline{\mathbf{R}}'}_{=H_{\mathrm{S}\varphi}^f(\hat{\underline{\mathbf{R}}},\omega)}\,\underline{\mathbf{e}}_\varphi,\end{aligned} \tag{13.225}$$

$$\begin{aligned}\underline{\mathbf{H}}_\mathrm{S}^h(\hat{\underline{\mathbf{R}}},\omega) &= -\frac{1}{2\pi c_\mathrm{S}}\iiint_{V_Q}\underline{\underline{\mathbf{h}}}(\underline{\mathbf{R}}',\omega)\cdot\hat{\underline{\mathbf{R}}}\,\mathrm{e}^{-\mathrm{j}k_\mathrm{S}\hat{\underline{\mathbf{R}}}\cdot\underline{\mathbf{R}}'}\,\mathrm{d}^3\underline{\mathbf{R}}'\cdot(\underline{\underline{\mathbf{I}}}-\hat{\underline{\mathbf{R}}}\hat{\underline{\mathbf{R}}}) \\ &= \underbrace{-\frac{1}{2\pi c_\mathrm{S}}\iiint_{V_Q}\underline{\underline{\mathbf{h}}}(\underline{\mathbf{R}}',\omega):\hat{\underline{\mathbf{R}}}\,\underline{\mathbf{e}}_\vartheta\,\mathrm{e}^{-\mathrm{j}k_\mathrm{S}\hat{\underline{\mathbf{R}}}\cdot\underline{\mathbf{R}}'}\,\mathrm{d}^3\underline{\mathbf{R}}'}_{=H_{\mathrm{S}\vartheta}^h(\hat{\underline{\mathbf{R}}},\omega)}\,\underline{\mathbf{e}}_\vartheta \\ &\quad - \underbrace{-\frac{1}{2\pi c_\mathrm{S}}\iiint_{V_Q}\underline{\underline{\mathbf{h}}}(\underline{\mathbf{R}}',\omega):\hat{\underline{\mathbf{R}}}\,\underline{\mathbf{e}}_\varphi\,\mathrm{e}^{-\mathrm{j}k_\mathrm{S}\hat{\underline{\mathbf{R}}}\cdot\underline{\mathbf{R}}'}\,\mathrm{d}^3\underline{\mathbf{R}}'}_{=H_{\mathrm{S}\varphi}^h(\hat{\underline{\mathbf{R}}},\omega)}\,\underline{\mathbf{e}}_\varphi.\end{aligned} \tag{13.226}$$

As expected, the integrals of the scalar S-radiation characteristics exhibit the S-point directivities (13.126) and (13.127) of the $\underline{\mathbf{f}}$-source and the S-point directivities (13.184) and (13.185) of the $\underline{\underline{\mathbf{h}}}$-source, where we must absolutely consider that the unit vectors $\underline{\mathbf{e}}_\vartheta$ and $\underline{\mathbf{e}}_\varphi$ belong to the orthogonal trihedron related to the observation point $\underline{\mathbf{R}}$ and not to the trihedron $\underline{\mathbf{e}}_{R'}, \underline{\mathbf{e}}_{\vartheta'}, \underline{\mathbf{e}}_{\varphi'}$ of the source point. Sketches of pertinent radiation patterns can be found in the

Figures 13.12 and 13.14, when cartesian $\underline{\mathbf{e}}_{x_i}$-, respectively, $\underline{\mathbf{e}}_{x_k}\underline{\mathbf{e}}_{x_l}$-components of the point-like sources are prescribed; if we equally decompose the extended sources $\underline{\mathbf{f}}(\underline{\mathbf{R}}', \omega)$ and $\underline{\underline{\mathbf{h}}}(\underline{\mathbf{R}}', \omega)$ according to

$$\underline{\mathbf{f}}(\underline{\mathbf{R}}', \omega) = f_{x_i}(\underline{\mathbf{R}}', \omega)\, \underline{\mathbf{e}}_{x_i}, \tag{13.227}$$

$$\underline{\underline{\mathbf{h}}}(\underline{\mathbf{R}}', \omega) = h_{x_k x_l}(\underline{\mathbf{R}}', \omega)\, \underline{\mathbf{e}}_{x_k}\underline{\mathbf{e}}_{x_l} \tag{13.228}$$

(using the summation convention), the scalar products $\underline{\mathbf{e}}_{x_i} \cdot \underline{\mathbf{e}}_{\vartheta}, \underline{\mathbf{e}}_{x_i} \cdot \underline{\mathbf{e}}_{\varphi}$, $\underline{\mathbf{e}}_{x_k}\underline{\mathbf{e}}_{x_l} : \hat{\underline{\mathbf{R}}}\,\underline{\mathbf{e}}_{\vartheta}, \underline{\mathbf{e}}_{x_k}\underline{\mathbf{e}}_{x_l} : \hat{\underline{\mathbf{R}}}\,\underline{\mathbf{e}}_{\varphi}$ $(i, k, l = 1, 2, 3)$ as point directivities of the cartesian components of the sources may come out of the integrals because they are equal for each source point $\underline{\mathbf{R}}'$ (the cartesian trihedron does not depend upon position). We obtain (using the summation convention):

$$H^f_{\mathrm{S}\vartheta,\varphi}(\hat{\underline{\mathbf{R}}}, \omega) = \frac{1}{4\pi c_{\mathrm{S}} Z_{\mathrm{S}}}\, \underline{\mathbf{e}}_{x_i} \cdot \underline{\mathbf{e}}_{\vartheta,\varphi} \iiint_{V_Q} f_{x_i}(\underline{\mathbf{R}}', \omega)\, \mathrm{e}^{-\mathrm{j}k_{\mathrm{S}}\hat{\underline{\mathbf{R}}}\cdot\underline{\mathbf{R}}'}\, \mathrm{d}^3\underline{\mathbf{R}}', \tag{13.229}$$

$$H^h_{\mathrm{S}\vartheta,\varphi}(\hat{\underline{\mathbf{R}}}, \omega) = -\frac{1}{2\pi c_{\mathrm{S}}}\, \underline{\mathbf{e}}_{x_i}\underline{\mathbf{e}}_{x_j} : \hat{\underline{\mathbf{R}}}\,\underline{\mathbf{e}}_{\vartheta,\varphi} \iiint_{V_Q} h_{x_i x_j}(\underline{\mathbf{R}}', \omega)\, \mathrm{e}^{-\mathrm{j}k_{\mathrm{S}}\hat{\underline{\mathbf{R}}}\cdot\underline{\mathbf{R}}'}\, \mathrm{d}^3\underline{\mathbf{R}}' \tag{13.230}$$

and similarly

$$H^f_{\mathrm{P}}(\hat{\underline{\mathbf{R}}}, \omega) = \frac{1}{4\pi c_{\mathrm{P}} Z_{\mathrm{P}}}\, \underline{\mathbf{e}}_{x_i} \cdot \hat{\underline{\mathbf{R}}} \iiint_{V_Q} f_{x_i}(\underline{\mathbf{R}}', \omega)\, \mathrm{e}^{-\mathrm{j}k_{\mathrm{P}}\hat{\underline{\mathbf{R}}}\cdot\underline{\mathbf{R}}'}\, \mathrm{d}^3\underline{\mathbf{R}}', \tag{13.231}$$

$$H^h_{\mathrm{P}}(\hat{\underline{\mathbf{R}}}, \omega) = -\frac{1}{4\pi c_{\mathrm{P}}^2 Z_{\mathrm{P}}}\, \underline{\mathbf{e}}_{x_i}\underline{\mathbf{e}}_{x_j} : (\lambda\underline{\underline{\mathbf{I}}} + 2\mu\hat{\underline{\mathbf{R}}}\,\hat{\underline{\mathbf{R}}}) \times \iiint_{V_Q} h_{x_i x_j}(\underline{\mathbf{R}}', \omega)\, \mathrm{e}^{-\mathrm{j}k_{\mathrm{P}}\hat{\underline{\mathbf{R}}}\cdot\underline{\mathbf{R}}'}\, \mathrm{d}^3\underline{\mathbf{R}}'. \tag{13.232}$$

Now the respective P- and S-radiation characteristic integrals look similar, and the point directivities appear as factors![240] Note: $H^f_{\mathrm{S}\vartheta,\varphi}$ and H^f_{P} consist of a sum of three and $H^h_{\mathrm{S}\vartheta,\varphi}$, H^h_{P} of a sum of nine integrals. Each of these integrals might be evaluated in those coordinates simplifying the calculation.

By the way: Even for the case of synchronous sources $\underline{\mathbf{f}}(\underline{\mathbf{R}}', \omega) = F(\omega)\underline{\mathbf{f}}(\underline{\mathbf{R}}')$, $\underline{\underline{\mathbf{h}}}(\underline{\mathbf{R}}', \omega) = H(\omega)\underline{\underline{\mathbf{h}}}(\underline{\mathbf{R}}')$, all radiation characteristics are frequency dependent, i.e., calculation of the pulsed sound field requires additional convolutions. Generally, it is not correct to impress the impulse $F(t)$ or $H(t)$ on the displacement calculated for a fixed frequency and a specific observation direction $\hat{\underline{\mathbf{R}}}$.

Based on (13.225) and (13.226), we immediately realize that

$$\underline{\mathbf{H}}^{f,h}_{\mathrm{S}}(\hat{\underline{\mathbf{R}}}, \omega) \cdot \hat{\underline{\mathbf{R}}} = 0 \tag{13.233}$$

[240] If we would like for some reason or another to decompose the vectorial $\underline{\mathbf{f}}$- and the tensorial $\underline{\underline{\mathbf{h}}}$-sources with regard to a trihedron of the source point—for example, $\underline{\mathbf{e}}_{R'}, \underline{\mathbf{e}}_{\vartheta'}, \underline{\mathbf{e}}_{\varphi'}$—the scalar products (the point directivities) $\underline{\mathbf{e}}_{R',\vartheta',\varphi} \cdot \underline{\mathbf{e}}_{R,\vartheta,\varphi}$, $\underline{\mathbf{e}}_{R',\vartheta',\varphi'}\underline{\mathbf{e}}_{R',\vartheta',\varphi'} : \underline{\mathbf{e}}_{R,\vartheta,\varphi}\underline{\mathbf{e}}_{R,\vartheta,\varphi}$ must remain under the integrals.

holds: In the far-field, shear waves from arbitrary $\underline{\mathbf{f}}$- and/or $\underline{\underline{\mathbf{h}}}$-sources are transversely polarized with regard to $\underline{\hat{\mathbf{R}}}$. This confirms that they locally behave like plane waves. This is also true for the corresponding stress tensor

$$\underline{\underline{\mathbf{T}}}_{\mathrm{S}}^{\mathrm{far}}(\underline{\mathbf{R}},\omega) = \mathrm{j}k_{\mathrm{S}}\mu \,\frac{\mathrm{e}^{\mathrm{j}k_{\mathrm{S}}R}}{R}\,\Big\{\underline{\hat{\mathbf{R}}}\Big[\underline{\mathbf{H}}_{\mathrm{S}}^{f}(\underline{\hat{\mathbf{R}}},\omega) + \underline{\mathbf{H}}_{\mathrm{S}}^{h}(\underline{\hat{\mathbf{R}}},\omega)\Big] + \Big[\underline{\mathbf{H}}_{\mathrm{S}}^{f}(\underline{\hat{\mathbf{R}}},\omega) + \underline{\mathbf{H}}_{\mathrm{S}}^{h}(\underline{\hat{\mathbf{R}}},\omega)\Big]\underline{\hat{\mathbf{R}}}\Big\} \tag{13.234}$$

if we convert according to

$$\begin{aligned}\underline{\mathbf{f}}\cdot(\underline{\hat{\mathbf{R}}}\,\underline{\underline{\mathbf{I}}}^{312} + \underline{\hat{\mathbf{R}}}\,\underline{\underline{\mathbf{I}}}^{321} - 2\underline{\hat{\mathbf{R}}}\,\underline{\hat{\mathbf{R}}}\,\underline{\hat{\mathbf{R}}}) &= \underline{\hat{\mathbf{R}}}\,\underline{\mathbf{f}} + \underline{\mathbf{f}}\,\underline{\hat{\mathbf{R}}} - 2\underline{\mathbf{f}}\cdot\underline{\hat{\mathbf{R}}}\,\underline{\hat{\mathbf{R}}}\,\underline{\hat{\mathbf{R}}}\\ &= \underline{\hat{\mathbf{R}}}\,\underline{\mathbf{f}}\cdot(\underline{\underline{\mathbf{I}}} - \underline{\hat{\mathbf{R}}}\,\underline{\hat{\mathbf{R}}}) + \underline{\mathbf{f}}\cdot(\underline{\underline{\mathbf{I}}} - \underline{\hat{\mathbf{R}}}\,\underline{\hat{\mathbf{R}}})\underline{\hat{\mathbf{R}}}\end{aligned}$$

and

$$\begin{aligned}&\underline{\underline{\mathbf{h}}} : (\underline{\hat{\mathbf{R}}}\,\underline{\hat{\mathbf{R}}}\,\underline{\underline{\mathbf{I}}}^{2314} + \underline{\hat{\mathbf{R}}}\,\underline{\hat{\mathbf{R}}}\,\underline{\underline{\mathbf{I}}}^{2341} + \underline{\hat{\mathbf{R}}}\,\underline{\hat{\mathbf{R}}}\,\underline{\underline{\mathbf{I}}}^{3214} + \underline{\hat{\mathbf{R}}}\,\underline{\hat{\mathbf{R}}}\,\underline{\underline{\mathbf{I}}}^{3241} - 4\underline{\hat{\mathbf{R}}}\,\underline{\hat{\mathbf{R}}}\,\underline{\hat{\mathbf{R}}}\,\underline{\hat{\mathbf{R}}})\\ &= 2(\underbrace{\underline{\underline{\mathbf{h}}} : \underline{\hat{\mathbf{R}}}\,\underline{\hat{\mathbf{R}}}\,\underline{\underline{\mathbf{I}}}^{2314}}_{=\underline{\hat{\mathbf{R}}}\,\underline{\underline{\mathbf{h}}}:\underline{\hat{\mathbf{R}}}\,\underline{\underline{\mathbf{I}}}} + \underbrace{\underline{\underline{\mathbf{h}}} : \underline{\hat{\mathbf{R}}}\,\underline{\hat{\mathbf{R}}}\,\underline{\underline{\mathbf{I}}}^{2341}}_{=\underline{\underline{\mathbf{h}}}:\underline{\hat{\mathbf{R}}}\,\underline{\underline{\mathbf{I}}}\,\underline{\hat{\mathbf{R}}}} - 2\underline{\underline{\mathbf{h}}} : \underline{\hat{\mathbf{R}}}\,\underline{\hat{\mathbf{R}}}\,\underline{\hat{\mathbf{R}}}\,\underline{\hat{\mathbf{R}}})\\ &= 2[\underline{\hat{\mathbf{R}}}\,\underline{\underline{\mathbf{h}}} : \underline{\hat{\mathbf{R}}}(\underline{\underline{\mathbf{I}}} - \underline{\hat{\mathbf{R}}}\,\underline{\hat{\mathbf{R}}}) + \underline{\underline{\mathbf{h}}} : \underline{\hat{\mathbf{R}}}(\underline{\underline{\mathbf{I}}} - \underline{\hat{\mathbf{R}}}\,\underline{\hat{\mathbf{R}}})\underline{\hat{\mathbf{R}}}].\end{aligned}$$

Sound pressure of elastodynamic source far-fields: The local similarity of elastodynamic source far-fields with plane waves suggests to give sound pressure definitions analogous to (8.123) and (8.124). With (13.221), we obtain the sound pressure

$$\begin{aligned}p_{\mathrm{P}}^{\mathrm{far}}(\underline{\mathbf{R}},\omega) &= -\,\underline{\underline{\mathbf{T}}}_{\mathrm{P}}^{\mathrm{far}}(\underline{\mathbf{R}},\omega) : \underline{\hat{\mathbf{R}}}\,\underline{\hat{\mathbf{R}}}\\ &= -\,\mathrm{j}\omega Z_{\mathrm{P}}\underbrace{\Big[H_{\mathrm{P}}^{f}(\underline{\hat{\mathbf{R}}},\omega) + H_{\mathrm{P}}^{h}(\underline{\hat{\mathbf{R}}},\omega)\Big]\frac{\mathrm{e}^{\mathrm{j}k_{\mathrm{P}}R}}{R}}_{=\,u_{\mathrm{P}}^{\mathrm{far}}(\underline{\mathbf{R}},\omega)}\end{aligned} \tag{13.235}$$

of the P-source field contribution (in the far-field!) and, with (13.234), the sound pressure of the S-source field contribution (in the far-field!):

$$\begin{aligned}p_{\mathrm{S}\vartheta}^{\mathrm{far}}(\underline{\mathbf{R}},\omega) &= -\,\underline{\underline{\mathbf{T}}}_{\mathrm{S}}^{\mathrm{far}}(\underline{\mathbf{R}},\omega) : \underline{\hat{\mathbf{R}}}\,\underline{\mathbf{e}}_{\vartheta}\\ &= -\,\mathrm{j}\omega Z_{\mathrm{S}}\underbrace{\Big[H_{\mathrm{S}\vartheta}^{f}(\underline{\hat{\mathbf{R}}},\omega) + H_{\mathrm{S}\vartheta}^{h}(\underline{\hat{\mathbf{R}}},\omega)\Big]\frac{\mathrm{e}^{\mathrm{j}k_{\mathrm{P}}R}}{R}}_{=\,u_{\mathrm{S}\vartheta}^{\mathrm{far}}(\underline{\mathbf{R}},\omega)},\end{aligned} \tag{13.236}$$

$$\begin{aligned}p_{\mathrm{S}\varphi}^{\mathrm{far}}(\underline{\mathbf{R}},\omega) &= -\,\underline{\underline{\mathbf{T}}}_{\mathrm{S}}^{\mathrm{far}}(\underline{\mathbf{R}},\omega) : \underline{\hat{\mathbf{R}}}\,\underline{\mathbf{e}}_{\varphi}\\ &= -\,\mathrm{j}\omega Z_{\mathrm{S}}\underbrace{\Big[H_{\mathrm{S}\varphi}^{f}(\underline{\hat{\mathbf{R}}},\omega) + H_{\mathrm{S}\varphi}^{h}(\underline{\hat{\mathbf{R}}},\omega)\Big]\frac{\mathrm{e}^{\mathrm{j}k_{\mathrm{P}}R}}{R}}_{=\,u_{\mathrm{S}\varphi}^{\mathrm{far}}(\underline{\mathbf{R}},\omega)}.\end{aligned} \tag{13.237}$$

Just as for plane waves, we state that an elastodynamic scattering problem can barely be solved with the sound pressure as field quantity because it does not satisfy any transition conditions at the boundary between the scatterer and the embedding material. However, we may define the sound pressure of the scattered far-field similar to (13.235) through (13.237) after we have settled the transition conditions for the particle displacement and the stress tensor.

13.3.3 Far-field approximations of two-dimensional elastodynamic source fields

Due to the reduced paper work, US-NDT often relies on two-dimensional simulations.[241] In Section 13.1, we already pointed out differences regarding the scalar Green function. In the present section, we want to look at the consequences for elastodynamic source fields if we assume $\partial/\partial y \equiv 0$, namely independence of all field quantities from the y-coordinate.

Two-dimensional elastodynamic $\underline{\mathbf{f}}$-source fields: At first, we calculate

$$\int_{-\infty}^{\infty} \boldsymbol{\nabla}'\boldsymbol{\nabla}' \frac{\mathrm{e}^{\mathrm{j}k\sqrt{(x-x')^2+(y-y')^2+(z-z')^2}}}{4\pi\sqrt{(x-x')^2+(y-y')^2+(z-z')^2}}\,\mathrm{d}y'$$

$$= \boldsymbol{\nabla}\boldsymbol{\nabla} \underbrace{\int_{-\infty}^{\infty} \frac{\mathrm{e}^{\mathrm{j}k\sqrt{(x-x')^2+(y-y')^2+(z-z')^2}}}{4\pi\sqrt{(x-x')^2+(y-y')^2+(z-z')^2}}\,\mathrm{d}y'}_{=\frac{\mathrm{j}}{4}\mathrm{H}_0^{(1)}\left(k\sqrt{(x-x')^2+(z-z')^2}\right)}$$

$$= \begin{pmatrix} \dfrac{\partial^2}{\partial x^2} & 0 & \dfrac{\partial^2}{\partial x \partial z} \\ 0 & 0 & 0 \\ \dfrac{\partial^2}{\partial z \partial x} & 0 & \dfrac{\partial^2}{\partial z^2} \end{pmatrix} \frac{\mathrm{j}}{4}\mathrm{H}_0^{(1)}\left(k\sqrt{(x-x')^2+(z-z')^2}\right) \tag{13.238}$$

using (13.24). With

$$\underline{\mathbf{r}} = x\,\underline{\mathbf{e}}_x + z\,\underline{\mathbf{e}}_z, \tag{13.239}$$

$$\underline{\mathbf{r}}' = x'\,\underline{\mathbf{e}}_x + z'\underline{\mathbf{e}}_z, \tag{13.240}$$

$$\mathrm{d}^2\underline{\mathbf{r}}' = \mathrm{d}x'\mathrm{d}z', \tag{13.241}$$

we write (13.213) for $\underline{\underline{\mathbf{h}}} \equiv \underline{\underline{\mathbf{0}}}$ assuming $\underline{\mathbf{f}}(x', y', z', \omega) = \underline{\mathbf{f}}(x', z', \omega) = \underline{\mathbf{f}}(\underline{\mathbf{r}}', \omega)$ and utilizing (13.91) and (13.238):

$$\underline{\mathbf{v}}(x, y, z, \omega) = -\mathrm{j}\omega \int\!\!\int_{V_Q} \int_{-\infty}^{\infty} \underline{\mathbf{f}}(x', z', \omega) \cdot \underline{\underline{\mathbf{G}}}(x - x', y - y', z - z', \omega)\,\mathrm{d}y'\,\mathrm{d}^2\underline{\mathbf{r}}'$$

[241] This is generally also true for our examples.

$$
\begin{aligned}
&= -\mathrm{j}\omega \iint_{V_Q} \underline{\mathbf{f}}(\underline{\mathbf{r}}', \omega) \cdot \underbrace{\int_{-\infty}^{\infty} \underline{\underline{\mathbf{G}}}(x - x', y - y', z - z', \omega)\, \mathrm{d}y'}_{\displaystyle = \frac{1}{\mu}\Big[\Big(\underline{\underline{\mathbf{I}}} + \frac{1}{k_{\mathrm{S}}^2}\nabla\nabla\Big)\frac{\mathrm{j}}{4}\,\mathrm{H}_0^{(1)}(k_{\mathrm{S}}|\underline{\mathbf{r}} - \underline{\mathbf{r}}'|) - \frac{1}{k_{\mathrm{S}}^2}\nabla\nabla\frac{\mathrm{j}}{4}\,\mathrm{H}_0^{(1)}(k_{\mathrm{P}}|\underline{\mathbf{r}} - \underline{\mathbf{r}}'|)\Big]} \mathrm{d}^2\underline{\mathbf{r}}' \\
&\stackrel{\text{def}}{=} \underline{\mathbf{v}}(x, z, \omega) \\
&= \underline{\mathbf{v}}(\underline{\mathbf{r}}, \omega). \qquad (13.242)
\end{aligned}
$$

The two-dimensionality of the source distribution yields the two-dimensionality of the Green function!

In Section 7.3, we showed that SH- and P-SV-waves are decoupled in the two-dimensional case; here, this should also be found. Therefore, we split (13.242) into cartesian components:

$$
\begin{aligned}
v_x(\underline{\mathbf{r}}, \omega) = -\mathrm{j}\omega \frac{1}{\mu} \iint_{V_Q} \underline{\mathbf{f}}(\underline{\mathbf{r}}', \omega) \cdot \Bigg[\left(\underline{\mathbf{e}}_x + \frac{1}{k_{\mathrm{S}}^2}\nabla\frac{\partial}{\partial x}\right)\frac{\mathrm{j}}{4}\,\mathrm{H}_0^{(1)}(k_{\mathrm{S}}|\underline{\mathbf{r}} - \underline{\mathbf{r}}'|) \\
- \frac{1}{k_{\mathrm{S}}^2}\nabla\frac{\partial}{\partial x}\frac{\mathrm{j}}{4}\,\mathrm{H}_0^{(1)}(k_{\mathrm{P}}|\underline{\mathbf{r}} - \underline{\mathbf{r}}'|)\Bigg]\, \mathrm{d}^2\underline{\mathbf{r}}',
\end{aligned}
\qquad (13.243)
$$

$$
v_y(\underline{\mathbf{r}}, \omega) = -\mathrm{j}\omega \frac{1}{\mu} \iint_{V_Q} \underline{\mathbf{f}}(\underline{\mathbf{r}}', \omega) \cdot \underline{\mathbf{e}}_y \frac{\mathrm{j}}{4}\,\mathrm{H}_0^{(1)}(k_{\mathrm{S}}|\underline{\mathbf{r}} - \underline{\mathbf{r}}'|)\, \mathrm{d}^2\underline{\mathbf{r}}', \qquad (13.244)
$$

$$
\begin{aligned}
v_z(\underline{\mathbf{r}}, \omega) = -\mathrm{j}\omega \frac{1}{\mu} \iint_{V_Q} \underline{\mathbf{f}}(\underline{\mathbf{r}}', \omega) \cdot \Bigg[\left(\underline{\mathbf{e}}_z + \frac{1}{k_{\mathrm{S}}^2}\nabla\frac{\partial}{\partial z}\right)\frac{\mathrm{j}}{4}\,\mathrm{H}_0^{(1)}(k_{\mathrm{S}}|\underline{\mathbf{r}} - \underline{\mathbf{r}}'|) \\
- \frac{1}{k_{\mathrm{S}}^2}\nabla\frac{\partial}{\partial z}\frac{\mathrm{j}}{4}\,\mathrm{H}_0^{(1)}(k_{\mathrm{P}}|\underline{\mathbf{r}} - \underline{\mathbf{r}}'|)\Bigg]\, \mathrm{d}^2\underline{\mathbf{r}}'.
\end{aligned}
\qquad (13.245)
$$

Obviously, the force density component f_y for a scalar SH-problem is only responsible for a v_y-component, whereas f_x and f_z radiate P- as well as SV-waves possessing only v_x- and v_z-components. The latter are combined to a PSV-vector:

$$
\begin{aligned}
\underline{\mathbf{v}}_{\mathrm{PSV}}(\underline{\mathbf{r}}, \omega) &= v_x(\underline{\mathbf{r}}, \omega)\,\underline{\mathbf{e}}_x + v_z(\underline{\mathbf{r}}, \omega)\,\underline{\mathbf{e}}_z \\
&= -\mathrm{j}\omega \iint_{V_Q} \underline{\mathbf{f}}(\underline{\mathbf{r}}', \omega) \cdot \underline{\underline{\mathbf{G}}}_{\mathrm{PSV}}(\underline{\mathbf{r}} - \underline{\mathbf{r}}', \omega)\, \mathrm{d}^2\underline{\mathbf{r}}', \qquad (13.246)
\end{aligned}
$$

where

$$
\begin{aligned}
\underline{\underline{\mathbf{G}}}_{\mathrm{PSV}}(\underline{\mathbf{r}} - \underline{\mathbf{r}}', \omega) = \frac{1}{\mu}\Bigg[\left(\underline{\underline{\mathbf{I}}} - \underline{\mathbf{e}}_y\underline{\mathbf{e}}_y + \frac{1}{k_{\mathrm{S}}^2}\nabla_{\underline{\mathbf{r}}}\nabla_{\underline{\mathbf{r}}}\right)\frac{\mathrm{j}}{4}\,\mathrm{H}_0^{(1)}(k_{\mathrm{S}}|\underline{\mathbf{r}} - \underline{\mathbf{r}}'|) \\
- \frac{1}{k_{\mathrm{S}}^2}\nabla_{\underline{\mathbf{r}}}\nabla_{\underline{\mathbf{r}}}\frac{\mathrm{j}}{4}\,\mathrm{H}_0^{(1)}(k_{\mathrm{P}}|\underline{\mathbf{r}} - \underline{\mathbf{r}}'|)\Bigg]
\end{aligned}
\qquad (13.247)
$$

with

$$\nabla_{\underline{\mathbf{r}}} = \underline{\mathbf{e}}_x \frac{\partial}{\partial x} + \underline{\mathbf{e}}_z \frac{\partial}{\partial z} \tag{13.248}$$

plays the role of a two-dimensional PSV-Green function;[242] the two-dimensionality is recognized by the argument $\underline{\mathbf{r}} - \underline{\mathbf{r}}'$.

Writing

$$v_y(\underline{\mathbf{r}}, \omega) = -\mathrm{j}\omega \iiint_{V_Q} f_y(\underline{\mathbf{r}}', \omega)\, G_{\mathrm{SH}}(\underline{\mathbf{r}} - \underline{\mathbf{r}}', \omega)\, \mathrm{d}^2\underline{\mathbf{r}}' \tag{13.249}$$

with the two-dimensional SH-Green function

$$G_{\mathrm{SH}}(\underline{\mathbf{r}} - \underline{\mathbf{r}}', \omega) = \frac{1}{\mu}\frac{\mathrm{j}}{4}\, \mathrm{H}_0^{(1)}(k_{\mathrm{S}}|\underline{\mathbf{r}} - \underline{\mathbf{r}}'|) \tag{13.250}$$

relates (13.244) to (13.246).

Two-dimensional elastodynamic $\underline{\underline{\mathbf{h}}}$-source fields: In the two-dimensionally spatially dependent $\underline{\underline{\mathbf{h}}}$-tensor, the h_{yx}-, h_{yy}-, and h_{yz}-components do not appear as sources because, in the two-dimensional (Fourier transformed) deformation rate equation

$$-\mathrm{j}\omega \underline{\underline{\mathbf{S}}}(\underline{\mathbf{r}}, \omega) = \frac{1}{2}\left\{\nabla_{\underline{\mathbf{r}}}\underline{\mathbf{v}}(\underline{\mathbf{r}}, \omega) + \left[\nabla_{\underline{\mathbf{r}}}\underline{\mathbf{v}}(\underline{\mathbf{r}}, \omega)\right]^{21}\right\} + \underline{\underline{\mathbf{h}}}(\underline{\mathbf{r}}, \omega), \tag{13.251}$$

the components $\nabla_{\underline{\mathbf{r}}}\underline{\mathbf{v}} : \underline{\mathbf{e}}_x\underline{\mathbf{e}}_y$, $\nabla_{\underline{\mathbf{r}}}\underline{\mathbf{v}} : \underline{\mathbf{e}}_y\underline{\mathbf{e}}_y$, $\nabla_{\underline{\mathbf{r}}}\underline{\mathbf{v}} : \underline{\mathbf{e}}_z\underline{\mathbf{e}}_y$ of the gradient dyadic $\nabla_{\underline{\mathbf{r}}}\underline{\mathbf{v}}$—and therefore of the deformation rate tensor according to the definition (3.73)—are equal to zero due to $\partial/\partial y \equiv 0$. Hence, in two dimensions, the injected deformation rate tensor $\underline{\underline{\mathbf{h}}}(\underline{\mathbf{r}}, \omega)$ has the component representation

$$\begin{aligned}\underline{\underline{\mathbf{h}}}(\underline{\mathbf{r}}, \omega) = {} & h_{xx}(\underline{\mathbf{r}}, \omega)\, \underline{\mathbf{e}}_x\underline{\mathbf{e}}_x + h_{xy}(\underline{\mathbf{r}}, \omega)\, \underline{\mathbf{e}}_x\underline{\mathbf{e}}_y + h_{xz}(\underline{\mathbf{r}}, \omega)\, \underline{\mathbf{e}}_x\underline{\mathbf{e}}_z + \\ & + h_{zx}(\underline{\mathbf{r}}, \omega)\, \underline{\mathbf{e}}_z\underline{\mathbf{e}}_x + h_{zy}(\underline{\mathbf{r}}, \omega)\, \underline{\mathbf{e}}_z\underline{\mathbf{e}}_y + h_{zz}(\underline{\mathbf{r}}, \omega)\, \underline{\mathbf{e}}_z\underline{\mathbf{e}}_z. \end{aligned} \tag{13.252}$$

With

$$\begin{aligned}\underline{\mathbf{v}}(x, y, z, \omega) &= \iint_{V_Q} \int_{-\infty}^{\infty} \underline{\underline{\mathbf{h}}}(x', z', \omega) : \underline{\underline{\underline{\mathbf{\Sigma}}}}(x - x', y - y', z - z', \omega)\, \mathrm{d}y'\, \mathrm{d}^2\underline{\mathbf{r}}' \\ &= \iint_{V_Q} \underline{\underline{\mathbf{h}}}(\underline{\mathbf{r}}', \omega) : \int_{-\infty}^{\infty} \underline{\underline{\underline{\underline{\mathbf{c}}}}} : \nabla \underline{\underline{\mathbf{G}}}(x - x', y - y', z - z', \omega)\, \mathrm{d}y'\, \mathrm{d}^2\underline{\mathbf{r}}'\end{aligned}$$

---

[242]Note: $\underline{\underline{\mathbf{G}}}_{\mathrm{PSV}}$ is a second rank tensor with nine components possessing a matrix scheme—compare (13.238)—with five zeroes.

$$
\begin{aligned}
&= \iint_{V_Q} \underline{\underline{\mathbf{h}}}(\underline{\mathbf{r}}', \omega) : \underline{\underline{\underline{\underline{\mathbf{c}}}}} : \nabla \underbrace{\int_{-\infty}^{\infty} \underline{\underline{\mathbf{G}}}(x - x', y - y', z - z', \omega)\, \mathrm{d}y'}_{\begin{subarray}{l} = \frac{1}{\mu}\left[\left(\underline{\underline{\mathbf{I}}} + \frac{1}{k_S^2}\nabla\nabla\right)\frac{j}{4}\,\mathrm{H}_0^{(1)}(k_S|\underline{\mathbf{r}} - \underline{\mathbf{r}}'|)\right. \\ \qquad \left. -\frac{1}{k_S^2}\nabla\nabla\frac{j}{4}\,\mathrm{H}_0^{(1)}(k_P|\underline{\mathbf{r}} - \underline{\mathbf{r}}'|)\right]\end{subarray}} \mathrm{d}^2\underline{\mathbf{r}}' \\
&\stackrel{\text{def}}{=} \underline{\mathbf{v}}(x, z, \omega) \\
&= \underline{\mathbf{v}}(\underline{\mathbf{r}}, \omega), \qquad (13.253)
\end{aligned}
$$

we now turn to the two-dimensional $\underline{\underline{\mathbf{h}}}$-sources (13.252).

At first, we calculate $v_y(\underline{\mathbf{r}}, \omega)$:

$$
\begin{aligned}
v_y(\underline{\mathbf{r}}, \omega) &= \frac{1}{\mu}\iint_{V_Q} \underline{\underline{\mathbf{h}}}(\underline{\mathbf{r}}', \omega) : \underline{\underline{\underline{\underline{\mathbf{c}}}}} : \nabla_{\underline{\mathbf{r}}}\left[\underline{\mathbf{e}}_y \frac{j}{4}\,\mathrm{H}_0^{(1)}(k_S|\underline{\mathbf{r}} - \underline{\mathbf{r}}'|)\right] \mathrm{d}^2\underline{\mathbf{r}}' \\
&= \frac{1}{\mu}\iint_{V_Q} \underline{\underline{\mathbf{h}}}(\underline{\mathbf{r}}', \omega) : \mu(\nabla_{\underline{\mathbf{r}}}\,\underline{\mathbf{e}}_y + \underline{\mathbf{e}}_y \nabla_{\underline{\mathbf{r}}})\frac{j}{4}\,\mathrm{H}_0^{(1)}(k_S|\underline{\mathbf{r}} - \underline{\mathbf{r}}'|)\, \mathrm{d}^2\underline{\mathbf{r}}' \\
&= 2\iint_{V_Q} \underline{\underline{\mathbf{h}}}(\underline{\mathbf{r}}', \omega) : \underline{\mathbf{e}}_y \nabla_{\underline{\mathbf{r}}} \frac{j}{4}\,\mathrm{H}_0^{(1)}(k_S|\underline{\mathbf{r}} - \underline{\mathbf{r}}'|)\, \mathrm{d}^2\underline{\mathbf{r}}' \\
&= 2\iint_{V_Q} \left[h_{xy}(\underline{\mathbf{r}}', \omega)\frac{\partial}{\partial x} + h_{zy}(\underline{\mathbf{r}}', \omega)\frac{\partial}{\partial z}\right] \frac{j}{4}\,\mathrm{H}_0^{(1)}(k_S|\underline{\mathbf{r}} - \underline{\mathbf{r}}'|)\, \mathrm{d}^2\underline{\mathbf{r}}'.
\end{aligned}
\qquad (13.254)
$$

Evidently, with (13.244) and (13.254), we have rediscovered the solution of the Fourier transformed two-dimensional differential equation (7.47) with the help of the two-dimensional scalar Green function.[243] As a matter of fact, the components $h_{xy}(\underline{\mathbf{r}}', \omega)$, $h_{zy}(\underline{\mathbf{r}}', \omega)$ of the two-dimensional $\underline{\underline{\mathbf{h}}}$-tensor only radiate SH-waves.

A column vector of $\underline{\underline{\mathbf{h}}}$ defined as

$$
\underline{\mathbf{h}}^y(\underline{\mathbf{r}}, \omega) = h_{xy}(\underline{\mathbf{r}}, \omega)\,\underline{\mathbf{e}}_x + h_{zy}(\underline{\mathbf{r}}, \omega)\,\underline{\mathbf{e}}_z \qquad (13.255)
$$

—the h_{yy}-component of $\underline{\underline{\mathbf{h}}}$ is not present according to (13.251)—allows for the short-hand notation

$$
v_y(\underline{\mathbf{r}}, \omega) = \iint_{V_Q} \underline{\mathbf{h}}^y(\underline{\mathbf{r}}', \omega) \cdot \underline{\mathbf{G}}_{\mathrm{SH}}(\underline{\mathbf{r}} - \underline{\mathbf{r}}', \omega)\, \mathrm{d}^2\underline{\mathbf{r}}' \qquad (13.256)
$$

of (13.254) with the vectorial SH-Green function

$$
\underline{\mathbf{G}}_{\mathrm{SH}}(\underline{\mathbf{r}} - \underline{\mathbf{r}}', \omega) = 2\nabla_{\underline{\mathbf{r}}} \frac{j}{4}\,\mathrm{H}_0^{(1)}(k_S|\underline{\mathbf{r}} - \underline{\mathbf{r}}'|). \qquad (13.257)
$$

[243] In (13.254), we differentiate the Green function, and in the differential equation (7.47), the h_{xy}, h_{zy}-sources are differentiated; with $\partial/\partial x, \partial/\partial z \Longrightarrow -\partial/\partial x', -\partial/\partial z'$ and partial integration, we realize that both the representations are identical.

With the definition of the two-dimensional third rank Green tensor

$$\begin{aligned}
\underline{\underline{\underline{\boldsymbol{\Sigma}}}}_{\mathrm{PSV}}(\underline{\mathbf{r}}, \omega) &= \underline{\underline{\underline{\underline{\mathbf{c}}}}} : \nabla_{\underline{\mathbf{r}}} \underline{\underline{\mathbf{G}}}_{\mathrm{PSV}}(\underline{\mathbf{r}}, \omega) \\
&= (\lambda \underline{\underline{\underline{\underline{\mathbf{I}}}}}^{\delta} + 2\mu \underline{\underline{\underline{\underline{\mathbf{I}}}}}^{+}) : \nabla_{\underline{\mathbf{r}}} \underline{\underline{\mathbf{G}}}_{\mathrm{PSV}}(\underline{\mathbf{r}}, \omega) \\
&= \lambda \underline{\underline{\mathbf{I}}}\, \nabla_{\underline{\mathbf{r}}} \cdot \underline{\underline{\mathbf{G}}}_{\mathrm{PSV}}(\underline{\mathbf{r}}, \omega) + \mu \left[\nabla_{\underline{\mathbf{r}}} \underline{\underline{\mathbf{G}}}_{\mathrm{PSV}}(\underline{\mathbf{r}}, \omega) + \nabla_{\underline{\mathbf{r}}} \underline{\underline{\mathbf{G}}}_{\mathrm{PSV}}^{213}(\underline{\mathbf{r}}, \omega) \right],
\end{aligned} \tag{13.258}$$

we may collect the x- and z-components of $\underline{\mathbf{v}}(\underline{\mathbf{r}}, \omega)$ in a PSV-vector

$$\underline{\mathbf{v}}_{\mathrm{PSV}}(\underline{\mathbf{r}}, \omega) = \iint_{V_Q} \underline{\underline{\mathbf{h}}}(\underline{\mathbf{r}}', \omega) : \underline{\underline{\underline{\boldsymbol{\Sigma}}}}_{\mathrm{PSV}}(\underline{\mathbf{r}} - \underline{\mathbf{r}}', \omega)\, \mathrm{d}^2\underline{\mathbf{r}}', \tag{13.259}$$

where the remaining components of the $\underline{\underline{\mathbf{h}}}$-tensors (13.251) appear.

Far-field approximations of two-dimensional elastodynamic source fields: The far-field approximation

$$\begin{aligned}
u_y^{\mathrm{far}}(\underline{\mathbf{r}}, \omega) &= \iint_{V_Q} f_y(\underline{\mathbf{r}}', \omega)\, G_{\mathrm{SH}}^{\mathrm{far}}(\underline{\mathbf{r}}, \underline{\mathbf{r}}', \omega)\, \mathrm{d}^2\underline{\mathbf{r}}' \\
&= \frac{\mathrm{e}^{\mathrm{j}k_{\mathrm{S}}r}}{\sqrt{r}}\, H_{\mathrm{SH}}^{f}(\underline{\hat{\mathbf{r}}}, \omega)
\end{aligned} \tag{13.260}$$

of the two-dimensional SH-f_y-source field of the particle displacement with the radiation characteristic

$$H_{\mathrm{SH}}^{f}(\underline{\hat{\mathbf{r}}}, \omega) = \frac{1}{\sqrt{\omega}} \frac{\mathrm{e}^{\mathrm{j}\frac{\pi}{4}}}{2Z_{\mathrm{S}}\sqrt{2\pi c_{\mathrm{S}}}} \iint_{V_Q} f_y(\underline{\mathbf{r}}', \omega)\, \mathrm{e}^{-\mathrm{j}k_{\mathrm{S}}\underline{\hat{\mathbf{r}}}\cdot\underline{\mathbf{r}}'}\, \mathrm{d}^2\underline{\mathbf{r}}' \tag{13.261}$$

is immediately obtained with (13.52). Besides the fact that (13.260) is recognized as the far-field approximation of a cylindrical wave, the scalar radiation characteristic (13.261) is different from the radiation characteristic components (13.225), in particular regarding the additional frequency dependence $1/\sqrt{\omega}$; for a pulse US-NDT, this is of considerable importance (compare Figures 13.4 and 13.6).

The far-field approximation

$$\begin{aligned}
\underline{\mathbf{u}}_{\mathrm{PSV}}^{\mathrm{far}}(\underline{\mathbf{r}}, \omega) &= \iint_{V_Q} \underline{\mathbf{f}}(\underline{\mathbf{r}}', \omega) \cdot \underline{\underline{\mathbf{G}}}_{\mathrm{PSV}}^{\mathrm{far}}(\underline{\mathbf{r}}, \underline{\mathbf{r}}', \omega)\, \mathrm{d}^2\underline{\mathbf{r}}' \\
&= \frac{\mathrm{e}^{\mathrm{j}k_{\mathrm{P}}r}}{\sqrt{r}}\, \underline{\mathbf{H}}_{\mathrm{P}}^{f}(\underline{\hat{\mathbf{r}}}, \omega) + \frac{\mathrm{e}^{\mathrm{j}k_{\mathrm{S}}r}}{\sqrt{r}}\, \underline{\mathbf{H}}_{\mathrm{SV}}^{f}(\underline{\hat{\mathbf{r}}}, \omega)
\end{aligned} \tag{13.262}$$

of the two-dimensional PSV-$\underline{\mathbf{f}}$-source field of the particle displacement with the radiation characteristics

$$\underline{\mathbf{H}}_{\mathrm{P}}^{f}(\underline{\hat{\mathbf{r}}}, \omega) = \underbrace{\frac{1}{\sqrt{\omega}} \frac{\mathrm{e}^{\mathrm{j}\frac{\pi}{4}}}{2Z_{\mathrm{P}}\sqrt{2\pi c_{\mathrm{P}}}} \iint_{V_Q} \underline{\mathbf{f}}(\underline{\mathbf{r}}', \omega) \cdot \underline{\hat{\mathbf{r}}}\, \mathrm{e}^{-\mathrm{j}k_{\mathrm{P}}\underline{\hat{\mathbf{r}}}\cdot\underline{\mathbf{r}}'}\, \mathrm{d}^2\underline{\mathbf{r}}'}_{= H_{\mathrm{P}}^{f}(\underline{\hat{\mathbf{r}}}, \omega)}\, \underline{\hat{\mathbf{r}}}, \tag{13.263}$$

$$\underline{\mathbf{H}}_{\mathrm{SV}}^{f}(\hat{\underline{\mathbf{r}}},\omega) = \underbrace{\frac{1}{\sqrt{\omega}}\frac{\mathrm{e}^{\mathrm{j}\frac{\pi}{4}}}{2Z_{\mathrm{S}}\sqrt{2\pi c_{\mathrm{S}}}}\int\!\!\int_{V_Q}\underline{\mathbf{f}}(\underline{\mathbf{r}}',\omega)\cdot\underline{\mathbf{e}}_{\theta}\,\mathrm{e}^{-\mathrm{j}k_{\mathrm{S}}\hat{\underline{\mathbf{r}}}\cdot\underline{\mathbf{r}}'}\,\mathrm{d}^2\underline{\mathbf{r}}'}_{=H_{\mathrm{SV}}^{f}(\hat{\underline{\mathbf{r}}},\omega)}\,\underline{\mathbf{e}}_{\theta} \tag{13.264}$$

is immediately obtained via derivation from (13.247) using

$$\nabla_{\underline{\mathbf{r}}} \Longrightarrow \mathrm{j}k_{\mathrm{P,S}}\hat{\underline{\mathbf{r}}} \tag{13.265}$$

and

$$\underline{\underline{\mathbf{I}}} = \hat{\underline{\mathbf{r}}}\hat{\underline{\mathbf{r}}} + \underline{\mathbf{e}}_{\theta}\,\underline{\mathbf{e}}_{\theta} + \underline{\mathbf{e}}_{y}\underline{\mathbf{e}}_{y} \tag{13.266}$$

as well as (13.52)

$$\begin{aligned}\underline{\underline{\mathbf{G}}}_{\mathrm{PSV}}^{\mathrm{far}}(\underline{\mathbf{r}},\underline{\mathbf{r}}',\omega) &= \frac{1}{\lambda+2\mu}\,\hat{\underline{\mathbf{r}}}\,\hat{\underline{\mathbf{r}}}\,\frac{1}{4}\,\mathrm{e}^{\mathrm{j}\frac{\pi}{4}}\sqrt{\frac{2}{\pi}}\,\frac{\mathrm{e}^{\mathrm{j}k_{\mathrm{P}}r}}{\sqrt{k_{\mathrm{P}}r}}\,\mathrm{e}^{-\mathrm{j}k_{\mathrm{P}}\hat{\underline{\mathbf{r}}}\cdot\underline{\mathbf{r}}'}\\&\quad+\frac{1}{\mu}\,\underline{\mathbf{e}}_{\theta}\,\underline{\mathbf{e}}_{\theta}\,\frac{1}{4}\,\mathrm{e}^{\mathrm{j}\frac{\pi}{4}}\sqrt{\frac{2}{\pi}}\,\frac{\mathrm{e}^{\mathrm{j}k_{\mathrm{S}}r}}{\sqrt{k_{\mathrm{S}}r}}\,\mathrm{e}^{-\mathrm{j}k_{\mathrm{S}}\hat{\underline{\mathbf{r}}}\cdot\underline{\mathbf{r}}'}.\end{aligned} \tag{13.267}$$

The respective three-dimensional radiation characteristics are numbered by (13.218) and (13.225). Equation 13.267 is the two-dimensional counterpart of (13.95) with (13.100) and (13.101).

The far-field approximation of the SH-particle displacement of two-dimensional $\underline{\underline{\mathbf{h}}}$-sources immediately results from (13.257) using (13.265) and (13.52):

$$u_y^{\mathrm{far}}(\underline{\mathbf{r}},\omega) = \frac{\mathrm{e}^{\mathrm{j}k_{\mathrm{P}}r}}{\sqrt{r}}\,H_{\mathrm{SH}}^{h}(\hat{\underline{\mathbf{r}}},\omega), \tag{13.268}$$

where

$$H_{\mathrm{SH}}^{h}(\hat{\underline{\mathbf{r}}},\omega) = -\frac{1}{\sqrt{\omega}}\frac{\mathrm{e}^{\mathrm{j}\frac{\pi}{4}}}{\sqrt{2\pi c_{\mathrm{S}}}}\int\!\!\int_{V_Q}\underline{\mathbf{h}}^{y}(\underline{\mathbf{r}}',\omega)\cdot\hat{\underline{\mathbf{r}}}\,\mathrm{e}^{-\mathrm{j}k_{\mathrm{S}}\hat{\underline{\mathbf{r}}}\cdot\underline{\mathbf{r}}'}\,\mathrm{d}^2\underline{\mathbf{r}}'. \tag{13.269}$$

For the PSV-contribution, we have to rely on (13.259) with (13.258); with (13.265) and (13.267), it follows

$$\begin{aligned}\underline{\underline{\boldsymbol{\Sigma}}}_{\mathrm{PSV}}^{\mathrm{far}}(\hat{\underline{\mathbf{r}}},\omega) &= \mathrm{j}k_{\mathrm{P}}\,\frac{1}{\lambda+2\mu}\,\frac{1}{4}\,\mathrm{e}^{\mathrm{j}\frac{\pi}{4}}\sqrt{\frac{2}{\pi}}\,\frac{\mathrm{e}^{\mathrm{j}k_{\mathrm{P}}r}}{\sqrt{k_{\mathrm{P}}r}}\,\underbrace{(\lambda\underline{\underline{\mathbf{I}}}^{\delta}+2\mu\underline{\underline{\mathbf{I}}}^{+}):\hat{\underline{\mathbf{r}}}\,\hat{\underline{\mathbf{r}}}\,\hat{\underline{\mathbf{r}}}}_{=(\lambda\underline{\underline{\mathbf{I}}}+2\mu\,\hat{\underline{\mathbf{r}}}\,\hat{\underline{\mathbf{r}}})\hat{\underline{\mathbf{r}}}}\\&\quad+\mathrm{j}k_{\mathrm{S}}\,\frac{1}{\mu}\,\frac{1}{4}\,\mathrm{e}^{\mathrm{j}\frac{\pi}{4}}\sqrt{\frac{2}{\pi}}\,\frac{\mathrm{e}^{\mathrm{j}k_{\mathrm{S}}r}}{\sqrt{k_{\mathrm{S}}r}}\,\underbrace{(\lambda\underline{\underline{\mathbf{I}}}^{\delta}+2\mu\underline{\underline{\mathbf{I}}}^{+}):\hat{\underline{\mathbf{r}}}\,\underline{\mathbf{e}}_{\theta}\,\underline{\mathbf{e}}_{\theta}}_{=\mu(\hat{\underline{\mathbf{r}}}\,\underline{\mathbf{e}}_{\theta}+\underline{\mathbf{e}}_{\theta}\hat{\underline{\mathbf{r}}})\underline{\mathbf{e}}_{\theta}}\end{aligned} \tag{13.270}$$

from (13.258), and hence

$$\underline{\mathbf{u}}_{\mathrm{PSV}}^{\mathrm{far}}(\underline{\mathbf{r}},\omega) = \frac{\mathrm{e}^{\mathrm{j}k_{\mathrm{P}}r}}{\sqrt{r}}\,\underline{\mathbf{H}}_{\mathrm{P}}^{h}(\hat{\underline{\mathbf{r}}},\omega) + \frac{\mathrm{e}^{\mathrm{j}k_{\mathrm{S}}r}}{\sqrt{r}}\,\underline{\mathbf{H}}_{\mathrm{SV}}^{h}(\hat{\underline{\mathbf{r}}},\omega) \tag{13.271}$$

with

$$\underline{\mathbf{H}}_{\mathrm{P}}^{h}(\hat{\underline{\mathbf{r}}}, \omega)$$
$$= -\frac{1}{\sqrt{\omega}} \frac{\mathrm{e}^{\mathrm{j}\frac{\pi}{4}}}{2c_{\mathrm{P}} Z_{\mathrm{P}} \sqrt{2\pi c_{\mathrm{P}}}} 2\mu \iint_{V_Q} \underline{\underline{\mathbf{h}}}(\underline{\mathbf{r}}', \omega) : (\lambda \underline{\underline{\mathbf{I}}} + 2\mu \hat{\underline{\mathbf{r}}}\,\hat{\underline{\mathbf{r}}})\, \mathrm{e}^{-\mathrm{j}k_{\mathrm{P}}\hat{\underline{\mathbf{r}}}\cdot\underline{\mathbf{r}}'}\, \mathrm{d}^2\underline{\mathbf{r}}'\, \hat{\underline{\mathbf{r}}}, \tag{13.272}$$

$$\underline{\mathbf{H}}_{\mathrm{SV}}^{h}(\hat{\underline{\mathbf{r}}}, \omega) = -\frac{1}{\sqrt{\omega}} \frac{\mathrm{e}^{\mathrm{j}\frac{\pi}{4}}}{\sqrt{2\pi c_{\mathrm{S}}}} \iint_{V_Q} \underline{\underline{\mathbf{h}}}(\underline{\mathbf{r}}', \omega) : \hat{\underline{\mathbf{r}}}\, \underline{\mathbf{e}}_{\theta}\, \mathrm{e}^{-\mathrm{j}k_{\mathrm{S}}\hat{\underline{\mathbf{r}}}\cdot\underline{\mathbf{r}}'}\, \mathrm{d}^2\underline{\mathbf{r}}'\, \underline{\mathbf{e}}_{\theta}. \tag{13.273}$$

The radiation characteristic (13.272) complements (13.263) and (13.273) complements (13.264).

13.3.4 Examples for two- and three-dimensional elastodynamic and acoustic source far-fields: Planar rectangular, planar circular, and planar strip-like force density distributions with constant amplitude

Planar rectangular force density distribution in an elastic full-space: Within the pressure and shear wave radiation characteristics (13.218) and (13.225) of the far-field (13.217) of the particle displacement, we may arbitrarily prescribe the force density vector $\underline{\mathbf{f}}(\underline{\mathbf{R}}, \omega)$ to subsequently obtain an intuitive idea of the resulting radiation field. We choose an example that may be considered as a preliminary US-NDT model[244] of a piezoelectric transducer with rectangular radiating surface:

$$\underline{\mathbf{f}}(\underline{\mathbf{R}}, \omega) = F(\omega) q_a(x) q_b(y) \delta(z) \underline{\mathbf{e}}_z. \tag{13.274}$$

Here, $q_a(x)$ and $q_b(y)$ are rectangular functions symmetric to the origin of width $2a$, respectively $2b$, according to (2.273), characterizing the rectangular geometry of the "transducer model" (13.274) and the constant force density within the aperture; with $\delta(z)$ in (13.274), we ensure a planar force density, and with $\underline{\mathbf{e}}_z$, we postulate the overall equal direction of the force density within the aperture; $F(\omega)$ stands for the spectrum of the excitation pulse of the synchronous source

$$\underline{\mathbf{f}}(\underline{\mathbf{R}}, t) = F(t) q_a(x) q_b(y) \delta(z) \underline{\mathbf{e}}_z \tag{13.275}$$

according to (13.61). Equation 13.274 is sketched in the coordinate system of Figure 13.16, where we count z downward into a virtual part.

The point directivities of the force density (13.274) have already been outlined in Figure 13.12(a) and (b), here we are concentrating on the directivities

[244]Preliminary, because at the moment, we can only calculate radiation fields of source distributions in full-space; physical effects resulting from the location of the transducer on a stress-free specimen surface are not yet accounted for.

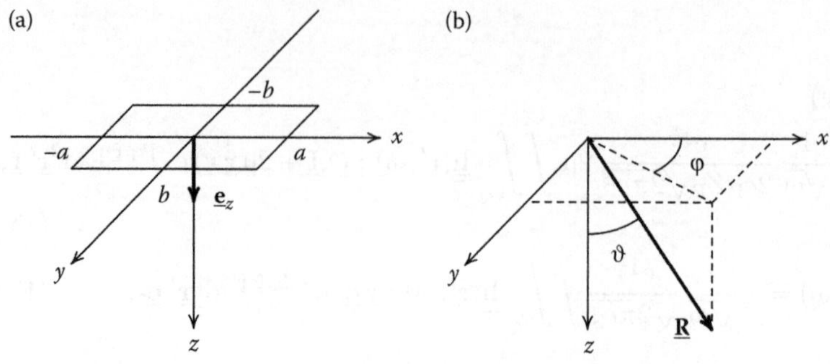

FIGURE 13.16
Planar rectangular force density distribution (a), and spherical coordinates for calculation of the radiation field (b).

of the finite aperture in an elastic full-space. We find them in terms of Equations 13.218 and 13.225, yet we may immediately shift the point directivities (13.231) and (13.229) in front of the integrals due to the $\underline{\mathbf{R}}'$-independent force density orientations:

$$H_{\mathrm{P}}^{f_z}(\hat{\underline{\mathbf{R}}}, \omega) = \frac{F(\omega)}{4\pi c_{\mathrm{P}} Z_{\mathrm{P}}} \underline{\mathbf{e}}_z \cdot \hat{\underline{\mathbf{R}}} \int_{-b}^{b} \int_{-a}^{a} \mathrm{e}^{-\mathrm{j}k_{\mathrm{P}} \hat{\underline{\mathbf{R}}} \cdot \underline{\mathbf{R}}'} \, \mathrm{d}x' \mathrm{d}y', \quad (13.276)$$

$$H_{\mathrm{S}\vartheta,\varphi}^{f_z}(\hat{\underline{\mathbf{R}}}, \omega) = \frac{F(\omega)}{4\pi c_{\mathrm{S}} Z_{\mathrm{S}}} \underline{\mathbf{e}}_z \cdot \underline{\mathbf{e}}_{\vartheta,\varphi} \int_{-b}^{b} \int_{-a}^{a} \mathrm{e}^{-\mathrm{j}k_{\mathrm{S}} \hat{\underline{\mathbf{R}}} \cdot \underline{\mathbf{R}}'} \, \mathrm{d}x' \mathrm{d}y'. \quad (13.277)$$

On behalf of the sifting property of the $\delta(z)$-distribution in (13.274), the expression $\hat{\underline{\mathbf{R}}} \cdot \underline{\mathbf{R}}'$ in (13.276) and (13.277) is given by

$$\begin{aligned} \hat{\underline{\mathbf{R}}} \cdot \underline{\mathbf{R}}' &= \hat{\underline{\mathbf{R}}} \cdot \underline{\mathbf{e}}_x \, x' + \hat{\underline{\mathbf{R}}} \cdot \underline{\mathbf{e}}_y \, y' \\ &= \sin\vartheta \cos\varphi \, x' + \sin\vartheta \sin\varphi \, y'. \end{aligned} \quad (13.278)$$

The resulting integrals can be easily calculated:

$$H_{\mathrm{P}}^{f_z}(\vartheta, \varphi, \omega) = \frac{abF(\omega)}{\pi c_{\mathrm{P}} Z_{\mathrm{P}}} \cos\vartheta \, \frac{\sin(k_{\mathrm{P}} a \sin\vartheta \cos\varphi)}{k_{\mathrm{P}} a \sin\vartheta \cos\varphi} \, \frac{\sin(k_{\mathrm{P}} b \sin\vartheta \sin\varphi)}{k_{\mathrm{P}} b \sin\vartheta \sin\varphi}, \quad (13.279)$$

$$H_{\mathrm{S}\vartheta}^{f_z}(\vartheta, \varphi, \omega) = -\frac{abF(\omega)}{\pi c_{\mathrm{S}} Z_{\mathrm{S}}} \sin\vartheta \, \frac{\sin(k_{\mathrm{S}} a \sin\vartheta \cos\varphi)}{k_{\mathrm{S}} a \sin\vartheta \cos\varphi} \, \frac{\sin(k_{\mathrm{S}} b \sin\vartheta \sin\varphi)}{k_{\mathrm{S}} b \sin\vartheta \sin\varphi}, \quad (13.280)$$

$$H_{\mathrm{S}\varphi}^{f_z}(\vartheta, \varphi, \omega) = 0. \quad (13.281)$$

In [Equation 13.217, respectively (13.224)]

$$\underline{\mathbf{u}}_{\mathrm{P}}^{\mathrm{far}}(R, \vartheta, \varphi, \omega) = \frac{\mathrm{e}^{\mathrm{j}k_{\mathrm{P}} R}}{R} H_{\mathrm{P}}^{f_z}(\vartheta, \varphi, \omega) \, \underline{\mathbf{e}}_R + \frac{\mathrm{e}^{\mathrm{j}k_{\mathrm{S}} R}}{R} H_{\mathrm{S}\vartheta}^{f_z}(\vartheta, \varphi, \omega) \, \underline{\mathbf{e}}_\vartheta, \quad (13.282)$$

the radiation characteristics determine the particle displacement far-field. We keep hold of the following: The radiation characteristics (13.279) and (13.280) are products of the respective point directivities [Figure 13.12(a) and (b)] and the directivity of the planar rectangular aperture with constant force distribution. For $\vartheta = 0, \pi$, we have

$$\left.\frac{\sin(k_{\mathrm{P,S}} a \sin\vartheta \cos\varphi)}{k_{\mathrm{P,S}} a \sin\vartheta \cos\varphi} \frac{\sin(k_{\mathrm{P,S}} b \sin\vartheta \sin\varphi)}{k_{\mathrm{P,S}} b \sin\vartheta \sin\varphi}\right|_{\vartheta=0,\pi} = 1, \tag{13.283}$$

yielding $H_{\mathrm{P}}^f(\vartheta = 0, \pi, \varphi, \omega)$ to be maximum due to the point directivity $\cos\vartheta$, whereas $H_{\mathrm{S}\vartheta}^f(\vartheta = 0, \pi, \varphi, \omega)$ has zeroes due to the point directivity $-\sin\vartheta$. Radiation patterns for (13.279) and (13.280) are displayed in Figure 13.17 for $y = 0$ as a function of ϑ in the coordinate system of Figure 13.16(b). These diagrams solely represent elastodynamics through appearance of the respective point directivity because the aperture diagram representing the geometry and the amplitude distribution of the planar source refers to every (rectangular) scalar aperture radiator. The essential parameters of the planar rectangular constant aperture distribution determining the radiation patterns are $k_{\mathrm{P,S}}a$ and $k_{\mathrm{P,S}}b$, hence $k_{\mathrm{P,S}}a$ for $y = 0$. For $k_{\mathrm{P,S}}a$-values of the order of 1 [Figure 13.17(a) and (b)], we are still close to a point source (Figure 13.12), the aperture directivity is barely recognizable. The chosen $k_{\mathrm{P,S}}a$-values for Figure 13.17(a) and (b) result, for example, for an aperture of x-extension $2a = 2\,\mathrm{mm}$ located in steel in combination with the frequency $f = 1\,\mathrm{MHz}$ ($\lambda_{\mathrm{P}} = 5.9\,\mathrm{mm}$, $\lambda_{\mathrm{S}} = 3.2\,\mathrm{mm}$). Increasing the x-aperture by a factor of 5 to $2a = 1\,\mathrm{cm}$, the diagrams of Figures 13.17(c) and (d) result, where in fact H_{P}^f already exhibits a significant directivity, while exactly this aperture main lobe in the $H_{\mathrm{S}\vartheta}^f$-diagram is suppressed by the multiplication with the zero of the point directivity $\sin\vartheta$ for $\vartheta = 0$ and simultaneously enhancing the amplitudes of the side lobes. This tendency—increasing directivity,[245] and an increasing number of side lobes in the H_{P}^f-diagram with increasing fanning out of the $H_{\mathrm{S}\vartheta}^f$-diagrams without a pronounced main lobe—continues with increasing $k_{\mathrm{P,S}}a$ (Figures 13.17(e) and (f); $2a = 2\,\mathrm{cm}$).

We want to make a remark on the "creeping wave transducer" using the present preliminary transducer model; therefore, we consider the non synchronous source

$$\underline{\mathbf{f}}(\underline{\mathbf{R}}, t) = F\left(t - \frac{x}{c_{\mathrm{A}}}\right) q_a(x) q_b(y) \delta(z) \underline{\mathbf{e}}_z \tag{13.284}$$

instead of (13.275). As before, this force density excitation impresses the same impulse $F(t)$ to the aperture line $x = 0, -b \leq y \leq b$ and also to the lines $-a \leq x < 0, -b \leq y \leq b$, however precipitated by x/c_{A}, and to the lines

[245]For a quantitative description of the directivity introducing the directivity factor, the respective literature on antennas should be consulted (for example, Balanis 1997; Langenberg 2005).

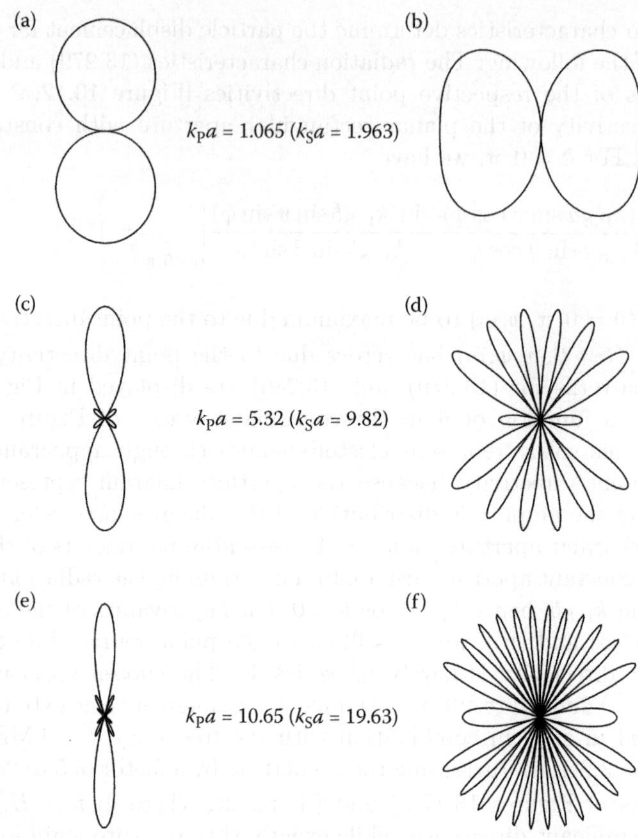

FIGURE 13.17
P- and S-radiation patterns [P: (a), (c), (e); S: (b), (d), (f)] of a planar rectangular f_z-force density aperture; $y = 0$ (steel: $c_P = 5900$ m/s, $c_S = 3200$ m/s, $\rho = 7.7 \cdot 10^3$ kg/m^3).

$0 < x \leq a, -b \leq y \leq b$ delayed by x/c_A, where c_A with the dimension of a velocity[246] denotes the parameter for the slope of this linear lead/delay. In the frequency domain, the nonsynchronous source (13.284) turns into the (linear) phase tapered source

$$\underline{\mathbf{f}}(\underline{\mathbf{R}}, \omega) = F(\omega)\mathrm{e}^{\mathrm{j}k_A x} q_a(x) q_b(y) \delta(z) \underline{\mathbf{e}}_z, \qquad (13.285)$$

where

$$k_A = \frac{\omega}{c_A} \qquad (13.286)$$

[246]For physically real angle transducers, c_A is the trace velocity of the P-wave from the piezoelectric crystal impinging on the material surface.

plays the role of a wave number for the trace velocity c_A of the tapering. With (13.285), we obtain the radiation characteristics

$$\begin{aligned} & H_\mathrm{P}^{f_z}(\vartheta,\varphi,\omega,k_\mathrm{A}) \\ & \quad = \frac{abF(\omega)}{\pi c_\mathrm{P} Z_\mathrm{P}} \cos\vartheta \, \frac{\sin(k_\mathrm{P} a \sin\vartheta\cos\varphi - k_\mathrm{A} a)}{k_\mathrm{P} a \sin\vartheta\cos\varphi - k_\mathrm{A} a} \, \frac{\sin(k_\mathrm{P} b \sin\vartheta\sin\varphi)}{k_\mathrm{P} b \sin\vartheta\sin\varphi}, \end{aligned} \tag{13.287}$$

$$\begin{aligned} & H_{\mathrm{S}\vartheta}^{f_z}(\vartheta,\varphi,\omega,k_\mathrm{A}) \\ & \quad = -\frac{abF(\omega)}{\pi c_\mathrm{S} Z_\mathrm{S}} \sin\vartheta \, \frac{\sin(k_\mathrm{S} a \sin\vartheta\cos\varphi - k_\mathrm{A} a)}{k_\mathrm{S} a \sin\vartheta\cos\varphi - k_\mathrm{A} a} \, \frac{\sin(k_\mathrm{S} b \sin\vartheta\sin\varphi)}{k_\mathrm{S} b \sin\vartheta\sin\varphi} \end{aligned} \tag{13.288}$$

instead of (13.279) and (13.280); the respective diagrams for $k_\mathrm{P} a = 10.65$ ($k_\mathrm{S} a = 19.63$) in Figure 13.18 illustrate that k_A turns into the parameter of a main lobe steering, where the factor α in $k_\mathrm{A} = \alpha k_\mathrm{P}$ equals the sine of the steering angle $\vartheta_{\mathrm{P}s}$ of the P-main lobe:

$$\sin\vartheta_{\mathrm{P}s} = \frac{k_\mathrm{A}}{k_\mathrm{P}} = \alpha. \tag{13.289}$$

This is obtained putting the argument $k_\mathrm{P} a \sin\vartheta\cos\varphi - k_\mathrm{A} a$ of the sinc function for $\varphi = 0$ equal to zero. Interesting enough, the steering of the P-main lobe simultaneously yields pronounced main lobes also for the S-diagrams. The steering prevents the "destruction" of the aperture main lobe by the point directivity $\sin\vartheta$. However, due to

$$\sin\vartheta_{\mathrm{S}s} = \frac{k_\mathrm{A}}{k_\mathrm{S}} = \frac{\alpha}{\kappa}, \tag{13.290}$$

the steering angle $\vartheta_{\mathrm{S}s}$ is smaller than $\vartheta_{\mathrm{P}s}$.

An interesting limiting case for beam steering results if $\alpha = 1$, i.e., $\vartheta_{\mathrm{P}s} = \pi/2$: The P-main lobe of the aperture itself—without point directivity (Figure 13.19)—points into the direction 90°, even though broadened as it is typical for such an end fire radiator (for electromagnetic waves: Balanis 1997); with the point directivity (with elastodynamics), the zero of $\cos\vartheta$ in this direction yields a completely different radiation pattern: The P-wave is now radiated under the angle(s) $\vartheta_{\mathrm{P}s} \simeq 64^\circ$ ($\vartheta_{\mathrm{P}s} \simeq 116^\circ$); it actually turns out to be the bulk pressure wave that accompanies the "creeping wave" of the accordingly named transducer (Erhard 1982; Langenberg et al. 1990). Figure 13.19 confirms that this P-wave is additionally accompanied by an S-main lobe under the angle $\vartheta_{\mathrm{S}s} \simeq 33^\circ$ (for steel). Apparently, the steering angle of the S-main lobe may be increased for increasing α as long as $\alpha/\kappa < 1$. As a consequence, the P-diagram fans out just like the S-diagram in Figure 13.17; it becomes useless.

Impulse radiation from a planar rectangular force density aperture: The frequency appears twice in the particle displacement far-field (13.282): in the exponential function and in the radiation characteristic. Therefore, if we are able to calculate $H_\mathrm{P}^{f_z}(\vartheta,\varphi,t)$ and $H_{\mathrm{S}\vartheta}^{f_z}(\vartheta,\varphi,t)$ via Fourier inversion

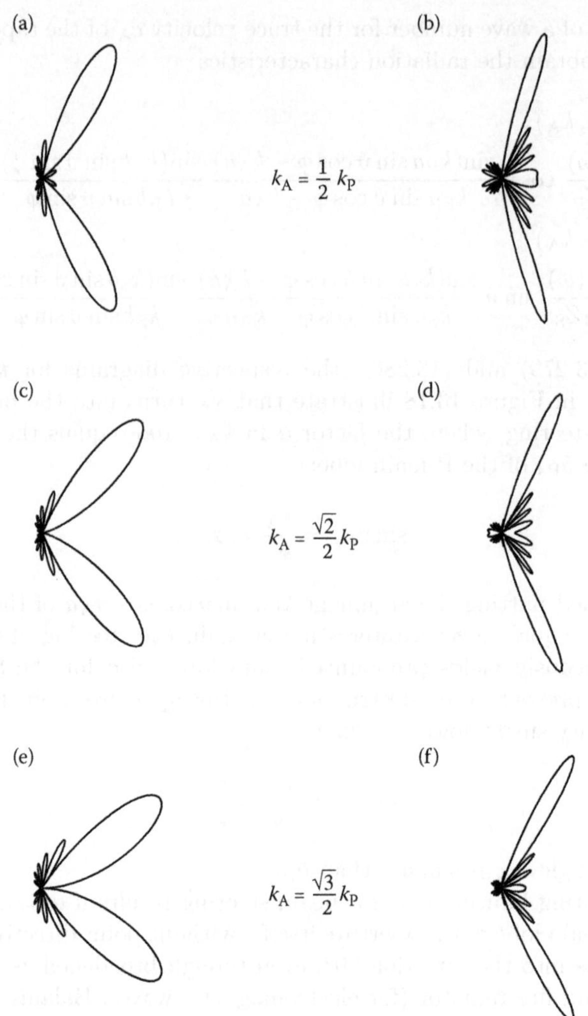

FIGURE 13.18
P- and S-radiation patterns [P: (a), (c), (e); S: (b), (d), (f)] of a planar rectangular f_z-force density aperture with linear x-phase distribution; $y = 0$ (steel: $c_{\mathrm{P}} = 5900$ m/s, $c_{\mathrm{S}} = 3200$ m/s, $\rho = 7.7 \cdot 10^3$ kg/m^3; $k_{\mathrm{P}}a = 10.65$, $k_{\mathrm{S}}a = 19.63$).

from $H_{\mathrm{P}}^f(\vartheta, \varphi, \omega)$ and $H_{\mathrm{S}\vartheta}^f(\vartheta, \varphi, \omega)$, the impulsive radiation field of the planar rectangular force density aperture becomes available according to

$$\underline{\mathbf{u}}_{\mathrm{P}}^{\mathrm{far}}(R, \vartheta, \varphi, t) = \frac{1}{R} H_{\mathrm{P}}^{f_z}\left(\vartheta, \varphi, t - \frac{R}{c_{\mathrm{P}}}\right) \underline{\mathbf{e}}_R + \frac{1}{R} H_{\mathrm{S}\vartheta}^{f_z}\left(\vartheta, \varphi, t - \frac{R}{c_{\mathrm{S}}}\right) \underline{\mathbf{e}}_\vartheta \tag{13.291}$$

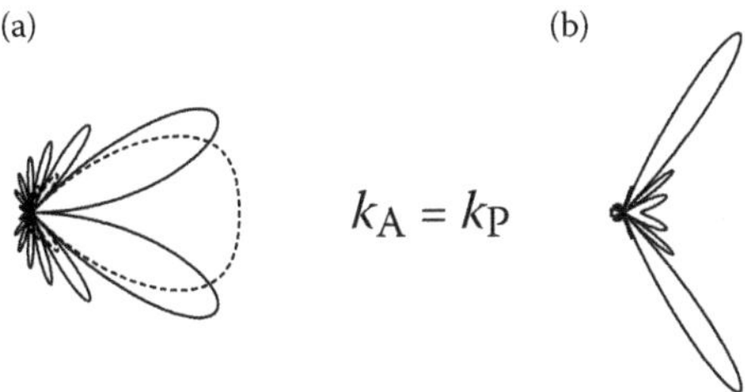

FIGURE 13.19
P- and S-radiation diagrams [P: (a); S: (b)] of a planar rectangular f_z-force density aperture with linear x-phase tapering (dashed: P-radiation diagram of the aperture without point directivity); $y = 0$ (steel: $c_P = 5900$ m/s, $c_S = 3200$ m/s, $\rho = 7.7 \cdot 10^3$ kg/m^3; $k_P a = 10.65$, $k_S a = 19.63$).

due to the shifting rule of the Fourier transform. We again restrict ourselves to the xz-plane, i.e., $y = 0$, and represent

$$H_P^{f_z}(\vartheta, \varphi = 0, t) = \frac{ab}{\pi c_P Z_P} \cos\vartheta \, F(t) * \mathcal{F}^{-1} \left\{ \frac{\sin\left(\dfrac{a}{c_P} \sin\vartheta \, \omega\right)}{\dfrac{a}{c_P} \sin\vartheta \, \omega} \right\} \tag{13.292}$$

as convolution integral of the excitation pulse $F(t)$ with the Fourier inversion of

$$\frac{\sin\left(\frac{a}{c_P} \sin\vartheta \, \omega\right)}{\frac{a}{c_P} \sin\vartheta \, \omega} = \begin{cases} 1 \text{ for } \vartheta = 0, \pi \\ \dfrac{c_P}{a \sin\vartheta} \dfrac{1}{\omega} \dfrac{1}{2j} \left(e^{j\frac{a}{c_P} \sin\vartheta \, \omega} - e^{-j\frac{a}{c_P} \sin\vartheta \, \omega} \right) \text{ for } 0 < \vartheta < \pi \end{cases}, \tag{13.293}$$

where we apparently have to distinguish the two cases in (13.293). The factor $1/\omega$ may be added to the $F(\omega)$-term to obtain

$$\begin{aligned} H_P^{f_z}(\vartheta, \varphi = 0, t) &= \begin{cases} \dfrac{ab}{\pi c_P Z_P} F(t) * \delta(t) \text{ für } \vartheta = 0, \pi \\ \dfrac{b}{2\pi Z_P} \dfrac{\cos\vartheta}{\sin\vartheta} \mathcal{F}^{-1} \left\{ \dfrac{F(\omega)}{-j\omega} \right\} * \left[-\delta\left(t - \dfrac{a}{c_P} \sin\vartheta\right) + \delta\left(t + \dfrac{a}{c_P} \sin\vartheta\right) \right] \\ \qquad \text{für } 0 < \vartheta < \pi \end{cases} \\ &= \begin{cases} \dfrac{ab}{\pi c_P Z_P} F(t) \text{ für } \vartheta = 0, \pi \\ \dfrac{b}{2\pi Z_P} \dfrac{\cos\vartheta}{\sin\vartheta} \displaystyle\int_{-\infty}^{t} \left[F\left(\tau + \dfrac{a}{c_P} \sin\vartheta\right) - F\left(\tau - \dfrac{a}{c_P} \sin\vartheta\right) \right] d\tau \\ \qquad \text{für } 0 < \vartheta < \pi \end{cases} \end{aligned} \tag{13.294}$$

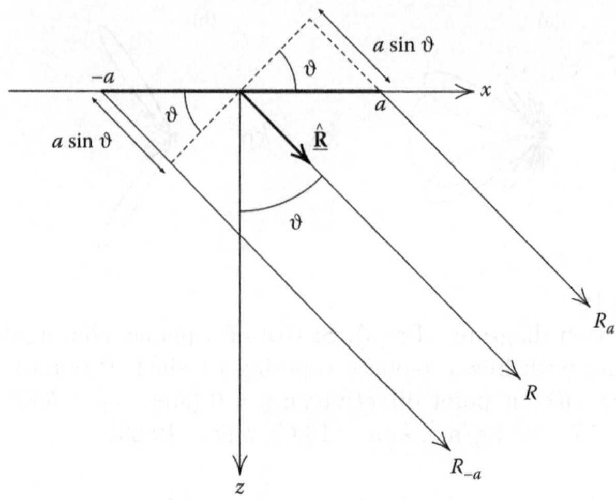

FIGURE 13.20
Far-field impulse radiation from a planar rectangular aperture in the plane $y = 0$.

with the integration rule (Equation 2.293) of the Fourier transform. For a given distance R satisfying $R \gg a$, all frequencies contained in $F(\omega)$ must fulfill the far-field condition $k_P R \gg 1$.

We state that: In the direction of the P-main lobes for $\vartheta = 0$ and $\vartheta = \pi$, the excitation impulse itself is radiated, while in all other directions, two timely separated pulses with different signs are radiated that are integrals of the excitation pulse. Figure 13.20 illustrates the time shift of both pulses identifying them as pulses emanating from the edges of the aperture: Per definition, in the far-field, the distances of all source points to a particular observation point are parallel yielding $R_a = R - a \sin\vartheta$ and $R_{-a} = R + a\sin\vartheta$; accordingly, the travel times of the pulses emanating from the edges differ by $2a\sin\vartheta/c_P$, and they alone compose the pulsed radiation far-field for $0 < \vartheta < \pi$ according to (13.294). However, for $\vartheta = 0, \pi$, a single impulse is observed that emanates from the center of the aperture because, in connection with (13.291), its travel time is given by R/c_P. Certainly, this impulse is also obtained in the limit superimposing the two edge pulses for $\vartheta \longrightarrow 0, \pi$; in other words, the pulsed far-field in the direction of the (time harmonic) main lobe(s) results from the isochronous superposition of both edge pulses. The transition from the near- to the far-field will be discussed with the help of AFIT/EFIT-simulations of piezoelectric transducers (Sections 14.1.1 and 14.2.1). Here, we refer to Figure 13.21 that displays the above facts of the pulsed far-field for an RC2(t)-excitation of the aperture [we have $\text{RC2}(\omega = 0) = 0$ as it is required for the application of the integration rule of the Fourier transform], where we have assumed $c_P T < 2a$; for $c_P T > 2a$, the two edge pulses can no longer be

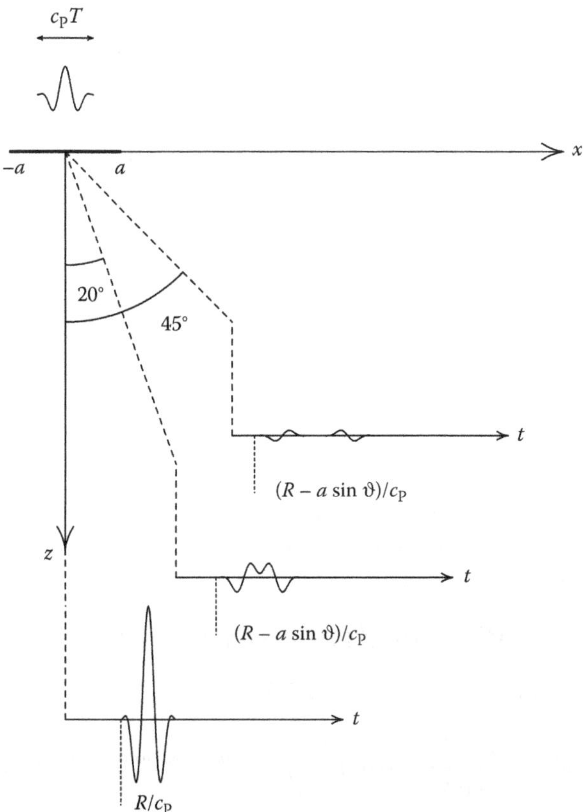

FIGURE 13.21
P-far-field particle displacement pulses from a planar rectangular f_z-force density aperture for RC2(t)-excitation; $y = 0$ (steel: $c_P = 5900$ m/s, $c_S = 3200$ m/s, $\rho = 7.7 \cdot 10^3$ kg/m^3).

separated, even for $\vartheta = \pi/2$. Note that the edge pulses in Figure 13.21 are actually integrated RC2(t)-pulses. The monochromatic beam steering of the P-main lobe manifests itself in the pulsed radiation field through the radiation of the single RC2(t)-pulse in the direction of the steered beam.

The above similarly holds for the pulsed S-radiation far-field.

Planar circular force density distribution in full-space: Instead of the planar rectangular force density distribution [Figure 13.16(a)] we now consider a planar circular force density distribution with radius a and constant amplitude [Figure 13.22(a)]:

$$\underline{\mathbf{f}}(\underline{\mathbf{R}}, \omega) = F(\omega) u(a - r) \delta(z) \underline{\mathbf{e}}_z; \tag{13.295}$$

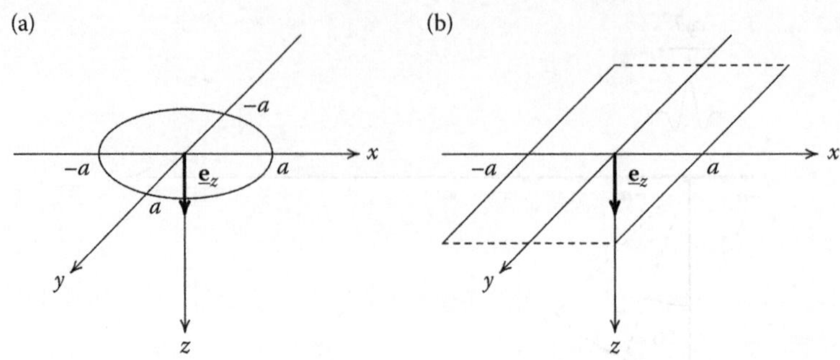

FIGURE 13.22
Planar circular force density distribution (a) and planar strip-like force density distribution (b).

$u(a-r)$ denotes the unit step-function that ensures the circular geometry and the constant amplitude of the force density distribution. Source point polar coordinates r', φ' in the xy-plane yield

$$\underline{\mathbf{R}}' = r' \cos\varphi' \, \underline{\mathbf{e}}_x + r' \sin\varphi' \, \underline{\mathbf{e}}_y + z' \, \underline{\mathbf{e}}_z, \tag{13.296}$$

and hence instead of (13.276) and (13.277):

$$H_{\mathrm{P}}^{f_z}(\hat{\underline{\mathbf{R}}}, \omega) = \frac{F(\omega)}{4\pi c_{\mathrm{P}} Z_{\mathrm{P}}} \cos\vartheta \int_0^{2\pi} \int_0^a \mathrm{e}^{-\mathrm{j}k_{\mathrm{P}} r' \sin\vartheta \cos(\varphi - \varphi')} r' \, \mathrm{d}r' \mathrm{d}\varphi', \tag{13.297}$$

$$H_{\mathrm{S}\vartheta}^{f_z}(\hat{\underline{\mathbf{R}}}, \omega) = -\frac{F(\omega)}{4\pi c_{\mathrm{S}} Z_{\mathrm{S}}} \sin\vartheta \int_0^{2\pi} \int_0^a \mathrm{e}^{-\mathrm{j}k_{\mathrm{S}} r' \sin\vartheta \cos(\varphi - \varphi')} r' \, \mathrm{d}r' \mathrm{d}\varphi'. \tag{13.298}$$

With the rotational symmetry of the source with regard to φ, we argue for the rotational symmetry of the radiation characteristics, i.e., their respective independence upon φ; therefore, we can insert any value for φ, for example, $\varphi = 0$ (or $\varphi = \pi$), into (13.297) and (13.298), and simultaneously, we interchange the order of integration:

$$H_{\mathrm{P}}^{f_z}(\vartheta, \omega) = \frac{F(\omega)}{4\pi c_{\mathrm{P}} Z_{\mathrm{P}}} \cos\vartheta \int_0^a r' \int_0^{2\pi} \mathrm{e}^{-\mathrm{j}k_{\mathrm{P}} r' \sin\vartheta \cos\varphi'} \, \mathrm{d}\varphi' \mathrm{d}r', \tag{13.299}$$

$$H_{\vartheta}^{f_z}(\mathrm{S}\vartheta, \omega) = -\frac{F(\omega)}{4\pi c_{\mathrm{S}} Z_{\mathrm{S}}} \sin\vartheta \int_0^a r' \int_0^{2\pi} \mathrm{e}^{-\mathrm{j}k_{\mathrm{S}} r' \sin\vartheta \cos\varphi'} \, \mathrm{d}\varphi' \mathrm{d}r'. \tag{13.300}$$

To evaluate the aperture integrals, we utilize the integral representation

$$\mathrm{J}_0(\zeta) = \frac{1}{2\pi} \int_0^{2\pi} \mathrm{e}^{\pm \mathrm{j}\zeta \cos\alpha} \, \mathrm{d}\alpha \tag{13.301}$$

for the Bessel function of zero order (Abramowitz and Stegun 1965), the recursion relation

$$\mathrm{J}_0(\zeta) = \mathrm{J}_1'(\zeta) + \frac{1}{\zeta}\,\mathrm{J}_1(\zeta) \tag{13.302}$$

and the partial integration:

$$H_{\mathrm{P}}^{f_z}(\vartheta, \omega) = \frac{a^2 F(\omega)}{2c_{\mathrm{P}} Z_{\mathrm{P}}} \cos\vartheta \, \frac{\mathrm{J}_1(k_{\mathrm{P}} a \sin\vartheta)}{k_{\mathrm{P}} a \sin\vartheta}, \tag{13.303}$$

$$H_{\mathrm{S}\vartheta}^{f_z}(\vartheta, \omega) = -\frac{a^2 F(\omega)}{2c_{\mathrm{S}} Z_{\mathrm{S}}} \sin\vartheta \, \frac{\mathrm{J}_1(k_{\mathrm{S}} a \sin\vartheta)}{k_{\mathrm{S}} a \sin\vartheta}. \tag{13.304}$$

As compared to (13.279) and (13.280), only the aperture radiation characteristic has changed! For $\vartheta = 0, \pi$, we have

$$\left. \frac{\mathrm{J}_1(k_{\mathrm{P,S}} a \sin\vartheta)}{k_{\mathrm{P,S}} a \sin\vartheta} \right|_{\vartheta=0,\pi} = \frac{1}{2} \tag{13.305}$$

basically suggesting radiation patterns as in Figure 13.17; however, the P-main lobe of the circular aperture is a little bit broader than the P-main lobe of the rectangular aperture (with the same linear dimension in the xz-plane) because the first zero of $\mathrm{J}_1(\zeta)$ occurs at $\zeta \simeq 3.83$, whereas the sine in (13.279) has its first zero already at $\pi \simeq 3.14$ (Figure 13.23).

Planar strip-like force density aperture in full-space: With

$$\underline{\mathbf{f}}(\underline{\mathbf{R}}, \omega) = F(\omega) q_a(x) \delta(z) \underline{\mathbf{e}}_z, \tag{13.306}$$

we describe the force density distribution of a planar strip-like source with constant amplitude [Figure 13.22(b)]. In the sense of Section 13.3.3, this is

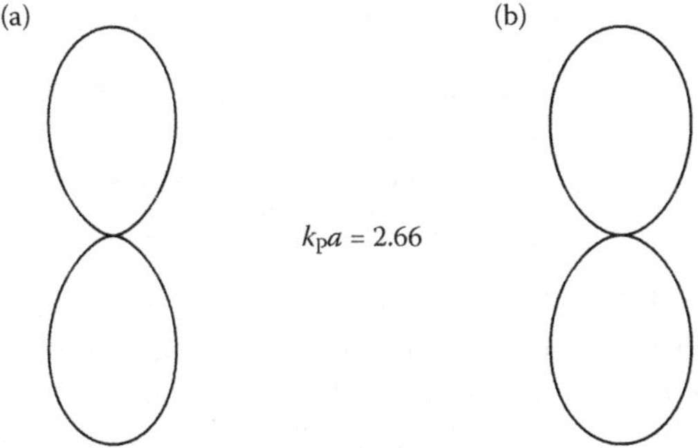

FIGURE 13.23
P-radiation pattern of the strip-like (a) and the circular aperture (b) (steel: $c_{\mathrm{P}} = 5900$ m/s, $c_{\mathrm{S}} = 3200$ m/s, $\rho = 7.7 \cdot 10^3$ kg/m^3).

a two-dimensional source distribution ($\partial/\partial y \equiv 0$) immediately allowing us to apply (13.262) to obtain its PSV-radiation field:

$$\underline{\mathbf{u}}_{\mathrm{PSV}}^{\mathrm{far}}(\underline{\mathbf{r}}, \omega) = \frac{\mathrm{e}^{\mathrm{j}k_{\mathrm{P}}r}}{\sqrt{r}} \underline{\mathbf{H}}_{\mathrm{P}}^{f_z}(\hat{\underline{\mathbf{r}}}, \omega) + \frac{\mathrm{e}^{\mathrm{j}k_{\mathrm{S}}r}}{\sqrt{r}} \underline{\mathbf{H}}_{\mathrm{SV}}^{f_z}(\hat{\underline{\mathbf{r}}}, \omega), \tag{13.307}$$

where we have

$$\underline{\mathbf{H}}_{\mathrm{P}}^{f_z}(\theta, \omega) = \frac{F(\omega)}{\sqrt{\omega}} \frac{\mathrm{e}^{\mathrm{j}\frac{\pi}{4}}}{2Z_{\mathrm{P}}\sqrt{2\pi c_{\mathrm{P}}}} \underline{\mathbf{e}}_z \cdot \hat{\underline{\mathbf{r}}} \int_{-a}^{a} \mathrm{e}^{-\mathrm{j}k_{\mathrm{P}}\hat{\underline{\mathbf{r}}}\cdot\underline{\mathbf{r}}'}\, \mathrm{d}x'\, \hat{\underline{\mathbf{r}}}, \tag{13.308}$$

$$\underline{\mathbf{H}}_{\mathrm{SV}}^{f_z}(\theta, \omega) = \frac{F(\omega)}{\sqrt{\omega}} \frac{\mathrm{e}^{\mathrm{j}\frac{\pi}{4}}}{2Z_{\mathrm{S}}\sqrt{2\pi c_{\mathrm{S}}}} \underline{\mathbf{e}}_z \cdot \underline{\mathbf{e}}_\theta \int_{-a}^{a} \mathrm{e}^{-\mathrm{j}k_{\mathrm{S}}\hat{\underline{\mathbf{r}}}\cdot\underline{\mathbf{r}}'}\, \mathrm{d}x'\, \underline{\mathbf{e}}_\theta \tag{13.309}$$

according to (13.263) and (13.264). Instead of the polar coordinate angle φ counted from the x-axis (Section 13.3.3), we have introduced the respective angle θ counted from the z-axis to achieve comparability with the polar angle counting in Figure 13.16(b); we have $\underline{\mathbf{e}}_\theta = -\underline{\mathbf{e}}_\varphi$. With

$$\hat{\underline{\mathbf{r}}} = \sin\theta\, \underline{\mathbf{e}}_x + \cos\theta\, \underline{\mathbf{e}}_z, \tag{13.310}$$

we obtain

$$\begin{aligned} \hat{\underline{\mathbf{r}}} \cdot \underline{\mathbf{r}}' &= \sin\theta\, x', \\ \underline{\mathbf{e}}_z \cdot \hat{\underline{\mathbf{r}}} &= \cos\theta, \\ \underline{\mathbf{e}}_z \cdot \underline{\mathbf{e}}_\theta &= -\sin\theta \end{aligned} \tag{13.311}$$

and, consequently,

$$\begin{aligned} H_{\mathrm{P}}^{f_z}(\theta, \omega) &= \frac{F(\omega)}{\sqrt{\omega}} \frac{\mathrm{e}^{\mathrm{j}\frac{\pi}{4}}}{2Z_{\mathrm{P}}\sqrt{2\pi c_{\mathrm{P}}}} \cos\theta \int_{-a}^{a} \mathrm{e}^{-\mathrm{j}k_{\mathrm{P}}\sin\theta x'}\, \mathrm{d}x' \\ &= \frac{F(\omega)}{\sqrt{\omega}} \frac{a\mathrm{e}^{\mathrm{j}\frac{\pi}{4}}}{2Z_{\mathrm{P}}\sqrt{2\pi c_{\mathrm{P}}}} \cos\theta\, \frac{2\sin(k_{\mathrm{P}}a\sin\theta)}{k_{\mathrm{P}}a\sin\theta}, \end{aligned} \tag{13.312}$$

$$\begin{aligned} H_{\mathrm{SV}}^{f_z}(\theta, \omega) &= -\frac{F(\omega)}{\sqrt{\omega}} \frac{\mathrm{e}^{\mathrm{j}\frac{\pi}{4}}}{2Z_{\mathrm{S}}\sqrt{2\pi c_{\mathrm{S}}}} \sin\theta \int_{-a}^{a} \mathrm{e}^{-\mathrm{j}k_{\mathrm{S}}\sin\theta x'}\, \mathrm{d}x' \\ &= -\frac{F(\omega)}{\sqrt{\omega}} \frac{a\mathrm{e}^{\mathrm{j}\frac{\pi}{4}}}{2Z_{\mathrm{S}}\sqrt{2\pi c_{\mathrm{S}}}} \sin\theta\, \frac{2\sin(k_{\mathrm{S}}a\sin\theta)}{k_{\mathrm{S}}a\sin\theta}. \end{aligned} \tag{13.313}$$

The comparison with (13.279) and (13.280) tells us that normalized monochromatic radiation patterns of the strip source are completely identical with the respective radiation patterns of the rectangular source for $y = 0$ ($\varphi = 0, \pi$): In this sense, the 2D-approximation of a 3D-strip source is perfect! It is not perfect regarding the behavior with distance and regarding pulsed radiation: Compared to (13.292), an additional convolution of the factor $1/\sqrt{\omega}$ appears as we already noted for the far-field approximation of the bandlimited two-dimensional (scalar) Green function (13.54).

Equally not perfect is the 2D-approximation of a circular aperture concerning pulsed radiation and radiation patterns: The Bessel function radiation patterns (13.303) and (13.304) differ in terms of their half-width from the sinc function radiation patterns (13.312) and (13.313); Figure 13.23 shows an example for $k_{\mathrm{P}}a = 2.66$, where $2a$ either denotes the width of the strip or the diameter of the circular aperture.

Acoustic source far-fields of planar dilatation rate distributions in full-space: Our preliminary model for piezoelectric transducers may also be used in immersion mode: We locate the planar source distribution in a liquid (water) to obtain a model for an immersion technique transducer. With the integral representations (5.63) and (5.73) of acoustic source fields, we are able to calculate their radiation field. Usually (Schmerr 1998), the dilatation rate $h(\underline{\mathbf{R}}, t)$ is prescribed for an immersion transducer (and not the force density), be it either planar rectangular or planar circular. With (13.46), we obtain the time harmonic far-field of the pressure as well as the particle velocity:

$$p^{\mathrm{far}}(\underline{\mathbf{R}}, \omega) = \mathrm{j}\omega\rho \frac{\mathrm{e}^{\mathrm{j}kR}}{4\pi R} \iiint_{V_Q} h(\underline{\mathbf{R}}', \omega)\, \mathrm{e}^{-\mathrm{j}k\hat{\underline{\mathbf{R}}}\cdot\underline{\mathbf{R}}'}\, \mathrm{d}^3\underline{\mathbf{R}}', \tag{13.314}$$

$$\underline{\mathbf{v}}^{\mathrm{far}}(\underline{\mathbf{R}}, \omega) = \mathrm{j}k \frac{\mathrm{e}^{\mathrm{j}kR}}{4\pi R} \iiint_{V_Q} h(\underline{\mathbf{R}}', \omega)\, \mathrm{e}^{-\mathrm{j}k\hat{\underline{\mathbf{R}}}\cdot\underline{\mathbf{R}}'}\, \mathrm{d}^3\underline{\mathbf{R}}'\ \hat{\underline{\mathbf{R}}}, \tag{13.315}$$

allowing for an equally elementary evaluation of the radiation characteristic integrals in (13.314) and (13.315) for planar h-sources, for example, the planar rectangular h-source:

$$h(\underline{\mathbf{R}}, \omega) = H(\omega) q_a(x) q_b(y) \delta(z). \tag{13.316}$$

The essential difference of the h-model of a time harmonic immersion transducer is the missing elastodynamic point directivity. For the pulsed radiation field, it is noteworthy that the factor ω yields a differentiation of the excitation pulse in the main lobe direction, whereas the edge pulses are replica of the excitation pulse. In two dimensions, ω is replaced by the square root of ω leading to a "partial" differentiation in the main lobe direction.[247]

Two facts should be noted:

- In the literature (for example, Schmerr 1998), the immersion transducer is modeled as an infinitely rigid baffled transducer with the so-called Rayleigh Sommerfeld integral. This implies the utilization of a Green function satisfying a Neumann boundary condition on the infinitely extended xy-plane containing the source distribution (13.316). We come back to this mentioning already that essentially only a factor 2 appears compared to (13.314) (Chapter 14).

[247]The necessary Fourier inversion of $\sqrt{|\omega|}$ to evaluate the convolution is given by Doetsch (1967). By the way: The resulting pulse structure is recognized in Figure 5.4 because the Huygens integral of the scattered pressure far-field also contains a factor $\sqrt{\omega}$ in two dimensions.

- The pulsed radiation field of a synchronous planar h-source may be analytically expressed even in the near-field; we refer to the literature (e.g.: Aulenbacher and Langenberg 1980; Schmerr 1998).

13.4 Elementary Spherical Waves and Plane Waves

For US-NDT, the concept of elastic plane waves to estimate the reflection, transmission, and mode conversion is equally important as the concept of elementary elastic spherical waves for a point source synthesis of elastodynamic source fields. Actually, both wave types may be converted into each other: A plane wave may be represented by an infinite series of (multipole) spherical waves, and a spherical wave by the integration of plane waves of arbitrary propagation directions: Instead of the point source synthesis of elastodynamic source fields, a spatial plane wave spectrum decomposition is obtained. This is considerably important for planar sources located on the surface of an elastic half-space and not, as discussed before, as planar "volume" sources in an elastic full-space; in fact, this is the actual transducer modeling problem (at least in an idealized manner); this is the reason why we called the results of Figures 13.17 through 13.19 and 13.21 results of preliminary transducer models.

13.4.1 Spatial plane wave spectrum of the three-dimensional scalar Green function: Weyl's integral representation

We write the differential equation (13.1) for the three-dimensional scalar Green function $G(\underline{\mathbf{R}} - \underline{\mathbf{R}}', \omega)$ in cartesian coordinates:

$$\begin{aligned} &\frac{\partial^2}{\partial x^2} G(x-x', y-y', z-z', \omega) + \frac{\partial^2}{\partial y^2} G(x-x', y-y', z-z', \omega) \\ &\quad + \frac{\partial^2}{\partial z^2} G(x-x', y-y', z-z', \omega) + k^2 G(x-x', y-y', z-z', \omega) \\ &\quad = -\delta(x-x')\delta(y-y')\delta(z-z'). \end{aligned} \tag{13.317}$$

Instead of the three-dimensional Fourier transform (13.12), we now introduce a two-dimensional Fourier transform with respect to x and y (according to Figure 13.16, the coordinate z should point into the specimen why we leave it untouched):

$$\hat{G}(K_x, K_y, z, \omega) = \int_{-\infty}^{\infty} \int_{-\infty}^{\infty} G(x, y, z, \omega)\, \mathrm{e}^{-\mathrm{j}K_x x - \mathrm{j}K_y y}\, \mathrm{d}x \mathrm{d}y, \tag{13.318}$$

where K_x, K_y denote the conjugate Fourier variables referring to x and y; as in Section 11.1, the transformed functions are characterized by a hat to distinguish them from (13.12). With the shifting rule, the differentiation rule, and the sifting property of the δ-distribution, the partial differential equation (13.317) turns into the ordinary differential equation

$$\begin{aligned}
&- K_x^2\hat{G}(K_x, K_y, z - z', \omega) - K_y^2\hat{G}(K_x, K_y, z - z', \omega)\\
&\quad + \frac{\mathrm{d}^2}{\mathrm{d}z^2}\hat{G}(K_x, K_y, z - z', \omega) + k^2\hat{G}(K_x, K_y, z - z', \omega)\\
&\quad = -\delta(z - z').
\end{aligned} \tag{13.319}$$

The conversion

$$\frac{\mathrm{d}^2}{\mathrm{d}z^2}\hat{G}(K_x, K_y, z - z', \omega) + \underbrace{(k^2 - K_x^2 - K_y^2)}_{\stackrel{\text{def}}{=} K_z^2}\hat{G}(K_x, K_y, z - z', \omega) = -\delta(z - z') \tag{13.320}$$

identifies this differential equation as defining equation of the one-dimensional Green function $\hat{G}(K_x, K_y, z - z', \omega)$ with respect to the coordinate z. If Equation 13.320 was a homogeneous equation [compare (11.2)] linearly independent solutions could be one-dimensional plane waves with wave number K_z propagating into positive as well as negative z-direction. Yet, here, we have to require that the waves propagate away from the source point z'. Consequently,

$$\hat{G}(K_x, K_y, z - z', \omega) = \hat{G}_0(K_x, K_y, \omega)\,\mathrm{e}^{\mathrm{j}K_z|z-z'|} \tag{13.321}$$

would be a suitable solution. As for the three-dimensional case (Section 13.1.1), we must calculate the amplitude factor $\hat{G}_0(K_x, K_y, \omega)$; for the one-dimensional case, this is comparatively simple because we only have to insert (13.321) into (13.320) differentiating in the distributional sense:[248]

$$\hat{G}_0(K_x, K_y, \omega) = \frac{\mathrm{j}}{2K_z}. \tag{13.322}$$

---

[248] Write

$$f(z, z') = |z - z'| = (z - z')\,\mathrm{sign}(z - z')$$

and differentiate

$$\begin{aligned}
\frac{\mathrm{d}f(z, z')}{\mathrm{d}z} &= \mathrm{sign}(z - z') + 2(z - z')\delta(z - z')\\
&= \mathrm{sign}(z - z'),\\
\frac{\mathrm{d}^2 f(z, z')}{\mathrm{d}z^2} &= 2\delta(z - z').
\end{aligned}$$

Furthermore, consider $\mathrm{sign}^2(z - z') = 1$ and

$$\mathrm{e}^{\mathrm{j}K_z|z-z'|}\delta(z - z') = \delta(z - z').$$

Thus, the three-dimensional scalar Green function with source point $x' = y' = z' = 0$ results as Weyl's integral representation in terms of the inverse Fourier integral (13.318):

$$
\begin{aligned}
G(x,y,z,\omega) &= \frac{\mathrm{e}^{\mathrm{j}k\sqrt{x^2+y^2+z^2}}}{4\pi\sqrt{x^2+y^2+z^2}} \\
&= \frac{1}{(2\pi)^2}\int_{-\infty}^{\infty}\int_{-\infty}^{\infty}\frac{\mathrm{j}}{2K_z}\,\mathrm{e}^{\mathrm{j}|z|K_z}\mathrm{e}^{\mathrm{j}K_x x+\mathrm{j}K_y y}\,\mathrm{d}K_x\mathrm{d}K_y \\
&= \frac{1}{(2\pi)^2}\,\frac{\mathrm{j}}{2}\int_{-\infty}^{\infty}\int_{-\infty}^{\infty}\frac{\mathrm{e}^{\mathrm{j}|z|\sqrt{k^2-K_x^2-K_y^2}}}{\sqrt{k^2-K_x^2-K_y^2}}\,\mathrm{e}^{\mathrm{j}K_x x+\mathrm{j}K_y y}\,\mathrm{d}K_x\mathrm{d}K_y.
\end{aligned}
\tag{13.323}
$$

The interpretation of Weyl's integral representation (13.323) as spatial spectrum of plane waves is achieved before the background of the considerations in Section 11.1 if we write (13.323) as

$$
G(x,y,z,\omega) = \frac{1}{(2\pi)^2}\,\frac{\mathrm{j}}{2}\int_{-\infty}^{\infty}\int_{-\infty}^{\infty}\frac{1}{K_z}\,\mathrm{e}^{\mathrm{j}\underline{\mathbf{K}}\cdot\underline{\mathbf{R}}}\,\mathrm{d}K_x\mathrm{d}K_y \tag{13.324}
$$

in the half-space $z > 0$ ($|z| = z$). Evidently, the "aperture" distribution $\hat{p}_0(K_x, K_y, \omega) \Longrightarrow \mathrm{j}/2K_z$ in the xy-plane yields the radiation field of a point source in terms of an elementary spherical wave for $z > 0$; the representation (13.324) is the spatial spectral plane wave decomposition of the Green function.

Since the coordinate z in (13.323) appears only as $|z|$, the above definition equally holds for the half-space $z < 0$, i.e., Figure 11.1 has to be symmetrically complemented for $z < 0$.

With

$$
\begin{aligned}
G(x,z,\omega) &= \frac{\mathrm{j}}{4}\,\mathrm{H}_0^{(1)}\left(k\sqrt{x^2+z^2}\right) \\
&= \frac{1}{2\pi}\,\frac{\mathrm{j}}{2}\int_{-\infty}^{\infty}\frac{\mathrm{e}^{\mathrm{j}|z|\sqrt{k^2-K_x^2}}}{\sqrt{k^2-K_x^2}}\,\mathrm{e}^{\mathrm{j}K_x x}\,\mathrm{d}K_x,
\end{aligned}
\tag{13.325}
$$

we may immediately write down the spectral plane wave decomposition of the two-dimensional scalar Green function.

An elementary spherical wave emanating from the origin is described by the Green function $\mathrm{e}^{\mathrm{j}kR}/4\pi R$ independent of the distance R; therefore, the application of the method of stationary phase to (13.323) according to (11.32) yields nothing but this:

$$
\begin{aligned}
G^{\mathrm{far}}(R,\omega) &= -\frac{\mathrm{j}}{2\pi}\,k\,\frac{\mathrm{j}}{2}\,\frac{1}{\sqrt{k^2-k^2\sin^2\vartheta\cos^2\varphi-k^2\sin^2\vartheta\sin^2\varphi}}\cos\vartheta\,\frac{\mathrm{e}^{\mathrm{j}kR}}{R} \\
&= \frac{1}{4\pi}\,\frac{\mathrm{e}^{\mathrm{j}kR}}{R}.
\end{aligned}
\tag{13.326}
$$

An elementary cylindrical wave is described by $\mathrm{j}/4\mathrm{H}_0^{(1)}(k\rho)$, where the Hankel function exhibits an explicit far-field behavior; application of the one-dimensional method of stationary phase according to (11.29) to the integral (13.325), therefore, yields exactly this (Equation 13.23 with Equation 13.22):

$$\begin{aligned} G^{\mathrm{far}}(\rho, \omega) &= \frac{\mathrm{j}}{2} k \cos\theta \frac{\mathrm{e}^{\mathrm{j}k\rho - \mathrm{j}\frac{\pi}{4}}}{\sqrt{2\pi k\rho}} \frac{1}{\sqrt{k^2 - k^2 \sin^2\theta}} \\ &= \frac{\mathrm{j}}{4} \sqrt{\frac{2}{\pi k}} \frac{\mathrm{e}^{\mathrm{j}k\rho - \mathrm{j}\frac{\pi}{4}}}{\sqrt{\rho}}. \end{aligned} \tag{13.327}$$

13.4.2 Spatial cylindrical wave spectrum of the three-dimensional scalar Green function: Sommerfeld integral

The formal transition to polar coordinates

$$\begin{aligned} x &= r\cos\varphi, \\ y &= r\sin\varphi \end{aligned} \tag{13.328}$$

in the xy-plane and polar coordinates

$$\begin{aligned} K_x &= K_r \cos\varphi_K, \\ K_y &= K_r \sin\varphi_K \end{aligned} \tag{13.329}$$

in the K_xK_y-plane turns (13.323) into

$$G(x, y, z, \omega) = \frac{1}{(2\pi)^2} \frac{\mathrm{j}}{2} \int_0^\infty \int_0^{2\pi} \frac{\mathrm{e}^{\mathrm{j}|z|\sqrt{k^2 - K_r^2}}}{\sqrt{k^2 - K_r^2}} \mathrm{e}^{\mathrm{j}K_r r\cos(\varphi - \varphi_K)} K_r \, \mathrm{d}\varphi_K \mathrm{d}K_r. \tag{13.330}$$

The rotational symmetry with respect to φ and the integral representation (13.301) of the Bessel function reveal (13.330) to be the spectral decomposition of the three-dimensional Green function into cylindrical waves:

$$\begin{aligned} G(x, y, z, \omega) &= \frac{\mathrm{e}^{\mathrm{j}k\sqrt{r^2 + z^2}}}{4\pi\sqrt{r^2 + z^2}} \\ &= \frac{\mathrm{j}}{4\pi} \int_0^\infty \frac{\mathrm{e}^{\mathrm{j}|z|\sqrt{k^2 - K_r^2}}}{\sqrt{k^2 - K_r^2}} K_r \, \mathrm{J}_0(K_r r) \, \mathrm{d}K_r. \end{aligned} \tag{13.331}$$

Again, we have to ensure that the square root in the exponential has the correct (positive) sign for $K_r > k$. The integral representation (13.331) is associated with the name of Sommerfeld.

The integral transform $\mathcal{B}\{\bullet\}$ according to

$$\begin{aligned} \Phi(r) &= \frac{1}{2\pi} \mathcal{B}\{\overline{\Phi}(K_r)\} \\ &= \frac{1}{2\pi} \int_0^\infty \overline{\Phi}(K_r) \, \mathrm{J}_0(rK_r) \, K_r \, \mathrm{d}K_r \end{aligned} \tag{13.332}$$

is called (inverse) Fourier–Bessel transform . It is symmetric with regard to the transform itself:

$$\begin{aligned}\overline{\Phi}(K_r) &= 2\pi\,\mathcal{B}\{\Phi(r)\}\\ &= 2\pi\int_0^\infty \Phi(r)\,\mathrm{J}_0(rK_r)\,r\,\mathrm{d}r.\end{aligned}\tag{13.333}$$

14

Force Density and Dilatation Rate Sources on Surfaces of Homogeneous Isotropic Half-Spaces, Radiation Fields of Piezoelectric Transducers

14.1 Acoustic Half-Spaces with Soft or Rigid Surfaces

14.1.1 AFIT-wavefronts of the line and strip-like rigidly baffled aperture radiator

AFIT—the Acoustic Finite Integration Technique—is the acronym for a numerical code to calculate the radiation, propagation, and scattering of acoustic waves (Wolter 1995; Marklein 1997): The propagation physics of acoustic waves, mathematically formulated with the acoustic governing equations, is literally visualized that way. Here, we calculate the pulsed wave field of an acoustic aperture radiator on the acoustically rigid surface of a half-space (infinitely rigid baffled transducer) using AFIT and display the field in terms of pulsed wavefronts (Figure 14.1). Afterward, we know what the respective analytical calculation should deliver, and we will see what it can deliver. Figure 14.1 (top) displays the pulsed wavefront of the pressure for a line dilatation rate source with RC2(t)-time dependence [more precisely: the time derivative of a dilatation rate source in order to account for the factor jω in (14.23)]; in the terminology of the next section, this is the RC2(ω)-bandlimited two-dimensional Green function satisfying a Neumann boundary condition: $G^{\mathrm{N}}(x, z, t) * \mathrm{RC2}(t)$. The result meets our expectations: We observe a (two-dimensional) semicircular wavefront that, according to our assumption, satisfies a Neumann boundary condition on the rigid surface, i.e., the normal component of the particle velocity vanishes. Figure 14.1 (bottom) shows the respective pulsed wavefronts of the pressure radiated from a strip-like dilatation rate aperture with constant amplitude: According to the scalar Huygens principle, each aperture point radiates semicircular wavefronts forming a plane pressure wavefront as envelope that is tangential to the semicircular wavefronts emanating from the aperture edges, it has the same lateral dimension as the aperture itself. Note: With increasing travel time, the radius of the aperture edge pulses increases, they nestle more and more against the aperture

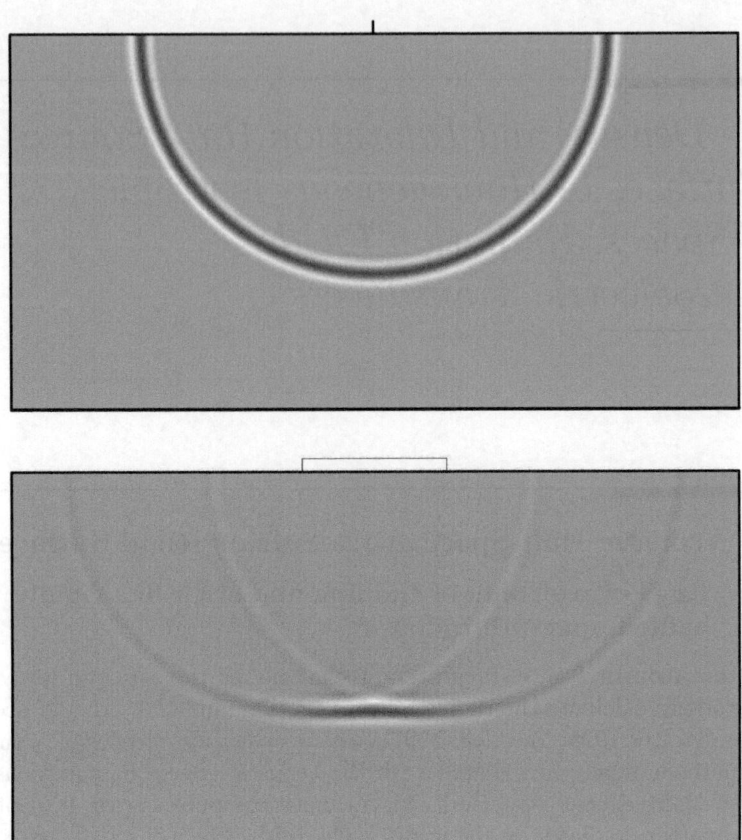

FIGURE 14.1
AFIT-wavefronts of the line and strip-like infinitely rigid baffled aperture radiator (acoustic pressure with pressure prescribed within the aperture).

wavefront; depending on the excitation pulse duration, they are "someday" no longer distinguishable from the aperture wavefront, and we will see later on that this leads to the transition near-field $\Longrightarrow$ far-field defining the near-field length.

14.1.2 Scalar half-space Green functions, Rayleigh–Sommerfeld integrals, plane wave spectral decomposition (integral representations of the Weyl type)

The theory of acoustic and elastodynamic source fields in Section 13.3 has been called a preliminary transducer model because the sources have been located in an infinite full-space. Yet, in general, US-NDT has to cope with finite sized parts that contain sources either in the interior [Figure 14.2(a)] or on the

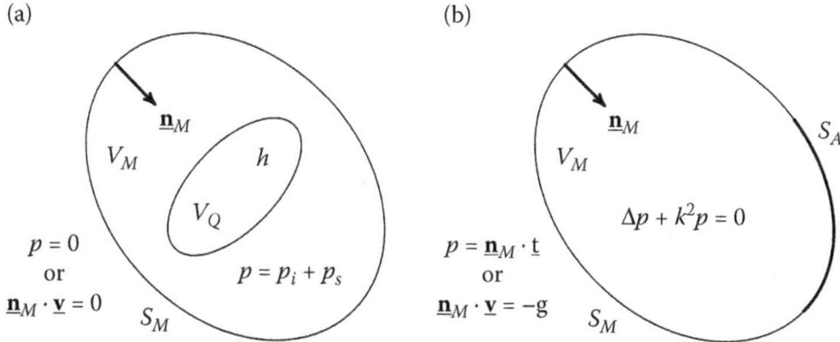

FIGURE 14.2
Volume sources in a "specimen" (a), and Surface sources on the "specimen" (b): Green functions satisfying boundary conditions are required.

surface within an aperture S_A [Figure 14.2(b)]. Normally, the surface S_M of the part volume V_M is also the measurement surface for ultrasonic signals, and in the exterior, we often assume vacuum. The above figures refer to the acoustic case because the solution principles become obvious without the mathematical complexity of elastodynamics.

Volume sources: scalar half-space Green functions: We refer to Figure 14.2(a) and consider for simplicity dilatation, hence h-sources, last but not least because they often serve to model immersion transducers. In full-space, the field of this source would be given by

$$p_i(\underline{\mathbf{R}}, \omega) = \iiint_{V_Q} \mathrm{j}\omega\rho\, h(\underline{\mathbf{R}}', \omega) G(\underline{\mathbf{R}} - \underline{\mathbf{R}}', \omega)\, \mathrm{d}^3\underline{\mathbf{R}}', \quad \underline{\mathbf{R}} \in V_M; \tag{14.1}$$

for $V_Q \subset V_M$, the pressure $p_i(\underline{\mathbf{R}}, \omega)$ should satisfy the boundary conditions specified for $\underline{\mathbf{R}} \in S_M$; this is not achieved by the incident field leading to the generation of secondary sources on S_M in such a way that they radiate a scattered field $p_s(\underline{\mathbf{R}}, \omega)$ superimposing to the incident field $p(\underline{\mathbf{R}}, \omega) = p_i(\underline{\mathbf{R}}, \omega) + p_s(\underline{\mathbf{R}}, \omega)$ to satisfy the boundary condition. According to the Helmholtz formulation of Huygens' principle, the field values $p(\underline{\mathbf{R}}, \omega)$, $\nabla p(\underline{\mathbf{R}}, \omega) \cdot \underline{\mathbf{n}}_M$ themselves are the secondary sources resulting in

$$p_s(\underline{\mathbf{R}}, \omega) = \iint_{S_M} \Big[p(\underline{\mathbf{R}}', \omega) \nabla' G(\underline{\mathbf{R}} - \underline{\mathbf{R}}', \omega) - G(\underline{\mathbf{R}} - \underline{\mathbf{R}}', \omega) \nabla' p(\underline{\mathbf{R}}', \omega) \Big] \cdot \underline{\mathbf{n}}_M'\, \mathrm{d}S' \tag{14.2}$$

as representation of the scattered field for $\underline{\mathbf{R}} \in V_M$ (Section 5.6, respectively Section 15.1.2; note: the normal $\underline{\mathbf{n}}_M$ points into the interior of V_M).

For a soft measurement surface S_M [Dirichlet boundary condition: $p(\underline{\mathbf{R}}', \omega) = 0$ for $\underline{\mathbf{R}}' \in S_M$], (14.2) reduces to

$$p_s^{\mathrm{D}}(\underline{\mathbf{R}}, \omega) = -\iint_{S_M} G(\underline{\mathbf{R}} - \underline{\mathbf{R}}', \omega) \nabla' p(\underline{\mathbf{R}}', \omega) \cdot \underline{\mathbf{n}}_M'\, \mathrm{d}S', \quad \underline{\mathbf{R}} \in V_M, \tag{14.3}$$

and for a rigid measurement surface S_M [Neumann boundary condition: $\nabla' p(\underline{\mathbf{R}}', \omega) \cdot \underline{\mathbf{n}}'_M = 0$ for $\underline{\mathbf{R}}' \in S_M$], we have

$$p_s^{\mathrm{N}}(\underline{\mathbf{R}}, \omega) = \iint_{S_M} p(\underline{\mathbf{R}}', \omega) \nabla' G(\underline{\mathbf{R}} - \underline{\mathbf{R}}', \omega) \cdot \underline{\mathbf{n}}'_M \, \mathrm{d}S', \quad \underline{\mathbf{R}} \in V_M. \tag{14.4}$$

To calculate the respective secondary sources, integral equations are resulting (Section 15.1.2) that have already been addressed in Section 5.6. After solving them, the superpositions $p^{\mathrm{D}} = p_s^{\mathrm{D}} + p_i$, $p^{\mathrm{N}} = p_s^{\mathrm{N}} + p_i$ are solutions of the respective problem.

An alternative to this approach uses the integral representations (14.1) and (14.2) with Green functions $G^{\mathrm{D}}(\underline{\mathbf{R}}', \underline{\mathbf{R}}, \omega)$, $G^{\mathrm{N}}(\underline{\mathbf{R}}', \underline{\mathbf{R}}, \omega)$ that satisfy the Dirichlet

$$G^{\mathrm{D}}(\underline{\mathbf{R}}', \underline{\mathbf{R}}, \omega) = 0, \quad \underline{\mathbf{R}}' \in S_M, \tag{14.5}$$

or the Neumann boundary condition

$$\nabla' G^{\mathrm{N}}(\underline{\mathbf{R}}', \underline{\mathbf{R}}, \omega) \cdot \underline{\mathbf{n}}'_M = 0, \quad \underline{\mathbf{R}}' \in S_M, \tag{14.6}$$

themselves besides the differential equation

$$\Delta' G^{\mathrm{D,N}}(\underline{\mathbf{R}}', \underline{\mathbf{R}}, \omega) + k^2 G^{\mathrm{D,N}}(\underline{\mathbf{R}}', \underline{\mathbf{R}}, \omega) = -\delta(\underline{\mathbf{R}}' - \underline{\mathbf{R}}), \quad \underline{\mathbf{R}}', \underline{\mathbf{R}} \in V_M. \tag{14.7}$$

Based on the reciprocity theorem, we may show (de Hoop 1995) that:

$$G^{\mathrm{D,N}}(\underline{\mathbf{R}}', \underline{\mathbf{R}}, \omega) = G^{\mathrm{D,N}}(\underline{\mathbf{R}}, \underline{\mathbf{R}}', \omega). \tag{14.8}$$

With (14.5), the Dirichlet scattered field (14.3) and, with (14.6), the Neumann scattered field, (14.4) is identically zero; hence, the incident fields

$$\begin{aligned} p_i(\underline{\mathbf{R}}, \omega) &= p^{\mathrm{D,N}}(\underline{\mathbf{R}}, \omega) \\ &= \iiint_{V_Q} \mathrm{j}\omega\rho\, h(\underline{\mathbf{R}}', \omega)\, G^{\mathrm{D,N}}(\underline{\mathbf{R}}', \underline{\mathbf{R}}, \omega)\, \mathrm{d}^3\underline{\mathbf{R}}' \end{aligned} \tag{14.9}$$

fulfill the required boundary conditions[249] for $\underline{\mathbf{R}} \in S_M$ due to (14.8). The problem is simply: How to obtain the Green functions $G^{\mathrm{D,N}}(\underline{\mathbf{R}}', \underline{\mathbf{R}}, \omega)$? After

[249] To show that $p^{\mathrm{D,N}}(\underline{\mathbf{R}}, \omega)$ according to (14.9) in fact satisfies the inhomogeneous differential equation

$$\Delta p^{\mathrm{D,N}}(\underline{\mathbf{R}}, \omega) + k^2 p^{\mathrm{D,N}}(\underline{\mathbf{R}}, \omega) = -\mathrm{j}\omega\rho\, h(\underline{\mathbf{R}}, \omega),$$

we also have to use (14.8). At first, we interchange $\underline{\mathbf{R}}$ and $\underline{\mathbf{R}}'$ in (14.9) according to

$$p^{\mathrm{D,N}}(\underline{\mathbf{R}}', \omega) = \iiint_{V_Q} \mathrm{j}\omega\rho\, h(\underline{\mathbf{R}}, \omega) G^{\mathrm{D,N}}(\underline{\mathbf{R}}, \underline{\mathbf{R}}', \omega)\, \mathrm{d}^3\underline{\mathbf{R}}',$$

and then we replace $G^{\mathrm{D,N}}(\underline{\mathbf{R}}, \underline{\mathbf{R}}', \omega)$ by $G^{\mathrm{D,N}}(\underline{\mathbf{R}}', \underline{\mathbf{R}}, \omega)$ and shift the operator $(\Delta' + k^2)p^{\mathrm{D,N}}(\underline{\mathbf{R}}', \omega)$ to $G^{\mathrm{D,N}}(\underline{\mathbf{R}}', \underline{\mathbf{R}}, \omega)$; with (14.7), we obtain $-\mathrm{j}\omega\rho\, h(\underline{\mathbf{R}}', \omega)$.

all, their calculation is nothing else but the solution of our problem for point-like h-sources. However, there are special cases of S_M-geometries offering a simple access to the calculation of (scalar) Green functions, for example, the planar surface of an (acoustic) half-space volume V_M: We image[250] the unit point source located at $\underline{\mathbf{R}}$ at this plane and construct

$$
\begin{aligned}
G^{\mathrm{D}}(\underline{\mathbf{R}}', \underline{\mathbf{R}}, \omega) &= \frac{\mathrm{e}^{\mathrm{j}k\sqrt{(x'-x)^2+(y'-y)^2+(z'-z)^2}}}{4\pi\sqrt{(x'-x)^2+(y'-y)^2+(z'-z)^2}} \\
&\quad - \frac{\mathrm{e}^{\mathrm{j}k\sqrt{(x'-x)^2+(y'-y)^2+(z'+z)^2}}}{4\pi\sqrt{(x'-x)^2+(y'-y)^2+(z'+z)^2}} \\
&= G^{\mathrm{D}}(x'-x, y'-y, z', z, \omega),
\end{aligned}
\tag{14.10}
$$

$$
\begin{aligned}
G^{\mathrm{N}}(\underline{\mathbf{R}}', \underline{\mathbf{R}}, \omega) &= \frac{\mathrm{e}^{\mathrm{j}k\sqrt{(x'-x)^2+(y'-y)^2+(z'-z)^2}}}{4\pi\sqrt{(x'-x)^2+(y'-y)^2+(z'-z)^2}} \\
&\quad + \frac{\mathrm{e}^{\mathrm{j}k\sqrt{(x'-x)^2+(y'-y)^2+(z'+z)^2}}}{4\pi\sqrt{(x'-x)^2+(y'-y)^2+(z'+z)^2}} \\
&= G^{\mathrm{N}}(x'-x, y'-y, z', z, \omega),
\end{aligned}
\tag{14.11}
$$

where

$$
\begin{aligned}
\underline{\mathbf{R}}' &= x'\underline{\mathbf{e}}_x + y'\underline{\mathbf{e}}_y + z'\underline{\mathbf{e}}_z, \\
\underline{\mathbf{R}} &= x\,\underline{\mathbf{e}}_x + y\,\underline{\mathbf{e}}_y + z\,\underline{\mathbf{e}}_z.
\end{aligned}
\tag{14.12}
$$

We immediately realize that G^{D} satisfies a Dirichlet and G^{N} a Neumann boundary condition for $z' = 0$.

Surface sources: Rayleigh–Sommerfeld integrals: In fact, the following is interesting: If we insert a surface dilatation source

$$
h(\underline{\mathbf{R}}', \omega) = g(x', y', \omega)\delta(z') \tag{14.13}
$$

into (14.9), we immediately have

$$
p^{\mathrm{D}}(\underline{\mathbf{R}}, \omega) \equiv 0, \quad \underline{\mathbf{R}} \in V_M, \tag{14.14}
$$

$$
p^{\mathrm{N}}(\underline{\mathbf{R}}, \omega) = 2\iint_{S_M} \mathrm{j}\omega\rho\, g(x', y', \omega)\, G(x-x', y-y', z, \omega)\, \mathrm{d}x'\mathrm{d}y', \quad \underline{\mathbf{R}} \in V_M, \tag{14.15}
$$

[250]This method of images results mathematically as follows: Complement the solution of the inhomogeneous equation (14.7) with a solution of the homogeneous equation, for example, the field of a point source emanating from a source point $\underline{\mathbf{R}}''$ in the exterior of V_M and determine $\underline{\mathbf{R}}''$ such that the respective boundary condition is satisfied (Langenberg 2005).

with the half-space Green functions, where $G(x-x', y-y', z-z', \omega)$ is the full-space (scalar) Green function! A surface dilatation source on the soft planar surface of a half-space is a nonradiating source (the source is short-circuited analogous to an electric current source on a perfectly conducting surface), while a respective source on the planar rigid surface of a half-space radiates a field twice as large as the one from a source in full-space: The source as well as its image are radiating!

If we want to create a nonzero surface source field with a Dirichlet boundary condition, we must turn to the complementary source, hence the (normal) surface force density. Again, the factor 2 appears as compared to the full-space:

$$p^{\mathrm{D}}(\underline{\mathbf{R}}, \omega) = 2 \iint_{S_M} f_z(x', y', \omega)\, \underline{\mathbf{n}}'_M \cdot \nabla' G(x-x', y-y', z-z', \omega)\Big|_{z'=0} \mathrm{d}x'\mathrm{d}y', \quad \underline{\mathbf{R}} \in V_M, \tag{14.16}$$

$$p^{\mathrm{N}}(\underline{\mathbf{R}}, \omega) \equiv 0, \quad \underline{\mathbf{R}} \in V_M. \tag{14.17}$$

The integrals (14.15) and (14.16) are called Rayleigh–Sommerfeld integrals. In particular, Equation 14.15 is relevant for practical applications because this integral characterizes the infinitely rigid baffled transducer.

Surface sources: spectral plane wave decomposition: Even though we already obtained the radiation field of surface source densities [Figure 14.2(b)] as a limiting case, we want to derive the Rayleigh–Sommerfeld integrals once more pursuing a different way. The reason: Applying the above method to the elastic case runs into problems: The method of images does no longer hold, and hence Green tensors satisfying boundary conditions may not simply be derived from the scalar Green functions (14.10) and (14.11) as it was true in full-space. Green tensors satisfying boundary conditions must be worked out, they result as integral representation of the Weyl type over plane wave spectra! Therefore, we want to practice that method for the scalar case.

For nonzero surface sources, the (time harmonic) acoustic pressure fulfills a homogeneous Helmholtz equation for $\underline{\mathbf{R}} \in V_M$:

$$\Delta p(\underline{\mathbf{R}}, \omega) + k^2 p(\underline{\mathbf{R}}, \omega) = 0, \quad \underline{\mathbf{R}} \in V_M, \tag{14.18}$$

and it must satisfy the inhomogeneous Dirichlet boundary condition (Equation 5.20)

$$p(\underline{\mathbf{R}}, \omega) = \underline{\mathbf{n}}_M \cdot \underline{\mathbf{t}}(\underline{\mathbf{R}}, \omega) \text{ for } \underline{\mathbf{R}} \in S_M, \tag{14.19}$$

or the inhomogeneous Neumann boundary condition (Equation 5.17)

$$\underline{\mathbf{n}}_M \cdot \underline{\mathbf{v}}(\underline{\mathbf{R}}, \omega) = -g(\underline{\mathbf{R}}, \omega) \text{ for } \underline{\mathbf{R}} \in S_M, \tag{14.20}$$

respectively,

$$\underline{\mathbf{n}}_M \cdot \nabla p(\underline{\mathbf{R}}, \omega) = -\mathrm{j}\omega\rho\, g(\underline{\mathbf{R}}, \omega) \text{ for } \underline{\mathbf{R}} \in S_M. \tag{14.21}$$

For a finite aperture $S_A \subset S_M$ that only extends over part of S_M—as indicated in Figure 14.2(b)—the inhomogeneities satisfy $\underline{\mathbf{n}}_M \cdot \underline{\mathbf{t}}(\underline{\mathbf{R}}, \omega) = 0$, $g(\underline{\mathbf{R}}, \omega) = 0$ for $\underline{\mathbf{R}} \in S_M \backslash S_A$.

A solution of (14.18) for $\underline{\mathbf{R}} \in V_M$ is the Helmholtz integral (14.2); utilizing Green functions satisfying the boundary condition G^{D}, respectively G^{N}, as well as, adequately, the inhomogeneous boundary condition (14.19), respectively (14.21), the explicit integral representations

$$p^{\mathrm{D}}(\underline{\mathbf{R}}, \omega) = \iint_{S_M} \underline{\mathbf{n}}'_M \cdot \underline{\mathbf{t}}(\underline{\mathbf{R}}', \omega) \boldsymbol{\nabla}' G^{\mathrm{D}}(\underline{\mathbf{R}}', \underline{\mathbf{R}}, \omega) \cdot \underline{\mathbf{n}}'_M \, \mathrm{d}S', \tag{14.22}$$

$$p^{\mathrm{N}}(\underline{\mathbf{R}}, \omega) = \iint_{S_M} \mathrm{j}\omega\rho \, g(\underline{\mathbf{R}}', \omega) G^{\mathrm{N}}(\underline{\mathbf{R}}', \underline{\mathbf{R}}, \omega) \, \mathrm{d}S' \tag{14.23}$$

result. We state once more—this time for arbitrary surfaces S_M: Prescribing the surface source as dilatation rate $g(\underline{\mathbf{R}}, \omega)$, $\underline{\mathbf{R}} \in S_M$ results in a rigid boundary condition of the radiated pressure (g is "short-circuited" on a soft surface). For planar surfaces S_M—as surfaces of half-spaces—we may again use the analytical expressions (14.10) and (14.11) turning (14.22) and (14.23) into the Rayleigh–Sommerfeld integrals (14.16) and (14.15).

Yet, the motivation for this paragraph was not to use Equations 14.10 and 14.11. Therefore, we do not write the solution of (14.18) for a half-space volume V_M $(z > 0)$ as a Huygens point source synthesis in terms of a Helmholtz integral but as a plane wave spectrum because, according to Section 13.4.1, we may represent the contribution of each point source within the aperture S_A for $z \geq 0$ by a respective spectrum whose arbitrary parameters—(14.18) is a homogeneous equation—must be determined through the appropriate boundary condition (14.19) or (14.20).

As in Section 11.1, we subject the differential equation (14.18) in cartesian coordinates to a two-dimensional spatial Fourier transform with respect to x and y:

$$\frac{\mathrm{d}^2}{\mathrm{d}z^2} \hat{p}^{\mathrm{N}}(K_x, K_y, z, \omega) + \underbrace{(k^2 - K_x^2 - K_y^2)}_{\stackrel{\mathrm{def}}{=} K_z} \hat{p}^{\mathrm{N}}(K_x, K_y, z, \omega) = 0, \tag{14.24}$$

where the upper index indicates that $p^{\mathrm{N}}(x, y, z, \omega)$ should satisfy the inhomogeneous Neumann boundary condition (14.20); for a planar surface S_M, it reads—we use the version (14.21)—

$$\underline{\mathbf{e}}_z \cdot \boldsymbol{\nabla} p^{\mathrm{N}}(x, y, z, \omega)\big|_{z=0} = -\mathrm{j}\omega\rho \, g(x, y, \omega), \quad x, y \in S_M. \tag{14.25}$$

As Fourier transform of (14.25), we obtain

$$\frac{\mathrm{d}}{\mathrm{d}z} \hat{p}^{\mathrm{N}}(K_x, K_y, z, \omega)\Big|_{z=0} = -\mathrm{j}\omega\rho \, \hat{g}(K_x, K_y, \omega). \tag{14.26}$$

Linearly independent solutions of (14.24) are the one-dimensional plane waves

$$\hat{p}^{\mathrm{N}}(K_x, K_y, z, \omega) = \hat{p}^{+}(K_x, K_y, \omega) \, \mathrm{e}^{\mathrm{j}K_z z} + \hat{p}^{-}(K_x, K_y, \omega) \, \mathrm{e}^{-\mathrm{j}K_z z} \tag{14.27}$$

propagating into positive and negative z-direction; the arbitrary amplitudes $\hat{p}^{\pm}(K_x, K_y, \omega)$ depending on the parameters K_x, K_y, ω of the differential equation (14.24) must be matched to the inhomogeneous boundary condition (14.26). Since the surface source density is located on the surface of the propagation half-space $z > 0$, physical intuition requires to put $\hat{p}^{-}(K_x, K_y, \omega)$ equal to zero. Subsequently, Equation 14.26 leads to

$$\hat{p}^{+}(K_x, K_y, \omega) = -\frac{\omega\rho}{K_z}\,\hat{g}(K_x, K_y, \omega). \tag{14.28}$$

The Fourier inversion of (14.27) with (14.28) results in

$$p^{\mathrm{N}}(x, y, z, \omega) = -\omega\rho\,\frac{1}{(2\pi)^2}\int_{-\infty}^{\infty}\int_{-\infty}^{\infty}\hat{g}(K_x, K_y, \omega)\,\frac{\mathrm{e}^{\mathrm{j}K_z z}}{K_z}\,\mathrm{e}^{\mathrm{j}K_x x+\mathrm{j}K_y y}\,\mathrm{d}K_x\mathrm{d}K_y, \tag{14.29}$$

namely the spectral plane wave decomposition of the sound field of a planar aperture radiator, where the two-dimensional Fourier transform of the aperture function $g(x, y, \omega)$ appears as spectral amplitude. For $z = 0$, the pressure $p^{\mathrm{N}}(x, y, z, \omega)$ satisfies the Neumann boundary condition (14.25), hence a rigid boundary condition in the exterior of a finite aperture S_A: The dilatation rate source is rigidly baffled. In Figure 14.1, this is clearly recognized: The pressure does not vanish on the excitation surface.

In Section 11.1, we put "aperture distribution" in quotes to indicate that the field distribution $p_0(x, y, \omega)$ as given in the xy-plane formally plays the same role in the integral representation (11.5) as solution of a homogeneous reduced wave equation as $g(x, y, \omega)$ in (13.302), yet it may not really be considered as a physically real aperture(function): A source is actually not present in a homogeneous equation. By means of the relation (14.28) between $\hat{g}(K_x, K_y, \omega)$ and $\hat{p}^{+}(K_x, K_y, \omega) \Longrightarrow \hat{p}_0(K_x, K_y, \omega)$, we immediately recognize that the factor $1/K_z$ may not lead to an identical physically real finite aperture $g(x, y, \omega)$ for a given finite sized "aperture distribution" $p_0(x, y, \omega)$. On the other hand, a finite aperture $g(x, y, \omega)$ yields an infinitely extended field distribution in the xy-plane—such that the pertinent normal derivative satisfies the inhomogeneous Neumann boundary condition (14.26)—which is understandable from a physical point of view: The surface source density $g(x, y, \omega)$ prescribed for $z = +0$, i.e., close to the rigid half-space surface at $z = 0$, illuminates this surface as an incident field!

Smuggling factors according to

$$p^{\mathrm{N}}(x, y, z, \omega) = 2\mathrm{j}\omega\rho\,\frac{1}{(2\pi)^2}\int_{-\infty}^{\infty}\int_{-\infty}^{\infty}\hat{g}(K_x, K_y, \omega)\,\frac{\mathrm{j}}{2K_z}\,\mathrm{e}^{\mathrm{j}K_z z}\mathrm{e}^{\mathrm{j}K_x x+\mathrm{j}K_y y}\,\mathrm{d}K_x\mathrm{d}K_y \tag{14.30}$$

into (14.29), we recognize the Fourier inversion of the product of the spectra $\hat{g}(K_x, K_y, \omega)$ and $\hat{G}(K_x, K_y, z, \omega)$ according to (13.321) and (13.322) for $z \geq 0$, i.e., the convolution of $g(x, y, \omega)$ with $G(x, y, z, \omega)$:

$$p^{\mathrm{N}}(x,y,z,\omega) = 2\mathrm{j}\omega\rho \int_{-\infty}^{\infty}\int_{-\infty}^{\infty} g(x',y',\omega)\, G(x-x',y-y',z,\omega)\, \mathrm{d}x'\mathrm{d}y'$$
$$= \mathrm{j}\,\frac{\omega\rho}{2\pi} \int_{-\infty}^{\infty}\int_{-\infty}^{\infty} g(x',y',\omega)\, \frac{\mathrm{e}^{\mathrm{j}k\sqrt{x-x')^2+(y-y')^2+z^2}}}{\sqrt{(x-x')^2+(y-y')^2+z^2}}\, \mathrm{d}x'\mathrm{d}y', \tag{14.31}$$

thus having rediscovered the Rayleigh–Sommerfeld integral (14.15) through explicit solution of a half-space boundary value problem.

Therefore, in case we know the Green function of the half-space explicitly, we may alternatively work with the Fourier integral (14.30) or with the convolution integral (14.31). In case we do not know the Green function—as it is true in elastodynamics at least for the time harmonic Green function—we have to take a back seat with the spectral plane wave decomposition (SPWD).

In Figure 14.3, we face alternative methods to calculate the sound field of an acoustic aperture radiator for the case of a strip source, where it is not quite irrespective that which integral is used for a particular distance z: The Weyl integral as spectral plane wave decomposition definitely has advantages if the sound field has to be calculated in planes parallel to the surface for distances not too large (within the near-field length); for larger z-values, the oscillations of the exponential function $\mathrm{e}^{\mathrm{j}K_z z}$—the propagator—complicate

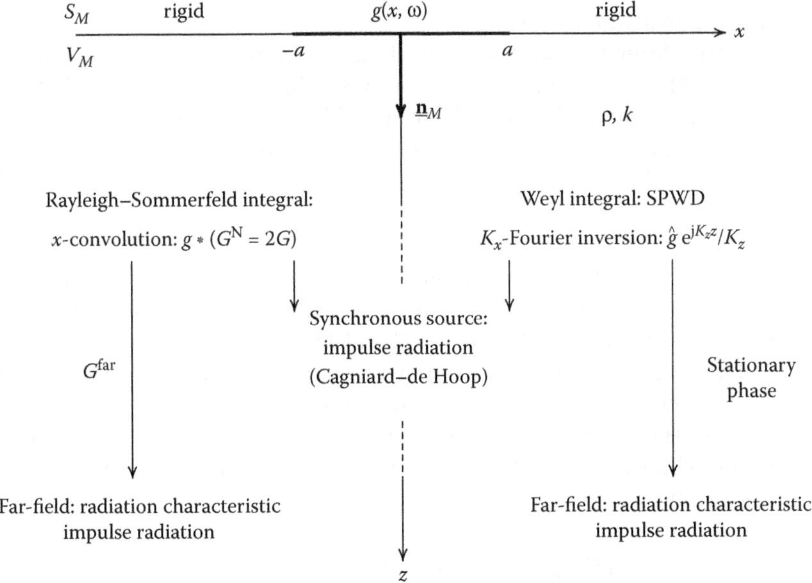

FIGURE 14.3
Alternative methods to calculate the sound field of an acoustic aperture radiator: strip-like rigidly baffled aperture.

numerical calculations. For those purposes of "parallel plane evaluation," the Rayleigh–Sommerfeld integral is not equally appropriate because the Green function is relatively slowly decaying with x and y; the power of the Rayleigh–Sommerfeld integral lies in the direct access of the far-field because the Weyl integral first requires the application of the method of stationary phase to arrive at the same radiation characteristic (Section 11.1.3). Yet both integrals lead to the same result independent of dimensionality and source geometry as we will show later on: In any case, the Fourier transform of the source distribution has to be calculated. If we would be interested in the pulsed radiation field of the aperture (for example, Figure 14.1), we should perform a numerical evaluation of the Rayleigh–Sommerfeld or the Weyl integral for each circular frequency within the spectrum of the excitation pulse with a subsequent inverse Fourier transform with regard to ω. However, for special source geometries—circular, strip-like, or rectangular apertures with constant amplitudes—and special excitation functions—generally step-function, Dirac impulse, or its derivatives—the pulsed radiation field may be analytically calculated for arbitrary distances z using a tricky integration of either one of the integral representations (Cagniard–de Hoop method: Aulenbacher 1988; Aulenbacher and Langenberg 1980; analytic integration: Schmerr 1998; Royer and Dieulesaint 2000); for other excitations, for example, RC$N(t)$-pulses, a subsequent convolution has to be performed with an increasingly unfriendly numerical effort the further we are in the far-field. After all, the transition near-field $\Longrightarrow$ far-field is most intuitively recognized from the pulsed radiation field (Figures 14.5 and 14.6).

The spatial filter interpretation as discussed in Section 11.1.2 equally applies to the field distribution $p^{\mathrm{N}}(x, y, z, \omega)$ according to (14.29), namely to the interpretation of the radiation field of a rigidly baffled acoustic transducer; only the propagation as spatial Fourier spectrum of the Green function becomes somewhat bulky because it is singular at the band limits $K_z = 0$ of the spatial low-pass filter.

Near-field length: Based on the numerical evaluation of the spatial plane wave spectrum of a strip-like "aperture," we could nicely illustrate the transition near-field $\Longrightarrow$ far-field (Figures 11.3 and 11.4); for another special case—the circular aperture—we may even perform analytic investigations if the Rayleigh–Sommerfeld is evaluated on the acoustic axis.

We write the Rayleigh–Sommerfeld integral (14.31) for $g(x', y', \omega) = u(a - r')$ in cylindrical coordinates and immediately choose $\varphi = 0$ due to the rotational symmetry:

$$p^{\mathrm{N}}(r, z, \omega) = 2\mathrm{j}\omega\rho \int_0^{2\pi} \int_0^{\infty} \frac{\mathrm{e}^{\mathrm{j}k\sqrt{r^2+r'^2-2rr'\cos\varphi'+z^2}}}{4\pi\sqrt{r^2+r'^2-2rr'\cos\varphi'+z^2}} r'\mathrm{d}r'\mathrm{d}\varphi'. \quad (14.32)$$

Per definition, we have $r = 0$ on the acoustic axis yielding an elementary integration after the substitution $\sqrt{r'^2+z^2} = \varrho$:

$$
\begin{aligned}
p^{\mathrm{N}}(0, z, \omega) &= \mathrm{j}\omega\rho \int_0^a \frac{\mathrm{e}^{\mathrm{j}k\sqrt{r'^2+z^2}}}{\sqrt{r'^2+z^2}}\, r'\mathrm{d}r' \\
&= \mathrm{j}\omega\rho \int_z^{\sqrt{z^2+a^2}} \mathrm{e}^{\mathrm{j}k\varrho}\, \mathrm{d}\varrho \\
&= Z \left[\mathrm{e}^{\mathrm{j}k\beta(z)} - 1\right] \mathrm{e}^{\mathrm{j}kz},
\end{aligned} \tag{14.33}
$$

where

$$
\beta(z) = \sqrt{z^2 + a^2} - z. \tag{14.34}
$$

Hence, we observe a one-dimensional amplitude and phase modulated wave propagating into positive z-direction; the magnitude

$$
|p^{\mathrm{N}}(0, z, \omega)| = 2Z \left| \sin \frac{k\beta(z)}{2} \right| \tag{14.35}
$$

yields the amplitude modulation that contains n zeroes according to

$$
\beta(z) = n\lambda, \quad n = 1, 2, 3, \ldots; \tag{14.36}
$$

their number is limited by the requirement $n\lambda < a$ because we have $\beta_{\max}(z) = \beta(0) = a$ (Figure 14.4). The physical reason for this oscillatory pressure behavior on the acoustic axis as displayed in Figure 14.4 is the interference of the aperture wavefront with the aperture edge pulses (Figure 14.1); the typical

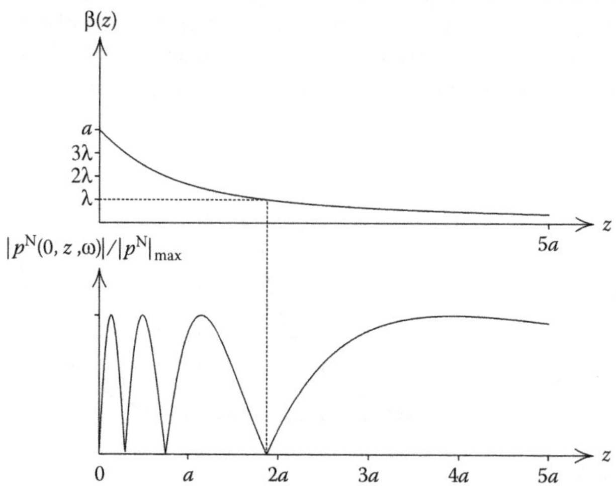

FIGURE 14.4
Phase and normalized magnitude of the sound pressure of a circular aperture radiator on the acoustic axis (for $a = 4\lambda$, as chosen here, the sound pressure indeed starts with a zero for $z = 0$).

far-field behavior as $1/R$-dependence ($1/z$ for $r = 0$) is only observed when these wavefronts become so close to each other that they superimpose due to their finite duration; namely, for $z \gg a$ we may approximate:

$$\begin{aligned}\beta(z) &= z\sqrt{1+\frac{a^2}{z^2}} - z \\ &\simeq z\left(1+\frac{a^2}{2z^2}\right) - z \\ &= \frac{a^2}{2z},\end{aligned} \tag{14.37}$$

yielding[251]

$$|p^{\mathrm{N}}(0, z, \omega)| \simeq \frac{\pi a^2 Z}{\lambda} \frac{1}{z} \tag{14.38}$$

if we approximate the sine function in (14.35) by its argument. We may even calculate the z-value where the far-field behavior becomes relevant; we define the near-field length N as the location of the last maximum of the sound pressure. We find the pertinent z-value requesting $\beta(z) = \lambda/2$, and we conclude for $a \gg \lambda$

$$N \simeq \frac{a^2}{\lambda}. \tag{14.39}$$

That way, we have constituted the near-field time harmonically, yet plausibility arguments have already been given based on the pulsed sound field as displayed in Figure 14.1. With a respective sketch of the pulsed sound field (Figure 14.5), we may also arrive at the expression (14.39). We define that travel distance $c_{\mathrm{P}}t$ of the aperture wavefront as near-field length N for which

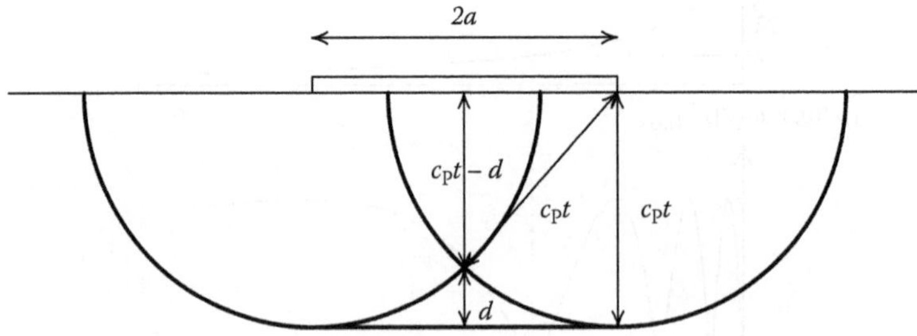

FIGURE 14.5
Calculation of the near-field length from the pulsed radiation field of the acoustic aperture radiator.

[251]Of course, the usual far-field approximation of the Rayleigh–Sommerfeld integral (14.32) with subsequent specialization to the acoustic axis yields the same result.

the distance d of the cross-points of the aperture edge pulses on the acoustic axis is approximately equal to a half of the impulse "duration length" $c_P T$. The geometry of Figure 14.5 tells us that

$$c_P t - d = \sqrt{c_P^2 t^2 - a^2} \tag{14.40}$$

holds, and consequently, we have according to the above definition

$$N - \sqrt{N^2 - a^2} \simeq \frac{c_P T}{2}, \tag{14.41}$$

respectively for $N \gg a$:

$$\frac{a^2}{N} \simeq c_P T. \tag{14.42}$$

With the definition of the pulse wavelength $c_P T \simeq \lambda$, (14.42) is equal to (14.39).

By the way: The near-field length of a pulsed sound field is causatively related to the band limitation of the aperture impulse, it does not exist for a Dirac pulse. To show that, we return to the monochromatic axis sound pressure (14.33) of the circular aperture that has been calculated for the spectrum of a Dirac pulse due to (14.32) and transform it into the time domain:

$$p^{\mathrm{N}}(0, z, t) = Z \left[\delta \left(t - \frac{1}{c} \sqrt{z^2 + a^2} \right) - \delta \left(t - \frac{z}{c} \right) \right]. \tag{14.43}$$

It obviously consists of two Dirac pulses with different signs and z-independent amplitude; they are assigned to the aperture wavefront and the superimposed edge pulses: The travel time difference of these pulses is given by $\beta(z)$, it is only zero at infinity! No superposition occurs for $z < \infty$ with the consequence that no "one impulse far-field behavior" with $1/R$-amplitude dependence is observed as we have derived it for the main lobe direction of an aperture in the far-field with bandlimited excitation (Equation 13.294 and Figure 13.21).

Figure 14.6(c) and (d) schematically display the far-field behavior for band limitation as given by (13.294)—the double circles may illustrate the finite impulse duration—by means of wavefronts; in contrast, Figure 14.6(a) and (b), similarly display the correct wave front behavior [numerically exact result: Figure 14.1(c) and (d)]: Evidently, according to the discussion of Figure 14.5, the differences become more and more marginal with increasing distance. Furthermore, we recognize the decreasing near-field length with increasing pulse duration according to (14.42). The limit "infinitely short (Dirac) impulse" yields an infinitely large near-field length. The validity condition for the monochromatic far-field approximation has been given as "distance large with regard to aperture size" and "distance large with regard to wavelength." Hence, we want to check the values of $2a$, R, and kR that have to be assigned to the wavefronts in Figure 14.6 if the wave number k is, for example, taken

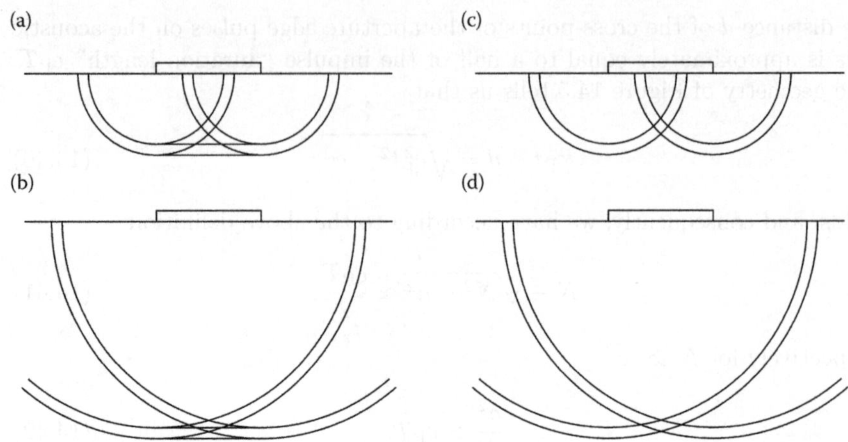

FIGURE 14.6
Bandlimited impulse wavefronts of a strip-like aperture radiator: transition near- to far-field; (a) and (b): near-field evaluation; (c) and (d): far-field evaluation.

from the center-frequency of an RC2(t)-pulse. The spatial extent of the wavefronts in Figure 14.6 yields $\lambda = 1\,\mathrm{mm}$ (wavelength, not Lamé's constant!) according to a center-frequency of ca. 600 kHz in steel. From that, we obtain $kR \simeq 80$ and $R/2a \simeq 0.7$ for the near-field sketches 14.6(a) and (c) as well as $kR \simeq 230$ and $R/2a \simeq 2$ for the far-field sketches 14.6(b) and (d). Even though the condition $kR \gg 1$ is satisfied for *all* sketches, the differences between 14.6(a) and 14.6(c) are still significant due to the too small $R/2a$-value. But the value $R/2a \simeq 2$ for the sketches (b) and (c) signals that we are not yet really in the far-field. The actual amplitudes are apparently not contained in this "validation" of the far-field approximation.

14.1.3 Far-field evaluation of Rayleigh–Sommerfeld and Weyl integrals

The far-field of the rigidly baffled aperture radiator is approximately obtained if we insert the far-field approximation (13.41) of the Green function into the Rayleigh–Sommerfeld integral (14.31):

$$p^{\mathrm{N,far}}(R,\vartheta,\varphi,\omega) = \frac{\mathrm{e}^{\mathrm{j}kR}}{R} \underbrace{\frac{\mathrm{j}\omega\rho}{2\pi} \int_{-\infty}^{\infty}\int_{-\infty}^{\infty} g(x',y',\omega)\,\mathrm{e}^{-\mathrm{j}k\sin\vartheta\cos\varphi x' - \mathrm{j}k\sin\vartheta\sin\varphi y'}\,\mathrm{d}x'\mathrm{d}y'}_{=H^{g}(\vartheta,\varphi,\omega)} . \tag{14.44}$$

Due to the rigid half-space surface, a factor 2 appears as compared to (13.65). The radiation characteristic $H^g(\vartheta, \varphi, \omega)$ of the surface dilatation rate source $g(x, y, \omega)$ turns out to be proportional to its two-dimensional spatial Fourier transform for $K_x = k \sin\vartheta \cos\varphi$, $K_y = k \sin\vartheta \sin\varphi$:

$$H^g(\vartheta, \varphi, \omega) = \frac{\mathrm{j}\omega\rho}{2\pi}\,\hat{g}(k \sin\vartheta \cos\varphi, k \sin\vartheta \sin\varphi, \omega). \tag{14.45}$$

Alternatively referring to the integral representation (14.29) of Weyl's type and calculating its respective far-field approximation with the method of stationary phase (Equation 11.32), we again obtain (14.44) and (14.45).

14.2 Strip-Like Normal and Tangential Force Density Distributions on the Stress-Free Surface of an Elastic Half-Space: Spectral Plane Wave Decomposition of the Two-Dimensional Second-Rank Green Tensor

14.2.1 EFIT-wavefronts of the linear and strip-like aperture radiator on the stress-free surface of an elastic half-space

Similar to Figure 14.1, we show respective EFIT-calculated pulsed wavefronts in Figure 14.7 for the case of an elastic half-space with stress-free surface excited by a normal force density with RC2(t)-time dependence. As output of EFIT, the pendant to AFIT for elastic waves (Fellinger 1991; Marklein 1997; Bihn 1998), we have displayed the magnitude of the particle velocity; insofar, it does not vanish on the excitation surface because it is stress-free. As compared to the particle displacement, the RC2(t)-excitation manifests itself as time derivative in the particle velocity.

The upper part of Figure 14.7 exhibits the wavefronts for the case of a line force density: We recognize semicircular pressure as well as shear wavefronts—the latter one with a zero in normal direction—that are connected by two head waves; their physical origin is the excitation of Huygens shear elementary waves through the pressure wave propagating along the surface with speed c_{P} that superimpose (in two dimensions) to a planar wavefront. It makes an angle with the surface ϑ_{K} that emerges as mode conversion angle for $\vartheta_{i\mathrm{P}} = \pi/2$ from (9.21). In the end, the head waves are needed because neither the pressure nor the shear wave can satisfy the stress-free boundary condition alone. Two Rayleigh surface waves finally complete the pulsed radiation field of a normal line source on a stress-free half-space.

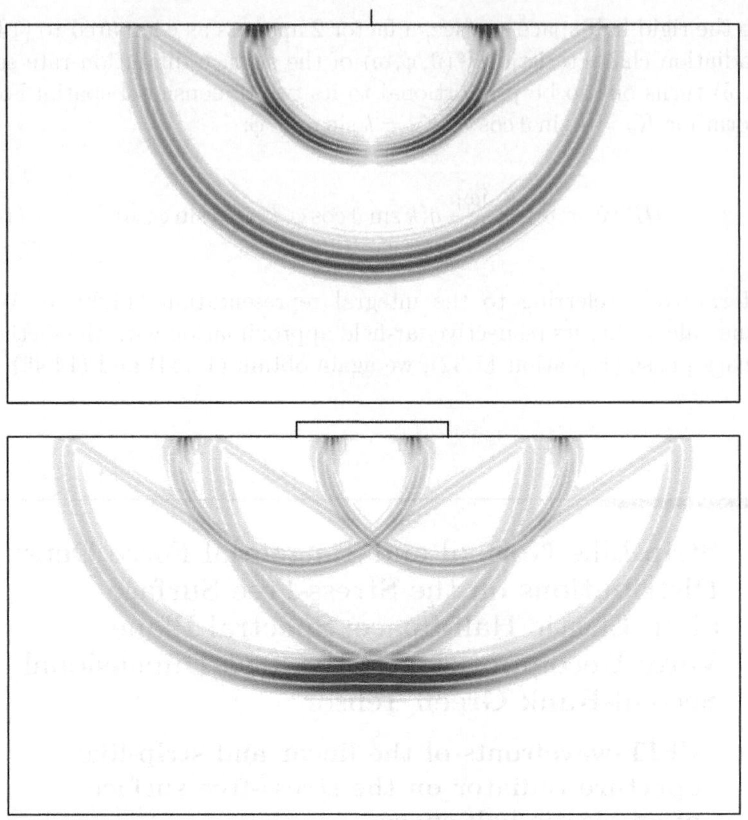

FIGURE 14.7
EFIT-wavefronts of the line source and the strip-like aperture radiator on the stress-free surface of an elastic half-space (magnitude of the particle velocity).

In the lower part of Figure 14.7, the excitation is given as a normal force density constantly distributed within an aperture (two-dimensional strip-like aperture radiator on a stress-free surface). According to Huygens' principle, each point within the aperture radiates semicircular pressure waves that superimpose to a planar (geometric optical) aperture wavefront, the "main lobe" of the strip "transducer." Equally recognizable are the semicircular pressure waves emanating from the edges of the aperture acting as line sources. Due to the zeroes of the Huygens elementary shear waves, no aperture wavefront with shear velocity can be formed; only the shear pulses from the aperture edges as well as, last but not least, the Rayleigh pulses from these line sources prevail. The comparison with Figure 14.1 reveals that the dominant EFIT-calculated elastic wavefronts may be satisfactorily approximated by acoustic waves.

14.2.2 Strip-like normal and tangential force density distributions on the stress-free surface of an elastic half-space

Normal force density: Again, we take Figure 13.22(b) as a basis, yet we assume a homogeneous isotropic half-space V_M for $z > 0$; its surface S_M—the xy-plane—be stress-free outside the strip-like aperture and inside be a constant synchronous surface force density independent of y, yielding the *ansatz*

$$\underline{\mathbf{t}}(x, \omega) = F(\omega) q_a(x) \underline{\mathbf{e}}_z \tag{14.46}$$

[compare the volume source (13.306)].

The complementary acoustic problem could be alternatively solved as an inhomogeneous boundary value problem either with the half-space scalar Green function or with the spectral plane wave decomposition. Here, we are forced to apply the second method because half-space Green tensors, especially the second rank half-space Green tensor, are not explicitly known[252] because the method of images does not hold: A line or point force density in front of a stress-free surface creates pressure and shear waves with different speeds, whose respective mode conversion at the surface may not be represented by image sources.

Therefore, we formulate the following two-dimensional inhomogeneous elastodynamic boundary value problem:[253] Solve the differential equation

$$\mu \Delta_{\underline{\mathbf{r}}} \underline{\mathbf{u}}(x, z, \omega) + (\lambda + \mu) \nabla_{\underline{\mathbf{r}}} \nabla_{\underline{\mathbf{r}}} \cdot \underline{\mathbf{u}}(x, z, \omega) + \omega^2 \rho \, \underline{\mathbf{u}}(x, z, \omega) = \underline{\mathbf{0}} \tag{14.47}$$

for the time harmonic particle displacement (Equation 7.24 for $\partial/\partial y \equiv 0$) for $z > 0$ under the condition

$$\underline{\underline{\mathbf{T}}}(x, z = 0, \omega) \cdot \underline{\mathbf{e}}_z = -\underline{\mathbf{t}}(x, \omega) \;\; \forall \; x \in S_M. \tag{14.48}$$

From Section 13.3.3, we know that the source (14.46) radiates only P-SV-waves, i.e., the particle displacement vector only has a component in the xz-plane, yet for both wave modes. The special solution "plane P-SV-waves with propagation direction $\hat{\underline{\mathbf{k}}}$" is found in terms of Equations 8.82 and 8.83 with $\hat{\underline{\mathbf{u}}}_{S1} = \underline{\mathbf{e}}_\theta$, where we once more introduce the "two-dimensional polar angle" θ counted from the z-axis. Now it becomes apparent that plane elastic waves are not just fiction, yet they are building blocks for elastodynamic

[252] As a result of the following calculation, we are at least able to give an integral representation of the Weyl type.

[253] The respective two-dimensional source field problem of full-space has the solution

$$\underline{\mathbf{u}}_{\mathrm{PSV}}(\underline{\mathbf{r}}, \omega) = \iint_{V_Q} \underline{\mathbf{f}}(\underline{\mathbf{r}}', \omega) \cdot \underline{\underline{\mathbf{G}}}_{\mathrm{PSV}}(\underline{\mathbf{r}} - \underline{\mathbf{r}}', \omega) \, \mathrm{d}^2 \underline{\mathbf{r}}'$$

according to (13.246) yielding the far-field approximation (13.312) and (13.313) with (13.306).

source fields in a spatial spectrum of plane elastic waves (compare Section 11.2 concerning theoretical arguments)

$$\underline{\mathbf{u}}_{\mathrm{PSV}}(x,z,\omega) = \frac{1}{2\pi}\int_{-\infty}^{\infty}\left[\hat{u}_{\mathrm{P}}(K_x,\omega)\,\mathrm{e}^{\mathrm{j}\underline{\mathbf{K}}_{\mathrm{P}}\cdot\underline{\mathbf{r}}}\,\hat{\underline{\mathbf{K}}}_{\mathrm{P}} + \hat{u}_{\mathrm{SV}}(K_x,\omega)\,\mathrm{e}^{\mathrm{j}\underline{\mathbf{K}}_{\mathrm{S}}\cdot\underline{\mathbf{r}}}\,\underline{\mathbf{e}}_{\theta_{K_{\mathrm{S}}}}\right]\mathrm{d}K_x; \tag{14.49}$$

as in (13.325), the two-dimensional case does not require a K_y-integration,[254] yet we continue to use the hat to characterize scalar spatial spectra $\hat{u}_{\mathrm{P}}(K_x,\omega)$, $\hat{u}_{\mathrm{SV}}(K_x,\omega)$. According to Figure 11.1, the phase vector of a spectral plane wave is given by the Fourier vector, hence

$$\begin{aligned}\underline{\mathbf{K}}_{\mathrm{P}} &= k_{\mathrm{P}}\hat{\underline{\mathbf{K}}}_{\mathrm{P}}\\ &= K_x\underline{\mathbf{e}}_x + \underbrace{\sqrt{k_{\mathrm{P}}^2 - K_x^2}}_{=K_{\mathrm{P}z}}\,\underline{\mathbf{e}}_z,\end{aligned} \tag{14.50}$$

$$\begin{aligned}\underline{\mathbf{K}}_{\mathrm{S}} &= k_{\mathrm{S}}\hat{\underline{\mathbf{K}}}_{\mathrm{S}}\\ &= K_x\underline{\mathbf{e}}_x + \underbrace{\sqrt{k_{\mathrm{S}}^2 - K_x^2}}_{=K_{\mathrm{S}z}}\,\underline{\mathbf{e}}_z;\end{aligned} \tag{14.51}$$

and consequently, the spectral P-contribution is polarized in the direction $\hat{\underline{\mathbf{K}}}_{\mathrm{P}}$, and the spectral SV-contribution in the direction $\underline{\mathbf{e}}_{\theta_{K_{\mathrm{S}}}}$ with $\hat{\underline{\mathbf{K}}}_{\mathrm{S}}\cdot\underline{\mathbf{e}}_{\theta_{K_{\mathrm{S}}}} = 0$, where (Figure 14.8)

$$\underline{\mathbf{e}}_{\theta_{K_{\mathrm{S}}}} = \cos\theta_{K_{\mathrm{S}}}\underline{\mathbf{e}}_x - \sin\theta_{K_{\mathrm{S}}}\underline{\mathbf{e}}_z. \tag{14.52}$$

Note: The polarization vectors $\hat{\underline{\mathbf{K}}}_{\mathrm{P}}$ and $\underline{\mathbf{e}}_{\theta_{K_{\mathrm{S}}}}$ are in fact spectral polarization vectors, they must remain under the integral! In the differential equation (14.47) K_x as conjugate Fourier variable with regard to x is—besides ω—a parameter, i.e., the spectral amplitudes $\hat{u}_{\mathrm{P}}(K_x,\omega)$ and $\hat{u}_{\mathrm{SV}}(K_x,\omega)$ may depend on these parameters. To actually consider (14.49) as the radiation field of the strip-like aperture, these spectral amplitudes must be matched to the inhomogeneous boundary condition[255] (14.48). Its Fourier transform reads as

$$\mathcal{F}_x\left\{\underline{\underline{\mathbf{T}}}(x,z=0,\omega)\right\}\cdot\underline{\mathbf{e}}_z = -F(\omega)\,\frac{2\sin aK_x}{K_x}\,\underline{\mathbf{e}}_z. \tag{14.53}$$

With the stress tensors for plane pressure and shear waves (Equations 8.121 and 8.122), we obtain the spectral decomposition

[254] The y-dependence of (14.46) may be represented by a factor 1; accordingly, the two-dimensional Fourier transform yields a $\delta(K_y)$-distribution, which makes the K_y-integral disappear due to the sifting property.

[255] For plane waves in full-space, these amplitudes may be arbitrary.

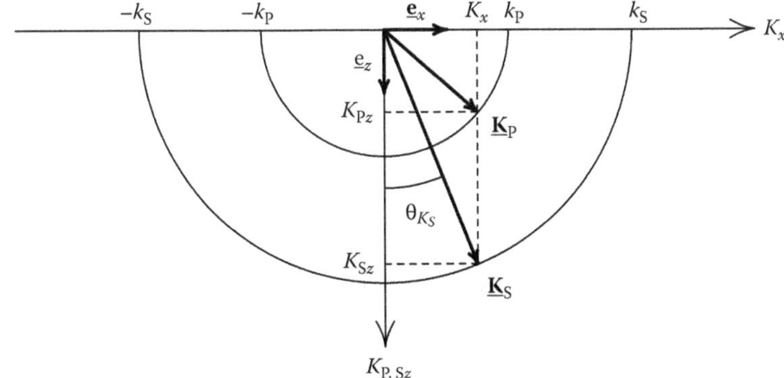

FIGURE 14.8
Propagating spectral components of plane P- and SV-waves.

$$\underline{\underline{\mathbf{T}}}_{\mathrm{PSV}}(x,z,\omega) = \frac{1}{2\pi}\int_{-\infty}^{\infty}\Big[\mathrm{j}k_{\mathrm{P}}\,\hat{u}_{\mathrm{P}}(K_x,\omega)\,\mathrm{e}^{\mathrm{j}\underline{\mathbf{K}}_{\mathrm{P}}\cdot\underline{\mathbf{r}}}\,(\lambda\underline{\underline{\mathbf{I}}} + 2\mu\underline{\hat{\mathbf{K}}}_{\mathrm{P}}\underline{\hat{\mathbf{K}}}_{\mathrm{P}})$$
$$+\,\mathrm{j}k_{\mathrm{S}}\mu\,\hat{u}_{\mathrm{SV}}(K_x,\omega)\,\mathrm{e}^{\mathrm{j}\underline{\mathbf{K}}_{\mathrm{S}}\cdot\underline{\mathbf{r}}}\,(\underline{\hat{\mathbf{K}}}_{\mathrm{S}}\underline{\mathbf{e}}_{\theta_{K_{\mathrm{S}}}} + \underline{\mathbf{e}}_{\theta_{K_{\mathrm{S}}}}\underline{\hat{\mathbf{K}}}_{\mathrm{S}})\Big]\,\mathrm{d}K_x \tag{14.54}$$

of the stress tensor into plane waves resulting in the requirement

$$\underline{\underline{\mathbf{T}}}_{\mathrm{PSV}}(x,z=0,\omega)\cdot\underline{\mathbf{e}}_z = -\frac{F(\omega)}{2\pi}\int_{-\infty}^{\infty}\frac{2\sin aK_x}{K_x}\,\mathrm{e}^{\mathrm{j}K_x x}\,\mathrm{d}K_x\,\underline{\mathbf{e}}_z \tag{14.55}$$

for the boundary condition (14.53), hence

$$\mathrm{j}k_{\mathrm{P}}\,\hat{u}_{\mathrm{P}}(K_x,\omega)\left(\lambda\underline{\underline{\mathbf{I}}} + \frac{2\mu}{k_{\mathrm{P}}^2}\,\underline{\mathbf{K}}_{\mathrm{P}}\underline{\mathbf{K}}_{\mathrm{P}}\right)\cdot\underline{\mathbf{e}}_z$$
$$+\,\mathrm{j}k_{\mathrm{S}}\mu\,\hat{u}_{\mathrm{SV}}(K_x,\omega)\left(\frac{\underline{\mathbf{K}}_{\mathrm{S}}}{k_{\mathrm{S}}}\,\underline{\mathbf{e}}_{\theta_{K_{\mathrm{S}}}} + \underline{\mathbf{e}}_{\theta_{K_{\mathrm{S}}}}\,\frac{\underline{\mathbf{K}}_{\mathrm{S}}}{k_{\mathrm{S}}}\right)\cdot\underline{\mathbf{e}}_z$$
$$= -F(\omega)\,\frac{2\sin aK_x}{K_x}\,\underline{\mathbf{e}}_z. \tag{14.56}$$

The vector equation (14.56) has the x- and z-components

$$\frac{2}{k_{\mathrm{P}}}\,K_x K_{\mathrm{P}z}\,\hat{u}_{\mathrm{P}}(K_x,\omega) + \left(K_x\,\underline{\mathbf{e}}_{\theta_{K_{\mathrm{S}}}}\cdot\underline{\mathbf{e}}_z + \underline{\mathbf{e}}_{\theta_{K_{\mathrm{S}}}}\cdot\underline{\mathbf{e}}_x K_{\mathrm{S}z}\right)\hat{u}_{\mathrm{SV}}(K_x,\omega) = 0, \tag{14.57}$$

$$\frac{1}{k_{\mathrm{P}}}\left(k_{\mathrm{S}}^2 - 2K_x^2\right)\hat{u}_{\mathrm{P}}(K_x,\omega) + 2K_{\mathrm{S}z}\,\underline{\mathbf{e}}_{\theta_{K_{\mathrm{S}}}}\cdot\underline{\mathbf{e}}_z\,\hat{u}_{\mathrm{SV}}(K_x,\omega)$$
$$= \mathrm{j}\,\frac{F(\omega)}{\mu}\,\frac{2\sin aK_x}{K_x}, \tag{14.58}$$

resulting in an inhomogeneous system of equations for the two unknowns $\hat{u}_{\mathrm{P}}(K_x, \omega)$ and $\hat{u}_{\mathrm{SV}}(K_x, \omega)$. We find

$$\hat{u}_{\mathrm{P}}(K_x, \omega) = \mathrm{j}k_{\mathrm{P}} \frac{F(\omega)}{\mu} \frac{2\sin aK_x}{K_x} \frac{k_{\mathrm{S}}\left(K_x \underline{\mathbf{e}}_{\theta_{K_{\mathrm{S}}}} \cdot \underline{\mathbf{e}}_z + K_{\mathrm{S}z} \underline{\mathbf{e}}_{\theta_{K_{\mathrm{S}}}} \cdot \underline{\mathbf{e}}_x\right)}{R(K_x)}, \tag{14.59}$$

$$\hat{u}_{\mathrm{SV}}(K_x, \omega) = -\mathrm{j} \frac{F(\omega)}{\mu} \frac{2\sin aK_x}{K_x} \frac{2k_{\mathrm{S}}K_x K_{\mathrm{P}z}}{R(K_x)}, \tag{14.60}$$

where

$$R(K_x) = k_{\mathrm{S}}\Big[\left(k_{\mathrm{S}}^2 - 2K_x^2\right)\left(K_x \underline{\mathbf{e}}_{\theta_{K_{\mathrm{S}}}} \cdot \underline{\mathbf{e}}_z + \underline{\mathbf{e}}_{\theta_{K_{\mathrm{S}}}} \cdot \underline{\mathbf{e}}_x K_{\mathrm{S}z}\right) - 4K_x K_{\mathrm{S}z} K_{\mathrm{P}z} \underline{\mathbf{e}}_{\theta_{K_{\mathrm{S}}}} \cdot \underline{\mathbf{e}}_z \Big]. \tag{14.61}$$

With (Figure 14.8)

$$\begin{aligned} K_x &= k_{\mathrm{S}} \sin\theta_{K_{\mathrm{S}}}, \\ K_{\mathrm{S}z} &= k_{\mathrm{S}} \cos\theta_{K_{\mathrm{S}}} \end{aligned} \tag{14.62}$$

and (14.52), we have

$$\underline{\mathbf{e}}_{\theta_{K_{\mathrm{S}}}} \cdot \underline{\mathbf{e}}_x = \frac{K_{\mathrm{S}z}}{k_{\mathrm{S}}}, \tag{14.63}$$

$$\underline{\mathbf{e}}_{\theta_{K_{\mathrm{S}}}} \cdot \underline{\mathbf{e}}_z = -\frac{K_x}{k_{\mathrm{S}}}; \tag{14.64}$$

hence,

$$R(K_x) = \left(k_{\mathrm{S}}^2 - 2K_x^2\right)^2 + 4K_x^2 \sqrt{k_{\mathrm{S}}^2 - K_x^2}\sqrt{k_{\mathrm{P}}^2 - K_x^2} \tag{14.65}$$

is the well-known Rayleigh function (Chapter 10: $k_{\mathrm{R}} \Longrightarrow K_x$). Additionally, noting

$$\underline{\hat{\mathbf{K}}}_{\mathrm{P}} \cdot \underline{\mathbf{e}}_x = \frac{K_x}{k_{\mathrm{P}}}, \tag{14.66}$$

$$\underline{\hat{\mathbf{K}}}_{\mathrm{P}} \cdot \underline{\mathbf{e}}_z = \frac{K_{\mathrm{P}z}}{k_{\mathrm{P}}}, \tag{14.67}$$

we finally obtain explicit representations of the cartesian components of the (two-dimensional) particle velocity field of a strip-like force density aperture with constant amplitude located on the elsewhere stress-free surface of an elastic half-space in terms of a spectral plane wave decomposition (Miller and

Pursey 1954; Achenbach 1973, 1988; Fellinger 1991; Kühnicke 2001; Harker 1988):

$$u^{t_z}_{\mathrm{PSV}x}(x,z,\omega) = \mathrm{j}\,\frac{F(\omega)}{2\pi\mu}\int_{-\infty}^{\infty}\frac{2\sin aK_x}{K_x}\,\frac{K_x}{R(K_x)}$$
$$\times\left[\left(k_{\mathrm{S}}^2-2K_x^2\right)\mathrm{e}^{\mathrm{j}\underline{\mathbf{K}}_{\mathrm{P}}\cdot\underline{\mathbf{r}}}-2\sqrt{k_{\mathrm{S}}^2-K_x^2}\sqrt{k_{\mathrm{P}}^2-K_x^2}\,\mathrm{e}^{\mathrm{j}\underline{\mathbf{K}}_{\mathrm{S}}\cdot\underline{\mathbf{r}}}\right]\mathrm{d}K_x, \tag{14.68}$$

$$u^{t_z}_{\mathrm{PSV}z}(x,z,\omega) = \mathrm{j}\,\frac{F(\omega)}{2\pi\mu}\int_{-\infty}^{\infty}\frac{2\sin aK_x}{K_x}\,\frac{\sqrt{k_{\mathrm{P}}^2-K_x^2}}{R(K_x)}$$
$$\times\left[\left(k_{\mathrm{S}}^2-2K_x^2\right)\mathrm{e}^{\mathrm{j}\underline{\mathbf{K}}_{\mathrm{P}}\cdot\underline{\mathbf{r}}}+2K_x^2\,\mathrm{e}^{\mathrm{j}\underline{\mathbf{K}}_{\mathrm{S}}\cdot\underline{\mathbf{r}}}\right]\mathrm{d}K_x. \tag{14.69}$$

The upper index characterizes the excitation "normal force" on a stress-free half-space.

For $K_x > k_{\mathrm{P}}$, respectively $K_x > k_{\mathrm{S}}$, the two vectors $\underline{\mathbf{K}}_{\mathrm{P}}$, respectively $\underline{\mathbf{K}}_{\mathrm{S}}$, become complex vectors (and, hence, also $\underline{\mathbf{e}}_{\theta_{K_{\mathrm{S}}}}$); the sign choice

$$\Im\underline{\mathbf{K}}_{\mathrm{P,S}} = \sqrt{k_{\mathrm{P,S}}^2-K_x^2}\,\underline{\mathbf{e}}_z \tag{14.70}$$

ensures the convergence of the integrals (14.68) and (14.69) for $z > 0$.

Tangential force density: For the tangential force density

$$\underline{\mathbf{t}}(x,\omega) = F(\omega)q_a(x)\underline{\mathbf{e}}_x, \tag{14.71}$$

the radiation field is obtained after a brief calculation:

$$u^{t_x}_{\mathrm{PSV}x}(x,z,\omega) = \mathrm{j}\,\frac{F(\omega)}{2\pi\mu}\int_{-\infty}^{\infty}\frac{2\sin aK_x}{K_x}\,\frac{\sqrt{k_{\mathrm{S}}^2-K_x^2}}{R(K_x)}$$
$$\times\left[2K_x^2\,\mathrm{e}^{\mathrm{j}\underline{\mathbf{K}}_{\mathrm{P}}\cdot\underline{\mathbf{r}}}+\left(k_{\mathrm{S}}^2-2K_x^2\right)\mathrm{e}^{\mathrm{j}\underline{\mathbf{K}}_{\mathrm{S}}\cdot\underline{\mathbf{r}}}\right]\mathrm{d}K_x, \tag{14.72}$$

$$u^{t_x}_{\mathrm{PSV}z}(x,z,\omega) = \mathrm{j}\,\frac{F(\omega)}{2\pi\mu}\int_{-\infty}^{\infty}\frac{2\sin aK_x}{K_x}\,\frac{K_x}{R(K_x)}$$
$$\times\left[2\sqrt{k_{\mathrm{S}}^2-K_x^2}\sqrt{k_{\mathrm{P}}^2-K_x^2}\,\mathrm{e}^{\mathrm{j}\underline{\mathbf{K}}_{\mathrm{P}}\cdot\underline{\mathbf{r}}}-\left(k_{\mathrm{S}}^2-2K_x^2\right)\mathrm{e}^{\mathrm{j}\underline{\mathbf{K}}_{\mathrm{S}}\cdot\underline{\mathbf{r}}}\right]\mathrm{d}K_x. \tag{14.73}$$

Methods to calculate the plane wave spectrum: Figure 14.3 tells us what we can do with the spectral plane wave decomposition of an acoustic sound field; Figure 14.9 converts this to the elastodynamic case. Beforehand we state that the integral representations (14.68) and (14.69), respectively (14.72) and (14.73), in contrast to a remark by Schmerr (1998), do not only

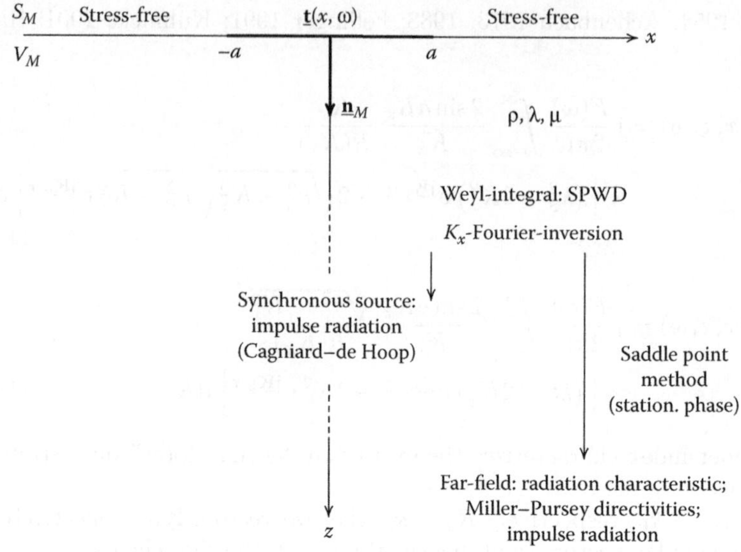

FIGURE 14.9
Alternative methods to calculate the radiation field of an elastic aperture radiator: strip-like aperture on a stress-free half-space.

contain independent P- and SV-bulk waves according to the two-dimensional second rank tensor (13.267), but the complete elastodynamic wave field inclusive head and Rayleigh waves (Figure 14.7). Insofar, the numerical evaluation of the K_x-Fourier transform for constant z-values (compare Figures 11.3 and 11.4 for the acoustic case) does not make much sense due to the inherent interferences; in addition, we even face numerical problems for $z = 0$ due to the zero of the Rayleigh function, in particular if a transform into the time domain is planned (Fellinger 1991). However, these difficulties can be avoided—others lurk around the corner—evaluating the Weyl-type integral representation, at least for synchronous sources with a special time dependence ($\delta'(t)$-dependence), directly in the time domain with the Cagniard–de Hoop method (de Hoop 1959; Achenbach 1973; Aulenbacher 1988). That way, the wavefronts depicted in Figure 14.7 are characteristically related to typical integration paths and poles in a complex plane: The semicircular (in two dimensions) pressure and shear wavefronts emerge from hyperbolic integration paths, whereas the head waves are due to a branch cut that is met by the SV-hyperbola for observation angles $\theta > \vartheta_{cmP}$ [ϑ_{cmP} according to (9.92)]. We will come back to the physical origin of the head waves in Section 14.2.4. Finally, the Rayleigh waves are related to a pole (a zero of the Rayleigh function), whose contribution becomes significant for $\theta \longrightarrow \pi/2$. One more evaluation method remains: For the acoustic case, we could immediately give a far-field approximation applying the method of stationary phase, yet for the present case, this procedure also faces problems, on one hand due to the zeroes of the Rayleigh function, and on the other hand due to the complex valued square root in the $\underline{\mathbf{K}}_\mathrm{S}$-integrals for $K_x > k_\mathrm{P}$. Instead, the method of steepest descent has to be applied in a complex K_x-plane (Miller and Pursey 1954; Achenbach 1973; Harker 1988), yet miraculously the same result is obtained if we unscrupulously apply the stationary phase formula (11.32). In the following, we will do that explicitly, but beforehand we want to point out that the complete left part of Figure 14.3 is missing in Figure 14.9: Green's second rank half-space tensor is not known explicitly impeding the presentation of an elastodynamic Rayleigh–Sommerfeld integral. We rather must apply the spectral representation in terms of plane waves to define the components of such a tensor and, hence, the far-field directivities of a line source on a stress-free half-space.

14.2.3 Spectral plane wave decomposition of the two-dimensional second-rank Green tensor

To assign a half-space Green tensor $\underline{\underline{\mathbf{G}}}^\mathrm{HS}_\mathrm{PSV}(\underline{\mathbf{r}}, \omega)$ satisfying a stress-free boundary condition to the above particle displacement components of the t_x-, respectively t_z-strip sources, similar to (13.116) according to

$$\underline{\mathbf{u}}^{\mathrm{LS}_t}_\mathrm{PSV}(\underline{\mathbf{r}}, \omega) = \underline{\underline{\mathbf{G}}}^\mathrm{HS}_\mathrm{PSV}(\underline{\mathbf{r}}, \omega) \cdot \hat{\underline{\mathbf{t}}}, \tag{14.74}$$

we have to switch from the strip source to the line source[256] and to assume $F(\omega) \equiv 1$:

$$\underline{\mathbf{t}}(x, \omega) = \delta(x)\, \hat{\underline{\mathbf{t}}}, \tag{14.75}$$

$$\mathcal{F}\{\underline{\mathbf{t}}(x, \omega)\} = \hat{\underline{\mathbf{t}}}. \tag{14.76}$$

With $\hat{\underline{\mathbf{t}}} = \underline{\mathbf{e}}_x$, the components of the particle displacement vector $\underline{\mathbf{u}}^{\mathrm{LS}_{t_x}}_\mathrm{PSV}(\underline{\mathbf{r}}, \omega)$ are "switched on" in (14.74) yielding per definition the cartesian components $\underline{\underline{\mathbf{G}}}^\mathrm{HS}_\mathrm{PSV}(\underline{\mathbf{r}}, \omega)$ as follows:

$$\begin{aligned}
\underline{\underline{\mathbf{G}}}^\mathrm{HS}_\mathrm{PSV}(\underline{\mathbf{r}}, \omega) : \underline{\mathbf{e}}_x\underline{\mathbf{e}}_x &\stackrel{\mathrm{def}}{=} \underline{\mathbf{u}}^{\mathrm{LS}_{t_x}}_\mathrm{PSV}(\underline{\mathbf{r}}, \omega) \cdot \underline{\mathbf{e}}_x \\
&= u^{\mathrm{LS}_{t_x}}_{\mathrm{PSV}x}(\underline{\mathbf{r}}, \omega),
\end{aligned} \tag{14.77}$$

$$\begin{aligned}
\underline{\underline{\mathbf{G}}}^\mathrm{HS}_\mathrm{PSV}(\underline{\mathbf{r}}, \omega) : \underline{\mathbf{e}}_x\underline{\mathbf{e}}_z &\stackrel{\mathrm{def}}{=} \underline{\mathbf{u}}^{\mathrm{LS}_{t_x}}_\mathrm{PSV}(\underline{\mathbf{r}}, \omega) \cdot \underline{\mathbf{e}}_z \\
&= u^{\mathrm{LS}_{t_x}}_{\mathrm{PSV}z}(\underline{\mathbf{r}}, \omega);
\end{aligned} \tag{14.78}$$

[256]Note: Miller and Pursey (1954) consider a narrow strip source with the limit

$$\hat{\underline{\mathbf{t}}}(K_x, \omega) = \lim_{a\to 0} \frac{2 \sin aK_x}{K_x} = 2a.$$

This, and the time dependence $\mathrm{e}^{+\mathrm{j}\omega t}$, is the origin of the somewhat different prefactors of their directivities.

analogously it follows for $\underline{\hat{\mathbf{t}}} = \underline{\mathbf{e}}_z$:

$$
\begin{aligned}
\underline{\underline{\mathbf{G}}}^{\text{HS}}_{\text{PSV}}(\underline{\mathbf{r}}, \omega) : \underline{\mathbf{e}}_z \underline{\mathbf{e}}_x &\stackrel{\text{def}}{=} \underline{\mathbf{u}}^{\text{LS}_{t_z}}_{\text{PSV}}(\underline{\mathbf{r}}, \omega) \cdot \underline{\mathbf{e}}_x \\
&= u^{\text{LS}_{t_z}}_{\text{PSV}x}(\underline{\mathbf{r}}, \omega), \qquad (14.79) \\
\underline{\underline{\mathbf{G}}}^{\text{HS}}_{\text{PSV}}(\underline{\mathbf{r}}, \omega) : \underline{\mathbf{e}}_z \underline{\mathbf{e}}_z &\stackrel{\text{def}}{=} \underline{\mathbf{u}}^{\text{LS}_{t_z}}_{\text{PSV}}(\underline{\mathbf{r}}, \omega) \cdot \underline{\mathbf{e}}_z \\
&= u^{\text{LS}_{t_z}}_{\text{PSV}z}(\underline{\mathbf{r}}, \omega). \qquad (14.80)
\end{aligned}
$$

With (14.76) and (14.68), (14.69), (14.72) and (14.73), we obviously have

$$
\begin{aligned}
\underline{\underline{\mathbf{G}}}^{\text{HS}}_{\text{PSV}}(\underline{\mathbf{r}}, \omega) : \underline{\mathbf{e}}_x \underline{\mathbf{e}}_x = \frac{\mathrm{j}}{2\pi\mu} \int_{-\infty}^{\infty} \frac{\sqrt{k_{\text{S}}^2 - K_x^2}}{R(K_x)} \\
\times \left[2K_x^2 \, \mathrm{e}^{\mathrm{j}\underline{\mathbf{K}}_{\text{P}} \cdot \underline{\mathbf{r}}} + \left(k_{\text{S}}^2 - 2K_x^2\right) \mathrm{e}^{\mathrm{j}\underline{\mathbf{K}}_{\text{S}} \cdot \underline{\mathbf{r}}} \right] \mathrm{d}K_x, \qquad (14.81)
\end{aligned}
$$

$$
\begin{aligned}
&\underline{\underline{\mathbf{G}}}^{\text{HS}}_{\text{PSV}}(\underline{\mathbf{r}}, \omega) : \underline{\mathbf{e}}_x \underline{\mathbf{e}}_z \\
&= \frac{\mathrm{j}}{2\pi\mu} \int_{-\infty}^{\infty} \frac{K_x}{R(K_x)} \qquad (14.82) \\
&\times \left[2\sqrt{k_{\text{S}}^2 - K_x^2}\sqrt{k_{\text{P}}^2 - K_x^2} \, \mathrm{e}^{\mathrm{j}\underline{\mathbf{K}}_{\text{P}} \cdot \underline{\mathbf{r}}} - \left(k_{\text{S}}^2 - 2K_x^2\right) \mathrm{e}^{\mathrm{j}\underline{\mathbf{K}}_{\text{S}} \cdot \underline{\mathbf{r}}} \right] \mathrm{d}K_x,
\end{aligned}
$$

$$
\begin{aligned}
&\underline{\underline{\mathbf{G}}}^{\text{HS}}_{\text{PSV}}(\underline{\mathbf{r}}, \omega) : \underline{\mathbf{e}}_z \underline{\mathbf{e}}_x \\
&= \frac{\mathrm{j}}{2\pi\mu} \int_{-\infty}^{\infty} \frac{K_x}{R(K_x)} \qquad (14.83) \\
&\times \left[\left(k_{\text{S}}^2 - 2K_x^2\right) \mathrm{e}^{\mathrm{j}\underline{\mathbf{K}}_{\text{P}} \cdot \underline{\mathbf{r}}} - 2\sqrt{k_{\text{S}}^2 - K_x^2}\sqrt{k_{\text{P}}^2 - K_x^2} \, \mathrm{e}^{\mathrm{j}\underline{\mathbf{K}}_{\text{S}} \cdot \underline{\mathbf{r}}} \right] \mathrm{d}K_x,
\end{aligned}
$$

$$
\begin{aligned}
\underline{\underline{\mathbf{G}}}^{\text{HS}}_{\text{PSV}}(\underline{\mathbf{r}}, \omega) : \underline{\mathbf{e}}_z \underline{\mathbf{e}}_z = \frac{\mathrm{j}}{2\pi\mu} \int_{-\infty}^{\infty} \frac{\sqrt{k_{\text{P}}^2 - K_x^2}}{R(K_x)} \\
\times \left[\left(k_{\text{S}}^2 - 2K_x^2\right) \mathrm{e}^{\mathrm{j}\underline{\mathbf{K}}_{\text{P}} \cdot \underline{\mathbf{r}}} + 2K_x^2 \, \mathrm{e}^{\mathrm{j}\underline{\mathbf{K}}_{\text{S}} \cdot \underline{\mathbf{r}}} \right] \mathrm{d}K_x. \qquad (14.84)
\end{aligned}
$$

Note: As a second rank tensor $\underline{\underline{\mathbf{G}}}^{\text{HS}}_{\text{PSV}}(\underline{\mathbf{r}}, \omega)$ is not symmetric, i.e., $\underline{\hat{\mathbf{t}}}$ may not be converted into a left factor of $\underline{\underline{\mathbf{G}}}^{\text{HS}}_{\text{PSV}}(\underline{\mathbf{r}}, \omega)$ in (14.74) [this would define $\underline{\underline{\mathbf{G}}}^{\text{HS}\,21}_{\text{PSV}}(\underline{\mathbf{r}}, \omega)$]. The reason is that the Green tensor for an inhomogeneous material—the half-space with stress-free surface is such a material—is only symmetric if the transpose is accompanied by an interchange of source and observation point[257] (de Hoop 1995). This is not possible for the present case because the integral representations (14.81) through (14.84) are only valid for a particular source point.

[257] It was not yet observed until now because homogeneous materials exhibit a $|\underline{\mathbf{R}} - \underline{\mathbf{R}}'|$-dependence as a special case of the $\underline{\mathbf{R}}, \underline{\mathbf{R}}'$-dependence.

14.2.4 Far-field radiation characteristics of normal and tangential line force densities on the surface of a stress-free half-space

We apparently suspect that, analogous to the respective volume source according to (13.262), the far-field of the strip source on the stress-free surface of a half-space will be given by independent P- and SV-waves with pertinent radiation characteristics. Being particularly interested in the radiation characteristics of normal and tangential line force densities, we apply formula (11.29) as asymptotic evaluation of a one-dimensional inverse Fourier integral (11.18) with the method of stationary phase—however, with mathematical concerns, yet with good hope—to the integral representations (14.81) through (14.84). We obtain[258]

$$u_{\mathrm{PSV}x}^{\mathrm{LS}_{t_z},\mathrm{far}}(r,\theta,\omega) = \frac{\mathrm{e}^{\mathrm{j}\frac{\pi}{4}}}{\mu}\frac{k_\mathrm{P}}{\sqrt{2\pi k_\mathrm{P}}}\frac{\mathrm{e}^{\mathrm{j}k_\mathrm{P}r}}{\sqrt{r}}\frac{k_\mathrm{P}\sin\theta\cos\theta(k_\mathrm{S}^2-2k_\mathrm{P}^2\sin^2\theta)}{R(k_\mathrm{P}\sin\theta)} + \frac{\mathrm{e}^{-\mathrm{j}\frac{3\pi}{4}}}{\mu}\frac{2k_\mathrm{S}}{\sqrt{2\pi k_\mathrm{S}}}\frac{\mathrm{e}^{\mathrm{j}k_\mathrm{S}r}}{\sqrt{r}}\frac{k_\mathrm{S}^2\sin\theta\cos^2\theta\sqrt{k_\mathrm{P}^2-k_\mathrm{S}^2\sin^2\theta}}{R(k_\mathrm{S}\sin\theta)}, \tag{14.85}$$

$$u_{\mathrm{PSV}z}^{\mathrm{LS}_{t_z},\mathrm{far}}(r,\theta,\omega) = \frac{\mathrm{e}^{\mathrm{j}\frac{\pi}{4}}}{\mu}\frac{k_\mathrm{P}}{\sqrt{2\pi k_\mathrm{P}}}\frac{\mathrm{e}^{\mathrm{j}k_\mathrm{P}r}}{\sqrt{r}}\frac{k_\mathrm{P}\cos^2\theta(k_\mathrm{S}^2-2k_\mathrm{P}^2\sin^2\theta)}{R(k_\mathrm{P}\sin\theta)} + \frac{\mathrm{e}^{\mathrm{j}\frac{\pi}{4}}}{\mu}\frac{2k_\mathrm{S}}{\sqrt{2\pi k_\mathrm{S}}}\frac{\mathrm{e}^{\mathrm{j}k_\mathrm{S}r}}{\sqrt{r}}\frac{k_\mathrm{S}^2\sin^2\theta\cos\theta\sqrt{k_\mathrm{P}^2-k_\mathrm{S}^2\sin^2\theta}}{R(k_\mathrm{S}\sin\theta)}; \tag{14.86}$$

$$u_{\mathrm{PSV}x}^{\mathrm{LS}_{t_x},\mathrm{far}}(r,\theta,\omega) = \frac{\mathrm{e}^{\mathrm{j}\frac{\pi}{4}}}{\mu}\frac{2k_\mathrm{P}}{\sqrt{2\pi k_\mathrm{P}}}\frac{\mathrm{e}^{\mathrm{j}k_\mathrm{P}r}}{\sqrt{r}}\frac{k_\mathrm{P}^2\sin^2\theta\cos\theta\sqrt{k_\mathrm{S}^2-k_\mathrm{P}^2\sin^2\theta}}{R(k_\mathrm{P}\sin\theta)} + \frac{\mathrm{e}^{\mathrm{j}\frac{\pi}{4}}}{\mu}\frac{k_\mathrm{S}}{\sqrt{2\pi k_\mathrm{S}}}\frac{\mathrm{e}^{\mathrm{j}k_\mathrm{S}r}}{\sqrt{r}}\frac{k_\mathrm{S}\cos^2\theta(k_\mathrm{S}^2-2k_\mathrm{S}^2\sin^2\theta)}{R(k_\mathrm{S}\sin\theta)}, \tag{14.87}$$

$$u_{\mathrm{PSV}z}^{\mathrm{LS}_{t_x},\mathrm{far}}(r,\theta,\omega) = \frac{\mathrm{e}^{\mathrm{j}\frac{\pi}{4}}}{\mu}\frac{2k_\mathrm{P}}{\sqrt{2\pi k_\mathrm{P}}}\frac{\mathrm{e}^{\mathrm{j}k_\mathrm{P}r}}{\sqrt{r}}\frac{k_\mathrm{P}^2\sin\theta\cos^2\theta\sqrt{k_\mathrm{S}^2-k_\mathrm{P}^2\sin^2\theta}}{R(k_\mathrm{P}\sin\theta)} + \frac{\mathrm{e}^{-\mathrm{j}\frac{\pi}{4}}}{\mu}\frac{k_\mathrm{S}}{\sqrt{2\pi k_\mathrm{S}}}\frac{\mathrm{e}^{\mathrm{j}k_\mathrm{S}r}}{\sqrt{r}}\frac{k_\mathrm{S}\sin\theta\cos\theta(k_\mathrm{S}^2-2k_\mathrm{S}^2\sin^2\theta)}{R(k_\mathrm{S}\sin\theta)}. \tag{14.88}$$

Note: In the $\underline{\mathbf{K}}_\mathrm{P}$-integral, we have to choose $K_x = k_\mathrm{P}\sin\theta$ and, in the $\underline{\mathbf{K}}_\mathrm{S}$-integral, $K_x = k_\mathrm{S}\sin\theta$. Combining

[258]The numerous indices characterizing the particle displacement precisely identify the relevant quantity as in (13.128) and (13.131).

$$u_r = \sin\theta\, u_x + \cos\theta\, u_z, \tag{14.89}$$

$$u_\theta = \cos\theta\, u_x - \sin\theta\, u_z \tag{14.90}$$

to r- and θ-components, we immediately find that the normal and the tangential forces only radiate a P-wave in the r-component and only an SV-wave in the θ-component of the far-field. We find

$$u_{\mathrm{P}r}^{\mathrm{LS}_{t_z},\mathrm{far}}(r,\theta,\omega) = \frac{\mathrm{e}^{\mathrm{j}k_\mathrm{P}r}}{\sqrt{r}}\, H_\mathrm{P}^{\mathrm{LS}_{t_z}}(\theta,\omega), \tag{14.91}$$

$$H_\mathrm{P}^{\mathrm{LS}_{t_z}}(\theta,\omega) = \frac{\mathrm{e}^{\mathrm{j}\frac{\pi}{4}}}{\mu\sqrt{2\pi k_\mathrm{P}}}\, \frac{\cos\theta(\kappa^2 - 2\sin^2\theta)}{(\kappa^2 - 2\sin^2\theta)^2 + 2\sin\theta\sin 2\theta\sqrt{\kappa^2 - \sin^2\theta}}; \tag{14.92}$$

$$u_{\mathrm{SV}\theta}^{\mathrm{LS}_{t_z},\mathrm{far}}(r,\theta,\omega) = \frac{\mathrm{e}^{\mathrm{j}k_\mathrm{S}r}}{\sqrt{r}}\, H_\mathrm{SV}^{\mathrm{LS}_{t_z}}(\theta,\omega), \tag{14.93}$$

$$H_\mathrm{SV}^{\mathrm{LS}_{t_z}}(\theta,\omega) = \frac{\mathrm{e}^{-\mathrm{j}\frac{3\pi}{4}}}{\mu\sqrt{2\pi k_\mathrm{S}}}\, \frac{\sin 2\theta\sqrt{1 - \kappa^2\sin^2\theta}}{\kappa(1 - 2\sin^2\theta)^2 + 2\sin\theta\sin 2\theta\sqrt{1 - \kappa^2\sin^2\theta}}; \tag{14.94}$$

$$u_{\mathrm{P}r}^{\mathrm{LS}_{t_x},\mathrm{far}}(r,\theta,\omega) = \frac{\mathrm{e}^{\mathrm{j}k_\mathrm{P}r}}{\sqrt{r}}\, H_\mathrm{P}^{\mathrm{LS}_{t_x}}(\theta,\omega), \tag{14.95}$$

$$H_\mathrm{P}^{\mathrm{LS}_{t_x}}(\theta,\omega) = \frac{\mathrm{e}^{\mathrm{j}\frac{\pi}{4}}}{\mu\sqrt{2\pi k_\mathrm{P}}}\, \frac{\sin 2\theta\sqrt{\kappa^2 - \sin^2\theta}}{(\kappa^2 - 2\sin^2\theta)^2 + 2\sin\theta\sin 2\theta\sqrt{\kappa^2 - \sin^2\theta}}; \tag{14.96}$$

$$u_{\mathrm{SV}\theta}^{\mathrm{LS}_{t_x},\mathrm{far}}(r,\theta,\omega) = \frac{\mathrm{e}^{\mathrm{j}k_\mathrm{S}r}}{\sqrt{r}}\, H_\mathrm{SV}^{\mathrm{LS}_{t_x}}(\theta,\omega), \tag{14.97}$$

$$H_\mathrm{SV}^{\mathrm{LS}_{t_x}}(\theta,\omega) = \frac{\mathrm{e}^{\mathrm{j}\frac{\pi}{4}}}{\mu\sqrt{2\pi k_\mathrm{S}}}\, \frac{\kappa\cos\theta\cos 2\theta}{\kappa(1 - 2\sin^2\theta)^2 + 2\sin\theta\sin 2\theta\sqrt{1 - \kappa^2\sin^2\theta}}. \tag{14.98}$$

With (14.92), (14.94), (14.96) and (14.98), we have given the (non-normalized) line force densities according to Miller and Pursey (1954) . Note: They are all different because the Green tensor is not symmetric. Figure 14.10 illustrates the line force density directivities as radiation patterns; the comparison with the full-space patterns in Figure 13.12 reveals that the stress-free surface is of great influence even in the far-field. Most relevant for US-NDT applications is the zero of $H_\mathrm{SV}^{\mathrm{LS}_{t_z}}$ for $\theta = 0$ (further zeroes are present for $\theta = \pm\vartheta_{cm\mathrm{P}}, \pm\pi/2$) as well as the maxima for $\theta \simeq \pm 35^\circ$ (steel); the performance of the 45°-angle transducer is based on that. Furthermore: Apart from the typical $1/\sqrt{\omega}$-dependence of two-dimensional fields, the line force density directivities are frequency independent.

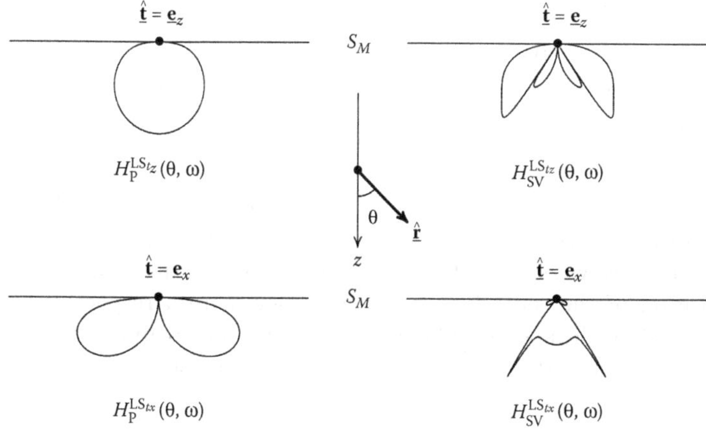

FIGURE 14.10
Radiation patterns for the particle displacement of a line force density (and point force density: Section 14.3) on the surface of an elastic half-space (magnitudes of the normalized radiation characteristics as function of θ; material of the half-space: steel).

For $K_x = k_P$, the spectrum of the plane P-waves becomes evanescent, yet in the stationary phase approximation, we then have $K_x = k_S \sin\theta = k_P$ for the SV-waves, hence

$$\theta = \arcsin\frac{c_S}{c_P}. \tag{14.99}$$

This is nothing but the critical angle ϑ_{cmP} (Equation 9.92) of a mode converted P-wave for an SV-wave incident on a stress-free surface. Consequently, the two $H_{SV}^{LS_{t_{z,x}}}$-directivities (14.94) and (14.98) for $\theta > \arcsin c_S/c_P$ become complex.[259] Similarly, $K_x = k_P$ implies that a P-wavefront moves along the surface with the (trace) velocity c_P; according to Section 9.1.2, this means that this "surface excitation" tails an SV-wavefront under the angle $\theta = \theta_{cmP}$, the plane wave spectrum must explicitly contain a plane SV-wave for $\vartheta_{cmP} < \theta \leq \pi/2$ (and symmetrically for respective negative angles), the so-called head or lateral wave (Section 14.2.1). The mathematical transition into the far-field applying the method of stationary phase according to (14.91) through (14.98) obviously suppresses the originally present head and Rayleigh waves (Figure 14.7). The line force density directivities exclusively represent direction-dependent amplitude distributions of the semicircular P- and SV-wavefronts—the zeroes of $H_{SV}^{LS_{t_z}}$ for $\pm\vartheta_{cmP}$ impede the excitation of head waves in case of normal force excitation.

[259]Therefore, if time domain wavefronts should be calculated via inverse Fourier transform using (14.93) and (14.94), respectively (14.97) and (14.98), we have to recognize (2.328). We come back to that in Section 14.3.2.

14.3 Circular Normal Force Density Distribution on the Stress-Free Surface of an Elastic Half-Space: Point Source Characteristic

In Section 13.3.4, we stated that radiation far-fields of planar force density distributions in full-space essentially differ with regard to the radiation characteristic of the aperture geometry; however, the point directivities of elastodynamics are always equal. There, we could conclude this from the analytic representation of the second rank Green tensor, here a respective assertion must be based on the spectral plane wave decomposition and its far-field evaluation. We explicitly deal with the circular aperture because it contains the point source as special case.

14.3.1 Integral representation of the Sommerfeld type

The circular (synchronous) normal force density aperture with constant amplitude

$$\underline{\mathbf{t}}(\underline{\mathbf{r}}, \omega) = F(\omega) u(a - r) \underline{\mathbf{e}}_z \tag{14.100}$$

is rotationally symmetric generally allowing for the application of a Fourier–Bessel transform according to (13.332) and (13.333) as an integral representation of the Sommerfeld type (13.331) to calculate its radiation field (Miller and Pursey 1954; Achenbach 1973; Schmerr 1998). The rotational symmetry[260] implies for the vector of the particle displacement $u_\varphi(r, \varphi, z, \omega) \equiv 0$ as well as $\partial/\partial\varphi \equiv 0$ for all field quantities. Forthrightly taking the homogeneous differential equation (7.24) for the particle velocity in cylindrical coordinates as basis, we observe, however, a coupling of its components through the remaining u_r, u_z-components suggesting to perform a decoupling in advance introducing Helmholtz-potentials $\Phi(r, z, \omega)$ and $\underline{\Psi}(r, z, \omega)$ (Section 7.2). Assuring the φ-component of

$$\underline{\mathbf{u}}(r, z, \omega) = \nabla \Phi(r, z, \omega) + \nabla \times \underline{\Psi}(r, z, \omega) \tag{14.101}$$

to be zero, we have to choose $\Psi_r = \Psi_z \equiv 0$ in addition to $\partial/\partial\varphi \equiv 0$; then, $\nabla \cdot \underline{\Psi} = 0$ is also assured, and we have

$$\underline{\mathbf{u}}(r, z, \omega) = \left[\frac{\partial \Phi(r, z, \omega)}{\partial r} - \frac{\partial \Psi_\varphi(r, z, \omega)}{\partial z} \right] \underline{\mathbf{e}}_r + \left[\frac{\partial \Phi(r, z, \omega)}{\partial z} + \frac{1}{r} \frac{\partial r \Psi_\varphi(r, z, \omega)}{\partial r} \right] \underline{\mathbf{e}}_z. \tag{14.102}$$

Consequently, the two resulting homogeneous scalar differential equations for $\Phi(r, z, \omega)$ and $\Psi_\varphi(r, z, \omega)$ read

[260] In each plane $\varphi = \text{const.}$, we expect P-SV-waves.

$$\frac{1}{r}\frac{\partial}{\partial r}\left[r\frac{\partial\Phi(r,z,\omega)}{\partial r}\right]+\frac{\partial^2\Phi(r,z,\omega)}{\partial z^2}+k_{\mathrm{P}}^2\Phi(r,z,\omega)=0, \tag{14.103}$$

$$\frac{1}{r}\frac{\partial}{\partial r}\left[r\frac{\partial\Psi_\varphi(r,z,\omega)}{\partial r}\right]-\frac{\Psi_\varphi(r,z,\omega)}{r^2}+\frac{\partial^2\Psi_\varphi(r,z,\omega)}{\partial z^2}+k_{\mathrm{S}}^2\Psi_\varphi(r,z,\omega)=0. \tag{14.104}$$

With the Fourier–Bessel transform *ansatz*[261]

$$\Phi(r,z,\omega)=\frac{1}{2\pi}\int_0^\infty\bar{\Phi}(K_r,z,\omega)\,\mathrm{J}_0(K_r r)\,K_r\mathrm{d}K_r, \tag{14.105}$$

$$\Psi_\varphi(r,z,\omega)=\frac{1}{2\pi}\int_0^\infty\bar{\Psi}_\varphi(K_r,z,\omega)\,\mathrm{J}_1(K_r r)\,K_r\mathrm{d}K_r, \tag{14.106}$$

we immediately find for $z\geq 0$

$$\bar{\Phi}(K_r,z,\omega)=\overline{\Phi}_0(K_r,\omega)\,\mathrm{e}^{\mathrm{j}zK_{\mathrm{P}z}}, \tag{14.107}$$

$$\bar{\Psi}_\varphi(r,z,\omega)=\overline{\Psi}_{\varphi 0}(K_r,\omega)\,\mathrm{e}^{\mathrm{j}zK_{\mathrm{S}z}} \tag{14.108}$$

with

$$K_{\mathrm{P,S}z}=\sqrt{k_{\mathrm{P,S}}^2-K_r^2}, \tag{14.109}$$

and, hence, for the components of the particle displacement[262]

$$u_r^{t_z}(r,z,\omega)=-\frac{1}{2\pi}\int_0^\infty\Big[K_r\overline{\Phi}_0(K_r,\omega)\,\mathrm{e}^{\mathrm{j}zK_{\mathrm{P}z}}+\mathrm{j}K_{\mathrm{S}z}\overline{\Psi}_{\varphi 0}(K_r,\omega)\,\mathrm{e}^{\mathrm{j}zK_{\mathrm{S}z}}\Big]K_r\mathrm{J}_1(K_r r)\,\mathrm{d}K_r, \tag{14.110}$$

$$u_z^{t_z}(r,z,\omega)=\frac{1}{2\pi}\int_0^\infty\Big[\mathrm{j}K_{\mathrm{P}z}\overline{\Phi}_0(K_r,\omega)\,\mathrm{e}^{\mathrm{j}zK_{\mathrm{P}z}}+K_r\overline{\Psi}_{\varphi 0}(K_r,\omega)\,\mathrm{e}^{\mathrm{j}zK_{\mathrm{S}z}}\Big]K_r\mathrm{J}_0(K_r r)\,\mathrm{d}K_r. \tag{14.111}$$

To calculate $\overline{\Phi}_0(K_r,\omega)$ and $\overline{\Psi}_{\varphi 0}(K_r,\omega)$ from the inhomogeneous boundary condition, we need the tangential and the normal components of $\underline{\underline{\mathbf{T}}}(r,z,\omega)\cdot\underline{\mathbf{e}}_z$:

$$\underline{\underline{\mathbf{T}}}(r,z,\omega):\underline{\mathbf{e}}_z\underline{\mathbf{e}}_r=\mu\left[\frac{\partial u_z(r,z,\omega)}{\partial r}+\frac{\partial u_r(r,z,\omega)}{\partial z}\right], \tag{14.112}$$

$$\underline{\underline{\mathbf{T}}}(r,z,\omega):\underline{\mathbf{e}}_z\underline{\mathbf{e}}_z=\lambda\,\frac{1}{r}\frac{\partial r u_r(r,z,\omega)}{\partial r}+(\lambda+2\mu)\,\frac{\partial u_z(r,z,\omega)}{\partial z}. \tag{14.113}$$

[261]Equation 14.104 is the differential equation of a cylindrical function of first order.

[262]We have (Abramowitz and Stegun 1965)

$$\frac{\mathrm{dJ}_0(K_r r)}{\mathrm{d}r}=-K_r\mathrm{J}_1(K_r r)$$

as well as

$$\frac{1}{r}\frac{\mathrm{d}}{\mathrm{d}r}\left[r\mathrm{J}_1(K_r r)\right]=K_r\mathrm{J}_0(K_r r).$$

With the Fourier–Bessel transform of (14.100) according to (Bracewell 1978)

$$\begin{aligned}\underline{\mathbf{t}}(K_r, \omega) &= 2\pi \int_0^\infty F(\omega) u(a-r)\, \mathrm{J}_0(K_r r)\, r\mathrm{d}r\, \underline{\mathbf{e}}_z \\ &= 2\pi a \frac{\mathrm{J}_1(K_r a)}{K_r} \underline{\mathbf{e}}_z,\end{aligned} \tag{14.114}$$

the tangential component (14.112) turns into a homogeneous and the normal component (14.113) into an inhomogeneous equation for $z = 0$. After resolving them, we have

$$\begin{aligned}&u_r^{t_z}(r, z, \omega) \\ &\quad = -\frac{F(\omega)}{2\pi\mu} \int_0^\infty \frac{2\pi a \mathrm{J}_1(K_r a)}{K_r} \frac{K_r}{R(K_r)} \\ &\quad \times \left[\left(k_\mathrm{S}^2 - 2K_r^2\right) \mathrm{e}^{\mathrm{j}zK_{\mathrm{P}z}} - 2\sqrt{k_\mathrm{S}^2 - K_r^2}\sqrt{k_\mathrm{P}^2 - K_r^2}\, \mathrm{e}^{\mathrm{j}zK_{\mathrm{S}z}} \right] K_r \mathrm{J}_1(K_r r)\, \mathrm{d}K_r,\end{aligned} \tag{14.115}$$

$$\begin{aligned}&u_z^{t_z}(r, z, \omega) \\ &\quad = \mathrm{j}\frac{F(\omega)}{2\pi\mu} \int_0^\infty \frac{2\pi a \mathrm{J}_1(K_r a)}{K_r} \frac{\sqrt{k_\mathrm{P}^2 - K_r^2}}{R(K_r)} \\ &\quad \times \left[\left(k_\mathrm{S}^2 - 2K_r^2\right) \mathrm{e}^{\mathrm{j}zK_{\mathrm{P}z}} + 2K_r^2\, \mathrm{e}^{\mathrm{j}zK_{\mathrm{S}z}} \right] K_r \mathrm{J}_0(K_r r)\, \mathrm{d}K_r.\end{aligned} \tag{14.116}$$

The comparison with (14.68) and (14.69) suggests that the (far-field) *point* directivities in each plane $\varphi = \mathrm{const}$ are principally equal to the *line* directivities.

14.3.2 Point source characteristics

For the Fourier-Bessel integral representations (14.115) and (14.116), we may immediately give the far-field approximations with the saddle point method (Miller and Pursey 1954). However, to be consistent with Section 14.2.4, we want to apply the method of stationary phase to two-dimensional (inverse) Fourier integrals (Equation 11.32), and therefore we replace

$$\mathrm{J}_0(K_r r) = \frac{1}{2\pi} \int_0^{2\pi} \mathrm{e}^{\mathrm{j}K_r r \cos\alpha}\, \mathrm{d}\alpha, \tag{14.117}$$

$$\mathrm{J}_1(K_r r) = -\frac{\mathrm{j}}{2\pi} \int_0^{2\pi} \mathrm{e}^{\mathrm{j}K_r r \cos\alpha} \cos\alpha\, \mathrm{d}\alpha; \tag{14.118}$$

in (14.115) and (14.116). Polar coordinates

$$\begin{aligned}K_x &= K_r \cos\varphi_K, \\ K_y &= K_r \sin\varphi_K; \\ x &= r\cos\varphi, \\ y &= r\sin\varphi\end{aligned} \tag{14.119}$$

imply $K_x x + K_y y = K_r r\cos(\varphi - \varphi_K)$ yielding the desired Fourier integrals for $\alpha = \varphi - \varphi_K$ (due to the rotational symmetry, we may choose $\varphi = 0$). Application of (11.32) to the resulting P- and S-terms and the combination

$$u_{\mathrm{P}R}^{t_z,\mathrm{far}}(R,\vartheta,\omega) = u_r^{t_z,\mathrm{far}}(r,z,\omega)\sin\vartheta + u_z^{t_z,\mathrm{far}}(r,z,\omega)\cos\vartheta, \tag{14.120}$$

$$u_{\mathrm{S}\vartheta}^{t_z,\mathrm{far}}(R,\vartheta,\omega) = u_r^{t_z,\mathrm{far}}(r,z,\omega)\cos\vartheta - u_z^{t_z,\mathrm{far}}(r,z,\omega)\sin\vartheta \tag{14.121}$$

result in the particle displacement far-field of the circular normal force density aperture on the stress-free surface of an elastic half-space decoupled into pressure and shear waves:

$$u_{\mathrm{P}R}^{t_z,\mathrm{far}}(R,\vartheta,\omega) = \frac{a^2F(\omega)}{2c_\mathrm{P}Z_\mathrm{P}}\cos\vartheta\,\frac{\mathrm{J}_1(k_\mathrm{P}a\sin\vartheta)}{k_\mathrm{P}a\sin\vartheta} \times \underbrace{\frac{2\kappa^2(\kappa^2-2\sin^2\vartheta)}{(\kappa^2-2\sin^2\vartheta)^2+2\sin\vartheta\sin2\vartheta\sqrt{\kappa^2-\sin^2\vartheta}}}_{=M_\mathrm{P}^{t_z}(\vartheta)}\frac{\mathrm{e}^{\mathrm{j}k_\mathrm{P}R}}{R}, \tag{14.122}$$

$$u_{\mathrm{S}\vartheta}^{t_z,\mathrm{far}}(R,\vartheta,\omega) = -\frac{a^2F(\omega)}{2c_\mathrm{S}Z_\mathrm{S}}\sin\vartheta\,\frac{\mathrm{J}_1(k_\mathrm{S}a\sin\vartheta)}{k_\mathrm{S}a\sin\vartheta} \times \underbrace{\frac{4\cos\vartheta\sqrt{1-\kappa^2\sin^2\vartheta}}{\kappa(1-2\sin^2\vartheta)^2+2\sin\vartheta\sin2\vartheta\sqrt{1-\kappa^2\sin^2\vartheta}}}_{=M_\mathrm{S}^{t_z}(\vartheta)}\frac{\mathrm{e}^{\mathrm{j}k_\mathrm{S}R}}{R}. \tag{14.123}$$

We purposely split off the factors $\cos\vartheta$ and $\sin\vartheta$ in (14.122) and (14.123) to allow for a direct comparison with the respective full-space source (Equations 13.303 and 13.304): The half-space (far-)field is obtained by a multiplication with the Miller–Pursey factors $M_\mathrm{P}^{t_z}(\vartheta)$ and $M_\mathrm{S}^{t_z}(\vartheta)$.

The Miller–Pursey point directivities are obtained if the circular aperture is replaced by a normal point force according to[263]

$$\underline{\mathbf{t}}(\underline{\mathbf{r}},\omega) = \frac{\delta(r)}{\pi r}\,\underline{\mathbf{e}}_z; \tag{14.124}$$

using the notation (13.125) and (13.126), we obtain

$$u_{\mathrm{P}R}^{\mathrm{PS}_{t_z},\mathrm{far}}(R,\vartheta,\omega) = \frac{\mathrm{e}^{\mathrm{j}k_\mathrm{P}R}}{R}\,H_\mathrm{P}^{\mathrm{PS}_{t_z}}(\vartheta), \tag{14.125}$$

---

[263] The πr-normalization of the δ-distribution is necessary for $\delta(r)/\pi r$ to be a unit source (Langenberg 2005):

$$\int_0^{2\pi}\int_0^\infty \frac{\delta(r)}{\pi r}\,\phi(r)\,r\mathrm{d}r\mathrm{d}\varphi = 2\underbrace{\int_0^\infty \delta(r)\phi(r)\,\mathrm{d}r}_{=1/2} = 1.$$

$$H_{\mathrm{P}}^{\mathrm{PS}_{t_z}}(\vartheta) = \frac{1}{4\pi(\lambda+2\mu)} \frac{2\kappa^2\cos\vartheta(\kappa^2 - 2\sin^2\vartheta)}{(\kappa^2 - 2\sin^2\vartheta)^2 + 2\sin\vartheta\sin 2\vartheta\sqrt{\kappa^2-\sin^2\vartheta}}; \tag{14.126}$$

$$u_{\mathrm{S}\vartheta}^{\mathrm{PS}_{t_z},\mathrm{far}}(R,\vartheta,\omega) = \frac{\mathrm{e}^{\mathrm{j}k_{\mathrm{S}}R}}{R} H_{\mathrm{S}}^{\mathrm{PS}_{t_z}}(\vartheta), \tag{14.127}$$

$$H_{\mathrm{S}}^{\mathrm{PS}_{t_z}}(\vartheta) = -\frac{2\sin 2\vartheta\sqrt{1-\kappa^2\sin^2\vartheta}}{\kappa(1-2\sin^2\vartheta)^2 + 2\sin\vartheta\sin 2\vartheta\sqrt{1-\kappa^2\sin^2\vartheta}}. \tag{14.128}$$

In connection with the line directivities (14.65) and (14.67), we had to point out the $1/\sqrt{\omega}$-frequency dependence, yet here we have to state the frequency independence of the point directivities. In case of a band limitation of the point source, for example, by an RC2(t)-impulse, a "quick look" would suggest, as in Figures 13.9 and 13.11 for the full-space, a similar appearance of RC2(t)-impulses for the particle displacement in both wavefronts. As compared to the full-space wavefronts in Figure 13.9, the direction dependent amplitude weights are given by the point directivities (14.126) and (14.128), and $H_{\mathrm{S}}^{\mathrm{PS}_{t_z}}(\vartheta)$ becomes complex for $\kappa\sin\vartheta > 1$, hence for $\vartheta > \vartheta_{cm\mathrm{P}}$! Similar to the derivation of the impulse equation (9.139), we have to complement

$$H_{\mathrm{S}}^{\mathrm{PS}_{t_z}}(\vartheta,\omega) \stackrel{\mathrm{def}}{=} |H_{\mathrm{S}}^{\mathrm{PS}_{t_z}}(\vartheta)|\mathrm{e}^{\mathrm{j}\phi_{H_{\mathrm{S}}}(\vartheta)\,\mathrm{sign}(\omega)} \tag{14.129}$$

for negative frequencies in the inverse Fourier transform with

$$\phi_{H_{\mathrm{S}}}(\vartheta) = \arg H_{\mathrm{S}}^{\mathrm{PS}_{t_z}}(\vartheta) \tag{14.130}$$

yielding an actual frequency dependence of the S-point directivity. With (2.328), it follows

$$u_{\mathrm{S},}^{\mathrm{PQ}_{t_z},\mathrm{far}}(R,\vartheta,t) = \begin{cases} H_{\mathrm{S}}^{\mathrm{PQ}_{t_z}}(\vartheta)\,\dfrac{F\left(t-\frac{R}{c_{\mathrm{S}}}\right)}{R}, \quad 0 \le \vartheta \le \vartheta_{cm\mathrm{P}} \\ |H_{\mathrm{S}}^{\mathrm{PQ}_{t_z}}(\vartheta)| \left[\cos\phi_{H_{\mathrm{S}}}(\vartheta)\,\dfrac{F\left(t-\frac{R}{c_{\mathrm{S}}}\right)}{R} - \sin\phi_{H_{\mathrm{S}}}(\vartheta)\,\dfrac{\mathcal{H}\left\{F\left(\tau-\frac{R}{c_{\mathrm{S}}}\right)\right\}}{R}\right], \\ \vartheta_{cm\mathrm{P}} < \vartheta \le \pi/2, \end{cases} \tag{14.131}$$

for the $F(\omega)$-bandlimited shear wavefront, while the bandlimited pressure wavefront is given by

$$u_{\mathrm{P}R}^{\mathrm{PS}_{t_z},\mathrm{far}}(R,\vartheta,t) = H_{\mathrm{P}}^{\mathrm{PS}_{t_z}}(\vartheta)\,\frac{F\left(t-\frac{R}{c_{\mathrm{P}}}\right)}{R} \tag{14.132}$$

for all ϑ. For an RC$N(t)$-time dependence of the point-like normal force density, the Hilbert transform in (14.131) with (2.306) may even be calculated—for $\mathcal{H}\{g(\tau)\} = f(t)$, we have $\mathcal{H}\{g(\tau - t_0)\} = f(t - t_0)$—:

$$u_{\mathrm{S},}^{\mathrm{PS}_{t_z},\mathrm{far}}(R,\vartheta,t) = \begin{cases} H_{\mathrm{S}}^{\mathrm{PS}_{t_z}}(\vartheta)\,\dfrac{eN\left(t-\frac{R}{c_{\mathrm{S}}}\right)\cos\omega_0\left(t-\frac{R}{c_{\mathrm{S}}}\right)}{R}, & 0 \le \vartheta \le \vartheta_{cm\mathrm{P}} \\ |H_{\mathrm{S}}^{\mathrm{PS}_{t_z}}(\vartheta)|\,\dfrac{eN\left(t-\frac{R}{c_{\mathrm{S}}}\right)\cos\omega_0\left[t-\frac{R}{c_{\mathrm{S}}}-\phi_{H_{\mathrm{S}}}(\vartheta)\right]}{R}, \\ \quad \vartheta_{cm\mathrm{P}} < \vartheta \le \pi/2, \end{cases} \tag{14.133}$$

where $eN(t)$ is the envelope of the RC$N(t)$-pulse (compare Equation 9.143). We recognize that the complex valued point directivity reflects itself in a phase shift of the carrier frequency of the RC$N(t)$-pulse for $\vartheta > \vartheta_{cm\mathrm{P}}$; therefore, a zero must exist in the S-point directivity for $\vartheta = \vartheta_{cm\mathrm{P}}$; as a consequence, the head wave is suppressed. Figure 14.11 displays the far-field approximation of the pulsed particle velocity field (this is the initial output of EFIT) of a normal point force with RC2(t)-time dependence in terms of wavefronts. The comparison with the approximation-free EFIT-result reveals the differences with respect to the near-field: The direct physical effects of the stress-free surface, head and Rayleigh waves, are missing.

If we want to calculate the respective time domain wavefronts in the far-field of a line source using (14.91) through (14.94), we have to observe the $1/\sqrt{\omega}$-frequency dependence of the line directivities (14.92) and (14.94). With the correspondence (13.53), we therefore obtain

$$H_{\mathrm{P}}^{\mathrm{LS}_{t_z}}(\theta,t) = \frac{1}{2\pi\mu}\sqrt{\frac{c_{\mathrm{P}}}{2}}\cos\theta\, M_{\mathrm{P}}^{t_z}(\theta)\,\frac{u(t)}{\sqrt{t}}, \tag{14.134}$$

(a) EFIT-result without approximations

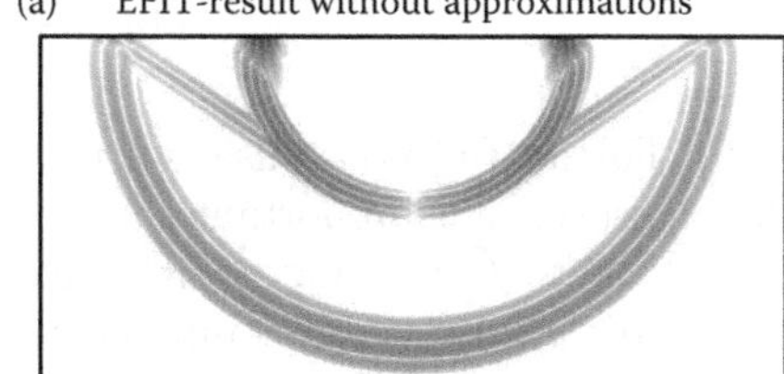

(b) Far-field: stationary phase approximation

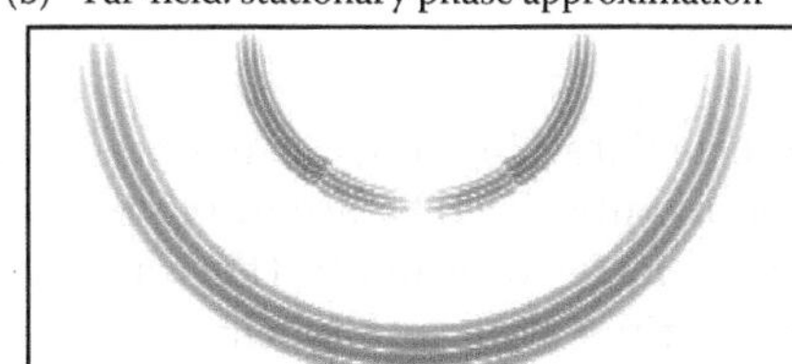

FIGURE 14.11
Wavefronts of a normal line force with RC2(t)-time dependence on a stress-free half-space (magnitude of the particle velocity): (a) EFIT-result without approximations and (b) wavefronts in the far-field with stationary phase approximation.

$$
H_{\mathrm{SV}}^{\mathrm{LS}_{t_z}}(\theta,t) = -\frac{1}{2\pi\mu}\sqrt{\frac{c_\mathrm{S}}{2}}\,\sin\theta
\begin{cases}
M_\mathrm{S}^{t_z}(\theta)\,\dfrac{u(t)}{\sqrt{t}}, & 0\le\theta\le\vartheta_{cm\mathrm{P}}\\[2ex]
|M_\mathrm{S}^{t_z}(\theta)|\left[\cos\phi_{M_\mathrm{S}^{t_z}}(\theta)\,\dfrac{u(t)}{\sqrt{t}} - \sin\phi_{M_\mathrm{S}^{t_z}}(\theta)\,\mathcal{H}\left\{\dfrac{u(\tau)}{\sqrt{\tau}}\right\}\right], & \\
\vartheta_{cm\mathrm{P}}<\theta\le\pi/2,
\end{cases}
\tag{14.135}
$$

where we have used the Miller–Pursey factors $M_\mathrm{P}^{t_z}(\theta)$ and $M_\mathrm{S}^{t_z}(\theta)$ with

$$
M_\mathrm{S}^{t_z}(\theta) = |M_\mathrm{S}^{t_z}(\theta)|\,\mathrm{e}^{\mathrm{j}\phi_{M_\mathrm{S}^{t_z}}(\theta)\,\mathrm{sign}(\omega)} \quad \text{for } \vartheta_{cm\mathrm{P}}<\theta\le\pi/2 \tag{14.136}
$$

as defined by (14.122) and (14.123). That way, we obtain $F(\omega)$-bandlimited time domain wavefronts of the particle displacement in the far-field:

$$
u_{\mathrm{P}r}^{\mathrm{LS}_{t_z},\mathrm{far}}(r,\theta,t) = \frac{1}{2\pi\mu}\sqrt{\frac{c_\mathrm{P}}{2}}\,\cos\theta\,M_\mathrm{P}^{t_z}(\theta)\,\frac{u(t)}{\sqrt{t}} * \frac{F\left(t-\frac{r}{c_\mathrm{P}}\right)}{\sqrt{r}}, \tag{14.137}
$$

$$
u_{\mathrm{SV}"}^{\mathrm{LS}_{t_z},\mathrm{far}}(r,\theta,t) = -\frac{1}{2\pi\mu}\sqrt{\frac{c_\mathrm{S}}{2}}\,\sin\theta
\begin{cases}
M_\mathrm{S}^{t_z}(\theta)\,\dfrac{u(t)}{\sqrt{t}} * \dfrac{F\left(t-\frac{r}{c_\mathrm{S}}\right)}{\sqrt{r}}, & 0\le\theta\le\vartheta_{cm\mathrm{P}}\\[2ex]
|M_\mathrm{S}^{t_z}(\theta)|\dfrac{u(t)}{\sqrt{t}} * \left[\cos\phi_{M_\mathrm{S}^{t_z}}(\theta)\,\dfrac{F\left(t-\frac{r}{c_\mathrm{S}}\right)}{\sqrt{r}} - \right. & \\
\qquad \left. - \sin\phi_{M_\mathrm{S}^{t_z}}(\theta)\,\dfrac{\mathcal{H}\left\{F\left(\tau-\frac{r}{c_\mathrm{S}}\right)\right\}}{\sqrt{r}}\right], & \\
\vartheta_{cm\mathrm{P}}<\theta\le\pi/2.
\end{cases}
\tag{14.138}
$$

To calculate (14.138), we have used (2.300). For an RC$N(t)$-pulse, we may even modify (14.138) according to (14.133); however, it is important that we always have to evaluate the convolution with $u(t)/\sqrt{t}$ yielding a difference in the time structure of the wavefronts in two dimensions as compared to Figure 14.11.

Due to the importance of the point directivities for practical applications—especially for the normal force[264]—we recap the respective part of Figure 14.10 in a matched terminology (Figure 14.12).

[264]The derivation of point directivities for the tangential force in the limit of a torque force is presented by Miller and Pursey (1954).

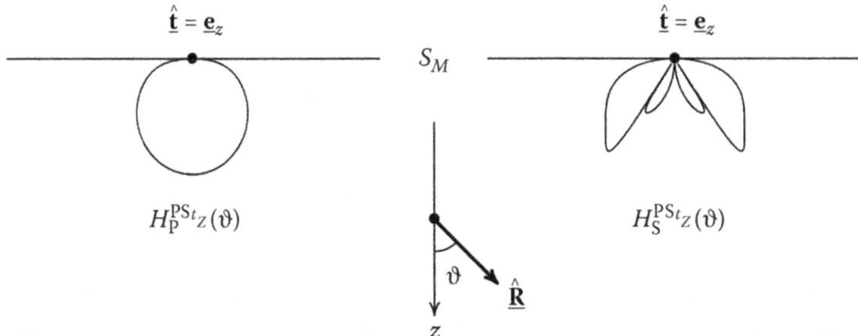

FIGURE 14.12
Radiation patterns of the particle displacement for a normal point force density on the surface of an elastic half-space (magnitudes of the normalized radiation characteristics as function of θ; material of the half-space: steel).

14.4 Radiation Fields of Piezoelectric Transducers

One of the basic tasks of US-NDT comprises modeling of the sound field of piezoelectric transducers with contact to the surface. An idealized model is sketched in Figure 14.13. On an infinitely large planar specimen surface S_M with vacuum on top, being often also the measurement surface for scattered ultrasonic signals, we prescribe a surface force density $\underline{\mathbf{t}}(x, y, t)$ within an aperture of arbitrary geometry S_A as function of the surface coordinates and time; through a Fourier transform with regard to t, the spectrum $\underline{\mathbf{t}}(x, y, \omega)$

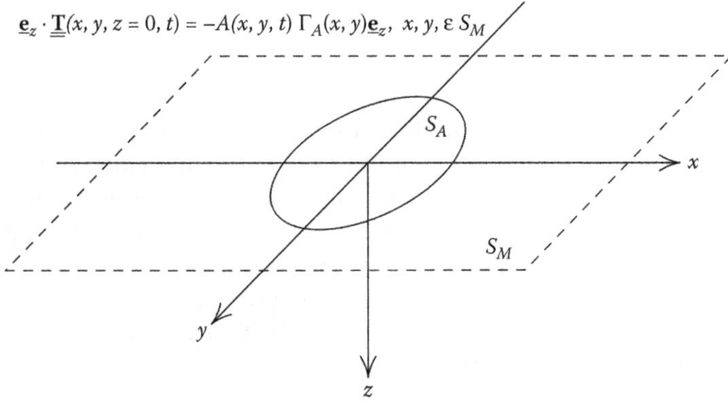

FIGURE 14.13
Basic task of US-NDT: sound field calculation of piezoelectric contact transducers prescribing an aperture distribution.

is specified for all frequencies, for the case of a monochromatic excitation, we may also select one particular frequency. As a consequence of the usual transducer fluid coupling, we exclusively prescribe the normal component of $\underline{\mathbf{t}}(x, y, t)$ according to

$$\underline{\mathbf{t}}(x, y, t) = A(x, y, t)\Gamma_A(x, y)\underline{\mathbf{e}}_z, \quad x, y \in S_M, \tag{14.139}$$

(Section 3.3.4), where $A(x, y, t)$ is a potentially nonsynchronous spatially and time-dependent aperture distribution within S_A; the restriction to S_A is ensured by the characteristic function $\Gamma_A(x, y)$ of S_A. A synchronous distribution would imply the splitting $A(x, y, t) = A(x, y)F(t)$.

Half-space Green tensor: In case the half-space $z > 0$ is homogeneous and isotropic, we know the Green tensors $\underline{\underline{\mathbf{G}}}$ and $\underline{\underline{\underline{\mathbf{\Sigma}}}}$ for the respective full-space allowing for an immediate citation of the time harmonic source field integral (13.213) for surface sources on S_M (we have $z > 0$):

$$\begin{aligned}\underline{\mathbf{v}}(x, y, z, \omega) = \int_{-\infty}^{\infty}\int_{-\infty}^{\infty} \Big[&- \mathrm{j}\omega\underline{\mathbf{t}}(x', y', \omega) \cdot \underline{\underline{\mathbf{G}}}(x - x', y - y', z, \omega) \\ &+ \underline{\underline{\mathbf{g}}}(x', y', \omega) : \underline{\underline{\underline{\mathbf{\Sigma}}}}(x - x', y - y', z, \omega)\Big]\, \mathrm{d}x'\mathrm{d}y'. \end{aligned} \tag{14.140}$$

It is consistent with the elastodynamic Huygens principle (Section 15.1.3) that both surface sources—surface force density $\underline{\mathbf{t}} = -\underline{\mathbf{e}}_z \cdot \underline{\underline{\mathbf{T}}}$ and surface deformation rate $\underline{\underline{\mathbf{g}}} = -\frac{1}{2}(\underline{\mathbf{e}}_z\underline{\mathbf{v}} + \underline{\mathbf{v}}\,\underline{\mathbf{e}}_z)$—appear in the integral. With (14.139), together with the symmetry (13.153) of $\underline{\underline{\underline{\mathbf{\Sigma}}}}$, (14.140) reduces to

$$\begin{aligned}\underline{\mathbf{v}}(x, y, z, \omega) = &- \mathrm{j}\omega \iint_{S_A} A(x', y', \omega)\, \underline{\mathbf{e}}_z \cdot \underline{\underline{\mathbf{G}}}(x - x', y - y', z, \omega)\, \mathrm{d}x'\mathrm{d}y' \\ &+ \int_{-\infty}^{\infty}\int_{-\infty}^{\infty} \underline{\mathbf{v}}(x', y', 0, \omega)\, \underline{\mathbf{e}}_z : \underline{\underline{\underline{\mathbf{\Sigma}}}}(x - x', y - y', z, \omega)\, \mathrm{d}x'\mathrm{d}y', \end{aligned} \tag{14.141}$$

where the second term is annoying because $\underline{\mathbf{v}}(x', y', 0, \omega)$ is not known beforehand, and it may not be prescribed independently of $A(x', y', \omega)$; in fact, we must perform the limit $z \longrightarrow 0$ in (14.141) mathematically very carefully resulting in an integral equation of the second kind for $\underline{\mathbf{v}}(x, y, z = 0, \omega)$. The solution of this integral equation is numerically expensive, yet it is advantageous because only known Green tensors are involved.

As an alternative to the solution of an integral equation, we may define a Green tensor[265] $\underline{\underline{\mathbf{G}}}^{\mathrm{N}}(x', y', z', x, y, z, \omega)$ of the half-space satisfying the differential equation (Equation 13.80)

[265] The upper index "N" stands for "Neumann": Within scalar acoustics, a (homogeneous) Dirichlet problem is defined by vanishing boundary values of the scalar pressure itself, whereas a Neumann problem is characterized by a vanishing normal derivative of the field quantity (Section 5.6). Here, the normal "component" of a "differentiated" tensor should similarly vanish.

$$\begin{aligned}&\left[(\mu\Delta' + \rho\omega^2)\,\underline{\underline{\mathbf{I}}} + (\lambda + \mu)\nabla'\nabla'\right]\\&\quad\cdot\,\underline{\underline{\mathbf{G}}}^{\mathrm{N}}(x', y', z', x, y, z, \omega) = -\delta(x' - x)\delta(y' - y)\delta(z' - z)\,\underline{\underline{\mathbf{I}}}\end{aligned} \tag{14.142}$$

as well as the stress-free boundary condition on S_M (of course, $\underline{\underline{\underline{\underline{\mathbf{c}}}}}$ is the stiffness tensor of a homogeneous isotropic material)

$$\underline{\mathbf{e}}_z \cdot \underline{\underline{\underline{\underline{\mathbf{c}}}}} : \nabla'\underline{\underline{\mathbf{G}}}^{\mathrm{N}}(x', y', z', x, y, z, \omega)\Big|_{z'=0} = \underline{\mathbf{0}}, \quad x', y' \in S_M. \tag{14.143}$$

Based on a reciprocity theorem, one can show (de Hoop 1995):

$$\underline{\underline{\mathbf{G}}}^{\mathrm{N}}(x', y', z', x, y, z, \omega) = \underline{\underline{\mathbf{G}}}^{\mathrm{N}\,21}(x, y, z, x', y', z', \omega). \tag{14.144}$$

With $\underline{\underline{\mathbf{G}}}^{\mathrm{N}}(x', y', z', x, y, z, \omega)$, we define $\underline{\underline{\boldsymbol{\Sigma}}}^{\mathrm{N}}(x', y', z', x, y, z, \omega)$ according to (Equation 13.142)

$$\underline{\underline{\underline{\boldsymbol{\Sigma}}}}^{\mathrm{N}}(x', y', z', x, y, z, \omega) = \underline{\underline{\underline{\underline{\mathbf{c}}}}} : \nabla'\underline{\underline{\mathbf{G}}}^{\mathrm{N}}(x', y', z', x, y, z, \omega), \tag{14.145}$$

and therefore we have

$$\underline{\mathbf{e}}_z \cdot \underline{\underline{\underline{\boldsymbol{\Sigma}}}}^{\mathrm{N}}(x', y', z' = 0, x, y, z, \omega) = \underline{\underline{\mathbf{0}}} \;\text{ for } x', y' \in S_M. \tag{14.146}$$

We conclude the dependence of $\underline{\underline{\mathbf{G}}}^{\mathrm{N}}$ and $\underline{\underline{\underline{\boldsymbol{\Sigma}}}}^{\mathrm{N}}$ from $x' - x$ and $y' - y$ via Fourier transforming (14.142); due to the nonsymmetry in z' and z as stated by the boundary condition (14.143), it remains the dependence "z' comma z":

$$\underline{\underline{\mathbf{G}}}^{\mathrm{N}}(x', y', z', x, y, z, \omega) \Longrightarrow \underline{\underline{\mathbf{G}}}^{\mathrm{N}}(x' - x, y' - y, z', z, \omega), \tag{14.147}$$

$$\underline{\underline{\underline{\boldsymbol{\Sigma}}}}^{\mathrm{N}}(x', y', z', x, y, z, \omega) \Longrightarrow \underline{\underline{\underline{\boldsymbol{\Sigma}}}}^{\mathrm{N}}(x' - x, y' - y, z', z, \omega). \tag{14.148}$$

With (14.146), the bothering integral in (14.141) disappears if we use the half-space Green tensor $\underline{\underline{\mathbf{G}}}^{\mathrm{N}}$ instead of the full-space Green tensor $\underline{\underline{\mathbf{G}}}$:

$$\begin{aligned}&\underline{\mathbf{v}}(x, y, z, \omega)\\&\quad = -\mathrm{j}\omega \iint_{S_A} A(x', y', \omega)\,\underline{\mathbf{e}}_z \cdot \underline{\underline{\mathbf{G}}}^{\mathrm{N}}(x' - x, y' - y, z' = 0, z, \omega)\,\mathrm{d}x'\mathrm{d}y'.\end{aligned} \tag{14.149}$$

Note: $\underline{\underline{\mathbf{G}}}^{\mathrm{N}}$ is symmetric on behalf of (14.144) allowing for an interchange of the factors in the dot product $\underline{\mathbf{e}}_z \cdot \underline{\underline{\mathbf{G}}}^{\mathrm{N}}$ only with a simultaneous interchange of the primed with the unprimed variables:

$$\underline{\mathbf{v}}(x, y, z, \omega) = -\mathrm{j}\omega \iint_{S_A} A(x', y', \omega)\,\underline{\underline{\mathbf{G}}}^{\mathrm{N}}(x - x', y - y', z, z' = 0, \omega) \cdot \underline{\mathbf{e}}_z\,\mathrm{d}x'\mathrm{d}y'. \tag{14.150}$$

For a unit point source $A(x', y', \omega) = \delta(x')\delta(y')$ at the origin, we in fact conclude according to

$$\begin{aligned}\underline{\mathbf{u}}^{\mathrm{PS}_{t_z}}(x, y, z, \omega) &= \underline{\underline{\mathbf{G}}}^{\mathrm{N}}(x, y, z, z' = 0, \omega) \cdot \underline{\mathbf{e}}_z \\ &\stackrel{\text{def}}{=} \underline{\underline{\mathbf{G}}}^{\mathrm{HS}}(x, y, z, \omega) \cdot \underline{\mathbf{e}}_z \end{aligned} \tag{14.151}$$

the sense of the Green tensor projected onto the direction of the exciting force density $\underline{\underline{\mathbf{G}}}^{\mathrm{N}} \cdot \underline{\mathbf{e}}_z$ as particle displacement of a normal force density on the stress-free surface of a homogeneous isotropic elastic half-space, because we have—in unprimed coordinates—

$$\begin{aligned}\underline{\mathbf{e}}_z \cdot \underline{\underline{\mathbf{T}}}^{\mathrm{PS}_{t_z}}(x, y, z = 0, \omega) &= \underline{\mathbf{e}}_z \cdot \underline{\underline{\underline{\underline{\mathbf{c}}}}} : \nabla \underline{\mathbf{u}}^{\mathrm{PS}_{t_z}}(x, y, z, \omega)\Big|_{z=0} \\ &= \underline{\mathbf{0}}, \quad x, y \in S_M, \end{aligned} \tag{14.152}$$

due to (14.143). With (14.74), we gave a respective definition of the two-dimensional half-space Green tensor.

In the limit $z \longrightarrow 0$, the integral representation (14.147) must actually assume the boundary values specified by

$$\underline{\mathbf{e}}_z \cdot \underline{\underline{\mathbf{T}}}(x, y, z = 0, \omega) = -A(x, y, \omega)\Gamma_A(x, y)\, \underline{\mathbf{e}}_z, \quad x, y \in S_M, \tag{14.153}$$

because it was accordingly constructed; however, the explicit proof is not at all trivial.

The integral representation (14.149) of the transducer sound field would be the elastodynamic counterpart to the Rayleigh–Sommerfeld integral[266] (14.15) if we would know $\underline{\underline{\mathbf{G}}}^{\mathrm{N}}(x' - x, y' - y, z', z, \omega)$ explicitly: Unfortunately, this is not true!

Strip-like aperture: However, we know integral representations of Weyl's or Sommerfeld's type for different alternatives of $\underline{\underline{\mathbf{G}}}^{\mathrm{N}}$. At first, we switch to the particle displacement and exploit the convolution structure of the aperture integral with regard to x' and y':

$$\begin{aligned}\underline{\mathbf{u}}(x, y, z, \omega) = \frac{1}{(2\pi)^2} \int_{-\infty}^{\infty} \int_{-\infty}^{\infty} & \hat{A}(K_x, K_y, \omega)\, \underline{\underline{\hat{\mathbf{G}}}}^{\mathrm{N}}(K_x, K_y, z, z' = 0, \omega) \\ & \cdot \underline{\mathbf{e}}_z\, \mathrm{e}^{\mathrm{j}K_x x + \mathrm{j}K_y y}\, \mathrm{d}K_x \mathrm{d}K_y, \end{aligned} \tag{14.154}$$

where $\hat{A}(K_x, K_y, \omega)$ and $\underline{\underline{\hat{\mathbf{G}}}}^{\mathrm{N}}(K_x, K_y, z, z' = 0, \omega)$ are spatial two-dimensional Fourier transforms of $A(x, y, \omega)$ and $\underline{\underline{\mathbf{G}}}^{\mathrm{N}}(x, y, z, z' = 0, \omega)$ with regard to x

[266] In acoustics, a free surface composes a Dirichlet problem for the pressure, yet in elastodynamics, a "*Neumann*" problem for the particle velocity because the pressure corresponds to the stress tensor.

and y. For a two-dimensional synchronous aperture distribution $A(x,y,\omega) = F(\omega)A(x)$ with

$$\hat{A}(K_x, K_y, \omega) = F(\omega)\,\hat{A}(K_x)\,2\pi\delta(K_y), \tag{14.155}$$

(14.154) specially reads

$$\underline{\mathbf{u}}(x,z,\omega) = \frac{F(\omega)}{2\pi}\int_{-\infty}^{\infty} \hat{A}(K_x)\hat{\underline{\underline{\mathbf{G}}}}^{\mathrm{N}}(K_x, K_y = 0, z, z' = 0, \omega)\cdot\underline{\mathbf{e}}_z\, \mathrm{e}^{\mathrm{j}K_x x}\,\mathrm{d}K_x, \tag{14.156}$$

resulting in

$$\hat{A}(K_x, K_y, \omega) = F(\omega)\,\frac{2\sin aK_x}{K_x}\,2\pi\delta(K_y), \tag{14.157}$$

and consequently

$$\underline{\mathbf{u}}(x,z,\omega) = \frac{F(\omega)}{2\pi}\int_{-\infty}^{\infty} \frac{2\sin aK_x}{K_x}\,\underbrace{\hat{\underline{\underline{\mathbf{G}}}}^{\mathrm{N}}(K_x, K_y = 0, z, z' = 0, \omega)\cdot\underline{\mathbf{e}}_z}_{\stackrel{\mathrm{def}}{=}\,\hat{\underline{\underline{\mathbf{G}}}}^{\mathrm{HS}}_{\mathrm{PSV}}(K_x, z, \omega)\cdot\underline{\mathbf{e}}_z}\, \mathrm{e}^{\mathrm{j}K_x x}\,\mathrm{d}K_x \tag{14.158}$$

for the strip-like synchronous aperture distribution with constant amplitude according to $A(x,y,\omega) = F(\omega)q_a(x)$; here, $\hat{\underline{\underline{\mathbf{G}}}}^{\mathrm{N}}\cdot\underline{\mathbf{e}}_z$ may be deduced from

$$\underline{\underline{\mathbf{G}}}^{\mathrm{HS}}_{\mathrm{PSV}}(x,z,\omega)\cdot\underline{\mathbf{e}}_z = \frac{1}{2\pi}\int_{-\infty}^{\infty} \hat{\underline{\underline{\mathbf{G}}}}^{\mathrm{N}}(K_x, K_y = 0, z, z' = 0, \omega)\cdot\underline{\mathbf{e}}_z\, \mathrm{e}^{\mathrm{j}K_x x}\,\mathrm{d}K_x \tag{14.159}$$

with recourse to (14.83) and (14.84). That way, Weyl's integral representation (14.156) is the extension of (14.68) and (14.69) to arbitrary two-dimensional aperture distributions.

Arbitrary geometry of the aperture S_A: To calculate $\underline{\underline{\mathbf{G}}}^{\mathrm{N}}\cdot\underline{\mathbf{e}}_z$ in

$$\underline{\mathbf{u}}(x,y,z,\omega) = \iint_{S_A} A(x',y',\omega)\,\underline{\underline{\mathbf{G}}}^{\mathrm{N}}(x-x', y-y', z, z'=0, \omega)\cdot\underline{\mathbf{e}}_z\,\mathrm{d}x'\mathrm{d}y' \tag{14.160}$$

for arbitrary aperture geometries and distributions, we have to modify the ansatz "spectrum of plane waves" (Equation 14.49) according to

$$\begin{aligned}\underline{\mathbf{u}}_{\mathrm{P,S}}(x,y,z,\omega) = \frac{1}{(2\pi)^2}\int_{-\infty}^{\infty}\int_{-\infty}^{\infty} \Big[&\hat{u}_{\mathrm{P}}(K_x,K_y,\omega)\,\mathrm{e}^{\mathrm{j}\underline{\mathbf{K}}_{\mathrm{P}}\cdot\underline{\mathbf{R}}}\,\hat{\underline{\mathbf{K}}}_{\mathrm{P}}\\ &+ \hat{u}_{\mathrm{S1}}(K_x,K_y,\omega)\,\mathrm{e}^{\mathrm{j}\underline{\mathbf{K}}_{\mathrm{S}}\cdot\underline{\mathbf{R}}}\,\underline{\mathbf{e}}_{\vartheta_{K_{\mathrm{S}}}}\\ &+ \hat{u}_{\mathrm{S2}}(K_x,K_y,\omega)\,\mathrm{e}^{\mathrm{j}\underline{\mathbf{K}}_{\mathrm{S}}\cdot\underline{\mathbf{R}}}\,\underline{\mathbf{e}}_{\varphi_{K_{\mathrm{S}}}}\Big]\,\mathrm{d}K_x\mathrm{d}K_y\end{aligned} \tag{14.161}$$

because we may no longer exclude a term with spectral $\underline{\mathbf{e}}_{\varphi_{K_S}}$-polarization due to symmetry considerations as for the circular aperture; therefore, its previously calculated sound field is useless solving the above task.

Sound fields of piezoelectric transducers in the time domain: Weyl's integral representation for the particle displacement field of a strip-like aperture with constant amplitude distribution (Equations 14.68 and 14.69) is the starting point for further conversions: As already mentioned, we obtain explicit expressions for the time domain field by applying the Cagniard–de Hoop method to the integral representation for the time harmonic field: $\underline{\mathbf{u}}(x, z, \omega) \Longrightarrow \underline{\mathbf{u}}(x, z, t)$, however, only for distributional excitation functions $F(t)$ (normal force density: Aulenbacher 1988) requiring subsequent convolutions for any arbitrary $F(t)$. Due to the "unlovely" structure of the scalar (full-space) Green function (13.36) with the square root singularity, the numerical evaluation is somewhat unpleasant leading at best to a mere basic insight. Nevertheless, we find point directivities for the near-field essentially revealing a shift of the maximum of the line source shear wave diagram to different angles.

An alternative to the above approach is the evaluation of (14.68) and (14.69) for those frequencies that—for example—are contained in an RC2(t)-impulse with a subsequent Fourier inversion with regard to ω. Especially for points close to the surface, this is not quite uncomplicated (Fellinger 1991).

Last but not least, we have the (two-dimensional) EFIT-code at hand to produce time domain wavefronts for the strip-like RC2(t)-excited aperture: In fact, the numerical tool implemented as EFIT visualizes $\underline{\mathbf{v}}(x, z, t)$ as function of space and time: Figure 14.7 already displayed examples. Advantageously, these numerical results can be validated against the above schemes (Fellinger 1991). Generally, EFIT also yields wavefronts for three-dimensional planar apertures; this is simply a question of available computer memory (and computation time). Again, for the point source, an analytic formulation is available (Achenbach 1973).

Sound fields of piezoelectric transducers in the far-field: If the planar (normal) force density (14.139) is located in full-space, we obtain the following particle displacement field (Equation 13.215 with 13.100, respectively 13.222 with 13.101):

$$\underline{\mathbf{u}}_{\mathrm{P}}^{t_z,\mathrm{far}}(\underline{\mathbf{R}}, \omega) = \frac{\mathrm{e}^{\mathrm{j}k_{\mathrm{P}}R}}{R} \frac{\underline{\underline{\mathbf{g}}}_{\mathrm{P}}(\hat{\underline{\mathbf{R}}}) \cdot \underline{\mathbf{e}}_z}{\lambda + 2\mu} \iint_{S_A} A(x', y', \omega) \times \mathrm{e}^{-\mathrm{j}k_{\mathrm{P}}x' \sin\vartheta\cos\varphi - \mathrm{j}k_{\mathrm{P}}y' \sin\vartheta\sin\varphi}\, \mathrm{d}x'\mathrm{d}y', \tag{14.162}$$

$$\underline{\mathbf{u}}_{\mathrm{S}}^{t_z,\mathrm{far}}(\underline{\mathbf{R}}, \omega) = \frac{\mathrm{e}^{\mathrm{j}k_{\mathrm{S}}R}}{R} \frac{\underline{\underline{\mathbf{g}}}_{\mathrm{S}}(\hat{\underline{\mathbf{R}}}) \cdot \underline{\mathbf{e}}_z}{\mu} \iint_{S_A} A(x', y', \omega) \times \mathrm{e}^{-\mathrm{j}k_{\mathrm{S}}x' \sin\vartheta\cos\varphi - \mathrm{j}k_{\mathrm{S}}y' \sin\vartheta\sin\varphi}\, \mathrm{d}x'\mathrm{d}y'. \tag{14.163}$$

Here, the aperture integral is nothing but the two-dimensional Fourier transform of the aperture distribution $A(x, y, \omega)\Gamma_A(x, y)$ for specially spatially depending values of the Fourier variables K_x and K_y:

$$
\begin{aligned}
\mathcal{F}_{xy}\{A(x,y,\omega)\Gamma_A(x,y)\}_{\substack{K_x = k_{\mathrm{P,S}}\sin\vartheta\cos\varphi \\ K_y = k_{\mathrm{P,S}}\sin\vartheta\sin\varphi}}
&= \iint_{S_A} A(x',y',\omega)\, \mathrm{e}^{-\mathrm{j}k_{\mathrm{P,S}}\sin\vartheta\cos\varphi - \mathrm{j}k_{\mathrm{P,S}}\sin\vartheta\sin\varphi}\, \mathrm{d}x'\mathrm{d}y' \\
&= \hat{A}(K_x = k_{\mathrm{P,S}}\sin\vartheta\cos\varphi, K_y = k_{\mathrm{P,S}}\sin\vartheta\sin\varphi, \omega).
\end{aligned}
\tag{14.164}
$$

With (Equations 13.102 and 13.103)

$$
\begin{aligned}
\underline{\underline{\mathbf{g}}}_{\mathrm{P}}(\hat{\underline{\mathbf{R}}}) \cdot \underline{\mathbf{e}}_z &= \frac{1}{4\pi} \hat{\underline{\mathbf{R}}}\hat{\underline{\mathbf{R}}} \cdot \underline{\mathbf{e}}_z \\
&= \frac{1}{4\pi} \cos\vartheta\, \hat{\underline{\mathbf{R}}},
\end{aligned}
\tag{14.165}
$$

$$
\begin{aligned}
\underline{\underline{\mathbf{g}}}_{\mathrm{S}}(\hat{\underline{\mathbf{R}}}) \cdot \underline{\mathbf{e}}_z &= \frac{1}{4\pi} (\underline{\underline{\mathbf{I}}} - \hat{\underline{\mathbf{R}}}\hat{\underline{\mathbf{R}}}) \cdot \underline{\mathbf{e}}_z \\
&= \frac{1}{4\pi} (\underline{\mathbf{e}}_\vartheta \underline{\mathbf{e}}_\vartheta - \underline{\mathbf{e}}_\varphi \underline{\mathbf{e}}_\varphi) \cdot \underline{\mathbf{e}}_z \\
&= \frac{1}{4\pi} \underline{\mathbf{e}}_\vartheta \underline{\mathbf{e}}_\vartheta \cdot \underline{\mathbf{e}}_z \\
&= -\frac{1}{4\pi} \sin\vartheta\, \underline{\mathbf{e}}_\vartheta,
\end{aligned}
\tag{14.166}
$$

we obtain

$$
\begin{aligned}
\underline{\mathbf{u}}_{\mathrm{P}}^{t_z,\mathrm{far}}(\underline{\mathbf{R}},\omega) &= \frac{1}{4\pi c_{\mathrm{P}} Z_{\mathrm{P}}} \frac{\mathrm{e}^{\mathrm{j}k_{\mathrm{P}}R}}{R} \\
&\quad \times \cos\vartheta\, \hat{A}(K_x = k_{\mathrm{P}}\sin\vartheta\cos\varphi, K_y = k_{\mathrm{P}}\sin\vartheta\sin\varphi, \omega)\, \underline{\mathbf{e}}_R,
\end{aligned}
\tag{14.167}
$$

$$
\begin{aligned}
\underline{\mathbf{u}}_{\mathrm{S}}^{t_z,\mathrm{far}}(\underline{\mathbf{R}},\omega) &= -\frac{1}{4\pi c_{\mathrm{S}} Z_{\mathrm{S}}} \frac{\mathrm{e}^{\mathrm{j}k_{\mathrm{S}}R}}{R} \\
&\quad \times \sin\vartheta\, \hat{A}(K_x = k_{\mathrm{S}}\sin\vartheta\cos\varphi, K_y = k_{\mathrm{S}}\sin\vartheta\sin\varphi, \omega)\, \underline{\mathbf{e}}_\vartheta.
\end{aligned}
\tag{14.168}
$$

We particularly emphasize that the far-field of a normal force density (the direction of the force density is parallel to a cartesian coordinate, here $\underline{\mathbf{e}}_z$) does not exhibit a φ-component of $\underline{\mathbf{u}}_{\mathrm{S}}^{t_z,\mathrm{far}}$. We have already given two examples for the general far-field equations of three-dimensionally planar force densities in full-space: the synchronous rectangular and circular aperture with constant amplitude distribution. For the rectangular aperture, we have (Equations 13.279 and 13.280)

$$
\begin{aligned}
&\hat{A}(K_x = k_{\mathrm{P,S}}\sin\vartheta\cos\varphi, K_y = k_{\mathrm{P,S}}\sin\vartheta\sin\varphi, \omega) \\
&\quad = F(\omega)\, \mathcal{F}_{xy}\{q_a(x)q_b(y)\}_{\substack{K_x = k_{\mathrm{P,S}}\sin\vartheta\cos\varphi \\ K_y = k_{\mathrm{P,S}}\sin\vartheta\sin\varphi}} \\
&\quad = F(\omega) \frac{2\sin(k_{\mathrm{P,S}}a\sin\vartheta\cos\varphi)}{k_{\mathrm{P,S}}\sin\vartheta\cos\varphi} \frac{2\sin(k_{\mathrm{P,S}}b\sin\vartheta\sin\varphi)}{k_{\mathrm{P,S}}\sin\vartheta\sin\varphi},
\end{aligned}
\tag{14.169}
$$

and for the circular aperture, we found (Equations 13.303 and 13.304)

$$
\begin{aligned}
\hat{A}(K_x &= k_{\mathrm{P,S}} \sin\vartheta \cos\varphi, K_y = k_{\mathrm{P,S}} \sin\vartheta \sin\varphi, \omega) \\
&= F(\omega)\, \mathcal{F}_{xy} \left\{ u\left(a - \sqrt{x^2+y^2}\right) \right\}_{\substack{K_x = k_{\mathrm{P,S}} \sin\vartheta \cos\varphi \\ K_y = k_{\mathrm{P,S}} \sin\vartheta \sin\varphi}} \\
&= F(\omega) \left. \frac{2\pi a \mathrm{J}_1\left(a\sqrt{K_x^2+K_y^2}\right)}{\sqrt{K_x^2+K_y^2}} \right|_{\substack{K_x = k_{\mathrm{P,S}} \sin\vartheta \cos\varphi \\ K_y = k_{\mathrm{P,S}} \sin\vartheta \sin\varphi}} \\
&= F(\omega) \left. \frac{2\pi a \mathrm{J}_1(aK_r)}{K_r} \right|_{K_r = k_{\mathrm{P,S}} \sin\vartheta} .
\end{aligned} \tag{14.170}
$$

For the two-dimensionally planar (strip-like) force density in full-space, the respective expressions for the particle displacement are also known (Equation 13.307 with 13.312 and 13.313):

$$
\underline{\mathbf{u}}_{\mathrm{P}}^{t_z,\mathrm{far}}(\underline{\mathbf{r}},\omega) = \frac{1}{\sqrt{\omega}} \frac{\mathrm{e}^{\mathrm{j}\frac{\pi}{4}}}{2Z_{\mathrm{P}}\sqrt{2\pi c_{\mathrm{P}}}} \frac{\mathrm{e}^{\mathrm{j}k_{\mathrm{P}}r}}{\sqrt{r}} \cos\theta\, \hat{A}(K_x = k_{\mathrm{P}} \sin\theta, \omega)\, \underline{\mathbf{e}}_r, \tag{14.171}
$$

$$
\underline{\mathbf{u}}_{\mathrm{S}}^{t_z,\mathrm{far}}(\underline{\mathbf{r}},\omega) = -\frac{1}{\sqrt{\omega}} \frac{\mathrm{e}^{\mathrm{j}\frac{\pi}{4}}}{2Z_{\mathrm{S}}\sqrt{2\pi c_{\mathrm{S}}}} \frac{\mathrm{e}^{\mathrm{j}k_{\mathrm{S}}r}}{\sqrt{r}} \sin\theta\, \hat{A}(K_x = k_{\mathrm{S}} \sin\theta, \omega)\, \underline{\mathbf{e}}_\theta, \tag{14.172}
$$

where

$$
\begin{aligned}
\hat{A}(K_x = k_{\mathrm{P,S}} \sin\theta, \omega) &= \mathcal{F}\{A(x,\omega) q_a(x)\}_{K_x = k_{\mathrm{P,S}}\sin\theta} \\
&= F(\omega)\, \frac{2\sin(k_{\mathrm{P,S}} a \sin\theta)}{k_{\mathrm{P,S}} \sin\theta}.
\end{aligned} \tag{14.173}
$$

On the other hand, we also calculated the particle displacement field of the strip-like and the circular synchronous aperture on a half-space. Therefore, a comparison is possible: At first, we refer to the half-space result for the strip-like synchronous aperture, namely the line source expressions multiplied by the radiation characteristic of the finite sized aperture:[267]

$$
\begin{aligned}
&\underline{\mathbf{u}}_{\mathrm{P}}^{t_z,\mathrm{far}}(r,\theta,\omega) \\
&= \underbrace{\underline{\mathbf{e}}_r \frac{1}{\sqrt{\omega}} \frac{\mathrm{e}^{\mathrm{j}\frac{\pi}{4}}}{2Z_{\mathrm{P}}\sqrt{2\pi c_{\mathrm{P}}}} \frac{\mathrm{e}^{\mathrm{j}k_{\mathrm{P}}r}}{\sqrt{r}} \cos\theta}_{\text{LS in full-space}} \underbrace{F(\omega) \frac{2\sin(k_{\mathrm{P}} a \sin\theta)}{k_{\mathrm{P}} \sin\theta}}_{\text{Scalar aperture}} \underbrace{M_{\mathrm{P}}^{t_z}(\theta)}_{\text{MP-factor}},
\end{aligned} \tag{14.174}
$$

$$
\begin{aligned}
&\underline{\mathbf{u}}_{\mathrm{SV}}^{t_z,\mathrm{far}}(r,\theta,\omega) \\
&= \underbrace{-\underline{\mathbf{e}}_\theta \frac{1}{\sqrt{\omega}} \frac{\mathrm{e}^{\mathrm{j}\frac{\pi}{4}}}{2Z_{\mathrm{S}}\sqrt{2\pi c_{\mathrm{S}}}} \frac{\mathrm{e}^{\mathrm{j}k_{\mathrm{S}}r}}{\sqrt{r}} \sin\theta}_{\text{LS in full-space}} \underbrace{F(\omega) \frac{2\sin(k_{\mathrm{S}} a \sin\theta)}{k_{\mathrm{S}} \sin\theta}}_{\text{Scalar aperture}} \underbrace{M_{\mathrm{S}}^{t_z}(\theta)}_{\text{MP-factor}} .
\end{aligned} \tag{14.175}
$$

[267]One has to evaluate the plane wave spectra (14.68) and (14.69) with the stationary phase method.

Certainly, the scalar aperture factor may be extended to a nonsynchronous aperture according to (with reference to Equations 13.287 and 13.288 with 13.286)

$$F(\omega)\,\frac{2\sin(k_{\mathrm{P,S}}a\sin\theta - k_A a)}{k_{\mathrm{P,S}}\sin\theta - k_A} \tag{14.176}$$

In (14.174) and (14.175), we may distinguish three different factors: line source in full-space with the respective radiation characteristics (Equations[268] 13.264 and 13.265), radiation characteristics of the scalar aperture (Equations 13.312 and 13.313), and the Miller–Pursey factors accounting for the half-space that have been defined with (14.122) and (14.123):

$$M_{\mathrm{P}}^{t_z}(\theta) = \frac{2\kappa^2(\kappa^2 - 2\sin^2\theta)}{(\kappa^2 - 2\sin^2\theta)^2 + 2\sin\theta\sin 2\theta\sqrt{\kappa^2 - \sin^2\theta}}, \tag{14.177}$$

$$M_{\mathrm{S}}^{t_z}(\theta) = \frac{4\cos\theta\sqrt{1 - \kappa^2\sin^2\theta}}{\kappa(1 - 2\sin^2\theta)^2 + 2\sin\theta\sin 2\theta\sqrt{1 - \kappa^2\sin^2\theta}}. \tag{14.178}$$

Next, we refer to the half-space result for the circular synchronous aperture (Equations 14.122 and 14.123):

$$\underline{\mathbf{u}}_{\mathrm{P}}^{t_z,\mathrm{far}}(R,\vartheta,\omega) = \underbrace{\underline{\mathbf{e}}_R\,\frac{1}{4\pi c_{\mathrm{P}} Z_{\mathrm{P}}}\,\frac{\mathrm{e}^{\mathrm{j}k_{\mathrm{P}}R}}{R}\,\cos\vartheta}_{\text{PS in full-space}}\;\underbrace{\frac{2\pi a \mathrm{J}_1(k_{\mathrm{P}}a\sin\vartheta)}{k_{\mathrm{P}}\sin\vartheta}}_{\text{Scalar aperture}}\;\underbrace{M_{\mathrm{P}}^{t_z}(\vartheta)}_{\text{MP-factor}}\,, \tag{14.179}$$

$$\underline{\mathbf{u}}_{\mathrm{S}}^{t_z,\mathrm{far}}(R,\vartheta,\omega) = \underbrace{-\underline{\mathbf{e}}_\vartheta\,\frac{1}{4\pi c_{\mathrm{S}} Z_{\mathrm{S}}}\,\frac{\mathrm{e}^{\mathrm{j}k_{\mathrm{S}}R}}{R}\,\sin\vartheta}_{\text{PS in full-space}}\;\underbrace{\frac{2\pi a \mathrm{J}_1(k_{\mathrm{S}}a\sin\vartheta)}{k_{\mathrm{S}}\sin\vartheta}}_{\text{Scalar aperture}}\;\underbrace{M_{\mathrm{S}}^{t_z}(\vartheta)}_{\text{MP-factor}}\,; \tag{14.180}$$

we discover exactly the same structure with the three characteristic factors [replace θ by ϑ in (14.177) and (14.178)]!

Instead of calculating the far-field for arbitrary three-dimensionally planar (non)synchronous normal force density aperture distributions with the *ansatz* (14.161), we generalize (14.179) and (14.180) accordingly:[269]

$$\underline{\mathbf{u}}_{\mathrm{P}}^{t_z,\mathrm{far}}(R,\vartheta,\omega) = \underbrace{\underline{\mathbf{e}}_R\,\frac{1}{4\pi c_{\mathrm{P}} Z_{\mathrm{P}}}\,\frac{\mathrm{e}^{\mathrm{j}k_{\mathrm{P}}R}}{R}\,\cos\vartheta}_{\text{PS in full-space}}$$
$$\underbrace{\times\,\mathcal{F}_{xy}\{A(x,y,\omega)\Gamma_A(x,y)\}_{\substack{K_x = k_{\mathrm{P}}\sin\vartheta\cos\varphi\\ K_y = k_{\mathrm{P}}\sin\vartheta\sin\varphi}}}_{\text{Scalar aperture}}\;\underbrace{M_{\mathrm{P}}^{t_z}(\vartheta)}_{\text{MP-factor}}\,, \tag{14.181}$$

[268]For $\underline{\mathbf{e}}_\varphi$ in (13.265), we have $\underline{\mathbf{e}}_\varphi = \underline{\mathbf{e}}_\theta$.

[269]Well understood, this is an obvious conclusion, not a mathematical proof.

$$\underline{\mathbf{u}}_{\mathrm{S}}^{t_z,\mathrm{far}}(R,\vartheta,\omega) = \underbrace{-\underline{\mathbf{e}}_\vartheta \frac{1}{4\pi c_{\mathrm{S}} Z_{\mathrm{S}}} \frac{\mathrm{e}^{\mathrm{j}k_{\mathrm{S}}R}}{R} \sin\vartheta}_{\text{PS in full-space}} \times \underbrace{\mathcal{F}_{xy}\{A(x,y,\omega)\Gamma_A(x,y)\}_{\substack{K_x = k_{\mathrm{S}} \sin\vartheta\cos\varphi \\ K_y = k_{\mathrm{S}}\sin\vartheta\sin\varphi}}}_{\text{Scalar aperture}} \underbrace{M_{\mathrm{S}}^{t_z}(\vartheta)}_{\text{MP-factor}} . \tag{14.182}$$

Note: A potential φ-dependence of the far-field of arbitrary transducer apertures may only, if at all, come from the scalar radiation characteristic of the aperture [compare (13.279) and (13.280)]. Furthermore: P,SV-waves propagate in each plane $\varphi = \text{const.}$

Now, we are able to accordingly modify the full-space radiation patterns of Figures 13.17 until 13.19 for the propagation into a half-space with a stress-free surface: They must only be multiplied with the Miller–Pursey factors. Figures 14.14 until 14.17 display respective results. We may generally state that: There is no big change as compared to the full-space diagrams; only the beam steering angles for the S-diagrams [Figure 13.18(b), (d), and (f) compared to Figure 14.15(b), (d), and (f)] are slightly different for the same k_A-values due to the difference in the point directivities. As compared to Figure 13.19, the S-main lobe of the "creeping wave transducer" is not as wide (Figure 14.16). By the way: To understand the operational mode of the creeping wave transducer as far as the accompanying bulk pressure wave is concerned, the stress-free surface contributes nothing, the full-space consideration is completely sufficient. However—as already multiply mentioned—head and Rayleigh wavefronts are missing after a transform into the time domain, as they have been observed in an EFIT-near-field simulation as a remarkable feature the above (time harmonic) far-field (Langenberg et al. 1990).

We should explicitly mention that far-field radiation patterns are not at all ultrasonic beams: They represent the sound field amplitude beyond the near-field length! A complete picture of the (time harmonic) sound field is only obtained through an approximation-free (numerical) evaluation of the spectral plane wave representation, through quantitative summation of ultrasonic beams (Shlivinski et al. 2004, 2005) or through EFIT-simulations (Fellinger 1991; Marklein 1997; Bihn 1998); Figure 14.18(a) and (b) exhibit respective EFIT-results for the 0°-pressure wave (P-$\perp$) and the 45°-shear wave transducer (SV-45°) replacing the pulse excitation by $\mathrm{e}^{\mathrm{j}\omega_0 t}u(t)$.

Figures 14.19 and 14.20 continue to show the comparison of approximation-free EFIT- and far-field approximated wavefronts for the strip-like aperture; in the beginning, there was Figure 14.11 for the line source. Figure 14.19 displays the logarithmically scaled versions of the wavefront amplitudes in order to magnify differences of both evaluation methods; in contrast, the linear representations in Figure 14.20 only accentuate the "NDT-relevant" features of the respective transducer field. The EFIT-calculated field of the 0°-pressure wave transducer (P-$\perp$) in Figure 14.19(a) teaches us that its acoustically

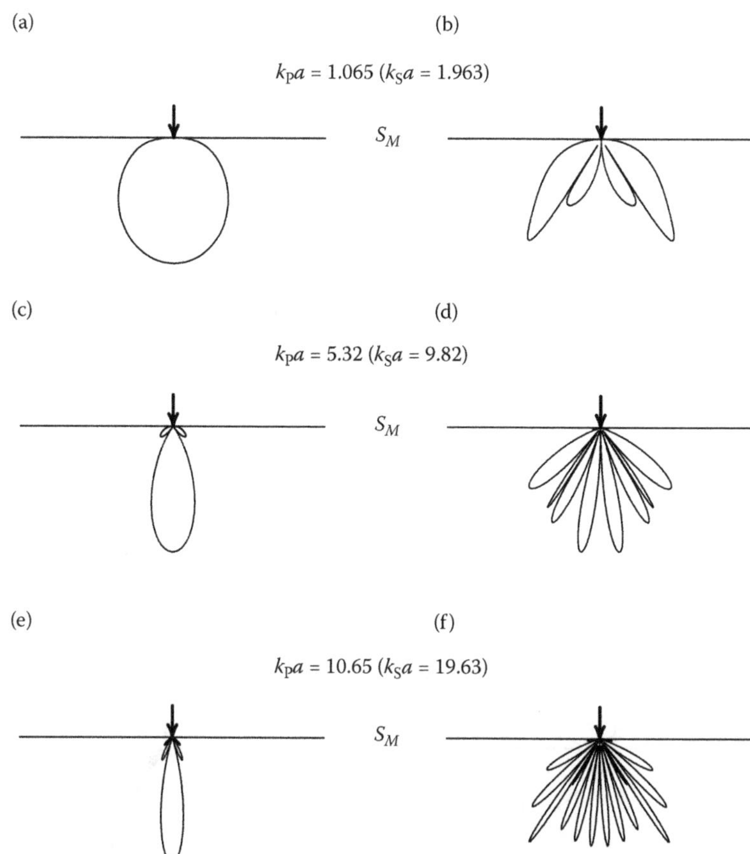

FIGURE 14.14
P- and S-radiation patterns [P: (a), (c), (e); S: (b), (d), (f)] of a planar rectangular t_z-force density aperture on the stress-free surface of a homogeneous isotropic half-space; $y = 0$ (steel: $c_P = 5900$ m/s, $c_S = 3200$ m/s, $\rho = 7.7 \cdot 10^3$ kg/m^3).

approximated pressure wave field according to Figure 14.1 must be complemented by the typical line source effects [Figure 14.11(a)] of both aperture edges, namely by shear, Rayleigh, and head waves. The corresponding far-field evaluation as given in Figure 14.19(d) only relies on the far-field equation (14.175) according to the prescription "P-$\perp$-transducer"; therefore, the shear wavefronts as described by Equation 14.175 are missing [compare Figure 14.11(b)]. However it is remarkable how well the exactly calculated sound field is approximated by the superposition of both edge pressure waves [also compare the sketch in Figure 14.6(d)]. This is also true for the 30°-pressure wave transducer [Figure 14.19(b) and (e)] and the 45°-shear wave transducer [Figure 14.19(c) and (f)], where the far-field of the latter results from the

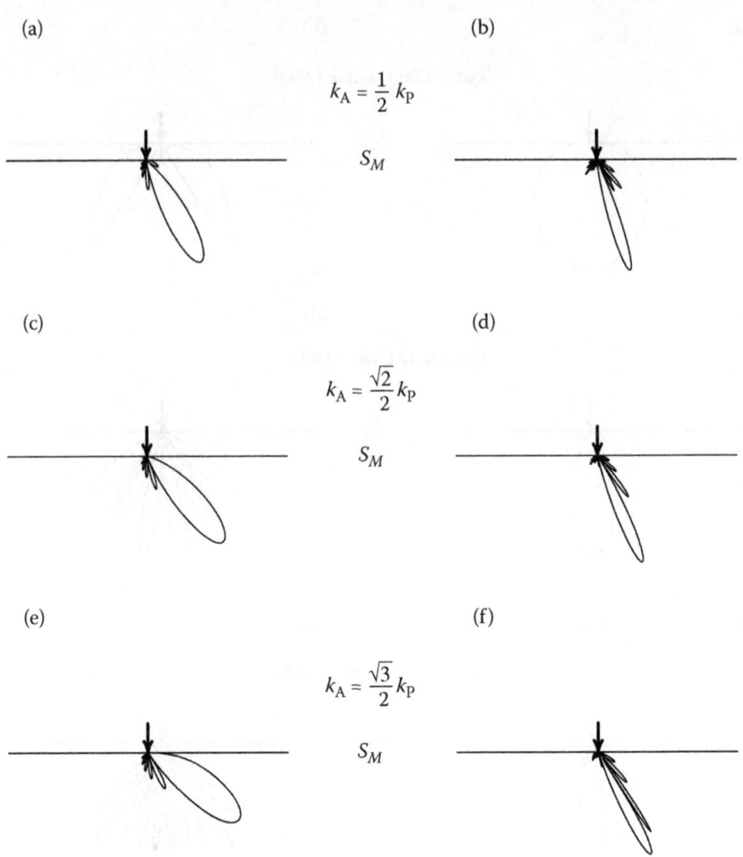

FIGURE 14.15
P- and S-radiation patterns [P: (a), (c), (e); S: (b), (d), (f)] of a planar rectangular t_z-force density aperture with linear x-phase tapering on the stress-free surface of a homogeneous isotropic half-space; $y = 0$ (steel: $c_P = 5900$ m/s, $c_S = 3200$ m/s, $\rho = 7.7 \cdot 10^3$ kg/m^3; $k_P a = 10.65$, $k_S a = 19.63$).

analysis of Equation 14.175. As compared to the P-⊥-field, the P-30°- and SV-45°-fields the far-field approximated wavefronts have already traveled a little bit further than the EFIT-calculated wavefronts. The explanation is given by the sketch in Figure 13.20. The EFIT-code pursues—wave theoretically correct—the superposition of Huygens-type elementary waves emanating from the aperture resulting in the spatial field distribution according to the chosen time, where the time origin corresponds to the first radiated elementary wave, in this case, to the one from the left aperture edge; in contrast, the far-field approximation replaces the finite spatial aperture by a point, here: line, source in the coordinate origin, namely the center of the aperture thus serving as the time origin. Hence, the right aperture edge is advanced, and the left one is retarded against the far-field time origin (Figure 13.20) resulting

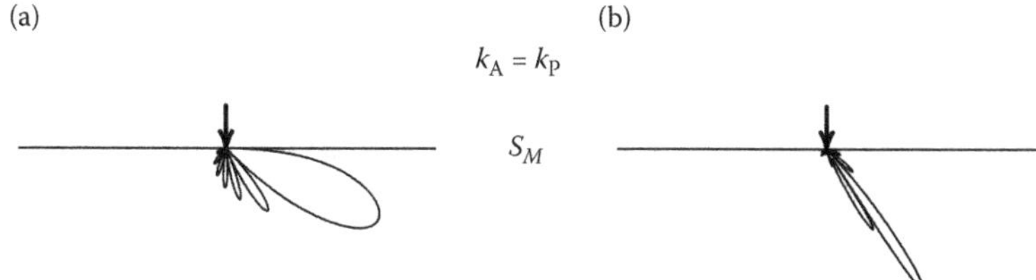

FIGURE 14.16
P- and S-radiation patterns [P: (a); S: (b)] of a planar rectangular t_z-force density aperture with linear x-phase tapering on the stress-free surface of a homogeneous isotropic half-space: bulk pressure wave of the creeping wave transducer; $y = 0$ (steel: $c_P = 5900$ m/s, $c_S = 3200$ m/s, $\rho = 7.7 \cdot 10^3$ kg/m^3; $k_P a = 10.65$, $k_S a = 19.63$)).

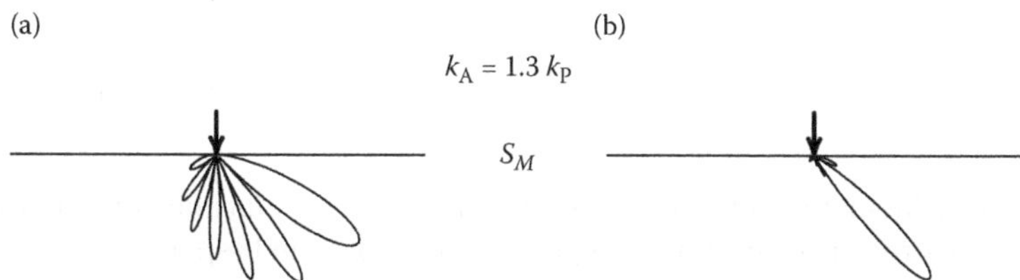

FIGURE 14.17
P- and S-radiation patterns [P: (a); S: (b)] of a planar rectangular t_z-force density aperture with linear x-phase tapering on the stress-free surface of a homogeneous isotropic half-space: 45°-shear wave transducer; $y = 0$ (steel: $c_P = 5900$ m/s, $c_S = 3200$ m/s, $\rho = 7.7 \cdot 10^3$ kg/m^3; $k_P a = 10.65$, $k_S a = 19.63$).

in a relatively earlier arrival of the pulse from the right aperture edge at a given observation point, respectively for the same time it has already traveled further. Therefore, matching the EFIT- and far-field time axes is trivial.

Far-field approximation of the half-space Green tensor $\underline{\underline{\mathbf{G}}}^{\mathrm{N}} \cdot \underline{\mathbf{e}}_z$: The sound field that has been heuristically generalized to arbitrary transducer apertures even allows for the presentation of the far-field component of the half-space Green tensor $\underline{\underline{\mathbf{G}}}^{\mathrm{N,far}}(\underline{\mathbf{R}}, \underline{\mathbf{R}}', \omega) \cdot \underline{\mathbf{e}}_z$: The sound field corresponding to the full-space t_z-source may be derived from (13.114) with $\underline{\underline{\mathbf{G}}}(\underline{\mathbf{R}} - \underline{\mathbf{R}}', \omega) \cdot \underline{\mathbf{e}}_z \Longrightarrow \underline{\underline{\mathbf{G}}}^{\mathrm{far}}(\underline{\mathbf{R}}, \underline{\mathbf{R}}', \omega) \cdot \underline{\mathbf{e}}_z$ [(13.95) $\Longrightarrow$ (13.100) and (13.101)]; yet exactly, this field explicitly appears in (14.181) and (14.182) multiplied by the Miller–Pursey factors. Consequently, we obtain the representation

$$\begin{aligned}\underline{\underline{\mathbf{G}}}^{\mathrm{N,far}}(\underline{\mathbf{R}}, \underline{\mathbf{R}}', \omega) \cdot \underline{\mathbf{e}}_z = {} & M_{\mathrm{P}}^{t_z}(\vartheta)\, \frac{1}{\lambda + 2\mu}\, \underline{\underline{\mathbf{g}}}_{\mathrm{P}}(\hat{\underline{\mathbf{R}}}) \cdot \underline{\mathbf{e}}_z\, \frac{\mathrm{e}^{\mathrm{j}k_{\mathrm{P}}R}}{R}\, \mathrm{e}^{-\mathrm{j}k_{\mathrm{P}}\hat{\underline{\mathbf{R}}}\cdot\underline{\mathbf{R}}'} \\ & + M_{\mathrm{S}}^{t_z}(\vartheta)\, \frac{1}{\mu}\, \underline{\underline{\mathbf{g}}}_{\mathrm{S}}(\hat{\underline{\mathbf{R}}}) \cdot \underline{\mathbf{e}}_z\, \frac{\mathrm{e}^{\mathrm{j}k_{\mathrm{S}}R}}{R}\, \mathrm{e}^{-\mathrm{j}k_{\mathrm{S}}\hat{\underline{\mathbf{R}}}\cdot\underline{\mathbf{R}}'} \qquad (14.183)\end{aligned}$$

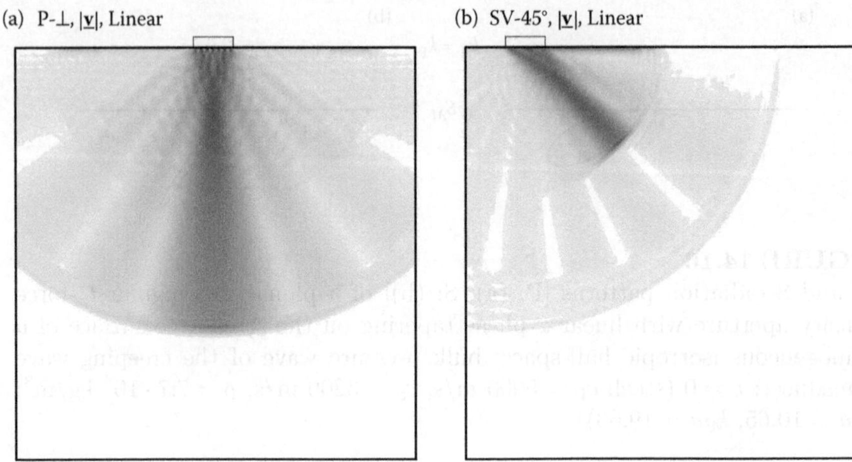

FIGURE 14.18
EFIT-simulated ultrasonic beams (linear scale); (a) P-⊥; (b) SV-45°.

as basis of a far-field point source synthesis for given normal force densities on the stress-free surface of a homogeneous isotropic half-space.

Heuristic approximation of the near-field of piezoelectric transducers: An obvious idea is the following: The terms

$$\phi_A^{\text{far}}(R, \vartheta, \varphi, \omega) = \frac{e^{jk_{P,S}R}}{4\pi R} \mathcal{F}\{A(x, y, \omega)\}_{\substack{K_x = k_{P,S} \sin\vartheta \cos\varphi \\ K_y = k_{P,S} \sin\vartheta \sin\varphi}} \tag{14.184}$$

are nothing else but the far-field approximation of the scalar field

$$\begin{aligned} &\phi_A(R, \vartheta, \varphi, \omega) \\ &\quad = \iint_{S_A} A(x', y', \omega) \\ &\quad \times \underbrace{\frac{e^{jk_{P,S}\sqrt{(R\sin\vartheta\cos\varphi - x')^2 + (R\sin\vartheta\sin\varphi - y')^2 + z^2\cos^2\vartheta}}}{4\pi\sqrt{(R\sin\vartheta\cos\varphi - x')^2 + (R\sin\vartheta\sin\varphi - y')^2 + z^2\cos^2\vartheta}}}_{= G_{P,S}(R, \vartheta, \varphi, x', y', z' = 0, \omega)} \, dx'dy', \end{aligned} \tag{14.185}$$

where $G_{P,S}(R, \vartheta, \varphi, x', y', z', \omega)$ is the three-dimensional scalar Green function of the acoustic full-space in "mixed" coordinates—spherical coordinates with regard to the observation point,[270] and cartesian coordinates with regard to the source point—; the index P or S determines the phase velocity of the

[270]Evidently, we could also choose cartesian coordinates with regard to the observation point.

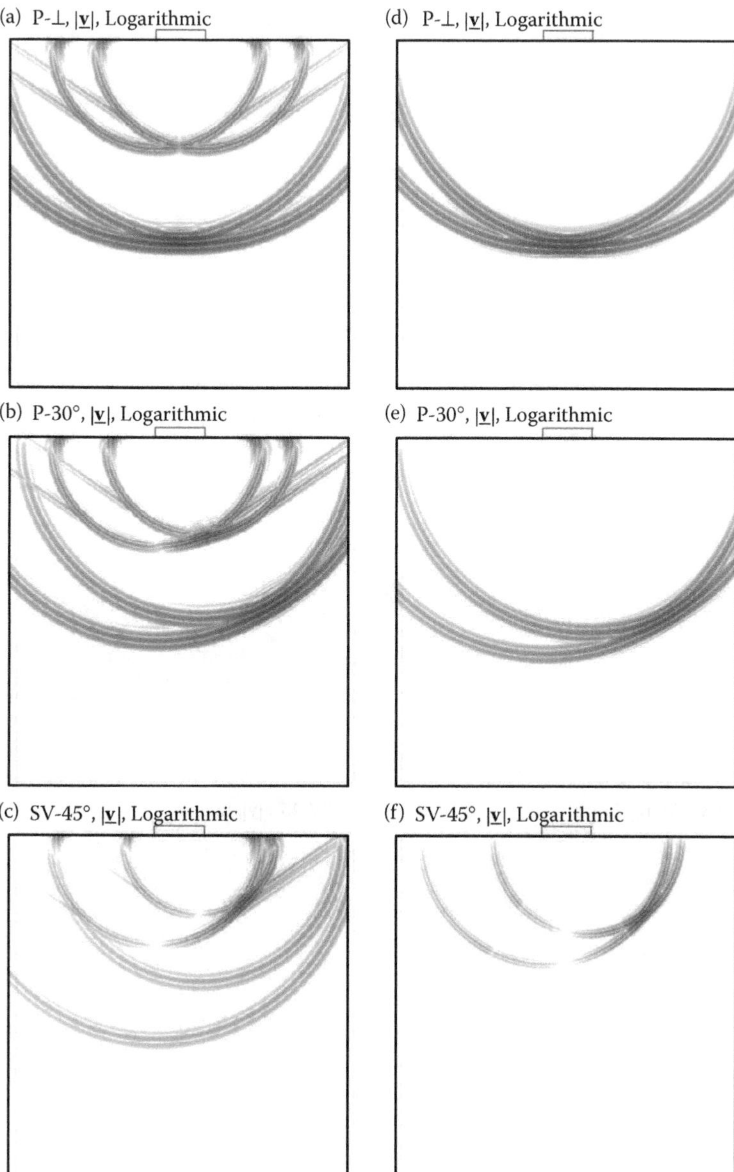

FIGURE 14.19
RC2(t)-impulse radiation form different piezoelectric transducers: EFIT-wavefronts [left column (a)–(c)] and far-field approximated wavefronts [right column (d)–(f)]; magnitude of the particle velocity: inverse Fourier-transform of the $-j\omega$ multiplied equations 14.174 and 14.175 with 14.176; logarithmic scale.

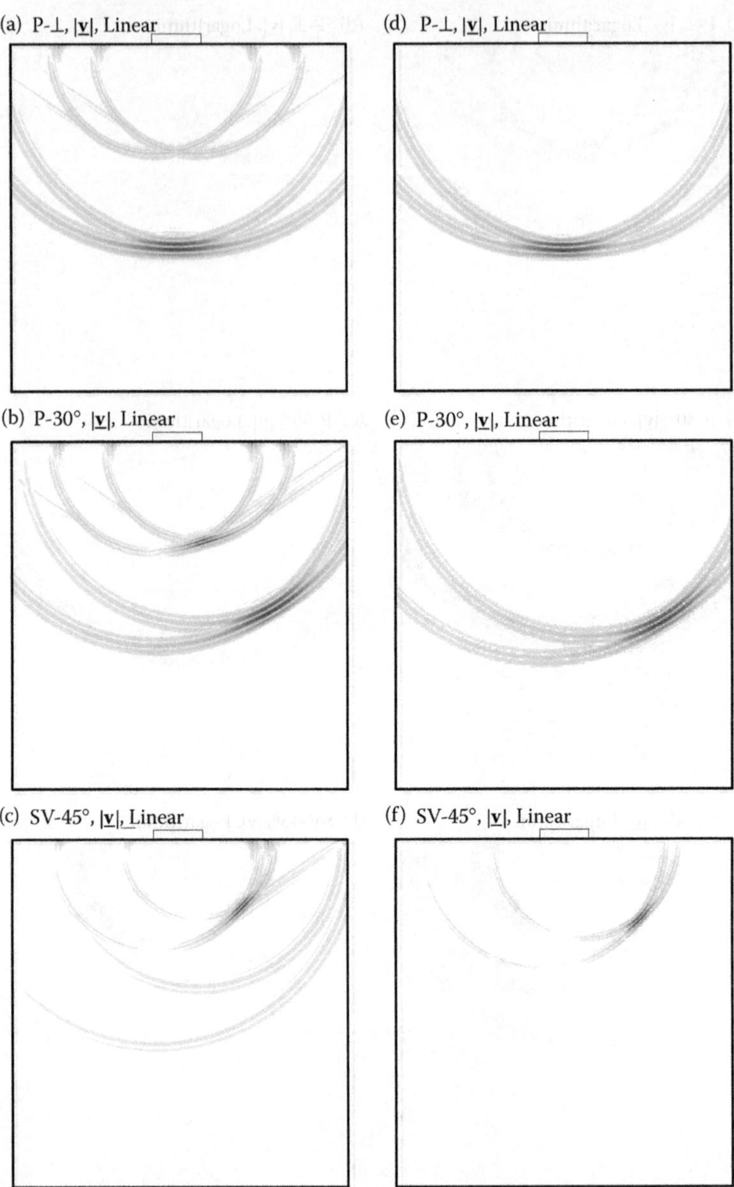

FIGURE 14.20
RC2(t)-impulse radiation from different piezoelectric transducers: EFIT-wavefronts [left column (a)–(c)] and far-field approximated wavefronts [right column (d)–(f)]; magnitude of the particle velocity: inverse Fourier-transform of the $-j\omega$ multiplied equations 14.174 and 14.175 with 14.176; linear scale.

acoustic full-space. The transition $\phi_A(R,\vartheta,\varphi,\omega) \Longrightarrow \phi_A^{\rm far}(R,\vartheta,\varphi,\omega)$ is tied to the assumptions $R/A \gg 1$ and $k_{\rm P,S}R \gg 1$, where A is the maximum linear dimension of the aperture geometry S_A. If these assumptions are not satisfied (e.g., for the wavefront sketches in Figure 14.6), we might think to extend the sound field equations (14.181) and (14.182) to the near-field[271]

$$\underline{\mathbf{u}}_{\rm P}^{t_z}(R,\vartheta,\varphi,\omega) = \underline{\mathbf{e}}_R \frac{1}{c_{\rm P} Z_{\rm P}} M_{\rm P}^{t_z}(\vartheta) \times \cos\vartheta \int\!\!\int_{S_A} A(x',y',\omega)\, G_{\rm P}(R,\vartheta,\varphi,x',y',z'=0,\omega)\, {\rm d}x'{\rm d}y', \tag{14.186}$$

$$\underline{\mathbf{u}}_{\rm S}^{t_z}(R,\vartheta,\varphi,\omega) = -\underline{\mathbf{e}}_\vartheta \frac{1}{c_{\rm S} Z_{\rm S}} M_{\rm S}^{t_z}(\vartheta) \times \sin\vartheta \int\!\!\int_{S_A} A(x',y',\omega)\, G_{\rm S}(R,\vartheta,\varphi,x',y',z'=0,\omega)\, {\rm d}x'{\rm d}y'. \tag{14.187}$$

That way, we have created a mixed near-far-field formulation: The scalar directivity of the aperture has been replaced by the correct near-field formula, i.e., the scalar point source synthesis uses the actual $A(x',y',\omega)$-weighted source points and not only the phase and amplitude correction of a spherical wave emanating from the origin; this means that even the aperture wavefront is present as sketched in Figures 14.6(a) and (b), respectively (two-dimensionally), calculated for Figures 14.1(c) and (d). On the other hand, the elastodynamic point directivities of the half-space are only available in the far-field in terms of the Miller–Pursey factors. Additionally, we could go that far to include the full-space point directivities $\cos\vartheta$ and $-\sin\vartheta$ in the point source synthesis according to

$$\underline{\mathbf{u}}_{\rm P}^{t_z}(R,\vartheta,\varphi,\omega) = \frac{1}{c_{\rm P} Z_{\rm P}} M_{\rm P}^{t_z}(\vartheta) \int\!\!\int_{S_A} A(x',y',\omega)\, \underline{\underline{\mathbf{G}}}_{\rm P}(R,\vartheta,\varphi,x',y',z'=0,\omega) \cdot \underline{\mathbf{e}}_z\, {\rm d}x'{\rm d}y', \tag{14.188}$$

$$\underline{\mathbf{u}}_{\rm S}^{t_z}(R,\vartheta,\varphi,\omega) = \frac{1}{c_{\rm S} Z_{\rm S}} M_{\rm S}^{t_z}(\vartheta) \int\!\!\int_{S_A} A(x',y',\omega)\, \underline{\underline{\mathbf{G}}}_{\rm S}(R,\vartheta,\varphi,x',y',z'=0,\omega) \cdot \underline{\mathbf{e}}_z\, {\rm d}x'{\rm d}y'; \tag{14.189}$$

In any case, $\underline{\mathbf{u}}_{\rm P}^{t_z}$ then exhibits $\underline{\mathbf{e}}_\vartheta$- and $\underline{\mathbf{e}}_\varphi$-components as well as $\underline{\mathbf{u}}_{\rm S}^{t_z}$ $\underline{\mathbf{e}}_R$- and $\underline{\mathbf{e}}_\varphi$-components that are not contained in (14.186) and (14.187). But: The particle displacement field (14.188) and (14.189) does not contain any head and Rayleigh waves because the Miller–Pursey factors characterizing the stress-free surface of the half-space appear as far-field approximation in front of the (elastodynamic) aperture integrals. This may not even be changed if we follow physical intuition and shift the Miller–Pursey factors under the integral

[271] An analogy: The P-half-space point directivity $M_{\rm P}^{t_z}(\vartheta)$ adopts the image value 2 for $\vartheta = 0$ just as the scalar Rayleigh–Sommerfeld integrals (14.15) and (14.16).

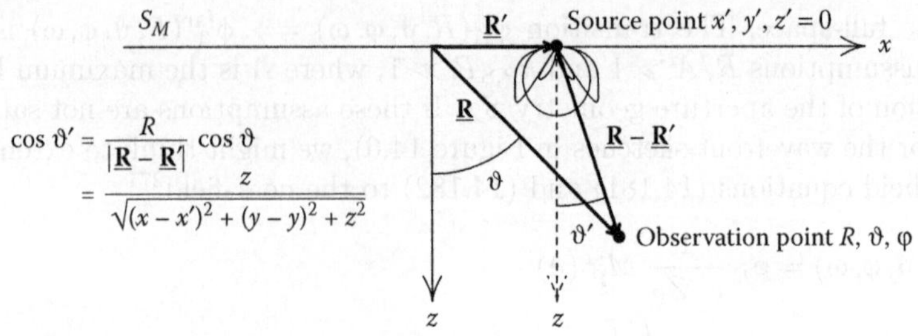

FIGURE 14.21
Point directivities for the source point.

replacing $\vartheta(x, y, z)$ by the polar angle $\vartheta'(x, y, z, x', y')$ of the source point (Figure 14.21):

$$\underline{\mathbf{u}}_{\mathrm{P}}^{t_z}(R, \vartheta, \varphi, \omega) = \frac{1}{c_{\mathrm{P}} Z_{\mathrm{P}}} \iint_{S_A} A(x', y', \omega)\, M_{\mathrm{P}}^{t_z}(\vartheta')\, \underline{\underline{\mathbf{G}}}_{\mathrm{P}}(R, \vartheta, \varphi, x', y', z' = 0, \omega) \cdot \underline{\mathbf{e}}_z\, \mathrm{d}x'\mathrm{d}y', \tag{14.190}$$

$$\underline{\mathbf{u}}_{\mathrm{S}}^{t_z}(R, \vartheta, \varphi, \omega) = \frac{1}{c_{\mathrm{S}} Z_{\mathrm{S}}} \iint_{S_A} A(x', y', \omega)\, M_{\mathrm{S}}^{t_z}(\vartheta')\, \underline{\underline{\mathbf{G}}}_{\mathrm{S}}(R, \vartheta, \varphi, x', y', z' = 0, \omega) \cdot \underline{\mathbf{e}}_z\, \mathrm{d}x'\mathrm{d}y'. \tag{14.191}$$

Having "risked" Equations 14.190 and 14.191, we can step back regarding complexity by complementing the Miller–Pursey factors by the (far-field) point directivities in full-space as contained in $\underline{\underline{\mathbf{G}}}_{\mathrm{P,S}} \cdot \underline{\mathbf{e}}_z$—yet with regard to the source point!—i.e., we shift the total point directivity under the integrals in (14.186) and (14.187):

$$\underline{\mathbf{u}}_{\mathrm{P}}^{t_z}(R, \vartheta, \varphi, \omega) = \underline{\mathbf{e}}_R \frac{1}{c_{\mathrm{P}} Z_{\mathrm{P}}} \iint_{S_A} A(x', y', \omega)\, M_{\mathrm{P}}^{t_z}(\vartheta') \times \cos\vartheta'\, G_{\mathrm{P}}(R, \vartheta, \varphi, x', y', z' = 0, \omega)\, \mathrm{d}x'\mathrm{d}y', \tag{14.192}$$

$$\underline{\mathbf{u}}_{\mathrm{S}}^{t_z}(R, \vartheta, \varphi, \omega) = -\underline{\mathbf{e}}_\vartheta \frac{1}{c_{\mathrm{S}} Z_{\mathrm{S}}} \iint_{S_A} A(x', y', \omega)\, M_{\mathrm{S}}^{t_z}(\vartheta') \times \sin\vartheta'\, G_{\mathrm{S}}(R, \vartheta, \varphi, x', y', z' = 0, \omega)\, \mathrm{d}x'\mathrm{d}y'. \tag{14.193}$$

These "point directivities of the source point" are illustrated in Figure 14.21. Figure 14.22 shows an example calculated according to (14.192) and (14.193); the respective far-field results are given in Figure 14.17: We recognize the formation of main lobes; however, Figure 14.22 is scaled with a length (here: m), whereas a far-field diagram does no longer contain the actual distance from the source. It is remarkable that the differences as compared to the far-field diagrams are only marginal.

We insist to point out that (14.190) and (14.191) do not represent the result of a calculation of the $\underline{\underline{\mathbf{G}}}^{\mathrm{N}} \cdot \underline{\mathbf{e}}_z$-component of the half-space Green tensor

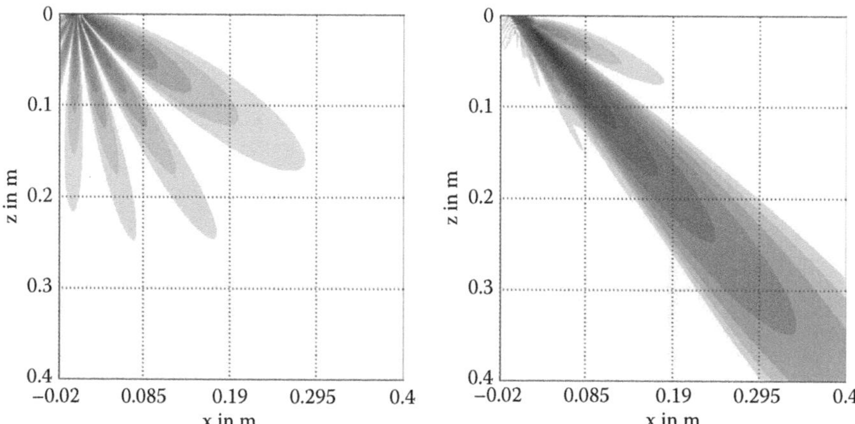

FIGURE 14.22
Near-field radiation pattern of a 45°-shear wave transducer in the plane $y = 0$ for $-20\,\mathrm{mm} \le x \le 200\,\mathrm{mm}$, $0 \le z \le 200\,\mathrm{mm}$; left: P-diagram, right: S-diagram; logarithmic scale (material: see Figure 14.17; frequency: 1 MHz; transducer width: $2a = 0.02$ m, i.e. $k_{\mathrm{P}}a = 10.65$).

in (14.149): All previous near-field formulas have been given based on physical intuition, yet heuristically; for each application, they have to prove their usefulness in comparison with an exact spectral plane wave representation as obtained with the *ansatz* (14.161). Alternatively, EFIT-results may be used for a validation. With regard to the numerical evaluation, we note that—as it is true for the scalar Rayleigh–Sommerfeld integrals (14.15) and (14.16)—we always face convolution integrals with regard to x and y provided the observation point is given in cartesian coordinates. Furthermore: Since the point directivities are frequency independent, we even may, similar to the scalar case, transform the near-field versions into the time domain with the Cagniard–de Hoop method to calculate wavefronts (Aulenbacher and Langenberg 1980).

15

Scatterers in Homogeneous Isotropic Nondissipative Infinite Spaces

The calculation of source fields, be it in full-space or on the surface of a half-space, certainly represents an important canonical problem of US-NDT; however, the real fundamental problem is sketched in Figure 15.1: A specimen volume V_M with surface S_M, often also being the stress-free measurement surface, contains (primary) sources of elastodynamic fields—source volumes V_Q—and "defects"—scattering bodies with, say, a volume V_c with surface S_c. If the specimen material is assumed to be linear, time invariant, and locally reacting (very little may be done analytically without this assumption) but nevertheless inhomogeneous, anisotropic, and dissipative, where dissipation is expressed by the frequency dependence of complex valued material tensors $\underline{\underline{\boldsymbol{\rho}}}_c^{(e)}(\underline{\mathbf{R}}, \omega)$, $\underline{\underline{\underline{\underline{\mathbf{c}}}}}_c^{(e)}(\underline{\mathbf{R}}, \omega)$ (Section 4.4). The scattering bodies may be characterized by material tensors $\underline{\underline{\boldsymbol{\rho}}}_c^{(i)}(\underline{\mathbf{R}}, \omega)$, $\underline{\underline{\underline{\underline{\mathbf{c}}}}}_c^{(i)}(\underline{\mathbf{R}}, \omega)$. The fundamental modeling[273] problem is the following: Calculate elastodynamic field quantities—generally $\underline{\mathbf{u}}(\underline{\mathbf{R}}, t)$—on S_M for given material tensors of V_M and V_c, and for given sources $\underline{\mathbf{f}}(\underline{\mathbf{R}}, t)$, $\underline{\underline{\mathbf{h}}}(\underline{\mathbf{R}}, t)$ in V_Q; due to the assumed linearity and time invariance of the materials, we may formulate the fundamental modeling problem as in Figure 15.1 for the Fourier spectra.

Analytical methods are not really available to solve this problem in general; for instance, going back to the point source synthesis involving Green functions, we mostly have to confine to homogeneous isotropic nondissipative specimen materials as well as a half-space as specimen volume, and even then, approximations have to be introduced regarding the source field representation (Section 14.4). Arguments for these approximations are formulated in the frequency domain allowing for frequency-dependent specimen materials; the Fourier inversion into the time domain must be anyway evaluated numerically in most cases. In this sense, the "defect" may be generally assumed to be inhomogeneous and anisotropic. Why? Because only those Green functions are needed that have already been used to calculate the source field. However, a complete arbitrariness of material properties and geometry of the defect first requires the numerical solution of integral equations for secondary

[273]In contrast to it, we face the assessment of ultrasonic signals: The sources are known, the specimen is known, and the field quantity is given on (part of) the surface; what can be deduced regarding V_c?

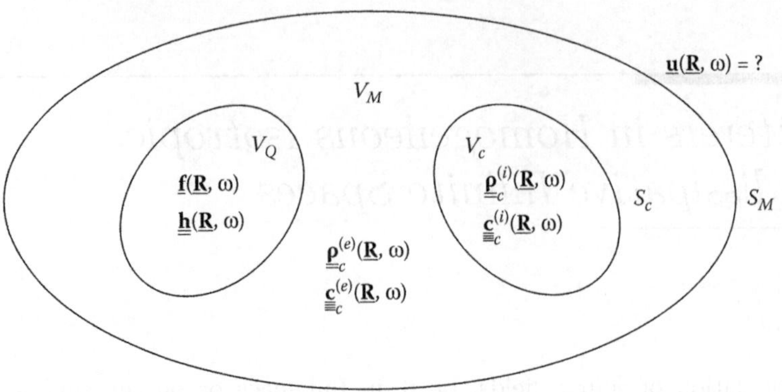

FIGURE 15.1
Modeling a US-NDT problem.

sources before inserting them into the integral representation for the scattered field. To avoid this, we may introduce the Kirchhoff or Born approximation for the secondary sources; the first one is relevant for voids, the latter one for inclusions. In contrast to the Green function based methods, we may rely on approximate analytical methods like ray tracing (Section 12.3.2) in inhomogeneous and/or anisotropic materials.

If a concrete situation does not allow for the above mentioned compromises, *ab initio* numerical methods—FE (finite elements), FDTD (finite difference time domain), FIT (finite integration technique) (EFIT: Fellinger 1991; Marklein 1997)—have to be applied. Yet, even relying on those does not solve all problems: New ones like discretization errors and numerical dispersion arise.[274]

15.1 Huygens' Principle

Christiaan Huygens (Blok et al. 1992) formulated a principle to be able to calculate certain problems of light wave propagation theoretically:

This principle postulates that each point on a wavefront acts as a point source for a spherical wave (elementary wave) propagating with the speed of light. For a later time, the field at a given point is the sum of all point source fields; the envelope of all elementary waves from all points represents the next wavefront.

For given source fields, we have already recognized this principle as the mathematical formulation of the point source synthesis, the elementary waves are described by the scalar Green function in the time domain according to

[274] There is no such thing as a free lunch.

(13.25). However, Huygens' principle claims more: Even if we know nothing about the sources, the knowledge of the wave field on an arbitrary surface is sufficient to calculate its further propagation. Furthermore, Huygens claims that: The waves only propagate forward, their backward propagation becomes extinct. We anticipate that a mathematical formulation of Huygens' principle will be found based on Green functions, and in fact, a completely formal application of Green's second integral formula yields the desired representation. In the time domain, this (integral) representation goes back to G. Kirchhoff, yet the frequency domain version was already given earlier by H. von Helmholtz generally leading to the notation as Helmholtz integral, sometimes Helmholtz–Kirchhoff integral: The Kirchhoff integral is nothing but the inverse Fourier transform of the Helmholtz integral.

Now we obviously expect that the elastodynamic version of such a Helmholtz integral simply emerges utilizing appropriate elastic elementary waves, i.e., their pertinent Green functions. The formal mathematical derivation—in particular, the backward extinction theorem—requires a tensorial counterpart of Green's second integral formula, and before the background of this quite challenging formalism, the physical content of the (elastodynamic) Huygens principle gets easily lost. Therefore, in the following, we first pursue two ways around the strict mathematical derivation:

- Using physical arguments postulating the extinction theorem;
- Deriving the scalar Huygens principle mathematically applying Green's second integral formula and interpreting it physically, because the physical content remains the same for the elastodynamic version.

15.1.1 Mathematical foundation of Huygens' principle of elastodynamics based on physical arguments

We argue with Fourier transformed fields in the frequency domain to find an integral representation of elastodynamic fields equivalent to the Helmholtz integral.

Huygens' principle as equivalence principle: The elastodynamic field (Equations 13.213 and 13.214)

$$\underline{\mathbf{v}}(\underline{\mathbf{R}}, \omega) = \iiint_{V_Q} \Big[-\mathrm{j}\omega \underline{\mathbf{f}}(\underline{\mathbf{R}}', \omega) \cdot \underline{\underline{\mathbf{G}}}(\underline{\mathbf{R}} - \underline{\mathbf{R}}', \omega) + \underline{\underline{\mathbf{h}}}(\underline{\mathbf{R}}', \omega) : \underline{\underline{\underline{\mathbf{\Sigma}}}}(\underline{\mathbf{R}} - \underline{\mathbf{R}}', \omega) \Big] \, \mathrm{d}^3\underline{\mathbf{R}}', \tag{15.1}$$

$$\underline{\underline{\mathbf{T}}}(\underline{\mathbf{R}}, \omega) = \iiint_{V_Q} \Big[\underline{\mathbf{f}}(\underline{\mathbf{R}}', \omega) \cdot \underline{\underline{\underline{\mathbf{\Sigma}}}}^{312}(\underline{\mathbf{R}}', \omega) - \frac{1}{\mathrm{j}\omega} \underline{\underline{\mathbf{h}}}(\underline{\mathbf{R}}', \omega) : \underline{\underline{\underline{\underline{\mathbf{\Pi}}}}}(\underline{\mathbf{R}} - \underline{\mathbf{R}}', \omega) \Big] \, \mathrm{d}^3\underline{\mathbf{R}}' \tag{15.2}$$

of given sources $\underline{\mathbf{f}}(\underline{\mathbf{R}}', \omega)$ and $\underline{\underline{\mathbf{h}}}(\underline{\mathbf{R}}', \omega)$ being zero in the exterior of V_Q satisfies the inhomogeneous elastodynamic governing equations

$$-\mathrm{j}\omega\varrho\,\underline{\mathbf{v}}(\underline{\mathbf{R}}, \omega) = \nabla \cdot \underline{\underline{\mathbf{T}}}(\underline{\mathbf{R}}, \omega) + \underline{\mathbf{f}}(\underline{\mathbf{R}}, \omega), \tag{15.3}$$

$$-\mathrm{j}\omega\,\underline{\underline{\mathbf{T}}}(\underline{\mathbf{R}}, \omega) = \underline{\underline{\underline{\underline{\mathbf{c}}}}} : \nabla\underline{\mathbf{v}}(\underline{\mathbf{R}}, \omega) + \underline{\underline{\underline{\underline{\mathbf{c}}}}} : \underline{\underline{\mathbf{h}}}(\underline{\mathbf{R}}, \omega) \tag{15.4}$$

in a homogeneous isotropic nondissipative full-space $\underline{\mathbf{R}} \in \mathbb{R}^3$, where $\underline{\underline{\underline{\underline{\mathbf{c}}}}} = \lambda\underline{\underline{\underline{\underline{\mathbf{I}}}}}^\delta + 2\mu\underline{\underline{\underline{\underline{\mathbf{I}}}}}^+$; the Green functions for this infinite space are given by (13.91), (13.154), and (13.208). In the exterior of a mathematically virtual closed surface S_g, i.e., for $\underline{\mathbf{R}} \in \mathbb{R}^3\backslash\overline{V}$, that encloses the source volume [Figure 15.2(a)], $\underline{\mathbf{v}}(\underline{\mathbf{R}}, \omega)$ and $\underline{\underline{\mathbf{T}}}(\underline{\mathbf{R}}, \omega)$ satisfy the homogeneous governing equations according to (15.1) and (15.2)

$$-\mathrm{j}\omega\varrho\,\underline{\mathbf{v}}(\underline{\mathbf{R}}, \omega) = \nabla \cdot \underline{\underline{\mathbf{T}}}(\underline{\mathbf{R}}, \omega), \tag{15.5}$$

$$-\mathrm{j}\omega\,\underline{\underline{\mathbf{T}}}(\underline{\mathbf{R}}, \omega) = \underline{\underline{\underline{\underline{\mathbf{c}}}}} : \nabla\underline{\mathbf{v}}(\underline{\mathbf{R}}, \omega). \tag{15.6}$$

Now we turn to the problem of Figure 15.2(b): The exterior of S_g should be composed of the same homogeneous isotropic nondissipative ϱ, λ, μ-material that hosts the source volume in Figure 15.2(a); however, in the interior of S_g, we postulate a null-field. How do we achieve the field exterior to S_g to be unchanged as compared to Figure 15.2(a)? Of course, we must locate sources somewhere, and the transition conditions (3.106) and (3.107) tell us that the discontinuous jump from a null-field inside S_g to the field $\underline{\mathbf{v}}, \underline{\underline{\mathbf{T}}}$ outside S_g may be sustained by surface sources

$$\underline{\mathbf{t}}(\underline{\mathbf{R}}', \omega) = -\,\underline{\mathbf{n}}' \cdot \underline{\underline{\mathbf{T}}}(\underline{\mathbf{R}}', \omega), \tag{15.7}$$

$$\underline{\underline{\mathbf{g}}}(\underline{\mathbf{R}}', \omega) = -\,\underline{\underline{\underline{\underline{\mathbf{I}}}}}^+ : \underline{\mathbf{n}}'\underline{\mathbf{v}}(\underline{\mathbf{R}}', \omega), \tag{15.8}$$

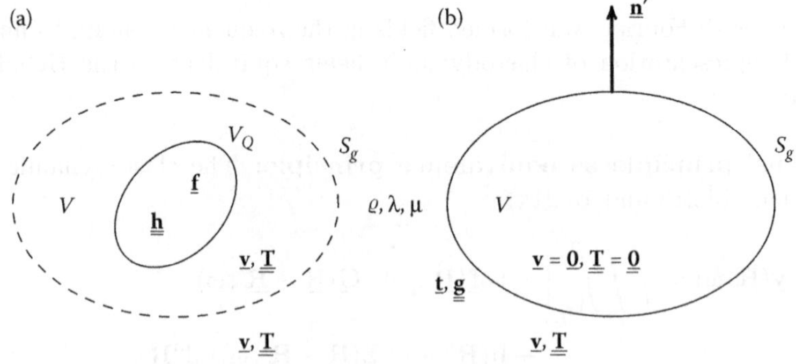

FIGURE 15.2
(a) Elastodynamic field of primary sources $\underline{\mathbf{f}}$ and $\underline{\underline{\mathbf{h}}}$; (b) "Huygens" sources $\underline{\mathbf{t}}$ and $\underline{\underline{\mathbf{g}}}$ as equivalent sources of the field in the exterior of S_g, and a null-field in the interior of S_g.

where $\underline{\mathbf{n}}'$ denotes the outward normal—pointing away from the null-field—at the point $\underline{\mathbf{R}}' \in S_g$. In (15.7) and (15.8), the field quantities $\underline{\mathbf{v}}(\underline{\mathbf{R}}', \omega)$ and $\underline{\underline{\mathbf{T}}}(\underline{\mathbf{R}}', \omega)$ are boundary values of the exterior field, i.e., the surface sources are field dependent. It seems that this is just a circular argument: A field is sustained by sources that contain the field itself! Nevertheless, a physical explanation is immediately at hand: With Figure 15.2(a), we actually assumed the excitation of the field by primary sources, and this evidently implies that also the field on the virtual surface S_g of Figure 15.2(a) originates from these sources, more precisely: the field components as given by (15.7) and (15.8). With Figure 15.2(b), these field components are kept, yet the interior of S_g is emptied, and interesting enough, in the exterior of S_g, nothing changes: Relating to the field in the exterior of S_g, the knowledge of the surface sources on S_g is equivalent to the knowledge of the primary volume sources in the interior of S_g. This is an equivalence principle (Langenberg 2005). It is also the Huygens principle because the knowledge of the field on a (closed) surface is sufficient for the knowledge of the field in the exterior of this surface. And: In the interior of S_g, the field of the equivalent sources is obviously canceled. Now only the mathematical representation of the field is missing; for that purpose, we invert (15.7), (15.8), and post: The field components given by

$$-\underline{\mathbf{n}}' \cdot \underline{\underline{\mathbf{T}}}(\underline{\mathbf{R}}', \omega) \stackrel{\text{def}}{=} \underline{\mathbf{t}}(\underline{\mathbf{R}}', \omega), \tag{15.9}$$

$$-\underline{\underline{\underline{\underline{\mathbf{I}}}}}^{+} : \underline{\mathbf{n}}'\underline{\mathbf{v}}(\underline{\mathbf{R}}', \omega) \stackrel{\text{def}}{=} \underline{\underline{\mathbf{g}}}(\underline{\mathbf{R}}', \omega) \tag{15.10}$$

define surface sources on S_g with $\underline{\mathbf{R}}' \in S_g$, and the vector singular function $\underline{\boldsymbol{\gamma}}(\underline{\mathbf{R}})$ of the surface S_g (Section 2.4.5) formally turns them into equivalent volume sources

$$\underline{\mathbf{f}}_c(\underline{\mathbf{R}}', \omega) = -\,\underline{\boldsymbol{\gamma}}(\underline{\mathbf{R}}') \cdot \underline{\underline{\mathbf{T}}}(\underline{\mathbf{R}}', \omega), \tag{15.11}$$

$$\underline{\underline{\mathbf{h}}}_c(\underline{\mathbf{R}}', \omega) = -\,\underline{\underline{\underline{\underline{\mathbf{I}}}}}^{+} : \underline{\boldsymbol{\gamma}}(\underline{\mathbf{R}}')\underline{\mathbf{v}}(\underline{\mathbf{R}}', \omega), \tag{15.12}$$

with[275] $\underline{\mathbf{R}}' \in \mathbb{R}^3$ that may be inserted into the integral representations (15.1) and (15.2):

$$\begin{aligned}\underline{\mathbf{v}}(\underline{\mathbf{R}}, \omega) = \iiint_{V_Q} \Big[&-\mathrm{j}\omega \underline{\mathbf{f}}_c(\underline{\mathbf{R}}', \omega) \cdot \underline{\underline{\mathbf{G}}}(\underline{\mathbf{R}} - \underline{\mathbf{R}}', \omega) \\ &+ \underline{\underline{\mathbf{h}}}_c(\underline{\mathbf{R}}', \omega) : \underline{\underline{\underline{\boldsymbol{\Sigma}}}}(\underline{\mathbf{R}} - \underline{\mathbf{R}}', \omega) \Big] \, \mathrm{d}^3\underline{\mathbf{R}}',\end{aligned} \tag{15.13}$$

$$\begin{aligned}\underline{\underline{\mathbf{T}}}(\underline{\mathbf{R}}, \omega) = \iiint_{V_Q} \Big[&\underline{\mathbf{f}}_c(\underline{\mathbf{R}}', \omega) \cdot \underline{\underline{\underline{\boldsymbol{\Sigma}}}}^{312}(\underline{\mathbf{R}}', \omega) \\ &- \frac{1}{\mathrm{j}\omega} \underline{\underline{\mathbf{h}}}_c(\underline{\mathbf{R}}', \omega) : \underline{\underline{\underline{\underline{\boldsymbol{\Pi}}}}}(\underline{\mathbf{R}} - \underline{\mathbf{R}}', \omega) \Big] \, \mathrm{d}^3\underline{\mathbf{R}}'.\end{aligned} \tag{15.14}$$

[275]Evidently, the volume sources (15.11) and (15.12) are only nonzero for $\underline{\mathbf{R}}' \in S_g$.

The sifting property (2.382) of the singular function transforms these volume integrals into surface integrals

$$\underline{\mathbf{v}}(\underline{\mathbf{R}}, \omega) = \iint_{S_g} \Big[\mathrm{j}\omega \underline{\mathbf{n}}' \cdot \underline{\underline{\mathbf{T}}}(\underline{\mathbf{R}}', \omega) \cdot \underline{\underline{\mathbf{G}}}(\underline{\mathbf{R}} - \underline{\mathbf{R}}', \omega) - \underline{\mathbf{n}}' \underline{\mathbf{v}}(\underline{\mathbf{R}}', \omega) : \underline{\underline{\underline{\mathbf{\Sigma}}}}(\underline{\mathbf{R}} - \underline{\mathbf{R}}', \omega) \Big] \, \mathrm{d}S', \tag{15.15}$$

$$\underline{\underline{\mathbf{T}}}(\underline{\mathbf{R}}, \omega) = \iint_{S_g} \Big[- \underline{\mathbf{n}}' \cdot \underline{\underline{\mathbf{T}}}(\underline{\mathbf{R}}', \omega) \cdot \underline{\underline{\underline{\mathbf{\Sigma}}}}^{312}(\underline{\mathbf{R}} - \underline{\mathbf{R}}', \omega) + \frac{1}{\mathrm{j}\omega} \underline{\mathbf{n}}' \underline{\mathbf{v}}(\underline{\mathbf{R}}', \omega) : \underline{\underline{\underline{\underline{\mathbf{\Pi}}}}}(\underline{\mathbf{R}} - \underline{\mathbf{R}}', \omega) \Big] \, \mathrm{d}S', \tag{15.16}$$

where we additionally recognized the symmetries (13.153) and (13.194) of the $\underline{\underline{\underline{\mathbf{\Sigma}}}}$- and $\underline{\underline{\underline{\underline{\mathbf{\Pi}}}}}$-tensor. The integral representations (15.15) and (15.16) do not only constitute an equivalence principle, they are the mathematical formulation of an elastodynamic Huygens principle, namely the elastodynamic counterpart to the scalar Helmholtz integral. For observation points $\underline{\mathbf{R}}$ in the exterior of S_g, they constitute a representation theorem as a solution of the homogeneous governing equations, and for observation points $\underline{\mathbf{R}}$ in the interior of S_g, they constitute an extinction theorem:[276] The mathematical formulation of Huygens' principle yields the field on one side of a surface with sources on the respective other side of the surface, and on the source side, a null-field is obtained. Yet, an explicit mathematical proof is still missing because the extinction theorem has only been postulated! We simply followed the ideas of Larmor (1903) to constitute an electromagnetic Huygens principle because (15.15) and (15.16) contain the correct boundary values of field components as equivalent sources with a significant probability due to their physical justification; viz., Equations (15.15) and (15.16) are solutions for arbitrary boundary values. Note: According to Larmor's arguments surface traction densities as well as surface deformation rates must be considered as equivalent surface sources being the essential reason for introducing both physical sources *ab initio* in the elastodynamic governing equations; else we could not have written down the correct elastodynamic equivalence principle on the basis of (15.1) and (15.2). As a trade-off, we had to derive all three Green tensors $\underline{\underline{\mathbf{G}}}$, $\underline{\underline{\underline{\mathbf{\Sigma}}}}$, and $\underline{\underline{\underline{\underline{\mathbf{\Pi}}}}}$.

Before we get more deeply involved in the mathematical derivation of (15.15) and (15.16), we want to sketch the benefit of an elastodynamic Huygens principle to solve the basic problem of US-NDT as displayed in Figure 15.1. Up to now, it is simply an equivalence principle.

Huygens' principle: integral representations of a scattered field: The following arguments have already been introduced in connection with acoustic and electromagnetic scattered fields; therefore, we simply copy Figures 5.2 and 6.3 elastodynamically (Figure 15.3). A source and a scattering volume may be

[276] For $\underline{\mathbf{R}} \in S_g$, they turn into the basis for BEM (boundary element method) as a numerical tool to calculate scattered fields.

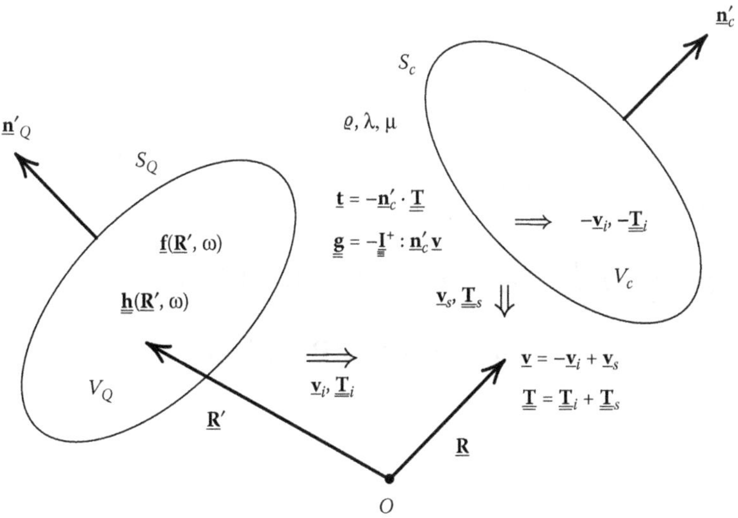

FIGURE 15.3
Elastodynamic scattering problem: surface sources of Huygens' principle.

embedded in a homogeneous isotropic nondissipative full-space; without the presence of the scatterer, the elastodynamic field would be a true source field of primary sources to be indexed as an incident field according to $\underline{\mathbf{v}}_i(\underline{\mathbf{R}}, \omega)$, $\underline{\underline{\mathbf{T}}}_i(\underline{\mathbf{R}}, \omega)$; it may be calculated with (Equations 15.1 and 15.2)

$$\underline{\mathbf{v}}_i(\underline{\mathbf{R}}, \omega) = \iiint_{V_Q} \Big[- \mathrm{j}\omega \underline{\mathbf{f}}(\underline{\mathbf{R}}', \omega) \cdot \underline{\underline{\mathbf{G}}}(\underline{\mathbf{R}} - \underline{\mathbf{R}}', \omega) + \underline{\underline{\mathbf{h}}}(\underline{\mathbf{R}}', \omega) : \underline{\underline{\underline{\mathbf{\Sigma}}}}(\underline{\mathbf{R}} - \underline{\mathbf{R}}', \omega) \Big] \, \mathrm{d}^3\underline{\mathbf{R}}', \tag{15.17}$$

$$\underline{\underline{\mathbf{T}}}_i(\underline{\mathbf{R}}, \omega) = \iiint_{V_Q} \Big[\underline{\mathbf{f}}(\underline{\mathbf{R}}', \omega) \cdot \underline{\underline{\underline{\mathbf{\Sigma}}}}^{312}(\underline{\mathbf{R}}', \omega) - \frac{1}{\mathrm{j}\omega} \underline{\underline{\mathbf{h}}}(\underline{\mathbf{R}}', \omega) : \underline{\underline{\underline{\underline{\mathbf{\Pi}}}}}(\underline{\mathbf{R}} - \underline{\mathbf{R}}', \omega) \Big] \, \mathrm{d}^3\underline{\mathbf{R}}'. \tag{15.18}$$

In the presence of a scattering object, this cannot be the only field because it does not necessarily satisfy the physically required boundary or transition conditions on S_c. Due to the linearity of the elastodynamic governing equations, we therefore superimpose a scattered field $\underline{\mathbf{v}}_s(\underline{\mathbf{R}}, \omega)$, $\underline{\underline{\mathbf{T}}}_s(\underline{\mathbf{R}}, \omega)$ to obtain the total field

$$\underline{\mathbf{v}}(\underline{\mathbf{R}}, \omega) = \underline{\mathbf{v}}_i(\underline{\mathbf{R}}, \omega) + \underline{\mathbf{v}}_s(\underline{\mathbf{R}}, \omega), \tag{15.19}$$

$$\underline{\underline{\mathbf{T}}}(\underline{\mathbf{R}}, \omega) = \underline{\underline{\mathbf{T}}}_i(\underline{\mathbf{R}}, \omega) + \underline{\underline{\mathbf{T}}}_s(\underline{\mathbf{R}}, \omega) \tag{15.20}$$

that should finally satisfy the boundary, respectively transition conditions on S_c. To ensure this, it would be helpful to have a mathematical representation of $\underline{\mathbf{v}}(\underline{\mathbf{R}},\omega)$, $\underline{\underline{\mathbf{T}}}(\underline{\mathbf{R}},\omega)$ that already contains the boundary values of the $\underline{\mathbf{v}}(\underline{\mathbf{R}},\omega)$, $\underline{\underline{\mathbf{T}}}(\underline{\mathbf{R}},\omega)$-field as parameters. The total field satisfies homogeneous elastodynamic governing equations in the exterior of V_c and V_Q—$\underline{\mathbf{R}} \in \mathrm{I\!R}^3 \backslash (\overline{V}_c \cup \overline{V}_Q)$—[the scattered field $\underline{\mathbf{v}}_s(\underline{\mathbf{R}},\omega)$, $\underline{\underline{\mathbf{T}}}_s(\underline{\mathbf{R}},\omega)$ even satisfies homogeneous equations for $\underline{\mathbf{R}} \in V_Q$, it is source-free in V_Q], and, therefore, we may rely on elastodynamic Huygens-type integral representations for the total field in terms of the integrals (15.15) and (15.16):

$$\begin{aligned}
\underline{\mathbf{v}}(\underline{\mathbf{R}},\omega) = &\iint_{S_Q} \Big[\mathrm{j}\omega \underline{\mathbf{n}}'_Q \cdot \underline{\underline{\mathbf{T}}}(\underline{\mathbf{R}}',\omega) \cdot \underline{\underline{\mathbf{G}}}(\underline{\mathbf{R}}-\underline{\mathbf{R}}',\omega) \\
&\qquad - \underline{\mathbf{n}}'_Q \underline{\mathbf{v}}(\underline{\mathbf{R}}',\omega) : \underline{\underline{\underline{\mathbf{\Sigma}}}}(\underline{\mathbf{R}}-\underline{\mathbf{R}}',\omega) \Big] \, \mathrm{d}S' \\
&+ \iint_{S_c} \Big[\mathrm{j}\omega \underline{\mathbf{n}}'_c \cdot \underline{\underline{\mathbf{T}}}(\underline{\mathbf{R}}',\omega) \cdot \underline{\underline{\mathbf{G}}}(\underline{\mathbf{R}}-\underline{\mathbf{R}}',\omega) \\
&\qquad - \underline{\mathbf{n}}'_c \underline{\mathbf{v}}(\underline{\mathbf{R}}',\omega) : \underline{\underline{\underline{\mathbf{\Sigma}}}}(\underline{\mathbf{R}}-\underline{\mathbf{R}}',\omega) \Big] \, \mathrm{d}S',
\end{aligned} \tag{15.21}$$

$$\begin{aligned}
\underline{\underline{\mathbf{T}}}(\underline{\mathbf{R}},\omega) = &\iint_{S_Q} \Big[-\underline{\mathbf{n}}'_Q \cdot \underline{\underline{\mathbf{T}}}(\underline{\mathbf{R}}',\omega) \cdot \underline{\underline{\underline{\mathbf{\Sigma}}}}^{312}(\underline{\mathbf{R}}',\omega) \\
&\qquad + \frac{1}{\mathrm{j}\omega} \underline{\mathbf{n}}'_Q \underline{\mathbf{v}}(\underline{\mathbf{R}}',\omega) : \underline{\underline{\underline{\underline{\mathbf{\Pi}}}}}(\underline{\mathbf{R}}-\underline{\mathbf{R}}',\omega) \Big] \, \mathrm{d}S' \\
&+ \iint_{S_c} \Big[-\underline{\mathbf{n}}'_c \cdot \underline{\underline{\mathbf{T}}}(\underline{\mathbf{R}}',\omega) \cdot \underline{\underline{\underline{\mathbf{\Sigma}}}}^{312}(\underline{\mathbf{R}}',\omega) \\
&\qquad + \frac{1}{\mathrm{j}\omega} \underline{\mathbf{n}}'_c \underline{\mathbf{v}}(\underline{\mathbf{R}}',\omega) : \underline{\underline{\underline{\underline{\mathbf{\Pi}}}}}(\underline{\mathbf{R}}-\underline{\mathbf{R}}',\omega) \Big] \, \mathrm{d}S'
\end{aligned} \tag{15.22}$$

for observation points in the exterior of S_Q and S_c; however, they still extend over both surfaces S_Q and S_c.

First of all, we concentrate on the S_Q-integrals in (15.21) and (15.22) and replace the total field, for example, in the integrand of (15.22) according to (15.5) and (15.6) by the superposition of the incident and scattered fields:

$$\begin{aligned}
&\iint_{S_Q} \Big[\mathrm{j}\omega \underline{\mathbf{n}}'_Q \cdot \underline{\underline{\mathbf{T}}}(\underline{\mathbf{R}}',\omega) \cdot \underline{\underline{\mathbf{G}}}(\underline{\mathbf{R}}-\underline{\mathbf{R}}',\omega) - \underline{\mathbf{n}}'_Q \underline{\mathbf{v}}(\underline{\mathbf{R}}',\omega) : \underline{\underline{\underline{\mathbf{\Sigma}}}}(\underline{\mathbf{R}}-\underline{\mathbf{R}}',\omega) \Big] \, \mathrm{d}S' \\
&= \underbrace{\iint_{S_Q} \Big[\mathrm{j}\omega \underline{\mathbf{n}}'_Q \cdot \underline{\underline{\mathbf{T}}}_i(\underline{\mathbf{R}}',\omega) \cdot \underline{\underline{\mathbf{G}}}(\underline{\mathbf{R}}-\underline{\mathbf{R}}',\omega) - \underline{\mathbf{n}}'_Q \underline{\mathbf{v}}_i(\underline{\mathbf{R}}',\omega) : \underline{\underline{\underline{\mathbf{\Sigma}}}}(\underline{\mathbf{R}}-\underline{\mathbf{R}}',\omega) \Big] \, \mathrm{d}S'}_{= \underline{\mathbf{v}}_i(\underline{\mathbf{R}},\omega) \text{ for } \underline{\mathbf{R}} \in \mathrm{I\!R}^3 \backslash \overline{V}_Q} \\
&+ \underbrace{\iint_{S_Q} \Big[\mathrm{j}\omega \underline{\mathbf{n}}'_Q \cdot \underline{\underline{\mathbf{T}}}_s(\underline{\mathbf{R}}',\omega) \cdot \underline{\underline{\mathbf{G}}}(\underline{\mathbf{R}}-\underline{\mathbf{R}}',\omega) - \underline{\mathbf{n}}'_Q \underline{\mathbf{v}}_s(\underline{\mathbf{R}}',\omega) : \underline{\underline{\underline{\mathbf{\Sigma}}}}(\underline{\mathbf{R}}-\underline{\mathbf{R}}',\omega) \Big] \, \mathrm{d}S'}_{= \underline{\mathbf{0}} \text{ for } \underline{\mathbf{R}} \in \mathrm{I\!R}^3 \backslash \overline{V}_Q} .
\end{aligned} \tag{15.23}$$

The underbrackets state that the S_Q-integrals in (15.21) and (15.22) yield nothing but the incident field, because the equivalence principle immediately tells us that the first integral on the right-hand of (15.23) is equal to $\underline{\mathbf{v}}_i$ for

$\underline{\mathbf{R}} \in \mathrm{IR}^3 \backslash \overline{V}_Q$, and the second integral is equal to zero; after all, the scattered field is source-free in the interior of S_Q. (A reminder: The Huygens integral is a representation of a field on one side of a surface for sources located on the other side.)

Hence, with (15.23), we may write the S_Q-equivalence integral in (15.21) as source field volume integral according to (15.17) that may also be evaluated for $\underline{\mathbf{R}} \in V_Q$. That way, we obtain for $\underline{\mathbf{R}} \in \mathrm{IR}^3 \backslash \overline{V}_c$:

$$\begin{aligned} \underline{\mathbf{v}}(\underline{\mathbf{R}}, \omega) &= \underline{\mathbf{v}}_i(\underline{\mathbf{R}}, \omega) \\ &\quad + \underbrace{\iint_{S_c} \left[\mathrm{j}\omega \underline{\mathbf{n}}_c' \cdot \underline{\underline{\mathbf{T}}}(\underline{\mathbf{R}}', \omega) \cdot \underline{\underline{\mathbf{G}}}(\underline{\mathbf{R}} - \underline{\mathbf{R}}', \omega) - \underline{\mathbf{n}}_c' \underline{\mathbf{v}}(\underline{\mathbf{R}}', \omega) : \underline{\underline{\underline{\boldsymbol{\Sigma}}}}(\underline{\mathbf{R}} - \underline{\mathbf{R}}', \omega) \right] \mathrm{d}S'}_{= \underline{\mathbf{v}}_s(\underline{\mathbf{R}}, \omega)}, \end{aligned} \tag{15.24}$$

and similarly

$$\begin{aligned} \underline{\underline{\mathbf{T}}}(\underline{\mathbf{R}}, \omega) &= \underline{\underline{\mathbf{T}}}_i(\underline{\mathbf{R}}, \omega) \\ &\quad + \underbrace{\iint_{S_c} \left[-\underline{\mathbf{n}}_c' \cdot \underline{\underline{\mathbf{T}}}(\underline{\mathbf{R}}', \omega) \cdot \underline{\underline{\underline{\boldsymbol{\Sigma}}}}^{312}(\underline{\mathbf{R}}', \omega) + \frac{1}{\mathrm{j}\omega} \underline{\mathbf{n}}_c' \underline{\mathbf{v}}(\underline{\mathbf{R}}', \omega) : \underline{\underline{\underline{\underline{\boldsymbol{\Pi}}}}}(\underline{\mathbf{R}} - \underline{\mathbf{R}}', \omega) \right] \mathrm{d}S'}_{= \underline{\underline{\mathbf{T}}}_s(\underline{\mathbf{R}}, \omega)}. \end{aligned} \tag{15.25}$$

With (15.24) and (15.25), we have found Huygens-type integral representations of the scattered field $\underline{\mathbf{v}}_s(\underline{\mathbf{R}}, \omega)$, $\underline{\underline{\mathbf{T}}}_s(\underline{\mathbf{R}}, \omega)$ for $\underline{\mathbf{R}} \in \mathrm{IR}^3 \backslash \overline{V}_c$; for $\underline{\mathbf{R}} \in V_c$, these representations yield $\underline{\mathbf{v}}_s(\underline{\mathbf{R}}, \omega) = -\underline{\mathbf{v}}_i(\underline{\mathbf{R}}, \omega)$, $\underline{\underline{\mathbf{T}}}_s(\underline{\mathbf{R}}, \omega) = -\underline{\underline{\mathbf{T}}}_i(\underline{\mathbf{R}}, \omega)$, because the integral representations (15.21) and (15.22) yield $\underline{\mathbf{v}}(\underline{\mathbf{R}}, \omega) = \underline{\mathbf{0}}$, $\underline{\underline{\mathbf{T}}}(\underline{\mathbf{R}}, \omega) = \underline{\underline{\mathbf{0}}}$ for $\underline{\mathbf{R}} \in V_c$ due to the extinction theorem.

From a physical point of view, the integral representations (15.24) and (15.25) are

- Elastodynamic Huygens integrals: The knowledge of well-defined field components on a closed surface S_c allows for the calculation of the (scattered) field outside the surface;

- Elastodynamic source field representations of secondary sources: With Larmor's arguments, we found the Huygens integrals of the scattered field through insertion of surface sources into volume integrals. The sources are secondary because they are excited—electromagnetically spoken: induced—by the incident field of the primary sources on behalf of the presence of the scatterer. We summarize that: The incident field has primary and the scattered field secondary sources.

There is an essential difference between primary and secondary sources: The former are prescribed independent of fields, the latter are field dependent, and, therefore, even dependent on the scattered field on S_c that has to be calculated from exactly these sources. Up to now, the Huygens integrals (15.24) and (15.25) are purely formal representations not yet ready to serve for an actual calculation of the scattered field. However, we have not yet introduced the possible physical properties of the scatterer, and it will turn out that the transition, respectively boundary, conditions will serve to calculate the secondary sources.

15.1.2 Mathematical derivation of Huygens' principle for scalar acoustic fields

Huygens and equivalence principle: We refer to Green's second integral formula

$$\iiint_V \{\Phi(\underline{\mathbf{R}}') \left[\Delta'\Psi(\underline{\mathbf{R}}') + k^2\Psi(\underline{\mathbf{R}}')\right] - \Psi(\underline{\mathbf{R}}') \left[\Delta'\Phi(\underline{\mathbf{R}}') + k^2\Phi(\underline{\mathbf{R}}')\right]\} \, \mathrm{d}V'$$

$$= \iint_{S_g} \left[\Phi(\underline{\mathbf{R}}')\underline{\mathbf{n}}' \cdot \nabla'\Psi(\underline{\mathbf{R}}') - \Psi(\underline{\mathbf{R}}')\underline{\mathbf{n}}' \cdot \nabla'\Phi(\underline{\mathbf{R}}')\right] \mathrm{d}S' \qquad (15.26)$$

slightly modifying (2.202): In the integrand of the volume integral, we add and subtract $k^2\Phi(\underline{\mathbf{R}}')\Psi(\underline{\mathbf{R}}')$, $k > 0$ yielding twice the Helmholtz operator $\Delta' + k^2$; in the volume as well as in the surface integral, we use $\underline{\mathbf{R}}'$ as integration variable (Figure 15.4), which becomes apparent in a minute. Within (15.26), $\Phi(\underline{\mathbf{R}}')$ and $\Psi(\underline{\mathbf{R}}')$ are initially arbitrary potential functions that will now be given a definite physical meaning: $\Phi(\underline{\mathbf{R}}') \Longrightarrow p(\underline{\mathbf{R}}', \omega)$ as a scalar field quantity may be considered to be the acoustic pressure that satisfies the homogeneous Helmholtz equation in V:

$$\Delta' p(\underline{\mathbf{R}}', \omega) + k^2 p(\underline{\mathbf{R}}', \omega) = 0, \quad \underline{\mathbf{R}}' \in V, \qquad (15.27)$$

with the wave number $k = \omega/c$ of a homogeneous isotropic nondissipative material (Equations 5.40, 5.42, and 5.50), and $\Psi(\underline{\mathbf{R}}') \Longrightarrow G(\underline{\mathbf{R}}' - \underline{\mathbf{R}}, \omega)$ may be the corresponding Green function

$$G(\underline{\mathbf{R}}' - \underline{\mathbf{R}}, \omega) = \frac{\mathrm{e}^{\mathrm{j}|\underline{\mathbf{R}}' - \underline{\mathbf{R}}|}}{4\pi|\underline{\mathbf{R}}' - \underline{\mathbf{R}}|} \qquad (15.28)$$

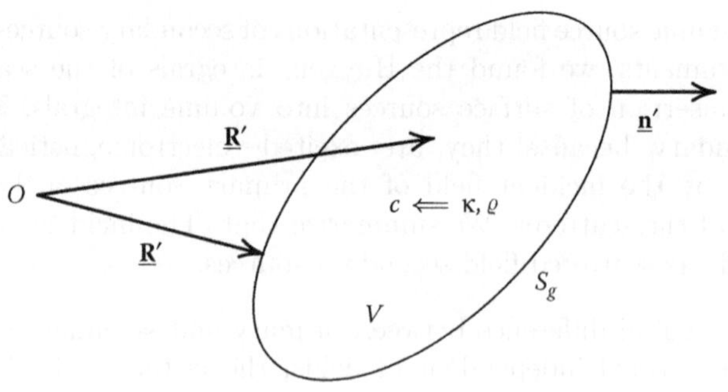

FIGURE 15.4
Mathematical derivation of Huygens' principle.

of infinite space that satisfies the differential equation for $\underline{\mathbf{R}}' \in \mathrm{IR}^3$ according to (5.58), respectively (13.1):

$$\Delta' G(\underline{\mathbf{R}}' - \underline{\mathbf{R}}, \omega) + k^2 G(\underline{\mathbf{R}}' - \underline{\mathbf{R}}, \omega) = -\delta(\underline{\mathbf{R}}' - \underline{\mathbf{R}}); \tag{15.29}$$

$\underline{\mathbf{R}}$ is an arbitrary position vector parameter that may point to the interior or to the exterior of V. Insertion of (15.27) and (15.29) into (15.26) yields

$$\left.\begin{array}{ll} \underline{\mathbf{R}} \in V: & p(\underline{\mathbf{R}}, \omega) \\ \underline{\mathbf{R}} \in \mathrm{IR}^3 \backslash \overline{V}: & 0 \end{array}\right\} = \iint_{S_g} \big[G(\underline{\mathbf{R}} - \underline{\mathbf{R}}', \omega)\underline{\mathbf{n}}' \cdot \nabla' p(\underline{\mathbf{R}}', \omega) - p(\underline{\mathbf{R}}', \omega)\underline{\mathbf{n}}' \cdot \nabla' G(\underline{\mathbf{R}} - \underline{\mathbf{R}}', \omega)\big] \, \mathrm{d}S' \tag{15.30}$$

depending[277] on $\underline{\mathbf{R}}$; with (15.28), we evidently have

$$G(\underline{\mathbf{R}}' - \underline{\mathbf{R}}, \omega) = G(\underline{\mathbf{R}} - \underline{\mathbf{R}}', \omega), \tag{15.31}$$

$$\underline{\mathbf{n}}' \cdot \nabla' G(\underline{\mathbf{R}}' - \underline{\mathbf{R}}, \omega) = \underline{\mathbf{n}}' \cdot \nabla' G(\underline{\mathbf{R}} - \underline{\mathbf{R}}', \omega). \tag{15.32}$$

Obviously, (15.30) constitutes the mathematical formulation of the scalar Huygens principle as a so-called Helmholtz integral: The knowledge of the field quantity "pressure" and its normal derivative[278] on a closed surface S_g is sufficient to calculate the field quantity on one side of the surface—here: in the interior—and on the other side, a null-field is obtained (representation theorem and extinction theorem). As a matter of fact, (15.30) is also an equivalence principle for $\underline{\mathbf{R}} \in V$ because this integral accounts for the contribution of sources located in the exterior of S_g to the interior of S_g. However, strictly speaking, we must no longer call them "principles" because (15.30) is a consequence of Helmholtz' reduced wave equation. Yet, we can close the circle of arguments in Section 15.1.1 if the acoustic equation of motion for homogeneous regions

$$\underline{\mathbf{v}}(\underline{\mathbf{R}}, \omega) = \frac{1}{\mathrm{j}\omega\varrho} \nabla p(\underline{\mathbf{R}}, \omega) \tag{15.33}$$

is recognized (Equation 5.75) and inserted into (15.30):

$$p(\underline{\mathbf{R}}, \omega) = \iint_{S_g} \big[\mathrm{j}\omega\varrho\, \underline{\mathbf{n}}' \cdot \underline{\mathbf{v}}(\underline{\mathbf{R}}', \omega) G(\underline{\mathbf{R}} - \underline{\mathbf{R}}', \omega) - p(\underline{\mathbf{R}}', \omega)\underline{\mathbf{n}}' \cdot \nabla' G(\underline{\mathbf{R}} - \underline{\mathbf{R}}', \omega)\big] \, \mathrm{d}S'. \tag{15.34}$$

Namely, additionally defining surface sources

$$\underline{\mathbf{t}}(\underline{\mathbf{R}}', \omega) = -\underline{\mathbf{n}}' p(\underline{\mathbf{R}}', \omega), \tag{15.35}$$

$$g(\underline{\mathbf{R}}', \omega) = \underline{\mathbf{n}}' \cdot \underline{\mathbf{v}}(\underline{\mathbf{R}}', \omega) \tag{15.36}$$

[277]This was the reason to choose $\underline{\mathbf{R}}'$ as integration variable in Green's second formula (15.26): With (15.30), $p(\underline{\mathbf{R}}, \omega)$ results as a function of the familiar variable $\underline{\mathbf{R}}$.

[278]Note: The scalar field quantity alone, as originally postulated by Huygens, is not sufficient.

for $\underline{\mathbf{R}}' \in S_g$—note: According to the definition (5.82) and (5.83), the normal should point away from the null-field, yet in Figure 15.4, it points toward the null-field as requested by Green's formula (15.26) yielding differing signs in (15.34) and (5.84)—then (15.34) turns into the surface source complement of the acoustic volume source integral (5.63):

$$p(\underline{\mathbf{R}}, \omega) = \iint_{S_g} \big[\mathrm{j}\omega\varrho\, g(\underline{\mathbf{R}}', \omega) G(\underline{\mathbf{R}} - \underline{\mathbf{R}}', \omega) + \underline{\mathbf{t}}(\underline{\mathbf{R}}', \omega) \cdot \boldsymbol{\nabla}' G(\underline{\mathbf{R}} - \underline{\mathbf{R}}', \omega) \big] \, \mathrm{d}S', \quad \underline{\mathbf{R}} \in V. \tag{15.37}$$

The Helmholtz integral (15.30) is a formal mathematical consequence of the Helmholtz wave equation for a scalar field, the version (15.34), respectively (15.37), additionally exhibits the physics of acoustic waves. Since (15.35) and (15.36) result from the transition conditions (5.11) and (5.12) and, hence, from the acoustic governing equations, we have proven Larmor's arguments with (15.37) for acoustics;[279] via insertion, we can immediately prove that (15.34) is a solution of the Helmholtz wave equation (15.27) for $\underline{\mathbf{R}} \in V$ (in unprimed coordinates). Therefore, we must not hesitate to insert the surface source densities (15.35) and (15.36) even into the integral representation (5.73) for the field quantity $\underline{\mathbf{v}}(\underline{\mathbf{R}}, \omega)$ that is complementary to $p(\underline{\mathbf{R}}, \omega)$ to obtain representations as (15.34), respectively (15.37):

$$\underline{\mathbf{v}}(\underline{\mathbf{R}}, \omega) = \iint_{S_g} \Big[\mathrm{j}\omega\kappa\, p(\underline{\mathbf{R}}', \omega) \underline{\mathbf{n}}' \cdot \underline{\underline{\mathbf{G}}}_v(\underline{\mathbf{R}} - \underline{\mathbf{R}}', \omega) - \underline{\mathbf{n}}' \cdot \underline{\mathbf{v}}(\underline{\mathbf{R}}', \omega) \boldsymbol{\nabla}' G(\underline{\mathbf{R}} - \underline{\mathbf{R}}', \omega) \Big] \mathrm{d}S' \tag{15.38}$$

$$= \iint_{S_g} \Big[-\mathrm{j}\omega\kappa\, \underline{\mathbf{t}}(\underline{\mathbf{R}}', \omega) \cdot \underline{\underline{\mathbf{G}}}_v(\underline{\mathbf{R}} - \underline{\mathbf{R}}', \omega) - g(\underline{\mathbf{R}}', \omega) \boldsymbol{\nabla}' G(\underline{\mathbf{R}} - \underline{\mathbf{R}}', \omega) \Big], \quad \underline{\mathbf{R}} \in V. \tag{15.39}$$

Here, $\underline{\underline{\mathbf{G}}}_v(\underline{\mathbf{R}} - \underline{\mathbf{R}}', \omega)$ is given by (5.68), respectively (5.71), where the distributional part as well as the principal value calculation may be disregarded for $\underline{\mathbf{R}} \in V$.

Integral representations of an acoustic scattered field: To utilize the Helmholtz integral (15.30), respectively (15.34), for calculation of an acoustic scattered field, we argue with Figure 15.5: To Figure 15.4, we add the mathematically virtual Huygens "volume" V to the volume V_Q with the primary sources and the scattering volume V_c embedded in a κ, ϱ-material (compare the elastodynamic counterpart in Figure 15.3) because the exterior of V is not

[279]Based on this physical understanding, it becomes clear why (15.30) must contain the scalar field quantity as well as its normal derivative; the physics of scalar acoustic wave propagation is primarily expressed in the acoustic governing equations—and the related transition conditions—and only subsequently in wave equations.

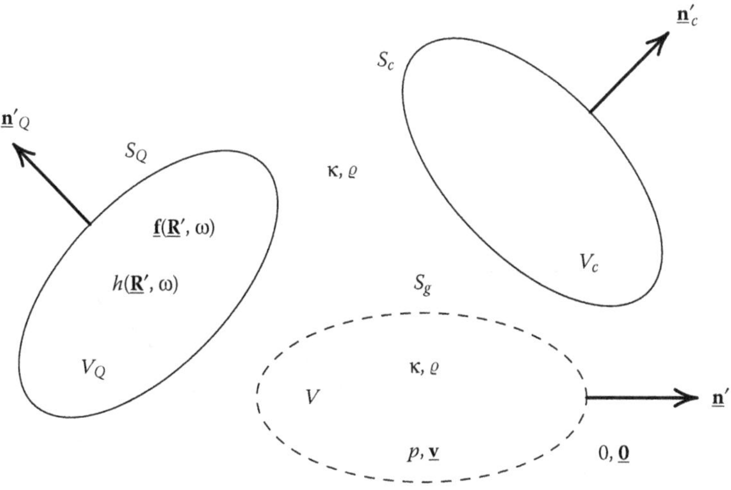

FIGURE 15.5
Acoustic scattering problem and Huygens "volume."

under concern in the derivation of the Helmholtz integral; hence, (15.30) still yields the field quantity $p(\underline{\mathbf{R}}, \omega)$ for $\underline{\mathbf{R}} \in V$—as it is true for (15.39) concerning $\underline{\mathbf{v}}(\underline{\mathbf{R}}, \omega)$—and both integrals yield a null-field for $\underline{\mathbf{R}} \in \mathbb{R}^3 \backslash \overline{V}$.

Figure 15.6 is topologically equivalent to the volume aggregation of Figure 15.5: The surface S_g is deformed in a way that it adapts the surfaces S_Q and S_c—note the different directions of normals—while being connected by two narrow "bridges" with a surface enwinding both volumes V_Q and V_c. While the integral contributions on adjacent "bridge rails" cancel due to the differing normal directions, we must require the contribution of the enwinding surface to vanish. From a physical view point, this is completely apparent: The Helmholtz integral interpreted as an equivalence integral represents the contribution of sources in the interior of the enwinding surface (yet in the exterior of V_Q and V_c) that are located in the exterior of this surface; but there are no sources! Mathematically, the vanishing of the enwinding integral is ensured by Sommerfeld's radiation condition[280] (Hönl et al. 1961; Langenberg 2005). Consequently, the following two integrals remain:

$$
\begin{aligned}
p(\underline{\mathbf{R}}, \omega) = & \iint_{S_Q} \left[\mathrm{j}\omega\varrho\, \underline{\mathbf{n}}' \cdot \underline{\mathbf{v}}(\underline{\mathbf{R}}', \omega) G(\underline{\mathbf{R}} - \underline{\mathbf{R}}', \omega) - p(\underline{\mathbf{R}}', \omega) \underline{\mathbf{n}}' \cdot \nabla' G(\underline{\mathbf{R}} - \underline{\mathbf{R}}', \omega) \right] \mathrm{d}S' \\
& + \iint_{S_c} \left[\mathrm{j}\omega\varrho\, \underline{\mathbf{n}}' \cdot \underline{\mathbf{v}}(\underline{\mathbf{R}}', \omega) G(\underline{\mathbf{R}} - \underline{\mathbf{R}}', \omega) - p(\underline{\mathbf{R}}', \omega) \underline{\mathbf{n}}' \cdot \nabla' G(\underline{\mathbf{R}} - \underline{\mathbf{R}}', \omega) \right] \mathrm{d}S'
\end{aligned}
$$

[280]The other way around: Sommerfeld's radiation condition ensures, together with transition, respectively boundary, conditions on S_c, the uniqueness of the solution of the scattering problem (Colton and Kress 1983).

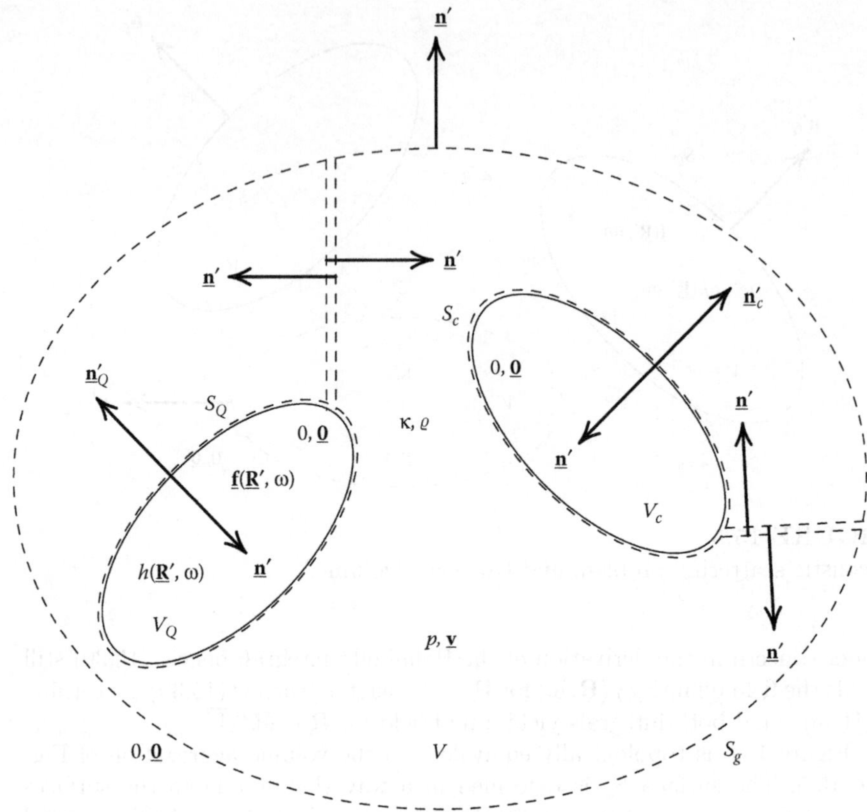

FIGURE 15.6
Acoustic scattering problem and distorted Huygens "volume."

$$= \underbrace{\iint_{S_Q} \left[-j\omega\varrho\, \underline{\mathbf{n}}'_Q \cdot \underline{\mathbf{v}}(\underline{\mathbf{R}}', \omega) G(\underline{\mathbf{R}} - \underline{\mathbf{R}}', \omega) + p(\underline{\mathbf{R}}', \omega)\underline{\mathbf{n}}'_Q \cdot \nabla' G(\underline{\mathbf{R}} - \underline{\mathbf{R}}', \omega)\right] \mathrm{d}S'}_{= p_i(\underline{\mathbf{R}}, \omega) \text{ for } \underline{\mathbf{R}} \in \mathbb{R}^3 \backslash \overline{V}_Q}$$
$$+ \underbrace{\iint_{S_c} \left[-j\omega\varrho\, \underline{\mathbf{n}}'_c \cdot \underline{\mathbf{v}}(\underline{\mathbf{R}}', \omega) G(\underline{\mathbf{R}} - \underline{\mathbf{R}}', \omega) + p(\underline{\mathbf{R}}', \omega)\underline{\mathbf{n}}'_c \cdot \nabla' G(\underline{\mathbf{R}} - \underline{\mathbf{R}}', \omega)\right] \mathrm{d}S'}_{= p_s(\underline{\mathbf{R}}, \omega) \text{ for } \underline{\mathbf{R}} \in \mathbb{R}^3 \backslash \overline{V}_c} \tag{15.40}$$

that represent the incident field and the scattered field according to (5.63) and per definition (Equations 5.84). With (15.39), we similarly obtain the particle velocity vector (Equations 5.73 and 5.86)

$$\underline{\mathbf{v}}(\underline{\mathbf{R}}, \omega) = \underline{\mathbf{v}}_i(\underline{\mathbf{R}}, \omega) + \iint_{S_c} \Big[- j\omega\kappa\, p(\underline{\mathbf{R}}', \omega)\underline{\mathbf{n}}'_c \cdot \underline{\underline{\mathbf{G}}}_v(\underline{\mathbf{R}} - \underline{\mathbf{R}}', \omega) + \underline{\mathbf{n}}'_c \cdot \underline{\mathbf{v}}(\underline{\mathbf{R}}', \omega) \nabla' G(\underline{\mathbf{R}} - \underline{\mathbf{R}}', \omega)\Big] \mathrm{d}S' \tag{15.41}$$

for $\underline{\mathbf{R}} \in \mathbb{R}^3 \backslash \overline{V}_c$. The comparison with the source field volume integrals (5.63) and (5.73) reveals that

$$\underline{\mathbf{t}}(\underline{\mathbf{R}}', \omega) = \underline{\mathbf{n}}'_c p(\underline{\mathbf{R}}', \omega), \tag{15.42}$$

$$g(\underline{\mathbf{R}}', \omega) = -\underline{\mathbf{n}}'_c \cdot \underline{\mathbf{v}}(\underline{\mathbf{R}}', \omega) \tag{15.43}$$

are in fact (secondary) surface sources, respectively,

$$\underline{\mathbf{f}}_c(\underline{\mathbf{R}}', \omega) = \underline{\boldsymbol{\gamma}}_c(\underline{\mathbf{R}}') p(\underline{\mathbf{R}}', \omega), \tag{15.44}$$

$$h_c(\underline{\mathbf{R}}', \omega) = -\underline{\boldsymbol{\gamma}}_c(\underline{\mathbf{R}}') \cdot \underline{\mathbf{v}}(\underline{\mathbf{R}}', \omega) \tag{15.45}$$

are formally (secondary) volume sources within the scattered field integral representations (15.40) and (15.41) (Figure 5.2).

15.1.3 Mathematical derivation of Huygens' principle for elastodynamic fields

Elastodynamic Huygens and equivalence principle: We started with the reduced wave equation (homogeneous in V) (15.27) as well as with the differential equation for the scalar Green function (15.29) to derive Huygens' principle for scalar fields mathematically, and both equations have been utilized in Green's second formula that fitted the Helmholtz operator in (15.27) and (15.29). For a similar approach in elastodynamics, we need the two differential equations

$$\nabla' \cdot \underline{\underline{\underline{\underline{\mathbf{c}}}}} : \nabla' \underline{\mathbf{v}}(\underline{\mathbf{R}}', \omega) + \omega^2 \varrho \, \underline{\mathbf{v}}(\underline{\mathbf{R}}', \omega) = \underline{\mathbf{0}}, \tag{15.46}$$

$$\nabla' \cdot \underline{\underline{\underline{\underline{\mathbf{c}}}}} : \nabla' \underline{\underline{\mathbf{G}}}(\underline{\mathbf{R}}' - \underline{\mathbf{R}}, \omega) + \omega^2 \varrho \, \underline{\underline{\mathbf{G}}}(\underline{\mathbf{R}}' - \underline{\mathbf{R}}, \omega) = -\delta(\underline{\mathbf{R}}' - \underline{\mathbf{R}}) \, \underline{\underline{\mathbf{I}}}, \tag{15.47}$$

where we imply homogeneity and isotropy of the material with the spatial dependence $\underline{\mathbf{R}}' - \underline{\mathbf{R}}$ of the second rank Green tensor, hence assuming $\underline{\underline{\underline{\underline{\mathbf{c}}}}} = \lambda \, \underline{\underline{\underline{\underline{\mathbf{I}}}}}^{\delta} + 2\mu \, \underline{\underline{\underline{\underline{\mathbf{I}}}}}^{+}$; nevertheless, for the sake of lucidity for the following derivation, we keep the short-hand notation (15.46) and (15.47) as compared to (13.78) and (13.80) without the specification of $\underline{\underline{\underline{\underline{\mathbf{c}}}}}$. Now we only need a suitable Green formula that contains the differential operator $\nabla' \cdot \underline{\underline{\underline{\underline{\mathbf{c}}}}} : \nabla'$ because we compose—and calculate with (15.46) and (15.47)—the expression

$$\begin{aligned} &\underline{\mathbf{v}}(\underline{\mathbf{R}}', \omega) \cdot \left[\nabla' \cdot \underline{\underline{\underline{\underline{\mathbf{c}}}}} : \nabla' \underline{\underline{\mathbf{G}}}(\underline{\mathbf{R}}' - \underline{\mathbf{R}}, \omega) + \omega^2 \varrho \, \underline{\underline{\mathbf{G}}}(\underline{\mathbf{R}}' - \underline{\mathbf{R}}, \omega) \right] \\ &\quad - \left[\nabla' \cdot \underline{\underline{\underline{\underline{\mathbf{c}}}}} : \nabla' \underline{\mathbf{v}}(\underline{\mathbf{R}}', \omega) + \omega^2 \varrho \, \underline{\mathbf{v}}(\underline{\mathbf{R}}', \omega) \right] \cdot \underline{\underline{\mathbf{G}}}(\underline{\mathbf{R}}' - \underline{\mathbf{R}}, \omega) \\ &\quad = -\underline{\mathbf{v}}(\underline{\mathbf{R}}', \omega) \, \delta(\underline{\mathbf{R}}' - \underline{\mathbf{R}}) \end{aligned} \tag{15.48}$$

analogous to the integrand (15.26), where we must adhere to the cancelation of the two terms $\omega^2 \varrho \, \underline{\mathbf{v}}(\underline{\mathbf{R}}', \omega) \cdot \underline{\underline{\mathbf{G}}}(\underline{\mathbf{R}}' - \underline{\mathbf{R}}, \omega)$ while taking the $\underline{\mathbf{v}}$- and

$\underline{\underline{\mathbf{G}}}$-contractions on the left-hand side of[281] (15.48). Quite clear: On the right-hand side, the volume integration of (15.48) either yields $\underline{\mathbf{v}}(\underline{\mathbf{R}}, \omega)$ or $\underline{\mathbf{0}}$ depending whether $\underline{\mathbf{R}}$ is located in the interior or exterior of V. Now, our goal must be to transform the volume integral of the left-hand side into a surface integral over S_g; that is why we need an appropriate Green formula. To derive it, we obviously have to pull out a ∇'-operator as divergence operator in both terms

$$\underline{\mathbf{v}} \cdot [\nabla' \cdot \underline{\underline{\underline{\underline{\mathbf{c}}}}} : \nabla' \underline{\underline{\mathbf{G}}}] \Longrightarrow \nabla' \cdot [\underline{\mathbf{v}} \cdot \underline{\underline{\underline{\underline{\mathbf{c}}}}} : \nabla' \underline{\underline{\mathbf{G}}}], \tag{15.49}$$

$$[\nabla' \cdot \underline{\underline{\underline{\underline{\mathbf{c}}}}} : \nabla' \underline{\mathbf{v}}] \cdot \underline{\underline{\mathbf{G}}} \Longrightarrow \nabla' \cdot [\underline{\underline{\underline{\underline{\mathbf{c}}}}} : \nabla' \underline{\mathbf{v}} \cdot \underline{\underline{\mathbf{G}}}] \tag{15.50}$$

—apparently yielding "disturbing" terms that hopefully cancel in the difference (15.48)—because afterward the application of Gauss' theorem results in a surface integral replacing the pulled out divergence operator by a contraction with the normal. In the formula collection, we find[282] differentiation formulas for the right-hand sides of (15.49) and (15.50):

$$\nabla' \cdot (\underline{\mathbf{A}} \cdot \underline{\underline{\underline{\mathbf{D}}}}) = \nabla' \underline{\mathbf{A}} : \underline{\underline{\underline{\mathbf{D}}}} + \underline{\mathbf{A}} \cdot \nabla' \cdot \underline{\underline{\underline{\mathbf{D}}}}^{213}, \tag{15.51}$$

$$\nabla' \cdot (\underline{\underline{\mathbf{B}}} \cdot \underline{\underline{\mathbf{C}}}) = (\nabla' \cdot \underline{\underline{\mathbf{B}}}) \cdot \underline{\underline{\mathbf{C}}} + \underline{\underline{\mathbf{B}}}^{21} : \nabla' \underline{\underline{\mathbf{C}}}. \tag{15.52}$$

With $\underline{\mathbf{A}} = \underline{\mathbf{v}}$, $\underline{\underline{\underline{\mathbf{D}}}} = \underline{\underline{\underline{\underline{\mathbf{c}}}}} : \nabla' \underline{\underline{\mathbf{G}}}$, $\underline{\underline{\mathbf{B}}} = \underline{\underline{\underline{\underline{\mathbf{c}}}}} : \nabla' \underline{\mathbf{v}}$, $\underline{\underline{\mathbf{C}}} = \underline{\underline{\mathbf{G}}}$, we are close to the desired transformations:

$$\underline{\mathbf{v}} \cdot \nabla' \cdot (\underline{\underline{\underline{\underline{\mathbf{c}}}}} : \nabla' \underline{\underline{\mathbf{G}}})^{213} = \nabla' \cdot (\underline{\mathbf{v}} \cdot \underline{\underline{\underline{\underline{\mathbf{c}}}}} : \nabla' \underline{\underline{\mathbf{G}}}) - \nabla' \underline{\mathbf{v}} : \underline{\underline{\underline{\underline{\mathbf{c}}}}} : \nabla' \underline{\underline{\mathbf{G}}}, \tag{15.53}$$

$$(\nabla' \cdot \underline{\underline{\underline{\underline{\mathbf{c}}}}} : \nabla' \underline{\mathbf{v}}) \cdot \underline{\underline{\mathbf{G}}} = \nabla' \cdot (\underline{\underline{\underline{\underline{\mathbf{c}}}}} : \nabla' \underline{\mathbf{v}} \cdot \underline{\underline{\mathbf{G}}}) - (\underline{\underline{\underline{\underline{\mathbf{c}}}}} : \nabla' \underline{\mathbf{v}})^{21} : \nabla' \underline{\underline{\mathbf{G}}}, \tag{15.54}$$

where "close" means that we have to take transposes left-sidedly in (15.53) and right-sidedly in (15.54), thus not yet realizing (15.49) with (15.53) completely, and not yet accounting for the cancelation of both terms on the right-hand sides of (15.53) and (15.54) in the difference (15.48). Both addressed transposes relate to the first two indices of $\underline{\underline{\underline{\underline{\mathbf{c}}}}}$, and $\underline{\underline{\underline{\underline{\mathbf{c}}}}}$ is respectively symmetric! Consequently, we have $(\underline{\underline{\underline{\underline{\mathbf{c}}}}} : \nabla' \underline{\underline{\mathbf{G}}})^{213} = \underline{\underline{\underline{\underline{\mathbf{c}}}}} : \nabla' \underline{\underline{\mathbf{G}}}$ and $(\underline{\underline{\underline{\underline{\mathbf{c}}}}} : \nabla' \underline{\mathbf{v}})^{21} = \underline{\underline{\underline{\underline{\mathbf{c}}}}} : \nabla' \underline{\mathbf{v}}$; due to the symmetry $\underline{\underline{\underline{\underline{\mathbf{c}}}}} = \underline{\underline{\underline{\underline{\mathbf{c}}}}}^{3412}$ (analogous to Equation 4.15), we further have $\underline{\underline{\underline{\underline{\mathbf{c}}}}} : \nabla' \underline{\mathbf{v}} = (\nabla' \underline{\mathbf{v}}) : \underline{\underline{\underline{\underline{\mathbf{c}}}}}$, and hence

[281] In a homogeneous isotropic material, $\underline{\underline{\mathbf{G}}}$ is symmetric also ensuring the cancelation of $\underline{\mathbf{v}} \cdot \underline{\underline{\mathbf{G}}}$ against $\underline{\underline{\mathbf{G}}} \cdot \underline{\mathbf{v}}$.

[282] In this collection, we actually find

$$\nabla \cdot (\underline{\mathbf{A}} \cdot \underline{\underline{\mathbf{D}}}) = \nabla \underline{\mathbf{A}} : \underline{\underline{\mathbf{D}}} + \underline{\mathbf{A}} \cdot \nabla \cdot \underline{\underline{\mathbf{D}}}^{21},$$

$$\nabla \cdot (\underline{\underline{\mathbf{D}}} \cdot \underline{\mathbf{A}}) = (\nabla \cdot \underline{\underline{\mathbf{D}}}) \cdot \underline{\mathbf{A}} + \underline{\underline{\mathbf{D}}}^{21} : \nabla \underline{\mathbf{A}},$$

yet in the first formula, we may enhance $\underline{\underline{\mathbf{D}}}$ to a third rank tensor, and in the second formula, $\underline{\mathbf{A}}$ may be enhanced to a second rank tensor because the additional indices are free indices in both formulas.

$$-\underline{\mathbf{v}}(\underline{\mathbf{R}}',\omega)\,\delta(\underline{\mathbf{R}}'-\underline{\mathbf{R}}) = \nabla'\cdot\left[\underline{\mathbf{v}}(\underline{\mathbf{R}}',\omega)\cdot\underline{\underline{\underline{\underline{\mathbf{c}}}}}:\nabla'\underline{\underline{\mathbf{G}}}(\underline{\mathbf{R}}'-\underline{\mathbf{R}},\omega)\right]$$
$$-\nabla'\cdot\left[\underline{\underline{\underline{\underline{\mathbf{c}}}}}:\underline{\mathbf{v}}(\underline{\mathbf{R}}',\omega)\cdot\underline{\underline{\mathbf{G}}}(\underline{\mathbf{R}}'-\underline{\mathbf{R}},\omega)\right]. \tag{15.55}$$

Now we integrate (15.55) over V, apply Gauss' theorem, and adhere to $\underline{\underline{\mathbf{G}}}(\underline{\mathbf{R}}'-\underline{\mathbf{R}},\omega) = \underline{\underline{\mathbf{G}}}(\underline{\mathbf{R}}-\underline{\mathbf{R}}',\omega)$ for homogeneous isotropic materials:

$$\left.\begin{array}{ll}\underline{\mathbf{R}}\in V: & \underline{\mathbf{v}}(\underline{\mathbf{R}},\omega)\\ \underline{\mathbf{R}}\in \mathrm{I\!R}^3\backslash\overline{V}: & \underline{\mathbf{0}}\end{array}\right\} = \iint_{S_g}\left[\underline{\mathbf{n}}'\cdot\underline{\underline{\underline{\underline{\mathbf{c}}}}}:\nabla'\underline{\mathbf{v}}(\underline{\mathbf{R}}',\omega)\cdot\underline{\underline{\mathbf{G}}}(\underline{\mathbf{R}}-\underline{\mathbf{R}}',\omega)\right.$$
$$\left.-\underline{\mathbf{n}}'\underline{\mathbf{v}}(\underline{\mathbf{R}}',\omega):\underline{\underline{\underline{\underline{\mathbf{c}}}}}:\nabla'\underline{\underline{\mathbf{G}}}(\underline{\mathbf{R}}-\underline{\mathbf{R}}',\omega)\right]\mathrm{d}S'. \tag{15.56}$$

This integral representation of the elastodynamic particle velocity vector is an elastodynamic representation theorem as well as an elastodynamic extinction theorem, hence, an elastodynamic Huygens as well as an equivalence integral in the same sense as the integral representation (15.30) of the scalar acoustic pressure. According to (13.151), we identify $\underline{\underline{\underline{\boldsymbol{\Sigma}}}}(\underline{\mathbf{R}}-\underline{\mathbf{R}}',\omega) = -\underline{\underline{\underline{\underline{\mathbf{c}}}}}:\nabla'\underline{\underline{\mathbf{G}}}(\underline{\mathbf{R}}-\underline{\mathbf{R}}',\omega)$ as a third rank Green tensor (of infinite space), and due to $\underline{\underline{\underline{\underline{\mathbf{c}}}}}:\nabla'\underline{\mathbf{v}}(\underline{\mathbf{R}}',\omega) = -\mathrm{j}\omega\underline{\underline{\mathbf{T}}}(\underline{\mathbf{R}}',\omega)$, the expressions

$$\underline{\mathbf{t}}(\underline{\mathbf{R}}',\omega) = \underline{\mathbf{n}}'\cdot\underline{\underline{\mathbf{T}}}(\underline{\mathbf{R}}',\omega), \tag{15.57}$$
$$\underline{\underline{\mathbf{g}}}(\underline{\mathbf{R}}',\omega) = \underline{\underline{\underline{\underline{\mathbf{I}}}}}^{+}:\underline{\mathbf{n}}'\underline{\mathbf{v}}(\underline{\mathbf{R}}',\omega) \tag{15.58}$$

define surface sources for $\underline{\mathbf{R}}'\in S_g$—note the differently oriented normal as compared to (15.7) and (15.8) that points toward the null-field—rendering (15.56) similar to (15.37) as a surface source counterpart for $\underline{\mathbf{R}}\in V$

$$\underline{\mathbf{v}}(\underline{\mathbf{R}},\omega) = \iint_{S_g}\left[-\mathrm{j}\omega\,\underline{\mathbf{n}}'\cdot\underline{\underline{\mathbf{T}}}(\underline{\mathbf{R}}',\omega)\cdot\underline{\underline{\mathbf{G}}}(\underline{\mathbf{R}}-\underline{\mathbf{R}}',\omega)\right.$$
$$\left.+\underline{\mathbf{n}}'\underline{\mathbf{v}}(\underline{\mathbf{R}}',\omega):\underline{\underline{\underline{\boldsymbol{\Sigma}}}}(\underline{\mathbf{R}}-\underline{\mathbf{R}}',\omega)\right]\mathrm{d}S' \tag{15.59}$$
$$= \iint_{S_g}\left[-\mathrm{j}\omega\,\underline{\mathbf{t}}(\underline{\mathbf{R}}',\omega)\cdot\underline{\underline{\mathbf{G}}}(\underline{\mathbf{R}}-\underline{\mathbf{R}}',\omega)\right.$$
$$\left.+\underline{\underline{\mathbf{g}}}(\underline{\mathbf{R}}',\omega):\underline{\underline{\underline{\boldsymbol{\Sigma}}}}(\underline{\mathbf{R}}-\underline{\mathbf{R}}',\omega)\right]\mathrm{d}S' \tag{15.60}$$

as compared to the elastodynamic volume source integral (15.13); vis-à-vis (15.15) the sign has changed in (15.59) due to the different normal orientation; in the version (15.60), this difference is not obvious due to the different definition of the surface sources according to (15.57) and (15.58).

With the proof of (15.56) and the interpretation (15.59), respectively (15.60), Larmor's arguments yielding (15.15) for electromagnetics have been validated. Therefore, we may complement the representations (15.59) and (15.60) by

$$\underline{\underline{\mathbf{T}}}(\underline{\mathbf{R}}, \omega) = \int\!\!\int_{S_g} \Big[\underline{\mathbf{n}}' \cdot \underline{\underline{\mathbf{T}}}(\underline{\mathbf{R}}', \omega) \cdot \underline{\underline{\underline{\mathbf{\Sigma}}}}^{312}(\underline{\mathbf{R}} - \underline{\mathbf{R}}', \omega) - \frac{1}{\mathrm{j}\omega} \underline{\mathbf{n}}' \underline{\mathbf{v}}(\underline{\mathbf{R}}', \omega) : \underline{\underline{\underline{\underline{\mathbf{\Pi}}}}}(\underline{\mathbf{R}} - \underline{\mathbf{R}}', \omega) \Big] \, \mathrm{d}S' \tag{15.61}$$

$$= \int\!\!\int_{S_g} \Big[\underline{\mathbf{t}}(\underline{\mathbf{R}}', \omega) \cdot \underline{\underline{\underline{\mathbf{\Sigma}}}}^{312}(\underline{\mathbf{R}} - \underline{\mathbf{R}}', \omega) - \frac{1}{\mathrm{j}\omega} \underline{\underline{\mathbf{g}}}(\underline{\mathbf{R}}', \omega) : \underline{\underline{\underline{\underline{\mathbf{\Pi}}}}}(\underline{\mathbf{R}} - \underline{\mathbf{R}}', \omega) \Big] \, \mathrm{d}S' \tag{15.62}$$

on the basis of (15.16). Mathematically, this follows applying Hooke's law to (15.59), respectively (15.60), similar to the calculation that yielded (13.190).

Integral representations of an elastodynamic scattered field: Now the integral representations (15.59) and (15.61) may be utilized to calculate elastodynamic scattered fields; to this end, we distort the integration surface S_g as in Figure 15.6, postulate the elastodynamic counterpart of a Sommerfeld radiation condition (Tan 1975b; Pao und Varatharajulu 1976; Alves and Kress 2002), and observe the different directions of the normals $\underline{\mathbf{n}}'$ and $\underline{\mathbf{n}}'_c$ to find

$$\underline{\mathbf{v}}(\underline{\mathbf{R}}, \omega) = \underline{\mathbf{v}}_i(\underline{\mathbf{R}}, \omega) + \underbrace{\int\!\!\int_{S_c} \Big[\mathrm{j}\omega \underline{\mathbf{n}}'_c \cdot \underline{\underline{\mathbf{T}}}(\underline{\mathbf{R}}', \omega) \cdot \underline{\underline{\mathbf{G}}}(\underline{\mathbf{R}} - \underline{\mathbf{R}}', \omega) - \underline{\mathbf{n}}'_c \underline{\mathbf{v}}(\underline{\mathbf{R}}', \omega) : \underline{\underline{\underline{\mathbf{\Sigma}}}}(\underline{\mathbf{R}} - \underline{\mathbf{R}}', \omega) \Big] \, \mathrm{d}S',}_{= \underline{\mathbf{v}}_s(\underline{\mathbf{R}}, \omega)} \tag{15.63}$$

$$\underline{\underline{\mathbf{T}}}(\underline{\mathbf{R}}, \omega) = \underline{\underline{\mathbf{T}}}_i(\underline{\mathbf{R}}, \omega) + \underbrace{\int\!\!\int_{S_c} \Big[-\underline{\mathbf{n}}'_c \cdot \underline{\underline{\mathbf{T}}}(\underline{\mathbf{R}}', \omega) \cdot \underline{\underline{\underline{\mathbf{\Sigma}}}}^{312}(\underline{\mathbf{R}}', \omega) + \frac{1}{\mathrm{j}\omega} \underline{\mathbf{n}}'_c \underline{\mathbf{v}}(\underline{\mathbf{R}}', \omega) : \underline{\underline{\underline{\underline{\mathbf{\Pi}}}}}(\underline{\mathbf{R}} - \underline{\mathbf{R}}', \omega) \Big] \, \mathrm{d}S'}_{= \underline{\underline{\mathbf{T}}}_s(\underline{\mathbf{R}}, \omega)} \tag{15.64}$$

for $\underline{\mathbf{R}} \in \mathrm{I\!R}^3 \backslash \overline{V}_c$ (Figure 15.3). As expected, these mathematically derived integral representations of the scattered field (de Hoop 1958, 1995; Tan 1975a; Pao and Varatharajulu 1976; Fellinger 1991) are identical with Equations 15.24 and 15.25 emerging from guess.

15.2 Integral Equations for Secondary Surface Deformation Sources on Scatterers with Stress-Free Surfaces: Displacement Field Integral Equation and Stress Field Integral Equation

The integral representation (15.63) for the scattered particle velocity vector holds for an arbitrary scatterer; it is a solution of the homogeneous reduced wave equation (15.46) in the exterior $\underline{\mathbf{R}} \in \mathrm{IR}^3 \backslash \overline{V}_c$ for arbitrary secondary sources. For a physically specified scatterer, the secondary sources have to be calculated with the help of the actual transition or boundary conditions. Specially, considering the NDT-relevant case of a scatterer with a stress-free surface—a vacuum "hole" in a solid, also modeling a crack for a vanishing volume with finite surface—the boundary condition enforces $\underline{\mathbf{n}}'_c \cdot \underline{\underline{\mathbf{T}}}(\underline{\mathbf{R}}', \omega) = \underline{\mathbf{0}}$ for $\underline{\mathbf{R}}' \in S_c$ yielding only the secondary surface deformation rate $\underline{\underline{\underline{\mathbf{I}}}}^+ : \underline{\mathbf{n}}'_c \underline{\mathbf{v}}(\underline{\mathbf{R}}', \omega)$ to be nonzero in the integral representation (15.63); yet, the latter is not specified by the boundary condition because the total field $\underline{\mathbf{v}}(\underline{\mathbf{R}}, \omega)$ defines these tensor components as secondary sources for $\underline{\mathbf{R}} \longrightarrow S_c$. However, exactly this fact is utilized for their calculation: The limit $\underline{\mathbf{R}} \longrightarrow S_c$ is explicitly performed with the integral representation (15.63) of the scattered field—we will see that this limit isolates a term $\underline{\mathbf{v}}(\underline{\mathbf{R}}, \omega)/2$ from the integral—hence an equation is obtained for $\underline{\mathbf{v}}(\underline{\mathbf{R}}, \omega)$, $\underline{\mathbf{R}} \in S_c$ containing the particle velocity vector explicitly, and also under the integral: Such an equation is called an integral equation of the second kind. This integral equation may be solved numerically for a given surface S_c—special cases also allow for an analytical solution—yielding numbers for the secondary deformation rate; afterward, these may be inserted into the scattered field representation (15.63) for $\underline{\mathbf{R}} \in \mathrm{IR}^3 \backslash \overline{V}_c$: The scattering problem has been solved! This exact approach is relatively expensive, and therefore approximate solutions of the integral equation are considered, and in many cases, applications indeed allow for the elastodynamically modified Kirchhoff or physical optics approximation that is well known in optics and electromagnetics. Many modeling codes developed for US-NDT rely on that approximation.

15.2.1 Integral equations relating secondary sources

At first, we perform the limit $\underline{\mathbf{R}} \longrightarrow S_c$ in (15.63) without specifying a boundary condition—for example: stress-free. Obviously, (15.63) may be immediately written down for the particle displacement resulting in a[283] displacement field integral equation (DFIE):

$$\begin{aligned} \underline{\mathbf{u}}(\underline{\mathbf{R}}, \omega) = \underline{\mathbf{u}}_i(\underline{\mathbf{R}}, \omega) - \iint_{S_c} \Big[& \underline{\mathbf{n}}'_c \cdot \underline{\underline{\mathbf{T}}}(\underline{\mathbf{R}}', \omega) \cdot \underline{\underline{\mathbf{G}}}(\underline{\mathbf{R}} - \underline{\mathbf{R}}', \omega) \\ & + \underline{\mathbf{u}}(\underline{\mathbf{R}}', \omega) \underline{\mathbf{n}}_c : \underline{\underline{\underline{\mathbf{\Sigma}}}}(\underline{\mathbf{R}} - \underline{\mathbf{R}}', \omega) \Big] \, \mathrm{d}S'; \end{aligned} \quad (15.65)$$

due to the symmetry of $\underline{\underline{\underline{\mathbf{\Sigma}}}}$ with regard to the first two indices, we may interchange $\underline{\mathbf{n}}'_c$ and $\underline{\mathbf{u}}(\underline{\mathbf{R}}', \omega)$. Insofar, the limit $\underline{\mathbf{R}} \longrightarrow S_c$ is not totally trivial due to the singularities of $\underline{\underline{\mathbf{G}}}(\underline{\mathbf{R}} - \underline{\mathbf{R}}', \omega)$ as well as $\underline{\underline{\underline{\mathbf{\Sigma}}}}(\underline{\mathbf{R}} - \underline{\mathbf{R}}', \omega)$ for $\underline{\mathbf{R}} \longrightarrow \underline{\mathbf{R}}'$, and integrals over singular functions are improper integrals whose existence must be proven for each special case. That is, exactly, our task in the present

[283]We follow the terminology of the theory of electromagnetic fields, where an electric field integral equation—EFIE—as well as a magnetic field integral equation—MFIE (Section 6.7; Langenberg 2005) are derived.

case; yet, fortunately, we must not start from the beginning because essential results are already at hand from the scalar and electromagnetic Huygens integrals (Langenberg 2005).

We introduce the short-hand notation

$$\underline{\mathbf{H}}^{\mathbf{u}}_{S_c}(\underline{\mathbf{R}}, \omega) = -\iint_{S_c} \Big[\underline{\mathbf{n}}'_c \cdot \underline{\underline{\mathbf{T}}}(\underline{\mathbf{R}}', \omega) \cdot \underline{\underline{\mathbf{G}}}(\underline{\mathbf{R}} - \underline{\mathbf{R}}', \omega) + \underline{\mathbf{u}}(\underline{\mathbf{R}}', \omega)\underline{\mathbf{n}}'_c : \underline{\underline{\underline{\mathbf{\Sigma}}}}(\underline{\mathbf{R}} - \underline{\mathbf{R}}', \omega) \Big] \, \mathrm{d}S' \tag{15.66}$$

of an elastodynamic Helmholtz integral for the particle displacement; we know already that

$$\underline{\mathbf{H}}^{\mathbf{u}}_{S_c}(\underline{\mathbf{R}}, \omega) = \begin{cases} \underline{\mathbf{u}}_s(\underline{\mathbf{R}}, \omega), & \underline{\mathbf{R}} \in \mathrm{I\!R}^3 \backslash \overline{V}_c \\ -\underline{\mathbf{u}}_i(\underline{\mathbf{R}}, \omega), & \underline{\mathbf{R}} \in V_c \end{cases} \tag{15.67}$$

holds; now, we are interested in $\underline{\mathbf{H}}^{\mathbf{u}}_{S_c}(\underline{\mathbf{R}}_0, \omega)$ with $\underline{\mathbf{R}}_0 \in S_c$, but we may not readily insert this position vector due to the above mentioned singularities; instead, we have to investigate the limit $\underline{\mathbf{H}}^{\mathbf{u}}_{S_c}(\underline{\mathbf{R}} \longrightarrow \underline{\mathbf{R}}_0, \omega)$. Doing so, we cut a small area S_ϵ out of S_c around the point $\underline{\mathbf{R}}_0$ and calculate

$$\underline{\mathbf{H}}^{\mathbf{u}}_{S_c}(\underline{\mathbf{R}} \longrightarrow \underline{\mathbf{R}}_0, \omega) = \underline{\mathbf{H}}^{\mathbf{u}}_{S_c \backslash \overline{S}_\epsilon}(\underline{\mathbf{R}}_0, \omega) + \lim_{\underline{\mathbf{R}} \to \underline{\mathbf{R}}_0} \underline{\mathbf{H}}^{\mathbf{u}}_{S_\epsilon}(\underline{\mathbf{R}}, \omega); \tag{15.68}$$

in the first term, we may offhand insert $\underline{\mathbf{R}}_0$ because this position vector is outside the integration surface $S_c \backslash \overline{S}_\epsilon$, yet in the second term, we must explicitly investigate the limit, and this is performed as follows: The integral $\underline{\mathbf{H}}^{\mathbf{u}}_{S_\epsilon}(\underline{\mathbf{R}}, \omega)$ is calculated for a point $\underline{\mathbf{R}}_0$ outside S_c for a very small area S_ϵ, and evaluating the limit $\underline{\mathbf{R}} \longrightarrow \underline{\mathbf{R}}_0$ subsequently. In order not to make it too cumbersome, we approximate S_ϵ by a planar circular disk K_ϵ centered at $\underline{\mathbf{R}}_0$ with radius ϵ and assume $\underline{\mathbf{R}}$ to move along the normal $\underline{\mathbf{n}}_0$ in $\underline{\mathbf{R}}_0$ toward $\underline{\mathbf{R}}_0$, i.e., along the constant normal on K_ϵ (Figure 15.7); we actually have

$$\underline{\mathbf{R}} = \underline{\mathbf{R}}_n = \underline{\mathbf{R}}_0 + h_n \underline{\mathbf{n}}_0 \tag{15.69}$$

finally resulting in the calculation of

$$\lim_{h_n \to 0} \underline{\mathbf{H}}^{\mathbf{u}}_{K_\epsilon}(\underline{\mathbf{R}}_n, \omega) = -\lim_{h_n \to 0} \iint_{K_\epsilon} \Big[\underline{\mathbf{n}}_0 \cdot \underline{\underline{\mathbf{T}}}(\underline{\mathbf{R}}', \omega) \cdot \underline{\underline{\mathbf{G}}}(\underline{\mathbf{R}}_n - \underline{\mathbf{R}}', \omega) + \underline{\mathbf{u}}(\underline{\mathbf{R}}', \omega)\underline{\mathbf{n}}_0 : \underline{\underline{\underline{\mathbf{\Sigma}}}}(\underline{\mathbf{R}}_n - \underline{\mathbf{R}}', \omega) \Big] \mathrm{d}S'. \tag{15.70}$$

For a sufficiently small disk K_ϵ, we surely may approximate (15.70) by

$$\lim_{h_n \to 0} \underline{\mathbf{H}}^{\mathbf{u}}_{K_\epsilon}(\underline{\mathbf{R}}_n, \omega) \simeq -\underline{\mathbf{n}}_0 \cdot \underline{\underline{\mathbf{T}}}(\underline{\mathbf{R}}_0, \omega) \cdot \lim_{h_n \to 0} \iint_{K_\epsilon} \underline{\underline{\mathbf{G}}}(\underline{\mathbf{R}}_n - \underline{\mathbf{R}}', \omega) \, \mathrm{d}S' - \underline{\mathbf{u}}(\underline{\mathbf{R}}_0, \omega)\underline{\mathbf{n}}_0 : \lim_{h_n \to 0} \iint_{K_\epsilon} \underline{\underline{\underline{\mathbf{\Sigma}}}}(\underline{\mathbf{R}}_n - \underline{\mathbf{R}}', \omega) \, \mathrm{d}S'. \tag{15.71}$$

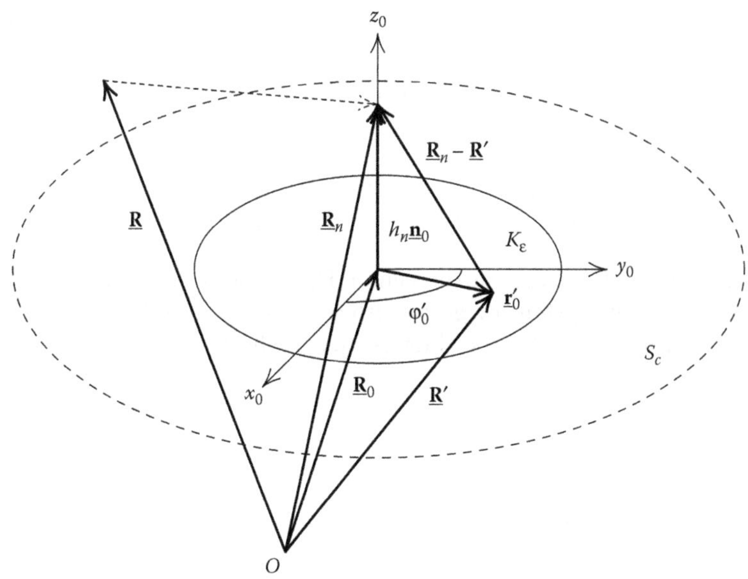

FIGURE 15.7
Calculation of the limit $\underline{\mathbf{R}} \longrightarrow \underline{\mathbf{R}}_0$ in the elastodynamic Helmholtz integral.

Now there is no further way around the integration of $\underline{\underline{\mathbf{G}}}$ and $\underline{\underline{\underline{\mathbf{\Sigma}}}}$; nevertheless, we may exploit that ϵ as well as h_n are small, in particular small with regard to the wavelength. Under this assumption—namely $k_{\mathrm{P,S}}|\underline{\mathbf{R}}_n - \underline{\mathbf{R}}'| \ll 1$—the exponential functions $\mathrm{e}^{\mathrm{j}k_{\mathrm{P,S}}|\underline{\mathbf{R}}_n - \underline{\mathbf{R}}'|}$ in $\underline{\underline{\mathbf{G}}}(\underline{\mathbf{R}}_n - \underline{\mathbf{R}}', \omega)$ and $\underline{\underline{\underline{\mathbf{\Sigma}}}}(\underline{\mathbf{R}}_n - \underline{\mathbf{R}}', \omega)$ may be replaced by the truncated Taylor series[284]

$$\mathrm{e}^{\mathrm{j}k_{\mathrm{P,S}}|\underline{\mathbf{R}}_n - \underline{\mathbf{R}}'|} \simeq 1 + \mathrm{j}k_{\mathrm{P,S}}|\underline{\mathbf{R}}_n - \underline{\mathbf{R}}'| - \frac{1}{2}\,k_{\mathrm{P,S}}^2|\underline{\mathbf{R}}_n - \underline{\mathbf{R}}'|^2. \tag{15.72}$$

With the approximation (15.72), we first investigate the $\underline{\underline{\mathbf{G}}}$-term in (15.71); with the representation (13.95) together with (13.98) and (13.99), we obtain[285]

$$\begin{aligned}\underline{\underline{\mathbf{G}}}(\underline{\mathbf{R}}_n - \underline{\mathbf{R}}', \omega) \simeq\; & \frac{1}{\lambda + 2\mu}\,\frac{1}{8\pi|\underline{\mathbf{R}}_n - \underline{\mathbf{R}}'|}\left[\underline{\underline{\mathbf{I}}} - (\widehat{\underline{\mathbf{R}}_n - \underline{\mathbf{R}}'})(\widehat{\underline{\mathbf{R}}_n - \underline{\mathbf{R}}'})\right] \\ & + \frac{1}{\mu}\,\frac{1}{8\pi|\underline{\mathbf{R}}_n - \underline{\mathbf{R}}'|}\left[\underline{\underline{\mathbf{I}}} + (\widehat{\underline{\mathbf{R}}_n - \underline{\mathbf{R}}'})(\widehat{\underline{\mathbf{R}}_n - \underline{\mathbf{R}}'})\right].\end{aligned} \tag{15.73}$$

---

[284] We must retain the quadratic term to get hold of all singular terms in $\underline{\underline{\mathbf{G}}}$ and $\underline{\underline{\underline{\mathbf{\Sigma}}}}$.

[285] This is not the $\underline{\underline{\mathbf{G}}}$-tensor of elastostatics.

Note the cancelation of the strongest singular $1/|\underline{\mathbf{R}}_n - \underline{\mathbf{R}}'|^3$-terms in $\underline{\underline{\mathbf{G}}}_{\mathrm{P,S}}$ for $k_{\mathrm{P,S}}|\underline{\mathbf{R}}_n - \underline{\mathbf{R}}'| \ll 1$; this is not true for electromagnetic fields yielding a particular problem (Langenberg 2002a). Hence, we are faced with the two integrals

$$\int\!\!\int_{K_\epsilon} \frac{1}{|\underline{\mathbf{R}}_n - \underline{\mathbf{R}}'|}\, \mathrm{d}S'; \quad \int\!\!\int_{K_\epsilon} \frac{(\underline{\mathbf{R}}_n - \underline{\mathbf{R}}')(\underline{\mathbf{R}}_n - \underline{\mathbf{R}}')}{|\underline{\mathbf{R}}_n - \underline{\mathbf{R}}'|^3}\, \mathrm{d}S' \tag{15.74}$$

in the first term of (15.71). In the center $\underline{\mathbf{R}}_0$ of the disk K_ϵ, we establish a cartesian coordinate system x_0, y_0, z_0 with $\underline{\mathbf{e}}_{z_0} = \underline{\mathbf{n}}_0$ (Figure 15.7) and perform the $\mathrm{d}S'$-integration in polar coordinates r_0', φ_0', where

$$\underline{\mathbf{R}}_0 - \underline{\mathbf{R}}' = -r_0' \cos\varphi_0'\, \underline{\mathbf{e}}_{x_0} - r_0' \sin\varphi_0'\, \underline{\mathbf{e}}_{y_0}. \tag{15.75}$$

This specifies the evaluation of the integrations (15.74):

$$\underbrace{\int_0^\epsilon \int_0^{2\pi} \frac{1}{(r_0'^2 + h_n^2)^{1/2}}\, r_0'\mathrm{d}r_0'\mathrm{d}\varphi_0'}_{= 2\pi(\sqrt{h_n^2+\epsilon^2} - h_n)}\ ; \quad \underbrace{\int_0^\epsilon \int_0^{2\pi} \frac{\underline{\underline{\mathbf{\Gamma}}}(r_0', \varphi_0')}{(r_0'^2 + h_n^2)^{3/2}}\, r_0'\mathrm{d}r_0'\mathrm{d}\varphi_0'}_{= \int_0^\epsilon \frac{r_0'}{(r_0'^2+h_n^2)^{3/2}} \int_0^{2\pi} \underline{\underline{\mathbf{\Gamma}}}(r_0', \varphi_0')\, \mathrm{d}\varphi_0'\mathrm{d}r_0'}, \tag{15.76}$$

where

$$\begin{aligned}
\underline{\underline{\mathbf{\Gamma}}}(r_0', \varphi_0') = {} & r_0'^2 \cos^2\varphi_0'\, \underline{\mathbf{e}}_{x_0}\underline{\mathbf{e}}_{x_0} + r_0'^2 \cos\varphi_0' \sin\varphi_0'\, \underline{\mathbf{e}}_{x_0}\underline{\mathbf{e}}_{y_0} - h_n r_0' \cos\varphi_0'\, \underline{\mathbf{e}}_{x_0}\underline{\mathbf{n}}_0 \\
& + r_0'^2 \cos\varphi_0' \sin\varphi_0'\, \underline{\mathbf{e}}_{y_0}\underline{\mathbf{e}}_{x_0} + r_0'^2 \sin^2\varphi_0'\, \underline{\mathbf{e}}_{y_0}\underline{\mathbf{e}}_{y_0} - h_n r_0' \sin\varphi_0'\, \underline{\mathbf{e}}_{y_0}\underline{\mathbf{n}}_0 \\
& - h_n r_0' \cos\varphi_0'\, \underline{\mathbf{n}}_0\underline{\mathbf{e}}_{x_0} - h_n r_0' \sin\varphi_0'\, \underline{\mathbf{n}}_0\underline{\mathbf{e}}_{y_0} + h_n^2\, \underline{\mathbf{n}}_0\underline{\mathbf{n}}_0.
\end{aligned} \tag{15.77}$$

It follows

$$\begin{aligned}
\int_0^\epsilon & \frac{r_0'}{(r_0'^2+h_n^2)^{3/2}} \int_0^{2\pi} \underline{\underline{\mathbf{\Gamma}}}(r_0', \varphi_0')\, \mathrm{d}\varphi_0'\mathrm{d}r_0' \\
& = \pi(\underline{\mathbf{e}}_{x_0}\underline{\mathbf{e}}_{x_0} + \underline{\mathbf{e}}_{y_0}\underline{\mathbf{e}}_{y_0}) \frac{2h_n(h_n - \sqrt{h_n^2+\epsilon^2}) + \epsilon^2}{\sqrt{h_n^2+\epsilon^2}} \\
& \quad - \pi\, \underline{\mathbf{n}}_0\underline{\mathbf{n}}_0 \frac{2h_n(h_n - \sqrt{h_n^2+\epsilon^2})}{\sqrt{h_n^2+\epsilon^2}}.
\end{aligned} \tag{15.78}$$

Obviously, the limit $h_n \longrightarrow 0$ yields a linear ϵ-term in the left as well as in the right integral (15.76); therefore, we have in (15.71)

$$\lim_{\epsilon \to 0} \lim_{h_n \to 0} \int\!\!\int_{K_\epsilon} \underline{\underline{\mathbf{G}}}(\underline{\mathbf{R}}_n - \underline{\mathbf{R}}', \omega)\, \mathrm{d}S' = \underline{\underline{\mathbf{0}}}. \tag{15.79}$$

Consequently, the $\underline{\underline{\mathbf{G}}}$-contribution in $\underline{\underline{\mathbf{H}}}^{\mathrm{u}}_{S_c\setminus\overline{S}_\epsilon}(\underline{\mathbf{R}}_0,\omega)$ reveals itself to be a convergent improper integral in the limit $\epsilon \longrightarrow 0$ (Colton and Kress 1983; Langenberg 2005).

Let us turn to the $\underline{\underline{\underline{\boldsymbol{\Sigma}}}}$-term in (15.71). Since $\underline{\underline{\underline{\boldsymbol{\Sigma}}}}$ contains another gradient as compared to $\underline{\underline{\mathbf{G}}}$, we must expect that the resulting stronger singularity does not lead to a result corresponding to (15.79). We refer to (13.157) and write this $\underline{\underline{\underline{\boldsymbol{\Sigma}}}}$-separation according to

$$\omega^2\varrho\,\underline{\underline{\underline{\boldsymbol{\Sigma}}}}(\underline{\mathbf{R}}-\underline{\mathbf{R}}',\omega) = k_{\mathrm{P}}^2\underline{\underline{\underline{\boldsymbol{\Sigma}}}}_{\mathrm{P}}(\underline{\mathbf{R}}-\underline{\mathbf{R}}',\omega) + k_{\mathrm{S}}^2\underline{\underline{\underline{\boldsymbol{\Sigma}}}}_{\mathrm{S}}(\underline{\mathbf{R}}-\underline{\mathbf{R}}',\omega) \tag{15.80}$$

immediately visualizing with the help of (13.160) and (13.161) the cancelation of the strongest singular $1/|\underline{\mathbf{R}}_n-\underline{\mathbf{R}}'|^4$-terms that result from the 1-term of the expansion (15.72). Yet, even with this finding, there is no way around to write down all single terms of (13.160) and (13.161) associated with the truncated Taylor series (15.72) in order to realize any compensations and to identify the dominant $1/|\underline{\mathbf{R}}_n-\underline{\mathbf{R}}'|^2$-terms

$$\begin{aligned}
&\omega^2\varrho\,\underline{\underline{\underline{\boldsymbol{\Sigma}}}}(\underline{\mathbf{R}}_n-\underline{\mathbf{R}}',\omega)\\
&\quad\simeq -\frac{k_{\mathrm{P}}^2}{4\pi|\underline{\mathbf{R}}_n-\underline{\mathbf{R}}'|^2}\Big\{\lambda\underline{\underline{\mathbf{I}}}(\widehat{\underline{\mathbf{R}}_n-\underline{\mathbf{R}}'}) + \mu(\widehat{\underline{\mathbf{R}}_n-\underline{\mathbf{R}}'})\underline{\underline{\mathbf{I}}} + \mu(\widehat{\underline{\mathbf{R}}_n-\underline{\mathbf{R}}'})\underline{\underline{\mathbf{I}}}^{213}\\
&\qquad\qquad + \mu\Big[\underline{\underline{\mathbf{I}}} - 3(\widehat{\underline{\mathbf{R}}_n-\underline{\mathbf{R}}'})(\widehat{\underline{\mathbf{R}}_n-\underline{\mathbf{R}}'})\Big](\widehat{\underline{\mathbf{R}}_n-\underline{\mathbf{R}}'})\Big\}\\
&\qquad + \frac{k_{\mathrm{S}}^2}{4\pi|\underline{\mathbf{R}}_n-\underline{\mathbf{R}}'|^2}\Big\{\mu\Big[\underline{\underline{\mathbf{I}}} - 3(\widehat{\underline{\mathbf{R}}_n-\underline{\mathbf{R}}'})(\widehat{\underline{\mathbf{R}}_n-\underline{\mathbf{R}}'})\Big](\widehat{\underline{\mathbf{R}}_n-\underline{\mathbf{R}}'})\Big\}
\end{aligned} \tag{15.81}$$

for $k_{\mathrm{P,S}}|\underline{\mathbf{R}}_n-\underline{\mathbf{R}}'| \ll 1$; we may ignore the terms that are constant with regard to $|\underline{\mathbf{R}}_n-\underline{\mathbf{R}}'|$ as well as the $1/|\underline{\mathbf{R}}_n-\underline{\mathbf{R}}'|$-terms, because their K_ϵ-integration has already been investigated and found to be not critical. In (15.81), we combine the P- and S-terms and add the $\underline{\mathbf{n}}_0$-contraction according (15.71):

$$\begin{aligned}
&\underline{\mathbf{n}}_0\cdot\underline{\underline{\underline{\boldsymbol{\Sigma}}}}(\underline{\mathbf{R}}_n-\underline{\mathbf{R}}',\omega)\\
&\quad\simeq \frac{1}{\lambda+2\mu}\,\frac{1}{4\pi|\underline{\mathbf{R}}_n-\underline{\mathbf{R}}'|^2}\Bigg[\mu\,\frac{\underline{\mathbf{n}}_0(\underline{\mathbf{R}}_n-\underline{\mathbf{R}}')}{|\underline{\mathbf{R}}_n-\underline{\mathbf{R}}'|} - \mu\,\frac{\underline{\mathbf{n}}_0\cdot(\underline{\mathbf{R}}_n-\underline{\mathbf{R}}')\underline{\underline{\mathbf{I}}}}{|\underline{\mathbf{R}}_n-\underline{\mathbf{R}}'|}\\
&\qquad - \mu\,\frac{(\underline{\mathbf{R}}_n-\underline{\mathbf{R}}')\underline{\mathbf{n}}_0}{|\underline{\mathbf{R}}_n-\underline{\mathbf{R}}'|} - 3(\lambda+\mu)\,\frac{\underline{\mathbf{n}}_0\cdot(\underline{\mathbf{R}}_n-\underline{\mathbf{R}}')(\underline{\mathbf{R}}_n-\underline{\mathbf{R}}')(\underline{\mathbf{R}}_n-\underline{\mathbf{R}}')}{|\underline{\mathbf{R}}_n-\underline{\mathbf{R}}'|^3}\Bigg].
\end{aligned} \tag{15.82}$$

With (15.69) and (15.75), we immediately realize the cancelation of the first and the third terms in the square brackets of (15.82) after the φ_0'-integration; it remains the calculation of

$$\begin{aligned}
\underline{\mathbf{n}}_0 \cdot \iint_{K_\epsilon} & \underline{\underline{\mathbf{\Sigma}}}(\underline{\mathbf{R}}_n - \underline{\mathbf{R}}', \omega)\, \mathrm{d}S' \\
&\simeq -\underline{\underline{\mathbf{I}}}\, \frac{\mu}{\lambda + 2\mu}\, \frac{h_n}{2} \underbrace{\int_0^\epsilon \frac{r_0'}{(r_0'^2 + h_n^2)^{3/2}}\, \mathrm{d}r_0'}_{= \frac{1}{h_n} - \frac{1}{(h_n^2 + \epsilon^2)^{1/2}}} \\
&\quad - \frac{3}{4} \frac{\lambda + \mu}{\lambda + 2\mu}\, h_n \Bigg[(\underline{\mathbf{e}}_{x0}\underline{\mathbf{e}}_{x0} + \underline{\mathbf{e}}_{y0}\underline{\mathbf{e}}_{y0}) \underbrace{\int_0^\epsilon \frac{r_0'^3}{(r_0'^2 + h_n^2)^{5/2}}\, \mathrm{d}r_0'}_{= -\frac{1}{3} \frac{\epsilon^2}{(h_n^2 + \epsilon^2)^{3/2}} + \frac{2}{3h_n} - \frac{2}{3} \frac{1}{(h_n^2 + \epsilon^2)^{1/2}}} \\
&\qquad + 2h_n^2\, \underline{\mathbf{n}}_0 \underline{\mathbf{n}}_0 \underbrace{\int_0^\epsilon \frac{r_0'}{(r_0'^2 + \epsilon^2)^{5/2}}\, \mathrm{d}r_0'}_{= \frac{1}{3h_n^3} - \frac{1}{3\,(h_n^2 + \epsilon^2)^{3/2}}} \Bigg]. \qquad (15.83)
\end{aligned}$$

In the limit $h_n \longrightarrow 0$, we finally find

$$\lim_{h_n \to 0} \underline{\mathbf{n}}_0 \cdot \iint_{K_\epsilon} \underline{\underline{\mathbf{\Sigma}}}(\underline{\mathbf{R}}_n - \underline{\mathbf{R}}', \omega)\, \mathrm{d}S' = -\frac{1}{2} \underline{\underline{\mathbf{I}}}; \qquad (15.84)$$

in contrast to (15.79), this limit is ϵ-independent unequal to zero resulting in

$$\lim_{\epsilon \to 0} \lim_{h_n \to 0} \underline{\mathbf{H}}^{\mathbf{u}}_{K_\epsilon}(\underline{\mathbf{R}}_n, \omega) = \frac{1}{2}\, \underline{\mathbf{u}}(\underline{\mathbf{R}}_0, \omega), \qquad (15.85)$$

as limit of (15.71) for $\epsilon \longrightarrow 0$.

According to the separation (15.68), we must now have a closer look at the limit

$$\lim_{\epsilon \to 0} \underline{\mathbf{H}}^{\mathbf{u}}_{S_c \setminus \overline{S}_\epsilon}(\underline{\mathbf{R}}_0, \omega); \qquad (15.86)$$

based on the same mathematical considerations as for the scalar acoustic case (Colton and Kress 1983; Langenberg 2005), we find (Chen and Zhou 1992) that this limit exists as a Cauchy principal value for the $\underline{\underline{\mathbf{\Sigma}}}$-contribution, i.e.,

$$\underline{\mathbf{H}}^{\mathbf{u}}_{S_c}(\underline{\mathbf{R}}_0, \omega) \stackrel{\mathrm{def}}{=} \lim_{\epsilon \to 0} \underline{\mathbf{H}}^{\mathbf{u}}_{S_c \setminus \overline{S}_\epsilon}(\underline{\mathbf{R}}_0, \omega) \qquad (15.87)$$

is an improper integral existing only in a special sense: Since $\underline{\mathbf{n}}_0 \cdot \underline{\underline{\mathbf{\Sigma}}}(\underline{\mathbf{R}}_0 - \underline{\mathbf{R}}', \omega)$ according to (15.82) exhibits a $|\underline{\mathbf{R}}_0 - \underline{\mathbf{R}}'|^{-2}$-singularity, we have to agree upon carrying out the φ'-integration before the r_0-integration while calculating $\underline{\mathbf{n}}_0 \cdot \iint_{S_c \setminus \overline{K}_\epsilon} \underline{\underline{\mathbf{\Sigma}}}(\underline{\mathbf{R}}_0 - \underline{\mathbf{R}}', \omega)\, \mathrm{d}S'$. In (15.68), the limit $\underline{\mathbf{H}}^{\mathbf{u}}_{S_c}(\underline{\mathbf{R}} \longrightarrow \underline{\mathbf{R}}_0, \omega)$ is nothing but

$$\begin{aligned}
\underline{\mathbf{H}}^{\mathbf{u}}_{S_c}(\underline{\mathbf{R}} \longrightarrow \underline{\mathbf{R}}_0, \omega) &= \underline{\mathbf{u}}_s(\underline{\mathbf{R}} \longrightarrow \underline{\mathbf{R}}_0, \omega) \\
&\stackrel{\mathrm{def}}{=} \underline{\mathbf{u}}_s(\underline{\mathbf{R}}_0, \omega), \qquad (15.88)
\end{aligned}$$

finally yielding

$$\underline{\mathbf{u}}_s(\underline{\mathbf{R}}, \omega) = \frac{1}{2}\,\underline{\mathbf{u}}(\underline{\mathbf{R}}, \omega) - \mathrm{PV}\int\!\!\int_{S_c} \Big[\underline{\mathbf{n}}_c' \cdot \underline{\underline{\mathbf{T}}}(\underline{\mathbf{R}}', \omega) \cdot \underline{\underline{\mathbf{G}}}(\underline{\mathbf{R}} - \underline{\mathbf{R}}', \omega) + \underline{\mathbf{u}}(\underline{\mathbf{R}}', \omega)\underline{\mathbf{n}}_c' : \underline{\underline{\underline{\boldsymbol{\Sigma}}}}(\underline{\mathbf{R}} - \underline{\mathbf{R}}', \omega)\Big]\,\mathrm{d}S' \tag{15.89}$$

for the "generalization" to the surface point $\underline{\mathbf{R}}_0 \Longrightarrow \underline{\mathbf{R}}$ with $\underline{\mathbf{R}} \in S_c$. Replacing the scattered field by the total field according to

$$\underline{\mathbf{u}}_s(\underline{\mathbf{R}}, \omega) = \underline{\mathbf{u}}(\underline{\mathbf{R}}, \omega) - \underline{\mathbf{u}}_i(\underline{\mathbf{R}}, \omega), \tag{15.90}$$

we obtain the integral relation for $\underline{\mathbf{R}} \in S_c$

$$\underline{\mathbf{u}}(\underline{\mathbf{R}}, \omega) = 2\underline{\mathbf{u}}_i(\underline{\mathbf{R}}, \omega) - 2\mathrm{PV}\int\!\!\int_{S_c} \Big[\underline{\mathbf{n}}_c' \cdot \underline{\underline{\mathbf{T}}}(\underline{\mathbf{R}}', \omega) \cdot \underline{\underline{\mathbf{G}}}(\underline{\mathbf{R}} - \underline{\mathbf{R}}', \omega) + \underline{\mathbf{u}}(\underline{\mathbf{R}}', \omega)\underline{\mathbf{n}}_c' : \underline{\underline{\underline{\boldsymbol{\Sigma}}}}(\underline{\mathbf{R}} - \underline{\mathbf{R}}', \omega)\Big]\,\mathrm{d}S' \tag{15.91}$$

as a composition of the secondary source "surface force density"

$$\underline{\mathbf{t}}_c(\underline{\mathbf{R}}', \omega) = -\underline{\mathbf{n}}_c' \cdot \underline{\underline{\mathbf{T}}}(\underline{\mathbf{R}}', \omega), \quad \underline{\mathbf{R}}' \in S_c, \tag{15.92}$$

with the secondary source "surface deformation"

$$\underline{\underline{\mathbf{g}}}_c^{\underline{\mathbf{u}}}(\underline{\mathbf{R}}', \omega) = -\underline{\underline{\underline{\underline{\mathbf{I}}}}}^+ : \underline{\mathbf{u}}(\underline{\mathbf{R}}', \omega)\underline{\mathbf{n}}_c', \quad \underline{\mathbf{R}}' \in S_c, \tag{15.93}$$

of the particle displacement field. Both sources are not independent upon each other. For example, prescribing a stress-free surface S_c, we must calculate $\underline{\mathbf{u}}$ on S_c and, hence, the secondary source $\underline{\underline{\mathbf{g}}}_c^{\underline{\mathbf{u}}}$ as a solution of an integral equation (three scalar integral equations).

15.2.2 Scatterers with stress-free surfaces: DFIE and Stress Field Integral Equation (SFIE)

Displacement field integral equation: The NDT-relevant stress-free boundary condition of S_c enforces $\underline{\mathbf{t}}_c$ to vanish (15.92) according to (15.92); therefore, the DFIE reads

$$\underline{\mathbf{u}}(\underline{\mathbf{R}}, \omega) = 2\underline{\mathbf{u}}_i(\underline{\mathbf{R}}, \omega) - 2\mathrm{PV}\int\!\!\int_{S_c} \underline{\mathbf{u}}(\underline{\mathbf{R}}', \omega)\underline{\mathbf{n}}_c' : \underline{\underline{\underline{\boldsymbol{\Sigma}}}}(\underline{\mathbf{R}} - \underline{\mathbf{R}}', \omega)\,\mathrm{d}S', \quad \underline{\mathbf{R}} \in S_c. \tag{15.94}$$

It is an integral equation of the second kind—the unknown quantity $\underline{\mathbf{u}}$ appears under the integral and outside—that basically holds for arbitrary scatterer geometries; yet, the fundamental NDT-problem of crack scattering is better formulated with an integral equation of the first kind (SFIE) to be discussed in a minute. Defining an integral operator

$$\begin{aligned}\mathcal{U}\{\underline{\mathbf{u}}\}(\underline{\mathbf{R}},\omega) &= 2\mathrm{PV}\iint_{S_c} \underline{\mathbf{u}}(\underline{\mathbf{R}}',\omega)\underline{\mathbf{n}}_c' : \underline{\underline{\underline{\boldsymbol{\Sigma}}}}(\underline{\mathbf{R}}-\underline{\mathbf{R}}',\omega)\,\mathrm{d}S' \\ &= 2\mathrm{PV}\iint_{S_c} \underline{\mathbf{u}}(\underline{\mathbf{R}}',\omega)\underline{\mathbf{n}}_c' : \underline{\underline{\underline{\underline{\mathbf{c}}}}} : \boldsymbol{\nabla}'\underline{\underline{\mathbf{G}}}(\underline{\mathbf{R}}-\underline{\mathbf{R}}',\omega)\,\mathrm{d}S', \quad \underline{\mathbf{R}}\in S_c,\end{aligned} \tag{15.95}$$

we may formally write (15.94) as

$$(\mathcal{I}+\mathcal{U})\{\underline{\mathbf{u}}\}(\underline{\mathbf{R}},\omega) = 2\underline{\mathbf{u}}_i(\underline{\mathbf{R}},\omega), \quad \underline{\mathbf{R}}\in S_c, \tag{15.96}$$

where $\mathcal{I}\{\underline{\mathbf{u}}\}(\underline{\mathbf{R}},\omega) = \underline{\underline{\mathbf{I}}}\cdot\underline{\mathbf{u}}(\underline{\mathbf{R}},\omega) = \underline{\mathbf{u}}(\underline{\mathbf{R}},\omega)$ denotes the identity operator. Therefore, it is clear how to calculate[286] $\underline{\mathbf{u}}(\underline{\mathbf{R}},\omega)$, $\underline{\mathbf{R}}\in S_c$: We have to invert $\mathcal{I}+\mathcal{U}$! Since $\mathcal{U}$ depends on the scattering geometry, there will be few possibilities to achieve that: As a matter of fact, an analytic inversion[287] only works for the circular cylinder and the sphere (Ying and Truell 1956; Klaholz 1992; Schubert 1999; Section 15.4), constraining us to numerical methods in all other cases. To solve the electromagnetic counterparts EFIE and MFIE—the MFIE corresponds to the DFIE—the method of moments has been proposed (Harrington 1968; Poggio and Miller 1987) and applied up to latest developments (Chew et al. 2002; Michielssen 2002; Wilton 2002). In elastodynamics, one also speaks of the boundary element method (Chen and Zhou 1992).

If the surface S_c consists of a front side S_c^+ and a back side S_c^- with an infinitesimal distance, the mathematical model of a crack emerges, and immediately, the above integral equation becomes useless. The integration over S_c decomposes into integrations over S_c^+ and S_c^- with the corresponding normals $\underline{\mathbf{n}}_c^+$ and $\underline{\mathbf{n}}_c^-$ and the particle displacements $\underline{\mathbf{u}}^+(\underline{\mathbf{R}}'\in S_c^+,\omega)$ and $\underline{\mathbf{u}}^-(\underline{\mathbf{R}}'\in S_c^-,\omega)$; according to the assumption $\underline{\mathbf{n}}_c^- = -\underline{\mathbf{n}}_c^+$, two identical integrals over S_c with different signs emerge, or, respectively, one integral results containing the unknown quantity as the crack opening displacement according to the difference $\underline{\mathbf{u}}^{\mathrm{cod}}(\underline{\mathbf{R}}',\omega) = \underline{\mathbf{u}}^+(\underline{\mathbf{R}}',\omega) - \underline{\mathbf{u}}^-(\underline{\mathbf{R}}',\omega)$. Yet, outside of this integral—it is an integral equation of the second kind!—the particle displacement $\underline{\mathbf{u}}^+(\underline{\mathbf{R}},\omega)$ appears for $\underline{\mathbf{R}}\in S_c^+$ and $\underline{\mathbf{u}}^-(\underline{\mathbf{R}},\omega)$ for $\underline{\mathbf{R}}\in S_c^-$! In acoustics and electromagnetics, we face the same dilemma having led to a way out: Formulate an integral equation of the first kind! Fortunately, we always have

[286] Investigating uniqueness—also in acoustics and electromagnetics—always relates to the nontrivial solutions of the interior (resonance problem regarding S_c) problem (Jones 1984; Colton and Kress 1983; Langenberg 2005). Therefore, all these wave theoretical integral operators exhibit a nonempty null-space for well-defined resonance frequencies, that is to say, for these frequencies, the pertinent homogeneous integral equations have nontrivial solutions.

[287] As a matter of fact, one does not solve the integral equation for eigenfunction expansions of the particle displacement on S_c, but the incident field as well as the scattered field (may be, even only the potentials) are expanded in terms of eigenfunctions resulting in the expansion coefficients via satisfying the boundary or transition conditions.

two related field quantities (after the definition of constitutive equations), here:[288] $\underline{\mathbf{u}}(\underline{\mathbf{R}}, \omega)$ and $\underline{\underline{\mathbf{T}}}(\underline{\mathbf{R}}, \omega)$. Therefore, instead of using (15.65), the integral representation (15.64) for the stress tensor is applied to derive an integral relation between secondary sources, and in that case, the stress-free boundary condition assures a vanishing term relating to this boundary condition under the integral as well as outside the integral: The result is an integral equation of the first kind for the secondary surface deformation. In the following paragraph, this integral equation will be explicitly given as the SFIE, and for reasons becoming obvious, we immediately specialize it to the problem of two-dimensional crack scattering, last but not least because we may rather easily obtain numerical results.

SFIE: two-dimensional crack scattering: Similar to (15.65), we introduce the short-hand notation

$$\underline{\underline{\mathbf{H}}}^{\mathbf{T}}_{S_c}(\underline{\mathbf{R}}, \omega) = -\iint_{S_c} \Big[\underline{\mathbf{n}}'_c \cdot \underline{\underline{\mathbf{T}}}(\underline{\mathbf{R}}', \omega) \cdot \underline{\underline{\underline{\mathbf{\Sigma}}}}^{312}(\underline{\mathbf{R}} - \underline{\mathbf{R}}', \omega) + \underline{\mathbf{u}}(\underline{\mathbf{R}}', \omega)\underline{\mathbf{n}}'_c : \underline{\underline{\underline{\mathbf{\Pi}}}}(\underline{\mathbf{R}} - \underline{\mathbf{R}}', \omega)\Big] \, \mathrm{d}S' \tag{15.97}$$

for the elastodynamic Huygens representation of the scattered stress tensor field; in this integral, we have to investigate the limit $\underline{\mathbf{R}} \longrightarrow \underline{\mathbf{R}}_0$ according to

$$\underline{\underline{\mathbf{H}}}^{\mathbf{T}}_{S_c}(\underline{\mathbf{R}} \longrightarrow \underline{\mathbf{R}}_0) = \lim_{\epsilon \to 0} \underline{\underline{\mathbf{H}}}^{\mathbf{T}}_{S_c \setminus \overline{S}_\epsilon}(\underline{\mathbf{R}}_0, \omega) + \lim_{\epsilon \to 0} \lim_{\underline{\mathbf{R}} \to \underline{\mathbf{R}}_0} \underline{\underline{\mathbf{H}}}^{\mathbf{T}}_{S_\epsilon}(\underline{\mathbf{R}}, \omega). \tag{15.98}$$

With the circular disk approximation of S_ϵ, we are able to calculate the double limit $\underline{\mathbf{R}} \longrightarrow \underline{\mathbf{R}}_0$ and $\epsilon \longrightarrow 0$ in the second term of (15.98) as before, where we advantageously contract with $\underline{\mathbf{n}}_0$, the normal in the surface point $\underline{\mathbf{R}}_0$, because we finally have to satisfy the boundary condition $\underline{\mathbf{n}}_0 \cdot \underline{\underline{\mathbf{T}}}(\underline{\mathbf{R}}_0, \omega) = \underline{\mathbf{0}}$ with (15.98). For the first term in the integrand of (15.97), we find

$$\begin{aligned} -\lim_{\epsilon \to 0} \lim_{\underline{\mathbf{R}} \to \underline{\mathbf{R}}_0} \underline{\mathbf{n}}_0 \cdot \iint_{K_\epsilon} \underline{\mathbf{n}}'_c \cdot \underline{\underline{\mathbf{T}}}(\underline{\mathbf{R}}', \omega) \cdot \underline{\underline{\underline{\mathbf{\Sigma}}}}^{312}(\underline{\mathbf{R}} - \underline{\mathbf{R}}', \omega) \, \mathrm{d}S' &= \frac{1}{2} \underline{\underline{\mathbf{I}}} \cdot \underline{\underline{\mathbf{T}}}(\underline{\mathbf{R}}_0, \omega) \cdot \underline{\mathbf{n}}_0 \\ &= \frac{1}{2} \underline{\underline{\mathbf{T}}}(\underline{\mathbf{R}}_0, \omega) \cdot \underline{\mathbf{n}}_0 \end{aligned} \tag{15.99}$$

in complete analogy to (15.84). Yet, with the second term in (15.97), we face problems because the del-operations in $\underline{\underline{\underline{\mathbf{\Pi}}}}(\underline{\mathbf{R}} - \underline{\mathbf{R}}', \omega)$ (one more than in $\underline{\underline{\underline{\mathbf{\Sigma}}}}$) lead to a singularity of the K_ϵ-integral in the limit $\epsilon \longrightarrow 0$. A similar problem for the electromagnetic case points to a solution (Langenberg 2005;

[288] In electromagnetics, one replaces the electric field strength for the case of perfectly conducting scatterers by the magnetic field strength, and instead of the MFIE as integral equation of the second kind, the EFIE as integral equation of the first kind results.

Morita et al. 1990): Sloppily spoken, the singularity emerges from the "upper boundary" of the K_ϵ-integral, namely from the boundary $r'_0 = \epsilon$ of the disk; yet, an exactly similar singularity is found in the limit

$$\underline{\underline{\mathbf{H}}}^{\mathbf{T}}_{S_c}(\underline{\mathbf{R}}_0, \omega) = \lim_{\epsilon \to 0} \underline{\underline{\mathbf{H}}}^{\mathbf{T}}_{S_c \setminus \overline{S}_\epsilon}(\underline{\mathbf{R}}_0, \omega) \tag{15.100}$$

at the lower boundary of the $S_c \setminus \overline{K}_\epsilon$-integral, i.e. the $\underline{\underline{\mathbf{H}}}^{\mathbf{T}}_{S_c}(\underline{\mathbf{R}}_0, \omega)$-integral is a PV-convergent improper integral. In such cases, help is found defining a special definition of the limit (15.87) that goes beyond the definition of a Cauchy PV—the explicit value of the singular term known from the calculation of the K_ϵ-integral is subtracted under the limes—and the result is called the ϵ-principal value (PV_ϵ) of the integral. We do not want to evaluate this further because our subsequent example of two-dimensional crack scattering allows for a circumvention via the hyper singularity of the integrand, thus avoiding the PV_ϵ-computation. Nevertheless, we write down the integral equation relation as it emerges from (15.97) through (15.99):

$$\begin{aligned} &\underline{\mathbf{n}}_c \cdot \underline{\underline{\mathbf{T}}}(\underline{\mathbf{R}}, \omega) \\ &\quad = 2\underline{\mathbf{n}}_c \cdot \underline{\underline{\mathbf{T}}}_i(\underline{\mathbf{R}}, \omega) - 2\,\mathrm{PV}_\epsilon\, \underline{\mathbf{n}}_c \cdot \iint_{S_c} \Big[\underline{\mathbf{n}}'_c \cdot \underline{\underline{\mathbf{T}}}(\underline{\mathbf{R}}', \omega) \cdot \underline{\underline{\boldsymbol{\Sigma}}}^{312}(\underline{\mathbf{R}} - \underline{\mathbf{R}}', \omega) \\ &\qquad\qquad + \underline{\mathbf{u}}(\underline{\mathbf{R}}', \omega)\underline{\mathbf{n}}'_c : \underline{\underline{\underline{\boldsymbol{\Pi}}}}(\underline{\mathbf{R}} - \underline{\mathbf{R}}', \omega)\Big]\,\mathrm{d}S' \end{aligned} \tag{15.101}$$

This is the counterpart to (15.91).

Equation 15.101 turns into the SFIE, an integral equation of the first kind for the secondary surface deformation, if the stress-free boundary condition of S_c is inserted:

$$\underline{\mathbf{n}}_c \cdot \underline{\underline{\mathbf{T}}}_i(\underline{\mathbf{R}}, \omega) = \mathrm{PV}_\epsilon\, \underline{\mathbf{n}}_c \cdot \iint_{S_c} \underline{\mathbf{u}}(\underline{\mathbf{R}}', \omega)\underline{\mathbf{n}}'_c : \underline{\underline{\underline{\boldsymbol{\Pi}}}}(\underline{\mathbf{R}} - \underline{\mathbf{R}}', \omega)\,\mathrm{d}S'. \tag{15.102}$$

Defining an integral operator

$$\mathcal{T}\{\underline{\mathbf{u}}\}(\underline{\mathbf{R}}, \omega) = 2\,\mathrm{PV}_\epsilon\, \underline{\mathbf{n}}_c \cdot \iint_{S_c} \underline{\mathbf{u}}(\underline{\mathbf{R}}', \omega)\underline{\mathbf{n}}'_c : \underline{\underline{\underline{\boldsymbol{\Pi}}}}(\underline{\mathbf{R}} - \underline{\mathbf{R}}', \omega)\,\mathrm{d}S' \tag{15.103}$$

similar to (15.95) yields the short-hand notation

$$\mathcal{T}\{\underline{\mathbf{u}}\}(\underline{\mathbf{R}}, \omega) = 2\,\underline{\mathbf{n}}_c \cdot \underline{\underline{\mathbf{T}}}(\underline{\mathbf{R}}, \omega). \tag{15.104}$$

As a hyper singular integral equation, the SFIE is not really suited for numerical evaluations; this corresponds to the EFIE in electromagnetics if the latter is written down in Franz's version (Langenberg 2002a); a way out either uses initially or after partial integration the Stratton–Chu version. Therefore, in the present case, we will also try to shift one of the del-operations in $\underline{\underline{\underline{\boldsymbol{\Pi}}}}$ to the unknown $\underline{\mathbf{u}}$ via partial integration. Below, following Tan (1977), this idea will be elaborated for two-dimensional crack scattering.

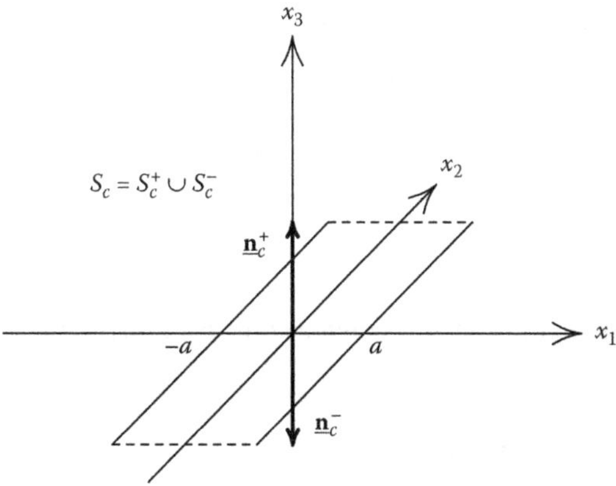

FIGURE 15.8
Strip-like two-dimensional crack.

Figure 15.8 displays the geometry of the (mathematically idealized) crack of width $2a$, i.e., the two-dimensional strip-like scatterer. The evaluation of various contractions in the SFIE suggests to number cartesian xyz-coordinates as x_j, $j = 1, 2, 3$. Due to the two-dimensionality of the problem, we have $\partial/\partial y = \partial/\partial x_2 \equiv 0$. We concentrate upon P-SV-scattering postulating $T_{ix_2x_j} \equiv 0$, $j = 1, 2, 3$ (Sections 9.1.1 and 9.1.2); since S_c^+ satisfies $\underline{\mathbf{n}}_c \stackrel{\text{def}}{=} \underline{\mathbf{n}}_c^+ = \underline{\mathbf{e}}_{x_3}$, only $T_{ix_2x_3} \equiv 0$ is relevant out of the three vanishing stress tensor components according to (15.102) allowing to restrict ourselves to the x_1- and the x_3-components of the SFIE. With $\underline{\mathbf{R}} = x_1\underline{\mathbf{e}}_{x_1} + x_3\underline{\mathbf{e}}_{x_3}$ and $\underline{\mathbf{R}}' = x_1'\underline{\mathbf{e}}_{x_1}$ and the definition of

$$\underline{\mathbf{u}}^{\text{cod}}(x_1', \omega) = \underline{\mathbf{u}}(x_1', x_3' = +0, \omega) - \underline{\mathbf{u}}(x_1', x_3' = -0, \omega) \tag{15.105}$$

as crack opening displacement, we first calculate the right-hand side of (15.102) applying (13.197) for $x_3 > 0$ thus ignoring the PV:[289]

$$\begin{aligned}
&\underline{\mathbf{n}}_c \cdot \iint_{S_c} \underline{\mathbf{u}}(\underline{\mathbf{R}}', \omega)\underline{\mathbf{n}}_c' : \underline{\underline{\underline{\mathbf{\Pi}}}}(\underline{\mathbf{R}} - \underline{\mathbf{R}}', \omega)\, \mathrm{d}S' \\
&\quad = \underline{\mathbf{e}}_{x_3} \cdot \int_{-a}^{a} \underline{\mathbf{u}}^{\text{cod}}(x_1', \omega)\underline{\mathbf{e}}_{x_3} : \underline{\underline{\underline{\mathbf{\Pi}}}}(x_1 - x_1', x_3, \omega)\, \mathrm{d}x_1' \\
&\quad = \underline{\mathbf{e}}_{x_3} \cdot \int_{-a}^{a} \underline{\mathbf{u}}^{\text{cod}}(x_1', \omega)\underline{\mathbf{e}}_{x_3} : \Big\{\lambda \boldsymbol{\nabla} \cdot \underline{\underline{\underline{\mathbf{\Sigma}}}}^{312}(x_1 - x_1', x_3, \omega)\underline{\underline{\mathbf{I}}} \\
&\qquad + \mu \left[\boldsymbol{\nabla}\underline{\underline{\underline{\mathbf{\Sigma}}}}^{2314}(x_1 - x_1', x_3, \omega) + \boldsymbol{\nabla}\underline{\underline{\underline{\mathbf{\Sigma}}}}^{2341}(x_1 - x_1', x_3, \omega)\right]\Big\}\mathrm{d}x_1'.
\end{aligned} \tag{15.106}$$

[289]On the left-hand side of the first line of (15.106), the three-dimensional scalar Green function appears in $\underline{\underline{\underline{\mathbf{\Pi}}}}$, yet on the right-hand side, we find the two-dimensional Green function due to the independence of the secondary source upon x_2' (compare Section 13.3.3).

For the x_1-components of the three terms in (15.106), we find

$$\underline{\mathbf{e}}_{x_3} \cdot \underline{\mathbf{u}}^{\text{cod}} \underline{\mathbf{e}}_{x_3} : \boldsymbol{\nabla} \cdot \underline{\underline{\underline{\boldsymbol{\Sigma}}}}^{312} \underline{\underline{\mathbf{I}}} \cdot \underline{\mathbf{e}}_{x_1} = 0, \tag{15.107}$$

$$\underline{\mathbf{e}}_{x_3} \cdot \underline{\mathbf{u}}^{\text{cod}} \underline{\mathbf{e}}_{x_3} : \boldsymbol{\nabla} \underline{\underline{\underline{\boldsymbol{\Sigma}}}}^{2314} \cdot \underline{\mathbf{e}}_{x_1} = u_{x_j}^{\text{cod}} \frac{\partial}{\partial x_3} \Sigma_{x_3 x_j x_1}, \tag{15.108}$$

$$\underline{\mathbf{e}}_{x_3} \cdot \underline{\mathbf{u}}^{\text{cod}} \underline{\mathbf{e}}_{x_3} : \boldsymbol{\nabla} \underline{\underline{\underline{\boldsymbol{\Sigma}}}}^{2341} \cdot \underline{\mathbf{e}}_{x_1} = u_{x_j}^{\text{cod}} \frac{\partial}{\partial x_1} \Sigma_{x_3 x_j x_3}; \tag{15.109}$$

we used Einstein's summation convention. Similarly, we obtain the x_3-components:

$$\underline{\mathbf{e}}_{x_3} \cdot \underline{\mathbf{u}}^{\text{cod}} \underline{\mathbf{e}}_{x_3} : \boldsymbol{\nabla} \cdot \underline{\underline{\underline{\boldsymbol{\Sigma}}}}^{312} \underline{\underline{\mathbf{I}}} \cdot \underline{\mathbf{e}}_{x_3} = u_{x_j}^{\text{cod}} \frac{\partial}{\partial x_k} \Sigma_{x_3 x_j x_k}, \tag{15.110}$$

$$\underline{\mathbf{e}}_{x_3} \cdot \underline{\mathbf{u}}^{\text{cod}} \underline{\mathbf{e}}_{x_3} : \boldsymbol{\nabla} \underline{\underline{\underline{\boldsymbol{\Sigma}}}}^{2314} \cdot \underline{\mathbf{e}}_{x_3} = u_{x_j}^{\text{cod}} \frac{\partial}{\partial x_3} \Sigma_{x_3 x_j x_3}, \tag{15.111}$$

$$\underline{\mathbf{e}}_{x_3} \cdot \underline{\mathbf{u}}^{\text{cod}} \underline{\mathbf{e}}_{x_3} : \boldsymbol{\nabla} \underline{\underline{\underline{\boldsymbol{\Sigma}}}}^{2341} \cdot \underline{\mathbf{e}}_{x_3} = u_{x_j}^{\text{cod}} \frac{\partial}{\partial x_3} \Sigma_{x_3 x_j x_3}. \tag{15.112}$$

Above all, our goal must be to transfer the derivatives of the $\underline{\underline{\underline{\boldsymbol{\Sigma}}}}$-components to derivatives of the $\underline{\mathbf{u}}^{\text{cod}}$-components via partial integration; the integration is with regard to x_1', therefore, the x_1- and essentially the x_3-derivatives must be transferred to x_1'-derivatives. The transfer of x_3- to x_1-derivatives is performed with the help of the differential equation (Equation 13.210)

$$\boldsymbol{\nabla} \cdot \underline{\underline{\underline{\boldsymbol{\Sigma}}}}(\underline{\mathbf{R}} - \underline{\mathbf{R}}', \omega) + \rho\omega^2 \underline{\underline{\mathbf{G}}}(\underline{\mathbf{R}} - \underline{\mathbf{R}}', \omega) = \underline{\underline{\mathbf{0}}}, \tag{15.113}$$

whose components—$j = 1, 2, 3$, $k = 1, 2, 3$—in the present case for $x_3 > 0$ explicitly read

$$\begin{aligned} &\frac{\partial}{\partial x_1} \Sigma_{x_1 x_j x_k}(x_1 - x_1', x_3, \omega) \\ &\quad + \frac{\partial}{\partial x_3} \Sigma_{x_3 x_j x_k}(x_1 - x_1', x_3, \omega) + \rho\omega^2 \, G_{x_j x_k}(x_1 - x_1', x_3, \omega) = 0. \end{aligned} \tag{15.114}$$

Now the sole x_1-derivatives may be transformed into x_1'-derivatives using $\partial/\partial x_1 = -\partial/\partial x_1'$ yielding the expressions

$$\underline{\mathbf{e}}_{x_3} \cdot \underline{\mathbf{u}}^{\text{cod}} \underline{\mathbf{e}}_{x_3} : \boldsymbol{\nabla} \underline{\underline{\underline{\boldsymbol{\Sigma}}}}^{2314} \cdot \underline{\mathbf{e}}_{x_1} = -u_{x_j}^{\text{cod}} \left(-\frac{\partial}{\partial x_1'} \Sigma_{x_1 x_j x_1} + \rho\omega^2 \, G_{x_j x_1} \right), \tag{15.115}$$

$$\underline{\mathbf{e}}_{x_3} \cdot \underline{\mathbf{u}}^{\text{cod}} \underline{\mathbf{e}}_{x_3} : \boldsymbol{\nabla} \underline{\underline{\underline{\boldsymbol{\Sigma}}}}^{2341} \cdot \underline{\mathbf{e}}_{x_1} = -u_{x_j}^{\text{cod}} \frac{\partial}{\partial x_1'} \Sigma_{x_3 x_j x_3}, \tag{15.116}$$

$$\begin{aligned} &\underline{\mathbf{e}}_{x_3} \cdot \underline{\mathbf{u}}^{\text{cod}} \underline{\mathbf{e}}_{x_3} : \boldsymbol{\nabla} \cdot \underline{\underline{\underline{\boldsymbol{\Sigma}}}}^{312} \underline{\underline{\mathbf{I}}} \cdot \underline{\mathbf{e}}_{x_3} \\ &\quad = -u_{x_j}^{\text{cod}} \left(\frac{\partial}{\partial x_1'} \Sigma_{x_3 x_j x_1} - \frac{\partial}{\partial x_1'} \Sigma_{x_1 x_j x_3} + \rho\omega^2 G_{x_j x_3} \right), \end{aligned} \tag{15.117}$$

$$\underline{\mathbf{e}}_{x_3} \cdot \underline{\mathbf{u}}^{\text{cod}} \underline{\mathbf{e}}_{x_3} : \nabla \underline{\underline{\underline{\mathbf{\Sigma}}}}^{2314} \cdot \underline{\mathbf{e}}_{x_3} = -u_{x_j}^{\text{cod}} \left(-\frac{\partial}{\partial x_1'} \Sigma_{x_1 x_j x_3} + \rho\omega^2 G_{x_j x_3} \right), \tag{15.118}$$

$$\underline{\mathbf{e}}_{x_3} \cdot \underline{\mathbf{u}}^{\text{cod}} \underline{\mathbf{e}}_{x_3} : \nabla \underline{\underline{\underline{\mathbf{\Sigma}}}}^{2341} \cdot \underline{\mathbf{e}}_{x_3} = -u_{x_j}^{\text{cod}} \left(-\frac{\partial}{\partial x_1'} \Sigma_{x_1 x_j x_3} + \rho\omega^2 G_{x_j x_3} \right) \tag{15.119}$$

instead of (15.108) through (15.112) to be inserted into (15.106):

$$\begin{aligned}
&\underline{\mathbf{e}}_{x_3} \cdot \int_{-a}^{a} \underline{\mathbf{u}}^{\text{cod}}(x_1', \omega) \underline{\mathbf{e}}_{x_3} : \underline{\underline{\underline{\mathbf{\Pi}}}}(x_1 - x_1', x_3, \omega)\, \mathrm{d}x_1 \cdot \underline{\mathbf{e}}_{x_1} \\
&\quad = -\mu\rho\omega^2 \int_{-a}^{a} u_{x_j}^{\text{cod}}(x_1', \omega) G_{x_j x_1}(x_1 - x_1', x_3, \omega)\, \mathrm{d}x_1' \\
&\qquad + \mu \int_{-a}^{a} u_{x_j}^{\text{cod}}(x_1', \omega) \frac{\partial}{\partial x_1'} \big[\Sigma_{x_1 x_j x_1}(x_1 - x_1', x_3, \omega) \\
&\qquad\qquad - \Sigma_{x_3 x_j x_3}(x_1 - x_1', x_3, \omega) \big] \mathrm{d}x_1',
\end{aligned} \tag{15.120}$$

$$\begin{aligned}
&\underline{\mathbf{e}}_{x_3} \cdot \int_{-a}^{a} \underline{\mathbf{u}}^{\text{cod}}(x_1', \omega) \underline{\mathbf{e}}_{x_3} : \underline{\underline{\underline{\mathbf{\Pi}}}}(x_1 - x_1', x_3, \omega)\, \mathrm{d}x_1 \cdot \underline{\mathbf{e}}_{x_3} \\
&\quad = -(\lambda + 2\mu)\rho\omega^2 \int_{-a}^{a} u_{x_j}^{\text{cod}}(x_1', \omega) G_{x_j x_3}(x_1 - x_1', x_3, \omega)\, \mathrm{d}x_1' \\
&\qquad + \int_{-a}^{a} u_{x_j}^{\text{cod}}(x_1', \omega) \frac{\partial}{\partial x_1'} \big[(\lambda + 2\mu) \Sigma_{x_1 x_j x_3}(x_1 - x_1', x_3, \omega) \\
&\qquad\qquad - \lambda \Sigma_{x_3 x_j x_1}(x_1 - x_1', x_3, \omega) \big] \mathrm{d}x_1'.
\end{aligned} \tag{15.121}$$

Under the self-evident side condition $u_{x_j}^{\text{cod}}(-a, \omega) = u_{x_j}^{\text{cod}}(a, \omega) = 0$, the x_1'-derivatives of the sigma-tensor components are shifted to the components of the crack opening displacement via partial integration resulting in a negative sign; afterward, the limit $x_3 \longrightarrow 0$ must be performed to obtain integral equations for the components of $\underline{\mathbf{u}}^{\text{cod}}(x_1', \omega)$. Due to this limit, the j-contraction in (15.120) is simultaneously constrained to $j = 1$, and the one in (15.121) to $j = 3$ because an odd number of x_3-derivatives in the sigma- and G-tensor components always produces a factor $x_3 - x_3'$ that vanishes for $x_3' = x_3 = 0$. In addition, we face the question whether the singularity of the sigma-tensor components is bothering; yet the—eventually—critical term (15.81) shows an asymmetry with regard to $x_1' = 0$ for $x_1 = 0$—it is $\underline{\mathbf{R}}_n = h_n \underline{\mathbf{e}}_{x_3}$—and $h_n \longrightarrow 0$ according to $x_3 \longrightarrow 0$—there is always an odd number of $(\underline{\mathbf{R}}_n - \underline{\mathbf{R}}')$-terms—yielding a vanishing $(-a, a)$-integral. Finally, we find the integral equations of the first kind—we return to the xyz-notation of cartesian coordinates—

$$\begin{aligned}
T_{izx}(x, 0, \omega) &= -\mu\rho\omega^2 \int_{-a}^{a} u_x^{\text{cod}}(x', \omega) G_{xx}(x - x', 0, \omega)\, \mathrm{d}x' \\
&\quad + \mu \int_{-a}^{a} \frac{\partial}{\partial x'} u_x^{\text{cod}}(x', \omega) \left[\Sigma_{zxz}(x - x', 0, \omega) - \Sigma_{xxx}(x - x', 0, \omega) \right] \mathrm{d}x',
\end{aligned} \tag{15.122}$$

$$T_{izz}(x,0,\omega) = -(\lambda+2\mu)\rho\omega^2 \int_{-a}^{a} u_z^{\mathrm{cod}}(x',\omega) G_{zz}(x-x',0,\omega)\,\mathrm{d}x'$$
$$+ \int_{-a}^{a} \frac{\partial}{\partial x'} u_z^{\mathrm{cod}}(x',\omega)\,[\lambda\Sigma_{zzx}(x-x',0,\omega) - (\lambda+2\mu)\Sigma_{zxz}(x-x',0,\omega)]\,\mathrm{d}x' \tag{15.123}$$

for both components of the crack opening displacement as they have been calculated[290] by Tan (1977); in both the equations, x is constrained to the interval $-a$ to a. Due to the two-dimensionality of the problem, the representation (13.248) has to be used to calculate $\underline{\underline{\mathbf{G}}}$ and $\underline{\underline{\underline{\mathbf{\Sigma}}}}$.

The SFIEs (15.122) and (15.123) are decoupled with regard to the components of the crack opening displacement, yet they contain the components themselves as well as their derivatives. But the numerical method to be applied solves this difficulty: The unknown components of the crack opening displacement are expanded into the so-called base functions—Tan chooses a trigonometric approximation—whose derivative can be easily calculated; afterward, the boundary condition is satisfied at discrete points x. This is one of the potential proposals of the method of moments (Harrington 1968; Poggio and Miller 1987; Wilton 2002); we have applied an EFIE-code developed by Wilton and collaborators (Langenberg et al. 1993a) to obtain the numerical results as discussed in Section 15.4.3, because the SFIE (15.122) and (15.123) exhibit the same mathematical structure as compared to their electromagnetic EFIE-counterparts in the Stratton–Chu version.

15.2.3 Kirchhoff approximation in elastodynamics

The formulation of Huygens' principle for scalar—optical—waves in terms of the Helmholtz integral as a solution of a homogeneous reduced wave equation provided a physical—in contrast to a geometrical—optics in the middle of the nineteenth century. Yet, possible solution methods were scarce; hence, approximate methods were investigated. The best known and most successful is the Kirchhoff approximation of physical optics giving a second name to the method, namely PO.[291] It may either be heuristically established or as a special solution of an integral equation of the second kind. For the examples of scalar acoustics and (two-dimensional) scalar SH-wave scattering, we will follow the second path just to introduce the Kirchhoff approximation of general elastodynamics heuristically based on this knowledge (the explicit mathematical derivation is rather confusing).

Kirchhoff approximation in acoustics: We investigate the DFIE (15.96) in the limit $\mu \longrightarrow 0$; with (13.154), it follows

[290] Note: For the $\tau^{\Gamma}_{\alpha;\beta;\gamma}$-tensor as defined by Tan, we have $\tau^{\Gamma}_{\alpha;\beta;\gamma} = -\Sigma_{\alpha\beta\gamma}$.

[291] With certain generalizations and augmented by other approximations, the Kirchhoff-PO is still the basis for the calculation of electromagnetic wave scattering by complex geometries (Greving 2000); it also plays a central role to model US-NDT problems (Spies 2000b; Boehm et al. 2002; Kühnicke 2001; Civa: www.civa.cea.fr; Schmitz et al. 2004b).

$$\underline{\underline{\underline{\Sigma}}}^{\mu\to 0}(\underline{\mathbf{R}}-\underline{\mathbf{R}}',\omega) = \lambda\,\underline{\underline{\mathbf{I}}}\nabla\cdot\underline{\underline{\mathbf{G}}}^{\mu\to 0}(\underline{\mathbf{R}}-\underline{\mathbf{R}}',\omega), \tag{15.124}$$

and the differential equation (13.80) for $\underline{\underline{\mathbf{G}}}^{\mu\to 0}$ reduces to

$$\nabla\nabla\cdot\underline{\underline{\mathbf{G}}}^{\mu\to 0}(\underline{\mathbf{R}}-\underline{\mathbf{R}}',\omega) + \frac{\rho\omega^2}{\lambda}\,\underline{\underline{\mathbf{G}}}^{\mu\to 0}(\underline{\mathbf{R}}-\underline{\mathbf{R}}',\omega) = -\frac{1}{\lambda}\,\delta(\underline{\mathbf{R}}-\underline{\mathbf{R}}')\underline{\underline{\mathbf{I}}}. \tag{15.125}$$

For $\underline{\mathbf{R}} \neq \underline{\mathbf{R}}'$, we obtain

$$\underline{\underline{\mathbf{G}}}^{\mu\to 0}(\underline{\mathbf{R}}-\underline{\mathbf{R}}',\omega) = -\frac{1}{\lambda}\frac{1}{k^2}\nabla\nabla\frac{\mathrm{e}^{\mathrm{j}k|\underline{\mathbf{R}}-\underline{\mathbf{R}}'|}}{4\pi|\underline{\mathbf{R}}-\underline{\mathbf{R}}'|} \tag{15.126}$$

as solution of (15.125) (Equation 5.69), where $k = \omega\sqrt{\rho/\lambda}$. Similarly, we have for $\underline{\mathbf{R}} \neq \underline{\mathbf{R}}'$

$$\begin{aligned}\nabla\cdot\underline{\underline{\mathbf{G}}}^{\mu\to 0}(\underline{\mathbf{R}}-\underline{\mathbf{R}}',\omega) &= \frac{1}{\lambda}\nabla\frac{\mathrm{e}^{\mathrm{j}k|\underline{\mathbf{R}}-\underline{\mathbf{R}}'|}}{4\pi|\underline{\mathbf{R}}-\underline{\mathbf{R}}'|}\\ &= \frac{1}{\lambda}\nabla G(\underline{\mathbf{R}}-\underline{\mathbf{R}}',\omega)\end{aligned} \tag{15.127}$$

under consideration of (5.58). Therefore, for $\underline{\mathbf{R}} \in \mathbb{R}^3\backslash\overline{V}_c$—$\underline{\mathbf{R}} \notin S_c$—we may rewrite the scattering integral based on (15.95):

$$\begin{aligned}&\iint_{S_c}\underline{\mathbf{u}}(\underline{\mathbf{R}}',\omega)\underline{\mathbf{n}}_c' : \underline{\underline{\underline{\Sigma}}}^{\mu\to 0}(\underline{\mathbf{R}}-\underline{\mathbf{R}}',\omega)\,\mathrm{d}S'\\ &\quad= \iint_{S_c}\underline{\mathbf{u}}(\underline{\mathbf{R}}',\omega)\cdot\underline{\mathbf{n}}_c'\nabla G(\underline{\mathbf{R}}-\underline{\mathbf{R}}',\omega)\,\mathrm{d}S'.\end{aligned} \tag{15.128}$$

For the limit $\underline{\mathbf{R}} \longrightarrow S_c$ we finally find (Colton and Kress 1983; Langenberg 2005):

$$\begin{aligned}&\lim_{\underline{\mathbf{R}}\to S_c}\underline{\mathbf{n}}_c\cdot\iint_{S_c}\underline{\mathbf{u}}(\underline{\mathbf{R}}',\omega)\cdot\underline{\mathbf{n}}_c'\nabla G(\underline{\mathbf{R}}-\underline{\mathbf{R}}',\omega)\,\mathrm{d}S'\\ &\quad= \frac{1}{2}\underline{\mathbf{u}}(\underline{\mathbf{R}},\omega)\cdot\underline{\mathbf{n}}_c + \iint_{S_c}\underline{\mathbf{u}}(\underline{\mathbf{R}}',\omega)\cdot\underline{\mathbf{n}}_c'\frac{\partial}{\partial n_c}G(\underline{\mathbf{R}}-\underline{\mathbf{R}}',\omega)\,\mathrm{d}S';\end{aligned} \tag{15.129}$$

hence, we obtain the integral equation of the second kind

$$(\mathcal{I}+\mathcal{K}')\{\underline{\mathbf{u}}\cdot\underline{\mathbf{n}}_c\}(\underline{\mathbf{R}},\omega) = 2\underline{\mathbf{u}}_i(\underline{\mathbf{R}},\omega)\cdot\underline{\mathbf{n}}_c, \quad \underline{\mathbf{R}} \in S_c, \tag{15.130}$$

with the operator

$$\mathcal{K}'\{\underline{\mathbf{u}}\cdot\underline{\mathbf{n}}_c\}(\underline{\mathbf{R}},\omega) = 2\iint_{S_c}\underline{\mathbf{u}}(\underline{\mathbf{R}}',\omega)\cdot\underline{\mathbf{n}}_c'\frac{\partial}{\partial n_c}G(\underline{\mathbf{R}}-\underline{\mathbf{R}}',\omega)\,\mathrm{d}S' \tag{15.131}$$

for the normal components of the particle displacement. On behalf of (5.74), this is an integral equation for the normal derivative of the field quantity "pressure" of an acoustic Dirichlet problem (soft boundary); its solution for a special case of S_c, namely an infinite planar (boundary) surface S_{xy} (of an acoustic half-space) may immediately be written down:

$$\underline{\mathbf{u}}(x,y,\omega)\cdot\underline{\mathbf{n}}_c = 2\underline{\mathbf{u}}_i(x,y,z=0,\omega)\cdot\underline{\mathbf{n}}_c, \quad x,y\in S_{xy}, \tag{15.132}$$

because due to

$$\frac{\partial}{\partial n_c}G(\underline{\mathbf{R}}-\underline{\mathbf{R}}',\omega) = \underbrace{\underline{\mathbf{n}}_c\cdot\frac{\underline{\mathbf{R}}-\underline{\mathbf{R}}'}{|\underline{\mathbf{R}}-\underline{\mathbf{R}}'|}}_{=0 \text{ for } \underline{\mathbf{R}},\underline{\mathbf{R}}'\in S_{xy}} \frac{\mathrm{e}^{\mathrm{j}k|\underline{\mathbf{R}}-\underline{\mathbf{R}}'|}}{4\pi|\underline{\mathbf{R}}-\underline{\mathbf{R}}'|}\left(\mathrm{j}k-\frac{1}{|\underline{\mathbf{R}}-\underline{\mathbf{R}}'|}\right), \tag{15.133}$$

we have

$$\mathcal{K}'\{\underline{\mathbf{u}}\cdot\underline{\mathbf{n}}_c\}(x,y,\omega) = 0, \quad x,y\in S_{xy}. \tag{15.134}$$

For an infinite planar "scattering" surface, the radiation interaction integral (15.134) is equal to zero, and hence, the secondary source is equal to twice the incident field: The surface is a perfect mirror!

Now Kirchhoff's approximation argues as follows: With the direction of the incident field, for example, the phase unit vector $\hat{\underline{\mathbf{k}}}_i$ of a plane wave, and the normal $\underline{\mathbf{n}}_c$ we define an illuminated side for convex scatterers by $\hat{\underline{\mathbf{k}}}_i\cdot\underline{\mathbf{n}}_c<0$ and a shadow side by $\hat{\underline{\mathbf{k}}}_i\cdot\underline{\mathbf{n}}_c>0$ (illustrated two-dimensionally in Figure 15.9). Then, the secondary source on the shadow side is set equal to zero, and on the illuminated side—following (15.132)—equal to twice the incident field, hence

$$\underline{\mathbf{u}}^{\mathrm{PO}}(\underline{\mathbf{R}},\omega)\cdot\underline{\mathbf{n}}_c = 2\underline{\mathbf{u}}_i(\underline{\mathbf{R}},\omega)\cdot\underline{\mathbf{n}}_c\,u(-\hat{\underline{\mathbf{k}}}_i\cdot\underline{\mathbf{n}}_c), \quad \underline{\mathbf{R}}\in S_c, \tag{15.135}$$

where the unit-step function $u(-\hat{\underline{\mathbf{k}}}_i\cdot\underline{\mathbf{n}}_c)$ accounts for the shadow side. Under the assumption of a smoothly curved surface S_c as compared to the wavelength—locally planar surface—there is some hope that this is an

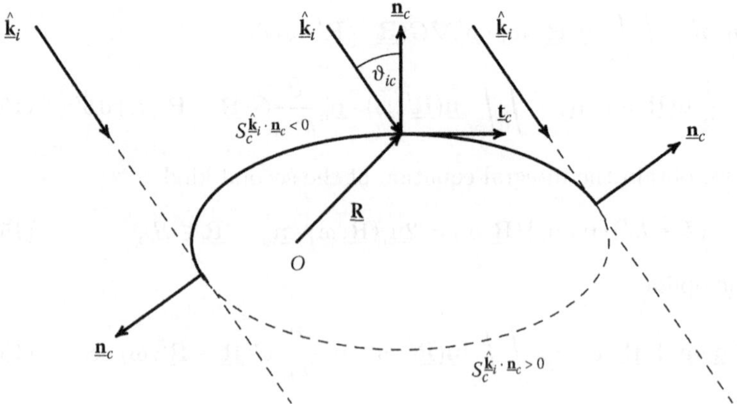

FIGURE 15.9
Kirchhoff approximation of elastodynamics ($S_c^{\hat{\underline{\mathbf{k}}}_i\cdot\underline{\mathbf{n}}_c<0}$ is the illuminated, $S_c^{\hat{\underline{\mathbf{k}}}_i\cdot\underline{\mathbf{n}}_c>0}$ the shadow side of S_c).

appropriate approximation; in fact, PO-approximated scattered far-fields are quite exact in the direction of specular reflection but not in the side lobe region (Figure 5.4); yet, a quantitative estimate of the error is not available.

Kirchhoff approximation for two-dimensional SH-wave scattering: According to (7.47), the y-component of the spectral particle displacement of two-dimensional ($\partial/\partial y \equiv 0$) SH-waves satisfies the homogeneous Helmholtz equation

$$\Delta u_y(\underline{\mathbf{r}}, \omega) + k_S^2 u_y(\underline{\mathbf{r}}, \omega) = 0 \tag{15.136}$$

in source-free space, where $\underline{\mathbf{r}} = x\underline{\mathbf{e}}_x + z\underline{\mathbf{e}}_z$. Following (7.49), the stress-free boundary condition results in the Neumann boundary condition:

$$\underline{\mathbf{n}}_c \cdot \nabla u_y(\underline{\mathbf{r}}, \omega) = 0, \quad \underline{\mathbf{r}} \in S_c, \tag{15.137}$$

where S_c is now a one-dimensional "surface," namely a closed contour in the xz-plane (Footnote 119). The Helmholtz integral

$$u_{sy}(\underline{\mathbf{r}}, \omega) = \int_{S_c} u_y(\underline{\mathbf{r}}', \omega) \frac{\partial}{\partial n_c'} G(\underline{\mathbf{r}} - \underline{\mathbf{r}}', \omega)\, \mathrm{d}S' \tag{15.138}$$

is the appropriate solution of (15.136) under the boundary condition (15.137) (Equation 5.85, respectively Equation 15.40, with 15.33) involving the two-dimensional scalar Green function (13.23). The limit $\underline{\mathbf{r}} \longrightarrow S_c$ in (15.138) yields

$$u_{sy}(\underline{\mathbf{r}}, \omega) = \frac{1}{2} u_y(\underline{\mathbf{r}}, \omega) + \int_{S_c} u_y(\underline{\mathbf{r}}', \omega) \frac{\partial}{\partial n_c'} G(\underline{\mathbf{r}} - \underline{\mathbf{r}}', \omega)\, \mathrm{d}S', \quad \underline{\mathbf{r}} \in S_c, \tag{15.139}$$

and consequently, the integral equation of the second kind

$$(\mathcal{I} - \mathcal{K})\{u_y\}(\underline{\mathbf{r}}, \omega) = 2u_{iy}(\underline{\mathbf{r}}, \omega), \quad \underline{\mathbf{r}} \in S_c, \tag{15.140}$$

with the operator

$$\mathcal{K}\{u_y\}(\underline{\mathbf{r}}, \omega) = 2 \int_{S_c} u_y(\underline{\mathbf{r}}', \omega) \frac{\partial}{\partial n_c'} G(\underline{\mathbf{r}} - \underline{\mathbf{r}}', \omega)\, \mathrm{d}S', \quad \underline{\mathbf{r}} \in S_c, \tag{15.141}$$

emerges for the secondary source $u_y(\underline{\mathbf{r}} \in S_c, \omega)$ of the scattered field. The integral equation of the second kind (15.140) for the Neumann problem complements the integral equation of the second kind (15.130) for the Dirichlet problem.[292]

Similar to (15.134), we find

$$\mathcal{K}\{u_y\}(x, \omega) = 0, \quad x \in S_x, \tag{15.142}$$

yielding once again (we have $\hat{\underline{\mathbf{k}}}_i \cdot \underline{\mathbf{e}}_y = 0$)

$$u_y^{\mathrm{PO}}(\underline{\mathbf{r}}, \omega) = 2u_{iy}(\underline{\mathbf{r}}, \omega)\, u(-\hat{\underline{\mathbf{k}}}_i \cdot \underline{\mathbf{n}}_c), \quad \underline{\mathbf{r}} \in S_c, \tag{15.143}$$

for the Kirchhoff approximated secondary source.

[292] The operators $\mathcal{K}$ and $\mathcal{K}'$ are adjoint to each other.

The typical structure "twice the incident field" (on the illuminated side) is also true for electromagnetic waves (Equation 6.157); again, the reason is that a perfectly conducting surface represents a perfect mirror for electromagnetic waves. This is not true for elastic waves!

Kirchhoff approximation for two-dimensional P-wave scattering: The reflection of a plane pressure wave at the planar stress-free boundary of an elastic half-space produces a reflected transversely polarized shear wave via mode conversion (Section 9.1.2), i.e., such a boundary is no perfect mirror for elastic waves. Therefore, we may not expect that the secondary source for the reflected pressure wave and the mode converted shear wave is just twice as large as the incident field at the boundary; certainly, reflection and mode conversion factors should come into play. In fact, we already calculated the relevant secondary source with (9.59). If we consider (9.58), we obtain for the spectral boundary particle displacement (for P-wave incidence)

$$\begin{aligned}\underline{\mathbf{u}}^{\mathrm{P}}(x,\omega,\vartheta_{i\mathrm{P}}) = {} & u_{i\mathrm{P}}(\omega)\,\mathrm{e}^{\mathrm{j}k_{\mathrm{P}}x\hat{\underline{\mathbf{k}}}_{i\mathrm{P}}\cdot\underline{\mathbf{e}}_x}\\ & \times\left[\hat{\underline{\mathbf{k}}}_{i\mathrm{P}} + R_{\mathrm{P}}(\vartheta_{i\mathrm{P}})\hat{\underline{\mathbf{k}}}_{r\mathrm{P}} + M_{\mathrm{S}}(\vartheta_{i\mathrm{P}})\hat{\underline{\mathbf{k}}}_{m\mathrm{S}}\times\underline{\mathbf{e}}_y\right], \quad x\in S_x.\end{aligned} \tag{15.144}$$

To prove explicitly that (15.144) is in fact the Huygens equivalent source for the reflected and mode converted waves, respectively secondary source of the scattered field in terms of reflected and mode converted waves, we either have to derive (15.144) as a solution of the (two-dimensional) DFIE (15.96) for $S_c = S_x$ or must at least show that (15.144) is a solution of (15.96). For both special cases of the previous paragraph, this proof was indeed simple due to the vanishing radiation interaction integral; here, it is extremely cumbersome: Following a spatial Fourier transform of (15.96) with regard to x, we managed to invert the operator $\boldsymbol{\mathcal{I}}+\boldsymbol{\mathcal{U}}$ yielding multiple combinations of sine and cosine functions, yet they did not allow us to bring the result into the explicit form (15.144); even symbolic math did not succeed. A nearly similar problem is encountered if (15.144) is inserted into (15.96): Here, numerical evaluation helps at least—with a discrepancy of about 10^{-10}—to prove that (15.144) is a solution of (15.96).

Due to these difficulties, we simply should accept (15.144) as secondary source based on physical intuition. Subsequently, it is only a small step to define a Kirchhoff approximated secondary source for P-wave incidence

$$\begin{aligned}\underline{\mathbf{u}}^{\mathrm{P,PO}}(\underline{\mathbf{r}},\omega,\vartheta_{ic}) = u_{i\mathrm{P}}(\omega)\,\mathrm{e}^{\mathrm{j}k_{\mathrm{P}}\hat{\underline{\mathbf{k}}}_{i\mathrm{P}}\cdot\underline{\mathbf{r}}}\left[\hat{\underline{\mathbf{k}}}_{i\mathrm{P}} + R_{\mathrm{P}}(\vartheta_{ic})\hat{\underline{\mathbf{k}}}_{r\mathrm{P}} + M_{\mathrm{S}}(\vartheta_{ic})\hat{\underline{\mathbf{k}}}_{m\mathrm{S}}\times\underline{\mathbf{e}}_y\right]\\ u(-\hat{\underline{\mathbf{k}}}_{i\mathrm{P}}\cdot\underline{\mathbf{n}}_c),\end{aligned} \tag{15.145}$$

$$\cos\vartheta_{ic} = -\hat{\underline{\mathbf{k}}}_{i\mathrm{P}}\cdot\underline{\mathbf{n}}_c, \tag{15.146}$$

$$\hat{\underline{\mathbf{k}}}_{i\mathrm{P}} = \sin\vartheta_{ic}\underline{\mathbf{t}}_c - \cos\vartheta_{ic}\underline{\mathbf{n}}_c, \tag{15.147}$$

$$\hat{\underline{\mathbf{k}}}_{r\mathrm{P}} = \sin\vartheta_{ic}\underline{\mathbf{t}}_c + \cos\vartheta_{ic}\underline{\mathbf{n}}_c, \tag{15.148}$$

$$\hat{\underline{\mathbf{k}}}_{mS} = \frac{k_P}{k_S}\sin\vartheta_{ic}\underline{\mathbf{t}}_c + \sqrt{1-\left(\frac{k_P}{k_S}\sin\vartheta_{ic}\right)^2}\,\underline{\mathbf{n}}_c, \tag{15.149}$$

$$\hat{\underline{\mathbf{k}}}_{mS}\times\underline{\mathbf{e}}_y = -\sqrt{1-\left(\frac{k_P}{k_S}\sin\vartheta_{ic}\right)^2}\,\underline{\mathbf{t}}_c + \frac{k_P}{k_S}\sin\vartheta_{ic}\underline{\mathbf{n}}_c \tag{15.150}$$

for $\underline{\mathbf{r}} \in S_c$. Note that normal $\underline{\mathbf{n}}_c$, tangential vector[293] $\underline{\mathbf{t}}_c = \underline{\mathbf{e}}_y \times \underline{\mathbf{n}}_c$, and angle of incidence ϑ_{ic} are locally defined on S_c (Figure 15.9) depending on $\underline{\mathbf{r}}$ and requesting to remain under the integral; as compared to (15.135) and (15.143), this "extended" spatial dependence—the phase of the incident field also yields a spatial dependence of the secondary sources (15.135) and (15.143)—is characterized in the argument of $\underline{\mathbf{u}}^{P,PO}(\underline{\mathbf{r}},\omega,\vartheta_{ic})$ through the explicit appearance of the angle ϑ_{ic}.

Equation 15.145 once again particularly exhibits—as compared to acoustics and electromagnetics—the additional problems of elastodynamics due to the existence of two wave modes with different velocities. We will see (Section 15.4) that this "extended" $\underline{\mathbf{r}}$-dependence even affects the radiation patterns of scattered fields.

Kirchhoff approximation for two-dimensional SV-wave scattering: From Section 9.1.2 (Equation 9.144), we extract the particle displacement of the secondary surface deformation source for SV-wave incidence:

$$\begin{aligned}\underline{\mathbf{u}}^{SV,PO}(\underline{\mathbf{r}},\omega,\vartheta_{ic}) = u_{iS}(\omega)\,e^{jk_S\hat{\underline{\mathbf{k}}}_{iS}\cdot\underline{\mathbf{r}}}\Big[\hat{\underline{\mathbf{k}}}_{iS}\times\underline{\mathbf{e}}_y + R_{SV}(\vartheta_{ic})\hat{\underline{\mathbf{k}}}_{rS}\times\underline{\mathbf{e}}_y \\ + M_P(\vartheta_{ic})\hat{\underline{\mathbf{k}}}_{mP}\Big]\,u(-\hat{\underline{\mathbf{k}}}_{iS}\cdot\underline{\mathbf{n}}_c),\end{aligned} \tag{15.151}$$

$$\cos\vartheta_{ic} = -\hat{\underline{\mathbf{k}}}_{iS}\cdot\underline{\mathbf{n}}_c, \tag{15.152}$$

$$\hat{\underline{\mathbf{k}}}_{iS} = \sin\vartheta_{ic}\underline{\mathbf{t}}_c - \cos\vartheta_{ic}\underline{\mathbf{n}}_c, \tag{15.153}$$

$$\hat{\underline{\mathbf{k}}}_{rS} = \sin\vartheta_{ic}\underline{\mathbf{t}}_c + \cos\vartheta_{ic}\underline{\mathbf{n}}_c, \tag{15.154}$$

$$\hat{\underline{\mathbf{k}}}_{mP} = \frac{k_S}{k_P}\sin\vartheta_{ic}\underline{\mathbf{t}}_c + \sqrt{1-\left(\frac{k_S}{k_P},\sin\vartheta_{ic}\right)^2}\,\underline{\mathbf{n}}_c. \tag{15.155}$$

Here, we have additionally to recognize complex valued expressions for $R_{SV}(\vartheta_{ic})$, $M_P(\vartheta_{ic})$, and $\hat{\underline{\mathbf{k}}}_{mP}$ for $\vartheta_{ic} > \vartheta_{cmP}$, where ϑ_{cmP} is the critical angle for the mode converted P-wave.

Kirchhoff approximation for three-dimensional pressure and shear wave scattering: In three dimensions, the separation of plane elastic shear waves into SH- and SV-polarizations has to be performed with regard to a given reference plane that is not yet defined by an arbitrary three-dimensional scatterer. Nevertheless, the direction of incidence $\hat{\underline{\mathbf{k}}}_{iP,S}$ defines a normal $\underline{\mathbf{n}}_c$ in each (illuminated) surface point $\underline{\mathbf{R}}$ and, hence, spatially dependent tangential

[293]Note: $\underline{\mathbf{t}}_c$ is defined in the direction $\hat{\underline{\mathbf{k}}}_{iP}$ yielding positive $\underline{\mathbf{t}}_c$-components of $\hat{\underline{\mathbf{k}}}_{iP}$, $\hat{\underline{\mathbf{k}}}_{rP}$, $\hat{\underline{\mathbf{k}}}_{mS}$.

and incidence planes as well as a tangential vector $\underline{\mathbf{t}}_c$ located in the respective plane of incidence. Therefore, for an incident pressure wave we may immediately choose (15.146) as Kirchhoff approximation, where ϑ_{ic}, $\underline{\mathbf{n}}_c$, and $\underline{\mathbf{t}}_c$ vary three-dimensionally for $\underline{\mathbf{R}} \in S_c$. The mode converted shear wave emerging from each of these "Kirchhoff reflections" is evidently SV-polarized with regard of the actually considered tangential, respectively incidence plane, but not SV-polarized with regard to all tangential planes (as in two dimensions); this must be recognized separating the scattered far-field into polarizations (Section 15.5). For an incident plane shear wave, an additional separation into SH- and SV-components with regard to the local tangential planes has to be performed; then, (15.143) (the y-component is the local tangential component perpendicular to $\underline{\mathbf{t}}_c$) and (15.151) may be applied.

For all cases of Kirchhoff approximated secondary sources being simply proportional to the incident field, the latter must not be just a plane wave; a real-life antenna or transducer field may also be inserted. Strictly speaking, for elastic waves, this is not true because the image principle does not hold: Reflection and mode conversion factors are only known for plane waves.

15.3 Integral Equations for the Equivalent Sources of Penetrable Scatterers

15.3.1 Lippmann–Schwinger integral equations for equivalent volume sources of inhomogeneous anisotropic scatterers

We refer to the elastodynamic scattering problem as sketched in Figure 15.10. With the reduced wave equations

$$\begin{aligned}\nabla \cdot \left[\underline{\underline{\underline{\underline{\mathbf{c}}}}}(\underline{\mathbf{R}}) : \nabla \underline{\mathbf{v}}(\underline{\mathbf{R}}, \omega)\right] + \omega^2 \rho(\underline{\mathbf{R}})\, \underline{\mathbf{v}}(\underline{\mathbf{R}}, \omega) &= j\omega \underline{\mathbf{f}}(\underline{\mathbf{R}}, \omega)\\ &\quad - \nabla \cdot \left[\underline{\underline{\underline{\underline{\mathbf{c}}}}}(\underline{\mathbf{R}}) : \underline{\underline{\mathbf{h}}}(\underline{\mathbf{R}}, \omega)\right],\end{aligned} \tag{15.156}$$

$$\begin{aligned}\underline{\underline{\mathbf{I}}}^+ : \nabla \left[\frac{1}{\rho(\underline{\mathbf{R}})} \nabla \cdot \underline{\underline{\mathbf{T}}}(\underline{\mathbf{R}}, \omega)\right] + \omega^2 \underline{\underline{\underline{\underline{\mathbf{s}}}}}(\underline{\mathbf{R}}) : \underline{\underline{\mathbf{T}}}(\underline{\mathbf{R}}, \omega)\\ = -\underline{\underline{\mathbf{I}}}^+ : \nabla \left[\frac{1}{\rho(\underline{\mathbf{R}})} \underline{\mathbf{f}}(\underline{\mathbf{R}}, \omega)\right] + j\omega \underline{\underline{\mathbf{h}}}(\underline{\mathbf{R}}, \omega)\end{aligned} \tag{15.157}$$

for the elastodynamic field quantities in inhomogeneous anisotropic materials as well as the definition[294]

$$\underline{\mathbf{f}}_c(\underline{\mathbf{R}}, \omega) \stackrel{\text{def}}{=} \underline{\mathbf{f}}_\rho(\underline{\mathbf{R}}, \omega) = -j\omega \Gamma_c(\underline{\mathbf{R}})\, [\rho - \rho(\underline{\mathbf{R}})]\, \underline{\mathbf{v}}(\underline{\mathbf{R}}, \omega), \tag{15.158}$$

[294]In $\underline{\underline{\mathbf{h}}}_c$, the italic "$c$" stands for the scattering volume V_c, and in $\underline{\underline{\mathbf{h}}}_{\mathbf{c}}$, the boldface "$\mathbf{c}$" stands for the inhomogeneity of the stiffness, respectively compliance tensor.

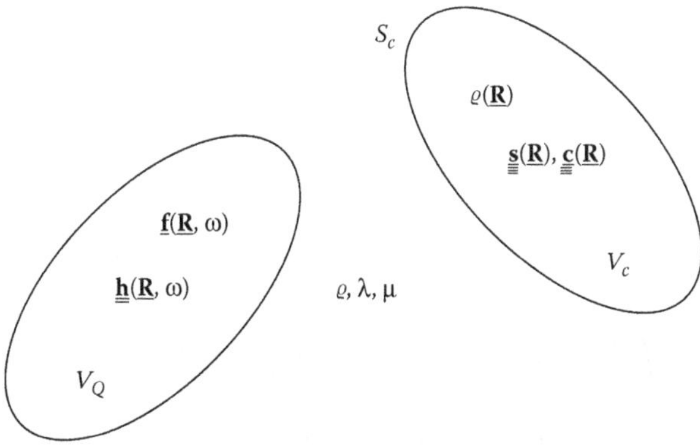

FIGURE 15.10
Elastodynamic scattering problem: penetrable inhomogeneous anisotropic scatterer in a homogeneous isotropic embedding material (equivalent volume sources).

$$\underline{\underline{\mathbf{h}}}_c(\underline{\mathbf{R}}, \omega) \stackrel{\text{def}}{=} \underline{\underline{\mathbf{h}}}_{\mathbf{c}}(\underline{\mathbf{R}}, \omega) = -\mathrm{j}\omega\Gamma_c(\underline{\mathbf{R}}) \left[\underline{\underline{\underline{\underline{\mathbf{s}}}}} - \underline{\underline{\underline{\underline{\mathbf{s}}}}}(\underline{\mathbf{R}})\right] : \underline{\underline{\mathbf{T}}}(\underline{\mathbf{R}}, \omega) \tag{15.159}$$

of equivalent sources of such a material inclusion (Section 7.1.1)—of a penetrable scatterer with the characteristic function $\Gamma_c(\underline{\mathbf{R}})$ of the scattering volume V_c—in a homogeneous isotropic material with known Green tensors, we may immediately write down the scattered field of a penetrable inhomogeneous anisotropic scatterer using the source field representations of point source synthesis (as so-called data equations):

$$\begin{aligned}\underline{\mathbf{v}}_s(\underline{\mathbf{R}}, \omega) = \iiint_{V_c} \Big[&-\mathrm{j}\omega \underline{\mathbf{f}}_c(\underline{\mathbf{R}}', \omega) \cdot \underline{\underline{\mathbf{G}}}(\underline{\mathbf{R}} - \underline{\mathbf{R}}', \omega) \\ &+ \underline{\underline{\mathbf{h}}}_c(\underline{\mathbf{R}}', \omega) : \underline{\underline{\underline{\boldsymbol{\Sigma}}}}(\underline{\mathbf{R}} - \underline{\mathbf{R}}', \omega)\Big] \mathrm{d}^3\underline{\mathbf{R}}', \end{aligned} \tag{15.160}$$

$$\begin{aligned}\underline{\underline{\mathbf{T}}}_s(\underline{\mathbf{R}}, \omega) = \iiint_{V_c} \Big[&\underline{\mathbf{f}}_c(\underline{\mathbf{R}}', \omega) \cdot \underline{\underline{\underline{\boldsymbol{\Sigma}}}}^{312}(\underline{\mathbf{R}}', \omega) \\ &- \frac{1}{\mathrm{j}\omega} \underline{\underline{\mathbf{h}}}_c(\underline{\mathbf{R}}', \omega) : \underline{\underline{\underline{\underline{\boldsymbol{\Pi}}}}}(\underline{\mathbf{R}} - \underline{\mathbf{R}}', \omega)\Big] \mathrm{d}^3\underline{\mathbf{R}}'. \end{aligned} \tag{15.161}$$

That way, the equivalent sources turn into secondary sources of the scattered field that depend upon the total field $\underline{\mathbf{v}}(\underline{\mathbf{R}}, \omega) = \underline{\mathbf{v}}_s(\underline{\mathbf{R}}, \omega) + \underline{\mathbf{v}}_i(\underline{\mathbf{R}}, \omega)$, $\underline{\underline{\mathbf{T}}}(\underline{\mathbf{R}}, \omega) = \underline{\underline{\mathbf{T}}}_s(\underline{\mathbf{R}}, \omega) + \underline{\underline{\mathbf{T}}}_i(\underline{\mathbf{R}}, \omega)$, where the incident field $\underline{\mathbf{v}}_i(\underline{\mathbf{R}}, \omega)$, $\underline{\underline{\mathbf{T}}}_i(\underline{\mathbf{R}}, \omega)$ comes from primary sources located outside V_c in the homogeneous isotropic background material.

Since (15.158) and (15.159) contain the total field, the scattered field point source synthesis is only a preliminary result: Beforehand, the total field in V_c must be calculated! Fortunately, the integral representations (15.160) and (15.161) also hold for $\underline{\mathbf{R}} \in V_c$—note: $\underline{\underline{\underline{\underline{\boldsymbol{\Pi}}}}}$ contains a δ-distributional term according to (13.208)—resulting in the system of coupled Lippmann–Schwinger integral equations

$$\underline{\mathbf{v}}(\underline{\mathbf{R}}, \omega) + \mathrm{j}\omega \iiint_{V_c} \Big\{ -\mathrm{j}\omega[\rho - \rho(\underline{\mathbf{R}}')]\, \underline{\mathbf{v}}(\underline{\mathbf{R}}', \omega) \cdot \underline{\underline{\mathbf{G}}}(\underline{\mathbf{R}} - \underline{\mathbf{R}}', \omega) + \Big[\underline{\underline{\underline{\underline{\mathbf{s}}}}} - \underline{\underline{\underline{\underline{\mathbf{s}}}}}(\underline{\mathbf{R}}')\Big] : \underline{\underline{\mathbf{T}}}(\underline{\mathbf{R}}', \omega) : \underline{\underline{\underline{\mathbf{\Sigma}}}}(\underline{\mathbf{R}} - \underline{\mathbf{R}}', \omega) \Big\}\, \mathrm{d}^3\underline{\mathbf{R}}' = \underline{\mathbf{v}}_i(\underline{\mathbf{R}}, \omega), \quad \underline{\mathbf{R}} \in V_c, \tag{15.162}$$

$$\underline{\underline{\mathbf{T}}}(\underline{\mathbf{R}}, \omega) + \mathrm{j}\omega \iiint_{V_c} \Big\{ [\rho - \rho(\underline{\mathbf{R}}')]\, \underline{\mathbf{v}}(\underline{\mathbf{R}}', \omega) \cdot \underline{\underline{\underline{\mathbf{\Sigma}}}}^{312}(\underline{\mathbf{R}} - \underline{\mathbf{R}}', \omega) - \frac{1}{\mathrm{j}\omega}\Big[\underline{\underline{\underline{\underline{\mathbf{s}}}}} - \underline{\underline{\underline{\underline{\mathbf{s}}}}}(\underline{\mathbf{R}}')\Big] : \underline{\underline{\mathbf{T}}}(\underline{\mathbf{R}}', \omega) : \underline{\underline{\underline{\underline{\mathbf{\Pi}}}}}(\underline{\mathbf{R}} - \underline{\mathbf{R}}', \omega) \Big\}\, \mathrm{d}^3\underline{\mathbf{R}}' = \underline{\underline{\mathbf{T}}}_i(\underline{\mathbf{R}}, \omega), \quad \underline{\mathbf{R}} \in V_c, \tag{15.163}$$

for the vector $\underline{\mathbf{v}}(\underline{\mathbf{R}}, \omega)$ and the tensor $\underline{\underline{\mathbf{T}}}(\underline{\mathbf{R}}, \omega)$ inside V_c: These are nine scalar equations for the three components of $\underline{\mathbf{v}}$ and the six components of $\underline{\underline{\mathbf{T}}}$ ($\underline{\underline{\mathbf{T}}}$ is symmetric); they are also called (coupled) object equations. As for the DFIE, the volume integrals in (15.162) and (15.163) stand for the radiation interaction inside V_c. Neglecting them results in the Born approximation for the secondary sources (Section 15.3.2). The explicit notation of the secondary sources in (15.162) and (15.163) immediately reveals the reason for the coupling of the object equations to be the inhomogeneity of *all* material parameters, i.e., density as well as elastic constants. Only for a sole density inhomogeneity with $\underline{\underline{\underline{\underline{\mathbf{s}}}}}(\underline{\mathbf{R}}) = \underline{\underline{\underline{\underline{\mathbf{s}}}}}$, a single Lippmann–Schwinger equation (three scalar equations) is obtained:[295]

$$\underline{\mathbf{v}}(\underline{\mathbf{R}}, \omega) + \omega^2\rho \iiint_{V_c} \chi_\rho(\underline{\mathbf{R}}')\underline{\underline{\mathbf{G}}}(\underline{\mathbf{R}} - \underline{\mathbf{R}}', \omega) \cdot \underline{\mathbf{v}}(\underline{\mathbf{R}}', \omega)\, \mathrm{d}^3\underline{\mathbf{R}}' = \underline{\mathbf{v}}_i(\underline{\mathbf{R}}, \omega), \quad \underline{\mathbf{R}} \in V_c; \tag{15.164}$$

here, $\chi_\rho(\underline{\mathbf{R}})$ is the density contrast of the penetrable scatterer according to (7.4). With the ($\underline{\mathbf{v}} = \underline{\mathbf{v}}_s + \underline{\mathbf{v}}_i$)-separation and the definition of the integral operator

$$\mathcal{V}\{\underline{\mathbf{v}}\}(\underline{\mathbf{R}}, \omega) = -\omega^2\rho \iiint_{V_c} \chi_\rho(\underline{\mathbf{R}}')\underline{\underline{\mathbf{G}}}(\underline{\mathbf{R}} - \underline{\mathbf{R}}', \omega) \cdot \underline{\mathbf{v}}(\underline{\mathbf{R}}', \omega)\, \mathrm{d}^3\underline{\mathbf{R}}', \quad \underline{\mathbf{R}} \in V_c, \tag{15.165}$$

we may write (15.164) as follows:

$$\underline{\mathbf{v}}_s(\underline{\mathbf{R}}, \omega) = (\mathcal{I} - \mathcal{V})^{-1}\{\mathcal{V}\{\underline{\mathbf{v}}_i\}\}(\underline{\mathbf{R}}, \omega), \quad \underline{\mathbf{R}} \in V_c, \tag{15.166}$$

where $\mathcal{I}\{\underline{\mathbf{v}}\}(\underline{\mathbf{R}}, \omega) = \underline{\underline{\mathbf{I}}} \cdot \underline{\mathbf{v}}(\underline{\mathbf{R}}, \omega) = \underline{\mathbf{v}}(\underline{\mathbf{R}}, \omega)$ denotes the identity operator. Specially choosing a plane pressure or shear wave according to

$$\underline{\mathbf{v}}_i(\underline{\mathbf{R}}, \omega, \underline{\mathbf{k}}_{i\mathrm{P,S}}) = v_{i\mathrm{P,S}}(\omega)\, \mathrm{e}^{\mathrm{j}k_{\mathrm{P,S}}\hat{\underline{\mathbf{k}}}_i \cdot \underline{\mathbf{R}}}\, \underline{\underline{\mathbf{I}}} \cdot \hat{\underline{\mathbf{v}}}_{i\mathrm{P,S}} \tag{15.167}$$

[295] Therefore, the solution of the inverse scattering problem starts with this simplifying assumption (Pelekanos et al. 2000).

as incident wave, the expression

$$\underline{\mathbf{v}}_s(\underline{\mathbf{R}}, \omega, \underline{\mathbf{k}}_{i\mathrm{P,S}}) = \underbrace{(\mathcal{I} - \mathcal{V})^{-1}\{\mathcal{V}\{v_{i\mathrm{P,S}}(\omega)\, \mathrm{e}^{\mathrm{j}k_{\mathrm{P,S}}\hat{\underline{\mathbf{k}}}_i\cdot\underline{\mathbf{R}}}\,\underline{\underline{\mathbf{I}}}\}\}(\underline{\mathbf{R}}, \omega)}_{=\underline{\underline{\boldsymbol{\Xi}}}_c^{\rho}(\underline{\mathbf{R}}, \omega, \underline{\mathbf{k}}_{i\mathrm{P,S}})} \cdot \hat{\underline{\mathbf{v}}}_{i\mathrm{P,S}} \tag{15.168}$$

reveals—as expected from the linearity of the governing equations—a linear dependence of the scattered field upon the polarization of the incident plane wave, thus defining a scattering tensor $\underline{\underline{\boldsymbol{\Xi}}}_c^{\rho}(\underline{\mathbf{R}}, \omega, \underline{\mathbf{k}}_{i\mathrm{P,S}})$ (for sole density variations).

To derive similar facts for $\underline{\mathbf{v}}_s(\underline{\mathbf{R}}, \omega)$ and $\underline{\underline{\mathbf{T}}}_s(\underline{\mathbf{R}}, \omega)$ in case of an additional inhomogeneity of the elastic constants, we use

$$\underline{\underline{\mathbf{T}}}_i(\underline{\mathbf{R}}, \omega, \underline{\mathbf{k}}_{i\mathrm{P,S}}) = -\frac{1}{\omega}\, v_{i\mathrm{P,S}}(\omega)\, \mathrm{e}^{\mathrm{j}k_{\mathrm{P,S}}\hat{\underline{\mathbf{k}}}_i\cdot\underline{\mathbf{R}}}\, (\lambda\, \underline{\underline{\mathbf{I}}}\, \underline{\mathbf{k}}_{i\mathrm{P,S}} + 2\mu\, \underline{\underline{\underline{\mathbf{I}}}}^{+} \cdot \underline{\mathbf{k}}_{i\mathrm{P,S}}) \cdot \hat{\underline{\mathbf{v}}}_{i\mathrm{P,S}} \tag{15.169}$$

as expression for the $\underline{\underline{\mathbf{T}}}_i$-wave associated with $\underline{\mathbf{v}}_i$ in the homogeneous isotropic embedding material [Equation 15.169 contains the two equations (8.121) and (8.122)]. The linear relation (15.168) may be generalized as follows and complemented by (15.171):

$$\underline{\mathbf{v}}_s(\underline{\mathbf{R}}, \omega, \underline{\mathbf{k}}_{i\mathrm{P,S}}) = \underline{\underline{\boldsymbol{\Xi}}}_c^{\rho,\mathbf{c}}(\underline{\mathbf{R}}, \omega, \underline{\mathbf{k}}_{i\mathrm{P,S}}) \cdot \hat{\underline{\mathbf{v}}}_{i\mathrm{P,S}}, \quad \underline{\mathbf{R}} \in V_c, \tag{15.170}$$

$$\underline{\underline{\mathbf{T}}}_s(\underline{\mathbf{R}}, \omega, \underline{\mathbf{k}}_{i\mathrm{P,S}}) = \underline{\underline{\underline{\boldsymbol{\Upsilon}}}}_c^{\rho,\mathbf{c}}(\underline{\mathbf{R}}, \omega, \underline{\mathbf{k}}_{i\mathrm{P,S}}) \cdot \hat{\underline{\mathbf{v}}}_{i\mathrm{P,S}}, \quad \underline{\mathbf{R}} \in V_c. \tag{15.171}$$

Here, we are only interested in the principal representation possibilities (15.170) and (15.171) and avoid to present the explicit (and complicated) expressions for $\underline{\underline{\boldsymbol{\Xi}}}_c^{\rho,\mathbf{c}}$ and $\underline{\underline{\underline{\boldsymbol{\Upsilon}}}}_c^{\rho,\mathbf{c}}$.

The comparison of the system of coupled Lippmann–Schwinger integral equations (15.162) and (15.163) with the DFIE (15.94) shows the essential complexity calculating the scattered field of an inclusion in a homogeneous isotropic elastic full-space as compared to a void (with a stress-free surface); this is apparently independent upon the homogeneity/inhomogeneity and/or the isotropy/anisotropy of the inclusion.

15.3.2 Born approximation for inhomogeneous anisotropic scatterers

The Kirchhoff approximation argues with a locally plane approximation of the scattering surface to linearize the void scattering problem.

The Born approximation argues with marginal differences for inclusion and embedding materials to linearize the inclusion scattering problem: It is assumed that the total field in the volume of the inclusion is not that different from the undisturbed incident field yielding $\underline{\mathbf{v}}_s(\underline{\mathbf{R}}, \omega) = \underline{\mathbf{0}}$ and $\underline{\underline{\mathbf{T}}}_s(\underline{\mathbf{R}}, \omega) = \underline{\underline{\mathbf{0}}}$ for $\underline{\mathbf{R}} \in V_c$. Therefore, Born secondary sources are given by

$$\underline{\mathbf{f}}_c^{\mathrm{Born}}(\underline{\mathbf{R}}, \omega) = -\mathrm{j}\omega\Gamma_c(\underline{\mathbf{R}})[\rho - \rho(\underline{\mathbf{R}})]\, \underline{\mathbf{v}}_i(\underline{\mathbf{R}}, \omega), \tag{15.172}$$

$$\underline{\underline{\mathbf{h}}}_c^{\mathrm{Born}}(\underline{\mathbf{R}}, \omega) = -\mathrm{j}\omega\Gamma_c(\underline{\mathbf{R}}) \left[\underline{\underline{\underline{\underline{\mathbf{s}}}}} - \underline{\underline{\underline{\underline{\mathbf{s}}}}}(\underline{\mathbf{R}})\right] : \underline{\underline{\mathbf{T}}}_i(\underline{\mathbf{R}}, \omega). \tag{15.173}$$

These sources result in a Born scattered field outside V_c, yet in the interior of V_c, they do not reproduce $\underline{\mathbf{v}}_i$, respectively $\underline{\underline{\mathbf{T}}}_i$, because the incident field comes from the primary sources in V_Q, even inside V_c.

The Born approximation ignores the scatterer with respect to its interior scattered field, hence, it will be the better the larger the wavelength as compared to V_c, i.e., in contrast to the Kirchhoff approximation, it is a low frequency approximation (for weak contrast). Chew (1990) gives a more precise assessment for scalar fields.

15.3.3 Coupled integral equations for equivalent surface sources of homogeneous isotropic scatterers

The Lippmann–Schwinger integral equations (15.162) and (15.163) for penetrable scatterers are coupled volume integral equations, yielding a costly numerical solution due to the required three-dimensional discretization. For homogeneous isotropic penetrable scatterers[296] (Figure 15.11), an alternative exists in terms of coupled surface integral equations; namely, in that case, Green's tensors are known for the isotropic material inside V_c—material parameters $\rho^{(i)}$, $\lambda^{(i)}$, $\mu^{(i)}$—as well as those for the outside material—material parameters $\rho^{(e)}$, $\lambda^{(e)}$, $\mu^{(e)}$—and we may write down elastodynamic Huygens integrals with the respective Green tensors $\underline{\underline{\mathbf{G}}}^{(i,e)}$, $\underline{\underline{\underline{\mathbf{\Sigma}}}}^{(i,e)}$ (of full-space) for the exterior scattered field, as well as for the interior scattered field. According to (15.63), the respective integral representation for the exterior particle velocity field reads ($\underline{\mathbf{R}} \in \mathbb{R}^3 \backslash \overline{V}_c$):

$$\begin{aligned}\underline{\mathbf{v}}^{(e)}(\underline{\mathbf{R}}, \omega) = \underline{\mathbf{v}}_i(\underline{\mathbf{R}}, \omega) + \iint_{S_c} \Big[&\mathrm{j}\omega\underline{\mathbf{n}}_c' \cdot \underline{\underline{\mathbf{T}}}^{(e)}(\underline{\mathbf{R}}', \omega) \cdot \underline{\underline{\mathbf{G}}}^{(e)}(\underline{\mathbf{R}} - \underline{\mathbf{R}}', \omega) \\ &- \underline{\mathbf{n}}_c'\underline{\mathbf{v}}^{(e)}(\underline{\mathbf{R}}', \omega) : \underline{\underline{\underline{\mathbf{\Sigma}}}}^{(e)}(\underline{\mathbf{R}} - \underline{\mathbf{R}}', \omega)\Big]\,\mathrm{d}S';\end{aligned} \tag{15.174}$$

of course, the secondary sources of the exterior scattered field exhibit the boundary values of the exterior total field. The counterpart to (15.174) for the interior total field reads ($\underline{\mathbf{R}} \in V_c$: No sources are located in V_c, hence, the scattered field is equal to the total field):

$$\begin{aligned}\underline{\mathbf{v}}^{(i)}(\underline{\mathbf{R}}, \omega) = -\iint_{S_c} \Big[&\mathrm{j}\omega\underline{\mathbf{n}}_c' \cdot \underline{\underline{\mathbf{T}}}^{(i)}(\underline{\mathbf{R}}', \omega) \cdot \underline{\underline{\mathbf{G}}}^{(i)}(\underline{\mathbf{R}} - \underline{\mathbf{R}}', \omega) \\ &- \underline{\mathbf{n}}_c'\underline{\mathbf{v}}^{(i)}(\underline{\mathbf{R}}', \omega) : \underline{\underline{\underline{\mathbf{\Sigma}}}}^{(i)}(\underline{\mathbf{R}} - \underline{\mathbf{R}}', \omega)\Big]\,\mathrm{d}S';\end{aligned} \tag{15.175}$$

[296] In case Green tensors are known for the inhomogeneous and/or anisotropic scatterer material, this alternative may also be pursued. Yet, the advantage of the Lippmann–Schwinger volume integral equation is the sole request for the tensors of the homogeneous isotropic embedding material.

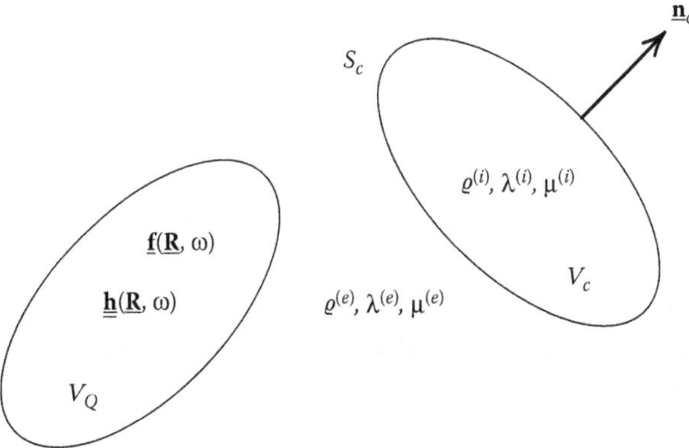

FIGURE 15.11
Elastodynamic scattering problem: penetrable homogeneous isotropic scatterer in a homogeneous isotropic embedding material (equivalent surface sources).

the minus sign goes back to the "wrong" direction of the normal $\underline{\mathbf{n}}_c$ for the secondary sources of the interior field (defined by the boundary values of the interior field). Based on the homogeneous transition conditions (3.88) and (3.93), we may eliminate the interior secondary sources:

$$\underline{\mathbf{v}}^{(i)}(\underline{\mathbf{R}}, \omega) = -\iint_{S_c} \Big[\mathrm{j}\omega \underline{\mathbf{n}}_c' \cdot \underline{\underline{\mathbf{T}}}^{(e)}(\underline{\mathbf{R}}', \omega) \cdot \underline{\underline{\mathbf{G}}}^{(i)}(\underline{\mathbf{R}} - \underline{\mathbf{R}}', \omega) \\ - \underline{\mathbf{n}}_c' \underline{\mathbf{v}}^{(e)}(\underline{\mathbf{R}}', \omega) : \underline{\underline{\underline{\mathbf{\Sigma}}}}^{(i)}(\underline{\mathbf{R}} - \underline{\mathbf{R}}', \omega) \Big] \, \mathrm{d}S'. \tag{15.176}$$

In (15.174) and (15.176), the six scalar components of $\underline{\mathbf{n}}_c \cdot \underline{\underline{\mathbf{T}}}^{(e)}(\underline{\mathbf{R}}, \omega)$ and $\underline{\mathbf{v}}^{(e)}(\underline{\mathbf{R}}, \omega)$, $\underline{\mathbf{R}} \in S_c$, appear as unknowns; hence, two vector (DFIE) surface integral equations are required. Therefore, we calculate the limit $\underline{\mathbf{R}} \longrightarrow S_c$ from the exterior of V_c in (15.174) and, in (15.176), the similar limit from the interior of V_c. For the first case, we may refer to (15.91):

$$\underline{\mathbf{v}}^{(e)}(\underline{\mathbf{R}}, \omega) = 2\, \underline{\mathbf{v}}_i(\underline{\mathbf{R}}, \omega) \\ + 2\,\mathrm{PV} \iint_{S_c} \Big[\mathrm{j}\omega \underline{\mathbf{n}}_c' \cdot \underline{\underline{\mathbf{T}}}^{(e)}(\underline{\mathbf{R}}', \omega) \cdot \underline{\underline{\mathbf{G}}}^{(e)}(\underline{\mathbf{R}} - \underline{\mathbf{R}}', \omega) \\ - \underline{\mathbf{n}}_c' \underline{\mathbf{v}}^{(e)}(\underline{\mathbf{R}}', \omega) : \underline{\underline{\underline{\mathbf{\Sigma}}}}^{(e)}(\underline{\mathbf{R}} - \underline{\mathbf{R}}', \omega) \Big] \, \mathrm{d}S'; \tag{15.177}$$

for the second case, the calculation in Section 15.2.1 has to be repeated for an interior point with the sole difference of an opposite normal; therefore, we anticipate the same basic result with a different sign:

$$\underline{\mathbf{v}}^{(i)}(\underline{\mathbf{R}}, \omega) = -2\,\mathrm{PV} \iint_{S_c} \Big[\mathrm{j}\omega \underline{\mathbf{n}}_c' \cdot \underline{\underline{\mathbf{T}}}^{(e)}(\underline{\mathbf{R}}', \omega) \cdot \underline{\underline{\mathbf{G}}}^{(i)}(\underline{\mathbf{R}} - \underline{\mathbf{R}}', \omega) \\ - \underline{\mathbf{n}}_c' \underline{\mathbf{v}}^{(e)}(\underline{\mathbf{R}}', \omega) : \underline{\underline{\underline{\mathbf{\Sigma}}}}^{(i)}(\underline{\mathbf{R}} - \underline{\mathbf{R}}', \omega) \Big] \, \mathrm{d}S'. \tag{15.178}$$

In (15.177) and (15.178), we now have $\underline{\mathbf{R}} \in S_c$ allowing to repeatedly apply the homogeneous transition condition (3.93) to finally eliminate $\underline{\mathbf{v}}^{(i)}(\underline{\mathbf{R}}, \omega)$:

$$\underline{\mathbf{v}}^{(e)}(\underline{\mathbf{R}}, \omega) = -2\,\mathrm{PV} \iint_{S_c} \Big[\mathrm{j}\omega \underline{\mathbf{n}}_c' \cdot \underline{\underline{\mathbf{T}}}^{(e)}(\underline{\mathbf{R}}', \omega) \cdot \underline{\underline{\mathbf{G}}}^{(i)}(\underline{\mathbf{R}} - \underline{\mathbf{R}}', \omega) \\ - \underline{\mathbf{n}}_c' \underline{\mathbf{v}}^{(e)}(\underline{\mathbf{R}}', \omega) : \underline{\underline{\underline{\mathbf{\Sigma}}}}^{(i)}(\underline{\mathbf{R}} - \underline{\mathbf{R}}', \omega) \Big] \mathrm{d}S'. \quad (15.179)$$

With (15.177) and (15.179), we have found the coupled system of surface integral equations we were looking for! The extension to piecewise homogeneous isotropic scatterers has been given by Tan (1975a).

15.4 Scattering Tensor; Far-Fields

15.4.1 Scattering tensor

Elastodynamic radiation fields are excited by primary sources, elastodynamic scattered fields are radiation fields of secondary sources. Therefore, we may come back to the results of Chapter 13 referring to the elastodynamic point source synthesis (Section 13.3.1) of scattered fields; note: The point source directivities of a point scatterer in full-space are identical to the ones of a point radiator, we have to use the Green functions of full-space according to (13.213) and (13.214). This is also true, for example, if the scatterer exhibits a stress-free boundary: Such a boundary condition is accounted for by the solution of an integral equation (DFIE: Equation 15.94) for the secondary source $\underline{\underline{\mathbf{h}}}_c$ or its Kirchhoff approximation and not using the Miller–Pursey factors based on the half-space Green tensor representation (14.183). The latter would be a contradiction by itself because the equivalent source of a stress-free scattering surface is a surface deformation (rate) and not a surface force density.

In the following, we explicitly present the scattered far-fields because we also give numerical examples for them; we only have to replace the primary sources $\underline{\mathbf{f}}$ and $\underline{\underline{\mathbf{h}}}$ in the vector radiation characteristics (13.217), (13.218) and (13.225), (13.226) by the secondary sources $\underline{\mathbf{f}}_c$ and $\underline{\underline{\mathbf{h}}}_c$ to obtain vector scattering amplitudes:

$$\underline{\mathbf{u}}_{s\mathrm{P}}^{\mathrm{far}}(\underline{\mathbf{R}}, \omega) = \frac{\mathrm{e}^{\mathrm{j}k_\mathrm{P} R}}{R} \left[\underline{\mathbf{H}}_{s\mathrm{P}}^{f_c}(\hat{\underline{\mathbf{R}}}, \omega) + \underline{\mathbf{H}}_{s\mathrm{P}}^{h_c}(\hat{\underline{\mathbf{R}}}, \omega) \right], \quad (15.180)$$

$$\underline{\mathbf{H}}_{s\mathrm{P}}^{f_c}(\hat{\underline{\mathbf{R}}}, \omega) = \frac{1}{4\pi c_\mathrm{P} Z_\mathrm{P}} \hat{\underline{\mathbf{R}}}\hat{\underline{\mathbf{R}}} \cdot \underbrace{\iiint_{V_c} \underline{\mathbf{f}}_c(\underline{\mathbf{R}}', \omega)\, \mathrm{e}^{-\mathrm{j}k_\mathrm{P} \hat{\underline{\mathbf{R}}} \cdot \underline{\mathbf{R}}'}\, \mathrm{d}^3\underline{\mathbf{R}}'}_{\substack{= \mathcal{F}_{3D}\{\underline{\mathbf{f}}_c(\underline{\mathbf{R}}, \omega)\}_{\underline{\mathbf{K}} = k_\mathrm{P}\hat{\underline{\mathbf{R}}}} \\ = \tilde{\underline{\mathbf{f}}}_c(\underline{\mathbf{K}} = k_\mathrm{P}\hat{\underline{\mathbf{R}}}, \omega)}}, \quad (15.181)$$

$$\underline{\mathbf{H}}_{s\mathrm{P}}^{h_c}(\hat{\underline{\mathbf{R}}}, \omega) = -\frac{1}{4\pi c_\mathrm{P}^2 Z_\mathrm{P}}\,\hat{\underline{\mathbf{R}}}(\lambda\underline{\underline{\mathbf{I}}} + 2\mu\hat{\underline{\mathbf{R}}}\hat{\underline{\mathbf{R}}}) : \underbrace{\iiint_{V_c} \underline{\underline{\mathbf{h}}}_c(\underline{\mathbf{R}}', \omega)\, \mathrm{e}^{-\mathrm{j}k_\mathrm{P}\hat{\underline{\mathbf{R}}}\cdot\underline{\mathbf{R}}'}\, \mathrm{d}^3\underline{\mathbf{R}}'}_{\substack{= \mathcal{F}_{3D}\{\underline{\underline{\mathbf{h}}}_c(\underline{\mathbf{R}}, \omega)\}_{\underline{\mathbf{K}}=k_\mathrm{P}\hat{\underline{\mathbf{R}}}} \\ = \tilde{\underline{\underline{\mathbf{h}}}}_c(\underline{\mathbf{K}} = k_\mathrm{P}\hat{\underline{\mathbf{R}}}, \omega)}}; \tag{15.182}$$

$$\underline{\mathbf{u}}_{s\mathrm{S}}^{\mathrm{far}}(\underline{\mathbf{R}}, \omega) = \frac{\mathrm{e}^{\mathrm{j}k_\mathrm{S}R}}{R}\left[\underline{\underline{\mathbf{H}}}_{s\mathrm{S}}^{f_c}(\hat{\underline{\mathbf{R}}}, \omega) + \underline{\underline{\mathbf{H}}}_{s\mathrm{S}}^{h_c}(\hat{\underline{\mathbf{R}}}, \omega)\right], \tag{15.183}$$

$$\underline{\underline{\mathbf{H}}}_{s\mathrm{S}}^{f_c}(\hat{\underline{\mathbf{R}}}, \omega) = \frac{1}{4\pi c_\mathrm{S} Z_\mathrm{S}}(\underline{\underline{\mathbf{I}}} - \hat{\underline{\mathbf{R}}}\hat{\underline{\mathbf{R}}}) \cdot \underbrace{\iiint_{V_c} \underline{\mathbf{f}}_c(\underline{\mathbf{R}}', \omega)\, \mathrm{e}^{-\mathrm{j}k_\mathrm{S}\hat{\underline{\mathbf{R}}}\cdot\underline{\mathbf{R}}'}\, \mathrm{d}^3\underline{\mathbf{R}}'}_{\substack{= \mathcal{F}_{3D}\{\underline{\mathbf{f}}_c(\underline{\mathbf{R}}, \omega)\}_{\underline{\mathbf{K}}=k_\mathrm{S}\hat{\underline{\mathbf{R}}}} \\ = \tilde{\underline{\mathbf{f}}}_c(\underline{\mathbf{K}} = k_\mathrm{S}\hat{\underline{\mathbf{R}}}, \omega)}}, \tag{15.184}$$

$$\underline{\underline{\mathbf{H}}}_{s\mathrm{S}}^{h_c}(\hat{\underline{\mathbf{R}}}, \omega) = -\frac{1}{2\pi c_\mathrm{S}}(\underline{\underline{\mathbf{I}}} - \hat{\underline{\mathbf{R}}}\hat{\underline{\mathbf{R}}})\hat{\underline{\mathbf{R}}} : \underbrace{\iiint_{V_c} \underline{\underline{\mathbf{h}}}_c(\underline{\mathbf{R}}', \omega)\, \mathrm{e}^{-\mathrm{j}k_\mathrm{S}\hat{\underline{\mathbf{R}}}\cdot\underline{\mathbf{R}}'}\, \mathrm{d}^3\underline{\mathbf{R}}'}_{\substack{= \mathcal{F}_{3D}\{\underline{\underline{\mathbf{h}}}_c(\underline{\mathbf{R}}, \omega)\}_{\underline{\mathbf{K}}=k_\mathrm{S}\hat{\underline{\mathbf{R}}}} \\ = \tilde{\underline{\underline{\mathbf{h}}}}_c(\underline{\mathbf{K}} = k_\mathrm{S}\hat{\underline{\mathbf{R}}}, \omega)}}. \tag{15.185}$$

Specifying secondary (volume) sources $\underline{\mathbf{f}}_c$ and $\underline{\underline{\mathbf{h}}}_c$ according to

$$\underline{\mathbf{f}}_c(\underline{\mathbf{R}}, \omega) = \underline{\mathbf{0}}, \tag{15.186}$$

$$\begin{aligned}\underline{\underline{\mathbf{h}}}_c(\underline{\mathbf{R}}, \omega) &= -\underline{\underline{\underline{\underline{\mathbf{I}}}}}^+ : \underline{\underline{\boldsymbol{\gamma}}}_c(\underline{\mathbf{R}})\underline{\mathbf{v}}(\underline{\mathbf{R}}, \omega) \\ &= \mathrm{j}\omega\,\underline{\underline{\underline{\underline{\mathbf{I}}}}}^+ : \underline{\underline{\boldsymbol{\gamma}}}_c(\underline{\mathbf{R}})\underline{\mathbf{u}}(\underline{\mathbf{R}}, \omega)\end{aligned} \tag{15.187}$$

yield (15.180) through (15.185) to be a scattered far-field of a scatterer with a stress-free surface, and specifying according to

$$\begin{aligned}\underline{\mathbf{f}}_c(\underline{\mathbf{R}}, \omega) &= -\mathrm{j}\omega\Gamma_c(\underline{\mathbf{R}})\,[\rho - \rho(\underline{\mathbf{R}})]\,\underline{\mathbf{v}}(\underline{\mathbf{R}}, \omega) \\ &= -\omega^2\Gamma_c(\underline{\mathbf{R}})\,[\rho - \rho(\underline{\mathbf{R}})]\,\underline{\mathbf{u}}(\underline{\mathbf{R}}, \omega),\end{aligned} \tag{15.188}$$

$$\underline{\underline{\mathbf{h}}}_c(\underline{\mathbf{R}}, \omega) = -\mathrm{j}\omega\Gamma_c(\underline{\mathbf{R}})\left[\underline{\underline{\underline{\underline{\mathbf{s}}}}} - \underline{\underline{\underline{\underline{\mathbf{s}}}}}(\underline{\mathbf{R}})\right] : \underline{\underline{\mathbf{T}}}(\underline{\mathbf{R}}, \omega) \tag{15.189}$$

yields a scattered far-field of a penetrable inhomogeneous anisotropic scatterer. The under brackets in (15.181) and (15.182), respectively (15.184) and (15.185), state the proportionality of scattered far-fields to the three-dimensionally Fourier transformed secondary sources on the respective Ewald spheres $\underline{\mathbf{K}} = k_{\mathrm{P,S}}\hat{\underline{\mathbf{R}}}$ as it is true for the far-fields of primary sources.[297] Hence,

[297] In these Fourier transforms $\underline{\mathbf{K}} = k_{\mathrm{P,S}}\hat{\underline{\mathbf{R}}}$ has to be inserted as Fourier variable; therefore, $\underline{\mathbf{K}}$ varies on so-called Ewald spheres with radii $K_{\mathrm{P,S}}$ for fixed wave numbers $k_{\mathrm{P,S}}$ if $\hat{\underline{\mathbf{R}}}$ varies.

monochromatic far-field measurements[298] of scattered fields, even though being recorded for all observation directions $\hat{\underline{\mathbf{R}}}$ (on the unit sphere), only contain the Ewald sphere information of the Fourier spectra, which is not sufficient to image the scatterer (the defect) because not even a partial volume of $\underline{\mathbf{K}}$-space is covered that way. One has to switch, for example, to a broadband impulse excitation providing spherical Ewald shell information. In the end, ultrasonic imaging techniques such as SAFT and its modifications use exactly this information (Chapter 16; Langenberg 1987; Langenberg et al. 1993a, 1999a, 2002; Langenberg 2002; Langenberg et al. 2004a, 2004b; Mayer et al. 1990; Marklein et al. 2002b; Kostka et al. 1998; Langenberg et al. 2006, 2007; Zimmer 2007).

Obviously, instead of (15.187), we may eventually use the Kirchhoff approximation and, instead of (15.188) and (15.189), the Born approximation; we then obtain Kirchhoff approximated, respectively Born approximated, scattered fields.

A particular compact notation of scattered far-fields for incident plane P,S-waves is obtained defining scattering tensors (of second rank). We write the secondary sources for these incident waves (15.188) and (15.189) utilizing (15.170) and (15.171) as well as (15.169) according to

$$\begin{aligned}\underline{\mathbf{f}}_c(\underline{\mathbf{R}}, \omega, \underline{\mathbf{k}}_{i\mathrm{P,S}}) = \mathrm{j}\omega\rho\, \chi_\rho(\underline{\mathbf{R}}) \Big[\underline{\underline{\boldsymbol{\Xi}}}_c^{\rho,\mathbf{c}}(\underline{\mathbf{R}}, \omega, \underline{\mathbf{k}}_{i\mathrm{P,S}})\, \mathrm{e}^{-\mathrm{j}k_{\mathrm{P,S}}\hat{\underline{\mathbf{k}}}_i\cdot\underline{\mathbf{R}}} \\ + v_{i\mathrm{P,S}}(\omega)\underline{\underline{\mathbf{I}}} \Big] \cdot \hat{\underline{\mathbf{v}}}_{i\mathrm{P,S}}\, \mathrm{e}^{\mathrm{j}k_{\mathrm{P,S}}\hat{\underline{\mathbf{k}}}_i\cdot\underline{\mathbf{R}}},\end{aligned} \tag{15.190}$$

$$\begin{aligned}\underline{\underline{\mathbf{h}}}_c(\underline{\mathbf{R}}, \omega, \underline{\mathbf{k}}_{i\mathrm{P,S}}) = -\mathrm{j}\omega\, \underline{\underline{\underline{\underline{\boldsymbol{\chi}}}}}_{\mathbf{c}}(\underline{\mathbf{R}}) : \underline{\underline{\underline{\underline{\mathbf{s}}}}}(\underline{\mathbf{R}}) : \Big[\underline{\underline{\underline{\boldsymbol{\Upsilon}}}}_c^{\mathrm{r},\mathbf{c}}(\underline{\mathbf{R}}, \omega, \underline{\mathbf{k}}_{i\mathrm{P,S}})\, \mathrm{e}^{-\mathrm{j}k_{\mathrm{P,S}}\hat{\underline{\mathbf{k}}}_i\cdot\underline{\mathbf{R}}} \\ - \frac{v_{i\mathrm{P,S}}(\omega)}{\omega} (\lambda\, \underline{\underline{\mathbf{I}}}\, \underline{\mathbf{k}}_{i\mathrm{P,S}} + 2\mu\, \underline{\underline{\underline{\underline{\mathbf{I}}}}}^+ \cdot \underline{\mathbf{k}}_{i\mathrm{P,S}}) \Big] \cdot \hat{\underline{\mathbf{v}}}_{i\mathrm{P,S}}\, \mathrm{e}^{\mathrm{j}k_{\mathrm{P,S}}\hat{\underline{\mathbf{k}}}_i\cdot\underline{\mathbf{R}}},\end{aligned} \tag{15.191}$$

and this results in

$$\underline{\mathbf{u}}_{s\mathrm{P,S}}^{\mathrm{far}}(\hat{\underline{\mathbf{R}}}, \omega, \underline{\mathbf{k}}_{i\mathrm{P,S}}) = \frac{\mathrm{e}^{\mathrm{j}k_{\mathrm{P,S}}R}}{R}\, \underline{\underline{\mathbf{U}}}_{\mathrm{P,S}}(\hat{\underline{\mathbf{R}}}, \omega, \underline{\mathbf{k}}_{i\mathrm{P,S}}) \cdot \hat{\underline{\mathbf{v}}}_{i\mathrm{P,S}} \tag{15.192}$$

with

$$\underline{\underline{\mathbf{U}}}_{\mathrm{P,S}}(\hat{\underline{\mathbf{R}}}, \omega, \underline{\mathbf{k}}_{i\mathrm{P,S}}) = \underline{\underline{\boldsymbol{\Xi}}}_{\mathrm{P,S}}^{\rho,\mathbf{c}}(\hat{\underline{\mathbf{R}}}, \omega, \underline{\mathbf{k}}_{i\mathrm{P,S}}) + \underline{\underline{\boldsymbol{\Upsilon}}}_{\mathrm{P,S}}^{\mathrm{r},\mathbf{c}}(\hat{\underline{\mathbf{R}}}, \omega, \underline{\mathbf{k}}_{i\mathrm{P,S}}), \tag{15.193}$$

$$\begin{aligned}\underline{\underline{\boldsymbol{\Xi}}}_{\mathrm{P}}^{\rho,\mathbf{c}}(\hat{\underline{\mathbf{R}}}, \omega, \underline{\mathbf{k}}_{i\mathrm{P,S}}) = \mathrm{j}\, \frac{k_{\mathrm{P}}}{4\pi c_{\mathrm{P}}}\, \hat{\underline{\mathbf{R}}}\hat{\underline{\mathbf{R}}} \cdot \iiint_{V_c} \chi_\rho(\underline{\mathbf{R}}') \\ \times \Big[\underline{\underline{\boldsymbol{\Xi}}}_c^{\rho,\mathbf{c}}(\underline{\mathbf{R}}', \omega, \underline{\mathbf{k}}_{i\mathrm{P,S}}) \mathrm{e}^{-\mathrm{j}\underline{\mathbf{k}}_{i\mathrm{P,S}}\cdot\underline{\mathbf{R}}'} + v_{i\mathrm{P,S}}(\omega)\underline{\underline{\mathbf{I}}} \Big] \\ \times\, \mathrm{e}^{-\mathrm{j}(k_{\mathrm{P}}\hat{\underline{\mathbf{R}}} - \underline{\mathbf{k}}_{i\mathrm{P,S}})\cdot\underline{\mathbf{R}}'}\, \mathrm{d}^3\underline{\mathbf{R}}',\end{aligned} \tag{15.194}$$

---

[298] This holds similarly for near-field measurements if the measurement surface is located outside the smallest sphere hosting the scatterer (Colton and Kress 1983; Dassios and Kleinman 2000).

$$\underline{\underline{\boldsymbol{\Upsilon}}}_{\mathrm{P}}^{\mathrm{r,c}}(\underline{\mathbf{R}}, \omega, \underline{\mathbf{k}}_{i\mathrm{P,S}}) = \mathrm{j}\, \frac{k_{\mathrm{P}}}{4\pi c_{\mathrm{P}} Z_{\mathrm{P}}}\, \hat{\underline{\mathbf{R}}}(\lambda\, \underline{\underline{\mathbf{I}}} + 2\mu\, \hat{\underline{\mathbf{R}}}\, \hat{\underline{\mathbf{R}}})$$
$$: \iiint_{V_c} \underline{\underline{\underline{\underline{\chi}}}}_{\mathbf{c}}(\underline{\mathbf{R}}') : \underline{\underline{\underline{\underline{\mathbf{s}}}}}(\underline{\mathbf{R}}') : \Bigg[\underline{\underline{\underline{\boldsymbol{\Upsilon}}}}_{c}^{\rho,\mathbf{c}}(\underline{\mathbf{R}}', \omega, \underline{\mathbf{k}}_{i\mathrm{P,S}}) \mathrm{e}^{-\mathrm{j}\underline{\mathbf{k}}_{i\mathrm{P,S}}\cdot\underline{\mathbf{R}}'}$$
$$- \frac{v_{i\mathrm{P,S}}(\omega)}{\omega} (\lambda\, \underline{\underline{\mathbf{I}}}\, \underline{\mathbf{k}}_{i\mathrm{P,S}} + 2\mu\, \underline{\underline{\underline{\underline{\mathbf{I}}}}}^{+} \cdot \underline{\mathbf{k}}_{i\mathrm{P,S}}) \Bigg] \mathrm{e}^{-\mathrm{j}(k_{\mathrm{P}}\hat{\underline{\mathbf{R}}} - \underline{\mathbf{k}}_{i\mathrm{P,S}})\cdot\underline{\mathbf{R}}'}\, \mathrm{d}^3\underline{\mathbf{R}}', \quad (15.195)$$

$$\underline{\underline{\boldsymbol{\Xi}}}_{\mathrm{S}}^{\rho,\mathbf{c}}(\underline{\mathbf{R}}, \omega, \underline{\mathbf{k}}_{i\mathrm{P,S}}) = \mathrm{j}\, \frac{k_{\mathrm{S}}}{4\pi c_{\mathrm{S}}}\, (\underline{\underline{\mathbf{I}}} - \hat{\underline{\mathbf{R}}}\, \hat{\underline{\mathbf{R}}})$$
$$\cdot \iiint_{V_c} \chi_\rho(\underline{\mathbf{R}}') \left[\underline{\underline{\boldsymbol{\Xi}}}_{c}^{\rho,\mathbf{c}}(\underline{\mathbf{R}}', \omega, \underline{\mathbf{k}}_{i\mathrm{P,S}}) \mathrm{e}^{-\mathrm{j}\underline{\mathbf{k}}_{i\mathrm{P,S}}\cdot\underline{\mathbf{R}}'} + v_{i\mathrm{P,S}}(\omega) \underline{\underline{\mathbf{I}}} \right]$$
$$\times\, \mathrm{e}^{-\mathrm{j}(k_{\mathrm{S}}\hat{\underline{\mathbf{R}}} - \underline{\mathbf{k}}_{i\mathrm{P,S}})\cdot\underline{\mathbf{R}}'}\, \mathrm{d}^3\underline{\mathbf{R}}', \quad (15.196)$$

$$\underline{\underline{\boldsymbol{\Upsilon}}}_{\mathrm{S}}^{\mathrm{r,c}}(\underline{\mathbf{R}}, \omega, \underline{\mathbf{k}}_{i\mathrm{P,S}}) = \mathrm{j}\, \frac{k_{\mathrm{S}}}{2\pi}\, (\underline{\underline{\mathbf{I}}} - \hat{\underline{\mathbf{R}}}\, \hat{\underline{\mathbf{R}}}) \hat{\underline{\mathbf{R}}} : \iiint_{V_c} \underline{\underline{\underline{\underline{\chi}}}}_{\mathbf{c}}(\underline{\mathbf{R}}') : \underline{\underline{\underline{\underline{\mathbf{s}}}}}(\underline{\mathbf{R}}')$$
$$: \left[\underline{\underline{\underline{\boldsymbol{\Upsilon}}}}_{c}^{\rho,\mathbf{c}}(\underline{\mathbf{R}}', \omega, \underline{\mathbf{k}}_{i\mathrm{P,S}}) \mathrm{e}^{-\mathrm{j}\underline{\mathbf{k}}_{i\mathrm{P,S}}\cdot\underline{\mathbf{R}}'} - \frac{v_{i\mathrm{P,S}}(\omega)}{\omega} (\lambda\, \underline{\underline{\mathbf{I}}}\, \underline{\mathbf{k}}_{i\mathrm{P,S}} + 2\mu\, \underline{\underline{\underline{\underline{\mathbf{I}}}}}^{+} \cdot \underline{\mathbf{k}}_{i\mathrm{P,S}}) \right]$$
$$\times\, \mathrm{e}^{-\mathrm{j}(k_{\mathrm{S}}\hat{\underline{\mathbf{R}}} - \underline{\mathbf{k}}_{i\mathrm{P,S}})\cdot\underline{\mathbf{R}}'}\, \mathrm{d}^3\underline{\mathbf{R}}', \quad (15.197)$$

if they are inserted into (15.181) and (15.182). Within the far-field approximation (15.192) of the particle displacement, the scattering tensor $\underline{\underline{\mathbf{U}}}_{\mathrm{P,S}}(\hat{\underline{\mathbf{R}}}, \omega, \underline{\mathbf{k}}_{i\mathrm{P,S}})$—its definition refers to the theory of electromagnetic waves (Langenberg 2005; Baum 2000)—relates the given polarization of the incident plane wave to the resulting polarization of the scattered field; it is composed of the terms $\underline{\underline{\boldsymbol{\Xi}}}_{\mathrm{P,S}}^{\rho,\mathbf{c}}(\hat{\underline{\mathbf{R}}}, \omega, \underline{\mathbf{k}}_{i\mathrm{P,S}})$ and $\underline{\underline{\boldsymbol{\Upsilon}}}_{\mathrm{P,S}}^{\rho,\mathbf{c}}(\hat{\underline{\mathbf{R}}}, \omega, \underline{\mathbf{k}}_{i\mathrm{P,S}})$ related to the density contrast $\chi_\rho(\underline{\mathbf{R}})$ and the stiffness contrast $\underline{\underline{\underline{\underline{\chi}}}}_{\mathbf{c}}(\underline{\mathbf{R}})$. Remote sensing with electromagnetic waves (Ulaby and Elachi 1990; Cloude 2002; Langenberg 2005) relies on a successful theory of object identification based on an algebraic analysis of the scattering tensor, respectively consecutively derived scattering matrices; these results still wait to be applied in US-NDT. Yet, fundamental facts for backscattering (pulse echo: $\hat{\underline{\mathbf{R}}} = -\hat{\underline{\mathbf{k}}}_i$) are immediately evident: The right-handed contraction of the scattering tensor with the polarization vector $\hat{\underline{\mathbf{v}}}_{i\mathrm{P,S}}$ (we could equally write $\hat{\underline{\mathbf{u}}}_{i\mathrm{P,S}}$ because we deal with a unit vector) results in a right-handed contraction of $\underline{\underline{\boldsymbol{\Xi}}}_{\mathrm{P,S}}^{\rho,\mathbf{c}}(\hat{\underline{\mathbf{R}}}, \omega, \underline{\mathbf{k}}_{i\mathrm{P,S}})$, respectively $\underline{\underline{\boldsymbol{\Upsilon}}}_{\mathrm{P,S}}^{\rho,\mathbf{c}}(\hat{\underline{\mathbf{R}}}, \omega, \underline{\mathbf{k}}_{i\mathrm{P,S}})$ with $\hat{\underline{\mathbf{v}}}_{i\mathrm{P,S}}$, and without further specification of theses tensors, no general result can be obtained; yet, in case of the Born approximation, $\underline{\underline{\boldsymbol{\Xi}}}_{\mathrm{P,S}}^{\rho,\mathbf{c}}(\hat{\underline{\mathbf{R}}}, \omega, \underline{\mathbf{k}}_{i\mathrm{P,S}})$ and $\underline{\underline{\boldsymbol{\Upsilon}}}_{\mathrm{P,S}}^{\rho,\mathbf{c}}(\hat{\underline{\mathbf{R}}}, \omega, \underline{\mathbf{k}}_{i\mathrm{P,S}})$ are deleted providing the following conclusions for isotropic scatterers:

- Under the Born approximation, there is no mode conversion in backscattering direction, i.e., an incident pressure wave is backscattered as a pure pressure wave and an incident shear wave as pure shear wave;

- Under the Born approximation, the scattered shear far-field in backscattering direction is equally polarized as the incident shear wave, there is no polarization rotation.

By the way, the same facts also hold for scatterers with a stress-free surface under the Kirchhoff approximation: The mirror point for backscattering is then given by $\underline{\mathbf{n}}_c = -\hat{\underline{\mathbf{k}}}_i$ (Figure 15.9), yielding the same direction of the surface particle displacement as the incident plane wave according to (15.145), respectively (15.151); this statement is obtained via the same calculation as above.

In reality, Born and Kirchhoff approaches may only be approximations identifying the deviation from the above statements as a measure for the validity of these approaches.

15.4.2 Two-dimensional scalar scattering problems: Pulsed SH-far-fields of circular cylindrical voids and strip-like cracks

Circular cylindrical void: SH-wave scattering is a scalar problem for two-dimensional scatterers, in particular, a scalar Neumann problem for voids with a stress-free surface (Section 7.3). We choose a circular cylindrical void with radius a (Figure 15.12) in steel ($\kappa = c_P/c_S = 1.827$)—model of a side wall drilled hole for US-NDT—and investigate the scattering of a plane SH-wave

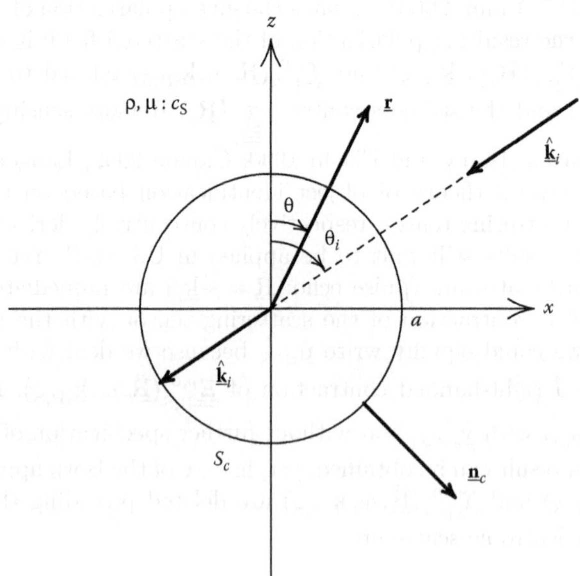

FIGURE 15.12
SH-wave scattering for a circular cylindrical void with a stress-free surface.

$$\underline{\mathbf{u}}_{i\mathrm{SH}}(\underline{\mathbf{r}}, \omega, \hat{\underline{\mathbf{k}}}_i) = -u_{i\mathrm{S}}(\omega)\, \mathrm{e}^{\mathrm{j}k_\mathrm{S}\hat{\underline{\mathbf{k}}}_i\cdot\underline{\mathbf{r}}}\, \underline{\mathbf{e}}_y, \tag{15.198}$$

with a polarization vector oriented in negative y-direction (parallel to the cylinder axis) as in Section 9.1.3. With

$$\hat{\underline{\mathbf{k}}}_i = -\sin\theta_i\, \underline{\mathbf{e}}_x - \cos\theta_i\, \underline{\mathbf{e}}_z, \tag{15.199}$$

$$\underline{\mathbf{r}} = r\sin\theta\, \underline{\mathbf{e}}_x + r\cos\theta\, \underline{\mathbf{e}}_z, \tag{15.200}$$

we obtain

$$\hat{\underline{\mathbf{k}}}_i \cdot \underline{\mathbf{r}} = -r\cos(\theta - \theta_i) \tag{15.201}$$

in polar coordinates in the xz-plane. For $r > a$, the only nonvanishing y-component of the scattered SH-field $\underline{\mathbf{u}}_{s\mathrm{SH}}(r, \theta, \omega) = u_{s\mathrm{SH}}(r, \theta, \omega)\underline{\mathbf{e}}_y$ satisfies the homogeneous reduced wave equation (7.47), namely

$$\Delta u_{s\mathrm{SH}}(r, \theta, \omega) + k_\mathrm{S}^2 u_{s\mathrm{SH}}(r, \theta, \omega) = 0, \tag{15.202}$$

and for $r = a$ the homogeneous Neumann boundary condition (Equation 7.53)

$$\left.\frac{\partial u_{s\mathrm{SH}}(r, \theta, \omega)}{\partial r}\right|_{r=a} = -\left.\frac{\partial u_{i\mathrm{SH}}(r, \theta, \omega)}{\partial r}\right|_{r=a} \tag{15.203}$$

of the total field. We conjecture that the scattered field can be represented as superposition of outgoing cylindrical waves and write (15.202) in polar coordinates:

$$\left[\frac{1}{r}\frac{\partial}{\partial r}\left(r\frac{\partial}{\partial r}\right) + \frac{1}{r^2}\frac{\partial^2}{\partial\theta^2} + k_\mathrm{S}^2\right] u_{s\mathrm{SH}}(r, \theta, \omega) = 0. \tag{15.204}$$

The so-called partial wave separation *ansatz*

$$u_{s\mathrm{SH}}(r, \theta, \omega) = \sum_{n=-\infty}^{\infty} a_n(\omega) R_n(r)\, \mathrm{e}^{\mathrm{j}n\theta} \tag{15.205}$$

turns (15.204) into a Bessel differential equation for the radial functions $R_n(r)$ (Schäfke 1967):

$$\left(\frac{\partial^2}{\partial r^2} + \frac{1}{r}\frac{\partial}{\partial r} - \frac{n^2}{r^2} + k_\mathrm{S}^2\right) R_n(r) = 0. \tag{15.206}$$

"Matching" solutions, i.e., those representing outgoing waves, are Hankel functions $\mathrm{H}_n^{(1)}(k_\mathrm{S}r)$ of the first kind of order n if the time dependence has been chosen according to $\mathrm{e}^{-\mathrm{j}\omega t}$, namely with the negative sign in the exponent of the inverse Fourier transform as always in this elaboration. That way, the partial wave amplitudes (the expansion coefficients) $a_n(\omega)$ in

$$u_{s\mathrm{SH}}(r, \theta, \omega) = \sum_{n=-\infty}^{\infty} a_n(\omega) \mathrm{H}_n^{(1)}(k_\mathrm{S}r)\, \mathrm{e}^{\mathrm{j}n\theta} \tag{15.207}$$

may be calculated with the help of the boundary condition (15.203) provided the respective partial wave separation for the incident plane wave (15.198) is available; we have (Schäfke 1967)

$$\mathrm{e}^{-\mathrm{j}k_{\mathrm{S}}r\cos(\theta-\theta_i)} = \sum_{n=-\infty}^{\infty} (-\mathrm{j})^n \mathrm{J}_n(k_{\mathrm{S}}r)\,\mathrm{e}^{\mathrm{j}n(\theta-\theta_i)}; \tag{15.208}$$

only the solutions of (15.206) free of singularities, the Bessel functions $\mathrm{J}_n(k_{\mathrm{S}}r)$ of order n appear because the plane wave has no singularities. With (15.207) and (15.208), we now obtain from (15.203)

$$a_n(\omega) = u_{i\mathrm{S}}(\omega)(-\mathrm{j})^n \frac{\mathrm{J}_n'(k_{\mathrm{S}}a)}{\mathrm{H}_n^{(1)'}(k_{\mathrm{S}}a)}\,\mathrm{e}^{-\mathrm{j}n\theta_i} \tag{15.209}$$

due to the orthogonality relation

$$\int_0^{2\pi} \mathrm{e}^{\mathrm{j}n\theta}\mathrm{e}^{-\mathrm{j}m\theta}\,\mathrm{d}\theta = 2\pi\delta_{nm}, \tag{15.210}$$

where the dash on the cylinder functions stands for a derivative with regard to the argument (not with regard to r!). With the asymptotics

$$\mathrm{H}_n^{(1)}(k_{\mathrm{S}}r) \simeq \sqrt{\frac{2}{\pi k_{\mathrm{S}}r}}\,\mathrm{e}^{\mathrm{j}k_{\mathrm{S}}r-\mathrm{j}n\frac{\pi}{2}-\mathrm{j}\frac{\pi}{4}} \tag{15.211}$$

of the Hankel functions (Abramowitz and Stegun 1965), we finally find

$$u_{s\mathrm{SH}}^{\mathrm{far}}(r,\theta,\omega) = \frac{\mathrm{e}^{\mathrm{j}k_{\mathrm{S}}r}}{\sqrt{r}}\,H_{\mathrm{SH}}(\theta,\omega), \tag{15.212}$$

$$H_{\mathrm{SH}}(\theta,\omega) = u_{i\mathrm{S}}(\omega)\mathrm{e}^{-\mathrm{j}\frac{\pi}{4}}\sqrt{\frac{2}{\pi k_{\mathrm{S}}}}\sum_{n=-\infty}^{\infty}(-1)^n\frac{\mathrm{J}_n'(k_{\mathrm{S}}r)}{\mathrm{H}_n^{(1)'}(k_{\mathrm{S}}a)}\,\mathrm{e}^{\mathrm{j}n(\theta-\theta_i)} \tag{15.213}$$

for the scattered far-field.

It should be noted that the partial wave separation method can also be applied to SH-wave scattering by a homogeneous penetrable cylinder: Only the emerging interior field must be expanded into standing wave cylinder functions, i.e., Bessel functions; its expansion coefficients as well as those for the exterior scattered field result from the transition conditions (7.51) and (7.52).

The above derivation of the SH-scattered field of a circular cylinder with a stress-free surface circumvents the field representation as a Huygens integral because the reduced wave equation is initially solved by a series expansion into the special functions of the underlying coordinate system (here: cylinder coordinates) (Langenberg 2005). This works only in so-called separable coordinate systems allowing for a representation of the scattering surface in terms of a coordinate surface and, hence, not for arbitrary scattering geometries; therefore, generally the integral equations resulting from Huygens' principle have to solved numerically. Nevertheless, we want briefly sketch how to deal with the Huygens integral in the case of the cylinder geometry even though this is a circumvention. Specializing (5.85) to SH-waves and to the actual geometry including the boundary condition results in

$$u_{s\mathrm{SH}}(r,\theta,\omega) = a\int_0^{2\pi} u_{\mathrm{SH}}(a,\theta',\omega)\,\frac{\partial}{\partial r'}G(\underline{\mathbf{r}}-\underline{\mathbf{r}}',\omega)\bigg|_{r'=a}\,\mathrm{d}\theta' \qquad (15.214)$$

as a Huygens-type scattered field representation, where $G(\underline{\mathbf{r}}-\underline{\mathbf{r}}',\omega)$ is the two-dimensional scalar Green function (13.23). The integral equation (of the second kind) for the secondary source $u_{\mathrm{SH}}(\theta,\omega) \stackrel{\mathrm{def}}{=} u_{\mathrm{SH}}(a,\theta,\omega) = u_{s\mathrm{SH}}(a,\theta,\omega) + u_{i\mathrm{SH}}(a,\theta,\omega)$ follows from (5.90):

$$u_{i\mathrm{SH}}(a,\theta,\omega) = \frac{1}{2}u_{\mathrm{SH}}(\theta,\omega) - a\int_0^{2\pi} u_{\mathrm{SH}}(\theta',\omega)\,\frac{\partial}{\partial r'}G(\underline{\mathbf{r}}-\underline{\mathbf{r}}',\omega)\bigg|_{r'=a}\,\mathrm{d}\theta',$$
$$0 \le \theta < 2\pi. \qquad (15.215)$$

The solution of (15.215) now follows the above separation scheme: The field $u_{\mathrm{SH}}(\theta,\omega)$ is expanded according to

$$u_{\mathrm{SH}}(\theta,\omega) = \sum_{n=-\infty}^{\infty} b_n(\omega)\,\mathrm{e}^{\mathrm{j}n\theta} \qquad (15.216)$$

as well as (Hönl et al. 1961)

$$\begin{aligned}\frac{\partial}{\partial r'}&G(\underline{\mathbf{r}}-\underline{\mathbf{r}}',\omega)\bigg|_{r'=a}\\ &= \frac{\mathrm{j}}{4}\frac{\partial}{\partial r'}\mathrm{H}_0^{(1)}\left(k_{\mathrm{S}}\sqrt{a^2+r'^2-2ar'\cos(\theta-\theta')}\right)\bigg|_{r'=a}\\ &= \frac{\mathrm{j}}{8}\,k_{\mathrm{S}}\sum_{m=-\infty}^{\infty}\left[\mathrm{H}_m^{(1)\prime}(k_{\mathrm{S}}a)\mathrm{J}_m(k_{\mathrm{S}}a) + \mathrm{H}_m^{(1)}(k_{\mathrm{S}}a)\mathrm{J}_m'(k_{\mathrm{S}}a)\right]\mathrm{e}^{\mathrm{j}m(\theta-\theta')}\end{aligned} \qquad (15.217)$$

into a Fourier series with regard to θ, respectively θ'; these series are inserted into (15.215), and the orthogonality relation (15.210) as well as the Wronski determinant between J_m and $\mathrm{H}_m^{(1)}$ (Schäfke 1967) are applied resulting in $b_n(\omega)$; now (15.216) and (15.217) are inserted into (15.214), and similarly as before, we find $u_{s\mathrm{SH}}$ according to (15.207) with (15.209).

To model a US-NDT problem with pulsed excitation, we have to evaluate (15.212) with (15.213) for all circular frequencies ω within the bandwidth of $u_{i\mathrm{S}}(\omega)$ applying a subsequent inverse Fourier transform. We choose $u_{i\mathrm{S}}(\omega)$ as spectrum of an origin symmetric, for the negative y-component of the particle displacement (15.198) positive RC2-pulse (Figure 2.20); likewise, the pulsed far-field in backscattering direction—this comes from the numerical evaluation—approximately exhibits the structure of a positive RC2-pulse in the negative y-component (compare the reflection of a plane SH-wave at a stress-free planar boundary in Section 9.1.3: The reflected pulse does not exhibit a sign change) stimulating us to plot $-u_{s\mathrm{SH}}^{\mathrm{RC2,far}}(r,\theta,t)$ in Figure 15.13;

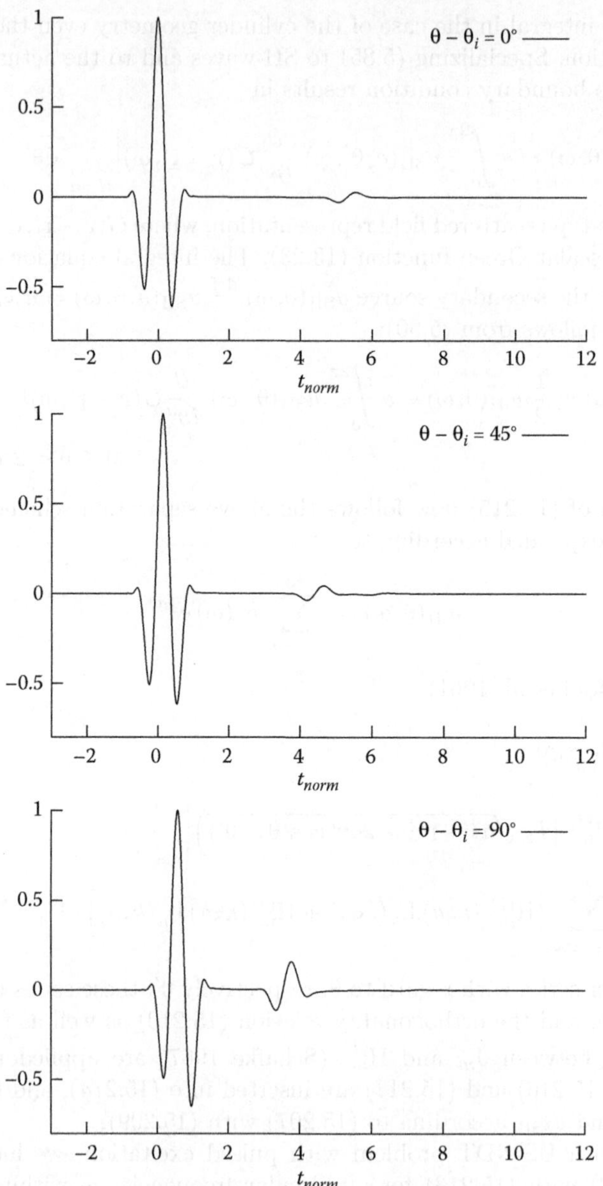

FIGURE 15.13
RC2-far-field scattered pulses $-u_{s\mathrm{SH}}^{\mathrm{RC2,far}}(r, \theta, t)$ for an incident plane SH-wave impinging on a circular cylinder with a stress-free surface (respective amplitudes normalized to the maximum value).

the total duration of the RC2-pulse is given by $T = 2a/c_S$ corresponding to the travel time of the shear wave along the diameter of the cylinder. The pulsed cylinder wavefront of the scattered far-field requires the travel time $t_r = r/c_S$ from the coordinate origin at the cylinder center to the observation point $\underline{\mathbf{R}}$ that is disregarded on a normalized time axis. The time origin adjusted to the pulsed incident plane wave coincides with the transmission of the RC2-pulse maximum through the plane $\hat{\underline{\mathbf{k}}}_i \cdot \underline{\mathbf{r}} = 0$; therefore, the pulse reflected at the cylinder front surface into backscattering direction arrives at time $t = t_r - 2t_a$ with $t_a = a/c_S$ at the observation point finally yielding

$$t_{\text{norm}} = \frac{c_S}{a} t - \frac{r}{a} + 2 \tag{15.218}$$

as normalized time axis for the scattered pulses in Figure 15.13.

Let us first consider the backscattered pulse for $\theta = \theta_i$: We recognize for $t_{\text{norm}} = 0$ the pulse being mirror reflected at the cylinder front surface exhibiting a certain asymmetry as compared to the symmetric RC2-pulse of the incident wave, which is not only related to the two-dimensionality of the problem (Figure 13.6); it essentially originates—even in the case of a strict validity of the Kirchhoff approximation (15.143)—from the adjacent points of the mirror point contributing to the backscattered pulse via Huygens integration (15.138). This first backscattered pulse is followed by a smaller so-called creeping wave pulse[299] for $t_{\text{norm}} \simeq 5$, whose physical nature is best explained by AFIT-simulations of the scattering process in Figures 15.14 through 15.17

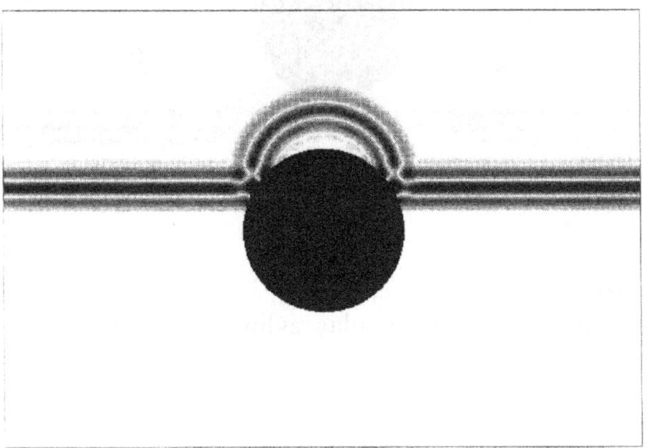

FIGURE 15.14
SH-RC2-pulse scattering for a circular cylinder with a stress-free surface: AFIT-simulation ($T = 2a/3c_S$).

[299]The Franz-type creeping waves circling a convex scatterer denote a precisely defined phenomenon in the literature (Hönl et al. 1961; Heyman and Felsen 1985); they have nothing to do with the radiation field of the "creeping wave transducer" (Langenberg et al. 1990).

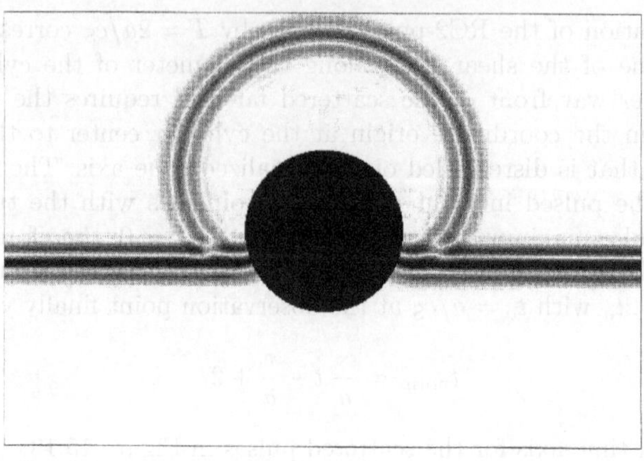

FIGURE 15.15
SH-RC2-pulse scattering for a circular cylinder with a stress-free surface: AFIT-simulation ($T = 2a/3c_S$).

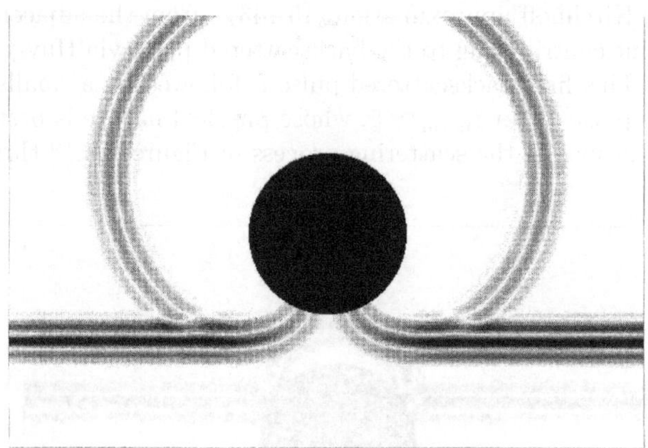

FIGURE 15.16
SH-RC2-pulse scattering for a circular cylinder with a stress-free surface: AFIT-simulation ($T = 2a/3c_S$).

(a scalar problem allows for the application of "AFIT"). Figure 15.14 initially illustrates the development of the primary scattered pulse; note that the superposition of this impulse with the incident plane wave on the surface of the cylinder is nonzero due to the Neumann boundary condition. After the incident wave has reached the time origin, it continues to travel straight ahead, yet it remains connected to the surface via the scattered wave (Figure 15.15) because the boundary condition must be satisfied for all times after the first contact with the scatterer. Figure 15.16 clearly shows the development

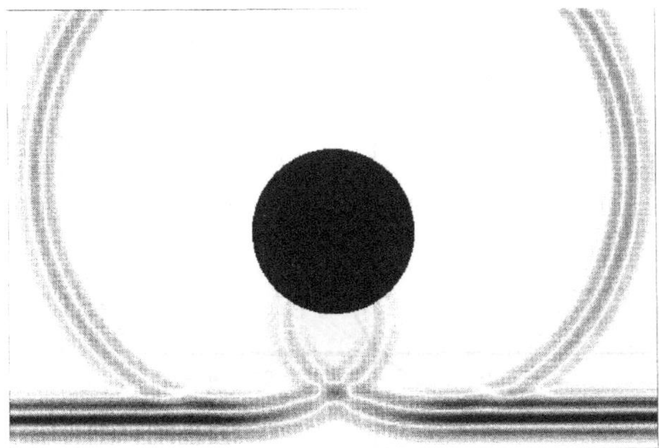

FIGURE 15.17
SH-RC2-pulse scattering for a circular cylinder with a stress-free surface: AFIT-simulation ($T = 2a/3c_S$).

of creeping waves resulting from the enforcement of the boundary condition: They stick to the surface of the scatterer while circling it, and on the rear surface, they cross each other (Figure 15.17); on the respective opposite side, they send wavefronts back to the observation point arriving there simultaneously. The travel time difference with regard to the mirror pulse is approximately calculated as follows: The mirror pulse is created at time $t_{\text{norm}} = 0$, both (symmetric) creeping waves exactly at $t_{\text{norm}} = 1$; then they circle the geometric shadow side of the circular cylinder during the time $t_{\text{norm}} = \pi$ and experience another delay $t_{\text{norm}} = 1$ compared to the mirror pulse while traveling to the (far-field) observation point resulting in a total delay of $t_{\text{norm}} = 1 + \pi + 1 \simeq 5$. Basically, an infinite series of creeping waves is created, yet their amplitudes decay rapidly leaving the second one already invisible within the chosen scale. Figure 15.17 also shows that the scatterer does not produce a geometric optical shadow.

It should be noted that the creeping wave phenomenon is a consequence of the exact calculation of the scattering problem: The Kirchhoff approximation does not contain creeping waves due to the zero secondary source on the shadow side. This jump discontinuity of the secondary source may lead to nonphysical "Kirchhoff signals" in the pulsed scattered field of a convex scatterer.

The AFIT-simulations obviously allow for an intuitive interpretation of the calculated A-scan in Figure 15.13, the single pulses can be uniquely allocated to physical phenomena. This is one basic advantage of such simulations. The two additional A-scans in Figure 15.13 have been calculated for $\theta - \theta_i = 45^\circ$ and $\theta - \theta_i = 90^\circ$; compared to backscattering, only the time delay and the amplitudes of the creeping waves change, and they do no longer arrive simultaneously at the observation point. The one with the smaller time delay

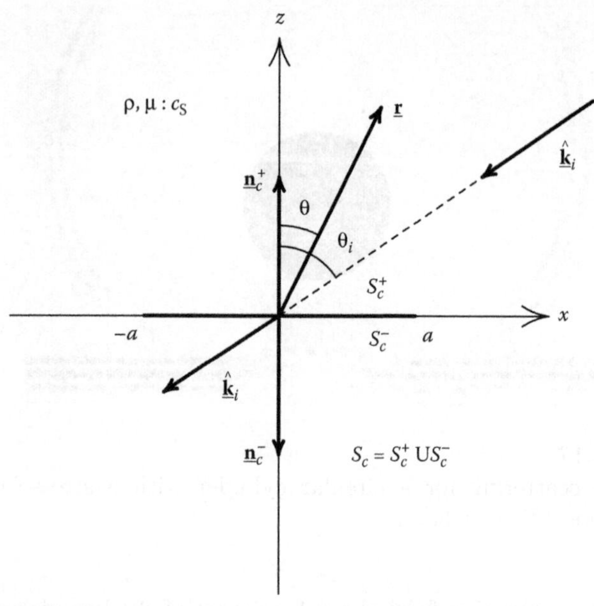

FIGURE 15.18
SH-wave scattering by a strip-like crack with a stress-free surface.

relative to the mirror pulse increases in visibility while the other one is already below the recording level due to the longer creeping path.

Strip-like crack: The strip-like crack may be considered as the limiting case of a cylinder with elliptical cross-section and stress-free surface. Advantage: The reduced wave equation is also separable in elliptical coordinates, the eigenfunctions are Mathieu functions (Schäfke 1967) to be calculated in terms of series involving Bessel functions. This method yielded the result cited in Section 5.6 for acoustic wave scattering by a rigid crack (Figure 5.3); here, we only have to reinterpret the physical quantities in terms of SH-wave scattering. Typically, in backscattering direction two scattered pulses are observed—again we have $T = 2a/c_S$—that emanate from the edges of the crack; the one from the farther edge is bigger in amplitude and exhibits a negative phase compared to the one from the closer edge. The edges may be considered as line sources, and, therefore, for an incident RC2-pulse, the time structure of the respectively bandlimited two-dimensional Green function according to Figure 13.5 is recovered.

Figures 15.19 through 15.20 illustrate the formation of the far-field pulses in Figure 5.3 with the help of AFIT-simulations, where Figure 15.20 illustrates the subsequent scattering of an edge pulse at the opposite edge; this basically leads to an infinite series of edge interaction pulses (with decreasing

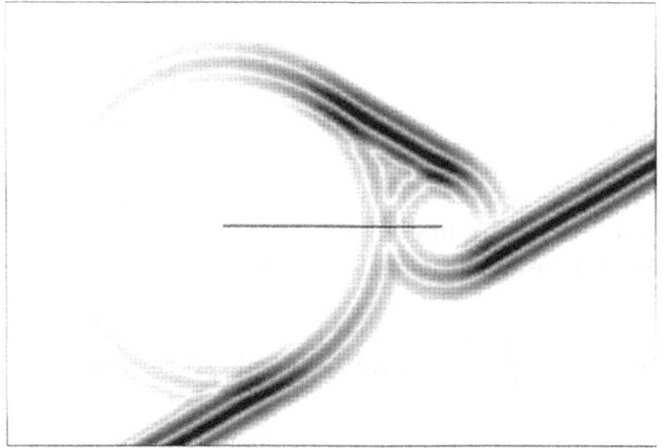

FIGURE 15.19
SH-RC2-pulse scattering by a strip-like crack with stress-free surface: AFIT-simulation: $\theta_i = -45^\circ$ ($T = 0.5a/c_S$).

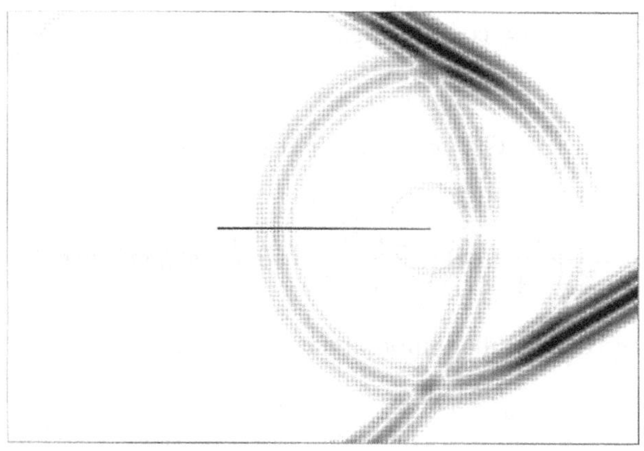

FIGURE 15.20
SH-RC2-pulse scattering by a strip-like crack with stress-free surface: AFIT-simulation: $\theta_i = -45^\circ$ ($T = 0.5a/c_S$).

amplitudes) appearing as resonances in the frequency spectrum; being a consequence of radiation interaction, they are not contained in the Kirchhoff approximation. Figures 15.32 and 15.33 additionally exhibit AFIT-simulations for perpendicular incidence.

In Section 5.6, we showed results in Figure 5.4 referring to acoustic crack scattering obtained with the Kirchhoff approximation. Here, we want to write down the respective equations for SH-wave scattering, last but not least to give the pulse structures in this figure, and, hence, in Figure 5.3 explicitly. First, we specify the scalar acoustic Huygens integral (5.85) to the actual physical situation:

$$u_{s\mathrm{SH}}(\underline{\mathbf{r}}, \omega) = \int_{S_c^+ \cup S_c^-} u_{\mathrm{SH}}(\underline{\mathbf{r}}', \omega) \nabla' G(\underline{\mathbf{r}} - \underline{\mathbf{r}}', \omega) \cdot \underline{\mathbf{n}}_c' \, \mathrm{d}s'; \tag{15.219}$$

here, $S_c = S_c^+ \cup S_c^-$ is the crack contour—the $\mathrm{d}s'$-integration extends along this line—composed of an upper and lower surface (Figure 15.18), and $G(\underline{\mathbf{r}} - \underline{\mathbf{r}}', \omega)$ is the two-dimensional Green function. In the far-field, (15.219) reduces to

$$u_{s\mathrm{SH}}^{\mathrm{far}}(\underline{\mathbf{r}}, \omega) = -\mathrm{j}k_{\mathrm{S}} \int_{S_c^+ \cup S_c^-} u_{\mathrm{SH}}(\underline{\mathbf{r}}', \omega) \, \hat{\underline{\mathbf{r}}} \cdot \underline{\mathbf{n}}_c' \, G^{\mathrm{far}}(\underline{\mathbf{r}}, \underline{\mathbf{r}}', \omega) \, \mathrm{d}s'. \tag{15.220}$$

According to (15.105), we define the crack opening displacement

$$u_{\mathrm{SH}}^{\mathrm{cod}}(x', \omega) = u_{\mathrm{SH}}(x', z' = +0, \omega) - u_{\mathrm{SH}}(x', z' = -0, \omega); \tag{15.221}$$

due to $\underline{\mathbf{n}}_c' = \underline{\mathbf{n}}_c^+ = \underline{\mathbf{e}}_z$ on S_c^+ and $\underline{\mathbf{n}}_c' = \underline{\mathbf{n}}_c^- = -\underline{\mathbf{e}}_z$ on S_c^-, we may afterward combine the S_c^--integral with the S_c^+-integral:

$$u_{s\mathrm{SH}}^{\mathrm{far}}(r, \theta, \omega) = -\mathrm{j}k_{\mathrm{S}} \cos\theta \int_{-a}^{a} u_{\mathrm{SH}}^{\mathrm{cod}}(x', \omega) \, G^{\mathrm{far}}(x, x', z, z' = 0, \omega) \, \mathrm{d}x'. \tag{15.222}$$

Explicit insertion of $G^{\mathrm{far}}(x, x', z, z' = 0, \omega) = G^{\mathrm{far}}(r, x', \omega)$ according to (13.52) yields

$$u_{s\mathrm{SH}}^{\mathrm{far}}(r, \theta, \omega) = -\mathrm{j}\omega \, \frac{1}{4c_{\mathrm{S}}} \, \mathrm{e}^{\mathrm{j}\frac{\pi}{4}} \sqrt{\frac{2}{\pi}} \frac{\mathrm{e}^{\mathrm{j}k_{\mathrm{S}}r}}{\sqrt{k_{\mathrm{S}}r}} \cos\theta \int_{-a}^{a} u_{\mathrm{SH}}^{\mathrm{cod}}(x', \omega) \, \mathrm{e}^{-\mathrm{j}k_{\mathrm{S}} \sin\theta x'} \, \mathrm{d}x'. \tag{15.223}$$

Introduction of the Kirchhoff approximation (Equation 15.143)

$$\begin{aligned} u_{\mathrm{SH}}^{\mathrm{cod,PO}}(x', \omega) &= 2u_{i\mathrm{SH}}(x', \omega) \\ &= 2u_{i\mathrm{S}}(\omega) \mathrm{e}^{-\mathrm{j}k_{\mathrm{S}} \sin\theta_i x'} \end{aligned} \tag{15.224}$$

leads to an immediate evaluation of the integral:

$$\begin{aligned} u_{s\mathrm{SH}}^{\mathrm{far,PO}}(r, \theta, \omega) = {} & -\mathrm{j}\omega u_{i\mathrm{S}}(\omega) \, \frac{1}{2c_{\mathrm{S}}} \, \mathrm{e}^{\mathrm{j}\frac{\pi}{4}} \sqrt{\frac{2}{\pi}} \frac{\mathrm{e}^{\mathrm{j}k_{\mathrm{S}}r}}{\sqrt{k_{\mathrm{S}}r}} \\ & \times \cos\theta \underbrace{\frac{\mathrm{e}^{\mathrm{j}k_{\mathrm{S}}a(\sin\theta + \sin\theta_i)} - \mathrm{e}^{-\mathrm{j}k_{\mathrm{S}}a(\sin\theta + \sin\theta_i)}}{\mathrm{j}k_{\mathrm{S}}(\sin\theta + \sin\theta_i)}}_{= \frac{2a \sin k_{\mathrm{S}}a(\sin\theta + \sin\theta_i)}{k_{\mathrm{S}}a \sin k_{\mathrm{S}}a(\sin\theta + \sin\theta_i)}} . \end{aligned} \tag{15.225}$$

Apart from the "Neumann factor" $\cos\theta$, the monochromatic PO-scattering diagram of the strip-like crack is given by the sinc-function with the argument $k_{\mathrm{S}}a(\sin\theta + \sin\theta_i)$ yielding the main lobe of width $k_{\mathrm{S}}a$ in the direction of specular reflection $\theta = -\theta_i$. Yet, the argument of the sinc-function is also equal to zero for $\theta = \theta_i + \pi$, defining the forward scattering direction; but this is a main lobe of the scattered field, it is required to create a shadow superimposing to the incident field, i.e., the secondary source of the strip scattered field must—as primary surface sources in full-space (Section 13.3.4)—also radiate in forward scattering direction. We obtain similar symmetric SH-scattering diagrams as for P-, respectively S, radiation diagrams in Figure 13.17; here, the local parameter $k_{\mathrm{A}}a$ is given by the angle of incidence[300] according to $-k_{\mathrm{S}}a\sin\theta_i$.

Having similar explicit expressions for the scattered SH-field as in Section 13.3.4 for radiated P-S-fields, we may also discuss the scattered pulsed SH-field via Fourier inversion. Yet we must likewise distinguish between the main lobe directions $\theta = -\theta_i$, respectively $\theta = \theta_i + \pi$, and all other directions because the frequency dependence of the sinc-function is dropped for the main lobe directions; therefore, an incident, for example, RC2-pulse must first be differentiated [factor $-\mathrm{j}\omega$ in (15.225)] and then convolved with the two-dimensional Green pulse (compare Equation 13.54); due to $\cos(\theta_i + \pi) = -\cos\theta_i$, it finally gets a negative sign in forward scattering direction (Figure 5.4). For all other scattering directions $\theta \neq -\theta_i, \theta_i + \pi$, $\theta_i \neq 0$—e.g., in backscattering direction $\theta = \theta_i \neq 0$ (as for the 45°-incidence in Figure 5.4)—the decomposition of the sinc function into two exponential functions reveals the superposition of two edge pulses in such a way as if the two edges would be line sources with opposite signs being switched on at times $t = \mp a\sin\theta_i/c_{\mathrm{S}}$ with an RC2-pulse (apart from the factor $\cot\theta_i$ that does not appear in Equation 13.54):

$$
\begin{aligned}
u_{s\mathrm{SH}}^{\mathrm{RC2,far,PO}}(r,\theta_i,t) = \cot\theta_i \Bigg[& G^{\mathrm{RC2,far}}\left(\underline{\mathbf{r}} = -r\hat{\underline{\mathbf{k}}}_i, \underline{\mathbf{r}}' = a\underline{\mathbf{e}}_x, t + \frac{a\sin\theta_i}{c_{\mathrm{S}}}\right) \\
& - G^{\mathrm{RC2,far}}\left(\underline{\mathbf{r}} = -r\hat{\underline{\mathbf{k}}}_i, \underline{\mathbf{r}}' = -a\underline{\mathbf{e}}_x, t - \frac{a\sin\theta_i}{c_{\mathrm{S}}}\right)\Bigg];
\end{aligned}
\tag{15.226}
$$

the edge initially illuminated by the incident wave radiates with a positive and the other one with a negative sign (Figure 5.4), where the equal amplitude magnitude is an immediate consequence of the Kirchhoff approximation; Figure 5.3, resulting from a calculation free of approximations, tells us that both edge pulses have indeed different signs yet nonequal amplitudes.

Let us go back to specular reflection, respectively forward scattering: As compared to the edge pulses the $(-\mathrm{j}\omega)$-factor approximately yields a reconstruction of the incident RC2-pulse; it is indeed required—with a negative sign—to produce a shadow of the scatterer; due to the above "approximately," this shadow is not geometric optical but scattered optical.

By the way, letting the strip not radiate as secondary but as primary SH-source, we have to prescribe a two-dimensional strip-like f_y-force density according to (7.47); hence, we have to solve

[300] A conventional angle transducer is actually realized by the angle of an incident wave!

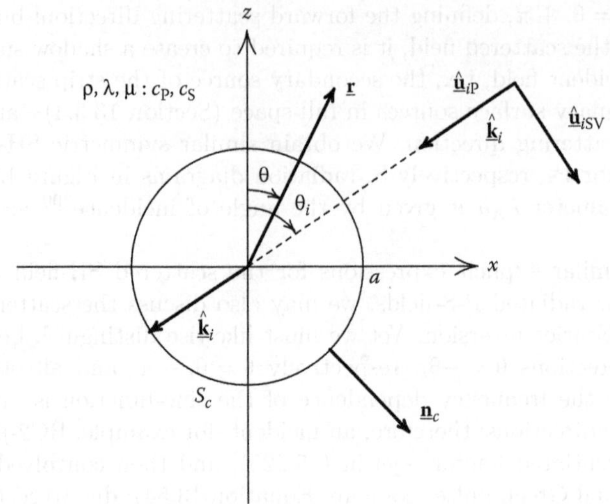

FIGURE 15.21
P-SV-wave scattering by a circular cylindrical void with stress-free surface.

$$\Delta u_y(\underline{\mathbf{r}}, \omega) + k_S^2 u_y(\underline{\mathbf{r}}, \omega) = -\frac{1}{\mu}\, f_y(\underline{\mathbf{r}}, \omega); \qquad (15.227)$$

with the two-dimensional Green function we immediately obtain

$$u_y(x, z, \omega) = \frac{1}{\mu} \int_{-a}^{a} f_y(x', \omega) G(x - x', z, z' = 0, \omega)\, dx'. \qquad (15.228)$$

According to Section 14.1.2, we may even embed the primary strip source in a stress-free surface—for SH-waves, this requires the enforcement of a Neumann boundary condition—utilizing the respective Green function $G^{\mathrm{N}} = 2G$ in (15.228):

$$u_y^{\mathrm{N}}(x, z, \omega) = \frac{2}{\mu} \int_{-a}^{a} f_y(x', \omega) G(x - x', z, z' = 0, \omega)\, dx'. \qquad (15.229)$$

We state that: Similar to the P-radiation of an f_z-source (Section 13.3.4; Figure 13.21), this primary SH-source does not exhibit a pulse differentiation in the far-field main lobe directions, and the edge pulses turn out to be integrals of the line source pulse (13.54).

15.4.3 Two-dimensional scattering problems: Pulsed P-SV-far-fields of circular cylindrical voids and strip-like cracks

Circular cylindrical void: Even for two-dimensional scatterers P-SV-wave scattering is not a scalar problem (compare reflection and mode conversion of a plane wave at a planar boundary: Chapter 9), yet only two scalar potentials are required, the scalar Helmholtz potential $\Phi(\underline{\mathbf{r}}, \omega)$ and the y-component $\Psi_y(\underline{\mathbf{r}}, \omega)$ of the vector potential $\underline{\mathbf{\Psi}}(\underline{\mathbf{r}}, \omega)$, because

$$\underline{\mathbf{u}}(\underline{\mathbf{r}}, \omega) = \nabla\Phi(\underline{\mathbf{r}}, \omega) + \nabla \times \Psi_y(\underline{\mathbf{r}}, \omega)\underline{\mathbf{e}}_y \tag{15.230}$$

solely yields particle displacement components located in the xz-plane to be allotted to P-, respectively SV-waves. Both potentials satisfy homogeneous scalar Helmholtz equations outside the scatterer

$$\Delta\Phi(\underline{\mathbf{r}}, \omega) + k_{\mathrm{P}}^2\Phi(\underline{\mathbf{r}}, \omega) = 0, \tag{15.231}$$

$$\Delta\Psi_y(\underline{\mathbf{r}}, \omega) + k_{\mathrm{S}}^2\Psi_y(\underline{\mathbf{r}}, \omega) = 0, \tag{15.232}$$

provided the incident field is assumed to be a (source-free) plane P- or SV-wave:

$$\Phi_i(\underline{\mathbf{r}}, \omega, \hat{\underline{\mathbf{k}}}_i) = \phi_i(\omega)\,\mathrm{e}^{\mathrm{j}k_{\mathrm{P}}\hat{\underline{\mathbf{k}}}_i\cdot\underline{\mathbf{r}}}, \tag{15.233}$$

$$\Psi_{iy}(\underline{\mathbf{r}}, \omega, \hat{\underline{\mathbf{k}}}_i) = \psi_i(\omega)\,\mathrm{e}^{\mathrm{j}k_{\mathrm{S}}\hat{\underline{\mathbf{k}}}_i\cdot\underline{\mathbf{r}}}. \tag{15.234}$$

With (15.230), the incident P-wave is obtained from (15.233):

$$\underline{\mathbf{u}}_{i\mathrm{P}}(\underline{\mathbf{r}}, \omega, \hat{\underline{\mathbf{k}}}_i) = \underbrace{\mathrm{j}k_{\mathrm{P}}\phi_i(\omega)}_{= u_{i\mathrm{P}}(\omega)}\,\mathrm{e}^{\mathrm{j}k_{\mathrm{P}}\hat{\underline{\mathbf{k}}}_i\cdot\underline{\mathbf{r}}}\,\hat{\underline{\mathbf{k}}}_i, \tag{15.235}$$

and the incident SV-wave from (15.234):

$$\underline{\mathbf{u}}_{i\mathrm{SV}}(\underline{\mathbf{r}}, \omega, \hat{\underline{\mathbf{k}}}_i) = \underbrace{\mathrm{j}k_{\mathrm{S}}\psi_i(\omega)}_{= u_{i\mathrm{S}}(\omega)}\,\mathrm{e}^{\mathrm{j}k_{\mathrm{S}}\hat{\underline{\mathbf{k}}}_i\cdot\underline{\mathbf{r}}}\,\hat{\underline{\mathbf{k}}}_i \times \underline{\mathbf{e}}_y, \tag{15.236}$$

where $\hat{\underline{\mathbf{k}}}_i$ is given by (15.199). For the circular cylindrical void, P-SV-scattering is sketched in Figure 15.21.

Apparently, it is now appropriate to solve the scalar Helmholtz equations (15.231) and (15.232) for the scattering potentials $\Phi_s(r, \theta, \omega)$ and $\Psi_{sy}(r, \theta, \omega)$ similar to (15.202) with partial wave separation:

$$\Phi_s(r, \theta, \omega) = \sum_{n=-\infty}^{\infty} \phi_n(\omega)\mathrm{H}_n^{(1)}(k_{\mathrm{P}}r)\,\mathrm{e}^{\mathrm{j}n\theta}, \tag{15.237}$$

$$\Psi_{sy}(r, \theta, \omega) = \sum_{n=-\infty}^{\infty} \psi_n(\omega)\mathrm{H}_n^{(1)}(k_{\mathrm{S}}r)\,\mathrm{e}^{\mathrm{j}n\theta}. \tag{15.238}$$

To calculate the expansion coefficients $\phi_n(\omega)$ and $\psi_n(\omega)$, we need two equations resulting from the stress-free boundary condition on the cylinder surface:

$$\underline{\underline{\mathbf{T}}}(a, \theta, \omega) \cdot \underline{\mathbf{e}}_r = \underline{\mathbf{0}}, \tag{15.239}$$

where

$$\underline{\underline{\mathbf{T}}}(a,\theta,\omega) = \lambda \underline{\underline{\mathbf{I}}} \nabla \cdot \underline{\mathbf{u}}(r,\theta,\omega)\Big|_{r=a} + \mu \left[\nabla \underline{\mathbf{u}}(r,\theta,\omega) + \nabla \underline{\mathbf{u}}^{21}(r,\theta,\omega)\right]\Big|_{r=a}. \tag{15.240}$$

With

$$\nabla \underline{\mathbf{u}}(r,\theta,\omega)\cdot \underline{\mathbf{e}}_r = \frac{\partial u_r(r,\theta,\omega)}{\partial r}\,\underline{\mathbf{e}}_r + \frac{1}{r}\left[\frac{\partial u_r(r,\theta,\omega)}{\partial \theta} - u_\theta(r,\theta,\omega)\right]\underline{\mathbf{e}}_\theta, \tag{15.241}$$

$$\nabla \underline{\mathbf{u}}^{21}(r,\theta,\omega)\cdot \underline{\mathbf{e}}_r = \frac{\partial u_r(r,\theta,\omega)}{\partial r}\,\underline{\mathbf{e}}_r + \frac{\partial u_\theta(r,\theta,\omega)}{\partial r}\,\underline{\mathbf{e}}_\theta, \tag{15.242}$$

where

$$u_r(r,\theta,\omega) = \frac{\partial \Phi(r,\theta,\omega)}{\partial r} + \frac{1}{r}\frac{\partial \Psi_y(r,\theta,\omega)}{\partial \theta}, \tag{15.243}$$

$$u_\theta(r,\theta,\omega) = \frac{1}{r}\frac{\partial \Phi(r,\theta,\omega)}{\partial \theta} - \frac{\partial \Psi_y(r,\theta,\omega)}{\partial r}, \tag{15.244}$$

we obtain using (15.231)

$$T_{rr}(r,\theta,\omega) = -k_{\mathrm{P}}^2\lambda\Phi(r,\theta,\omega) + 2\mu\left\{\frac{\partial^2\Phi(r,\theta,\omega)}{\partial r^2} + \frac{\partial}{\partial r}\left[\frac{1}{r}\frac{\partial \Psi_y(r,\theta,\omega)}{\partial \theta}\right]\right\}, \tag{15.245}$$

$$\begin{aligned}\frac{1}{\mu}\,T_{\theta r}(r,\theta,\omega) = {} & 2\frac{\partial}{\partial r}\left[\frac{1}{r}\frac{\partial \Phi(r,\theta,\omega)}{\partial \theta}\right] - r\frac{\partial}{\partial r}\left[\frac{1}{r}\frac{\partial \Psi_y(r,\theta,\omega)}{\partial r}\right] \\ & + \frac{1}{r^2}\frac{\partial^2 \Psi_y(r,\theta,\omega)}{\partial \theta^2}.\end{aligned} \tag{15.246}$$

For $r = a$, we get both nonzero components of (15.239) as the required equations. Note: The potentials Φ and Ψ_y in (15.245) and (15.246) are total potentials $\Phi = \Phi_i + \Phi_s$, $\Psi_y = \Psi_{iy} + \Psi_{sy}$ allowing for the calculation of the expansion coefficients for P-, respectively SV-wave, incidence depending on the choice of $\psi_i(\omega) = 0$, $\phi_i(\omega) \neq 0$, respectively $\psi_i(\omega) \neq 0$, $\phi_i(\omega) = 0$, and using (15.208).

The r- and θ-components of the scattered field are calculated with the help of (15.243) and (15.244), where we know that in the far-field pressure wavefronts are given by u_{sr}^{far}, and shear wavefronts are given by $u_{s\theta}^{\text{far}}$; therefore, for P-wave incidence, u_{sr}^{far} is the directly scattered and $u_{s\theta}^{\text{far}}$ the mode converted wavefront, and vice versa. Furthermore, we have to take care of a differentiating ($\mathrm{j}\omega$)-factor between the potential spectra and the particle displacement spectra when calculating pulsed wavefronts (Equations 15.235 and 15.236).

Figure 15.22 displays a matrix representation of pulsed far-fields as function of the normalized time

$$t_{\text{norm}}^{\mathrm{P,S}} = \frac{c_{\mathrm{P}}}{a}\,t - \frac{c_{\mathrm{P}}}{c_{\mathrm{P,S}}}\frac{r}{a} + 2\frac{c_{\mathrm{P}}}{c_{\mathrm{P,S}}} \tag{15.247}$$

prescribing $u_{iP}(\omega)$ (top), respectively $u_{iS}(\omega)$ (bottom), as RC2-spectra; for all cases, the total duration of the RC2-pulse with $T = 2a/c_P$ is approximately as large as the pressure wave travel time over the diameter of the cylinder. The origin of the normalized time axes always coincides with the arrival of the first (specularly reflected) pressure or shear pulse. As compared to the scalar case, we notice—e.g., in backscattering direction—a differentiation of the incident pulse (similar to the specular reflection in case of the strip-like crack: Figure 5.3); the reason is the (jω)-factor appearing in the secondary $\underline{\underline{\mathbf{h}}}_c$-source (15.187) that penetrates the scattering amplitudes (15.182) and (15.185). Since all time axes are normalized with the pressure wave velocity, we observe the (SV$\Longrightarrow$SV)-creeping wave pulse in backscattering direction later than in Figure 15.13 because in the former case, the time axis was normalized with the shear wave velocity. In backscattering direction, the clockwise and counter-clockwise traveling creeping waves arrive simultaneously, in other directions, they can be timely separated. Due to the lower attenuation in the (SV$\Longrightarrow$SV)-case, the creeping wave pulse with the larger travel distance becomes also visible.

Figures 15.23 and 15.24 display EFIT-calculated wavefronts for P-wave incidence and Figures 15.25 and 15.26 for SV-wave incidence for two different times: As compared to SH-wavefronts (Figures 15.14 through 15.17), we observe mode conversion together with creeping waves in both modes, and the scattered wavefronts exhibit a characteristic amplitude structure due to the elastodynamic point directivity of the secondary sources (consult Figure 13.14 for the $\underline{\mathbf{e}}_x\underline{\mathbf{e}}_z$-source to understand, for example, the amplitude structure for SV-wave incidence).

Strip-like crack: In contrast to the partial wave separation in cylindrical coordinates (Equation 15.207), the coordinates of the elliptical cylinder—including the limiting case of the strip-like crack—exhibit wave numbers in all separation functions (Mathieu functions), even in the generalized angular functions. If only a single wave number has to be considered as for SH-scattering, orthogonality can still be exploited to calculate the expansion coefficients. Yet, P-SV-scattering has to cope with two wave numbers foreclosing a respective approach. Therefore, the present case requires the numerical solution of integral equations (15.122) and (15.123) as derived in Section 15.2.2 for the components of the crack opening displacement, and this may be achieved with the method of moments[301] (Harrington 1968).

Figures 15.27 through 15.29 exemplify magnitudes of far-field radiation patterns for a fixed frequency and crack width $2a$; throughout we have $k_Pa = 11$, and $\kappa = 1.827$ for the pressure shear wave ratio yields $k_Sa = 20.1$ (this κ-value is slightly different from $\kappa = 1.844$ used elsewhere in this elaboration). The figures vary with the angle of incidence; each of them exhibits diagrams

[301] We could apply a code developed by D.R. Wilton and coworkers from the University of Houston, USA.

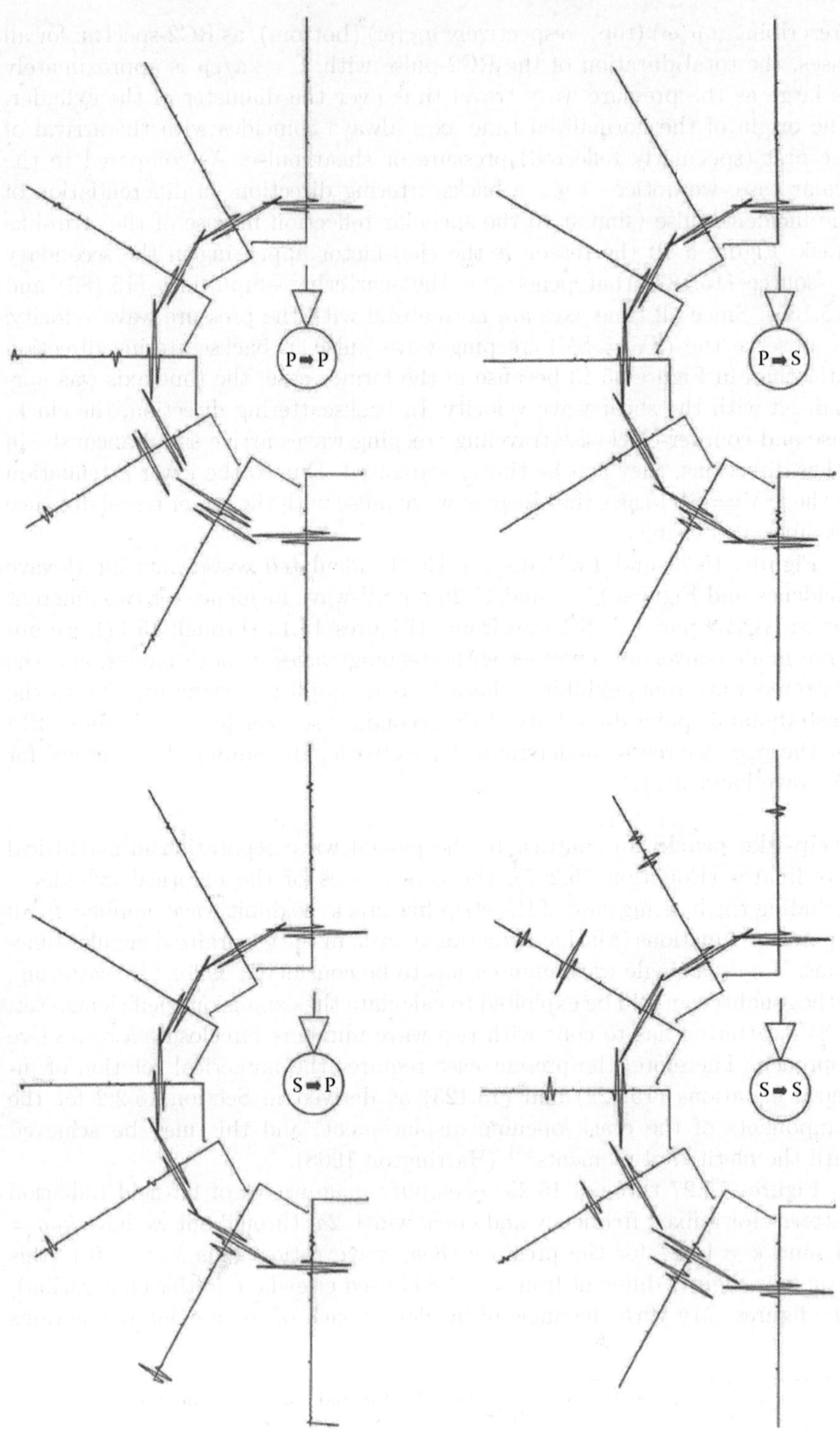
P ➡ P
P ➡ S
S ➡ P
S ➡ S

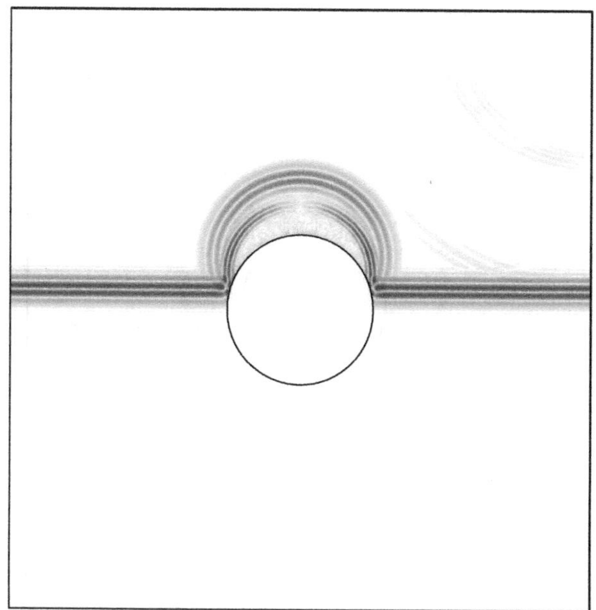

FIGURE 15.23
P-RC2-pulse scattering by a circular cylinder with a stress-free surface: EFIT-simulation.

for P-, respectively SV-wave, incidence in a matrix representation as in Figure 15.22. We know from Section 15.4 that scattered fields are described by similar equations as radiation fields in full-space, only the primary sources have to be replaced by secondary sources. Admittedly, primary sources are usually realized prescribing the force density,[302] while secondary sources on stress-free surfaces are given by deformation rates. Point directivities of Figure 13.12 are relevant for the first ones, and those of Figure 13.14 for the latter ones. Yet, for perpendicular P-wave incidence (Figure 15.27: top)—the deformation

---

FIGURE 15.22
RC2-pulsed scattered far-field $\underline{\mathbf{u}}_{s\mathrm{P}}^{\mathrm{RC2,far}}(r,\theta,t)\cdot\underline{\mathbf{e}}_r$, respectively $\underline{\mathbf{u}}_{s\mathrm{SV}}^{\mathrm{RC2,far}}$ $(r,\theta,t)\cdot\underline{\mathbf{e}}_\theta$ for plane P- (top), respectively SV-wave (bottom), incidence on a circular cylinder with a stress-free surface (amplitudes normalized to their respective maximum value resulting in nonzero pulses for (P$\Longrightarrow$S)- and (S$\Longrightarrow$P)-backscattering).

[302]Because we consider primary force density sources on stress-free surfaces as models for piezoelectric transducers; prescription of a deformation rate yields a "short circuit."

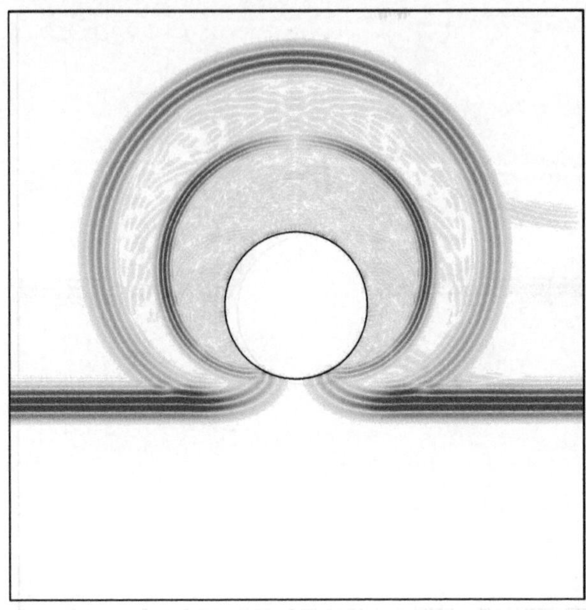

FIGURE 15.24
P-RC2-pulse scattering by a circular cylinder with a stress-free surface: EFIT-simulation.

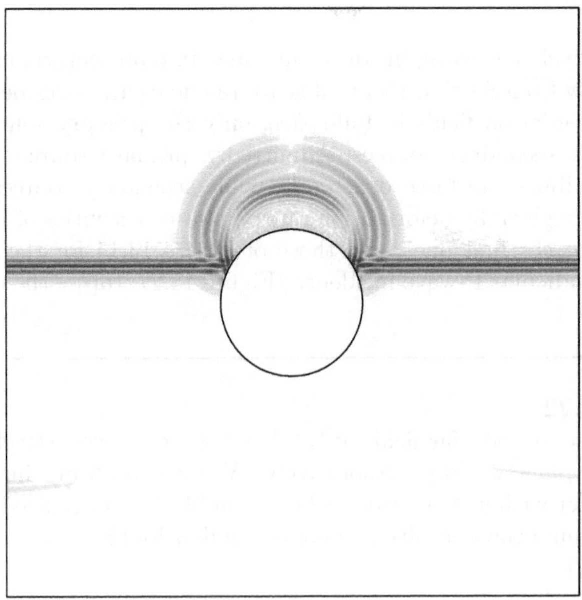

FIGURE 15.25
SV-RC2-pulse scattering by a circular cylinder with a stress-free surface: EFIT-simulation.

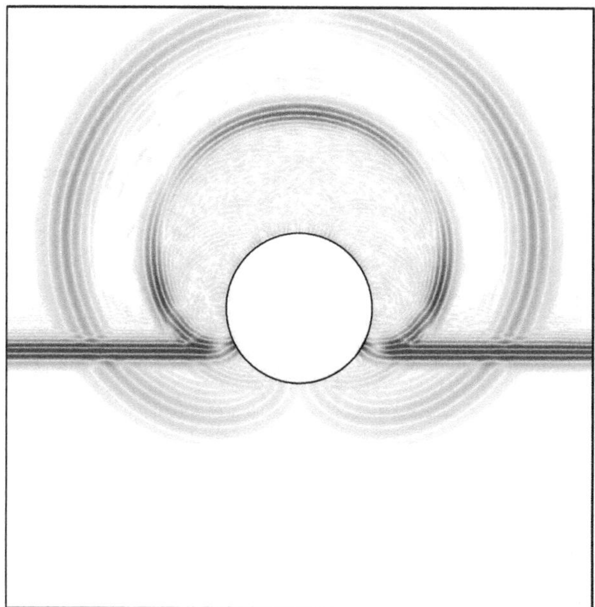

FIGURE 15.26
SV-RC2-pulse scattering by a circular cylinder with a stress-free surface: EFIT-simulation.

rate is a $\underline{\mathbf{e}}_z\underline{\mathbf{e}}_z$-source—the differences compared to the $\underline{\mathbf{e}}_z$-force density source are covered by the scalar radiation diagram of the strip leading in fact to a congruence of the (P$\Longrightarrow$P)- and (P$\Longrightarrow$SV)-scattering diagrams of Figure 15.27 with the P- and S-radiation diagrams of Figures 13.17(e) and (f) (apart from the slightly different $k_{\mathrm{P,S}}a$-values). The same is true for the $\underline{\mathbf{e}}_x\underline{\mathbf{e}}_z$-deformation rate source (Figure 13.14) for perpendicular SV-wave incidence as compared to the $\underline{\mathbf{e}}_x$-force density source (Figure 13.12). Vis-à-vis P-wave incidence only the main lobes are narrower due to the larger $k_{\mathrm{S}}a$-value. As for the scalar SH-wave incidence on the strip (Section 15.4.2), we observe respective main lobes in the direction of specular reflection (in this case in backscattering direction) and, in forward scattering direction, to yield a shadow.

Figure 15.28 displays scattering diagrams for $\theta_i = 210°$, an angle of incidence of 30° with regard to the crack normal. First, we recognize clear main lobes even for the mode converted scattered fields (P$\Longrightarrow$SV; SV$\Longrightarrow$P), yet only in the direction of "specular" mode conversion because no shadow has to be formed. Furthermore, we state an asymmetry within the (P$\Longrightarrow$P)- and (SV$\Longrightarrow$SV)-diagrams obviously foreclosing to argue with the superposition of the magnitude-symmetric deformation rate point directivities of Figure 13.14 to understand the structure of crack scattered fields, and by no means, we may refer to the primary f_z-full-space source [Figures 13.17(a) and (b)] illustration.

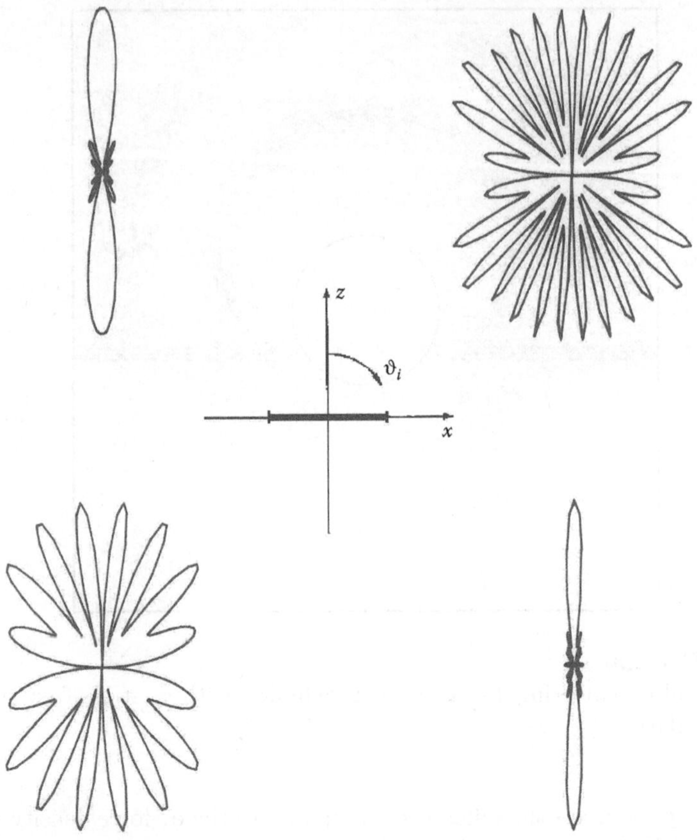

FIGURE 15.27
P-SV-far-field radiation patterns of a strip-like crack with a stress-free surface for plane P- (top: P$\Longrightarrow$P; P$\Longrightarrow$SV), respectively SV-wave, incidence (bottom: SV$\Longrightarrow$P; SV$\Longrightarrow$SV) (respective amplitudes normalized to their maximum value: $k_{\mathrm{P}}a = 11$, $k_{\mathrm{S}}a = 20.1$, $\theta_i = 180^{\mathrm{o}}$).

The asymmetry of the (P$\Longrightarrow$P)- and (SV$\Longrightarrow$SV)-main lobe diagrams in the directions of specular reflection and forward scattering for nonperpendicular incidence is in fact a specialty of elastodynamics as compared to acoustics and electromagnetics: Cause are the elastodynamic point directivities (13.174) and (13.176) of deformation rate sources that essentially codetermine the angular dependence of P-SV-scattered fields via the scattering amplitude representations (15.182) and (15.185); with the secondary source (Equation 15.187)

$$\underline{\underline{\mathbf{h}}}_c^{\mathrm{cod}}(x, y, z, \omega) = \mathrm{j}\omega \underline{\underline{\underline{\underline{\mathbf{I}}}}}^{+} : \underline{\mathbf{e}}_z \underline{\mathbf{u}}^{\mathrm{cod}}(x, \omega)\delta(z) q_a(x), \tag{15.248}$$

the scalar radiation characteristics of the crack "aperture" turn out to be magnitude-symmetric for the transition from the scattering direction θ to the

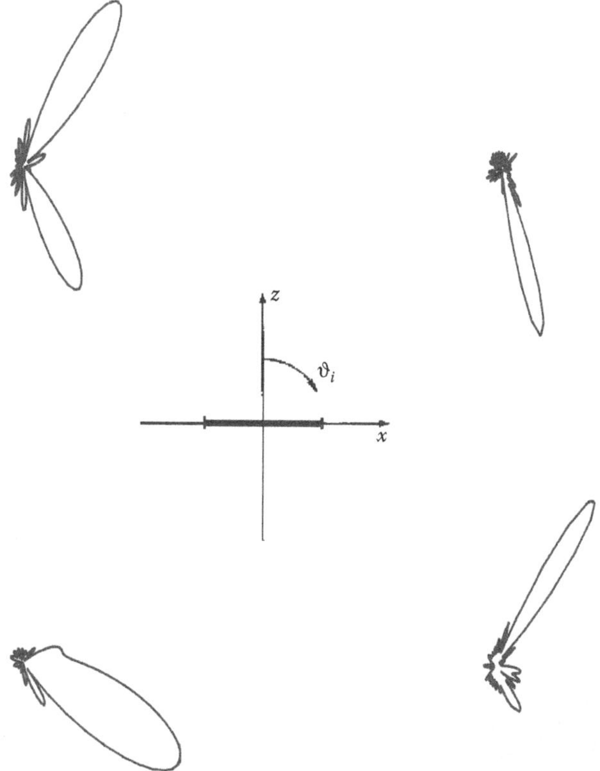

FIGURE 15.28
P-SV-far-field radiation patterns of a strip-like crack with a stress-free surface for plane P- (top: P$\Longrightarrow$P; P$\Longrightarrow$SV), respectively SV-wave, incidence (bottom: SV$\Longrightarrow$P; SV$\Longrightarrow$SV) (respective amplitudes normalized to their maximum value: $k_\text{P}a = 11$, $k_\text{S}a = 20.1$, $\theta_i = 210^\text{o}$).

scattering direction $\pi - \theta$, but not the superpositions of $\underline{\mathbf{e}}_x\underline{\mathbf{e}}_z$- and $\underline{\mathbf{e}}_z\underline{\mathbf{e}}_z$-point directivities because of the combined appearance of $\cos\theta$- and $\sin\theta$-functions.

Figure 15.29 finally shows the magnitudes of the scattered far-fields for an incidence angle of 45° with regard to the crack normal, i.e., for $\theta_i = 225^\text{o}$. For the case of reflection and mode conversion of a plane SV-wave at an infinite stress-free surface, we would already find ourselves beyond the critical angle for P-mode conversion; evidently, the same is true for the crack of finite width as can be concluded from the (SV$\Longrightarrow$P)-diagram: For $\theta_i = \pi + \vartheta_{\text{cmP}}$, the main lobe of the scalar crack "aperture" points into the direction $\theta = 90^\text{o}$ allowing the point directivities to visualize mainly side lobes.

As with Figures 15.22 through 15.26, we now turn to pulsed scattered fields first showing 2D-EFIT-wavefronts for a long-$\perp$-, respectively a shear-$\perp$, transducer model on a stress-free surface (Figures 15.30 and 15.31, respectively

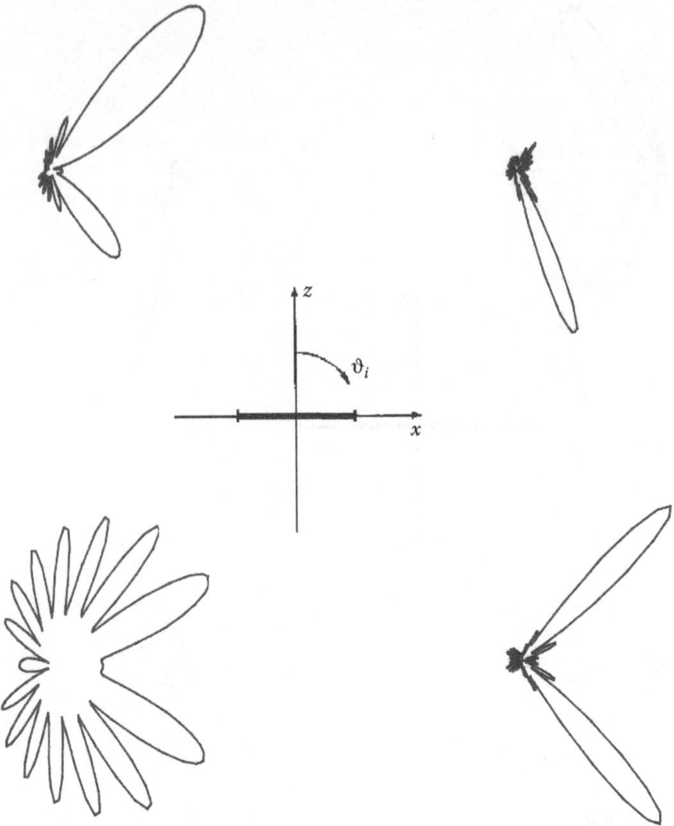

FIGURE 15.29
P-SV-far-field radiation patterns of a strip-like crack with a stress-free surface for plane P- (top: P⟹P; P⟹SV), respectively SV-wave, incidence (bottom: SV⟹P; SV⟹SV) (respective amplitudes normalized to their maximum value: $k_{\mathrm{P}}a = 11$, $k_{\mathrm{S}}a = 20.1$, $\theta_i = 225°$).

15.32 and 15.33). Mode conversion and Rayleigh waves traveling along the crack faces are nicely recognized exciting the crack edges—basically infinitely often—as line sources for cylindrical elastic waves. They become visible in A-scans as resonance pulses (e.g., P⟹P in Figure 15.34, backscattering). The EFIT-simulations in Figures 15.30 and 15.31 are complemented by AFIT-simulations assuming a fictitious μ-free material with the same pressure wave velocity as for steel; then the stress-free boundary condition on the specimen and the crack surface manifests itself as a Dirichlet boundary condition for the pressure due to $\underline{\underline{\mathbf{T}}} = -p\underline{\underline{\mathbf{I}}}$ (Equation 5.3) (soft surfaces), and prescribing a normal force density within the transducer aperture, i.e., the stress tensor component T_{zz}, manifests itself as pressure prescription. From this confrontation,

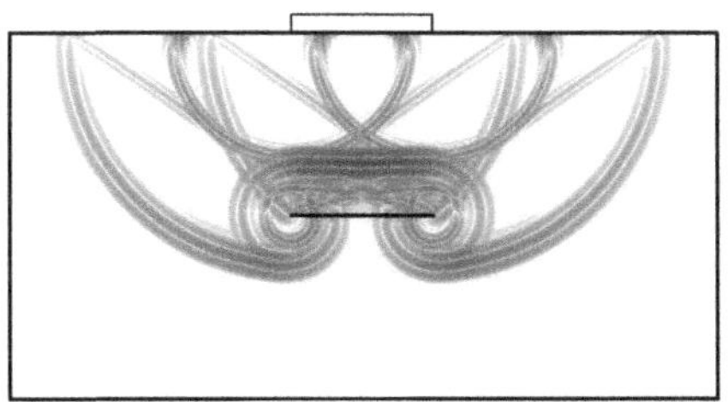
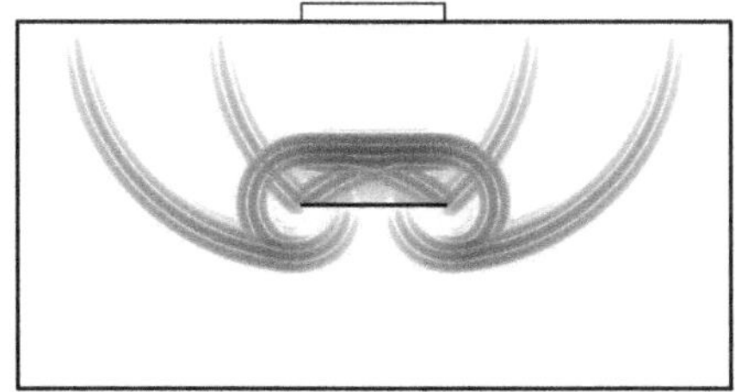

FIGURE 15.30
P-RC2-pulse scattering ($t = t_1$) by a strip-like crack with a stress-free surface (left: EFIT-simulation; magnitude of the particle velocity), respectively by a strip-like "Crack" with a soft surface (right: AFIT-simulation; for a direct comparison magnitude of the particle velocity and not pressure as in Figure 14.1).

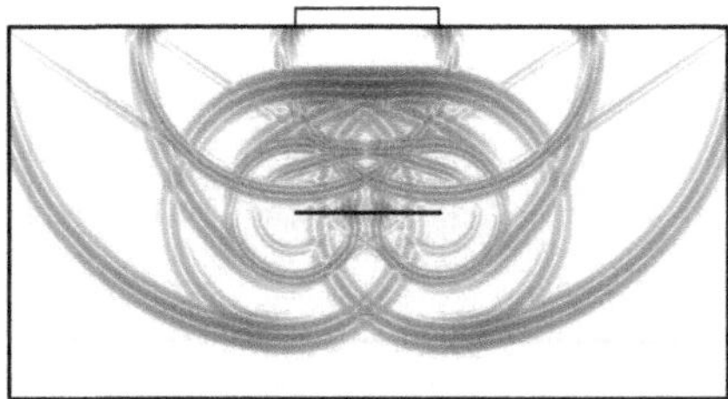
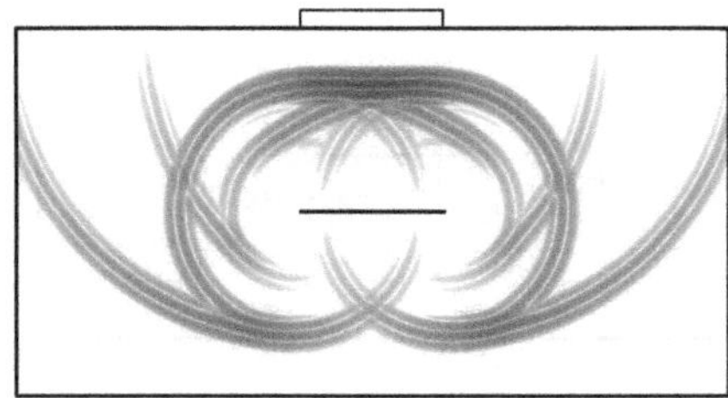

FIGURE 15.31
P-RC2-pulse scattering ($t = t_2 > t_1$) by a strip-like crack with a stress-free surface (left: EFIT-simulation; magnitude of the particle velocity), respectively by a strip-like "crack" with a soft surface (right: AFIT-simulation; for a direct comparison magnitude of the particle velocity and not pressure as in Figure 14.1).

we learn a lot about the complexity of elastodynamic wave fields as compared to their acoustic counterparts. In Figures 15.32 and 15.33, the AFIT-simulation stands for SH-wave scattering (we have $c_{\rm SH} = c_{\rm SV}$), i.e., as scalar boundary condition the Neumann boundary condition replaces the Dirichlet boundary condition of Figures 15.30 and 15.31. Insofar, Figures 15.32 and 15.33 complement Figures 15.19 and 15.20.

The already cited Figure 15.34 shows, comparable to Figure 15.22, RC2-pulse A-scans for the special incidence angle of 30°, counted from the normal, that yielded distinct scattering diagrams in all four cases according to Figure 15.28. The impulses for the (P$\Longrightarrow$P)-case look very similar as the scalar acoustic pulses in Figure 5.3 (i.e., the SH-case), yet augmented by pronounced resonance pulses; the differentiation in the main lobe directions comes from the (jω)-factor of the secondary deformation rate source (15.248).

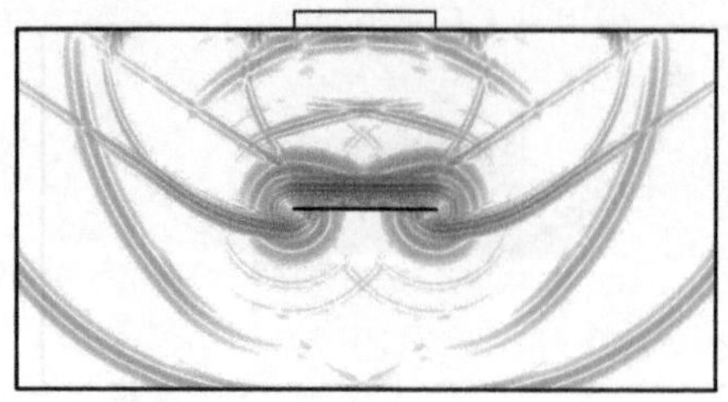
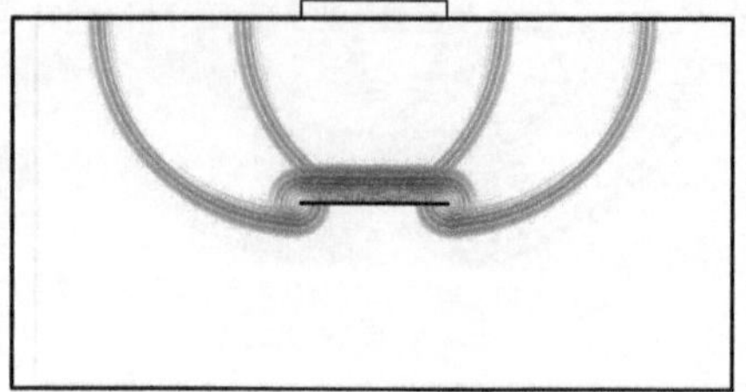

FIGURE 15.32
SV-RC2-pulse scattering ($t = t_1$) by a strip-like crack with a stress-free surface (left: EFIT-simulation; magnitude of the particle velocity), respectively SH-RC2-pulse scattering (right: AFIT-simulation; magnitude of the y-component of the particle velocity).

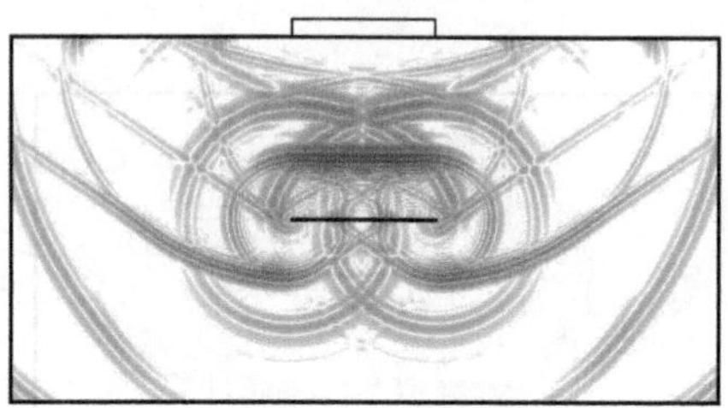
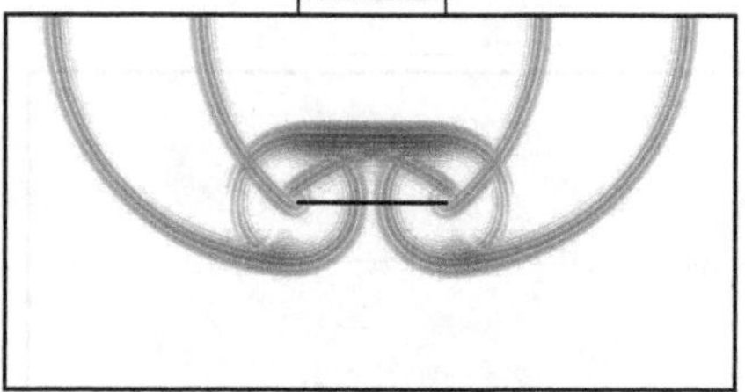

FIGURE 15.33
SV-RC2-pulse scattering ($t = t_2 > t_1$) by a strip-like crack with a stress-free surface (left: EFIT-simulation; magnitude of the particle velocity), respectively SH-RC2-pulse scattering (right: AFIT-simulation; magnitude of the y-component of the particle velocity).

Due to the c_P-time axis normalization in all cases, the (SV$\Longrightarrow$SV)-edge pulses are further apart. As compared to (P$\Longrightarrow$P)-scattering, we state a much lower amplitude of the pulse from the closer edge with respect to the one from the farther edge; this weakens the depth assessment of a back wall breaking crack in pulse-echo mode as well as the time of flight diffraction (TOFD)-technique.

15.4.4 Three-dimensional scattering problems: Pulsed P-S-far-fields of spherical voids

In the case of the two-dimensional circular cylinder, we were able to apply the same mathematical calculus for either pressure or shear wave scattering (Section 15.4.3), in case of the three-dimensional sphere, shear wave scattering is much more elaborate, while pressure wave scattering follows the beaten path.

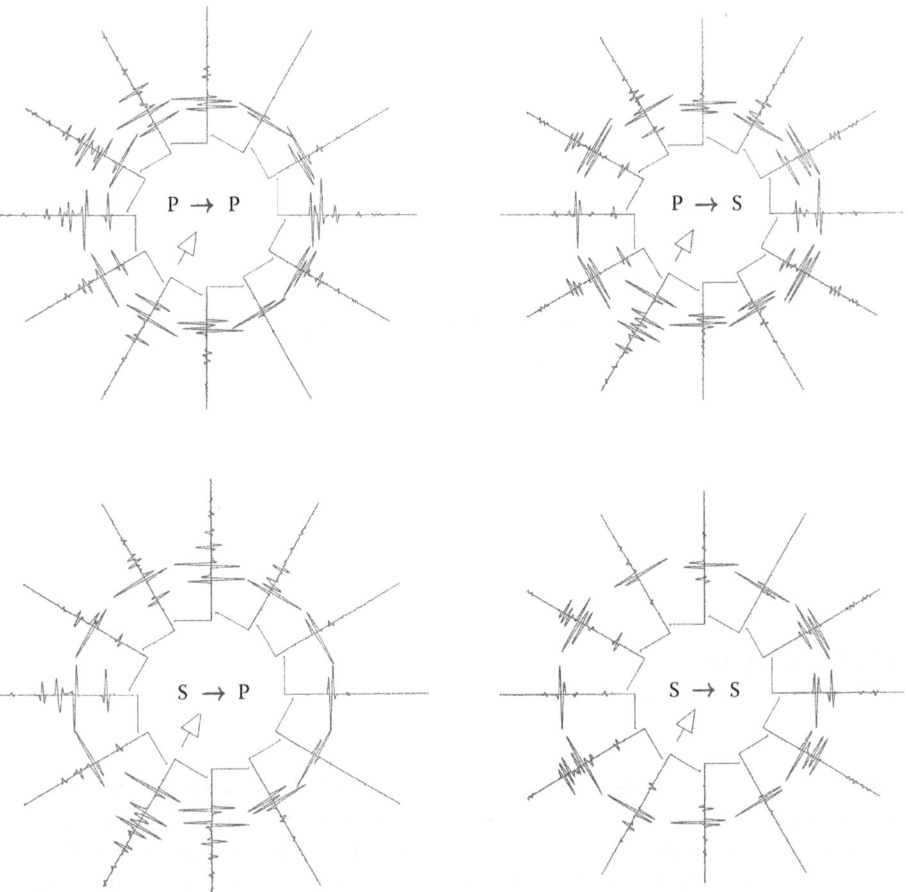

FIGURE 15.34
RC2-pulsed far-field $\underline{\mathbf{u}}_{sP}^{\mathrm{RC2,far}}(r, \theta, t) \cdot \underline{\mathbf{e}}_r$, respectively $\underline{\mathbf{u}}_{sSV}^{\mathrm{RC2,far}}(r, \theta, t) \cdot \underline{\mathbf{e}}_\theta$ by a strip-like crack with a stress-free surface for plane P- (top), respectively SV-wave incidence (bottom) (respective amplitudes normalized to maximum value); $\theta_i = 210^\mathrm{o}$.

P-wave incidence: Due to the spherical symmetry, we may orient a cartesian coordinate system always with the negative z-direction coinciding with the direction of incidence of the P-wave (Figure 15.35). The *ansatz*

$$\underline{\mathbf{u}}_{i\mathrm{P}}(\underline{\mathbf{R}}, \omega, \hat{\underline{\mathbf{k}}}_i) = u_{i\mathrm{P}}(\omega)\, \mathrm{e}^{\mathrm{j}k_\mathrm{P}\hat{\underline{\mathbf{k}}}_i \cdot \underline{\mathbf{R}}}\, \hat{\underline{\mathbf{k}}}_i \tag{15.249}$$

then yields φ-independent R- and ϑ-components of the incident P-wave with

$$\begin{aligned} \hat{\underline{\mathbf{k}}}_i &= -\underline{\mathbf{e}}_z \\ &= -\cos\vartheta\, \underline{\mathbf{e}}_R + \sin\vartheta\, \underline{\mathbf{e}}_\vartheta; \end{aligned} \tag{15.250}$$

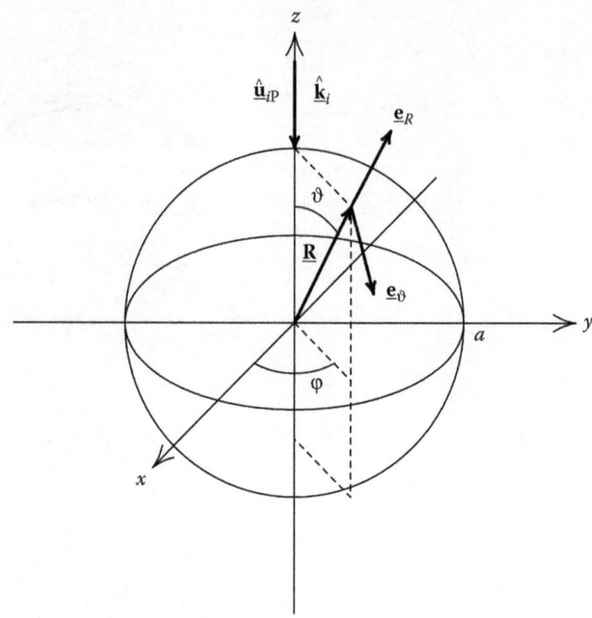

FIGURE 15.35
P-wave scattering by a spherical void with a stress-free surface.

therefore, we expect that $\partial/\partial\varphi \equiv 0$ holds and that the scattered field—also in the mode converted part—does not contain φ-components: In each plane $\varphi = \text{const}$, it is P-SV-polarized meaning that the reference plane for SV-polarization is φ-dependent: We deal with P-S-wave scattering. In the far-field, we obviously have $\underline{\mathbf{u}}_{sP}^{\text{far}} = u_{sP}^{\text{far}}\,\underline{\mathbf{e}}_R$ and $\underline{\mathbf{u}}_{sS}^{\text{far}} = u_{sS}^{\text{far}}\,\underline{\mathbf{e}}_\vartheta$ (Figure 15.35). With the plane "potential wave"

$$\Phi_i(\underline{\mathbf{R}}, \omega, \hat{\underline{\mathbf{k}}}_i) = \phi_i(\omega)\,\mathrm{e}^{\mathrm{j}k_\mathrm{P}\hat{\underline{\mathbf{k}}}_i\cdot\underline{\mathbf{R}}} \tag{15.251}$$

and

$$\begin{aligned}\underline{\mathbf{u}}_{i\mathrm{P}}(\underline{\mathbf{R}}, \omega, \hat{\underline{\mathbf{k}}}_i) &= \nabla\Phi_i(\underline{\mathbf{R}}, \omega, \hat{\underline{\mathbf{k}}}_i)\\ &= \underline{\mathbf{e}}_R\,\frac{\partial\Phi_i(\underline{\mathbf{R}}, \omega, \hat{\underline{\mathbf{k}}}_i)}{\partial R} + \underline{\mathbf{e}}_\vartheta\,\frac{1}{R}\,\frac{\partial\Phi_i(\underline{\mathbf{R}}, \omega, \hat{\underline{\mathbf{k}}}_i)}{\partial\vartheta},\end{aligned} \tag{15.252}$$

we again obtain (15.249) with (15.250) provided we put

$$u_{i\mathrm{P}}(\omega) = \mathrm{j}k_\mathrm{P}\phi_i(\omega). \tag{15.253}$$

The desired R- and ϑ-components of the scattered field $\underline{\mathbf{u}}_s(R, \vartheta, \omega)$ follow according to

$$
\begin{aligned}
\underline{\mathbf{u}}_s(R,\vartheta,\omega) &= \nabla\Phi_s(R,\vartheta,\omega) + \nabla\times\left[\Psi_{s\varphi}(R,\vartheta,\omega)\,\underline{\mathbf{e}}_\varphi\right] \\
&= \left\{\frac{\partial\Phi_s(R,\vartheta,\omega)}{\partial R} + \frac{1}{R\sin\vartheta}\frac{\partial}{\partial\vartheta}\left[\sin\vartheta\,\Psi_{s\varphi}(R,\vartheta,\omega)\right]\right\}\underline{\mathbf{e}}_R \\
&\quad + \left\{\frac{1}{R}\frac{\partial\Phi_s(R,\vartheta,\omega)}{\partial\vartheta} - \frac{1}{R}\frac{\partial}{\partial R}\left[R\,\Psi_{s\varphi}(R,\vartheta,\omega)\right]\right\}\underline{\mathbf{e}}_\vartheta
\end{aligned} \tag{15.254}
$$

from the Helmholtz potentials $\Phi_s(R,\omega,\vartheta)$ and $\underline{\mathbf{\Psi}}_s(R,\vartheta,\omega) = \Psi_{s\varphi}(R,\vartheta,\omega)\,\underline{\mathbf{e}}_\varphi$, where these potentials must satisfy the differential equation

$$\Delta\Phi_s(R,\vartheta,\omega) + k_{\mathrm{P}}^2\Phi_s(R,\vartheta,\omega) = 0, \tag{15.255}$$

$$\Delta\Psi_{s\varphi}(R,\vartheta,\omega) + \left(k_{\mathrm{S}}^2 - \frac{1}{R^2\sin^2\vartheta}\right)\Psi_{s\varphi}(R,\vartheta,\omega) = 0; \tag{15.256}$$

the additional term in (15.256) is due to the fact that $\underline{\mathbf{\Psi}}_s$ satisfies a vector Helmholtz equation, i.e., in $\Delta\Psi_{s\varphi}\underline{\mathbf{e}}_\varphi$, we also have to differentiate $\underline{\mathbf{e}}_\varphi$.

As in Sections 15.4.2 and 15.4.3, we solve (15.255) and (15.256) separating into partial waves, yet this time partial spherical waves. With the delta-operator

$$\Delta = \frac{\partial^2}{\partial R^2} + \frac{2}{R}\frac{\partial}{\partial R} + \frac{1}{R^2\sin\vartheta}\frac{\partial}{\partial\vartheta}\sin\vartheta\frac{\partial}{\partial\vartheta} + \frac{1}{R^2\sin^2\vartheta}\frac{\partial^2}{\partial\varphi^2} \tag{15.257}$$

in spherical coordinates, we find the general solution of

$$\Delta W(R,\vartheta,\varphi,\omega) + k^2 W(R,\vartheta,\varphi) = 0 \tag{15.258}$$

in terms of an eigenfunction expansion:

$$W(R,\vartheta,\varphi,\omega) = \sum_{n=0}^{\infty}\sum_{m=-n}^{n} w_{nm}(\omega)\mathrm{h}_n^{(1)}(kR)\mathrm{P}_n^m(\cos\vartheta)\mathrm{e}^{\mathrm{j}m\varphi}, \tag{15.259}$$

where $\mathrm{h}_n^{(1)}(kR)$ denote spherical Hankel functions according to

$$\mathrm{h}_n^{(1)}(kR) = \sqrt{\frac{\pi}{2kR}}\,\mathrm{H}_{n+\frac{1}{2}}^{(1)}(kR), \tag{15.260}$$

and $\mathrm{P}_n^m(\zeta)$ associated Legendre functions of the first kind according to (Schäfke 1967)

$$\mathrm{P}_n^m(\zeta) = (-1)^m(1-\zeta^2)^{\frac{m}{2}}\frac{\mathrm{d}^m\mathrm{P}_n(\zeta)}{\mathrm{d}\zeta^m}, \tag{15.261}$$

$$\mathrm{P}_n(\zeta) = \frac{1}{2^n n!}\frac{\mathrm{d}^n}{\mathrm{d}\zeta^n}(\zeta^2-1)^n; \tag{15.262}$$

they satisfy the differential equations

$$\frac{\partial^2\mathrm{h}_n^{(1)}(kR)}{\partial R^2} + \frac{2}{R}\frac{\partial\mathrm{h}_n^{(1)}(kR)}{\partial R} + \left(k^2 - \frac{n(n+1)}{R^2}\right)\mathrm{h}_n^{(1)}(kR) = 0, \tag{15.263}$$

$$\frac{1}{\sin\vartheta}\frac{\partial}{\partial\vartheta}\left[\sin\vartheta\,\frac{\partial\mathrm{P}_n^m(\cos\vartheta)}{\partial\vartheta}\right] + \left[n(n+1) - \frac{m^2}{\sin^2\vartheta}\right]\mathrm{P}_n^m(\cos\vartheta) = 0. \tag{15.264}$$

Since the solutions of (15.255) and (15.256) should be rotationally symmetric with regard to φ, only the index $m = 0$ is relevant in the eigenfunction expansion (15.259) allowing to write down the series

$$\Phi_s(R, \vartheta, \omega) = \sum_{n=0}^{\infty} \phi_n(\omega) \mathrm{h}_n^{(1)}(k_\mathrm{P} R) \mathrm{P}_n(\cos\vartheta) \tag{15.265}$$

for the scattering potential $\Phi_s(R, \vartheta, \omega)$; here, $\mathrm{P}_n(\cos\vartheta) = \mathrm{P}_n^0(\cos\vartheta)$ are Legendre polynomials in $\cos\vartheta$. Furthermore, we state for $m = \pm 1$ that (15.264) is just the ϑ-separated part of the differential equation (15.256) in case $\partial/\partial\varphi = 0$ resulting in

$$\Psi_{s\varphi}(R, \vartheta, \omega) = \sum_{n=0}^{\infty} \psi_n(\omega) \mathrm{h}_n^{(1)}(k_\mathrm{S} R) \mathrm{P}_n^1(\cos\vartheta) \tag{15.266}$$

as (rotationally symmetric) series expansion for the scattering potential[303] $\Psi_{s\varphi}(R, \vartheta, \omega)$; since P_n^{-1} according to $\mathrm{P}_n^{-1} = -\mathrm{P}_n^1/n(n+1)$ is proportional to P_n^1, we must not consider this term separately. We take the eigenfunction expansion for the incident plane pressure wave potential from the literature (Schäfke 1967; Langenberg 2005):

$$\begin{aligned}\Phi_i(R, \vartheta, \omega, \hat{\underline{\mathbf{k}}}_i = -\underline{\mathbf{e}}_z) &= \phi_i(\omega)\, \mathrm{e}^{-\mathrm{j}k_\mathrm{P} R\cos\vartheta} \\ &= \phi_i(\omega) \sum_{n=0}^{\infty} (-\mathrm{j})^n (2n+1) \mathrm{j}_n(k_\mathrm{P} R) \mathrm{P}_n(\cos\vartheta),\end{aligned} \tag{15.267}$$

where j_n denote spherical Bessel functions.

As always, the boundary condition

$$\begin{aligned}\underline{\underline{\mathbf{T}}}(a, \vartheta, \omega) \cdot \underline{\mathbf{e}}_R &= \underline{\underline{\mathbf{T}}}_s(a, \vartheta, \omega) \cdot \underline{\mathbf{e}}_R + \underline{\underline{\mathbf{T}}}_{i\mathrm{P}}(a, \vartheta, \omega, -\underline{\mathbf{e}}_z) \cdot \underline{\mathbf{e}}_R \\ &= \underline{\mathbf{0}}\end{aligned} \tag{15.268}$$

with

$$\begin{aligned}&\underline{\underline{\mathbf{T}}}(R, \vartheta, \omega) \cdot \underline{\mathbf{e}}_R \\ &\quad = \lambda\, \underline{\mathbf{e}}_R \nabla \cdot \underline{\mathbf{u}}(R, \vartheta, \omega) + \mu \left[\nabla \underline{\mathbf{u}}(R, \vartheta, \omega) \cdot \underline{\mathbf{e}}_R + \nabla \underline{\mathbf{u}}^{21}(R, \vartheta, \omega) \cdot \underline{\mathbf{e}}_R\right] \\ &\quad = -\lambda k_\mathrm{P}^2 \Phi(R, \vartheta, \omega)\, \underline{\mathbf{e}}_R + \mu \left[\nabla \underline{\mathbf{u}}(R, \vartheta, \omega) \cdot \underline{\mathbf{e}}_R + \nabla \underline{\mathbf{u}}^{21}(R, \vartheta, \omega) \cdot \underline{\mathbf{e}}_R\right],\end{aligned} \tag{15.269}$$

$$\nabla \underline{\mathbf{u}}(R, \vartheta, \omega) \cdot \underline{\mathbf{e}}_R = \frac{\partial u_R(R, \vartheta, \omega)}{\partial R} \underline{\mathbf{e}}_R + \frac{1}{R} \left[\frac{\partial u_R(R, \vartheta, \omega)}{\partial \vartheta} - u_\vartheta(R, \vartheta, \omega)\right] \underline{\mathbf{e}}_\vartheta, \tag{15.270}$$

$$\nabla \underline{\mathbf{u}}^{21}(R, \vartheta, \omega) \cdot \underline{\mathbf{e}}_R = \frac{\partial u_R(R, \vartheta, \omega)}{\partial R} \underline{\mathbf{e}}_R + \frac{\partial u_\vartheta(R, \vartheta, \omega)}{\partial R} \underline{\mathbf{e}}_\vartheta, \tag{15.271}$$

---

[303] Due to (Equation 15.261)

$$\mathrm{P}_n^1(\cos\vartheta) = \frac{\partial \mathrm{P}_n(\cos\vartheta)}{\partial\vartheta}$$

the literature (Mow 1965) initially starts motivation-less with $\partial\Psi_{s\varphi}/\partial\vartheta$ as potential.

serves to calculate the expansion coefficients $\phi_n(\omega)$ and $\psi_n(\omega)$ because (15.268) has exactly *two* scalar components $\underline{\mathbf{e}}_R \cdot \underline{\underline{\mathbf{T}}} \cdot \underline{\mathbf{e}}_R = T_{RR}$ and $\underline{\mathbf{e}}_\vartheta \cdot \underline{\underline{\mathbf{T}}} \cdot \underline{\mathbf{e}}_R = T_{\vartheta R}$:

$$T_{RR}(R,\vartheta,\omega) = -\lambda k_P^2 \Phi(R,\vartheta,\omega) + 2\mu \frac{\partial^2 \Phi(R,\vartheta,\omega)}{\partial R^2} + 2\mu \frac{1}{\sin\vartheta} \frac{\partial}{\partial R} \left\{ \frac{1}{R} \frac{\partial}{\partial \vartheta} [\sin\vartheta\, \Psi_\varphi(R,\vartheta,\omega)] \right\}, \tag{15.272}$$

$$\frac{1}{\mu} T_{\vartheta R}(R,\vartheta,\omega) = 2\frac{\partial}{\partial R}\left[\frac{1}{R}\frac{\partial \Phi(R,\vartheta,\omega)}{\partial \vartheta}\right] - \frac{\partial^2 \Psi_\varphi(R,\vartheta,\omega)}{\partial R^2} + \frac{2}{R^2}\Psi_\varphi(R,\vartheta,\omega) + \frac{1}{R^2}\frac{\partial}{\partial\vartheta}\left\{\frac{1}{\sin\vartheta}\frac{\partial}{\partial\vartheta}[\sin\vartheta\,\Psi_\varphi(R,\vartheta,\omega)]\right\}. \tag{15.273}$$

Insertion of the series expansions (15.265), (15.266), and (15.267) and considering Footnote 303 and the differential equations (15.263) and (15.264) together with the orthogonality relation

$$\int_0^\pi \mathrm{P}_n^m(\cos\vartheta)\mathrm{P}_{n'}^m(\cos\vartheta)\sin\vartheta\,\mathrm{d}\vartheta = \frac{2}{2n+1}\frac{(n+m)!}{(n-m)!}\delta_{nn'} \tag{15.274}$$

yields the following system of equations for $\phi_n(\omega)$ and $\psi_n(\omega)$:

$$\begin{aligned}&\left\{\left[2n(n+1) - k_S^2 a^2\right]\mathrm{h}_n^{(1)}(k_P a) - 4k_P a\,\mathrm{h}_n^{(1)'}(k_P a)\right\}\phi_n(\omega)\\ &+ 2n(n+1)\left[k_S a\,\mathrm{h}_n^{(1)'}(k_S a) - \mathrm{h}_n^{(1)}(k_S a)\right]\psi_n(\omega)\\ &= -(-\mathrm{j})^n(2n+1)\left\{\left[2n(n+1) - k_S^2 a^2\right]\mathrm{j}_n(k_P a) - 4k_P a\,\mathrm{j}_n'(k_P a)\right\}\phi_i(\omega),\end{aligned} \tag{15.275}$$

$$\begin{aligned}&\left[k_P a\,\mathrm{h}_n^{(1)'}(k_P a) - \mathrm{h}_n^{(1)}(k_P a)\right]\phi_n(\omega)\\ &- \left\{\left[n(n+1) - 1 - \frac{k_S^2 a^2}{2}\right]\mathrm{h}_n^{(1)}(k_S a) - k_S a\,\mathrm{h}_n^{(1)'}(k_S a)\right\}\psi_n(\omega)\\ &= -(-\mathrm{j})^n(2n+1)\left[k_P a\,\mathrm{j}_n'(k_P a) - \mathrm{j}_n(k_P a)\right]\phi_i(\omega).\end{aligned} \tag{15.276}$$

With (15.254), we finally obtain the components of the scattered particle velocity field:

$$u_{sR}(R,\vartheta,\omega) = \sum_{n=0}^\infty \left[k_P \phi_n(\omega)\mathrm{h}_n^{(1)'}(k_P R) - n(n+1)\psi_n(\omega)\frac{\mathrm{h}_n^{(1)}(k_S R)}{R}\right]\mathrm{P}_n(\cos\vartheta), \tag{15.277}$$

$$u_{s\vartheta}(R,\vartheta,\omega) = \sum_{n=0}^\infty \left\{\phi_n(\omega)\frac{\mathrm{h}_n^{(1)}(k_P R)}{R} - \psi_n(\omega)\left[k_S \mathrm{h}_n^{(1)'}(k_S R) + \frac{\mathrm{h}_n^{(1)}(k_S R)}{R}\right]\right\}\mathrm{P}_n^1(\cos\vartheta). \tag{15.278}$$

With the asymptotic

$$\mathrm{h}_n^{(1)}(k_{\mathrm{P,S}}R) \simeq \frac{\mathrm{j}^{-(n+1)}}{k_{\mathrm{P,S}}} \frac{\mathrm{e}^{\mathrm{j}k_{\mathrm{P,S}}R}}{R}, \tag{15.279}$$

$$\mathrm{h}_n^{(1)'}(k_{\mathrm{P,S}}R) \simeq \mathrm{j}^{-n} \frac{\mathrm{e}^{\mathrm{j}k_{\mathrm{P,S}}R}}{R} \tag{15.280}$$

of the spherical Hankel functions the far-field approximation emerges:

$$u_{sR}^{\mathrm{far}}(R, \vartheta, \omega) = \frac{\mathrm{e}^{\mathrm{j}k_{\mathrm{P}}R}}{R} k_{\mathrm{P}} \sum_{n=0}^{\infty} \mathrm{j}^{-n} \phi_n(\omega) \mathrm{P}_n(\cos\vartheta), \tag{15.281}$$

$$u_{s\vartheta}^{\mathrm{far}}(R, \vartheta, \omega) = -\frac{\mathrm{e}^{\mathrm{j}k_{\mathrm{S}}R}}{R} k_{\mathrm{S}} \sum_{n=0}^{\infty} \mathrm{j}^{-n} \psi_n(\omega) \mathrm{P}_n^1(\cos\vartheta). \tag{15.282}$$

That way, the scattering amplitudes (15.181) and (15.185) are explicitly given as eigenfunction expansions: As expected, in the far-field, we observe exclusively a pressure wave in the longitudinal R-component, and a shear wave in the transverse ϑ-component, the scattered wave modes are decoupled through polarization.

S-wave incidence: Choosing the (negative) z-axis of a cartesian coordinate system as direction of incidence of a plane shear wave on a spherical void, we may rotate the coordinate system around the z-axis until the x-axis coincides with the polarization vector of the shear wave (Figure 15.36). Hence, we assume

$$\underline{\mathbf{u}}_{i\mathrm{S}}(\underline{\mathbf{R}}, \omega, \hat{\underline{\mathbf{k}}}_i) = u_{i\mathrm{S}}(\omega)\, \mathrm{e}^{\mathrm{j}k_{\mathrm{S}}\hat{\underline{\mathbf{k}}}_i \cdot \underline{\mathbf{R}}}\, \underline{\mathbf{e}}_x \tag{15.283}$$

with $\hat{\underline{\mathbf{k}}}_i = -\underline{\mathbf{e}}_z$ as incident wave. Since $\underline{\mathbf{e}}_x$ defines a preference direction in the xy-plane, we may no longer count on the rotational symmetry of the pressure wave incidence, we must admit $\partial/\partial\varphi \not\equiv 0$. Furthermore, we may choose a surface element on the sphere where $\hat{\underline{\mathbf{k}}}_i$ and $\underline{\mathbf{e}}_R$ span a plane of incidence, and it is obvious that $\hat{\underline{\mathbf{u}}}_{i\mathrm{S}} = \underline{\mathbf{e}}_x$ is generally not located in this plane: In three dimensions, we can no longer speak of SV-scattering, we have to cope with S-scattering without decoupling into SV and SH. This requires two components of the vector potential $\underline{\boldsymbol{\Psi}}$ besides the scalar potential Φ, namely its ϑ- and φ-components. Yet, we may easily show that the resulting ϑ- and φ-components of the vector Helmholtz equation $\Delta\underline{\boldsymbol{\Psi}}(\underline{\mathbf{R}}, \omega) + k_{\mathrm{S}}^2\underline{\boldsymbol{\Psi}}(\underline{\mathbf{R}}, \omega) = \underline{\mathbf{0}}$ are not decoupled.[304] But from the theory of electromagnetic waves, we borrow a "trick" to decouple a vector wave equation (Hönl et al. 1961; Langenberg 2005) that also works for the present problem[305] (Brill and Gaunaurd 1987): We put $\underline{\boldsymbol{\Psi}}(\underline{\mathbf{R}}, \omega)$ divergence-free according to

[304]In fact, due to the divergence condition, the R-component is equal to zero, but in the ϑ- as well as in the φ-component both components of $\underline{\boldsymbol{\Psi}}$ appear.

[305]Alternatively, we may work with vector wave functions (Stratton 1941; Einspruch et al. 1960).

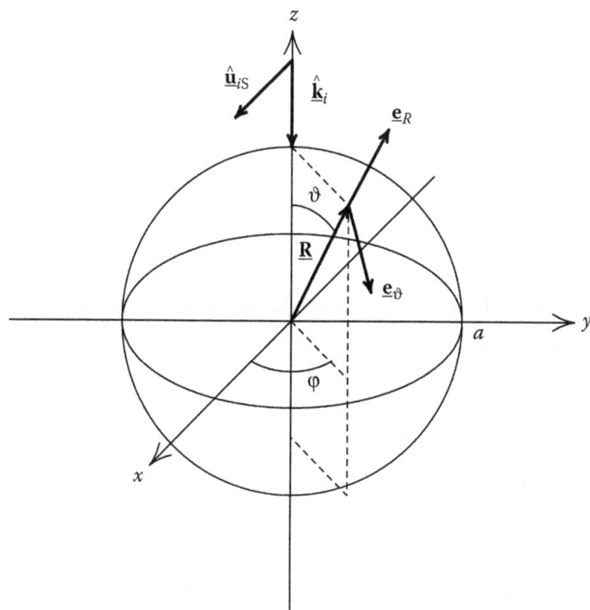

FIGURE 15.36
S-wave scattering by a spherical void with a stress-free surface.

$$\underline{\mathbf{\Psi}}(\underline{\mathbf{R}}, \omega) = \nabla \times [\underline{\mathbf{R}}\, U(\underline{\mathbf{R}}, \omega)] + \nabla \times \nabla \times [\underline{\mathbf{R}}\, V(\underline{\mathbf{R}}, \omega)] \tag{15.284}$$

introducing two scalar potentials $U(\underline{\mathbf{R}}, \omega)$ and $V(\underline{\mathbf{R}}, \omega)$, the so-called Debye potentials. Applying the Helmholtz operator $\Delta + k_\mathrm{S}^2$ to (15.284), we obtain after some calculus

$$\begin{aligned}(\Delta + k_\mathrm{S}^2)\underline{\mathbf{\Psi}}(\underline{\mathbf{R}}, \omega) = &-(\underline{\mathbf{R}} \times \nabla)\left[\Delta U(\underline{\mathbf{R}}, \omega) + k_\mathrm{S}^2 U(\underline{\mathbf{R}}, \omega)\right] \\ &+ (\underline{\mathbf{R}} \cdot \nabla\nabla + 2\nabla - \underline{\mathbf{R}}\,\Delta)\left[\Delta V(\underline{\mathbf{R}}, \omega) + k_\mathrm{S}^2 V(\underline{\mathbf{R}}, \omega)\right],\end{aligned} \tag{15.285}$$

and hence

$$\Delta U(\underline{\mathbf{R}}, \omega) + k_\mathrm{S}^2 U(\underline{\mathbf{R}}, \omega) = 0, \tag{15.286}$$

$$\Delta V(\underline{\mathbf{R}}, \omega) + k_\mathrm{S}^2 V(\underline{\mathbf{R}}, \omega) = 0 \tag{15.287}$$

as a collection of sufficient conditions to assure that $\underline{\mathbf{\Psi}}(\underline{\mathbf{R}}, \omega)$ satisfies a homogeneous vector Helmholtz equation. In addition, we have

$$\Delta \Phi(\underline{\mathbf{R}}, \omega) + k_\mathrm{P}^2 \Phi(\underline{\mathbf{R}}, \omega) = 0 \tag{15.288}$$

for the scalar pressure wave potential. Evidently, the solutions of all scalar Helmholtz equations have to be expanded into partial waves in spherical coordinates; without relying on rotational symmetry, we have to assume the general eigenfunction expansion (15.259) according to

$$\Phi_s(R,\vartheta,\varphi,\omega)=\sum_{n=0}^{\infty}\sum_{m=-n}^{n}\phi_{nm}^{(s)}(\omega)\mathrm{h}_n^{(1)}(k_\mathrm{P}R)\mathrm{P}_n^m(\cos\vartheta)\mathrm{e}^{\mathrm{j}m\varphi}, \tag{15.289}$$

$$U_s(R,\vartheta,\varphi,\omega)=\sum_{n=0}^{\infty}\sum_{m=-n}^{n}u_{nm}^{(s)}(\omega)\mathrm{h}_n^{(1)}(k_\mathrm{S}R)\mathrm{P}_n^m(\cos\vartheta)\mathrm{e}^{\mathrm{j}m\varphi}, \tag{15.290}$$

$$V_s(R,\vartheta,\varphi,\omega)=\sum_{n=0}^{\infty}\sum_{m=-n}^{n}v_{nm}^{(s)}(\omega)\mathrm{h}_n^{(1)}(k_\mathrm{S}R)\mathrm{P}_n^m(\cos\vartheta)\mathrm{e}^{\mathrm{j}m\varphi}. \tag{15.291}$$

In particular, we have to find the expansions for the potentials $U_i(\underline{\mathbf{R}},\omega,\hat{\underline{\mathbf{k}}}_i)$, $V_i(\underline{\mathbf{R}},\omega,\hat{\underline{\mathbf{k}}}_i)$ of the incident shear wave in a way that

$$\begin{aligned}\underline{\mathbf{u}}_{i\mathrm{S}}(\underline{\mathbf{R}},\omega,-\underline{\mathbf{e}}_z)&=\nabla\times\nabla\times[\underline{\mathbf{R}}\,U_i(\underline{\mathbf{R}},\omega,-\underline{\mathbf{e}}_z)]\\&\quad+\nabla\times\nabla\times\nabla\times[\underline{\mathbf{R}}\,V_i(\underline{\mathbf{R}},\omega,-\underline{\mathbf{e}}_z)],\end{aligned} \tag{15.292}$$

hence,

$$\begin{aligned}&u_{i\mathrm{S}}(\omega)\,\mathrm{e}^{-\mathrm{j}k_\mathrm{S}R\cos\vartheta}\,\sin\vartheta\cos\varphi\\&\quad=\nabla\times\nabla\times[\underline{\mathbf{R}}\,U_i(R,\vartheta,\varphi,\omega,-\underline{\mathbf{e}}_z)]\cdot\underline{\mathbf{e}}_R\\&\qquad+\nabla\times\nabla\times\nabla\times[\underline{\mathbf{R}}\,V_i(R,\vartheta,\varphi,\omega,-\underline{\mathbf{e}}_z)]\cdot\underline{\mathbf{e}}_R,\end{aligned} \tag{15.293}$$

$$\begin{aligned}&u_{i\mathrm{S}}(\omega)\,\mathrm{e}^{-\mathrm{j}k_\mathrm{S}R\cos\vartheta}\,\cos\vartheta\cos\varphi\\&\quad=\nabla\times\nabla\times[\underline{\mathbf{R}}\,U_i(R,\vartheta,\varphi,\omega,-\underline{\mathbf{e}}_z)]\cdot\underline{\mathbf{e}}_\vartheta\\&\qquad+\nabla\times\nabla\times\nabla\times[\underline{\mathbf{R}}\,V_i(R,\vartheta,\varphi,\omega,-\underline{\mathbf{e}}_z)]\cdot\underline{\mathbf{e}}_\vartheta,\end{aligned} \tag{15.294}$$

$$\begin{aligned}&-u_{i\mathrm{S}}(\omega)\,\mathrm{e}^{-\mathrm{j}k_\mathrm{S}R\cos\vartheta}\,\sin\varphi\\&\quad=\nabla\times\nabla\times[\underline{\mathbf{R}}\,U_i(R,\vartheta,\varphi,\omega,-\underline{\mathbf{e}}_z)]\cdot\underline{\mathbf{e}}_\varphi\\&\qquad+\nabla\times\nabla\times\nabla\times[\underline{\mathbf{R}}\,V_i(R,\vartheta,\varphi,\omega,-\underline{\mathbf{e}}_z)]\cdot\underline{\mathbf{e}}_\varphi,\end{aligned} \tag{15.295}$$

holds. Explicitly differentiating the right-hand side of (15.292) yields

$$\begin{aligned}&2\nabla U_i(\underline{\mathbf{R}},\omega,-\underline{\mathbf{e}}_z)+\underline{\mathbf{R}}\cdot\nabla\nabla U_i(\underline{\mathbf{R}},\omega,-\underline{\mathbf{e}}_z)+k_\mathrm{S}^2\,\underline{\mathbf{R}}\,U_i(\underline{\mathbf{R}},\omega,-\underline{\mathbf{e}}_z)\\&\quad-k_\mathrm{S}^2\,\underline{\mathbf{R}}\times\nabla V_i(\underline{\mathbf{R}},\omega,-\underline{\mathbf{e}}_z)\end{aligned} \tag{15.296}$$

allowing to write down the components required in (15.293) through (15.295):

$$\begin{aligned}u_{i\mathrm{S}}(\omega)\,\mathrm{e}^{-\mathrm{j}k_\mathrm{S}R\cos\vartheta}\,\sin\vartheta\cos\varphi&=R\frac{\partial^2U_i(R,\vartheta,\varphi,\omega,-\underline{\mathbf{e}}_z)}{\partial R^2}\\&\quad+2\frac{\partial U_i(R,\vartheta,\varphi,\omega,-\underline{\mathbf{e}}_z)}{\partial R}\\&\quad+k_\mathrm{S}^2\,R\,U_i(R,\vartheta,\varphi,\omega,-\underline{\mathbf{e}}_z),\end{aligned} \tag{15.297}$$

$$\begin{aligned}u_{i\mathrm{S}}(\omega)\,\mathrm{e}^{-\mathrm{j}k_\mathrm{S}R\cos\vartheta}\,\cos\vartheta\cos\varphi&=\frac{1}{R}\frac{\partial^2}{\partial\vartheta\partial R}[R\,U_i(R,\vartheta,\varphi,\omega,-\underline{\mathbf{e}}_z)]\\&\quad+k_\mathrm{S}^2\,\frac{1}{\sin\vartheta}\frac{\partial V_i(R,\vartheta,\varphi,\omega,-\underline{\mathbf{e}}_z)}{\partial\varphi},\end{aligned} \tag{15.298}$$

$$-u_{iS}(\omega)\,\mathrm{e}^{-\mathrm{j}k_S R\cos\vartheta}\sin\varphi = \frac{1}{R\sin\vartheta}\frac{\partial^2}{\partial\varphi\partial R}\left[R\,U_i(R,\vartheta,\varphi,\omega,-\underline{\mathbf{e}}_z)\right] - k_S^2\,\frac{\partial V_i(R,\vartheta,\varphi,\omega,-\underline{\mathbf{e}}_z)}{\partial\vartheta}. \tag{15.299}$$

Of course, we have to assure a correct calculation of the components of the dyadic differential operator $\boldsymbol{\nabla\nabla}$ in spherical coordinates (alternatively, we may check the collection of formulas). Fortunately, (15.297) contains only U_i, and, hence, only u_{nm}-coefficients; however, with (15.267), we know the eigenfunction expansion for $\mathrm{e}^{-\mathrm{j}kR\cos\vartheta}$ but not the one for $\sin\vartheta\,\mathrm{e}^{-\mathrm{j}kR\cos\vartheta}$. This can be readily fixed observing

$$u_{iS}(\omega)\,\mathrm{e}^{-\mathrm{j}k_S R\cos\vartheta}\sin\vartheta\cos\varphi = \frac{u_{iS}(\omega)}{\mathrm{j}k_S R}\cos\varphi\frac{\partial}{\partial\vartheta}\,\mathrm{e}^{-\mathrm{j}k_S R\cos\vartheta} \tag{15.300}$$

because then we have

$$u_{iS}(\omega)\,\mathrm{e}^{-\mathrm{j}k_S R\cos\vartheta}\sin\vartheta\cos\varphi = \frac{u_{iS}(\omega)}{\mathrm{j}k_S R}\cos\varphi\sum_{n=0}^{\infty}(-\mathrm{j})^n(2n+1)\mathrm{j}_n(k_S R)\mathrm{P}_n^1(\cos\vartheta). \tag{15.301}$$

Since Equations 15.297 through 15.299 must hold for all values of R, ϑ, and φ, the factor $\cos\varphi$ on the left-hand side of (15.299) enforces a similar factor on the right-hand side giving the eigenfunction expansion of $U_i(R,\vartheta,\varphi,\omega,-\underline{\mathbf{e}}_z)$ undoubtedly the following appearance:

$$U_i(R,\vartheta,\varphi,\omega,-\underline{\mathbf{e}}_z) = \cos\varphi\sum_{n=0}^{\infty}u_n^{(i)}(\omega)\mathrm{j}_n(k_S R)\mathrm{P}_n^1(\cos\vartheta). \tag{15.302}$$

Applied to the spherical Bessel functions in (15.302), the differential operator on the right-hand side of (15.297) yields $n(n+1)\mathrm{j}_n(k_S R)/R$ finally resulting in the eigenfunction expansion

$$U_i(R,\vartheta,\varphi,\omega,-\underline{\mathbf{e}}_z) = \frac{u_{iS}(\omega)}{\mathrm{j}k_S}\cos\varphi\sum_{n=0}^{\infty}(-\mathrm{j})^n\frac{2n+1}{n(n+1)}\mathrm{j}_n(k_S R)\mathrm{P}_n^1(\cos\vartheta). \tag{15.303}$$

Equations 15.298 and 15.299 exhibit the same R-differential operator applied to U_i, once with an additional ϑ-, and once with an additional φ-differentiation; therefore, the following approach is offered: We differentiate (15.298) with respect to φ, multiply (15.299) with $\sin\vartheta$, and differentiate the resulting equation with respect to ϑ; afterward, both equations are subtracted: The terms with U_i are canceled! It remains

$$\frac{1}{k_S^2}\,u_{iS}(\omega)\sin\varphi\frac{\partial}{\partial\vartheta}\left(\mathrm{e}^{-\mathrm{j}k_S R\cos\vartheta}\right) = \frac{1}{\sin^2\vartheta}\frac{\partial^2 V_i(R,\vartheta,\varphi,\omega,-\underline{\mathbf{e}}_z)}{\partial\varphi^2} + \frac{1}{\sin\vartheta}\frac{\partial}{\partial\vartheta}\left[\sin\vartheta\,\frac{\partial V_i(R,\vartheta,\varphi,\omega,-\underline{\mathbf{e}}_z)}{\partial\vartheta}\right]. \tag{15.304}$$

With the differential equation (15.264), the *ansatz*

$$V_i(R,\vartheta,\varphi,\omega,-\underline{\mathbf{e}}_z)=\sin\varphi\sum_{n=0}^{\infty}v_n^{(i)}(\omega)\mathrm{j}_n(k_{\mathrm{S}}R)\mathrm{P}_n^1(\cos\vartheta) \tag{15.305}$$

leads to

$$V_i(R,\vartheta,\varphi,\omega,-\underline{\mathbf{e}}_z)=-\frac{u_{i\mathrm{S}}(\omega)}{k_{\mathrm{S}}^2}\sin\varphi\sum_{n=0}^{\infty}(-\mathrm{j})^n\frac{2n+1}{n(n+1)}\mathrm{j}_n(k_{\mathrm{S}}R)\mathrm{P}_n^1(\cos\vartheta). \tag{15.306}$$

Equation 15.269 is now also φ-dependent according to

$$\begin{aligned}\underline{\underline{\mathbf{T}}}(R,\vartheta,\varphi,\omega)\cdot\underline{\mathbf{e}}_R=&-\lambda k_{\mathrm{P}}^2\Phi(R,\vartheta,\varphi,\omega)\,\underline{\mathbf{e}}_R\\&+\mu\left[\nabla\underline{\mathbf{u}}(R,\vartheta,\varphi,\omega)\cdot\underline{\mathbf{e}}_R+\nabla\underline{\mathbf{u}}^{21}(R,\vartheta,\varphi,\omega)\cdot\underline{\mathbf{e}}_R\right];\end{aligned} \tag{15.307}$$

hence, in contrast to (15.270) and (15.271), we also need the φ-components in $\nabla\underline{\mathbf{u}}\cdot\underline{\mathbf{e}}_R$ and $\nabla\underline{\mathbf{u}}^{21}\cdot\underline{\mathbf{e}}_R$:

$$\begin{aligned}&\nabla\underline{\mathbf{u}}(R,\vartheta,\varphi,\omega)\cdot\underline{\mathbf{e}}_R\\&=\frac{\partial u_R(R,\vartheta,\varphi,\omega)}{\partial R}\underline{\mathbf{e}}_R+\frac{1}{R}\left[\frac{\partial u_R(R,\vartheta,\varphi,\omega)}{\partial\vartheta}-u_\vartheta(R,\vartheta,\varphi,\omega)\right]\underline{\mathbf{e}}_\vartheta\\&+\frac{1}{R\sin\vartheta}\left[\frac{\partial u_R(R,\vartheta,\varphi,\omega)}{\partial\varphi}-\sin\vartheta\,u_\varphi(R,\vartheta,\varphi,\omega)\right]\underline{\mathbf{e}}_\varphi,\end{aligned} \tag{15.308}$$

$$\begin{aligned}&\nabla\underline{\mathbf{u}}^{21}(R,\vartheta,\varphi,\omega)\cdot\underline{\mathbf{e}}_R\\&=\frac{\partial u_R(R,\vartheta,\varphi,\omega)}{\partial R}\underline{\mathbf{e}}_R+\frac{\partial u_\vartheta(R,\vartheta,\varphi,\omega)}{\partial\vartheta}\underline{\mathbf{e}}_\vartheta+\frac{\partial u_\varphi(R,\vartheta,\varphi,\omega)}{\partial\varphi}\underline{\mathbf{e}}_\varphi.\end{aligned} \tag{15.309}$$

With the differentiation prescriptions

$$\begin{aligned}u_R(R,\vartheta,\varphi,\omega)=&\frac{\partial\Phi(R,\vartheta,\varphi,\omega)}{\partial R}+\frac{1}{R}\frac{\partial}{\partial R}\left[R^2\frac{\partial U(R,\vartheta,\varphi,\omega)}{\partial R}\right]\\&+k_{\mathrm{S}}^2RU(R,\vartheta,\varphi,\omega),\end{aligned} \tag{15.310}$$

$$\begin{aligned}u_\vartheta(R,\vartheta,\varphi,\omega)=&\frac{1}{R}\frac{\partial\Phi(R,\vartheta,\varphi,\omega)}{\partial\vartheta}+\frac{1}{R}\frac{\partial}{\partial R}\left[R\frac{\partial U(R,\vartheta,\varphi,\omega)}{\partial\vartheta}\right]\\&+\frac{k_{\mathrm{S}}^2}{\sin\vartheta}\frac{\partial V(R,\vartheta,\varphi,\omega)}{\partial\varphi},\end{aligned} \tag{15.311}$$

$$\begin{aligned}u_\varphi(R,\vartheta,\varphi,\omega)=&\frac{1}{R\sin\vartheta}\frac{\partial\Phi(R,\vartheta,\varphi,\omega)}{\partial\varphi}+\frac{1}{R\sin\vartheta}\frac{\partial}{\partial R}\left[R\frac{\partial U(R,\vartheta,\varphi,\omega)}{\partial\varphi}\right]\\&-k_{\mathrm{S}}^2\frac{\partial V(R,\vartheta,\varphi,\omega)}{\partial\vartheta},\end{aligned} \tag{15.312}$$

we obtain

$$\begin{aligned} T_{RR}(R,\vartheta,\varphi,\omega) &= -\lambda k_{\mathrm{P}}^2 \Phi(R,\vartheta,\varphi,\omega) + 2\mu \,\frac{\partial^2 \Phi(R,\vartheta,\varphi,\omega)}{\partial R^2} \\ &\quad + 2\mu \,\frac{\partial}{\partial R}\left\{\frac{1}{R}\frac{\partial}{\partial R}\left[R^2 \frac{\partial U(R,\vartheta,\varphi,\omega)}{\partial R}\right]\right\} \\ &\quad + 2\mu k_{\mathrm{S}}^2 \,\frac{\partial}{\partial R}\left[R\, U(R,\vartheta,\varphi,\omega)\right], \end{aligned} \tag{15.313}$$

$$\begin{aligned} \frac{1}{\mu}\, T_{\vartheta R}(R,\vartheta,\varphi,\omega) &= 2\frac{\partial}{\partial R}\left[\frac{1}{R}\,\frac{\partial \Phi(R,\vartheta,\varphi,\omega)}{\partial \vartheta}\right] \\ &\quad + 2\left(\frac{\partial^2}{\partial R^2} + \frac{1}{R}\frac{\partial}{\partial R} - \frac{1}{R^2}\right)\frac{\partial U(R,\vartheta,\varphi,\omega)}{\partial \vartheta} \\ &\quad + k_{\mathrm{S}}^2\,\frac{\partial U(R,\vartheta,\varphi,\omega)}{\partial \vartheta} + \frac{k_{\mathrm{S}}^2}{\sin\vartheta}\, R \frac{\partial}{\partial R}\left[\frac{1}{R}\,\frac{\partial V(R,\vartheta,\varphi,\omega)}{\partial \varphi}\right], \end{aligned} \tag{15.314}$$

$$\begin{aligned} \frac{1}{\mu}\, T_{\varphi R}(R,\vartheta,\varphi,\omega) &= \frac{2}{\sin\vartheta}\,\frac{\partial}{\partial R}\left[\frac{1}{R}\,\frac{\partial \Phi(R,\vartheta,\varphi,\omega)}{\partial \varphi}\right] \\ &\quad + \frac{2}{\sin\vartheta}\left(\frac{\partial^2}{\partial R^2} + \frac{1}{R}\frac{\partial}{\partial R} - \frac{1}{R^2}\right)\frac{\partial U(R,\vartheta,\varphi,\omega)}{\partial \varphi} \\ &\quad + \frac{k_{\mathrm{S}}^2}{\sin\vartheta}\,\frac{\partial U(R,\vartheta,\varphi,\omega)}{\partial \varphi} - k_{\mathrm{S}}^2 R \frac{\partial}{\partial R}\left[\frac{1}{R}\,\frac{\partial V(R,\vartheta,\varphi,\omega)}{\partial \vartheta}\right]. \end{aligned} \tag{15.315}$$

In (15.313), we have $\Phi = \Phi_s$ and $U = U_i + U_s$; since U_i contains $\cos\varphi$ and P_n^1, the following specializations of the scattering potential expansion Φ_s and U_s are mandatory:

$$\Phi_s(R,\vartheta,\varphi,\omega) = \cos\varphi \sum_{n=0}^{\infty} \phi_n^{(s)}(\omega)\mathrm{h}_n^{(1)}(k_{\mathrm{P}}R)\mathrm{P}_n^1(\cos\vartheta), \tag{15.316}$$

$$U_s(R,\vartheta,\varphi,\omega) = \cos\varphi \sum_{n=0}^{\infty} u_n^{(s)}(\omega)\mathrm{h}_n^{(1)}(k_{\mathrm{S}}R)\mathrm{P}_n^1(\cos\vartheta), \tag{15.317}$$

because the differential operators in (15.313) only apply to the spherical cylinder functions, and, hence, the orthogonality relation (15.274) is applicable for $m = 1$. In (15.314) and (15.315) also appear φ- and ϑ-differentiations apart from the R-differentiations; at first: In (15.314), we have $V = V_i + V_s$; so, if we specialize

$$V_s(R,\vartheta,\varphi,\omega) = \sin\varphi \sum_{n=0}^{\infty} v_n^{(s)}(\omega)\mathrm{h}_n^{(1)}(k_{\mathrm{S}}R)\mathrm{P}_n^1(\cos\vartheta), \tag{15.318}$$

then all terms of (15.314) contain $\cos\varphi$, and all terms of (15.315) contain $\sin\varphi$, and, hence, the φ-dependence can be eliminated from the homogeneous boundary condition equations (15.313) through (15.315) for $R = a$. However, the orthogonality relation (15.274) may not immediately be applied because the ϑ-differentiations in both equations (15.314) and (15.315) yield "friendly" $P_n^1(\cos\vartheta)$-terms as well as "unfriendly" $\partial P_n^1(\cos\vartheta)/\partial\vartheta$-terms, and the latter ones may "only" be changed into P_n- and P_n^2-terms according to—keeping the lower index—(Stratton 1941, however, with a wrong sign)

$$\frac{\partial P_n^1(\cos\vartheta)}{\partial\vartheta} = -\frac{1}{2}\left[n(n+1)P_n(\cos\vartheta) - P_n^2(\cos\vartheta)\right]. \tag{15.319}$$

Yet, the following reasoning is successful: The short-hand notation

$$\begin{aligned}\frac{1}{\mu}T_{\vartheta R}(R,\vartheta,\varphi,\omega) &= \partial_R^\Phi \frac{\partial\Phi(R,\vartheta,\varphi,\omega)}{\partial\vartheta} + \partial_R^U \frac{\partial U(R,\vartheta,\varphi,\omega)}{\partial\vartheta}\\ &\quad + \frac{k_S^2}{\sin\vartheta}\partial_R^V \frac{\partial V(R,\vartheta,\varphi,\omega)}{\partial\varphi},\end{aligned} \tag{15.320}$$

$$\begin{aligned}\frac{1}{\mu}T_{\varphi R}(R,\vartheta,\varphi,\omega) &= \frac{1}{\sin\vartheta}\partial_R^\Phi \frac{\partial\Phi(R,\vartheta,\varphi,\omega)}{\partial\varphi} + \frac{1}{\sin\vartheta}\partial_R^U \frac{\partial U(R,\vartheta,\varphi,\omega)}{\partial\varphi}\\ &\quad - k_S^2\partial_R^V \frac{\partial V(R,\vartheta,\varphi,\omega)}{\partial\vartheta}\end{aligned} \tag{15.321}$$

of (15.314) and (15.315) using Φ, U, V-specific R-differential operators ∂_R^Φ, $\partial_R^U, \partial_R^V$ particularly enlightens the appearance of similar differential operators in both equations allowing to introduce respectively similar coefficients $\alpha_n(a,\omega)$ and $\beta_n(a,\omega)$

$$\begin{aligned}\alpha_n(a,\omega) &= \phi_n^{(s)}(\omega)\partial_R^\Phi h_n^{(1)}(k_P R)\Big|_{R=a} + u_n^{(s)}(\omega)\partial_R^U h_n^{(1)}(k_S R)\Big|_{R=a}\\ &\quad + u_n^{(i)}(\omega)\partial_R^U j_n(k_S R)\Big|_{R=a},\end{aligned} \tag{15.322}$$

$$\beta_n(a,\omega) = k_S^2 v_n^{(s)}(\omega)\partial_R^V h_n^{(1)}(k_S R)\Big|_{R=a} + k_S^2 v_n^{(i)}(\omega)\partial_R^V j_n(k_S R)\Big|_{R=a} \tag{15.323}$$

after insertion of the expansions (15.316) through (15.318) for $R = a$ yielding the notation:

$$\sum_{n=0}^{\infty}\left[\alpha_n(a,\omega)\sin^2\vartheta\frac{\partial P_n^1(\cos\vartheta)}{\partial\vartheta} + \beta_n(a,\omega)\sin\vartheta P_n^1(\cos\vartheta)\right] = 0, \tag{15.324}$$

$$\sum_{n=0}^{\infty}\left[\alpha_n(a,\omega)\sin\vartheta P_n^1(\cos\vartheta) - \beta_n(a,\omega)\sin^2\vartheta\frac{\partial P_n^1(\cos\vartheta)}{\partial\vartheta}\right] = 0; \tag{15.325}$$

the $u_n^{(i)}$- and $v_n^{(i)}$-coefficients are known from (15.303), respectively (15.306). Multiplication of both equations with $P_{n'}^1(\cos\vartheta)$ and integration with respect to ϑ from 0 to π yield

$$\beta_{n'}(a,\omega) = -\sum_{n=0}^{\infty}\alpha_n(a,\omega)\underbrace{\frac{2n'+1}{2n'(n'+1)}\int_0^\pi \sin^2\vartheta\,\frac{\partial P_n^1(\cos\vartheta)}{\partial\vartheta}\,P_{n'}^1(\cos\vartheta)\,d\vartheta}_{=\gamma_{nn'}}, \tag{15.326}$$

$$\alpha_{n'}(a,\omega) = \sum_{n=0}^{\infty} \beta_n(a,\omega) \underbrace{\frac{2n'+1}{2n'(n'+1)} \int_0^{\pi} \sin^2\vartheta \, \frac{\partial \mathrm{P}_n^1(\cos\vartheta)}{\partial\vartheta} \, \mathrm{P}_{n'}^1(\cos\vartheta) \, d\vartheta}_{=\gamma_{nn'}} \tag{15.327}$$

according to the orthogonality relation (15.274) for $m = 1$. Inserting, for example, (15.327) into (15.326)—we denote the summation index n'' in (15.326)—we obtain

$$\beta_{n'}(a,\omega) = -\sum_{n''=0}^{\infty} \gamma_{n''n'} \sum_{n=0}^{\infty} \beta_n(a,\omega)\gamma_{nn''} \tag{15.328}$$

$$= -\sum_{n=0}^{\infty} \beta_n(a,\omega) \sum_{n''=0}^{\infty} \gamma_{nn''}\gamma_{n''n'}; \tag{15.329}$$

hence, we must have

$$\sum_{n''=0}^{\infty} \gamma_{nn''}\gamma_{n''n'} = -\delta_{nn'}. \tag{15.330}$$

A numerical calculation requires a truncation of the series expansions, say at $n = n' = n'' = N$; then the $\gamma_{nn'}$ fill an $N \times N$-matrix $\underline{\underline{\Gamma}}$, and (15.330) can be written:

$$\underline{\underline{\Gamma}} \cdot \underline{\underline{\Gamma}} = -\underline{\underline{\boldsymbol{I}}}, \tag{15.331}$$

where $\underline{\underline{\boldsymbol{I}}}$ denotes the $N \times N$-unit matrix; as usual, the dot stands for the contraction of adjacent indices. With the calculation rules $\det(\underline{\underline{\Gamma}} \cdot \underline{\underline{\Gamma}}) = \det^2\underline{\underline{\Gamma}}$ and $\det(-\underline{\underline{\boldsymbol{I}}}) = -\det\underline{\underline{\boldsymbol{I}}} = -1$, we immediately see that this cannot be true. Consequently, in (15.324) as well as in (15.325), the summations must separately be zero:

$$\sum_{n=0}^{\infty} \beta_n(a,\omega) \sin\vartheta \, \mathrm{P}_n^1(\cos\vartheta) = 0, \tag{15.332}$$

$$\sum_{n=0}^{\infty} \alpha_n(a,\omega) \sin\vartheta \, \mathrm{P}_n^1(\cos\vartheta) = 0, \tag{15.333}$$

$$\sum_{n=0}^{\infty} \alpha_n(a,\omega) \sin^2\vartheta \, \frac{\partial \mathrm{P}_n^1(\cos\vartheta)}{\partial\vartheta} = 0, \tag{15.334}$$

$$\sum_{n=0}^{\infty} \beta_n(a,\omega) \sin^2\vartheta \, \frac{\partial \mathrm{P}_n^1(\cos\vartheta)}{\partial\vartheta} = 0. \tag{15.335}$$

Now the orthogonality relation (15.274) yields for $m = 1$:

$$\beta_n(a,\omega) = 0, \tag{15.336}$$

$$\alpha_n(a,\omega) = 0 \tag{15.337}$$

if we use (15.332) and (15.333); with (15.336) and (15.337), even the requirements (15.334) and (15.335) are correct. Due to (15.323), Equation 15.336 states the decoupling of the V-potential from the U-potential; we find explicitly

$$\begin{aligned} v_n^{(s)}(\omega) &= -v_n^{(i)}(\omega) \left. \frac{\partial_R^V \mathrm{j}_n(k_S R)}{\partial_R^V \mathrm{h}_n^{(1)}(k_S R)} \right|_{R=a} \\ &= -v_n^{(i)}(\omega) \frac{k_S a \, \mathrm{j}_n'(k_S a) - \mathrm{j}_n(k_S a)}{k_S a \, \mathrm{h}_n^{(1)\prime}(k_S a) - \mathrm{h}_n^{(1)}(k_S a)}, \end{aligned} \qquad (15.338)$$

where the dashes on the spherical cylinder functions denote derivatives with regard to the argument. The $v_n^{(i)}(\omega)$ may be read off from (15.306):

$$v_n^{(i)}(\omega) = -\frac{u_{iS}(\omega)}{k_S^2} (-\mathrm{j})^n \frac{2n+1}{n(n+1)}. \qquad (15.339)$$

To calculate the expansion coefficients $\phi_n^{(s)}(\omega)$ and $u_n^{(s)}(\omega)$, we must combine[306] (15.337) according to

$$\begin{aligned} &\left[k_P a \, \mathrm{h}_n^{(1)\prime}(k_P a) - \mathrm{h}_n^{(1)}(k_P a)\right] \phi_n^{(s)}(\omega) \\ &\quad + \left\{\left[n(n+1) - 1 - \frac{k_S^2 a^2}{2}\right] \mathrm{h}_n^{(1)}(k_S a) - k_S a \, \mathrm{h}_n^{(1)\prime}(k_S a)\right\} u_n^{(s)}(\omega) \\ &\quad = -\left\{\left[n(n+1) - 1 - \frac{k_S^2 a^2}{2}\right] \mathrm{j}_n(k_S a) - k_S a \, \mathrm{j}_n'(k_S a)\right\} u_n^{(i)}(\omega) \end{aligned} \qquad (15.340)$$

with the boundary condition $T_{RR}(a, \vartheta, \varphi, \omega) = 0$ resulting from (15.313):

$$\begin{aligned} &\left\{\left[2n(n+1) - k_S^2 a^2\right] \mathrm{h}_n^{(1)}(k_P a) - 4 k_P a \, \mathrm{h}_n^{(1)\prime}(k_P a)\right\} \phi_n^{(s)}(\omega) \\ &\quad + 2n(n+1) \left[k_S a \, \mathrm{h}_n^{(1)\prime}(k_S a) - \mathrm{h}_n^{(1)}(k_S a)\right] u_n^{(s)}(\omega) \\ &\quad = -2n(n+1) \left[k_S a \, \mathrm{j}_n'(k_S a) - \mathrm{j}_n(k_S a)\right] u_n^{(i)}(\omega). \end{aligned} \qquad (15.341)$$

The $u_n^{(i)}(\omega)$ are read off from (15.303):

$$u_n^{(i)}(\omega) = \frac{u_{iS}(\omega)}{\mathrm{j} k_S} (-\mathrm{j})^n \frac{2n+1}{n(n+1)}. \qquad (15.342)$$

With the expansion coefficients of the potentials, we may finally calculate the components of the particle displacement according to (15.310) through (15.312):

[306] The expansion coefficients are similarly given by Brill and Gaunaurd (1987), yet without calculation.

$$u_{sR}(R,\vartheta,\varphi,\omega) = \cos\varphi \sum_{n=0}^{\infty} \left[k_{\mathrm{P}} \phi_n^{(s)}(\omega) \mathrm{h}_n^{(1)\prime}(k_{\mathrm{P}}R) + n(n+1) u_n^{(s)}(\omega) \frac{\mathrm{h}_n^{(1)}(k_{\mathrm{S}}R)}{R} \right] \mathrm{P}_n^1(\cos\vartheta), \quad (15.343)$$

$$u_{s\vartheta}(R,\vartheta,\varphi,\omega) = \cos\varphi \sum_{n=0}^{\infty} \left\{ \phi_n^{(s)}(\omega) \frac{\mathrm{h}_n^{(1)}(k_{\mathrm{P}}R)}{R} \frac{\partial \mathrm{P}_n^1(\cos\vartheta)}{\partial\vartheta} + u_n^{(s)}(\omega) \left[k_{\mathrm{S}} \mathrm{h}_n^{(1)\prime}(k_{\mathrm{S}}R) + \frac{\mathrm{h}_n^{(1)}(k_{\mathrm{S}}R)}{R} \right] \frac{\partial \mathrm{P}_n^1(\cos\vartheta)}{\partial\vartheta} + k_{\mathrm{S}}^2 v_n^{(s)}(\omega) \mathrm{h}_n^{(1)}(k_{\mathrm{S}}R) \frac{\mathrm{P}_n^1(\cos\vartheta)}{\sin\vartheta} \right\}, \quad (15.344)$$

$$u_{s\varphi}(R,\vartheta,\varphi,\omega) = -\sin\varphi \sum_{n=0}^{\infty} \left\{ \phi_n^{(s)}(\omega) \frac{\mathrm{h}_n^{(1)}(k_{\mathrm{P}}R)}{R} \frac{\mathrm{P}_n^1(\cos\vartheta)}{\sin\vartheta} + u_n^{(s)}(\omega) \left[k_{\mathrm{S}} \mathrm{h}_n^{(1)\prime}(k_{\mathrm{S}}R) + \frac{\mathrm{h}_n^{(1)}(k_{\mathrm{S}}R)}{R} \right] \frac{\mathrm{P}_n^1(\cos\vartheta)}{\sin\vartheta} + k_{\mathrm{S}}^2 v_n^{(s)}(\omega) \mathrm{h}_n^{(1)}(k_{\mathrm{S}}R) \frac{\partial \mathrm{P}_n^1(\cos\vartheta)}{\partial\vartheta} \right\}. \quad (15.345)$$

Obviously, these representation only hold for $R \geq a$. It is interesting to note the agreement of the components with those for P-wave incidence with respect to the R-dependence (Equations 15.277 and 15.278). With the asymptotic expansions (15.277) and (15.278), we find the far-field approximations:

$$u_{sR}^{\mathrm{far}}(R,\vartheta,\varphi,\omega) = \frac{\mathrm{e}^{\mathrm{j}k_{\mathrm{P}}R}}{R} k_{\mathrm{P}} \cos\varphi \sum_{n=0}^{\infty} \mathrm{j}^{-n} \phi_n^{(s)}(\omega) \mathrm{P}_n^1(\cos\vartheta), \quad (15.346)$$

$$u_{s\vartheta}^{\mathrm{far}}(R,\vartheta,\varphi,\omega) = \frac{\mathrm{e}^{\mathrm{j}k_{\mathrm{S}}R}}{R} k_{\mathrm{S}} \cos\varphi \sum_{n=0}^{\infty} \mathrm{j}^{-n} \left[u_n^{(s)}(\omega) \frac{\partial \mathrm{P}_n^1(\cos\vartheta)}{\partial\vartheta} - \mathrm{j}k_{\mathrm{S}} v_n^{(s)}(\omega) \frac{\mathrm{P}_n^1(\cos\vartheta)}{\sin\vartheta} \right], \quad (15.347)$$

$$u_{s\varphi}^{\mathrm{far}}(R,\vartheta,\varphi,\omega) = -\frac{\mathrm{e}^{\mathrm{j}k_{\mathrm{S}}R}}{R} k_{\mathrm{S}} \sin\varphi \sum_{n=0}^{\infty} \mathrm{j}^{-n} \left[u_n^{(s)}(\omega) \frac{\mathrm{P}_n^1(\cos\vartheta)}{\sin\vartheta} - \mathrm{j}k_{\mathrm{S}} v_n^{(s)}(\omega) \frac{\partial \mathrm{P}_n^1(\cos\vartheta)}{\partial\vartheta} \right]. \quad (15.348)$$

To calculate the ϑ-derivative of $\mathrm{P}_n^1(\cos\vartheta)$, we can use (15.319), and with (15.261) and (15.262), we can show that $\mathrm{P}_n^1(\cos\vartheta)/\sin\vartheta$ is nonsingular for $\vartheta = 0, \pi$. As in the P-case, the mode decoupling is once more explicitly given by longitudinal and transverse polarizations. Yet, we should note that both shear wave expansion coefficients enter the transverse components, i.e., their

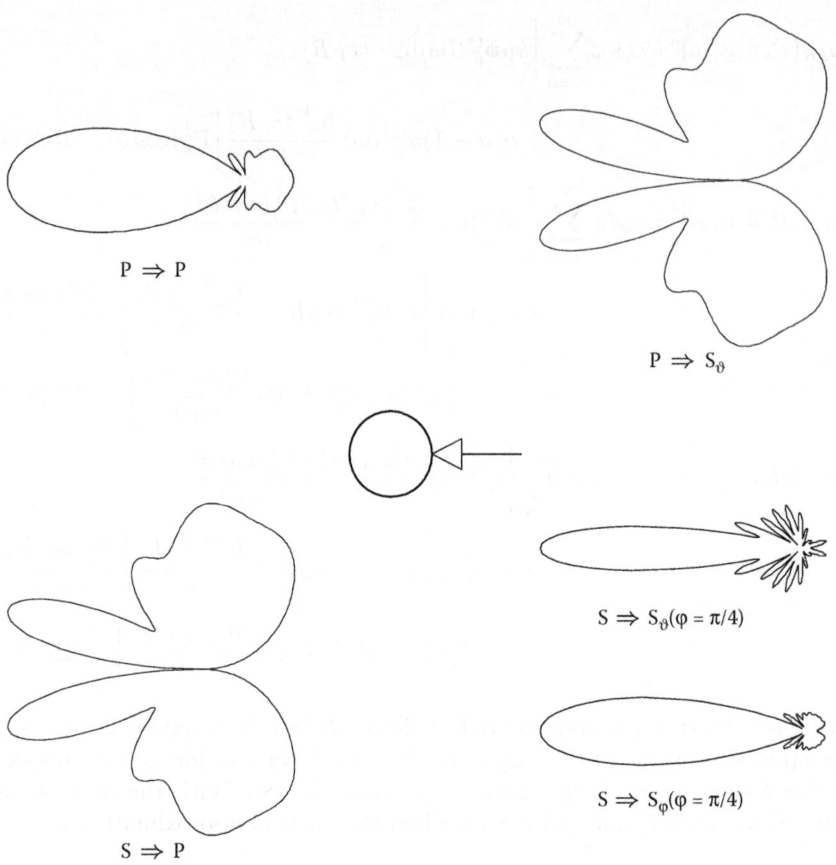

FIGURE 15.37
Far-field radiation patterns of the spherical void in a homogeneous isotropic material (steel) for plane wave incidence; $k_{\mathrm{P}}a = 10.39$, $k_{\mathrm{S}}a = 19.18$: $\lambda_{\mathrm{P}} = a$.

decoupling is not reflected by the particle displacement components in the spherical coordinate system: We observe S- and not SV-, respectively, SH-scattering. By the way: The far-field approximation according to (13.47), namely via the substitution $\boldsymbol{\nabla} \stackrel{\mathrm{far}}{\Longrightarrow} \mathrm{j}k_{\mathrm{P,S}}\underline{\hat{\mathbf{R}}}$, may not immediately be performed with the Debye *ansatz* (15.284) because the vector of position $\underline{\mathbf{R}}$ explicitly appears in addition to the differential operators: For example, defining a vector potential $\underline{\mathbf{U}} = U\,\underline{\mathbf{R}}$ we have $\underline{\mathbf{U}}^{\mathrm{far}} \neq U^{\mathrm{far}}\underline{\mathbf{R}}$.

Figures 15.37 through 15.42 display results of the numerical evaluation of (15.346) through (15.348), once in the frequency domain as scattering diagrams and once as scattered pulses in the time domain following an inverse Fourier transform. For steel as embedding material of a spherical void Figure 15.37 shows scattering diagrams for plane P-, respectively, S-wave

($\underline{\mathbf{e}}_x$-polarization) incidence for the directly scattered as well as for the mode converted part ($k_{\mathrm{P}}a = 6.28$, $k_{\mathrm{S}}a = 11.58$); for direct shear wave scattering, the plane ($\varphi = \pi/4$) has been chosen exhibiting both nonvanishing transverse components because the respective diagrams are different. As usual, we state that these diagrams feature more side lobes than the (P$\Longrightarrow$P)-diagram due to the smaller wavelength also resulting in a stronger penciling of forward scattering; by the way, superimposed to the incident field, the latter serves to form a "shadow" of the void; here, only the scattered field is displayed. Therefore, the—identical—mode conversion diagrams have zeroes for forward scattering, and additionally—according to the point directivities of the third rank Green tensor (Figure 13.14)—in back scattering direction.

Prescribing $u_{i\mathrm{P}}(\omega)$, respectively $u_{i\mathrm{S}}(\omega)$, as RC2-pulse spectra with the center frequency 200 kHz, evaluating (15.346) through (15.348) within the relevant RC2-frequency band, and subsequently Fourier inverting, we calculate far-field scattered pulses in each $\hat{\underline{\mathbf{R}}}$-scattering direction as displayed in Figures[307] 15.38 through 15.42. As compared to the scattered pulses for the cylindrical void (Figure 15.22), an absolute time axis scaled in microseconds has been chosen; only the travel time from the origin in the center of the sphere to the far-field observation point has been subtracted precipitating, for example, the (P$\Longrightarrow$P)- and (S$\Longrightarrow$S)-backscattered pulses (Figure 15.38, respectively, Figures 15.41 and 15.42) by $2t_a^{\mathrm{P}} = 2a/c_{\mathrm{P}}$, respectively $2t_a^{\mathrm{S}} = 2a/c_{\mathrm{S}}$, with regard to the time origin because they originate at the front surface of the sphere, while the time origin is allocated to the passage of the maximum of the incident RC2-pulse through the xy-plane. Due to the mode conversion scattering diagrams in Figure 15.37, there are no mode converted pulses in backscattering direction (Figures 15.39 and 15.40). Figure 15.38 shows the excitation of creeping waves yet with considerable larger relative amplitudes as compared to the cylindrical void (Figure 15.22) because they are able to circle the sphere in each $R\varphi$-plane. Moving with the observation point from the backscattering to the forward scattering direction (as displayed in the picture sequence of Figures 15.38 through 15.42) causes the creeping wave with the shorter travel distance to approach the directly scattered pulse with increasing amplitude—the other one is already too much attenuated to be still visible—superimposing it in forward scattering direction. For the (S$\Longrightarrow$S)-case (Figure 15.41: ϑ-component for $\varphi = 0$; Figure 15.42: φ-component for $\varphi = \pi/2$), the creeping wave amplitude in backscattering direction is nearly as large as the amplitude of the directly scattered pulse, and in the ($\varphi = 0$)-plane—the ϑ-component in this plane is quasi-SV with regard to the incident wave—we nicely recognize both circling creeping waves with increasing ϑ (*both* means: clockwise and counterclockwise). Elastodynamic creeping waves for various geometries (e.g., spheres, spheroids) have been under concern for defect shape recognition (Bollig and Langenberg 1983).

[307]Note (Figure 8.3): As a function of time, the smaller shear wavelength is not visible because the P- and S-RC2-pulses have equal duration for the same center frequency of the RC2-pulse.

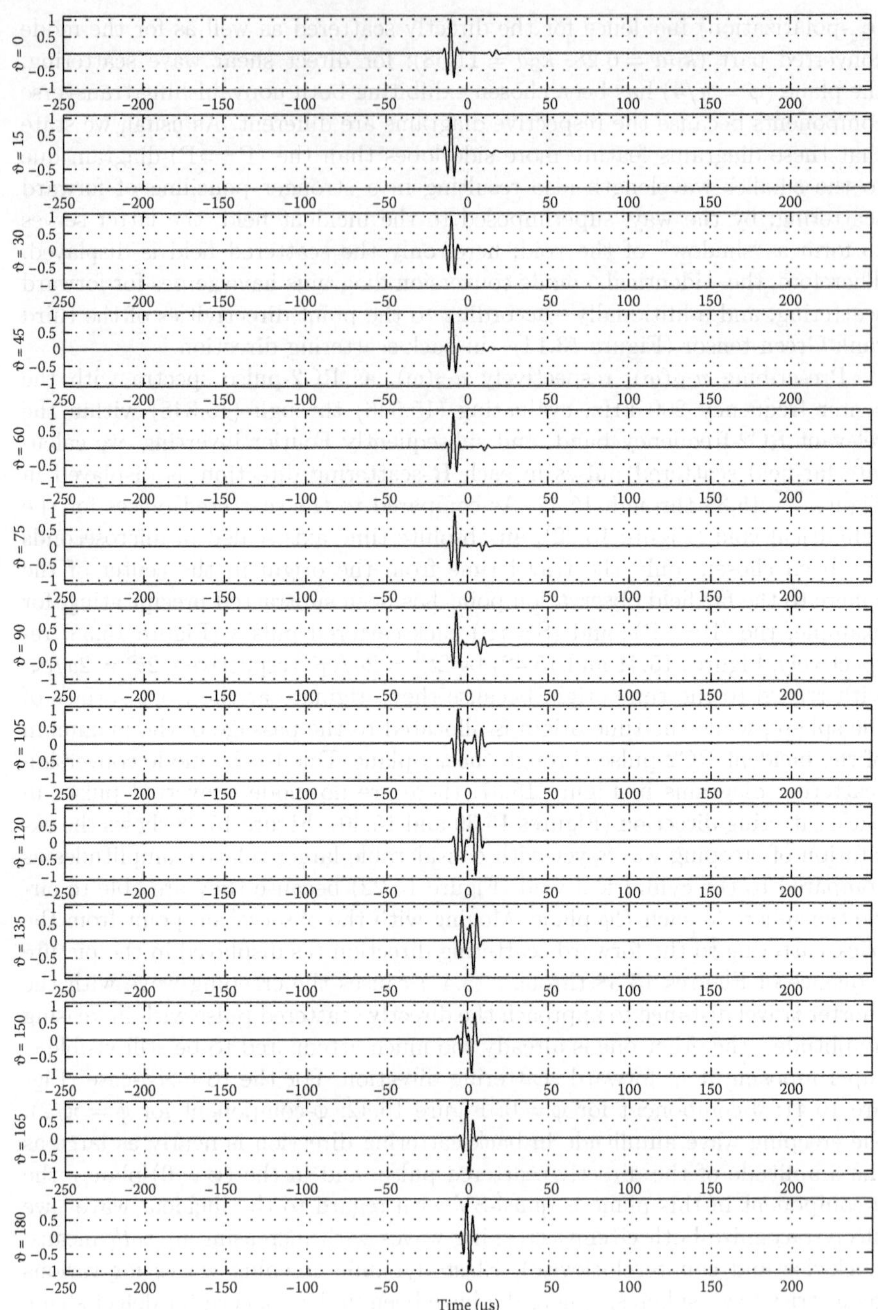

FIGURE 15.38

Pulsed scattered P-far-field of the spherical void in a homogeneous isotropic material (steel) for plane P-wave incidence for various observation angles; $k_{\text{P}}a = 6.28$, $k_{\text{S}}a = 11.58$ for the center frequency 200 kHz of the RC2-pulse spectrum; $a = 29.5 \cdot 10^{-3}$ m ($\lambda_{\text{P}} = a$); horizontal axis: time in μs.

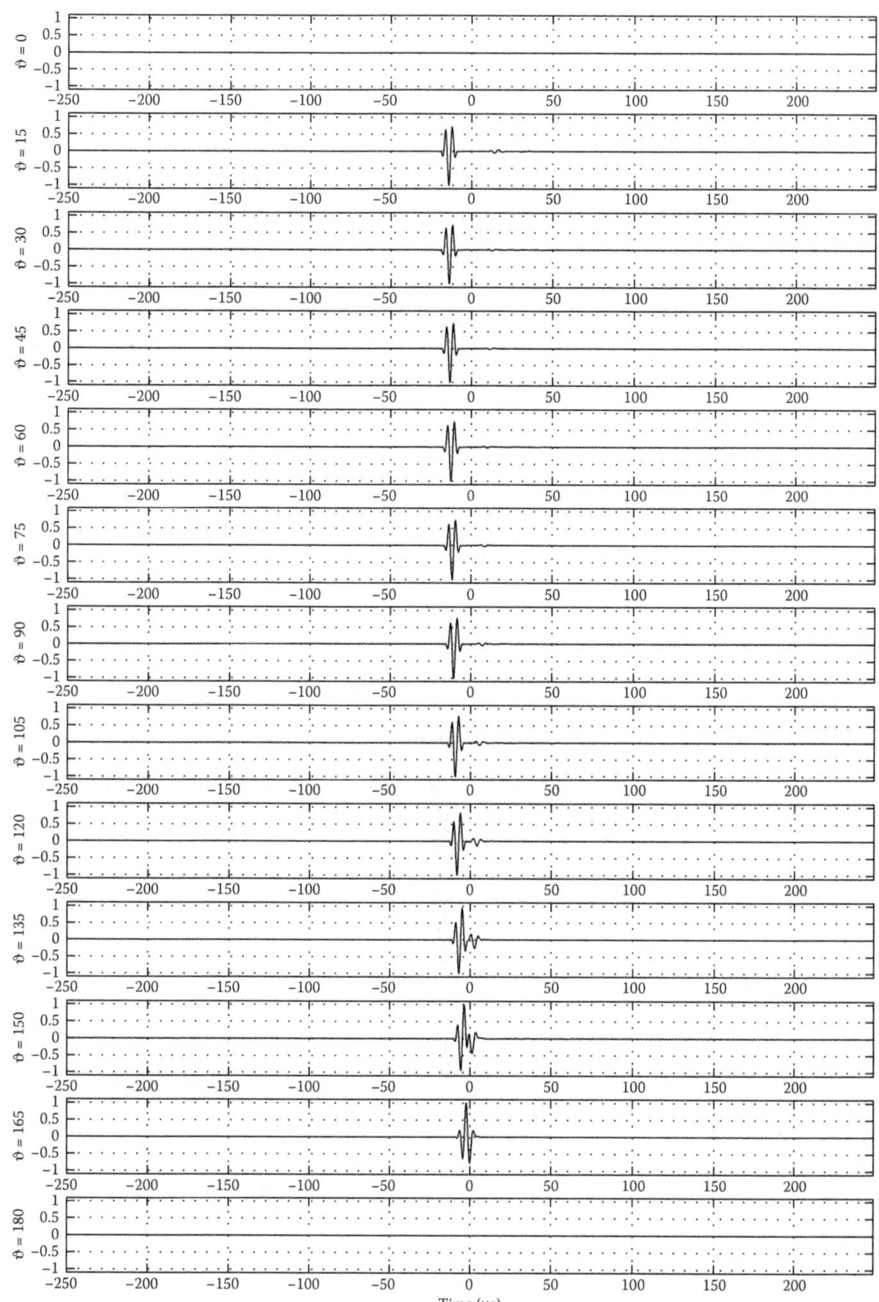

FIGURE 15.39
Pulsed scattered S-far-field (ϑ-component for $\varphi = 0$) of the spherical void in a homogeneous isotropic material (steel) for plane P-wave incidence for various observation angles; $k_P a = 6.28$, $k_S a = 11.58$ for the center frequency 200 kHz of the RC2-pulse spectrum; $a = 29.5 \cdot 10^{-3}$ m ($\lambda_P = a$); horizontal axis: time in μs.

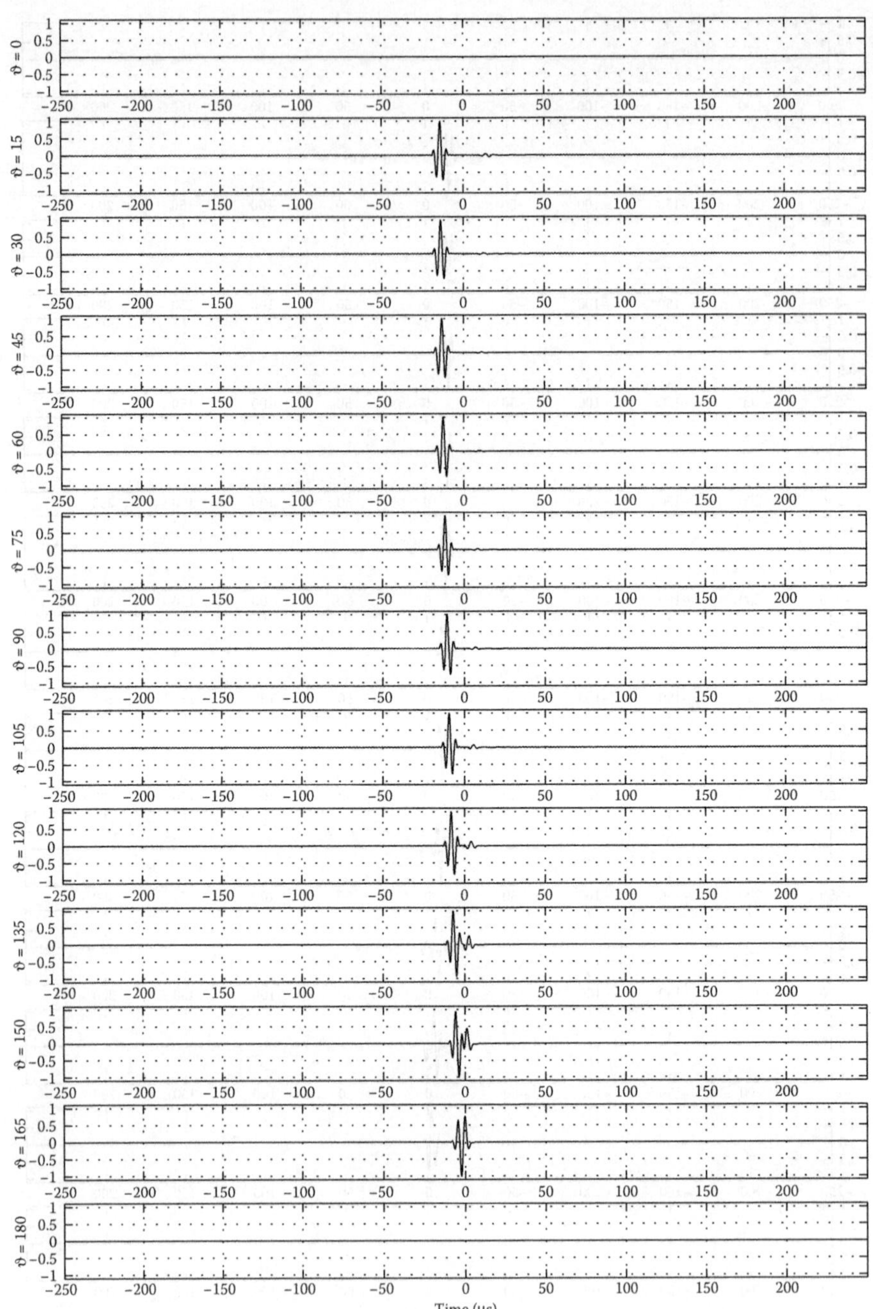

FIGURE 15.40
Pulsed scattered P-far-field of the spherical void in a homogeneous isotropic material (steel) for plane S-wave incidence for various observation angles; $k_{\mathrm{P}}a = 6.28$, $k_{\mathrm{S}}a = 11.58$ for the center frequency 200 kHz of the RC2-pulse spectrum; $a = 29.5 \cdot 10^{-3}$ m ($\lambda_{\mathrm{P}} = a$); horizontal axis: time in μs.

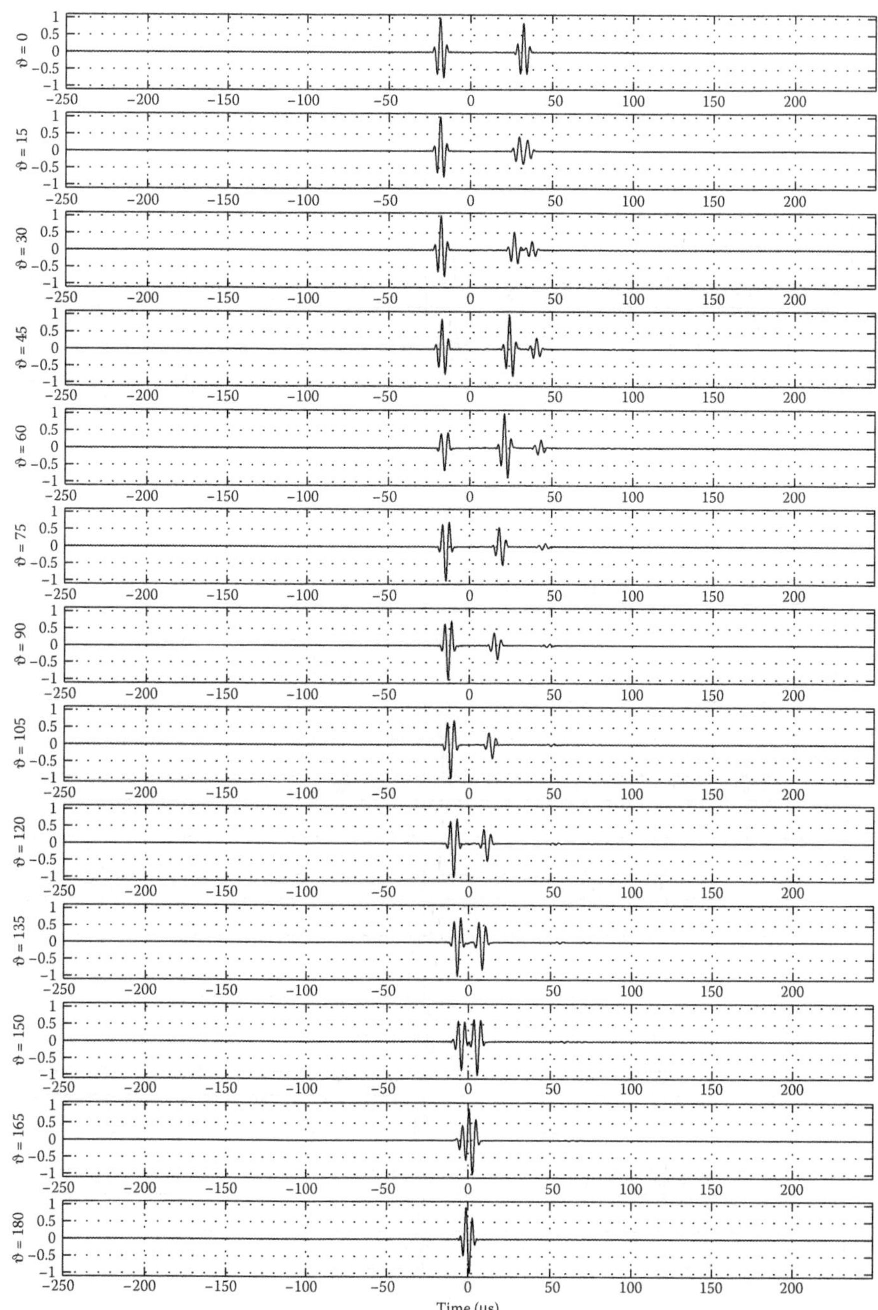

FIGURE 15.41

Pulsed scattered S-far-field (ϑ-component for $\varphi = 0$) of the spherical void in a homogeneous isotropic material (steel) for plane S-wave incidence for various observation angles; $k_{\mathrm{P}}a = 6.28$, $k_{\mathrm{S}}a = 11.58$ for the center frequency 200 kHz of the RC2-pulse spectrum; $a = 29.5 \cdot 10^{-3}$ m ($\lambda_{\mathrm{P}} = a$); horizontal axis: time in μs.

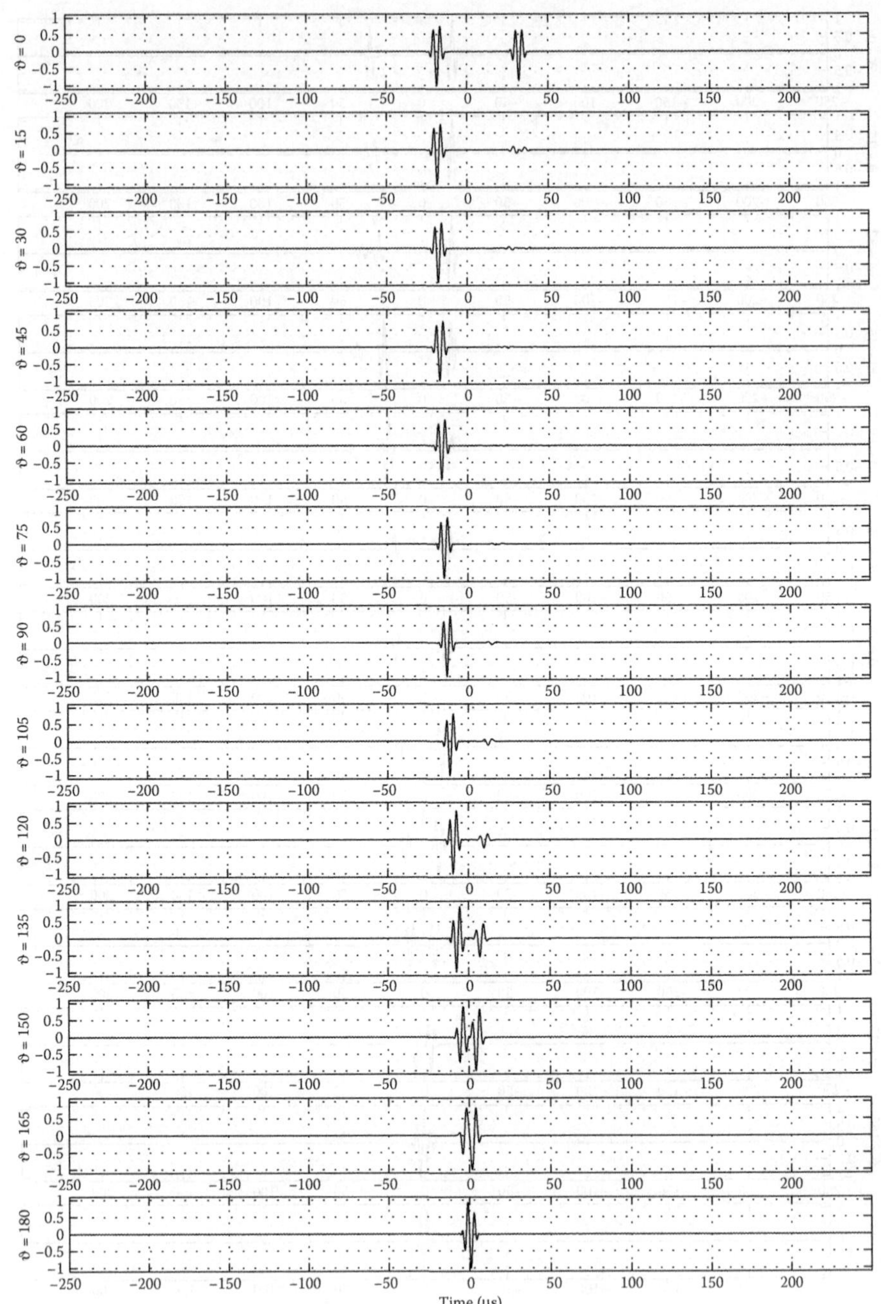

FIGURE 15.42
Pulsed scattered S-far-field (φ-component for $\varphi = \pi/2$) of the spherical void in a homogeneous isotropic material (steel) for plane S-wave incidence for various observation angles; $k_{\mathrm{P}}a = 6.28$, $k_{\mathrm{S}}a = 11.58$ foe the center frequency 200 kHz of the RC2-pulse spectrum; $a = 29.5 \cdot 10^{-3}$ m ($\lambda_{\mathrm{P}} = a$); horizontal axis: time in μs.

15.5 3D System Model of Pulsed Ultrasonic Scattering within Kirchhoff's Approximation

The first step in a system model construction for ultrasonic scattering consists of transducer modeling [the electronic equipment is included by Schmerr and Song (2007)]; apparently, we will utilize the results of Section 14.4 yet with citation of the relevant equations for the sake of completeness.

Figure 15.43 illustrates the respective coordinate system for transducer modeling (index i for incident field). Within an aperture S_A as part of a stress-free planar measurement surface S_M ($x_i y_i$-plane of a cartesian coordinate system: Figure 14.13), we prescribe a perpendicular (Fourier-transformed) force density[308]

$$\underline{\mathbf{t}}(x_i, y_i, \omega) = u_i(\omega) A(x_i, y_i, \omega) \Gamma_A(x_i, y_i) \, \underline{\mathbf{e}}_z, \tag{15.349}$$

where $\Gamma_A(x_i, y_i)$ as characteristic function of the aperture describes its geometry, and $u_i(\omega)$ is the given pulse spectrum. With the prescription of (15.349), we introduce the first approximation[309] into our system model; there are more to come.

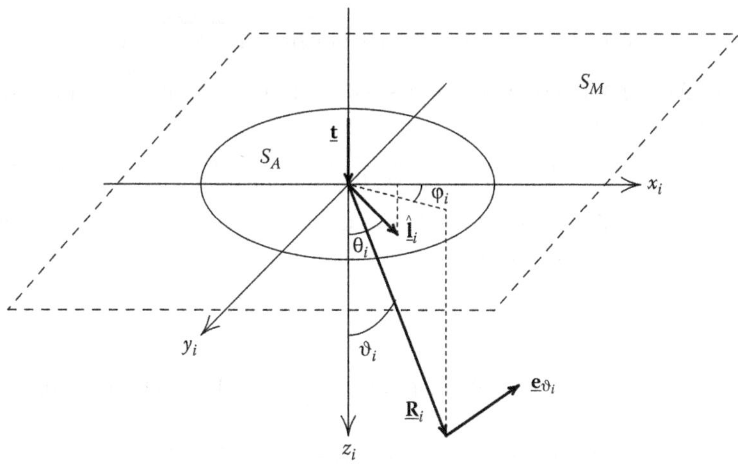

FIGURE 15.43
Transducer modeling for a 3D-US-scattering model.

[308]With respect to (14.153), we explicitly indicate the given pulse spectrum $u_i(\omega)$; yet $A(x_i, y_i, \omega)$ remains frequency dependent due to the phase tapering (15.350).

[309]To go beyond requires a detailed model of the total transducer, respectively, a precise measurement of the particle displacement amplitude and pulse structure on the transducer radiation surface (Marklein 1997).

With[310]

$$A(x_i, y_i, \omega) = q_a(x_i) q_b(y_i)\, \mathrm{e}^{\mathrm{j}k_\mathrm{S} \sin\theta_i x_i}, \tag{15.350}$$

we specialize to a rectangular shear wave angle transducer and define a respective main lobe direction $\hat{\underline{\mathbf{l}}}_i$ in the $x_i z_i$-plane through θ_i (maximum of the shear wave beam according to Figure 14.14). As a second approximation,[311] we use the far-field formula (14.182) to obtain the particle displacement spectrum

$$\begin{aligned}
&\underline{\mathbf{u}}_{i\mathrm{S}}^{\mathrm{far}}(R_i, \vartheta_i, \varphi_i, \omega)\\
&\quad = -\underline{\mathbf{e}}_{\vartheta_i} \sin\vartheta_i\, M_\mathrm{S}^{t_z}(\vartheta_i)\, \frac{u_i(\omega)}{4\pi c_\mathrm{S} Z_\mathrm{S}} \frac{\mathrm{e}^{\mathrm{j}k_\mathrm{S} R_i}}{R_i} \mathcal{F}_{x_i y_i}\{A(x_i, y_i, \omega)\}_{\substack{K_{x_i} = k_\mathrm{S}\sin\vartheta_i\cos\varphi_i \\ K_{y_i} = k_\mathrm{S}\sin\vartheta_i\sin\varphi_i}}\\
&\quad \stackrel{\mathrm{def}}{=} H_{i\mathrm{S}}(\hat{\underline{\mathbf{R}}}_{ik}, \omega) \frac{\mathrm{e}^{\mathrm{j}k_\mathrm{S} R_{ik}}}{R_i} \underline{\mathbf{e}}_{\vartheta_i}
\end{aligned} \tag{15.351}$$

of the incident SV-wave, where

$$\begin{aligned}
\mathcal{F}_{x_i y_i}\{A(x_i, y_i, \omega)\} &= \hat{A}(K_{x_i}, K_{y_i}, \omega)\\
&= \frac{2\sin a(K_{x_i} - k_\mathrm{S}\sin\theta_i)}{K_{x_i} - k_\mathrm{S}\sin\theta_i} \frac{2\sin bK_{y_i}}{K_{y_i}},
\end{aligned} \tag{15.352}$$

and (Equation 14.178)

$$M_\mathrm{S}^{t_z}(\vartheta_i) = \frac{4\cos\vartheta_i\sqrt{1-\kappa^2\sin^2\vartheta_i}}{\kappa(1-2\sin^2\vartheta_i)^2 + 2\sin\vartheta_i\sin 2\vartheta_i\sqrt{1-\kappa^2\sin^2\vartheta_i}}. \tag{15.353}$$

Note that $M_\mathrm{S}^{t_z}(\vartheta_i)$ is complex for $\kappa\sin\vartheta_i > 1$, i.e., for $\vartheta_i > \vartheta_{cm\mathrm{P}}$, and frequency dependent according to

$$M_\mathrm{S}^{t_z}(\vartheta_i) = |M_\mathrm{S}^{t_z}(\vartheta_i)|\, \mathrm{e}^{\mathrm{j}\phi_{M_\mathrm{S}}(\vartheta_i)\,\mathrm{sign}(\omega)} \tag{15.354}$$

—for example, this is exactly the domain for a 45°-shear wave angle transducer—yielding the respective particle displacement pulse corresponding to (15.351) as

$$\begin{aligned}
&\underline{\mathbf{u}}_{i\mathrm{S}}^{\mathrm{far}}(R_i, \vartheta_i, \varphi_i, t)\\
&= -\underline{\mathbf{e}}_{\vartheta_i} \frac{\sin\vartheta_i}{4\pi c_\mathrm{S} Z_\mathrm{S}}
\begin{cases}
M_\mathrm{S}^{t_z}(\vartheta_i)\, \hat{A}(\vartheta_i, \varphi_i, t) * u_i\left(t - \frac{R_i}{c_\mathrm{S}}\right) \quad \text{for } 0 \le \vartheta_i \le \vartheta_{cm\mathrm{P}}\\
|M_\mathrm{S}^{t_z}(\vartheta_i)| \left[\cos\phi_{M_\mathrm{S}}(\vartheta_i)\, \hat{A}(\vartheta_i, \varphi_i, t) * u_i\left(t - \frac{R_i}{c_\mathrm{S}}\right)\right.\\
\qquad \left. - \sin\phi_{M_\mathrm{S}}(\vartheta_i)\, \mathcal{H}\left\{\hat{A}(\vartheta_i, \varphi_i, t) * u_i\left(t - \frac{R_i}{c_\mathrm{S}}\right)\right\}\right]\\
\qquad\qquad \text{for } \vartheta_{cm\mathrm{P}} < \vartheta_i < \pi/2
\end{cases}
\end{aligned} \tag{15.355}$$

---

[310]The product of the rectangular functions $q_a(x_i)$, $q_b(y_i)$ in (15.350) is the characteristic function $\Gamma_A(x_i, y_i)$ of the aperture.

[311]Here, we could be more general according to Equations 14.187, respectively 14.193, but then we would have to calculate an integral for each beam direction $\underline{\mathbf{R}}_i$ and each frequency.

(Equation 14.132), where $u_i(t) = \mathcal{F}_{\omega}^{-1}\{u_i(\omega)\}$. With $\hat{A}(\vartheta_i, \varphi_i, t)$, we denote the inverse Fourier transform

$$\hat{A}(\vartheta_i, \varphi_i, t) = \mathcal{F}_{\omega}^{-1}\left\{\hat{A}\left(K_{x_i} = \frac{\omega}{c_S}\sin\vartheta_i\cos\varphi_i, K_{y_i} = \frac{\omega}{c_S}\sin\vartheta_i\sin\varphi_i, \omega\right)\right\} \tag{15.356}$$

of the scalar aperture directivity with regard to ω requiring a case-by-case analysis. For $\underline{\hat{\mathbf{R}}}_i = \underline{\hat{\mathbf{l}}}_i$, i.e., in the main beam direction $\varphi_i = 0$, $\vartheta_i = \theta_i$ of the transducer, we obtain (in the far-field!) only a single pulse:

$$\hat{A}(\vartheta_i = \theta_i, \varphi_i = 0, t) * u_i\left(t - \frac{R_i}{c_S}\right) = 4ab\, u_i\left(t - \frac{R_i}{c_S}\right). \tag{15.357}$$

For observation points in the x_iz_i-plane outside the main beam, i.e., for $\varphi_i = 0$, $\vartheta_i \neq \theta_i$, we certainly observe, as shown in Section 13.3.4 (Figure 13.21), two pulses:

$$\begin{aligned}\hat{A}(\vartheta_i \neq \theta_i, \varphi_i = 0, t) &* u_i\left(t - \frac{R_i}{c_S}\right)\\ = \frac{2c_S b}{\sin\vartheta_i - \sin\theta_i}\int_{-\infty}^{t}&\left[u_i\left(\tau - \frac{R_i}{c_S} + \frac{a}{c_S}(\sin\vartheta_i - \sin\theta_i)\right)\right.\\ &\left. - u_i\left(\tau - \frac{R_i}{c_S} - \frac{a}{c_S}(\sin\vartheta_i - \sin\theta_i)\right)\right] d\tau\end{aligned} \tag{15.358}$$

that emanate from the $\pm a$-edges of the rectangular aperture; the $\pm b$-edges are not visible in this plane. Two pulses coming solely from the $\pm b$-edges are observed for $\vartheta_i\varphi_i$-combinations satisfying $\sin\vartheta_i\cos\varphi_i = \sin\theta_i$. Finally, we obtain four pulses for arbitrary ϑ_i outside the x_iz_i-plane:

$$\begin{aligned}\hat{A}(\vartheta_i, \varphi_i \neq 0, t) &* u_i\left(t - \frac{R_i}{c_S}\right)\\ = &\frac{c_S^2}{(\sin\vartheta_i\cos\varphi_i - \sin\theta_i)\sin\vartheta_i\sin\varphi_i}\\ &\times\int_{-\infty}^{t}\left[u_i\left(\tau - \frac{R_i}{c_S} - \frac{a}{c_S}(\sin\vartheta_i\cos\varphi_i - \sin\theta_i)\right)\right.\\ &\left. - u_i\left(\tau - \frac{R_i}{c_S} + \frac{a}{c_S}(\sin\vartheta_i\cos\varphi_i - \sin\theta_i)\right)\right] d\tau\\ &*\int_{-\infty}^{t}\left[u_i\left(\tau - \frac{R_i}{c_S} - \frac{b}{c_S}\sin\vartheta_i\sin\varphi_i\right)\right.\\ &\left. - u_i\left(\tau - \frac{R_i}{c_S} + \frac{b}{c_S}\sin\vartheta_i\sin\varphi_i\right)\right] d\tau.\end{aligned} \tag{15.359}$$

The existence of a defect V_c, namely a scatterer, excites a scattered field that may be calculated with the elastodynamic Huygens integral within the

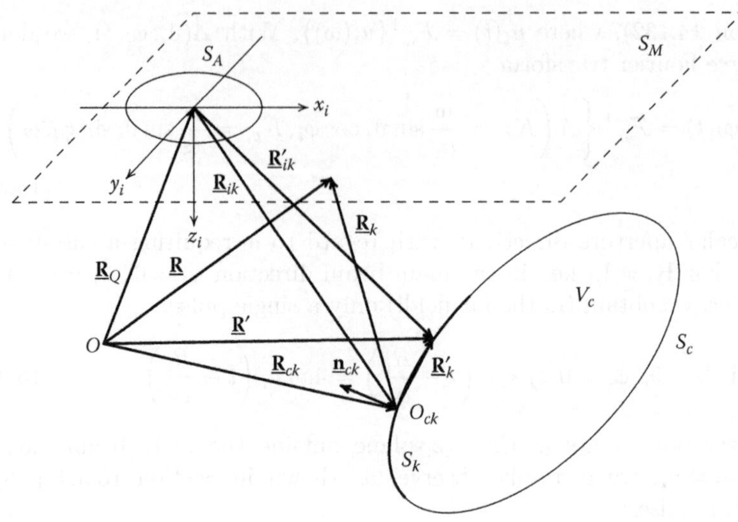

FIGURE 15.44
3D-US-system model: transducer and scatterer.

Kirchhoff approximation (Section 15.2.3). We need observation and source point coordinates $\underline{\mathbf{R}}$, respectively $\underline{\mathbf{R}}'$, with coordinate origin O (Figure 15.44): Per definition $\underline{\mathbf{R}}'$ points toward S_c, $\underline{\mathbf{R}}$ is generally located on S_M. With respect to this coordinate origin, the origin of our $x_i y_i z_i$-coordinate system is identified by $\underline{\mathbf{R}}_Q$. For a numerical calculation of the elastodynamic Huygens integral for arbitrary scattering surfaces S_c, we have to discretize S_c via tessellation into two-dimensional patches; the kth patch be characterized by the fixed position vector $\underline{\mathbf{R}}_{ck}$. With regard to the endpoint O_{ck} of $\underline{\mathbf{R}}_{ck}$, we define a Dupin-coordinate system with surface normal $\underline{\mathbf{n}}_{ck}$. (After introducing the Kirchhoff approximation, S_k turns into a planar surface patch, and the Dupin coordinates degenerate into a local cartesian $x_k y_k z_k$-coordinate system with $\underline{\mathbf{e}}_{z_k} = \underline{\mathbf{n}}_{ck}$.) A source point $\underline{\mathbf{R}}' \in S_k$ for the scattered field may finally be characterized by $\underline{\mathbf{R}}'_k$ with respect to O_{ck}, just as $\underline{\mathbf{R}}_k$ characterizes an observation point of the scattered field.

We assume the defect to be a void with stress-free surface and come back to (15.65) in terms of

$$\underline{\mathbf{u}}_s(\underline{\mathbf{R}}, \omega) = -\iint_{S_c} \underline{\mathbf{u}}(\underline{\mathbf{R}}', \omega)\underline{\mathbf{n}}'_c : \underline{\underline{\mathbf{\Sigma}}}(\underline{\mathbf{R}} - \underline{\mathbf{R}}', \omega)\, \mathrm{d}S' \tag{15.360}$$

to calculate the scattered field. Utilization of the full-space Green tensor implies the third approximation neglecting the retroaction of the measurement surface on the scattered field (multiple reflections between S_M and S_c are not considered), i.e., the total field $\underline{\mathbf{u}}_{i\mathrm{S}}(\underline{\mathbf{R}}, \omega) + \underline{\mathbf{u}}_s(\underline{\mathbf{R}}, \omega)$ as superposition of incident and scattered fields does not satisfy the stress-free boundary condition for $\underline{\mathbf{R}} \in S_M \backslash \overline{S}_A$. (It should already be satisfied by the incident field alone, which is not true for the far-field approximation.)

Due to the discretization of S_c, we may represent the scattered field as sum of the scattered fields of the single patches:

$$\underline{\mathbf{u}}_s(\underline{\mathbf{R}}, \omega) = \sum_k \underline{\mathbf{u}}_{sk}(\underline{\mathbf{R}}_{ck} + \underline{\mathbf{R}}_k, \omega), \tag{15.361}$$

where

$$\underline{\mathbf{u}}_{sk}(\underline{\mathbf{R}}_{ck} + \underline{\mathbf{R}}_k, \omega) = - \iint_{S_k} \underline{\mathbf{u}}(\underline{\mathbf{R}}_{ck} + \underline{\mathbf{R}}'_k, \omega)\underline{\mathbf{n}}'_{ck} : \underline{\underline{\underline{\mathbf{\Sigma}}}}(\underline{\mathbf{R}}_k - \underline{\mathbf{R}}'_k, \omega)\, \mathrm{d}^2\underline{\mathbf{R}}'_k. \tag{15.362}$$

The radiation interaction of single patches is still contained in the globally—for S_c, and not locally for S_k—calculated secondary source as solution of the integral equation (15.2.2); only after the introduction of the Kirchhoff approximation independent secondary sources of the patches are postulated. Before doing this, we apply the far-field approximation of each patch scattered field with regard to the local origin O_{ck} as the forth approximation:[312]

$$\underline{\mathbf{u}}^{\mathrm{far}}_{sk}(\underline{\mathbf{R}}_{ck} + \underline{\mathbf{R}}_k, \omega) = - \iint_{S_k} \underline{\mathbf{u}}(\underline{\mathbf{R}}_{ck} + \underline{\mathbf{R}}'_k, \omega)\underline{\mathbf{n}}'_{ck} : \underline{\underline{\underline{\mathbf{\Sigma}}}}^{\mathrm{far}}(\underline{\mathbf{R}}_k - \underline{\mathbf{R}}'_k, \omega)\, \mathrm{d}^2\underline{\mathbf{R}}'_k, \tag{15.363}$$

where (Equation 13.157 with 13.162 and 13.163)

$$\begin{aligned}\underline{\underline{\underline{\mathbf{\Sigma}}}}^{\mathrm{far}}(\underline{\mathbf{R}}_k, \underline{\mathbf{R}}'_k, \omega) &= \mathrm{j}\omega \frac{\mathrm{e}^{\mathrm{j}k_\mathrm{P} R_k}}{4\pi R_k} \frac{1}{\varrho c_\mathrm{P}^3} \left(\lambda\, \underline{\underline{\mathbf{I}}}\, \hat{\underline{\mathbf{R}}}_k + 2\mu\, \hat{\underline{\mathbf{R}}}_k \hat{\underline{\mathbf{R}}}_k \hat{\underline{\mathbf{R}}}_k\right) \mathrm{e}^{-\mathrm{j}k_\mathrm{P}\hat{\underline{\mathbf{R}}}_k \cdot \underline{\mathbf{R}}'_k} \\ &\quad + \mathrm{j}\omega \frac{\mathrm{e}^{\mathrm{j}k_\mathrm{S} R_k}}{4\pi R_k} \frac{1}{c_\mathrm{S}} \left(\hat{\underline{\mathbf{R}}}_k \underline{\underline{\mathbf{I}}} + \hat{\underline{\mathbf{R}}}_k \underline{\underline{\mathbf{I}}}^{213} - 2\hat{\underline{\mathbf{R}}}_k \hat{\underline{\mathbf{R}}}_k \hat{\underline{\mathbf{R}}}_k\right) \mathrm{e}^{-\mathrm{j}k_\mathrm{S}\hat{\underline{\mathbf{R}}}_k \cdot \underline{\mathbf{R}}'_k}.\end{aligned} \tag{15.364}$$

The result is the separation of the patch scattered field into a directly scattered S- and a mode converted P-part:

$$\underline{\mathbf{u}}^{\mathrm{far}}_{sk}(\underline{\mathbf{R}}_{ck} + \underline{\mathbf{R}}_k, \omega) = \underline{\mathbf{u}}^{\mathrm{far}}_{sk\mathrm{S}}(\underline{\mathbf{R}}_{ck} + \underline{\mathbf{R}}_k, \omega) + \underline{\mathbf{u}}^{\mathrm{far}}_{sk\mathrm{P}}(\underline{\mathbf{R}}_{ck} + \underline{\mathbf{R}}_k, \omega) \tag{15.365}$$

with

$$\begin{aligned}&\underline{\mathbf{u}}^{\mathrm{far}}_{sk\mathrm{S}}(\underline{\mathbf{R}}_{ck} + \underline{\mathbf{R}}_k, \omega) \\ &\quad = \frac{\mathrm{e}^{\mathrm{j}k_\mathrm{S} R_k}}{R_k} \left(-\frac{\mathrm{j}\omega}{4\pi c_\mathrm{S}}\right) \iint_{S_k} \underline{\mathbf{u}}(\underline{\mathbf{R}}_{ck} + \underline{\mathbf{R}}'_k, \omega)\underline{\mathbf{n}}'_{ck} \mathrm{e}^{-\mathrm{j}k_\mathrm{S}\hat{\underline{\mathbf{R}}}_k \cdot \underline{\mathbf{R}}'_k}\, \mathrm{d}^2\underline{\mathbf{R}}'_k \\ &\quad : \left(\hat{\underline{\mathbf{R}}}_k \underline{\underline{\mathbf{I}}} + \hat{\underline{\mathbf{R}}}_k \underline{\underline{\mathbf{I}}}^{213} - 2\hat{\underline{\mathbf{R}}}_k \hat{\underline{\mathbf{R}}}_k \hat{\underline{\mathbf{R}}}_k\right),\end{aligned} \tag{15.366}$$

[312]We could avoid this at the expense of a considerable calculation effort, yet the precision of the Kirchhoff approximation in the near-field is not known.

$$\begin{aligned}
&\underline{\mathbf{u}}_{sk\mathrm{P}}^{\mathrm{far}}(\underline{\mathbf{R}}_{ck}+\underline{\mathbf{R}}_k,\omega)\\
&\quad=\frac{\mathrm{e}^{\mathrm{j}k_\mathrm{P}R_k}}{R_k}\left(-\frac{\mathrm{j}\omega}{4\pi\varrho c_\mathrm{P}^3}\right)\int\!\!\int_{S_k}\underline{\mathbf{u}}(\underline{\mathbf{R}}_{ck}+\underline{\mathbf{R}}_k',\omega)\underline{\mathbf{n}}_{ck}'\mathrm{e}^{-\mathrm{j}k_\mathrm{P}\hat{\underline{\mathbf{R}}}_k\cdot\underline{\mathbf{R}}_k'}\,\mathrm{d}^2\underline{\mathbf{R}}_k'\\
&\qquad:\left(\lambda\,\underline{\underline{\mathbf{I}}}\,\hat{\underline{\mathbf{R}}}_k+2\mu\,\hat{\underline{\mathbf{R}}}_k\hat{\underline{\mathbf{R}}}_k\hat{\underline{\mathbf{R}}}_k\right).
\end{aligned}\tag{15.367}$$

Now, these integral representations of the scattered field require the secondary surface deformation source $\underline{\mathbf{u}}(\underline{\mathbf{R}}_{ck}+\underline{\mathbf{R}}_k',\omega)\underline{\mathbf{n}}_{ck}'$, and consequently

$$\underline{\mathbf{u}}(\underline{\mathbf{R}}_{ck}+\underline{\mathbf{R}}_k',\omega)\stackrel{\mathrm{def}}{=}\underline{\mathbf{u}}(\underline{\mathbf{R}}_k',\omega)\tag{15.368}$$

as function of the integration variable $\underline{\mathbf{R}}_k'$; to apply the Kirchhoff approximation, we initially need the incident field at this point. Yet, $\underline{\mathbf{R}}_{ck}$ defines not only the "center point" of a patch but also a ray (Section 12.3.2) of the incident field in the direction of $\underline{\mathbf{R}}_{ik}$ (Figure 15.44); accordingly, $\underline{\mathbf{R}}_k'$ isolates a ray $\underline{\mathbf{R}}_{ik}'$, and therefore we need

$$\underline{\mathbf{u}}_{ik\mathrm{S}}^{\mathrm{far}}(\underline{\mathbf{R}}_{ik}',\omega)=H_{ik\mathrm{S}}(\hat{\underline{\mathbf{R}}}_{ik}',\omega)\,\frac{\mathrm{e}^{\mathrm{j}k_\mathrm{S}R_{ik}'}}{R_{ik}'}\,\underline{\mathbf{e}}_{\vartheta_{ik}'}\tag{15.369}$$

as an incident spherical wave. For acoustic and electromagnetic waves, this knowledge is sufficient for the Kirchhoff approximation, but for elastic waves, we additionally need the reflected as well as the mode converted spherical wave with the respective amplitude factors; yet simple expressions are only known for plane waves (and planar surfaces S_c). Consequently, as a fifth approximation, we consider the spherical wave to be a plane wave; Figure 15.45 serves as illustration. Namely, for $R_{ik} \gg R_k'$, we have

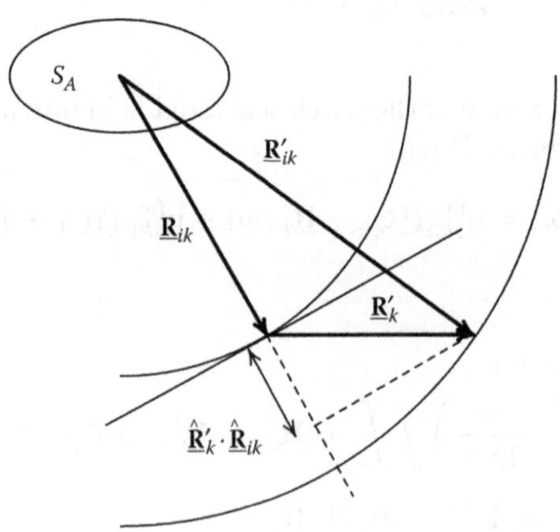

FIGURE 15.45
Approximation of the incident spherical wave by a plane wave.

$$\begin{aligned}
\vartheta'_{ik} &\simeq \vartheta_{ik}, \\
\varphi'_{ik} &\simeq \varphi_{ik}, \\
R'_{ik} &\simeq R_{ik} + \underline{\mathbf{R}}'_k \cdot \hat{\underline{\mathbf{R}}}_{ik},
\end{aligned} \tag{15.370}$$

allowing for the approximation

$$\begin{aligned}
\underline{\mathbf{u}}^{\text{far}}_{ik\text{S}}(\underline{\mathbf{R}}'_{ik}, \omega) &\simeq H_{ik\text{S}}(\hat{\underline{\mathbf{R}}}_{ik}, \omega) \frac{\mathrm{e}^{\mathrm{j}k_\text{S} R_{ik}}}{R_{ik}} \,\mathrm{e}^{\mathrm{j}k_\text{S}\hat{\underline{\mathbf{R}}}_{ik}\cdot\underline{\mathbf{R}}'_k}\, \underline{\mathbf{e}}_{\vartheta_{ik}} \\
&\stackrel{\text{def}}{=} \underline{\mathbf{u}}_{ik\text{S}}(\underline{\mathbf{R}}'_k, \omega)
\end{aligned} \tag{15.371}$$

for $k_\text{S} R_{ik} \gg 1$. Now, the term $\mathrm{e}^{\mathrm{j}k_\text{S}\hat{\underline{\mathbf{R}}}_{ik}\cdot\underline{\mathbf{R}}'_k}\, \underline{\mathbf{e}}_{\vartheta_{ik}}$ in (15.371) defines the desired transverse—we have $\underline{\mathbf{e}}_{\vartheta_{ik}} \cdot \hat{\underline{\mathbf{R}}}_{ik} = 0$—shear wave at $\underline{\mathbf{R}}'_k \in S_k$ with patch-dependent phase, amplitude, and polarization propagating into $\hat{\underline{\mathbf{R}}}_{ik}$-direction.

With the Kirchhoff approximation as sixth approximation, we calculate $\underline{\mathbf{u}}(\underline{\mathbf{R}}'_k, \omega)$ so as if the patch S_k would be a planar (Kirchhoff) patch $S_{k\text{K}}$ illuminated by a plane wave $\mathrm{e}^{\mathrm{j}k_\text{S}\hat{\underline{\mathbf{R}}}_{ik}\cdot\underline{\mathbf{R}}_k}\, \underline{\mathbf{e}}_{\vartheta_{ik}}$; moreover, its contribution to the scattered field is only considered if $\hat{\underline{\mathbf{R}}}_{ik} \cdot \underline{\mathbf{n}}_{ck} < 0$ holds, because for $\hat{\underline{\mathbf{R}}}_{ik} \cdot \underline{\mathbf{n}}_{ck} > 0$, it is located in the shadow of the incident wave. For a planar patch, the Dupin coordinates degenerate to a cartesian $x_k y_k z_k$-coordinate system with the orthonormal trihedron $\underline{\mathbf{e}}_{x_k}, \underline{\mathbf{e}}_{y_k}, \underline{\mathbf{e}}_{z_k}$ (Figure 15.46), where the direction of, for example $\underline{\mathbf{e}}_{x_k}$ and, hence, the direction of $\underline{\mathbf{e}}_{y_k}$, must still be defined. As sketched in Figure 15.46, we choose $\underline{\mathbf{e}}_{x_k}$ "in the direction" of $\underline{\mathbf{R}}_{ik}$, i.e.,

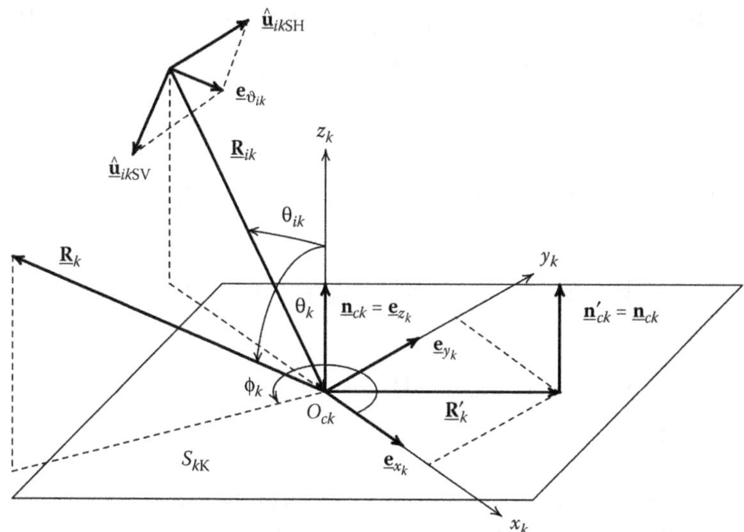

FIGURE 15.46
Illustration of the Kirchhoff approximation.

$$\underline{\mathbf{e}}_{x_k} = \frac{(\underline{\underline{\mathbf{I}}} - \underline{\mathbf{n}}_{ck}\underline{\mathbf{n}}_{ck}) \cdot \hat{\underline{\mathbf{R}}}_{ik}}{|(\underline{\underline{\mathbf{I}}} - \underline{\mathbf{n}}_{ck}\underline{\mathbf{n}}_{ck}) \cdot \hat{\underline{\mathbf{R}}}_{ik}|}, \tag{15.372}$$

and consequently, we obtain $\underline{\mathbf{e}}_{y_k} = \underline{\mathbf{n}}_{ck} \times \underline{\mathbf{e}}_{x_k}$. That way, the propagation vector $\hat{\underline{\mathbf{R}}}_{ik}$ of the incident plane wave is located in the $x_k z_k$-plane, and we may immediately adopt—according to the Kirchhoff approximation—the formulas from Sections 9.1.2 and 9.1.3 for reflection and mode conversion of transversely polarized plane shear waves. Yet, we must bear the matter in mind that $\underline{\mathbf{e}}_{\vartheta_{ik}}$ stands for an SV-polarized shear wave with regard to S_M, with regard to $S_{k\mathrm{K}}$, the vector $\underline{\mathbf{e}}_{\vartheta_{ik}}$ generally contains SV- and SH-components that have to be identified beforehand. Therefore, we decompose

$$\underline{\mathbf{e}}_{\vartheta_{ik}} = \underline{\mathbf{e}}_{\vartheta_{ik}} \cdot \hat{\underline{\mathbf{u}}}_{ik\mathrm{SH}}\, \hat{\underline{\mathbf{u}}}_{ik\mathrm{SH}} + \underline{\mathbf{e}}_{\vartheta_{ik}} \cdot \hat{\underline{\mathbf{u}}}_{ik\mathrm{SV}}\, \hat{\underline{\mathbf{u}}}_{ik\mathrm{SV}} \tag{15.373}$$

in an SH-SV-polarization basis with regard to $S_{k\mathrm{K}}$ consisting of the unit vectors $\hat{\underline{\mathbf{u}}}_{ik\mathrm{SH}} = \underline{\mathbf{e}}_{y_k}$ and $\hat{\underline{\mathbf{u}}}_{ik\mathrm{SV}} = \underline{\mathbf{e}}_{y_k} \times \hat{\underline{\mathbf{R}}}_{ik}$ (Figure 15.46). In terms of the incidence angle θ_{ik}, the vectors $\hat{\underline{\mathbf{R}}}_{ik}$ and $\hat{\underline{\mathbf{u}}}_{ik\mathrm{SV}}$ have components

$$\hat{\underline{\mathbf{R}}}_{ik} = \sin\theta_{ik}\underline{\mathbf{e}}_{x_k} - \cos\theta_{ik}\underline{\mathbf{e}}_{z_k}, \tag{15.374}$$

$$\hat{\underline{\mathbf{u}}}_{ik\mathrm{SV}} = -\cos\theta_{ik}\underline{\mathbf{e}}_{x_k} - \sin\theta_{ik}\underline{\mathbf{e}}_{z_k} \tag{15.375}$$

in the $x_k y_k z_k$-coordinate system. Yet, given are the vectors $\hat{\underline{\mathbf{R}}}_{ik}$ and $\underline{\mathbf{e}}_{\vartheta_{ik}}$—and also $\hat{\underline{\mathbf{u}}}_{ik\mathrm{SV}}$—in the $x_i y_i z_i$-coordinate system according to

$$\hat{\underline{\mathbf{R}}}_{ik} = \hat{\underline{\mathbf{R}}}_{ik} \cdot \underline{\mathbf{e}}_{x_i}\, \underline{\mathbf{e}}_{x_i} + \hat{\underline{\mathbf{R}}}_{ik} \cdot \underline{\mathbf{e}}_{y_i}\, \underline{\mathbf{e}}_{y_i} + \hat{\underline{\mathbf{R}}}_{ik} \cdot \underline{\mathbf{e}}_{z_i}\, \underline{\mathbf{e}}_{z_i}, \tag{15.376}$$

$$\underline{\mathbf{e}}_{\vartheta_{ik}} = \underline{\mathbf{e}}_{\vartheta_{ik}} \cdot \underline{\mathbf{e}}_{x_i}\, \underline{\mathbf{e}}_{x_i} + \underline{\mathbf{e}}_{\vartheta_{ik}} \cdot \underline{\mathbf{e}}_{y_i}\, \underline{\mathbf{e}}_{y_i} + \underline{\mathbf{e}}_{\vartheta_{ik}} \cdot \underline{\mathbf{e}}_{z_i}\, \underline{\mathbf{e}}_{z_i} \tag{15.377}$$

requiring the transformation:

$$\begin{aligned}\hat{\underline{\mathbf{R}}}_{ik} \cdot \underline{\mathbf{e}}_{x_k} &= \hat{\underline{\mathbf{R}}}_{ik} \cdot \underline{\mathbf{e}}_{x_i}\, \underline{\mathbf{e}}_{x_i} \cdot \underline{\mathbf{e}}_{x_k} + \hat{\underline{\mathbf{R}}}_{ik} \cdot \underline{\mathbf{e}}_{y_i}\, \underline{\mathbf{e}}_{y_i} \cdot \underline{\mathbf{e}}_{x_k} + \hat{\underline{\mathbf{R}}}_{ik} \cdot \underline{\mathbf{e}}_{z_i}\, \underline{\mathbf{e}}_{z_i} \cdot \underline{\mathbf{e}}_{x_k}\\ &\overset{\mathrm{def}}{=} \sin\theta_{ik},\end{aligned} \tag{15.378}$$

$$\begin{aligned}\hat{\underline{\mathbf{R}}}_{ik} \cdot \underline{\mathbf{e}}_{y_k} &= \hat{\underline{\mathbf{R}}}_{ik} \cdot \underline{\mathbf{e}}_{x_i}\, \underline{\mathbf{e}}_{x_i} \cdot \underline{\mathbf{e}}_{y_k} + \hat{\underline{\mathbf{R}}}_{ik} \cdot \underline{\mathbf{e}}_{y_i}\, \underline{\mathbf{e}}_{y_i} \cdot \underline{\mathbf{e}}_{y_k} + \hat{\underline{\mathbf{R}}}_{ik} \cdot \underline{\mathbf{e}}_{z_i}\, \underline{\mathbf{e}}_{z_i} \cdot \underline{\mathbf{e}}_{y_k}\\ &\overset{\mathrm{def}}{=} 0,\end{aligned} \tag{15.379}$$

$$\begin{aligned}\hat{\underline{\mathbf{R}}}_{ik} \cdot \underline{\mathbf{e}}_{z_k} &= \hat{\underline{\mathbf{R}}}_{ik} \cdot \underline{\mathbf{e}}_{x_i}\, \underline{\mathbf{e}}_{x_i} \cdot \underline{\mathbf{e}}_{z_k} + \hat{\underline{\mathbf{R}}}_{ik} \cdot \underline{\mathbf{e}}_{y_i}\, \underline{\mathbf{e}}_{y_i} \cdot \underline{\mathbf{e}}_{z_k} + \hat{\underline{\mathbf{R}}}_{ik} \cdot \underline{\mathbf{e}}_{z_i}\, \underline{\mathbf{e}}_{z_i} \cdot \underline{\mathbf{e}}_{z_k}\\ &\overset{\mathrm{def}}{=} -\cos\theta_{ik};\end{aligned} \tag{15.380}$$

the nine scalar products between the unit vectors of the i- and the k-coordinate system trihedron stand for the relative orientation of these two coordinate systems with respect to each other and must be calculated for each patch.

The SH-SV-decomposition of our transversely polarized plane shear wave (15.371) with regard to the kth patch now reads as

$$\begin{aligned}
\underline{\mathbf{u}}_{ik\mathrm{S}}(\underline{\mathbf{R}}'_k, \omega) &= H_{ik\mathrm{S}}(\hat{\underline{\mathbf{R}}}_{ik}, \omega) \frac{\mathrm{e}^{\mathrm{j}k_\mathrm{S} R_{ik}}}{R_{ik}} \mathrm{e}^{\mathrm{j}k_\mathrm{S}\hat{\underline{\mathbf{R}}}_{ik}\cdot\underline{\mathbf{R}}'_k} \underline{\mathbf{e}}_{\vartheta_{ik}} \cdot \underline{\mathbf{e}}_{y_k} \underline{\mathbf{e}}_{y_k} \\
&\quad + H_{ik\mathrm{S}}(\hat{\underline{\mathbf{R}}}_{ik}, \omega) \frac{\mathrm{e}^{\mathrm{j}k_\mathrm{S} R_{ik}}}{R_{ik}} \mathrm{e}^{\mathrm{j}k_\mathrm{S}\hat{\underline{\mathbf{R}}}_{ik}\cdot\underline{\mathbf{R}}'_k} \underline{\mathbf{e}}_{\vartheta_{ik}} \cdot \hat{\underline{\mathbf{u}}}_{ik\mathrm{SV}} \hat{\underline{\mathbf{u}}}_{ik\mathrm{SV}} \\
&\stackrel{\mathrm{def}}{=} A_{ik\mathrm{SH}}(\underline{\mathbf{R}}_{ik}, \omega) \, \mathrm{e}^{\mathrm{j}k_\mathrm{S}\hat{\underline{\mathbf{R}}}_{ik}\cdot\underline{\mathbf{R}}'_k} \underline{\mathbf{e}}_{y_k} \\
&\quad + A_{ik\mathrm{SV}}(\underline{\mathbf{R}}_{ik}, \omega) \, \mathrm{e}^{\mathrm{j}k_\mathrm{S}\hat{\underline{\mathbf{R}}}_{ik}\cdot\underline{\mathbf{R}}'_k} \underline{\mathbf{e}}_{y_k} \times \hat{\underline{\mathbf{R}}}_{ik}
\end{aligned} \tag{15.381}$$

with

$$\underline{\mathbf{e}}_{\vartheta_{ik}} = -\cos\theta_{ik} \, \underline{\mathbf{e}}_{\vartheta_{ik}} \cdot \underline{\mathbf{e}}_{x_k} - \sin\theta_{ik} \, \underline{\mathbf{e}}_{\vartheta_{ik}} \cdot \underline{\mathbf{e}}_{z_k}. \tag{15.382}$$

According to the combination of the Kirchhoff approximations (15.143) and (15.151), we postulate[313]

$$\begin{aligned}
\underline{\mathbf{u}}(\underline{\mathbf{R}}'_k, \omega) = \underline{\mathbf{u}}_{ik\mathrm{S}}(\underline{\mathbf{R}}'_k, \omega) &+ A_{ik\mathrm{SH}}(\underline{\mathbf{R}}_{ik}, \omega) R_\mathrm{SH}(\theta_{ik}) \, \mathrm{e}^{\mathrm{j}k_\mathrm{S}\hat{\underline{\mathbf{R}}}_{kr\mathrm{S}}\cdot\underline{\mathbf{R}}'_k} \underline{\mathbf{e}}_{y_k} \\
&+ A_{ik\mathrm{SV}}(\underline{\mathbf{R}}_{ik}, \omega) R_\mathrm{SV}(\theta_{ik}) \, \mathrm{e}^{\mathrm{j}k_\mathrm{S}\hat{\underline{\mathbf{R}}}_{kr\mathrm{S}}\cdot\underline{\mathbf{R}}'_k} \underline{\mathbf{e}}_{y_k} \times \hat{\underline{\mathbf{R}}}_{kr\mathrm{S}} \\
&+ A_{ik\mathrm{SV}}(\underline{\mathbf{R}}_{ik}, \omega) M_\mathrm{P}(\theta_{ik}) \, \mathrm{e}^{\mathrm{j}k_\mathrm{S}\hat{\underline{\mathbf{R}}}_{km\mathrm{P}}\cdot\underline{\mathbf{R}}'_k} \hat{\underline{\mathbf{R}}}_{km\mathrm{P}},
\end{aligned} \tag{15.383}$$

where

$$\underline{\mathbf{R}}'_k = x'_k \underline{\mathbf{e}}_{x_k} + y'_k \underline{\mathbf{e}}_{y_k}, \tag{15.384}$$

$$\hat{\underline{\mathbf{R}}}_{kr\mathrm{S}} = \sin\theta_{ik} \underline{\mathbf{e}}_{x_k} + \cos\theta_{ik} \underline{\mathbf{e}}_{z_k}, \tag{15.385}$$

$$\hat{\underline{\mathbf{R}}}_{km\mathrm{P}} = \kappa \sin\theta_{ik} \underline{\mathbf{e}}_{x_k} + \cos\theta_{km\mathrm{P}} \underline{\mathbf{e}}_{z_k}, \tag{15.386}$$

$$\cos\theta_{km\mathrm{P}} = \begin{cases} \sqrt{1 - \kappa^2 \sin^2\theta_{ik}} & \text{for } \theta_{ik} \le \vartheta_{cm\mathrm{P}} \\ \mathrm{j}\sqrt{\kappa^2 \sin^2\theta_{ik} - 1} & \text{for } \vartheta_{ik} > \theta_{cm\mathrm{P}} \end{cases}, \tag{15.387}$$

$$\vartheta_{cm\mathrm{P}} = \arcsin\frac{1}{\kappa}, \tag{15.388}$$

$$R_\mathrm{SH}(\theta_{ik}) = -1, \tag{15.389}$$

$$R_\mathrm{SV}(\theta_{ik}) = \frac{\sin 2\theta_{ik} \sin 2\theta_{km\mathrm{P}} - \kappa^2 \cos^2\theta_{ik}}{\sin 2\theta_{ik} \sin 2\theta_{km\mathrm{P}} + \kappa^2 \cos^2\theta_{ik}}, \tag{15.390}$$

$$M_\mathrm{P}(\theta_{ik}) = -\kappa \frac{\sin 4\theta_{ik}}{\sin 2\theta_{ik} \sin 2\theta_{km\mathrm{P}} + \kappa^2 \cos^2\theta_{ik}}; \tag{15.391}$$

(15.392)

consequently, we obtain

$$\begin{aligned}
\underline{\mathbf{u}}(\underline{\mathbf{R}}'_k, \omega) &= [A_{ik\mathrm{SV}}(\underline{\mathbf{R}}_{ik}, \omega)(\underline{\mathbf{e}}_{y_k} \times \hat{\underline{\mathbf{R}}}_{ik} + R_\mathrm{SV}(\theta_{ik}) \underline{\mathbf{e}}_{y_k} \times \hat{\underline{\mathbf{R}}}_{kr\mathrm{S}} \\
&\qquad + M_\mathrm{P}(\theta_{ik}) \hat{\underline{\mathbf{R}}}_{km\mathrm{P}}) + 2A_{ik\mathrm{SH}}(\underline{\mathbf{R}}_{ik}, \omega) \underline{\mathbf{e}}_{y_k}] \, \mathrm{e}^{\mathrm{j}k_\mathrm{S} \sin\theta_{ik} x'_k} \\
&\stackrel{\mathrm{def}}{=} \underline{\mathbf{u}}(\underline{\mathbf{R}}_{ik}, \omega) \, \mathrm{e}^{\mathrm{j}k_\mathrm{S} \sin\theta_{ik} x'_k}
\end{aligned} \tag{15.393}$$

[313]Due to the abundance of indices, we omit the explicit PO-characterization for the Kirchhoff approximation.

from (15.382). Now, we insert the Kirchhoff approximation (15.393) into (15.366) and (15.367):

$$\begin{aligned}\underline{\mathbf{u}}_{sk\mathrm{S}}^{\mathrm{far}}(\underline{\mathbf{R}}_k,\omega) &= -\frac{\mathrm{j}\omega}{4\pi c_\mathrm{S}}\, I_{sk\mathrm{S}}(\hat{\underline{\mathbf{R}}}_k,\theta_{ik},\omega)\,\frac{\mathrm{e}^{\mathrm{j}k_\mathrm{S}R_k}}{R_k}\,\underline{\mathbf{u}}(\underline{\mathbf{R}}_{ik},\omega)\underline{\mathbf{e}}_{z_k}\\ &\quad :\left(\hat{\underline{\mathbf{R}}}_k\underline{\underline{\mathbf{I}}}+\hat{\underline{\mathbf{R}}}_k\underline{\underline{\mathbf{I}}}^{213}-2\hat{\underline{\mathbf{R}}}_k\hat{\underline{\mathbf{R}}}_k\hat{\underline{\mathbf{R}}}_k\right), \qquad (15.394)\\ \underline{\mathbf{u}}_{sk\mathrm{P}}^{\mathrm{far}}(\underline{\mathbf{R}}_k,\omega) &= -\frac{\mathrm{j}\omega}{4\pi\varrho c_\mathrm{P}^3}\, I_{sk\mathrm{P}}(\hat{\underline{\mathbf{R}}}_k,\theta_{ik},\omega)\,\frac{\mathrm{e}^{\mathrm{j}k_\mathrm{P}R_k}}{R_k}\,\underline{\mathbf{u}}(\underline{\mathbf{R}}_{ik},\omega)\underline{\mathbf{e}}_{z_k}\\ &\quad :\left(\lambda\,\underline{\underline{\mathbf{I}}}\,\hat{\underline{\mathbf{R}}}_k+2\mu\,\hat{\underline{\mathbf{R}}}_k\hat{\underline{\mathbf{R}}}_k\hat{\underline{\mathbf{R}}}_k\right). \qquad (15.395)\end{aligned}$$

We have

$$\begin{aligned}I_{sk\mathrm{S}}(\hat{\underline{\mathbf{R}}}_k,\theta_{ik},\omega) &= \iint_{S_{k\mathrm{K}}} \mathrm{e}^{-\mathrm{j}k_\mathrm{S}(\hat{\underline{\mathbf{R}}}_k\cdot\underline{\mathbf{R}}_k'-\sin\theta_{ik}x_k')}\,\mathrm{d}x_k'\mathrm{d}y_k'\\ &= \mathcal{F}_{x_k'y_k'}\{\Gamma_{k\mathrm{K}}(x_k',y_k')\mathrm{e}^{\mathrm{j}k_\mathrm{S}\sin\theta_{ik}x_k'}\}_{\substack{K_{x_k'}=k_\mathrm{S}\sin\theta_k\cos\phi_k\\ K_{y_k'}=k_\mathrm{S}\sin\theta_k\sin\phi_k}}\\ &= \hat{\Gamma}_{k\mathrm{K}}(k_\mathrm{S}\sin\theta_k\cos\phi_k-k_\mathrm{S}\sin\theta_{ik},k_\mathrm{S}\sin\theta_k\sin\phi_k), \qquad (15.396)\\ I_{sk\mathrm{P}}(\hat{\underline{\mathbf{R}}}_k,\theta_{ik},\omega) &= \iint_{S_{k\mathrm{K}}} \mathrm{e}^{-\mathrm{j}k_\mathrm{P}(\hat{\underline{\mathbf{R}}}_k\cdot\underline{\mathbf{R}}_k'-\kappa\sin\theta_{ik}x_k')}\,\mathrm{d}x_k'\mathrm{d}y_k'\\ &= \mathcal{F}_{x_k'y_k'}\{\Gamma_{k\mathrm{K}}(x_k',y_k')\mathrm{e}^{\mathrm{j}k_\mathrm{S}\sin\theta_{ik}x_k'}\}_{\substack{K_{x_k'}=k_\mathrm{P}\sin\theta_k\cos\phi_k\\ K_{y_k'}=k_\mathrm{P}\sin\theta_k\sin\phi_k}}\\ &= \hat{\Gamma}_{k\mathrm{K}}(k_\mathrm{P}\sin\theta_k\cos\phi_k-k_\mathrm{S}\sin\theta_{ik},k_\mathrm{S}\sin\theta_k\sin\phi_k) \qquad (15.397)\end{aligned}$$

with

$$\hat{\underline{\mathbf{R}}}_k = \sin\theta_k\cos\phi_k\,\underline{\mathbf{e}}_{x_k}+\sin\theta_k\sin\phi_k\,\underline{\mathbf{e}}_{y_k}+\cos\theta_k\,\underline{\mathbf{e}}_{z_k} \qquad (15.398)$$

as frequency-dependent scalar radiation characteristics of the kth patch with the characteristic function $\Gamma_{k\mathrm{K}}(x_k',y_k')$ for the directly scattered S- and for the mode converted P-part. Obviously, we observe the S-main lobes—more precisely: their scalar aperture factors—for angles θ_k and ϕ_k given by[314]

$$\begin{aligned}\sin\theta_k\cos\phi_k-\sin\theta_{ik} &= 0,\\ \sin\theta_k\sin\phi_k &= 0; \qquad (15.399)\end{aligned}$$

we find $\phi_k=0$ and $\theta_k=\theta_{ik}$ according to the reflection law. To determine the polar and azimuth angle of the (scalar) P-main lobe, we obtain the equations

$$\begin{aligned}k_\mathrm{P}\sin\theta_k\cos\phi_k-k_\mathrm{S}\sin\theta_{ik} &= 0,\\ \sin\theta_k\sin\phi_k &= 0 \qquad (15.400)\end{aligned}$$

[314]The maximum of a spatial spectrum $\tilde{\Gamma}(\underline{\mathbf{K}})$ of a three-dimensional characteristic function $\Gamma(\underline{\mathbf{R}})$ is directed toward $\underline{\mathbf{K}}=\underline{\mathbf{0}}$; here, this is applied to the two-dimensional characteristic function of the patch $S_{k\mathrm{K}}$ modulated by the incident wave.

resulting in $\phi_k = 0$ and $\theta_k = \arcsin(\kappa \sin\theta_{ik})$, namely in the mode conversion law. For $\kappa\sin\theta_{ik} = 1$ we have $\theta_k = \pi/2$, and the lobe angle cannot become larger, i.e. for $\theta_{ik} > \vartheta_{cm\mathrm{P}}$ the P-main lobe disappears form the visible range in the terminology of antenna theory (Balanis 1997; Langenberg 2005); consequently, the P-part is then ignored in the numerical evaluation. For an illustration of main lobes of scattered far-fields of a primary surface force density rectangular patch (in full-space), the reader is referred to Figures 13.17 and 13.18 as well as 13.19 for the limit of the visible range.

As in Section 13.2.4, we must now calculate the point directivities of the secondary surface deformation $\underline{\mathbf{u}}(\underline{\mathbf{R}}_{ik}, \omega)\underline{\mathbf{e}}_{z_k}$ with formulas (13.183) through (13.185) according to (15.394) and (15.395). With (15.394), we find the vector components of the S-point directivity:

$$\begin{aligned}
&\underline{\mathbf{u}}(\underline{\mathbf{R}}_{ik}, \omega)\underline{\mathbf{e}}_{z_k} : \left(\hat{\underline{\mathbf{R}}}_k \underline{\underline{\mathbf{I}}} + \hat{\underline{\mathbf{R}}}_k \underline{\underline{\mathbf{I}}}^{213} - 2\hat{\underline{\mathbf{R}}}_k\hat{\underline{\mathbf{R}}}_k\hat{\underline{\mathbf{R}}}_k\right)\\
&\quad = \cos\theta_k\, \underline{\mathbf{u}}(\underline{\mathbf{R}}_{ik}, \omega) - \underline{\mathbf{u}}(\underline{\mathbf{R}}_{ik}, \omega)\cdot\hat{\underline{\mathbf{R}}}_k(\cos\theta_k\hat{\underline{\mathbf{R}}}_k + \sin\theta_k\underline{\mathbf{e}}_{\theta_k}),
\end{aligned} \tag{15.401}$$

and state—as always—that $\underline{\mathbf{u}}_{sk\mathrm{S}}^{\mathrm{far}}(\underline{\mathbf{R}}_k, \omega)$ has only $\underline{\mathbf{e}}_{\theta_k}$- and $\underline{\mathbf{e}}_{\phi_k}$-components in the $x_k y_k z_k$-coordinate system:

$$\underline{\mathbf{u}}_{sk\mathrm{S}}^{\mathrm{far}}(\underline{\mathbf{R}}_k, \omega)\cdot\hat{\underline{\mathbf{R}}}_k = 0, \tag{15.402}$$

$$\underline{\mathbf{u}}_{sk\mathrm{S}}^{\mathrm{far}}(\underline{\mathbf{R}}_k, \omega)\cdot\underline{\mathbf{e}}_{\theta_k} \sim \cos\theta_k\,\underline{\mathbf{u}}(\underline{\mathbf{R}}_{ik}, \omega)\cdot\underline{\mathbf{e}}_{\theta_k} - \sin\theta_k\,\underline{\mathbf{u}}(\underline{\mathbf{R}}_{ik}, \omega)\cdot\hat{\underline{\mathbf{R}}}_k, \tag{15.403}$$

$$\underline{\mathbf{u}}_{sk\mathrm{S}}^{\mathrm{far}}(\underline{\mathbf{R}}_k, \omega)\cdot\underline{\mathbf{e}}_{\phi_k} \sim \cos\theta_k\,\underline{\mathbf{u}}(\underline{\mathbf{R}}_{ik}, \omega)\cdot\underline{\mathbf{e}}_{\phi_k}, \tag{15.404}$$

which have to be found explicitly. With (15.393), we obtain after some calculus:

$$\begin{aligned}
&\underline{\mathbf{u}}(\underline{\mathbf{R}}_{ik}, \omega)\cdot\hat{\underline{\mathbf{R}}}_k\\
&\quad = A_{ik\mathrm{SV}}(\underline{\mathbf{R}}_{ik}, \omega)\Big\{\sin\theta_k\cos\phi_k\left[\cos\theta_{ik}\left(R_{SV}(\theta_{ik}) - 1\right) + \kappa\,\sin\theta_{ik}M_{\mathrm{P}}(\theta_{ik})\right]\\
&\qquad - \cos\theta_k\left[\sin\theta_{ik}\left(R_{SV}(\theta_{ik}) + 1\right) - \sqrt{1 - \kappa^2\sin^2\theta_{ik}}\,M_{\mathrm{P}}(\theta_{ik})\right]\Big\}\\
&\qquad + 2A_{ik\mathrm{SH}}(\underline{\mathbf{R}}_{ik}, \omega)\sin\theta_k\sin\phi_k,
\end{aligned} \tag{15.405}$$

$$\begin{aligned}
&\underline{\mathbf{u}}(\underline{\mathbf{R}}_{ik}, \omega)\cdot\underline{\mathbf{e}}_{\theta_k}\\
&\quad = A_{ik\mathrm{SV}}(\underline{\mathbf{R}}_{ik}, \omega)\Big\{\cos\theta_k\cos\phi_k\left[\cos\theta_{ik}\left(R_{\mathrm{SV}}(\theta_{ik}) - 1\right) + \kappa\,\sin\theta_{ik}M_{\mathrm{P}}(\theta_{ik})\right]\\
&\qquad + \sin\theta_k\left[\sin\theta_{ik}\left(R_{\mathrm{SV}}(\theta_{ik}) + 1\right) - \sqrt{1 - \kappa^2\sin^2\theta_{ik}}\,M_{\mathrm{P}}(\theta_{ik})\right]\Big\}\\
&\qquad + 2A_{ik\mathrm{SH}}(\underline{\mathbf{R}}_{ik}, \omega)\cos\theta_k\sin\phi_k,
\end{aligned} \tag{15.406}$$

$$
\begin{aligned}
&\underline{\mathbf{u}}(\underline{\mathbf{R}}_{ik},\omega)\cdot\underline{\mathbf{e}}_{\phi_k}\\
&\quad= A_{ik\mathrm{SV}}(\underline{\mathbf{R}}_{ik},\omega)\left\{\sin\phi_k\left[\cos\theta_{ik}\left(1-R_{\mathrm{SV}}(\theta_{ik})\right)-\kappa\,\sin\theta_{ik}M_{\mathrm{P}}(\theta_{ik})\right]\right\}\\
&\qquad+2A_{ik\mathrm{SH}}(\underline{\mathbf{R}}_{ik},\omega)\cos\phi_k. \qquad (15.407)
\end{aligned}
$$

With these projections of the Kirchhoff approximated secondary surface deformation source, we are able to calculate the S-scattered field of the kth patch:

$$
\begin{aligned}
\underline{\mathbf{u}}^{\mathrm{far}}_{sk\mathrm{S}}(\underline{\mathbf{R}}_k,\omega) = &-\frac{\mathrm{j}\omega}{4\pi c_{\mathrm{S}}}\,I_{sk\mathrm{S}}(\hat{\underline{\mathbf{R}}}_k,\theta_{ik},\omega)\,\frac{\mathrm{e}^{\mathrm{j}k_{\mathrm{S}}R_k}}{R_k}\\
&\times\left\{\left[\cos\theta_k\,\underline{\mathbf{u}}(\underline{\mathbf{R}}_{ik},\omega)\cdot\underline{\mathbf{e}}_{\theta_k}-\sin\theta_k\,\underline{\mathbf{u}}(\underline{\mathbf{R}}_{ik},\omega)\cdot\hat{\underline{\mathbf{R}}}_k\right]\underline{\mathbf{e}}_{\theta_k}\right.\\
&\left.\qquad+\cos\theta_k\,\underline{\mathbf{u}}(\underline{\mathbf{R}}_{ik},\omega)\cdot\underline{\mathbf{e}}_{\phi_k}\,\underline{\mathbf{e}}_{\phi_k}\right\}. \qquad (15.408)
\end{aligned}
$$

In fact, for $\theta_{ik}=0$ and $A_{ik\mathrm{SH}}(\underline{\mathbf{R}}_{ik},\omega)=0$, we obtain the S-point directivity of an $\underline{\mathbf{e}}_x\underline{\mathbf{e}}_z$-deformation rate as displayed in Figure 13.14.

We turn to the mode converted P-scattered field and calculate

$$
\begin{aligned}
&\underline{\mathbf{u}}(\underline{\mathbf{R}}_{ik},\omega)\underline{\mathbf{e}}_{z_k}:\left(\lambda\,\underline{\underline{\mathbf{I}}}\hat{\underline{\mathbf{R}}}_k+2\mu\,\hat{\underline{\mathbf{R}}}_k\hat{\underline{\mathbf{R}}}_k\hat{\underline{\mathbf{R}}}_k\right)\\
&\quad=\left[\lambda\,\underline{\mathbf{u}}(\underline{\mathbf{R}}_{ik},\omega)\cdot\underline{\mathbf{e}}_{z_k}+2\mu\,\cos\theta_k\,\underline{\mathbf{u}}(\underline{\mathbf{R}}_{ik},\omega)\cdot\hat{\underline{\mathbf{R}}}_k\right]\hat{\underline{\mathbf{R}}}_k \qquad (15.409)
\end{aligned}
$$

analogous to (15.401); as expected, it follows

$$
\begin{aligned}
&\underline{\mathbf{u}}^{\mathrm{far}}_{sk\mathrm{P}}(\underline{\mathbf{R}}_k,\omega)\cdot\hat{\underline{\mathbf{R}}}_k\sim\lambda\,\underline{\mathbf{u}}(\underline{\mathbf{R}}_{ik},\omega)\cdot\underline{\mathbf{e}}_{z_k}+2\mu\,\cos\theta_k\,\underline{\mathbf{u}}(\underline{\mathbf{R}}_{ik},\omega)\cdot\hat{\underline{\mathbf{R}}}_k, \qquad (15.410)\\
&\underline{\mathbf{u}}^{\mathrm{far}}_{sk\mathrm{P}}(\underline{\mathbf{R}}_k,\omega)\cdot\underline{\mathbf{e}}_{\theta_k}=0, \qquad (15.411)\\
&\underline{\mathbf{u}}^{\mathrm{far}}_{sk\mathrm{P}}(\underline{\mathbf{R}}_k,\omega)\cdot\underline{\mathbf{e}}_{\phi_k}=0. \qquad (15.412)
\end{aligned}
$$

The requested term $\underline{\mathbf{u}}(\underline{\mathbf{R}}_{ik},\omega)\cdot\hat{\underline{\mathbf{R}}}_k$ is already given with (15.405) requiring only the calculation of

$$
\begin{aligned}
\underline{\mathbf{u}}(\underline{\mathbf{R}}_{ik},\omega)\cdot\underline{\mathbf{e}}_{z_k}=A_{ik\mathrm{SV}}(\underline{\mathbf{R}}_{ik},\omega)&\Big[-\sin\theta_{ik}\left(1+R_{\mathrm{SV}}(\theta_{ik})\right)\\
&+\sqrt{1-\kappa^2\sin^2\theta_{ik}}\,M_{\mathrm{P}}(\theta_{ik})\Big]. \qquad (15.413)
\end{aligned}
$$

We have agreed to put the mode converted P-scattered field equal to zero for $\theta_{ik}>\vartheta_{cm\mathrm{P}}$; yet for $\theta_{ik}\leq\vartheta_{cm\mathrm{P}}$, we calculate

$$
\begin{aligned}
\underline{\mathbf{u}}^{\mathrm{far}}_{sk\mathrm{P}}(\underline{\mathbf{R}}_k,\omega)=&-\frac{\mathrm{j}\omega}{4\pi\varrho c_{\mathrm{P}}^3}\,I_{sk\mathrm{P}}(\hat{\underline{\mathbf{R}}}_k,\omega)\,\frac{\mathrm{e}^{\mathrm{j}k_{\mathrm{P}}R_k}}{R_k}\\
&\times\left[\lambda\,\underline{\mathbf{u}}(\underline{\mathbf{R}}_{ik},\omega)\cdot\underline{\mathbf{e}}_{z_k}+2\mu\,\cos\theta_k\,\underline{\mathbf{u}}(\underline{\mathbf{R}}_{ik},\omega)\cdot\hat{\underline{\mathbf{R}}}_k\right]\hat{\underline{\mathbf{R}}}_k. \qquad (15.414)
\end{aligned}
$$

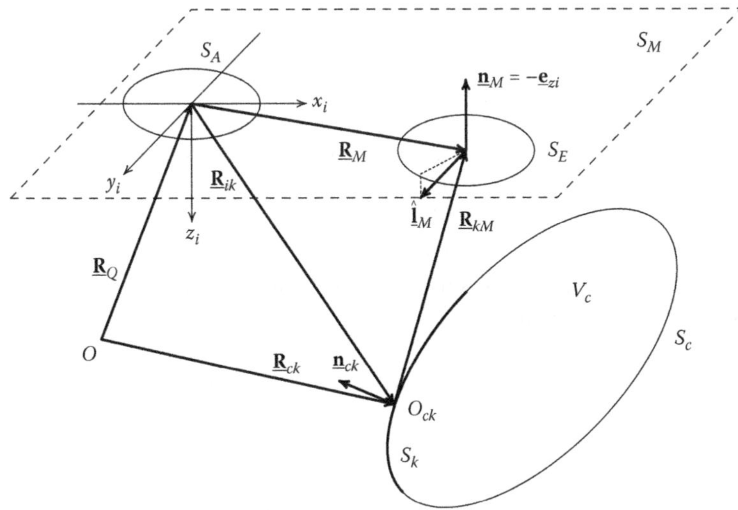

FIGURE 15.47
3D-US-system model: reception.

Again, the point directivity of an $\underline{\mathbf{e}}_x\underline{\mathbf{e}}_z$-deformation rate source resulting for $\theta_{ik} = 0$ and $A_{ik\text{SH}}(\underline{\mathbf{R}}_{ik}, \omega) = 0$ is displayed in Figure 13.14.

Now we turn to reception simulation definitely locating the point of observation $\underline{\mathbf{R}}_k \Longrightarrow \underline{\mathbf{R}}_{kM}$ on the measurement surface S_M (Figure 15.47). In the $x_iy_iz_i$-coordinate system, we characterize the end point of $\underline{\mathbf{R}}_{kM}$ by $\underline{\mathbf{R}}_M$, hence

$$\underline{\mathbf{R}}_{kM} = \underline{\mathbf{R}}_M - \underline{\mathbf{R}}_{ik} \tag{15.415}$$

holds. The normal component of the sum[315] of all patch scattered fields that originate from illuminated patches—for a planar measurement surface S_M, we have $\underline{\mathbf{n}}_M = -\underline{\mathbf{e}}_{z_i}$—

$$u_M(\underline{\mathbf{R}}_M, \omega) = -\underline{\mathbf{e}}_{z_i} \cdot \sum_k^{\hat{\underline{\mathbf{R}}}_{ik}\cdot\underline{\mathbf{n}}_{ck}<0} \left[\underline{\mathbf{u}}_{sk\text{S}}^{\text{far}}(\underline{\mathbf{R}}_{kM}, \omega) + \underline{\mathbf{u}}_{sk\text{P}}^{\text{far}}(\underline{\mathbf{R}}_{kM}, \omega)\right] \tag{15.416}$$

then defines a scalar[316] "point reception expression" $u_M(\underline{\mathbf{R}}_M, \omega)$ for SV-P-waves (SV with regard to S_M); for its explicit calculation, we once more need the direction cosines between the trihedrons of the $x_iy_iz_i$- and the $x_ky_kz_k$-coordinate system for $\hat{\underline{\mathbf{R}}}_{kM} \cdot \underline{\mathbf{e}}_{z_i}$, $\underline{\mathbf{e}}_{\theta_{kM}} \cdot \underline{\mathbf{e}}_{z_i}$, $\underline{\mathbf{e}}_{\phi_{kM}} \cdot \underline{\mathbf{e}}_{z_i}$. For $\underline{\mathbf{R}}_M \neq \underline{\mathbf{0}}$, the reception quantity $u_M(\underline{\mathbf{R}}_M, \omega)$ stands for a pitch-catch mode, and for $\underline{\mathbf{R}}_M = \underline{\mathbf{0}}$,

[315]The Kirchhoff approximation allows for a simple summation because radiation interaction is neglected.

[316]Comparable to the open circuit voltage in antenna theory (Balanis 1997; Langenberg 2005).

we have a pulse-echo-mode. Yet it should be noted that, due to the utilization of full-space Green tensors in the elastodynamic Huygens integral, the scattered field does not satisfy a boundary condition on S_M; to account for that, we would have to introduce partial fields reflected and mode converted at S_M to be calculated with the approximation of incident plane waves (else the reflection and mode conversion factors are not known).

In addition, a receiving transducer may be approximately modeled as a receiving aperture S_E (Figure 15.47). Let the receiving transducer have the main beam direction $\hat{\underline{\mathbf{l}}}_M$ in transmission mode—either for SV- or P-waves (SV with respect to S_M)—then

$$u_E^{\mathrm{P,SV}}(\omega) \iint_{S_E} u_M(\underline{\mathbf{R}}_M, \omega)\, \mathrm{e}^{\mathrm{j}k_{\mathrm{P,S}}\hat{\underline{\mathbf{l}}}_M \cdot \underline{\mathbf{R}}_M}\, \mathrm{d}^2\underline{\mathbf{R}}_M \tag{15.417}$$

yields a receiving transducer quantity for these P-, respectively SV-waves. This is only an approximation since the scattered field does not satisfy a transition condition on the transducer aperture, and, hence, there is no extraction of energy. This is not even changed incorporating the surface S_M to be stress-free. Nevertheless, (15.417) contains the mode-dependent direction selectivity of reception, a fact that is absolutely necessary to assess A- or B-scans properly (e.g., Shlivinski et al. 2004b); to produce such data, an inverse Fourier transform of $u_E(\omega)$ according to the modification

$$u_E^{\mathrm{P,SV}}(\omega) = u_i(\omega) \iint_{S_E} u_M(\underline{\mathbf{R}}_M, \omega)\, \mathrm{e}^{\mathrm{j}k_{\mathrm{P,S}}\hat{\underline{\mathbf{l}}}_M \cdot \underline{\mathbf{R}}_M}\, \mathrm{d}^2\underline{\mathbf{R}}_M \tag{15.418}$$

must be performed for a given pulse spectrum $u_i(\omega)$. As an example, we evaluated (15.418) for the "defects" in the specimen as displayed in Figure 15.48 for 45°-shear wave incidence (2 MHz) without consideration of its actual surface, and not only for a single reception point but also for a linear pulse-echo scan path defined on the surface of the specimen (opposite to the display in Figure 15.48 from left to right). The result is given in Figure 15.49 in terms of

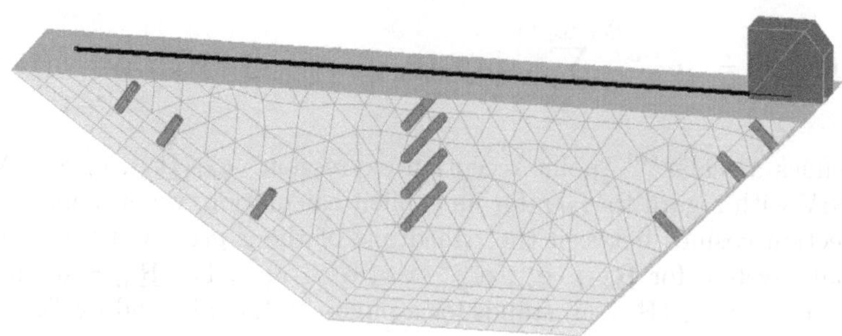

FIGURE 15.48
CAD-model of a specimen with side wall drilled holes and flat bottom holes (V. Schmitz, Fraunhofer IZFP).

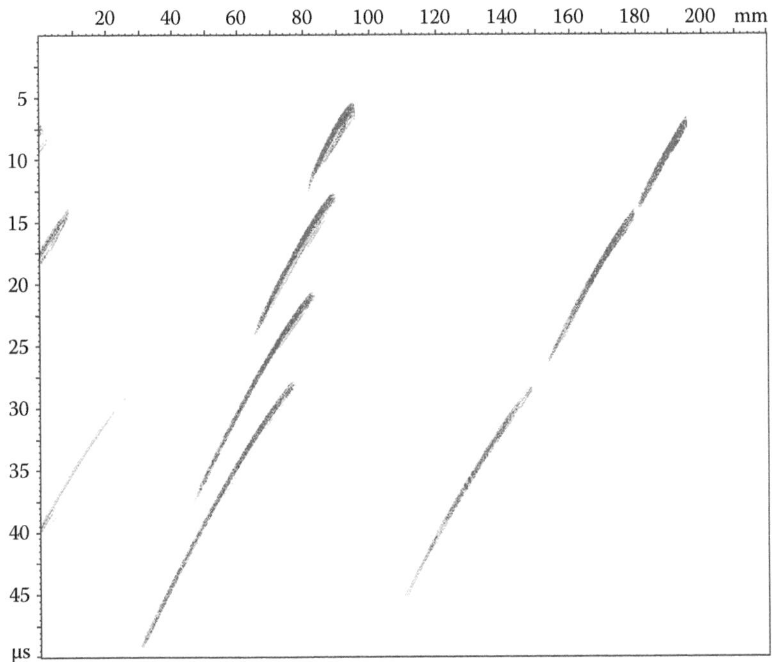

FIGURE 15.49
Calculated B-scan for a linear scan path from left to right on the top surface of the specimen in Figure 15.48: 45°-2 MHz-shear wave transducer, RC3-pulse; horizontal axis = scan coordinate, vertical axis = time counted downward (V. Schmitz, Fraunhofer IZFP).

a B-scan. Based on Figure 16.4, we will discuss the SAFT-processing of these synthetic data in comparison with an experiment (Figure 16.3).

Two general statements concerning the structure of scattered far-fields in backscattering direction under Kirchhoff or Born approximation are established in Section 15.4: In backscattering direction, there is no mode conversion and no shear wave polarization rotation.

16

Inverse Scattering: US-NDT Imaging

We meet a direct scattering problem if scatterer, embedding material, and incident field are known and the scattered field has to be calculated. In the sense of Huygens' principle superimposing elementary waves—a point source synthesis—we have to move forward in time requiring the knowledge of Green functions of the embedding material. As a special problem the radiation interaction of the scatterer with itself arises rendering the direct scattering problem as a nonlinear problem with regard to the geometry of the scatterer. Therefore, linearizations such as the Born or Kirchhoff approximation neglect this radiation interaction.

An inverse scattering problem is met if embedding material, incident field, and scattered field (on a measurement surface) are known, and the location, geometry, and material composition of the scatterer has to be calculated. Replacing "scatterer" by "defect," this is the classical problem of US-NDT! Considering the measured values of the scattered field as "sources," we apparently have to invert the scattered wave propagation with the knowledge of the embedding material Green functions, i.e., a temporal backward oriented point source synthesis, with these sources has to be performed. For this back propagation, the above mentioned nonlinearity is debilitating because the radiation interaction of the scatterer with itself must be "feazed." To avoid this—thus considerably simplifying the inverse scattering problem—linearizations like the Born or Kirchhoff approximation are introduced.

The simplest case of a scalar scattering problem is met for an acoustic point scatterer in a homogeneous isotropic material because there is no radiation interaction. The scattering data are immediately given by the scalar full-space Green function allowing for a direct back propagation of these "measured" data—diffraction curves (or surfaces) in B-scans—with the knowledge of exactly this Green function: Applying the resulting algorithm to arbitrary scatterers (linearizingly) assuming that they are composed of noninteracting point scatterers, we have invented the imaging scheme SAFT.

16.1 SAFT: Synthetic Aperture Focusing Technique

16.1.1 Integration along diffraction curves (surfaces) and back propagation

We execute the following "experiment" (Figure 16.1): A point scatterer in an acoustic homogeneous isotropic nondissipative material is hit by an

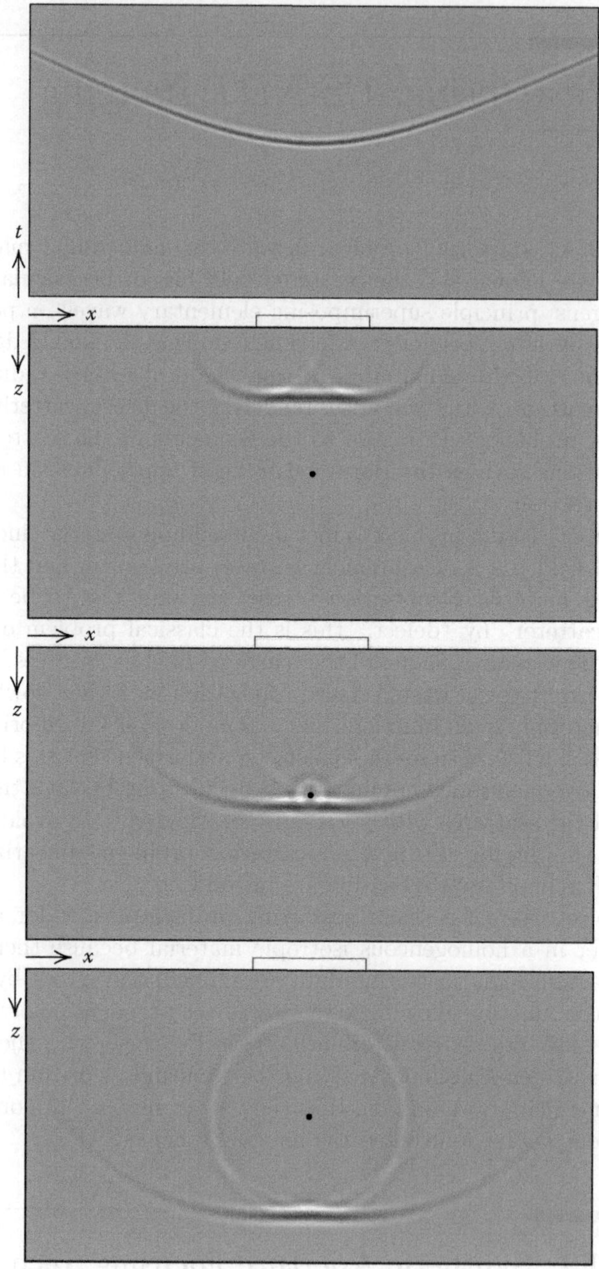

FIGURE 16.1
2D-AFIT-simulation: a point scatterer generates "measurement" data as a diffraction curve.

RC2-bandlimited geometric "optical" (two-dimensional) aperture wavefront (Figure 16.1: second picture from top) from a "transducer" (Figure 16.1: third picture from top); that way, it is turned into a secondary source, and generates per definition a scattered field in terms of a two-dimensional bandlimited Green function (Equation 13.38) that becomes visible in the bottom picture of Figure 16.1. If this scattered wavefront reaches the "measurement" surface, it can be principally pointwise detected to build up a B-scan (of course, reflection takes place at the rigid surface): A diffraction curve emerges as a B-scan data space (Figure 16.1: top). For a planar reception surface, the diffraction curve is a hyperbola, in three dimensions a rotationally symmetric hyperboloid. This is readily shown if we identify those times for which the δ-distribution of the three-dimensional full-space scalar Green function (Equation 13.25)

$$G(\underline{\mathbf{R}} - \underline{\mathbf{R}}', t - t') = \frac{\delta\left(t - t' - \frac{|\underline{\mathbf{R}} - \underline{\mathbf{R}}'|}{c}\right)}{4\pi|\underline{\mathbf{R}} - \underline{\mathbf{R}}'|} \tag{16.1}$$

is singular for given t' (the incident wave hits the point scatterer, here $t' = 0$) and given position $\underline{\mathbf{R}}'$ varying $\underline{\mathbf{R}}$ on the (finite sized) measurement surface S_M:

$$\begin{aligned} t &= \frac{|\underline{\mathbf{R}} - \underline{\mathbf{R}}'|}{c} \\ &= \frac{1}{c}\sqrt{(x - x')^2 + (y - y')^2 + (z - z')^2}; \end{aligned} \tag{16.2}$$

according to the planar measurement surface, we assume cartesian coordinates and position the point scatterer at $x' = x'_0, y' = y'_0, z' = z'_0$. The measurement surface should have distance d from the xy-plane, hence, we have

$$t^2 - \frac{(x - x'_0)^2}{c^2} - \frac{(y - y'_0)^2}{c^2} = \frac{(d - z'_0)^2}{c^2}, \tag{16.3}$$

respectively

$$t = \frac{1}{c}\sqrt{(d - z'_0)^2 + (x - x'_0)^2 + (y - y'_0)^2} > 0, \tag{16.4}$$

as a hyperboloid equation in xyt-data space; this (single shell) hyperboloid has its apex at $x = x'_0, y = y'_0$ exhibiting the shortest travel time $t = (d - z'_0)/c$. Having in mind this geometry information of the diffraction surface, we may postulate the following inversion algorithm: Choose a point x', y', z' in xyz-"reconstruction space" and calculate the respective hyperboloid as if there would reside a point scatterer; then integrate in xyt-data space along this surface as two-dimensionally sketched in Figure 16.2; it is anticipated that the resulting value would be rather low according to the intersection point of the dashed with the solid hyperbola if the point scatterer actually resides at x'_0, y'_0, z'_0. The same holds for all fictitious point scatterers $x' \neq x'_0, y' \neq y'_0, z' \neq z'_0$, only if we come to the voxel hosting the actually existing point scatterer the integration yields a "high" value because it is along the real diffraction

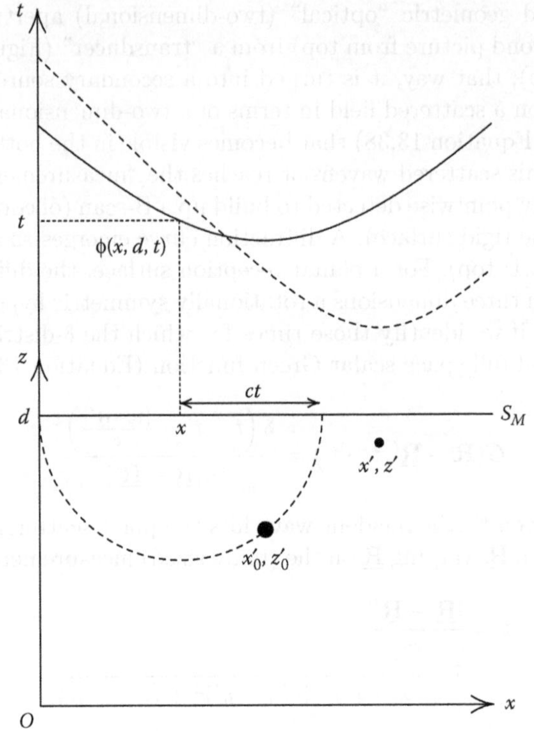

FIGURE 16.2
SAFT: integration of scalar scattered field data along hyperbolic diffraction curves, respectively back propagating them.

curve (surface). The result is a "reconstruction" of the point scatterer, an inverse scattering problem has been solved for this particular point scatterer via focusing of its diffraction curve (surface). Now, assuming that an arbitrary scatterer is composed of independent point scatterers—implying a linearization of scattering and inverse scattering—and integrating its scalar scattered field $\phi(x, y, d, t)$ according to

$$o(x', y', z') = \int\!\!\int_{S_M} \phi\left(x, y, d, t = \frac{1}{c}\sqrt{(x - x')^2 + (y - y')^2 + (d - z')^2}\right) \mathrm{d}x\mathrm{d}y \tag{16.5}$$

for all points x', y', z' in reconstruction space (region of interest) over all points $x, y \in S_M$ hoping that this focusing of scattering data within a synthetic aperture S_M—therefore: SAFT—would yield a suitable image $o(x', y', z')$ of the scatterer. Since the integration (16.5) starts with voxels it is called the voxel, driven approach of SAFT (in two dimensions: pixel driven approach). Even though Figure 16.1 actually stands for an acoustic scattering problem, we denote the scalar scattering data with $\phi(\underline{\mathbf{R}}, t)$ because application in US-NDT is not for an acoustic pressure.

An alternative to the voxel driven approach, is the A-scan driven approach, which is also sketched in Figure 16.2: Each data point $\phi(x, y, d, t)$ is back propagated to its $x'y'z'$-isochrone surface $(0 < z' < d)$

$$z' - d = -\sqrt{c^2t^2 - (x' - x)^2 - (y' - y)^2} < 0, \tag{16.6}$$

i.e., equally distributed on the half-sphere with midpoint x, y, d and radius ct as given by (16.6). If the respective data point lies on the diffraction surface of the $x_0'y_0'z_0'$-point scatterer, the pertinent half-sphere contains exactly this point scatterer due to (16.3)—this situation is depicted in Figure 16.2—i.e., all respective half-spheres intersect at x_0', y_0', z_0' and yield a "high" amplitude via superposition, the location of the point scatterer has been reconstructed. The generalization to arbitrary scatterers that are composed of independent point scatterers results in the linear algorithm "A-scan driven SAFT." As a point data source synthesis that is back oriented in time—we synthesize not a field but an image—the advanced Green function

$$G_a(\underline{\mathbf{R}} - \underline{\mathbf{R}}', t) = \frac{\delta\left(t + \frac{|\underline{\mathbf{R}} - \underline{\mathbf{R}}'|}{c}\right)}{4\pi|\underline{\mathbf{R}} - \underline{\mathbf{R}}'|} \tag{16.7}$$

may be used to formulate this Huygens-like referring to the time domain version of (15.37) (we ignore the distance dependence of the elementary wave amplitudes):

$$\begin{aligned} o(x', y', z') = \iint_{S_M} \int_{-\infty}^{\infty} & \phi(x, y, d, t')\,\delta \\ & \times \left(t - t' + \frac{1}{c}\sqrt{(x - x')^2 + (y - y')^2 + (d - z')^2}\right) \mathrm{d}t'\mathrm{d}x\mathrm{d}y\Bigg|_{t=0} \quad ; \end{aligned} \tag{16.8}$$

we have to choose $t = 0$ after having calculated the t'-convolution integral, because per definition, each single point scatterer has been excited for this particular time. Obviously, the result (16.8) is identical to (16.5) due to the relation $f(t) * \delta(t - t_0) = f(t - t_0)$.

We explicitly point out that algorithms to solve the inverse scattering problem require *a priori* information about the embedding material; it is the Green function, respectively the phase-, and energy velocity of the elementary waves in the homogeneous isotropic embedding material that has to be known to be inserted into (16.5), respectively (16.8).

16.1.2 Pitch-catch and pulse-echo versions of SAFT

At the beginning of the chapter about inverse scattering, we assumed the knowledge of the incident field to define an inverse scattering problem, yet it does not appear in the inversion formula (16.5). The reason is the agreement $t' = 0$: A δ-pulsed plane wave in the direction $\underline{\hat{\mathbf{k}}}_i$ in terms of $\delta(t - \underline{\hat{\mathbf{k}}}_i \cdot \underline{\mathbf{R}}/c)$ meets a point scatterer located in the coordinate origin for $t = 0$ (respectively, on the plane $\underline{\hat{\mathbf{k}}}_i \cdot \underline{\mathbf{R}} = 0$); to meet the point scatterer at $\underline{\mathbf{R}}_0'$ for time $t = 0$, it must be modified according to $\phi_i(t - \underline{\hat{\mathbf{k}}}_i \cdot (\underline{\mathbf{R}} - \underline{\mathbf{R}}_0')/c)$. This is not practical because in the inversion formula (16.5), we face the transition $\underline{\mathbf{R}}_0' \Longrightarrow \underline{\mathbf{R}}'$ as variable in reconstruction space; it is more convenient to choose $t' = \underline{\hat{\mathbf{k}}}_i \cdot \underline{\mathbf{R}}_0'/c$ in (16.1) because then we have

$$\begin{aligned} o(x', y', z') &= \iint_{S_M} \phi\left(x, y, d, t = \frac{1}{c}\sqrt{(x-x')^2 + (y-y')^2 + (d-z')^2} + \frac{\underline{\hat{\mathbf{k}}}_i \cdot \underline{\mathbf{R}}'}{c}\right) \\ &\quad \times dxdy \end{aligned} \tag{16.9}$$

as inversion formula allowing for a voxel-dependent time normalization given by the direction of the incident plane wave.

Instead of a nonphysical incident plane wave coming from infinity, we may as well accomplish the illumination of the x_0', y_0', z_0'-point scatterer by an x_Q, y_Q, z_Q-localized point source according to

$$t' = \frac{1}{c}\sqrt{(x_Q - x_0')^2 + (y_Q - y_0')^2 + (d - z_0')^2}. \tag{16.10}$$

The respective inversion formula then reads

$$\begin{aligned} o(x', y', z') = \iint_{S_M} \phi\Bigg(x, y, d, t &= \frac{1}{c}\sqrt{(x-x')^2 + (y-y')^2 + (d-z')^2} \\ &+ \frac{1}{c}\sqrt{(x_Q-x')^2 + (y_Q-y')^2 + (d-z')^2}\Bigg)\, dxdy. \end{aligned} \tag{16.11}$$

Specialization to the practically relevant pulse-echo operational mode may then be realized through $x_Q = x$, $y_Q = y$:

$$\begin{aligned} o(x', y', z') = \iint_{S_M} \phi\Bigg(x, y, d, t \\ = \frac{2}{c}\sqrt{(x-x')^2 + (y-y')^2 + (d-z')^2}\Bigg) dxdy. \end{aligned} \tag{16.12}$$

Note: The integrations in the SAFT inversion formulas (16.9), (16.11), and (16.12) do not yield the same results; so, strictly speaking, the resulting images $o(x', y', z')$ should be denoted differently.

16.1.3 SAFT with Hilbert transformed pulse data

Regrettably, practical US-NDT may not rely on a δ-pulse as incident transducer field; generally, a bandlimited pulse with, for example, RC2-dependence serves as an appropriate model. Therefore, the scattered pulse reflects exactly this oscillatory behavior (compare Figure 16.1) that is rediscovered in a B-scan data field as diffraction curves (surfaces); the SAFT-inversion integrals back propagate these oscillations into reconstruction space yielding, according to the pulse duration, to a limited resolution as well as to oscillations of the image function $o(x', y', z')$. Yet, they may be eliminated via SAFT signal processing.[317] For that purpose, a complex valued signal is defined applying a Hilbert transform (Section 2.3.4) with regard to time according to

$$\phi_c(\underline{\mathbf{R}}, t) = \phi(\underline{\mathbf{R}}, t) + j\mathcal{H}_\tau\{\phi(\underline{\mathbf{R}}, \tau)\} \tag{16.13}$$

inserting it into the inversion formulas (16.9), (16.11), and (16.12); the result is a complex valued image function $o_c(\underline{\mathbf{R}}')$ with a magnitude representing the envelope of the oscillatory image function; this works for arbitrary geometries of measurement surfaces (Langenberg et al. 1993b). Alternatively, for planar measurement surfaces, we may create a complex valued image function according to[318]

$$o_c(x', y', z') = o(x', y', z') - j\mathcal{H}_{\zeta'}\{o(x', y', \zeta')\} \tag{16.14}$$

calculating the magnitude afterward; it turns out to be equivalent to the previous approach (Langenberg 1987; Section 16.2.5). Yet, the calculation of the magnitude of the complex valued scattered signal with subsequent SAFT-inversion is not equivalent; it yields a deterioration of resolution.

Since SAFT is a heuristically proposed imaging algorithm, be it either voxel driven or A-scan driven, we may equally establish heuristically based modifications.

- First: US-NDT meets scalar wave fields only for SH-waves. Nevertheless, the application to P- as well as SV-waves (in three dimensions: P- and S-waves) is possible with two assumptions: Any scalar measurement quantity is selected, for example, the electric voltage delivered by the transducer, hoping to be proportional to the normal component of the particle displacement on the specimen surface (for fluid coupling, the horizontal component is not transmitted); however, due to the appearance of the wave speed c in (16.5), it must be known beforehand, i.e., the assignment of diffraction surfaces to pressure or shear waves is mandatory.

- Nonplanar measurement surfaces may be considered through $d \Longrightarrow z(x, y)$.

- A further physical understanding of wave propagation suggests the implementation of a depth adjustment under the SAFT-integral (16.5) because data from point scatterers in greater depth definitely exhibit smaller amplitudes than those close to the surface.

[317] Various signal processing techniques, e.g., a deconvolution of the scattered signal with the incident pulse, allow for an improvement of resolution (Mayer 1989).

[318] Regarding the negative sign, Section 16.2.5 should be consulted.

- In practical applications, the incident field is neither a plane wave nor does it come from a point source; yet transducer radiation characteristics may easily be incorporated in a simple manner through spatial angle limitation of the isochrone surface (e.g., for the A-scan driven approach).
- For homogeneous anisotropic materials, the isochrone surfaces are given by energy velocity surfaces; therefore, a modification of SAFT for those embedding materials is immediately at hand (Langenberg et al. 1997). Accounting for refraction at boundaries of a piecewise (anisotropic) inhomogeneous embedding material (defect imaging in anisotropic welds), a ray geometric SAFT algorithm may be formulated and implemented for such applications (Hannemann 2001; Marklein et al. 2002b; Shlivinski et al. 2004b).

Finally: Even though knowledge concerning wave propagation has entered the formulation of the SAFT imaging algorithm, it is not a rigorous mathematical solution of the inverse scattering problem, which becomes apparent embedding it into an inverse scattering theory (Langenberg 1987; Section 16.2.5); only the inherent linearization based on the negligence of radiation interaction is obvious, yet at this point we do not know anything about further implicit assumptions, and, hence, about precision and resolution, and about the factual physical meaning of the image $o(x', y', z')$. These questions must be answered on the basis of theoretical investigations (Langenberg 1987) or applying the algorithms to test specimens, respectively using synthetic data (Langenberg et al. 2004a,b; Mayer et al. 2003; Schmitz 2002; Langenberg et al. 1999a, 1993a). In the following, we give an example: For the test specimen displayed in Figure 15.48, pulse-echo B-scan data have been recorded along a linear scan path—in Figure 16.3, from left to right—with a 45°-2 MHz-shear wave transducer (MWB45N2) and processed with the SAFT algorithm. The result displayed in Figure 16.3 shows in fact a focusing of diffraction curves to "defect" surfaces as they are recorded by the transducer, where the axial resolution is given by the pulse duration and the lateral resolution is given by the transducer, respectively the (synthetic) measurement aperture. Clearly,

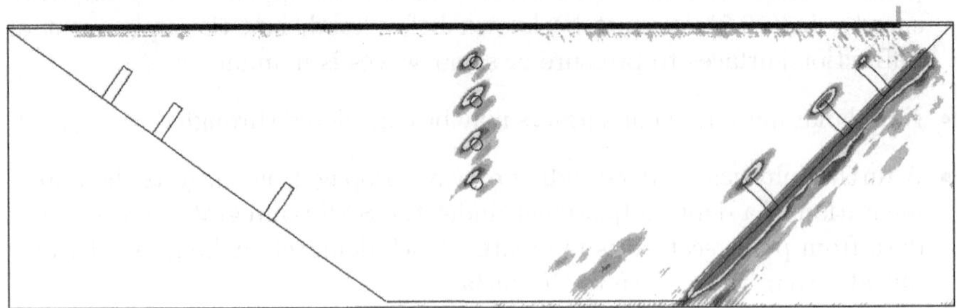

FIGURE 16.3
SAFT imaging with experimental data (V. Schmitz: Fraunhofer IZFP).

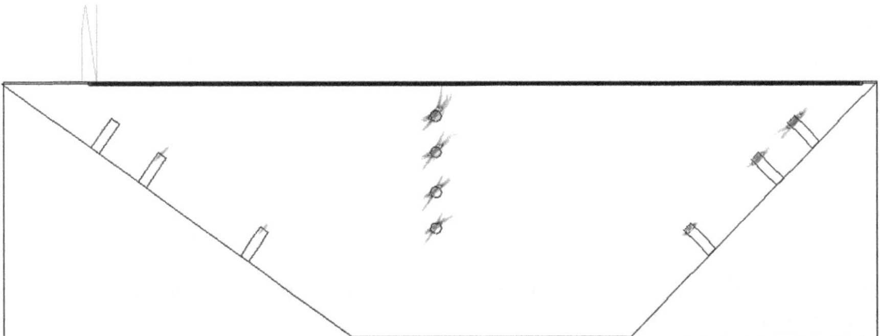

FIGURE 16.4
SAFT image with simulation data (V. Schmitz: Fraunhofer IZFP).

artifacts (ghost images) are recognizable that may be explained by multiple scattering, respectively creeping waves for the cylindrical drills.

Figure 16.4 shows SAFT results for the same test specimen and the same scan path if synthetic data calculated with the Kirchhoff approximation (Figure 15.49) are inserted. First: Due to the shorter simulation pulse, the axial resolution is better, and then: multiply scattered and creeping wave pulses are missing, because they are not considered within the Kirchhoff approximation. By the way: The images of the side wall drills exhibit the axially resolved edge pulses of the transducer aperture as source of the incident field (Section 13.3.4).

16.2 FT-SAFT: Fourier Transform Synthetic Aperture Focusing Technique

The SAFT algorithms are based on heuristic arguments: Even though their "derivation" uses the knowledge of the time domain Green function of the embedding material, more precisely: the geometry of elementary wavefronts—potential radiation characteristics of elementary waves are not incorporated—the mathematical relation between scatterer and its scattered field is not considered. However, this relation is available in terms of volume integrals over secondary sources that are equivalent to the scatterer. Since the scattered field is the quantity to be measured, it should in principle be possible to invert these volume integrals, i.e., to formulate an inverse scattering theory. For scalar wave fields, this theory has been widely finalized as a linear (Langenberg 1987, 2002a) as well as a nonlinear theory (Belkebir and Saillard 2001). On one hand, the linear theory contains the SAFT algorithms as special cases after introducing several approximations (Langenberg 1987;

pulse-echo version: Section 16.2.5), and on the other hand, it allows for the implementation of algorithmic alternatives, most effectively: FT-SAFT for planar measurement surfaces (Mayer 1989; Mayer et al. 1990; Langenberg et al. 1999a; Mayer et al. 2003; Langenberg et al. 2004a, 2004b) because essentially Fourier transforms are applied, hence the acronym "FT" stands for.

16.2.1 Scalar secondary sources: Contrast sources

For acoustically penetrable scatterers, the Fourier spectrum of the scattered pressure is calculated as

$$p_s(\underline{\mathbf{R}}, \omega) = \int_{-\infty}^{\infty}\int_{-\infty}^{\infty}\int_{-\infty}^{\infty} \Big[\mathrm{j}\omega\rho\, h_\kappa(\underline{\mathbf{R}}', \omega) G(\underline{\mathbf{R}} - \underline{\mathbf{R}}', \omega) \\ + \underline{\mathbf{f}}_\rho(\underline{\mathbf{R}}', \omega) \cdot \nabla' G(\underline{\mathbf{R}} - \underline{\mathbf{R}}', \omega)\Big] \, \mathrm{d}^3\underline{\mathbf{R}}' \quad (16.15)$$

in terms of volume integration (5.106) of the Fourier spectra of the secondary sources (5.100) and (5.101):

$$\begin{aligned} h_\kappa(\underline{\mathbf{R}}, \omega) &= \mathrm{j}\omega\, \Gamma_c(\underline{\mathbf{R}}) \left[\kappa - \kappa^{(i)}(\underline{\mathbf{R}})\right] p(\underline{\mathbf{R}}, \omega) \\ &= -\,\mathrm{j}\omega\kappa\, \chi_\kappa(\underline{\mathbf{R}}) p(\underline{\mathbf{R}}, \omega), \qquad (16.16) \\ \underline{\mathbf{f}}_\rho(\underline{\mathbf{R}}, \omega) &= -\,\mathrm{j}\omega\, \Gamma_c(\underline{\mathbf{R}}) \left[\rho - \rho^{(i)}(\underline{\mathbf{R}})\right] \underline{\mathbf{v}}(\underline{\mathbf{R}}, \omega) \\ &= \mathrm{j}\omega\rho\, \chi_\rho(\underline{\mathbf{R}}) \underline{\mathbf{v}}(\underline{\mathbf{R}}, \omega) \\ &= \chi_\rho(\underline{\mathbf{R}}) \nabla p(\underline{\mathbf{R}}, \omega); \qquad (16.17) \end{aligned}$$

here, $\kappa^{(i)}(\underline{\mathbf{R}})$ and $\rho^{(i)}(\underline{\mathbf{R}})$ denote compressibility and mass density in the scattering volume V_c with the characteristic function $\Gamma_c(\underline{\mathbf{R}})$ that resides in the homogeneous isotropic embedding material κ, ρ with the Green function $G(\underline{\mathbf{R}} - \underline{\mathbf{R}}', \omega)$; $\chi_\rho(\underline{\mathbf{R}})$ and $\chi_\kappa(\underline{\mathbf{R}})$ are contrast functions defined by (5.94) and (5.95). For perfect acoustic scatterers with a soft or rigid surface S_c, we conveniently start with the Helmholtz formulation (5.84) of Huygens' principle

$$p_s(\underline{\mathbf{R}}, \omega) = \iint_{S_c} \left[\mathrm{j}\omega\rho\, g(\underline{\mathbf{R}}', \omega) G(\underline{\mathbf{R}} - \underline{\mathbf{R}}', \omega) + \underline{\mathbf{t}}(\underline{\mathbf{R}}', \omega) \cdot \nabla' G(\underline{\mathbf{R}} - \underline{\mathbf{R}}', \omega)\right] \mathrm{d}S', \quad (16.18)$$

where

$$\begin{aligned} g(\underline{\mathbf{R}}, \omega) &= -\,\underline{\mathbf{n}}_c \cdot \underline{\mathbf{v}}(\underline{\mathbf{R}}, \omega) \\ &= -\,\frac{1}{\mathrm{j}\omega\rho}\, \underline{\mathbf{n}}_c \cdot \nabla p(\underline{\mathbf{R}}, \omega), \qquad (16.19) \\ \underline{\mathbf{t}}(\underline{\mathbf{R}}, \omega) &= \underline{\mathbf{n}}_c p(\underline{\mathbf{R}}, \omega) \qquad (16.20) \end{aligned}$$

denote the secondary surface sources (5.82) and (5.83).

A formulation of (16.18) that is equivalent to (16.15) and (16.18) reads

$$p_s(\underline{\mathbf{R}}, \omega) = \int_{-\infty}^{\infty}\int_{-\infty}^{\infty}\int_{-\infty}^{\infty} \big[\mathrm{j}\omega\rho\, h_c(\underline{\mathbf{R}}', \omega) G(\underline{\mathbf{R}} - \underline{\mathbf{R}}', \omega) + \underline{\mathbf{f}}_c(\underline{\mathbf{R}}', \omega) \cdot \nabla' G(\underline{\mathbf{R}} - \underline{\mathbf{R}}', \omega) \big] \, \mathrm{d}^3\underline{\mathbf{R}}' \tag{16.21}$$

via the definition of secondary volume sources (15.44) and (15.45):

$$\begin{aligned} h_c(\underline{\mathbf{R}}, \omega) &= -\underline{\boldsymbol{\gamma}}_c(\underline{\mathbf{R}}) \cdot \underline{\mathbf{v}}(\underline{\mathbf{R}}, \omega) \\ &= -\frac{1}{\mathrm{j}\omega\rho} \underline{\boldsymbol{\gamma}}_c(\underline{\mathbf{R}}) \cdot \nabla p(\underline{\mathbf{R}}, \omega), \end{aligned} \tag{16.22}$$

$$\underline{\mathbf{f}}_c(\underline{\mathbf{R}}, \omega) = \underline{\boldsymbol{\gamma}}_c(\underline{\mathbf{R}}) p(\underline{\mathbf{R}}, \omega). \tag{16.23}$$

For the secondary sources (16.22) and (16.23), the specializations

$$\underline{\mathbf{f}}_c^{\mathrm{s}}(\underline{\mathbf{R}}, \omega) = \underline{\mathbf{0}}, \tag{16.24}$$

$$h_c^{\mathrm{s}}(\underline{\mathbf{R}}, \omega) = -\frac{1}{\mathrm{j}\omega\rho} \underline{\boldsymbol{\gamma}}_c(\underline{\mathbf{R}}) \cdot \nabla p(\underline{\mathbf{R}}, \omega) \tag{16.25}$$

to a soft scatterer: Dirichlet boundary condition ($p(\underline{\mathbf{R}}, \omega) = 0$, $\underline{\mathbf{R}} \in S_c$), respectively to a rigid scatterer, is appropriate: Neumann boundary condition ($\underline{\mathbf{n}}_c \cdot \nabla p(\underline{\mathbf{R}}, \omega) = 0$, $\underline{\mathbf{R}} \in S_c$):

$$\underline{\mathbf{f}}_c^{\mathrm{r}}(\underline{\mathbf{R}}, \omega) = \underline{\boldsymbol{\gamma}}_c(\underline{\mathbf{R}}) p(\underline{\mathbf{R}}, \omega), \tag{16.26}$$

$$h_c(\underline{\mathbf{R}}, \omega) = 0. \tag{16.27}$$

Since the rigid scatterer exhibits the gradient of the Green function in (16.21), we may produce the representation

$$p_s^{\mathrm{r}}(\underline{\mathbf{R}}, \omega) = -\int_{-\infty}^{\infty}\int_{-\infty}^{\infty}\int_{-\infty}^{\infty} G(\underline{\mathbf{R}} - \underline{\mathbf{R}}', \omega) \nabla' \cdot \underline{\mathbf{f}}_c^{\mathrm{r}}(\underline{\mathbf{R}}', \omega) \, \mathrm{d}^3\underline{\mathbf{R}}' \tag{16.28}$$

according to (5.63) $\Longrightarrow$ (5.61), where we may replace

$$\nabla \cdot \underline{\mathbf{f}}_c^{\mathrm{r}}(\underline{\mathbf{R}}, \omega) = p(\underline{\mathbf{R}}, \omega) \nabla \cdot \underline{\boldsymbol{\gamma}}_c(\underline{\mathbf{R}}) \tag{16.29}$$

using (16.26) and applying the Neumann boundary condition one more time. Further specialization to a penetrable scatterer with $\chi_\rho(\underline{\mathbf{R}}) \equiv 0$, we obtain a similar representation

$$p_s^{\mathrm{pen,s,r}}(\underline{\mathbf{R}}, \omega) = \int_{-\infty}^{\infty}\int_{-\infty}^{\infty}\int_{-\infty}^{\infty} q_c^{\mathrm{pen,s,r}}(\underline{\mathbf{R}}', \omega) G(\underline{\mathbf{R}} - \underline{\mathbf{R}}', \omega) \, \mathrm{d}^3\underline{\mathbf{R}}' \tag{16.30}$$

for all three canonical scatterers if we define

$$q_c^{\mathrm{pen,s,r}}(\underline{\mathbf{R}}, \omega) = \begin{cases} k^2 \chi_\kappa(\underline{\mathbf{R}}) p(\underline{\mathbf{R}}, \omega) & \text{penetrable (only } \kappa\text{-contrast)} \\ -\underline{\boldsymbol{\gamma}}_c(\underline{\mathbf{R}}) \cdot \nabla p(\underline{\mathbf{R}}, \omega) & \text{soft} \\ -\left[\nabla \cdot \underline{\boldsymbol{\gamma}}_c(\underline{\mathbf{R}}) \right] p(\underline{\mathbf{R}}, \omega) & \text{rigid} \end{cases} . \tag{16.31}$$

In each case, we meet a contrast source $w(\underline{\mathbf{R}}, \omega)$

$$q_c^{\text{pen,s,r}}(\underline{\mathbf{R}}, \omega) \Longrightarrow w(\underline{\mathbf{R}}, \omega) = \chi(\underline{\mathbf{R}})\phi(\underline{\mathbf{R}}, \omega) \tag{16.32}$$

that contains the geometry/material properties of the scatterer with $\chi(\underline{\mathbf{R}})$; it is field dependent via $\phi(\underline{\mathbf{R}}, \omega)$; now we generally write $\phi(\underline{\mathbf{R}}, \omega)$ (for the penetrable scatterer $\phi(\underline{\mathbf{R}}, \omega)$ contains the factor k^2)—consequently $\phi_s(\underline{\mathbf{R}}, \omega)$ for the scattered field—to indicate that the integral representation

$$\phi_s(\underline{\mathbf{R}}, \omega) = \int_{-\infty}^{\infty}\int_{-\infty}^{\infty}\int_{-\infty}^{\infty} w(\underline{\mathbf{R}}', \omega) G(\underline{\mathbf{R}} - \underline{\mathbf{R}}', \omega)\, \mathrm{d}^3\underline{\mathbf{R}}' \tag{16.33}$$

is a basis of an inversion theory not only for scalar acoustic waves but also for any scalar wave field satisfying the differential equation

$$\Delta\phi_s(\underline{\mathbf{R}}, \omega) + k^2\phi_s(\underline{\mathbf{R}}, \omega) = -w(\underline{\mathbf{R}}, \omega). \tag{16.34}$$

We would like to note that the distributional contrast sources of perfect scatterers may be represented by strongly lossy penetrable scatterers. To this end, we specialize the Maxwell model (4.78) for the compliance tensor to the acoustic case:

$$\frac{\partial S(\underline{\mathbf{R}}, t)}{\partial t} = -\kappa(\underline{\mathbf{R}}) \frac{\partial p(\underline{\mathbf{R}}, t)}{\partial t} - \Gamma(\underline{\mathbf{R}}) p(\underline{\mathbf{R}}, t). \tag{16.35}$$

For the Fourier spectra, we obtain the complex compressibility

$$\kappa_c(\underline{\mathbf{R}}) = \kappa(\underline{\mathbf{R}}) + \mathrm{j}\,\frac{\Gamma(\underline{\mathbf{R}})}{\omega} \tag{16.36}$$

yielding

$$\kappa_c^{(i)} = \kappa + \mathrm{j}\,\frac{\Gamma}{\omega} \tag{16.37}$$

for a homogeneous κ-lossy scatterer, whose complex compressibility should have the same real part as the embedding material; consequently, we obtain a purely imaginary contrast

$$\chi_\kappa = \mathrm{j}\,\frac{\Gamma}{\omega\kappa}, \tag{16.38}$$

whose magnitude can be arbitrarily *increased* choosing Γ (for fixed frequency) accordingly. The question is now: What kind of boundary condition may be approximately realized with $|\chi_\kappa| \gg 1$. We calculate the reflection of a plane wave

$$p_i(\underline{\mathbf{R}}, \omega) = p_0(\omega)\, \mathrm{e}^{-\mathrm{j}k \sin\vartheta_i y - \mathrm{j}k\cos\vartheta_i z} \tag{16.39}$$

with $k = \omega\sqrt{\rho\kappa}$ at a lossy half-space with the wave number $k_c = \omega\sqrt{\rho(\kappa + \mathrm{j}\Gamma/\omega)}$ (surface $\equiv xy$-plane) and obtain for the reflected wave

$$p_r(\underline{\mathbf{R}}, \omega) = R(\vartheta_i, \omega)\, p_i(\omega)\, \mathrm{e}^{-\mathrm{j}k\sin\vartheta_i y + \mathrm{j}\Re k_{tz}^{(c)} z} \tag{16.40}$$

due to the transition conditions (5.13) and (5.14); the reflection coefficient is given by

$$R(\vartheta_i, \omega) = \frac{k \cos\vartheta_i - k_{tz}^{(c)}}{k \cos\vartheta_i + k_{tz}^{(c)}}, \tag{16.41}$$

and the complex z-component

$$k_{tz}^{(c)} = k\sqrt{\cos^2\vartheta_i + \mathrm{j}\,\frac{\Gamma}{\omega\kappa}} \tag{16.42}$$

of the complex phase vector by $\underline{\mathbf{k}}_t^{(c)}$ of the accordingly attenuated $\mathrm{e}^{\Im k_{tz}^{(c)} z}$, $z < 0$, transmitted plane wave. For $\Gamma/\omega\kappa \gg 1$, we have $R \simeq -1$, i.e., the acoustic pressure approximately satisfies a Dirichlet boundary condition. Therefore, contrast source inversion (CSI) (Section 16.2.2) implements the soft scatterer in terms of a purely imaginary contrast (16.38); a Maxwell model similar to (16.35) for the density analogously approximates the rigid scatterer.

To linearize the direct as well as the inverse scattering problems, the back coupling of the scattered field to the contrast source must be canceled; this is basically achieved if the total field in (16.31) is set proportional to the incident field, more precisely: if the penetrable scatterer is Born and the perfect scatterer Kirchhoff approximated:

$$w^{\text{lin}}(\underline{\mathbf{R}}, \omega) = \begin{cases} k^2 \chi_\kappa(\underline{\mathbf{R}}) p_i(\underline{\mathbf{R}}, \omega) & \text{penetrable (only } \kappa\text{−contrast)} \\ -2\underline{\boldsymbol{\gamma}}_u(\underline{\mathbf{R}}) \cdot \boldsymbol{\nabla} p_i(\underline{\mathbf{R}}, \omega) & \text{soft} \\ -2\left[\boldsymbol{\nabla} \cdot \underline{\boldsymbol{\gamma}}_u(\underline{\mathbf{R}})\right] p_i(\underline{\mathbf{R}}, \omega) & \text{rigid.} \end{cases} \tag{16.43}$$

Here, $\underline{\boldsymbol{\gamma}}_u(\underline{\mathbf{R}})$ stands for the illuminated scattering surface, i.e., considering the Kirchhoff shadow boundary. With this linearized version, the FT-SAFT algorithm is essentially based on the contrast sources. Before we come to that the nonlinear CSI will be briefly addressed because the necessary equations are available anyway.

16.2.2 Contrast source inversion

In fact, Equation 16.33 can be used twice, once as the data equation

$$\phi_s(\underline{\mathbf{R}}, \omega) = \int_{-\infty}^{\infty}\int_{-\infty}^{\infty}\int_{-\infty}^{\infty} w(\underline{\mathbf{R}}', \omega) G(\underline{\mathbf{R}} - \underline{\mathbf{R}}', \omega)\, \mathrm{d}^3\underline{\mathbf{R}}', \quad \underline{\mathbf{R}} \in S_M, \tag{16.44}$$

and once more as the object equation

$$\phi(\underline{\mathbf{R}}, \omega) = \phi_i(\underline{\mathbf{R}}, \omega) + \int_{-\infty}^{\infty}\int_{-\infty}^{\infty}\int_{-\infty}^{\infty} w(\underline{\mathbf{R}}', \omega) G(\underline{\mathbf{R}} - \underline{\mathbf{R}}', \omega)\, \mathrm{d}^3\underline{\mathbf{R}}', \quad \underline{\mathbf{R}} \in V_c. \tag{16.45}$$

For known contrast $\chi(\underline{\mathbf{R}})$, Equation 16.45 is the Lippmann–Schwinger integral equation (5.108) for the interior total field $\phi(\underline{\mathbf{R}}, \omega)$, $\underline{\mathbf{R}} \in V_c$, for vanishing ρ-contrast; hence, it is the basis to solve the direct scattering problem.[319] In case of the inverse problem, the contrast source is the desired unknown quantity yielding the following iterative solution of (16.44) and (16.45) without linearization:

- An initial guess $w^{(0)}(\underline{\mathbf{R}}, \omega)$ is calculated with a linear back propagation method yielding an initial guess of the interior total field $\phi^{(0)}(\underline{\mathbf{R}}, \omega)$ with (16.45), and hence with the modified equation (16.32)

$$\chi(\underline{\mathbf{R}}) = \frac{w(\underline{\mathbf{R}}, \omega)\phi^*(\underline{\mathbf{R}}, \omega)}{|\phi(\underline{\mathbf{R}}, \omega)|^2} \tag{16.46}$$

  a zero-order approximation of the contrast.

- Multiplying (16.45) with $\chi(\underline{\mathbf{R}})$ results in the modified object equation

$$\begin{aligned} w(\underline{\mathbf{R}}, \omega) &= \chi(\underline{\mathbf{R}})\phi_i(\underline{\mathbf{R}}, \omega) \\ &+ \chi(\underline{\mathbf{R}}) \int_{-\infty}^{\infty}\int_{-\infty}^{\infty}\int_{-\infty}^{\infty} w(\underline{\mathbf{R}}', \omega) G(\underline{\mathbf{R}} - \underline{\mathbf{R}}', \omega)\, \mathrm{d}^3\underline{\mathbf{R}}', \quad \underline{\mathbf{R}} \in V_c, \end{aligned} \tag{16.47}$$

  which represents a system of integral equations together with (16.44) to calculate $w^{(1)}(\underline{\mathbf{R}}, \omega)$ if $\chi^{(0)}(\underline{\mathbf{R}})$ is used.

- With (16.45), we obtain $\phi^{(1)}(\underline{\mathbf{R}}, \omega)$, with (16.46) $\chi^{(1)}(\underline{\mathbf{R}})$ allowing for the calculation of $w^{(2)}(\underline{\mathbf{R}}, \omega)$, and so on.

The flow chart in Figure 16.5 clearly arranges the single steps of the CSI algorithm.

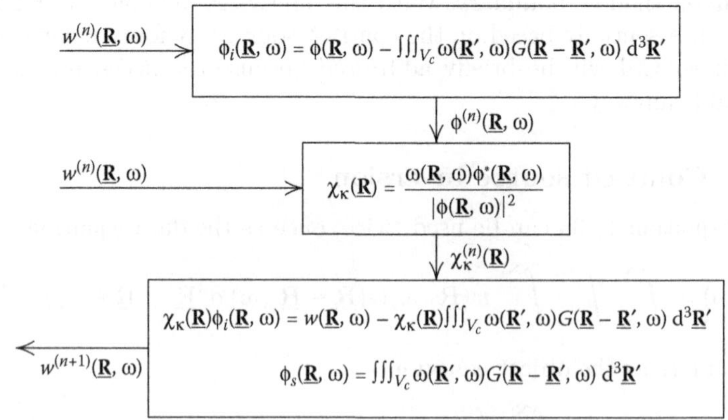

FIGURE 16.5
Flow chart for nonlinear CSI.

[319] For perfect scatterers, it reduces to integral equations (5.89), respectively (5.90).

This nonlinear iterative algorithm has been published as CSI (including modifications) (Kleinman and van den Berg 1997; van den Berg 1999; Haak 1999), and we applied it successfully to electromagnetic data (Marklein et al. 2001, 2002a). However, it is numerically costly, and, hence, it is legitimate to ask what kind of improvements may be obtained as compared to linear inversion. It is completely clear: If the (non-Born) contrast of a penetrable scatterer has to be calculated quantitatively, there is no way around a nonlinear inversion algorithm. On the other hand, if one is only interested in the surface contour of a perfect (non-Kirchhoff) scatterer, results for ultrasonic data seem to indicate (Marklein et al. 2002b; Schmitz et al. 2004a) only a marginal gain. However: For typical measured electromagnetic data, the gain may be significant (Marklein et al. 2001).

16.2.3 Generalized holography

In connection with the CSI, we referred to a zero-order approximation of the contrast source as solution of a linear back propagation scheme: It is available in various versions of generalized holography (Langenberg 1987). We consider—as for the A-scan driven approach of SAFT—the time harmonic scattered field data (the Fourier spectra of the time-dependent scattered field) on a closed measurement surface S_M surrounding the scatterer completely as point "sources" that have to be back propagated into the embedding material in terms of elementary waves; hence,

$$\theta_H(\underline{\mathbf{R}}, \omega) = \iint_{S_M} \left[G^*(\underline{\mathbf{R}} - \underline{\mathbf{R}}', \omega)\nabla'\phi_s(\underline{\mathbf{R}}', \omega) \right. \\ \left. -\phi_s(\underline{\mathbf{R}}', \omega)\nabla' G^*(\underline{\mathbf{R}} - \underline{\mathbf{R}}', \omega)\right] \cdot \underline{\mathbf{n}}'_M \,\mathrm{d}S' \qquad (16.48)$$

defines the (single frequency) "image" of the Huygens-type back propagation for $\underline{\mathbf{R}} \in V_M$, because $G^*(\underline{\mathbf{R}} - \underline{\mathbf{R}}', \omega)$ is nothing but the Fourier spectrum of the advanced Green function (16.7); $\underline{\mathbf{n}}'_M$ is the outer normal on S_M. Since the secondary sources of the scattered field reside in the interior of V_M, the volume enclosed by S_M, the "original" Huygens integral (16.48) with G instead of G^* would yield a null-field for $\underline{\mathbf{R}} \in V_M$. In contrast, (16.48) yields the nonvanishing generalized holographic field, even though at this point, it is not yet known how it is related to the contrast sources of the scatterer. However, application of Green's theorem to V_M results in the scattered field equation (16.34) and its solution (16.33) in the Porter–Bojarski integral equation (of the first kind)

$$\theta_H(\underline{\mathbf{R}}, \omega) = 2\mathrm{j} \int_{-\infty}^{\infty}\int_{-\infty}^{\infty}\int_{-\infty}^{\infty} w(\underline{\mathbf{R}}', \omega)\, \Im G(\underline{\mathbf{R}} - \underline{\mathbf{R}}', \omega)\,\mathrm{d}^3\underline{\mathbf{R}}' \qquad (16.49)$$

for the contrast source, where the kernel is the imaginary part of Green's function of the embedding material. A more precise investigation reveals (Langenberg 1987) that θ_H is already a solution for $w(\underline{\mathbf{R}}, \omega)$, namely the solution of minimal norm (minimal energy). Under certain circumstances, e.g., for strip-like scatterers (US-NDT: cracks) this is already something (Langenberg 1987). If not, one must try harder "to put more energy" into the solution of (16.49), e.g., integrating over frequency (within the bandwidth of the ultrasonic pulse). Yet, since the contrast source is a function of both variables $\underline{\mathbf{R}}$ *and* ω, in general not in the synchronous form $F(\omega)w(\underline{\mathbf{R}})$, the spatial distribution of $w(\underline{\mathbf{R}}, \omega)$ is prone to vary for a different frequency, and this so-called frequency diversity will only yield a result if the frequency dependence of the contrast source is controlled from outside by the operator. This is achieved via the linearization $w \Longrightarrow w^{\text{lin}}$ (Equation 16.43) and prescription of the incident field, for example, as plane wave with direction $\hat{\underline{\mathbf{k}}}_i$. That way, the door for an angular diversity is opened, i.e. (16.49) is integrated over a spatial angle interval of $\hat{\underline{\mathbf{k}}}_i$ (with single frequency excitation). Both diversities lead to explicit inversion algorithms for the contrast (Langenberg 1987, 2002). For US-NDT, the frequency diversity is primarily relevant, and in fact, if the resulting formula for contrast inversion is transformed into the time domain, we obtain—after several subsequent approximations—SAFT! That way, we have provided a field theoretical derivation of SAFT! For the special case of planar measurement surfaces, we will actually derive the field theoretical exact pulse-echo version of SAFT in Section 16.2.5.

The numerical evaluation of the back propagation integral (16.48) is especially effective if we can apply—for planar measurement surfaces—spatial Fourier transforms: The result is the FT-SAFT algorithm that is, due to the same field theoretical foundations, the result equivalent to SAFT. However, the direct derivation of FT-SAFT without using generalized holography is simpler favoring this approach in the following section.

16.2.4 FT-SAFT

Fourier diffraction slice theorem: Similar to Figure 16.2, we configure a planar measurement surface in a cartesian coordinate system as xy-plane for $z = d$. The scattered field representation (16.43) reveals itself according to

$$\phi_s(x, y, d, \omega) = \int_{-\infty}^{\infty}\int_{-\infty}^{\infty}\int_{-\infty}^{\infty} w(x', y', z', \omega)G(x - x', y - y', d - z', \omega) \times \mathrm{d}x'\mathrm{d}y'\mathrm{d}z' \qquad (16.50)$$

as two-dimensional convolution integral with respect to x and y; it is resolved via the Fourier transform convolution theorem to be transformed into a product of the spatial spectra:

$$\hat{\phi}_s(K_x, K_y, d, \omega) = \int_{-\infty}^{\infty} \hat{w}(K_x, K_y, z', \omega)\hat{G}(K_x, K_y, d - z', \omega)\,\mathrm{d}z', \qquad (16.51)$$

where the spectra—for example, $\hat{\phi}_s(K_x, K_y, d, \omega)$—are given by

$$\hat{\phi}_s(K_x, K_y, d, \omega) = \int_{-\infty}^{\infty}\int_{-\infty}^{\infty} \phi_s(x, y, d, \omega)\,\mathrm{e}^{-\mathrm{j}K_x x - \mathrm{j}K_y y}\,\mathrm{d}x\mathrm{d}y; \qquad (16.52)$$

the spectrum

$$\hat{G}(K_x, K_y, d - z', \omega) = \frac{\mathrm{j}}{2\sqrt{k^2 - K_x^2 - K_y^2}} \, \mathrm{e}^{\mathrm{j}|d-z'|\sqrt{k^2-K_x^2-K_y^2}} \tag{16.53}$$

is explicitly known from Weyl's integral representation (13.323). With the short-hand notation

$$K_z = \sqrt{k^2 - K_x^2 - K_y^2} \tag{16.54}$$

and under the given assumption $d > z'$—then we have $|d - z'| = d - z'$—(16.51) is finally revealed as spatial Fourier integral with respect to z', where K_z plays the role of the Fourier variable:

$$\hat{\phi}_s(K_x, K_y, d, \omega) = \frac{\mathrm{j}}{2K_z} \, \mathrm{e}^{\mathrm{j}dK_z} \int_{-\infty}^{\infty} \hat{w}(K_x, K_y, z', \omega) \, \mathrm{e}^{-\mathrm{j}K_z z'} \, \mathrm{d}z'. \tag{16.55}$$

Hence, the so-called Fourier diffraction slice theorem holds:

$$\hat{\phi}_s(K_x, K_y, d, \omega) = \frac{\mathrm{j}}{2K_z} \, \mathrm{e}^{\mathrm{j}dK_z} \, \tilde{w}(K_x, K_y, K_z = \sqrt{k^2 - K_x^2 - K_y^2}, \omega). \tag{16.56}$$

In words: For $K_x^2 + K_y^2 \leq k^2$, the two-dimensionally Fourier transformed scattered field $\hat{\phi}_s(K_x, K_y, d, \omega)$ with respect to the measurement coordinates x and y is proportional to the three-dimensional spatial Fourier spectrum $\tilde{w}(K_x, K_y, K_z, \omega)$ on the Ewald sphere $K_z = \sqrt{k^2 - K_x^2 - K_y^2}$. The mapping prescription (16.54) of the $K_x K_y \omega$-space into $K_x K_y K_z$-space distributes according to $K_x^2 + K_y^2 \leq k^2$ bandlimited Fourier-transformed measured data[320] depending upon variables K_x, K_y, ω on a half-sphere surface with radius $k = \omega/c$ in $\underline{\mathbf{K}}$-space, where $\underline{\mathbf{K}} = K_x \underline{\mathbf{e}}_x + K_y \underline{\mathbf{e}}_y + K_z \underline{\mathbf{e}}_z$. The radius parameter k inter alia permits to cover a partial volume of $\underline{\mathbf{K}}$-space varying frequency requiring—as for the integration of the Porter–Bojarski equation (16.49)—the linearization of the contrast source. We again distinguish between frequency diversity of a multibistatic (multipitch-catch) setup, angular diversity of a multibistatic (multipitch-catch) single frequency setup, as well as frequency diversity of a multimonostatic (pulse-echo) setup.

FT-SAFT: multibistatic frequency and multibistatic angular diversity: We assume the incident wave to be a plane wave

$$\phi_i(\underline{\mathbf{R}}, \omega, \hat{\underline{\mathbf{k}}}_i) = \phi_0(\omega, \hat{\underline{\mathbf{k}}}_i) \, \mathrm{e}^{\mathrm{j}\underline{\mathbf{k}}_i \cdot \underline{\mathbf{R}}} \tag{16.57}$$

with the diversity parameters ω and $\hat{\underline{\mathbf{k}}}_i = \underline{\mathbf{k}}_i / k$: For varying frequency and $\underline{\mathbf{R}} \in S_M$, an arbitrarily fixed illumination direction $\hat{\underline{\mathbf{k}}}_i$ corresponds to a broadband multiple pitch-catch experiment, hence to multibistatic frequency diversity

[320]Figure 11.3 depicts intuitively that this band limitation is actually realized for a sufficiently large measurement distance d.

(radar terminology: $\underline{\mathbf{R}} = \underline{\mathbf{R}}_0 \in S_M$ is a bistatic experiment, hence for arbitrary $\underline{\mathbf{R}} \in S_M$ a multibistatic experiment), and for varying angle of incidence and $\underline{\mathbf{R}} \in S_M$, an arbitrarily fixed frequency corresponds to a multibistatic single frequency angular diversity. However, in each case, (16.57) yields the linearized contrast source

$$w^{\mathrm{lin}}(\underline{\mathbf{R}}, \omega, \hat{\underline{\mathbf{k}}}_i) = \chi(\underline{\mathbf{R}})\phi_i(\underline{\mathbf{R}}, \omega, \hat{\underline{\mathbf{k}}}_i), \tag{16.58}$$

where we hide the factor k^2 for the penetrable scatterer in $\phi_0(\omega, \hat{\underline{\mathbf{k}}}_i)$. Inserting (16.58) into (16.56), the Fourier transform of a modulated contrast function has to be calculated due to the exponential function in (16.57), i.e., the three-dimensional Fourier spectrum $\chi(\underline{\mathbf{R}})$ is displaced in $\underline{\mathbf{K}}$-space. Yet, it is advantageous to displace the Ewald sphere instead; to achieve this, we multiply the scattered field with $\mathrm{e}^{-\mathrm{j}\underline{\mathbf{k}}_i \cdot \underline{\mathbf{R}}}$ and combine the resulting exponential term $\mathrm{e}^{-\mathrm{j}\underline{\mathbf{k}}_i \cdot (\underline{\mathbf{R}} - \underline{\mathbf{R}}')}$ with the Green function (16.44) before applying the two-dimensional Fourier transform:

$$\begin{aligned} &\mathrm{e}^{-\mathrm{j}\underline{\mathbf{k}}_i \cdot \underline{\mathbf{R}}} \phi_s(\underline{\mathbf{R}}, \omega) \\ &\quad = \phi_0(\omega, \hat{\underline{\mathbf{k}}}_i) \int_{-\infty}^{\infty}\int_{-\infty}^{\infty}\int_{-\infty}^{\infty} \chi(\underline{\mathbf{R}}') \frac{\mathrm{e}^{\mathrm{j}k|\underline{\mathbf{R}} - \underline{\mathbf{R}}'|}}{4\pi|\underline{\mathbf{R}} - \underline{\mathbf{R}}'|} \mathrm{e}^{-\mathrm{j}\underline{\mathbf{k}}_i \cdot (\underline{\mathbf{R}} - \underline{\mathbf{R}}')} \, \mathrm{d}^3\underline{\mathbf{R}}'; \end{aligned} \tag{16.59}$$

due to this modulation function the Fourier variables K_x and K_y are now displaced by the components k_{ix}, respectively k_{iy}:

$$\begin{aligned} &\mathrm{e}^{-\mathrm{j}k_{iz}d} \hat{\phi}_s(K_x + k_{ix}, K_y + k_{iy}, d, z) \\ &\quad = \phi_0(\omega, \hat{\underline{\mathbf{k}}}_i) \frac{\mathrm{j}}{2\sqrt{k^2 - (K_x + k_{ix})^2 - (K_y + k_{iy})^2}} \\ &\qquad \times \int_{-\infty}^{\infty} \hat{\chi}(K_x, K_y, z') \, \mathrm{e}^{-\mathrm{j}k_{iz}(d - z')} \, \mathrm{e}^{\mathrm{j}|d - z'|\sqrt{k^2 - (K_x + k_{ix})^2 - (K_y + k_{iy})^2}} \, \mathrm{d}z'. \end{aligned} \tag{16.60}$$

With the assumption $d > z'$, we have

$$\begin{aligned} &\hat{\phi}_s(K_x + k_{ix}, K_y + k_{iy}, d, \omega) \\ &\quad = \phi_0(\omega, \hat{\underline{\mathbf{k}}}_i) \frac{\mathrm{j}\mathrm{e}^{\mathrm{j}d\sqrt{k^2 - (K_x + k_{ix})^2 - (K_y + k_{iy})^2}}}{2\sqrt{k^2 - (K_x + k_{ix})^2 - (K_y + k_{iy})^2}} \\ &\qquad \times \int_{-\infty}^{\infty} \hat{\chi}(K_x, K_y, z') \, \mathrm{e}^{-\mathrm{j}z'(\sqrt{k^2 - (K_x + k_{ix})^2 - (K_y + k_{iy})^2} - k_{iz})} \, \mathrm{d}z' \end{aligned} \tag{16.61}$$

resulting in the Fourier diffraction slice theorem

$$\begin{aligned} &\hat{\phi}_s(K_x + k_{ix}, K_y + k_{iy}, d, \omega) \\ &\quad = \phi_0(\omega, \hat{\underline{\mathbf{k}}}_i) \frac{\mathrm{j}}{2(K_z + k_{iz})} \, \mathrm{e}^{\mathrm{j}d(K_z + k_{iz})} \, \tilde{\chi}(K_x, K_y, K_z) \end{aligned} \tag{16.62}$$

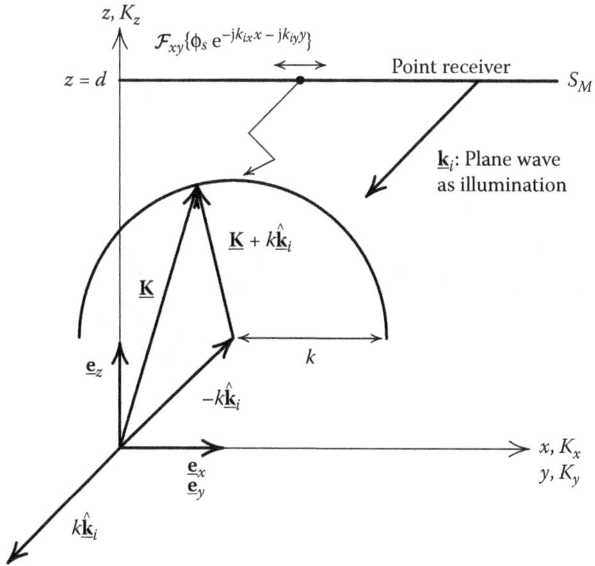

FIGURE 16.6
Illustration of the multibistatic FT-SAFT mapping prescription.

with the mapping prescription

$$K_z = \sqrt{k^2 - (K_x + k_{ix})^2 - (K_y + k_{iy})^2} - k_{iz}, \tag{16.63}$$

respectively,

$$|\underline{\mathbf{K}} + \underline{\mathbf{k}}_i| = k \tag{16.64}$$

with $K_z + k_{iz} \geq 0$. Figure 16.6 illustrates this mapping of the Fourier transformed data into $\underline{\mathbf{K}}$-space, where once again only propagating spectral components $(K_x + k_{ix})^2 + (K_y + k_{iy})^2 \leq k^2$ of $\hat{\phi}_s$ are considered: The midpoint of the Ewald sphere is now located at $-\underline{\mathbf{k}}_i$, and the transformed data are placed on that part of the hemisphere that is oriented toward the measurement surface[321]; hence, the parameters of the mapping are $\hat{\underline{\mathbf{k}}}_i$ and k yielding typical $\underline{\mathbf{K}}$-space coverings for frequency-, respectively angular diversity,[322] as displayed in Figure 16.7 that are different in both cases consequently yielding different spatially bandlimited results after Fourier inversion into reconstruction space. By the way: Even for infinite frequency bandwidth $(0 \leq k < \infty)$,

[321]Measuring the scattered field in transmission on a plane orthogonal to the propagation direction, the Fourier diffraction slice theorem is immediately recognized as wave theoretical counterpart of the X-ray tomography Fourier slice theorem (Langenberg 1987). Therefore, FT-SAFT is also called diffraction tomography (Devaney 1986).

[322]Obviously, for angular diversity intersection lines (surfaces) of Ewald hemispheres are obtained yielding superpositions of (Fourier transformed) data; this must be accounted for by a filter operation that has been derived by Langenberg (2002a) for the case of far-field inversion.

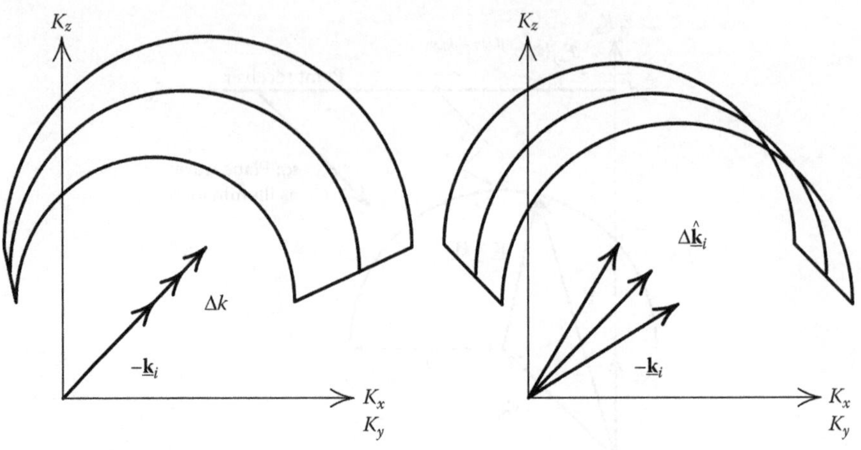

FIGURE 16.7
FT-SAFT $\underline{\mathbf{K}}$-space covering for multibistatic frequency- (left), respectively multibistatic angular diversity (right).

respectively complete covering of the $\hat{\underline{\mathbf{k}}}_i$-unit sphere, we obtain at most bandlimited contrast functions, a fact that is also true for SAFT as time domain version of frequency diversity, yet it is not apparent in the heuristic motivation of this algorithm. This fact should always be in mind if US-imaging is addressed as defect reconstruction. Evidently, this is especially true if bandwidth, respectively illumination interval, is restricted; for example, neither operational mode reaches the origin of $\underline{\mathbf{K}}$-space in that case, i.e., the Fourier inversion of the data covered limited $\underline{\mathbf{K}}$-space regions initially always results in—as SAFT with bandlimited pulses—oscillatory images of scatterers; yet in general, these are complex valued allowing for a simple magnitude operation to get rid of the oscillations having the same effect as the additional SAFT processing of Hilbert transformed pulsed data.

We have already emphasized that within the framework of a linear inverse scattering theory SAFT and FT-SAFT are just two sides of the same medal; for the multibistatic frequency diversity, this is illustrated in the flow chart of Figure 16.8 based on synthetic EFIT-data, whereas the mathematical proof has been given by Langenberg (1987). Starting point—in two dimensions—is an xt-data field (the A-scans in a B-scan), here: the elastodynamic field of a pulsed plane pressure wave scattered by a circular cylinder with a stress-free surface; the pixel-driven SAFT algorithm requires the calculation of diffraction hyperbolas $t(x', z')$ for each $x'z'$-pixel, and subsequent data integration along the hyperbolas. The result is an image of the illuminated surface contour of the scatterer with an axial resolution corresponding to the pulse duration (through simultaneous SAFT processing of Hilbert transformed data and magnitude calculation the pulse oscillations have been suppressed). Alternatively,

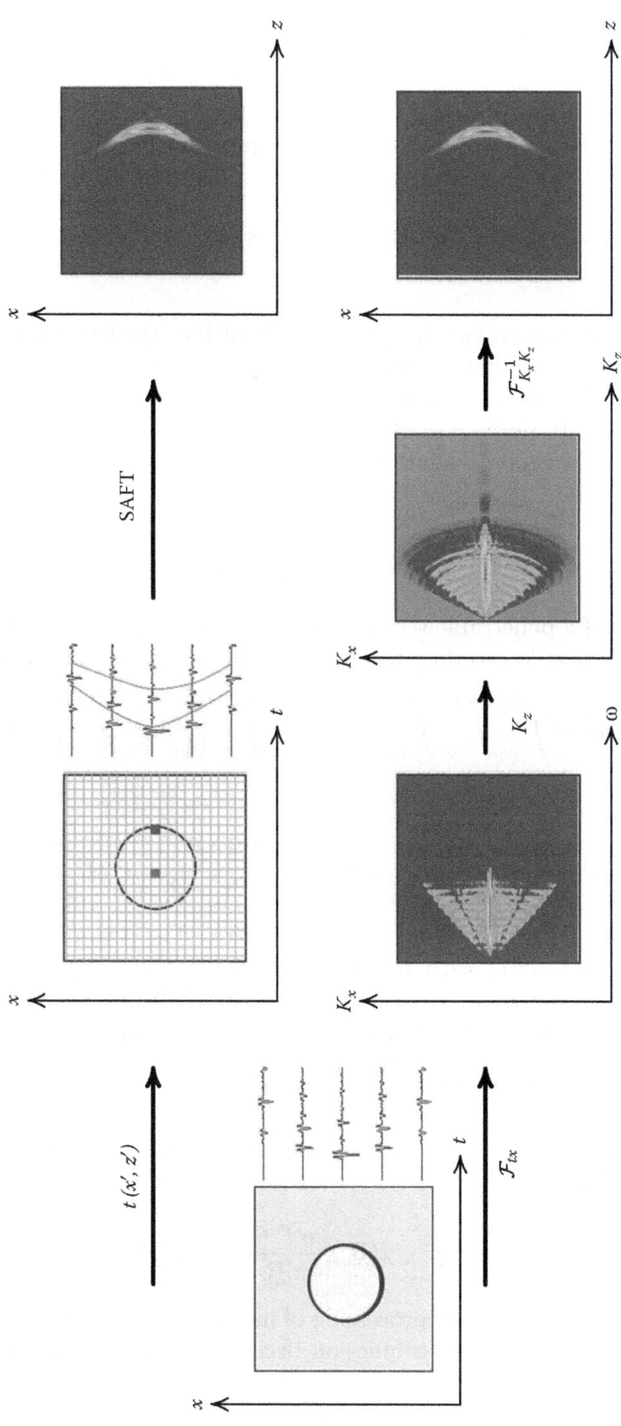

FIGURE 16.8
SAFT and FT-SAFT (multibistatic frequency diversity).

the xt-data are Fourier transformed with regard to t and x resulting in an ωK_x-space (exhibiting very nicely the spatial band limitation of Fourier transformed data to $k_x \leq k$); the mapping $K_z = \sqrt{\omega^2/c_{\mathrm{P}}^2 - K_x^2}$ (c_{P} is the pressure wave velocity) turns it into $\underline{\mathbf{K}}$-space with coordinates K_x and K_z, and a two-dimensional inverse Fourier transform with subsequent magnitude calculation yields the FT-SAFT result that is practically nondistinguishable from the SAFT result.

The equivalence of pulse-echo SAFT and FT-SAFT will be proved in the last section of this elaboration.

FT-SAFT: multimonostatic frequency diversity (pulse-echo version): Due to the simple data acquisition, the pulse-echo mode plays an outstanding role in US-NDT; a corresponding SAFT algorithm has been formulated with (16.12) to derive a multimonstatic FT-SAFT version the measurements must first be preprocessed. As for (16.12), we start from the incident field of a point-source at $\underline{\mathbf{R}}_0 \in S_M$, hence we modify (16.57) according to

$$\phi_i(\underline{\mathbf{R}}, \omega, \underline{\mathbf{R}}_0) = \phi_0(\omega)\,\frac{\mathrm{e}^{\mathrm{j}k|\underline{\mathbf{R}}-\underline{\mathbf{R}}_0|}}{4\pi|\underline{\mathbf{R}}-\underline{\mathbf{R}}_0|}; \tag{16.65}$$

Note: For the case of a penetrable scatterer $\phi_0(\omega)$ contains the factor k^2. The contrast source linearized according to (16.65) results in the scattered field

$$\phi_s(\underline{\mathbf{R}}, \omega, \underline{\mathbf{R}}_0) = \phi_0(\omega)\int_{-\infty}^{\infty}\int_{-\infty}^{\infty}\int_{-\infty}^{\infty}\chi(\underline{\mathbf{R}}')\,\frac{\mathrm{e}^{\mathrm{j}k|\underline{\mathbf{R}}'-\underline{\mathbf{R}}_0|}}{4\pi|\underline{\mathbf{R}}'-\underline{\mathbf{R}}_0|}\,\frac{\mathrm{e}^{\mathrm{j}k|\underline{\mathbf{R}}-\underline{\mathbf{R}}'|}}{4\pi|\underline{\mathbf{R}}-\underline{\mathbf{R}}'|}\,\mathrm{d}^3\underline{\mathbf{R}}' \tag{16.66}$$

that reduces to

$$\phi_s^{\mathrm{m}}(\underline{\mathbf{R}}, \omega) = \phi_0(\omega)\int_{-\infty}^{\infty}\int_{-\infty}^{\infty}\int_{-\infty}^{\infty}\chi(\underline{\mathbf{R}}')\,\frac{\mathrm{e}^{2\mathrm{j}k|\underline{\mathbf{R}}-\underline{\mathbf{R}}'|}}{(4\pi)^2|\underline{\mathbf{R}}-\underline{\mathbf{R}}'|^2}\,\mathrm{d}^3\underline{\mathbf{R}}' \tag{16.67}$$

in pulse-echo mode $\underline{\mathbf{R}}_0 = \underline{\mathbf{R}}$ ("m" for monostatic). The two-dimensional Fourier transform of (16.67) with respect to x and y is no longer offhand possible because the integrand contains the square of the Green function. With the definition of the modified monostatic scattered field

$$\phi_s^{\mathrm{mo}}(\underline{\mathbf{R}}, \omega) = \frac{2\pi}{\mathrm{j}}\,\frac{\partial}{\partial k}\,\frac{\phi_s^{\mathrm{m}}(\underline{\mathbf{R}}, \omega)}{\phi_0(\omega)}, \tag{16.68}$$

we achieve the appearance of a "monostatic Green function" $G^{\mathrm{mo}}(\underline{\mathbf{R}} - \underline{\mathbf{R}}', \omega) = G(\underline{\mathbf{R}} - \underline{\mathbf{R}}', 2\omega)$ in

$$\phi_s^{\mathrm{mo}}(\underline{\mathbf{R}}, \omega) = \int_{-\infty}^{\infty}\int_{-\infty}^{\infty}\int_{-\infty}^{\infty}\chi(\underline{\mathbf{R}}')\,\frac{\mathrm{e}^{2\mathrm{j}k|\underline{\mathbf{R}}-\underline{\mathbf{R}}'|}}{4\pi|\underline{\mathbf{R}}-\underline{\mathbf{R}}'|}\,\mathrm{d}^3\underline{\mathbf{R}}'; \tag{16.69}$$

(16.68) is the above mentioned preprocessing of monostatic data. As a matter of fact, $G^{\mathrm{mo}}(\underline{\mathbf{R}} - \underline{\mathbf{R}}', \omega)$ is a Green function because $\phi_s^{\mathrm{mo}}(\underline{\mathbf{R}}, \omega)$ satisfies the differential equation

$$\Delta\phi_s^{\mathrm{mo}}(\underline{\mathbf{R}}, \omega) + 4k^2\phi_s^{\mathrm{mo}}(\underline{\mathbf{R}}, \omega) = -\chi(\underline{\mathbf{R}}) \tag{16.70}$$

with the Green function $G^{\mathrm{mo}}(\underline{\mathbf{R}} - \underline{\mathbf{R}}', \omega)$. Yet it should be pointed out that, in contrast to (16.34), this only holds within the framework of linearization.

The Fourier diffraction slice theorem corresponding to (16.69) with the mapping prescription

$$K_z = \sqrt{4k^2 - K_x^2 - K_y^2} \geq 0, \tag{16.71}$$

respectively

$$|\underline{\mathbf{K}}| = 2k, \tag{16.72}$$

can now immediately be written down:

$$\hat{\phi}_s^{\mathrm{mo}}(K_x, K_y, d, \omega) = \frac{\mathrm{j}}{2K_z} \, \mathrm{e}^{\mathrm{j}dK_z} \, \tilde{\chi}(K_x, K_y, K_z). \tag{16.73}$$

The mapping (16.71) is sketched in Figure 16.9; obviously, we obtain origin-centered hemispheres with radii $2k$. Frequency diversity is immediately at hand with (16.73) and (16.71), because (16.73) is already linearized; in Figure 16.9, the diversity is indicated by the dashed mapping hemispheres (circles). Again, it is apparent that finite frequency bandwidth yields oscillatory images being fixed with magnitude calculation.

As in Figure 16.8 for multibistatic frequency diversity, we illustrate for multimonostatic frequency diversity the equivalence of FT-SAFT and pulse-echo SAFT, yet this time for experimental US-data (V. Schmitz: Fraunhofer IZFP). Figure 16.10 displays images of a crack orthogonal to the surface of a steel specimen; illumination occurred with (SV-) shear waves under 45°, hence with elastic vector waves. We consider the transducer receiving voltage as scalar "wave field"; apparently, it does not satisfy the scalar wave equation (16.34) consequently, the defect images in Figure 16.10 may not be considered

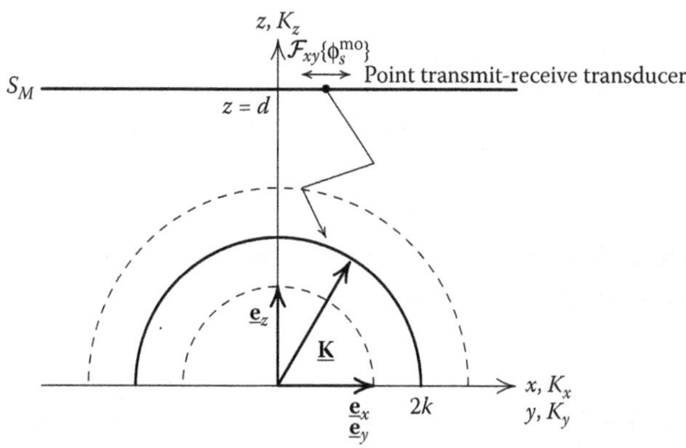

FIGURE 16.9
Illustration of multimonostatic FT-SAFT mapping.

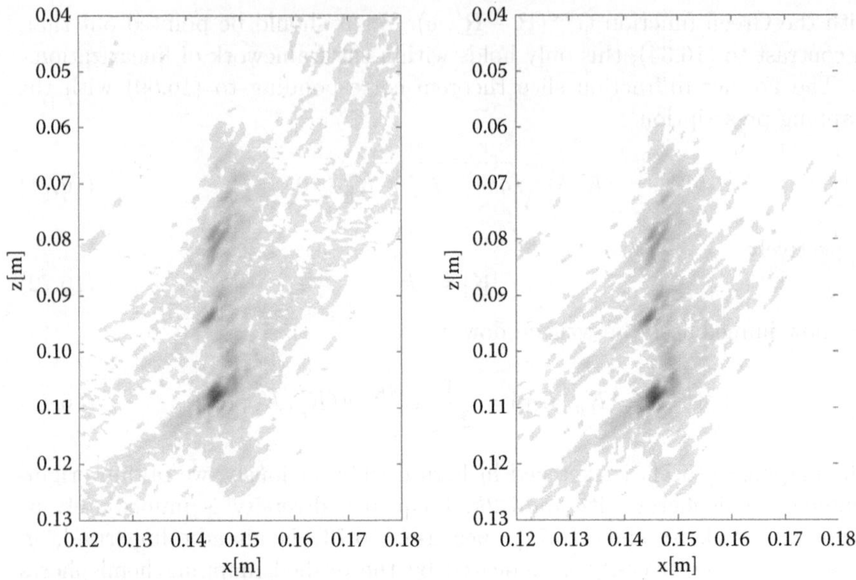

FIGURE 16.10
Pulse-echo SAFT- (left) and multimonostatic FT-SAFT imaging (right) of an actual crack for 45°-shear wave incidence.

to be defect reconstructions. A generalization of FT-SAFT to elastic waves utilizing mode conversion has also been formulated (Kostka et al. 1998; Langenberg et al. 2006, 2007; Zimmer 2007).

To what extent multimonostatic frequency diversity of FT-SAFT is in fact equivalent to the pulse-echo version of SAFT will be analytically shown in the next section. Based on simulations and analysis of experimental data, we could show that in fact slowness surfaces must be used for the mapping prescription within FT-SAFT in anisotropic materials in contrast to the utilization of wave (energy) surfaces within SAFT (Zimmer 2007); after all we superimpose spectra of plane waves and do not back propagate elementary waves as within SAFT.

16.2.5 Exact derivation of pulse-echo SAFT for planar measurement surfaces

The notation

$$\tilde{\chi}(K_x, K_y, K_z) = \frac{2}{\mathrm{j}} K_z \, \mathrm{e}^{-\mathrm{j}dK_z} \, \hat{\phi}_s^{\mathrm{mo}} \left(K_x, K_y, d, \omega = \frac{c}{2} \sqrt{K_x^2 + K_y^2 + K_z^2} \right) \tag{16.74}$$

of the monostatic Fourier diffraction slice theorem directly reveals that the contrast function $\chi(x, y, z)$ should be available with the help of a three-dimensional inverse Fourier transform with regard to K_x, K_y, K_z. Yet the mapping prescription (16.71) tells us—as illustrated in Figure 16.9—that only the upper K_z-half-space[323]—$K_z \geq 0$ is accessible with transformed data for $k \geq 0$, i.e., at most

$$\frac{1}{(2\pi)^3} \int_{-\infty}^{\infty} \int_{-\infty}^{\infty} \int_{-\infty}^{\infty} \tilde{\chi}(K_x, K_y, K_z)\, u(K_z)\, \mathrm{d}K_x \mathrm{d}K_y \mathrm{d}K_z \tag{16.75}$$

can be calculated with transformed data appending $u(K_z)$ as unit step function to $\tilde{\chi}(K_x, K_y, K_z)$ excluding the lower half-space. Yet, this three-dimensional inverse Fourier integral yields the Hilbert transform with regard to z besides the real valued contrast function $\chi(x, y, z)$

$$\begin{aligned} &\frac{1}{(2\pi)^3} \int_{-\infty}^{\infty} \int_{-\infty}^{\infty} \int_{-\infty}^{\infty} \tilde{\chi}(K_x, K_y, K_z)\, u(K_z)\, \mathrm{d}K_x \mathrm{d}K_y \mathrm{d}K_z \\ &\quad = \frac{1}{2} \left[\chi(x, y, z) - \mathrm{j}\mathcal{H}_\zeta\{\chi(x, y, \zeta)\}\right], \end{aligned} \tag{16.76}$$

considering (2.287) and (2.298), where the negative sign of the Hilbert transform is determined by the kernel of the spatial Fourier transform (Equations 2.329 and 2.330). Hence, $\chi(x, y, z)$ is found from (16.76) via magnitude calculation considering (16.74):

$$\begin{aligned} &\chi(x, y, z) \\ &\quad = \Re\Bigg\{ \frac{4}{\mathrm{j}(2\pi)^3} \int_{-\infty}^{\infty} \int_{-\infty}^{\infty} \int_{-\infty}^{\infty} K_z\, \mathrm{e}^{-\mathrm{j}K_z(d-z)}\, \hat{\phi}_s^{\mathrm{mo}} \\ &\qquad\qquad \times \left(K_x, K_y, d, \omega = \frac{c}{2}\sqrt{K_x^2 + K_y^2 + K_z^2}\right) \\ &\qquad\qquad \times\, u(K_z)\, \mathrm{e}^{\mathrm{j}K_x x + \mathrm{j}K_y y}\, \mathrm{d}K_x \mathrm{d}K_y \mathrm{d}K_z \Bigg\}. \end{aligned} \tag{16.77}$$

Comparison with (16.53)—there we have $\mathrm{e}^{\mathrm{j}(d-z)K_z}$ for $d > z$—shows that the resulting wave propagation from the scatterer to the measurement surface in (16.77)—we had $\mathrm{e}^{-\mathrm{j}(d-z)K_z}$—has been transformed in a back propagation from the measurement surface; hence, the SAFT idea is already recognizable. However, SAFT is a time domain algorithm, here the K_z-integration reflects frequency diversity suggesting to transform the K_z-Integral into a k-, respectively ω-integral, to interpret the latter as Fourier integral. Via the mapping prescription (16.71), we substitute

$$K_z\, \mathrm{d}K_z = 4k\, \mathrm{d}k, \quad k \geq 0, \tag{16.78}$$

[323]The lower half-space could be accessible with a second measurement surface at $z = -d$.

and obtain

$$\begin{aligned}\chi(x,y,z) &= \Re\Bigg\{\frac{16}{\mathrm{j}(2\pi)^3}\int_{-\infty}^{\infty} u(k)\,k\int_{-\infty}^{\infty}\int_{-\infty}^{\infty} \mathrm{e}^{-\mathrm{j}(d-z)\sqrt{4k^2-K_x^2-K_y^2}} \\ &\quad\times \hat{\phi}_s^{\mathrm{mo}}(K_x,K_y,d,\omega=ck)\,\mathrm{e}^{\mathrm{j}K_x x+\mathrm{j}K_y y}\,\mathrm{d}K_x\mathrm{d}K_y\;\mathrm{d}k\Bigg\}. \qquad (16.79)\end{aligned}$$

Now, it is time to critically inspect the K_xK_y-integrations, because $\sqrt{4k^2-K_x^2-K_y^2}$ becomes purely imaginary for $K_x^2+K_y^2>4k^2$, and only a proper sign choice of the complex square root determines convergence of the K_xK_y-integrals. This sign, namely $\Im\sqrt{4k^2-K_x^2-K_y^2}>0$, is already prescribed in the two-dimensional Fourier transform

$$\hat{G}^{\mathrm{mo}}(K_x,K_y,d-z,\omega) = \begin{cases}\dfrac{\mathrm{j}}{2\sqrt{4k^2-K_x^2-K_y^2}}\,\mathrm{e}^{\mathrm{j}(d-z)\sqrt{4k^2-K_x^2-K_y^2}} & \text{for } K_x^2+K_y^2\le 4k^2\\[2ex] \dfrac{1}{2\sqrt{K_x^2+K_y^2-4k^2}}\,\mathrm{e}^{-(d-z)\sqrt{K_x^2+K_y^2-4k^2}} & \text{for } K_x^2+K_y^2> 4k^2\end{cases} \qquad (16.80)$$

of the monostatic Green function ensuring the convergence of Weyl's integral representation, or, physically expressed, ensuring the evanescent waves emanating from the source at z—here: the scatterer—are in fact attenuated reaching the measurement surface at $d>z$. Yet, in (16.79), we find $\mathrm{e}^{-\mathrm{j}(d-z)\sqrt{4k^2-K_x^2-K_y^2}}$ leading to

$$\begin{array}{ll}\mathrm{e}^{-\mathrm{j}(d-z)\sqrt{4k^2-K_x^2-K_y^2}} & \text{for } K_x^2+K_y^2\le 4k^2,\\ \mathrm{e}^{(d-z)\sqrt{K_x^2+K_y^2-4k^2}} & \text{for } K_x^2+K_y^2> 4k^2\end{array} \qquad (16.81)$$

with the same sign choice of the imaginary part. Consequence: The K_xK_y-integrals in (16.79) would not converge! Yet, on the other side, we know (compare Figure 11.3) that $\hat{\phi}_s^{\mathrm{mo}}(K_x,K_y,d,\omega=ck)$ contains only "few" spectral components for $K_x^2+K_y^2>4k^2$ due to the attenuation of the evanescent waves allowing for an explicit articulation via a multiplication with a circular disc filter $u\left(2k-\sqrt{K_x^2+K_y^2}\right)$ in the K_xK_y-plane.[324] In the integral representation

[324] Within the $K_x\omega$-diagram of Figure 16.8, this filter—the radius of the circular disc increases linearly with frequency—is clearly visible.

$$\chi_\circ(\underline{\mathbf{R}}) = \Re\left\{\frac{16}{\mathrm{j}(2\pi)^3}\int_{-\infty}^{\infty} u(k)\,k \int_{-\infty}^{\infty}\int_{-\infty}^{\infty} u\left(2k-\sqrt{K_x^2+K_y^2}\right) \times \mathrm{e}^{-\mathrm{j}(d-z)\sqrt{4k^2-K_x^2-K_y^2}} \times \hat{\phi}_s^{\mathrm{mo}}(K_x,K_y,d,\omega=ck)\,\mathrm{e}^{\mathrm{j}K_x x+\mathrm{j}K_y y}\,\mathrm{d}K_x\mathrm{d}K_y\;\mathrm{d}k\right\}, \tag{16.82}$$

we do no longer face any convergence problems; however, as a consequence, we only obtain a filtered contrast function to be indicated by a respective index, i.e., the final loss of evanescent partial waves in $\hat{\phi}_s^{\mathrm{mo}}$ reduces the spatial resolution.[325] Obviously, we have

$$u\left(2k-\sqrt{K_x^2+K_y^2}\right)\,\mathrm{e}^{-\mathrm{j}(d-z)\sqrt{4k^2-K_x^2-K_y^2}} = 2\frac{\partial}{\partial z}\hat{G}^{\mathrm{mo}*}(K_x,K_y,d-z,\omega)\,u\left(2k-\sqrt{K_x^2+K_y^2}\right); \tag{16.83}$$

with the definition of the z-derivative of a circular disc bandlimited complex conjugate Green function $\hat{G}_\circ^{\mathrm{mo}*}(K_x,K_y,d-z,\omega)$ according to

$$\hat{G}_\circ^{\mathrm{mo}*}(K_x,K_y,d-z,\omega) = \hat{G}^{\mathrm{mo}*}(K_x,K_y,d-z,\omega)\,u\left(2k-\sqrt{K_x^2+K_y^2}\right) \tag{16.84}$$

K_xK_y-integrations in (16.82) may be written as a two-dimensional convolution integral:

$$\chi_\circ(x,y,z) = \Re\left\{\frac{32}{2\pi\mathrm{j}}\int_{-\infty}^{\infty} u(k)\,k \int_{-\infty}^{\infty}\int_{-\infty}^{\infty}\frac{\partial}{\partial z}G_\circ^{\mathrm{mo}*}(x-x',y-y',d-z,\omega=ck) \times \phi_s^{\mathrm{mo}}(x',y',d,\omega=ck)\,\mathrm{d}x'\mathrm{d}y'\;\mathrm{d}k\right\}, \tag{16.85}$$

where $G_\circ^{\mathrm{mo}*}(x,y,z,\omega)$, according to a correspondence given by Bracewell (1978), may be represented by the respective two-dimensional convolution integral

$$G_\circ^{\mathrm{mo}*}(x,y,d-z,\omega) = \frac{\mathrm{e}^{-2\mathrm{j}k\sqrt{x^2+y^2+(d-z)^2}}}{4\pi\sqrt{x^2+y^2+(d-z)^2}} \overset{xy}{**} \frac{k}{\pi}\,\frac{\mathrm{J}_1(2k\sqrt{x^2+y^2})}{\sqrt{x^2+y^2}}. \tag{16.86}$$

[325] To obtain a so-called super resolution, the partial evanescent waves have to be enhanced exponentially while back propagating from the measurement surface according to (16.81); yet this is very noise sensitive requiring special algorithmic care (Bertero and De Mol 1996).

Now, the first approximation on the way to the SAFT algorithm comes into play: For large values of $k\sqrt{x^2+y^2}$—the US-pulse technique is a high-frequency approximation—the Bessel function term is approximated by a δ-pulse with the consequence:

$$G_\circ^{\mathrm{mo}*}(x,y,d-z,\omega) \simeq \frac{\mathrm{e}^{-2\mathrm{j}k\sqrt{x^2+y^2+(d-z)^2}}}{4\pi\sqrt{x^2+y^2+(d-z)^2}}. \tag{16.87}$$

This approximation is of interest because in the time domain,[326] we have

$$\mathcal{F}_\omega^{-1}\{G_\circ^{\mathrm{mo}*}(x,y,d-z,\omega)\} \simeq \frac{\delta\left(t+\frac{2}{c}\sqrt{x^2+y^2+(d-z)^2}\right)}{4\pi\sqrt{x^2+y^2+(d-z)^2}}, \tag{16.88}$$

and the time domain is our final goal to establish SAFT mathematically. With the short-hand notation

$$\begin{aligned} F(x,y,z,\omega) \\ = \int_{-\infty}^{\infty}\int_{-\infty}^{\infty} \frac{\partial}{\partial z} G_\circ^{\mathrm{mo}*}(x-x',y-y',d-z,\omega)\phi_s^{\mathrm{mo}}(x',y',d,\omega)\,\mathrm{d}x'\mathrm{d}y', \end{aligned} \tag{16.89}$$

we write (16.85)

$$\chi_\circ(x,y,z) = \Re\left\{\frac{32}{c^2}\frac{1}{2\pi}\int_{-\infty}^{\infty} u(\omega)(-\mathrm{j}\omega)F(x,y,z,\omega)\,\mathrm{d}\omega\right\}; \tag{16.90}$$

now, we find an inverse Fourier integral with regard to ω for $t=0$ on the right-hand side; therefore, the following calculation

$$\begin{aligned} &\frac{1}{2\pi}\int_{-\infty}^{\infty} u(\omega)(-\mathrm{j}\omega)F(x,y,z,\omega)\,\mathrm{d}\omega \\ &= \frac{1}{2\pi}\int_{-\infty}^{\infty} u(\omega)(-\mathrm{j}\omega)F(x,y,z,\omega)\,\mathrm{e}^{-\mathrm{j}\omega t}\,\mathrm{d}\omega\Bigg|_{t=0} \\ &= \mathcal{F}_\omega^{-1}\{u(\omega)(-\mathrm{j}\omega)F(x,y,z,\omega)\}\Big|_{t=0} \\ &= \frac{\partial}{\partial t}f(x,y,z,t) * \left[\frac{1}{2}\delta(t) - \frac{\mathrm{j}}{2\pi}\,\mathrm{pf}\frac{1}{t}\right]\Bigg|_{t=0} \\ &= \frac{1}{2}\left[\frac{\partial}{\partial t}f(x,y,z,t) + \mathrm{j}\mathcal{H}_\tau\left\{\frac{\partial}{\partial \tau}f(x,y,z,\tau)\right\}\right]\Bigg|_{t=0} \end{aligned} \tag{16.91}$$

---

[326] That way, we approximate the circular disc bandlimited Green function by the Green function itself; yet, the latter contains those (evanescent) spectral components that we got rid of beforehand. In fact, the Green function is only complete with these components (Tygel and Hubral 1987).

yields in fact a time domain algorithm:

$$\chi_{\circ}(x,y,z) = \Re\left\{\frac{16}{c^2}\left[\frac{\partial}{\partial t}f(x,y,z,t) + \mathrm{j}\mathcal{H}_{\tau}\left\{\frac{\partial}{\partial \tau}f(x,y,z,\tau)\right\}\right]\Bigg|_{t=0}\right\}$$
$$= \frac{16}{c^2}\frac{\partial}{\partial t}f(x,y,z,t)\Bigg|_{t=0} . \qquad (16.92)$$

It becomes intuitive if we explicitly calculate

$$f(x,y,z,t) \simeq \int_{-\infty}^{\infty}\int_{-\infty}^{\infty}\frac{\partial}{\partial z}\frac{\delta\left(t+\frac{2}{c}\sqrt{(x-x')^2+(y-y')^2+(d-z)^2}\right)}{4\pi\sqrt{(x-x')^2+(y-y')^2+(d-z)^2}}$$
$$\overset{t}{*}\ \phi_s^{\mathrm{mo}}(x',y',d,t)\,\mathrm{d}x'\mathrm{d}y'$$
$$\simeq \frac{1}{4\pi}\int_{-\infty}^{\infty}\int_{-\infty}^{\infty}\frac{d-z}{(x-x')^2+(y-y')^2+(d-z)^2}$$
$$\times\left[\frac{\delta\left(t+\frac{2}{c}\sqrt{(x-x')^2+(y-y')^2+(d-z)^2}\right)}{\sqrt{(x-x')^2+(y-y')^2+(d-z)^2}}\right.$$
$$\left.-\frac{2}{c}\frac{\partial}{\partial t}\delta\left(t+\frac{2}{c}\sqrt{(x-x')^2+(y-y')^2+(d-z)^2}\right)\right]$$
$$\overset{t}{*}\ \phi_s^{\mathrm{mo}}(x',y',d,t)\,\mathrm{d}x'\mathrm{d}y' \qquad (16.93)$$

with the help of (16.88) based on (16.89). Due to the relations $\delta(t+t_0)*f(t) = f(t+t_0)$, $\delta'(t+t_0)*f(t) = f'(t+t_0)$, it follows

$$f(x,y,z,t) \simeq \frac{1}{4\pi}\int_{-\infty}^{\infty}\int_{-\infty}^{\infty}\frac{d-z}{(x-x')^2+(y-y')^2+(d-z)^2}$$
$$\times\left[\frac{\phi_s^{\mathrm{mo}}(x',y',d,t')}{\sqrt{(x-x')^2+(y-y')^2+(d-z)^2}}\right.$$
$$\left.-\frac{2}{c}\frac{\partial}{\partial t'}\phi_s^{\mathrm{mo}}(x',y',d,t')\right]\Bigg|_{t'=t+\frac{2}{c}\sqrt{(x-x')^2+(y-y')^2+(d-z)^2}}\mathrm{d}x'\mathrm{d}y'$$
$$(16.94)$$

finally resulting in the time domain algorithm

$$\chi_{\circ}(x,y,z) \simeq \frac{4}{\pi c^2}\int_{-\infty}^{\infty}\int_{-\infty}^{\infty}\frac{d-z}{(x-x')^2+(y-y')^2+(d-z)^2}\times$$
$$\times\left[\frac{\frac{\partial}{\partial t'}\phi_s^{\mathrm{mo}}(x',y',d,t')}{\sqrt{(x-x')^2+(y-y')^2+(d-z)^2}}-\right.$$
$$\left.-\frac{2}{c}\frac{\partial^2}{\partial t'^2}\phi_s^{\mathrm{mo}}(x',y',d,t')\right]\Bigg|_{t'=\frac{2}{c}\sqrt{(x-x')^2+(y-y')^2+(d-z)^2}}\mathrm{d}x'\mathrm{d}y'$$
$$(16.95)$$

with (16.92) to retrieve the contrast function. The $\simeq$-character and the o-index in (16.95) illustrate that the contrast function is not "reconstructed." In addition, the inverse problem has been linearized with Born's approximation, and last but not least, formula (16.95) is based on the model of acoustic wave propagation without density contrast. Furthermore, the upper index "mo" on the data declares the deconvolution with $\phi_0(\omega)$ of the actual monostatic data according to (16.68), where $\phi_0(\omega)$ is not even the spectrum of the exciting pulse, but it also contains the factor ω^2. Hence, it is surely legitimate to simply define an "image" of the scatterer

$$o(x, y, z) = \int_{-\infty}^{\infty}\int_{-\infty}^{\infty} \phi_s^{\mathrm{m}}\left(x', y', d, t = \frac{2}{c}\sqrt{(x-x')^2 + (y-y')^2 + (d-z)^2}\right) \mathrm{d}x'\mathrm{d}y', \tag{16.96}$$

while keeping the idea of time domain back propagation according to (16.95) that is nothing more than the result of the pulse-echo SAFT algorithm according to (16.12)! However, we know now that this algorithm has to satisfy well defined assumptions.

The theoretical considerations of this section additionally yield another result taking not the real part but the imaginary part of (16.76), and considering (16.91):

$$\mathcal{H}_\zeta\{\chi_\mathrm{o}(x, y, \zeta)\} = -\frac{16}{c^2}\mathcal{H}_\tau\left\{\frac{\partial}{\partial\tau} f(x, y, z, \tau)\right\}\bigg|_{t=0}. \tag{16.97}$$

For planar measurement surfaces, the back propagation of Hilbert-transformed data is in fact equivalent to the Hilbert transform of the beforehand calculated SAFT image with respect to the ζ-(depth)coordinate. In case of SAFT processing of spectrally bandlimited data, namely data without a (complete) deconvolution, the oscillations occurring in the image may be deleted either before or after back propagation:

$$|\chi_\mathrm{o}(x, y, z) - \mathcal{H}_\zeta\{\chi_\mathrm{o}(x, y, \zeta)\}| = \frac{16}{c^2}\left|\frac{\partial}{\partial t} f(x, y, z, t) + \mathrm{j}\mathcal{H}_\tau\left\{\frac{\partial}{\partial\tau} f(x, y, z, \tau)\right\}\right|_{t=0}. \tag{16.98}$$

Note: Back propagation of the magnitude of data being complex complemented by a Hilbert transform does not yield the same result!

The explicit Hilbert transform—either with regard to ζ or τ (depth coordinate or time)—may be avoided taking the magnitude of the anyway complex valued FT-SAFT result.

Utilizing the generalized holographic field allows for the derivation of an exact time domain SAFT algorithm even for arbitrary geometries of measurement surfaces (Langenberg 1987).

Appendix

Collection of Mathematical Definitions and Identities

The following formula collection for vector/tensor algebra and analysis especially contains the substantial collection of Chen (1993) and the identities published by van Bladel (1985) and Ben-Menahem and Singh (1981), various others have been contributed by ourselves together with some corrections in the collections we referred to; hopefully no new errors have entered.

A.1 Vector Identities

$$
\begin{aligned}
\underline{\mathbf{A}} \cdot (\underline{\mathbf{B}} \times \underline{\mathbf{C}}) &= \underline{\mathbf{C}} \cdot (\underline{\mathbf{A}} \times \underline{\mathbf{B}}) \\
&= \underline{\mathbf{B}} \cdot (\underline{\mathbf{C}} \times \underline{\mathbf{A}}) \\
&= [\underline{\mathbf{A}}\,\underline{\mathbf{B}}\,\underline{\mathbf{C}}] \quad \text{(triple product)} \\
&= [\underline{\mathbf{C}}\,\underline{\mathbf{A}}\,\underline{\mathbf{B}}] \\
&= [\underline{\mathbf{B}}\,\underline{\mathbf{C}}\,\underline{\mathbf{A}}]
\end{aligned}
$$

$$
\begin{aligned}
\underline{\mathbf{A}} \times \underline{\mathbf{B}} &= -\underline{\mathbf{B}} \times \underline{\mathbf{A}} \\
&= (\underline{\underline{\mathbf{I}}} \times \underline{\mathbf{A}}) \cdot \underline{\mathbf{B}} \\
&= \underline{\mathbf{A}} \cdot (\underline{\underline{\mathbf{I}}} \times \underline{\mathbf{B}})
\end{aligned}
$$

$$
\begin{aligned}
\underline{\mathbf{A}} \times (\underline{\mathbf{B}} \times \underline{\mathbf{C}}) &= \underline{\mathbf{B}}(\underline{\mathbf{A}} \cdot \underline{\mathbf{C}}) - \underline{\mathbf{C}}(\underline{\mathbf{A}} \cdot \underline{\mathbf{B}}) \\
&= \underline{\mathbf{B}}\,\underline{\mathbf{A}} \cdot \underline{\mathbf{C}} - \underline{\mathbf{C}}\,\underline{\mathbf{A}} \cdot \underline{\mathbf{B}} \\
&= (\underline{\mathbf{B}}\,\underline{\mathbf{C}} - \underline{\mathbf{C}}\,\underline{\mathbf{B}}) \cdot \underline{\mathbf{A}}
\end{aligned}
$$

$$
\begin{aligned}
(\underline{\mathbf{A}} \times \underline{\mathbf{B}}) \times \underline{\mathbf{C}} &= \underline{\mathbf{B}}(\underline{\mathbf{C}} \cdot \underline{\mathbf{A}}) - \underline{\mathbf{A}}(\underline{\mathbf{C}} \cdot \underline{\mathbf{B}}) \\
&= (\underline{\mathbf{B}}\,\underline{\mathbf{A}} - \underline{\mathbf{A}}\,\underline{\mathbf{B}}) \cdot \underline{\mathbf{C}}
\end{aligned}
$$

$$
\begin{aligned}
\underline{\mathbf{A}} \times (\underline{\mathbf{B}} \times \underline{\mathbf{C}}) - (\underline{\mathbf{A}} \times \underline{\mathbf{B}}) \times \underline{\mathbf{C}} &= \underline{\mathbf{B}} \times (\underline{\mathbf{A}} \times \underline{\mathbf{C}}) \\
&= \underline{\mathbf{A}}(\underline{\mathbf{C}} \cdot \underline{\mathbf{B}}) - \underline{\mathbf{C}}(\underline{\mathbf{A}} \cdot \underline{\mathbf{B}})
\end{aligned}
$$

$$(\underline{\mathbf{A}} \times \underline{\mathbf{B}}) \cdot (\underline{\mathbf{C}} \times \underline{\mathbf{D}}) = (\underline{\mathbf{A}} \cdot \underline{\mathbf{C}})(\underline{\mathbf{B}} \cdot \underline{\mathbf{D}}) - (\underline{\mathbf{A}} \cdot \underline{\mathbf{D}})(\underline{\mathbf{B}} \cdot \underline{\mathbf{C}})$$
$$= \underline{\mathbf{A}} \cdot [\underline{\mathbf{B}} \times (\underline{\mathbf{C}} \times \underline{\mathbf{D}})]$$

$$(\underline{\mathbf{A}} \times \underline{\mathbf{B}}) \times (\underline{\mathbf{C}} \times \underline{\mathbf{D}}) = [(\underline{\mathbf{A}} \times \underline{\mathbf{B}}) \cdot \underline{\mathbf{D}}]\underline{\mathbf{C}} - [(\underline{\mathbf{A}} \times \underline{\mathbf{B}}) \cdot \underline{\mathbf{C}}]\underline{\mathbf{D}}$$
$$= [\underline{\mathbf{A}}\,\underline{\mathbf{B}}\,\underline{\mathbf{D}}]\underline{\mathbf{C}} - [\underline{\mathbf{A}}\,\underline{\mathbf{B}}\,\underline{\mathbf{C}}]\underline{\mathbf{D}}$$

$$\underline{\mathbf{A}} \times [\underline{\mathbf{B}} \times (\underline{\mathbf{C}} \times \underline{\mathbf{D}})] = (\underline{\mathbf{B}} \cdot \underline{\mathbf{D}})(\underline{\mathbf{A}} \times \underline{\mathbf{C}}) - (\underline{\mathbf{B}} \cdot \underline{\mathbf{C}})(\underline{\mathbf{A}} \times \underline{\mathbf{D}})$$
$$(\underline{\mathbf{A}} \times \underline{\mathbf{B}}) \cdot [(\underline{\mathbf{B}} \times \underline{\mathbf{C}}) \times (\underline{\mathbf{C}} \times \underline{\mathbf{A}})] = [\underline{\mathbf{A}} \cdot (\underline{\mathbf{B}} \times \underline{\mathbf{C}})]^2.$$

The following double products are defined according to Ben-Menahem and Singh (1981); van Bladel (1985), Lindell (1992), or Gibbs (1913) using different definitions.

$$(\underline{\mathbf{A}}\,\underline{\mathbf{B}}) : (\underline{\mathbf{C}}\,\underline{\mathbf{D}}) = (\underline{\mathbf{A}} \cdot \underline{\mathbf{D}})(\underline{\mathbf{B}} \cdot \underline{\mathbf{C}})$$
$$(\underline{\mathbf{A}}\,\underline{\mathbf{B}}) \overset{\times}{\times} (\underline{\mathbf{C}}\,\underline{\mathbf{D}}) = (\underline{\mathbf{A}} \times \underline{\mathbf{D}})(\underline{\mathbf{B}} \times \underline{\mathbf{C}})$$
$$(\underline{\mathbf{A}}\,\underline{\mathbf{B}}) \overset{\cdot}{\times} (\underline{\mathbf{C}}\,\underline{\mathbf{D}}) = (\underline{\mathbf{A}} \times \underline{\mathbf{D}})(\underline{\mathbf{B}} \cdot \underline{\mathbf{C}})$$
$$(\underline{\mathbf{A}}\,\underline{\mathbf{B}}) \overset{\times}{\cdot} (\underline{\mathbf{C}}\,\underline{\mathbf{D}}) = (\underline{\mathbf{A}} \cdot \underline{\mathbf{D}})(\underline{\mathbf{B}} \times \underline{\mathbf{C}}).$$

A.2 Tensor Identities

Summation convention: If an index on one side of an equation appears twice or multiply (and not on the other side), it is subject to a summation from 1 to 3; no summation is involved if the index also appears on the other side.

A.2.1 Permutation tensor

$$\underline{\underline{\underline{\boldsymbol{\varepsilon}}}} = \sum_{i=1}^{3}\sum_{j=1}^{3}\sum_{k=1}^{3} \varepsilon_{ijk}\underline{\mathbf{e}}_{x_i}\underline{\mathbf{e}}_{x_j}\underline{\mathbf{e}}_{x_k}$$
$$= \varepsilon_{ijk}\underline{\mathbf{e}}_{x_i}\underline{\mathbf{e}}_{x_j}\underline{\mathbf{e}}_{x_k} \quad \text{(summation convention)}$$
$$= \underline{\underline{\boldsymbol{\varepsilon}}}_i\underline{\mathbf{e}}_{x_i} \quad \text{(summation convention)}$$

$$\varepsilon_{ijk} = \begin{cases} 0 & \text{, if two indices are equal} \\ 1 & \text{, if } ijk \text{ is an even permutation of 123 (e.g., 231)} \\ -1 & \text{, if } ijk \text{ is an odd permutation of 123 (e.g., 213)} \end{cases}$$

$$\underline{\underline{\boldsymbol{\varepsilon}}}_1 = \begin{pmatrix} 0 & 0 & 0 \\ 0 & 0 & 1 \\ 0 & -1 & 0 \end{pmatrix}$$

$$\underline{\underline{\boldsymbol{\varepsilon}}}_2 = \begin{pmatrix} 0 & 0 & -1 \\ 0 & 0 & 0 \\ 1 & 0 & 0 \end{pmatrix}$$

$$\underline{\underline{\boldsymbol{\varepsilon}}}_3 = \begin{pmatrix} 0 & 1 & 0 \\ -1 & 0 & 0 \\ 0 & 0 & 0 \end{pmatrix}$$

$$\underline{\underline{\underline{\boldsymbol{\varepsilon}}}}^{213} = -\underline{\underline{\underline{\boldsymbol{\varepsilon}}}}$$
$$\underline{\underline{\underline{\boldsymbol{\varepsilon}}}}^{132} = -\underline{\underline{\underline{\boldsymbol{\varepsilon}}}}$$
$$\underline{\underline{\underline{\boldsymbol{\varepsilon}}}}^{321} = -\underline{\underline{\underline{\boldsymbol{\varepsilon}}}}$$
$$\underline{\underline{\underline{\boldsymbol{\varepsilon}}}}^{312} = \underline{\underline{\underline{\boldsymbol{\varepsilon}}}}$$
$$\underline{\underline{\underline{\boldsymbol{\varepsilon}}}}^{231} = \underline{\underline{\underline{\boldsymbol{\varepsilon}}}}$$

$$\underline{\mathbf{e}}_{x_i} \times \underline{\mathbf{e}}_{x_j} = \varepsilon_{kij}\underline{\mathbf{e}}_{x_k}\ ,\ i, j = 1, 2, 3$$
$$\varepsilon_{kij} = \underline{\mathbf{e}}_{x_k} \cdot (\underline{\mathbf{e}}_{x_i} \times \underline{\mathbf{e}}_{x_j})$$

$$\begin{aligned}\underline{\underline{\underline{\boldsymbol{\varepsilon}}}} \cdot \underline{\underline{\underline{\boldsymbol{\varepsilon}}}} &= \varepsilon_{ijk}\varepsilon_{klm}\underline{\mathbf{e}}_{x_i}\underline{\mathbf{e}}_{x_j}\underline{\mathbf{e}}_{x_l}\underline{\mathbf{e}}_{x_m} \\ &= (\delta_{il}\delta_{jm} - \delta_{im}\delta_{jl})\underline{\mathbf{e}}_{x_i}\underline{\mathbf{e}}_{x_j}\underline{\mathbf{e}}_{x_l}\underline{\mathbf{e}}_{x_m} \\ &= \underline{\underline{\mathbf{I}}}\,\underline{\underline{\mathbf{I}}}^{1324} - \underline{\underline{\mathbf{I}}}\,\underline{\underline{\mathbf{I}}}^{1342}\end{aligned}$$

$$\underline{\underline{\underline{\boldsymbol{\varepsilon}}}} : \underline{\underline{\underline{\boldsymbol{\varepsilon}}}} = -2\underline{\underline{\mathbf{I}}}$$
$$\underline{\underline{\underline{\boldsymbol{\varepsilon}}}} \,\vdots\, \underline{\underline{\underline{\boldsymbol{\varepsilon}}}} = -6$$
$$\underline{\underline{\underline{\boldsymbol{\varepsilon}}}} : \underline{\mathbf{B}}\,\underline{\mathbf{A}} = \underline{\mathbf{A}} \times \underline{\mathbf{B}}$$
$$\underline{\underline{\underline{\boldsymbol{\varepsilon}}}} : \underline{\underline{\mathbf{D}}}^{21} = \langle \underline{\underline{\mathbf{D}}} \rangle$$
$$\underline{\mathbf{A}} \times \underline{\mathbf{B}} = \langle \underline{\mathbf{A}}\,\underline{\mathbf{B}} \rangle$$
$$\langle \underline{\underline{\mathbf{D}}}^{21} \rangle = -\langle \underline{\underline{\mathbf{D}}} \rangle$$
$$\langle \underline{\underline{\mathbf{D}}} \rangle = \underline{\mathbf{0}} \text{ if } \underline{\underline{\mathbf{D}}} \text{ is symmetric}$$

$$\begin{aligned}\underline{\underline{\mathbf{I}}} &= \sum_{i=1}^{3}\sum_{j=1}^{3} \delta_{ij}\underline{\mathbf{e}}_{x_i}\underline{\mathbf{e}}_{x_j} \\ &= \delta_{ij}\underline{\mathbf{e}}_{x_i}\underline{\mathbf{e}}_{x_j} \quad \text{(summation convention)} \\ &= \sum_{i=1}^{3} \underline{\mathbf{e}}_{x_i}\underline{\mathbf{e}}_{x_i} \\ &= \underline{\mathbf{e}}_{x_i}\underline{\mathbf{e}}_{x_i} \quad \text{(summation convention)}\end{aligned}$$

$$\underline{\underline{\mathbf{I}}} \cdot \underline{\mathbf{A}} = \underline{\mathbf{A}} \cdot \underline{\underline{\mathbf{I}}} = \underline{\mathbf{A}}$$
$$\underline{\underline{\mathbf{I}}} \cdot \underline{\underline{\mathbf{D}}} = \underline{\underline{\mathbf{D}}} \cdot \underline{\underline{\mathbf{I}}} = \underline{\underline{\mathbf{D}}}$$
$$\underline{\underline{\mathbf{I}}} : \underline{\underline{\mathbf{D}}} = \underline{\underline{\mathbf{D}}} : \underline{\underline{\mathbf{I}}} = \text{trace}\,\underline{\underline{\mathbf{D}}}$$
$$\underline{\underline{\mathbf{I}}} : \underline{\underline{\mathbf{I}}} = 3$$

$$
\begin{aligned}
\underline{\underline{\underline{\underline{\mathbf{I}}}}}^{\delta} = \underline{\underline{\mathbf{I}}}\,\underline{\underline{\mathbf{I}}} &= \delta_{ij}\underline{\mathbf{e}}_{x_i}\underline{\mathbf{e}}_{x_j}\delta_{kl}\underline{\mathbf{e}}_{x_k}\underline{\mathbf{e}}_{x_l}\\
&= \underline{\mathbf{e}}_{x_i}\underline{\mathbf{e}}_{x_i}\underline{\mathbf{e}}_{x_k}\underline{\mathbf{e}}_{x_k}
\end{aligned}
$$

$$
\underline{\underline{\underline{\underline{\mathbf{I}}}}}^{\delta} : \underline{\underline{\mathbf{D}}} = \underline{\underline{\mathbf{D}}} : \underline{\underline{\underline{\underline{\mathbf{I}}}}}^{\delta} = \underline{\underline{\mathbf{I}}}\,\mathrm{trace}\,\underline{\underline{\mathbf{D}}}
$$

$$
\begin{aligned}
\underline{\underline{\mathbf{I}}}\,\underline{\underline{\mathbf{I}}}^{1324} &= \delta_{ij}\delta_{kl}\underline{\mathbf{e}}_{x_i}\underline{\mathbf{e}}_{x_k}\underline{\mathbf{e}}_{x_j}\underline{\mathbf{e}}_{x_l}\\
&= \delta_{ik}\delta_{jl}\underline{\mathbf{e}}_{x_i}\underline{\mathbf{e}}_{x_j}\underline{\mathbf{e}}_{x_k}\underline{\mathbf{e}}_{x_l}\\
&= \underline{\mathbf{e}}_{x_i}\underline{\mathbf{e}}_{x_j}\underline{\mathbf{e}}_{x_i}\underline{\mathbf{e}}_{x_j}
\end{aligned}
$$

$$
\begin{aligned}
\underline{\underline{\mathbf{I}}}\,\underline{\underline{\mathbf{I}}}^{1342} &= \delta_{ij}\delta_{kl}\underline{\mathbf{e}}_{x_i}\underline{\mathbf{e}}_{x_k}\underline{\mathbf{e}}_{x_l}\underline{\mathbf{e}}_{x_j}\\
&= \delta_{il}\delta_{jk}\underline{\mathbf{e}}_{x_i}\underline{\mathbf{e}}_{x_j}\underline{\mathbf{e}}_{x_k}\underline{\mathbf{e}}_{x_l}\\
&= \underline{\mathbf{e}}_{x_i}\underline{\mathbf{e}}_{x_j}\underline{\mathbf{e}}_{x_j}\underline{\mathbf{e}}_{x_i}
\end{aligned}
$$

$$
\begin{aligned}
\underline{\underline{\mathbf{I}}}\,\underline{\underline{\mathbf{I}}}^{1324} : \underline{\underline{\mathbf{D}}} &= \underline{\underline{\mathbf{D}}} : \underline{\underline{\mathbf{I}}}\,\underline{\underline{\mathbf{I}}}^{1324} = \underline{\underline{\mathbf{D}}}^{21}\\
\underline{\underline{\mathbf{I}}}\,\underline{\underline{\mathbf{I}}}^{1342} : \underline{\underline{\mathbf{D}}} &= \underline{\underline{\mathbf{D}}} : \underline{\underline{\mathbf{I}}}\,\underline{\underline{\mathbf{I}}}^{1342} = \underline{\underline{\mathbf{D}}}\\
\underline{\underline{\underline{\underline{\mathbf{I}}}}}^{+} &= \frac{1}{2}\left(\underline{\underline{\mathbf{I}}}\,\underline{\underline{\mathbf{I}}}^{1342} + \underline{\underline{\mathbf{I}}}\,\underline{\underline{\mathbf{I}}}^{1324}\right)\\
\underline{\underline{\underline{\underline{\mathbf{I}}}}}^{-} &= \frac{1}{2}\left(\underline{\underline{\mathbf{I}}}\,\underline{\underline{\mathbf{I}}}^{1342} - \underline{\underline{\mathbf{I}}}\,\underline{\underline{\mathbf{I}}}^{1324}\right)\\
\underline{\underline{\mathbf{D}}}_{\mathrm{s}} &= \underline{\underline{\underline{\underline{\mathbf{I}}}}}^{+} : \underline{\underline{\mathbf{D}}} = \underline{\underline{\mathbf{D}}} : \underline{\underline{\underline{\underline{\mathbf{I}}}}}^{+}\\
\underline{\underline{\mathbf{D}}}_{\mathrm{a}} &= \underline{\underline{\underline{\underline{\mathbf{I}}}}}^{-} : \underline{\underline{\mathbf{D}}} = \underline{\underline{\mathbf{D}}} : \underline{\underline{\underline{\underline{\mathbf{I}}}}}^{-}\\
\underline{\underline{\mathbf{I}}}\,\underline{\underline{\mathbf{I}}}^{1342} : \underline{\underline{\mathbf{I}}}\,\underline{\underline{\mathbf{I}}}^{1342} &= \underline{\underline{\mathbf{I}}}\,\underline{\underline{\mathbf{I}}}^{1342}\\
\underline{\underline{\mathbf{I}}}\,\underline{\underline{\mathbf{I}}}^{1324} : \underline{\underline{\mathbf{I}}}\,\underline{\underline{\mathbf{I}}}^{1324} &= \underline{\underline{\mathbf{I}}}\,\underline{\underline{\mathbf{I}}}^{1342}\\
\underline{\underline{\mathbf{I}}}\,\underline{\underline{\mathbf{I}}}^{1342} : \underline{\underline{\mathbf{I}}}\,\underline{\underline{\mathbf{I}}}^{1324} &= \underline{\underline{\mathbf{I}}}\,\underline{\underline{\mathbf{I}}}^{1324}\\
\underline{\underline{\mathbf{I}}}\,\underline{\underline{\mathbf{I}}}^{1324} : \underline{\underline{\mathbf{I}}}\,\underline{\underline{\mathbf{I}}}^{1342} &= \underline{\underline{\mathbf{I}}}\,\underline{\underline{\mathbf{I}}}^{1324}.
\end{aligned}
$$

A.2.2 Products

$$
\begin{aligned}
(\underline{\mathbf{A}} \cdot \underline{\underline{\mathbf{D}}}) \cdot \underline{\underline{\mathbf{B}}} &= \underline{\mathbf{A}} \cdot (\underline{\underline{\mathbf{D}}} \cdot \underline{\underline{\mathbf{B}}})\\
&= \underline{\mathbf{A}} \cdot \underline{\underline{\mathbf{D}}} \cdot \underline{\underline{\mathbf{B}}}
\end{aligned}
$$

$$
\begin{aligned}
(\underline{\underline{\mathbf{D}}} \cdot \underline{\underline{\mathbf{B}}}) \cdot \underline{\mathbf{A}} &= \underline{\underline{\mathbf{D}}} \cdot (\underline{\underline{\mathbf{B}}} \cdot \underline{\mathbf{A}})\\
&= \underline{\underline{\mathbf{D}}} \cdot \underline{\underline{\mathbf{B}}} \cdot \underline{\mathbf{A}}
\end{aligned}
$$

$$
\begin{aligned}
(\underline{\mathbf{A}} \times \underline{\underline{\mathbf{D}}}) \cdot \underline{\underline{\mathbf{B}}} &= \underline{\mathbf{A}} \times (\underline{\underline{\mathbf{D}}} \cdot \underline{\underline{\mathbf{B}}})\\
&= \underline{\mathbf{A}} \times \underline{\underline{\mathbf{D}}} \cdot \underline{\underline{\mathbf{B}}}
\end{aligned}
$$

$$\begin{aligned}(\underline{\underline{\mathbf{D}}} \cdot \underline{\underline{\mathbf{B}}}) \times \underline{\mathbf{A}} &= \underline{\underline{\mathbf{D}}} \cdot (\underline{\underline{\mathbf{B}}} \times \underline{\mathbf{A}}) \\ &= \underline{\underline{\mathbf{D}}} \cdot \underline{\underline{\mathbf{B}}} \times \underline{\mathbf{A}}\end{aligned}$$

$$\underline{\mathbf{A}}\,\underline{\mathbf{B}}^{21} = \underline{\mathbf{B}}\,\underline{\mathbf{A}}$$

$$\underline{\mathbf{A}} \cdot \underline{\underline{\mathbf{D}}} = \underline{\underline{\mathbf{D}}}^{21} \cdot \underline{\mathbf{A}}$$

$$\underline{\mathbf{A}} \cdot \underline{\underline{\mathbf{D}}} \cdot \underline{\mathbf{B}} = \underline{\mathbf{B}} \cdot \underline{\underline{\mathbf{D}}}^{21} \cdot \underline{\mathbf{A}}$$

$$(\underline{\underline{\mathbf{B}}} \cdot \underline{\underline{\mathbf{A}}})^{21} = \underline{\underline{\mathbf{A}}}^{21} \cdot \underline{\underline{\mathbf{B}}}^{21}$$

$$(\underline{\underline{\mathbf{A}}} \cdot \underline{\underline{\mathbf{B}}} \cdot \underline{\underline{\mathbf{C}}})^{21} = \underline{\underline{\mathbf{C}}}^{21} \cdot \underline{\underline{\mathbf{B}}}^{21} \cdot \underline{\underline{\mathbf{A}}}^{21}$$

$$\underline{\mathbf{A}} \times \underline{\underline{\mathbf{I}}} = \underline{\underline{\mathbf{I}}} \times \underline{\mathbf{A}}$$

$$(\underline{\mathbf{A}} \times \underline{\underline{\mathbf{I}}})^{21} = -\underline{\mathbf{A}} \times \underline{\underline{\mathbf{I}}}$$

$$\underline{\underline{\mathbf{D}}} = \underline{\mathbf{A}} \times \underline{\underline{\mathbf{I}}} \text{ (most general antisymmetric tensor)}$$

$$\text{with } \underline{\mathbf{A}} = \frac{1}{2}\underline{\underline{\underline{\boldsymbol{\varepsilon}}}} : \underline{\underline{\mathbf{D}}} = \frac{1}{2}\langle\underline{\underline{\mathbf{D}}}^{21}\rangle = -\frac{1}{2}\langle\underline{\underline{\mathbf{D}}}\rangle$$

$$\begin{aligned}\underline{\mathbf{B}} \times \underline{\mathbf{A}} &= \underline{\mathbf{B}} \cdot (\underline{\mathbf{A}} \times \underline{\underline{\mathbf{I}}}) \\ &= \underline{\mathbf{B}} \cdot (\underline{\underline{\mathbf{I}}} \times \underline{\mathbf{A}})\end{aligned}$$

$$\begin{aligned}\underline{\mathbf{A}} \times \underline{\mathbf{B}} &= (\underline{\underline{\mathbf{I}}} \times \underline{\mathbf{A}}) \cdot \underline{\mathbf{B}} \\ &= -\underline{\mathbf{B}} \cdot (\underline{\underline{\mathbf{I}}} \times \underline{\mathbf{A}}) \\ &= \underline{\underline{\underline{\boldsymbol{\varepsilon}}}} : \underline{\mathbf{B}}\,\underline{\mathbf{A}} \\ &= -\underline{\mathbf{A}} \cdot \underline{\underline{\underline{\boldsymbol{\varepsilon}}}} \cdot \underline{\mathbf{B}}\end{aligned}$$

$$\begin{aligned}\underline{\mathbf{A}} \times \underline{\underline{\mathbf{D}}} &= (\underline{\mathbf{A}} \times \underline{\underline{\mathbf{I}}}) \cdot \underline{\underline{\mathbf{D}}} \\ &= (\underline{\underline{\mathbf{I}}} \times \underline{\mathbf{A}}) \cdot \underline{\underline{\mathbf{D}}} \\ &= \underline{\underline{\underline{\boldsymbol{\varepsilon}}}} : (\underline{\underline{\mathbf{D}}}\,\underline{\mathbf{A}})^{132} \\ &= -\left(\underline{\underline{\mathbf{D}}}^{21} \times \underline{\mathbf{A}}\right)^{21} \\ &= -\underline{\mathbf{A}} \cdot \underline{\underline{\underline{\boldsymbol{\varepsilon}}}} \cdot \underline{\underline{\mathbf{D}}}\end{aligned}$$

$$\underline{\mathbf{A}} \times \underline{\mathbf{B}}\,\underline{\mathbf{C}} = (\underline{\mathbf{A}} \times \underline{\mathbf{B}})\underline{\mathbf{C}}$$

$$\begin{aligned}\underline{\underline{\mathbf{D}}} \times \underline{\mathbf{A}} &= \underline{\underline{\mathbf{D}}} \cdot (\underline{\underline{\mathbf{I}}} \times \underline{\mathbf{A}}) \\ &= \underline{\underline{\mathbf{D}}} \cdot (\underline{\mathbf{A}} \times \underline{\underline{\mathbf{I}}}) \\ &= -[\underline{\underline{\underline{\boldsymbol{\varepsilon}}}} : (\underline{\underline{\mathbf{D}}}\,\underline{\mathbf{A}})^{231}]^{21} \\ &= -(\underline{\mathbf{A}} \times \underline{\underline{\mathbf{D}}}^{21})^{21}\end{aligned}$$

$$\underline{\mathbf{B}}\,\underline{\mathbf{C}} \times \underline{\mathbf{A}} = \underline{\mathbf{B}}(\underline{\mathbf{C}} \times \underline{\mathbf{A}})$$

$$\begin{aligned}(\underline{\mathbf{A}} \times \underline{\underline{\mathbf{D}}}) \times \underline{\mathbf{B}} &= \underline{\mathbf{A}} \times (\underline{\underline{\mathbf{D}}} \times \underline{\mathbf{B}}) \\ &= \underline{\mathbf{A}} \times \underline{\underline{\mathbf{D}}} \times \underline{\mathbf{B}}\end{aligned}$$

$$(\underline{\mathbf{A}}\,\underline{\mathbf{B}} - \underline{\mathbf{B}}\,\underline{\mathbf{A}}) \cdot \underline{\mathbf{C}} = (\underline{\mathbf{B}} \times \underline{\mathbf{A}}) \times \underline{\mathbf{C}}$$

$$\begin{aligned}(\underline{\mathbf{A}} \times \underline{\underline{\mathbf{D}}}) \cdot \underline{\mathbf{B}} &= \underline{\underline{\underline{\boldsymbol{\varepsilon}}}} : (\underline{\underline{\mathbf{D}}}\,\underline{\mathbf{A}})^{132} \cdot \underline{\mathbf{B}} \\ &= \underline{\mathbf{A}} \times (\underline{\underline{\mathbf{D}}} \cdot \underline{\mathbf{B}})\end{aligned}$$

$$\begin{aligned}\underline{\mathbf{B}} \cdot (\underline{\mathbf{A}} \times \underline{\underline{\mathbf{D}}}) &= (\underline{\mathbf{B}} \times \underline{\mathbf{A}}) \cdot \underline{\underline{\mathbf{D}}} \\ &= -\underline{\mathbf{A}} \cdot (\underline{\mathbf{B}} \times \underline{\underline{\mathbf{D}}})\end{aligned}$$

$$\underline{\underline{\mathbf{B}}} \cdot (\underline{\mathbf{A}} \times \underline{\underline{\mathbf{D}}}) = -(\underline{\mathbf{A}} \times \underline{\underline{\mathbf{B}}}^{21})^{21} \cdot \underline{\underline{\mathbf{D}}}$$

$$\begin{aligned}(\underline{\underline{\mathbf{D}}} \times \underline{\mathbf{A}}) \cdot \underline{\mathbf{B}} &= -\underline{\underline{\mathbf{D}}} \cdot \underline{\underline{\underline{\boldsymbol{\varepsilon}}}} \cdot \underline{\mathbf{A}} \cdot \underline{\mathbf{B}} \\ &= -\underline{\underline{\mathbf{D}}} \cdot \underline{\underline{\underline{\boldsymbol{\varepsilon}}}} : \underline{\mathbf{A}}\,\underline{\mathbf{B}} \\ &= -(\underline{\underline{\mathbf{D}}} \times \underline{\mathbf{B}}) \cdot \underline{\mathbf{A}} \\ &= \underline{\underline{\mathbf{D}}} \cdot (\underline{\mathbf{A}} \times \underline{\mathbf{B}})\end{aligned}$$

$$(\underline{\underline{\mathbf{D}}} \times \underline{\mathbf{A}}) \cdot \underline{\underline{\mathbf{B}}} = \underline{\underline{\mathbf{D}}} \cdot (\underline{\mathbf{A}} \times \underline{\underline{\mathbf{B}}})$$

$$\begin{aligned}\underline{\mathbf{B}} \cdot (\underline{\underline{\mathbf{D}}} \times \underline{\mathbf{A}}) &= (\underline{\mathbf{B}} \cdot \underline{\underline{\mathbf{D}}}) \times \underline{\mathbf{A}} \\ &= \underline{\mathbf{B}} \cdot \underline{\underline{\mathbf{D}}} \times \underline{\mathbf{A}}\end{aligned}$$

$$\begin{aligned}(\underline{\mathbf{A}} \times \underline{\mathbf{B}})(\underline{\mathbf{C}} \times \underline{\mathbf{D}}) = (\underline{\mathbf{A}} \times \underline{\mathbf{B}}) \cdot (\underline{\mathbf{C}} \times \underline{\mathbf{D}})\,\underline{\underline{\mathbf{I}}} + (\underline{\mathbf{A}} \cdot \underline{\mathbf{D}})\underline{\mathbf{C}}\,\underline{\mathbf{B}} + \\ + (\underline{\mathbf{B}} \cdot \underline{\mathbf{C}})\underline{\mathbf{D}}\,\underline{\mathbf{A}} - (\underline{\mathbf{A}} \cdot \underline{\mathbf{C}})\underline{\mathbf{D}}\,\underline{\mathbf{B}} - (\underline{\mathbf{B}} \cdot \underline{\mathbf{D}})\underline{\mathbf{C}}\,\underline{\mathbf{A}}\end{aligned}$$

$$\begin{aligned}(\underline{\mathbf{A}} \times \underline{\underline{\mathbf{I}}}) \cdot (\underline{\mathbf{B}} \times \underline{\underline{\mathbf{I}}}) &= \underline{\mathbf{A}} \times (\underline{\mathbf{B}} \times \underline{\underline{\mathbf{I}}}) \\ &= \underline{\mathbf{B}}\,\underline{\mathbf{A}} - (\underline{\mathbf{A}} \cdot \underline{\mathbf{B}})\,\underline{\underline{\mathbf{I}}}\end{aligned}$$

$$\underline{\mathbf{A}} \times (\underline{\mathbf{B}} \times \underline{\underline{\mathbf{D}}}) = \underline{\mathbf{B}}(\underline{\mathbf{A}} \cdot \underline{\underline{\mathbf{D}}}) - \underline{\underline{\mathbf{D}}}(\underline{\mathbf{A}} \cdot \underline{\mathbf{B}})$$

$$\begin{aligned}(\underline{\mathbf{A}} \times \underline{\underline{\mathbf{I}}})^2 &= \underline{\mathbf{A}}\,\underline{\mathbf{A}} - (\underline{\mathbf{A}} \cdot \underline{\mathbf{A}})\,\underline{\underline{\mathbf{I}}} \\ &= \underline{\mathbf{A}}\,\underline{\mathbf{A}} - A^2\,\underline{\underline{\mathbf{I}}}\end{aligned}$$

$$(\underline{\mathbf{A}} \times \underline{\mathbf{B}}) \times \underline{\underline{\mathbf{I}}} = \underline{\mathbf{B}}\,\underline{\mathbf{A}} - \underline{\mathbf{A}}\,\underline{\mathbf{B}}$$

$$\underline{\underline{\mathbf{D}}} \cdot (\underline{\mathbf{A}} \times \underline{\underline{\mathbf{I}}}) + (\underline{\mathbf{A}} \times \underline{\underline{\mathbf{I}}}) \cdot \underline{\underline{\mathbf{D}}}^{21} = (\underline{\mathbf{A}}\,\mathrm{trace}\,\underline{\underline{\mathbf{D}}} - \underline{\underline{\mathbf{D}}}^{21} \cdot \underline{\mathbf{A}}) \times \underline{\underline{\mathbf{I}}}$$

$$\underline{\underline{\mathbf{I}}} \overset{\times}{\times} \underline{\underline{\mathbf{D}}} = \underline{\underline{\mathbf{I}}}\,\mathrm{trace}\,\underline{\underline{\mathbf{D}}} - \underline{\underline{\mathbf{D}}}$$

$$\underline{\underline{\mathbf{D}}} \overset{\times}{\times} \underline{\underline{\mathbf{I}}} = (\underline{\underline{\mathbf{I}}} \overset{\times}{\times} \underline{\underline{\mathbf{D}}})^{21}$$

$$\underline{\underline{\mathbf{D}}} \overset{\cdot}{\times} \underline{\underline{\mathbf{I}}} = \langle \underline{\underline{\mathbf{D}}} \rangle$$

$$\underline{\underline{\mathbf{D}}} \overset{\times}{\times} \underline{\underline{\mathbf{E}}} = \underline{\underline{\mathbf{D}}} : (\underline{\underline{\underline{\boldsymbol{\varepsilon}}}}\,\underline{\underline{\underline{\boldsymbol{\varepsilon}}}})^{412536} : \underline{\underline{\mathbf{E}}}.$$

A.2.3 Traces

$$
\begin{aligned}
\operatorname{trace} \underline{\underline{\mathbf{A}}} &= A_{11} + A_{22} + A_{33} \\
&= A_{ii} \\
&= \underline{\underline{\mathbf{I}}} : \underline{\underline{\mathbf{A}}} = \underline{\underline{\mathbf{A}}} : \underline{\underline{\mathbf{I}}}
\end{aligned}
$$

$$
\begin{aligned}
\operatorname{trace} \underline{\underline{\mathbf{I}}} &= 3 \\
\operatorname{trace} (\underline{\mathbf{A}}\,\underline{\mathbf{B}}) &= \underline{\mathbf{A}} \cdot \underline{\mathbf{B}} \\
\operatorname{trace} (\underline{\underline{\mathbf{A}}} + \underline{\underline{\mathbf{B}}}) &= \operatorname{trace} \underline{\underline{\mathbf{A}}} + \operatorname{trace} \underline{\underline{\mathbf{B}}} \\
\operatorname{trace} (\underline{\underline{\mathbf{A}}} + \underline{\mathbf{C}}\,\underline{\mathbf{D}}) &= \operatorname{trace} \underline{\underline{\mathbf{A}}} + \underline{\mathbf{C}} \cdot \underline{\mathbf{D}} \\
\operatorname{trace} (\underline{\underline{\mathbf{A}}} \pm \alpha \underline{\underline{\mathbf{I}}}) &= \operatorname{trace} \underline{\underline{\mathbf{A}}} \pm 3\alpha \\
\operatorname{trace} (\alpha \underline{\underline{\mathbf{I}}} + \underline{\mathbf{C}}\,\underline{\mathbf{D}}) &= 3\alpha + \underline{\mathbf{C}} \cdot \underline{\mathbf{D}} \\
\operatorname{trace} (\alpha \underline{\underline{\mathbf{I}}} + \underline{\mathbf{A}}_1 \underline{\mathbf{C}}_1 + \underline{\mathbf{A}}_2 \underline{\mathbf{C}}_2) &= 3\alpha + \underline{\mathbf{A}}_1 \cdot \underline{\mathbf{C}}_1 + \underline{\mathbf{A}}_2 \cdot \underline{\mathbf{C}}_2 \\
\operatorname{trace} \alpha \underline{\underline{\mathbf{A}}} &= \alpha \operatorname{trace} \underline{\underline{\mathbf{A}}} \\
\operatorname{trace} \underline{\underline{\mathbf{A}}}^{21} &= \operatorname{trace} \underline{\underline{\mathbf{A}}} \\
\operatorname{trace} (\underline{\underline{\mathbf{A}}} \cdot \underline{\underline{\mathbf{B}}}) &= \operatorname{trace} (\underline{\underline{\mathbf{B}}} \cdot \underline{\underline{\mathbf{A}}}) \\
\operatorname{trace} (\underline{\mathbf{A}}\,\underline{\mathbf{B}} \cdot \underline{\underline{\mathbf{D}}}) &= \underline{\mathbf{B}} \cdot \underline{\underline{\mathbf{D}}} \cdot \underline{\mathbf{A}} \\
\operatorname{trace} (\alpha \underline{\underline{\mathbf{I}}} + \underline{\mathbf{A}}\,\underline{\mathbf{B}} \cdot \underline{\underline{\mathbf{D}}}) &= 3\alpha + \underline{\mathbf{B}} \cdot \underline{\underline{\mathbf{D}}} \cdot \underline{\mathbf{A}}
\end{aligned}
$$

$$
\begin{aligned}
\operatorname{trace} (\underline{\underline{\mathbf{A}}} \cdot \underline{\underline{\mathbf{B}}} \cdot \underline{\underline{\mathbf{C}}}) &= \operatorname{trace} (\underline{\underline{\mathbf{B}}} \cdot \underline{\underline{\mathbf{C}}} \cdot \underline{\underline{\mathbf{A}}}) \\
&= \operatorname{trace} (\underline{\underline{\mathbf{C}}} \cdot \underline{\underline{\mathbf{A}}} \cdot \underline{\underline{\mathbf{B}}})
\end{aligned}
$$

$$
\begin{aligned}
\operatorname{trace} (\underline{\mathbf{A}} \times \underline{\underline{\mathbf{I}}}) &= 0 \\
\operatorname{trace} (\underline{\underline{\mathbf{D}}} + \underline{\mathbf{A}} \times \underline{\underline{\mathbf{I}}}) &= \operatorname{trace} \underline{\underline{\mathbf{D}}} \\
\operatorname{trace} (\underline{\mathbf{A}}\,\underline{\mathbf{B}} + \underline{\mathbf{C}} \times \underline{\underline{\mathbf{I}}}) &= \underline{\mathbf{A}} \cdot \underline{\mathbf{B}} \\
\operatorname{trace} (\alpha \underline{\underline{\mathbf{I}}} + \underline{\mathbf{A}}\,\underline{\mathbf{B}} + \underline{\mathbf{C}} \times \underline{\underline{\mathbf{I}}}) &= 3\alpha + \underline{\mathbf{A}} \cdot \underline{\mathbf{B}} \\
\operatorname{trace} [(\underline{\mathbf{A}} \times \underline{\underline{\mathbf{I}}}) \cdot (\underline{\mathbf{B}} \times \underline{\underline{\mathbf{I}}})] &= -2\underline{\mathbf{A}} \cdot \underline{\mathbf{B}} \\
\operatorname{trace} (\underline{\mathbf{A}} \times \underline{\underline{\mathbf{I}}})^2 &= -2A^2 \\
\operatorname{trace} \operatorname{adj} \underline{\underline{\mathbf{A}}} &= \frac{1}{2}(\operatorname{trace}^2 \underline{\underline{\mathbf{A}}} - \operatorname{trace} \underline{\underline{\mathbf{A}}}^2) \\
\operatorname{trace} \operatorname{adj} (\underline{\underline{\mathbf{A}}} + \underline{\underline{\mathbf{B}}}) &= \operatorname{trace} \operatorname{adj} \underline{\underline{\mathbf{A}}} + \operatorname{trace} \operatorname{adj} \underline{\underline{\mathbf{B}}} - \operatorname{trace} (\underline{\underline{\mathbf{A}}} \cdot \underline{\underline{\mathbf{B}}}) \\
&\quad + \operatorname{trace} \underline{\underline{\mathbf{A}}} \operatorname{trace} \underline{\underline{\mathbf{B}}}
\end{aligned}
$$

$$
\begin{aligned}
\operatorname{trace} \operatorname{adj} (\underline{\underline{\mathbf{A}}} \pm \alpha \underline{\underline{\mathbf{I}}}) &= 3\alpha^2 \pm 2\alpha \operatorname{trace} \underline{\underline{\mathbf{A}}} + \operatorname{trace} \operatorname{adj} \underline{\underline{\mathbf{A}}} \\
\operatorname{trace} \operatorname{adj} (\underline{\underline{\mathbf{A}}} + \underline{\mathbf{C}}\,\underline{\mathbf{D}}) &= \operatorname{trace} \operatorname{adj} \underline{\underline{\mathbf{A}}} + (\underline{\mathbf{C}} \cdot \underline{\mathbf{D}}) \operatorname{trace} \underline{\underline{\mathbf{A}}} - \underline{\mathbf{D}} \cdot \underline{\underline{\mathbf{A}}} \cdot \underline{\mathbf{C}} \\
\operatorname{trace} \operatorname{adj} (\alpha \underline{\underline{\mathbf{I}}} + \underline{\mathbf{C}}\,\underline{\mathbf{D}}) &= \alpha(3\alpha + 2\underline{\mathbf{C}} \cdot \underline{\mathbf{D}}) \\
\operatorname{trace} \operatorname{adj} (\alpha \underline{\underline{\mathbf{I}}} + \underline{\mathbf{A}}_1 \underline{\mathbf{C}}_1 + \underline{\mathbf{A}}_2 \underline{\mathbf{C}}_2) &= \alpha[3\alpha + 2(\underline{\mathbf{A}}_1 \cdot \underline{\mathbf{C}}_1 + \underline{\mathbf{A}}_2 \cdot \underline{\mathbf{C}}_2)] \\
&\quad + (\underline{\mathbf{C}}_1 \times \underline{\mathbf{C}}_2) \cdot (\underline{\mathbf{A}}_1 \times \underline{\mathbf{A}}_2)
\end{aligned}
$$

$$\text{trace adj}\,(\alpha\underline{\underline{\mathbf{I}}} + \underline{\mathbf{A}}\,\underline{\mathbf{B}} \cdot \underline{\underline{\mathbf{D}}}) = \alpha(3\alpha + 2\underline{\mathbf{B}} \cdot \underline{\underline{\mathbf{D}}} \cdot \underline{\mathbf{A}})$$

$$\begin{aligned} \text{trace adj}\,(\underline{\underline{\mathbf{D}}} + \underline{\mathbf{A}} \times \underline{\underline{\mathbf{I}}}) &= \text{trace adj}\,\underline{\underline{\mathbf{D}}} + A^2 - \text{trace}\,[\underline{\underline{\mathbf{D}}} \cdot (\underline{\mathbf{A}} \times \underline{\underline{\mathbf{I}}})] \\ &= \text{trace adj}\,\underline{\underline{\mathbf{D}}} + A^2, \text{if } \underline{\underline{\mathbf{D}}} \text{ is symmetric} \end{aligned}$$

$$\text{trace adj}\,(\alpha\underline{\underline{\mathbf{I}}} + \underline{\mathbf{A}} \times \underline{\underline{\mathbf{I}}}) = 3\alpha^2 + A^2$$

$$\text{trace adj}\,(\underline{\mathbf{A}}\,\underline{\mathbf{B}} + \underline{\mathbf{C}} \times \underline{\underline{\mathbf{I}}}) = C^2 - \underline{\mathbf{C}} \cdot (\underline{\mathbf{A}} \times \underline{\mathbf{B}})$$

$$\text{trace adj}\,(\alpha\underline{\underline{\mathbf{I}}} + \underline{\mathbf{A}}\,\underline{\mathbf{B}} + \underline{\mathbf{C}} \times \underline{\underline{\mathbf{I}}}) = 3\alpha^2 + 2\alpha(\underline{\mathbf{A}} \cdot \underline{\mathbf{B}}) + C^2 - \underline{\mathbf{C}} \cdot (\underline{\mathbf{A}} \times \underline{\mathbf{B}}).$$

A.2.4 Determinants

$$\begin{aligned} \det \underline{\underline{\mathbf{D}}} &= \begin{vmatrix} D_{11} & D_{12} & D_{13} \\ D_{21} & D_{22} & D_{23} \\ D_{31} & D_{32} & D_{33} \end{vmatrix} \\ &= D_{11}D_{22}D_{33} + D_{12}D_{23}D_{31} + D_{21}D_{32}D_{13} - D_{13}D_{22}D_{31} \\ &\quad - D_{12}D_{21}D_{33} - D_{11}D_{32}D_{23} \\ &= \begin{vmatrix} \underline{\mathbf{D}}_1 \\ \underline{\mathbf{D}}_2 \\ \underline{\mathbf{D}}_3 \end{vmatrix} \\ &= \underline{\underline{\underline{\boldsymbol{\varepsilon}}}} \,\vdots\, \underline{\mathbf{D}}_3\underline{\mathbf{D}}_2\underline{\mathbf{D}}_1 \\ &= \left|\underline{\mathbf{D}}^1\;\underline{\mathbf{D}}^2\;\underline{\mathbf{D}}^3\right| \\ &= \underline{\underline{\underline{\boldsymbol{\varepsilon}}}} \,\vdots\, \underline{\mathbf{D}}^3\underline{\mathbf{D}}^2\underline{\mathbf{D}}^1 \end{aligned}$$

with the row vectors $\underline{\mathbf{D}}_i$, and the column vectors $\underline{\mathbf{D}}^i$ of $\underline{\underline{\mathbf{D}}}$

$$\det(\underline{\mathbf{A}} \times \underline{\underline{\mathbf{I}}}) = 0$$

$$\det(\underline{\underline{\mathbf{A}}} \cdot \underline{\underline{\mathbf{B}}}) = \det \underline{\underline{\mathbf{A}}} \det \underline{\underline{\mathbf{B}}}$$

$$\det \underline{\underline{\mathbf{A}}}^{21} = \det \underline{\underline{\mathbf{A}}}$$

$$\det \underline{\underline{\mathbf{A}}}^{-1} = \frac{1}{\det \underline{\underline{\mathbf{A}}}}$$

$$\det(\alpha\,\underline{\underline{\mathbf{A}}}) = \alpha^3 \det \underline{\underline{\mathbf{A}}}$$

$$\det \text{adj}\,\underline{\underline{\mathbf{A}}} = \det{}^2 \underline{\underline{\mathbf{A}}}$$

$$\det \underline{\underline{\mathbf{A}}} = \frac{1}{6}\left[\text{trace}^3 \underline{\underline{\mathbf{A}}} - 3\,\text{trace}\,\underline{\underline{\mathbf{A}}}\;\text{trace}\,\underline{\underline{\mathbf{A}}}^2 + 2\,\text{trace}\,\underline{\underline{\mathbf{A}}}^3\right]$$

$$\det(\underline{\underline{\mathbf{A}}} + \underline{\underline{\mathbf{B}}}) = \det \underline{\underline{\mathbf{A}}} + \det \underline{\underline{\mathbf{B}}} + \text{trace}\,(\text{adj}\,\underline{\underline{\mathbf{A}}} \cdot \underline{\underline{\mathbf{B}}}) + \text{trace}\,(\underline{\underline{\mathbf{A}}} \cdot \text{adj}\,\underline{\underline{\mathbf{B}}})$$

$$\det(\underline{\mathbf{A}}\,\underline{\mathbf{B}}) = 0$$

$$\det(\underline{\mathbf{A}}_1\underline{\mathbf{C}}_1 + \underline{\mathbf{A}}_2\underline{\mathbf{C}}_2) = 0$$

$$\det(\underline{\mathbf{A}}_1\underline{\mathbf{C}}_1 + \underline{\mathbf{A}}_2\underline{\mathbf{C}}_2 + \underline{\mathbf{A}}_3\underline{\mathbf{C}}_3) = (\underline{\mathbf{A}}_1 \cdot \underline{\mathbf{A}}_2 \times \underline{\mathbf{A}}_3)(\underline{\mathbf{C}}_1 \cdot \underline{\mathbf{C}}_2 \times \underline{\mathbf{C}}_3)$$

$$\det(\underline{\underline{\mathbf{A}}} \pm \alpha\underline{\underline{\mathbf{I}}}) = \pm\alpha^3 + \alpha^2 \operatorname{trace} \underline{\underline{\mathbf{A}}} \pm \alpha \operatorname{trace} \operatorname{adj} \underline{\underline{\mathbf{A}}} + \det \underline{\underline{\mathbf{A}}}$$

$$\det(\underline{\underline{\mathbf{A}}} + \underline{\mathbf{C}}\,\underline{\mathbf{D}}) = \det \underline{\underline{\mathbf{A}}} + \underline{\mathbf{D}} \cdot (\operatorname{adj} \underline{\underline{\mathbf{A}}}) \cdot \underline{\mathbf{C}}$$

$$\det(\alpha\underline{\underline{\mathbf{I}}} + \underline{\mathbf{C}}\,\underline{\mathbf{D}}) = \alpha^2(\alpha + \underline{\mathbf{C}} \cdot \underline{\mathbf{D}})$$

$$\det(\alpha\underline{\underline{\mathbf{I}}} + \underline{\mathbf{A}}\,\underline{\mathbf{B}} \cdot \underline{\underline{\mathbf{D}}}) = \alpha^2(\alpha + \underline{\mathbf{B}} \cdot \underline{\underline{\mathbf{D}}} \cdot \underline{\mathbf{A}})$$

$$\begin{aligned}\det(\alpha\underline{\underline{\mathbf{I}}} + \underline{\mathbf{A}}_1\underline{\mathbf{C}}_1 + \underline{\mathbf{A}}_2\underline{\mathbf{C}}_2) &= \alpha[\alpha^2 + \alpha(\underline{\mathbf{A}}_1 \cdot \underline{\mathbf{C}}_1 + \underline{\mathbf{A}}_2 \cdot \underline{\mathbf{C}}_2) \\ &\quad + (\underline{\mathbf{C}}_1 \times \underline{\mathbf{C}}_2) \cdot (\underline{\mathbf{A}}_1 \times \underline{\mathbf{A}}_2)]\end{aligned}$$

$$\begin{aligned}\det(\underline{\underline{\mathbf{D}}} + \underline{\mathbf{A}} \times \underline{\underline{\mathbf{I}}}) &= \det \underline{\underline{\mathbf{D}}} + \operatorname{trace}[\operatorname{adj} \underline{\underline{\mathbf{D}}} \cdot (\underline{\mathbf{A}} \times \underline{\underline{\mathbf{I}}})] + \underline{\mathbf{A}} \cdot \underline{\underline{\mathbf{D}}} \cdot \underline{\mathbf{A}} \\ &= \det \underline{\underline{\mathbf{D}}} + \underline{\mathbf{A}} \cdot \underline{\underline{\mathbf{D}}} \cdot \underline{\mathbf{A}}, \text{ if } \underline{\underline{\mathbf{D}}} \text{ is symmetric}\end{aligned}$$

$$\det(\alpha\underline{\underline{\mathbf{I}}} + \underline{\mathbf{A}} \times \underline{\underline{\mathbf{I}}}) = \alpha(\alpha^2 + A^2)$$

$$\det(\underline{\mathbf{A}}\,\underline{\mathbf{B}} + \underline{\mathbf{C}} \times \underline{\underline{\mathbf{I}}}) = (\underline{\mathbf{A}} \cdot \underline{\mathbf{C}})(\underline{\mathbf{B}} \cdot \underline{\mathbf{C}})$$

$$\begin{aligned}\det(\alpha\underline{\underline{\mathbf{I}}} + \underline{\mathbf{A}}\,\underline{\mathbf{B}} + \underline{\mathbf{C}} \times \underline{\underline{\mathbf{I}}}) &= \alpha^3 + \alpha^2\underline{\mathbf{A}} \cdot \underline{\mathbf{B}} + \alpha(C^2 - \underline{\mathbf{A}} \times \underline{\mathbf{B}} \cdot \underline{\mathbf{C}}) \\ &\quad + (\underline{\mathbf{A}} \cdot \underline{\mathbf{C}})(\underline{\mathbf{B}} \cdot \underline{\mathbf{C}}).\end{aligned}$$

A.2.5 Adjoints and inverses

$$\operatorname{adj} \underline{\underline{\mathbf{D}}} = \frac{1}{2}\underline{\underline{\boldsymbol{\varepsilon}}} : (\underline{\underline{\mathbf{D}}}\,\underline{\underline{\mathbf{D}}})^{4231} : \underline{\underline{\boldsymbol{\varepsilon}}}$$

$$\begin{aligned}(\operatorname{adj} \underline{\underline{\mathbf{D}}})_{nk} &= \underline{\mathbf{e}}_{x_n} \cdot \operatorname{adj} \underline{\underline{\mathbf{D}}} \cdot \underline{\mathbf{e}}_{x_k} \\ &= \frac{1}{2}\varepsilon_{ijk}\varepsilon_{lmn}D_{il}D_{jm}\end{aligned}$$

$$\operatorname{adj} \underline{\underline{\mathbf{D}}} = \begin{pmatrix} D_{22}D_{33} - D_{23}D_{32} & D_{32}D_{13} - D_{33}D_{12} & D_{12}D_{23} - D_{13}D_{22} \\ D_{31}D_{23} - D_{21}D_{33} & D_{11}D_{33} - D_{13}D_{31} & D_{13}D_{21} - D_{11}D_{23} \\ D_{21}D_{32} - D_{31}D_{22} & D_{12}D_{31} - D_{11}D_{32} & D_{11}D_{22} - D_{12}D_{21} \end{pmatrix}$$

$$\operatorname{adj}(\underline{\mathbf{A}}\,\underline{\mathbf{B}}) = \underline{\underline{\mathbf{0}}}$$

$$\operatorname{adj}(\underline{\mathbf{A}} \times \underline{\underline{\mathbf{I}}}) = \underline{\mathbf{A}}\,\underline{\mathbf{A}}$$

$$\operatorname{adj}(\underline{\underline{\mathbf{A}}} \cdot \underline{\underline{\mathbf{B}}}) = \operatorname{adj} \underline{\underline{\mathbf{B}}} \cdot \operatorname{adj} \underline{\underline{\mathbf{A}}}$$

$$\operatorname{adj} \underline{\underline{\mathbf{A}}}^{21} = (\operatorname{adj} \underline{\underline{\mathbf{A}}})^{21}$$

$$\begin{aligned}\operatorname{adj} \underline{\underline{\mathbf{A}}}^{-1} &= (\operatorname{adj} \underline{\underline{\mathbf{A}}})^{-1} \\ &= \frac{\underline{\underline{\mathbf{A}}}}{\det \underline{\underline{\mathbf{A}}}}\end{aligned}$$

$$\operatorname{adj}(\alpha\,\underline{\underline{\mathbf{A}}}) = \alpha^2 \operatorname{adj} \underline{\underline{\mathbf{A}}}$$

$$\operatorname{adj}(\alpha\,\underline{\underline{\mathbf{I}}}) = \alpha^2\underline{\underline{\mathbf{I}}}$$

$$\text{adj}\,\text{adj}\,\underline{\underline{\mathbf{A}}} = \underline{\underline{\mathbf{A}}}\,\det\underline{\underline{\mathbf{A}}}$$

$$\begin{aligned}\text{adj}\,\underline{\underline{\mathbf{A}}} &= \underline{\underline{\mathbf{A}}}^2 - \underline{\underline{\mathbf{A}}}\,\text{trace}\,\underline{\underline{\mathbf{A}}} + \frac{1}{2}\left[\text{trace}^2\underline{\underline{\mathbf{A}}} - \text{trace}\,\underline{\underline{\mathbf{A}}}^2\right]\underline{\underline{\mathbf{I}}}\\ &= \underline{\underline{\mathbf{A}}}^2 - \underline{\underline{\mathbf{A}}}\,\text{trace}\,\underline{\underline{\mathbf{A}}} + \underline{\underline{\mathbf{I}}}\,\text{trace}\,\text{adj}\,\underline{\underline{\mathbf{A}}}\end{aligned}$$

$$\begin{aligned}\underline{\underline{\mathbf{A}}}\cdot\text{adj}\,\underline{\underline{\mathbf{A}}} &= (\text{adj}\,\underline{\underline{\mathbf{A}}})\cdot\underline{\underline{\mathbf{A}}}\\ &= \underline{\underline{\mathbf{I}}}\det\underline{\underline{\mathbf{A}}}\end{aligned}$$

$$\begin{aligned}\text{adj}\,\underline{\underline{\mathbf{A}}}\cdot(\underline{\mathbf{B}}\times\underline{\mathbf{C}}) &= (\underline{\underline{\mathbf{A}}}^{21}\cdot\underline{\mathbf{B}})\times(\underline{\underline{\mathbf{A}}}^{21}\times\underline{\mathbf{C}})\\ &= (\underline{\mathbf{B}}\cdot\underline{\underline{\mathbf{A}}})\times(\underline{\mathbf{C}}\cdot\underline{\underline{\mathbf{A}}})\end{aligned}$$

$$(\underline{\mathbf{B}}\times\underline{\underline{\mathbf{I}}})\cdot\text{adj}\,\underline{\underline{\mathbf{A}}}\cdot(\underline{\mathbf{C}}\times\underline{\underline{\mathbf{I}}}) = \underline{\underline{\mathbf{A}}}^{21}\cdot\underline{\mathbf{C}}\,\underline{\mathbf{B}}\cdot\underline{\underline{\mathbf{A}}}^{21} - (\underline{\mathbf{C}}\cdot\underline{\underline{\mathbf{A}}}\cdot\underline{\mathbf{B}})\underline{\underline{\mathbf{A}}}^{21}$$

$$\begin{aligned}\text{adj}\,(\underline{\underline{\mathbf{A}}}+\underline{\underline{\mathbf{B}}}) = {}&\text{adj}\,\underline{\underline{\mathbf{A}}} + \text{adj}\,\underline{\underline{\mathbf{B}}} + \underline{\underline{\mathbf{A}}}\cdot\underline{\underline{\mathbf{B}}} + \underline{\underline{\mathbf{B}}}\cdot\underline{\underline{\mathbf{A}}} -\\ &- \underline{\underline{\mathbf{B}}}\,\text{trace}\,\underline{\underline{\mathbf{A}}} - \underline{\underline{\mathbf{A}}}\,\text{trace}\,\underline{\underline{\mathbf{B}}} + \underline{\underline{\mathbf{I}}}\,\text{trace}\,\underline{\underline{\mathbf{A}}}\,\text{trace}\,\underline{\underline{\mathbf{B}}} - \underline{\underline{\mathbf{I}}}\,\text{trace}\,(\underline{\underline{\mathbf{A}}}\cdot\underline{\underline{\mathbf{B}}})\end{aligned}$$

$$\text{adj}\,(\underline{\mathbf{A}}_1\underline{\mathbf{C}}_1 + \underline{\mathbf{A}}_2\underline{\mathbf{C}}_2) = (\underline{\mathbf{C}}_1\times\underline{\mathbf{C}}_2)(\underline{\mathbf{A}}_1\times\underline{\mathbf{A}}_2)$$

$$\begin{aligned}\text{adj}\,(\underline{\mathbf{A}}_1\underline{\mathbf{C}}_1 + \underline{\mathbf{A}}_2\underline{\mathbf{C}}_2 + \underline{\mathbf{A}}_3\underline{\mathbf{C}}_3) = {}&(\underline{\mathbf{C}}_2\times\underline{\mathbf{C}}_3)(\underline{\mathbf{A}}_2\times\underline{\mathbf{A}}_3) + (\underline{\mathbf{C}}_3\times\underline{\mathbf{C}}_1)(\underline{\mathbf{A}}_3\times\underline{\mathbf{A}}_1)\\ &+ (\underline{\mathbf{C}}_1\times\underline{\mathbf{C}}_2)(\underline{\mathbf{A}}_1\times\underline{\mathbf{A}}_2)\end{aligned}$$

$$\text{adj}\,(\underline{\underline{\mathbf{A}}}\pm\alpha\underline{\underline{\mathbf{I}}}) = \alpha^2\underline{\underline{\mathbf{I}}} \pm \alpha(\underline{\underline{\mathbf{I}}}\,\text{trace}\,\underline{\underline{\mathbf{A}}} - \underline{\underline{\mathbf{A}}}) + \text{adj}\,\underline{\underline{\mathbf{A}}}$$

$$\begin{aligned}\text{adj}\,(\underline{\underline{\mathbf{A}}} + \underline{\mathbf{C}}\,\underline{\mathbf{D}}) &= \text{adj}\,\underline{\underline{\mathbf{A}}} + (\underline{\underline{\mathbf{A}}} - \underline{\underline{\mathbf{I}}}\,\text{trace}\,\mathbf{A})\cdot(\underline{\mathbf{D}}\times\underline{\underline{\mathbf{I}}})\cdot(\underline{\mathbf{C}}\times\underline{\underline{\mathbf{I}}})\\ &\quad + [(\underline{\mathbf{D}}\cdot\mathbf{A})\times\underline{\underline{\mathbf{I}}}]\cdot(\underline{\mathbf{C}}\times\underline{\underline{\mathbf{I}}})\\ &= \text{adj}\,\underline{\underline{\mathbf{A}}} - (\underline{\mathbf{D}}\times\underline{\underline{\mathbf{I}}})\cdot\underline{\underline{\mathbf{A}}}^{21}\cdot(\underline{\mathbf{C}}\times\underline{\underline{\mathbf{I}}})\end{aligned}$$

$$\text{adj}\,(\alpha\underline{\underline{\mathbf{I}}} + \underline{\mathbf{C}}\,\underline{\mathbf{D}}) = \alpha[(\alpha + \underline{\mathbf{C}}\cdot\underline{\mathbf{D}})\underline{\underline{\mathbf{I}}} - \underline{\mathbf{C}}\,\underline{\mathbf{D}}]$$

$$\text{adj}\,(\alpha\underline{\underline{\mathbf{I}}} + \underline{\mathbf{A}}\,\underline{\mathbf{B}}\cdot\underline{\underline{\mathbf{D}}}) = \alpha[(\alpha + \underline{\mathbf{B}}\cdot\underline{\underline{\mathbf{D}}}\cdot\underline{\mathbf{A}})\underline{\underline{\mathbf{I}}} - \underline{\mathbf{A}}\,\underline{\mathbf{B}}\cdot\underline{\underline{\mathbf{D}}}]$$

$$\begin{aligned}\text{adj}\,(\alpha\underline{\underline{\mathbf{I}}} + \underline{\mathbf{A}}_1\underline{\mathbf{C}}_1 + \underline{\mathbf{A}}_2\underline{\mathbf{C}}_2) = {}&\alpha[(\alpha + \underline{\mathbf{A}}_1\cdot\underline{\mathbf{C}}_1 + \underline{\mathbf{A}}_2\cdot\underline{\mathbf{C}}_2)\underline{\underline{\mathbf{I}}} - \underline{\mathbf{A}}_1\underline{\mathbf{C}}_1 - \underline{\mathbf{A}}_2\underline{\mathbf{C}}_2]\\ &+ (\underline{\mathbf{C}}_1\times\underline{\mathbf{C}}_2)(\underline{\mathbf{A}}_1\times\underline{\mathbf{A}}_2)\end{aligned}$$

$$\begin{aligned}\text{adj}\,(\underline{\underline{\mathbf{D}}} + \underline{\mathbf{A}}\times\underline{\underline{\mathbf{I}}}) &= \text{adj}\,\underline{\underline{\mathbf{D}}} + \underline{\mathbf{A}}\,\underline{\mathbf{A}} + (\underline{\underline{\mathbf{D}}} - \underline{\underline{\mathbf{I}}}\,\text{trace}\,\underline{\underline{\mathbf{D}}})\cdot(\underline{\mathbf{A}}\times\underline{\underline{\mathbf{I}}}) + (\underline{\mathbf{A}}\times\underline{\underline{\mathbf{I}}})\cdot\underline{\underline{\mathbf{D}}}\\ &\quad + \underline{\underline{\mathbf{I}}}\,\text{trace}\,[\underline{\underline{\mathbf{D}}}\cdot(\underline{\mathbf{A}}\times\underline{\underline{\mathbf{I}}})]\\ &= \text{adj}\,\underline{\underline{\mathbf{D}}} + \underline{\mathbf{A}}\,\underline{\mathbf{A}} - (\underline{\underline{\mathbf{D}}}\cdot\underline{\mathbf{A}})\times\underline{\underline{\mathbf{I}}}, \text{ if } \underline{\underline{\mathbf{D}}} \text{ is symmetric}\end{aligned}$$

$$\text{adj}\,(\alpha\underline{\underline{\mathbf{I}}} + \underline{\mathbf{A}}\times\underline{\underline{\mathbf{I}}}) = \alpha(\alpha\underline{\underline{\mathbf{I}}} - \underline{\mathbf{A}}\times\underline{\underline{\mathbf{I}}}) + \underline{\mathbf{A}}\,\underline{\mathbf{A}}$$

$$\text{adj}\,(\underline{\mathbf{A}}\,\underline{\mathbf{B}} + \underline{\mathbf{C}}\times\underline{\underline{\mathbf{I}}}) = \underline{\mathbf{C}}\,\underline{\mathbf{C}} - (\underline{\mathbf{B}}\cdot\underline{\mathbf{C}})(\underline{\mathbf{A}}\times\underline{\underline{\mathbf{I}}}) - \underline{\mathbf{C}}(\underline{\mathbf{A}}\times\underline{\mathbf{B}})$$

$$\begin{aligned}\text{adj}\,(\alpha\underline{\underline{\mathbf{I}}} + \underline{\mathbf{A}}\,\underline{\mathbf{B}} + \underline{\mathbf{C}}\times\underline{\underline{\mathbf{I}}}) = {}&\alpha^2\underline{\underline{\mathbf{I}}} - \alpha[\underline{\mathbf{A}}\,\underline{\mathbf{B}} - (\underline{\mathbf{A}}\cdot\underline{\mathbf{B}})\underline{\underline{\mathbf{I}}} + \underline{\mathbf{C}}\times\underline{\underline{\mathbf{I}}}] + \underline{\mathbf{A}}\,\underline{\mathbf{A}}\\ &- (\underline{\mathbf{B}}\cdot\underline{\mathbf{C}})(\underline{\mathbf{A}}\times\underline{\underline{\mathbf{I}}}) - \underline{\mathbf{C}}(\underline{\mathbf{A}}\times\underline{\mathbf{B}})\end{aligned}$$

$$\underline{\underline{\mathbf{A}}}^{-1} = \frac{\operatorname{adj} \underline{\underline{\mathbf{A}}}}{\det \underline{\underline{\mathbf{A}}}}$$
$$(\alpha\, \underline{\underline{\mathbf{A}}})^{-1} = \frac{1}{\alpha} \underline{\underline{\mathbf{A}}}^{-1}$$
$$(\underline{\underline{\mathbf{A}}}^{-1})^{-1} = \underline{\underline{\mathbf{A}}}$$
$$(\underline{\underline{\mathbf{A}}} \cdot \underline{\underline{\mathbf{B}}})^{-1} = \underline{\underline{\mathbf{B}}}^{-1} \cdot \underline{\underline{\mathbf{A}}}^{-1}$$
$$(\underline{\underline{\mathbf{A}}}^{-1})^{21} = (\underline{\underline{\mathbf{A}}}^{21})^{-1}.$$

A.3 Coordinate Systems

A.3.1 Cartesian coordinates

$$x, y, z \text{ or } x_i\ ,\ i = 1, 2, 3$$

with the orthonormal trihedron $\underline{\mathbf{e}}_x, \underline{\mathbf{e}}_y, \underline{\mathbf{e}}_z$ or $\underline{\mathbf{e}}_{x_i}\ ,\ i = 1, 2, 3$

$$\begin{aligned}\underline{\mathbf{R}} &= x\underline{\mathbf{e}}_x + y\underline{\mathbf{e}}_y + z\underline{\mathbf{e}}_z\\ &= x_i \underline{\mathbf{e}}_{x_i} \text{ (summation convention)}\end{aligned}$$

$$|\underline{\mathbf{R}} - \underline{\mathbf{R}}'| = \sqrt{(x-x')^2 + (y-y')^2 + (z-z')^2}$$

vector components: $A_{x_i} = \underline{\mathbf{A}} \cdot \underline{\mathbf{e}}_{x_i}\ ,\ i = 1, 2, 3$

$$\begin{aligned}\underline{\mathbf{A}} &= A_x\underline{\mathbf{e}}_x + A_y\underline{\mathbf{e}}_y + A_z\underline{\mathbf{e}}_z\\ &= \sum_{i=1}^{3} A_{x_i}\underline{\mathbf{e}}_{x_i}\\ &= A_{x_i}\underline{\mathbf{e}}_{x_i} \text{ (summation convention)}\\ &= A_i\underline{\mathbf{e}}_{x_i}\end{aligned}$$

$$\begin{aligned}\text{tensor components: } D_{x_i x_j} &= \underline{\mathbf{e}}_{x_i} \cdot \underline{\underline{\mathbf{D}}} \cdot \underline{\mathbf{e}}_{x_j}\\ &= \underline{\underline{\mathbf{D}}} : \underline{\mathbf{e}}_{x_j}\underline{\mathbf{e}}_{x_i}\ ,\ i, j = 1, 2, 3\end{aligned}$$

$$\begin{aligned}\underline{\underline{\mathbf{D}}} &= D_{xx}\underline{\mathbf{e}}_x\underline{\mathbf{e}}_x + D_{xy}\underline{\mathbf{e}}_x\underline{\mathbf{e}}_y + D_{xz}\underline{\mathbf{e}}_x\underline{\mathbf{e}}_z\\ &\quad + D_{yx}\underline{\mathbf{e}}_y\underline{\mathbf{e}}_x + D_{yy}\underline{\mathbf{e}}_y\underline{\mathbf{e}}_y + D_{yz}\underline{\mathbf{e}}_y\underline{\mathbf{e}}_z +\\ &\quad + D_{zx}\underline{\mathbf{e}}_z\underline{\mathbf{e}}_x + D_{zy}\underline{\mathbf{e}}_z\underline{\mathbf{e}}_y + D_{zz}\underline{\mathbf{e}}_z\underline{\mathbf{e}}_z\\ &= \sum_{i=1}^{3}\sum_{j=1}^{3} D_{x_i x_j}\underline{\mathbf{e}}_{x_i}\underline{\mathbf{e}}_{x_j}\\ &= D_{x_i} D_{x_j}\underline{\mathbf{e}}_{x_i}\underline{\mathbf{e}}_{x_j} \text{ (summation convention)}\end{aligned}$$

$$= D_{ij}\underline{\mathbf{e}}_{x_i}\underline{\mathbf{e}}_{x_j}$$
$$= \underline{\mathbf{e}}_x\underline{\mathbf{D}}_x + \underline{\mathbf{e}}_y\underline{\mathbf{D}}_y + \underline{\mathbf{e}}_z\underline{\mathbf{D}}_z$$

$$= \underline{\mathbf{D}}^x\underline{\mathbf{e}}_x + \underline{\mathbf{D}}^y\underline{\mathbf{e}}_y + \underline{\mathbf{D}}^z\underline{\mathbf{e}}_z$$

with row vectors $\underline{\mathbf{D}}_{x_i}$ and column vectors $\underline{\mathbf{D}}^{x_j}$ of $\underline{\underline{\mathbf{D}}}$, e.g.

$$\underline{\mathbf{D}}_x = D_{xx}\underline{\mathbf{e}}_x + D_{xy}\underline{\mathbf{e}}_y + D_{xz}\underline{\mathbf{e}}_z$$
$$\underline{\mathbf{D}}^x = D_{xx}\underline{\mathbf{e}}_x + D_{yx}\underline{\mathbf{e}}_y + D_{zx}\underline{\mathbf{e}}_z$$

$$\text{trace}\,\underline{\underline{\mathbf{D}}} = D_{xx} + D_{yy} + D_{zz}$$

$$\text{scalar products: } \underline{\mathbf{A}}\cdot\underline{\mathbf{B}} = A_xB_x + A_yB_y + A_zB_z$$
$$= A_iB_i$$

$$\mathrm{d}s^2 = \mathrm{d}x^2 + \mathrm{d}y^2 + \mathrm{d}z^2$$

$$\text{vector product: } \underline{\mathbf{A}}\times\underline{\mathbf{B}} = (A_yB_z - A_zB_y)\underline{\mathbf{e}}_x + (A_zB_x - A_xB_z)\underline{\mathbf{e}}_y$$
$$+ (A_xB_y - A_yB_x)\underline{\mathbf{e}}_z$$
$$= \varepsilon_{ijk}A_jB_k\underline{\mathbf{e}}_{x_i}$$
$$= \underline{\underline{\underline{\boldsymbol{\varepsilon}}}} : \underline{\mathbf{B}}\,\underline{\mathbf{A}}$$
$$= -\underline{\mathbf{A}}\cdot\underline{\underline{\underline{\boldsymbol{\varepsilon}}}}\cdot\underline{\mathbf{B}}$$
$$= \langle\underline{\mathbf{A}}\,\underline{\mathbf{B}}\rangle$$

$$\langle\underline{\underline{\mathbf{D}}}\rangle = \underline{\underline{\underline{\boldsymbol{\varepsilon}}}} : \underline{\underline{\mathbf{D}}}^{21} \quad \text{(rotation vector of } \underline{\underline{\mathbf{D}}}\text{)}$$
$$= \varepsilon_{ijk}D_{jk}\underline{\mathbf{e}}_{x_i}$$
$$= -\langle\underline{\underline{\mathbf{D}}}^{21}\rangle$$

$$\mathrm{d}V = \mathrm{d}x\mathrm{d}y\mathrm{d}z$$

$$\boldsymbol{\nabla} = \underline{\mathbf{e}}_x\frac{\partial}{\partial x} + \underline{\mathbf{e}}_y\frac{\partial}{\partial y} + \underline{\mathbf{e}}_z\frac{\partial}{\partial z}$$

$$\boldsymbol{\nabla}\Phi = \text{grad}\,\Phi$$
$$= \frac{\partial\Phi}{\partial x}\underline{\mathbf{e}}_x + \frac{\partial\Phi}{\partial y}\underline{\mathbf{e}}_y + \frac{\partial\Phi}{\partial z}\underline{\mathbf{e}}_z$$

$$\boldsymbol{\nabla}\boldsymbol{\nabla}\Phi = \frac{\partial^2\Phi}{\partial x^2} + \frac{\partial^2\Phi}{\partial x\partial y} + \frac{\partial^2\Phi}{\partial x\partial z}$$
$$+ \frac{\partial^2\Phi}{\partial y\partial x} + \frac{\partial^2\Phi}{\partial y^2} + \frac{\partial^2\Phi}{\partial y\partial z}$$
$$+ \frac{\partial^2\Phi}{\partial z\partial x} + \frac{\partial^2\Phi}{\partial z\partial y} + \frac{\partial^2\Phi}{\partial z^2}$$

$$\begin{aligned}\Delta\Phi &= \boldsymbol{\nabla}\cdot\boldsymbol{\nabla}\Phi \\ &= \frac{\partial^2\Phi}{\partial x^2}+\frac{\partial^2\Phi}{\partial y^2}+\frac{\partial^2\Phi}{\partial z^2}\end{aligned}$$

$$\operatorname{trace}(\boldsymbol{\nabla}\boldsymbol{\nabla}\Phi) = \Delta\Phi$$

$$\begin{aligned}\boldsymbol{\nabla}\cdot\underline{\mathbf{A}} &= \operatorname{div}\underline{\mathbf{A}} \\ &= \frac{\partial A_x}{\partial x}+\frac{\partial A_y}{\partial y}+\frac{\partial A_z}{\partial z}\end{aligned}$$

$$\begin{aligned}\boldsymbol{\nabla}\times\underline{\mathbf{A}} &= \operatorname{curl}\underline{\mathbf{A}} \\ &= \left(\frac{\partial A_z}{\partial y}-\frac{\partial A_y}{\partial z}\right)\underline{\mathbf{e}}_x+\left(\frac{\partial A_x}{\partial z}-\frac{\partial A_z}{\partial x}\right)\underline{\mathbf{e}}_y+\left(\frac{\partial A_y}{\partial x}-\frac{\partial A_x}{\partial y}\right)\underline{\mathbf{e}}_z\end{aligned}$$

$$\begin{aligned}\boldsymbol{\nabla}\underline{\mathbf{A}} &= \operatorname{grad}\underline{\mathbf{A}} \\ &= \frac{\partial A_x}{\partial x}\underline{\mathbf{e}}_x\underline{\mathbf{e}}_x+\frac{\partial A_y}{\partial x}\underline{\mathbf{e}}_x\underline{\mathbf{e}}_y+\frac{\partial A_z}{\partial x}\underline{\mathbf{e}}_x\underline{\mathbf{e}}_z \\ &\quad+\frac{\partial A_x}{\partial y}\underline{\mathbf{e}}_y\underline{\mathbf{e}}_x+\frac{\partial A_y}{\partial y}\underline{\mathbf{e}}_y\underline{\mathbf{e}}_y+\frac{\partial A_z}{\partial y}\underline{\mathbf{e}}_y\underline{\mathbf{e}}_z \\ &\quad+\frac{\partial A_x}{\partial z}\underline{\mathbf{e}}_z\underline{\mathbf{e}}_x+\frac{\partial A_y}{\partial z}\underline{\mathbf{e}}_z\underline{\mathbf{e}}_y+\frac{\partial A_z}{\partial z}\underline{\mathbf{e}}_z\underline{\mathbf{e}}_z \\ &= \boldsymbol{\nabla}A_x\underline{\mathbf{e}}_x+\boldsymbol{\nabla}A_y\underline{\mathbf{e}}_y+\boldsymbol{\nabla}A_z\underline{\mathbf{e}}_z\end{aligned}$$

$$\begin{aligned}(\boldsymbol{\nabla}\cdot\underline{\underline{\mathbf{D}}})\cdot\underline{\mathbf{e}}_x &= \frac{\partial D_{xx}}{\partial x}+\frac{\partial D_{yx}}{\partial y}+\frac{\partial D_{zx}}{\partial z} \\ (\boldsymbol{\nabla}\cdot\underline{\underline{\mathbf{D}}})\cdot\underline{\mathbf{e}}_y &= \frac{\partial D_{xy}}{\partial x}+\frac{\partial D_{yy}}{\partial y}+\frac{\partial D_{zy}}{\partial z} \\ (\boldsymbol{\nabla}\cdot\underline{\underline{\mathbf{D}}})\cdot\underline{\mathbf{e}}_z &= \frac{\partial D_{xz}}{\partial x}+\frac{\partial D_{yz}}{\partial y}+\frac{\partial D_{zz}}{\partial z}\end{aligned}$$

$$\begin{aligned}\operatorname{trace}(\boldsymbol{\nabla}\underline{\mathbf{A}}) &= \boldsymbol{\nabla}\cdot\underline{\mathbf{A}} \\ \langle\boldsymbol{\nabla}\underline{\mathbf{A}}\rangle &= \boldsymbol{\nabla}\times\underline{\mathbf{A}}\end{aligned}$$

$$\begin{aligned}\Delta\underline{\mathbf{A}} &= \boldsymbol{\nabla}\cdot\boldsymbol{\nabla}\underline{\mathbf{A}} \\ &= \boldsymbol{\nabla}\boldsymbol{\nabla}\cdot\underline{\mathbf{A}}-\boldsymbol{\nabla}\times\boldsymbol{\nabla}\times\underline{\mathbf{A}} \\ &= \left(\frac{\partial^2 A_x}{\partial x^2}+\frac{\partial^2 A_x}{\partial y^2}+\frac{\partial^2 A_x}{\partial z^2}\right)\underline{\mathbf{e}}_x \\ &\quad+\left(\frac{\partial^2 A_y}{\partial x^2}+\frac{\partial^2 A_y}{\partial y^2}+\frac{\partial^2 A_y}{\partial z^2}\right)\underline{\mathbf{e}}_y \\ &\quad+\left(\frac{\partial^2 A_z}{\partial x^2}+\frac{\partial^2 A_z}{\partial y^2}+\frac{\partial^2 A_z}{\partial z^2}\right)\underline{\mathbf{e}}_z \\ &= \underline{\mathbf{e}}_x\Delta A_x+\underline{\mathbf{e}}_y\Delta A_y+\underline{\mathbf{e}}_z\Delta A_z.\end{aligned}$$

A.4 Curvilinear Orthogonal Coordinates

$$\xi_1, \xi_2, \xi_3$$

with the orthonormal trihedron $\underline{\mathbf{e}}_{\xi_1}, \underline{\mathbf{e}}_{\xi_2}, \underline{\mathbf{e}}_{\xi_3}$

$$\begin{aligned} x &= x(\xi_1, \xi_2, \xi_3) \\ y &= y(\xi_1, \xi_2, \xi_3) \\ z &= z(\xi_1, \xi_2, \xi_3) \end{aligned}$$

$$\underline{\mathbf{R}} = x(\xi_1, \xi_2, \xi_3)\underline{\mathbf{e}}_x + y(\xi_1, \xi_2, \xi_3)\underline{\mathbf{e}}_y + z(\xi_1, \xi_2, \xi_3)\underline{\mathbf{e}}_z$$

$$h_{\xi_i} = \sqrt{\left(\frac{\partial x}{\partial \xi_i}\right)^2 + \left(\frac{\partial y}{\partial \xi_i}\right)^2 + \left(\frac{\partial z}{\partial \xi_i}\right)^2}$$

$$\begin{aligned} \underline{\mathbf{e}}_{\xi_i} &= \frac{1}{h_{\xi_i}} \frac{\partial \underline{\mathbf{R}}}{\partial \xi_i} \\ &= \frac{1}{h_{\xi_i}} \frac{\partial x_j}{\partial \xi_i} \underline{\mathbf{e}}_{x_j} \\ &= \gamma_{ij} \underline{\mathbf{e}}_{x_j} \end{aligned}$$

$$\begin{aligned} \gamma_{ij} &= \underline{\mathbf{e}}_{\xi_i} \cdot \underline{\mathbf{e}}_{x_j} \\ \gamma_{ij}\gamma_{kj} &= \delta_{ik} \\ \gamma_{ij}\gamma_{ik} &= \delta_{jk} \end{aligned}$$

vector components: $A_{\xi_i} = \underline{\mathbf{A}} \cdot \underline{\mathbf{e}}_{\xi_i}$, $i = 1, 2, 3$

$$\begin{aligned} \underline{\mathbf{A}} &= A_{\xi_1}\underline{\mathbf{e}}_{\xi_1} + A_{\xi_2}\underline{\mathbf{e}}_{\xi_2} + A_{\xi_3}\underline{\mathbf{e}}_{\xi_3} \\ &= \sum_{i=1}^{3} A_{\xi_i}\underline{\mathbf{e}}_{\xi_i} \\ &= A_i\underline{\mathbf{e}}_{\xi_i} \text{ (summation convention)} \end{aligned}$$

$$\begin{aligned} A_{\xi_i}\underline{\mathbf{e}}_{\xi_i} &= A_{x_j}\underline{\mathbf{e}}_{x_j} \\ A_{\xi_i} &= \gamma_{ij} A_{x_j} \\ A_{x_i} &= \gamma_{ji} A_{\xi_j} \end{aligned}$$

tensor components: $D_{\xi_i\xi_j} = \underline{\mathbf{e}}_{\xi_i} \cdot \underline{\underline{\mathbf{D}}} \cdot \underline{\mathbf{e}}_{\xi_j} = \underline{\underline{\mathbf{D}}} : \underline{\mathbf{e}}_{\xi_j}\underline{\mathbf{e}}_{\xi_i}$, $i, j = 1, 2, 3$

$$\begin{aligned} \underline{\underline{\mathbf{D}}} = {} & D_{\xi_1\xi_1}\underline{\mathbf{e}}_{\xi_1}\underline{\mathbf{e}}_{\xi_1} + D_{\xi_1\xi_2}\underline{\mathbf{e}}_{\xi_1}\underline{\mathbf{e}}_{\xi_2} + D_{\xi_1\xi_3}\underline{\mathbf{e}}_{\xi_1}\underline{\mathbf{e}}_{\xi_3} \\ & + D_{\xi_2\xi_1}\underline{\mathbf{e}}_{\xi_2}\underline{\mathbf{e}}_{\xi_1} + D_{\xi_2\xi_2}\underline{\mathbf{e}}_{\xi_2}\underline{\mathbf{e}}_{\xi_2} + D_{\xi_2\xi_3}\underline{\mathbf{e}}_{\xi_2}\underline{\mathbf{e}}_{\xi_3} \\ & + D_{\xi_3\xi_1}\underline{\mathbf{e}}_{\xi_3}\underline{\mathbf{e}}_{\xi_1} + D_{\xi_3\xi_2}\underline{\mathbf{e}}_{\xi_3}\underline{\mathbf{e}}_{\xi_2} + +D_{\xi_3\xi_3}\underline{\mathbf{e}}_{\xi_3}\underline{\mathbf{e}}_{\xi_3} \end{aligned}$$

$$= \sum_{i=1}^{3}\sum_{j=1}^{3} D_{\xi_i\xi_j}\underline{\mathbf{e}}_{\xi_i}\underline{\mathbf{e}}_{\xi_j}$$

$$= D_{ij}\underline{\mathbf{e}}_{\xi_i}\underline{\mathbf{e}}_{\xi_j} \text{ (summation convention)}$$

$$\underline{\underline{\mathbf{I}}} = \delta_{ij}\underline{\mathbf{e}}_{\xi_i}\underline{\mathbf{e}}_{\xi_j}$$

$$\underline{\underline{\underline{\boldsymbol{\varepsilon}}}} = \varepsilon_{ijk}\underline{\mathbf{e}}_{\xi_i}\underline{\mathbf{e}}_{\xi_j}\underline{\mathbf{e}}_{\xi_k}$$

$$D_{\xi_i\xi_j}\underline{\mathbf{e}}_{\xi_i}\underline{\mathbf{e}}_{\xi_j} = D_{x_k x_l}\underline{\mathbf{e}}_{x_k}\underline{\mathbf{e}}_{x_l}$$

$$D_{\xi_k\xi_l} = \gamma_{ki}\gamma_{lj}D_{x_i x_j}$$

$$D_{x_k x_l} = \gamma_{ik}\gamma_{jl}D_{\xi_i\xi_j}$$

$$\text{trace}\,\underline{\underline{\mathbf{D}}} = D_{\xi_1\xi_1} + D_{\xi_2\xi_2} + D_{\xi_3\xi_3}$$

$$= D_{xx} + D_{yy} + D_{zz}$$

$$\text{scalar product: } \underline{\mathbf{A}}\cdot\underline{\mathbf{B}} = A_{\xi_1}B_{\xi_1} + A_{\xi_2}B_{\xi_2} + A_{\xi_3}B_{\xi_3}$$

$$= A_i B_i$$

$$= A_x B_x + A_y B_y + A_z B_z$$

$$\text{double contraction: } \underline{\underline{\mathbf{C}}} : \underline{\underline{\mathbf{D}}} = C_{\xi_i\xi_j}D_{\xi_j\xi_i} = C_{x_i x_j}D_{x_j x_i}$$

$$\mathrm{d}s^2 = h_{\xi_1}^2\mathrm{d}\xi_1^2 + h_{\xi_2}^2\mathrm{d}\xi_2^2 + h_{\xi_3}^2\mathrm{d}\xi_3^2$$

$$= \mathrm{d}x^2 + \mathrm{d}y^2 + \mathrm{d}z^2$$

$$\text{vector product: } \underline{\mathbf{A}}\times\underline{\mathbf{B}} = (A_{\xi_2}B_{\xi_3} - A_{\xi_3}B_{\xi_2})\underline{\mathbf{e}}_{\xi_1} + (A_{\xi_3}B_{\xi_1} - A_{\xi_1}B_{\xi_3})\underline{\mathbf{e}}_{\xi_2}$$

$$+ (A_{\xi_1}B_{\xi_2} - A_{\xi_2}B_{\xi_1})\underline{\mathbf{e}}_{\xi_3}$$

$$= \varepsilon_{ijk}A_j B_k\underline{\mathbf{e}}_{\xi_i}$$

$$= \underline{\underline{\underline{\boldsymbol{\varepsilon}}}} : \underline{\mathbf{B}}\,\underline{\mathbf{A}}$$

$$= -\underline{\mathbf{A}}\cdot\underline{\underline{\underline{\boldsymbol{\varepsilon}}}}\cdot\underline{\mathbf{B}}$$

$$= \langle\underline{\mathbf{A}}\,\underline{\mathbf{B}}\rangle$$

$$\mathrm{d}V = h_{\xi_1}h_{\xi_2}h_{\xi_3}\mathrm{d}\xi_1\mathrm{d}\xi_2\mathrm{d}\xi_3$$

$$\boldsymbol{\nabla} = \underline{\mathbf{e}}_{\xi_i}\frac{1}{h_{\xi_i}}\frac{\partial}{\partial\xi_i}$$

$$\boldsymbol{\nabla}\Phi = \text{grad }\Phi$$

$$= \frac{1}{h_{\xi_1}}\frac{\partial\Phi}{\partial\xi_1}\underline{\mathbf{e}}_{\xi_1} + \frac{1}{h_{\xi_2}}\frac{\partial\Phi}{\partial\xi_2}\underline{\mathbf{e}}_{\xi_2} + \frac{1}{h_{\xi_3}}\frac{\partial\Phi}{\partial\xi_3}\underline{\mathbf{e}}_{\xi_3}$$

$$\boldsymbol{\nabla}\xi_i = \frac{1}{h_{\xi_i}}\underline{\mathbf{e}}_{\xi_i}$$

$$
\begin{aligned}
\nabla \cdot \underline{\mathbf{A}} &= \operatorname{div} \underline{\mathbf{A}} \\
&= \frac{1}{h_{\xi_i}} \underline{\mathbf{e}}_{\xi_i} \cdot \frac{\partial}{\partial \xi_i} \left(A_{\xi_j} \underline{\mathbf{e}}_{\xi_j} \right) \\
&= \frac{1}{h_{\xi_i}} \underline{\mathbf{e}}_{\xi_i} \cdot \left(\frac{\partial A_{\xi_j}}{\partial \xi_i} \underline{\mathbf{e}}_{\xi_j} + A_{\xi_j} \frac{\partial \underline{\mathbf{e}}_{\xi_j}}{\partial \xi_i} \right) \\
&= \frac{1}{h_{\xi_1} h_{\xi_2} h_{\xi_3}} \left(\frac{\partial A_{\xi_1} h_{\xi_2} h_{\xi_3}}{\partial \xi_1} + \frac{\partial A_{\xi_2} h_{\xi_1} h_{\xi_3}}{\partial \xi_2} + \frac{\partial A_{\xi_3} h_{\xi_1} h_{\xi_2}}{\partial \xi_3} \right),
\end{aligned}
$$

where

$$
\frac{\partial \underline{\mathbf{e}}_{\xi_j}}{\partial \xi_i} = \frac{1}{h_{\xi_j}} \underline{\mathbf{e}}_{\xi_i} \frac{\partial h_{\xi_i}}{\partial \xi_j} (1 - \delta_{ij}) - \frac{1}{h_{\xi_k}} \underline{\mathbf{e}}_{\xi_k} \frac{\partial h_{\xi_j}}{\partial \xi_k} \delta_{ij} (1 - \delta_{kj}).
$$

In the second term, summation is over k, where the value $k = j$ does not appear in the sum due to the factor $1 - \delta_{kj}$.

Christoffel symbols of the second kind:

$$
\alpha_k(j, i) = \begin{pmatrix} & k & \\ j & & i \end{pmatrix}
$$

$$
\frac{\partial \underline{\mathbf{e}}_{\xi_j}}{\partial \xi_i} = \alpha_k(j, i) \underline{\mathbf{e}}_{\xi_k}
$$

$$
\begin{pmatrix} & k & \\ j & & j \end{pmatrix} = -\frac{1}{h_{\xi_k}} \frac{\partial h_{\xi_j}}{\partial \xi_k}, \quad \begin{pmatrix} & j & \\ j & & j \end{pmatrix} = 0
$$

$$
\begin{pmatrix} & i & \\ j & & i \end{pmatrix} = \frac{1}{h_{\xi_j}} \frac{\partial h_{\xi_i}}{\partial \xi_j}
$$

$$
\begin{pmatrix} & j & \\ j & & i \end{pmatrix} = 0, \quad \begin{pmatrix} & k & \\ j & & i \end{pmatrix} = 0 \text{ for } i \neq j \neq k
$$

$$
\begin{aligned}
\Delta \Phi &= \nabla \cdot \nabla \Phi \\
&= \frac{1}{h_{\xi_i}} \underline{\mathbf{e}}_{\xi_i} \cdot \frac{\partial}{\partial \xi_i} \left(\underline{\mathbf{e}}_{\xi_j} \frac{1}{h_{\xi_j}} \frac{\partial}{\partial \xi_j} \right) \\
&= \frac{1}{h_{\xi_1} h_{\xi_2} h_{\xi_3}} \left[\frac{\partial}{\partial \xi_1} \left(\frac{h_{\xi_2} h_{\xi_3}}{h_{\xi_1}} \frac{\partial \Phi}{\partial \xi_1} \right) + \frac{\partial}{\partial \xi_2} \left(\frac{h_{\xi_3} h_{\xi_1}}{h_{\xi_2}} \frac{\partial \Phi}{\partial \xi_2} \right) \right. \\
&\qquad \left. + \frac{\partial}{\partial \xi_3} \left(\frac{h_{\xi_1} h_{\xi_2}}{h_{\xi_3}} \frac{\partial \Phi}{\partial \xi_3} \right) \right]
\end{aligned}
$$

$$
\nabla \underline{\mathbf{A}} = \frac{1}{h_{\xi_i}} \left(\frac{\partial A_{\xi_j}}{\partial \xi_i} \underline{\mathbf{e}}_{\xi_i} \underline{\mathbf{e}}_{\xi_j} + A_{\xi_j} \underline{\mathbf{e}}_{\xi_i} \frac{\partial \underline{\mathbf{e}}_{\xi_j}}{\partial \xi_i} \right)
$$

$$
\operatorname{trace} (\nabla \underline{\mathbf{A}}) = \nabla \cdot \underline{\mathbf{A}}
$$

$$
\langle \nabla \underline{\mathbf{A}} \rangle = \nabla \times \underline{\mathbf{A}}
$$

$$
\begin{aligned}
\nabla \times \underline{\mathbf{A}} &= \operatorname{curl} \underline{\mathbf{A}} \\
&= \underline{\underline{\underline{\boldsymbol{\varepsilon}}}} : (\nabla \underline{\mathbf{A}})^{21} \\
&= \frac{1}{h_{\xi_i}} \underline{\mathbf{e}}_{\xi_i} \times \frac{\partial}{\partial \xi_i} (A_{\xi_j} \underline{\mathbf{e}}_{\xi_j}) \\
&= \frac{1}{h_{\xi_i}} \underline{\mathbf{e}}_{\xi_i} \times \left[\frac{\partial A_{\xi_j}}{\partial \xi_i} \underline{\mathbf{e}}_{\xi_j} + A_{\xi_j} \begin{pmatrix} & k & \\ j & & i \end{pmatrix} \underline{\mathbf{e}}_{\xi_k} \right] \\
&= \frac{1}{h_{\xi_2} h_{\xi_3}} \left[\frac{\partial A_{\xi_3} h_{\xi_3}}{\partial \xi_2} - \frac{\partial A_{\xi_2} h_{\xi_2}}{\partial \xi_3} \right] \underline{\mathbf{e}}_{\xi_1} \\
&\quad + \frac{1}{h_{\xi_1} h_{\xi_3}} \left[\frac{\partial A_{\xi_1} h_{\xi_1}}{\partial \xi_3} - \frac{\partial A_{\xi_3} h_{\xi_3}}{\partial \xi_1} \right] \underline{\mathbf{e}}_{\xi_2} \\
&\quad + \frac{1}{h_{\xi_1} h_{\xi_2}} \left[\frac{\partial A_{\xi_2} h_{\xi_2}}{\partial \xi_1} - \frac{\partial A_{\xi_1} h_{\xi_1}}{\partial \xi_2} \right] \underline{\mathbf{e}}_{\xi_3}
\end{aligned}
$$

$$
\begin{aligned}
\Delta \underline{\mathbf{A}} &= \nabla \cdot \nabla \underline{\mathbf{A}} \\
&= \nabla \nabla \cdot \underline{\mathbf{A}} - \nabla \times \nabla \times \underline{\mathbf{A}}.
\end{aligned}
$$

A.5 Cylindrical Coordinates

$$r, \varphi, z$$

with the orthonormal trihedron $\underline{\mathbf{e}}_r, \underline{\mathbf{e}}_\varphi, \underline{\mathbf{e}}_z$

$$
\begin{aligned}
x &= r \cos \varphi \qquad & r &= \sqrt{x^2 + y^2} \\
y &= r \sin \varphi \qquad & \varphi &= \arctan \tfrac{y}{x} \\
z &= z \qquad & z &= z
\end{aligned}
$$

$$
\begin{aligned}
\underline{\mathbf{R}} &= r \cos \varphi \, \underline{\mathbf{e}}_x + r \sin \varphi \, \underline{\mathbf{e}}_y + z \underline{\mathbf{e}}_z \\
&= r \underline{\mathbf{e}}_r + z \underline{\mathbf{e}}_z
\end{aligned}
$$

$$|\underline{\mathbf{R}} - \underline{\mathbf{R}}'| = \sqrt{r^2 + r'^2 - 2rr' \cos(\varphi - \varphi') + (z - z')^2}$$

$$
\begin{aligned}
h_r &= 1 \\
h_\varphi &= r \\
h_z &= 1
\end{aligned}
$$

$$
\begin{aligned}
\underline{\mathbf{e}}_r &= \cos \varphi \, \underline{\mathbf{e}}_x + \sin \varphi \, \underline{\mathbf{e}}_y \qquad & \underline{\mathbf{e}}_x &= \cos \varphi \, \underline{\mathbf{e}}_r - \sin \varphi \, \underline{\mathbf{e}}_\varphi \\
\underline{\mathbf{e}}_\varphi &= -\sin \varphi \, \underline{\mathbf{e}}_x + \cos \varphi \, \underline{\mathbf{e}}_y \qquad & \underline{\mathbf{e}}_y &= \sin \varphi \, \underline{\mathbf{e}}_r + \cos \varphi \, \underline{\mathbf{e}}_\varphi \\
\underline{\mathbf{e}}_z &= \underline{\mathbf{e}}_z \qquad & \underline{\mathbf{e}}_z &= \underline{\mathbf{e}}_z
\end{aligned}
$$

$$\text{vector components: } A_r = \underline{\mathbf{A}} \cdot \underline{\mathbf{e}}_r$$
$$A_\varphi = \underline{\mathbf{A}} \cdot \underline{\mathbf{e}}_\varphi$$
$$A_z = \underline{\mathbf{A}} \cdot \underline{\mathbf{e}}_z$$

$$\underline{\mathbf{A}} = A_r \underline{\mathbf{e}}_r + A_\varphi \underline{\mathbf{e}}_\varphi + A_z \underline{\mathbf{e}}_z$$

$$\begin{pmatrix} A_r \\ A_\varphi \\ A_z \end{pmatrix} = \begin{pmatrix} \cos\varphi & \sin\varphi & 0 \\ -\sin\varphi & \cos\varphi & 0 \\ 0 & 0 & 1 \end{pmatrix} \begin{pmatrix} A_x \\ A_y \\ A_z \end{pmatrix}$$

$$\begin{pmatrix} A_x \\ A_y \\ A_z \end{pmatrix} = \begin{pmatrix} \cos\varphi & -\sin\varphi & 0 \\ \sin\varphi & \cos\varphi & 0 \\ 0 & 0 & 1 \end{pmatrix} \begin{pmatrix} A_r \\ A_\varphi \\ A_z \end{pmatrix}$$

$$\text{tensor components: } D_{rr} = \underline{\mathbf{e}}_r \cdot \underline{\underline{\mathbf{D}}} \cdot \underline{\mathbf{e}}_r$$
$$= \underline{\underline{\mathbf{D}}} : \underline{\mathbf{e}}_r \underline{\mathbf{e}}_r$$
$$D_{r\varphi} = \underline{\mathbf{e}}_r \cdot \underline{\underline{\mathbf{D}}} \cdot \underline{\mathbf{e}}_\varphi$$
$$= \underline{\underline{\mathbf{D}}} : \underline{\mathbf{e}}_\varphi \underline{\mathbf{e}}_r$$

etc.

$$\begin{aligned} \underline{\underline{\mathbf{D}}} = {} & D_{rr} \underline{\mathbf{e}}_r \underline{\mathbf{e}}_r + D_{r\varphi} \underline{\mathbf{e}}_r \underline{\mathbf{e}}_\varphi + D_{rz} \underline{\mathbf{e}}_r \underline{\mathbf{e}}_z \\ & + D_{\varphi r} \underline{\mathbf{e}}_\varphi \underline{\mathbf{e}}_r + D_{\varphi\varphi} \underline{\mathbf{e}}_\varphi \underline{\mathbf{e}}_\varphi + D_{\varphi z} \underline{\mathbf{e}}_\varphi \underline{\mathbf{e}}_z \\ & + D_{zr} \underline{\mathbf{e}}_z \underline{\mathbf{e}}_r + D_{z\varphi} \underline{\mathbf{e}}_z \underline{\mathbf{e}}_\varphi + D_{zz} \underline{\mathbf{e}}_z \underline{\mathbf{e}}_z \end{aligned}$$

$$\underline{\underline{\mathbf{I}}} = \underline{\mathbf{e}}_r \underline{\mathbf{e}}_r + \underline{\mathbf{e}}_\varphi \underline{\mathbf{e}}_\varphi + \underline{\mathbf{e}}_z \underline{\mathbf{e}}_z$$

$$\begin{pmatrix} D_{rr} & D_{r\varphi} & D_{rz} \\ D_{\varphi r} & D_{\varphi\varphi} & D_{\varphi z} \\ D_{zr} & D_{z\varphi} & D_{zz} \end{pmatrix} = \begin{pmatrix} \cos\varphi & \sin\varphi & 0 \\ -\sin\varphi & \cos\varphi & 0 \\ 0 & 0 & 1 \end{pmatrix} \begin{pmatrix} D_{xx} & D_{xy} & D_{xz} \\ D_{yx} & D_{yy} & D_{yz} \\ D_{zx} & D_{zy} & D_{zz} \end{pmatrix}$$
$$\times \begin{pmatrix} \cos\varphi & -\sin\varphi & 0 \\ \sin\varphi & \cos\varphi & 0 \\ 0 & 0 & 1 \end{pmatrix}$$

for example:

$$D_{rr} = D_{xx} \cos^2\varphi + (D_{xy} + D_{yx}) \cos\varphi \sin\varphi + D_{yy} \sin^2\varphi$$
$$D_{\varphi\varphi} = D_{xx} \sin^2\varphi - (D_{xy} + D_{yx}) \cos\varphi \sin\varphi + D_{yy} \cos^2\varphi$$

$$\begin{aligned} \operatorname{trace} \underline{\underline{\mathbf{D}}} &= D_{rr} + D_{\varphi\varphi} + D_{zz} \\ &= D_{xx} + D_{yy} + D_{zz} \end{aligned}$$

$$\begin{aligned} \text{scalar product: } \underline{\mathbf{A}} \cdot \underline{\mathbf{B}} &= A_r B_r + A_\varphi B_\varphi + A_z B_z \\ &= A_x B_x + A_y B_y + A_z B_z \end{aligned}$$

$$\mathrm{d}s^2 = \mathrm{d}r^2 + r^2 \mathrm{d}\varphi^2 + \mathrm{d}z^2$$

$$\text{vector product: } \underline{\mathbf{A}} \times \underline{\mathbf{B}} = (A_\varphi B_z - A_z B_\varphi)\underline{\mathbf{e}}_r + (A_z B_r - A_r B_z)\underline{\mathbf{e}}_\varphi + (A_r B_\varphi - A_\varphi B_r)\underline{\mathbf{e}}_z = \underline{\underline{\underline{\boldsymbol{\varepsilon}}}} : \underline{\mathbf{B}}\,\underline{\mathbf{A}} = -\underline{\mathbf{A}} \cdot \underline{\underline{\underline{\boldsymbol{\varepsilon}}}} \cdot \underline{\mathbf{B}} = \langle \underline{\mathbf{A}}\,\underline{\mathbf{B}} \rangle$$

$$\mathrm{d}V = r\mathrm{d}r\mathrm{d}\varphi\mathrm{d}z$$

$$\boldsymbol{\nabla} = \underline{\mathbf{e}}_r \frac{\partial}{\partial r} + \frac{1}{r}\underline{\mathbf{e}}_\varphi \frac{\partial}{\partial \varphi} + \underline{\mathbf{e}}_z \frac{\partial}{\partial z}$$

$$\boldsymbol{\nabla}\Phi = \operatorname{grad}\Phi = \frac{\partial \Phi}{\partial r}\underline{\mathbf{e}}_r + \frac{1}{r}\frac{\partial \Phi}{\partial \varphi}\underline{\mathbf{e}}_\varphi + \frac{\partial \Phi}{\partial z}\underline{\mathbf{e}}_z$$

$$\boldsymbol{\nabla}\boldsymbol{\nabla}\Phi = \frac{\partial^2 \Phi}{\partial r^2}\underline{\mathbf{e}}_r\underline{\mathbf{e}}_r + \frac{1}{r}\left(\frac{\partial^2 \Phi}{\partial r \partial \varphi} - \frac{1}{r}\frac{\partial \Phi}{\partial \varphi}\right)\underline{\mathbf{e}}_r\underline{\mathbf{e}}_\varphi + \frac{\partial^2 \Phi}{\partial r \partial z}\underline{\mathbf{e}}_r\underline{\mathbf{e}}_z + \frac{1}{r}\left(\frac{\partial^2 \Phi}{\partial r \partial \varphi} - \frac{1}{r}\frac{\partial \Phi}{\partial \varphi}\right)\underline{\mathbf{e}}_\varphi\underline{\mathbf{e}}_r + \frac{1}{r^2}\left(\frac{\partial^2 \Phi}{\partial \varphi^2} + r\frac{\partial \Phi}{\partial r}\right)\underline{\mathbf{e}}_\varphi\underline{\mathbf{e}}_\varphi + \frac{1}{r}\frac{\partial^2 \Phi}{\partial \varphi \partial z}\underline{\mathbf{e}}_\varphi\underline{\mathbf{e}}_z + \frac{\partial^2 \Phi}{\partial r \partial z}\underline{\mathbf{e}}_z\underline{\mathbf{e}}_r + \frac{1}{r}\frac{\partial^2 \Phi}{\partial \varphi \partial z}\underline{\mathbf{e}}_z\underline{\mathbf{e}}_\varphi + \frac{\partial^2 \Phi}{\partial z^2}\underline{\mathbf{e}}_z\underline{\mathbf{e}}_z.$$

Christoffel symbols of the second kind:

$$\frac{\partial \underline{\mathbf{e}}_r}{\partial r} = \underline{\mathbf{0}} \qquad \frac{\partial \underline{\mathbf{e}}_\varphi}{\partial r} = \underline{\mathbf{0}} \qquad \frac{\partial \underline{\mathbf{e}}_z}{\partial r} = \underline{\mathbf{0}}$$

$$\frac{\partial \underline{\mathbf{e}}_r}{\partial \varphi} = \underline{\mathbf{e}}_\varphi = \alpha_2(1,2)\underline{\mathbf{e}}_\varphi \qquad \frac{\partial \underline{\mathbf{e}}_\varphi}{\partial \varphi} = -\underline{\mathbf{e}}_r = \alpha_1(2,2)\underline{\mathbf{e}}_r \qquad \frac{\partial \underline{\mathbf{e}}_z}{\partial \varphi} = \underline{\mathbf{0}}$$

$$\frac{\partial \underline{\mathbf{e}}_r}{\partial z} = \underline{\mathbf{0}} \qquad \frac{\partial \underline{\mathbf{e}}_\varphi}{\partial z} = \underline{\mathbf{0}} \qquad \frac{\partial \underline{\mathbf{e}}_z}{\partial z} = \underline{\mathbf{0}}$$

$$\Delta\Phi = \frac{1}{r}\frac{\partial}{\partial r}\left(r\frac{\partial \Phi}{\partial r}\right) + \frac{1}{r^2}\frac{\partial^2 \Phi}{\partial \varphi^2} + \frac{\partial^2 \Phi}{\partial z^2} = \frac{\partial^2 \Phi}{\partial r^2} + \frac{1}{r}\frac{\partial \Phi}{\partial r} + \frac{1}{r^2}\frac{\partial^2 \Phi}{\partial \varphi^2} + \frac{\partial^2 \Phi}{\partial z^2}$$

$$\operatorname{trace}(\boldsymbol{\nabla}\boldsymbol{\nabla}\Phi) = \Delta\Phi \tag{A.1}$$

$$\boldsymbol{\nabla} \cdot \underline{\mathbf{A}} = \operatorname{div}\underline{\mathbf{A}} = \frac{1}{r}\frac{\partial r A_r}{\partial r} + \frac{1}{r}\frac{\partial A_\varphi}{\partial \varphi} + \frac{\partial A_z}{\partial z}$$

$$\boldsymbol{\nabla} \cdot \underline{\mathbf{e}}_r = \frac{1}{r}$$

$$\boldsymbol{\nabla} \cdot \underline{\mathbf{e}}_\varphi = 0$$

$$\nabla \cdot \underline{\mathbf{e}}_z = 0$$

$$\begin{aligned}\nabla \times \underline{\mathbf{A}} &= \operatorname{curl} \underline{\mathbf{A}}\\ &= \left(\frac{1}{r}\frac{\partial A_z}{\partial \varphi} - \frac{\partial A_\varphi}{\partial z}\right)\underline{\mathbf{e}}_r + \left(\frac{\partial A_r}{\partial z} - \frac{\partial A_z}{\partial r}\right)\underline{\mathbf{e}}_\varphi + \left(\frac{1}{r}\frac{\partial r A_\varphi}{\partial r} - \frac{1}{r}\frac{\partial A_r}{\partial \varphi}\right)\underline{\mathbf{e}}_z\end{aligned}$$

$$\nabla \times \underline{\mathbf{e}}_r = \underline{\mathbf{0}}$$

$$\nabla \times \underline{\mathbf{e}}_\varphi = \frac{1}{r}\underline{\mathbf{e}}_z$$

$$\nabla \times \underline{\mathbf{e}}_z = \underline{\mathbf{0}}$$

$$\begin{aligned}\nabla \underline{\mathbf{A}} &= \operatorname{grad} \underline{\mathbf{A}}\\ &= \frac{\partial A_r}{\partial r}\underline{\mathbf{e}}_r\underline{\mathbf{e}}_r + \frac{\partial A_\varphi}{\partial r}\underline{\mathbf{e}}_r\underline{\mathbf{e}}_\varphi + \frac{\partial A_z}{\partial r}\underline{\mathbf{e}}_r\underline{\mathbf{e}}_z\\ &\quad + \frac{1}{r}\left(\frac{\partial A_r}{\partial \varphi} - A_\varphi\right)\underline{\mathbf{e}}_\varphi\underline{\mathbf{e}}_r + \frac{1}{r}\left(\frac{\partial A_\varphi}{\partial \varphi} + A_r\right)\underline{\mathbf{e}}_\varphi\underline{\mathbf{e}}_\varphi + \frac{1}{r}\frac{\partial A_z}{\partial \varphi}\underline{\mathbf{e}}_\varphi\underline{\mathbf{e}}_z\\ &\quad + \frac{\partial A_r}{\partial z}\underline{\mathbf{e}}_z\underline{\mathbf{e}}_r + \frac{\partial A_\varphi}{\partial z}\underline{\mathbf{e}}_z\underline{\mathbf{e}}_\varphi + \frac{\partial A_z}{\partial z}\underline{\mathbf{e}}_z\underline{\mathbf{e}}_z\end{aligned}$$

$$(\nabla \cdot \underline{\underline{\mathbf{D}}}) \cdot \underline{\mathbf{e}}_r = \frac{1}{r}\frac{\partial r D_{rr}}{\partial r} + \frac{1}{r}\frac{\partial D_{\varphi r}}{\partial \varphi} + \frac{\partial D_{zr}}{\partial z} - \frac{D_{\varphi\varphi}}{r}$$

$$(\nabla \cdot \underline{\underline{\mathbf{D}}}) \cdot \underline{\mathbf{e}}_\varphi = \frac{1}{r}\frac{\partial r D_{r\varphi}}{\partial r} + \frac{1}{r}\frac{\partial D_{\varphi\varphi}}{\partial \varphi} + \frac{\partial D_{z\varphi}}{\partial z} + \frac{D_{\varphi r}}{r}$$

$$(\nabla \cdot \underline{\underline{\mathbf{D}}}) \cdot \underline{\mathbf{e}}_z = \frac{1}{r}\frac{\partial r D_{rz}}{\partial r} + \frac{1}{r}\frac{\partial D_{\varphi z}}{\partial \varphi} + \frac{\partial D_{zz}}{\partial z}$$

$$\operatorname{trace}(\nabla \underline{\mathbf{A}}) = \nabla \cdot \underline{\mathbf{A}}$$

$$\langle \nabla \underline{\mathbf{A}} \rangle = \nabla \times \underline{\mathbf{A}}$$

$$\begin{aligned}\nabla\nabla \cdot \underline{\mathbf{A}} = &\left(\frac{\partial^2 A_r}{\partial r^2} + \frac{\partial^2 A_z}{\partial r \partial z} + \frac{1}{r}\frac{\partial^2 A_\varphi}{\partial r \partial \varphi} + \frac{1}{r}\frac{\partial A_r}{\partial r} - \frac{1}{r^2}\frac{\partial A_\varphi}{\partial \varphi} - \frac{A_r}{r^2}\right)\underline{\mathbf{e}}_r\\ &+ \left(\frac{1}{r}\frac{\partial^2 A_z}{\partial \varphi \partial z} + \frac{1}{r^2}\frac{\partial^2 A_\varphi}{\partial \varphi^2} + \frac{1}{r}\frac{\partial^2 A_r}{\partial r \partial \varphi} + \frac{1}{r^2}\frac{\partial A_r}{\partial \varphi}\right)\underline{\mathbf{e}}_\varphi\\ &+ \left(\frac{\partial^2 A_z}{\partial z^2} + \frac{1}{r}\frac{\partial^2 A_\varphi}{\partial \varphi \partial z} + \frac{\partial^2 A_r}{\partial r \partial z} + \frac{1}{r}\frac{\partial A_r}{\partial z}\right)\underline{\mathbf{e}}_z\end{aligned}$$

$$\begin{aligned}&\nabla \times \nabla \times \underline{\mathbf{A}}\\ &= \left(-\frac{1}{r^2}\frac{\partial^2 A_r}{\partial \varphi^2} - \frac{\partial^2 A_r}{\partial z^2} + \frac{\partial^2 A_z}{\partial r \partial z} + \frac{1}{r}\frac{\partial^2 A_\varphi}{\partial r \partial \varphi} + \frac{1}{r^2}\frac{\partial A_\varphi}{\partial \varphi}\right)\underline{\mathbf{e}}_r\\ &+ \left(-\frac{\partial^2 A_\varphi}{\partial z^2} + \frac{1}{r}\frac{\partial^2 A_z}{\partial \varphi \partial z} - \frac{\partial^2 A_\varphi}{\partial r^2} - \frac{1}{r}\frac{\partial A_\varphi}{\partial r} + \frac{A_\varphi}{r^2} - \frac{1}{r^2}\frac{\partial A_r}{\partial \varphi} + \frac{1}{r}\frac{\partial^2 A_r}{\partial \varphi \partial r}\right)\underline{\mathbf{e}}_\varphi\\ &+ \left(-\frac{\partial^2 A_z}{\partial r^2} - \frac{1}{r^2}\frac{\partial^2 A_z}{\partial \varphi^2} + \frac{\partial^2 A_r}{\partial r \partial z} + \frac{1}{r}\frac{\partial^2 A_\varphi}{\partial \varphi \partial z} + \frac{1}{r}\frac{\partial A_r}{\partial z} - \frac{1}{r}\frac{\partial A_z}{\partial r}\right)\underline{\mathbf{e}}_z\end{aligned}$$

$$\begin{aligned}
\Delta\underline{\mathbf{A}} &= \boldsymbol{\nabla}\cdot\boldsymbol{\nabla}\underline{\mathbf{A}} \\
&= \boldsymbol{\nabla}\boldsymbol{\nabla}\cdot\underline{\mathbf{A}} - \boldsymbol{\nabla}\times\boldsymbol{\nabla}\times\underline{\mathbf{A}} \\
&= \left(\Delta A_r - \frac{A_r}{r^2} - \frac{2}{r^2}\frac{\partial A_\varphi}{\partial\varphi}\right)\underline{\mathbf{e}}_r \\
&\quad + \left(\Delta A_\varphi - \frac{A_\varphi}{r^2} + \frac{2}{r^2}\frac{\partial A_r}{\partial\varphi}\right)\underline{\mathbf{e}}_\varphi \\
&\quad + \Delta A_z\underline{\mathbf{e}}_z.
\end{aligned}$$

A.6 Spherical Coordinates

$$R, \vartheta, \varphi$$

with the orthonormal trihedron $\underline{\mathbf{e}}_R, \underline{\mathbf{e}}_\vartheta, \underline{\mathbf{e}}_\varphi$

$$\begin{aligned}
x &= R\sin\vartheta\cos\varphi & R &= \sqrt{x^2+y^2+z^2} \\
y &= R\sin\vartheta\sin\varphi & \vartheta &= \arctan\frac{\sqrt{x^2+y^2}}{z} \\
z &= R\cos\vartheta & \varphi &= \arctan\frac{y}{x}
\end{aligned}$$

$$\begin{aligned}
\underline{\mathbf{R}} &= R\sin\vartheta\cos\varphi\,\underline{\mathbf{e}}_x + R\sin\vartheta\sin\varphi\,\underline{\mathbf{e}}_y + R\cos\vartheta\,\underline{\mathbf{e}}_z \\
&= R\underline{\mathbf{e}}_R
\end{aligned}$$

$$|\underline{\mathbf{R}} - \underline{\mathbf{R}}'| = \sqrt{R^2 + R'^2 - 2RR'[\sin\vartheta\sin\vartheta'\cos(\varphi-\varphi') + \cos\vartheta\cos\vartheta']}$$

$$\begin{aligned}
h_R &= 1 \\
h_\vartheta &= R \\
h_\varphi &= R\sin\vartheta
\end{aligned}$$

$$\begin{aligned}
\underline{\mathbf{e}}_R &= \sin\vartheta\cos\varphi\,\underline{\mathbf{e}}_x + \sin\vartheta\sin\varphi\,\underline{\mathbf{e}}_y & \underline{\mathbf{e}}_x &= \sin\vartheta\cos\varphi\,\underline{\mathbf{e}}_R + \cos\vartheta\cos\varphi\,\underline{\mathbf{e}}_\vartheta \\
&\quad + \cos\vartheta\,\underline{\mathbf{e}}_z & &\quad - \sin\varphi\,\underline{\mathbf{e}}_\varphi \\
\underline{\mathbf{e}}_\vartheta &= \cos\vartheta\cos\varphi\,\underline{\mathbf{e}}_x + \cos\vartheta\sin\varphi\,\underline{\mathbf{e}}_y & \underline{\mathbf{e}}_y &= \sin\vartheta\sin\varphi\,\underline{\mathbf{e}}_R + \cos\vartheta\sin\varphi\,\underline{\mathbf{e}}_\vartheta \\
&\quad - \sin\vartheta\,\underline{\mathbf{e}}_z & &\quad + \cos\varphi\,\underline{\mathbf{e}}_\varphi \\
\underline{\mathbf{e}}_\varphi &= -\sin\varphi\,\underline{\mathbf{e}}_x + \cos\varphi\,\underline{\mathbf{e}}_y & \underline{\mathbf{e}}_z &= \cos\vartheta\,\underline{\mathbf{e}}_R - \sin\vartheta\,\underline{\mathbf{e}}_\vartheta
\end{aligned}$$

vector components: $A_R = \underline{\mathbf{A}}\cdot\underline{\mathbf{e}}_R$

$$\begin{aligned}
A_\vartheta &= \underline{\mathbf{A}}\cdot\underline{\mathbf{e}}_\vartheta \\
A_\varphi &= \underline{\mathbf{A}}\cdot\underline{\mathbf{e}}_\varphi
\end{aligned}$$

$$\underline{\mathbf{A}} = A_R\underline{\mathbf{e}}_R + A_\vartheta\underline{\mathbf{e}}_\vartheta + A_\varphi\underline{\mathbf{e}}_\varphi$$

$$\begin{pmatrix} A_R \\ A_\vartheta \\ A_\varphi \end{pmatrix} = \begin{pmatrix} \sin\vartheta\cos\varphi & \sin\vartheta\sin\varphi & \cos\vartheta \\ \cos\vartheta\cos\varphi & \cos\vartheta\sin\varphi & -\sin\vartheta \\ -\sin\varphi & \cos\varphi & 0 \end{pmatrix} \begin{pmatrix} A_x \\ A_y \\ A_z \end{pmatrix}$$

$$\begin{pmatrix} A_x \\ A_y \\ A_z \end{pmatrix} = \begin{pmatrix} \sin\vartheta\cos\varphi & \cos\vartheta\cos\varphi & -\sin\varphi \\ \sin\vartheta\sin\varphi & \cos\vartheta\sin\varphi & \cos\varphi \\ \cos\vartheta & -\sin\vartheta & 0 \end{pmatrix} \begin{pmatrix} A_R \\ A_\vartheta \\ A_\varphi \end{pmatrix}$$

$$\begin{aligned} \text{tensor components: } D_{RR} &= \underline{\mathbf{e}}_R \cdot \underline{\underline{\mathbf{D}}} \cdot \underline{\mathbf{e}}_R \\ &= \underline{\underline{\mathbf{D}}} : \underline{\mathbf{e}}_R \underline{\mathbf{e}}_R \\ D_{R\vartheta} &= \underline{\mathbf{e}}_R \cdot \underline{\underline{\mathbf{D}}} \cdot \underline{\mathbf{e}}_\vartheta \\ &= \underline{\underline{\mathbf{D}}} : \underline{\mathbf{e}}_\vartheta \underline{\mathbf{e}}_R \\ &\text{etc.} \end{aligned}$$

$$\begin{aligned} \underline{\underline{\mathbf{D}}} = {} & D_{RR} \underline{\mathbf{e}}_R \underline{\mathbf{e}}_R + D_{R\vartheta} \underline{\mathbf{e}}_R \underline{\mathbf{e}}_\vartheta + D_{R\varphi} \underline{\mathbf{e}}_R \underline{\mathbf{e}}_\varphi \\ & + D_{\vartheta R} \underline{\mathbf{e}}_\vartheta \underline{\mathbf{e}}_R + D_{\vartheta\vartheta} \underline{\mathbf{e}}_\vartheta \underline{\mathbf{e}}_\vartheta + D_{\vartheta\varphi} \underline{\mathbf{e}}_\vartheta \underline{\mathbf{e}}_\varphi \\ & + D_{\varphi R} \underline{\mathbf{e}}_\varphi \underline{\mathbf{e}}_R + D_{\varphi\vartheta} \underline{\mathbf{e}}_\varphi \underline{\mathbf{e}}_\vartheta + D_{\varphi\varphi} \underline{\mathbf{e}}_\varphi \underline{\mathbf{e}}_\varphi \end{aligned}$$

$$\underline{\underline{\mathbf{I}}} = \underline{\mathbf{e}}_R \underline{\mathbf{e}}_R + \underline{\mathbf{e}}_\vartheta \underline{\mathbf{e}}_\vartheta + \underline{\mathbf{e}}_\varphi \underline{\mathbf{e}}_\varphi$$

$$\begin{aligned} & \begin{pmatrix} D_{RR} & D_{R\vartheta} & D_{R\varphi} \\ D_{\vartheta R} & D_{\vartheta\vartheta} & D_{\vartheta\varphi} \\ D_{\varphi R} & D_{\varphi\vartheta} & D_{\varphi\varphi} \end{pmatrix} \\ & = \begin{pmatrix} \sin\vartheta\cos\varphi & \sin\vartheta\sin\varphi & \cos\vartheta \\ \cos\vartheta\cos\varphi & \cos\vartheta\sin\varphi & -\sin\vartheta \\ -\sin\varphi & \cos\varphi & 0 \end{pmatrix} \begin{pmatrix} D_{xx} & D_{xy} & D_{xz} \\ D_{yx} & D_{yy} & D_{yz} \\ D_{zx} & D_{zy} & D_{zz} \end{pmatrix} \\ & \quad \begin{pmatrix} \sin\vartheta\cos\varphi & \cos\vartheta\cos\varphi & -\sin\varphi \\ \sin\vartheta\sin\varphi & \cos\vartheta\sin\varphi & \cos\varphi \\ \cos\vartheta & -\sin\vartheta & 0 \end{pmatrix} \end{aligned}$$

for example:

$$\begin{aligned} D_{RR} = {} & (D_{xx}\cos^2\varphi + D_{yy}\sin^2\varphi)\sin^2\vartheta + D_{zz}\cos^2\vartheta \\ & + (D_{xy} + D_{yx})\sin^2\vartheta\cos\varphi\sin\varphi + (D_{xz} + D_{zx})\sin\vartheta\cos\vartheta\cos\varphi \\ & + (D_{yz} + D_{zy})\sin\vartheta\cos\vartheta\sin\varphi \\ D_{\vartheta\vartheta} = {} & (D_{xx}\cos^2\varphi + D_{yy}\sin^2\varphi)\cos^2\vartheta + D_{zz}\sin^2\vartheta \\ & + (D_{xy} + D_{yx})\cos^2\vartheta\cos\varphi\sin\varphi - (D_{xz} + D_{zx})\sin\vartheta\cos\vartheta\cos\varphi \\ & - (D_{yz} + D_{zy})\sin\vartheta\cos\vartheta\sin\varphi \\ D_{\varphi\varphi} = {} & D_{xx}\sin^2\varphi + D_{yy}\cos^2\varphi \\ & - (D_{xy} + D_{yx})\cos\varphi\sin\varphi \end{aligned}$$

$$\begin{aligned} \text{trace } \underline{\underline{\mathbf{D}}} &= D_{RR} + D_{\vartheta\vartheta} + D_{\varphi\varphi} \\ &= D_{xx} + D_{yy} + D_{zz} \\ \text{scalar product: } \underline{\mathbf{A}} \cdot \underline{\mathbf{B}} &= A_R B_R + A_\vartheta B_\vartheta + A_\varphi B_\varphi \\ &= A_x B_x + A_y B_y + A_z B_z \end{aligned}$$

$$\mathrm{d}s^2 = \mathrm{d}R^2 + R^2\mathrm{d}\vartheta^2 + R^2\sin^2\vartheta\mathrm{d}\varphi^2$$

$$\text{vector product: } \underline{\mathbf{A}} \times \underline{\mathbf{B}} = (A_\vartheta B_\varphi - A_\varphi B_\vartheta)\underline{\mathbf{e}}_R + (A_\varphi B_R - A_R B_\varphi)\underline{\mathbf{e}}_\vartheta + (A_R B_\vartheta - A_\vartheta B_R)\underline{\mathbf{e}}_\varphi = \underline{\underline{\underline{\boldsymbol{\varepsilon}}}} : \underline{\mathbf{B}}\,\underline{\mathbf{A}} = -\underline{\mathbf{A}} \cdot \underline{\underline{\underline{\boldsymbol{\varepsilon}}}} \cdot \underline{\mathbf{B}} = \langle \underline{\mathbf{A}}\,\underline{\mathbf{B}} \rangle$$

$$\mathrm{d}V = R^2\sin\vartheta\mathrm{d}R\mathrm{d}\vartheta\mathrm{d}\varphi$$

$$\boldsymbol{\nabla} = \underline{\mathbf{e}}_R\frac{\partial}{\partial R} + \frac{1}{R}\underline{\mathbf{e}}_\vartheta\frac{\partial}{\partial\vartheta} + \frac{1}{R\sin\vartheta}\underline{\mathbf{e}}_\varphi\frac{\partial}{\partial\varphi}$$

$$\boldsymbol{\nabla}\Phi = \text{grad}\,\Phi = \frac{\partial\Phi}{\partial R}\underline{\mathbf{e}}_R + \frac{1}{R}\frac{\partial\Phi}{\partial\vartheta}\underline{\mathbf{e}}_\vartheta + \frac{1}{R\sin\vartheta}\frac{\partial\Phi}{\partial\varphi}\underline{\mathbf{e}}_\varphi$$

$$\begin{aligned}\boldsymbol{\nabla\nabla}\Phi = {} & \frac{\partial^2\Phi}{\partial R^2}\underline{\mathbf{e}}_R\underline{\mathbf{e}}_R + \frac{1}{R}\left(\frac{\partial^2\Phi}{\partial R\partial\vartheta} - \frac{1}{R}\frac{\partial\Phi}{\partial\vartheta}\right)\underline{\mathbf{e}}_R\underline{\mathbf{e}}_\vartheta \\ & + \frac{1}{R\sin\vartheta}\left(\frac{\partial^2\Phi}{\partial R\partial\varphi} - \frac{1}{R}\frac{\partial\Phi}{\partial\varphi}\right)\underline{\mathbf{e}}_R\underline{\mathbf{e}}_\varphi \\ & + \frac{1}{R}\left(\frac{\partial^2\Phi}{\partial R\partial\vartheta} - \frac{1}{R}\frac{\partial\Phi}{\partial\vartheta}\right)\underline{\mathbf{e}}_\vartheta\underline{\mathbf{e}}_R + \frac{1}{R^2}\left(\frac{\partial^2\Phi}{\partial\vartheta^2} + R\frac{\partial\Phi}{\partial R}\right)\underline{\mathbf{e}}_\vartheta\underline{\mathbf{e}}_\vartheta \\ & + \frac{1}{R^2\sin\vartheta}\left(\frac{\partial^2\Phi}{\partial\vartheta\partial\varphi} - \cot\vartheta\frac{\partial\Phi}{\partial\varphi}\right)\underline{\mathbf{e}}_\vartheta\underline{\mathbf{e}}_\varphi \\ & + \frac{1}{R\sin\vartheta}\left(\frac{\partial^2\Phi}{\partial R\partial\varphi} - \frac{1}{R}\frac{\partial\Phi}{\partial\varphi}\right)\underline{\mathbf{e}}_\varphi\underline{\mathbf{e}}_R \\ & + \frac{1}{R^2\sin\vartheta}\left(\frac{\partial^2\Phi}{\partial\vartheta\partial\varphi} - \cot\vartheta\frac{\partial\Phi}{\partial\varphi}\right)\underline{\mathbf{e}}_\varphi\underline{\mathbf{e}}_\vartheta \\ & + \frac{1}{R^2}\left(\frac{1}{\sin^2\vartheta}\frac{\partial^2\Phi}{\partial\varphi^2} + \cot\vartheta\frac{\partial\Phi}{\partial\vartheta} + R\frac{\partial\Phi}{\partial R}\right)\underline{\mathbf{e}}_\varphi\underline{\mathbf{e}}_\varphi\end{aligned}$$

Christoffel symbols of the second kind:

$$\frac{\partial\underline{\mathbf{e}}_R}{\partial R} = \underline{\mathbf{0}} \qquad \frac{\partial\underline{\mathbf{e}}_\vartheta}{\partial R} = \underline{\mathbf{0}} \qquad \frac{\partial\underline{\mathbf{e}}_\varphi}{\partial R} = \underline{\mathbf{0}}$$

$$\frac{\partial\underline{\mathbf{e}}_R}{\partial\vartheta} = \underline{\mathbf{e}}_\vartheta = \alpha_2(1,2)\underline{\mathbf{e}}_\vartheta \qquad \frac{\partial\underline{\mathbf{e}}_\vartheta}{\partial\vartheta} = -\underline{\mathbf{e}}_R = \alpha_1(2,2)\underline{\mathbf{e}}_R \qquad \frac{\partial\underline{\mathbf{e}}_\varphi}{\partial\vartheta} = \underline{\mathbf{0}}$$

$$\frac{\partial\underline{\mathbf{e}}_R}{\partial\varphi} = \sin\vartheta\,\underline{\mathbf{e}}_\varphi = \alpha_3(1,3)\underline{\mathbf{e}}_\varphi \qquad \frac{\partial\underline{\mathbf{e}}_\vartheta}{\partial\varphi} = \cos\vartheta\,\underline{\mathbf{e}}_\varphi = \alpha_3(2,3)\underline{\mathbf{e}}_\varphi \qquad \frac{\partial\underline{\mathbf{e}}_\varphi}{\partial\varphi} = -\sin\vartheta\,\underline{\mathbf{e}}_R - \cos\vartheta\,\underline{\mathbf{e}}_\vartheta = \alpha_1(3,3)\underline{\mathbf{e}}_R + \alpha_2(3,3)\underline{\mathbf{e}}_\vartheta$$

$$\begin{aligned}
\Delta\Phi &= \frac{1}{R^2}\frac{\partial}{\partial R}\left(R^2\frac{\partial\Phi}{\partial R}\right) + \frac{1}{R^2\sin\vartheta}\frac{\partial}{\partial\vartheta}\left(\sin\vartheta\frac{\partial\Phi}{\partial\vartheta}\right) + \frac{1}{R^2\sin^2\vartheta}\frac{\partial^2\Phi}{\partial\varphi^2}\\
&= \frac{\partial^2\Phi}{\partial R^2} + \frac{2}{R}\frac{\partial\Phi}{\partial R} + \frac{1}{R^2\sin\vartheta}\frac{\partial^2\Phi}{\partial\vartheta^2} + \frac{\cot\vartheta}{R^2}\frac{\partial\Phi}{\partial\vartheta} + \frac{1}{R^2\sin^2\vartheta}\frac{\partial^2\Phi}{\partial\varphi^2}\\
&= \frac{1}{R^2}\frac{\partial}{\partial R}\left(R^2\frac{\partial\Phi}{\partial R}\right) + (\underline{\mathbf{e}}_R\times\mathbf{\nabla})^2\Phi\\
&= \frac{1}{R^2}\frac{\partial}{\partial R}\left(R^2\frac{\partial\Phi}{\partial R}\right) + \mathcal{B}\{\Phi\}\ ;\ \mathcal{B}\text{ is the Beltrami operator}
\end{aligned}$$

$$\operatorname{trace}(\mathbf{\nabla}\mathbf{\nabla}\Phi) = \Delta\Phi \tag{A.2}$$

$$\begin{aligned}
\mathbf{\nabla}\cdot\underline{\mathbf{A}} &= \operatorname{div}\underline{\mathbf{A}}\\
&= \frac{1}{R^2}\frac{\partial}{\partial R}(R^2A_R) + \frac{1}{R\sin\vartheta}\frac{\partial}{\partial\vartheta}(\sin\vartheta A_\vartheta) + \frac{1}{R\sin\vartheta}\frac{\partial A_\varphi}{\partial\varphi}
\end{aligned}$$

$$\mathbf{\nabla}\cdot\underline{\mathbf{e}}_R = \frac{2}{R}$$

$$\mathbf{\nabla}\cdot\underline{\mathbf{e}}_\vartheta = \frac{1}{R\tan\vartheta}$$

$$\mathbf{\nabla}\cdot\underline{\mathbf{e}}_\varphi = 0$$

$$\begin{aligned}
\mathbf{\nabla}\times\underline{\mathbf{A}} &= \operatorname{curl}\underline{\mathbf{A}}\\
&= \frac{1}{R\sin\vartheta}\left[\frac{\partial}{\partial\vartheta}(\sin\vartheta A_\varphi) - \frac{\partial A_\vartheta}{\partial\varphi}\right]\underline{\mathbf{e}}_R\\
&\quad + \frac{1}{R}\left[\frac{1}{\sin\vartheta}\frac{\partial A_R}{\partial\varphi} - \frac{\partial}{\partial R}(RA_\varphi)\right]\underline{\mathbf{e}}_\vartheta\\
&\quad + \frac{1}{R}\left[\frac{\partial}{\partial R}(RA_\vartheta) - \frac{\partial A_R}{\partial\vartheta}\right]\underline{\mathbf{e}}_\varphi
\end{aligned}$$

$$\mathbf{\nabla}\times\underline{\mathbf{e}}_R = \underline{\mathbf{0}}$$

$$\mathbf{\nabla}\times\underline{\mathbf{e}}_\vartheta = \frac{1}{R}\underline{\mathbf{e}}_\varphi$$

$$\mathbf{\nabla}\times\underline{\mathbf{e}}_\varphi = \frac{1}{R\tan\vartheta}\underline{\mathbf{e}}_R - \frac{1}{R}\underline{\mathbf{e}}_\vartheta$$

$$\begin{aligned}
\mathbf{\nabla}\underline{\mathbf{A}} &= \frac{\partial A_R}{\partial R}\underline{\mathbf{e}}_R\underline{\mathbf{e}}_R + \frac{\partial A_\vartheta}{\partial R}\underline{\mathbf{e}}_R\underline{\mathbf{e}}_\vartheta + \frac{\partial A_\varphi}{\partial R}\underline{\mathbf{e}}_R\underline{\mathbf{e}}_\varphi\\
&\quad + \frac{1}{R}\left(\frac{\partial A_R}{\partial\vartheta} - A_\vartheta\right)\underline{\mathbf{e}}_\vartheta\underline{\mathbf{e}}_R + \frac{1}{R}\left(\frac{\partial A_\vartheta}{\partial\vartheta} + A_R\right)\underline{\mathbf{e}}_\vartheta\underline{\mathbf{e}}_\vartheta + \frac{1}{R}\frac{\partial A_\varphi}{\partial\vartheta}\underline{\mathbf{e}}_\vartheta\underline{\mathbf{e}}_\varphi\\
&\quad + \frac{1}{R\sin\vartheta}\left(\frac{\partial A_R}{\partial\varphi} - \sin\vartheta A_\varphi\right)\underline{\mathbf{e}}_\varphi\underline{\mathbf{e}}_R + \frac{1}{R\sin\vartheta}\left(\frac{\partial A_\vartheta}{\partial\varphi} - \cos\vartheta A_\varphi\right)\underline{\mathbf{e}}_\varphi\underline{\mathbf{e}}_\vartheta\\
&\quad + \frac{1}{R\sin\vartheta}\left(\frac{\partial A_\varphi}{\partial\varphi} + \sin\vartheta A_R + \cos\vartheta A_\vartheta\right)\underline{\mathbf{e}}_\varphi\underline{\mathbf{e}}_\varphi
\end{aligned}$$

$$
\begin{aligned}
(\nabla\cdot\underline{\underline{\mathbf{D}}})\cdot\underline{\mathbf{e}}_R &= \frac{1}{R^2}\frac{\partial R^2 D_{RR}}{\partial R}+\frac{1}{R\sin\vartheta}\frac{\partial\sin\vartheta D_{\vartheta R}}{\partial\vartheta}+\frac{1}{R\sin\vartheta}\frac{\partial D_{\varphi R}}{\partial\varphi}\\
&\quad -\frac{1}{R}D_{\vartheta\vartheta}-\frac{1}{R}D_{\varphi\varphi}\\
(\nabla\cdot\underline{\underline{\mathbf{D}}})\cdot\underline{\mathbf{e}}_\vartheta &= \frac{1}{R^2}\frac{\partial R^2 D_{R\vartheta}}{\partial R}+\frac{1}{R\sin\vartheta}\frac{\partial\sin\vartheta D_{\vartheta\vartheta}}{\partial\vartheta}\\
&\quad +\frac{1}{R\sin\vartheta}\frac{\partial D_{\varphi\vartheta}}{\partial\varphi}+\frac{1}{R}D_{\vartheta R}-\frac{\cot\vartheta}{R}D_{\varphi\varphi}\\
(\nabla\cdot\underline{\underline{\mathbf{D}}})\cdot\underline{\mathbf{e}}_\varphi &= \frac{1}{R^2}\frac{\partial R^2 D_{R\varphi}}{\partial R}+\frac{1}{R\sin\vartheta}\frac{\partial\sin\vartheta D_{\vartheta\varphi}}{\partial\vartheta}\\
&\quad +\frac{1}{R\sin\vartheta}\frac{\partial D_{\varphi\varphi}}{\partial\varphi}+\frac{1}{R}D_{\varphi R}+\frac{\cot\vartheta}{R}D_{\varphi\vartheta}
\end{aligned}
$$

$$
\operatorname{trace}(\nabla\underline{\mathbf{A}}) = \nabla\cdot\underline{\mathbf{A}}
$$

$$
\langle\underline{\mathbf{A}}\,\underline{\mathbf{B}}\rangle = \nabla\times\underline{\mathbf{A}}
$$

$$
\begin{aligned}
\nabla\nabla\cdot\underline{\mathbf{A}} = &\left(\frac{\partial^2 A_R}{\partial R^2}+\frac{2}{R}\frac{\partial A_R}{\partial R}-\frac{2A_R}{R^2}-\frac{A_\vartheta}{R^2\tan\vartheta}+\frac{1}{R\tan\vartheta}\frac{\partial A_\vartheta}{\partial R}+\frac{1}{R}\frac{\partial^2 A_\vartheta}{\partial\vartheta\partial R}\right.\\
&\left.-\frac{1}{R^2}\frac{\partial A_\vartheta}{\partial\vartheta}+\frac{1}{R\sin\vartheta}\frac{\partial^2 A_\varphi}{\partial\varphi\partial R}-\frac{1}{R^2\sin\vartheta}\frac{\partial A_\varphi}{\partial\varphi}\right)\underline{\mathbf{e}}_R\\
&+\left(\frac{1}{R}\frac{\partial^2 A_R}{\partial R\partial\vartheta}+\frac{2}{R^2}\frac{\partial A_R}{\partial\vartheta}-\frac{A_\vartheta}{R^2\sin^2\vartheta}+\frac{1}{R^2\tan\vartheta}\frac{\partial A_\vartheta}{\partial\vartheta}+\frac{1}{R^2}\frac{\partial^2 A_\vartheta}{\partial\vartheta^2}\right.\\
&\left.+\frac{1}{R^2\sin\vartheta}\frac{\partial^2 A_\varphi}{\partial\varphi\partial\vartheta}-\frac{\cot\vartheta}{R^2\sin\vartheta}\frac{\partial A_\varphi}{\partial\varphi}\right)\underline{\mathbf{e}}_\vartheta\\
&+\left(\frac{1}{R\sin\vartheta}\frac{\partial^2 A_R}{\partial R\partial\varphi}+\frac{2}{R^2\sin\vartheta}\frac{\partial A_R}{\partial\varphi}+\frac{\cot\vartheta}{R^2\sin\vartheta}\frac{\partial A_\vartheta}{\partial\varphi}+\frac{1}{R^2\sin\vartheta}\frac{\partial^2 A_\varphi}{\partial\varphi\partial\vartheta}\right.\\
&\left.+\frac{1}{R^2\sin^2\vartheta}\frac{\partial^2 A_\varphi}{\partial\varphi^2}\right)\underline{\mathbf{e}}_\varphi
\end{aligned}
$$

$$
\begin{aligned}
\nabla\times\nabla\times\underline{\mathbf{A}} = &\left(\frac{1}{R}\frac{\partial^2 A_\vartheta}{\partial R\partial\vartheta}+\frac{1}{R^2}\frac{\partial A_\vartheta}{\partial\vartheta}-\frac{1}{R^2}\frac{\partial^2 A_R}{\partial\vartheta^2}+\frac{1}{R\tan\vartheta}\frac{\partial A_\vartheta}{\partial R}+\frac{1}{R\tan\vartheta}\frac{A_\vartheta}{R}\right.\\
&-\frac{1}{R^2\tan\vartheta}\frac{\partial A_R}{\partial\vartheta}-\frac{1}{R^2\sin^2\vartheta}\frac{\partial^2 A_R}{\partial\varphi^2}+\frac{1}{R\sin\vartheta}\frac{\partial^2 A_\varphi}{\partial R\partial\varphi}\\
&\left.+\frac{1}{R^2\sin\vartheta}\frac{\partial A_\varphi}{\partial\varphi}\right)\underline{\mathbf{e}}_R\\
&+\left(\frac{1}{R^2\sin^2\vartheta}\frac{\partial^2 A_\varphi}{\partial\varphi\partial\vartheta}+\frac{\cot\vartheta}{R^2\sin\vartheta}\frac{\partial A_\varphi}{\partial\varphi}-\frac{1}{R^2\sin^2\vartheta}\frac{\partial^2 A_\vartheta}{\partial\varphi^2}-\frac{2}{R}\frac{\partial A_\vartheta}{\partial R}\right.\\
&\left.+\frac{1}{R}\frac{\partial^2 A_R}{\partial R\partial\vartheta}-\frac{\partial^2 A_\vartheta}{\partial R^2}\right)\underline{\mathbf{e}}_\vartheta
\end{aligned}
$$

$$+\left(\frac{1}{R\sin\vartheta}\frac{\partial^2 A_R}{\partial\varphi\partial R}-\frac{2}{R}\frac{\partial A_\varphi}{\partial R}-\frac{1}{R^2}\frac{\partial^2 A_\varphi}{\partial\vartheta^2}-\frac{\partial^2 A_\varphi}{\partial R^2}\right.$$
$$-\frac{1}{R^2\tan\vartheta}\frac{\partial A_\varphi}{\partial\vartheta}+\frac{A_\varphi}{R^2\sin^2\vartheta}+\frac{1}{R^2\sin^2\vartheta}\frac{\partial^2 A_\vartheta}{\partial\vartheta\partial\varphi}$$
$$\left.-\frac{\cot\vartheta}{R^2\sin\vartheta}\frac{\partial A_\vartheta}{\partial\varphi}\right)\underline{\mathbf{e}}_\varphi$$

$$\begin{aligned}\Delta\underline{\mathbf{A}} &= \nabla\cdot\nabla\underline{\mathbf{A}}\\ &= \nabla\nabla\cdot\underline{\mathbf{A}}-\nabla\times\nabla\times\underline{\mathbf{A}}\\ &= \left(\Delta A_R-\frac{2A_R}{R^2}-\frac{2\cot\vartheta}{R^2}A_\vartheta-\frac{2}{R^2}\frac{\partial A_\vartheta}{\partial\vartheta}-\frac{2}{R^2\sin\vartheta}\frac{\partial A_\varphi}{\partial\varphi}\right)\underline{\mathbf{e}}_R\\ &\quad+\left(\Delta A_\vartheta+\frac{2}{R^2}\frac{\partial A_R}{\partial\vartheta}-\frac{A_\vartheta}{R^2\sin^2\vartheta}-\frac{2\cot\vartheta}{R^2\sin\vartheta}\frac{\partial A_\varphi}{\partial\varphi}\right)\underline{\mathbf{e}}_\vartheta\\ &\quad+\left(\Delta A_\varphi+\frac{2}{R^2\sin\vartheta}\frac{\partial A_R}{\partial\varphi}-\frac{1}{R^2\sin^2\vartheta}A_\varphi+\frac{2\cot\vartheta}{R^2\sin\vartheta}\frac{\partial A_\vartheta}{\partial\varphi}\right)\underline{\mathbf{e}}_\varphi\end{aligned}$$

A.7 Identities for the Del Operator

A.7.1 General scalar, vector, and tensor fields

Single del-operations

- Scalar fields

$$\nabla\Phi[\phi_1(\underline{\mathbf{R}}),\ldots,\phi_n(\underline{\mathbf{R}})]=\sum_{i=1}^{n}\frac{\partial\Phi(\phi_1,\ldots,\phi_n)}{\partial\phi_i}\nabla\phi_i(\underline{\mathbf{R}})$$
$$\nabla(\Phi\Psi)=\Phi\nabla\Psi+\Psi\nabla\Phi$$

- Scalar and vector fields

$$\nabla(\Phi\underline{\mathbf{A}})=\Phi\nabla\underline{\mathbf{A}}+(\nabla\Phi)\underline{\mathbf{A}}$$
$$\nabla\cdot(\Phi\underline{\mathbf{A}})=\Phi\nabla\cdot\underline{\mathbf{A}}+\underline{\mathbf{A}}\cdot\nabla\Phi$$
$$\nabla\times(\Phi\underline{\mathbf{A}})=\Phi\nabla\times\underline{\mathbf{A}}-\underline{\mathbf{A}}\times\nabla\Phi$$

- Scalar and tensor fields

$$\nabla(\Phi\underline{\underline{\mathbf{D}}})=(\nabla\Phi)\underline{\underline{\mathbf{D}}}+\Phi\nabla\underline{\underline{\mathbf{D}}}$$
$$\nabla(\Phi\,\underline{\underline{\mathbf{I}}})=(\nabla\Phi)\,\underline{\underline{\mathbf{I}}}$$
$$\nabla\cdot(\Phi\underline{\underline{\mathbf{D}}})=\nabla\Phi\cdot\underline{\underline{\mathbf{D}}}+\Phi\nabla\cdot\underline{\underline{\mathbf{D}}}$$

$$\nabla\cdot(\Phi\underline{\underline{\mathbf{I}}}) = \nabla\Phi$$
$$\nabla\times(\Phi\underline{\underline{\mathbf{D}}}) = \nabla\Phi\times\underline{\underline{\mathbf{D}}} + \Phi\nabla\times\underline{\underline{\mathbf{D}}}$$
$$\nabla\times(\Phi\underline{\underline{\mathbf{I}}}) = \nabla\Phi\times\underline{\underline{\mathbf{I}}}$$

- Vector fields

$$\nabla\cdot\underline{\mathbf{A}}[\phi_1(\underline{\mathbf{R}}),\ldots,\phi_n(\underline{\mathbf{R}})] = \sum_{i=1}^{n}\frac{\partial\underline{\mathbf{A}}(\phi_1,\ldots,\phi_n)}{\partial\phi_i}\cdot\nabla\phi_i(\underline{\mathbf{R}})$$
$$\nabla\times\underline{\mathbf{A}}[\phi_1(\underline{\mathbf{R}}),\ldots,\phi_n(\underline{\mathbf{R}})] = \sum_{i=1}^{n}\nabla\phi_i(\underline{\mathbf{R}})\times\frac{\partial\underline{\mathbf{A}}(\phi_1,\ldots,\phi_n)}{\partial\phi_i}$$
$$\begin{aligned}\nabla(\underline{\mathbf{A}}\cdot\underline{\mathbf{B}}) &= (\nabla\underline{\mathbf{A}})\cdot\underline{\mathbf{B}} + (\nabla\underline{\mathbf{B}})\cdot\underline{\mathbf{A}}\\ &= \underline{\mathbf{B}}\cdot(\nabla\underline{\mathbf{A}})^{21} + \underline{\mathbf{A}}\cdot(\nabla\underline{\mathbf{B}})^{21}\\ &= \underline{\mathbf{A}}\times(\nabla\times\underline{\mathbf{B}}) + \underline{\mathbf{A}}\cdot\nabla\underline{\mathbf{B}} + \underline{\mathbf{B}}\times(\nabla\times\underline{\mathbf{A}}) + \underline{\mathbf{B}}\cdot\nabla\underline{\mathbf{A}},\end{aligned}$$

weil

$$\begin{aligned}\underline{\mathbf{A}}\times(\nabla\times\underline{\mathbf{B}}) &= -\underline{\mathbf{A}}\cdot\underline{\underline{\boldsymbol{\varepsilon}}}\cdot\underline{\underline{\boldsymbol{\varepsilon}}}:(\nabla\underline{\mathbf{B}})^{21}\\ &= -\underline{\mathbf{A}}\cdot(\nabla\underline{\mathbf{B}}) + \underline{\mathbf{A}}\cdot(\nabla\underline{\mathbf{B}})^{21}\\ &= (\nabla\underline{\mathbf{B}})\cdot\underline{\mathbf{A}} - \underline{\mathbf{A}}\cdot(\nabla\underline{\mathbf{B}})\end{aligned}$$
$$\nabla(\underline{\mathbf{A}}\times\underline{\mathbf{B}}) = (\nabla\underline{\mathbf{A}})\times\underline{\mathbf{B}} - (\nabla\underline{\mathbf{B}})\times\underline{\mathbf{A}}$$
$$\nabla(\underline{\mathbf{A}}\,\underline{\mathbf{B}}) = (\nabla\underline{\mathbf{A}})\underline{\mathbf{B}} + (\underline{\mathbf{A}}\nabla\underline{\mathbf{B}})^{213}$$
$$\nabla\cdot(\underline{\mathbf{A}}\times\underline{\mathbf{B}}) = (\nabla\times\underline{\mathbf{A}})\cdot\underline{\mathbf{B}} - \underline{\mathbf{A}}\cdot(\nabla\times\underline{\mathbf{B}})$$
$$\nabla\cdot(\underline{\mathbf{A}}\,\underline{\mathbf{B}}) = (\nabla\cdot\underline{\mathbf{A}})\underline{\mathbf{B}} + \underline{\mathbf{A}}\cdot(\nabla\underline{\mathbf{B}})$$
$$\begin{aligned}\nabla\times(\underline{\mathbf{A}}\times\underline{\mathbf{B}}) &= \underline{\mathbf{B}}\cdot\nabla\underline{\mathbf{A}} - \underline{\mathbf{A}}\cdot\nabla\underline{\mathbf{B}} + (\nabla\cdot\underline{\mathbf{B}})\underline{\mathbf{A}} - (\nabla\cdot\underline{\mathbf{A}})\underline{\mathbf{B}}\\ &= \nabla\cdot(\underline{\mathbf{B}}\,\underline{\mathbf{A}} - \underline{\mathbf{A}}\,\underline{\mathbf{B}})\end{aligned}$$
$$\nabla\times(\underline{\mathbf{A}}\,\underline{\mathbf{B}}) = (\nabla\times\underline{\mathbf{A}})\underline{\mathbf{B}} - \underline{\mathbf{A}}\times(\nabla\underline{\mathbf{B}})$$
$$\underline{\underline{\mathbf{I}}}:\nabla\underline{\mathbf{A}} = \nabla\cdot\underline{\mathbf{A}}$$
$$\underline{\underline{\mathbf{I}}}\stackrel{\cdot}{\times}\nabla\underline{\mathbf{A}} = \nabla\times\underline{\mathbf{A}}$$
$$\underline{\underline{\mathbf{I}}}\stackrel{\times}{\cdot}\nabla\underline{\mathbf{A}} = -\nabla\times\underline{\mathbf{A}}$$
$$\underline{\underline{\mathbf{I}}}\stackrel{\times}{\times}\nabla\underline{\mathbf{A}} = \underline{\underline{\mathbf{I}}}\,\nabla\cdot\underline{\mathbf{A}} - \nabla\underline{\mathbf{A}}$$
$$(\nabla\boldsymbol{A})_{\mathrm{a}} = \frac{1}{2}(\underline{\underline{\boldsymbol{\varepsilon}}}:\nabla\underline{\mathbf{A}})\times\underline{\underline{\mathbf{I}}} = -\frac{1}{2}(\nabla\times\underline{\mathbf{A}})\times\underline{\underline{\mathbf{I}}}$$

- Vector and tensor fields

$$\nabla(\underline{\mathbf{A}}\,\underline{\underline{\mathbf{D}}}) = (\nabla\underline{\mathbf{A}})\underline{\underline{\mathbf{D}}} + (\underline{\mathbf{A}}\nabla\underline{\underline{\mathbf{D}}})^{2134}$$
$$\nabla(\underline{\underline{\mathbf{D}}}\cdot\underline{\mathbf{A}}) = (\nabla\underline{\underline{\mathbf{D}}})\cdot\underline{\mathbf{A}} + (\nabla\underline{\mathbf{A}})\cdot\underline{\underline{\mathbf{D}}}^{21}$$
$$\nabla(\underline{\mathbf{A}}\cdot\underline{\underline{\mathbf{D}}}) = (\nabla\underline{\mathbf{A}})\cdot\underline{\underline{\mathbf{D}}} + \underline{\mathbf{A}}\cdot(\nabla\underline{\underline{\mathbf{D}}})^{213}$$

$$\nabla(\underline{\mathbf{A}}\times\underline{\underline{\mathbf{I}}}) = \nabla\underline{\mathbf{A}}\times\underline{\underline{\mathbf{I}}}$$
$$\nabla\cdot(\underline{\underline{\mathbf{D}}}\,\underline{\mathbf{A}}) = (\nabla\cdot\underline{\underline{\mathbf{D}}})\underline{\mathbf{A}} + \underline{\underline{\mathbf{D}}}^{21}\cdot\nabla\underline{\mathbf{A}}$$
$$\nabla\cdot(\underline{\underline{\mathbf{I}}}\,\underline{\mathbf{A}}) = \nabla\underline{\mathbf{A}}$$
$$\nabla\cdot(\underline{\mathbf{A}}\,\underline{\underline{\mathbf{D}}}) = (\nabla\cdot\underline{\mathbf{A}})\underline{\underline{\mathbf{D}}} + \underline{\mathbf{A}}\cdot\nabla\underline{\underline{\mathbf{D}}}$$
$$\nabla\cdot(\underline{\mathbf{A}}\,\underline{\underline{\mathbf{I}}}) = \underline{\underline{\mathbf{I}}}\,\nabla\cdot\underline{\mathbf{A}}$$

$$\begin{aligned}\nabla\cdot(\underline{\underline{\mathbf{D}}}\cdot\underline{\mathbf{A}}) &= (\nabla\cdot\underline{\underline{\mathbf{D}}})\cdot\underline{\mathbf{A}} + \mathrm{trace}\,(\underline{\underline{\mathbf{D}}}^{21}\cdot\nabla\underline{\mathbf{A}})\\ &= (\nabla\cdot\underline{\underline{\mathbf{D}}})\cdot\underline{\mathbf{A}} + \underline{\underline{\mathbf{D}}}^{21} : \nabla\underline{\mathbf{A}}\end{aligned}$$

$$\begin{aligned}\nabla\cdot(\underline{\mathbf{A}}\cdot\underline{\underline{\mathbf{D}}}) &= \nabla\underline{\mathbf{A}} : \underline{\underline{\mathbf{D}}} + \underline{\mathbf{A}}\cdot\nabla\cdot\underline{\underline{\mathbf{D}}}^{21}\\ &= \mathrm{trace}\,(\nabla\underline{\mathbf{A}}\cdot\underline{\underline{\mathbf{D}}}) + \underline{\mathbf{A}}\cdot\nabla\cdot\underline{\underline{\mathbf{D}}}^{21}\end{aligned}$$

$$\begin{aligned}\nabla\cdot(\underline{\mathbf{A}}\times\underline{\underline{\mathbf{D}}}) &= (\nabla\times\underline{\mathbf{A}})\cdot\underline{\underline{\mathbf{D}}} - \underline{\mathbf{A}}\cdot(\nabla\times\underline{\underline{\mathbf{D}}})\\ &= -\nabla\cdot(\underline{\underline{\mathbf{D}}}^{21}\times\underline{\mathbf{A}})\\ &= -(\nabla\times\underline{\underline{\mathbf{D}}})^{21}\cdot\underline{\mathbf{A}} + \underline{\underline{\mathbf{D}}}^{21}\cdot(\nabla\times\underline{\mathbf{A}})\end{aligned}$$

$$\nabla\cdot(\underline{\underline{\mathbf{D}}}\times\underline{\mathbf{A}}) = (\nabla\cdot\underline{\underline{\mathbf{D}}})\times\underline{\mathbf{A}} + \underline{\underline{\mathbf{D}}}^{21}\,\dot{\times}\,\nabla\underline{\mathbf{A}}$$
$$\nabla\cdot(\underline{\underline{\mathbf{I}}}\times\underline{\mathbf{A}}) = \nabla\times\underline{\mathbf{A}}$$
$$\nabla\cdot(\underline{\mathbf{A}}\times\underline{\underline{\mathbf{I}}}) = \nabla\times\underline{\mathbf{A}}$$
$$\nabla\times(\underline{\underline{\mathbf{D}}}\cdot\underline{\mathbf{A}}) = (\nabla\times\underline{\underline{\mathbf{D}}})\cdot\underline{\mathbf{A}} - \underline{\underline{\mathbf{D}}}^{21}\overset{\times}{\cdot}\nabla\underline{\mathbf{A}}$$
$$\nabla\times(\underline{\underline{\mathbf{D}}}\times\underline{\mathbf{A}}) = (\nabla\times\underline{\underline{\mathbf{D}}})\times\underline{\mathbf{A}} - \nabla\underline{\mathbf{A}}\overset{\times}{\times}\underline{\underline{\mathbf{D}}}^{21}$$
$$\nabla\times(\underline{\underline{\mathbf{I}}}\times\underline{\mathbf{A}}) = (\nabla\underline{\mathbf{A}})^{21} - \underline{\underline{\mathbf{I}}}\,\nabla\cdot\underline{\mathbf{A}}$$

- Tensor fields

$$\nabla\cdot(\underline{\underline{\mathbf{D}}}\cdot\underline{\underline{\mathbf{E}}}) = (\nabla\cdot\underline{\underline{\mathbf{D}}})\cdot\underline{\underline{\mathbf{E}}} + \underline{\underline{\mathbf{D}}}^{21} : \nabla\underline{\underline{\mathbf{E}}}$$
$$\nabla\cdot(\underline{\underline{\mathbf{D}}}^{21}\times\underline{\underline{\mathbf{E}}}) = (\nabla\times\underline{\underline{\mathbf{D}}})^{21}\cdot\underline{\underline{\mathbf{E}}} - \underline{\underline{\mathbf{D}}}^{21}\cdot(\nabla\times\underline{\underline{\mathbf{E}}})$$

Double del-operations

$$\nabla\cdot\nabla\Phi = \Delta\Phi$$
$$\nabla\cdot\nabla\underline{\mathbf{A}} = \Delta\underline{\mathbf{A}}$$
$$\nabla\cdot(\nabla\underline{\mathbf{A}})^{21} = \nabla\nabla\cdot\underline{\mathbf{A}}$$
$$\nabla\cdot(\nabla\times\underline{\mathbf{A}}) = 0$$
$$\nabla\cdot(\nabla\times\underline{\underline{\mathbf{D}}}) = \underline{\mathbf{0}}$$
$$\nabla\times(\nabla\Phi) = \underline{\mathbf{0}}$$
$$\nabla\times(\nabla\underline{\mathbf{A}}) = \underline{\underline{\mathbf{0}}}$$
$$\nabla\times(\nabla\underline{\mathbf{A}})^{21} = (\nabla\nabla\times\underline{\mathbf{A}})^{21}$$

$$\nabla \times \nabla \times \underline{\mathbf{A}} = \nabla\nabla \cdot \underline{\mathbf{A}} - \nabla \cdot \nabla \underline{\mathbf{A}}$$

$$\Delta(\Phi\Psi) = \Phi\Delta\Psi + 2\nabla\Phi \cdot \nabla\Psi + \Psi\Delta\Phi$$

$$\Delta(\Phi\underline{\mathbf{A}}) = \Phi\Delta\underline{\mathbf{A}} + 2\nabla\Phi \cdot \nabla\underline{\mathbf{A}} + \underline{\mathbf{A}}\Delta\Phi$$

$$\Delta(\underline{\mathbf{A}}\,\underline{\mathbf{B}}) = \underline{\mathbf{A}}\Delta\underline{\mathbf{B}} + 2(\nabla\underline{\mathbf{A}})^{21} \cdot \nabla\underline{\mathbf{B}}) + (\Delta\underline{\mathbf{A}})\underline{\mathbf{B}}$$

$$\Delta(\Phi\underline{\underline{\mathbf{D}}}) = \Phi\Delta\underline{\underline{\mathbf{D}}} + (\Delta\Phi)\underline{\underline{\mathbf{D}}} + 2\nabla\Phi \cdot \nabla\underline{\underline{\mathbf{D}}}$$

$$\Delta(\underline{\mathbf{A}} \cdot \underline{\mathbf{B}}) = \underline{\mathbf{A}} \cdot \Delta\underline{\mathbf{B}} + 2(\nabla\underline{\mathbf{A}})^{21} : \nabla\underline{\mathbf{B}} + (\Delta\underline{\mathbf{A}}) \cdot \underline{\mathbf{B}}$$

$$\Delta(\underline{\mathbf{A}} \cdot \underline{\underline{\mathbf{D}}}) = \underline{\mathbf{A}} \cdot \Delta\underline{\underline{\mathbf{D}}} + 2(\nabla\underline{\mathbf{A}})^{21} : (\nabla\underline{\underline{\mathbf{D}}}) + (\Delta\underline{\mathbf{A}}) \cdot \underline{\underline{\mathbf{D}}}$$

$$\Delta(\underline{\underline{\mathbf{D}}} \cdot \underline{\mathbf{A}}) = \underline{\underline{\mathbf{D}}} \cdot \Delta\underline{\mathbf{A}} + 2(\nabla\underline{\underline{\mathbf{D}}})^{312} : \nabla\underline{\mathbf{A}} + (\Delta\underline{\underline{\mathbf{D}}}) \cdot \underline{\mathbf{A}}$$

$$\Delta(\underline{\mathbf{A}} \times \underline{\underline{\mathbf{I}}}) = (\Delta\underline{\mathbf{A}}) \times \underline{\underline{\mathbf{I}}}$$

$$\nabla\nabla \cdot (\Phi\underline{\mathbf{A}}) = (\nabla\Phi)\nabla \cdot \underline{\mathbf{A}} + \Phi\nabla\nabla \cdot \underline{\mathbf{A}} + \nabla\Phi \times (\nabla \times \underline{\mathbf{A}}) + \underline{\mathbf{A}} \cdot \nabla\nabla\Phi + \nabla\Phi \cdot \nabla\underline{\mathbf{A}}$$

$$\nabla \times \nabla \times (\Phi\underline{\mathbf{A}}) = \nabla\Phi \times (\nabla \times \underline{\mathbf{A}}) - \underline{\mathbf{A}}\Delta\Phi + \underline{\mathbf{A}} \cdot \nabla\nabla\Phi + \Phi\nabla \times \nabla \times \underline{\mathbf{A}} + (\nabla\Phi)\nabla \cdot \underline{\mathbf{A}} - \nabla\Phi \cdot \nabla\underline{\mathbf{A}}$$

$$\nabla \times \nabla \times (\Phi\,\underline{\underline{\mathbf{I}}}) = \nabla\nabla\Phi - \underline{\underline{\mathbf{I}}}\,\Delta\Phi$$

A.8 Special Vector Fields Depending on the Vector of Position

$$\nabla R = \hat{\underline{\mathbf{R}}}$$

$$\nabla\frac{1}{R} = -\frac{1}{R^2}\hat{\underline{\mathbf{R}}}$$

$$\nabla R^n = nR^{n-1}\hat{\underline{\mathbf{R}}}\,,\ n = 0, \pm 1, \pm 2, \ldots$$

$$\nabla\Phi(R) = \Phi'(R)\hat{\underline{\mathbf{R}}}$$

$$\nabla\frac{\mathrm{e}^{\,\mathrm{j}kR}}{R} = \left(\mathrm{j}k - \frac{1}{R}\right)\frac{\mathrm{e}^{\,\mathrm{j}kR}}{R}\hat{\underline{\mathbf{R}}}$$

$$\nabla|\underline{\mathbf{R}} - \underline{\mathbf{R}}'| = \frac{\underline{\mathbf{R}} - \underline{\mathbf{R}}'}{|\underline{\mathbf{R}} - \underline{\mathbf{R}}'|}$$

$$\nabla'|\underline{\mathbf{R}} - \underline{\mathbf{R}}'| = -\nabla|\underline{\mathbf{R}} - \underline{\mathbf{R}}'|$$

$$\nabla\frac{1}{|\underline{\mathbf{R}} - \underline{\mathbf{R}}'|} = -\frac{\underline{\mathbf{R}} - \underline{\mathbf{R}}'}{|\underline{\mathbf{R}} - \underline{\mathbf{R}}'|^3}$$

$$\nabla \cdot \underline{\mathbf{R}} = 3$$

$$\nabla \cdot \hat{\underline{\mathbf{R}}} = \frac{2}{R}$$

$$\nabla \cdot \frac{\underline{\mathbf{R}} - \underline{\mathbf{R}}'}{|\underline{\mathbf{R}} - \underline{\mathbf{R}}'|} = \frac{2}{|\underline{\mathbf{R}} - \underline{\mathbf{R}}'|}$$

$$\nabla\cdot[\Phi(R)\underline{\hat{\mathbf{R}}}] = \frac{2\Phi(R)}{R} + \Phi'(R)$$

$$\nabla\cdot(R^n\underline{\hat{\mathbf{R}}}) = (n+2)R^{n-1}\,,\ n = 0, \pm 1, \pm 2, \ldots$$

$$\nabla\cdot\left(\frac{\mathrm{e}^{\mathrm{j}kR}}{R}\underline{\hat{\mathbf{R}}}\right) = \left(\frac{1}{R} + \mathrm{j}k\right)\frac{\mathrm{e}^{\mathrm{j}kR}}{R}$$

$$\nabla\times\underline{\mathbf{R}} = \underline{\mathbf{0}}$$

$$\nabla\times[\Phi(R)\underline{\hat{\mathbf{R}}}] = \underline{\mathbf{0}}$$

$$\nabla\underline{\mathbf{R}} = \underline{\underline{\mathbf{I}}}$$

$$\nabla\cdot\nabla\underline{\mathbf{R}} = \underline{\mathbf{0}}$$

$$\nabla(\underline{\mathbf{R}}\,\underline{\mathbf{R}}) = \underline{\underline{\mathbf{I}}}\,\underline{\mathbf{R}} + (\underline{\mathbf{R}}\,\underline{\underline{\mathbf{I}}})^{213}$$

$$\nabla\underline{\hat{\mathbf{R}}} = \nabla\cdot(\underline{\underline{\mathbf{I}}}\,\underline{\hat{\mathbf{R}}})$$

$$= \frac{1}{R}(\underline{\underline{\mathbf{I}}} - \underline{\hat{\mathbf{R}}}\,\underline{\hat{\mathbf{R}}})$$

$$\nabla[\Phi(R)\underline{\hat{\mathbf{R}}}] = \left[\Phi'(R) - \frac{\Phi(R)}{R}\right]\underline{\hat{\mathbf{R}}}\,\underline{\hat{\mathbf{R}}} + \frac{\Phi(R)}{R}\underline{\underline{\mathbf{I}}}$$

$$\nabla\cdot(\underline{\hat{\mathbf{R}}}\,\underline{\underline{\mathbf{I}}}) = \frac{2}{R}\underline{\underline{\mathbf{I}}}$$

$$\nabla\cdot(\underline{\hat{\mathbf{R}}}\,\underline{\hat{\mathbf{R}}}\,\underline{\hat{\mathbf{R}}}) = \frac{2}{R}\underline{\hat{\mathbf{R}}}\,\underline{\hat{\mathbf{R}}}$$

$$\nabla\cdot[\Phi(R)\underline{\underline{\mathbf{I}}}] = \nabla\Phi(R)$$

$$= \Phi'(R)\underline{\hat{\mathbf{R}}}$$

$$\nabla\cdot[\Phi(R)\underline{\hat{\mathbf{R}}}\,\underline{\hat{\mathbf{R}}}] = \left[\Phi'(R) + \frac{2}{R}\Phi(R)\right]\underline{\hat{\mathbf{R}}}$$

$$\nabla\cdot[\underline{\underline{\mathbf{D}}}(\underline{\mathbf{R}})\times\underline{\mathbf{R}}] = [\nabla\cdot\underline{\underline{\mathbf{D}}}(\underline{\mathbf{R}})]\times\underline{\mathbf{R}} - \langle\underline{\underline{\mathbf{D}}}(\underline{\mathbf{R}})\rangle$$

$$\Delta R^n = n(n+1)R^{n-2}\,,\ n = 0, \pm 1, \pm 2, \ldots$$

$$\Delta\Phi(R) = \frac{2\Phi'(R)}{R} + \Phi''(R)$$

$$\Delta\frac{1}{R} = -4\pi\delta(x)\delta(y)\delta(z)$$

$$\Delta\frac{\mathrm{e}^{\mathrm{j}kR}}{R} = -k^2\frac{\mathrm{e}^{\mathrm{j}kR}}{R} - 4\pi\delta(x)\delta(y)\delta(z)$$

$$\nabla\nabla\frac{1}{R} = -\frac{1}{R^3}(\underline{\underline{\mathbf{I}}} - 3\underline{\hat{\mathbf{R}}}\,\underline{\hat{\mathbf{R}}})$$

für $R \neq 0$

$$\nabla\nabla\frac{\mathrm{e}^{\mathrm{j}kR}}{R} = \left[-k^2\underline{\hat{\mathbf{R}}}\,\underline{\hat{\mathbf{R}}} + \mathrm{j}k\frac{1}{R}(\underline{\underline{\mathbf{I}}} - 3\underline{\hat{\mathbf{R}}}\,\underline{\hat{\mathbf{R}}}) - \frac{1}{R^2}(\underline{\underline{\mathbf{I}}} - 3\underline{\hat{\mathbf{R}}}\,\underline{\hat{\mathbf{R}}})\right]\frac{\mathrm{e}^{\mathrm{j}kR}}{R}$$

for $R \neq 0$.

For $R = 0$, both above formulas have to be understood as pseudofunctions in a distributional sense, and they must be complemented by a δ-singular term;

hence we have for all $R \geq 0$:

$$\nabla\nabla\frac{\mathrm{e}^{\,\mathrm{j}kR}}{R} = \mathrm{pf}_{\delta}\left[-k^2\underline{\hat{\mathbf{R}}}\,\underline{\hat{\mathbf{R}}} + \mathrm{j}k\frac{1}{R}(\underline{\underline{\mathbf{I}}} - 3\underline{\hat{\mathbf{R}}}\,\underline{\hat{\mathbf{R}}}) - \frac{1}{R^2}(\underline{\underline{\mathbf{I}}} - 3\underline{\hat{\mathbf{R}}}\,\underline{\hat{\mathbf{R}}})\right]\frac{\mathrm{e}^{\,\mathrm{j}kR}}{R} + 4\pi k^2\underline{\underline{\mathbf{L}}}^{V_\delta}\,\delta(\underline{\mathbf{R}})$$

$$\text{mit } \underline{\underline{\mathbf{L}}}^{V_\delta} = -\frac{1}{4\pi k^2}\lim_{V_\delta\to 0}\iint_{S_\delta}\frac{\underline{\mathbf{n}}\,\underline{\hat{\mathbf{R}}}}{R^2}\,\mathrm{d}S$$

V_δ is an exclusion volume with surface S_δ.
For a spherical exclusion volume we have $V_\delta = V_K$:
$\mathrm{pf}_\delta \Longrightarrow \mathrm{PV}$ (Cauchy principal value)

$$\underline{\underline{\mathbf{L}}}^{V_K} = -\frac{1}{3k^2}\,\underline{\underline{\mathbf{I}}}$$

For a constant vector $\underline{\mathbf{a}}$ we have

$$\nabla(\underline{\mathbf{R}}\cdot\underline{\mathbf{a}}) = \underline{\mathbf{a}}$$

$$\nabla\cdot(R\,\underline{\mathbf{a}}) = \underline{\hat{\mathbf{R}}}\cdot\underline{\mathbf{a}}$$

$$\nabla\times(R\,\underline{\mathbf{a}}) = \underline{\hat{\mathbf{R}}}\times\underline{\mathbf{a}}$$

$$\nabla\times(\underline{\mathbf{a}}\times\underline{\mathbf{R}}) = 2\underline{\mathbf{a}}.$$

References

Abramowitz, M. and Stegun, I.A. (1965) *Handbook of mathematical functions.* Dover Publications: New York

Achenbach, J.D. (1973) *Wave propagation in elastic solids.* North-Holland: Amsterdam

Achenbach, J.D., Gautesen, A.K. and McMaken, H. (1982) *Ray methods for elastic waves in solids.* Pitman Advanced Publishing Program: Boston

Alves, C.J.S. and Kress, R. (2002) On the far field operator in elastic obstacle scattering. *IMA J Appl Math* **67**:1

Auld, B.A. (1973) *Acoustic fields and waves in solids, Vol. I and Vol. II.* John Wiley & Sons: New York

Aulenbacher, U. (1988) *Breitbandige und monochromatische Ultraschallfelder piezoelektrischer Winkel- und Array-Prüfköpfe bei Ankopplung an elastische Medien.* Dissertation, Universität des Saarlandes: Saarbrücken

Aulenbacher, U. and Langenberg, K.J. (1980) Analytical representation of transient ultrasonic phased-array near- and far-fields. *J Nondestr Evaluation* **1**:53

Balanis, C.A. (1997) *Antenna theory.* John Wiley & Sons: New York

Bamler, R. (1989) *Mehrdimensionale lineare Systeme.* Springer-Verlag: Berlin

Baum, C.E. (2000) Target symmetry and the scattering dyadic. In: D.H. Werner, and R. Mittra (Eds.): *Frontiers in electromagnetics.* IEEE Press: Piscataway

Becker, K.D. (1974) *Ausbreitung elektromagnetischer Wellen.* Springer-Verlag: Berlin

Behnke, H. and Sommer, F. (1965) *Theorie der analytischen Funktionen einer komplexen Veränderlichen.* Springer-Verlag: Berlin

Belkebir, K. and Saillard, M. (Eds.) (2001) Special section: Testing inversion algorithms against experimental data. *Inverse Problems* **17**:1565

Ben-Menahem, A. and Singh, S.J. (1981) *Seismic waves and sources.* Springer-Verlag: New York

Bertero, M. and De Mol, C. (1996) Super-resolution by data inversion. In: E. Wolf (Ed.): *Progress in optics.* Elsevier: Amsterdam

Bihn, M. (1998) *Zur numerischen Berechnung elastischer Wellen im Zeitbereich.* Dissertation, TU Darmstadt, Shaker Verlag: Aachen

Bleistein, N. (1984) *Mathematical methods for wave phenomena.* Academic Press: Orlando

Blok, H., Ferweda, H.A. and Kuiken, H.K. (Eds.) (1992) *Huygens' principle 1690–1990.* North-Holland, Elsevier: Amsterdam

Blume, S. (1991) *Theorie elektromagnetischer Felder.* Hüthig: Heidelberg

Boehm, R., Erhard, A., Wüstenberg, H. and Rehfeldt, T. (2002) Dreidimensionale Berechnung von Schallfeldern unter dem Einfluss zylindrischer Bauteilkrümmungen für fokussierende Prüfköpfe und Gruppenstrahler. Berichtsband **80-CD**, Deutsche Gesellschaft für Zerstörungsfreie Prüfung: Berlin

Bollig, G. and Langenberg, K.J. (1983) The Singularity Expansion Method as applied to the elastodynamic scattering problem. *Wave Motion* **5**:331

Born, M. and Wolf, E. (1975) *Principles of optics*. Pergamon Press: Oxford

Bowman, J.J., Senior T.B.A. and Uslenghi, P.L.E. (Eds.) (1987) *Electromagnetic and acoustic scattering by simple shapes*. Hemisphere Publ. Corp.: New York

Bracewell, R.N. (1978) *The fourier transform and its applications*. McGraw-Hill: New York

Brill, D. and Gaunaurd, G. (1987) Resonance theory of elastic waves ultrasonically scattered from an elastic sphere. *J Acoust Soc Am* **81**:1

Brillouin, L. (1914) Über die Fortpflanzung des Lichts in dispergierenden Medien. Annalen der Physik **44**:203

Burg, K., Haf, H. and Wille, F. (1990) *Höhere Mathematik für Ingenieure, Band IV: Vektoranalysis und Funktionentheorie*. B.G. Teubner: Stuttgart

Červený, V. (2001) *Seismic ray theory*. Cambridge University Press: Cambridge

Červený, V., Popov, M.M. and Pšenčík, I. (1982) Computation of wave fields in inhomogeneous media—Gaussian beam approach. *Geophys J R Astr Soc* **70**:109

Chen, H.C. (1983) *Theory of electromagnetic waves*. McGraw-Hill: New York

Chen, G. and Zhou, J. (1992) *Boundary element methods*. Academic Press: London

Chew, W.C. (1990) *Waves and fields in inhomogeneous media*. Van Nostrand Reinhold: New York

Chew, W.C., Song, J.M., Velamparambil, S., Cui, T.J., Zhao, S., Pan, Y.C., Hu, B. and Chao, H.Y. (2002) Recent developments in fast frequency-domain integral-equation solvers in large-scale computational electromagnetics. In: W.R. Stone (Ed.): *Review of radio science 1999–2002*. IEEE Press/John Wiley: New York

Cloude, S.R. (2002) Polarimetry in wave scattering applications. In: P.C. Sabatier and E.R. Pike (Eds.): *Scattering*. Academic Press: London

Colton, D. and Kress, R. (1983) *Integral equation methods in scattering theory*. John Wiley & Sons: New York

Dassios, G. and Kleinman, R. (2000) *Low frequency scattering*. Clarendon Press: Oxford

de Hoop, A.T. (1958) *Representation theorems for the displacement in an elastic solid and their application to elastodynamic diffraction theory*. Ph.D. Thesis, Technical University, Delft

de Hoop, A.T. (1959) A modification of Cagniard's method of solving seismic pulse problems. *Appl Sci Res* **B8**:349

de Hoop, A.T. (1995) *Handbook of radiation and scattering of waves.* Academic Press: London (2008) Electronic Reproduction with Corrections http://www.atdehoop.com/Books/DeHoopAT book 2008/DeHoop AT1995Handbook.index.html

DeSanto, J.A. (1992) *Scalar wave theory.* Springer-Series on Wave Phenomena **12** (Ed.: L.B. Felsen), Springer-Verlag: Berlin

Devaney, A.J. (1986) Reconstructive tomography with diffracting wavefields. *Inv Probl* **2**:161

Doetsch, G. (1967) Funktionaltransformationen. In: R. Sauer and I. Szabó (Eds.): *Mathematische Hilfsmittel des Ingenieurs, Teil I.* Springer-Verlag: Berlin

Dudley, D.G. (1994) *Mathematical foundations for electromagnetic theory.* IEEE Press: Piscataway

Einspruch, N.G., Witterholt, E.J. and Truell, R. (1960) Scattering of a plane transverse wave by a spherical obstacle in an elastic medium. *J Appl Phys* **31**:806

Erdélyi, A. (Ed.) (1954) *Tables of integral transforms, Vol. I and Vol. II.* McGraw-Hill: New York

Erhard, A. (1982) *Untersuchungen zur Ausbreitung von Longitudinalwellen an Oberflächen bei der Materialprüfung mit Ultraschall.* Dissertation, TU Berlin

Fellinger, P. (1991) *Verfahren zur numerischen Lösung elastischer Wellenausbreitungsprobleme im Zeitbereich durch direkte Diskretisierung der elastodynamischen Grundgleichungen.* Dissertation, Universität Kassel: Kassel

Fellinger, P., Marklein, R., Langenberg, K.J. and Klaholz, S. (1995) Numerical modeling of elastic wave propagation and scattering with EFIT—Elastodynamic Finite Integration Technique. *Wave Motion* **21**:47

Fink, M., Sandrin, L., Tanter, M., Catheline, S., Chaffai, S., Bercoff, J. and Gennison, J.L. (2002) Ultra high speed imaging of elasticity. Proc IEEE-UFFC Symposium, München

Gabor, D. (1946) Theory of communication. *J Inst Elec Eng* **93**:429

Galdi, V., Felsen, L.B. and Castanon, D.A. (2001) Narrow-waisted Gaussian beam discretization for short-pulse radiation from one-dimensional large apertures. *IEEE Trans Ant Propagat* **AP-49**:1322

——— (2003) Time-domain radiation from large two-dimensional apertures via narrow-waisted Gaussian beams. *IEEE Trans Ant Propagat* **AP-51**:78

Gibbs, J.W. (1913) *Vector analysis.* Yale University Press: New Haven

Greving, G. (2000) Numerical system-simulations including antennas and propagation exemplified for a radio navigation system. *Int J Electron Comm (AEÜ)* **54**:1

Haak, K.F.I. (1999) *Multi-frequency non-linear profile inversion methods.* Delft University Press: Delft

Hahn, S.L. (1997) *Hilbert transforms in signal processing.* Artech House: Boston

Hannemann, R. (2001) *Modeling and imaging of elastodynamic wave fields in inhomogeneous anisotropic media.* Dissertation (www.dissertation.de), Universität Kassel: Kassel

Harker, A.H. (1988) *Elastic waves in solids.* Adam Hilger: Bristol

Harker, A.H., Ogilvy, J.A. and Temple J.A.G. (1990) Modeling ultrasonic inspection of austenitic welds. *J Nondestr Eval* **9**:155

Harrington, R.F. (1968) *Field computation by moment methods.* The Macmillan Comp.: New York

Helbig, K. (1994) *Foundations of anisotropy for exploration seismics.* Pergamon, Elsevier: Oxford

Hertz, H. (1890) Über die Grundgleichungen der Elektrodynamik für ruhende Körper. *Annalen der Physik* **40**:577

Heyman, E. (1994) Pulsed beam propagation in inhomogeneous medium. *IEEE Trans Ant Propagat* **AP-42**:311

Heyman, E. (2002) Pulsed beam solutions for propagation and scattering problems. In: P.C. Sabatier and E.R. Pike (Eds.): *Scattering*. Academic Press: London

Heyman, E. and Felsen, L.B. (1985) A wavefront interpretation of the singularity expansion method. *IEEE Trans Ant Propagat* **AP-33**:706

——— (2001) Gaussian beam and pulsed beam dynamics: complex-source and complex-spectrum formulations within and beyond paraxial asymptotics. *J Opt Soc Am A* **18**:1588

Hönl, H., Maue, A.W. and Westpfahl, K. (1961) Theorie der Beugung. In: S. Flügge (Ed.): *Handbuch der Physik*, XXV/1. Springer-Verlag: Berlin

IEEE Committee (1973) Transducers and Resonators Committee of the IEEE Group on Sonics and Ultrasonics (Sponsor): IEEE Standard on Magnetostrictive Materials: Piezomagnetic Nomenclature. *IEEE Trans Sonics Ultrasonics* **SU-20**:67

Ishimaru, A. (1991) *Electromagnetic wave propagation, radiation, and scattering.* Prentice Hall, Englewood Cliffs

Jackson, J.D. (1975) *Classical electrodynamics*. John Wiley & Sons: New York

Jones, D.S. (1984) A uniqueness theorem in elastodynamics. *Q J Mech Appl Math* **37**, Pt. 1:121

Karlsson, A. and Kristensson, G. (1992) Constitutive relations, dissipation and reciprocity for the Maxwell equations in the time domain. *J Electrom Waves Applic* **6**:537

King, R.W.P. and Harrison, C.W. (1969) *Antennas and waves: A modern approach.* The M.I.T. Press: Cambridge/MA

Klaholz, S. (1992) *Die Streuung elastischer Wellenfelder an der kreiszylindrischen Pore—analytische und numerische Lösung im Vergleich.* Diplomarbeit im Fachgebiet Theoretische Elektrotechnik, Universität Kassel: Kassel

Kleinman R.E. and van den Berg, P.M. (1997) A contrast source inversion method. *Inv Probl* **13**:1607

Klimeš, L. (1989) Gaussian packets in the computation of seismic wavefields. *Geophys J Int* **99**:421

Kostka, J., Langenberg, K.J., Mayer, K. and Krause, M. (1998) Improved flaw imaging applying elastodynamic far-field Fourier inversion (EL-FT-SAFT). In: *Computer methods and inverse problems in nondestructive testing and diagnostics.* Berichtsband **64**, Deutsche Gesellschaft für Zerstörungsfreie Prüfung: Berlin

Krautkrämer, J. and Krautkrämer, H. (1986) *Werkstoffprüfung mit Ultraschall.* Springer-Verlag: Berlin

Krieger, J., Krause, M. and Wiggenhauser, H. (1998) *Erprobung und Bewertung zerstörungsfreier Prüfmethoden für Betonbrücken.* Berichte der Bundesanstalt für Straßenwesen, Brücken- und Ingenieurbau **18**: Bergisch-Gladbach

Kristensson, G., Karlsson, A. and Rikte, S. (2000) Electromagnetic wave propagation in dispersive and complex material with time domain techniques. In: P.C. Sabatier and E.R. Pike (Eds.): *Scattering.* Academic Press: London

Kühnicke, E. (2001) *Elastische Wellen in geschichteten Festkörpersystemen.* Habilitationsschrift, TU Dresden, TIMUG Wissenschaftliche Vereinigung: Bonn

Kutzner, J. (1983) Grundlagen der Ultraschallphysik. B.G. Teubner: Stuttgart

Landau, L.D., Lifshitz, E.M. and Pitaevskii, L.P. (1984) *Electrodynamics of continuous media.* Pergamon Press: Oxford

Langenberg, K.J. (1983) *Vorlesungen über theoretische Grundlagen der zerstörungsfreien Werkstoffprüfung mit Ultraschall. Teil II: Ebene elastische Wellen.* Bericht 830215-TW des Fraunhofer-Instituts für Zerstörungsfreie Prüfverfahren: Saarbrücken

Langenberg, K.J. (1987) Applied inverse problems for acoustic, electromagnetic and elastic waves. In: P.C. Sabatier (Ed.): *Basic methods of tomography and inverse problems.* Adam Hilger: Bristol

Langenberg, K.J. (2002) Linear scalar inverse scattering. In: P.C. Sabatier and E.R. Pike (Eds.): *Scattering.* Academic Press: London

Langenberg, K.J. (2005) *Theorie elektromagnetischer Wellen.* Vorlesungsmanuskript Universität Kassel: Kassel

Langenberg, K.J., Fellinger, P. and Marklein, R. (1990) On the nature of the so-called subsurface longitudinal wave and/or the surface longitudinal "creeping" wave. *Res Nondestr Eval* **2**:59

Langenberg, K.J., Fellinger, P., Marklein, R., Zanger, P., Mayer, K. and Kreutter, T. (1993a) Inverse methods and imaging. In: J.D. Achenbach (Ed.): *Evaluation of materials and structures by quantitative ultrasonics.* Springer-Verlag: Wien

Langenberg, K.J., Brandfaß, M., Mayer, K., Kreutter, T., Brüll, A., Fellinger, P. and Huo, D. (1993b) Principles of microwave imaging and inverse scattering. *EARSeL Advances in Remote Sensing* **2**:163

Langenberg, K.J., Brandfaß, M., Fellinger, P., Gurke, T. and Kreutter, T. (1994) A unified theory of multidimensional electromagnetic vector inverse scattering within the Kirchhoff or Born approximation. In: W.-M. Boerner and H. Überall (Eds.) *Radar target imaging.* Springer-Verlag: Berlin

Langenberg, K.J., Brandfaß, M., Klaholz, S., Marklein, R., Mayer, K., Pitsch, A. and Schneider, R. (1997) Applied inversion in nondestructive testing. In: H.W. Engl, A.K. Louis and W. Rundell (Eds.): *Inverse problems in medical imaging and nondestructive testing*. Springer: Wien

Langenberg, K.J., Brandfaß, M., Hannemann, R., Hofmann, C., Kaczorowski, T., Kostka, J., Marklein, R., Mayer, K. and Pitsch, A. (1999a) Inverse scattering with acoustic, electromagnetic, and elastic waves as applied in nondestructive evaluation. In: A. Wirgin (Ed.): *Wavefield inversion*. Springer-Verlag: Wien

Langenberg, K.J., Brandfaß, M., Fritsch, A. and Potzkai, B. (1999b) Linearized 3-D electromagnetic vector wave inversion. In: M. Oristaglio and B. Spies (Eds.): *Three-dimensional electromagnetics*. Society of Exploration Geophysicists: Tulsa

Langenberg, K.J. and Fellinger, P. (1995) Singularity evaluation of the dyadic Green function utilizing distributional analysis. Proceedings of the 1995 U.R.S.I. International Symposium on Electromagnetic Theory, St. Petersburg, Russia, May 23–26

Langenberg, K.J., Hannemann, R., Kaczorowski, T., Marklein, R., Koehler, B., Schurig, C. and Walte, F. (2000) Application of modeling techniques for ultrasonic austenitic weld inspection. *NDT&E International* **33**:465

Langenberg, K.J., Marklein, R. and Mayer, K. (2002) Applications in nondestructive testing. In: P.C. Sabatier and E.R. Pike (Eds.): *Scattering*. Academic Press: London

Langenberg, K.J., Marklein, R., Mayer, K., Krylow, T., Ampha, P., Krause, M. and Streicher, D. (2004a) Wavefield inversion in nondestructive testing. In: I.M. Pinto, V. Galdi and L.B. Felsen (Eds.): *Electromagnetics in a complex world: challenges and perspectives*. Springer *Proceedings in Physics* **97**, Springer: Berlin

Langenberg, K.J., Marklein, R., Mayer, K., Wiggenhauser, H. and Krause, M. (2004b) Multimode modeling and imaging in nondestructive testing. In: P. Russer and M. Mongiardo (Eds.): *Fields, networks, computational methods, and systems in modern electrodynamics*. Springer Proceedings in Physics **96**, Springer: Berlin

Langenberg, K.J. and Marklein, R. (2005) Transient elastic waves applied to nondestructive testing of transversely isotropic lossless materials: A coordinate-free approach. *Wave Motion* **41**:247

Langenberg, K.J., Mayer, K. and Marklein, R. (2006) Nondestructive testing of concrete with electromagnetic and elastic waves: Modeling and imaging. *Cement and Concrete Composites* **28**:370

Langenberg, K.J., Mayer, K. and Marklein, R. (2007) Zerstörungsfreie Prüfung von Beton: Modellierung und Abbildung. In: Reinhardt, H.-W. et al. (Eds.): *Echo-Verfahren in der zerstörungsfreien Zustandsuntersuchung von Betonbauteilen*. Beton Kalender 2007: Verkehrsbauten—Flächentragwerke, Ernst & Sohn: Berlin

Larmor, J. (1903) On the mathematical expression of the principle of Huygens. *London Math Soc Proc* **1**:1

Lindell, I.V. (1992) *Methods for Electromagnetic Field Analysis.* Clarendon Press: Oxford

Lindell, I.V. and Kiselev, A.P. (2000) Polyadic methods in elastodynamics. *PIERS 2000 Proceedings*, The Electromagnetics Academy: Cambridge/MA

Maciel, J.J. and Felsen, L.B. (1990) Gaussian beam analysis of propagation from an extended plane aperture distribution through dielectric layers. I. Plane layer. *IEEE Trans Ant Propagat* **38**:1607

Marklein, R. (1997) *Numerische Verfahren zur Modellierung von akustischen, elektromagnetischen, elastischen und piezoelektrischen Wellenausbreitungsproblemen im Zeitbereich basierend auf der Finiten Integrationstechnik.* Dissertation, Universität Kassel, Shaker Verlag: Aachen

Marklein, R. (2002) The Finite Integration Technique as a general tool to compute acoustic, electromagnetic, elastodynamic, and coupled wave fields. In: W.R. Stone (Ed.): *Review of radio science 1999–2002.* IEEE Press: Piscataway

Marklein, R., Balasubramanian, K., Qing, A. and Langenberg, K.J. (2001) Linear and non-linear iterative scalar inversion of multi-frequency multi-bistatic experimental electromagnetic scattering data. *Inv Probl* **17**:1597

Marklein, R., Balasubramanian, K., Langenberg, K.J., Miao, J. and Sinaga, S.M. (2002a) Applied linear and non-linear scalar inverse scattering: A comparative study. Proc. XXVIIth General Assembly of the International Union of Radio Science (URSI), August 17–24: Maastricht/The Netherlands

Marklein, R., Mayer, K., Hannemann, R., Krylow, T., Balasubramanian, K., Langenberg, K.J. and Schmitz, V. (2002b) Linear and non-linear inversion algorithms applied in nondestructive evaluation. *Inv Probl* **18**:1733

Martensen, E. (1968) *Potentialtheorie.* B.G. Teubner: Stuttgart

Mayer, K. (1989) *Ultraschallabbildungsverfahren: Algorithmen, Methoden der Signalverarbeitung und Realisierung.* Dissertation, Universität Kassel: Kassel

Mayer, K., Marklein, R., Langenberg, K.J. and Kreutter, T. (1990) Three-dimensional imaging system based on Fourier transform synthetic aperture focusing technique. *Ultrasonics* **28**:241

Mayer, K., Zimmer, A., Langenberg, K.J., Kohl, C. and Maierhofer, C. (2003) Nondestructive evaluation of embedded structures in concrete: Modeling and imaging. Proc. Int. Symp. Non-Destructive Testing in Civil Engineering (NDT-CE), Sept. 16–19: Berlin

Michielssen, E., Shanker, B., Aygun, K., Lu, M. and Ergin, A. (2002) Plane-wave time-domain algorithms and fast time-domain integral-equation solvers. In: W.R. Stone (Ed.) *Review of radio science 1999–2002*. IEEE Press/John Wiley: New York

Miller, G.F. and Pursey, H. (1954) The field and radiation impedance of mechanical radiators on the free surface of a semi-infinite isotropic solid. Proc R Soc **A 223**:521

Morita, N., Kumagai, N. and Mautz, J.R. (1990) *Integral equation methods for electromagnetics*. Artech House: Boston

Morse, P.M. and Feshbach, H. (1953) *Methods of theoretical physics, Part I and Part II*. McGraw-Hill: New York

Morse, P.M. and Ingard, K.U. (1968) *Theoretical acoustics*. McGraw-Hill: New York

Mow, C.C. (1965) Transient response of a rigid spherical inclusion in an elastic medium. *J Appl Mech* **32**:637

Nayfeh, A.H. (1995) *Wave propagation in layered anisotropic media with applications to composites*. Elsevier: Amsterdam

Neumann, E., Hirsekorn, S., Hübschen, G., Just, T. and Schmid, R. (1995) *Ultraschallprüfung von austenitischen Plattierungen, Mischnähten und austenitischen Schweißnähten*. Expert-Verlag: Reveringen-Malmsheim

Nevels, R. and Shin, C. (2001) Lorenz, Lorentz, and the gauge. *IEEE Ant Propagat Mag* **43**

Norris, A.N. (1987) A theory of pulsed propagation in anisotropic elastic solids. *Wave Motion* **9**:509

Norris, A.N. (1988) Elastic Gaussian wave packets in isotropic media. *Acta Mech* **71**:95

Ogilvy, J.A. (1986) Ultrasonic beam profiles and beam propagation in an austenitic weld using a theoretical ray tracing model. *Ultrasonics* **24**:337

Pao, Y.H. and Varatharajulu, V. (1976) Huygens' principle, radiation conditions, and integral formulas for the scattering of elastic waves. *J Acoust Soc Am* **59**

Pelekanos, G., Kleinman, R.E. and van den Berg, P.M. (2000) Inverse scattering in elasticity—a modified gradient approach. *Wave Motion* **32**:57

Poggio, A.J. and Miller, E.K. (1987) Integral equation solutions of three-dimensional scattering problems. In: R. Mittra (Ed.): *Computer techniques for electromagnetics*. Hemisphere Publ. Corp.: Washington

Popov, M.M. (1982) A new method of computation of wave fields using Gaussian beams. *Wave Motion* **4**:85

Popov, M.M. (2004) Gaussian beam method for propagation of wave packets modulated both in frequency and amplitude. *Proc URSI Int Symp on Electromagnetic Field Theory*, 337: Pisa/Italy

Rose, J.L. (1999) *Ultrasonic waves in solid media*. Cambrigde University Press: Cambridge

Royer, D. and Dieulesaint, E. (2000) *Elastic waves in solids, I, II*. Springer-Verlag: Berlin

Schäfke, F.W. (1967) Spezielle Funktionen. In: R. Sauer and I. Szabó (Eds.): *Mathematische Hilfsmittel des Ingenieurs, Teil I*. Springer-Verlag: Berlin

Schleichert, U. (1989) *Die Theorie der optischen Erzeugung elastischer Wellen in Festkörpern*. Dissertation, Universität Kassel: Kassel

Schleichert, U., Paul, M., Hoffmann, B., Langenberg, K.J. and Arnold, W. (1988) Theoretical and experimental investigations of broadband thermoelastically generated ultrasonic pulses. In: P. Hess and J. Pelzl (Eds.): *Photoacoustic and photothermal phenomena*. Springer Series in Optical Sciences **58**, Springer: Berlin

Schmerr, L.W. (1998) *Fundamentals of ultrasonic nondestructive evaluation.* Plenum Press: New York

Schmerr, L.W. and Song, J.-S. (2007) *Ultrasonic nondestructive evaluation systems. Models and measurements.* Springer-Verlag: Berlin

Schmitz, V. (2002) Nondestructive acoustic imaging techniques. In: M. Fink, W.A. Kuperman, J.-P. Montagner and A. Turin (Eds.): *Imaging of complex media with acoustic and seismic waves.* Topics in Applied Physics **84**, Springer: Berlin

Schmitz, V., Müller, W., Maisl, M., Gondrom, S., Langenberg, K.J., Marklein, R., Miao, J. and Julisch, P. (2004a) Grundlagen für eine beanspruchungsorientierte zerstörungsfreie Prüfung von druckführenden Komponenten. Abschlussbericht des Reaktorsicherheitsforschungsvorhabens 1501220, Fraunhofer IZFP: Saarbrücken

Schmitz, V., Langenberg, K.J. and Chaklov, S. (2004b) Calculation of high frequency ultrasonic signals for shear wave insonification in solid material. *Ultrasonics* **42**:249

Schubert, F. (1999) *Ausbreitungsverhalten von Ultraschallimpulsen in Beton und Schlussfolgerungen für die zerstörungsfreie Prüfung.* Dissertation, Technische Universität, Dresden

Shlivinski, A. (2004) Personal Communication

Shlivinski, A., Heyman, E., Boag, A. and Letrou, C. (2004) A phase-space beam summation formulation for ultra wideband radiation. *IEEE Trans Ant Propagat* **AP-52**:2042

Shlivinski, A., Heyman, E. and Boag, A. (2005) A phase-space beam summation formulation for ultra wideband radiation. Part II–A multi-band scheme. *IEEE Trans Ant Propagat* **AP-53**:948

Shlivinski, A., Langenberg, K.J. and Marklein, R. (2004b) Ultrasonic modeling and imaging in dissimilar welds. Beitrag Nr. 49 im 2. Band des 30. MPA-Seminars, Stuttgart

Siegman, A.E. (1986) *Lasers.* University Science Books, Mill Valley/CA

Sihvola, A. (1991) Lorenz-Lorentz or Lorentz-Lorenz. *IEEE Trans Ant Propagat Mag* **33**:56

Snieder, R. (2002) General theory of elastic wave scattering. In: P.C. Sabatier and E.R. Pike (Eds.): *Scattering.* Academic Press: London

Sommerfeld, A. (1914) Über die Fortpflanzung des Lichts in dispergierenden Medien. *Annalen der Physik* **44**:177

Sommerfeld, A. (bearbeitet von F. Bopp and J. Meixner) (1964) *Vorlesungen über Theoretische Physik, Band III: Elektrodynamik.* Akademische Verlagsgesellschaft: Leipzig

Spies, M. (1992) *Elastische Wellen in transversal-isotropen Medien: ebene Wellen, Gauß'sche Wellenpakete, Green'sche Funktionen, elastische Holographie.* Dissertation, Universität des Saarlandes: Saarbrücken

Spies, M. (1994) Elastic waves in homogeneous and layered transversely isotropic media: Plane waves and Gaussian wave packets. A general approach. *J Acoust Soc Am* **95**:1748

Spies, M. (1998) Transducer field modeling in anisotropic media by superposition of Gaussian base functions. *J Acoust Soc Am* **105**: 633

Spies, M. (2000a) Modeling of transducer fields in inhomogeneous ansisotropic materials using Gaussian beam superposition. *NDT&E International* **33**:155

Spies, M. (2000b) Kirchhoff evaluation of scattered elastic wavefields in anisotropic media. *J Acoust Soc Am* **107**:2755

Stratton, J.A. (1941) *Electromagnetic theory*. McGraw-Hill: New York

Tan, T.H. (1975a) Scattering of elastic waves by elastically transparent obstacles (integral-equation method). *Appl Sci Res* **31**:29

Tan, T.H. (1975b) Far-field radiation characteristics of elastic waves and the elastodynamic radiation condition. *Appl Sci Res* **31**:363

Tan, T.H. (1977) Scattering of plane, elastic waves by a plane crack of finite width. *Appl Sci Res* **33**:75

Tygel, M. and Hubral, P. (1987) *Transient waves in layered media*. Elsevier: Amsterdam

Ulaby, F.T. and Elachi, C. (Eds.) (1990) *Radar polarimetry for geoscience applications*. Artech House: Norwood/MA

van Bladel, J. (1961) Some remarks on Green's dyadic for infinite space. *IRE Trans Ant Propagat* **AP-9**:563

van Bladel, J. (1985) *Electromagnetic Fields*. Hemisphere Publishing Corporation: New York

van Bladel, J. (1991) *Singular Electromagnetic Fields and Sources*. Clarendon Press: Oxford

van den Berg, P.M. (1999) Reconstruction of media posed as an optimization problem. In: A. Wirgin (Ed.): *Wavefield inversion*. Springer-Verlag: Wien

Vinh, P.C. and R.W. Ogden (2004) On formulas for the Rayleigh wave speed. *Wave Motion* **39**:191

Wang, C.-Y. (2002) New expressions for cylindrical waves in a transversely isotropic solid and cuspidal borehole modes and dual arrivals. *Proceedings IEEE-UFFC Symposium*, München

Wilton, D.R. (2002) Computational methods. In: P.C. Sabatier and E.R. Pike (Eds.): *Scattering*. Academic Press: London

Wilbrand, A. (1989) *Theoretische und experimentelle Untersuchungen zu einem quantitativen Modell für elektromagnetische Ultraschallprüfköpfe*. Dissertation, Universität des Saarlandes: Saarbrücken

Wolf, H. (1976) *Spannungsoptik*. Springer-Verlag: Berlin

Wolter, H. (1995) *Berechnung akustischer Wellen und Resonatoren mit der FIT-Methode*. Dissertation, TU Darmstadt

Ying, C.F. and Truell, R. (1956) Scattering of a plane longitudinal wave by a spherical obstacle in an isotropically elastic solid. *J Appl Phys* **27**:1086

Zeroug, S., Stanke, F.E. and Burridge R. (1996) A complex-transducer-point model for finite emitting and receiving ultrasonic transducers. *Wave Motion* **24**:21

Zimmer, A. (2007) *Abbildende zerstörungsfreie Prüfverfahren mit elastischen und elektromagnetischen Wellen*. Dissertation, Universität Kassel: Kassel

Index